Normal Curve Areas

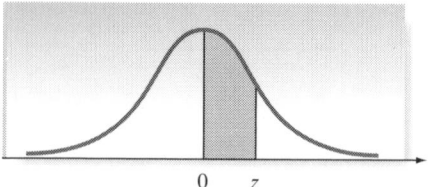

z	.00	.01	.02	.03	.04	.05	.06	.07	.08	.09
.0	.0000	.0040	.0080	.0120	.0160	.0199	.0239	.0279	.0319	.0359
.1	.0398	.0438	.0478	.0517	.0557	.0596	.0636	.0675	.0714	.0753
.2	.0793	.0832	.0871	.0910	.0948	.0987	.1026	.1064	.1103	.1141
.3	.1179	.1217	.1255	.1293	.1331	.1368	.1406	.1443	.1480	.1517
.4	.1554	.1591	.1628	.1664	.1700	.1736	.1772	.1808	.1844	.1879
.5	.1915	.1950	.1985	.2019	.2054	.2088	.2123	.2157	.2190	.2224
.6	.2257	.2291	.2324	.2357	.2389	.2422	.2454	.2486	.2517	.2549
.7	.2580	.2611	.2642	.2673	.2704	.2734	.2764	.2794	.2823	.2852
.8	.2881	.2910	.2939	.2967	.2995	.3023	.3051	.3078	.3106	.3133
.9	.3159	.3186	.3212	.3238	.3264	.3289	.3315	.3340	.3365	.3389
1.0	.3413	.3438	.3461	.3485	.3508	.3531	.3554	.3577	.3599	.3621
1.1	.3643	.3665	.3686	.3708	.3729	.3749	.3770	.3790	.3810	.3830
1.2	.3849	.3869	.3888	.3907	.3925	.3944	.3962	.3980	.3997	.4015
1.3	.4032	.4049	.4066	.4082	.4099	.4115	.4131	.4147	.4162	.4177
1.4	.4192	.4207	.4222	.4236	.4251	.4265	.4279	.4292	.4306	.4319
1.5	.4332	.4345	.4357	.4370	.4382	.4394	.4406	.4418	.4429	.4441
1.6	.4452	.4463	.4474	.4484	.4495	.4505	.4515	.4525	.4535	.4545
1.7	.4554	.4564	.4573	.4582	.4591	.4599	.4608	.4616	.4625	.4633
1.8	.4641	.4649	.4656	.4664	.4671	.4678	.4686	.4693	.4699	.4706
1.9	.4713	.4719	.4726	.4732	.4738	.4744	.4750	.4756	.4761	.4767
2.0	.4772	.4778	.4783	.4788	.4793	.4798	.4803	.4808	.4812	.4817
2.1	.4821	.4826	.4830	.4834	.4838	.4842	.4846	.4850	.4854	.4857
2.2	.4861	.4864	.4868	.4871	.4875	.4878	.4881	.4884	.4887	.4890
2.3	.4893	.4896	.4898	.4901	.4904	.4906	.4909	.4911	.4913	.4916
2.4	.4918	.4920	.4922	.4925	.4927	.4929	.4931	.4932	.4934	.4936
2.5	.4938	.4940	.4941	.4943	.4945	.4946	.4948	.4949	.4951	.4952
2.6	.4953	.4955	.4956	.4957	.4959	.4960	.4961	.4962	.4963	.4964
2.7	.4965	.4966	.4967	.4968	.4969	.4970	.4971	.4972	.4973	.4974
2.8	.4974	.4975	.4976	.4977	.4977	.4978	.4979	.4979	.4980	.4981
2.9	.4981	.4982	.4982	.4983	.4984	.4984	.4985	.4985	.4986	.4986
3.0	.4987	.4987	.4987	.4988	.4988	.4989	.4989	.4989	.4990	.4990

Source: Abridged from Table I of A. Hald, *Statistical Tables and Formulas* (New York: Wiley), 1952. Reproduced by permission of A. Hald.

Critical Values of t

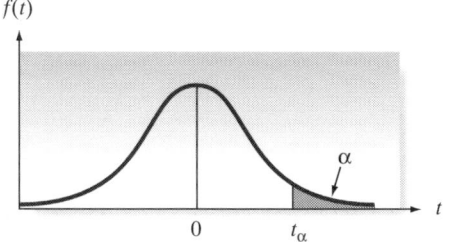

v	$t_{.100}$	$t_{.050}$	$t_{.025}$	$t_{.010}$	$t_{.005}$	$t_{.001}$	$t_{.0005}$
1	3.078	6.314	12.706	31.821	63.657	318.31	636.62
2	1.886	2.920	4.303	6.965	9.925	22.326	31.598
3	1.638	2.353	3.182	4.541	5.841	10.213	12.924
4	1.533	2.132	2.776	3.747	4.604	7.173	8.610
5	1.476	2.015	2.571	3.365	4.032	5.893	6.869
6	1.440	1.943	2.447	3.143	3.707	5.208	5.959
7	1.415	1.895	2.365	2.998	3.499	4.785	5.408
8	1.397	1.860	2.306	2.896	3.355	4.501	5.041
9	1.383	1.833	2.262	2.821	3.250	4.297	4.781
10	1.372	1.812	2.228	2.764	3.169	4.144	4.587
11	1.363	1.796	2.201	2.718	3.106	4.025	4.437
12	1.356	1.782	2.179	2.681	3.055	3.930	4.318
13	1.350	1.771	2.160	2.650	3.012	3.852	4.221
14	1.345	1.761	2.145	2.624	2.977	3.787	4.140
15	1.341	1.753	2.131	2.602	2.947	3.733	4.073
16	1.337	1.746	2.120	2.583	2.921	3.686	4.015
17	1.333	1.740	2.110	2.567	2.898	3.646	3.965
18	1.330	1.734	2.101	2.552	2.878	3.610	3.922
19	1.328	1.729	2.093	2.539	2.861	3.579	3.883
20	1.325	1.725	2.086	2.528	2.845	3.552	3.850
21	1.323	1.721	2.080	2.518	2.831	3.527	3.819
22	1.321	1.717	2.074	2.508	2.819	3.505	3.792
23	1.319	1.714	2.069	2.500	2.807	3.485	3.767
24	1.318	1.711	2.064	2.492	2.797	3.467	3.745
25	1.316	1.708	2.060	2.485	2.787	3.450	3.725
26	1.315	1.706	2.056	2.479	2.779	3.435	3.707
27	1.314	1.703	2.052	2.473	2.771	3.421	3.690
28	1.313	1.701	2.048	2.467	2.763	3.408	3.674
29	1.311	1.699	2.045	2.462	2.756	3.396	3.659
30	1.310	1.697	2.042	2.457	2.750	3.385	3.646
40	1.303	1.684	2.021	2.423	2.704	3.307	3.551
60	1.296	1.671	2.000	2.390	2.660	3.232	3.460
120	1.289	1.658	1.980	2.358	2.617	3.160	3.373
	1.282	1.645	1.960	2.326	2.576	3.090	3.291

Source: This table is reproduced with the kind permission of the Trustees of Biometrika from E. S. Pearson and H. O. Hartley (eds.), *The Biometrika Tables for Statisticians,* Vol. 1, 3d ed., Biometrika, 1966.

Statistics

Statistics

Tenth Edition

James T. McClave

Info Tech, Inc.
University of Florida

Terry Sincich

University of South Florida

PEARSON

Prentice
Hall

PEARSON EDUCATION INTERNATIONAL

Editor-in-Chief: *Sally Yagan*
Executive Acquisition Editor: *Petra Recter*
Project Manager: *Jacquelyn Riotto Zupic*
Production Editors: *Jeanne Audino/Bayani Mendoza de Leon*
Assistant Managing Editor: *Bayani Mendoza de Leon*
Senior Managing Editor: *Linda Mihatov Behrens*
Executive Managing Editor: *Kathleen Schiaparelli*
Assistant Manufacturing Manager/Buyer: *Michael Bell*
Media Production Editor: *Donna Crilly*
Managing Editor, Digital Supplements: *Nicole M. Jackson*
Marketing Assistant: *Rebecca Alimena*
Director of Creative Services: *Paul Belfanti*
Art Director: *Maureen Eidea*
Art Editor: *Thomas Benfatti*
Interior Designer: *Joseph Sengotta*
Cover Designer: *Kiwi Design*
Director, Image Resource Center: *Melinda Reo*
Manager, Rights and Permissions: *Zina Arabia*
Manager, Visual Research: *Beth Brenzel*
Image Permission Coordinator: *Carolyn Gauntt*
Photo Researcher: *Kathy Weisbrod*
Composition: *Progressive*
Art Studio: *Precision Graphics*
Editorial Assistant/Supplement Editor: *Joanne Wendelken*
Cover photos: *Getty Images*

© 2006, 2003 by Pearson Education, Inc.
Pearson Prentice Hall
Pearson Education, Inc.
Upper Saddle River, New Jersey 07458

Printed in the United States of America

10 9 8 7 6 5 4 3 2 1

ISBN 0-13-200302-3

Pearson Education LTD., *London*
Pearson Education Australia PTY, Limited, *Sydney*
Pearson Education Singapore, Pte. Ltd
Pearson Education North Asia Ltd, *Hong Kong*
Pearson Education Canada, Ltd., *Toronto*
Pearson Educación de Mexico, S.A. de C.V.
Pearson Education—Japan, *Tokyo*
Pearson Education Malaysia, Pte. Ltd
Pearson Education, Upper Saddle River, *New Jersey*

Contents

Preface xi
Applications Index xviii

APPENDICES

Appendix A Tables   881

Preface

This 10th edition of *Statistics* is an introductory text emphasizing inference, with extensive coverage of data collection and analysis as needed to evaluate the reported results of statistical studies and make good decisions. As in earlier editions, the text stresses the development of statistical thinking, the assessment of credibility and value of the inferences made from data, both by those who consume and those who produce them. It assumes a mathematical background of basic algebra.

A briefer version of the book, *A First Course in Statistics*, is available for single semester courses that include minimal coverage of regression analysis, analysis of variance, and categorical data analysis.

NEW IN THE 10TH EDITION

- ***Over 1,200 Exercises, with Revisions and Updates to 50%*** Revised to provide a greater variety in level of difficulty. In addition to "*Understanding the Principles*" and "*Learning the Mechanics*" exercises, "*Applied Exercises*" are categorized into "*Basic*," "*Intermediate*," and "*Advanced*" at the end of each section. Many of these exercises foster and promote critical thinking skills.

- ***Critical Thinking Challenges*** At the end of the "Chapter Supplementary Exercises," students are asked to apply their critical thinking skills to solve one or two challenging real-life problems.

- ***Over 125 Examples*** All examples now have three components: (1) "problem"; (2) "solution"; and (3) "look back". This step-by-step process provides students with a defined structure by which to approach problems and enhances their problem solving skills. The "look back" feature often gives helpful hints to solving the problem.

- ***Now Work*** A "Now Work" exercise follows each example. "Now Work" suggests an end-of-section exercise that is similar in style and concept to the text example. This gives the student the opportunity to test and confirm their understanding.

- ***7 New Statistics in Action Cases*** Each chapter now begins with a description of an actual, engaging, case study on a contemporary, controversial or high-profile issue ("Statistics in Action") and the accompanying data. We revisit the case throughout the chapter in order to demonstrate how to analyze the data and interpret the results ("Statistics in Action Revisited"). Our goal is to show students the importance of applying sound statistical techniques in order to evaluate the findings and to think through the statistical issues involved.

- ***MINITAB Software Tutorials*** At the end of each chapter are "*Using Technology*" tutorials with point-and-click instructions and with screen shots for MINITAB. These tutorials are easily located and provide students with useful information on how to best use and maximize MINITAB statistical software.

- ***Statistical Software Printouts*** These appear throughout the text in examples and exercises and include MINITAB, as well as SPSS and SAS printouts. Students are exposed to the computer printouts they will encounter in the hi-tech world.

- ***Profiles of Statisticians in History*** Side boxes feature within each of them a photo of a famous statistician and a brief description of his/her achievements.

With these profiles, students will develop an appreciation of the statistician's efforts and the discipline of statistics as a whole.

- *Chapter Summary Notes*   Now provided with end of chapter material. These notes help the student summarize and reinforce the important points from the chapter, and are useful for study tools.

- *Student Data CD*   The text is accompanied by a CD that contains files for all of the data sets marked with a CD icon in the text. These include data sets for text examples, exercises, and *Statistics in Action* cases. All data files are saved in four different formats: MINITAB, SAS, SPSS, and ASCII for easy importing into other statistical software packages.

Chapter Specific

- *Chapter 3 (Probability)*   An optional section on Bayes's Rule has been added (Section 3.9).

- *Chapter 7 (Confidence Intervals for a Single Sample)*   An introductory section on identifying the parameter of interest (a population mean or proportion) has been added (Section 7.1).

- *Chapter 9 (Inferences for Two Samples)*   An introductory section on identifying the target parameter (difference between means, difference between proportions, or ratio of variances) has been added (Section 9.1).

- *Chapter 11 (Simple Linear Regression)*   Several sections from the previous edition have been combined and streamlined to shorten the chapter.

- *Chapter 12 (Multiple Regression and Model Building)*   Several sections from the previous edition have been combined and streamlined to shorten the chapter.

TRADITIONAL STRENGTHS

We have maintained the pedagogical features of *Statistics* that we believe make it unique among introductory statistics texts. These features, which assist the student in achieving an overview of statistics and an understanding of its relevance in everyday life, are as follows:

- *Use of Examples as a Teaching Device*   Almost all new ideas are introduced and illustrated by data-based applications and examples. We believe that students better understand definitions, generalizations, and theoretical concepts after seeing an application.

- *Statistics in Action*   Each chapter begins with a case study on an actual contemporary, controversial, or high-profile issue. Relevant research questions and data from the study are presented and the proper analysis demonstrated in short "Statistics in Action Revisited" sections throughout the chapter. These motivate students to critically evaluate the findings and think through the statistical issues involved. Additional cases appear on the text Web site.

- *Real Data Exercises*   The text includes more than 1,200 exercises based on applications in a variety of disciplines and research areas. All the applied exercises employ the use of current real data extracted from a wide variety of publications (e.g., newspapers, magazines, journals, and the Internet). Some students have difficulty learning the mechanics of statistical techniques when all problems are couched in terms of realistic applications. For this reason, all exercise sections are divided into five parts:

    Understanding the Principles. Short answer exercises on definitions, concepts, and assumptions.

Learning the Mechanics. Designed as straightforward applications of new concepts, these exercises allow students to test their ability to comprehend a mathematical concept or a definition.

Applying the Concepts—Basic. Based on applications taken from a wide variety of journals, newspapers, and other sources, these short exercises help the student to begin developing the skills necessary to diagnose and analyze real-world problems.

Applying the Concepts—Intermediate. Based on more detailed real-world applications, these exercises require the student to apply critical thinking and their knowledge of the technique presented in the section.

Applying the Concepts—Advanced. These more challenging real-data exercises require the student to utilize their critical thinking skills.

- ***Exploring Data with Statistical Computer Software and the Graphing Calculator*** Each statistical analysis method presented is demonstrated using output from three leading Windows-based statistical software packages: SPSS, MINITAB, and SAS. In addition, output and keystroke instructions for the TI-83 Graphing Calculator are covered in optional boxes which are easy to locate throughout the text.

End of Chapter Materials Include:

- ***Quick Review*** Each chapter ends with a list of key terms and formulas, with reference to the page number where they first appear, as well as a brief summary of the concepts covered.

- ***Language Lab*** Following the Quick Review is a pronunciation guide to Greek letters and other special terms. Usage notes are also provided. This lab gives students a quick reference to the myriad of symbols encountered in statistics.

- ***Student Projects*** Presented at the end of each chapter, these projects emphasize gathering data, analyzing data, and/or report writing.

FLEXIBILITY IN COVERAGE

The text is written to allow the instructor flexibility in coverage of topics through sections marked "optional" in relevant chapters. Suggestions for covering two topics, probability and regression, are given below.

- ***Probability and Counting Rules*** One of the most troublesome aspects of an introductory business statistics course is the study of probability. Probability poses a challenge for instructors because they must decide on the level of presentation, and students find it a difficult subject to comprehend. We believe that one cause of these problems is the mixture of probability and counting rules that occur in most introductory texts. Consequently, we have included the counting rules (with examples) in a separate optional section at the end of Chapter 3 (Probability). In addition, all exercises that require the use of counting rules are marked with an asterisk (*). Thus, the instructor can control the level of coverage of probability.

- ***Multiple Regression and Model Building*** This topic represents one of the most useful statistical tools for the solution of applied problems. Although an entire text could be devoted to regression modeling, we feel that we have presented coverage that is understandable, usable, and much more comprehensive that the presentations in other introductory statistics texts. We have devoted two full chapters to discussing the major types of inferences that can be derived from a regression analysis, showing how these results appear in the output from statistical software, and, most importantly, selecting multiple regression models to be used in an analysis. Thus, the instructor has the choice of a one-chapter coverage of simple linear regression (Chapter 11), a two-chapter treatment of simple and multiple regression (excluding the optional sections in Chapter 13 on model building), or complete coverage of regression analysis, including model building and regression diagnostics. This extensive coverage of such useful statistical tools will provide added evidence to the student of the relevance of statistics to real-world business problems.

- ***Footnotes***   Although the text is designed for students with a non-calculus background, footnotes explain the role of calculus in various derivations. Footnotes are also used to inform the student about some of the theory underlying certain methods of analysis. These footnotes allow additional flexibility in the mathematical and theoretical level at which the material is presented.

ACKNOWLEDGMENTS

This book reflects the efforts of a great many people over a number of years. First, we would like to thank the following professors, whose reviews and comments on this and prior editions have contributed to the tenth edition:

Reviewers Involved with the 10th Edition of *Statistics*

Sneh Gulati, *Florida International University*
Golan Kibria, *Florida International University*
Ina Parks Howell, *Florida International University*
Barbara Wainwright, *Salisbury University*
Roddy Akbari, *Guilford Technical Community College*
Kazemi Mohammed, *UNC-Charlotte*
Susan Nolan, *Seton Hall University*
Sean Simpson, *Westchester CC*
Marvin Bishop, *Manhattan College*
Khadiga Gamgoum, *Northern Virginia CC*

Reviewers of Previous Editions of *Statistics*

Bill Adamson, South Dakota State; Ibrahim Ahmad, Northern Illinois University; David Atkinson, Olivet Nazarene University; Mary Sue Beersman, Northeast Missouri State University; William H. Beyer, University of Akron; Patricia M. Buchanan, Pennsylvania State University; Dean S. Burbank, Gulf Coast Community College; Kathryn Chaloner, University of Minnesota; Hanfeng Chen, Bowling Green State University; Gerardo Chin-Leo, The Everygreen State College; Linda Brant Collins, Iowa State University; Brant Deppa, Winona State University; John Dirkse, California State University—Bakersfield; N.B. Ebrahimi, Northern Illinois University; John Egenolf University of Alaska—Anchorage; Dale Everson, University of Idaho; Christine Franklin, University of Georgia; Rudy Gideon, University of Montana; Victoria Marie Gribshaw, Seton Hill College; Larry Griffey, Florida Community College; David Groggel, Miami University at Oxford; John E. Groves, California Polytechnic State

University—San Luis Obispo; Dale K. Hathaway, Olivet Nazarene University; Shu-ping Hodgson, Central Michigan University; Jean L. Holton, Virginia Commonwealth University; Soon Hong, Grand Valley; Ina Parks S. Howell, Florida International University; Gary Itzkowitz, Rowan College of New Jersey; John H. Kellermeier, State University College at Plattsburgh; Timothy J. Killeen, University of Connecticut; William G. Koellner, Montclair State University; James R. Lackritz, San Diego State University; Diane Lambert, AT&T/Bell Laboratories; Edwin G. Landauer, Clackamas Community College; James Lang, Valencia Junior College; Glenn Larson, University of Regina; Kjohn J. Lefante, Jr., University of South Alabama; Pi-Erh Lin, Florida State University; R. Bruce Lind, University of Puget Sound; Rhonda Magel, North Dakota State University; Linda C. Malone, University of Central Florida; Allen E. Martin, California State University—Los Angeles; Rick Martinez, Foothill College; Brenda Masters, Oklahoma State University; Leslie Matekaitis, Cal Genetics; E. Donice McCune, Stephen F. Austin State University; Mark M. Mdeerschaert, University of Nevada—Reno; Greg Miller, Steven F. Austin State University; Satya Narayan Mishra, University of South Alabama; Christopher Morrell, Loyola College in Maryland; A. Mukherjea, University of South Florida; Thomas O'Gorman, Northern Illinois University; Bernard Ostle, University of Central Florida; William B. Owen, Central Washington University; Won J. Park, Wright State University; John J. Peterson, Smith Kline & French Laboratories; Ronald Pierce, Eastern Kentucky University; Betty Rehfuss, North Dakota State University—Bottineau; Andrew Rosalsky, University of Florida; C. Bradley Russell, Clemson University; Rita Schillaber, University of Alberta; James R. Schott, University of Central Florida; Susan C. Schott, University of Central Florida; George Schultz, St. Petersburg Junior College; Carl James Schwarz. University of Manitoba; Mike Seyfried, Shippensburg University; Arvind K. Shah. University of South Alabama; Lewis Shoemaker, Millersville University; Charles W. Sinclair, Portland State University; Robert K. Smidt, California Polytechnic State University—San Luis Obispo; Vasanth B. Solomon, Drake University; W. Robert Stephenson, Iowa State University; Thaddeus Tarpey, Wright State University; Kathy Taylor, Clackamas Community College; Barbara Treadwell, Western Michigan University; Dan Voss, Wright State University; Augustin Vukov, University of Toronto; Dennis D. Wackerly, University of Florida; Matthew Wood; University of Missouri—Columbia; Ann Cascarelle, St. Petersburg College; Dale K. Hathaway, Olivet Nazarene University; Gary S. Itzkowitz, Rowan College of New Jersey; Mir Mortazavi, Eastern New Mexico University; Steve Nimmo, Morningside College (Iowa); James R. Schott, University of Central Florida

Other Contributors

Special thanks are due to our ancillary authors, Nancy Boudreau, Mark Dummeldinger, and to typist Kelly Barber, several of whom have worked with us for many years. Sudhir Goel has done an excellent job of accuracy checking the 10th edition and has helped us to ensure a highly accurate, clean text. Finally, the Prentice Hall staff of Petra Recter, Joanne Wendelken, Patrice Jones, Jeanne Audino, Jacquelyn Riotto Zupic, Sally Yagan, Bayani Mendoza de Leon, Thomas Benfatti, Rebecca Alimena, Michael Bell and Linda Behrens helped greatly with all phases of the text development, production, and marketing effort.

STUDENT RESOURCES

Student Study Pack

Everything a student needs to succeed in one place. Free packaged with the book, or available stand-alone. (The Student Study Pack is not available in the High School Channel.)

The **Student Study Pack** contains:

- *Student Solutions Manual*—includes complete worked out solutions to the odd-numbered text exercises.

- *Technology Manual*—Provides tutorial instruction and worked out examples for the TI-83 Calculator, Excel and Minitab Version 14. Includes a CD with PHStat (Excel Plug-in).

- *Pearson Tutor Center*—Tutors provide one-on-one tutoring for any problem with an answer At the back of the book. Students access the Tutor Center via toll-free phone, fax, or email. For more information, see www.prenhall.com/tutorcenter.

INSTRUCTOR RESOURCES

Content Distribution Center

All instructor resources can be downloaded from www.prenhall.com (then search for McClave, *Statistics 10E*). Once there, click the "instructor" link in the left menu bar. This is a password-protected site that will require registration. Alternatively, instructor resources can be ordered individually from your Prentice Hall sales representative.

- *Annotated Instructor's Edition (AIE) (ISBN: 0-13-149818-5)*—Marginal notes placed next to discussions of essential teaching concepts include:

  Teaching Tips—suggest alternative presentations or point out common student errors

  Exercises—reference specific section and chapter exercises that reinforce the concept

  Short Answers—section and chapter exercises are provided next to the selected exercises.

- *Instructor Solutions Manual (ISBN: 0-13-149819-3)*—Solutions to all of the even-numbered exercises are provided. Careful attention has been paid to ensure that all methods of solution and notation are consistent with those used in the text.

- *TestGen (ISBN: 0-13-149822-3)*—Test-generating software—creates tests from textbook section objectives. Over 1,000 questions are contained and algorithmic generation allows for unlimited test versions. Instructors may also edit questions or create their own.

- *Test-Item File (ISBN: 0-13-149820-7)*—A printed text bank derived from TestGen.

- *Power Point Lecture Slides*—Fully editable and printable slides that include many examples from the text as well as important pieces of art. Important concepts and definitions are included (only available for download).

INTERNET RESOURCES

MyMathLab™

Provides a rich and flexible set of course materials for your textbook, along with course-management tools so that instructors can easily customize an online course. MyMathLab gives instructors everything they need to deliver all or a portion of their course online, whether their students are in a lab setting or working from home.

MyMathLab offers the following features:

- Comprehensive gradebook tracking
- Complete online course content and customization tools
- Ability to copy courses and share them with other instructors
- Guided mathematical instruction for students with MathXL
- Multimedia learning aids for students
- Student study plan for self-paced learning
- Free tutoring for students from the Math Tutor Center
- All instructor and student resources (i.e., Solutions Manual, Power Point Slides, Technology Manual, etc.)

For more information, go to **www.mymathlab.com**

MathXL (Internet)

A powerful online homework, tutorial, and assessment system designed specifically for Pearson Education textbooks in mathematics and statistics. With MathXL, instructors can create, edit, and assign online homework and tests and track all student work in the online class gradebook. MathXL creates personalized study plans for your students based on their test results and offers unlimited (algorithmically generated) practice exercises correlated directly to the exercises in the textbook. MathXL is designed to be used in a lecture, self-paced, or distance-learning course. It does not contain a homework manager or gradebook.

For more information, go to **www.mathxl.com**

MyMathLab or MathXL can be bundled with a new textbook free of charge or can be purchased as a stand-alone.

Applications Index

Designed for Your Success…

Over 1,000 Interesting and Diverse Applications

- Represent a wide and diverse array of relevant business and decision making applications.

- Feature unique and sourced data that illustrates the role that statistics play in decision making.

- Using real-world data to emphasize that statistics is important to understanding the world around us.

- Emphasis is placed on the interpretation of real problems and adapting them to standard methods of analysis.

2.35 College protests of labor exploitation. The United Students Against Sweatshops (USAS) was formed by students on American college campuses in 1999 to protest labor exploitation in the apparel industry. Clark University sociologist Robert Ross analyzed the USAS movement in the *Journal of World-Systems Research* (Winter 2004). Between 1999 and 2000 there were 18 student "sit-ins" for a "sweat free campus" organized at several universities. The table gives the duration (in days) of each sit-in as well as the number of student arrests.

SITIN

Sit-in	Year	University	Duration (days)	Number of Arrests	Tier Ranking
1	1999	Duke	1	0	1st
2	1999	Georgetown	4	0	1st
3	1999	Wisconsin	1	0	1st
4	1999	Michigan	1	0	1st
5	1999	Fairfield	1	0	1st
6	1999	North Carolina	1	0	1st
7	1999	Arizona	10	0	1st
8	2000	Toronto	11	0	1st
9	2000	Pennsylvania	9	0	1st
10	2000	Macalester	2	0	1st
11	2000	Michigan	3	0	1st
12	2000	Wisconsin	4	54	1st
13	2000	Tulane	12	0	1st

Source: Ross, R. J. S. "From antisweatshop to global justice to antiwar: How the new new left is the same and different from the old new left." *Journal of Word-Systems Research,* Vol. X, No. 1, Winter 2004 (Tables 1 and 3).

Page 52

Section 2.8   Methods for Detecting Outliers (Optional)   91

EXAMPLE 2.18   INFERENCE USING Z-SCORES

Problem   Suppose a female bank employee believes that her salary is low as a result of sex discrimination. To substantiate her belief, she collects information on the salaries of her male counterparts in the banking business. She finds that their salaries have a mean of $54,000 and a standard deviation of $2,000. Her salary is $47,000. Does this information support her claim of sex discrimination?

Solution   The analysis might proceed as follows: First, we calculate the z-score for the woman's salary with respect to those of her male counterparts. Thus,

$$z = \frac{\$47,000 - \$54,000}{\$2,000} = -3.5$$

The implication is that the woman's salary is 3.5 standard deviations *below* the mean of the male salary distribution. Furthermore, if a check of the male salary data shows that the frequency distribution is mound shaped, we can infer that very few salaries in this distribution should have a z-score less than -3, as shown in Figure 2.33. Clearly, a z-score of -3.5 represents an outlier. Either her salary is from a distribution different from the male salary distribution or it is a very unusual (highly improbable) measurement from a salary distribution no different from the male distribution.

Figure 2.33
Male Salary Distribution

Look Back Which of the two situations do you think prevails? Statistical thinking would lead us to conclude that her salary does not come from the male salary distribution, lending support to the female bank employee's claim of sex discrimination. However, the careful investigator should require more information before inferring sex discrimination as the cause. We would want to know more about the data collection technique the woman used and more about her competence at her job. Also, perhaps other factors such as length of employment should be considered in the analysis.

Now Work   *Exercise 2.123*

■ ■ ■

More than 125 Clear and Interesting Examples with Solutions

- A step-by-step approach to reaching the solution with thorough explanations built into each step. The problem is identified, the solution is presented in a step-by-step manner, then the "Insight" is given which offers a look back at what has just been learned.

- Solutions are carefully explained to better prepare students for the end-of-section exercises.

- **"Now Work"** appears after each example and directs students to an exercise at the end of section that allows them to practice the skill and concept presented in the example.

Thorough and Extensive Exercise Sets—Over 1,200 Exercises Total!

- End-of-Section Exercises are numerous and divided into three parts: **Understanding the Principles, Learning the Mechanics** and **Applying the Concepts**. The *Applying the Concepts* word problems are divided into three levels of difficulty (*Basic, Intermediate, Advanced*), emphasizing critical thinking skills and requiring students to apply statistical techniques in solving real-world problems.

- Many exercises contain data and information taken from newspaper articles, magazines and journals. Inclusion of this data serves to illustrate the relevancy of this material. Each data set is saved in a file on a CD for the student to analyze using statistical software.

- Includes, where appropriate, computer output screens to give students practice in interpretation of data.

Exercises 2.103–2.117

Understanding the Principles

2.103 Give the percentage of measurements in a data set that are above and below each of the following percentiles:
   a. 75th percentile
   b. 50th percentile
   c. 20th percentile
   d. 84th percentile

2.104 What is the 50th percentile of a quantitative data set called?

2.105 For mound-shaped data, what percentage of measurements have a z-score between -2 and 2?

Learning the Mechanics

2.106 Compute the z-score corresponding to each of the following values of x:
   a. $x = 40$, $s = 5$, $\bar{x} = 30$
   b. $x = 90$, $\mu = 89$, $\sigma = 2$
   c. $\mu = 50$, $\sigma = 5$, $x = 50$
   d. $s = 4$, $x = 20$, $\bar{x} = 30$
   e. In parts **a–d**, state whether the z-score locates x within a sample or a population.
   f. In parts **a–d**, state whether each value of x lies above or below the mean and by how many standard deviations.

2.107 Compare the z-scores to decide which of the following x values lie the greatest distance above the mean and the greatest distance below the mean.
   a. $x = 100$, $\mu = 50$, $\sigma = 25$
   b. $x = 1$, $\mu = 4$, $\sigma = 1$
   c. $x = 0$, $\mu = 200$, $\sigma = 100$
   d. $x = 10$, $\mu = 5$, $\sigma = 3$

2.108 Suppose that 40 and 90 are two elements of a population data set and that their z-scores are -2 and 3, respectively. Using only this information, is it possible to determine the population's mean and standard deviation? If so, find them. If not, explain why it's not possible.

Applying the Concepts—Basic

2.109 Drivers stopped by police. According to the Bureau of Justice Statistics (March 2002), 73.5% of all licensed drivers stopped by police are 25 years or older. Give a [...] or the age of 25 years in the distri[...] licensed drivers stopped by police.

[...]th graders. According to the Na[...] Education Statistics (2000), scores [...] assessment test for United States [...] e a mean of 500, a 5th percentile [...] entile of 435, a 75th percentile [...] ercentile of 653. Interpret each of [...] scriptive measures.

SHIPSANIT

2.111 Sanitation inspection of cruise ships. Refer to the sanitation levels of cruise ships, Exercise 2.93 (p. 79), saved in the SHIPSANIT file.
   a. Give a measure of relative standing for the Nautilus Explorer's score of 74. Interpret the result.
   b. Give a measure of relative standing for the Rotterdam's score of 93. Interpret the result.

JAPANESE

2.112 Reading Japanese books. Refer to the *Reading in a Foreign Language* (Apr. 2004) experiment to improve the Japanese reading comprehension levels of 14 University of Hawaii students, Exercises 2.57 and 2.81 (p. 71). The data on number of books read and grade for each student are saved in the JAPANESE file.
   a. Find the mean and standard deviation of the number of books read by students who earned an A grade. Find the z-score for an A student who read 40 books. Interpret the result.
   b. Find the mean and standard deviation of the number of books read by students who either earned either a B or C grade. Find the z-score for a B or C student who read 40 books. Interpret the result.
   c. Refer to parts **a** and **b**. Which of the two groups of students is more likely to have read 40 books? Explain.

Applying the Concepts—Intermediate

NZBIRDS

2.113 Extinct New Zealand birds. Refer to the *Evolutionary Ecology Research* (July 2003) study of the patterns of extinction in the New Zealand bird population, Exercise 2.96 (p. 80). Again, consider the data on the egg length (measured in millimeters) for the 132 bird species saved in the NZBIRDS file.
   a. Find the 10th percentile for the egg length distribution and interpret its value.
   b. The *Moas, P. australis* bird species has an egg length of 205 millimeters. Find the z-score for this species of bird and interpret its value.

Applying the Concepts—Intermediate

2.114 Lead in drinking water. The US. Environmental Protection Agency (EPA) sets a limit on the amount of lead permitted in drinking water. The EPA *Action Level* for lead is .015 milligrams per liter (mg/L) of water. Under EPA guidelines, if 90% of a water system's study samples have a lead concentration less than .015 mg/L, the water is considered safe for drinking. I (co-author Sincich) received a report on a study of lead levels in the drinking water of homes in my subdivision. The 90th percentile of the study sample had a lead concentration of .00372 mg/L. Are water customers in my subdivision at risk of drinking water with unhealthy lead levels? Explain.

Page 84

Statistics in *ACTION*

The "Eye Cue" Test: Does Experience Improve Performance?

In 1948, famous child psychologist Jean Piaget devised a test of basic perceptual and conceptual skills dubbed the "water-level task." Subjects were shown a drawing of a glass being held perfectly still (at a 45° angle) by an invisible hand so that any water in it had to be at rest (see Figure SIA2.1).

The task for the subject was to draw a line representing the surface of the water – a line that would touch the black dot pictured on the right side of the glass. Piaget found that young children typically failed the test. Fifty years later, research psychologists still use the water-level task to test the perception of both adults and children. Surprisingly, about 40% of the adult population also fail. In addition, males tend to do better than females, and younger adults tend to do better than older adults.

Will people with experience handling liquid-filled containers perform better on this "eye cue" test? This question was the focus of research conducted by psychologists Heiko Hecht (NASA) and Dennis R. Proffitt (University of Virginia) and published in *Psychological Science* (Mar. 1995). The researchers presented the task to each of six different groups: (1) male college students, (2) female college students, (3) professional waitresses, (4) housewives, (5) male bartenders, and (6) male bus drivers. A total of 120 subjects (20 per group) participated in the study. Two of the groups, waitresses and bartenders, were assumed to have considerable experience handling liquid-filled glasses.

After each subject completed the drawing task, the researchers recorded the deviation* (in angle degrees) of the judged line from the true line. If the deviation was within 5° of the true water surface angle, the answer was considered correct. Deviations of more than 5° in either direction were considered incorrect answers.

The data for the water-level task (simulated based on summary results presented in the journal article) are provided in the **EYECUE** file. For each of the 120 subjects in the experiment, the following variables (in the order they appear on the data file) were measured:

EYECUE

Variables

GENDER (F or M)
GROUP (Student, Waitress, Wife, Bartender, or Busdriver)
DEVIATION (angle, in degrees, of the judged line from the true line)
JUDGE (Within5, More5Above, More5Below)

The researchers are interested in testing several theories concerning performance on the water-level task, including (1) males perform better than females, (2) younger adults perform better than older adults, and (3) experience improves task performance.

In the following *Statistics in Action Revisited* sections, we apply the graphical and numerical descriptive techniques of this chapter to the **EYECUE** data to answer some of the researchers questions.

Statistics in Action Revisited for Chapter 2

- Interpreting a Pie Chart (p. 35)
- Interpreting a Histogram (p. 49)
- Interpreting Descriptive Statistics (p. 78)
- Detecting Outliers (p. 93)

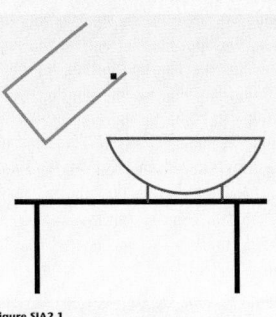

Figure SIA2.1
Drawing of the Water-Level Task

*The true surface line is perfectly parallel to the table top.

Page 29

Statistics in Action

- Each chapter opens with a **Statistics in Action** case study that highlights controversial, contemporary issues that involve statistics.

- The themes and issues are revisited and built upon throughout the chapter, by way of easily identifiable boxes labeled *Statistics in Action Revisited*.

Technology

Optional technology boxes appear throughout the text, both in the discussion and the exercises, facilitating computer-based analysis.

* They include output from EXCEL, MINITAB, and SPSS.

* Includes step-by-step instructions on how to use the TI-83 calculator in a variety of applications.

* **Technology Tutorials ("Using Technology")** appear at the end of chapters. They include point & click instructions and screen shots for EXCEL, MINITAB, and SPSS.

* These technology tutorials help students better understand what statistical tools to use, how to apply them and how to interpret the results.

Page 104

Box Plots

Using the T1-83 Graphing Calculator

Making a Box Plot

Step 1   Enter the data
Press STAT and select 1:Edit
Note: If the list already contains data, clear the old data. Use the up arrow to highlight 'L1'. Press CLEAR ENTER.
Use the arrow and ENTER keys to enter the data set into L1.

Step 2   Set up the box plot

Press 2nd **Y =** for STAT PLOT
Press 1 for Plot 1

Set the cursor so that 'ON' is flashing.
For TYPE, use the right arrow to scroll through the plot icons and select the boxplot in the middle of the second row.
For XLIST, choose L1.
Set FREQ to 1.

```
Plot1  Plot2  Plot3
On  Off
Type:
Xlist:L1
Freq:L2
```

Step 3   View the graph
Press ZOOM and select 9:ZoomStat

Optional   *Read the five number summary*

Page 92

End of Chapter Review

Each chapter concludes with information designed to help check your understanding of the material and study for tests. These resources include:

- **Quick Review**, a list of key terms and formulas with page number references for easy look up.

- **Language Lab**, that helps students learn the language of statistics through pronunciation guides, descriptions of symbols, names, etc.

- **Chapter Summary,** that provides a concise overview of key concepts.

- **Supplementary Exercises,** a review of the important topics introduced in the chapter.

Includes **Critical Thinking challenges** in which students are asked to apply their critical thinking skills to solve 1-2 challenging real-life problems.

Page 105

Quick Review

Key Terms

Note: Starred (*) terms are from the optional sections to this chapter

Bar graph 32
Bivariate relationship* 95
Box plots* 88
Central tendency 55
Chebyshev's Rule 73

Histogram 43
Inner fences* 87
Interquartile range* 86
Lower quartile* 86
Mean 56
Measures of central tendency 55
Measures of relative standing 71
Measures of variation or spread 67
Median 57

Percentile 81
Pie chart* 86
Quartiles* 86
Range 67
Rare-event approach* 91
Scattergram* 95
Scatterplot* 95
Skewness 59
Standard deviation 60

Language Lab

Symbol	Pronunciation	Description
Σ	sum of	Summation notation; $\sum_{i=1}^{n} x_i$ represents the sum of the measurements $x_1, x_2, \ldots, x_n$,
μ	mu	Population mean
$\bar{x}$	x-bar	Sample mean
σ^2	sigma-squared	Population variance
σ	sigma	Population standard deviation

Chapter Summary

- Graphical methods for qualitative data: **pie chart, bar graph,** and **Pareto diagram**
- Graphical methods for quantitative data: **dot plot, stem-and-leaf display,** and **histogram**
- Numerical measures of central tendency: **mean, median,** and **mode**
- Numerical measures of variation: **range, variance,** and **standard deviation**

- Rules for determining the percentage of measurements in the interval (mean) ± 2(std. dev.): **Chebyshev's Rule** (at least 75%) and **Empirical Rule** (approximately 95%)
- Measures of relative standing: **percentile score** and **z-score**
- Methods for detecting outliers: box plots and z-scores
- Method for graphing the relationship between two quantitative variables: **scatterplot**

Supplementary Exercises 2.149–2.187

Note: Starred (*) exercises refer to the optional sections in this chapter.

Understanding the Principles

2.149 Discuss conditions under which the median is preferred to the mean as a measure of central tendency.

2.150 Explain why we generally prefer the standard deviation to the range as a measure of variation.

2.151 Give a situation where we will prefer using a stem-leaf display over a histogram when graphically describing quantitative data.

2.152 Give a situation where we will prefer using a box plot over z-scores to detect an outlier.

2.153 Give a technique that is used to distort information shown on a graph.

Learning the Mechanics

2.154 Construct a relative frequency histogram for the data summarized in the table below.

Measurement Class	Relative Frequency	Interval Class	Relative Frequency
.00–.75	.02	5.25–6.00	.15
.75–1.50	.01	6.00–6.75	.12
1.50–2.25	.03	6.75–7.50	.09
2.25–3.00	.05	7.50–8.25	.05
3.00–3.75	.10	8.25–9.00	.04

b. $\sum_{i=1}^{n} x_i^2 = 666, \sum_{i=1}^{n} x_i = 106, n = 25$

c. $\sum_{i=1}^{n} x_i^2 = 76, \sum_{i=1}^{n} x_i = 11, n = 7$

2.158 If the range of a set of data is 20, find a rough approximation to the standard deviation of the data set.

2.159 For each of the following data sets, compute $\bar{x}$, s^2, and s. If appropriate, specify the units in which your answers are expressed.
a. 4, 6, 6, 5, 6, 7
b. −$1, $4, −$3, $0, −$3, −$6
c. 3/5%, 4/5%, 2/5%, 1/5%, 1/16%
d. Calculate the range of each data set in parts **a–c**.

2.160 For each of the following data sets, compute $\bar{x}$, s^2, and s:
a. 13, 1, 10, 3, 3
b. 13, 6, 6, 0
c. 1, 0, 1, 10, 11, 11, 15
d. 3, 3, 3, 3
e. For each data set in parts **a–d**, form the interval $\bar{x} \pm 2s$ and calculate the percentage of the measurements that fall in the interval.

*2.161 Construct a scatterplot for the data listed here. Do you detect any trends?

Variable #1:	174	268	345	119	400	520	190	448	307	252
Variable #2:	8	10	15	7	22	31	15	20	11	9

Page 106

Statistics

1

Statistics, Data, and Statistical Thinking

Contents

Statistics in Action

USA Weekend Teen Survey:
Are Boys Really from Mars and
Girls from Venus?

Using Technology

Creating and Listing Data
Using MINITAB

Where We're Going

- Introduce the field of statistics.
- Demonstrate how statistics applies to real-world problems.
- Establish the link between statistics and data.
- Identify the different types of data and data collection methods.
- Differentiate between population and sample data.
- Differentiate between descriptive and inferential statistics.

Statistics in *ACTION*

USA WEEKEND Teen Survey: Are Boys Really from Mars and Girls from Venus?

Welcome to USA WEEKEND, the magazine that makes a difference.

www.usaweekend.com

USA Weekend is a magazine supplement in approximately 600 United States newspapers, with an estimated 50 million readers. Each year, the magazine conducts a Teen Survey in which America's teenagers give opinions on a wide variety of topics, including their views of the opposite sex (2003 survey) and their use of newspapers (2004 survey). The surveys are made available in print in the magazine supplement, at the www.usaweekend.com Web site, and via the mail using newspaper subscriber lists.

Each survey is typically sponsored by a special interest group. For example, the American Society of Newspaper Editors and the Newspaper Association of America Foundation were sponsors of the 2004 survey, while *YM Magazine* and John Gray (author of the best-selling book *Men are from Mars, Women are from Venus*) sponsored the 2003 survey. The number of teens responding to the survey varies as well (37,000 teenagers in 2003 and 65,000 in 2004).

Some of the questions (and corresponding results) for the 2003 and 2004 Teen Surveys are shown in the following table.

2004 Survey: Teens & Newspapers

1. Where do you get most of your news?

Television:	48%
Newspaper:	18%
Word of Mouth:	14%
Online:	9%
Radio:	7%
Magazines:	4%

2. How relevant is the newspaper to your life?

Very relevant:	18%
Somewhat relevant:	53%
Not very relevant:	23%
Not at all relevant:	6%

3. How many days a week do you read the newspaper?

Average = 3.1

4. Do you think newspapers will become obsolete because of other media?

Yes:	24%
No:	50%
Don't know:	26%

2003 Survey: Teens & the Opposite Sex

1. Which quality first catches your attention in the opposite sex?

	Boys	Girls
Looks:	58%	40%
Personality:	27%	36%
Athleticism:	1%	2%
Confidence:	2%	3%
Sense of Humor:	9%	16%
Intelligence:	2%	2%
Popularity:	1%	1%

2. Which quality of your sex first catches the attention of the opposite sex?

	Boys	Girls
Looks:	56%	79%
Personality:	16%	11%
Athleticism:	6%	0%
Confidence:	3%	3%
Sense of Humor:	12%	3%
Intelligence:	2%	1%
Popularity:	5%	3%

3. Do you think it's OK for girls in your grade to call guys?

	Boys	Girls
Yes:	96%	95%
No:	4%	5%

4. Do you think it's OK for girls in your grade to ask guys out?

	Boys	Girls
Yes:	90%	80%
No:	10%	20%

5. What part of the newspaper do you read first?

Front page:	36%
Comics:	19%
Sports:	15%
Other:	30%

5. Is it possible to have a good friend of the opposite sex?

	Boys	*Girls*
No:	6%	3%
Yes, but only if part of a group:	8%	7%
Yes, anytime:	86%	90%

In the following *Statistics in Action Revisited* sections, we discuss several key statistical concepts covered in this chapter that are relevant to the *USA Weekend* Annual Teen Surveys.

- Identifying the data collection method and data type (p. 16)
- Critically assessing the ethics of a statistical study (p. 18)

Statistics in Action Revisited for Chapter 1

- Identifying the population, sample, and inference (p. 12)

1.1 The Science of Statistics

What does statistics mean to you? Does it bring to mind batting averages, Gallup polls, unemployment figures, numerical distortions of facts (lying with statistics!)? Or is it simply a college requirement you have to complete? We hope to persuade you that statistics is a meaningful, useful science whose broad scope of applications to business, government, and the physical and social sciences is almost limitless. We also want to show that statistics can lie only when they are misapplied. Finally, we wish to demonstrate the key role statistics plays in critical thinking—whether in the classroom, on the job, or in everyday life. Our objective is to leave you with the impression that the time you spend studying this subject will repay you in many ways.

The *Random House College Dictionary* defines *statistics* as "the science that deals with the collection, classification, analysis, and interpretation of information or data." Thus, a statistician isn't just someone who calculates batting averages at baseball games or tabulates the results of a Gallup Poll. Professional statisticians are trained in *statistical science*. That is, they are trained in collecting numerical information in the form of **data**, evaluating it, and drawing conclusions from it. Furthermore, statisticians determine what information is relevant in a given problem and whether the conclusions drawn from a study are to be trusted.

> **DEFINITION 1.1**
>
> **Statistics** is the science of data. This involves collecting, classifying, summarizing, organizing, analyzing, and interpreting numerical information.

In the next section, you'll see several real-life examples of statistical applications that involve making decisions and drawing conclusions.

1.2 Types of Statistical Applications

Statistics means "numerical descriptions" to most people. Monthly housing starts, the failure rate of liver transplants, and the proportion of African-Americans who feel brutalized by local police all represent statistical descriptions of large sets of data collected on some phenomenon. Often the data are selected from

Biography

FLORENCE NIGHTINGALE (1820–1910)— The Passionate Statistician

In Victorian England, the "Lady of the Lamp" had a mission to improve the squalid field hospital conditions of the British army during the Crimean War. Today, most historians consider Florence Nightingale to be the founder of the nursing profession. To convince members of the British Parliament of the need for supplying nursing and medical care to soldiers in the field, Nightingale compiled massive amounts of data from the army files. Through a remarkable series of graphs (which included the first pie chart), she demonstrated that most of the deaths in the war were due to illnesses contracted outside the battlefield or long after battle action from wounds that went untreated. Florence Nightingale's compassion and self-sacrificing nature, coupled with her ability to collect, arrange, and present large amounts of data, led some to call her the Passionate Statistican.

some larger set of data whose characteristics we wish to estimate. We call this selection process *sampling*. For example, you might collect the ages of a sample of customers at a video store to estimate the average age of *all* customers of the store. Then you could use your estimate to target the store's advertisements to the appropriate age group. Notice that statistics involves two different processes: (1) describing sets of data and (2) drawing conclusions (making estimates, decisions, predictions, etc.) about the sets of data based on sampling. So, the applications of statistics can be divided into two broad areas: *descriptive statistics* and *inferential statistics*.

DEFINITION 1.2

Descriptive statistics utilizes numerical and graphical methods to look for patterns in a data set, to summarize the information revealed in a data set, and to present that information in a convenient form.

DEFINITION 1.3

Inferential statistics utilizes sample data to make estimates, decisions, predictions, or other generalizations about a larger set of data.

Although we'll discuss both descriptive and inferential statistics in the following chapters, the primary theme of the text is **inference**.

Let's begin by examining some studies that illustrate applications of statistics.

Study 1 "Speed Training Program for High School Football Players" (*The Sport Journal*, Winter 2004)

Michael Gray and Jessica Sauerbeck, researchers at Northern Kentucky University, designed and tested a speed training program for junior varsity and varsity high school football players. Each participant was timed in a 40-yard sprint prior to the start of the training program and timed again after completing the program. Based on these sprint times, each participant was classified as having an "improved" time, "no change" in time, or a "decrease" in time. The results are summarized in Figure 1.1. From the graph, you can clearly see that nearly 87% of the players improved their sprint times and only about 3% had decreased times.

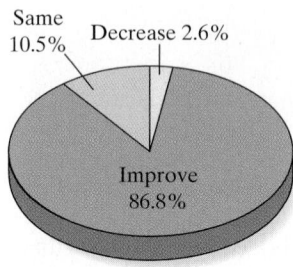

Figure 1.1
Success Rate of a Speed
Training Program for High
School Football Players

Since Figure 1.1 describes the categories of improvement in sprint times for players who participate in a speed training program, the graphic is an example of descriptive statistics.

Study 2   "Does Height Influence Progression through Primary School Grades?" (*The Archives of Disease in Childhood*, April 2000)

Researchers from the University of Melbourne (Australia) analyzed the heights of over 2,800 students in primary school (grades 1 through 6). After dividing the students into three equal groups based on age (youngest third, middle third, and oldest third) within each grade level, the researchers found that the oldest group of students were the shortest in height, on average — and that this phenomenon was mostly due to the 133 children (mostly boys) who had been held back a grade level. From the analysis, the researchers inferred "that older boys within grades were relatively shorter than their younger peers" and that "when boys experienced school difficulties, height [is] one factor influencing the final decision to [hold back the student a grade level]." Thus, inferential statistics was applied to arrive at this conclusion.

Study 3   "The Significance of Belief and Expectancy Within the Spiritual Healing Encounter" (*Social Science & Medicine*, July 1995)

"Spiritual healing techniques have been a fundamental component of the healing rituals of virtually all societies since the advent of man," says researcher Daniel P. Wirth. What effect, if any, do spiritual healing encounters have on the mental and physical health of patients in today's society? And do the effects depend on the patient's expectations? To answer these questions, Wirth conducted a scientific study of 48 individuals residing in a northern California suburb. The mental and physical health of each subject was determined via a battery of standardized tests. Each individual then underwent a spiritual healing encounter that involved a "magnetic laying-on-of-hands approach." One week following the healing session, subjects were retested.

An analysis of the data revealed a significant difference between the pretest and posttest health scores for the subjects. Further analysis led Wirth to conclude that "high healer and patient expectancy [are] important . . . predictors as well as facilitators of the healing process."

Like Study 2, this study is an example of the use of inferential statistics. Wirth used data from 48 individuals to make inferences about the beliefs and health improvements of all patients who experience a spiritual healing encounter.

These studies provide three real-life examples of the uses of statistics. Notice that each involves an analysis of data, either for the purpose of describing the data set (Study 1) or for making inferences about a data set (Studies 2 and 3).

1.3 Fundamental Elements of Statistics

Statistical methods are particularly useful for studying, analyzing, and learning about *populations* of *experimental units*.

DEFINITION 1.4

An **experimental unit** is an object (e.g., person, thing, transaction, or event) upon which we collect data.

> **DEFINITION 1.5**
>
> A **population** is a set of units (usually people, objects, transactions, or events) that we are interested in studying.

For example, populations may include (1) *all* employed workers in the United States, (2) *all* registered voters in California, (3) *everyone* who is afflicted with AIDS, (4) *all* the cars produced last year by a particular assembly line, (5) the entire stock of spare parts available at United Airlines' maintenance facility, (6) *all* sales made at the drive-in window of a McDonald's restaurant during a given year, and (7) the set of *all* accidents occurring on a particular stretch of interstate highway during a holiday period. Notice that the first three population examples (1–3) are sets (groups) of people, the next two (4–5) are sets of objects, the next (6) is a set of transactions, and the last (7) is a set of events. Also notice that *each set includes all the units in the population.*

In studying a population, we focus on one or more characteristics or properties of the units in the population. We call such characteristics *variables*. For example, we may be interested in the variables age, gender, and/or the number of years of education of the people currently unemployed in the United States.

> **DEFINITION 1.6**
>
> A **variable** is a characteristic or property of an individual population unit.

The name *variable* is derived from the fact that any particular characteristic may vary among the units in a population.

In studying a particular variable it is helpful to be able to obtain a numerical representation for it. Often, however, numerical representations are not readily available, so the process of measurement plays an important supporting role in statistical studies. **Measurement** is the process we use to assign numbers to variables of individual population units. We might, for instance, measure the performance of the President by asking a registered voter to rate it on a scale from 1 to 10. Or we might measure workforce age simply by asking each worker, "How old are you?" In other cases, measurement involves the use of instruments such as stopwatches, scales, and calipers.

If the population you wish to study is small, it is possible to measure a variable for every unit in the population. For example, if you are measuring the GPA for all incoming first-year students at your university, it is at least feasible to obtain every GPA. When we measure a variable for every unit of a population, it is called a **census** of the population. Typically, however, the populations of interest in most applications are much larger, involving perhaps many thousands or even an infinite number of units. Examples of large populations are those following Definition 1.5, as well as all graduates of your university or college, all potential buyers of a new fax machine, and all pieces of first-class mail handled by the U.S. Post Office. For such populations conducting a census would be prohibitively time-consuming and/or costly. A reasonable alternative would be to select and study a *subset* (or portion) of the units in the population.

> **DEFINITION 1.7**
>
> A **sample** is a subset of the units of a population.

For example, instead of polling all 120 million registered voters in the United States during a presidential election year, a pollster might select and question a sample of just 1,500 voters (see Figure 1.2). If he is interested in the variable "presidential preference," he would record (measure) the preference of each sampled vote.

After the variable(s) of interest for every unit in the sample (or population) are measured, the data are analyzed, either by descriptive or inferential statistical methods. The pollster, for example, may be interested only in *describing* the voting patterns of the sample of 1,500 voters. More likely, however, he will want to use the information in the sample to make inferences about the population of all 120 million voters.

> ### DEFINITION 1.8
>
> A **statistical inference** is an estimate, prediction, or some other generalization about a population based on information contained in a sample.

That is, *we use the information contained in the smaller sample to learn about the larger population.** Thus, from the sample of 1,500 voters, the pollster may estimate

Figure 1.2
A Sample of Voter
Registration Cards for
All Registered Voters

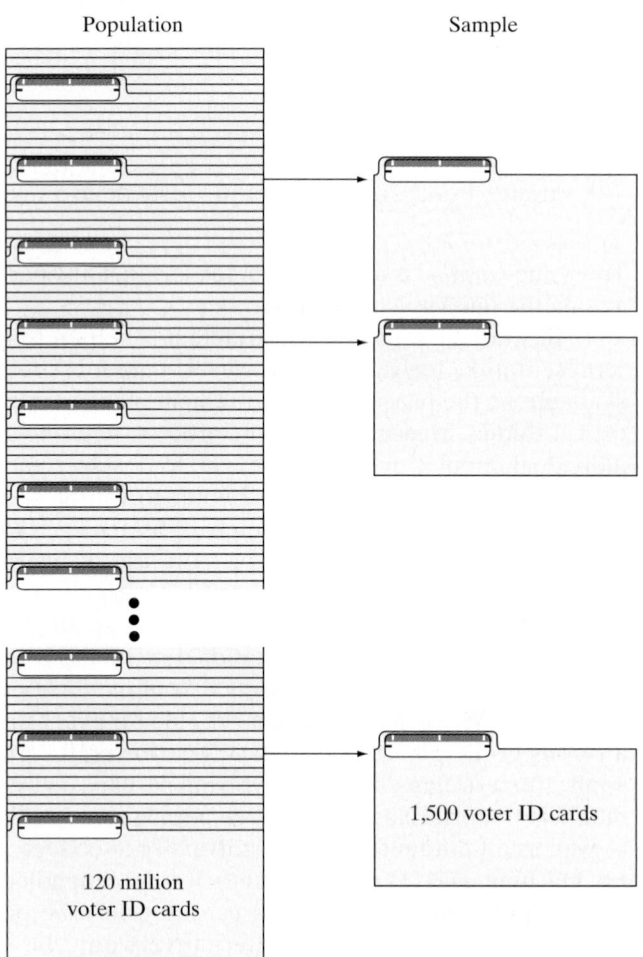

Population   Sample

1,500 voter ID cards

120 million
voter ID cards

*The terms *population* and *sample* are often used to refer to the sets of measurements themselves, as well as to the units on which the measurements are made. When a single variable of interest is being measured, this usage causes little confusion. But when the terminology is ambiguous, we'll refer to the measurements as *population data sets* and sample data sets, respectively.

the percentage of all the voters who would vote for each presidential candidate if the election were held on the day the poll was conducted, or he might use the results to predict the outcome on election day.

EXAMPLE 1.1

KEY ELEMENTS OF A STATISTICAL PROBLEM

Problem According to the *Atlanta Journal-Constitution* (Nov. 4, 2003), the average age of viewers of MSNBC cable television news programming is 58 years. Suppose a rival network executive hypothesizes that the average age of MSNBC news viewers is less than 58. To test her hypothesis, she samples 500 MSNBC news viewers and determines the age of each.

   a. Describe the population.

   b. Describe the variable of interest.

   c. Describe the sample.

   d. Describe the inference.

Solution
   a. The population is the set of units of interest to the cable TV executive, which is the set of all MSNBC news viewers.

   b. The age (in years) of each viewer is the variable of interest.

   c. The sample must be a subset of the population. In this case, it is the 500 MSNBC news viewers selected by the executive.

   d. The inference of interest involves the *generalization* of the information contained in the sample of 500 viewers to the population of all MSNBC viewers. In particular, the executive wants to *estimate* the average age of the viewers in order to determine whether it is less than 58 years. She might accomplish this by calculating the average age in the sample and using the sample average to estimate the population average.

Look Back A key to diagnosing a statistical problem is to identify the data set collected (in this example, the ages of the 500 cable viewers) as a population or sample.

■ ■ ■

EXAMPLE 1.2

KEY ELEMENTS OF A STATISTICAL PROBLEM

Problem "Cola wars" is the popular term for the intense competition between Coca-Cola and Pepsi displayed in their marketing campaigns. Their campaigns have featured movie and television stars, rock videos, athletic endorsements, and claims of consumer preference based on taste tests. Suppose, as part of a Pepsi marketing campaign, 1,000 cola consumers are given a blind taste test (i.e., a taste test in which the two brand names are disguised). Each consumer is asked to state a preference for brand A or brand B.

   a. Describe the population.

   b. Describe the variable of interest.

   c. Describe the sample.

   d. Describe the inference.

Solution

a. Since we are interested in the responses of cola consumers in a taste test, a cola consumer is the experimental unit. Thus, the population of interest is the collection or set of all consumers.

b. The characteristic that Pepsi wants to measure is the consumer's cola preference as revealed under the conditions of a blind taste test, so *cola preference* is the variable of interest.

c. The sample is the 1,000 cola consumers selected from the population of all cola consumers.

d. The inference of interest is the *generalization* of the cola preferences of the 1,000 sampled consumers to the population of all cola consumers. In particular, the preferences of the consumers in the sample can be used to *estimate* the percentage of all cola consumers who prefer each brand.

Look Back In determining whether the study is inferential or descriptive, we assess whether Pepsi is interested in the responses of only the 1,000 sampled customers (descriptive statistics) or in the responses for the entire population of consumers (inferential statistics).

Now Work   *Exercise 1.14b*

■ ■ ■

The preceding definitions and examples identify four of the five elements of an inferential statistical problem: a population, one or more variables of interest, a sample, and an inference. But making the inference is only part of the story. We also need to know its **reliability** — that is, how good the inference is. The only way we can be certain that an inference about a population is correct is to include the entire population in our sample. However, because of *resource constraints* (i.e., insufficient time and/or money), we usually can't work with whole populations, so we base our inferences on just a portion of the population (a sample). Thus, we introduce an element of *uncertainty* into our inferences. Consequently, whenever possible, it is important to determine and report the reliability of each inference made. Reliability, then, is the fifth element of inferential statistical problems.

The measure of reliability that accompanies an inference separates the science of statistics from the art of fortune-telling. A palm reader, like a statistician, may examine a sample (your hand) and make inferences about the population (your life). However, unlike statistical inferences, the palm reader's inferences include no measure of reliability.

Suppose, like the cable TV executive in Example 1.1, we are interested in the *error of estimation* (i.e., the difference between the average age of a population of cable TV viewers and the average age of a sample of viewers). Using statistical methods, we can determine a *bound on the estimation error*. This bound is simply a number that our estimation error is not likely to exceed. We'll see in later chapters that this bound is a measure of the uncertainty of our inference. The reliability of statistical inferences is discussed throughout this text. For now, we simply want you to realize that an inference is incomplete without a measure of its reliability.

DEFINITION 1.9

A **measure of reliability** is a statement (usually quantified) about the degree of uncertainty associated with a statistical inference.

Let's conclude this section with a summary of the elements of both descriptive and inferential statistical problems and an example to illustrate a measure of reliability.

Four Elements of Descriptive Statistical Problems

1. The population or sample of interest
2. One or more variables (characteristics of the population or sample units) that are to be investigated
3. Tables, graphs, or numerical summary tools
4. Identification of patterns in the data

Five Elements of Inferential Statistical Problems

1. The population of interest
2. One or more variables (characteristics of the population units) that are to be investigated
3. The sample of population units
4. The inference about the population based on information contained in the sample
5. A measure of reliability for the inference

EXAMPLE 1.3   RELIABILITY OF AN INFERENCE

Problem   Refer to Example 1.2, in which the cola preferences of 1,000 consumers were indicated in a taste test. Describe how the reliability of an inference concerning the preferences of all cola consumers in the Pepsi bottler's marketing region could be measured.

Solution   When the preferences of 1,000 consumers are used to estimate those of all consumers in the region, the estimate will not exactly mirror the preferences of the population. For example, if the taste test shows that 56% of the 1,000 consumers preferred Pepsi, it does not follow (nor is it likely) that exactly 56% of all cola drinkers in the region prefer Pepsi. Nevertheless, we can use sound statistical reasoning (which we'll explore later in the text) to ensure that the sampling procedure will generate estimates that are almost certainly within a specified limit of the true percentage of all consumers who prefer Pepsi. For example, such reasoning might assure us that the estimate of the preference for Pepsi is almost certainly within 5% of the actual population preference. The implication is that the actual preference for Pepsi is between 51% [i.e., $(56 - 5)$%] and 61% [i.e., $(56 + 5)$%] — that is, (56 ± 5)%. This interval represents a measure of reliability for the inference.

Look Back   The interval 56 ± 5 is called a *confidence interval*, since we are confident that the true percentage of customers that prefer Pepsi in a taste test falls into the range (51, 61). In Chapter 7, we learn how to assess the degree of confidence (e.g., 90% or 95% level of confidence) in the interval.

■ ■ ■

Statistics in Action Revisited
Identifying the Population, Sample, and Inference

Consider the 2003 Teen Survey conducted by *USA Weekend* magazine. One of the survey questions is, "Which quality of your sex first catches the attention of the opposite sex?" The experimental unit for the study is a teenager (the person answering the question), and the variable measured is the response ("looks," "personality," "athleticism," "confidence," "sense of humor," "intelligence," or "popularity") to the question. The magazine reported that approximately 37,000 teens participated in the study. Obviously, this is not all of the teenagers in the United States. Consequently, the 37,000 responses represent a sample selected from the much larger population of all American teenagers.

USA Weekend reported that 79% of the girls felt that their "Looks" is the quality that first catches the attention of teenage boys, while only 56% of the boys felt that their "Looks" was the first thing noticed by girls. This large discrepancy in percentages (as well as some others) led the magazine to infer that "boys are from Mars, girls are from Venus."

1.4 Types of Data

You have learned that statistics is the science of data and that data are obtained by measuring the values of one or more variables on the units in the sample (or population). All data (and hence the variables we measure) can be classified as one of two general types: *quantitative data* and *qualitative data*.

Quantitative data are data that are measured on a naturally occurring numerical scale.* The following are examples of quantitative data:

1. The temperature (in degrees Celsius) at which each in a sample of 20 pieces of heat-resistant plastic begins to melt

2. The current unemployment rate (measured as a percentage) for each of the 50 states

3. The scores of a sample of 150 law school applicants on the LSAT, a standardized law school entrance exam administered nationwide

4. The number of convicted murderers who receive the death penalty each year over a 10-year period

> **DEFINITION 1.10**
>
> **Quantitative data** are measurements that are recorded on a naturally occurring numerical scale.

In contrast, qualitative data cannot be measured on a natural numerical scale; they can only be classified into categories.† Examples of qualitative data include

1. The political party affiliation (Democrat, Republican, or Independent) in a sample of 50 voters

2. The defective status (defective or not) of each of 100 computer chips manufactured by Intel

---

*Quantitative data can be subclassified as either *interval data* or *ratio data*. For ratio data, the origin (i.e., the value 0) is a meaningful number. But the origin has no meaning with interval data. Consequently, we can add and subtract interval data, but we can't multiply and divide them. Of the four quantitative data sets listed above, (1) and (3) are interval data, while (2) and (4) are ratio data.
†Qualitative data can be subclassified as either *nominal data* or *ordinal data*. The categories of an ordinal data set can be ranked or meaningfully ordered, but the categories of a nominal data set can't be ordered. Of the four qualitative data sets listed above, (1) and (2) are nominal and (3) and (4) are ordinal.

3. The size of a car (subcompact, compact, mid-size, or full-size) rented by each of a sample of 30 business travelers

4. A taste-tester's ranking (best, worst, etc.) of four brands of barbecue sauce for a panel of 10 testers

Often, we assign arbitrary numerical values to qualitative data for ease of computer entry and analysis. But these assigned numerical values are simply codes: They cannot be meaningfully added, subtracted, multiplied, or divided. For example, we might code Democrat = 1, Republican = 2, and Independent = 3. Similarly, a taste tester might rank the barbecue sauces from 1 (best) to 4 (worst). These are simply arbitrarily selected numerical codes for the categories and have no utility beyond that.

DEFINITION 1.11

Qualitative data are measurements that cannot be measured on a natural numerical scale; they can only be classified into one of a group of categories.

EXAMPLE 1.4   DATA TYPES

Problem   Chemical and manufacturing plants often discharge toxic-waste materials such as DDT into nearby rivers and streams. These toxins can adversely affect the plants and animals inhabiting the river and the river bank. The U.S. Army Corps of Engineers conducted a study of fish in the Tennessee River (in Alabama) and its three tributary creeks: Flint Creek, Limestone Creek, and Spring Creek. A total of 144 fish were captured, and the following variables were measured for each:

1. River/creek where each fish was captured
2. Species (channel catfish, largemouth bass, or smallmouth buffalofish)
3. Length (centimeters)
4. Weight (grams)
5. DDT concentration (parts per million)

Classify each of the five variables measured as quantitative or qualitative.

Solution   The variables length, weight, and DDT concentration are quantitative because each is measured on a numerical scale: length in centimeters, weight in grams, and DDT in parts per million. In contrast, river/creek and species cannot be measured quantitatively: They can only be classified into categories (e.g., channel catfish, largemouth bass, and smallmouth buffalofish for species). Consequently, data on river/creek and species are qualitative.

Look Back   It is essential that you understand whether data is quantitative or qualitative in nature, since the statistical method appropriate for describing, reporting, and analyzing the data depends on the data type (quantitative or qualitative).

Now Work   *Exercise 1.12*

---   ■ ■ ■   ---

We demonstrate many useful methods for analyzing quantitative and qualitative data in the remaining chapters of the text. But first, we discuss some important ideas on data collection.

1.5 Collecting Data

Once you decide on the type of data — quantitative or qualitative — appropriate for the problem at hand, you'll need to collect the data. Generally, you can obtain data in four different ways:

1. Data from a *published source*
2. Data from a *designed experiment*
3. Data from a *survey*
4. Data from an *observational study*

Sometimes, the data set of interest has already been collected for you and is available in a **published source**, such as a book, journal, or newspaper. For example, you may want to examine and summarize the divorce rates (i.e., number of divorces per 1,000 population) in the 50 states of the United States. You can find this data set (as well as numerous other data sets) at your library in the *Statistical Abstract of the United States*, published annually by the U.S. government. Similarly, someone who is interested in monthly mortgage applications for new home construction would find this data set in the *Survey of Current Business*, another government publication. Other examples of published data sources include *The Wall Street Journal* (financial data) and *The Sporting News* (sports information). The Internet (World Wide Web) now provides a medium by which data from published sources are readily obtained.*

A second method of collecting data involves conducting a **designed experiment**, in which the researcher exerts strict control over the units (people, objects, or things) in the study. For example, a recent medical study investigated the potential of aspirin in preventing heart attacks. Volunteer physicians were divided into two groups — the *treatment* group and the *control* group. In the treatment group, each physician took one aspirin tablet a day for one year, while each physician in the control group took an aspirin-free placebo made to look like an aspirin tablet. The researchers, not the physicians under study, controlled who received the aspirin (the treatment) and who received the placebo. As you'll learn in Chapter 10, a properly designed experiment allows you to extract more information from the data than is possible with an uncontrolled study.

Surveys are a third source of data. With a **survey**, the researcher samples a group of people, asks one or more questions, and records the responses. Probably the most familiar type of survey is the political poll conducted by any one of a number of organizations (e.g., Harris, Gallup, Roper, and CNN) and designed to predict the outcome of a political election. Another familiar survey is the Nielsen survey, which provides the major networks with information on the most watched programs on television. Surveys can be conducted through the mail, with telephone interviews, or with in-person interviews. Although in-person surveys are more expensive than mail or telephone surveys, they may be necessary when complex information is to be collected.

Finally, observational studies can be employed to collect data. In an **observational study**, the researcher observes the experimental units in their natural setting and records the variable(s) of interest. For example, a child psychologist might observe and record the level of aggressive behavior of a sample of fifth graders playing on a school playground. Similarly, a zoologist may observe and measure the weights of newborn elephants born in captivity. Unlike a designed experiment, an observational study is one in which the researcher makes no attempt to control any aspect of the experimental units.

*With published data, we often make a distinction between the *primary source* and a *secondary source*. If the publisher is the original collector of the data, the source is primary. Otherwise, the data is secondary source data.

Regardless of which data collection method is employed, it is likely that the data will be a sample from some population. And if we wish to apply inferential statistics, we must obtain a *representative sample*.

> **DEFINITION 1.12**
>
> A **representative sample** exhibits characteristics typical of those possessed by the target population.

For example, consider a political poll conducted during a presidential election year. Assume the pollster wants to estimate the percentage of all 120 million registered voters in the United States who favor the incumbent President. The pollster would be unwise to base the estimate on survey data collected for a sample of voters from the incumbent's own state. Such an estimate would almost certainly be *biased* high; consequently, it would not be very reliable.

The most common way to satisfy the representative sample requirement is to select a random sample. A **random sample** ensures that every subset of fixed size in the population has the same chance of being included in the sample. If the pollster samples 1,500 of the 120 million voters in the population so that every subset of 1,500 voters has an equal chance of being selected, he has devised a random sample. The procedure for selecting a random sample is discussed in Chapter 3. Here, we look at two examples involving actual sampling studies.

> **DEFINITION 1.13**
>
> A **random sample** of n experimental units is a sample selected from the population in such a way that every different sample of size n has an equal chance of selection.

EXAMPLE 1.5   METHOD OF DATA COLLECTION

Problem   What percentage of Web users are addicted to the Internet? To find out, a psychologist designed a series of 10 questions based on a widely used set of criteria for gambling addiction and distributed them through the website *ABCNews.com*. (A sample question: "Do you use the Internet to escape problems?") A total of 17,251 Web users responded to the questionnaire. If participants answered "yes" to at least half of the questions, they were viewed as addicted. The findings, released at the 1999 annual meeting of the American Psychological Association, revealed that 990 respondents, or 5.7%, are addicted to the Internet (*Tampa Tribune*, Aug. 23, 1999).

    a. Identify the data collection method.

    b. Identify the target population.

    c. Are the sample data representative of the population?

Solution   **a.** The data collection method is a survey: 17,251 Internet users responded to the questions posed at the *ABCNews.com* website.

    b. Since the website can be accessed by anyone surfing the Internet, presumably the target population is *all* Internet users.

    c. Because the 17,251 respondents clearly make up a subset of the target population, they do form a sample. Whether or not the sample is representative is unclear, since we are given no information on the 17,251 respondents. However, a survey like this one in which the respondents are *self-selected* (i.e., each Internet user who saw the survey chose whether or not to respond to it) often suffers from *nonresponse bias*. It is possible that many Internet users who chose not to respond (or who never saw the survey) would have answered the questions differently, leading to a higher (or lower) percentage of affirmative answers.

Look Back Any inferences based on survey samples that employ self-selection are suspect due to potential nonresponse bias.

■ ■ ■

EXAMPLE 1.6   METHOD OF DATA COLLECTION

Problem Psychologists at the University of Tennessee carried out a study of the susceptibility of people to hypnosis (*Psychological Assessment*, Mar. 1995). In a random sample of 130 undergraduate psychology students at the university, each experienced both traditional hypnosis and computer-assisted hypnosis. Approximately half were randomly assigned to undergo the traditional procedure first, followed by the computer-assisted procedure. The other half were randomly assigned to experience computer-assisted hypnosis first, then traditional hypnosis. Following the hypnosis episodes, all students filled out questionnaires designed to measure a student's susceptibility to hypnosis. The susceptibility scores of the two groups of students were compared.

a. Identify the data collection method.

b. Is the sample data representative of the target population?

Solution **a.** Here, the experimental units are the psychology students. Since the researchers controlled which type of hypnosis — traditional or computer assisted — the students experienced first (through random assignment), a designed experiment was used to collect the data.

b. The sample of 130 psychology students was randomly selected from all psychology students at the University of Tennessee. If the target population is *all University of Tennessee psychology students*, it is likely that the sample is representative. However, the researchers warn that the sample data should not be used to make inferences about other, more general, populations.

Look Back By using randomization in a designed experiment, the researcher is attempting to eliminate different types of bias, including self-selection bias.

Now Work   *Exercise 1.15*

■ ■ ■

Statistics in Action Revisited
Identifying the Data Collection Method and Data Type

In the *USA Weekend* Annual Teen Survey, American teenagers are asked to respond to a variety of questions on topics such as newspapers usage (2004 survey) and the opposite sex (2003 survey). The title of the study implies that the data are collected using a survey of teenagers. The *USA Weekend* surveys were made available using printed forms in the magazine, in an online survey form, and through the mail using newspaper subscriber lists.

Both quantitative and qualitative data were collected in the surveys. For example, the survey question "Do you think its OK for girls in your grade to call guys?" is phrased to elicit a "yes" or "no" response. Since the responses produced for this question are categorical in nature, the data are qualitative. However, the question "How many days a week do you read the newspaper?" will give meaningful numerical responses such as one, two, or three days. Thus, this data is quantitative.

1.6 The Role of Statistics in Critical Thinking

According to H. G. Wells, author of such science-fiction classics as *The War of the Worlds* and *The Time Machine*, "*Statistical thinking* will one day be as necessary for efficient citizenship as the ability to read and write." Written more than a hundred years ago, Wells's prediction is proving true today.

Biography

**H. G. WELLS
(1866–1946)—
Writer and Novelist**

English-born Herbert George Wells published his first novel, *The Time Machine*, in 1895 as a parody of the English class division and as a satirical warning that human progress is inevitable. Although most famous as a science-fiction novelist, Wells was a prolific writer as a journalist, sociologist, historian, and philosopher. Wells's prediction about statistical thinking (see Definition 1.14) is just one of a plethora of observations he made about life on this world. Here are a few more of H. G. Wells's more famous quotes:

"Advertising is legalized lying."

"Crude classification and false generalizations are the curse of organized life."

"The crisis of today is the joke of tomorrow."

"Fools make researchers and wise men exploit them."

"The only true measure of success is the ration between what we might have done and what we might have been on the one hand, and the thing we have made and the things we have made of ourselves on the other."

The growth in data collection associated with scientific phenomena, business operations, and government activities (quality control, statistical auditing, forecasting, etc.) has been remarkable in the past several decades. Every day the media present us with published results of political, economic, and social surveys. In increasing government emphasis on drug and product testing, for example, we see vivid evidence of the need to be able to evaluate data sets intelligently. Consequently, each of us has to develop a discerning sense — an ability to use rational thought to interpret the meaning of data. This ability can help you make intelligent decisions, inferences, and generalizations; that is, it helps you *think critically* using statistics.

> **DEFINITION 1.14**
>
> **Statistical thinking** involves applying rational thought and the science of statistics to critically assess data and inferences.

To gain some insight into the role statistics plays in critical thinking, let's look at a *New York Times* (June 17, 1995) article titled "The Case for No Helmets." The article, written by the editor of a magazine for Harley-Davidson bikers, lists several reasons why motorcyclists should not be required by law to wear helmets. First, he argued that helmets may actually kill, since in collisions at speeds greater than 15 miles an hour, the heavy helmet may protect the head but snap the spine. Second, the editor cited a "study" that claimed "nine states without helmet laws had a lower fatality rate (3.05 deaths per 10,000 motorcycles) than those that mandated helmets (3.38)." Finally, he reported that "in a survey of 2,500 [at a rally], 98% of the respondents opposed such laws."

You can use "statistical thinking" to help you critically evaluate the arguments presented in this article. For example, before you accept the 98% estimate, you would want to know how the data were collected for the study cited by the editor of the biker magazine. If a survey was conducted, it's possible that the 2,500 bikers in the sample were not selected at random from the target population of all bikers, but rather were "self-selected." (Remember, they were all attending a rally — maybe even a rally for bikers who oppose the law.) If the respondents were likely to have strong opinions regarding the helmet law (e.g., strongly oppose the law), the resulting estimate is probably biased high. Also, if the biased sample was intentional, with the sole purpose to mislead the public, the researcher would be guilty of **unethical statistical practice**.

You'd also want more information about the study comparing the motorcycle fatality rate of the nine states without a helmet law to those states that mandate helmets: Were the data obtained from a published source? Were all 50 states included in the study? That is, are you seeing sample data or population data? Furthermore, do the helmet laws vary among states? If so, can you really compare the fatality rates?

These questions actually led to the discovery of two scientific and statistically sound studies on helmets. The first, a UCLA study of nonfatal injuries, disputed the charge that helmets shift injuries to the spine. The second study reported a dramatic *decline* in motorcycle crash deaths after California passed its helmet law.

As in the motorcycle helmet study, many statistical studies are based on survey data. Most of the problems with these surveys result from the use of *nonrandom* samples. These samples are subject to potential errors, such as *selection bias, nonresponse bias* (recall Example 1.6), and *measurement error*. Researchers who are aware of these problems and continue to use the sample data to make inferences are practicing unethical statistics.

DEFINITION 1.15

Selection bias results when a subset of the experimental units in the population is excluded so that these units have no chance of being selected in the sample.

DEFINITION 1.16

Nonresponse bias results when the researchers conducting a survey or study are unable to obtain data on all experimental units selected for the sample.

DEFINITION 1.17

Measurement error refers to inaccuracies in the values of the data recorded. In surveys, the error may be due to ambiguous or leading questions and the interviewer's effect on the respondent.

Statistics in Action Revisited
Critically Assessing the Ethics of a Statistical Study

The results from the *USA Weekend* Teen Surveys led the magazine to make conclusions such as "boys are from Mars, girls are from Venus" when relating to the opposite sex (2003 survey) and "America's teens read newspapers ... and find them relevant and reliable" (2004 survey). Both surveys were based on a large sample of teen respondents — 37,000 in 2003 and 65,000 in 2004. However, several problems lead a critical thinker to cast doubt on the validity of the inferences.

First, are the samples representative of all American teens? *USA Weekend* admits that its annual teen survey is "unscientific," meaning that the teens are not randomly selected from all teens in the United States. In fact, the teens who responded are "self-selected," choosing to respond to the survey via the printed copy in the magazine, the Internet survey, or the copy mailed to their home. Therefore, there is a strong selection bias in the survey results. Second, because not all American teens receive the magazine, were aware of the Internet survey, or were on the magazine subscriber mailing list, the potential for non-response bias is extremely high.

Finally, the wording in several of the survey questions is ambiguous or leading. This can occur when the survey is designed to support a specific point of view or organization (such as the Newspaper Association of America Foundation). For example, the question "How relevant is the news-paper to your life?" is likely to lead to the responses "very relevant" or "somewhat relevant" since the phrase "relevant … to your life" has different meanings to different people. (A teenager who delivers papers every afternoon after school but never reads the paper would likely answer "very relevant.") Ambiguities such as these will likely result in measurement error in the data.

In the remaining chapters of the text, you'll become familiar with the tools essential for building a firm foundation in statistics and statistical thinking.

Quick Review

Key Terms

Census   8
Data   4
Descriptive statistics   5
Designed experiment   14
Experimental unit   7
Inference   5
Inferential statistics   5
Measure of reliability   11
Measurement   7

Measurement error   18
Nonresponse bias   18
Observational study   14
Population   7
Published source   14
Qualitative data   13
Quantitative data   12
Random sample   15
Reliability   10
Representative sample   15

Sample   8
Selection bias   18
Statistical inference   9
Statistical thinking   17
Statistics   4
Survey   14
Unethical Statistical Practice   18
Variable   7

Chapter Summary

- Two types of statistical applications: **descriptive** and **inferential**
- Fundamental elements of statistics: **population, experimental units, variable, sample, inference, measure of reliability**

- Two types of data: **quantitative** and **qualitative**
- Data collection methods: **published source, designed experiment, survey**, and **observationally**
- Potential problems with nonrandom samples: **selection bias, nonresponse bias, measurement error**.

Exercises 1.1–1.32

Understanding the Principles

1.1   What is statistics?

1.2   Explain the difference between descriptive and inferential statistics.

1.3   List and define the five elements of an inferential statistical analysis.

1.4   List the four major methods of collecting data and explain their differences.

1.5   Explain the difference between quantitative and qualitative data.

1.6   Explain how populations and variables differ.

1.7   Explain how populations and samples differ.

1.8   What is a representative sample? What is its value?

1.9   Why would a statistician consider an inference incomplete without an accompanying measure of its reliability?

1.10   Define statistical thinking.

1.11   Suppose you're given a data set that classifies each sample unit into one of four categories: A, B, C, or D. You plan to create a computer database consisting of these data, and you decide to code the data as A = 1, B = 2, C = 3, and D = 4. Are the data consisting of the classifications A, B, C, and D qualitative or quantitative? After the data are input as 1, 2, 3, or 4, are they qualitative or quantitative? Explain your answers.

Applying the Concepts—Basic

1.12 College application. Colleges and universities are
NW requiring an increasing amount of information about
applicants before making acceptance and financial
aid decisions. Classify each of the following types of
data required on a college application as quantitative
or qualitative.

a. High school GPA
b. High school class rank
c. Applicant's score on the SAT or ACT
d. Gender of applicant
e. Parents' income
f. Age of applicant

1.13 Miscellaneous data. Classify the following examples of
data as either qualitative or quantitative:

a. The bacteria count in the water at each of 30 city
swimming pools
b. The occupation of each of 200 shoppers at a supermarket
c. The marital status of each person living on a city
block
d. The time (in months) between auto maintenance for
each of 100 used cars

1.14 Sprint speed training. Refer to *The Sport Journal* (Winter, 2004) study of a speed training program for high
school football players in Section 1.2 (Study 1, p. 5).
Recall that each participant was timed in a 40-yard
sprint both before and after training. The researchers
measured two variables: (1) the difference between the
before and after sprint times (in seconds), and (2) the
category of improvement ("improved," "no change,"
and "worse") for each player.

a. Identify the type (quantitative or qualitative) of each
variable measured.
NW b. A total of 14 high school football players participated in the speed training program. Does the data
set collected represent a population or a sample?
Explain.

1.15 Iraq War poll. A poll is to be conducted in which 2,000
NW individuals are asked whether the United States was
justified in invading Iraq in 2003. The 2,000 individuals
are selected by random-digit telephone dialing and
asked the question over the phone.

a. What is the relevant population?
b. What is the variable of interest? Is it quantitative or
qualitative?
c. What is the sample?
d. What is the inference of interest to the pollster?
e. What method of data collection is employed?

f. How likely is the sample to be representative?

1.16 Student GPAs. Consider the set of all students enrolled
in your statistics course this term. Suppose you're
interested in learning about the current grade point
averages (GPAs) of this group.

a. Define the population and variable of interest.
b. Is the variable qualitative or quantitative?

c. Suppose you determine the GPA of every member
of the class. Would this represent a census or a
sample?
d. Suppose you determine the GPA of 10 members of
the class. Would this represent a census or a sample?
e. If you determine the GPA of every member of the
class and then calculate the average, how much reliability does this have as an "estimate" of the class
average GPA?
f. If you determine the GPA of 10 members of the class
and then calculate the average, will the number you
get necessarily be the same as the average GPA for
the whole class? On what factors would you expect
the reliability of the estimate to depend?
g. What must be true in order for the sample of 10 students you select from your class to be considered a
random sample?

1.17 The Executive Compensation Scoreboard. Each year,
Business Week publishes its "Executive Compensation
Scoreboard." Data are collected for a sample of chief
executive officers and the following variables are measured for each CEO: (1) the industry type of the CEO's
company (e.g., banking, consumer products, etc.), (2) the
CEO's total compensation (in $ millions) for the year,
(3) the CEO's total compensation (in $ millions) over
the previous three years, and (4) the return (in dollars)
on a $100 investment in the CEO's company made in
three years earlier.

a. Describe the population of interest to *Business Week*.
b. Identify the type (quantitative or qualitative) of each
variable measured.
c. In 2003, *Business Week* surveyed 370 CEOs from the
top-ranked companies listed in the *Business Week
1000*. Discuss any biases that may exist in the sample.

1.18 Extinct birds. Biologists at the University of California
(Riverside) are studying the patterns of extinction in the
New Zealand bird population. (*Evolutionary Ecology
Research*, July 2003.) At the time of the Maori coloniza-
tion of New Zealand (prior to European contact), the
following variables were measured for each bird species.
Identify each variable as quantitative or qualitative.

a. Flight capability (volant or flightless)
b. Habitat type (aquatic, ground terrestrial, or aerial
terrestrial)
c. Nesting site (ground, cavity within ground, tree, cavity above ground)
d. Nest density (high or low)
e. Diet (fish, vertebrates, vegetables, or invertebrates)

f. Body mass (grams)
g. Egg length (millimeters)
h. Extinct status (extinct, absent from island, present)

1.19 E-commerce survey. The Cutter Consortium surveyed
154 U.S. companies to determine the extent of their
involvement in electronic commerce (called *e-commerce*).
Four of the questions they asked follow. (*Internet Week*,
Sept. 6, 1999.)

1. Do you have an overall e-commerce strategy?
2. If you don't already have an e-commerce plan, when will you implement one?
3. Are you delivering products over the Internet?
4. What was your company's total revenue in the last fiscal year?
   a. For each question, determine the variable of interest and classify it as quantitative or qualitative.
   b. Do the data collected for the 154 companies represent a sample or a population? Explain.

Applying the Concepts—Intermediate

1.20 Herbal Medicines. *The American Association of Nurse Anesthetists Journal* (Feb. 2000) published the results of a study on the use of herbal medicines before surgery. Of 500 surgical patients that were randomly selected for the study, 51% used herbal or alternative medicines (e.g., garlic, ginkgo, kava, fish oil) against their doctor's advice prior to surgery.
   a. Do the 500 surgical patients represent a population or a sample? Explain.
   b. If your answer was sample in part **a**, is the sample representative of the population? If you answered population in part **a**, explain how to obtain a representative sample from the population.
   c. For each patient, what variable is measured? Is the data collected quantitative or qualitative?

1.21 National Bridge Inventory. All highway bridges in the United States are inspected periodically for structural deficiency by the Federal Highway Administration (FHWA). Data from the FHWA inspections are compiled into the National Bridge Inventory (NBI). Several of the nearly 100 variables maintained by the NBI are listed below. Classify each variable as quantitative or qualitative.
   a. Length of maximum span (feet)
   b. Number of vehicle lanes
   c. Toll bridge (yes or no)
   d. Average daily traffic
   e. Condition of deck (good, fair, or poor)
   f. Bypass or detour length (miles)
   g. Route type (interstate, U.S., state, county, or city)

1.22 Structurally deficient bridges. Refer to Exercise 1.21. The most recent NBI data were analyzed, and the results published in the *Journal of Infrastructure Systems* (June 1995). Using the FHWA inspection ratings, each of the 470,515 highway bridges in the United States was categorized as structurally deficient, functionally obsolete, or safe. About 26% of the bridges were found to be structurally deficient, while 19% were functionally obsolete.
   a. What is the variable of interest to the researchers?
   b. Is the variable of part **a** quantitative or qualitative?
   c. Is the data set analyzed a population or a sample? Explain.
   d. How did the researchers obtain the data for their study?

1.23 CT scanning for lung cancer. According to the American Lung Association, lung cancer accounts for 28% of all cancer deaths in the United States. A new type of screening for lung cancer, computed tomography (CT), has been developed. Medical researchers believe CT scans are more sensitive than regular X-rays in pinpointing small tumors. The H. Lee Moffitt Cancer Center at the University of South Florida is currently conducting a clinical trial of 50,000 smokers nationwide to compare the effectiveness of CT scans with X-rays for detecting lung cancer. (*Todays' Tomorrows*, Fall 2002.) Each participating smoker is randomly assigned to one of two screening methods, CT or chest X-ray, and their progress tracked over time. The age at which the scanning method first detects a tumor is the variable of interest.
   a. Identify the data collection method used by the cancer researchers.
   b. Identify the experimental units of the study.
   c. Identify the type (quantitative or qualitative) of the variable measured.
   d. Identify the population and sample.
   e. What is the inference that will ultimately be drawn from the clinical trial?

1.24 Gallup poll of teenagers. A Gallup Youth Poll was conducted to determine the topics that teenagers most want to discuss with their parents. The findings show that 46% would like more discussion about the family's financial situation, 37% would like to talk about school, and 30% would like to talk about religion. The survey was based on a national sampling of 505 teenagers, selected at random from all U.S. teenagers.
   a. Describe the sample.
   b. Describe the population from which the sample was selected.
   c. Is the sample representative of the population?
   d. What is the variable of interest?
   e. How is the inference expressed?
   f. Newspaper accounts of most polls usually give a *margin of error* (i.e., plus or minus 3%) for the survey result. What is the purpose of the margin of error and what is its interpretation?

1.25 Massage vs. rest in boxing. Does a massage enable the muscles of tired athletes to recover from exertion faster than usual? To answer this question, researchers recruited eight amateur boxers to participate in an experiment. (*British Journal of Sports Medicine*, April 2000.) After a 10-minute workout in which each boxer threw 400 punches, half the boxers were given a 20-minute massage and half just rested for 20 minutes. Before returning to the ring for a second workout, the heart rate (beats per minute) and blood lactate level (micromoles) were recorded for each boxer. The researchers found no difference in the means of the two groups of boxers for either variable.
   a. Identify the data collection method used by the researchers.
   b. Identify the experimental units of the study.

c. Identify the variables measured and their type (quantitative or qualitative).

d. What is the inference drawn from the analysis?

e. Comment on whether this inference can be made about all athletes.

1.26 Acid neutralizer experiment. A first-year chemistry student conducts an experiment to determine the amount of hydrochloric acid necessary to neutralize 2 milliliters (ml) of a basic solution. The student prepares five 2-ml portions of the solution and adds a known concentration of hydrochloric acid to each. The amount of acid necessary to achieve neutrality of the solution is recorded for each of the five portions.

a. Describe the population of interest to the student.

b. What is the variable of interest? Is it quantitative or qualitative?

c. Describe the sample.

d. Describe the data collection method.

1.27 Blood loss in burn patients. A group of University of South Florida surgery researchers believe that the drug aprotinin is effective in reducing the blood loss of burn patients who undergo skin replacement surgery. (*USF Office of Research Annual Report, 1997–1998.*) In a clinical trial of 14 burn patients, half were randomly assigned to receive the drug and half a placebo (no drug). A preliminary analysis of the patients' blood loss revealed that the group with the drug had a significant reduction of bleeding.

a. Identify the data collection method used for this study.

b. Does the study involve descriptive or inferential statistics? Explain.

c. What is the population (or sample) of interest to the researchers?

Applying the Concepts—Advanced

1.28 Object recall study. Are men or women more adept at remembering where they leave misplaced items (like car keys)? According to University of Florida psychology professor Robin West, women show greater competence in actually finding these objects (*Explore*, Fall 1998). Approximately 300 men and women from Gainesville, Florida, participated in a study in which each person placed 20 common objects in a 12-room "virtual" house represented on a computer screen. Thirty minutes later, the subjects were asked to recall where they put each of the objects. For each object, a recall variable was measured as "yes" or "no."

a. Identify the population of interest to the psychology professor.

b. Identify the sample.

c. Does the study involve descriptive or inferential statistics? Explain.

d. Are the variables measured in the study quantitative or qualitative?

1.29 Soccer "headers" and IQ. *USA Today* (Aug. 14, 1995) reported on a study that suggests "frequently 'heading' the ball in soccer lowers players' IQs." A psychologist tested 60 male soccer players, ages 14–29, who played up to five times a week. Players who averaged 10 or more headers a game had an average IQ of 103, while players who headed one or fewer times per game had an average IQ of 112.

a. Describe the population of interest to the psychologist.

b. Identify the variables of interest.

c. Identify the type (qualitative or quantitative) of the variables, part **b**.

d. Describe the sample.

e. What is the inference made by the psychologist?

f. Discuss possible reasons why the inference, part **e**, may be misleading.

Critical Thinking Challenges

1.30 "20/20" survey exposés. The popular prime-time ABC television program "20/20" presented several misleading (and possibly unethical) surveys in a segment titled "Facts or Fiction? — Exposés of So-Called Surveys" (March 31, 1995). The information reported from four of these surveys, conducted by businesses or special interest groups with specific objectives in mind, are given. (Actual survey facts are provided in parentheses.)

Quaker Oats Study: Eating oat bran is a cheap and easy way to reduce your cholesterol count. (*Fact*: Diet must consist of nothing but oat bran to achieve a slightly lower cholesterol count.)

March of Dimes Report: Domestic violence causes more birth defects than all medical issues combined. (*Fact*: No study — false report.)

American Association of University Women (AAUW) study: Only 29% of high school girls are happy with themselves, compared to 66% of elementary school girls. (*Fact*: Of 3,000 high school girls, 29% responded "Always true" to the statement, "I am happy the way I am." Most answered, "Sort of true" and "Sometimes true.")

Food Research and Action Center study: One in four American children under age 12 is hungry or at risk of hunger. (*Fact*: Based on responses to questions: "Do you ever cut the size of meals?", "Do you ever eat less than you feel you should?", "Did you ever rely on limited numbers of foods to feed your children because you were running out of money to buy food for a meal?")

a. Refer to the Quaker Oats study relating oat bran to cholesterol levels. Discuss why it is unethical to report the results as stated.

b. Consider the false March of Dimes report on domestic violence and birth defects. Discuss the type of data required to investigate the impact of domestic violence on birth defects. What data collection method would you recommend?

c. Refer to the AAUW study of self-esteem of high school girls. Explain why the results of the study are likely to be misleading. What data might be appropriate for assessing the self-esteem of high school girls?

d. Refer to the Food Research and Action Center study of hunger in America. Explain why the results of the study are likely to be misleading. What data would provide insight into the proportion of hungry American children?

1.31 Roper poll on the Holocaust. The Holocaust — the Nazi Germany extermination of the Jews during World War II — has been well documented through film, books, and interviews with the concentration camp survivors. But a recent Roper poll found that one in five U.S. residents doubted the Holocaust really occurred. A national sample of more than 1,000 adults and high school students were asked, "Does it seem possible or does it seem impossible to you that the Nazi extermination of the Jews never happened?" Twenty-two percent of the adults and 20% of the high school students responded "Yes." (*New York Times*, May 16, 1994.) Discuss reasons for doubting the reliability of the inference.

1.32 Poll on alien spacecraft. "Have you ever seen anything that you believe was a spacecraft from another planet?" This was the question put to 1,500 American adults in a national poll conducted by ABC News and *The Washington Post*. The pollsters used random-digit telephone dialing to contact adult Americans, until 1,500 responded. Ten percent (i.e., 150) of the respondents answered that they had, in fact, seen an alien spacecraft. (*Chance*, Summer 1997.) No information was provided on how many adults were called and, for one reason or another, did not answer the question.
a. Identify the data collection method.
b. Identify the target population.
c. Comment on the validity of the survey results.

Student Projects

Scan your daily newspaper or weekly news magazine or search the Internet for articles that contain numerical data. The data might be a summary of the results of a public opinion poll, the results of a vote by the U.S. Senate, or a list of crime rates, birth or death rates, etc. For each article you find, answer the following questions:

a. Do the data constitute a sample or an entire population? If a sample has been taken, clearly identify both the sample and the population; otherwise, identify the population.

b. What type of data (quantitative or qualitative) has been collected?

c. What is the data source?

d. If a sample has been observed, is it likely to be representative of the population?

e. If a sample has been observed, does the article present an explicit (or implied) inference about the population of interest? If so, state the inference made in the article.

f. If an inference has been made, has a measure of reliability been included? What is it?

g. Use your answers to questions d–f to critically evaluate the article.

REFERENCES

Brochures about Survey Research, Section on Survey Research Methods, American Statistical Association, 2004. (www.amstat.org)

Careers in Statistics, American Statistical Association, Biometric Society, Institute of Mathematical Statistics and Statistical Society of Canada, 2004. (www.amstat.org)

Cochran, W. G. *Sampling Techniques*, 3rd ed. New York: Wiley, 1977.

Deming, W. E. *Sample Design in Business Research*. New York: Wiley, Inc., 1963.

Ethical Guidelines for Statistical Practice. American Statistical Association, 1995.

Hansen, M. H., Hurwitz, W. N., and Madow, W. G. *Sample Survey Methods and Theory*, Vol. 1. New York: Wiley, 1953.

Kirk, R. E., ed. *Statistical Issues: A Reader for the Behavioral Sciences*. Monterey, Calif.: Brooks/Cole, 1972.

Kish, L. *Survey Sampling*. New York: Wiley, 1965.

Scheaffer, R., Mendenhall, W., and Ott, R. L. *Elementary Survey Sampling*, 2nd ed. Boston: Duxbury, 1979.

Tanur, J. M., Mosteller, F., Kruskal, W. H., Link, R. E., Pieters, R. S., and Rising, G. R. *Statistics: A Guide to the Unknown* (E. L. Lehmann, special editor). San Francisco: Holden-Day, 1989.

Yamane, T. *Elementary Sampling Theory*, 3rd ed. Englewood Cliffs, N.J.: Prentice Hall, 1967.

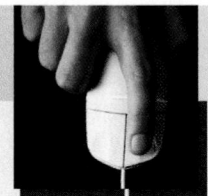

Using Technology

Creating and Listing Data Using MINITAB

Upon entering into a MINITAB session, you will see a screen similar to Figure 1.M.1. The bottom portion of the screen is an empty spreadsheet — called a MINITAB worksheet — with columns representing variables and rows representing observations (or cases). The very top of the screen is the MINITAB main menu bar, with buttons for the different functions and procedures available in MINITAB. Once you have entered data into the spreadsheet, you can analyze the data by clicking the appropriate menu buttons. The results will appear in the Session window.

Figure 1.M.1
Initial Screen Viewed by MINITAB User

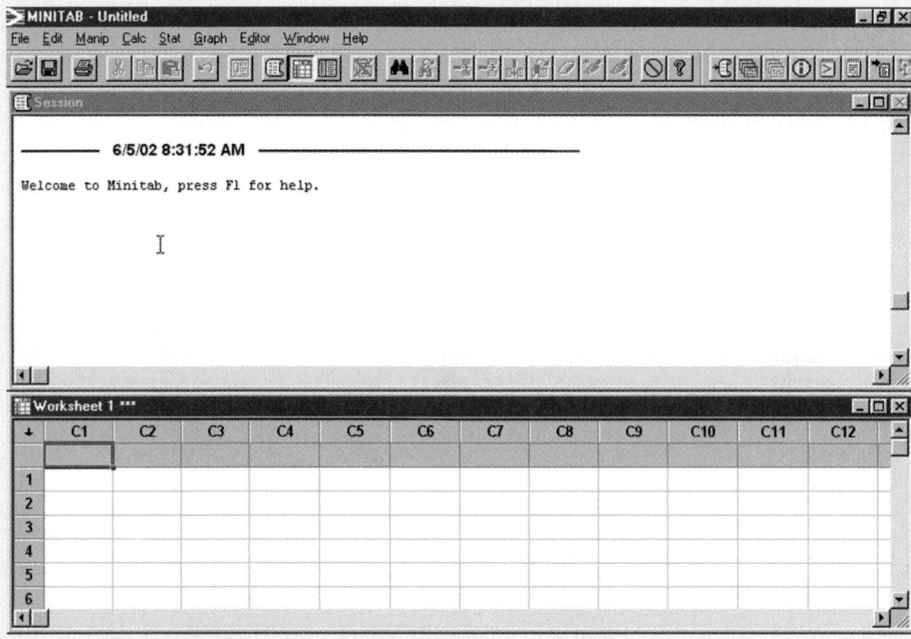

You can create a MINITAB data file by entering data directly into the worksheet. Figure 1.M.2 shows data entered for a variable called "Ratio". The variables (columns) can be named by typing in the name of each variable in the box below the column number.

If the data are saved in an external data file, you can access it using the options available in MINITAB. Click the "File" button on the menu bar, then click "Open Worksheet", as shown in Figure 1.M.3. The dialog box shown in Figure 1.M.4 will appear.

Specify the disk drive and folder that contain the external data file and the file type, then click on the file name, as shown in Figure 1.M.4. If the data set contains qualitative data or data with special characters, click on the "Options" button as shown in Figure 1.M.4. The Options dialog box, shown in Figure 1.M.5, will appear. Specify the appropriate options for the data set, then click "OK" to return

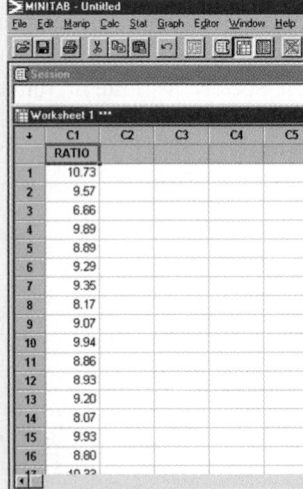

Figure 1.M.2
Data Entered into the MINITAB Worksheet

Figure 1.M.3

MINITAB Options for
Reading Data from an
External File

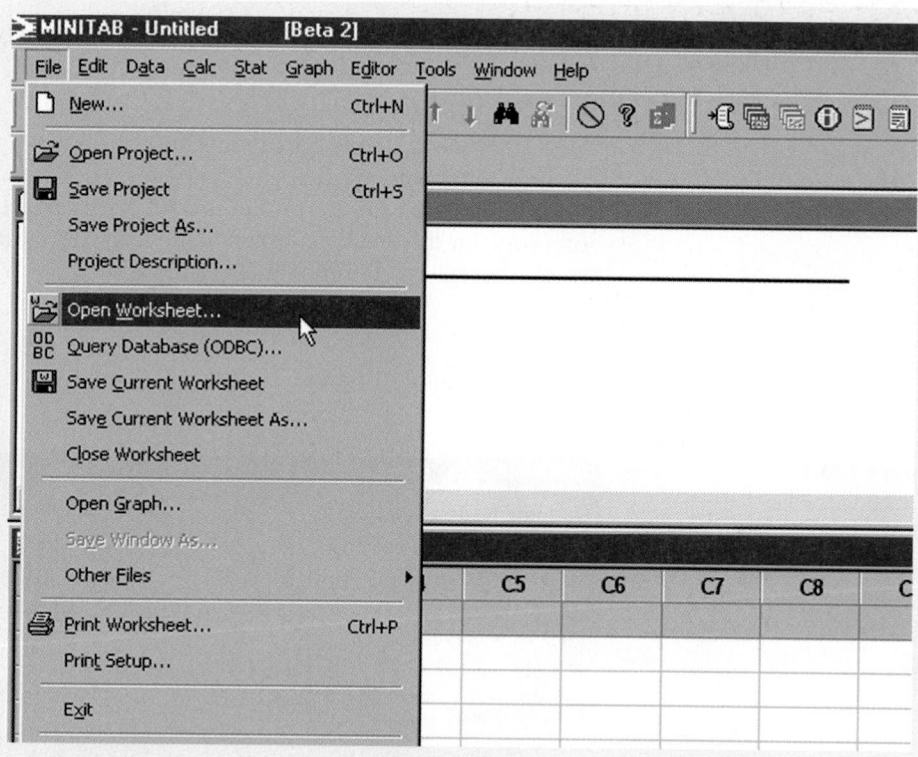

to the "Open Worksheet" dialog box (Figure 1.M.4). Click "Open" and the
MINITAB worksheet will appear with the data from the external data file, as
shown in Figure 1.M.6.

Figure 1.M.4

Selecting the External
Data File

Reminder. The variables (columns) can be named by typing in the name of each variable in the box under the column number.

To obtain a listing (printout) of your data, click on the "Data" button on the MINITAB main menu bar, then click on "Display Data" (see Figure 1.M.7). The resulting menu, or dialog box, appears as in Figure 1.M.8. Enter the names of the variables you want to print in the "Columns constants and matrices to display" box (you can do this by simply clicking on the variables), then click "OK". The printout will show up on your MINITAB session screen.

Figure 1.M.5
Selecting the Data
Input Options

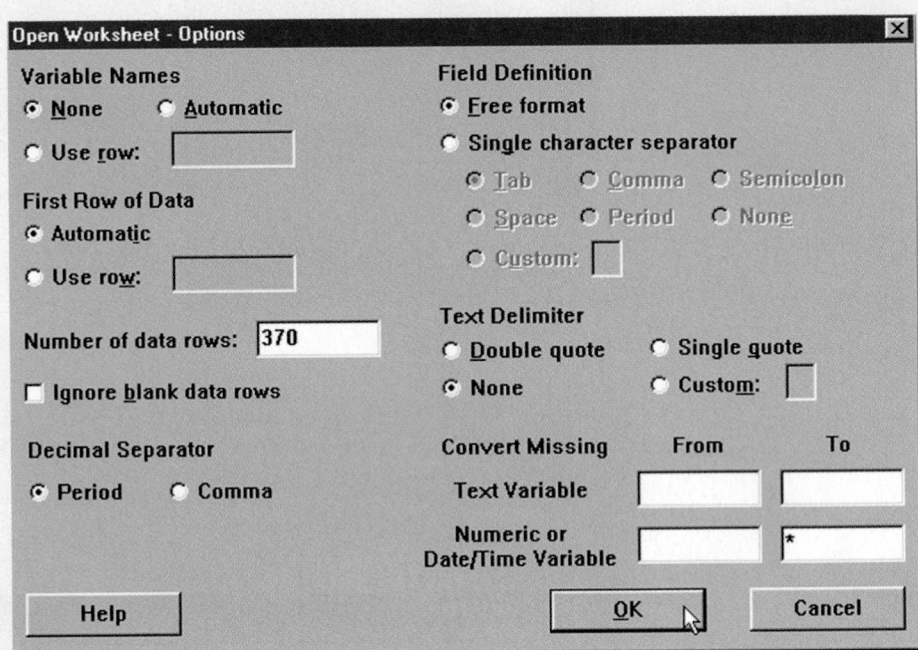

Figure 1.M.6
The MINITAB Worksheet
with the Imported Data

↓	C1-T	C2-T	C3-T	C4-T	C5	C6	C7	C8	C9	C10
1	AEROSPACE	IHT	Boeing	P.M.Condit	4145	27794	83	3		
2	AEROSPACE	IHT	GeneralDynamics	N.D.Chabraja	15245	31367	157	4		
3	AEROSPACE	IHT	Goodrich	D.L.Burner	2884	7449	77	4		
4	AEROSPACE	IHT	LockheedMartin	V.D.Coffman	25337	47702	274	4		
5	AEROSPACE	IHT	NorthropGrumman	K.Kresa	9222	23927	190	4		
6	AEROSPACE	IHT	Raytheon	D.P.Burnham	8922	19591	125	4		
7	AEROSPACE	IHT	UnitedTech.	G.David	9664	51136	99	3		
8	AIRLINES	TRAN	DeltaAirLines	L.F.Mullin	4688	9931	25	1		
9	APPLIANCES	CP	Leggett&Platt	F.E.Wright	1096	5017	112	4		
10	APPLIANCES	CP	Maytag	R.F.Hake	1703	0	63	0		
11	APPLIANCES	CP	Whirlpool	D.R.Whitwam	4819	11381	86	3		
12	AUTOMOTIVE	CP	Dana	J.M.Magliochetti	1433	4416	44	3		

bwecstest.dat ***

Figure 1.M.7
MINITAB Menu Options
for Obtaining a List of
Your Data

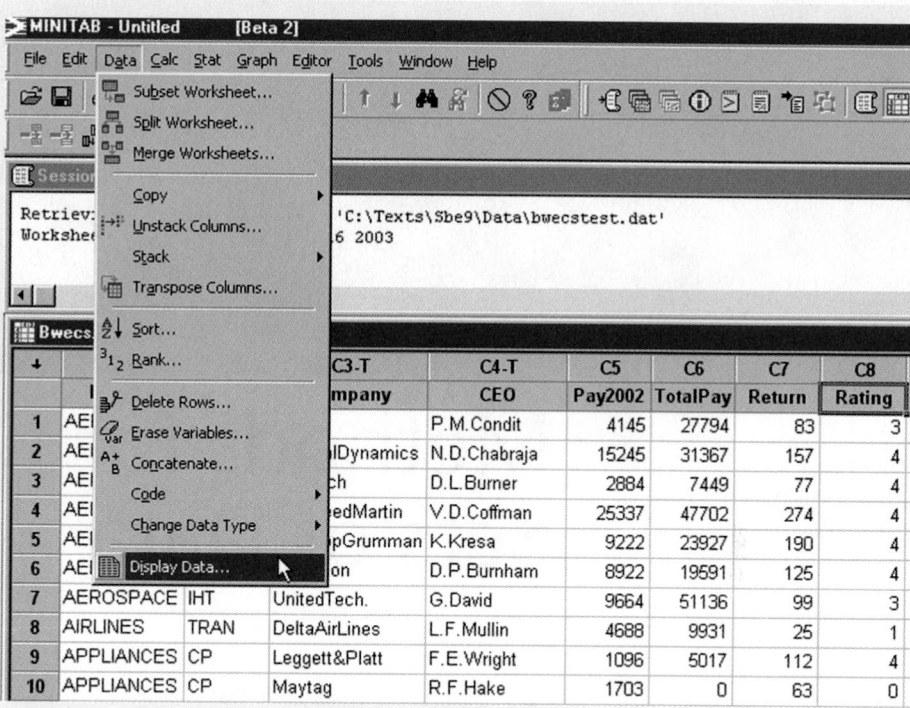

Figure 1.M.8
Display Data Dialog Box

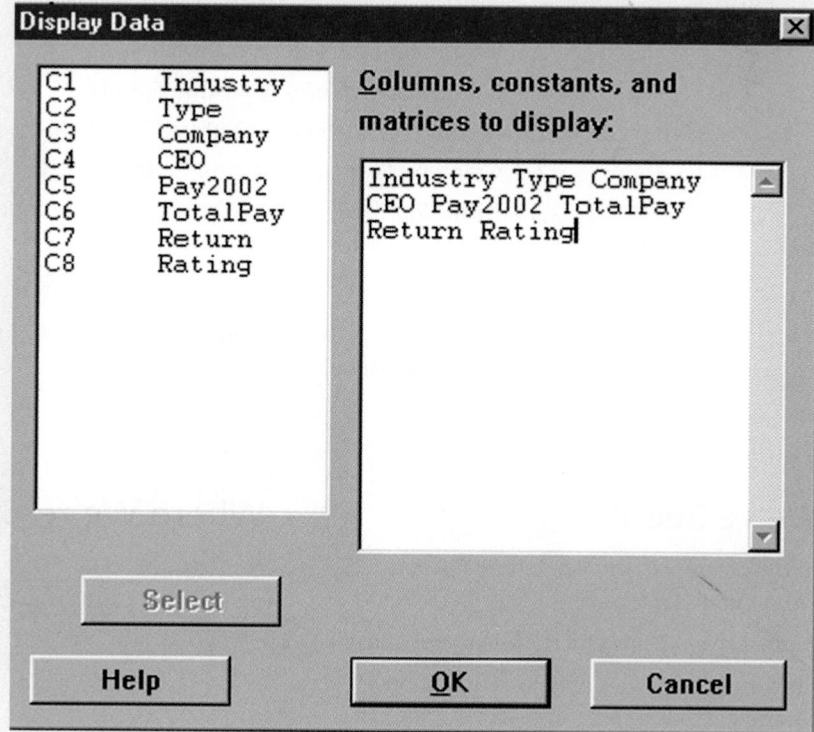

2

Methods for Describing Sets of Data

Contents

Statistics in Action

The "Eye Cue" Test: Does Experience Improve Performance?

Using Technology

Describing Data Using MINITAB

Where We've Been

- Examined the difference between inferential and descriptive statistics
- Described the key elements of a statistical problem
- Learned about the two types of data—quantitative and qualitative
- Discussed the role of statistical thinking in managerial decision making

Where We're Going

- Describe data using graphs.
- Describe data using numerical measures.

Statistics in *ACTION*

The "Eye Cue" Test: Does Experience Improve Performance?

In 1948, famous child psychologist Jean Piaget devised a test of basic perceptual and conceptual skills dubbed the "water-level task." Subjects were shown a drawing of a glass being held perfectly still (at a 45° angle) by an invisible hand so that any water in it had to be at rest (see Figure SIA2.1).

The task for the subject was to draw a line representing the surface of the water — a line that would touch the black dot pictured on the right side of the glass. Piaget found that young children typically failed the test. Fifty years later, research psychologists still use the water-level task to test the perception of both adults and children. Surprisingly, about 40% of the adult population also fail. In addition, males tend to do better than females, and younger adults tend to do better than older adults.

Will people with experience handling liquid-filled containers perform better on this "eye cue" test? This question was the focus of research conducted by psychologists Heiko Hecht (NASA) and Dennis R. Proffitt (University of Virginia) and published in *Psychological Science* (Mar. 1995). The researchers presented the task to each of six different groups: (1) male college students, (2) female college students, (3) professional waitresses, (4) housewives, (5) male bartenders, and (6) male bus drivers. A total of 120 subjects

(20 per group) participated in the study. Two of the groups, waitresses and bartenders, were assumed to have considerable experience handling liquid-filled glasses.

After each subject completed the drawing task, the researchers recorded the deviation* (in angle degrees) of the judged line from the true line. If the deviation was within 5° of the true water surface angle, the answer was considered correct. Deviations of more than 5° in either direction were considered incorrect answers.

The data for the water-level task (simulated based on summary results presented in the journal article) are provided in the **EYECUE** file. For each of the 120 subjects in the experiment, the following variables (in the order they appear on the data file) were measured:

💿 EYECUE

Variables

GENDER (F or M)
GROUP (Student, Waitress, Wife, Bartender, or Busdriver)
DEVIATION (angle, in degrees, of the judged line from the true line)
JUDGE (Within5, More5Above, More5Below)

The researchers are interested in testing several theories concerning performance on the water-level task, including (1) males perform better than females, (2) younger adults perform better than older adults, and (3) experience improves task performance.

In the following *Statistics in Action Revisited* sections, we apply the graphical and numerical descriptive techniques of this chapter to the **EYECUE** data to answer some of the researchers questions.

Statistics in Action Revisited for Chapter 2

- Interpreting a Pie Chart (p. 35)
- Interpreting a Histogram (p. 49)
- Interpreting Descriptive Statistics (p. 78)
- Detecting Outliers (p. 93)

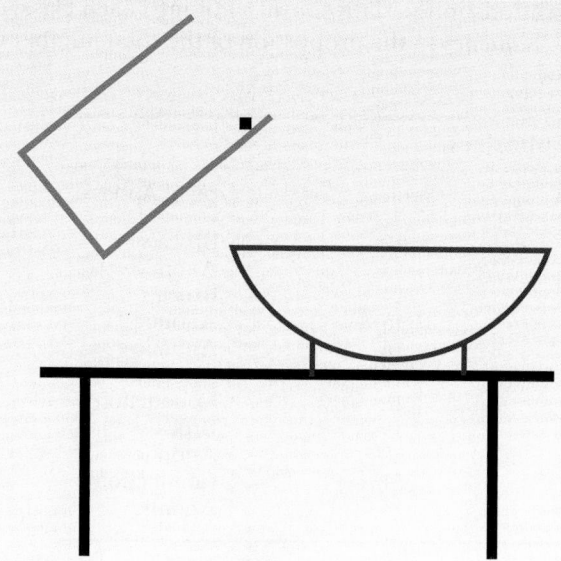

Figure SIA2.1
Drawing of the Water-Level Task

*The true surface line is perfectly parallel to the table top.

Suppose you wish to evaluate the mathematical capabilities of a class of 1,000 first-year college students based on their quantitative Scholastic Aptitude Test (SAT) scores. How would you describe these 1,000 measurements? Characteristics of interest include the typical or most frequent SAT score, the variability in the scores, the highest and lowest scores, the "shape" of the data, and whether or not the data set contains any unusual scores. Extracting this information isn't easy. The 1,000 scores provide too many bits of information for our minds to comprehend. Clearly we need some method for summarizing and characterizing the information in such a data set. Methods for describing data sets are also essential for statistical inference. Most populations are large data sets. Consequently, we need methods for describing a data set that let us make descriptive statements (inferences) about a population based on information contained in a sample.

Two methods for describing data are presented in this chapter, one *graphical* and the other *numerical*. Both play an important role in statistics. Section 2.1 presents both graphical and numerical methods for describing qualitative data. Graphical methods for describing quantitative data are illustrated in Sections 2.2, 2.8, and 2.9; numerical descriptive methods for quantitative data are presented in Sections 2.3–2.7. We end this chapter with a section on the *misuse* of descriptive techniques.

2.1 Describing Qualitative Data

Consider a study of aphasia published in the *Journal of Communication Disorders* (Mar. 1995). Aphasia is the "impairment or loss of the faculty of using or understanding spoken or written language." Three types of aphasia have been identified by researchers: Broca's, conduction, and anomic. They wanted to determine whether one type of aphasia occurs more often than any other, and, if so, how often. Consequently, they measured aphasia type for a sample of 22 adult aphasiacs. Table 2.1 gives the type of aphasia diagnosed for each aphasiac in the sample.

For this study, the variable of interest, aphasia type, is qualitative in nature. Qualitative data are nonnumerical in nature; thus, the value of a qualitative variable can only be classified into categories called *classes*. The possible aphasia types — Broca's, conduction, and anomic — represent the classes for this qualitative variable. We can summarize such data numerically in two ways: (1) by computing the *class frequency* — the number of observations in the data set that fall into each class; or (2) by computing the *class relative frequency* — the proportion of the total number of observations falling into each class.

 **APHASIA**

TABLE 2.1   **Data on 22 Adult Aphasiacs**

Subject	Type of Aphasia	Subject	Type of Aphasia
1	Broca's	12	Broca's
2	Anomic	13	Anomic
3	Anomic	14	Broca's
4	Conduction	15	Anomic
5	Broca's	16	Anomic
6	Conduction	17	Anomic
7	Conduction	18	Conduction
8	Anomic	19	Broca's
9	Conduction	20	Anomic
10	Anomic	21	Conduction
11	Conduction	22	Anomic

Source: Li, E. C., Williams, S. E., and Volpe, R. D., "The effects of topic and listener familiarity of discourse variables in procedural and narrative discourse tasks." *Journal of Communication Disorders*, Vol. 28, No. 1, Mar. 1995, p. 44 (Table 1).

> **DEFINITION 2.1**
>
> A **class** is one of the categories into which qualitative data can be classified.

> **DEFINITION 2.2**
>
> The **class frequency** is the number of observations in the data set falling in a particular class.

> **DEFINITION 2.3**
>
> The **class relative frequency** is the class frequency divided by the total number of observations in the data set; that is,
>
> $$\text{class relative frequency} = \frac{\text{class frequency}}{n}$$

> **DEFINITION 2.4**
>
> The **class percentage** is the class relative frequency multiplied by 100; that is,
>
> $$\text{class percentage} = (\text{class relative frequency}) \times 100$$

Examining Table 2.1, we observe that 5 aphasiacs in the study were diagnosed as suffering from Broca's aphasia, 7 from conduction aphasia, and 10 from anomic aphasia. These numbers — 5, 7, and 10 — represent the class frequencies for the three classes and are shown in the summary table, Figure 2.2, produced using SPSS.

Figure 2.2 also gives the relative frequency of each of the three aphasia classes. From Definition 2.3, we know that we calculate the relative frequency by dividing the class frequency by the total number of observations in the data set. Thus, the relative frequencies for the three types of aphasia are

$$\text{Broca's:} \quad \frac{5}{22} = .227$$

$$\text{Conduction:} \quad \frac{7}{22} = .318$$

$$\text{Anomic:} \quad \frac{10}{22} = .455$$

These values, expressed as a percent, are shown in the SPSS summary table, Figure 2.1. From these relative frequencies we observe that nearly half (45.5%) of the 22 subjects in the study are suffering from anomic aphasia.

Although the summary table of Figure 2.1 adequately describes the data of Table 2.1, we often want a graphical presentation as well. Figures 2.2 and 2.3 show two of the most widely used graphical methods for describing qualitative data — bar

Figure 2.1

SPSS Summary Table for Types of Aphasia

TYPE

		Frequency	Percent	Valid Percent	Cumulative Percent
Valid	Anomic	10	45.5	45.5	45.5
	Brocas	5	22.7	22.7	68.2
	Conduction	7	31.8	31.8	100.0
	Total	22	100.0	100.0	

FREQUENCY

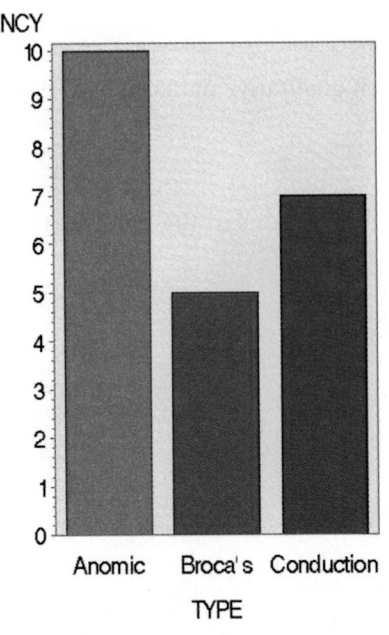

Figure 2.2
SAS Bar Graph for
Aphasia Type

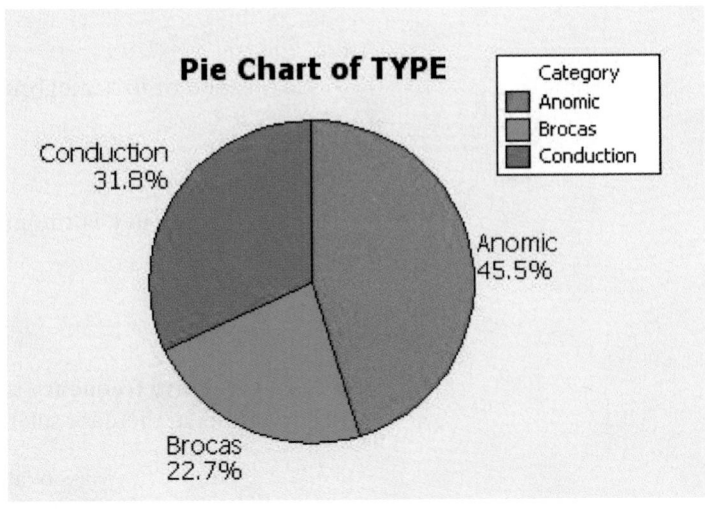

Figure 2.3
MINITAB Pie Chart for Aphasia Type

graphs and pie charts. Figure 2.2 shows the frequencies of aphasia types in a **bar graph** produced with SAS. Note that the height of the rectangle, or "bar," over each class is equal to the class frequency. (Optionally, the bar heights can be proportional to class relative frequencies.)

In contrast, Figure 2.3 shows the relative frequencies of the three types of aphasia in a **pie chart** generated with MINITAB. Note that the pie is a circle (spanning 360°) and the size (angle) of the "pie slice" assigned to each class is proportional to the class relative frequency. For example, the slice assigned to anomic aphasia is 45.5% of 360°, or $(.455)(360°) = 163.8°$.

Before leaving the data set in Table 2.1, consider the bar graph shown in Figure 2.4, produced using SPSS. Note that the bars for types of aphasia are arranged in descending order of height from left to right across the horizontal axis. That is, the tallest bar (Anomic) is positioned at the far left and the shortest bars (Broca's) are

Figure 2.4
SPSS Pareto Diagram for
Type of Aphasia

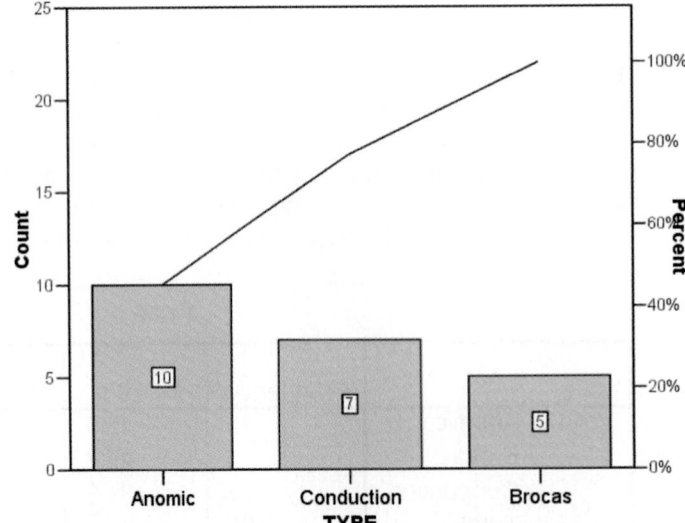

at the far right. This rearrangement of the bars in a bar graph is called a **Pareto diagram**. One goal of a Pareto diagram (named for the Italian economist, Vilfredo Pareto) is to make it easy to locate the "most important" categories — those with the largest frequencies.

Biography

VILFREDO PARETO (1843–1923)—The Pareto Principle

Born in Paris to an Italian aristocratic family, Vilfredo Pareto was educated at the University of Turin, where he studied engineering and mathematics. After the death of his parents, Pareto quit his job as an engineer and began writing and lecturing on the evils of the economic policies of the Italian government. While at the University of Lausanne in Switzerland in 1896, he published his first paper, *Cours d'economie politique*. In the paper, Pareto derived a complicated mathematical formula to prove that the distribution of income and wealth in society is not random but that a consistent pattern appears throughout history in all societies. Essentially, Pareto showed that approximately 80% of the total wealth in a society lies with only 20% of the families. This famous law about the "vital few and the trivial many" is widely known as the Pareto principle in economics.

Summary of Graphical Descriptive Methods for Qualitative Data

Bar Graph: The categories (classes) of the qualitative variable are represented by bars, where the height of each bar is either the class frequency, class relative frequency, or class percentage.

Pie Chart: The categories (classes) of the qualitative variable are represented by slices of a pie (circle). The size of each slice is proportional to the class relative frequency.

Pareto Diagram: A bar graph with the categories (classes) of the qualitative variable (i.e., the bars) arranged by height in descending order from left to right.

Now Work   *Exercise 2.4*

Let's look at a practical example that requires interpretation of the graphical results.

EXAMPLE 2.1   GRAPHING AND SUMMARIZING QUALITATIVE DATA

Problem   A group of cardiac physicians in southwest Florida have been studying a new drug designed to reduce blood loss in coronary bypass operations. Blood loss data for 114 coronary bypass patients (some who received a dosage of the drug and others who did not) are saved in the **BLOODLOSS** file. Although the drug shows promise in reducing blood loss, the physicians are concerned about possible side effects and complications. So their data set includes not only the qualitative variable DRUG, which indicates whether or not the patient received the drug, but also the qualitative variable COMP, which specifies the type (if any) of complication experienced by the patient. The four values of COMP are (1) redo surgery, (2) post-op infection, (3) both, or (4) none.

BLOODLOSS

a. Figure 2.5, generated using SAS, shows summary tables for the two qualitative variables, DRUG and COMP. Interpret the results.

b. Interpret the MINITAB and SPSS printouts shown in Figures 2.6 and 2.7, respectively.

Figure 2.5
SAS Summary Tables for
DRUG and COMP

The FREQ Procedure

DRUG	Frequency	Percent	Cumulative Frequency	Cumulative Percent
NO	57	50.00	57	50.00
YES	57	50.00	114	100.00

COMP	Frequency	Percent	Cumulative Frequency	Cumulative Percent
BOTH	6	5.26	6	5.26
INFECT	15	13.16	21	18.42
NONE	79	69.30	100	87.72
REDO	14	12.28	114	100.00

Figure 2.6
MINITAB Side-by-Side Bar
Graphs for COMP by Value
of DRUG

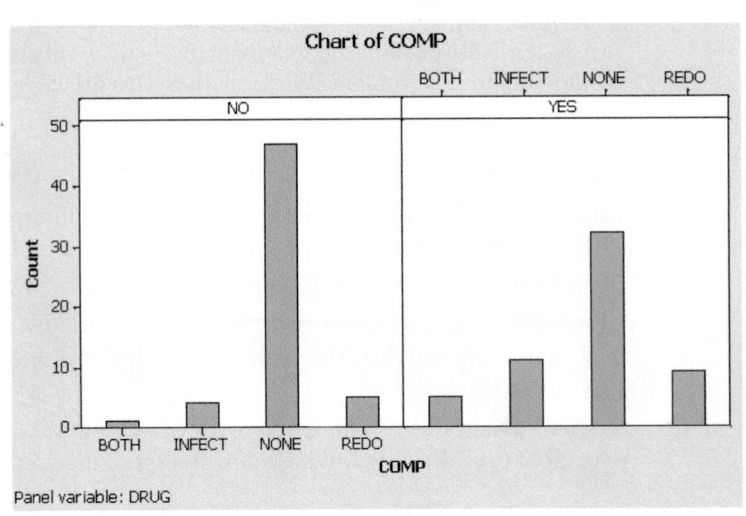

Figure 2.7
SPSS Summary Tables for
COMP by Value of Drug

COMP

DRUG			Frequency	Percent	Valid Percent	Cumulative Percent
NO	Valid	BOTH	1	1.8	1.8	1.8
		INFECT	4	7.0	7.0	8.8
		NONE	47	82.5	82.5	91.2
		REDO	5	8.8	8.8	100.0
		Total	57	100.0	100.0	
YES	Valid	BOTH	5	8.8	8.8	8.8
		INFECT	11	19.3	19.3	28.1
		NONE	32	56.1	56.1	84.2
		REDO	9	15.8	15.8	100.0
		Total	57	100.0	100.0	

Solution   a. The top table in Figure 2.5 is a summary frequency table for DRUG. Note that exactly half (57) of the 114 coronary bypass patients received the drug and half did not. The bottom table in Figure 2.5 is a summary frequency table for COMP. The class relative frequencies are given in the Percent column. We see that about 69% of the 114 patients had no complications, leaving about 31% who experienced either a redo surgery, a post-op infection, or both.

b.   Figure 2.6 is a MINITAB side-by-side bar graph for the data. The four bars in the left-side graph represent the frequencies of COMP for the 57 patients who did not receive the drug; the four bars in the right-side graph represent the frequencies of COMP for the 57 patients who did receive a dosage of the drug. The graph clearly shows that patients who did not receive the drug suffered fewer complications. The exact percentages are displayed in the SPSS summary tables of Figure 2.8. Over 56% of the patients who got the drug had no complications, compared to about 83% for the patients who got no drug.

Look Back   Although these results show that the drug may be effective in reducing blood loss, Figures 2.6 and 2.7 imply that patients on the drug may have a higher risk of complications. But before using this information to make a decision about the drug, the physicians will need to provide a measure of reliability for the inference. That is, the physicians will want to know whether the difference between the percentages of patients with complications observed in this sample of 114 patients is generalizable to the population of all coronary bypass patients.

Now Work   *Exercise 2.17*

■ ■ ■

Statistics in Action Revisited

Interpreting Pie Charts

In the *Psychological Science* "water-level task" experiment, the researchers measured three qualitative variables: *Gender* (F or M), *Subject Group* (student, waitress, wife, bartender, or bus driver), and *Judged Task Performance* (within 5° of the line, more than 5° above the line, or more than 5° below the line). Pie charts and bar graphs can be used to summarize and describe the responses in these variables are categories. Recall that the data is saved in the **EYECUE** file. These variables are named GENDER, GROUP, and JUDGE in the data file. We created pie charts for these variables using MINITAB.

Figure SIA2.2 is a pie chart for the JUDGE variable. Clearly, the large slice for "Within5" indicates that a majority of subjects (52.5%) were judged to be within 5° of the correct line. The researchers want to know if the water-level task

performance will vary across gender. Figure SIA2.3 shows side-by-side pie charts of the JUDGE variable for each level of GENDER. You can see that 65% of the male subjects (right-side chart) were judged to be within 5° of the line, compared to only 40% for female subjects (left-side chart). These graphs support the prevailing theory that men will perform better than women on the task.

We produced a similar set of side-by-side pie charts to compare the performances of the different groups of subjects in Figure SIA2.4. These charts again show that bus drivers (75% judged to be within 5° of the correct line) and college students (72.5%) perform the best, while bartenders (40%), wives (30%) and (surprisingly) waitresses (25%) perform the worst. These graphs do not seem to support the researchers' theory that experience improves task performance.

Figure SIA2.2

MINITAB Pie Chart for
Judged Task Performance

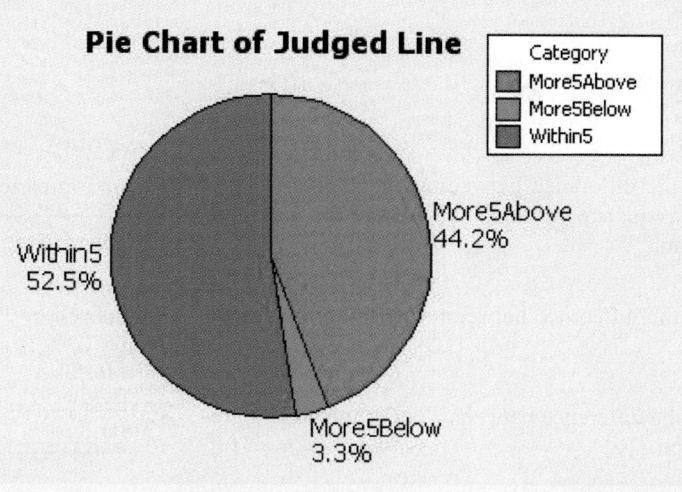

Figure SIA2.3
MINITAB Pie Charts for
Judged Task Performance —
Females versus Males

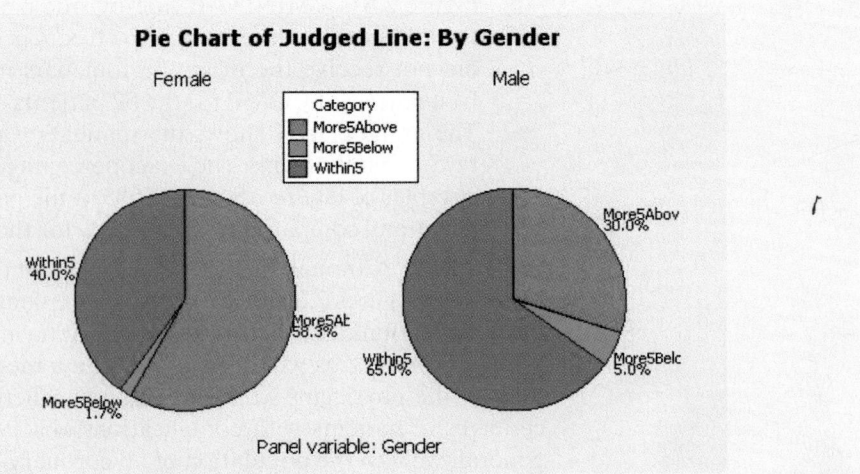

Figure SIA2.4
MINITAB Pie Charts for
Judged Task Performance —
Group Comparisons

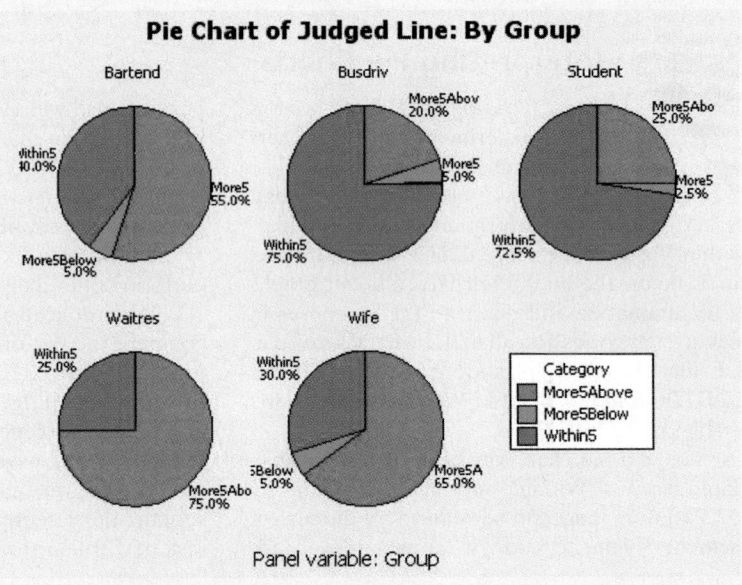

Exercises 2.1–2.20

Understanding the Principles

2.1 Explain the difference between class frequency, class relative frequency, and class percentage for a qualitative variable.

2.2 Explain the difference between a bar graph and a pie chart.

2.3 Explain the difference between a bar graph and a Pareto diagram.

Learning the Mechanics

2.4 Complete the following table.

Grade on Statistics Exam	Frequency	Relative Frequency
A: 90–100		.08
B: 80–89	36	
C: 65–79	90	
D: 50–64	30	
F: Below 50	28	
Total	200	1.00

2.5   A qualitative variable with three classes (X, Y, and Z) is measured for each of 20 units randomly sampled from a target population. The data (observed class for each unit) are listed below.

Y   X   X   Z   X   Y   Y   Y   X   X   Z   X
Y   Y   X   Z   Y   Y   Y   X

a. Compute the frequency for each of the three classes.
b. Compute the relative frequency for each of the three classes.
c. Display the results, part **a**, in a frequency bar graph.
d. Display the results, part **b**, in a pie chart.

Applying the Concepts — Basic

2.6   **Types of cancer treated.** The Moffitt Cancer Center at the University of South Florida treats over 25,000 patients a year. The graphic below describes the types of cancer treated in Moffitt's patients during fiscal year 2002.

[NW] a. What type of graph is portrayed?
b. Which type of cancer is treated most often at Moffitt?

c. What percentage of Moffitt's patients are treated for melanoma, lymphoma, or leukemia?

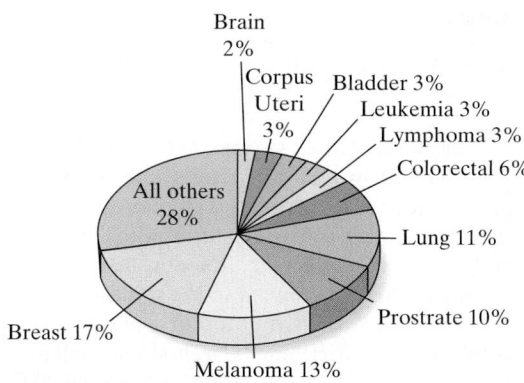

Source: Today's Tomorrows, Annual Report, H. Lee Moffitt Cancer Center & Research Institute, Winter 2002–2003.

2.7   **Japanese reading levels.** University of Hawaii language professors C. Hitosugi and R. Day incorporated a 10-week extensive reading program into a second semester Japanese course in an effort to improve students' Japanese reading comprehension. (*Reading in a Foreign Language,* Apr. 2004.) The professors collected 266 books originally written for Japanese children and required their students to read at least 40 of them as part of the grade in the course. The books were categorized into reading levels (color coded for easy selection) according to length and complexity. The reading levels for the 266 books are summarized in the table.

Reading Level	Number
Level 1 (Red)	39
Level 2 (Blue)	76
Level 3 (Yellow)	50
Level 4 (Pink)	87
Level 5 (Orange)	11
Level 6 (Green)	3
Total	266

Source: Hitosugi, C. I. and Day, R. R. "Extensive reading in Japanese," *Reading in a Foreign Language,* Vol. 16, No. 1. Apr. 2004 (Table 2).

a. Calculate the proportion of books at reading level 1 (red).
b. Repeat part **a** for each of the remaining reading levels.
c. Verify that the proportions in parts **a** and **b** sum to 1.
d. Use the previous results to form a bar graph for the reading levels.
e. Construct a Pareto diagram for the data. Use the diagram to identify the reading level that occurs most often.

2.8   **Dates of pennies.** *Chance* (Spring 2000) reported on a study to estimate the number of pennies required to fill a coin collectors album. The data used in the study were obtained by noting the mint date on each in a sample of 2,000 pennies. The distribution of mint dates are summarized in the following table.

Mint Date	Number
Pre-1960	18
1960s	125
1970s	330
1980s	727
1990s	800

Source: Lu, S., and Skiena, S. "Filling a penny album," *Chance,* Vol. 13, No. 2, Spring 2000, p. 26 (Table 1).

a. Identify the experimental unit for the study.
b. Identify the variable measured.
c. What proportion of pennies in the sample have mint dates in the 1960s?
d. Construct a pie chart to describe the distribution of mint dates for the 2,000 sampled pennies.

2.9   **Estimating the rhino population.** The International Rhino Federation estimates that there are 17,470 rhinoceroses living in the wild in Africa and Asia. A breakdown of the number of rhinos of each species is reported in the accompanying table.

Rhino Species	Population Estimate
African Black	3,610
African White	11,100
(Asian) Sumatran	300
(Asian) Javan	60
(Asian) Indian	2,400
Total	17,470

Source: International Rhino Federation, Sep. 2004.

a. Construct a relative frequency table for the data.
b. Display the relative frequencies in a bar graph.
c. What proportion of the 17,470 rhinos are African rhinos? Asian?

2.10 Role importance for the elderly. In *Psychology and Aging* (Dec. 2000), University of Michigan School of Public Health researchers studied the roles that elderly people feel are the most important to them in late life. The accompanying table summarizes the most salient roles identified by each in a national sample of 1,102 adults, 65 years or older.

Most Salient Role	Number
Spouse	424
Parent	269
Grandparent	148
Other relative	59
Friend	73
Homemaker	59
Provider	34
Volunteer, club, church member	36
Total	1,102

Source: Krause, N., and Shaw, B. A. "Role-specific feelings of control and mortality," *Psychology and Aging*, Vol. 15, No. 4, Dec. 2000 (Table 2).

a. Describe the qualitative variable summarized in the table. Give the categories associated with the variable.
b. Are the numbers in the table frequencies or relative frequencies?
c. Display the information in the table in a bar graph.
d. Which role is identified by the highest percentage of elderly adults? Interpret the relative frequency associated with this role.

2.11 Switching off air bags. Driver-side and passenger-side air bags are installed in all new cars to prevent serious or fatal injury in an automobile crash. However, air bags have been found to cause deaths in children and small people or people with handicaps in low-speed crashes. Consequently, in 1998 the federal government began allowing vehicle owners to request installation of an on-off switch for air bags. The table describes the reasons for requesting the installation of passenger-side on-off switches given by car owners in 1998 and 1999.

Reason	Number of Requests
Infant	1,852
Child	17,148
Medical	8,377
Infant & Medical	44
Child & Medical	903
Infant & Child	1,878
Infant & Child & Medical	135
Total	30,337

Source: National Highway Transportation Safety Administration. Sept. 2000.

a. What type of variable, quantitative or qualitative, is summarized in the table? Give the values that the variable could assume.
b. Calculate the relative frequencies for each reason.
c. Display the information in the table in an appropriate graph.
d. What proportion of the car owners who requested on-off air bag switches gave Medical as one of the reasons?

Applying the Concepts—Intermediate

2.12 Excavating ancient pottery. Archaeologists excavating the ancient Greek settlement at Phylakopi classified the pottery found in trenches (*Chance*, Fall 2000). The accompanying table describes the collection of 837 pottery pieces uncovered in a particular layer at the excavation site. Construct and interpret a graph that will aid the archaeologists in understanding the distribution of the pottery types found at the site.

Pot Category	Number Found
Burnished	133
Monochrome	460
Slipped	55
Painted in curvilinear decoration	14
Painted in geometric decoration	165
Painted in naturalistic decoration	4
Cycladic white clay	4
Conical cup clay	2
Total	837

Source: Berg, I., and Bliedon, S. "The Pots of Phylakopi: Applying Statistical Techniques to Archaeology," *Chance*, Vol. 13, No. 4, Fall 2000.

2.13 "Made in the USA" survey. "Made in the USA" is a claim stated in many product advertisements or on product labels. Advertisers want consumers to believe that the product is manufactured with 100% U.S. labor and materials — which is often not the case. What does "Made in the USA" mean to the typical consumer? To answer this question, a group of marketing professors conducted an experiment at a shopping mall in Muncie, Indiana. (*Journal of Global Business*, Spring 2002.) They asked every fourth adult entrant to the mall to participate in the study. A total of 106 shoppers agreed to answer the question, "'Made in the USA' means what percentage of US labor and materials?" The responses of the 106 shoppers are summarized in the table.

Response to "Made in the USA"	Number of Shoppers
100%	64
75 to 99%	20
50 to 74%	18
Less than 50%	4

Source: "'Made in the USA': Consumer perceptions, deception and policy alternatives," *Journal of Global Business*, Vol. 13, No. 24, Spring 2002 (Table 3).

a. What type of data collection method was used?

b. What type of variable, quantitative or qualitative, is measured?

c. Present the data in the table in graphical form. Use the graph to make a statement about the percentage of consumers who believe "Made in the USA" means 100% U.S. labor and materials.

2.14 Hearing loss in senior citizens. Audiologists have recently developed a rehabilitation program for hearing-impaired patients in a Canadian home for senior citizens. (*Journal of the Academy of Rehabilitative Audiology*, 1994.) Each of the 30 residents of the home were diagnosed for degree and type of sensorineural hearing loss, coded as follows: 1 = hear within normal limits, 2 = high-frequency hearing loss, 3 = mild loss, 4 = mild-to-moderate loss, 5 = moderate loss, 6 = moderate-to-severe loss, and 7 = severe-to-profound loss. The data are listed in the next table. Use a graph to portray the results. Which type of hearing loss appears to be the most prevalent among nursing home residents?

HEARLOSS

6	7	1	1	2	6	4	6	4	2	5	2	5
1	5	4	6	6	5	5	5	2	5	3	6	4
6	6	4	2									

Source: Jennings, M. B., and Head, B. G. "Development of an ecological audiologic rehabilitation program in a home-for-the-aged." *Journal of the Academy of Rehabilitative Audiology*, Vol. 27, 1994, p. 77 (Table 1).

2.15 Best-paid CEOs. *Forbes* magazine periodically conducts a salary survey of chief executive officers. In addition to salary information, *Forbes* collects and reports personal data on the CEOs, including level of education and age. Do most CEOs have advanced degrees, such as masters degrees or doctorates? The data in the table below represent the highest degree obtained for each of the top 25 best-paid CEOs of 2003. Use a graphical method to summarize the highest degree obtained for these CEOs. What is your opinion about whether most CEOs have advanced degrees?

DDT

2.16 Species of Contaminated Fish. Refer to Example 1.4 (p. 13) and the U.S. Army Corps of Engineers data on fish contaminated from the toxic discharges of a chemical plant located on the banks of the Tennessee River in Alabama. The engineers determined the species (channel catfish, largemouth bass, or smallmouth buffalofish) for each of the 144 captured fish. The data on species are saved in the DDT file. Use a graphical method to describe the frequency of occurrence of the three fish species in the 144 captured fish.

FORBES25

CEO	Company	Degree	Age
1. Jeffry C. Barbakow	Tenet Healthcare	Masters	59
2. Dwight C. Schar	NVR	Bachelors	61
3. Michael S. Dell	Dell Computer	None	38
4. Irwin M. Jacobs	Qualcomm	Masters	69
5. Barry Diller	USA Interactive	None	61
6. Dan M. Palmer	Concord EFS	Bachelors	60
7. Charles T. Fote	First Data	None	54
8. Orin C. Smith	Starbucks	Masters	60
9. Richard S. Fuld, Jr.	Lehman Bros Holding	Masters	57
10. Maurice R. Greenberg	American Int Group	Law	78
11. Charles M. Cawley	MBNA	Bachelors	62
12. James E. Cayne	Bear Stearns	None	69
13. Scott G. McNealy	Sun Microsystems	Masters	48
14. Philip J. Purcell	Morgan Stanley	Masters	59
15. Vance D. Coffman	Lockheed Martin	PhD	59
16. Lee R. Raymond	ExxonMobil	PhD	64
17. Richard J. Kogan	Schering-Plough	Masters	61
18. Kenneth W. Freeman	Quest Diagnostics	Masters	52
19. Leonard D. Schaeffer	WellPoint Health	Bachelors	57
20. Stuart A. Miller	Lennar	Law	45
21. Robert L. Tillman	Lowe's	Bachelors	59
22. Sumner M. Redstone	Viacom	Law	79
23. Peter Cartwright	Calpine	Masters	73
24. David D. Halbert	Advance PCS	Bachelors	47
25. Craig R. Barrett	Intel	PhD	63

Source: Forbes, April 23, 2003.

2.17 Unauthorized computer use. The Computer Security Institute (CSI) conducts an annual survey of computer crime at United States businesses. CSI sends survey questionnaires to computer security personnel at all U.S. corporations and government agencies. In 2001, 64% of the respondents admitted unauthorized use of computer systems at their firms during the year. (*Computer Security Issues & Trends*, Spring 2001.) One survey question asked, "If your business website suffered unauthorized use, where did the attack come from, inside or outside the company?" The responses for those business websites that did, in fact, experience unauthorized use are summarized in the table for two survey years, 1999 (125 reported attacks) and 2001 (163 reported attacks). Compare the responses for the two years using side-by-side bar charts. What inference can be made from the charts?

WWW Site Attack	Percentage in 1999	Percentage in 2001
Inside	7	4
Outside	38	47
Both	41	22
Don't Know	14	26
Totals	100	100

Source: "2001 CSI/FBI Computer Crime and Security Survey," *Computer Security Issues & Trends,* Vol. 7, No. 1, Spring 2001.

Applying the Concepts—Advanced

NZBIRDS

2.18 Extinct New Zealand birds. Refer to the *Evolutionary Ecology Research* (July 2003) study of the patterns of extinction in the New Zealand bird population, Exercise 1.18 (p. 20). Data on flight capability (volant or flightless), habitat (aquatic, ground terrestrial, or aerial terrestrial), nesting site (ground, cavity within ground, tree, cavity above ground), nest density (high or low), diet (fish, vertebrates, vegetables, or invertebrates), body mass (grams), egg length (millimeters), and extinct status (extinct, absent from island, present) for 132 bird species at the time of the Maori colonization of New Zealand are saved in the NZBIRDS file. Use a graphical method to investigate the theory that extinct status is related to flight capability, habitat, and nest density.

2.19 Whistling dolphins. Marine scientists who study dolphin communication have discovered that bottlenose dolphins exhibit an individualized whistle contour known as the *signature whistle*. A study was conducted to categorize the signature whistles of ten captive adult bottlenose dolphins in socially interactive contexts. (*Ethology*, July 1995.) A total of 185 whistles were recorded during the study period; each whistle contour was analyzed and assigned to a category using a contour similarity (CS) technique. The results are reported in the accompanying table. Use a graphical method to summarize the results. Do you detect any patterns in the data that might be helpful to marine scientists?

DOLPHIN

Whistle Category	Number of Whistles
Type a	97
Type b	15
Type c	9
Type d	7
Type e	7
Type f	2
Type g	2
Type h	2
Type i	2
Type j	4
Type k	13
Other types	25

Source: McCowan, B., and Reiss, D. "Quantitative comparison of whistle repertoires from captive adult bottlenose dolphins (Delphiniae, Tursiops truncatus): A re-evaluation of the signature whistle hypothesis." *Ethology,* Vol. 100, No. 3, July 1995, p. 200 (Table 2).

2.20 Benford's Law of Numbers. According to *Benford's Law*, certain digits $(1, 2, 3, \ldots, 9)$ are more likely to occur as the first significant digit in a randomly selected number than other digits. For example, the law predicts that the number "1" is the most likely to occur (30% of the time) as the first digit. In a study reported in the *American Scientist* (July–Aug. 1998) to test Benford's Law, 743 first-year college students were asked to write down a six-digit number at random. The first significant digit of each number was recorded and its distribution summarized in the following table.

DIGITS

First Digit	Number of Occurrences
1	109
2	75
3	77
4	99
5	72
6	117
7	89
8	62
9	43
Total	743

Source: Hill, T. P. "The First Digit Phenomenon." *American Scientist,* Vol. 86, No. 4, July–Aug. 1998, p. 363 (Figure 5).

a. Describe the first digit of the "random guess" data with an appropriate graph.

b. Does the graph support Benford's Law? Explain.

2.2 Graphical Methods for Describing Quantitative Data

Recall from Section 1.4 that quantitative data sets consist of data that are recorded on a meaningful numerical scale. For describing, summarizing, and detecting patterns in such data, we can use three graphical methods: *dot plots, stem-and-leaf displays,* and *histograms*. Since most statistical software packages can be used to construct these displays, we'll focus here on their interpretation rather than their construction.

For example, the Environmental Protection Agency (EPA) performs extensive tests on all new car models to determine their mileage ratings. Suppose that the 100 measurements in Table 2.2 represent the results of such tests on a certain new car model. How can we summarize the information in this rather large sample?

 EPAGAS

TABLE 2.2   EPA Mileage Ratings on 100 Cars

36.3	41.0	36.9	37.1	44.9	36.8	30.0	37.2	42.1	36.7
32.7	37.3	41.2	36.6	32.9	36.5	33.2	37.4	37.5	33.6
40.5	36.5	37.6	33.9	40.2	36.4	37.7	37.7	40.0	34.2
36.2	37.9	36.0	37.9	35.9	38.2	38.3	35.7	35.6	35.1
38.5	39.0	35.5	34.8	38.6	39.4	35.3	34.4	38.8	39.7
36.3	36.8	32.5	36.4	40.5	36.6	36.1	38.2	38.4	39.3
41.0	31.8	37.3	33.1	37.0	37.6	37.0	38.7	39.0	35.8
37.0	37.2	40.7	37.4	37.1	37.8	35.9	35.6	36.7	34.5
37.1	40.3	36.7	37.0	33.9	40.1	38.0	35.2	34.8	39.5
39.9	36.9	32.9	33.8	39.8	34.0	36.8	35.0	38.1	36.9

A visual inspection of the data indicates some obvious facts. For example, most of the mileages are in the 30s, with a smaller fraction in the 40s. But it is difficult to provide much additional information on the 100 mileage ratings without resorting to some method of summarizing the data. One such method is a dot plot.

Dot Plots

A MINITAB **dot plot** for the 100 EPA mileage ratings is shown in Figure 2.8. The horizontal axis of Figure 2.8 is a scale for the quantitative variable in miles per gallon (mpg). The numerical value of each measurement in the data set is located on the

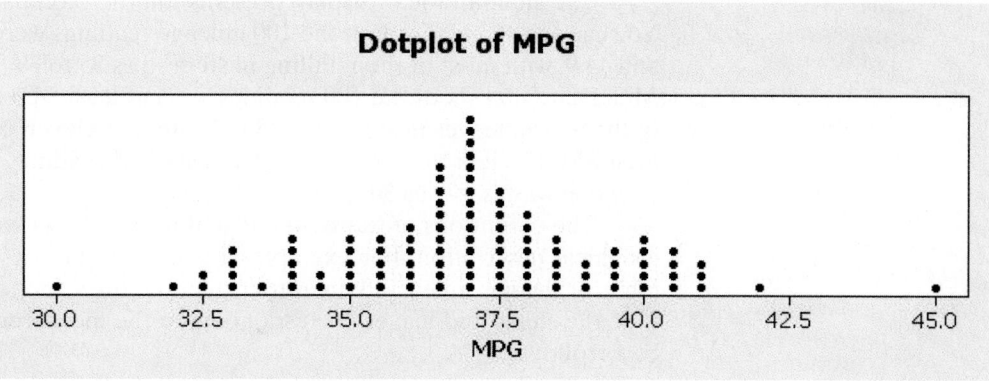

Figure 2.8
MINITAB Dot Plot for 100 EPA Mileage Ratings

horizontal scale by a dot. When data values repeat, the dots are placed above one another, forming a pile at that particular numerical location. As you can see, this dot plot verifies that almost all of the mileage ratings are in the 30s, with most falling between 35 and 40 miles per gallon.

Stem-and-Leaf Display

Another graphical representation of these same data, a MINITAB **stem-and-leaf display**, is shown in Figure 2.9. In this display the *stem* is the portion of the measurement (mpg) to the left of the decimal point, while the remaining portion to the right of the decimal point is the *leaf*.

Figure 2.9

MINITAB Stem-and-Leaf Display for 100 Mileage Ratings

```
Stem-and-leaf of MPG   N  = 100
Leaf Unit = 0.10

   1      30   0
   2      31   8
   6      32   5799
  12      33   126899
  18      34   024588
  29      35   01235667899
  49      36   01233445566777888999
 (21)     37   000011122334456677899
  30      38   0122345678
  20      39   00345789
  12      40   0123557
   5      41   002
   2      42   1
   1      43
   1      44   9
```

In Figure 2.9, the stems for the data set are listed in the second column from the smallest (30) to the largest (44). Then the leaf for each observation is listed to the right in the row of the display corresponding to the observation's stem. For example, the leaf 3 of the first observation (36.3) in Table 2.2 appears in the row corresponding to the stem 36. Similarly, the leaf 7 for the second observation (32.7) in Table 2.2 appears in the row corresponding to the stem 32, while the leaf 5 for the third observation (40.5) appears in the row corresponding to the stem 40. (The stems and leaves for these first three observations are highlighted in Figure 2.9.) Typically, the leaves in each row are ordered as shown in the MINITAB stem-and-leaf display.

The stem-and-leaf display presents another compact picture of the data set. You can see at a glance that the 100 mileage readings were distributed between 30.0 and 44.9, with most of them falling in stem rows 35 to 39. The six leaves in stem row 34 indicate that six of the 100 readings were at least 34.0 but less than 35.0. Similarly, the eleven leaves in stem row 35 indicate that eleven of the 100 readings were at least 35.0 but less than 36.0. Only five cars had readings equal to 41 or larger, and only one was as low as 30.

The definitions of the stem and leaf for a data set can be modified to alter the graphical description. For example, suppose we had defined the stem as the tens digit for the gas mileage data, rather than the ones and tens digits. With this definition, the stems and leaves corresponding to the measurements 36.3 and 32.7 would be as follows:

Stem	Leaf		Stem	Leaf
3	6		3	2

Note that the decimal portion of the numbers has been dropped. Generally, only one digit is displayed in the leaf.

    If you look at the data, you'll see why we didn't define the stem this way. All the mileage measurements fall in the 30s and 40s, so all the leaves would fall into just two stem rows in this display. The resulting picture would not be nearly as informative as Figure 2.9.

Now Work   ***Exercise 2.25***

Biography

JOHN TUKEY (1915–2000)— The Picasso of Statistics

Like the legendary artist Pablo Picasso, who mastered and revolutionized a variety of art forms during his lifetime, John Tukey is recognized for his contributions to many subfields of statistics. Born in Massachusetts, Tukey was home-schooled, graduated with his bachelor's and master's degrees in chemistry from Brown University, and received his Ph.D. in mathematics from Princeton University. While at Bell Telephone Laboratories in the 1960s and early 1970s, Tukey developed exploratory data analysis, a set of graphical descriptive methods for summarizing and presenting huge amounts of data. Many of these tools, including the stem-and-leaf display and the box plot, are now standard features of modern statistical software packages. (In fact, it was Tukey himself who coined the term *software* for computer programs.)

Histograms

An SPSS **histogram** for these 100 EPA mileage readings is shown in Figure 2.10. The horizontal axis of Figure 2.10, which gives the miles per gallon for a given automobile, is divided into **class intervals** commencing with the interval from 30.0–31.5 and proceeding in intervals of equal size to 43.5–45.0 mpg. The vertical axis gives the

Figure 2.10
SPSS Histogram for
100 EPA Mileage Ratings

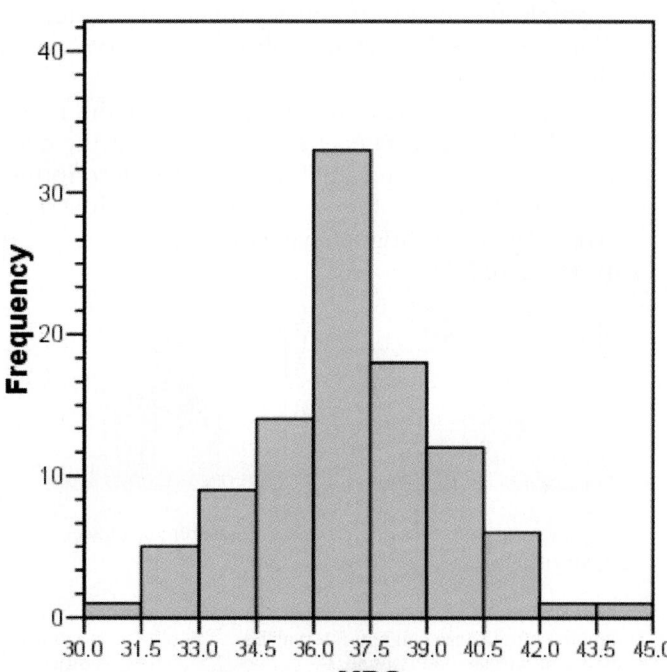

number (or *frequency*) of the 100 readings that fall in each interval. It appears that about 33 of the 100 cars, or 33%, obtained a mileage between 36.0 and 37.5. This class interval contains the highest frequency, and the intervals tend to contain a smaller number of the measurements as the mileages get smaller or larger.

Histograms can be used to display either the frequency or relative frequency of the measurements falling into the class intervals. The class intervals, frequencies, and relative frequencies for the EPA car mileage data are shown in the summary table, Table 2.3.*

TABLE 2.3 Class Intervals, Frequencies, and Relative Frequencies for the Car Mileage Data

Class Interval	Frequency	Relative Frequency
30.0–31.5	1	.01
31.5–33.0	5	.05
33.0–34.5	9	.09
34.5–36.0	14	.14
36.0–37.5	33	.33
37.5–39.0	18	.18
39.0–40.5	12	.12
40.5–42.0	6	.06
42.0–43.5	1	.01
43.5–45.0	1	.01
Totals	100	1.00

By summing the relative frequencies in the intervals 34.5–36.0, 36.0–37.5, and 37.5–39.0, you can see that 65% of the mileages are between 34.5 and 39.0. Similarly, only 2% of the cars obtained a mileage rating over 42.0. Many other summary statements can be made by further study of the histogram and accompanying summary table. Note that the sum of all class frequencies will always equal the sample size, n.

When interpreting a histogram consider two important facts. First, the proportion of the total area under the histogram that falls above a particular interval on the x-axis is equal to the relative frequency of measurements falling in the interval. For example, the relative frequency for the class interval 36.0–37.5 in Figure 2.10 is .33. Consequently, the rectangle above the interval contains .33 of the total area under the histogram.

Second, imagine the appearance of the relative frequency histogram for a very large set of data (say, a population). As the number of measurements in a data set is increased, you can obtain a better description of the data by decreasing the width of the class intervals. When the class intervals become small enough, a relative frequency histogram will (for all practical purposes) appear as a smooth curve (see Figure 2.11).

Figure 2.11

The Effect of the Size of a Data Set on the Outline of a Histogram

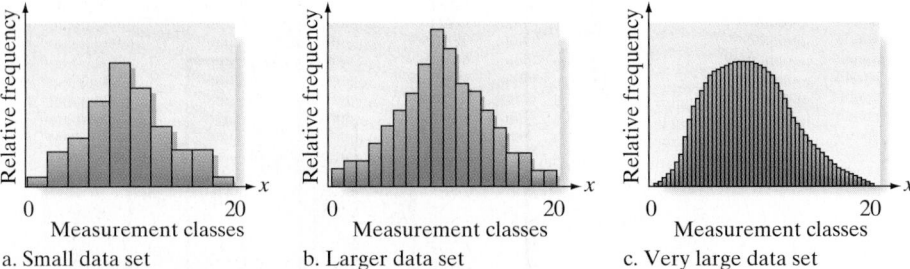

a. Small data set b. Larger data set c. Very large data set

*SPSS, like many software packages, will classify an observation that falls on the borderline of a class interval into the next highest interval. For example, the gas mileage of 37.5, which falls on the border between the class intervals 36.0–37.5 and 37.5–39.0, is classified into the 37.5–39.0 class. The frequencies in Table 2.3 reflect this convention.

Some recommendations for selecting the number of intervals in a histogram for smaller data sets are given in the following box.

Determining the Number of Classes in a Histogram

Number of Observations in Data Set	Number of Classes
Less than 25	5–6
25–50	7–14
More than 50	15–20

While histograms provide good visual descriptions of data sets — particularly very large ones — they do not let us identify individual measurements. In contrast, each of the original measurements is visible to some extent in a dot plot and clearly visible in a stem-and-leaf display. The stem-and-leaf display arranges the data in ascending order, so it's easy to locate the individual measurements. For example, in Figure 2.9 we can easily see that two of the gas mileage measurements are equal to 36.3, but can't see that fact by inspecting the histogram in Figure 2.10. However, stem-and-leaf displays can become unwieldy for very large data sets. A very large number of stems and leaves causes the vertical and horizontal dimensions of the display to become cumbersome, diminishing the usefulness of the visual display.

EXAMPLE 2.2

GRAPHING A QUANTITATIVE VARIABLE

Problem The data in Table 2.4 give, by state, the percentages of the total number of college or university student loans that are in default.

NW

a. Create a histogram for these data. Where do most of the default rates lie?

b. Create a stem-and-leaf display for these data. Locate Wyoming's default rate of 2.7 on the graph.

Solution

a. We used SAS to generate a relative frequency histogram in Figure 2.12. Note that 7 classes were formed. The classes are identified by their *midpoints* rather than their endpoints. Thus, the first interval has a midpoint of 1.5, the second of 4.5, and so on. The corresponding class intervals based on these midpoints are therefore 1.0–3.0, 3.0–5.0, etc. Note that the classes with midpoints of 7.5, and 10.5 (ranging from 6.0 to 12.0) contain approximately 60% of the 51 default measurements.

LOANDEFAULT

TABLE 2.4   Percentage of Student Loans (per State) in Default

State	%	State	%	State	%	State	%
Ala.	12.0	Ill.	9.3	Mont.	6.4	R.I.	8.8
Alaska	19.7	Ind.	6.7	Nebr.	4.9	S.C.	14.1
Ariz.	12.1	Iowa	6.2	Nev.	10.1	S. Dak.	5.5
Ark.	12.9	Kans.	5.7	N.H.	7.9	Tenn.	12.3
Calif.	11.4	Ky.	10.3	N.J.	12.0	Tex.	15.2
Colo.	9.5	La.	13.5	N. Mex.	7.5	Utah	6.0
Conn.	8.8	Maine	9.7	N.Y.	11.3	Vt.	8.3
Del.	10.9	Md.	16.6	N.C.	15.5	Va.	14.4
D.C.	14.7	Mass.	8.3	N. Dak.	4.8	Wash.	8.4
Fla.	11.8	Mich.	11.4	Ohio	10.4	W Va.	9.5
Ga.	14.8	Minn.	6.6	Okla.	11.2	Wis.	9.0
Hawaii	12.8	Miss.	15.6	Oreg.	7.9	Wyo.	2.7
Idaho	7.1	Mo.	8.8	Pa.	8.7		

Source: National Direct Student Loan Coalition.

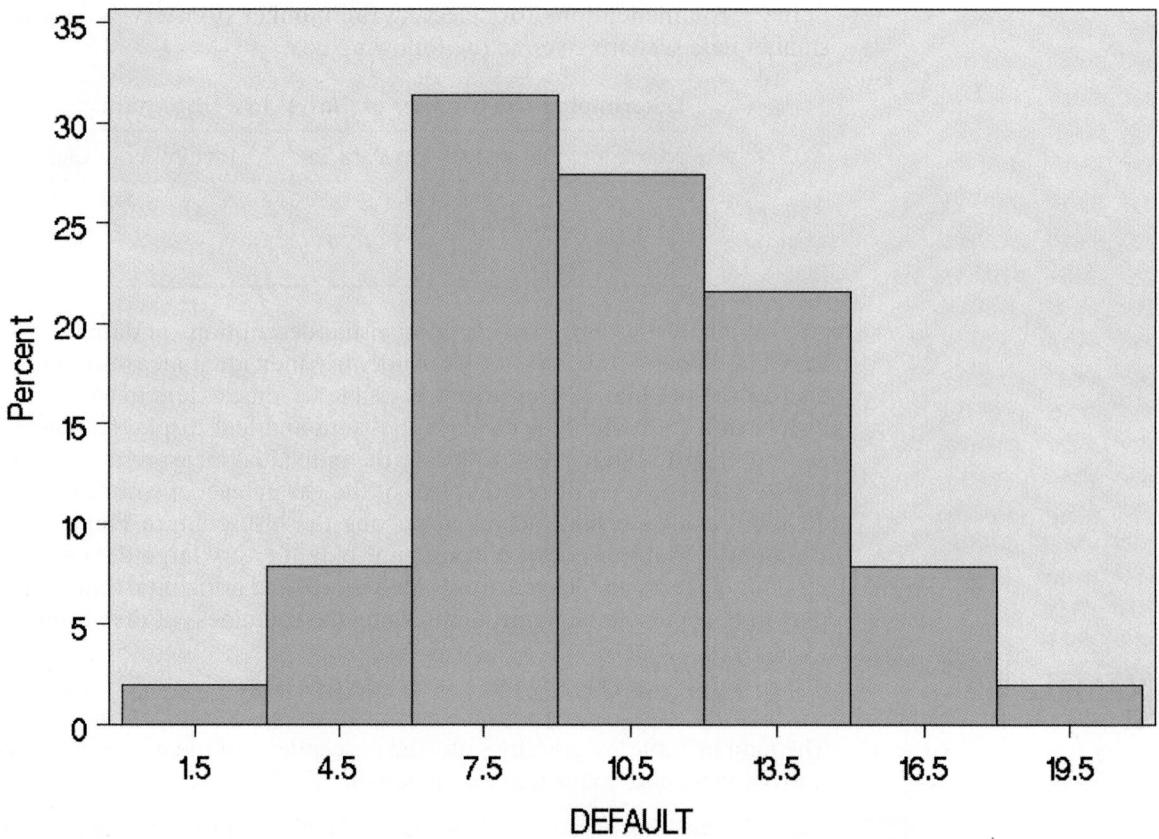

Figure 2.12
SAS Relative Frequency Histogram for Student Loan Default Rate Data

b. We used MINITAB to generate the stem-and-leaf display in Figure 2.13. Note that the stem, (the *second* column in the printout) has been defined as the number *two* places to the left of the decimal. The leaf (the third column in the printout) is the number one place to the left of the decimal.* (The digit to the right of the decimal is not shown.) Thus, the leaf 2 in the stem 0 row (the first row of the printout) represents the default rate of 2.7 for Wyoming.

Figure 2.13
MINITAB Stem-and-Leaf
Display for Student Loan
Default Rate Data

```
Stem-and-leaf of DEFAULT  N  = 51
Leaf Unit = 1.0


    1    0  2
    5    0  4455
   14    0  666667777
  (12)   0  888888899999
   25    1  000011111
   16    1  2222223
    9    1  4444555
    2    1  6
    1    1  9
```

*The first column of the MINITAB stem-and-leaf display represents the cumulative number of measurements from the class interval to the nearest extreme class interval.

Look Back As is usually the case for data sets that are not too large (say, fewer than 100 measurements), the stem-and-leaf display provides more detail than the histogram without being unwieldy. For instance, the stem-and-leaf display in Figure 2.13 clearly indicates the values of the individual measurements in the data set. For example, the highest default rate (representing the measurement 19.7 for Alaska) is shown in the last stem row. Histograms are most useful for displaying very large data sets when the overall shape of the distribution of measurements is more important than the identification of individual measurements.

Now Work	*Exercise 2.30*

■ ■ ■

Most statistical software packages can be used to generate histograms, stem-and-leaf displays, and dot plots. All three are useful tools for graphically describing data sets. We recommend that you generate and compare the displays whenever you can. You'll find that histograms are generally more useful for very large data sets, while stem-and-leaf displays and dot plots provide useful detail for smaller data sets.

Summary of Graphical Descriptive Methods for Quantitative Data

Dot Plot: The numerical value of each quantitative measurement in the data set is represented by a dot on a horizontal scale. When data values repeat, the dots are placed above one another vertically.

Stem-and-Leaf Display: The numerical value of the quantitative variable is partitioned into a "stem" and a "leaf." The possible stems are listed in order in a column. The leaf for each quantitative measurement in the data set is placed in the corresponding stem row. Leaves for observations with the same stem value are listed in increasing order horizontally.

Histogram: The possible numerical values of the quantitative variable are partitioned into class intervals, where each interval has the same width. These intervals form the scale of the horizontal axis. The frequency or relative frequency of observations in each class interval is determined. A vertical bar is placed over each class interval with height equal to either the class frequency or class relative frequency.

Histograms

Using the TI-83 Graphing Calculator

Making a Histogram from Raw Data

Step 1   *Enter the data*
Press **STAT** and select **1:Edit**
Note: If the list already contains data, clear the old data. Use the up arrow to highlight '**L1**'. Press **CLEAR ENTER**.
Use the arrow and **ENTER** keys to enter the data set into **L1**.

Step 2   *Set up the histogram plot*
Press **2nd** and Press **Y =** for **STAT PLOT**
Press **1** for **Plot 1**

Set the cursor so that **ON** is flashing.
For **Type**, use the arrow and Enter keys to highlight and select the histogram.
For **Xlist**, choose the column containing the data (in most cases, L1).
Note: Press **2nd 1** for **L1**
Freq should be set to 1.

Step 3 *Select your window settings*
Press **WINDOW** and adjust the settings as follows:
Xmin = lowest class boundary
Xmax = highest class boundary
Xscl = class width
Ymin = 0
Ymax ≥ greatest class frequency
Yscl = 1
Xres = 1

Step 4 *View the graph*
Press **GRAPH**

Optional *Read class frequencies and class boundaries*

Step You can press **TRACE** to read the class frequencies and class boundaries. Use the arrow keys to move between bars.

Example The figures below show TI-83 window settings and histogram for the following sample data:
86, 70, 62, 98, 73, 56, 53, 92, 86, 37, 62, 83, 78, 49, 78, 37, 67, 79, 57

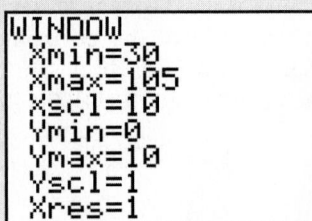
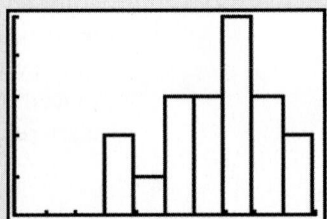

II Making a Histogram from a Frequency Table

Step 1 *Enter the data*
Press **STAT** and select **1:Edit**
Note: If a list already contains data, clear the old data. Use the up arrow to highlight the list name, 'L1' or 'L2'.
Press **CLEAR ENTER**.
Enter the midpoint of each class into **L1**
Enter the class frequencies or relative frequencies into **L2**

Step 2 *Set up the histogram plot*
Press **2nd** and **Y** = for **STAT PLOT**
Press **1** for **Plot 1**
Set the cursor so that **ON** is flashing.

For **Type**, use the arrow and Enter keys to highlight and select the histogram.
For **Xlist**, choose the column containing the midpoints.
For **Freq**, choose the column containing the frequencies or relative frequencies.

Step 3–4 *Follow steps 3–4 given above.*
   Note: To set up the Window for relative frequencies, be sure to set **Ymax**
   to a value that is greater than or equal to the largest relative frequency.

Statistics in Action Revisited

Interpreting Histograms

A quantitative variable measured in the *Psychological Science* "water-level task" experiment was *Deviation angle* (measured in degrees) of the judged line from the true surface line (parallel to the table top). The smaller the deviation, the better the task performance. Recall that the researchers want to test the prevailing theory that males will do better than females on judging the correct water level. To check this theory, we accessed the **EYECUE** data file in MINITAB and created two frequency histograms for deviation angle — one for male subjects and one for female subjects. These side-by-side histograms are displayed in Figure SIA2.5.

From the histograms, there is some support for the theory. The histogram for males (the histogram on the right in Figure SIA2.5) shows the center of the distribution at about 0°, with most (about 70%) of the deviation values falling between −10° and 10°. In general, the male subjects tended to perform fairly well on the water-level task. Alternatively, the histogram for females (the histogram on the left in Figure SIA2.5) is centered at about 10°, with most (about 70%) of the deviation values falling above 0°. Thus, the histogram for females shows a tendency for female subjects to have greater deviations (and, thus, poorer performances) in the water-level task than males. In later chapters, we'll learn how to attach a measure of reliability to such an inference.

Figure SIA2.5

MINITAB Histograms for Deviation Angle in the Water-Level Task — Males versus Females

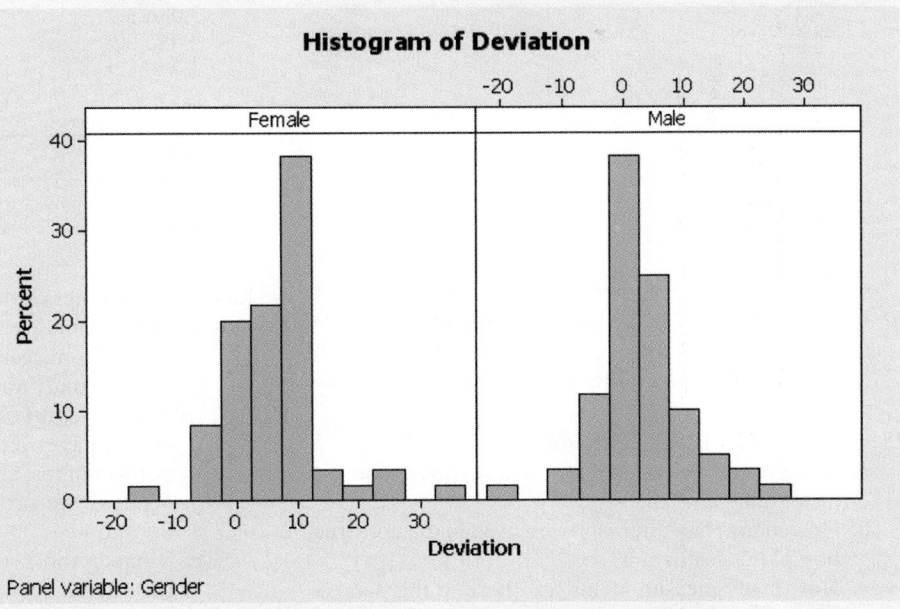

Exercises 2.21–2.41

Understanding the Principles

2.21 Explain the difference between a dot plot and a stem-and-leaf display.

2.22 Explain the difference between the stem and the leaf in a stem-and-leaf display.

2.23 In a histogram, what are the class intervals?

2.24 How many classes are recommended in a histogram for a data set with more than 50 observations?

Learning the Mechanics

2.25 Consider the stem-and-leaf display shown here.
NW

Stem	Leaf
5	1
4	457
3	00036
2	1134599
1	2248
0	012

a. How many observations were in the original data set?

b. In the bottom row of the stem-and-leaf display, identify the stem, the leaves, and the numbers in the original data set represented by this stem and its leaves.

c. Re-create all the numbers in the data set and construct a dot plot.

Learning the Mechanics

2.26 Graph the relative frequency histogram for the 500 measurements summarized in the accompanying relative frequency table.

Class Interval	Relative Frequency
.5–2.5	.10
2.5–4.5	.15
4.5–6.5	.25
6.5–8.5	.20
8.5–10.5	.05
10.5–12.5	.10
12.5–14.5	.10
14.5–16.5	.05

2.27 Refer to Exercise 2.26. Calculate the number of the 500 measurements falling into each of the measurement classes. Then graph a frequency histogram for these data.

2.28 Consider the following histogram:

a. Is this a frequency histogram or a relative frequency histogram? Explain.

b. How many class intervals were used in the construction of this histogram?

c. How many measurements are there in the data set described by this histogram?

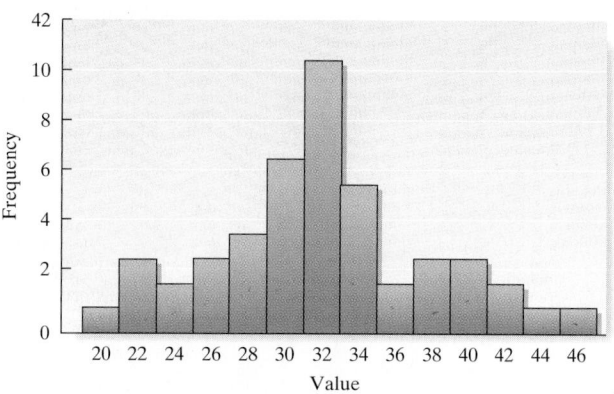

Applying the Concepts—Basic

2.29 **Earthquake magnitudes.** The strength of an earthquake — called magnitude — is measured quantitatively using the Richter scale. A magnitude below 3.0 is considered a very minor earthquake, while a magnitude of 7.0 or above is considered a major earthquake. The National Earthquake Information Center located 31,366 earthquakes worldwide in 2003. The magnitudes (on the Richter scale) of these earthquakes are summarized in the accompanying table. Display the data in a graph.

Magnitude	Number of Earthquakes
8.0 to 9.9	1
7.0 to 7.9	14
6.0 to 6.9	140
5.0 to 5.9	1,160
4.0 to 4.9	8,419
3.0 to 3.9	7,705
2.0 to 2.9	7,723
1.0 to 1.9	2,471
0.1 to 0.9	128
No Magnitude	3,605
Total	31,366

Source: National Earthquake Information Center, Department of Interior, U.S. Geological Survey.

2.30 **Aggressiveness scores.** The graph on p. 25 summarizes the scores obtained by 100 students on a questionnaire designed to measure aggressiveness. (Scores are integer values that range from 0 to 20. A high score indicates a high level of aggression.)

a. Which measurement class contains the highest proportion of test scores?

b. What is the proportion of scores that lie between 3.5 and 5.5?

c. What is the proportion of scores that are higher than 11.5?

d. How many students scored less than 5.5?

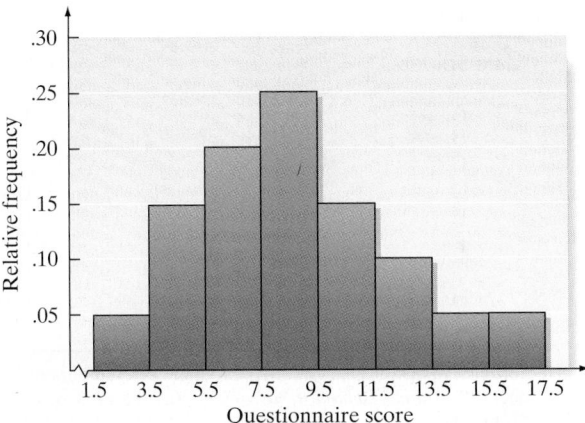

2.31 Reading Japanese books. Refer to the *Reading in a Foreign Language* (Apr. 2004) experiment to improve the Japanese reading comprehension levels of University of Hawaii students, Exercise 2.7 (p. 37). Fourteen students participated in a 10-week extensive reading program in a second semester Japanese course. The number of books read by each student and the student's course grade are listed in the table.

🔵 **JAPANESE**

Number of Books	Course Grade	Number of Books	Course Grade
53	A	30	A
42	A	28	B
40	A	24	A
40	B	22	C
39	A	21	B
34	A	20	B
34	A	16	B

Source: Hitosugi, C. I., and Day, R. R. "Extensive reading in Japanese," *Reading in a Foreign Language*, Vol. 16, No. 1, Apr. 2004 (Table 4).

a. Construct a stem-and-leaf display for the number of books read by the students.
b. Highlight (or circle) the leaves in the display that correspond to students who earned an A grade in the course. What inference can you make about these students?

2.32 Ratings of chess players. The United States Chess Federation (USCF) establishes a numerical rating for each competitive chess player. The USCF rating is a number between 0 and 4,000 that changes over time depending on the outcome of tournament games. The higher the rating, the better (more successful) the player. The next graph describes the rating distribution of 27,563 players who were active competitors in 1997.
a. What type of graph is displayed?
b. Is the variable displayed on the graph quantitative or qualitative?
c. What percentage of players has a USCF rating above 1,000?

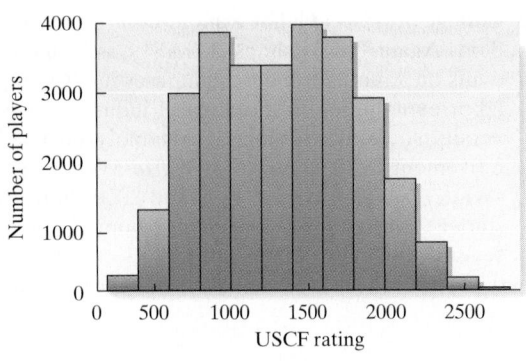

Source: United States Chess Federation, Jan. 1998.

Applying the Concepts—Intermediate

🔵 **DDT**

2.33 Contaminated fish. Refer to Exercise 2.16 (p. 39) and the U.S. Army Corps of Engineers data on contaminated fish saved in the DDT file. In addition to species (channel catfish, largemouth bass, or smallmouth buffalofish), the length (in centimeters), weight (in grams), and DDT level (in parts per million) were measured for each of the 144 captured fish.
a. Use a graphical method to describe the distribution of the 144 fish lengths.
b. Use a graphical method to describe the distribution of the 144 fish weights.
c. Use a graphical method to describe the distribution of the 144 DDT measurements.

2.34 Radioactive lichen. Lichen has a high absorbance capacity for radiation fallout from nuclear accidents. Since lichen is a major food source for Alaskan caribou, and caribou are, in turn, a major food source for many Alaskan villagers, it is important to monitor the level of radioactivity in lichen. Researchers at the University of Alaska, Fairbanks collected data on 9 lichen specimens at various locations for this purpose. The amount of the radioactive element cesium-137 was measured (in microcuries per milliliter) for each specimen. The data values, converted to logarithms, are given in the table. (Note that the closer the value is to zero, the greater the amount of cesium in the specimen.)

🔵 **LICHEN**

Location			
Bethel	−5.50	−5.00	
Eagle Summit	−4.15	−4.85	
Moose Pass	−6.05		
Turnagain Pass	−5.00		
Wickersham Dome	−4.10	−4.50	−4.60

Source: Lichen Radionuclide Baseline Research Project, 2003.

a. Construct a dot plot for the nine measurements.
b. Construct a stem-and-leaf display for the nine measurements.
c. Construct a histogram plot for the nine measurements.
d. Which of the three graphs, parts **a–c**, is more informative?
e. What proportion of the measurements have a radioactivity level of −5.00 or lower?

2.35 College protests of labor exploitation. The United Students Against Sweatshops (USAS) was formed by students on American college campuses in 1999 to protest labor exploitation in the apparel industry. Clark University sociologist Robert Ross analyzed the USAS movement in the *Journal of World-Systems Research* (Winter 2004). Between 1999 and 2000 there were 18 student "sit-ins" for a "sweat free campus" organized at several universities. The table gives the duration (in days) of each sit-in as well as the number of student arrests.

⚙ **SITIN**

Sit-in	Year	University	Duration (days)	Number of Arrests	Tier Ranking
1	1999	Duke	1	0	1st
2	1999	Georgetown	4	0	1st
3	1999	Wisconsin	1	0	1st
4	1999	Michigan	1	0	1st
5	1999	Fairfield	1	0	1st
6	1999	North Carolina	1	0	1st
7	1999	Arizona	10	0	1st
8	2000	Toronto	11	0	1st
9	2000	Pennsylvania	9	0	1st
10	2000	Macalester	2	0	1st
11	2000	Michigan	3	0	1st
12	2000	Wisconsin	4	54	1st
13	2000	Tulane	12	0	1st
14	2000	SUNY Albany	1	11	2nd
15	2000	Oregon	3	14	2nd
16	2000	Purdue	12	0	2nd
17	2000	Iowa	4	16	2nd
18	2000	Kentucky	1	12	2nd

Source: Ross, R. J. S. "From antisweatshop to global justice to antiwar: How the new new left is the same and different from the old new left." *Journal of Word-Systems Research,* Vol. X, No. 1, Winter 2004 (Tables 1 and 3).

a. Summarize the data on sit-in duration using a stem-and-leaf display.

b. Highlight (or circle) the leaves in the display that correspond to sit-ins where at least one arrest was made. Does the pattern revealed support the theory that sit-ins of longer duration are more likely to lead to arrests?

2.36 Fluid loss in spiders. A group of University of Virginia biologists studied nectivory (nectar drinking) in crab spiders to determine if adult males were feeding on nectar to prevent fluid loss. (*Animal Behavior,* June 1995.) Nine male spiders were weighed and then placed on the flowers of Queen Anne's lace. One hour later, the spiders were removed and reweighed. The evaporative fluid loss (in milligrams) of each of the nine male spiders is given in the next table.

⚙ **SPIDERS**

Male Spider	Fluid Loss
A	.018
B	.020
C	.017
D	.024
E	.020
F	.024
G	.003
H	.001
I	.009

Source: Pollard, S. D., et al. "Why do male crab spiders drink nectar?" *Animal Behavior,* Vol. 49, No. 6, June 1995, p.1445 (Table II).

a. Summarize the fluid losses of male crab spiders with a stem-and-leaf display.

b. Of the nine spiders, only three drank any nectar from the flowers of Queen Anne's lace. These three spiders are identified as G, H, and I in the table. Locate and circle these three fluid losses on the stem-and-leaf display. Does the pattern depicted in the graph give you any insight into whether feeding on flower nectar reduces evaporative fluid loss for male crab spiders? Explain.

2.37 Research on brain specimens. *Postmortem interval* (PMI) is defined as the elapsed time between death and an autopsy. Knowledge of PMI is considered essential when conducting medical research on human cadavers. The data in the table are the PMIs of 22 human brain specimens obtained at autopsy in a recent study. (*Brain and Language*, June 1995.) Graphically describe the PMI data with a dot plot. Based on the plot, make a summary statement about the PMI of the 22 human brain specimens.

⚙ **BRAINPMI**

Postmortem Intervals for 22 Human Brain Specimens

5.5	14.5	6.0	5.5	5.3	5.8	11.0	6.1
7.0	14.5	10.4	4.6	4.3	7.2	10.5	6.5
3.3	7.0	4.1	6.2	10.4	4.9		

Source: Hayes, T. L., and Lewis, D. A. "Anatomical specialization of the anterior motor speech area: Hemispheric differences in magnopyramidal neurons." *Brain and Language,* Vol. 49, No. 3, June 1995, p. 292 (Table 1).

2.38 Research on eating disorders. Data from a psychology experiment were reported and analyzed in *The American Statistician* (May 2001). Two samples of female students participated in the experiment. One sample consisted of 11 students known to suffer from the eating disorder bulimia; the other sample consisted of 14 students with normal eating habits. Each student completed a questionnaire from which a "fear of negative evaluation" (FNE) score was produced. (The higher the score, the greater the fear of negative evaluation.) The data are displayed in the table on p. 53.

a. Construct a dot plot or stem-and-leaf display for the FNE scores of all 25 female students.
b. Highlight the bulimic students on the graph, part **a**. Does it appear that bulimics tend to have a greater fear of negative evaluation? Explain.
c. Why is it important to attach a measure of reliability to the inference made in part **b**?

🔵 BULIMIA

Bulimic
students: 21 13 10 20 25 19 16 21 24 13 14
Normal
students: 13 6 16 13 8 19 23 18 11 19 7 10 15 20

Source: Randles, R. H. "On neutral responses (zeros) in the sign test and ties in the Wilcoxon-Mann-Whitney test," *The American Statistician,* Vol. 55, No. 2, May 2001 (Figure 3).

2.39 Eclipses and occults. Saturn has five satellites that rotate around the planet. *Astronomy* (Aug. 1995) lists 19 different events involving eclipses or occults of Saturnian satellites during the month of August. For each event, the percent of light lost by the eclipsed or occulted satellite at midevent is recorded in the table.

🔵 SATURN

Date	Event	Light Loss (%)
Aug. 2	Eclipse	65
4	Eclipse	61
5	Occult	1
6	Eclipse	56
8	Eclipse	46
8	Occult	2
9	Occult	9
11	Occult	5
12	Occult	39
14	Occult	1
14	Eclipse	100
15	Occult	5
15	Occult	4
16	Occult	13
20	Occult	11
23	Occult	3
23	Occult	20
25	Occult	20
28	Occult	12

Source: ASTRONOMY magazine, Aug. 1995, p. 60.

a. Construct a stem-and-leaf display for light loss percentage of the 19 events.
b. Locate on the stem-and-leaf plot, part **a**, the light losses associated with eclipses of Saturnian satellites. (Circle the light losses on the plot.)
c. Based on the marked stem-and-leaf display, part **b**, make an inference about which event type (eclipse or occult) is more likely to lead to a greater light loss.

Applying the Concepts—Advanced

2.40 Comparing SAT scores. Educators are constantly evaluating the efficacy of public schools in the education and training of American students. One quantitative assessment of change over time is the difference in scores on the SAT, which has been used for decades by colleges and universities as one criterion for admission. The SATSCORES file contains the average SAT scores for each of the 50 states and District of Columbia for 1990 and 2000. Selected observations are shown below.

🔵 SATSCORES (First five and last two deservations)

State	1990	2000
Alabama	1079	1114
Alaska	1015	1034
Arizona	1041	1044
Arkansas	1077	1117
California	1002	1015
⋮	⋮	⋮
Wisconsin	1111	1181
Wyoming	1072	1090

Source: College Entrance Examination Board, 2001.

a. Use graphs to display the two SAT score distributions. How have the distributions of average state scores changed over the last decade?
b. As another method of comparing the 1990 and 2000 average SAT scores, compute the **paired difference** by subtracting the 1990 score from the 2000 score for each state. Summarize these differences with a graph.
c. Interpret the graph, part **b**. How do your conclusions compare to those you reached when comparing the two graphs in part **a**?
d. What is the largest improvement in the average score between 1990 and 2000 as indicated on the graph, part **b**? With which state is this improvement associated in the table of data?

2.41 Speech listening study. The role that listener knowledge plays in the perception of imperfectly articulated speech was investigated in the *American Journal of Speech–Language Pathology* (Feb. 1995). Thirty female college students, randomly divided into three groups of ten, participated as listeners in the study. All subjects were required to listen to a 48-sentence audiotape of a Korean woman with cerebral palsy and asked to transcribe her entire speech. For the first group of students (the *control group*), the speaker used her normal manner of speaking. For the second group (the *treatment group*), the speaker employed a learned breathing pattern (called breath-group strategy) to improve speech efficiency. The third group (the *familiarity group*) also listened to the tape with the breath-group strategy, but only after they had practiced listening twice to another tape in which they were told exactly what the speaker was saying. At the end of the listening/transcribing session, two quantitative variables were measured for each listener: (1) the total number of words transcribed (called the rate of response) and (2) the percentage of words correctly transcribed (called the accuracy score). The data for all 30 subjects are provided in the table on p. 54.

a. Use a graphical method to describe the differences in the distributions of response rates among the three groups.
b. Use a graphical method to describe the distribution of accuracy scores for the three listener groups.

LISTEN

Control Group		Treatment Group		Familiarization Group	
Rate of Response	Percent Correct	Rate of Response	Percent Correct	Rate of Response	Percent Correct
250	23.6	254	26.0	193	36.0
230	26.0	178	32.6	223	41.0
197	26.0	139	32.6	232	43.0
238	26.7	249	33.0	214	44.0
174	29.2	236	34.4	269	44.0
263	29.5	231	36.5	256	46.0
275	31.3	161	38.5	224	46.0
193	32.9	255	40.0	225	48.0
204	35.4	275	41.7	288	49.0
168	29.2	181	44.8	244	52.0

Source: Tjaden, K., and Liss, J. M. "The influence of familiarity on judgments of treated speech." *American Journal of Speech–Language Pathology,* Vol. 4, No. 1, Feb. 1995, p. 43 (Table 1).

2.3 Summation Notation

Now that we've examined some graphical techniques for summarizing and describing quantitative data sets, we turn to numerical methods for accomplishing this objective. Before giving the formulas for calculating numerical descriptive measures, let's look at some shorthand notation that will simplify our calculation instructions. Remember that such notation is used for one reason only — to avoid repeating the same verbal descriptions over and over. If you mentally substitute the verbal definition of a symbol each time you read it, you'll soon get used to it.

We denote the measurements of a quantitative data set as

$$x_1, x_2, x_3, \ldots, x_n$$

where x_1 is the first measurement in the data set, x_2 is the second measurement in the data set, x_3 is the third measurement in the data set, ..., and x_n is the nth (and last) measurement in the data set. Thus, if we have five measurements in a set of data, we will write x_1, x_2, x_3, x_4, x_5 to represent the measurements. If the actual numbers are 5, 3, 8, 5, and 4, we have $x_1 = 5$, $x_2 = 3$, $x_3 = 8$, $x_4 = 5$, and $x_5 = 4$.

Most of the formulas we use require a summation of numbers. For example, one sum we'll need to obtain is the sum of all the measurements in the data set, or $x_1 + x_2 + x_3 + \cdots + x_n$. To shorten the notation, we use the symbol $\sum$ for the summation. That is, $x_1 + x_2 + x_3 + \cdots + x_n = \sum_{i=1}^{n} x_i$. Verbally translate $\sum_{i=1}^{n} x_i$ as follows: "The sum of the measurements, whose typical member is x_i, beginning with the member x_1 and ending with the member x_n."

Suppose, as in our earlier example, $x_1 = 5$, $x_2 = 3$, $x_3 = 8$, $x_4 = 5$, and $x_5 = 4$.

Then the sum of the five measurements, denoted $\sum_{i=1}^{5} x_i$ is obtained as follows:

$$\sum_{i=1}^{5} x_i = x_1 + x_2 + x_3 + x_4 + x_5$$

$$= 5 + 3 + 8 + 5 + 4 = 25$$

Another important calculation requires that we square each measurement and then sum the squares. The notation for this sum is $\sum_{i=1}^{n} x_i^2$. For the five measurements, we have

$$\sum_{i=1}^{5} x_i^2 = x_1^2 + x_2^2 + x_3^2 + x_4^2 + x_5^2$$

$$= 5^2 + 3^2 + 8^2 + 5^2 + 4^2$$

$$= 25 + 9 + 64 + 25 + 16 = 139$$

In general, the symbol following the summation sign $\sum$ represents the variable (or function of the variable) that is to be summed.

> **The Meaning of Summation Notation** $\sum_{i=1}^{n} x_i$
>
> Sum the measurements on the variable that appears to the right of the summation symbol, beginning with the 1st measurement and ending with the nth measurement.

Exercises 2.42–2.45

Learning the Mechanics

Note: In all exercises, $\sum$ represents $\sum_{i=1}^{n}$.

2.42 A data set contains the observations 5, 1, 3, 2, 1. Find
a. $\sum x = 12$ b. $\sum x^2 = 40$ c. $\sum (x - 1) = 7$
d. $\sum (x - 1)^2$ e. $\left(\sum x\right)^2 = 144$

2.43 Suppose a data set contains the observations 3, 8, 4, 5, 3, 4, 6. Find
a. $\sum x$ b. $\sum x^2$ c. $\sum (x - 5)^2$
d. $\sum (x - 2)^2$ e. $\left(\sum x\right)^2$

2.44 Refer to Exercise 2.42. Find
a. $\sum x^2 - \dfrac{\left(\sum x\right)^2}{5}$ b. $\sum (x - 2)^2$ c. $\sum x^2 - 10$

2.45 A data set contains the observations 6, 0, −2, −1, 3. Find
a. $\sum x$ b. $\sum x^2$ c. $\sum x^2 - \dfrac{\left(\sum x\right)^2}{5}$

(handwritten)
a) $40 - \dfrac{144}{5}$

b) $(5-2)^2 + (1-2)^2 + (3-2)^2 + (2-2)^2 + (1-2)^2$
$\quad 9 + 1 + 1 + 0 + 1 = 12$

2.4 Numerical Measures of Central Tendency

When we speak of a data set, we refer to either a sample or to a population. If statistical inference is our goal, we'll wish ultimately to use sample **numerical descriptive measures** to make inferences about the corresponding measures for a population.

As you'll see, a large number of numerical methods are available to describe quantitative data sets. Most of these methods measure one of two data characteristics:

1. The **central tendency** of the set of measurements — that is, the tendency of the data to cluster, or center, about certain numerical values (see Figure 2.14a).

2. The **variability** of the set of measurements — that is, the spread of the data (see Figure 2.14b).

In this section we concentrate on **measures of central tendency**. In the next section, we discuss measures of variability.

The most popular and best understood measure of central tendency for a quantitative data set is the *arithmetic mean* (or simply the mean) of a data set.

Figure 2.14
Numerical Descriptive
Measures

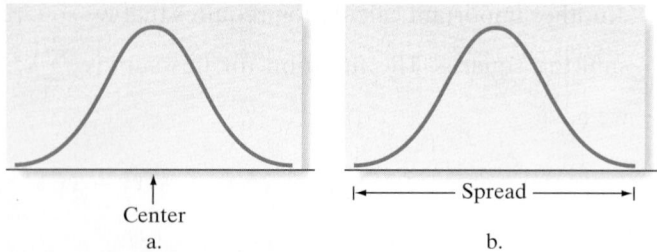

Center
a.

Spread
b.

DEFINITION 2.5

The **mean** of a set of quantitative data is the sum of the measurements divided by the number of measurements contained in the data set.

In everyday terms, the mean is the average value of the data set and is often used to represent a "typical" value. We denote the **mean of a sample** of measurements by $\bar{x}$ (read "x-bar"), and represent the formula for its calculation as shown in the box:

Formula for a Sample Mean

$$\bar{x} = \frac{\sum_{i=1}^{n} x_i}{n}$$

EXAMPLE 2.3 COMPUTING THE SAMPLE MEAN

Problem Calculate the mean of the following five sample measurements: 5, 3, 8, 5, 6.

Solution Using the definition of sample mean and the summation notation, we find

$$\bar{x} = \frac{\sum_{i=1}^{5} x_i}{5} = \frac{5 + 3 + 8 + 5 + 6}{5} = \frac{27}{5} = 5.4$$

Thus, the mean of this sample is 5.4.

Look Back There is no specific rule for rounding when calculating $\bar{x}$ because $\bar{x}$ is specifically defined to be the sum of all measurements divided by n; that is, it is a specific fraction. When $\bar{x}$ is used for descriptive purposes, it is often convenient to round the calculated value of $\bar{x}$ to the number of significant figures used for the original measurements. When $\bar{x}$ is to be used in other calculations, however, it may be necessary to retain more significant figures.

Now Work *Exercise 2.55*

EXAMPLE 2.4 FINDING THE MEAN ON A PRINTOUT

Problem Calculate the sample mean for the 100 EPA mileages given in Table 2.2.

Solution The mean gas mileage for the 100 cars is denoted

$$\bar{x} = \frac{\sum_{i=1}^{100} x_i}{100}$$

Rather than compute $\bar{x}$ by hand (or calculator), we employed SAS to compute the mean. The SAS printout is shown in Figure 2.15. The sample mean, highlighted on the printout, is $\bar{x} = 36.9940$.

The MEANS Procedure

Analysis Variable : MPG

Mean	Std Dev	Variance	N	Minimum	Maximum	Median
36.9940000	2.4178971	5.8462263	100	30.0000000	44.9000000	37.0000000

Figure 2.15

SAS Numerical Descriptive Measures for 100 EPA Gas Mileages

Look Back Given this information, you can visualize a distribution of gas mileage readings centered in the vicinity of $\bar{x} \approx 37$. An examination of the relative frequency histogram (Figure 2.10) confirms that $\bar{x}$ does in fact fall near the center of the distribution.

■ ■ ■

The sample mean $\bar{x}$ will play an important role in accomplishing our objective of making inferences about populations based on sample information. For this reason we need to use a different symbol for the *mean of a population* — the mean of the set of measurements on every unit in the population. We use the Greek letter μ (mu) for the population mean.

> ### Symbols for the Sample Mean and the Population Mean
>
> In this text, we adopt a general policy of using Greek letters to represent numerical descriptive measures for the population and Roman letters to represent corresponding descriptive measures for the sample. The symbols for the mean are
>
> $\bar{x}$ = Sample mean μ = Population mean

We'll often use the sample mean, $\bar{x}$, to estimate (make an inference about) the population mean, μ. For example, the EPA mileages for the population consisting of *all* cars has a mean equal to some value, μ. Our sample of 100 cars yielded mileages with a mean of $\bar{x} = 36.9940$. If, as is usually the case, we don't have access to the measurements for the entire population, we could use $\bar{x}$ as an estimator or approximator for μ. Then we'd need to know something about the reliability of our inference. That is, we'd need to know how accurately we might expect $\bar{x}$ to estimate μ. In Chapter 7, we'll find that this accuracy depends on two factors:

1. The *size of the sample*. The larger the sample, the more accurate the estimate will tend to be.

2. The *variability, or spread, of the data*. All other factors remaining constant, the more variable the data, the less accurate the estimate.

Another important measure of central tendency is the **median**.

> ### DEFINITION 2.6
>
> The **median** of a quantitative data set is the middle number when the measurements are arranged in ascending (or descending) order.

The median is of most value in describing large data sets. If the data set is characterized by a relative frequency histogram (Figure 2.16), the median is the point on the x-axis such that half the area under the histogram lies above the medi-

Figure 2.16
Location of the Median

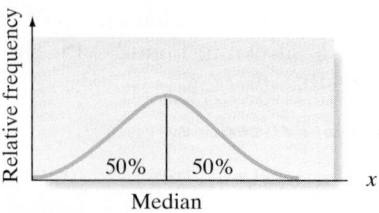

an and half lies below. [*Note*: In Section 2.2 we observed that the relative frequency associated with a particular interval on the *x*-axis is proportional to the amount of area under the histogram that lies above the interval.] We denote the *median* of a *sample* by M.

Calculating a Sample Median, M

Arrange the *n* measurements from the smallest to the largest.

1. If *n* is odd, *M* is the middle number.
2. If *n* is even, *M* is the mean of the middle two numbers.

EXAMPLE 2.5   COMPUTING THE MEDIAN

Problem   Consider the following sample of $n = 7$ measurements: 5, 7, 4, 5, 20, 6, 2

    a. Calculate the median *M* of this sample.

    b. Eliminate the last measurement (the 2) and calculate the median of the remaining $n = 6$ measurements.

Solution   **a.** The seven measurements in the sample are ranked in ascending order: 2, 4, 5, 5, 6, 7, 20. Because the number of measurements is odd, the median is the middle measurement. Thus, the median of this sample is $M = 5$.

    b. After removing the 2 from the set of measurements, we rank the sample measurements in ascending order as follows: 4, 5, 5, 6, 7, 20. Now the number of measurements is even, so we average the middle two measurements. The median is $M = (5 + 6)/2 = 5.5$.

Look Back When the sample size *n* is even (as in part **b**), exactly half of the measurements will fall below the calculated median *M*. However, when *n* is odd (as in part **a**), the percentage of measurements that fall below *M* is approximately 50%. The approximation improves as *n* increases.

    Now Work   *Exercise 2.52*

    In certain situations, the median may be a better measure of central tendency than the mean. In particular, the median is less sensitive than the mean to extremely large or small measurements. Note, for instance, that all but one of the measurements in part **a** of Example 2.5 center about $x = 5$. The single relatively large measurement, $x = 20$, does not affect the value of the median, 5, but it causes the mean, $\bar{x} = 7$, to lie to the right of most of the measurements.

    As another example of data for which the central tendency is better described by the median than the mean, consider the household incomes of a community being studied by a sociologist. The presence of just a few households with very high incomes will affect the mean more than the median. Thus, the median will provide a more accurate picture of the typical income for the community. The mean could exceed the vast majority of the sample measurements (household incomes), making it a misleading measure of central tendency.

EXAMPLE 2.6   FINDING THE MEDIAN ON A PRINTOUT

Problem Calculate the median for the 100 EPA mileages given in Table 2.2. Compare the median to the mean computed in Example 2.4.

Solution For this large data set, we again resort to a computer analysis. The median is highlighted on the SAS printout displayed in Figure 2.15 (p. 57). You can see that the median is 37.0. This value implies that half of the 100 mileages in the data set fall below 37.0 and half lie above 37.0. Note that the median, 37.0, and the mean, 36.9940, are almost equal. This fact indicates a lack of **skewness** in the data — that is, the tendency to have as many measurements in the left tail of the distribution as in the right tail (recall the histogram, Figure 2.10).

Look Back In general, extreme values (large or small) affect the mean more than the median since these values are used explicitly in the calculation of the mean. On the other hand, the median is not affected directly by extreme measurements since only the middle measurement (or two middle measurements) is explicitly used to calculate the median. Consequently, if measurements are pulled toward one end of the distribution, the mean will shift toward that tail more than the median.

DEFINITION 2.7

A data set is said to be **skewed** if one tail of the distribution has more extreme observations than the other tail.

A comparison of the mean and median gives us a general method for detecting skewness in data sets, as shown in the next box.

Detecting Skewness by Comparing the Mean and the Median

If the data set is skewed to the right, then the median is less than the mean.

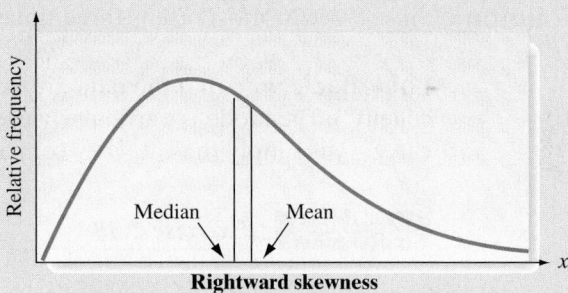

Rightward skewness

If the data set is symmetric, the mean equals the median.

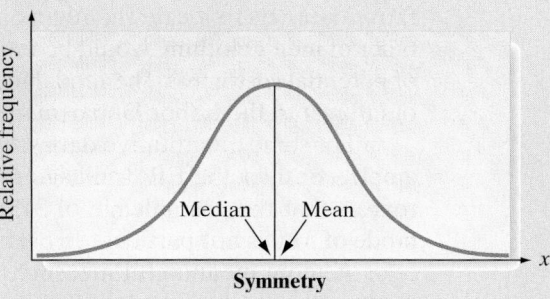

Symmetry

If the data set is skewed to the left, the mean is less than (to the left of) the median.

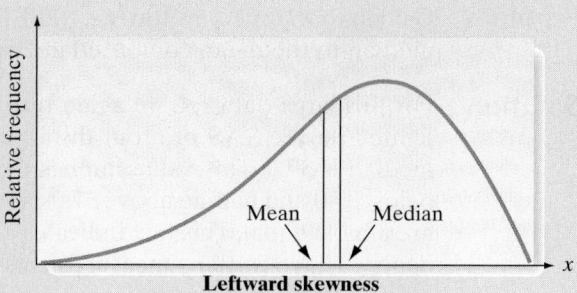

Leftward skewness

Now Work   *Exercise 2.51*

A third measure of central tendency is the **mode** of a set of measurements.

DEFINITION 2.8
The **mode** is the measurement that occurs most frequently in the data set.

Therefore, the mode shows where the data tend to concentrate.

EXAMPLE 2.7   FINDING THE MODE

Problem   Each of ten taste testers rated a new brand of barbecue sauce on a 10-point scale, where 1 = awful and 10 = excellent. Find the mode for the ten ratings shown below.

$$8 \quad 7 \quad 9 \quad 6 \quad 8 \quad 10 \quad 9 \quad 9 \quad 5 \quad 7$$

Solution   Since 9 occurs most often (three times), the mode of the ten taste ratings is 9.

Look Back Note that the data are actually qualitative in nature (e.g., "awful," "excellent"). The mode is particularly useful for describing qualitative data. The modal category is simply the category (or class) that occurs most often.

Now Work   *Exercise 2.53*

∎ ∎ ∎

Because it emphasizes data concentration, the mode is also used with quantitative data sets to locate the region in which much of the data is concentrated. A retailer of men's clothing would be interested in the modal neck size and sleeve length of potential customers. The modal income class of the laborers in the United States is of interest to the Labor Department.

For some quantitative data sets, the mode may not be very meaningful. For example, consider the EPA mileage ratings in Table 2.2. A reexamination of the data reveals that the gas mileage of 37.0 occurs most often (four times). However, the mode of 37.0 is not particularly useful as a measure of central tendency.

A more meaningful measure can be obtained from a relative frequency histogram for quantitative data. The measurement class containing the largest relative frequency is called the **modal class**. Several definitions exist for locating the position of the mode within a modal class, but the simplest is to define the mode as the midpoint of the modal class. For example, examine the frequency histogram for the EPA mileage ratings, in Figure 2.10 (p. 43). You can see that the modal class is the interval 36.0–37.5. The mode (the midpoint) is 36.75. This modal class (and the mode itself) identifies the area in which the data are most concentrated, and in that sense it is a measure of central tendency. However, for most applications involving quantitative data, the mean and median provide more descriptive information than the mode.

EXAMPLE 2.8   COMPARING THE MEAN, MEDIAN, AND MODE

Problem   Each year, *Business Week* magazine compiles its "Executive Compensation Scoreboard" based on a survey of approximately 350 executives at U.S. companies. The 2003 scoreboard contains the total annual pay for each CEO as well as the dollar value of a $100 investment in the CEO's company stock made three years earlier. If the ratio of shareholder return to total pay is high, *Business Week* deems the CEO "worth his/her pay." Based on the return-to-pay ratio, each CEO was assigned a rating of 1 to 5, where 1 represents a "low" ratio and 5 represents a "high" ratio. (*Business Week*, April 21, 2003.) The data for the 2003 scoreboard, saved in the **BWECS** file, includes the variables "shareholder return" and "rating." Find the mean, median, and mode for both of these variables. Which measure of central tendency is better for describing the distribution of shareholder returns? Ratings?

Solution   Measures of central tendency for the two variables were obtained using SPSS. The means, medians, and modes are displayed at the top of the SPSS printout, Figure 2.17.

For "shareholder return," the mean, median, and mode are $111.80, $103.50, and $125, respectively. Note that the mean is greater than the median, indicating that the data is skewed right. This rightward skewness (graphically shown on the histogram for the returns in the middle of Figure 2.17) is due to a few exceptionally high shareholder return values. Consequently, we would probably want to use the median, $103.50, as the "typical" value for shareholder return. The mode of $125 is the return value that occurs most often in the data set, but it is not very descriptive of the "center" of the return distribution.

For "Rating," the mean, median, and mode are 3.16, 3, and 3, respectively. The rating value assigned by *Business Week* is really categorical in nature (low return-to-pay ratios are assigned the "low" rating of 1, high return-to-pay ratios are assigned the "high" rating of 5). Consequently, the more meaningful measure of central tendency would be the mode of 3—it is the most frequently occurring category, as shown in the histogram at the bottom of Figure 2.17. The mean of 3.16 would not be very descriptive of this type of data.

Look Back The choice of which measure of central tendency to use will depend on the properties of the data set analyzed and the application. Consequently, it is vital that you understand how the mean, median, and mode are computed.

Now Work   *Exercise 2.66 a,b*

■ ■ ■

Figure 2.17

SPSS Analysis of Return-on-
Investment and Rating for
CEOs in Executive
Compensation Scoreboard

Statistics

		return	rating
N	Valid	350	330
	Missing	13	33
Mean		111.80	3.16
Median		103.50	3.00
Mode		125	3

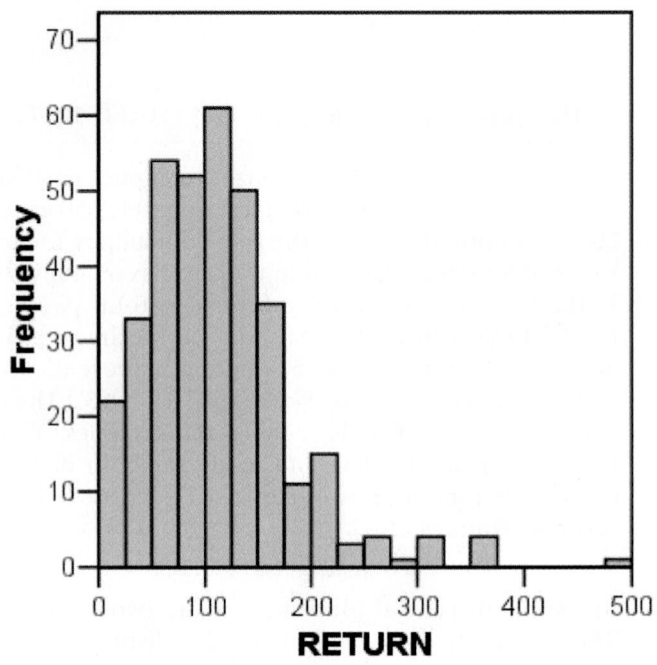

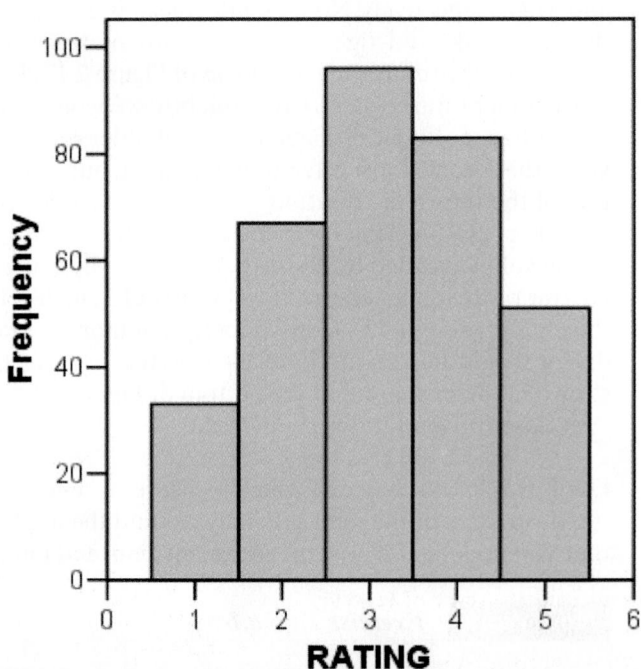

One-Variable Descriptive Statistics

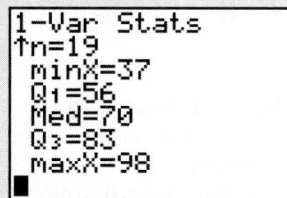

Using the TI-83 Graphing Calculator

Step 1   Enter the data
Press STAT and select 1:Edit
Note: If the list already contains data, clear the old data. Use the up arrow to highlight 'L1'. Press CLEAR ENTER.
Use the arrow and ENTER keys to enter the data set into L1.

Step 2   Calculate descriptive statistics
Press STAT
Press the right arrow key to highlight CALC
Press ENTER for 1-Var Stats
Enter the name of the list containing your data.
Press 2nd 1 for L1 (or 2nd 2 for L2 etc.)
Press ENTER

You should see the statistics on your screen. Some of the statistics are off the bottom of the screen. Use the down arrow to scroll through to see the remaining statistics. Use the up arrow to scroll back up.

Example   The descriptive statistics for the sample data set

86, 70, 62, 98, 73, 56, 53, 92, 86, 37, 62, 83, 78, 49, 78, 37, 67, 79, 57

The output screens for this example are shown below.

```
1-Var Stats
x̄=68.57894737
Σx=1303
Σx²=94897
Sx=17.54142966
σx=17.07357389
↓n=19
```

```
1-Var Stats
↑n=19
 minX=37
 Q₁=56
 Med=70
 Q₃=83
 maxX=98
■
```

Sorting
Data   The descriptive statistics do not include the mode. To find the mode, sort your data as follows:

Press STAT
Press 2 for SORTA(
Enter the name of the list your data is in. If your data is in L1, press 2nd 1

Press ENTER
The screen will say: DONE
To see the sorted data, press STAT and select 1:Edit
Scroll down through the list and locate the data value that occurs most frequently.

Exercises 2.46–2.68

Understanding the Principles:

2.46 Explain the difference between a measure of central tendency and a measure of variability.

2.47 Give three different measures of central tendency.

2.48 What is the symbol used to represent the sample mean? The population mean?

2.49 What two factors impact the accuracy of the sample mean as an estimate of the population mean?

2.50 Explain the concept of a skewed distribution.

2.51 Describe how the mean compares to the median for a distribution as follows:
  a. Skewed to the left
  b. Skewed to the right
  c. Symmetric

Learning the Mechanics

2.52 Calculate the mean and median of the following grade
NW point averages:

$$3.2 \quad 2.5 \quad 2.1 \quad 3.7 \quad 2.8 \quad 2.0$$

2.53 Calculate the mode, mean, and median of the following
NW data:

$$18 \quad 10 \quad 15 \quad 13 \quad 17 \quad 15 \quad 12 \quad 15 \quad 18 \quad 16 \quad 11$$

2.54 Construct one data set consisting of five measurements and another consisting of six measurements for which the medians are equal.

2.55 Calculate the mean for samples where
  a. $n = 10$, $\sum x = 85$
  b. $n = 16$, $\sum x = 400$
  c. $n = 45$, $\sum x = 35$
  d. $n = 18$, $\sum x = 242$

2.56 Calculate the mean, median, and mode for each of the following samples:
  a. $7, -2, 3, 3, 0, 4$
  b. $2, 3, 5, 3, 2, 3, 4, 3, 5, 1, 2, 3, 4$
  c. $51, 50, 47, 50, 48, 41, 59, 68, 45, 37$

Applying the Concepts

2.57 **Reading Japanese Books.** Refer to the *Reading in a Foreign Language* (Apr. 2004) experiment to improve the Japanese reading comprehension levels of 14 University of Hawaii students, Exercise 2.31 (p. 51). The number of books read by each student and the student's course grade are repeated in the table.

JAPANESE

Number of Books	Course Grade	Number of Books	Course Grade
53	A	30	A
42	A	28	B
40	A	24	A
40	B	22	C
39	A	21	B
34	A	20	B
34	A	16	B

Source: Hitosugi, C. I., and Day, R. R. "Extensive reading in Japanese," *Reading in a Foreign Language,* Vol. 16, No. 1, Apr. 2004 (Table 4).

  a. Find the mean, median, and mode of the number of books read. Interpret these values.

  b. What do the mean and median indicate about the skewness of the distribution of the data?

2.58 **Most powerful women in American** *Fortune* (Oct. 14, 2002) published a list of the 50 most powerful women in America. The data on age (in years) and title of each of these 50 women are stored in the **WPOWER50** file. The first five and last two observations of the data are listed in the accompanying table.
  a. Find the mean, median, and modal age of these 50 women.
  b. What do the mean and median indicate about the skewness of the age distribution?
  c. Construct a relative frequency histogram for the age data. What is the modal age class?

WPOWER50 (Selected observations)

Rank	Name	Age	Company	Title
1	Carly Fiorina	48	Hewlet-Packard	CEO
2	Betsy Holden	46	Kraft Foods	CEO
3	Meg Whitman	46	eBay	CEO
4	Indra Nooyi	46	PepsiCo	CFO
5	Andrea Jung	44	Avon Products	CEO
.	.	.		
49	Fran Keeth	56	Royal Dutch Petrol.	CEO
50	Heidi Miller	49	Bank One	EVP

Source: Fortune, Oct. 14, 2002.

2.59 **Ammonia in car exhaust.** Three-way catalytic converters have been installed in new vehicles in order to reduce pollutants from motor vehicle exhaust emissions. However, these converters unintentionally increase the level of ammonia in the air. *Environmental Science & Technology* (Sept. 1, 2000) published a study on the ammonia levels near the exit ramp of a San Francisco highway tunnel. The data in the table represent daily ammonia concentrations (parts per million) on eight randomly selected days during afternoon drive-time in the summer of a recent year,

AMMONIA

1.53	1.50	1.37	1.51	1.55	1.42	1.41	1.48

  a. Find the mean daily ammonia level in air in the tunnel.

  b. Find the median ammonia level.
  c. Interpret the values obtained in parts **a** and **b**.

DDT

2.60 **Contaminated fish.** Refer to Exercise 2.33 (p. 51) and the U.S. Army Corps of Engineers data on contaminated fish saved in the **DDT** file. Consider the quantitative variables length (in centimeters), weight (in grams), and DDT level (in parts per million).
  a. Find three numerical measures of central tendency for the 144 fish lengths. Interpret these values.

  b. Find three numerical measures of central tendency for the 144 fish weights. Interpret these values.

c. Find three numerical measures of central tendency for the 144 DDT measurements. Interpret these values.

d. Use the results, part **a**, and the graph of the data from Exercise 2.33a to make a statement about the type of skewness in the fish length distribution.

e. Use the results, part **b**, and the graph of the data from Exercise 2.33b to make a statement about the type of skewness in the fish weight distribution.

f. Use the results, part **c**, and the graph of the data from Exercise 2.33c to make a statement about the type of skewness in the fish DDT distribution.

2.61 Radioactive lichen. Refer to the University of Alaska study to monitor the level of radioactivity in lichen, Exercise 2.34 (p. 51). The amount of the radioactive element cesium-137 (measured in microcuries per milliliter) for each of nine lichen specimens is repeated in the table.

LICHEN

Location			
Bethel	−5.50	−5.00	
Eagle Summit	−4.15	−4.85	
Moose Pass	−6.05		
Turnagain Pass	−5.00		
Wickersham Dome	−4.10	−4.50	−4.60

Source: Lichen Radionuclide Baseline Research Project, 2003.

a. Find the mean, median, and mode of the radioactivity levels.

b. Interpret the value of each measure of central tendency, part **a**.

Applying the Concepts—Intermediate

2.62 Recommendation letters for professors. Applicants for an academic position (e.g., assistant professor) at a college or university are usually required to submit at least three letters of recommendation. A study of 148 applicants for an entry-level position in experimental psychology at the University of Alaska Anchorage revealed that many did not meet the three-letter requirement. (*American Psychologist*, July 1995.) Summary statistics for the number of recommendation letters in each application are given below. Interpret these summary measures.

Mean = 2.28   Median = 3   Mode = 3

2.63 Training zoo animals. "The Training Game" is an activity used in psychology in which one person shapes an arbitrary behavior by selectively reinforcing the movements of another person. A group of 15 psychology students at Georgia Institute of Technology played "The Training Game" at Zoo Atlanta while participating in an experimental psychology laboratory in which they assisted in the training of animals. (*Teaching of Psychology*, May 1998.) At the end of the session, each student was asked to rate the statement: "'The Training Game' is a great way for students to understand the animal's perspective during training." Responses were recorded on a 7-point scale ranging from 1 (strongly disagree) to 7 (strongly agree). The 15 responses were summarized as follows:

mean = 5.87, mode = 6.

a. Interpret the measures of central tendency in the words of the problem.

b. What type of skewness (if any) is likely to be present in the distribution of student responses? Explain.

2.64 Symmetric or skewed? Would you expect the data sets described below to possess relative frequency distributions that are symmetric, skewed to the right, or skewed to the left? Explain.

a. The salaries of all persons employed by a large university

b. The grades on an easy test

c. The grades on a difficult test

d. The amounts of time students in your class studied last week

e. The ages of automobiles on a used-car lot

f. The amounts of time spent by students on a difficult examination (maximum time is 50 minutes)

2.65 Children's use of pronouns. Clinical observations suggest that specifically language-impaired (SLI) children have great difficulty with the proper use of pronouns. This phenomenon was investigated and reported in the *Journal of Communication Disorders* (Mar. 1995). Thirty children, all from low-income families, participated in the study. Ten were 5-year-old SLI children, ten were younger (3-year-old) normally developing (YND) children, and ten were older (5-year-old) normally developing (OND) children. The table contains the gender, deviation intelligence quotient (DIQ), and percentage of pronoun errors observed for each of the 30 subjects.

SLI

Subject	Gender	Group	DIQ	Pronoun Errors (%)
1	F	YND	110	94.40
2	F	YND	92	19.05
3	F	YND	92	62.50
4	M	YND	100	18.75
5	F	YND	86	0
6	F	YND	105	55.00
7	F	YND	90	100.00
8	M	YND	96	86.67
9	M	YND	90	32.43
10	F	YND	92	0
11	F	SLI	86	60.00
12	M	SLI	86	40.00
13	M	SLI	94	31.58
14	M	SLI	98	66.67
15	F	SLI	89	42.86
16	F	SLI	84	27.27

SLI (Continued)

Subject	Gender	Group	DIQ	Pronoun Errors (%)
17	M	SLI	110	33.33
18	F	SLI	107	0
19	F	SLI	87	0
20	M	SLI	95	0
21	M	OND	110	0
22	M	OND	113	0
23	M	OND	113	0
24	F	OND	109	0
25	M	OND	92	0
26	F	OND	108	0
27	M	OND	95	0
28	F	OND	87	0
29	F	OND	94	0
30	F	OND	98	0

Source: Moore, M. E. "Error analysis of pronouns by normal and language-impaired children." *Journal of Communication Disorders,* Vol. 28, No. 1, Mar. 1995, p. 62 (Table 2), p. 67 (Table 5).

a. Identify the variables in the data set as quantitative or qualitative.

b. Why is it nonsensical to compute numerical descriptive measures for qualitative variables?

c. Compute measures of central tendency for DIQ for the ten SLI children.

d. Compute measures of central tendency for DIQ for the ten YND children.

e. Compute measures of central tendency for DIQ for the ten OND children.

f. Use the results, parts **c–e**, to compare the DIQ central tendencies of the three groups of children. Is it reasonable to use a single number (e.g., mean or median) to describe the center of the DIQ distribution? Or should three "centers" be calculated, one for each of the three groups of children? Explain.

g. Repeat parts **c–f** for the percentage of pronoun errors.

2.66 Mongolian desert ants. The *Journal of Biogeography* (Dec. 2003) published an article on the first comprehensive study of ants in Mongolia (Central Asia). [NW]

Botanists placed seed baits at 11 study sites and observed the ant species attracted to each site. Some of the data recorded at each study site are provided in the table at the bottom of the page.

a. Find the mean, median, and mode for the number of ant species discovered at the 11 sites. Interpret each of these values.

b. Which measure of central tendency would you recommend to describe the center of the number of ant species distribution? Explain.

c. Find the mean, median, and mode for the total plant cover percentage at the 5 Dry Steppe sites only.

d. Find the mean, median, and mode for the total plant cover percentage at the 6 Gobi Desert sites only.

e. Based on the results, parts **c** and **d**, does the center of the total plant cover percentage distribution appear to be different at the two regions?

Applying the Concepts—Advanced

2.67 Eye refractive study. The conventional method of measuring the refractive status of an eye involves three quantities: (1) sphere power, (2) cylinder power, and (3) axis. Optometric researchers studied the variation in these three measures of refraction. (*Optometry and Vision Science*, June 1995.) Twenty-five successive refractive measurements were obtained on the eyes of over 100 university students. The cylinder power measurements for the left eye of one particular student (ID #11) are listed in the table. [*Note:* All measurements are negative values.]

LEFTEYE

.08	.08	1.07	.09	.16	.04	.07	.17	.11
.06	.12	.17	.20	.12	.17	.09	.07	.16
.15	.16	.09	.06	.10	.21	.06		

Source: Rubin, A., and Harris, W. F. "Refractive variation during autorefraction: Multivariate distribution of refractive status." *Optometry and Vision Science,* Vol. 72, No. 6, June 1995, p. 409 (Table 4).

GOBIANTS

Site	Region	Annual Rainfall (mm)	Max. Daily Temp. (°C)	Total Plant Cover (%)	Number of Ant Species	Species Diversity Index
1	Dry Steppe	196	5.7	40	3	.89
2	Dry Steppe	196	5.7	52	3	.83
3	Dry Steppe	179	7.0	40	52	1.31
4	Dry Steppe	197	8.0	43	7	1.48
5	Dry Steppe	149	8.5	27	5	.97
6	Gobi Desert	112	10.7	30	49	.46
7	Gobi Desert	125	11.4	16	5	1.23
8	Gobi Desert	99	10.9	30	4	
9	Gobi Desert	125	11.4	56	4	.76
10	Gobi Desert	84	11.4	22	5	1.26
11	Gobi Desert	115	11.4	14	4	.69

Source: Pfeiffer, M., et al. "Community organization and species richness of ants in Mongolia along an ecological gradient from steppe to Gobi desert," *Journal of Biogeography,* Vol. 30, No. 12, Dec. 2003 (Tables 1 and 2).

a. Find measures of central tendency for the data and interpret their values.

b. Note that the data contains one unusually large (negative) cylinder power measurement relative to the other measurements in the data set. Find this measurement. (In Section 2.8, we call this value an **outlier**).

c. Delete the outlier, part **b**, from the data set and recalculate the measures of central tendency. Which measure is most affected by the deletion of the outlier?

2.68 **Active nuclear power plants.** The U.S. Energy Information Administration monitors all nuclear power plants operating in the United States. The table lists the number of active nuclear power plants operating in each of a sample of 20 states.

a. Find the mean, median, and mode of this data set.

b. Eliminate the largest value from the data set and repeat part **a**. What effect does dropping this measurement have on the measures of central tendency found in part **a**?

c. Arrange the 20 values in the table from lowest to highest. Next, eliminate the lowest two values and the highest two values from the data set and find the mean of the remaining data values. The result is called a *10% trimmed mean*, since it is calculated after removing the highest 10% and the lowest 10% of the data values. What advantages does a trimmed mean have over the regular arithmetic mean?

NUCLEAR

State	Number of Power Plants
Alabama	5
Arizona	3
California	4
Florida	5
Georgia	4
Illinois	13
Kansas	1
Louisiana	2
Massachusetts	1
Mississippi	1
New Hampshire	1
New York	6
North Carolina	5
Ohio	2
Pennsylvania	9
South Carolina	7
Tennessee	3
Texas	4
Vermont	1
Wisconsin	3

Source: Statistical Abstract of the United States, 2000 (Table 966). U.S. Energy Information Administration, Electric Power Annual.

2.5 Numerical Measures of Variability

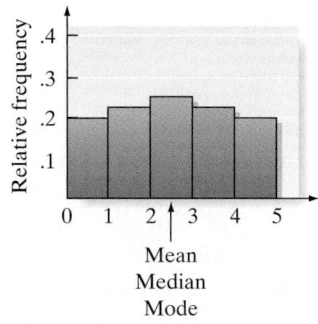

a. Data set 1

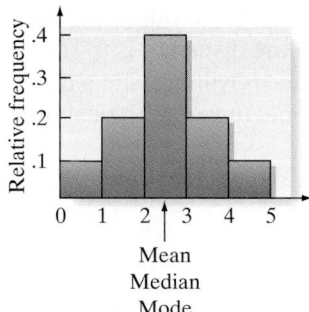

b. Data set 2

Figure 2.18
Hypothetical Data Sets

Measures of central tendency provide only a partial description of a quantitative data set. The description is incomplete without a **measure of the variability**, or **spread**, of the data set. Knowledge of the data set's variability along with its center can help us visualize the shape of a data set as well as its extreme values.

If you examine the two histograms in Figure 2.18, you'll notice that both hypothetical data sets are symmetric with equal modes, medians, and means. However, data set 1 (Figure 2.18a) has measurements spread with almost equal relative frequency over the measurement classes, while data set 2 (Figure 2.18b) has most of its measurements clustered about its center. Thus, data set 2 is *less variable* than data set 1. Consequently, you can see that we need a measure of variability as well as a measure of central tendency to describe a data set.

Perhaps the simplest measure of the variability of a quantitative data set is its *range*.

DEFINITION 2.9

The **range** of a quantitative data set is equal to the largest measurement minus the smallest measurement.

The range is easy to compute and easy to understand, but it is a rather insensitive measure of data variation when the data sets are large. This is because two data sets can have the same range and be vastly different with respect to data variation. This phenomenon is demonstrated in Figure 2.18. Both distributions of data shown in the figure have the same range, but most of the measurements in data set 2 tend to concentrate near the center of the distribution. Consequently, the data are much less variable than the data in set 1. Thus, you can see that the range does not always detect differences in data variation for large data sets.

Let's see if we can find a measure of data variation that is more sensitive than the range. Consider the two samples in Table 2.5: Each has five measurements. (We have ordered the numbers for convenience.) Note that both samples have a mean of 3 and that we have also calculated the distance between each measurement and the mean. What information do these distances contain? If they tend to be large in magnitude, as in sample 1, the data are spread out, or highly variable. If the distances are mostly small, as in sample 2, the data are clustered around the mean, $\bar{x}$, and therefore do not exhibit much variability. You can see that these distances, displayed graphically in Figure 2.19, provide information about the variability of the sample measurements.

TABLE 2.5   Two Hypothetical Data Sets

	Sample 1	Sample 2
Measurements	$1, 2, 3, 4, 5$	$2, 3, 3, 3, 4$
Mean	$\bar{x} = \dfrac{1 + 2 + 3 + 4 + 5}{5} = \dfrac{15}{5} = 3$	$\bar{x} = \dfrac{2 + 3 + 3 + 3 + 4}{5} = \dfrac{15}{5} = 3$
Distances of measurement values from $\bar{x}$	$(1 - 3), (2 - 3), (4 - 3),$ $(5 - 3)$ or $-2, -1, 0, 1, 2$	$(2 - 3), (3 - 3), (3 - 3), (3 - 3),$ $(4 - 3)$ or $-1, 0, 0, 0, 1$

Figure 2.19
Dot Plots for Two Data Sets

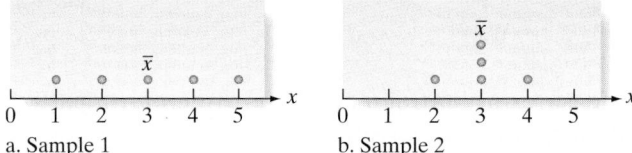

a. Sample 1   b. Sample 2

The next step is to condense the information in these distances into a single numerical measure of variability. Averaging the distances from $\bar{x}$ won't help because the negative and positive distances cancel; that is, the sum of the deviations (and thus the average deviation) is always equal to zero.

Two methods come to mind for dealing with the fact that positive and negative distances from the mean cancel. The first is to treat all the distances as though they were positive, ignoring the sign of the negative distances. We won't pursue this line of thought because the resulting measure of variability (the mean of the absolute values of the distances) presents analytical difficulties beyond the scope of this text. A second method of eliminating the minus signs associated with the distances is to square them. The quantity we can calculate from the squared distances will provide a meaningful description of the variability of a data set and presents fewer analytical difficulties in inference making.

To use the squared distances calculated from a data set, we first calculate the *sample variance*.

DEFINITION 2.10

The **sample variance** for a sample of n measurements is equal to the sum of the squared distances from the mean divided by $(n - 1)$. The symbol, s^2 is used to represent the sample variance.

Formula for the Sample Variance:

$$s^2 = \frac{\sum_{i=1}^{n}(x_i - \bar{x})^2}{n - 1}$$

Note: A shortcut formula for calculating s^2 is

$$s^2 = \frac{\sum_{i=1}^{n} x_i^2 - \dfrac{\left(\sum_{i=1}^{n} x_i\right)^2}{n}}{n - 1}$$

Referring to the two samples in Table 2.5, you can calculate the variance for sample 1 as follows:

$$s^2 = \frac{(1 - 3)^2 + (2 - 3)^2 + (3 - 3)^2 + (4 - 3)^2 + (5 - 3)^2}{5 - 1}$$

$$= \frac{4 + 1 + 0 + 1 + 4}{4} = 2.5$$

The second step in finding a meaningful measure of data variability is to calculate the *standard deviation* of the data set:

DEFINITION 2.11

The **sample standard deviation**, s, is defined as the positive square root of the sample variance, s^2. Thus,

$$s = \sqrt{s^2}$$

The population variance, denoted by the symbol σ^2 (sigma squared), is the average of the squared distances of the measurements on *all* units in the population from the mean, μ, and σ (sigma) is the square root of this quantity.

Symbols for Variance and Standard Deviation

s^2 = Sample variance

s = Sample standard deviation

σ^2 = Population variance

σ = Population standard deviation

Notice that, unlike the variance, the standard deviation is expressed in the original units of measurement. For example, if the original measurements are in dollars, the variance is expressed in the peculiar units "dollar squared," but the standard deviation is expressed in dollars.

You may wonder why we use the divisor $(n - 1)$ instead of n when calculating the sample variance. Wouldn't using n seem more logical, so that the sample variance would be the average squared distance from the mean? The trouble is, using n tends to produce an underestimate of the population variance, σ^2. So we use $(n - 1)$ in the denominator to provide the appropriate correction for this tendency.* Since sample statistics like s^2 are primarily used to estimate population parameters like σ^2, $(n - 1)$ is preferred to n when defining the sample variance.

**Appropriate here means that s^2 with a divisor of $(n - 1)$ is an *unbiased estimator* of σ^2. We define and discuss unbiasedness of estimators in Chapter 6.*

EXAMPLE 2.9   **COMPUTING MEASURES OF VARIATION**

Problem   Calculate the variance and standard deviation of the following sample: 2, 3, 3, 3, 4.

Solution   As the number of measurements increases, calculating s^2 and s becomes very tedious. Fortunately, as we show in Example 2.10, we can use a statistical software package (or calculator) to find these values. If you must calculate these quantities by hand, it is advantageous to use the shortcut formula provided in Definition 2.10.

To do this, we need two summations: $\sum x$ and $\sum x^2$. These can easily be obtained from the following type of tabulation:

x	x^2
2	4
3	9
3	9
3	9
4	16
$\sum x = 15$	$\sum x^2 = 47$

Then we use*

$$s^2 = \frac{\sum_{i=1}^{n} x_i^2 - \frac{\left(\sum_{i=1}^{n} x_i\right)^2}{n}}{n-1} = \frac{47 - \frac{(15)^2}{5}}{5-1} = \frac{2}{4} = .5$$

$$s = \sqrt{.5} = .71$$

Look Back As the sample size n increases, these calculations can become very tedious. As the next example shows, we can use the computer to find s^2 and s.

> Now Work   *Exercise 2.76a*

EXAMPLE 2.10   **FINDING MEASURES OF VARIATION ON A PRINTOUT**

Problem   Use the computer to find the sample variance s^2 and the sample standard deviation s for the 100 gas mileage readings given in Table 2.2.

Solution   The SAS printout describing the gas mileage data is reproduced in Figure 2.20. The variance and standard deviation, highlighted on the printout, are (rounded) $s^2 = 5.85$ and $s = 2.42$.

■ ■ ■

You now know that the standard deviation measures the variability of a set of data and how to calculate it. The larger the standard deviation, the more variable the data are. The smaller the standard deviation, the less variation in the data. But how can we practically interpret the standard deviation and use it to make inferences? This is the topic of Section 2.6.

*When calculating s^2, how many decimal places should you carry? Although there are no rules for the rounding procedure, it is reasonable to retain twice as many decimal places in s^2 as you ultimately wish to have in s. If you wish to calculate s to the nearest hundredth (two decimal places), for example, you should calculate s^2 to the nearest ten-thousandth (four decimal places).

```
                        The MEANS Procedure

                      Analysis Variable : MPG

      Mean        Std Dev        Variance     N      Minimum       Maximum       Median

   36.9940000    2.4178971      5.8462263    100    30.0000000    44.9000000    37.0000000
```

Figure 2.20
Reproduction of SAS Numerical Descriptive Measures for 100 EPA Mileages

Exercises 2.69–2.85

Understanding the Principles

2.69 What is the range of a data set?

2.70 What is the primary disadvantage of using the range to compare the variability of data sets?

2.71 Describe the sample variance using words rather than a formula. Do the same with the population variance.

2.72 Can the variance of a data set ever be negative? Explain. Can the variance ever be smaller than the standard deviation? Explain.

2.73 If the standard deviation increases, does this imply that the data are more variable or less variable?

Learning the Mechanics

2.74 Calculate the variance and standard deviation for samples where
   a. $n = 10$, $\sum x^2 = 84$, $\sum x = 20$
   b. $n = 40$, $\sum x^2 = 380$, $\sum x = 100$
   c. $n = 20$, $\sum x^2 = 18$, $\sum x = 17$

2.75 Calculate the range, variance, and standard deviation for the following samples:
   a. 39, 42, 40, 37, 41
   b. 100, 4, 7, 96, 80, 3, 1, 10, 2
   c. 100, 4, 7, 30, 80, 30, 42, 2

2.76 Calculate the range, variance, and standard deviation for the following samples:
   a. 4, 2, 1, 0, 1
   b. 1, 6, 2, 2, 3, 0, 3
   c. 8, −2, 1, 3, 5, 4, 4, 1, 3
   d. 0, 2, 0, 0, −1, 1, −2, 1, 0, −1, 1, −1, 0, −3, −2, −1, 0, 1

2.77 Using only integers between 0 and 10, construct two data sets with at least 10 observations each so that the two sets have the same mean but different variances. Construct dot plots for each of your data sets, and mark the mean of each data set on its dot plot.

2.78 Using only integers between 0 and 10, construct two data sets with at least 10 observations each that have the same range but different means. Construct a dot plot for each of your data sets, and mark the mean of each data set on its dot plot.

2.79 Consider the following sample of five measurements: 2, 1, 1, 0, 3.

   a. Calculate the range, s^2, and s.
   b. Add 3 to each measurement and repeat part **a**.
   c. Subtract 4 from each measurement and repeat part **a**.
   d. Considering your answers to parts **a**, **b**, and **c**, what seems to be the effect on the variability of a data set by adding the same number to or subtracting the same number from each measurement?

2.80 Compute s^2, and s for each of the following data sets. If appropriate, specify the units in which your answer is expressed.
   a. 3, 1, 10, 10, 4
   b. 8 feet, 10 feet, 32 feet, 5 feet
   c. −1, −4, −3, 1, −4, −4
   d. 1/5 ounce, 1/5 ounce, 1/5 ounce, 2/5 ounce, 1/5 ounce, 4/5 ounce

Applying the Concepts—Basic

JAPANESE
2.81 Reading Japanese books. Refer to the *Reading in a Foreign Language* (Apr. 2004) experiment to improve the Japanese reading comprehension levels of 14 University of Hawaii students, Exercises 2.31 and 2.57 (p. 64). The data on number of books read and grade for each student is saved in the JAPANESE file.
   a. Find the range, variance, and standard deviation of the number of books read by students who earned an A grade.
   b. Find the range, variance, and standard deviation of the number of books read by students who earned either a B or C grade.
   c. Refer to parts **a** and **b**. Which of the two groups of students has a more variable distribution for number of books read?

DDT
2.82 Contaminated fish. Refer to Exercise 2.60 (p. 64) and the U.S. Army Corps of Engineers data on contaminated fish saved in the DDT file. Consider the quantitative variables length (in centimeters), weight (in grams), and DDT level (in parts per million).
   a. Find three different measures of variation for the 144 fish lengths. Give the units of measurement for each.

b. Find three different measures of variation for the 144 fish weights. Give the units of measurement for each.

c. Find three different measures of variation for the 144 DDT values. Give the units of measurement for each.

2.83 Ammonia in car exhaust. Refer to the *Environmental Science & Technology* (Sept. 1, 2000) study on the ammonia levels near the exit ramp of a San Francisco highway tunnel, Exercise 2.59 (p. 54). The data (in parts per million) for 8 days during afternoon drive-time are reproduced in the table.

⊙ **AMMONIA**

1.53	1.50	1.37	1.51	1.55	1.42	1.41	1.48

a. Find the range of the ammonia levels.
b. Find the variance of the ammonia levels.
c. Find the standard deviation of the ammonia levels.
d. Suppose the standard deviation of the daily ammonia levels during morning drive-time at the exit ramp is 1.45 ppm. Which time, morning or afternoon drive-time, has more variable ammonia levels?

Applying the Concepts—Intermediate

⊙ **WPOWER50**

2.84 Most powerful women in America. Refer to Exercise 2.58 (p. 54) and *Fortune's* (Oct. 14, 2002) list of the 50 most powerful women in America. The data are stored in the WPOWER50 file.

a. Find the range of the ages for these 50 women.
b. Find the variance of the ages for these 50 women.
c. Find the standard deviation of the ages for these 50 women.
d. Suppose the standard deviation of the ages of the most powerful women in Europe is 10 years. For which location, the United States or Europe, is the age data more variable?
e. If the largest age in the data set is omitted, would the standard deviation increase or decrease? Verify your answer.

⊙ **NUCLEAR**

2.85 Active nuclear power plants. Refer to Exercise 2.68 (p. 67) and the U.S. Energy Information Administration's data on the number of nuclear power plants operating in each of 20 states. The data are saved in the NUCLEAR file.

a. Find the range, variance, and standard deviation of this data set.
b. Eliminate the largest value from the data set and repeat part **a**. What effect does dropping this measurement have on the measures of variation found in part **a**?
c. Eliminate the smallest and largest value from the data set and repeat part **a**. What effect does dropping both of these measurements have on the measures of variation found in part **a**?

2.6 Interpreting the Standard Deviation

We've seen that if we are comparing the variability of two samples selected from a population, the sample with the larger standard deviation is the more variable of the two. Thus, we know how to interpret the standard deviation on a relative or comparative basis, but we haven't explained how it provides a measure of variability for a single sample.

To understand how the standard deviation provides a measure of variability of a data set, consider a specific data set and answer the following questions: How many measurements are within 1 standard deviation of the mean? How many measurements are within 2 standard deviations? For example, look at the 100 mileage per gallon readings given in Table 2.2. Recall that $\bar{x} = 36.99$ and $s = 2.42$. Then

$$\bar{x} - s = 34.57 \qquad \bar{x} + s = 39.41$$

$$\bar{x} - 2s = 32.15 \qquad \bar{x} + 2s = 41.83$$

If we examine the data, we find that 68 of the 100 measurements, or 68%, are in the interval

$$\bar{x} - s \text{ to } \bar{x} + s$$

Similarly, we find that 96, or 96%, of the 100 measurements are in the interval

$$\bar{x} - 2s \text{ to } \bar{x} + 2s$$

We usually write these intervals as

$$(\bar{x} - s, \bar{x} + s) \text{ and } (\bar{x} - 2s, \bar{x} + 2s)$$

Such observations identify criteria for interpreting a standard deviation that apply to *any* set of data, whether a population or a sample. The criteria, expressed as a mathematical theorem and as a rule of thumb, are presented in Tables 2.6 and 2.7. In these tables we give two sets of answers to the questions of how many measurements fall within 1, 2, and 3 standard deviations of the mean. The first, which applies to *any* set of data, is derived from a theorem proved by the Russian mathematician P. L. Chebyshev (1821–1894). The second, which applies to **mound-shaped, symmetric, distributions** of data (where the mean, median, and mode are all about the same), is based upon empirical evidence that has accumulated over the years. However, the percentages given for the intervals in Table 2.7 provide remarkably good approximations even when the distribution of the data is slightly skewed or asymmetric. Note that the rules apply to either population or sample data sets.

TABLE 2.6   Interpreting the Standard Deviation: Chebyshev's Rule

---

Chebyshev's Rule applies to any data set, regardless of the shape of the frequency distribution of the data.

a. It is possible that very few of the measurements will fall within 1 standard deviation of the mean, i.e., within the interval $(\bar{x} - s, \bar{x} + s)$ for samples and $(\mu - \sigma, \mu + \sigma)$ for populations.

b. At least $\frac{3}{4}$ of the measurements will fall within 2 standard deviations of the mean, i.e., within the interval $(\bar{x} - 2s, \bar{x} + 2s)$ for samples and $(\mu - 2\sigma, \mu + 2\sigma)$ for populations.

c. At least $\frac{8}{9}$ of the measurements will fall within 3 standard deviations of the mean, i.e., within the interval $(\bar{x} - 3s, \bar{x} + 3s)$ for samples and $(\mu - 3\sigma, \mu + 3\sigma)$ for populations.

d. Generally, for any number k greater than 1, at least $(1 - 1/k^2)$ of the measurements will fall within k standard deviations of the mean, i.e., within the interval $(\bar{x} - ks, \bar{x} + ks)$ for samples and $(\mu - k\sigma, \mu + k\sigma)$ for populations.

Biography

PAFNUTY L. CHEBYSHEV (1821–1894)—
The Splendid Russian Mathematician

P. L. Chebyshev was educated in mathematical science at Moscow University, eventually earning his master's degree. Following his graduation, Chebyshev joined St. Petersburg (Russia) University as a professor, becoming part of the well-known "Petersburg mathematical school." It was here that Chebyshev proved his famous theorem about the probability of a measurement being within k standard deviations of the mean (Table 2.6). His fluency in French allowed him to gain international recognition in probability theory. In fact, Chebyshev once objected to being described as a "splendid Russian mathematician," saying he surely was a "world-wide mathematician." One student remembered Chebyshev as "a wonderful lecturer" who "was always prompt for class," and "as soon as the bell sounded, he immediately dropped the chalk, and, limping, left the auditorium."

TABLE 2.7   Interpreting the Standard Deviation: The Empirical Rule

The **Empirical Rule** is a rule of thumb that applies to data sets with frequency distributions that are mound shaped and symmetric, as shown below.

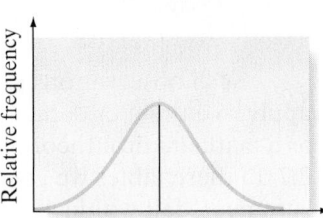

a. Approximately 68% of the measurements will fall within 1 standard deviation of the mean, i.e., within the interval $(\bar{x} - s, \bar{x} + s)$ for samples and $(\mu - \sigma, \mu + \sigma)$ for populations.
b. Approximately 95% of the measurements will fall within 2 standard deviations of the mean, i.e., within the interval $(\bar{x} - 2s, \bar{x} + 2s)$ for samples and $(\mu - 2\sigma, \mu + 2\sigma)$ for populations.
c. Approximately 99.7% (essentially all) of the measurements will fall within 3 standard deviations of the mean, i.e., within the interval $(\bar{x} - 3s, \bar{x} + 3s)$ for samples and $(\mu - 3\sigma, \mu + 3\sigma)$ for populations.

EXAMPLE 2.11

INTERPRETING THE STANDARD DEVIATION

Problem   Thirty students in an experimental psychology class use various techniques to train a rat to move through a maze. At the end of the course, each student's rat is timed through the maze. The results (in minutes) are listed in Table 2.8. Determine the fraction of the 30 measurements in the intervals $\bar{x} \pm s$, $\bar{x} \pm 2s$, and $\bar{x} \pm 3s$, and compare the results with those predicted in Tables 2.6 and 2.7.

⊙ **RATMAZE**

TABLE 2.8   Times (in Minutes) of 30 Rats Running Through a Maze

1.97	.60	4.02	3.20	1.15	6.06	4.44	2.02	3.37	3.65
1.74	2.75	3.81	9.70	8.29	5.63	5.21	4.55	7.60	3.16
3.77	5.36	1.06	1.71	2.47	4.25	1.93	5.15	2.06	1.65

Solution   First, we entered the data into the computer and used MINITAB to produce summary statistics. The mean and standard deviation of the sample data, highlighted on the printout shown in Figure 2.21, are (rounded)

$$\bar{x} = 3.74 \text{ minutes} \qquad s = 2.20 \text{ minutes}$$

Figure 2.21
MINITAB Descriptive
Statistics for Rat Maze
Times

Descriptive Statistics: RUNTIME

Variable	N	Mean	StDev	Minimum	Median	Maximum
RUNTIME	30	3.744	2.198	0.600	3.510	9.700

Now, we form the interval

$$(\bar{x} - s, \bar{x} + s) = (3.74 - 2.20, 3.74 + 2.20) = (1.54, 5.94)$$

A check of the measurements shows that 23 of the times are within this 1 standard deviation interval around the mean. This number represents 23/30, or ≈77% of the sample measurements.

The next interval of interest is

$$(\bar{x} - 2s, \bar{x} + 2s) = (3.74 - 4.40, 3.74 + 4.40) = (-.66, 8.14)$$

All but two of the times are within this interval, so 28/30, or approximately 93%, are within 2 standard deviations of $\bar{x}$.

Finally, the 3-standard-deviation interval around $\bar{x}$ is

$$(\bar{x} - 3s, \bar{x} + 3s) = (3.74 - 6.60, 3.74 + 6.60) = (-2.86, 10.34)$$

All of the times fall within 3 standard deviations of the mean.

These 1-, 2-, and 3-standard-deviation percentages (77%, 93%, and 100%) agree fairly well with the approximations of 68%, 95%, and 100% given by the Empirical Rule (Table 2.7).

Look Back If you look at the MINITAB frequency histogram for this data set in Figure 2.22, you'll note that the distribution is not really mound shaped, nor is it extremely skewed. Thus, we get reasonably good results from the mound-shaped approximations. Of course, we know from Chebyshev's Rule (Table 2.6) that no matter what the shape of the distribution, we would expect at least 75% and at least 89% of the measurements to lie within 2 and 3 standard deviations of $\bar{x}$, respectively.

Figure 2.22
MINITAB Histogram of Rat Maze Times

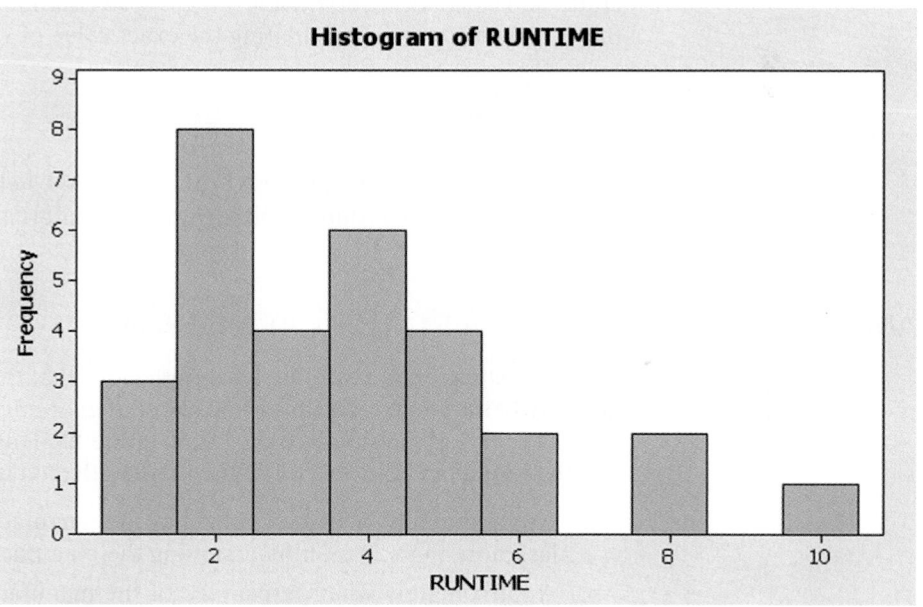

Now Work   *Exercise 2.90*

EXAMPLE 2.12   CHECKING THE CALCULATION OF THE SAMPLE STANDARD DEVIATION

Problem   Chebyshev's Rule and the Empirical Rule are useful as a check on the calculation of the standard deviation. For example, suppose we calculated the standard deviation for the gas mileage data (Table 2.2) to be 5.85. Are there any "clues" in the data that enable us to judge whether this number is reasonable?

Solution The range of the mileage data in Table 2.2 is $44.0 - 30.0 = 14.9$. From Chebyshev's Rule and the Empirical Rule we know that most of the measurements (approximately 95% if the distribution is mound shaped) will be within 2 standard deviations of the mean. And, regardless of the shape of the distribution and the number of measurements, almost all of them will fall within 3 standard deviations of the mean. Consequently, we would expect the range of the measurements to be between 4 (i.e., $\pm 2s$) and 6 (i.e., $\pm 3s$) standard deviations in length (see Figure 2.23). For the car mileage data, this means that s should fall between

$$\frac{\text{Range}}{6} = \frac{14.9}{6} = 2.48 \quad \text{and} \quad \frac{\text{Range}}{4} = \frac{14.9}{4} = 3.73$$

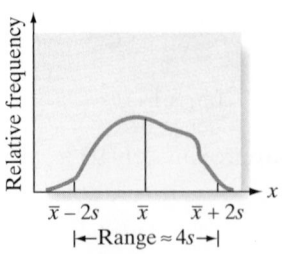

Figure 2.23
The Relation Between the Range and the Standard Deviation

In particular, the standard deviation should not be much larger than 1/4 of the range, particularly for the data set with 100 measurements. Thus, we have reason to believe that the calculation of 5.85 is too large. A check of our work reveals that 5.85 is the variance s^2, not the standard deviation s (see Example 2.10). We "forgot" to take the square root (a common error); the correct value is $s = 2.42$. Note that this value is slightly smaller than the range divided by 6 (2.48). The larger the data set, the greater the tendency for very large or very small measurements (extreme values) to appear, and when they do, the range may exceed 6 standard deviations.

Look Back In examples and exercises we'll sometimes use $s \approx$ range/4 to obtain a crude, and usually conservatively large, approximation for s. However, we stress that this is no substitute for calculating the exact value of s when possible.

| Now Work | *Exercise 2.91* |

In the next example, we use the concepts in Chebyshev's Rule and the Empirical Rule to build the foundation for statistical inference-making.

EXAMPLE 2.13   MAKING A STATISTICAL INFERENCE

Problem A manufacturer of automobile batteries claims that the average length of life for its grade A battery is 60 months. However, the guarantee on this brand is for just 36 months. Suppose the standard deviation of the life length is known to be 10 months, and the frequency distribution of the life-length data is known to be mound shaped.

a. Approximately what percentage of the manufacturer's grade A batteries will last more than 50 months, assuming the manufacturer's claim is true?

b. Approximately what percentage of the manufacturer's batteries will last less than 40 months, assuming the manufacturer's claim is true?

c. Suppose your battery lasts 37 months. What could you infer about the manufacturer's claim?

Solution If the distribution of life length is assumed to be mound shaped with a mean of 60 months and a standard deviation of 10 months, it would appear as shown in Figure 2.24. Note that we can take advantage of the fact that mound-shaped distributions are (approximately) symmetric about the mean, so that the percentages given by the Empirical Rule can be split equally between the halves of the distribution on each side of the mean.

Figure 2.24

Battery Life-Length
Distribution: Manufacturer's
Claim Assumed True

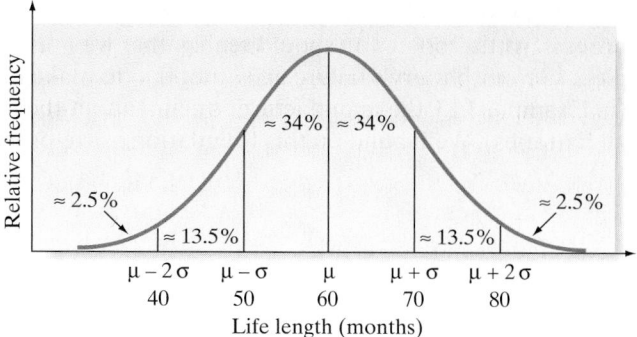

For example, since approximately 68% of the measurements will fall within 1 standard deviation of the mean, the distribution's symmetry implies that approximately $(1/2)(68\%) = 34\%$ of the measurements will fall between the mean and 1 standard deviation on each side. This concept is illustrated in Figure 2.24. The figure also shows that 2.5% of the measurements lie beyond 2 standard deviations in each direction from the mean. This result follows from the fact that if approximately 95% of the measurements fall within 2 standard deviations of the mean, then about 5% fall outside 2 standard deviations; if the distribution is approximately symmetric, then about 2.5% of the measurements fall beyond 2 standard deviations on each side of the mean.

a.   It is easy to see in Figure 2.24 that the percentage of batteries lasting more than 50 months is approximately 34% (between 50 and 60 months) plus 50% (greater than 60 months). Thus, approximately 84% of the batteries should have life length exceeding 50 months.

b.   The percentage of batteries that last less than 40 months can also be easily determined from Figure 2.24. Approximately 2.5% of the batteries should fail prior to 40 months, assuming the manufacturer's claim is true.

c.   If you are so unfortunate that your grade A battery fails at 37 months, you can make one of two inferences: Either your battery was one of the approximately 2.5% that fail prior to 40 months, or something about the manufacturer's claim is not true. Because the chances are so small that a battery fails before 40 months, you would have good reason to have serious doubts about the manufacturer's claim. A mean smaller than 60 months and/or a standard deviation longer than 10 months would both increase the likelihood of failure prior to 40 months.*

Look Back The approximations given in Figure 2.24 are more dependent on the assumption of a mound-shaped distribution than those given by the Empirical Rule (Table 2.7), because the approximations in Figure 2.24 depend on the (approximate) symmetry of the mound-shaped distribution. We saw in Example 2.11 that the Empirical Rule can yield good approximations even for skewed distributions. This will *not* be true of the approximations in Figure 2.24; the distribution *must* be mound shaped and (approximately) symmetric.

---

*The assumption that the distribution is mound shaped and symmetric may also be incorrect. However, if the distribution were skewed to the right, as life-length distributions often tend to be, the percentage of measurements more than 2 standard deviations below the mean would be even less than 2.5%.

Example 2.13 is our initial demonstration of the statistical inference-making process. At this point you should realize that we'll use sample information (in Example 2.13, your battery's failure at 37 months) to make inferences about the population (in Example 2.13, the manufacturer's claim about the life length for the population of all batteries). We'll build on this foundation as we proceed.

Statistics in Action Revisited
Interpreting Descriptive Statistics

We return to the analysis of data from the *Psychological Science* "water-level task" experiment. The quantitative variable of interest is *Deviation angle* (measured in degrees) of the judged line from the true surface line (parallel to the table top). Recall that the researchers want to test the theory that males will do better than females on judging the correct water level (i.e., that males, in general, have deviation angles closer to 0 than females). The MINITAB descriptive statistics printout for the EYECUE data is displayed in Figure SIA2.6, with the means and standard deviations highlighted.

The sample mean for females is 6.57 degrees and the mean for males is 2.92 degrees. Our interpretation is that males' judged lines have smaller deviation angles from the true surface line (2.92 degrees, on average) than females' judged lines (6.57 degrees, on average), supporting the theory stated by the researchers.

To interpret the standard deviation, we substitute into the formula, Mean $\pm$ 2(Standard deviation), to obtain the intervals:

Females: $6.57 \pm 2(7.82) = 6.57 \pm 15.64 = (-9.07, 22.21)$
Males: $2.92 \pm 2(7.54) = 2.92 \pm 15.08 = (-12.16, 18.00)$

From the Chebyshev's Rule (Table 2.6), we know that at least 75% of the females who perform the water task will have deviation angles anywhere between -9.07 and 22.21 degrees. Similarly, we know that at least 75% of the males will have deviation angles anywhere from -12.16 to 18.00 degrees. Note that these ranges indicate that there is very little difference in the spread of the deviation angle distributions of the two groups. However, the intervals indicate that a male subject is more likely to judge the water line below the actual surface (i.e., with a greater negative deviation angle) than a female, while a female subject is more likely to judge the water line above the actual surface (i.e., with a greater positive deviation angle) than a male.

Descriptive Statistics: Deviation

Variable	Gender	N	Mean	StDev	Variance	Minimum	Median	Maximum
Deviation	Female	60	6.57	7.82	61.10	-15.00	7.00	35.00
	Male	60	2.917	7.543	56.891	-18.000	2.000	27.000

Figure SIA2.6
MINITAB Analysis of Deviation Angle in the Water-Level Task — Males versus Females

Exercises 2.86–2.102

Understanding the Principles

2.86 To what kind of data sets can Chebyshev's Rule be applied? The Empirical Rule?

2.87 The output from a statistical computer program indicates that the mean and standard deviation of a data set consisting of 200 measurements are $1,500 and $300, respectively.

a. What are the units of measurement of the variable of interest? Based on the units, what type of data is this: quantitative or qualitative?

b. What can be said about the number of measurements between $900 and $2,100? Between $600 and $2,400? Between $1,200 and $1,800? Between $1,500 and $2,100?

2.88 For any set of data, what can be said about the percentage of the measurements contained in each of the following intervals?
  a. $\bar{x} - s$ to $\bar{x} + s$
  b. $\bar{x} - 2s$ to $\bar{x} + 2s$
  c. $\bar{x} - 3s$ to $\bar{x} + 3s$

2.89 For a set of data with a mound-shaped relative frequency distribution, what can be said about the percentage of the measurements contained in each of the intervals specified in Exercise 2.88?

Learning the Mechanics

2.90 The following is a sample of 25 measurements:

 LM 2-90

7	6	6	11	8	9	11	9	10	8	7	7	5
9	10	7	7	7	7	9	12	10	10	8	6	

  a. Compute $\bar{x}$, s^2, and s for this sample.

  b. Count the number of measurements in the intervals $\bar{x} \pm s$, $\bar{x} \pm 2s$, and $\bar{x} \pm 3s$. Express each count as a percentage of the total number of measurements.
  c. Compare the percentages found in part **b** to the percentages given by the Empirical Rule and Chebyshev's Rule.
  d. Calculate the range and use it to obtain a rough approximation for s. Does the result compare favorably with the actual value for s found in part **a**?

2.91 Given a data set with a largest value of 760 and a smallest value of 135, what would you estimate the standard deviation to be? Explain the logic behind the procedure you used to estimate the standard deviation. Suppose the standard deviation is reported to be 25. Is this feasible? Explain.

Applying the Concepts—Basic

 WPOWER50

2.92 Most powerful women in America. Refer to the *Fortune* (Oct. 14, 2002) list of the 50 most powerful women in America, saved in the WPOWER file. In Exercise 2.58 (p. 64) you found the mean age of the 50 women in the data set and in Exercise 2.84 (p. 72) you found the standard deviation. Use the mean and standard deviation to form an interval that will contain at least 75% of the ages in the data set.

2.93 Sanitation inspection of cruise ships. To minimize the potential for gastrointestinal disease outbreaks, all passenger cruise ships arriving at U.S. ports are subject to unannounced sanitation inspections. Ships are rated on a 100-point scale by the Centers for Disease Control and Prevention. A score of 86 or higher indicates that the ship is providing an accepted standard of sanitation. The May 2004 sanitation scores for 174 cruise ships are saved in the SHIPSANIT file. The first five and last five observations in the data set are listed in the accompanying table.

  a. Find the mean and standard deviation of the sanitation scores.
  b. Calculate the intervals $\bar{x} \pm s$, $\bar{x} \pm 2s$, $\bar{x} \pm 3s$.
  c. Find the percentage of measurements in the data set that fall within each of the intervals, part **b**. Do these percentages agree with either Chebyshev's Theorem or the Empirical Rule?

 SHIPSANIT (Selected observations)

Ship Name	Sanitation Score
Adonia	99
Adventure of the Seas	97
AIDA Aura	99
AIDA Avita	98
Albatross	96
.	.
.	.
.	.
Volendam	97
Voyager of the Seas	97
Wind Spirit	97
Wind Surf	98
World Discoverer	89

Source: National Center for Environmental Health, Centers for Disease Control and Prevention, May 24, 2004.

2.94 Recommendation letters for professors. Refer to the *American Psychologist* (July 1995) study of 148 applicants for a position in experimental psychology, Exercise 2.62 (p. 65). Recall that the mean number of recommendation letters included in each application packet was $\bar{x} = 2.28$. The standard deviation was also reported in the article; it's value was $s = 1.48$.
  a. Sketch the relative frequency distribution for the number of recommendation letters included in each application for the experimental psychology position. (Assume the distribution is mound-shaped and relatively symmetric.)
  b. Locate an interval on the distribution, part **a**, that captures approximately 95% of the measurements in the sample.
  c. Locate an interval on the distribution, part **a**, that captures almost all the sample measurements.

2.95 Dentists' use of anesthetics. A study published in *Current Allergy & Clinical Immunology* (Mar. 2004) investigated allergic reactions of dental patients to local anesthetics (LA). Based on a survey of dental practitioners, the study reported that the mean number of units (ampoules) of LA used per week by dentists was 79 with a standard deviation of 23. Suppose we want to determine the percentage of dentists that use less than 102 units of LA per week.
  a. Assuming nothing is known about the shape of the distribution for the data, what percentage of dentists use less than 102 units of LA per week?
  b. Assuming the data has a mound-shaped distribution, what percentage of dentists use less than 102 units of LA per week?

Applying the Concepts — Intermediate

⊘ NZBIRDS

2.96 Extinct New Zealand birds. Refer to the *Evolutionary Ecology Research* (July 2003) study of the patterns of extinction in the New Zealand bird population, Exercise 2.18 (p. 40). Consider the data on the egg length (measured in millimeters) for the 132 bird species saved in the NZBIRDS file.
  a. Find the mean and standard deviation of the egg lengths.
  b. Form an interval that can be used to predict the egg length of a bird species found in New Zealand.

2.97 Handwashing versus handrubbing. In hospitals, handwashing with soap is emphasized as the single most important measure to prevent infections. As an alternative to handwashing, some hospitals allow health workers to rub their hands with an alcohol-based antiseptic. The *British Medical Journal* (Aug. 17, 2002) reported on a study to compare the effectiveness of handwashing with soap and handrubbing with alcohol. One group of health care workers used handrubbing, while a second group used handwashing to clean their hands. The bacterial count (number of colony forming units) on the hand of each worker was recorded. The table gives descriptive statistics on bacteria counts for the two groups of health care workers.

	Mean	Standard Deviation
Handrubbing	35	59
Handwashing	69	106

  a. For handrubbers, form an interval that contains about 95% of the bacterial counts. (*Note*: The bacterial count cannot be less than 0.)
  b. Repeat part **a** for handwashers.
  c. Based on the results, parts **a** and **b**, make an inference about the effectiveness of the two hand cleaning methods.

2.98 Sentence complexity study. A study published in *Applied Psycholinguistics* (June 1998) compared the language skills of young children (16–30 months old) from low income and middle income families. A total of 260 children — 65 in the low income and 195 in the middle income group — completed the Communicative Development Inventory (CDI) exam. One of the variables measured on each child was sentence complexity score. Summary statistics for the scores of the two groups are reproduced in the table. Use this information to sketch a graph of the sentence complexity score distribution for each income group. (Assume the distributions are mound-shaped and symmetric.) Compare the distributions. What can you infer?

	Low Income	Middle Income
Sample Size	65	195
Mean	7.62	15.55
Median	4	14
Standard Deviation	8.91	12.24
Minimum	0	0
Maximum	36	37

Source: Arriaga, R. I. et al. "Scores on the MacArthur Communicative Development Inventory of children from low-income and middle-income families." *Applied Psycholinguistics,* Vol. 19, No. 2, June 1998, p. 217 (Table 7).

2.99 Velocity of Winchester bullets. The *American Rifleman* (June 1993) reported on the velocity of ammunition fired from the FEG P9R pistol, a 9mm gun manufactured in Hungary. Field tests revealed that Winchester bullets fired from the pistol had a mean velocity (at 15 feet) of 936 feet per second and a standard deviation of 10 feet per second. Tests were also conducted with Uzi and Black Hills ammunition.
  a. Describe the velocity distribution of Winchester bullets fired from the FEG P9R pistol.
  b. A bullet, brand unknown, is fired from the FEG P9R pistol. Suppose the velocity (at 15 feet) of the bullet is 1,000 feet per second. Is the bullet likely to be manufactured by Winchester? Explain.

2.100 Speed of light from galaxies. Astronomers theorize that cold dark matter (CDM) caused the formation of galaxies and clusters of galaxies in the universe. The theoretical CDM model requires an estimate of the velocity of light emitted from the galaxy cluster. *The Astronomical Journal* (July 1995) published a study of observed velocities for galaxies in four different galaxy clusters. Galaxy velocity was measured in kilometers per second (km/s) using a spectrograph and high-power telescope.
  a. The observed velocities of 103 galaxies located in the cluster named A2142 are summarized in the

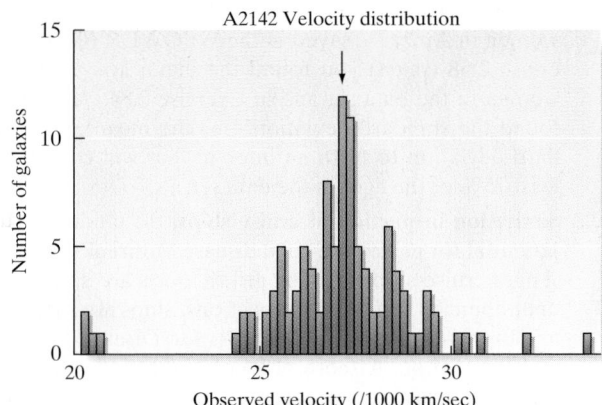

Source: Oegerle, W. R., Hill, J. M., and Fitchett, M. J. "Observations of high dispersion clusters of galaxies: Constraints on cold dark matter." *The Astronomical Journal*, Vol. 110, No. 1, July 1995, p. 37 (Figure 1).

accompanying histogram. Comment on whether the Empirical Rule is applicable for describing the velocity distribution for this cluster.

b. The mean and standard deviation of the 103 velocities observed in galaxy cluster A2142 were reported as $\bar{x} = 27{,}117$ km/s and $s = 1{,}280$ km/s, respectively. Use this information to construct an interval that captures approximately 95% of the galaxy velocities in the cluster.

c. Recommend a single velocity value to be used in the CDM model for galaxy cluster A2142. Explain your reasoning

Applying the Concepts—Advanced

2.101 Improving SAT scores. The National Education Longitudinal Survey (NELS) tracks a nationally representative sample of U.S. students from eighth grade through high school and college. Research published in *Chance* (Winter 2001) examined the Standardized Admission Test (SAT) scores of 265 NELS students who paid a private tutor to help them improve their scores. The table summarizes the changes in both the SAT-Mathematics and SAT-Verbal scores for these students.

	SAT-Math	SAT-Verbal
Mean change in score	19	7
Standard deviation of score changes	65	49

a. Suppose one of the 265 students who paid a private tutor is selected at random. Give an interval that is likely to contain this student's change in the SAT-Math score.

b. Repeat part **a** for the SAT-Verbal score.

c. Suppose the selected student's score increased on one of the SAT tests by 140 points. Which test, the SAT-Math or SAT-Verbal, is the one most likely to have the 140-point increase? Explain.

2.102 Land purchase decision. A buyer for a lumber company must decide whether to buy a piece of land containing 5,000 pine trees. If 1,000 of the trees are at least 40 feet tall, the buyer will purchase the land; otherwise, he won't. The owner of the land reports that the height of the trees has a mean of 30 feet and a standard deviation of 3 feet. Based on this information, what is the buyer's decision?

2.7 Numerical Measures of Relative Standing

We've seen that numerical measures of central tendency and variability describe the general nature of a quantitative data set (either a sample or a population). In addition, we may also be interested in describing the *relative* quantitative location of a particular measurement within a data set. Descriptive measures of the relationship of a measurement to the rest of the data are called **measures of relative standing**.

One measure of the relative standing of a measurement is its *percentile ranking*. For example, suppose you scored an 80 on a test and you want to know how you fared in comparison with others in your class. If the instructor tells you that you scored at the 90th percentile, it means that 90% of the grades were lower than yours and 10% were higher. Thus, if the scores were described by the relative frequency histogram in Figure 2.25, the 90th percentile would be located at a point such that 90% of the total area under the relative frequency histogram lies below the 90th percentile and 10% lies above. If the instructor tells you that you scored in the 50th percentile (the median of the data set), 50% of the test grades would be lower than yours and 50% would be higher.

Percentile rankings are of practical value only for large data sets. Finding them involves a process similar to the one used in finding a median. The measurements are ranked in order and a rule is selected to define the location of each percentile. Since we are primarily interested in interpreting the percentile rankings of measurements (rather than finding particular percentiles for a data set), we define the *pth percentile* of a data set as shown in Definition 2.12.

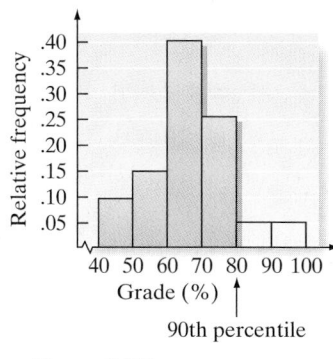

Figure 2.25
Location of 90th Percentile for Test Grades

DEFINITION 2.12

For any set of *n* measurements (arranged in ascending or descending order), the **ptb percentile** is a number such that $p\%$ of the measurements fall below the *p*th percentile and $(100 - p)\%$ fall above it.

EXAMPLE 2.14   FINDING AND INTERPRETING PERCENTILES

Problem   Refer to the student default rates of the 50 states (and District of Columbia) in Table 2.4. An SPSS printout describing the data is shown in Figure 2.26. Locate the 25th percentile and 95th percentile on the printout and interpret these values.

Figure 2.26
SPSS Percentiles for Student
Default Rate Data

Statistics

DEFAULT

N	Valid	51
	Missing	0
Percentiles	5	4.860
	10	5.760
	25	7.900
	50	9.700
	75	12.300
	90	15.120
	95	16.000

Solution   Both the 25th percentile and 95th percentile are highlighted on the SPSS printout, Figure 2.26. These values are 7.9 and 16.0, respectively. Our interpretations are as follows: 25% of the 51 default rates fall below 7.9 and 95% of the default rates fall below 16.0.

Look Back   The method for computing percentiles with small data sets varies according to the software used. As the sample size increases, the percentiles from the different software packages will converge to a single number.

Now Work	*Exercise 2.103*

■ ■ ■

Another measure of relative standing in popular use is the *z-score*. As you can see in Definition 2.13, the *z*-score makes use of the mean and standard deviation of the data set in order to specify the relative location of the measurement.

Note that the *z*-score is calculated by subtracting $\bar{x}$ (or μ) from the measurement x and then dividing the result by s (or σ). The final result, the **z-score**, represents the distance between a given measurement x and the mean, expressed in standard deviations.

DEFINITION 2.13

The **sample z-score** for a measurement x is

$$z = \frac{x - \bar{x}}{s}$$

The **population z-score** for a measurement x is

$$z = \frac{x - \mu}{\sigma}$$

EXAMPLE 2.15   FINDING A Z-SCORE

Problem   Suppose a sample of 2,000 high school seniors' verbal SAT scores is selected. The mean and standard deviation are

$$\bar{x} = 550 \quad s = 75$$

Suppose Joe Smith's score is 475. What is his sample *z*-score?

Solution Joe Smith's Verbal SAT score lies below the mean score of the 2,000 seniors; as shown in Figure 2.27.

Figure 2.27
Verbal SAT Scores of High School Seniors

325	475	550	775
$\bar{x} - 3s$	Joe Smith's score	$\bar{x}$	$\bar{x} + 3s$

We compute

$$z = \frac{x - \bar{x}}{s} = \frac{475 - 550}{75} = -1.0$$

which tells us that Joe Smith's score is 1.0 standard deviation *below* the sample mean; in short, his sample z-score is -1.0.

Look Back The numerical value of the z-score reflects the relative standing of the measurement. A large positive z-score implies that the measurement is larger than almost all other measurements, whereas a large negative z-score indicates that the measurement is smaller than almost every other measurement. If a z-score is 0 or near 0, the measurement is located at or near the mean of the sample or population.

Now Work	*Exercise 2.106*

■ ■ ■

We can be more specific if we know that the frequency distribution of the measurements is mound shaped. In this case, the following interpretation of the z-score can be given:

Interpretation of z-Scores for Mound-Shaped Distributions of Data

1. Approximately 68% of the measurements will have a z-score between -1 and 1.

2. Approximately 95% of the measurements will have a z-score between -2 and 2.

3. Approximately 99.7% (almost all) of the measurements will have a z-score between -3 and 3.

Note that this interpretation of z-scores is identical to that given by the Empirical Rule for mound-shaped distributions (Table 2.7). The statement that a measurement falls in the interval $(\mu - \sigma)$ to $(\mu + \sigma)$ is equivalent to the statement that a measurement has a population z-score between -1 and 1, since all measurements between $(\mu - \sigma)$ and $(\mu + \sigma)$ are within 1 standard deviation of μ. These z-scores are displayed in Figure 2.28.

Figure 2.28
Population z-Scores for a Mound-Shaped Distribution

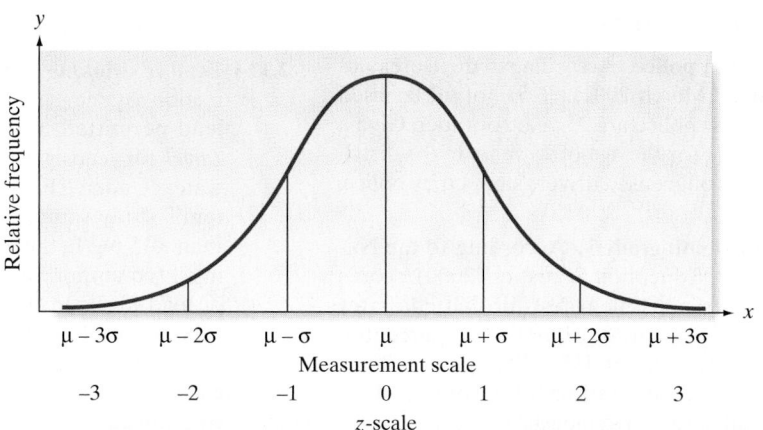

Exercises 2.103–2.117

Understanding the Principles

2.103 Give the percentage of measurements in a data set
NW that are above and below each of the following percentiles:
   a. 75th percentile
   b. 50th percentile
   c. 20th percentile
   d. 84th percentile

2.104 What is the 50th percentile of a quantitative data set called?

2.105 For mound-shaped data, what percentage of measurements have a z-score between -2 and 2?

Learning the Mechanics

2.106 Compute the z-score corresponding to each of the following values of x:
NW
   a. $x = 40$, $s = 5$, $\bar{x} = 30$
   b. $x = 90$, $\mu = 89$, $\sigma = 2$
   c. $\mu = 50$, $\sigma = 5$, $x = 50$
   d. $s = 4$, $x = 20$, $\bar{x} = 30$
   e. In parts **a–d**, state whether the z-score locates x within a sample or a population.
   f. In parts **a–d**, state whether each value of x lies above or below the mean and by how many standard deviations.

2.107 Compare the z-scores to decide which of the following x values lie the greatest distance above the mean and the greatest distance below the mean.
   a. $x = 100$, $\mu = 50$, $\sigma = 25$
   b. $x = 1$, $\mu = 4$, $\sigma = 1$
   c. $x = 0$, $\mu = 200$, $\sigma = 100$
   d. $x = 10$, $\mu = 5$, $\sigma = 3$

2.108 Suppose that 40 and 90 are two elements of a population data set and that their z-scores are -2 and 3, respectively. Using only this information, is it possible to determine the population's mean and standard deviation? If so, find them. If not, explain why it's not possible.

Applying the Concepts—Basic

2.109 **Drivers stopped by police.** According to the Bureau of Justice Statistics (March 2002), 73.5% of all licensed drivers stopped by police are 25 years or older. Give a percentile ranking for the age of 25 years in the distribution of all ages of licensed drivers stopped by police.

2.110 **Math scores of eighth graders.** According to the National Center for Education Statistics (2000), scores on a mathematics assessment test for United States eighth graders have a mean of 500, a 5th percentile of 356, a 25th percentile of 435, a 75th percentile of 563, and a 95th percentile of 653. Interpret each of these numerical descriptive measures.

SHIPSANIT

2.111 **Sanitation inspection of cruise ships.** Refer to the sanitation levels of cruise ships, Exercise 2.93 (p. 79), saved in the SHIPSANIT file.
   a. Give a measure of relative standing for the Nautilus Explorer's score of 74. Interpret the result.
   b. Give a measure of relative standing for the Rotterdam's score of 93. Interpret the result.

JAPANESE

2.112 **Reading Japanese books.** Refer to the *Reading in a Foreign Language* (Apr. 2004) experiment to improve the Japanese reading comprehension levels of 14 University of Hawaii students, Exercises 2.57 and 2.81 (p. 71). The data on number of books read and grade for each student are saved in the JAPANESE file.
   a. Find the mean and standard deviation of the number of books read by students who earned an A grade. Find the z-score for an A student who read 40 books. Interpret the result.
   b. Find the mean and standard deviation of the number of books read by students who either earned either a B or C grade. Find the z-score for a B or C student who read 40 books. Interpret the result.
   c. Refer to parts **a** and **b**. Which of the two groups of students is more likely to have read 40 books? Explain.

Applying the Concepts—Intermediate

NZBIRDS

2.113 **Extinct New Zealand birds.** Refer to the *Evolutionary Ecology Research* (July 2003) study of the patterns of extinction in the New Zealand bird population, Exercise 2.96 (p. 80). Again, consider the data on the egg length (measured in millimeters) for the 132 bird species saved in the NZBIRDS file.
   a. Find the 10th percentile for the egg length distribution and interpret its value.
   b. The *Moas, P. australis* bird species has an egg length of 205 millimeters. Find the z-score for this species of bird and interpret its value.

Applying the Concepts—Intermediate

2.114 **Lead in drinking water.** The US. Environmental Protection Agency (EPA) sets a limit on the amount of lead permitted in drinking water. The EPA *Action Level* for lead is .015 milligrams per liter (mg/L) of water. Under EPA guidelines, if 90% of a water system's study samples have a lead concentration less than .015 mg/L, the water is considered safe for drinking. I (co-author Sincich) received a report on a study of lead levels in the drinking water of homes in my subdivision. The 90th percentile of the study sample had a lead concentration of .00372 mg/L. Are water customers in my subdivision at risk of drinking water with unhealthy lead levels? Explain.

⊘ **LEFTEYE**

2.115 Eye refractive study. Refer to the *Optometry and Vision Science* (June 1995) study of refractive variation in eyes, Exercise 2.67 (p. 66). The 25 cylinder power measurements are saved in the LEFTEYE file.

   a. Find the 10th percentile of cylinder power measurements. Interpret the result.

   b. Find the 95th percentile of cylinder power measurements. Interpret the result.

   c. Calculate the *z*-score for the cylinder power measurement of 1.07. Interpret the result.

2.116 Blue versus red exam study. In a study of how external clues influence performance, psychology professors at the University of Alberta and Pennsylvania State University gave two different forms of a midterm examination to a large group of introductory psychology students. The questions on the exam were identical and in the same order, but one exam was printed on blue paper and the other on red paper. (*Teaching Psychology*, May 1998.) Grading only the difficult questions on the exam, the researchers found that scores on the blue exam had a distribution with a mean of 53% and a standard deviation of 15%, while scores on the red exam had a distribution with a mean of 39% and a standard deviation of 12%. (Assume that both distributions are approximately mound-shaped and symmetric.)

   a. Give an interpretation of the standard deviation for the students who took the blue exam.

   b. Give an interpretation of the standard deviation for the students who took the red exam.

   c. Suppose a student is selected at random from the group of students who participated in the study and the student's score on the difficult questions is 20%. Which exam form is the student more likely to have taken, the blue or the red exam? Explain.

Applying the Concepts—Advanced

2.117 GPAs of students. At one university, the students are given *z*-scores at the end of each semester rather than the traditional GPAs. The mean and standard deviation of all students' cumulative GPAs, on which the *z*-scores are based, are 2.7 and .5, respectively.

   a. Translate each of the following *z*-scores to corresponding GPA scores: $z = 2.0$, $z = -1.0$, $z = .5$, $z = -2.5$.

   b. Students with *z*-scores below -1.6 are put on probation. What is the corresponding probationary GPA?

   c. The president of the university wishes to graduate the top 16% of the students with *cum laude* honors and the top 2.5% with *summa cum laude* honors. Where (approximately) should the limits be set in terms of *z*-scores? In terms of GPAs? What assumption, if any, did you make about the distribution of the GPAs at the university?

2.8 Methods for Detecting Outliers (Optional)

Sometimes it is important to identify inconsistent or unusual measurements in a data set. An observation that is unusually large or small relative to the data values we want to describe is called an **outlier**.

     Outliers are often attributable to one of several causes. First, the measurement associated with the outlier may be invalid. For example, the experimental procedure used to generate the measurement may have malfunctioned, the experimenter may have misrecorded the measurement, or the data might have been coded incorrectly in the computer. Second, the outlier may be the result of a misclassified measurement. That is, the measurement belongs to a population different from that from which the rest of the sample was drawn. Finally, the measurement associated with the outlier may be recorded correctly and from the same population as the rest of the sample, but represents a rare (chance) event. Such outliers occur most often when the relative frequency distribution of the sample data is extremely skewed, because such a distribution has a tendency to include extremely large or small observations relative to the others in the data set.

> **DEFINITION 2.14**
>
> An observation (or measurement) that is unusually large or small relative to the other values in a data set is called an **outlier**. Outliers typically are attributable to one of the following causes:
>
> **1.** The measurement is observed, recorded, or entered into the computer incorrectly.
>
> **2.** The measurement comes from a different population.
>
> **3.** The measurement is correct, but represents a rare (chance) event.

Two useful methods for detecting outliers, one graphical and one numerical, are **box plots** and z-scores. The box plot is based on the *quartiles* of a data set. **Quartiles** are values that partition the data set into four groups, each containing 25% of the measurements. The *lower quartile* Q_L is the 25th percentile, the *middle quartile* is the median M (the 50th percentile), and the *upper quartile* Q_U is the 75th percentile (see Figure 2.29).

Figure 2.29

The Quartiles for a Data Set

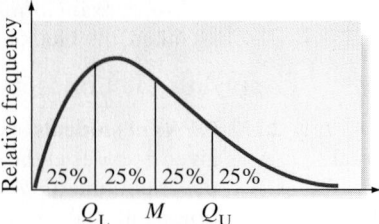

DEFINITION 2.15

The **lower quartile** Q_L is the 25th percentile of a data set. The **middle quartile** M is the median. The **upper quartile** Q_U is the 75th percentile.

A box plot is based on the *interquartile range (IQR)*, the distance between the lower and upper quartiles:

$$IQR = Q_U - Q_L$$

DEFINITION 2.16

The **interquartile range (IQR)** is the distance between the lower and upper quartiles:
$$IQR = Q_U - Q_L$$

An annotated MINITAB box plot for the gas mileage data (Table 2.2) is shown in Figure 2.30.* Note that a rectangle (the *box*) is drawn, with the bottom and top of the rectangle (the **hinges**) drawn at the quartiles Q_L and Q_U, respectively.

Figure 2.30

Annotated MINITAB Box Plot for EPA Gas Mileages

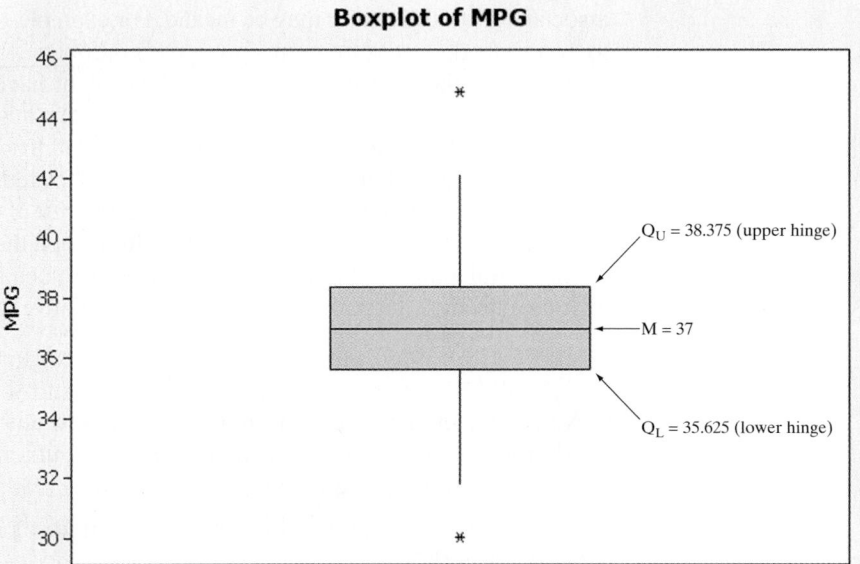

*Although box plots can be generated by hand, the amount of detail required makes them particularly well suited for computer generation. We use computer software to generate the box plots in this section.

By definition, then, the "middle" 50% of the observations — those between Q_L and Q_U — fall inside the box. For the gas mileage data, these quartiles are at 35.625 and 38.375. Thus,

$$IQR = 38.375 - 35.625 = 2.75$$

The median is shown at 37 by a horizontal line within the box.

To guide the construction of the "tails" of the box plot, two sets of limits, called **inner fences** and **outer fences**, are used. Neither set of fences actually appears on the box plot. Inner fences are located at a distance of 1.5(IQR) from the hinges. Emanating from the hinges of the box are vertical lines called the **whiskers**. The two whiskers extend to the most extreme observation inside the inner fences. For example, the inner fence on the lower side of the gas mileage box plot is

$$\text{Lower inner fence} = \text{Lower hinge} - 1.5(\text{IQR})$$
$$= 35.625 - 1.5(2.75)$$
$$= 35.625 - 4.125 = 31.5$$

The smallest measurement *inside* this fence is the second smallest measurement, 31.8. Thus, the lower whisker extends to 31.8. Similarly, the upper whisker extends to 42.1, the largest measurement inside the upper inner fence

$$\text{Upper inner fence} = \text{Upper hinge} + 1.5(\text{IQR})$$
$$= 38.375 + 1.5(2.75)$$
$$= 38.375 + 4.125 = 42.5$$

Values that are beyond the inner fences are deemed *potential outliers* because they are extreme values that represent relatively rare occurrences. In fact, for mound-shaped distributions, fewer than 1% of the observations are expected to fall outside the inner fences. Two of the 100 gas mileage measurements, 30.0 and 44.9, fall beyond the inner fences, one on each end of the distribution. Each of these potential outliers is represented by a common symbol (an asterisk in MINITAB).

The other two imaginary fences, the outer fences, are defined at a distance 3(IQR) from each end of the box. Measurements that fall beyond the outer fences (also represented by an asterisk in MINITAB) are very extreme measurements that require special analysis. Since less than one-hundredth of 1% (.01% or .0001) of the measurements from mound-shaped distributions are expected to fall beyond the outer fences, these measurements are considered to be *outliers*. Since there are no measurements of gas mileage beyond the outer fences, there are no outliers.

Recall that outliers may be incorrectly recorded observations, members of a population different from the rest of the sample, or, at the least, very unusual measurements from the same population. The box plot of Figure 2.30 detected two potential outliers — the two gas mileage measurements beyond the inner fences. When we analyze these measurements, we find that they are correctly recorded. Perhaps they represent mileages that correspond to exceptional models of the car being tested or to unusual gas mixtures. Outlier analysis often reveals useful information of this kind and therefore plays an important role in the statistical inference-making process.

In addition to detecting outliers, box plots provide useful information on the variation in a data set. The elements (and nomenclature) of box plots are summarized in the next box. Some aids to the interpretation of box plots are also given.

Elements of a Box Plot

1. A rectangle (the **box**) is drawn with the ends (the **hinges**) drawn at the lower and upper quartiles (Q_L and Q_U). The median of the data is shown in the box, usually by a line.

2. The points at distances 1.5(IQR) from each hinge mark the **inner fences** of the data set. Lines (the **whiskers**) are drawn from each hinge to the most extreme measurement inside the inner fence.

$$\text{Lower inner fence} = Q_L - 1.5(\text{IQR})$$

$$\text{Upper inner fence} = Q_U + 1.5(\text{IQR})$$

3. A second pair of fences, the **outer fences**, appear at a distance of 3 interquartile ranges, 3(IQR), from the hinges. One symbol (e.g., "*") is used to represent measurements falling between the inner and outer fences, and another (e.g., "0") is used to represent measurements beyond the outer fences. Thus, outer fences are not shown unless one or more measurements lie beyond them.

$$\text{Lower outer fence} = Q_L - 3(\text{IQR})$$

$$\text{Upper outer fence} = Q_U + 3(\text{IQR})$$

4. The symbols used to represent the median and the extreme data points (those beyond the fences) will vary depending on the software you use to construct the box plot. (You may use your own symbols if you are constructing a box plot by hand.) You should consult the program's documentation to determine exactly which symbols are used.

Aids to the Interpretation of Box Plots

1. Examine the length of the box. The IQR is a measure of the sample's variability and is especially useful for the comparison of two samples (see Example 2.16).

2. Visually compare the lengths of the whiskers. If one is clearly longer, the distribution of the data is probably skewed in the direction of the longer whisker.

3. Analyze any measurements that lie beyond the fences. Fewer than 5% should fall beyond the inner fences, even for very skewed distributions. Measurements beyond the outer fences are probably outliers, with one of the following explanations:
   a. The measurement is incorrect. It may have been observed, recorded, or entered into the computer incorrectly.
   b. The measurement belongs to a population different from the population that the rest of the sample was drawn from (see Example 2.17).
   c. The measurement is correct *and* from the same population as the rest. Generally, we accept this explanation only after carefully ruling out all others.

EXAMPLE 2.16   BOX PLOTS USING THE COMPUTER

Problem   Use a statistical software package to draw a box plot for the student loan default data, Table 2.4. Identify any outliers in the data set.

Solution   The MINITAB box plot for the student loan default rates is shown in Figure 2.32. Note that the median appears to be about 9.5, and, with the exception of a single extreme observation, the distribution appears to be symmetrically distributed between approximately 3% and 17%. The single outlier is beyond the inner fence but inside the outer fence. Examination of the data reveals that this observation corresponds to Alaska's default rate of 19.7%.

Figure 2.31
MINITAB Box Plot for
Student Loan Default Rates

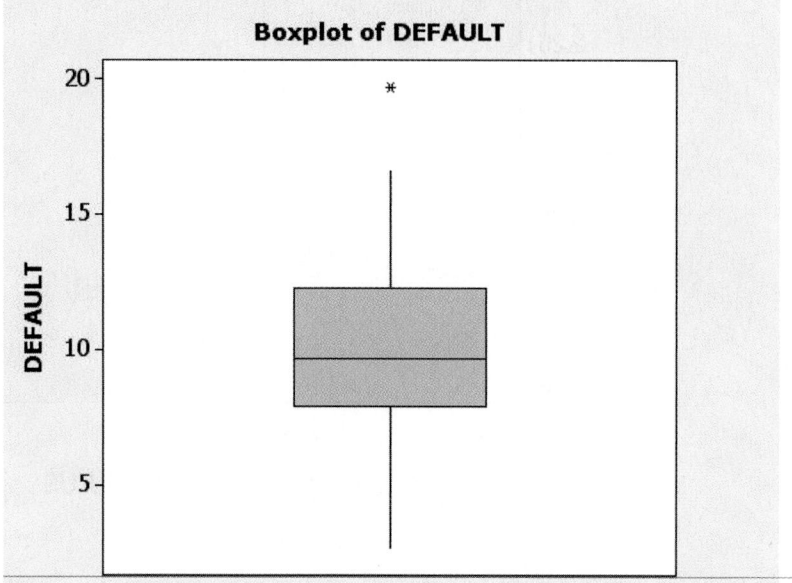

Look Back Before removing the outlier from the data set, a good analyst will make a concerted effort to find the cause of the outlier.

Now Work   *Exercise 2.125*

EXAMPLE 2.17   COMPARING BOX PLOTS

Problem   A Ph.D. student in psychology conducted a stimulus reaction experiment as a part of her dissertation research. She subjected 50 subjects to a threatening stimulus and 50 to a nonthreatening stimulus. The reaction times of all 100 students, recorded to the nearest tenth of a second, are listed in Table 2.9. Box plots of the two resulting samples of reaction times, generated using SAS, are shown in Figure 2.32. Interpret the box plots.

REACTION

TABLE 2.9   Reaction Times of Students

Nonthreatening Stimulus									
2.0	1.8	2.3	2.1	2.0	2.2	2.1	2.2	2.1	2.1
2.0	2.0	1.8	1.9	2.2	2.0	2.2	2.4	2.1	2.0
2.2	2.1	2.2	1.9	1.7	2.0	2.0	2.3	2.1	1.9
2.0	2.2	1.6	2.1	2.3	2.0	2.0	2.0	2.2	2.6
2.0	2.0	1.9	1.9	2.2	2.3	1.8	1.7	1.7	1.8

Threatening Stimulus									
1.8	1.7	1.4	2.1	1.3	1.5	1.6	1.8	1.5	1.4
1.4	2.0	1.5	1.8	1.4	1.7	1.7	1.7	1.4	1.9
1.9	1.7	1.6	2.5	1.6	1.6	1.8	1.7	1.9	1.9
1.5	1.8	1.6	1.9	1.3	1.5	1.6	1.5	1.6	1.5
1.3	1.7	1.3	1.7	1.7	1.8	1.6	1.7	1.7	1.7

Figure 2.32
SAS Box Plots for Reaction
Time Data

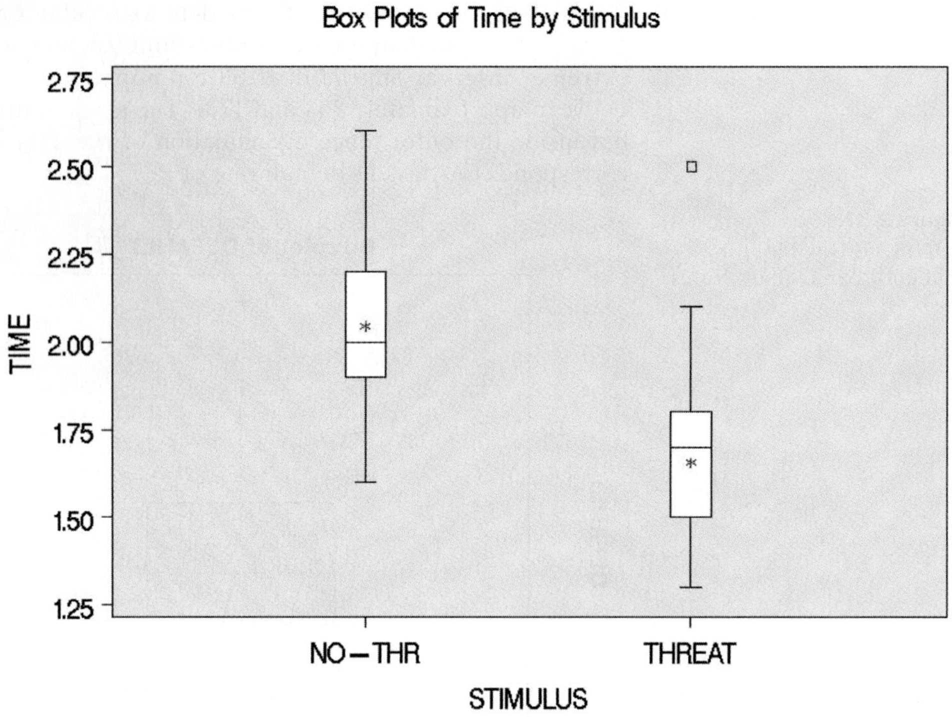

Box Plots of Time by Stimulus

Solution In SAS, the median is represented by the horizontal line through the box, while the asterisk (*) symbol represents the mean. Analysis of the box plots on the same numerical scale reveals that the distribution of times corresponding to the threatening stimulus lies below that of the nonthreatening stimulus. The implication is that the reaction times tend to be faster to the threatening stimulus. Note, too, that the upper whiskers of both samples are longer than the lower whiskers, indicating that the reaction times are positively skewed.

No observations in the two samples fall between the inner and outer fences. However, there is one outlier — the observation of 2.5 seconds corresponding to the threatening stimulus that is beyond the outer fence (denoted by the square symbol in SAS). When the researcher examined her notes from the experiments, she found that the subject whose time was beyond the outer fence had mistakenly been given the nonthreatening stimulus. You can see in Figure 2.32 that his time would have been within the upper whisker if moved to the box plot corresponding to the nonthreatening stimulus. The box plots should be reconstructed since they will both change slightly when this misclassified reaction time is moved from one sample to the other.

Look Back The researcher concluded that the reactions to the threatening stimulus were faster than those to the nonthreatening stimulus. However, she was asked by her Ph.D. committee whether the results were *statistically significant*. Their question addresses the issue of whether the observed difference between the samples might be attributable to chance or sampling variation rather than to real differences between the populations. To answer their question, the researcher must use inferential statistics rather than graphical descriptions. We discuss how to compare two samples using inferential statistics in Chapter 9.

■ ■ ■

The following example illustrates how z-scores can be used to detect outliers and make inferences.

EXAMPLE 2.18   INFERENCE USING Z-SCORES

Problem   Suppose a female bank employee believes that her salary is low as a result of sex dis-crimination. To substantiate her belief, she collects information on the salaries of her male counterparts in the banking business. She finds that their salaries have a mean of $54,000 and a standard deviation of $2,000. Her salary is $47,000. Does this infor-mation support her claim of sex discrimination?

Solution   The analysis might proceed as follows: First, we calculate the z-score for the woman's salary with respect to those of her male counterparts. Thus,

$$z = \frac{\$47,000 - \$54,000}{\$2,000} = -3.5$$

The implication is that the woman's salary is 3.5 standard deviations *below* the mean of the male salary distribution. Furthermore, if a check of the male salary data shows that the frequency distribution is mound shaped, we can infer that very few salaries in this distribution should have a z-score less than -3, as shown in Figure 2.33. Clearly, a z-score of -3.5 represents an outlier. Either her salary is from a distribution different from the male salary distribution or it is a very unusual (highly improbable) measurement from a salary distribution no different from the male distribution.

Figure 2.33
Male Salary Distribution

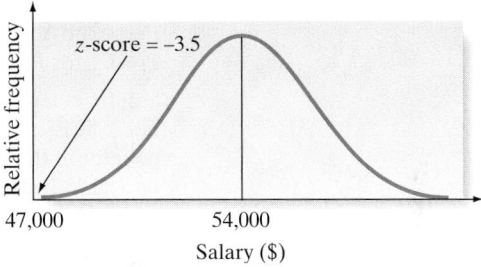

Look Back Which of the two situations do you think prevails? Statistical thinking would lead us to conclude that her salary does not come from the male salary distri-bution, lending support to the female bank employee's claim of sex discrimination. However, the careful investigator should require more information before inferring sex discrimination as the cause. We would want to know more about the data collec-tion technique the woman used and more about her competence at her job. Also, per-haps other factors such as length of employment should be considered in the analysis.

Now Work   *Exercise 2.123*

■ ■ ■

Examples 2.17 and 2.18 exemplify an approach to statistical inference that might be called the **rare-event approach**. An experimenter hypothesizes a specific frequency distribution to describe a population of measurements. Then a sample of measurements is drawn from the population. If the experimenter finds it unlikely that the sample came from the hypothesized distribution, the hypothesis is conclud-ed to be false. Thus, in Example 2.18 the woman believes her salary reflects discrim-ination. She hypothesizes that her salary should be just another measurement in the distribution of her male counterparts' salaries if no discrimination exists. However, it is so unlikely that the sample (in this case, her salary) came from the male frequency distribution that she rejects that hypothesis, concluding that the distribution from which her salary was drawn is different from the distribution for the men.

This rare-event approach to inference-making is discussed further in later chap-ters. Proper application of the approach requires a knowledge of probability, the subject of our next chapter.

We conclude this section with some rules of thumb for detecting outliers.

Rules of Thumb for Detecting Outliers*

	Suspect Outliers	Highly Suspect Outliers
Box Plots:	Data points between inner & outer fences	Data points beyond outer fences
z-Scores:	$2 < |z| < 3$	$|z| > 3$

Box Plots

Using the T1-83 Graphing Calculator

Making a Box Plot

Step 1   Enter the data
Press STAT and select 1:Edit
Note: If the list already contains data, clear the old data. Use the up arrow to highlight 'L1'. Press CLEAR ENTER.
Use the arrow and ENTER keys to enter the data set into L1.

Step 2   Set up the box plot

Press 2nd $Y =$ for STAT PLOT
Press 1 for Plot 1

Set the cursor so that 'ON' is flashing.
For TYPE, use the right arrow to scroll through the plot icons and select the boxplot in the middle of the second row.
For XLIST, choose L1.
Set FREQ to 1.

Step 3   View the graph
Press ZOOM and select 9:ZoomStat

Optional   *Read the five number summary*
Step Press TRACE
Use the left and right arrow keys to move between minX, Q1, Med, Q3, and maxX.

Example   Make a box plot for the given data,
86, 70, 62, 98, 73, 56, 53, 92, 86, 37, 62, 83, 78, 49, 78, 37, 67, 79, 57

The output screen for this example is shown below.

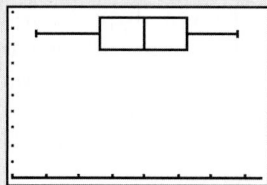

*The z-score and box plot methods both establish rule-of-thumb limits outside of which a measurement is deemed to be an outlier. Usually, the two methods produce similar results. However, the presence of one or more outliers in a data set can inflate the computed value of s. Consequently, it will be less likely that an errant observation would have a z-score larger than 3 in absolute value. In contrast, the values of the quartiles used to calculate the intervals for a box plot are not affected by the presence of outliers.

Statistics in Action Revisited

Detecting Outliers

In the *Psychological Science* "water-level task" experiment, the quantitative variable of interest is *Deviation angle* (measured in degrees) of the judged line from the true surface line (parallel to the table top). Are there any unusual values of this variable in the **EYECUE** data set? We will employ both the box plot and *z*-score methods to aid in identifying outliers in the data.

Descriptive statistics for the deviation angles of all 120 experimental subjects, produced using MINITAB, are shown in Figure SIA2.7. To employ the *z*-score method, we need the mean and standard deviation. These values are highlighted on Figure SIA2.7. Then, the 3-standard deviation interval is

$$4.742 \pm 3(7.865) = 4.742 \pm 23.595 = (-18.853, 28.337)$$

If you examine the deviation angles in the **EYECUE** data set, you find that only one of these values, 35, falls beyond the 3-standard deviation interval. Thus, the data for the subject whose judged line had a deviation angle of 35° (*z*-score = 3.85) is considered a highly suspect outlier using the *z*-score approach.

A box plot for the data is shown in Figure SIA2.8. Although there are several suspect outliers (asterisks) shown on the box plot, there are no highly suspect outliers (zeros) shown. That is, there are no data points that fall beyond the outer fences. Thus, the box plot method does not identify the value of 35 as a highly suspect outlier, only as a suspect outlier.

Before any type of inference is made concerning the population of deviation angles for subjects performing the water-level task, we should consider whether this potential outlier is a legitimate observation (in which case it will remain in the data set for analysis) or is associated with a subject that is not a member of the population of interest (in which case it will be removed from the data set).

Descriptive Statistics: Deviation

Variable	N	Mean	StDev	Minimum	Q1	Median	Q3	Maximum	IQR
Deviation	120	4.742	7.865	-18.000	0.000	4.000	9.000	35.000	9.000

Figure SIA2.7
MINITAB Descriptive Statistics for Deviation Angle of Judged Water Level

Figure SIA2.8
MINITAB Box Plot for Deviation Angle of Judged Water Level

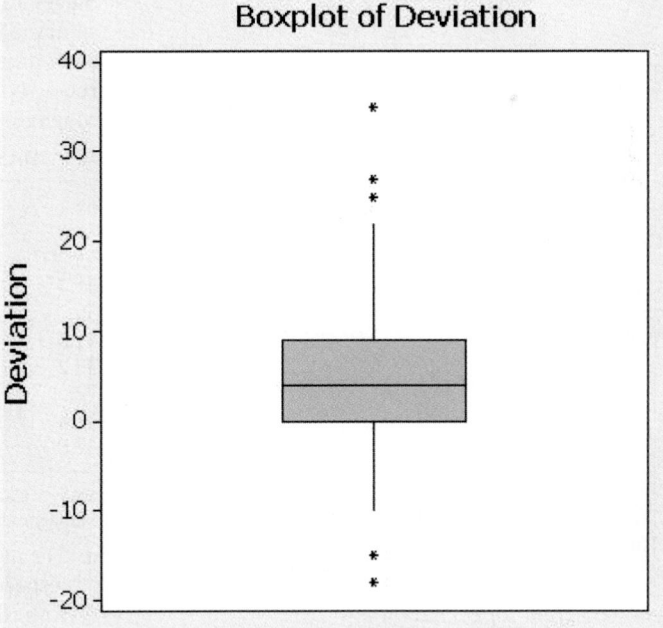

Exercises 2.118–2.135

Understanding the Principles:

2.118 Define an outlier.

2.119 Define the 25th, 50th, and 75th percentiles of a data set.

2.120 What is the interquartile range?

2.121 What are the hinges of a box plot?

2.122 With mound-shaped data, what proportion of the measurements have z-scores between -3 and 3?

Learning the Mechanics

2.123 A sample data set has a mean of 57 and a standard
NW deviation of 11. Determine whether each of the following sample measurements are outliers.
   a. 65   **b.** 21   **c.** 72   **d.** 98

2.124 Suppose a data set consisting of exam scores has a lower quartile $Q_L = 60$, a median $M = 75$, and an upper quartile $Q_U = 85$. The scores on the exam range from 18 to 100. Without having the actual scores available to you, construct as much of the box plot as possible.

2.125 Consider the horizontal box plot shown below.
NW **a.** What is the median of the data set (approximately)?

   b. What are the upper and lower quartiles of the data set (approximately)?
   c. What is the interquartile range of the data set (approximately)?
   d. Is the data set skewed to the left, skewed to the right, or symmetric?
   e. What percentage of the measurements in the dataset lie to the right of the median? To the left of the upper quartile?
   f. Identify any outliers in the data.

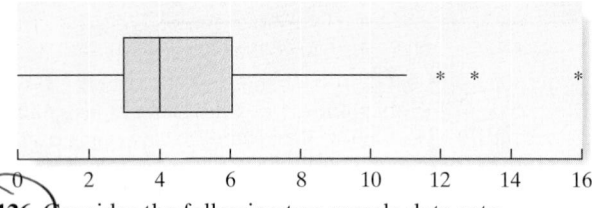

2.126 Consider the following two sample data sets:

LM2_126

	Sample A			Sample B	
121	171	158	171	152	170
173	184	163	168	169	171
157	85	145	190	183	185
165	172	196	140	173	206
170	159	172	172	174	169
161	187	100	199	151	180
142	166	171	167	170	188

   a. Construct a box plot for each data set.
   b. Identify any outliers that may exist in the two data sets.

Applying the Concepts—Basic

PMI

2.127 **Research on brain specimens.** Refer to the *Brain and Language* data on postmortem intervals (PMIs) of 22 human brain specimens, Exercise 2.37 (p. 52). The mean and standard deviation of the PMI values are 7.3 and 3.18, respectively.
   a. Find the z-score for the PMI value of 3.3.
   b. Is the PMI value of 3.3 considered an outlier? Explain.

2.128 **Dentists' use of anesthetics.** Refer to the *Current Allergy & Clinical Immunology* (Mar. 2004) study of the use of local anesthetics (LA) in dentistry, Exercise 2.95 (p. 79). Recall that the mean number of units (ampoules) of LA used per week by dentists was 79 with a standard deviation of 23. Consider a dentist who used 175 units of LA in a week.
   a. Find the z-score for this measurement.
   b. Would you consider this measurement to be an outlier? Explain.
   c. Give several reasons why the outlier may occur.

SATSCORES

2.129 **Comparing SAT Scores.** Refer to Exercise 2.40 (p. 53) in which we compared states' average SAT scores in 1990 and 2000. The data is saved in the SATSCORES file.
   a. Construct side-by-side box plots of the SAT scores for the two years.
   b. Compare the variability of the SAT scores for the two years.
   c. Are any states' SAT scores outliers in either year? If so, identify them.

2.130 **Salary offers to MBAs.** The table contains the top salary offer (in thousands of dollars) received by each member of a sample of 50 MBA students who recently graduated from the Graduate School of Management at Rutgers University.

MBASAL

61.1	48.5	47.0	49.1	43.5
50.8	62.3	50.0	65.4	58.0
53.2	39.9	49.1	75.0	51.2
41.7	40.0	53.0	39.6	49.6
55.2	54.9	62.5	35.0	50.3
41.5	56.0	55.5	70.0	59.2
39.2	47.0	58.2	59.0	60.8
72.3	55.0	41.4	51.5	63.0
48.4	61.7	45.3	63.2	41.5
47.0	43.2	44.6	47.7	58.6

Source: Career Services Office, Graduate School of Management, Rutgers University.

   a. The mean and standard deviation are 52.33 and 9.22, respectively. Find and interpret the z-score associated with the highest salary offer, the lowest salary offer, and the mean salary offer. Would you consider the highest offer to be unusually high? Why or why not?

b. Construct a box plot for this data set and identify any outliers.

Applying the Concepts—Intermediate

2.131 Speech listening study. Refer to the *American Journal of Speech–Language Pathology* (Feb. 1995) study, Exercise 2.41 (p. 53). Recall that three groups of college students listened to an audio tape of a woman with imperfect speech. The groups were called *control*, *treatment*, and *familiarity*. At the end of the session, each student transcribed the spoken words.

a. Construct box plots for the percentage of words correctly transcribed (i.e., accuracy rate) for each group.

b. How do the median accuracy rates compare for the three groups?

c. How do the variabilities of the accuracy rates compare for the three groups?

d. The standard deviations of the accuracy rates are 3.56 for the control group, 5.45 for the treatment group, and 4.46 for the familiarity group. Do the standard deviations agree with the interquartile ranges (part **c**) with regard to the comparison of the variabilities of the accuracy rates?

e. Is there evidence of outliers in any of the three distributions?

WPOWER50

2.132 Most powerful woman in America. Refer to the *Fortune* (Oct. 14, 2002) ranking of the 50 most powerful women in America, Exercise 2.92 (p. 79). Create box plots to compare the ages of the women in three groups based on their position within the firm: Group 1 (CEO, CFO, CIO, or COO); Group 2 (Chairman, President, or Director); and Group 3 (Vice President, Vice Chairman, or General Manager).

SHIPSANIT

2.133 Sanitation inspection of cruise ships. Refer to the data on sanitation levels of cruise ships, Exercise 2.93 (p. 53).

a. Use the box plot method to detect any outliers in the data.

b. Use the z-score method to detect any outliers in the data.

c. Do the two methods agree? If not, explain why.

Applying the Concepts—Advanced

2.134 Library book checkouts. A city librarian claims that books have been checked out an average of seven (or more) times in the last year. You suspect he has exaggerated the checkout rate (book usage) and that the mean number of checkouts per book per year is, in fact, less than seven. Using the computerized card catalog, you randomly select one book and find that it has been checked out four times in the last year. Assume that the standard deviation of the number of checkouts per book per year is approximately 1.

a. If the mean number of checkouts per book per year really is seven, what is the z-score corresponding to four?

b. Considering your answer to part **a**, do you have reason to believe that the librarian's claim is incorrect?

c. If you knew that the distribution of the number of checkouts were mound shaped, would your answer to part **b** change? Explain.

d. If the standard deviation of the number of checkouts per book per year were 2 (instead of 1), would your answers to parts **b** and **c** change? Explain.

2.135 Zinc phosphide in sugarcane. A chemical company produces a substance composed of 98% cracked corn particles and 2% zinc phosphide for use in controlling rat populations in sugarcane fields. Production must be carefully controlled to maintain the 2% zinc phosphide because too much zinc phosphide will cause damage to the sugarcane and too little will be ineffective in controlling the rat population. Records from past production indicate that the distribution of the actual percentage of zinc phosphide present in the substance is approximately mound-shaped, with a mean of 2.0% and a standard deviation of .08%. Suppose one batch chosen randomly actually contains 1.80% zinc phosphide. Does this indicate that there is too little zinc phosphide in today's production? Explain your reasoning.

2.9 Graphing Bivariate Relationships (Optional)

The claim is often made that the crime rate and the unemployment rate are "highly correlated." Another popular belief is that smoking and lung cancer are "related." Some people even believe that the Dow Jones Industrial Average and the lengths of fashionable skirts are "associated." The words *correlated*, *related*, and *associated* imply a relationship between two variables — in the examples above, two *quantitative* variables.

One way to describe the relationship between two quantitative variables — called a **bivariate relationship** — is to plot the data in a **scattergram** (or **scatterplot**). A scattergram is a two-dimensional plot, with one variable's values plotted along the vertical axis and the other along the horizontal axis. For example, Figure 2.34 is a scattergram relating (1) the cost of mechanical work (heating, ventilating, and

Figure 2.34

Scattergram of cost vs.
floor area

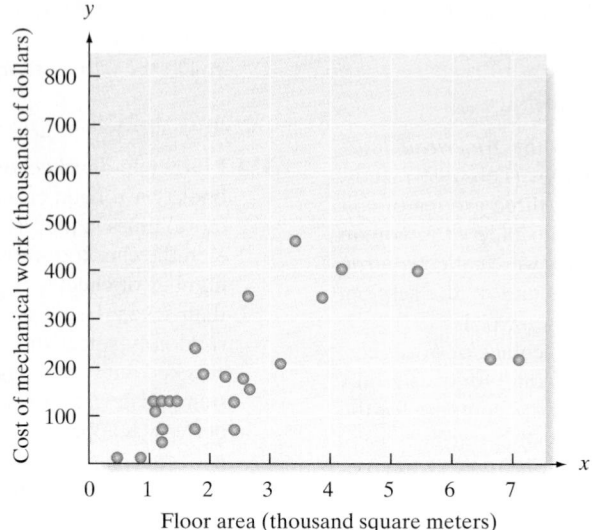

plumbing) to (2) the floor area of the building for a sample of 26 factory and ware-house buildings. Note that the scattergram suggests a general tendency for mechan-ical cost to increase as building floor area increases.

When an increase in one variable is generally associated with an increase in the second variable, we say that the two variables are "positively related" or "positively correlated."* Figure 2.34 implies that mechanical cost and floor area are positively correlated. Alternatively, if one variable has a tendency to decrease as the other in-creases, we say the variables are "negatively correlated." Figure 2.35 shows several hypothetical scattergrams that portray a positive bivariate relationship (Figure 2.35a), a negative bivariate relationship (Figure 2.35b), and a situation where the two variables are unrelated (Figure 2.35c).

Figure 2.35

Hypothetical Bivariate
Relationship

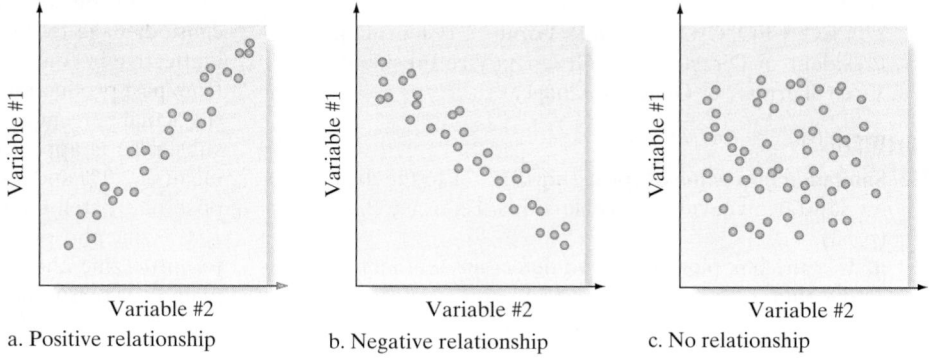

a. Positive relationship   b. Negative relationship   c. No relationship

EXAMPLE 2.19   GRAPHING BIVARIATE DATA

Problem   A medical item used to administer to a hospital patient is called a *factor*. For exam-ple, factors can be intravenous (IV) tubing, IV fluid, needles, shave kits, bedpans, di-apers, dressings, medications, and even code carts. The coronary care unit at Bayonet Point Hospital (St. Petersburg, Florida) recently investigated the relationship be-tween the number of factors administered per patient and the patient's length of stay (in days). Data on these two variables for a sample of 50 coronary care patients are given in Table 2.10. Use a scattergram to describe the relationship between the two variables of interest, number of factors, and length of stay.

*A formal definition of correlation is given in Chapter 11. We will learn that correlation measures the strength of the linear (or straight-line) relationship between two quantitative variables.

◎ **MEDFACTORS**

TABLE 2.10   Data on Patient's Factors and Length of Stay

Number of Factors	Length of Stay (days)	Number of Factors	Length of Stay (days)
231	9	354	11
323	7	142	7
113	8	286	9
208	5	341	10
162	4	201	5
117	4	158	11
159	6	243	6
169	9	156	6
55	6	184	7
77	3	115	4
103	4	202	6
147	6	206	5
230	6	360	6
78	3	84	3
525	9	331	9
121	7	302	7
248	5	60	2
233	8	110	2
260	4	131	5
224	7	364	4
472	12	180	7
220	8	134	6
383	6	401	15
301	9	155	4
262	7	338	8

Source: Bayonet Point Hospital, Coronary Care Unit.

Solution   Rather than construct the plot by hand, we resort to a statistical software package. The SPSS plot of the data in Table 2.10, with length of stay (LOS) on the horizontal axis and number of factors (FACTORS) on the vertical axis, is shown in Figure 2.36.

Figure 2.36
SPSS Scatterplot of Data in Table 2.10

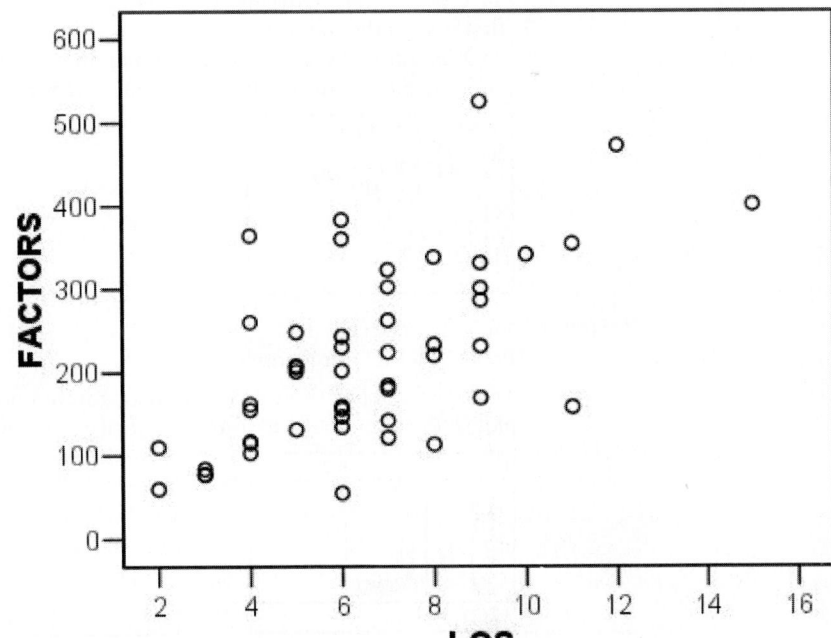

Although the plotted points exhibit a fair amount of variation, the scattergram clearly shows an increasing trend. It appears that a patient's length of stay is positively correlated with the number of factors administered to the patient.

Look Back If hospital administrators can be confident that the sample trend shown in Figure 2.36 accurately describes the trend in the population, then they may use this information to improve their forecasts of lengths of stay for future patients.

> Now Work   *Exercise 2.139*

The scattergram is a simple but powerful tool for describing a bivariate relationship. However, keep in mind that it is only a graph. No measure of reliability can be attached to inferences made about bivariate populations based on scattergrams of sample data. The statistical tools that enable us to make inferences about bivariate relationships are presented in Chapter 11.

Scatterplots

Using the T1-83 Graphing Calculator

Making Scatterplots

Step 1   Enter the data
Press STAT and select 1:Edit
Note: If a list already contains data, clear the old data. Use the up arrow to highlight the list name, 'L1' or 'L2'.
Press CLEAR ENTER
Enter your x-data in L1 and your y-data in L2.

Step 2   Set up the scatterplot
Press 2nd **Y =** for STAT PLOT
Press 1 for Plot1
Set the cursor so that ON is flashing.
For Type, use the arrow and Enter keys to highlight and select the scatterplot (first icon in the first row).
For Xlist, choose the column containing the x-data.
For Freq, choose the column containing the y-data.

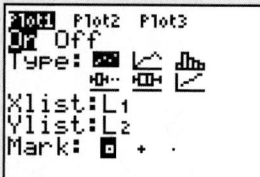

Step 3   View the scatterplot
Press ZOOM 9 for ZoomStat

Example   The figures below show a table of data entered on the T1-83 and the scatterplot of the data obtained using the steps given above.

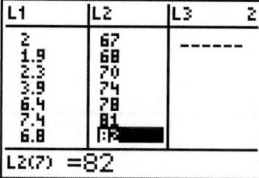   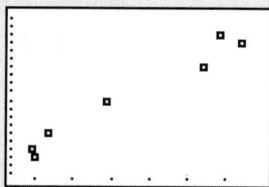

Exercises 2.136–2.148

Understanding the Principles

2.136 For what types of variables, quantitative or qualitative, are scatterplots useful?

2.137 Define a bivariate relationship.

2.138 What is the difference between positive association and negative association when describing the relationship between two variables?

Learning the Mechanics

2.139 Construct a scatterplot for the data in the following
[NW] table. Do you detect a trend?

Variable #1:	5	3	−1	2	7	6	4	0	8
Variable #2:	14	3	10	1	8	5	3	2	12

2.140 Construct a scatterplot for the data in the following table. Do you detect a trend?

Variable #1:	.5	1	1.5	2	2.5	3	3.5	4	4.5	5
Variable #2:	2	1	3	4	6	10	9	12	17	17

Applying the Concepts—Basic

2.141 Baseball batting averages versus wins. Baseball wisdom says if you can't hit, you can't win. Is the number of games won by a major league baseball team in a season related to the team's batting average? The information in the table, found in the *Baseball Almanac* (2003), shows the number of games won and the batting averages for the 14 teams in the American League for the 2002 Major League Baseball season. Construct a scatterplot for the data. Do you observe a trend?

⊙ ALWINS

Team	Games Won	Batting Ave.
New York	103	.275
Toronto	78	.261
Baltimore	67	.246
Boston	93	.277
Tampa Bay	55	.253
Cleveland	74	.249
Detroit	55	.248
Chicago	81	.268
Kansas City	62	.256
Minnesota	94	.272
Anaheim	99	.282
Texas	72	.269
Seattle	93	.275
Oakland	103	.261

Source: Baseball Almanac, 2003.

2.142 Feeding behavior of fish. Zoologists at the University of Western Australia conducted a study of the feeding behavior of blackbream fish spawned in aquariums. (*Brain and Behavior Evolution*, April 2000.) In one experiment, the zoologists recorded the number of aggressive strikes of two blackbream fish feeding at the bottom of the aquarium in the 10-minute period following the addition of food. The number of strikes and age of the fish (in days) was recorded approximately each week for nine weeks, as shown in the table.

⊙ BLACKBREAM

Week	Number of Strikes	Age of Fish (days)
1	85	120
2	63	136
3	34	150
4	39	155
5	58	162
6	35	169
7	57	178
8	12	184
9	15	190

Source: J. Shand, et al. "Variability in the location of the retinal ganglion cell area centralis is correlated with ontogenetic changes in feeding behavior in the Blackbream, Acanthopagrus 'butcher'," *Brain and Behavior,* Vol. 55, No. 4, April 2000 (Figure H).

a. Construct a scattergram for the data, with number of strikes on the *y*-axis and age of the fish on the *x*-axis.

b. Examine the scattergram of part **a**. Do you detect a trend?

2.143 Walking study. A "self-avoiding walk" describes a path in which you never retrace your steps or cross your own path. An "unrooted walk" is a path in which it is impossible to distinguish between the starting point and ending point of the path. The *American Scientist* (July–Aug. 1998) investigated the relationship between self-avoiding and unrooted walks. The table gives the number of unrooted walks and possible number of self-avoiding walks of various lengths, where length is measured as number of steps.

⊙ WALK

Walk Length (Number of Steps)	Unrooted Walks	Self-Avoiding Walks
1	1	4
2	2	12
3	4	36
4	9	100
5	22	284
6	56	780
7	147	2,172
8	388	5,916

Source: Hayes, B. "How to avoid yourself." *American Scientist,* Vol. 86, No. 4, July–Aug.1998, p. 317 (Figure 5).

a. Construct a plot to investigate the relationship between total possible number of self-avoiding walks and walk length. What pattern (if any) do you observe?

b. Repeat part **a** for unrooted walks.

DDT

2.144 Contaminated fish. Refer to the U.S. Army Corps of Engineers data on contaminated fish saved in the DDT file. Three quantitative variables are measured for each of the 144 captured fish: length (in centimeters), weight (in grams), and DDT level (in parts per million). Form a scatterplot for each pair of these variables. What trends, if any, do you detect?

Applying the Concepts—Intermediate

SITIN

2.145 College protests of labor exploitation. Refer to the *Journal of World-Systems Research* (Winter 2004) study of 14 student "sit-ins" for a "sweat free campus" at universities in 1999 and 2000, Exercise 2.35 (p. 52). The SITIN file contains data on the duration (in days) of each sit-in as well as the number of student arrests.

a. Use a scatterplot to graph the relationship between duration and number of arrests. Do you detect a trend?

b. Repeat part **a**, but only graph the data for sit-ins where there was at least one arrest. Do you detect a trend?

c. Comment on the reliability of the trend you detected in part **b**.

2.146 Short-term memory study. Research published in the *American Journal of Psychiatry* (July 1995) attempted to establish a link between hippocampal (brain) volume and short-term verbal memory of patients with combat-related post-traumatic stress disorder (PTSD). A sample of 21 Vietnam veterans with a history of combat-related PTSD participated in the study. Magnetic resonance imaging was used to measure the volume of the right hippocampus (in cubic millimeters) of each subject. Verbal memory retention of each subject was measured by the percent retention subscale of the Wechsler Memory Scale. The data for the 21 patients are plotted in the scattergram. The researchers "hypothesized that smaller hippocampal volume would be associated with deficits in short-term verbal memory in patients with PTSD." Does the scattergram provide visual evidence to support this theory?

GOBIANTS

2.147 Mongolian desert ants. Refer to the *Journal of Biogeography* (Dec. 2003) study of ants in Mongolia, Exercise 2.66 (p. 66). Data on annual rainfall, maximum daily temperature, plant cover percentage, number of ant species, and species diversity index recorded at each of 11 study sites are saved in the GOBIANTS file.

a. Construct a scatterplot to investigate the relationship between annual rainfall and maximum daily temperature. What type of trend (if any) do you detect?

b. Use scatterplots to investigate the relationship that annual rainfall has with each of the other four variables in the data set. Are the other variables positively or negatively related to rainfall?

2.148 Forest fragmentation study. Ecologists classify the cause of forest fragmentation as either anthropogenic (i.e., due to human development activities such as road construction or logging) or natural in origin (e.g., due to wetlands or wildfire). *Conservation Ecology* (Dec. 2003) published an article on the causes of fragmentation for 54 South American forests. Using advanced high resolution satellite imagery, the researchers developed two fragmentation indices for each forest — one index for anthropogenic fragmentation and one for fragmentation from natural causes. The values of these two indices (where higher values indicate more fragmentation) for five of the forests in the sample are shown in the accompanying table. The data for all 54 forests are saved in the FORFRAG file.

FORFRAG (Selected observations)

Ecoregion (forest)	Anthropogenic Index	Natural Origin Index
Araucaria moist forests	34.09	30.08
Atlantic Coast restingas	40.87	27.60
Bahia coastal forests	44.75	28.16
Bahia interior forests	37.58	27.44
Bolivian Yungas	12.40	16.75

Source: Wade, T. G., et al. "Distribution and causes of global forest fragmentation," *Conservation Ecology,* Vol. 72, No. 2, Dec. 2003 (Table 6).

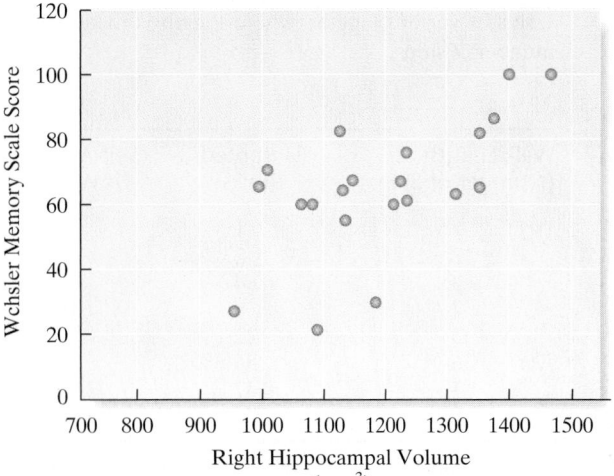

a. Ecologists theorize that an approximately linear (straight-line) relationship exists between the two fragmentation indices. Graph the data for all 54 forests. Does the graph support the theory?

b. Delete the data for the three forests with the largest anthropogenic indices and reconstruct the graph, part **a**. Comment on the ecologists' theory.

2.10 Distorting the Truth with Descriptive Techniques

A picture may be "worth a thousand words," but pictures can also color messages or distort them. In fact, the pictures in statistics — histograms, bar charts, and other graphical descriptions — are susceptible to distortion, so we have to examine each of them with care. We'll mention a few of the pitfalls to watch for when interpreting a chart or graph.

One common way to change the impression conveyed by a graph is to change the scale on the vertical axis, the horizontal axis, or both. For example, Figure 2.37 is a bar graph that shows the market share of sales for a company for each of the years 2000 to 2005. If you want to show that the change in firm A's market share over time is moderate, you should pack in a large number of units per inch on the vertical axis. That is, make the distance between successive units on the vertical scale small, as shown in Figure 2.37. You can see that a change in the firm's market share over time is barely apparent.

Figure 2.37

Firm A's Market Share: Packed Vertical Axis

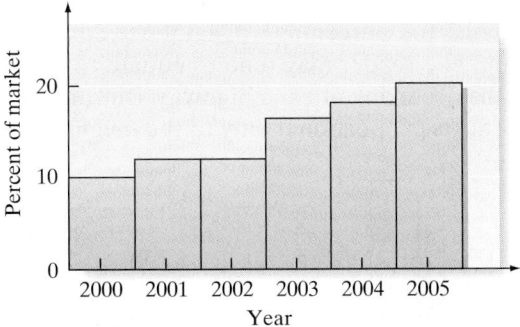

If you want to use the same data to make the changes in firm A's market share appear large, you should increase the distance between successive units on the vertical axis. That is, stretch the vertical axis by graphing only a few units per inch as in Figure 2.38. A telltale sign of stretching is a long vertical axis, but this is often hidden by starting the vertical axis at some point above 0. The same effect can be achieved by using a broken line — called a *scale break* — for the vertical axis, as shown in Figure 2.39.

Figure 2.38

Firm A's Market Share: Stretched Vertical Axis

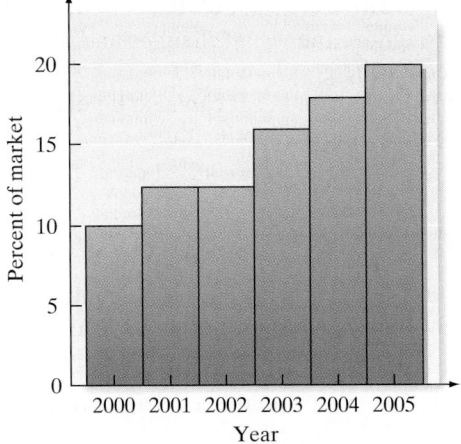

Figure 2.39
Firm A's Market Share:
Scale Break

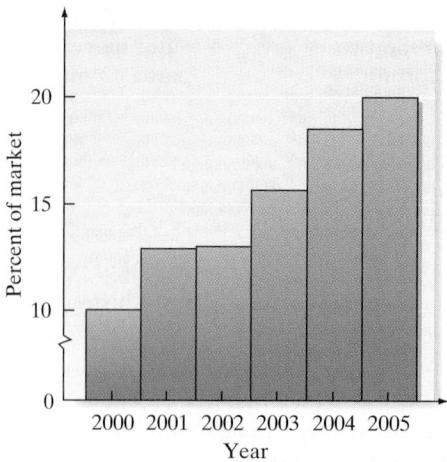

The changes in categories indicated by a bar graph can also be emphasized or deemphasized by stretching or shrinking the vertical axis. Another method of achieving visual distortion with bar graphs is by making the width of the bars proportional to height. For example, look at the bar chart in Figure 2.40a, which depicts the percentage of the total number of motor vehicle deaths in a year that occurred on each of four major highways. Now suppose we make both the width and the height grow as the percentage of fatal accidents grows. This change is shown in Figure 2.40b. The reader may tend to equate the *area* of the bars with the percentage of deaths occurring at each highway. But in fact, the true relative frequency of fatal accidents is proportional only to the *height* of the bars.

Figure 2.40
Relative Frequency of
Fatal Motor Vehicle
Accidents on Each of Four
Major Highways

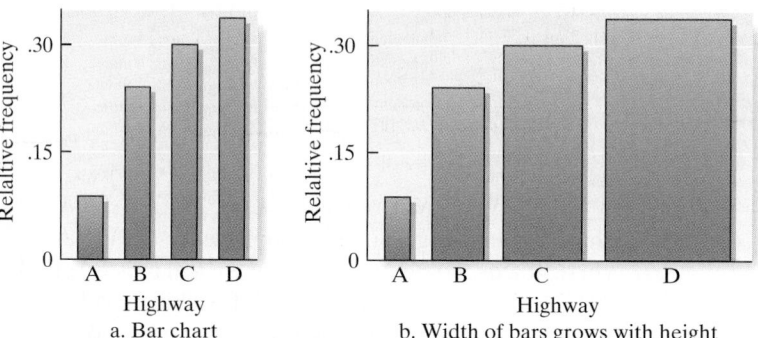

Although we've discussed only a few of the ways that graphs can be used to convey misleading pictures of phenomena, the lesson is clear. Look at all graphical descriptions of data with a critical eye. Particularly, check the axes and the size of the units on each axis. Ignore the visual changes and concentrate on the actual numerical changes indicated by the graph or chart.

The information in a data set can also be distorted by using numerical descriptive measures, as Example 2.20 indicates.

EXAMPLE 2.20   MISLEADING DESCRIPTIVE STATISTICS

Problem Suppose you're considering working for a small law firm — one that currently has a senior member and three junior members. You inquire about the salary you could expect to earn if you join the firm. Unfortunately, you receive two answers:

> *Answer A:* The senior member tells you that an "average employee" earns $87,500

Answer B: One of the junior members later tells you that an "average employee" earns $75,000

Which answer can you believe?

Solution   The confusion exists because the phrase "average employee" has not been clearly defined. Suppose the four salaries paid are $75,000 for each of the three junior members and $125,000 for the senior member. Thus,

$$\bar{x} = \frac{3(\$75,000) + \$125,000}{4} = \frac{\$350,000}{4} = \$87,500$$

Median $= \$75,000$

You can now see how the two answers were obtained. The senior member reported the mean of the four salaries, and the junior member reported the median. The information you received was distorted because neither person stated which measure of central tendency was being used.

Look Back Based on our earlier discussion of the mean and median, we would probably prefer the median as the number that best describes the salary of the "average" employee.

Another distortion of information in a sample occurs when *only* a measure of central tendency is reported. Both a measure of central tendency and a measure of variability are needed to obtain an accurate mental image of a data set.

Suppose you want to buy a new car and are trying to decide which of two models to purchase. Since energy and economy are both important issues, you decide to purchase model A because its EPA mileage rating is 32 miles per gallon in the city, whereas the mileage rating for model B is only 30 miles per gallon in the city.

However, you may have acted too quickly. How much variability is associated with the ratings? As an extreme example, suppose that further investigation reveals that the standard deviation for model A mileages is 5 miles per gallon, whereas that for model B is only 1 mile per gallon. If the mileages form a mound-shaped distribution, they might appear as shown in Figure 2.41. Note that the larger amount of variability associated with model A implies that more risk is involved in purchasing model A. That is, the particular car you purchase is more likely to have a mileage rating that will greatly differ from the EPA rating of 32 miles per gallon if you purchase model A, while a model B car is not likely to vary from the 30 miles per gallon rating by more than 2 miles per gallon.

We conclude this section with another example on distorting the truth with numerical descriptive measures.

Figure 2.41

Mileage Distributions for Two Car Models

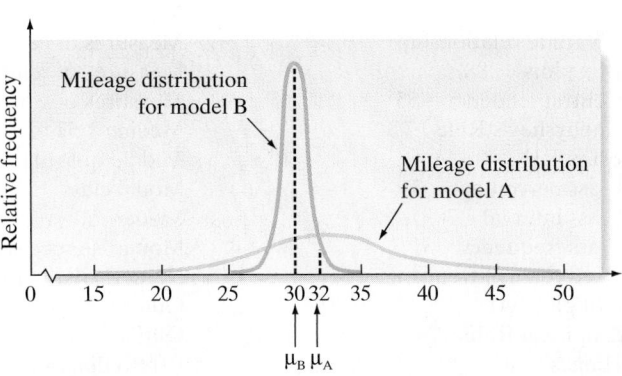

EXAMPLE 2.21   MORE MISLEADING DESCRIPTIVE STATISTICS

Problem   *Children Out of School in America* is a report on delinquency of school-age children by the Children's Defense Fund (CDF). Consider the following three reported results of the CDF survey.

- Reported result: 25% of the 16- and 17-year-olds in the Portland, Maine, Bayside East Housing Project were out of school. Actual data: *Only eight children were surveyed; two were found to be out of school.*

- Reported result: Of all the secondary school students who had been suspended more than once in census tract 22 in Columbia, South Carolina, 33% had been suspended two times and 67% had been suspended three or more times. Actual data: *CDF found only three children in that entire census tract who had been suspended; one child was suspended twice and the other two children, three or more times.*

- Reported result: In the Portland Bayside East Housing Project, 50% of all the secondary school children who had been suspended more than once had been suspended three or more times. Actual data: *The survey found two secondary school children had been suspended in that area; one of them had been suspended three or more times.*

Identify the potential distortions in the results reported by the CDF.

Solution   In each of these examples the reporting of percentages (i.e., relative frequencies) instead of the numbers themselves is misleading. No inference we might draw from the cited examples would be reliable. (We'll see how to measure the reliability of estimated percentages in Chapter 7.) In short, either the report should state the numbers alone instead of percentages, or, better yet, it should state that the numbers were too small to report by region. If several regions were combined, the numbers (and percentages) would be more meaningful.

Look Back If several regions were combined, the numbers (and percentages) would be more meaningful.

Quick Review

Key Terms

Note: Starred (*) terms are from the optional sections to this chapter

Bar graph   32
Bivariate relationship*   95
Box plots*   88
Central tendency   55
Chebyshev's Rule   73
Class   31
Class percentage   31
Class interval   38
Class frequency   31
Class relative frequency   31
Dot plot   41
Empirical Rule   74
Hinges*   87

Histogram   43
Inner fences*   87
Interquartile range*   86
Lower quartile*   86
Mean   56
Measures of central tendency   55
Measures of relative standing   71
Measures of variation or spread   67
Median   57
Middle quartile*   86
Modal class   51
Mode   50
Mound-shaped distribution   73
Numerical descriptive measures   55
Outer fences*   87
Outlier*   85
Pareto diagram   33

Percentile   81
Pie chart   86
Quartiles*   86
Range   67
Rare-event approach*   91
Scattergram*   95
Scatterplot*   95
Skewness   59
Standard deviation   69
Stem-and-leaf display   52
Symmetric distribution   59
Upper quartile*   86
Variability   55
Variance   68
Whiskers*   88
z-score   82

Key Formulas

$$\frac{\text{Class frequency}}{n}$$
Class relative frequency   31

$$\left(\frac{\text{Class frequency}}{n}\right) \times 100$$
Class percentage   31

$$\bar{x} = \frac{\sum\limits_{i=1}^{n} x_i}{n}$$
Sample mean   56

$$s^2 = \frac{\sum\limits_{i=1}^{n}(x_i - \bar{x})^2}{n-1} = \frac{\sum\limits_{i=1}^{n} x_i^2 - \dfrac{\left(\sum\limits_{i=1}^{n} x_i\right)^2}{n}}{n-1}$$
Sample variance   69

$$s = \sqrt{s^2}$$
Sample standard deviation   69

$$z = \frac{x - \bar{x}}{s}$$
Sample z-score   82

$$z = \frac{x - \mu}{\sigma}$$
Population z-score   82

$$\text{IQR} = Q_U - Q_L$$
Interquartile range   86

$$Q_L - 1.5(\text{IQR})$$
*Lower inner fence   88

$$Q_U + 1.5(\text{IQR})$$
*Upper inner fence   88

$$Q_L - 3(\text{IQR})$$
*Lower outer fence   88

$$Q_U + 3(\text{IQR})$$
*Upper outer fence   88

Chapter Summary

- Graphical methods for qualitative data: **pie chart, bar graph,** and **Pareto diagram**
- Graphical methods for quantitative data: **dot plot, stem-and-leaf display,** and **histogram**
- Numerical measures of central tendency: **mean, median,** and **mode**
- Numerical measures of variation: **range, variance,** and **standard deviation**

- Rules for determining the percentage of measurements in the interval (mean) $\pm$ 2(std. dev.): **Chebyshev's Rule** (at least 75%) and **Empirical Rule** (approximately 95%)
- Measures of relative standing: **percentile score** and **z-score**
- Methods for detecting outliers: box plots and z-scores
- Method for graphing the relationship between two quantitative variables: **scatterplot**

Language Lab

Symbol	Pronunciation	Description
$\sum$	sum of	Summation notation; $\sum\limits_{i=1}^{n} x_i$ represents the sum of the measurements $x_1, x_2, \ldots, x_n$,
μ	mu	Population mean
$\bar{x}$	x-bar	Sample mean
σ^2	sigma-squared	Population variance
σ	sigma	Population standard deviation
s^2		Sample variance
s		Sample standard deviation
z		z-score for a measurement

M	Median (middle quartile) of a sample data set
Q_L	Lower quartile (25th percentile)
Q_U	Upper quartile (75th percentile)
IQR	Interquartile range

Supplementary Exercises 2.149–2.187

Note: Starred (*) exercises refer to the optional sections in this chapter.

Understanding the Principles

2.149 Discuss conditions under which the median is preferred to the mean as a measure of central tendency.

2.150 Explain why we generally prefer the standard deviation to the range as a measure of variation.

2.151 Give a situation where we will prefer using a stem-and-leaf display over a histogram when graphically describing quantitative data.

2.152 Give a situation where we will prefer using a box plot over z-scores to detect an outlier.

2.153 Give a technique that is used to distort information shown on a graph.

Learning the Mechanics

2.154 Construct a relative frequency histogram for the data summarized in the table below.

Measurement Class	Relative Frequency	Interval Class	Relative Frequency
.00–.75	.02	5.25–6.00	.15
.75–1.50	.01	6.00–6.75	.12
1.50–2.25	.03	6.75–7.50	.09
2.25–3.00	.05	7.50–8.25	.05
3.00–3.75	.10	8.25–9.00	.04
3.75–4.50	.14	9.00–9.75	.01
4.50–5.25	.19		

2.155 Consider the following three measurements: 50, 70, 80. Find the z-score for each measurement if they are from a population with a mean and standard deviation equal to
  a. $\mu = 60, \sigma = 10$
  b. $\mu = 60, \sigma = 5$
  c. $\mu = 40, \sigma = 10$
  d. $\mu = 40, \sigma = 100$

*2.156 Refer to Exercise 2.155. For parts **a–d**, determine whether the three measurements 50, 70, and 80 are outliers.

2.157 Compute s^2 for data sets with the following characteristics:
  a. $\sum_{i=1}^{n} x_i^2 = 246, \sum_{i=1}^{n} x_i = 63, n = 22$

  b. $\sum_{i=1}^{n} x_i^2 = 666, \sum_{i=1}^{n} x_i = 106, n = 25$

  c. $\sum_{i=1}^{n} x_i^2 = 76, \sum_{i=1}^{n} x_i = 11, n = 7$

2.158 If the range of a set of data is 20, find a rough approximation to the standard deviation of the data set.

2.159 For each of the following data sets, compute $\bar{x}, s^2$, and s. If appropriate, specify the units in which your answers are expressed.
  a. 4, 6, 6, 5, 6, 7
  b. −\$1, \$4, −\$3, \$0, −\$3, −\$6
  c. 3/5%, 4/5%, 2/5%, 1/5%, 1/16%
  d. Calculate the range of each data set in parts **a–c**.

2.160 For each of the following data sets, compute $\bar{x}, s^2$, and s:
  a. 13, 1, 10, 3, 3
  b. 13, 6, 6, 0
  c. 1, 0, 1, 10, 11, 11, 15
  d. 3, 3, 3, 3
  e. For each data set in parts **a–d**, form the interval $\bar{x} \pm 2s$ and calculate the percentage of the measurements that fall in the interval.

*2.161 Construct a scatterplot for the data listed here. Do you detect any trends?

Variable #1:	174	268	345	119	400	520	190	448	307	252
Variable #2:	8	10	15	7	22	31	15	20	11	9

2.162 The following data sets have been invented to demonstrate that the lower bounds given by Chebyshev's Rule are appropriate. Notice that the data are contrived and would not be encountered in a real-life problem.
  a. Consider a data set that contains ten 0s, two 1s, and ten 2s. Calculate $\bar{x}, s^2$, and s. What percentage of the measurements are in the interval $\bar{x} \pm s$? Compare this result to Chebyshev's Rule.
  b. Consider a data set that contains five 0s, thirty-two 1s, and five 2s. Calculate $\bar{x}, s^2$, and s. What percentage of the measurements are in the interval $\bar{x} \pm 2s$? Compare this result to Chebyshev's Rule.
  c. Consider a data set that contains three 0s, fifty 1s, and three 2s. Calculate $\bar{x}, s^2$, and s. What percentage of the measurements are in the interval $\bar{x} \pm 3s$? Compare this result to Chebyshev's Rule.

d. Draw a histogram for each of the data sets in parts **a**, **b**, and **c**. What do you conclude from these graphs and the answers to parts **a**, **b**, and **c**?

Applying the Concepts—Basic

2.163 Gallup poll on smoking. *USA Today* (Dec. 31, 1999) reported on a Gallup Survey which asked: "Do you think cigarette smoking is one of the causes of lung cancer?" The accompanying table shows the results of the 1999 survey compared to the results of a similar survey conducted 45 years earlier.

Is Smoking a Cause of Lung Cancer?	1954 Survey	1999 Survey
Yes	41%	92%
No	31%	6%
No opinion	28%	2%

a. Construct a pie chart for the 1954 survey results.
b. Construct a pie chart for the 1999 survey results.
c. Compare the two graphs. Do the graphs show a dramatic shift in people's opinion on whether smoking causes lung cancer?

CRASH

2.164 Crash tests on new cars. Each year, the National Highway Traffic Safety Administration (NHTSA) crash tests new car models to determine how well they protect the driver and front-seat passenger in a head-on collision. The NHTSA has developed a "star" scoring system for the frontal crash test, with results ranging from one star (*) to five stars (*****). The more stars in the rating, the better the level of crash protection in a head-on collision. The NHTSA crash test results for 98 cars in a recent model year are stored in the data file named CRASH. The driver-side star ratings for the 98 cars are summarized in the accompanying MINITAB printout. Use the information in the printout to form a pie chart. Interpret the graph.

Tally for Discrete Variables: DRIVSTAR

```
DRIVSTAR   Count   Percent
       2       4      4.08
       3      17     17.35
       4      59     60.20
       5      18     18.37
     N=       98
```

2.165 Crash Tests on New Cars. Refer to Exercise 2.164 and the NHTSA crash test data. One quantitative variable recorded by the NHTSA is driver's severity of head injury (measured on a scale from 0 to 1,500). The mean and standard deviation for the 98 driver head-injury ratings in the CRASH file are displayed in the

MINITAB printout at the bottom of the page. Use these values to find the z-score for a driver head-injury rating of 408. Interpret the result.

2.166 Competition for bird nest holes. *The Condor* (May 1995) published a study of competition for nest holes among collared flycatchers, a bird species. The authors collected the data for the study by periodically inspecting nest boxes located on the island of Gotland in Sweden. The nest boxes were grouped into 14 discrete locations (called plots). The accompanying table gives the number of flycatchers killed and the number of flycatchers breeding at each plot.

CONDOR

Plot Number	Number Killed	Number of Breeders
1	5	30
2	4	28
3	3	38
4	2	34
5	2	26
6	1	124
7	1	68
8	1	86
9	1	32
10	0	30
11	0	46
12	0	132
13	0	100
14	0	6

Source: Merilä, J., and Wiggins, D. A. "Interspecific competition for nest holes causes adult mortality in the collared flycatcher." *The Condor,* Vol. 97, No. 2, May 1995, p. 447 (Table 4). Cooper Ornithological Society.

a. Calculate the mean, median, and mode for the number of flycatchers killed at the 14 study plots.
b. Interpret the measures of central tendency, part **a**.
*c. Graphically examine the relationship between number killed and number of breeders. Do you detect a trend?

2.167 Beanie Babies. Beanie Babies are toy stuffed animals that have become valuable collector's items. *Beanie World Magazine* provided the age, retired status, and value of 50 Beanie Babies. The data one saved in the BEANIE file, with several of the observations shown in the table on p. 82.
a. Summarize the retried/current status of the 50 Beanie Babies with an appropriate graph. Interpret the graph.
b. Summarize the values of the 50 Beanie Babies with an appropriate graph. Interpret the graph.
*c. Use a graph to portray the relationship between a Beanie Baby's value and its age. Do you detect a trend?

Descriptive Statistics: DRIVHEAD

```
Variable    N    Mean   StDev   Minimum     Q1   Median      Q3   Maximum
DRIVHEAD   98   603.7   185.4     216.0  475.0    605.0   724.3    1240.0
```

🖸 **BEANIE (Selected observations)**

Name	Age (Months) as of Sept. 1998	Retired (R) Current (C)	Value ($)
1. Ally the Alligator	52	R	55.00
2. Batty the Bat	12	C	12.00
3. Bongo the Brown Monkey	28	R	40.00
4. Blackie the Bear	52	C	10.00
5. Bucky the Beaver	40	R	45.00
⋮	⋮	⋮	⋮
46. Stripes the Tiger (Gold/Black)	40	R	400.00
47. Teddy the 1997 Holiday Bear	12	R	50.00
48. Tuffy the Terrier	17	C	10.00
49. Tracker the Basset Hound	5	C	15.00
50. Zip the Black Cat	28	R	40.00

Source: Beanie World Magazine, Sept. 1998.

d. According to Chebyshev's Rule, what percentage of the age measurements would you expect to find in the intervals $\bar{x} \pm .75s$, $\bar{x} \pm 2.5s$, $\bar{x} \pm 4s$?

e. What percentage of the age measurements actually fall in the intervals of part **e?** Compare your results with those of part

f. Repeat parts **d** and **e** for value.

2.168 Public image of mental illness. A survey was conducted to investigate the impact of the mass media on the public's perception of mental illness. (*Health Education Journal,* Sept. 1994.) The media coverage for each of 562 items related to mental health in Scotland was classified into one of five categories: violence to others, sympathetic coverage, harm to self, comic images, and criticism of accepted definitions of mental illness. A summary of the results is provided in the accompanying table.

Media Coverage	Number of Items
Violence to others	373
Sympathetic	102
Harm to self	71
Comic images	12
Criticism of definitions	4
Total	**562**

Source: Philo, G., et al. "The impact of the mass media on public images of mental illness: Media content and audience belief." *Health Education Journal,* Vol. 53, No. 3, Sept. 1994, p.274 (Table 1).

a. Construct a relative frequency table for the data.

b. Display the relative frequencies in a graph.

c. Discuss the findings.

2.169 Computer anxiety of teachers. The extent of computer anxiety among secondary technical education teachers in West Virginia was examined in the *Journal of Studies in Technical Careers* (Vol. 15, 1995). Level of computer anxiety was measured by the Computer Anxiety Scale (COMPAS), with scores ranging from 10 to 50. (Lower scores reflect a higher degree of anxiety.) One objective was to compare the computer anxiety levels of male and female secondary technical education teachers in West Virginia. The data for the 116 teachers who participated in the survey are summarized in the following table.

	Male Teachers	Female Teachers
n	68	48
$\bar{x}$	26.4	24.5
s	10.6	11.2

Source: Gordon, H. R. D. "Analysis of the computer anxiety levels of secondary technical education teachers in West Virginia." *Journal of Studies in Technical Careers,* Vol. 15, No. 2, 1995, pp. 26–27 (Table 1).

a. Interpret the mean COMPAS score for male teachers.

b. Repeat part **a** for female teachers.

c. Give an interval that contains about 95% of the COMPAS scores for male teachers. (Assume the distribution of scores for males is symmetric and mound shaped.)

d. Repeat part **c** for female teachers.

2.170 Achievement test scores. The distribution of scores on a nationally administered college achievement test has a median of 520 and a mean of 540.

a. Explain why it is possible for the mean to exceed the median for this distribution of measurements.

b. Suppose you are told that the 90th percentile is 660. What does this mean?

c. Suppose you are told that you scored at the 94th percentile. Interpret this statement.

Applying the Concepts—Intermediate

2.171 New book reviews. *Choice* magazine provides new-book reviews each issue. A random sample of 375 *Choice* book reviews in American history, geography, and area studies was selected and the "overall opinion" of the book stated in each review was ascertained. (*Library Acquisitions: Practice and Theory*, Vol. 19, 1995.) Overall opinion was coded as follows: 1 = would not recommend, 2 = cautious or very little recommendation, 3 = little or no preference, 4 = favorable/recommended, 5 = outstanding/significant contribution. A summary table for the data is provided below.

Opinion Code	Number of Reviews
1	19
2	37
3	35
4	238
5	46
Total	375

Source: Carlo, P. W., and Natowitz, A. "Choice book reviews in American history, geography, and area studies: An analysis for 1988–1993." *Library Acquisitions: Practice & Theory,* Vol. 19, No. 2, 1995, p. 159 (Figure 1).

a. Use a Pareto diagram to find the opinion that occurred most often. What proportion of the books reviewed had this opinion?

b. Do you agree with the following statement extracted from the study: "A majority (more than 75%) of books reviewed are evaluated favorably and recommended for purchase."?

2.172 Archaeologists study of ring diagrams. Archaeologists gain insight into the social life of ancient tribes by measuring the distance (in centimeters) of each artifact found at a site from the "central hearth" (i.e., the middle of an artifact scatter). A graphical summary of these distances, in the form of a histogram, is called a *ring diagram*. (*World Archaeology*, Oct. 1997.) Ring diagrams for two archaeological sites (A and G) in Europe follow.

a. Identify the type of skewness (if any) present in the data collected at the two sites.

b. Archaeologists have associated unimodal ring diagrams with open air hearths and multimodal ring diagrams with hearths inside dwellings. Can you identify the type of hearth (*open air or inside dwelling*) that was most likely present at each site?

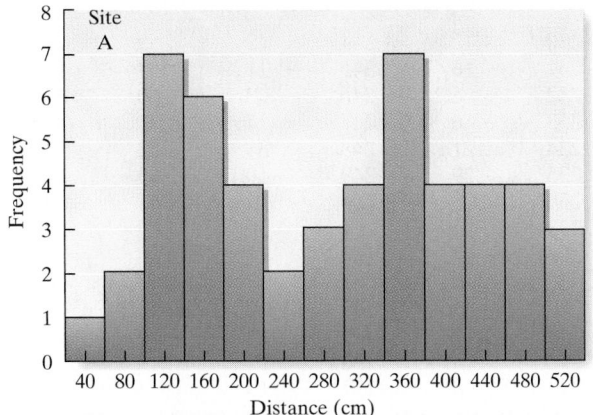

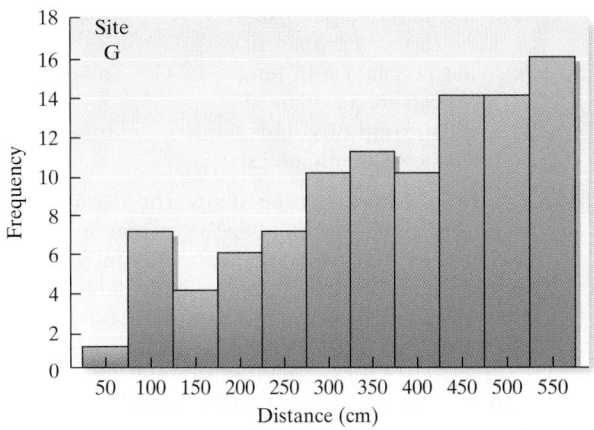

2.173 Golf handicaps. The United States Golf Association (USGA) Handicap System is designed to allow golfers of differing abilities to enjoy fair competition. The handicap index is a measure of a player's potential scoring ability on an 18-hole golf course of standard difficulty. For example, on a par-72 course, a golfer with a handicap of 7 will typically have a score of 79 (seven strokes over par). Over 4.5 million golfers have an official USGA Handicap index. The handicap indexes for both male and female golfers were obtained from the USGA and are summarized in the two histograms shown below.

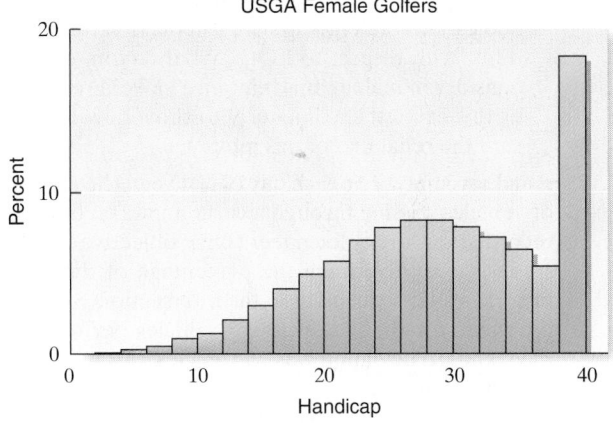

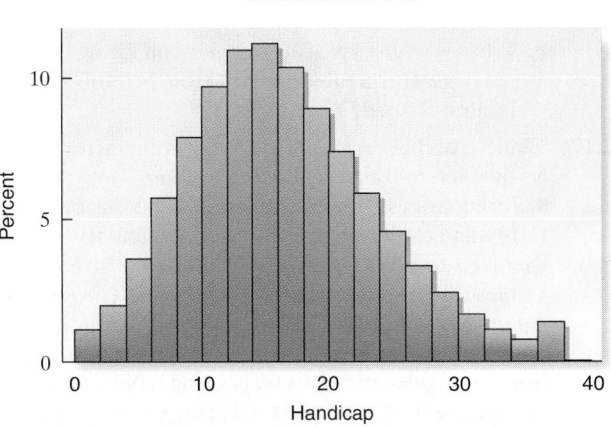

a. What percentage of male USGA golfers have a handicap greater than 20?

b. What percentage of female USGA golfers have a handicap greater than 20?

c. Which group of golfers, male or female, tend to have the greater handicap?

2.174 Control of life perception study. The Locus of Control (LOC) is a measure of one's perception of control over factors affecting one's life. In one study, the LOC was measured for two groups of individuals undergoing weight-reduction for treatment of obesity. (*Journal of Psychology*, Mar. 1991.) The LOC mean and standard deviation for a sample of 46 adults were 6.45 and 2.89, respectively, while the LOC mean and standard deviation for a sample of 19 adolescents were 10.89 and 2.48, respectively. A lower score on the LOC scale indicates a perception that internal factors are in control, while a higher score indicates a perception that external factors are in control.

a. Calculate the 1- and 2-standard-deviation intervals around the means for each group. Plot these intervals on a line graph using different colors or symbols to represent each group.

b. Assuming that the distributions of LOC scores are approximately mound-shaped, estimate the numbers of individuals within each interval.

c. Based on your answers to parts **a** and **b**, do you think an inference can be made that *all* adults and adolescents undergoing weight-reduction treatment differ with respect to LOC? What factors did you consider in making this inference? (We'll reconsider this exercise in Chapter 9 to show how to measure the reliability of this inference.)

2.175 Road use study. For each day of last year, the number of vehicles passing through a certain intersection was recorded by a city engineer. One objective of this study was to determine the percentage of days that more than 425 vehicles used the intersection. Suppose the mean for the data was 375 vehicles per day and the standard deviation was 25 vehicles.

a. What can you say about the percentage of days that more than 425 vehicles used the intersection? Assume you know nothing about the shape of the relative frequency distribution for the data.

b. What is your answer to part **a** if you know that the relative frequency distribution for the data is mound-shaped?

2.176 Chinese herbal drugs. Platelet-activating factor (PAF) is a potent chemical that occurs in patients suffering from shock, inflammation, hypotension, respiratory, and cardiovascular disorders. A bioassay was undertaken to investigate the potential of 17 traditional Chinese herbal drugs in PAF inhibition. (*Progress in Natural Science*, June 1995.) The prevention of the PAF binding process, measured as a percentage, for each drug is provided in the accompanying table.

a. Construct a stem-and-leaf display for the data.

🔘 **PAF**

Drug	PAF Inhibition (%)
Hai-feng-teng (Fuji)	77
Hai-feng-teng (Japan)	33
Shan-ju	75
Zhang-yiz-hu-jiao	62
Shi-nan-teng	70
Huang-hua-hu-jiao	12
Hua-nan-hu-jiao	0
Xiao-yie-pa-ai-xiang	0
Mao-ju	0
Jia-ju	15
Xie-yie-ju	25
Da-yie-ju	0
Bian-yie-hu-jiao	9
Bi-bo	24
Duo-mai-hu-jiao	40
Yan-sen	0
Jiao-guo-hu-jiao	31

Source: Guiqiu, H. "PAF receptor antagonistic principles from Chinese traditional drugs." *Progress in Natural Science*, Vol. 5, No. 3, June 1995, p. 301 (Table 1).

b. Compute the mean, median, and mode of the inhibition percentages for the 17 herbal drugs. Interpret the results.

c. Locate the median, mean, and mode on the stem-and-leaf display, part **a**. Do these measures of central tendency appear to locate the center of the data?

d. Compute the range, variance, and standard deviation for the inhibition percentages.

e. Form an interval that is highly likely to capture the inhibition percentage of a Chinese herbal drug.

*2.177 Computer system down time study. A manufacturer of minicomputers is investigating the down time of its systems. The 40 most recent customers were surveyed to determine the amount of down time (in hours) they had experienced during the previous month. These data are listed in the table.

🔘 **DOWNTIME**

Customer Number	Down Time	Customer Number	Down Time	Customer Number	Down Time
230	12	244	2	258	28
231	16	245	11	259	19
232	5	246	22	260	34
233	16	247	17	261	26
234	21	248	31	262	17
235	29	249	10	263	11
236	38	250	4	264	64
237	14	251	10	265	19
238	47	252	15	266	18
239	0	253	7	267	24
240	24	254	20	268	49
241	15	255	9	269	50
242	13	256	22		
243	8	257	18		

a. Use a statistical software package to construct a box plot for these data. Use the information reflected in the box plot to describe the frequency distribution of the data set. Your description should address central tendency, variation, and skewness.

b. Use your box plot to determine which customers are having unusually lengthy down times.

c. Find and interpret the z-scores associated with customers you identified in part **b**.

🔿 SLI

*2.178 **Childrens use of pronouns.** Refer to the *Journal of Communication Disorders* data on deviation intelligence quotient (DIQ) and percent pronoun errors for 30 children, Exercise 2.65 (p. 65). The data are saved in the SLI file.

a. Plot all the data to investigate a possible trend between DIQ and proper use of pronouns. What do you observe?

b. Plot the data for the ten SLI children only. Is there a trend between DIQ and proper use of pronouns?

2.179 **Oil spill impact on seabirds.** The *Journal of Agricultural, Biological, and Environmental Statistics* (Sept. 2000) published a study on the impact of the *Exxon Valdez* tanker oil spill on the seabird population in Prince William Sound, Alaska. A subset of the data analyzed is stored in the EVOS file. Data were collected on 96 shoreline locations (called transects) of constant width but variable length. For each transect, the number of seabirds found is recorded as well as the length (in kilometers) of the transect and whether or not the transect was in an oiled area. (The first five and last five observations in the EVOS file are listed in the table.)

a. Identify the variables measured as quantitative or qualitative.

🔿 **EVOS (Selected observations)**

Transect	Seabirds	Length	Oil
1	0	4.06	No
2	0	6.51	No
3	54	6.76	No
4	0	4.26	No
5	14	3.59	No
⋮	⋮	⋮	⋮
92	7	3.40	Yes
93	4	6.67	Yes
94	0	3.29	Yes
95	0	6.22	Yes
96	27	8.94	Yes

Source: McDonald, T. L., Erickson, W. P., and McDonald, L. L. "Analysis of count data from before–after control-impact studies," *Journal of Agricultural, Biological, and Environmental Statistics*, Vol. 5, No. 3, Sept. 2000, pp. 277–8 (Table A.1).

b. Identify the experimental unit.

c. Use a pie chart to describe the percentage of transects in oiled and unoiled areas.

d. Use a graphical method to examine the relationship between observed number of seabirds and transect length.

e. Observed seabird density is defined as the observed count divided by the length of the transect. MINITAB descriptive statistics for seabird densities in unoiled and oiled transects are displayed in the printout shown at the bottom of the page. Assess whether the distribution of seabird densities differs for transects in oiled and unoiled areas.

f. For unoiled transects, give an interval of values that is likely to contain at least 75% of the seabird densities.

g. For oiled transects, give an interval of values that is likely to contain at least 75% of the seabird densities.

h. Which type of transect, an oiled or unoiled one, is more likely to have a seabird density of 16? Explain.

Applying the Concepts—Advanced

2.180 **Standardized test "average."** *US News & World Report* reported on many factors contributing to the breakdown of public education. One study mentioned in the article found that over 90% of the nation's school districts reported that their students were scoring "above the national average" on standardized tests. Using your knowledge of measures of central tendency, explain why the schools' reports are incorrect. Does your analysis change if the term "average" refers to the mean? To the median? Explain what effect this misinformation might have on the perception of the nation's schools.

2.181 **Speed of light from galaxies.** Refer to *The Astronomical Journal* study of galaxy velocities, Exercise 2.100 (p. 80). A second cluster of galaxies, named A1775, is thought to be a *double cluster*; that is, two clusters of galaxies in close proximity. Fifty-one velocity observations (in kilometers per second, km/s) from cluster A1775 are listed in the table.

🔿 **GALAXY2**

22922	20210	21911	19225	18792	21993	23059
20785	22781	23303	22192	19462	19057	23017
20186	23292	19408	24909	19866	22891	23121
19673	23261	22796	22355	19807	23432	22625
22744	22426	19111	18933	22417	19595	23408
22809	19619	22738	18499	19130	23220	22647
22718	22779	19026	22513	19740	22682	19179
19404	22193					

Source: Oegerle, W. R., Hill, J. M., and Fitchett, M. J. "Observations of high dispersion clusters of galaxies: Constraints on cold dark matter." *The Astronomical Journal*, Vol. 110, No. 1, July 1995, p. 34 (Table 1), p. 37 (Figure 1).

Descriptive Statistics: Density

Variable	Oil	N	Mean	StDev	Minimum	Q1	Median	Q3	Maximum
Density	no	36	3.27	6.70	0.000	0.000	0.890	3.87	36.23
	yes	60	3.495	5.968	0.0000	0.000	0.700	5.233	32.836

a. Use a graphical method to describe the velocity distribution of galaxy cluster A1775.

b. Examine the graph, part **a**. Is there evidence to support the double cluster theory? Explain.

c. Calculate numerical descriptive measures (e.g., mean and standard deviation) for galaxy velocities in cluster A1775. Depending on your answer to part **b**, you may need to calculate two sets of numerical descriptive measures, one for each of the clusters (say, A1775A and A1775B) within the double cluster.

d. Suppose you observe a galaxy velocity of 20,000 km/s. Is this galaxy likely to belong to cluster A1775A or A1775B? Explain.

OILSPILL

2.182 Hull failures of oil tankers. Owing to several major ocean oil spills by tank vessels, Congress passed the 1990 Oil Pollution Act, which requires all tankers to be designed with thicker hulls. Further improvements in the structural design of a tank vessel have been proposed since then, each with the objective of reducing the likelihood of an oil spill and decreasing the amount of outflow in the event of a hull puncture. To aid in this development, *Marine Technology* (Jan. 1995) reported on the spillage amount (in thousands of metric tons) and cause of puncture for 50 recent major oil spills from tankers and carriers. [*Note*: Cause of puncture is classified as either collision (C), fire/explosion (FE), hull failure (HF) or grounding (G).] The data are saved in the OILSPILL file.

a. Use a graphical method to describe the cause of oil spillage for the 50 tankers. Does the graph suggest that any one cause is more likely to occur than any other? How is this information of value to the design engineers?

b. Find and interpret descriptive statistics for the 50 spillage amounts. Use this information to form an interval that can be used to predict the spillage amount of the next major oil spill.

2.183 Risk of jail suicides. Suicide is the leading cause of death of Americans incarcerated in correctional facilities. To determine what factors increase the risk of suicide in urban jails, a group of researchers collected data on all 37 suicides that occurred over a 15-year period in the Wayne County Jail, Detroit, Michigan (*American Journal of Psychiatry*, July 1995). The data for each suicide victim are saved in the SUICIDE file. Selected observations are shown in the accompanying table.

a. Identify the type (quantitative or qualitative) of each variable measured.

b. Are suicides at the jail more likely to be committed by inmates charged with murder/manslaughter or with lesser crimes? Illustrate with a graph.

c. Are suicides at the jail more likely to be committed at night? Illustrate with a graph.

d. What is the mean length of time an inmate is in jail before committing suicide? What is the median? Interpret these two numbers.

e. Is it likely that a future suicide at the jail will occur after 200 days? Explain.

*f. Have suicides at the jail declined over the years? Support your answer with a graph.

Critical Thinking Challenges

2.184 Grades in statistics. The final grades given by two professors in introductory statistics courses have been carefully examined. The students in the first professor's class had a grade point average of 3.0 and a standard deviation of .2. Those in the second professor's class had grade points with an average of 3.0 and a standard deviation of 1.0. If you had a choice, which professor would you take for this course?

2.185 Salaries of professional athletes. The salaries of superstar professional athletes receive much attention in the media. The multimillion-dollar long-term contract is now commonplace among this elite group. Nevertheless, rarely does a season pass without negotiations between one or more of the players' associations and team owners for additional salary and fringe benefits for *all* players in their particular sports.

a. If a players' association wanted to support its argument for higher "average" salaries, which measure of central tendency do you think it should use? Why?

b. To refute the argument, which measure of central tendency should the owners apply to the players' salaries? Why?

2.186 No Child Left Behind Act. According to the government, federal spending on K–12 education has increased dramatically over the past 20 years, but student performance has essentially stayed the same. Hence, in 2002, President George Bush signed into law the No

SUICIDE (Selected Observations)

Victim	Days in Jail Before Suicide	Marital Status	Race	Murder/Manslaughter Charge	Time of Suicide	Year
1	3	Married	W	Yes	Night	1972
2	4	Single	W	Yes	Night	1987
3	5	Single	NW	Yes	Afternoon	1975
4	7	Widowed	NW	Yes	Night	1981
5	10	Single	NW	Yes	Afternoon	1982
⋮	⋮	⋮	⋮	⋮	⋮	⋮
36	41	Single	NW	No	Night	1985
37	86	Married	W	No	Night	1968

Source: DuRand, C. J., et al. "A quarter century of suicide in a major urban jail: Implications for community psychiatry." *American Journal of Psychiatry*, Vol. 152, No. 7, July 1995, p. 1078 (Table 1).

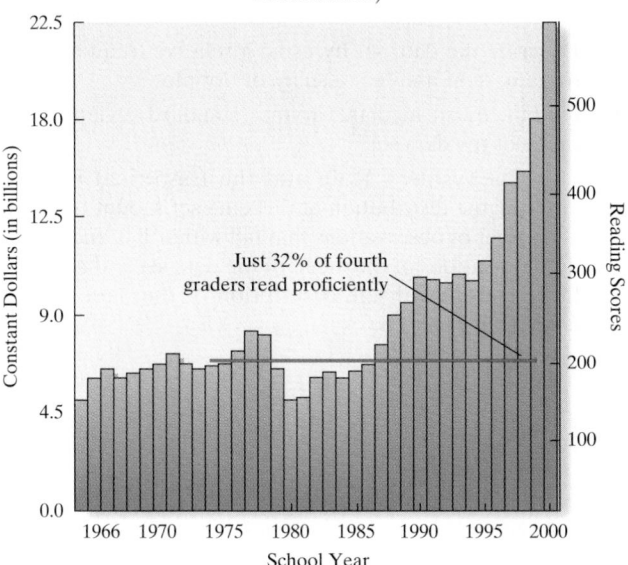

Federal Spending on
K-12 Education
(Elementary and Secondary
Education Act)

Just 32% of fourth
graders read proficiently

School Year

U.S. Department of Education $$$

Child Left Behind Act, a bill that promised improved student achievement for all U.S. children. *Chance* (Fall 2003) reported on a graphic obtained from the U.S. Department of Education Web site (www.ed.gov) that was designed to support the new legislation. The graphic is reproduced above. The bars in the graph represent annual federal spending on education, in billions of dollars (left-side vertical axis). The horizontal line represents the annual average fourth grade children's reading ability score (right-side vertical axis). Critically assess the information portrayed in the graph. Does it, in fact, support the government's position that our children are not making classroom improvements despite federal spending on education? Use the following facts (divulged in the *Chance* article) to help you frame your answer: (1) The U.S. student population has also increased dramatically over the past 20 years, (2) fourth-grade reading test scores are designed to have an average of 250 with a standard deviation of 50, and (3) the reading test scores of seventh and twelfth grades and the mathematics scores of fourth graders did improve substantially over the past 20 years.

2.187 The new Hite Report. In 1968 researcher Shere Hite shocked conservative America with her famous Hite Report on the permissive sexual attitudes of American men and women. Some 20 years later, Hite was surrounded by controversy again with her book, *Women and Love: A Cultural Revolution in Progress* (Knopf Press, 1988). In this book, Hite reveals some startling statistics describing how women feel about contemporary relationships:

- Eighty-four percent are not emotionally satisfied with their relationship.
- Ninety-five percent report "emotional and psychological harassment" from their men.
- Seventy percent of those married 5 years or more are having extramarital affairs.
- Only 13% of those married more than 2 years are "in love."

Hite conducted the survey by mailing out 100,000 questionnaires to women across the country over a 7-year period. These questionnaires were mailed to a wide variety of organizations, including church groups, women's voting and political groups, women's rights organizations, and counseling and walk-in centers for women. Organizational leaders were asked to circulate the questionnaires to their members. Hite also relied on volunteer respondents who wrote in for copies of the questionnaire. Each questionnaire consisted of 127 open-ended questions, many with numerous subquestions and follow-ups. Hite's instructions read, "It is not necessary to answer every question! Feel free to skip around and answer those questions you choose." Approximately 4,500 completed questionnaires were returned for a response rate of 4.5%, and they form the data set from which the percentages above were determined. Hite claims that these 4,500 women are a representative sample of all women in the United States and, therefore, the survey results imply that vast numbers of women are "suffering a lot of pain in their love relationships with men." Critically assess the survey results. Do you believe they are reliable?

Student Project

We list here several sources of real-life data sets (many in the list have been obtained from Wasserman and Bernero's *Statistics Sources* and many can be accessed via the Internet). This index of data sources is very complete and is a useful reference for anyone interested in finding almost any type of data. First we list some almanacs:

CBS News Almanac
Information Please Almanac
World Almanac and Book of Facts

United States Government publications are also rich sources of data:

Agricultural Statistics
Digest of Educational Statistics
Handbook of Labor Statistics
Housing and Urban Development Yearbook
Social Indicators
Uniform Crime Reports for the United States

Vital Statistics of the United States
Business Conditions Digest
Economic Indicators
Monthly Labor Review
Survey of Current Business
Bureau of the Census Catalog
Statistical Abstract of the United States

Main data sources are published on an annual basis:

Commodity Yearbook
Facts and Figures on Government Finance
Municipal Yearbook
Standard and Poor's Corporation, Trade and Securities: Statistics

Some sources contain data that are international in scope:

Compendium of Social Statistics
Demographic Yearbook
United Nations Statistical Yearbook
World Handbook of Political and Social Indicators

Utilizing the data sources listed, sources suggested by your instructor, or your own resourcefulness, find one real-life quantitative data set that stems from an area of particular interest to you.

a. Describe the data set by using a relative frequency histogram, stem-and-leaf display, or dot plot.

b. Find the mean, median, variance, standard deviation, and range of the data set.

c. Use Chebyshev's Rule and the Empirical Rule to describe the distribution of this data set. Count the actual number of observations that fall within 1, 2, and 3 standard deviations of the mean of the data set and compare these counts with the description of the data set you developed in part **b**.

References

Hoff, D. *How to Lie with Statistics*. New York: Norton, 1954.

Mendenhall, W., Beaver, R. J., and Beaver, B. M. *Introduction to Probability and Statistics*, 10th ed. North Scituate, Mass.: Duxbury, 1999.

Tufte, E. R. *Envisioning Information*. Cheshire, Conn.: Graphics Press, 1990.

Tufte, E.R. *Visual Explanations*. Cheshire, Conn.: Graphics Press, 1997.

Tufte, E. R. *Visual Display of Quantitative Information*. Cheshire, Conn.: Graphics Press, 1983.

Tukey, J. *Exploratory Data Analysis*. Reading, Mass.: Addison-Wesley, 1977.

Using Technology

Describing Data Using MINITAB

Graphing Data

To obtain graphical descriptions of your data, click on the "Graph" button on the MINITAB menu bar. The resulting menu list appears as shown in Figure 2.M.1. Several of the options covered in this text are "Bar Chart," "Pie Chart", "Scatter Plot," "Histogram," "Dotplot," and "Stem-and-Leaf (display)." Click on the graph of your choice to view the appropriate dialog box. For example, the dialog box for a histogram is shown in Figure 2.M.2. Make the appropriate variable selections and click "OK" to view the graph.

Numerical Descriptive Statistics

To obtain numerical descriptive measures for a quantitative variable (e.g., mean, standard deviation, etc.), click on the "Stat" button on the main menu bar, then click on "Basic Statistics, then click on "Display Descriptive Statistics" (see Figure 2.M.3). The resulting dialog box appears in Figure 2.M.4.

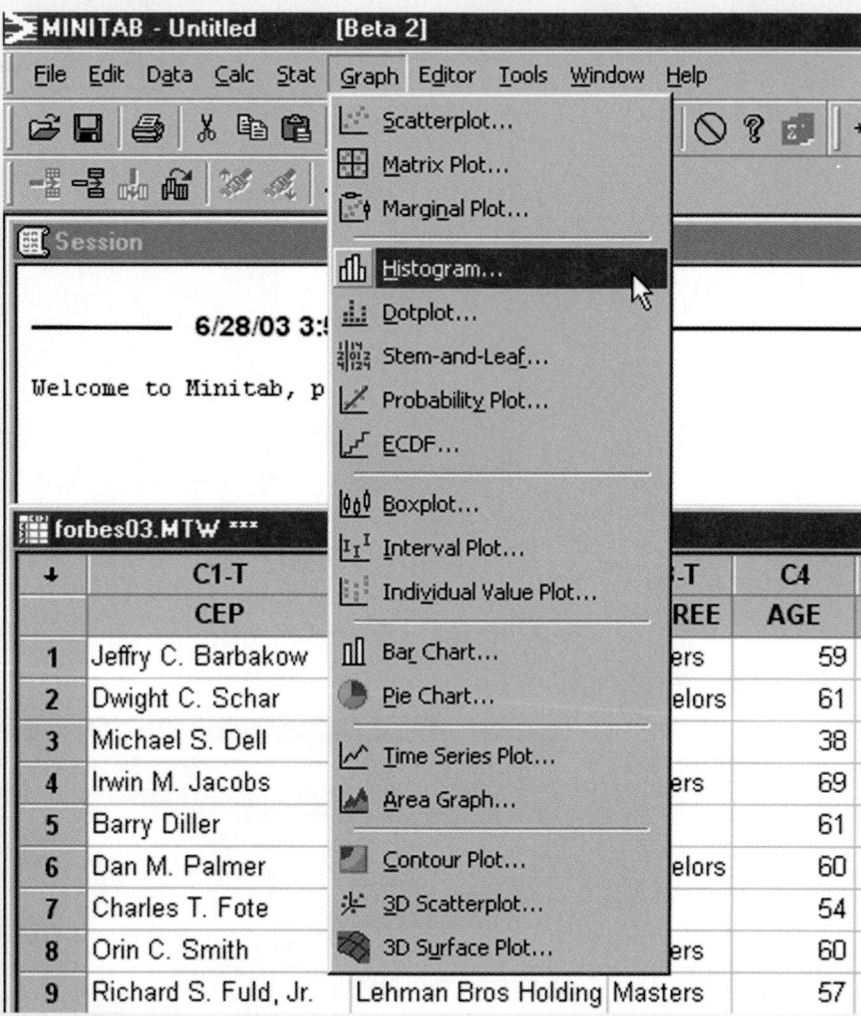

Figure 2.M.1
MINITAB Menu Options for Graphing Your Data

Select the quantitative variables you want to analyze and place them in the "Variables" box. You can control which particular descriptive statistics appear by clicking the "Statistics" button on the dialog box and making your selections. (As an option, you can create histograms and dot plots for the data by clicking the "Graphs" button and making the appropriate selections.) Click on "OK" to view the descriptive statistics printout.

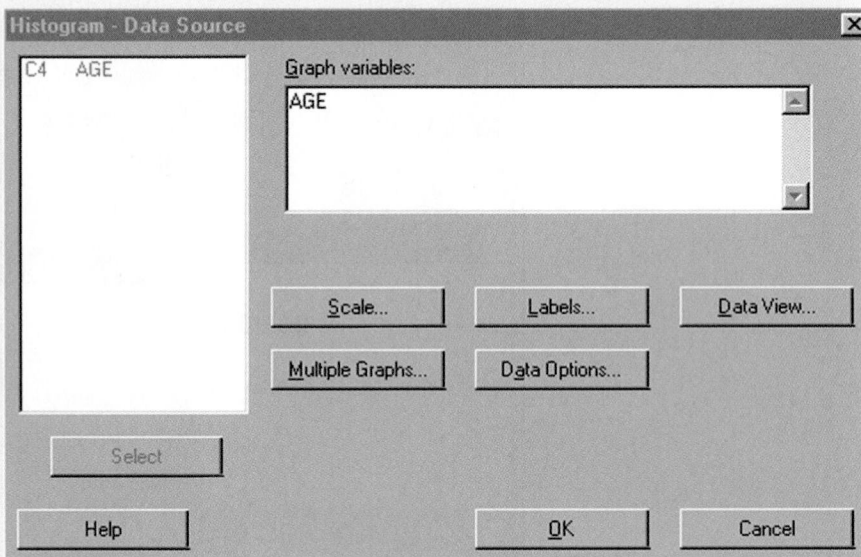

Figure 2.M.2
Histogram Dialog Box

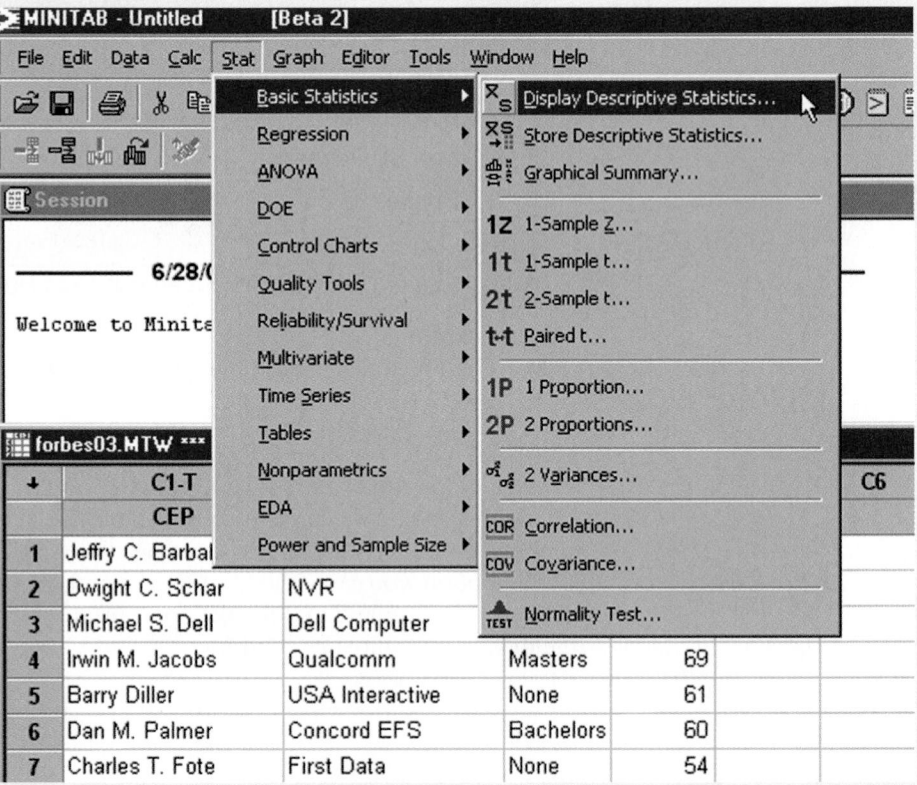

Figure 2.M.3
MINITAB Options for Obtaining Descriptive Statistics

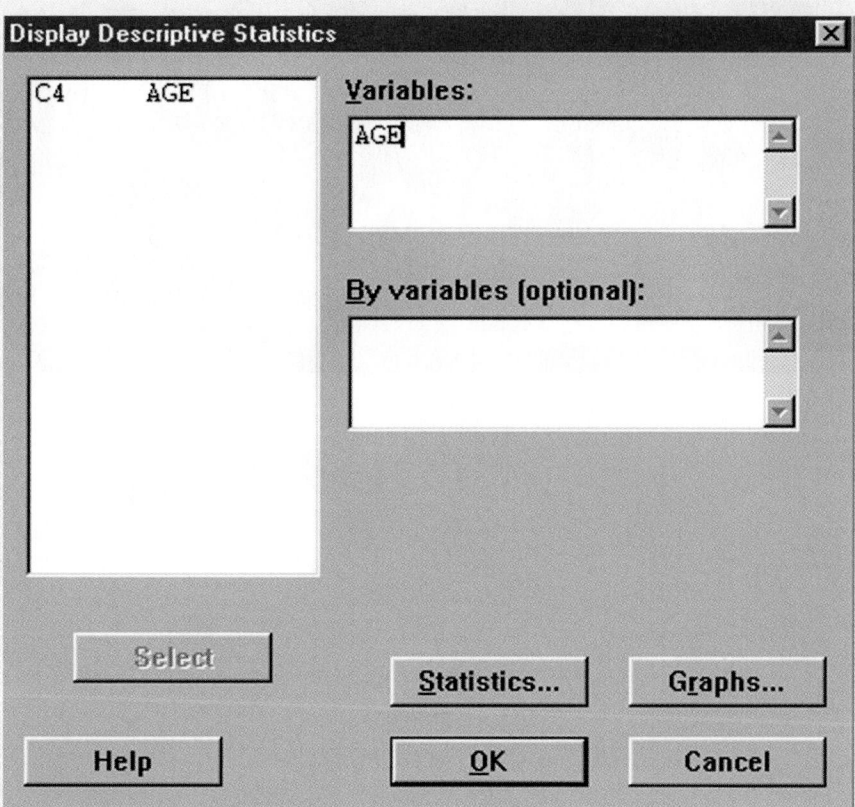

Figure 2.M.4
Descriptive Statistics Dialog Box

3

Probability

Contents

Statistics in Action

Lotto Buster!

Using Technology

Generating a Random Sample
Using MINITAB

Where We've Been

*   Identified the objective of inferential statistics—to make inferences about a population based on information in a sample

*   Introduced graphical and numerical descriptive measures for both quantitative and qualitative data

Where We're Going

*   Develop probability as a measure of uncertainty.

*   Introduce basic rules for finding probabilities.

*   Use a probability as a measure of reliability for an inference.

Statistics in *ACTION*

Lotto Buster!

"Welcome to the Wonderful World of Lottery Bus$ters." So began the premier issue of *Lottery Buster*, a monthly publication for players of the state lottery games. *Lottery Buster* provides interesting facts and figures on the 37 state lotteries currently operating in the United States and, more importantly, tips on how to increase a player's odds of winning the lottery.

New Hampshire, in 1963, was the first state in modern times to authorize a state lottery as an alternative to increasing taxes. (Prior to this time, beginning in 1895, lotteries were banned in America because of corruption.) Since then, lotteries have become immensely popular for two reasons. First, they lure you with the opportunity to win millions of dollars with a $1 investment, and second, when you lose, at least you believe your money is going to a good cause. Many state lotteries, like Florida, designate a high percentage of lottery revenues to fund state education.

The popularity of the state lottery has brought with it an avalanche of "experts" and "mathematical wizards" (such as the editors of *Lottery Buster*) who provide advice on how to win the lottery — for a fee, of course! Many offer guaranteed "systems" of winning through computer software products with catchy names such as Lotto Wizard, Lottorobics, Win4D, and Loto-Luck.

For example, most knowledgeable lottery players would agree that the "golden rule" or "first rule" in winning lotteries is *game selection*. State lotteries generally offer three types of games: Instant (scratch-off tickets or online) games, Daily Numbers (Pick-3 or Pick-4), and the weekly Pick-6 Lotto game.

One version of the Instant game involves scratching off the thin opaque covering on a ticket with the edge of a coin to determine whether you have won or lost. The cost of a ticket ranges from 50¢ to $1, and the amount won ranges from $1 to $100,000 in most states and to as much as $1 million in others. *Lottery Buster* advises against playing the Instant game because it is "a pure chance play, and you can win only by dumb luck. No skill can be applied to this game."

The Daily Numbers game permits you to choose either a three-digit (Pick-3) or four-digit (Pick-4) number at a cost of $1 per ticket. Each night, the winning number is drawn. If your number matches the winning number, you win a large sum of money, usually $100,000. You do have some control over the Daily Numbers game (since you pick the numbers that you play) and, consequently, there are strategies available to increase your chances of winning. However, the Daily Numbers game, like the Instant game, is not available for out-of-state play.

To play Pick-6 Lotto, you select six numbers of your choice from a field of numbers ranging from 1 to N, where N depends on which state's game you are playing. For example, Florida's current Lotto game involves picking six numbers ranging from 1 to 53. (See Figure SIA3.1.) The cost of a ticket is $1, and the payoff, if your six numbers match the winning numbers drawn, is $7 million or more, depending on the number of tickets purchased. (To date, Florida has had the largest state weekly payoff of over $200 million.) In addition to the grand prize, you can win second-, third-, and fourth-prize payoffs by matching five, four, and three of the six numbers drawn, respectively. And you don't have to be a resident of the state to play the state's Lotto game.

In this chapter, several Statistics In Action Revisited examples demonstrate how to use the basic concepts of probability to compute the odds of winning a state lottery game and to assess the validity of the strategies suggested by lottery "experts."

Figure SIA3.1

Recall that one branch of statistics is concerned with decisions about a population based on sample information. You can see how this is accomplished more easily if you understand the relationship between population and sample — a relationship that becomes clearer if we reverse the statistical procedure of making inferences from sample to population. In this chapter we assume the population is *known* and calculate the chances of obtaining various samples from the population. Thus, we show that probability is the reverse of statistics: In probability, we use the population information to infer the probable nature of the sample.

Probability plays an important role in inference making. Suppose, for example, you have an opportunity to invest in an oil exploration company. Past records show that out of ten previous oil drillings (a sample of the company's experiences), all ten came up dry. What do you conclude? Do you think the chances are better than 50 : 50 that the company will hit a gusher? Should you invest in this company? Chances are, your answer to these questions will be an emphatic No. If the company's exploratory prowess is sufficient to hit a producing well 50% of the time, a record of ten dry wells out of ten drilled is an event that is just too improbable.

Or suppose you're playing poker with what your opponents assure you is a well-shuffled deck of cards. In three consecutive five-card hands, the person on your right is dealt four aces. Based on this sample of three deals, do you think the cards are being adequately shuffled? Again, your answer is likely to be No because dealing three hands of four aces is just too improbable if the cards were properly shuffled.

Note that the decisions concerning the potential success of the oil drilling company and the adequacy of card shuffling both involve knowing the chance — or probability — of a certain sample result. Both situations were contrived so that you could easily conclude that the probabilities of the sample results were small. Unfortunately, the probabilities of many observed sample results aren't so easy to evaluate intuitively. For these cases we need the assistance of a theory of probability.

3.1 Events, Sample Spaces, and Probability

Let's begin our treatment of probability with straightforward examples that are easily described. With the aid of simple examples, we can introduce important definitions that will help us develop the notion of probability more easily.

Suppose a coin is tossed once and the up face is recorded. The result we see is called an *observation*, or *measurement*, and the process of making an observation is called an *experiment*. Notice that our definition of experiment is broader than the one used in the physical sciences, where you tend to picture test tubes, microscopes, and other laboratory equipment. Among other things, statistical experiments may include recording an Internet user's preference for a Web browser, recording a voter's opinion on an important political issue, measuring the amount of dissolved oxygen in a polluted river, observing the level of anxiety of a test taker, counting the number of errors in an inventory, and observing the fraction of insects killed by a new insecticide. The point is that a statistical experiment can be almost any act of observation as long as the outcome is uncertain.

> **DEFINITION 3.1**
>
> An **experiment** is an act or process of observation that leads to a single outcome that cannot be predicted with certainty.

Consider another simple experiment consisting of tossing a die and observing the number on the up face. The six possible outcomes to this experiment are:

1. Observe a 1

2. Observe a 2

3. Observe a 3

4. Observe a 4

5. Observe a 5

6. Observe a 6

Note that if this experiment is conducted once, *you can observe one and only one of these six basic outcomes, and the outcome cannot be predicted with certainty.* Also, these possibilities cannot be decomposed into more basic outcomes. Because observing the outcome of an experiment is similar to selecting a sample from a population, the basic possible outcomes to an experiment are called **sample points**.*

> **DEFINITION 3.2**
>
> A **sample point** is the most basic outcome of an experiment.

EXAMPLE 3.1   LISTING SAMPLE POINTS

Problem   Two coins are tossed, and their up faces are recorded. List all the sample points for this experiment.

Solution   Even for a seemingly trivial experiment, we must be careful when listing the sample points. At first glance, we might expect three basic outcomes: Observe two heads; Observe two tails; or Observe one head and one tail. However, further reflection reveals that the last of these, Observe one head and one tail, can be decomposed into two outcomes: Head on coin 1, Tail on coin 2; and Tail on coin 1, Head on coin 2. Thus, we have four sample points:

1. Observe *HH*

2. Observe *HT*

3. Observe *TH*

4. Observe *TT*

where *H* in the first position means "Head on coin 1," *H* in the second position means "Head on coin 2," and so on.

Look Back   Even if the coins are identical in appearance they are, in fact, two distinct coins. Thus, the sample points must account for this distinction.

| Now Work | *Exercise 3.15a*

■ ■ ■

---

*Alternatively, the term *simple event* can be used.

TABLE 3.1 Experiments and Their Sample Spaces

Experiment: Observe the up face on a coin.
Sample Space: 1. Observe a head
2. Observe a tail

This sample space can be represented in set notation as a set containing two sample points:

$$S: \quad \{H, T\}$$

where H represents the sample point Observe a head and T represents the sample point Observe a tail.

Experiment: Observe the up face on a die.
Sample Space: 1. Observe a 1
2. Observe a 2
3. Observe a 3
4. Observe a 4
5. Observe a 5
6. Observe a 6

This sample space can be represented in set notation as a set of six sample points:

$$S: \quad \{1, 2, 3, 4, 5, 6\}$$

Experiment: Observe the up faces on two coins.
Sample Space: 1. Observe HH
2. Observe HT
3. Observe TH
4. Observe TT

This sample space can be represented in set notation as a set of four sample points:

$$S: \quad \{HH, HT, TH, TT\}$$

We often wish to refer to the collection of all the sample points of an experiment. This collection is called the *sample space* of the experiment. For example, there are six sample points in the sample space associated with the die-toss experiment. The sample spaces for the experiments discussed thus far are shown in Table 3.1.

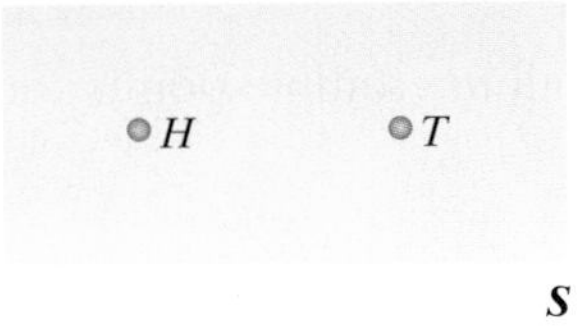

> **DEFINITION 3.3**
>
> The **sample space** of an experiment is the collection of all its sample points.

Just as graphs are useful in describing sets of data, a pictorial method for presenting the sample space will often be useful. Figure 3.1 shows such a representation for each of the experiments in Table 3.1. In each case, the sample space is shown as a closed figure, labeled **S**, containing all possible sample points. Each sample point is represented by a solid dot (i.e., a "point") and labeled accordingly. Such graphical representations are called **Venn diagrams**.

Figure sidebar (left column):

∘ *H*   ∘ *T*

S

a. Experiment: Observe the up face on a coin

∘ 1   ∘ 2   ∘ 3
∘ 4   ∘ 5   ∘ 6

S

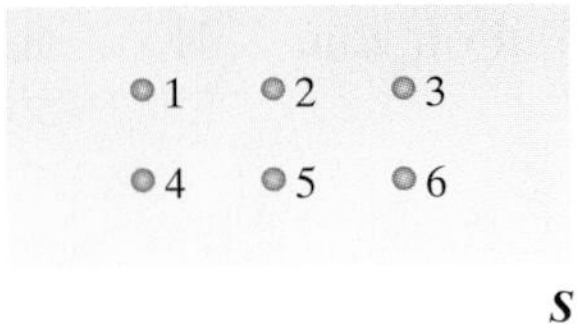

b. Experiment: Observe the up face on a die

∘ *HH*   ∘ *HT*
∘ *TH*   ∘ *TT*

S

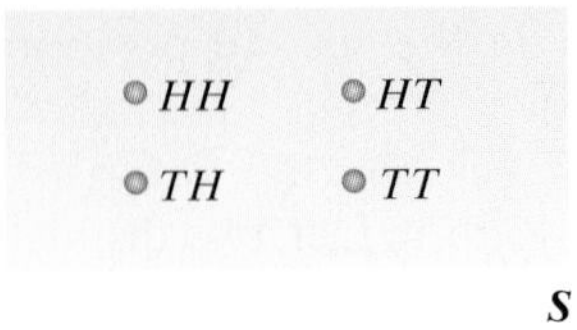

c. Experiment: Observe the up faces on two coins

Figure 3.1
Venn Diagrams for the Three Experiments from Table 3.1

Biography

JOHN VENN (1834–1923)— The English Logician

Born in Hull, England, John Venn is probably best known for his pictorial representation of unions and intersections (the Venn diagram). While lecturing in Moral Science at Cambridge University, Venn wrote *The Logic of Chance* and two other treatises on logic. In these works, Venn probably had his greatest contribution to the field of probability and statistics—the notion that the probability of an event is simply the long-run proportion of times the event occurs. Besides being a well-known mathematician, Venn was also a historian, philosopher, priest, and skilled machine builder. (His machine for bowling cricket balls once beat a top star on the Australian cricket team.)

Now that we know that an experiment will result in *only one* basic outcome — called a sample point — and that the sample space is the collection of all possible sample points, we're ready to discuss the probabilities of the sample points. You have undoubtedly used the term *probability* and have some intuitive idea about its meaning. Probability is generally used synonymously with "chance," "odds," and similar concepts. For example, if a fair coin is tossed, we might reason that both the sample points, Observe a head and Observe a tail, have the same *chance* of occurring. Thus, we might state that "the probability of observing a head is 50%" or "the odds of seeing a head are 50 : 50." Both these statements are based on an informal knowledge of probability. We'll begin our treatment of probability by using such informal concepts and then solidify what we mean later.

The probability of a sample point is a number between 0 and 1 that measures the likelihood that the outcome will occur when the experiment is performed. This number is usually taken to be the relative frequency of the occurrence of a sample point in a very long series of repetitions of an experiment.* For example, if we are assigning probabilities to the two sample points in the coin-toss experiment (Observe a head and Observe a tail), we might reason that if we toss a balanced coin a very large number of times, the sample points Observe a head and Observe a tail will occur with the same relative frequency of .5.

Our reasoning is supported by Figure 3.2. The figure plots the relative frequency of the number of times that a head occurs when simulating (by computer) the toss of a coin N times, where N ranges from as few as 25 tosses to as many as 1,500 tosses of the coin. You can see that when N is large (i.e., $N = 1,500$), the relative frequency is converging to .5. Thus, the probability of each sample point in the coin tossing experiment is .5.

For some experiments, we may have little or no information on the relative frequency of occurrence of the sample points; consequently, we must assign probabilities

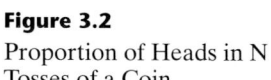
Figure 3.2
Proportion of Heads in N Tosses of a Coin

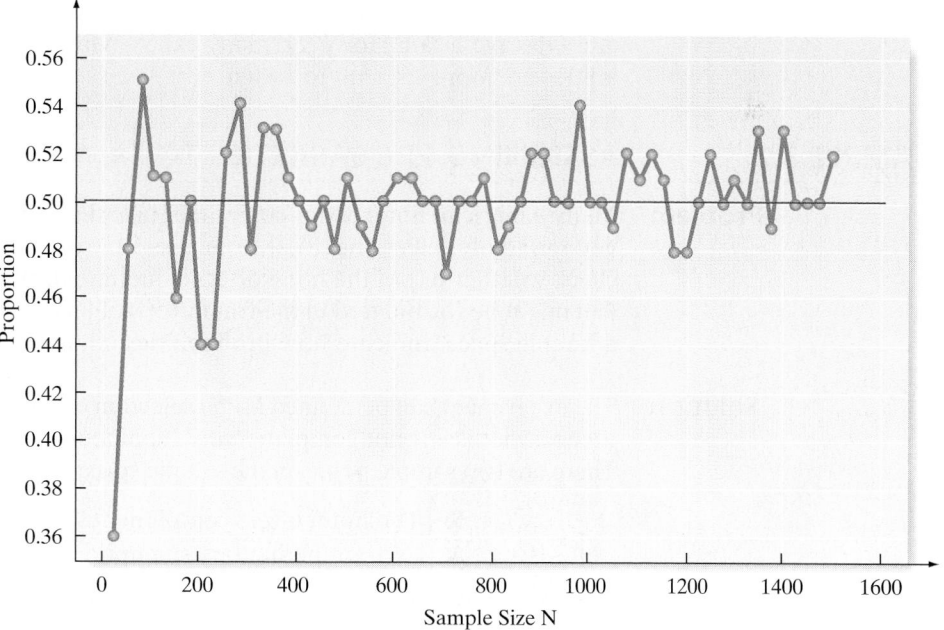

*The result derives from an axiom in probability theory called the *Law of Large Numbers*. Phrased informally, the law states that the relative frequency of the number of times that an outcome occurs when an experiment is replicated over and over again (i.e., a large number of times) approaches the true (or theoretical) probability of the outcome.

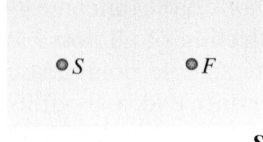

Figure 3.3
Experiment: Invest in a Business Venture and Observe Whether It Succeeds (S) or Fails (F)

to the sample points based on general information about the experiment. For example, if the experiment is to invest in a business venture and to observe whether it succeeds or fails, the sample space would appear as in Figure 3.3.

We are unlikely to be able to assign probabilities to the sample points of this experiment based on a long series of repetitions since unique factors govern each performance of this kind of experiment. Instead, we may consider factors such as the personnel managing the venture, the general state of the economy at the time, the rate of success of similar ventures, and any other information deemed pertinent. If we finally decide that the venture has an 80% chance of succeeding, we assign a probability of .8 to the sample point Success. This probability can be interpreted as a measure of our degree of belief in the outcome of the business venture; that is, it is a subjective probability. Notice, however, that such probabilities should be based on expert information that is carefully assessed. If not, we may be misled on any decisions based on these probabilities or based on any calculations in which they appear. [*Note:* For a text that deals in detail with the subjective evaluation of probabilities, see Winkler (1972) or Lindley (1985).]

No matter how you assign the probabilities to sample points, the probabilities assigned must obey two rules:

> **Probability Rules for Sample Points**
>
> Let p_i represent the probability of sample point i.
>
> **1.** All sample point probabilities *must* lie between 0 and 1 (i.e., $0 \leq p_i \leq 1$).
> **2.** The probabilities of all the sample points within a sample space *must* sum to 1 (i.e., $\Sigma p_i = 1$).

Assigning probabilities to sample points is easy for some experiments. For example, if the experiment is to toss a fair coin and observe the face, we would probably all agree to assign a probability of $\frac{1}{2}$ to the two sample points, Observe a head and Observe a tail. However, many experiments have sample points whose probabilities are more difficult to assign.

EXAMPLE 3.2   ASSIGNING PROBABILITIES TO SAMPLE POINTS

Problem   Many American hotels offer complimentary shampoo in their guest rooms. Suppose you randomly select one hotel from a registry of all hotels in the United States and check whether or not the hotel offers complimentary shampoo. Show how this problem might be formulated in the framework of an experiment with sample points and a sample space. Indicate how probabilities might be assigned to the sample points.

Solution   The experiment can be defined as the selection of an American hotel and the observation of whether or not complimentary shampoo is offered in the hotel's guest rooms. There are two sample points in the sample space corresponding to this experiment:

S: {The hotel offers complimentary shampoo}
N: {No complimentary shampoo is offered by the hotel}

The difference between this and the coin-toss experiment becomes apparent when we attempt to assign probabilities to the two sample points. What probability should we assign to the sample point S? If you answer .5, you are assuming that the events S and N should occur with equal likelihood, just like the sample points Heads and Tails in the coin-toss experiment. But assignment of sample point probabilities for the hotel-shampoo experiment is not so easy. In fact, a recent survey of American hotels found that 80% now offer complimentary shampoo to guests. Then it might be reasonable to approximate the probability of the sample point S as .8 and that of the sample point N as .2.

Look Back Here we see that the sample points are not always equally likely, so assigning probabilities to them can be complicated — particularly for experiments that represent real applications (as opposed to coin- and die-toss experiments).

<u>Now Work</u> *Exercise 3.20*

■ ■ ■

Although the probabilities of sample points are often of interest in their own right, it is usually probabilities of collections of sample points that are important. Example 3.3 demonstrates this point.

EXAMPLE 3.3

FINDING THE PROBABILITY OF A COLLECTION OF SAMPLE POINTS

Problem   A fair die is tossed, and the up face is observed. If the face is even, you win $1. Otherwise, you lose $1. What is the probability that you win?

Solution   Recall that the sample space for this experiment contains six sample points:

$$S: \{1, 2, 3, 4, 5, 6\}$$

Since the die is balanced, we assign a probability of $\frac{1}{6}$ to each of the sample points in this sample space. An even number will occur if one of the sample points, Observe a 2, Observe a 4, or Observe a 6, occurs. A collection of sample points such as this is called an *event*, which we denote by the letter A. Since the event A contains three sample points — all with probability $\frac{1}{6}$ — and since no sample points can occur simultaneously, we reason that the probability of A is the sum of the probabilities of the sample points in A. Thus, the probability of A is $\frac{1}{6} + \frac{1}{6} + \frac{1}{6} = \frac{1}{2}$.

Look Back Based on our notion of probability, this implies that, *in the long run*, you will win $1 half the time and lose $1 half the time.

<u>Now Work</u> *Exercise 3.14*

■ ■ ■

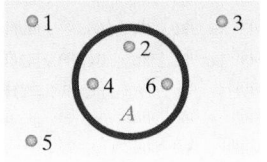

Figure 3.4
Die-Toss Experiment with Event A: Observe an Even Number

Figure 3.4 is a Venn diagram depicting the sample space associated with a die-toss experiment and the event A, Observe an even number. The event A is represented by the closed figure inside the sample space S. This closed figure A contains all the sample points that comprise it.

To decide which sample points belong to the set associated with an event A, test each sample point in the sample space S. If event A occurs, then that sample point is in the event A. For example, the event A, Observe an even number, in the die-toss experiment will occur if the sample point Observe a 2 occurs. By the same reasoning, the sample points Observe a 4 and Observe a 6 are also in event A.

To summarize, we have demonstrated that an event can be defined in words, or it can be defined as a specific set of sample points. This leads us to the following general definition of an *event*:

DEFINITION 3.4

An **event** is a specific collection of sample points.

EXAMPLE 3.4   FINDING THE PROBABILITY OF AN EVENT

Problem   Consider the experiment of tossing two *unbalanced* coins. Because the coins are *not* balanced, their outcomes (*H* or *T*) are not equiprobable. Suppose the correct probabilities associated with the sample points are given in the accompanying table. [*Note:* The necessary properties for assigning probabilities to sample points are satisfied.]

Consider the events:

$$A: \{\text{Observe exactly one head}\}$$
$$B: \{\text{Observe at least one head}\}$$

Calculate the probability of *A* and the probability of *B*.

Sample Point	Probability
HH	$\frac{4}{9}$
HT	$\frac{2}{9}$
TH	$\frac{2}{9}$
TT	$\frac{1}{9}$

Solution   Event *A* contains the sample points *HT* and *TH*. Since two or more sample points cannot occur at the same time, we can easily calculate the probability of event *A* by summing the probabilities of the two sample points. Thus, the probability of observing exactly one head (event *A*), denoted by the symbol $P(A)$, is

$$P(A) = P(\text{Observe } HT) + P(\text{Observe } TH) = \frac{2}{9} + \frac{2}{9} = \frac{4}{9}$$

Similarly, since *B* contains the sample points *HH, HT,* and *TH,*

$$P(B) = \frac{4}{9} + \frac{2}{9} + \frac{2}{9} = \frac{8}{9}$$

Look Back   Again, these probabilities should be interpreted *in the long run.* For example, $P(B) = \frac{8}{9} \approx .89$ implies that if we were to toss two coins an infinite number of times, we would observe at least two heads on about 89% of the tosses.

Now Work   *Exercise 3.11*

◾ ◾ ◾

The preceding example leads us to a general procedure for finding the probability of an event *A*:

> **Probability of an Event**
>
> The probability of an event *A* is calculated by summing the probabilities of the sample points in the sample space for *A*.

Thus, we can summarize the steps for calculating the probability of any event, as indicated in the next box.

> **Steps for Calculating Probabilities of Events**
> 1. Define the experiment, that is, describe the process used to make an observation and the type of observation that will be recorded.
> 2. List the sample points.
> 3. Assign probabilities to the sample points.
> 4. Determine the collection of sample points contained in the event of interest.
> 5. Sum the sample point probabilities to get the event probability.

EXAMPLE 3.5

APPLYING THE FIVE STEPS

Problem The American Association for Marriage and Family Therapy (AAMFT) is a group of professional therapists and family practitioners that treats many of the nation's couples and families. The AAMFT released the findings of a study that tracked the post-divorce history of 100 pairs of former spouses with children. Each divorced couple was classified into one of four groups, nicknamed "perfect pals (PP)," "cooperative colleagues (CC)," "angry associates (AA)," and "fiery foes (FF)." The proportions classified into each group are shown in Table 3.2.

Suppose one of the 100 couples is selected at random.

a. Define the experiment that generated the data in Table 3.2, and list the sample points.

b. Assign probabilities to the sample points.

c. What is the probability that the former spouses are "fiery foes?"

d. What is the probability that the former spouses have at least some conflict in their relationship?

TABLE 3.2   Results of AAMFT Study of Divorced Couples

Group	Proportion
Perfect Pals (PP) (Joint-custody parents who get along well)	.12
Cooperative Colleagues (CC) (Occasional conflict, likely to be remarried)	.38
Angry Associates (AA) (Cooperate on children-related issues only, conflicting otherwise)	.25
Fiery Foes (FF) (Communicate only through children, hostile toward each other)	.25

Solution

a. The experiment is the act of classifying the randomly selected couple. The sample points, the simplest outcomes of the experiment, are the four groups (categories) listed in Table 3.2. They are shown in the Venn diagram in Figure 3.5.

b. If, as in Example 3.1, we were to assign equal probabilities in this case, each of the response categories would have a probability of one-fourth (1/4), or .25. But, by examining Table 3.2 you can see that equal probabilities are not reasonable in this case because the response percentages are not all the same in the four categories. It is more reasonable to assign a probability equal to the response proportion in each class, as shown in Table 3.3.*

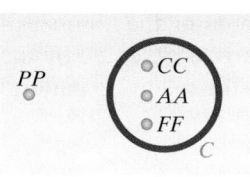

Figure 3.5
Venn Diagram for
AAMFT Survey

*Since the response percentages were based on a sample of divorced couples, these assigned probabilities are estimates of the true population response percentages. You will learn how to measure the reliability of probability estimates in Chapter 7.

TABLE 3.3   **Sample Point Probabilities for AAMFT Survey**

Sample Point	Probability
PP	.12
CC	.38
AA	.25
FF	.25

c. The event that the former spouses are "fiery foes" corresponds to the sample point FF. Consequently, the probability of the event is the probability of the sample point. From Table 3.3, we find $P(FF) = .25$. Therefore, there is a .25 probability (or one-fourth chance) that the couple we select are "fiery foes."

d. The event that the former spouses have at least some conflict in their relationship, call it event C, is not a sample point because it consists of more than one of the response classifications (the sample points). In fact, as shown in Figure 3.5, the event C consists of three sample points: CC, AA, and FF. The probability of C is defined to be the sum of the probabilities of the sample points in C:

$$P(C) = P(CC) + P(AA) + P(FF) = .38 + .25 + .25 = .88$$

Thus, the chance that we observe a divorced couple with some degree of conflict in their relationship is .88 — a fairly high probability.

Look Back The key to solving this problem is to follow the steps outlined in the box. We defined the experiment (Step 1) and listed the sample points (Step 2) in part **a**. The assignment of probabilities to the sample points (Step 3) was done in part **b**. For each probability in parts c and d, we identified the collection of sample points in the event (Step 4) and summed their probabilities (Step 5).

Now Work   *Exercise 3.23*

■ ■ ■

The preceding examples have one thing in common: The number of sample points in each of the sample spaces was small; hence, the sample points were easy to identify and list. How can we manage this when the sample points run into the thousands or millions? For example, suppose you wish to select five marines for a dangerous mission from a division of 1,000. Then each different group of five marines would represent a sample point. How can you determine the number of sample points associated with this experiment?

One method of determining the number of sample points for a complex experiment is to develop a counting system. Start by examining a simple version of the experiment. For example, see if you can develop a system for counting the number of ways to select two marines from a total of four. If the marines are represented by the symbols M_1, M_2, M_3, and M_4, the sample points could be listed in the following pattern:

(M_1, M_2) (M_2, M_3) (M_3, M_4)
(M_1, M_3) (M_2, M_4)
(M_1, M_4)

Note the pattern and now try a more complex situation — say, sampling three marines out of five. List the sample points and observe the pattern. Finally, see if you can deduce the pattern for the general case. Perhaps you can program a computer to produce the matching and counting for the number of samples of 5 selected from a total of 1,000.

A second method of determining the number of sample points for an experiment is to use **combinatorial mathematics**. This branch of mathematics is concerned with developing counting rules for given situations. For example, there is a simple rule for finding the number of different samples of five marines selected from 1,000. This rule, called the **Combinations Rule**, is given in the box.

Combinations Rule

A sample of n elements is to be drawn from a set of N elements. Then, the number of different samples possible is denoted by $\dbinom{N}{n}$ and is equal to

$$\binom{N}{n} = \frac{N!}{n!(N-n)!}$$

where the factorial symbol (!) means that

$$n! = n(n-1)(n-2)\cdots(3)(2)(1)$$

For example, $5! = 5 \cdot 4 \cdot 3 \cdot 2 \cdot 1$. [*Note:* The quantity 0! is defined to be equal to 1.]

EXAMPLE 3.6

USING THE COMBINATIONS RULE

Problem   Consider the task of choosing 2 marines from a platoon of 4 to send on a dangerous mission. Use the combinations counting rule to determine how many different selections can be made.

Solution   For this example, $N = 4$, $n = 2$, and

$$\binom{4}{2} = \frac{4!}{2!2!} = \frac{4 \cdot 3 \cdot 2 \cdot 1}{(2 \cdot 1)(2 \cdot 1)} = 6$$

Look Back   You can see that this agrees with the number of sample points listed on p. 128.

> **Now Work**   *Exercise 3.13*

■ ■ ■

EXAMPLE 3.7

USING THE COMBINATIONS RULE

Problem   Suppose you plan to invest equal amounts of money in each of five business ventures. If you have 20 ventures from which to make the selection, how many different samples of five ventures can be selected from the 20?

Solution   For this example, $N = 20$ and $n = 5$. Then the number of different samples of 5 that can be selected from the 20 ventures is

$$\binom{20}{5} = \frac{20!}{5!(20-5)!} = \frac{20!}{5!15!}$$

$$= \frac{20 \cdot 19 \cdot 18 \cdot \cdots \cdot 3 \cdot 2 \cdot 1}{(5 \cdot 4 \cdot 3 \cdot 2 \cdot 1)(15 \cdot 14 \cdot 13 \cdot \cdots \cdot 3 \cdot 2 \cdot 1)} = 15{,}504$$

Look Back   You can see that attempting to list all the sample points for this experiment would be an extremely tedious and time consuming, if not practically impossible, task.

■ ■ ■

The Combinations Rule is just one of a large number of counting rules that have been developed by combinatorial mathematicians. This counting rule applies to situations in which the experiment calls for selecting n elements from a total of N elements, without replacing each element before the next is selected. Several other basic counting rules are presented in optional Section 3.8.

Statistics in Action Revisited

Computing and Understanding the Probability of Winning Lotto

In Florida's state lottery game, called Pick-6 Lotto, you select six numbers of your choice from a set of numbers ranging from 1 to 53. We can apply the Combinations Rule to determine the total number of combinations of 6 numbers selected from 53 (i.e., the total number of sample points [or possible winning tickets]). Here, $N = 53$ and $n = 6$; therefore, we have

$$\binom{N}{n} = \frac{N!}{n!(N-n)!} = \frac{53!}{6!47!}$$
$$= \frac{(53)(52)(51)(50)(49)(48)(47!)}{(6)(5)(4)(3)(2)(1)(47!)}$$
$$= 22{,}957{,}480$$

Now, since the Lotto balls are selected at random, each of these 22,957,480 combinations is equally likely to occur. Therefore, the probability of winning Lotto is

$$P(\text{Win } 6/53 \text{ Lotto}) = 1/(22{,}957{,}480) = .00000004356$$

This probability is often stated as follows: The odds of winning the game with a single ticket are 1 in 22,957,480, or, 1

in approximately 23 million. For all practical purposes, this probability is 0, implying that you have almost no chance of winning the lottery with a single ticket. Yet each week there is almost always a winner in the Florida Lotto. This apparent contradiction can be explained with the following analogy.

Suppose there is a line of minivans, front-to-back, from New York City to Los Angeles, California. Based on the distance between the two cities and the length of a standard minivan, there would be approximately 23 million minivans in line. Lottery officials will select, at random, one of the minivans and put a check for $10 million dollars in the glove compartment. For a cost of $1, you may roam the country and select one (and only one) minivan and check the glove compartment. Do you think you will find $10 million in the minivan you choose? You can be almost certain that you won't. But now permit anyone to enter the lottery for $1 and suppose that 50 million people do so. With such a large number of participants, it is very likely that someone will find the minivan with the $10 million — but it almost certainly won't be you! (This example illustrates an axiom in statistics called the Law of Large Numbers. See the footnote at the bottom of p. 123.)

Exercises 3.1–3.32

Understanding the Principles

3.1 What is an experiment?

3.2 What are the most basic outcomes of an experiment called?

3.3 Define the sample space.

3.4 What is a Venn diagram?

3.5 Give two probability rules for sample points.

3.6 What is an event?

3.7 How do you find the probability of an event made up of several sample points?

3.8 Give a scenario where the combinations rule is appropriate for counting the number of sample points.

Learning the Mechanics

3.9 An experiment results in one of the following sample points: E_1, E_2, E_3, E_4, or E_5.
   a. Find $P(E_3)$ if $P(E_1) = .1$, $P(E_2) = .2$, $P(E_4) = .1$, and $P(E_5) = .1$
   b. Find $P(E_3)$ if $P(E_1) = P(E_3)$, $P(E_2) = .1$, $P(E_4) = .2$, and $P(E_5) = .1$

   c. Find $P(E_3)$ if $P(E_1) = P(E_2) = P(E_4) = P(E_5) = .1$

3.10 The accompanying Venn diagram describes the sample space of a particular experiment and events A and B.

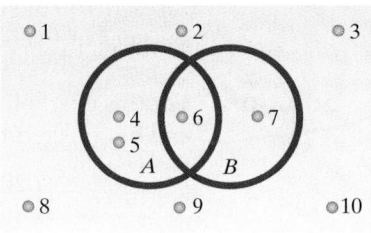

   a. Suppose the sample points are equally likely. Find $P(A)$ and $P(B)$
   b. Suppose $P(1) = P(2) = P(3) = P(4) = P(5) = \frac{1}{20}$ and $P(6) = P(7) = P(8) = P(9) = P(10) = \frac{3}{20}$. Find $P(A)$ and $P(B)$.

3.11 The sample space for an experiment contains five sample points with probabilities as shown in the table. Find the probability of each of the following events:

A: {Either 1, 2, or 3 occurs}
B: {Either 1, 3, or 5 occurs}
C: {4 does not occur}

Sample Points	Probabilities
1	.05
2	.20
3	.30
4	.30
5	.15

3.12 Compute each of the following:

a. $\binom{9}{4}$ **b.** $\binom{7}{2}$ **c.** $\binom{4}{4}$

d. $\binom{5}{0}$ **e.** $\binom{6}{5}$

3.13 Compute the number of ways you can select *n* elements from *N* elements for each of the following:
a. $n = 2, N = 5$ **b.** $n = 3, N = 6$
c. $n = 5, N = 20$

3.14 Two fair dice are tossed, and the up face on each die is recorded.
a. List the 36 sample points contained in the sample space.
b. Assign probabilities to the sample points.
c. Find the probability of observing each of the following events:

A: {A 3 appears on each of the two dice}
B: {The sum of the numbers is even}
C: {The sum of the numbers is equal to 7}
D: {A 5 appears on at least one of the dice}
E: {The sum of the numbers is 10 or more}

3.15 Two marbles are drawn at random and without replacement from a box containing two blue marbles and three red marbles.
a. List the sample points.
b. Assign probabilities to the sample points.
c. Determine the probability of observing each of the following events:

A: {Two blue marbles are drawn}
B: {A red and a blue marble are drawn}
C: {Two red marbles are drawn}

3.16 Simulate the experiment described in Exercise 3.15 using any five identically shaped objects, two of which are one color and three, another. Mix the objects, draw two, record the results, and then replace the objects. Repeat the experiment a large number of times (at least 100). Calculate the proportion of times events *A, B,* and *C* occur. How do these proportions compare with the probabilities you calculated in Exercise 3.8? Should these proportions equal the probabilities? Explain.

Applying the Concepts—Basic

3.17 Post office violence. *The Wall Street Journal* (Sept. 1, 2000) reported on an independent study of postal workers and violence at post offices. In a sample of 12,000 postal workers, 600 of them were physically assaulted on the job in the past year. Use this information to estimate the probability that a randomly selected postal worker will be physically assaulted on the job during the year.

3.18 International Nanny Association. According to *USA Today* (Sept. 19, 2000), there are 650 members of the International Nanny Association (INA). Of these, only three are men. Find the probability that a randomly selected member of the INA is a man.

3.19 USDA chicken inspection. The United States Department of Agriculture (USDA) reports that, under its standard inspection system, one in every 100 slaughtered chickens pass inspection with fecal contamination. (*Tampa Tribune*, Mar. 31, 2000.)
a. If a slaughtered chicken is selected at random, what is the probability that it passes inspection with fecal contamination?
b. The probability of part **a** was based on a USDA study that found that 306 of 32,075 chicken carcasses passed inspection with fecal contamination. Do you agree with the USDA's statement about the likelihood of a slaughtered chicken passing inspection with fecal contamination?

3.20 African rhinos. Two species of rhinoceros native to Africa are black rhinos and white rhinos. The International Rhino Federation estimates that the African rhinoceros population consists of 2,600 white rhinos and 8,400 black rhinos. Suppose one rhino is selected at random from the African rhino population and its species (black or white) is observed.
a. List the sample points for this experiment.
b. Assign probabilities to the sample points based on the estimates made by the International Rhino Federation.

3.21 Fungi in beech forest trees. Beechwood forests in East Central Europe are being threatened by dynamic changes in land ownership and economic upheaval. The current status of the beech tree species in this area was evaluated by Hungarian university professors in *Applied Ecology and Environmental Research* (Vol. 1, 2003). Of 188 beech trees surveyed, 49 of the trees had been damaged by fungi. Depending on the species of fungi, damage will occur either on the trunk, branches, or leaves of the tree. In the damaged trees, the trunk was affected 85% of the time, the leaves 10% of the time, and the branches 5% of the time.
a. Give a reasonable estimate of the probability of a beech tree in East Central Europe being damaged by fungi.

b. A fungi-damaged beech tree is selected at random and the area (trunk, leaf, or branch) affected is observed. List the sample points for this experiment and assign a reasonable probability to each sample point.

3.22 Chance of rain. Answer the following question posed in the *Atlanta Journal-Constitution* (Feb. 7, 2000): When a meteorologist says that "the probability of rain this afternoon is .4," does it mean that 40% of the time during the afternoon, it will be raining?

Applying the Concepts—Intermediate

3.23 Benford's Law of numbers. Refer to the *American Scientist* (July–Aug.1998) study of *Benford's Law*, Exercise 2.14 (p. 51). Recall that *Benford's Law* states that the integer 1 is more likely to occur as the first significant digit in a randomly selected number than the other integers. The table summarizes the results of a study where 743 college freshmen selected a six-digit number at random. Assume that one of these 743 numbers is selected and the first significant digit is observed.

DIGITS

First Digit	Number of Occurrences
1	109
2	75
3	77
4	99
5	72
6	117
7	89
8	62
9	43
Total	743

Source: Hill, T. P. "The first digit phenomenon." *American Scientist,* Vol. 86, No. 4, July–Aug. 1998, p. 363 (Figure 5).

a. List the sample points for this experiment.
b. Explain why the sample points are not equally likely to occur.
c. Assign reasonable probabilities to the sample points. Verify that the probabilities sum to 1.
d. What is the probability that the first significant digit is 1 or 2?
e. What is the probability that the first significant digit is greater than 5?

3.24 Sex composition patterns of children in families. When having children, is there a genetic factor that causes some families to favor one sex over the other? That is, does having boys or girls "run in the family?" This was the question of interest in *Chance* (Fall 2001). Using data collected on children's sex for over 4,000 American families that had at least 2 children, the researchers compiled the following table. An American family with at least 2 children is selected, and the sex composition of the first two children is observed.

Sex Composition of First Two Children	Frequency
Boy-Boy	1,085
Boy-Girl	1,086
Girl-Boy	1,111
Girl-Girl	926
TOTAL	4,208

Source: Rodgers, J. L., and Doughty, D. "Does having boys or girls run in the family?" *Chance,* Vol. 14, No. 4, Fall 2001, Table 3.

a. List the sample points for this experiment.
b. If having a boy is no more likely than having a girl and vice versa, assign a probability to each sample point.
c. Use the information in the table to estimate the sample point probabilities. Do these estimates agree (to a reasonable degree of approximation) with the probabilities, part **b**?
d. Make an inference about whether having boys or girls "runs in the family."

3.25 Choosing portable grill displays. University of Maryland marketing professor R. W. Hamilton studied how people attempt to influence the choices of others by offering undesirable alternatives. (*Journal of Consumer Research,* Mar. 2003.) Such a phenomenon typically occurs when family members propose a vacation spot, friends recommend a restaurant for dinner, and realtors show the buyer potential home sites. In one phase of the study, the researcher had each of 124 college students select showroom displays for portable grills. Five different displays (representing five different sized grills) were available, but only three displays would be selected. The students were instructed to select the displays to maximize purchases of Grill #2 (a smaller-sized grill).

a. In how many possible ways can the 3-grill displays be selected from the 5 displays? List the possibilities.
b. The table shows the grill display combinations and number of each selected by the 124 students. Use this information to assign reasonable probabilities to the different display combinations.
c. Find the probability that a student who participated in the study selected a display combination involving Grill #1.

Grill Display Combination	Number of Students
1-2-3	35
1-2-4	8
1-2-5	42
2-3-4	4
2-3-5	1
2-4-5	34

Source: Hamilton, R. W. "Why do people suggest what they do not want? Using context effects to influence others' choices," *Journal of Consumer Research,* Vol. 29, Mar. 2003, Table 1.

3.26 Perfect SAT scores. The maximum score possible on the Standardized Admission Test (SAT) is 1600. According to the test developers, the chance that a student scores a perfect 1600 on the SAT is 5 in 10,000.

a. Find the probability that a randomly selected student scores a 1600 on the SAT.

b. In a recent year, 545 students scored a perfect 1600 on the SAT. How is this possible given the probability calculated in part **a**?

3.27 Jai-alai quinella bet. The Quinella bet at the paramutual game of jai-alai consists of picking the jai-alai players that will place first and second in a game *irrespective* of order. In jai-alai, eight players (numbered 1, 2, 3, . . . , 8) compete in every game.

a. How many different Quinella bets are possible?

b. Suppose you bet the Quinella combination of 2-7. If the players are of equal ability, what is the probability that you win the bet?

3.28 Alzheimer's gene study. Scientists have already discovered two genes (on chromosomes 19 and 21) that mutate to cause the early onset of Alzheimer's disease. *Science News* (July 8, 1995) reported on a search for a third gene that causes Alzheimer's. An international team of scientists gathered genetic information from 21 families afflicted by the early-onset form of the disease. In six of these families, mutations in the S182 gene on chromosome 14 accounted for the early onset of Alzheimer's.

a. Consider a family afflicted by the early onset of Alzheimer's disease. Find the approximate probability that the researchers will find mutations in the S182 gene on chromosome 14 for this family

b. Discuss the reliability of the probability, part **a**. How could a more accurate estimate of the probability be obtained?

Applying the Concepts—Advanced

3.29 Odd Man Out. Three people play a game called "Odd Man Out." In this game, each player flips a fair coin until the outcome (heads or tails) for one of the players is not the same as the other two players. This player is then "the odd man out" and loses the game. Find the probability that the game ends (i.e., either exactly one of the coins will fall heads or exactly one of the coins will fall tails) after only one toss by each player. Suppose one of the players, hoping to reduce his chances of being the odd man, uses a two-headed coin. Will this ploy be successful? Solve by listing the sample points in the sample space.

3.30 Matching socks. Consider the following question posed to Marilyn vos Savant in her weekly newspaper column "Ask Marilyn":

I have two pairs of argyle socks, and they look nearly identical—one navy blue and the other black. [When doing the laundry] my wife matches the socks incorrectly much more often than she does correctly. . . . If all four socks are in front of her, it seems to me that her chances are 50% for a wrong match and 50% for a right match. What do you think?

Source: *Parade Magazine*, Feb. 27, 1994.

Use your knowledge of probability to answer this question. [*Hint:* List the sample points in the experiment.]

3.31 Post-op nausea study. Nausea and vomiting after surgery are common side effects of anesthesia and painkillers. Six different drugs, varying in cost, were compared for their effectiveness in preventing nausea and vomiting. (*New England Journal of Medicine*, June 10, 2004.) The medical researchers looked at all possible combinations of the drugs as treatments, including a single drug, as well as 2-drug, 3-drug, 4-drug, 5-drug, and 6-drug combinations.

a. How many 2-drug combinations of the six drugs are possible?

b. How many 3-drug combinations of the six drugs are possible?

c. How many 4-drug combinations of the six drugs are possible?

d. How many 5-drug combinations of the six drugs are possible?

e. The researchers stated that a total of 64 drug combinations were tested as treatments for nausea. Verify that there are 64 ways that the six drugs can be combined. (Remember to include the 1-drug and 6-drug combinations, as well as the control treatment of no drugs.)

3.32 Dominant versus recessive traits. An individual's genetic makeup is determined by the genes obtained from each parent. For every genetic trait, each parent possesses a gene pair, and each contributes one-half of this gene pair, with equal probability, to their offspring, forming a new gene pair. The offspring's traits (eye color, baldness, etc.) come from this new gene pair, where each gene in this pair possesses some characteristic.

For the gene pair that determines eye color, each gene trait may be one of two types: dominant brown (*B*) or recessive blue (*b*). A person possessing the gene pair *BB* or *Bb* has brown eyes, whereas the gene pair *bb* produces blue eyes.

a. Suppose both parents of an individual are brown-eyed, each with a gene pair of the type *Bb*. What is the probability that a randomly selected child of this couple will have blue eyes?

b. If one parent has brown eyes, type *Bb*, and the other has blue eyes, what is the probability that a randomly selected child of this couple will have blue eyes?

c. Suppose one parent is brown-eyed, type *BB*. What is the probability that a child has blue eyes?

3.2 Unions and Intersections

An event can often be viewed as a composition of two or more other events. Such events, which are called **compound events**, can be formed (composed) in two ways, as defined and illustrated here.

> **DEFINITION 3.5**
>
> The **union** of two events A and B is the event that occurs if either A or B or both occur on a single performance of the experiment. We denote the union of events A and B by the symbol $A \cup B$. $A \cup B$ consists of all the sample points that belong to A or B or both. (See Figure 3.6a.)

> **DEFINITION 3.6**
>
> The **intersection** of two events A and B is the event that occurs if both A and B occur on a single performance of the experiment. We write $A \cap B$ for the intersection of A and B. $A \cap B$ consists of all the sample points belonging to *both A and B*. (See Figure 3.6b.)

Figure 3.6
Venn Diagrams for Union
and Intersection

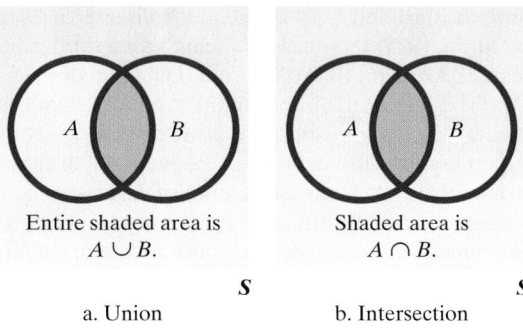

EXAMPLE 3.8   UNION AND INTERSECTION OF TWO EVENTS

Problem   Consider the die-toss experiment. Define the following events:

A: {Toss an even number}
B: {Toss a number less than or equal to 3}

  a. Describe $A \cup B$ for this experiment.
  b. Describe $A \cap B$ for this experiment.
  c. Calculate $P(A \cup B)$ and $P(A \cap B)$ assuming the die is fair.

Solution   Draw the Venn diagram as shown in Figure 3.7.

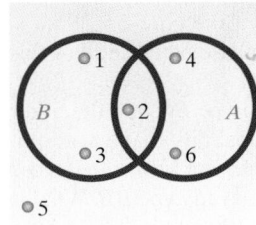

Figure 3.7
Venn Diagram for
Die Toss

  a. The union of A and B is the event that occurs if we observe either an even number, a number less than or equal to 3, or both on a single throw of the die. Consequently, the sample points in the event $A \cup B$ are those for which A occurs, B occurs, or both A and B occur. Checking the sample points in the entire sample space, we find that the collection of sample points in the union of A and B is

$$A \cup B = \{1, 2, 3, 4, 6\}$$

  b. The intersection of A and B is the event that occurs if we observe *both* an even number and a number less than or equal to 3 on a single throw of the die. Checking the sample points to see which imply the occurrence of *both* events A and B, we see that the intersection contains only one sample point:

$$A \cap B = \{2\}$$

In other words, the intersection of A and B is the sample point Observe a 2.

c. Recalling that the probability of an event is the sum of the probabilities of the sample points of which the event is composed, we have

$$P(A \cup B) = P(1) + P(2) + P(3) + P(4) + P(6)$$

$$= \frac{1}{6} + \frac{1}{6} + \frac{1}{6} + \frac{1}{6} + \frac{1}{6} = \frac{5}{6}$$

and

$$P(A \cap B) = P(2) = \frac{1}{6}$$

Look Back Since the 6 sample points are equally likely, then the probabilities in part c are simply the number of sample points in the event of interest divided by 6.

| Now Work | *Exercise 3.43a–d*

— ∎ ∎ ∎ —

Unions and intersections can be defined for more than two events. For example, the event $A \cup B \cup C$ represents the union of three events — A, B, and C. This event, which includes the set of sample points in A, B, or C, will occur if any one or more of the events A, B, or C occurs. Similarly, the intersection $A \cap B \cap C$ is the event that all three of the events A, B, and C occur. Therefore, $A \cap B \cap C$ is the set of sample points that are in all three of the events A, B, and C.

EXAMPLE 3.9 UNION AND INTERSECTION OF THREE EVENTS

Problem Refer to Example 3.6 and define the event

$$C: \{\text{Toss a number greater than 1}\}$$

Find the sample points in

a. $A \cup B \cup C$
b. $A \cap B \cap C$

where

$$A: \{\text{Toss an even number}\}$$
$$B: \{\text{Toss a number less than or equal to 3}\}$$

Solution a. Event C contains the sample points corresponding to tossing a 2, 3, 4, 5, or 6; event A contains the sample points 2, 4, and 6; and event B contains the sample points 1, 2, and 3. Therefore, the event that either A, B, or C occurs contains all six sample points in S — that is, those corresponding to tossing a 1, 2, 3, 4, 5, or 6.

b. You will observe all of the events A, B, and C only if you observe a 2. Therefore, the intersection $A \cap B \cap C$ contains the single sample point Toss a 2.

Look Back Again, the probability of the two events can be found by dividing the number of sample points in the event by 6.

| Now Work | *Exercise 3.46a*

— ∎ ∎ ∎ —

EXAMPLE 3.10 — FINDING PROBABILITIES FROM A TWO-WAY TABLE

Problem   *Family Planning Perspectives* reported on a study of over 200,000 births in New Jersey over a recent two-year period. The study investigated the link between the mother's race and the age at which she gave birth (called maternal age). The percentages of the total number of births in New Jersey are given by the maternal age and race classifications in Table 3.4.

This table is called a **two-way table** since responses are classified according to two variables, maternal age (rows) and race (columns).

TABLE 3.4   Percentage of New Jersey Birth
Mothers in Age–Race Classes

	Race	
Maternal Age (years)	White	Black
≤17	2%	2%
18–19	3%	2%
20–29	41%	12%
≥30	33%	5%

Source: Reichman, N. E., and Pagnini, D. L. "Maternal age and birth outcomes: Data from New Jersey." *Family Planning Perspectives*, Vol. 29, No. 6, Nov./Dec. 1997, p. 269 (adapted from Table 1).

Define the following events:

A: {A New Jersey birth mother is white}
B: {A New Jersey mother was a teenager when giving birth}

a. Find $P(A)$ and $P(B)$.
b. Find $P(A \cup B)$.
c. Find $P(A \cap B)$.

Solution   Following the steps for calculating probabilities of events, we first note that the objective is to characterize the race and maternal age distribution of New Jersey birth mothers. To accomplish this, we define the experiment to consist of selecting a birth mother from the collection of all New Jersey birth mothers during the two-year period and observing her race and maternal age class. The sample points are the eight different age-race classifications:

E_1: {≤17 yrs., white} E_5: {≤17 yrs., black}
E_2: {18–19 yrs., white} E_6: {18–19 yrs., black}
E_3: {20–29 yrs., white} E_7: {20–29 yrs., black}
E_4: {≥30 yrs., white} E_8: {≥30 yrs., black}

Next, we assign probabilities to the sample points. If we blindly select one of the birth mothers, the probability that she will occupy a particular age–race classification is just the proportion, or relative frequency, of birth mothers in the classification. These proportions (as percentages) are given in Table 3.4. Thus,

$P(E_1)$ = Relative frequency of birth mothers in age–race class {≤17 yrs., white} = .02
$P(E_2)$ = .03
$P(E_3)$ = .41
$P(E_4)$ = .33

$$P(E_5) = .02$$
$$P(E_6) = .02$$
$$P(E_7) = .12$$
$$P(E_8) = .05$$

You may verify that the sample point probabilities add to 1.

a. To find $P(A)$, we first determine the collection of sample points contained in event A. Since A is defined as {white}, we see from Table 3.4 that A contains the four sample points represented by the first column of the table. In words, the event A consists of the race classification {white} in all four age classifications. The probability of A is the sum of the probabilities of the sample points in A:

$$P(A) = P(E_1) + P(E_2) + P(E_3) + P(E_4) = .02 + .03 + .41 + .33 = .79$$

Similarly, $B = $ {teenage mother, age ≤ 19 years} consists of the four sample points in the first and second rows of Table 3.4:

$$P(B) = P(E_1) + P(E_2) + P(E_5) + P(E_6) = .02 + .03 + .02 + .02 = .09$$

b. The union of events A and B, $A \cup B$, consists of all sample points in *either A or B or both*. That is, the union of A and B consists of all birth mothers who are white *or* who gave birth as a teenager. In Table 3.4 this is any sample point found in the first column *or* the first two rows. Thus,

$$P(A \cup B) = .02 + .03 + .41 + .33 + .02 + .02 = .83$$

c. The intersection of events A and B, $A \cap B$, consists of all sample points in *both A and B*. That is, the intersection of A and B consists of all birth mothers who are white *and* who gave birth as a teenager. In Table 3.4 this is any sample point found in the first column *and* the first two rows. Thus,

$$P(A \cap B) = .02 + .03 = .05.$$

Look Back As with previous problems, the key to finding the probabilities of parts **b** and **c** is to identify the sample points that comprise the event of interest. In a two-way table like Table 3.4, the total number of sample points will be equal to the number of rows times the number of columns.

Now Work *Exercise 3.45f–g*

■ ■ ■

3.3 Complementary Events

A very useful concept in the calculation of event probabilities is the notion of **complementary events**:

> **DEFINITION 3.7**
>
> The **complement** of an event A is the event that A does *not* occur — that is, the event consisting of all sample points that are not in event A. We denote the complement of A by A^c.

An event A is a collection of sample points, and the sample points included in A^c are those not in A. Figure 3.8 demonstrates this idea. Note from the figure that all sample points in S are included in *either* A or A^c and that *no* sample point is in both A

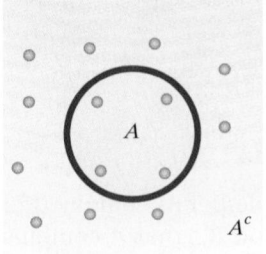

Figure 3.8
Venn Diagram of
Complementary Events

and A^c. This leads us to conclude that the probabilities of an event and its complement *must sum to 1*:

> **Rule of Complements**
>
> The sum of the probabilities of complementary events equals 1; that is,
> $P(A) + P(A^c) = 1$.

In many probability problems calculating the probability of the complement
of the event of interest is easier than calculating the event itself. Then, because

$$P(A) + P(A^c) = 1$$

we can calculate $P(A)$ by using the relationship

$$P(A) = 1 - P(A^c).$$

EXAMPLE 3.11   FINDING THE PROBABILITY OF A COMPLEMENTARY EVENT

Problem   Consider the experiment of tossing two fair coins. (The sample space for this experiment is shown in Figure 3.9.) Calculate the probability of event A: {Observing
at least one head}.

Solution   We know that the event A: {Observing at least one head} consists of the sample
points

$$A: \{HH, HT, TH\}$$

The complement of A is defined as the event that occurs when A does not occur.
Therefore,

$$A^c: \{\text{Observe no heads}\} = \{TT\}$$

This complementary relationship is shown in Figure 3.9. Assuming the coins are
balanced,

$$P(A^c) = P(TT) = \frac{1}{4}$$

and

$$P(A) = 1 - P(A^c) = 1 - \frac{1}{4} = \frac{3}{4}.$$

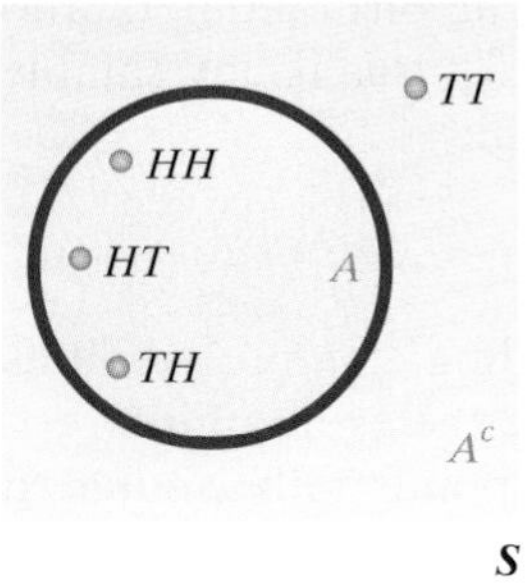

Figure 3.9
Complementary Events
in the Toss of Two Coins

Look Back   Note that we can find $P(A)$ by summing the probabilities of the sample
points, HH, HT, and TH, in A. Many times, it is easier to find $P(A^c)$ using the Rule
of Complements.

| Now Work | *Exercise 3.43e–f* |

■ ■ ■

EXAMPLE 3.12   APPLYING THE RULE OF COMPLEMENTS

Problem   A fair coin is tossed 10 times, and the up face is recorded after each toss. What is the
probability of event A: {Observe at least one head}?

Solution   We solve this problem by following the five steps for calculating probabilities of
events (see Section 3.1).

Step 1 Define the experiment. The experiment is to record the results of the 10 tosses of the coin.

Step 2 List the sample points. A sample point consists of a particular sequence of ten heads and tails. Thus, one sample point is *HHTTTHTHTT*, which denotes head on first toss, head on second toss, tail on third toss, etc. Others are *HTHHHTTTTT* and *THHTHTHTTH*. Obviously, the number of sample points is very large — too many to list. It can be shown (see Section 3.8) that there are $2^{10} = 1{,}024$ sample points for this experiment.

Step 3 Assign probabilities. Since the coin is fair, each sequence of heads and tails has the same chance of occurring, and therefore all the sample points are equally likely. Then

$$P(\text{Each sample point}) = \frac{1}{1{,}024}$$

Step 4 Determine the sample points in event *A*. A sample point is in *A* if at least one *H* appears in the sequence of 10 tosses. However, if we consider the complement of *A*, we find that

$$A^c = \{\text{No heads are observed in 10 tosses}\}$$

Thus, A^c contains only one sample point:

$$A^c: \quad \{TTTTTTTTTT\}$$

and $P(A^c) = \dfrac{1}{1{,}024}$

Step 5 Now we use the relationship of complementary events to find $P(A)$:

$$P(A) = 1 - P(A^c) = 1 - \frac{1}{1{,}024} = \frac{1{,}023}{1{,}024} = .999$$

Look Back With such a high probability, we are virtually certain of observing at least one head in 10 tosses of the coin.

■ ■ ■

3.4 The Additive Rule and Mutually Exclusive Events

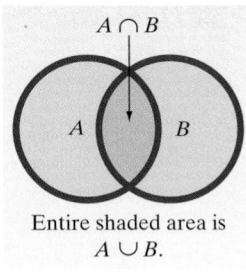

Entire shaded area is $A \cup B$.

Figure 3.10
Venn Diagram of Union

In Section 3.2 we saw how to determine which sample points are contained in a union and how to calculate the probability of the union by adding the probabilities of the sample points in the union. It is also possible to obtain the probability of the union of two events by using the *additive rule of probability*.

By studying the Venn diagram in Figure 3.10, you can see that the probability of the union of two events, *A* and *B*, can be obtained by summing $P(A)$ and $P(B)$ and subtracting $P(A \cap B)$. We must subtract $P(A \cap B)$ because the sample point probabilities in $A \cap B$ have been included twice — once in $P(A)$ and once in $P(B)$.

The formula for calculating the probability of the union of two events is given in the next box.

Additive Rule of Probability

The probability of the union of events *A* and *B* is the sum of the probability of events *A* and *B* minus the probability of the intersection of events *A* and *B*, that is

$$P(A \cup B) = P(A) + P(B) - P(A \cap B)$$

EXAMPLE 3.13   APPLYING THE ADDITIVE RULE

Problem   Hospital records show that 12% of all patients are admitted for surgical treatment, 16% are admitted for obstetrics, and 2% receive both obstetrics and surgical treatment. If a new patient is admitted to the hospital, what is the probability that the patient will be admitted either for surgery, obstetrics, or both?

Solution   Consider the following events:

> A: {A patient admitted to the hospital receives surgical treatment}
> B: {A patient admitted to the hospital receives obstetrics treatment}

Then, from the given information

$$P(A) = .12$$
$$P(B) = .16$$

and the probability of the event that a patient receives both obstetrics and surgical treatment is

$$P(A \cap B) = .02$$

The event that a patient admitted to the hospital receives either surgical treatment, obstetrics treatment, or both is the union $A \cup B$. The probability of $A \cup B$ is given by the additive rule of probability:

$$P(A \cup B) = P(A) + P(B) - P(A \cap B)$$
$$= .12 + .16 - .02 = .26$$

Thus, 26% of all patients admitted to the hospital receive either surgical treatment, obstetrics treatment, or both.

Look Back   From the information given, it is not possible to list and assign probabilities to all the sample points. Consequently, we cannot proceed through the 5-step process (p. 127) for finding $P(A \cup B)$ and must use the Additive Rule.

> **Now Work**   *Exercise 3.41*

--- ■ ■ ■ ---

A very special relationship exists between events A and B when $A \cap B$ contains no sample points. In this case we call the events A and B *mutually exclusive events*.

DEFINITION 3.8

Events A and B are **mutually exclusive** if $A \cap B$ contains no sample points, that is, if A and B have no sample points in common. For mutually exclusive events,

$$P(A \cap B) = 0$$

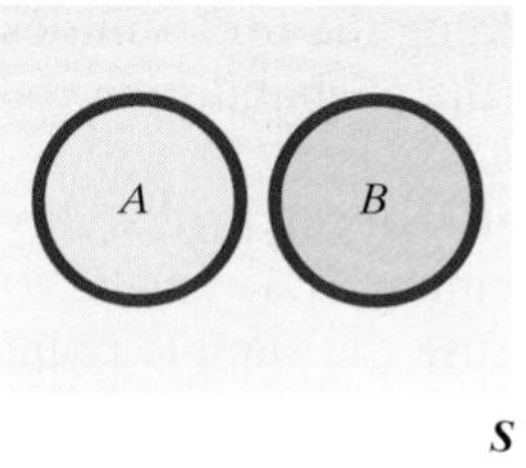

Figure 3.11
Venn Diagram of
Mutually Exclusive
Events

Figure 3.11 shows a Venn diagram of two mutually exclusive events. The events A and B have no sample points in common, that is, A and B cannot occur simultaneously and $P(A \cap B) = 0$. Thus, we have the important relationship given in the box.

Probability of Union of Two Mutually Exclusive Events

If two events A and B are *mutually exclusive*, the probability of the union of A and B equals the sum of the probabilities of A and B; that is, $P(A \cup B) = P(A) + P(B)$.

Caution

The formula shown above is *false* if the events are *not* mutually exclusive. In this case (i.e., two nonmutually exclusive events), you must apply the general additive rule of probability.

EXAMPLE 3.14

UNION OF TWO MUTUALLY EXCLUSIVE EVENTS

Problem Consider the experiment of tossing two balanced coins. Find the probability of observing *at least* one head.

Solution Define the events

A: {Observe at least one head}
B: {Observe exactly one head}
C: {Observe exactly two heads}

Note that

$$A = B \cup C$$

and that $B \cap C$ contains no sample points (see Figure 3.12). Thus, B and C are mutually exclusive, so that

$$P(A) = P(B \cup C) = P(B) + P(C) = \frac{1}{2} + \frac{1}{4} = \frac{3}{4}$$

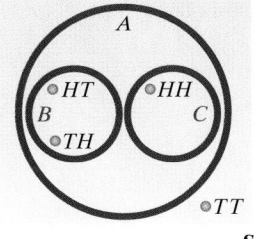

Figure 3.12
Venn Diagram for Coin
Toss Experiment

Look Back Although this example is very simple, it shows us that writing events with verbal descriptions that include the phrases "at least" or "at most" as unions of mutually exclusive events is very useful. This practice enables us to find the probability of the event by adding the probabilities of the mutually exclusive events.

■ ■ ■

Statistics in Action Revisited

The Probability of Winning Lotto with a Wheel System

Refer to Florida's Pick-6 Lotto game in which you select six numbers of your choice from a field of numbers ranging from 1 to 53. In Section 3.2, we learned that the probability of winning Lotto on a single ticket is only 1 in approximately 23 million. The "experts" at Lotto Buster recommend many strategies for increasing the odds of winning the lottery. One strategy is to employ a *wheeling system*. In a complete wheeling system, you select more than six numbers, say, seven, and play every combination of six of those seven numbers.

Suppose you choose to "wheel" the following seven numbers: 2, 7, 18, 23, 30, 32, and 51. Every combination of six of these seven numbers is listed in Table SIA3.1. You can see that there are 7 different possibilities. (Use the Combinations Rule with $N = 7$ and $n = 6$ to verify this.) Thus, we would purchase seven tickets (at a cost of $7) corresponding to these different combinations in a complete wheeling system.

To determine if this strategy does, in fact, increase our odds of winning, we need to find the probability that one of these seven combinations occurs during the 6/53 Lotto

TABLE SIA3.1   Wheeling the Six Numbers 2, 7, 18, 23, 30, 32, and 51

Ticket #1	2	7	18	23	30	32
Ticket #2	2	7	18	23	30	51
Ticket #3	2	7	18	23	32	51
Ticket #4	2	7	18	30	32	51
Ticket #5	2	7	23	30	32	51
Ticket #6	2	18	23	30	32	51
Ticket #7	7	18	23	30	32	51

draw. That is, we need to find the probability that either Ticket #1 or Ticket #2 or Ticket #3 or Ticket #4 or Ticket #5 or Ticket #6 or Ticket #7 is the winning combination. Note that this probability is stated using the word *or*, implying a union of seven events. Letting T_1 represent the event that Ticket #1 wins, and defining $T_2, T_3, \ldots, T_7$ in a similar fashion, we want to find

$$P(T_1 \text{ or } T_2 \text{ or } T_3 \text{ or } T_4 \text{ or } T_5 \text{ or } T_6 \text{ or } T_7)$$

Recall (Section 3.2) that the 22,957,480 possible combinations in Pick-6 Lotto are mutually exclusive and equally likely to occur. Consequently, the probability of the union of the seven events is simply the sum of the probabilities of the individual events, where each event has probability of $1/(22{,}957{,}480)$:

$P(\text{win Lotto with 7 Wheeled Numbers})$
$$= P(T_1 \text{ or } T_2 \text{ or } T_3 \text{ or } T_4 \text{ or } T_5 \text{ or } T_6 \text{ or } T_7)$$
$$= 7/(22{,}957{,}480) = .0000003$$

In terms of odds, we now have 3 chances in 10 million of winning the Lotto with the complete wheeling system. The "experts" are correct—our odds of winning Lotto have increased (from 1 in 23 million). However, the probability of winning is so close to 0 we question whether the $7 spent on lottery tickets is worth the negligible increase in odds. In fact, it can be shown that to increase your chance of winning the 6/53 Lotto to 1 chance in 100 (i.e., .01) using a complete wheeling system, you would have to wheel 26 of your favorite numbers—a total of 230,230 combinations at a cost of $230,230!

Exercises 3.33–3.56

Understanding the Principles

3.33 Define the union of two events, in words.

3.34 Define the intersection of two events, in words.

3.35 Define the complement of an event, in words.

3.36 State the Rule of Complements.

3.37 State the Additive Rule of Probability for any two events.

3.38 Define mutually exclusive events, in words.

3.39 State the Additive Rule of Probability for mutually exclusive events.

Learning the Mechanics

3.40 Suppose $P(A) = .4$, $P(B) = .7$, and $P(A \cap B) = .3$. Find the following probabilities:
  a. $P(B^c)$ **b.** $P(A^c)$ **c.** $P(A \cup B)$

3.41 A fair coin is tossed three times and the events A and B are defined as follows:

  A: {At least one head is observed}
  B: {The number of heads observed is odd}

  a. Identify the sample points in the events A, B, $A \cup B$, A^c, and $A \cap B$.
  b. Find $P(A)$, $P(B)$, $P(A \cup B)$, $P(A^c)$, and $P(A \cap B)$ by summing the probabilities of the appropriate sample points.
  c. Find $P(A \cup B)$ using the additive rule. Compare your answer to the one you obtained in part **b**.
  d. Are the events A and B mutually exclusive? Why?

3.42 A pair of fair dice is tossed. Define the following events:

  A: {You will roll a 7 (i.e., the sum of the numbers of dots on the upper faces of the two dice is equal to 7)}
  B: {At least one of the two dice is showing a 4}

  a. Identify the sample points in the events A, B, $A \cap B$, $A \cup B$, and A^c.
  b. Find $P(A)$, $P(B)$, $P(A \cap B)$, $P(A \cup B)$, and $P(A^c)$ by summing the probabilities of the appropriate sample points.
  c. Find $P(A \cup B)$ using the additive rule. Compare your answer to that for the same event in part **b**.
  d. Are A and B mutually exclusive? Why?

3.43 Consider the Venn diagram, where $P(E_1) = P(E_2) =$ **NW** $P(E_3) = \frac{1}{5}$, $P(E_4) = P(E_5) = \frac{1}{20}$, $P(E_6) = \frac{1}{10}$, and $P(E_7) = \frac{1}{5}$. Find each of the following probabilities.

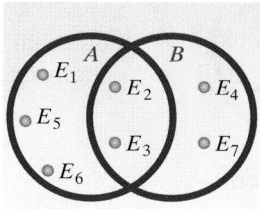

  a. $P(A)$ **b.** $P(B)$ **c.** $P(A \cup B)$
  d. $P(A \cap B)$ **e.** $P(A^c)$ **f.** $P(B^c)$
  g. $P(A \cup A^c)$ **h.** $P(A^c \cap B)$

3.44 Consider the accompanying Venn diagram, where $P(E_1) = .10$, $P(E_2) = .05$, $P(E_3) = P(E_4) = .2$, $P(E_5) = .06$, $P(E_6) = .3$, $P(E_7) = .06$, and $P(E_8) = .03$. Find the following probabilities:

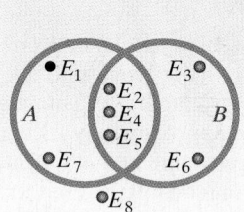

  a. $P(A^c)$ **b.** $P(B^c)$ **c.** $P(A^c \cap B)$
  d. $P(A \cup B)$ **e.** $P(A \cap B)$
  f. $P(A^c \cup B^c)$
  g. Are events A and B mutually exclusive? Why?

3.45 The outcomes of two variables are (Low, Medium, **NW** High) and (On, Off), respectively. An experiment is conducted in which the outcomes of each of the two variables are observed. The probabilities associated with each of the six possible outcome pairs are given in the accompanying table.

	Low	Medium	High
On	.50	.10	.05
Off	.25	.07	.03

Consider the following events:

A: {On}
B: {Medium or On}
C: {Off and Low}
D: {High}

a. Find $P(A)$.   **b.** Find $P(B)$.
c. Find $P(C)$.   **d.** Find $P(D)$.
e. Find $P(A^c)$.   **f.** Find $P(A \cup B)$.
g. Find $P(A \cap B)$.
h. Consider each possible pair of events A, B, C, and D. List the pairs of events that are mutually exclusive. Justify your choices.

3.46 Three fair coins are tossed. We wish to find the probability of the event A: {Observe at least one head}.
NW
    a. Express A as the union of three mutually exclusive events. Find the probability of A using this expression.
    b. Express A as the complement of an event. Find the probability of A using this expression.

Applying the Concepts—Basic

3.47 Mathematics achievement test. According to the National Center for Education Statistics (2002), only 5% of United States eighth graders score above 655 on a mathematics assessment test. Make a probability statement about the event that a randomly selected eighth grader has a score of 655 or below on the mathematics assessment test.

3.48 Binge alcohol drinking. A study of binge alcohol drinking by college students was published in the *American Journal of Public Health* (July 1995). Suppose an experiment consists of randomly selecting one of the undergraduate students who participated in the study. Consider the following events:

A: {The student is a binge drinker}
B: {The student is a male}
C: {The student lives in a coed dorm}

Describe each of the following events in terms of unions, intersections, and complements ($A \cup B$, $A \cap B$, A^c, etc.):
a. The student is male and a binge drinker.
b. The student is not a binge drinker.
c. The student is male or lives in a coed dorm.
d. The student is female and not a binge drinker.

3.49 Odds of winning roulette. *Roulette* is a very popular game in many American casinos. In Roulette, a ball spins on a circular wheel that is divided into 38 arcs of equal length, bearing the numbers $00, 0, 1, 2, \ldots, 35, 36$. The number of the arc on which the ball stops is the outcome of one play of the game. The numbers are also colored in the manner shown in the table.

Red: 1, 3, 5, 7, 9, 12, 14, 16, 18, 19, 21, 23, 25, 27, 30, 32, 34, 36
Black: 2, 4, 6, 8, 10, 11, 13, 15, 17, 20, 22, 24, 26, 28, 29, 31, 33, 35
Green: 00, 0

Players may place bets on the table in a variety of ways, including bets on odd, even, red, black, high, low, etc. Consider the following events:

A: {Outcome is an odd number (00 and 0 are considered neither odd nor even)}
B: {Outcome is a black number}
C: {Outcome is a low number (1–18)}

a. Define the event $A \cap B$ as a specific set of sample points.
b. Define the event $A \cup B$ as a specific set of sample points.
c. Find $P(A)$, $P(B)$, $P(A \cap B)$, $P(A \cup B)$, and $P(C)$ by summing the probabilities of the appropriate sample points.
d. Define the event $A \cap B \cap C$ as a specific set of sample points.
e. Find $P(A \cup B)$ using the additive rule. Are events A and B mutually exclusive? Why?
f. Find $P(A \cap B \cap C)$ by summing the probabilities of the sample points given in part **d**.
g. Define the event $(A \cup B \cup C)$ as a specific set of sample points.
h. Find $P(A \cup B \cup C)$ by summing the probabilities of the sample points given in part **g**.

3.50 Abortion Provider Survey. The Alan Guttmacher Institute (AGI) Abortion Provider Survey is a survey of all 258 known nonhospital abortion providers in the United States (*Perspectives on Sexual and Reproductive Health*, Jan./Feb. 2003.) For one part of the survey, the 358 providers were classified according to caseload (number of abortions performed per year) and whether they permit their patient to take the abortion drug, misoprostol, at home or return to the abortion facility to receive the drug. The responses are summarized in the accompanying table. Suppose we select, at random, one of the 358 providers and observe the provider's caseload (less than 50, or, 50 or more) and home use of the drug (yes or no).

	Number of Abortions		
Permit Drug at Home	Less than 50	50 or More	Totals
Yes	170	130	300
No	48	10	58
Totals	218	140	358

Source: Henshaw, S. K., and Finer, L. B. "The accessibility of abortion services in the United States, 2001," *Perspectives on Sexual and Reproductive Health*, Vol. 35, No. 1, Jan./Feb. 2003 (Table 4).

a. List all the possible outcomes for this provider.
b. Based on the table, assign reasonable probabilities to the outcomes, part **a**.
c. Find the probability that the provider permits home use of the abortion drug.

d. Find the probability that the provider permits home use of the drug or has a caseload of less than 50 abortions.

e. Find the probability that the provider permits home use of the drug and has a caseload of less than 50 abortions.

Applying the Concepts—Intermediate

3.51 Federal civil trial appeals. The *Journal of the American Law and Economics Association* (Vol. 3, 2001) published the results of a study of appeals of Federal civil trials. The following table, extracted from the article, gives a break-down of 2,143 civil cases that were appealed by either the plaintiff or defendant. The outcome of the appeal, as well as the type of trial (judge or jury), was determined for each civil case. Suppose one of the 2,143 cases is selected at random and both the outcome of the appeal and type of trial are observed.

	Jury	Judge	Totals
Plaintiff trial win — reversed	194	71	265
Plaintiff trial win — affirmed/dismissed	429	240	669
Defendant trial win — reversed	111	68	179
Defendant trial win — affirmed/dismissed	731	299	1,030
Totals	1,465	678	2,143

a. List the sample points for this experiment.
b. Find $P(A)$, where $A = \{$jury trial$\}$.
c. Find $P(B)$, where $B = \{$plaintiff trial win is reversed$\}$.
d. Are A and B mutually exclusive events?
e. Find $P(A^c)$.
f. Find $P(A \cup B)$.
g. Find $P(A \cap B)$.

3.52 Stacking in the NBA. In professional sports, *stacking* is a term used to describe the practice of African-American players being excluded from certain positions because of race. To illustrate the stacking phenomenon, the *Sociology of Sport Journal* (Vol. 14, 1997) presented the table shown here. The table summarizes the race and positions of 368 National Basketball Association (NBA) players in 1993. Suppose an NBA player is selected at random from that year's player pool.

	POSITION			
	Guard	Forward	Center	Totals
White	26	30	28	84
Black	128	122	34	284
Totals	154	152	62	368

a. What is the probability that the player is white?
b. What is the probability that the player is a center?
c. What is the probability that the player is African-American and plays guard?

d. What is the probability that the player is not a guard?
e. What is the probability that the player is white or a center?

3.53 Elderly wheelchair user study. The *American Journal of Public Health* (Jan. 2002) reported on a study of elderly wheelchair users who live at home. A sample of 306 wheelchair users, age 65 or older, were surveyed about whether they had an injurious fall during the year and whether their home features any one of five struc-tural modifications: bathroom modifications, widened doorways/hallways, kitchen modifications, installed rail-ings, and easy-open doors. The responses are summa-rized in the accompanying table. Suppose we select, at random, one of the 306 surveyed wheelchair users.

Home Features	Injurious Fall(s)	No Falls	Totals
All 5	2	7	9
At least 1, but not all	26	162	188
None	20	89	109
Totals	48	258	306

Source: Berg, K., Hines, M., and Allen, S. "Wheelchair users at home: Few home modifications and many injurious falls," *American Journal of Public Health,* Vol. 92, No. 1, Jan. 2002 (Table 1).

a. Find the probability that the wheelchair user had an injurious fall.
b. Find the probability that the wheelchair user had all five features installed in the home.
c. Find the probability that the wheelchair user had no falls and none of the features installed in the home.

3.54 Gang research study. The National Gang Crime Re-search Center (NGCRC) has developed a six-level gang classification system for both adults and juveniles. The NGCRC collected data on approximately 7,500 confined offenders and assigned each a score using the gang classification system. (*Journal of Gang Research,* Winter 1997.) One of several other variables measured by the NGCRC was whether or not the offender has ever carried a homemade weapon (e.g., knife) while in custody. The table on p. 145 gives the number of con-fined offenders in each of the gang score and home-made weapon categories. Assume one of the confined offenders is randomly selected.
a. Find the probability that the offender has a gang score of 5.
b. Find the probability that the offender has carried a homemade weapon.
c. Find the probability that the offender has a gang score below 3.
d. Find the probability that the offender has a gang score of 5 and has carried a homemade weapon.
e. Find the probability that the offender has a gang score of 0 or has never carried a homemade weapon.
f. Are the events described in parts **a** and **b** mutually exclusive? Explain.
g. Are the events described in parts **a** and **c** mutually exclusive? Explain.

Applying the Concepts—Advanced

3.55 Buying men's dress pants. A buyer for a large metropolitan department store must choose two firms from the four available to supply the store's fall line of men's dress pants. The buyer has not dealt with any of the four firms before and considers their products equally attractive. Unknown to the buyer, two of the four firms are having serious financial problems that may result in their not being able to deliver the fall line of slacks as soon as promised. The four firms are identified as G_1 and G_2 (firms in good financial condition) and P_1 and P_2 (firms in poor financial condition). If the buyer selects the firms at random, find

a. The probability that firm P_1 is selected.
b. The probability that at least one of the selected firms is in good financial condition.

3.56 Galileo's Passedix game. Passedix is a game of chance played with three fair dice. Players bet whether the sum of the faces shown on the dice will be above or below 10. During the late sixteenth century, the astronomer and mathematician Galileo Galilei was asked by the Grand Duke of Tuscany to explain why "the chance of throwing a total 9 with three fair dice was less than that of throwing a total of 10." (*Interstat*, Jan. 2004.) The Grand Duke believed that the chance should be the same since "there are an equal number of partitions of the numbers 9 and 10." Find the flaw in the Grand Duke's reasoning and answer the question posed to Galileo.

Gang Classification Score	Weapon		Totals
	Yes	No	
0 (Never joined a gang, no close friends in a gang)	255	2,551	2,806
1 (Never joined a gang, 1–4 close friends in a gang)	110	560	670
2 (Never joined a gang, 5 or more friends in a gang)	151	636	787
3 (Inactive gang member)	271	959	1,230
4 (Active gang member, no position of rank)	175	513	688
5 (Active gang member, holds position of rank)	476	831	1,307
Totals	**1,438**	**6,050**	**7,488**

Source: Knox, G. W. et al. "A gang classification system for corrections," *Journal of Gang Research*, Vol. 4, No. 2, Winter 1997, p. 54 (Table 4).

3.5 Conditional Probability

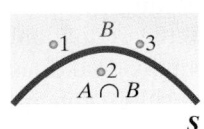

Figure 3.13
Reduced Sample Space for the Die Toss Experiment: Given that Event B Has Occurred

The event probabilities we've been discussing give the relative frequencies of the occurrences of the events when the experiment is repeated a very large number of times. Such probabilities are often called **unconditional probabilities** because no special conditions are assumed, other than those that define the experiment.

Often, however, we have additional knowledge that might affect the outcome of an experiment, so we may need to alter the probability of an event of interest. A probability that reflects such additional knowledge is called the **conditional probability** of the event. For example, we've seen that the probability of observing an even number (event A) on a toss of a fair die is $\frac{1}{2}$. But suppose we're given the information that on a particular throw of the die the result was a number less than or equal to 3 (event B). Would the probability of observing an even number on that throw of the die still be equal to $\frac{1}{2}$? It can't be, because making the assumption that B has occurred reduces the sample space from six sample points to three sample points (namely, those contained in event B). This reduced sample space is as shown in Figure 3.13.

Because the sample points for the die-toss experiment are equally likely, each of the three sample points in the reduced sample space is assigned an equal *conditional probability* of $\frac{1}{3}$. Since the only even number of the three in the reduced sample space B is the number 2 and the die is fair, we conclude that the probability that A occurs *given that B occurs* is $\frac{1}{3}$. We use the symbol $P(A|B)$ to represent the probability of event A given that event B occurs. For the die-toss example

$$P(A|B) = \frac{1}{3}$$

To get the probability of event A given that event B occurs, we proceed as follows. We divide the probability of the part of A that falls within the reduced sample space B, namely $P(A \cap B)$, by the total probability of the reduced sample space, namely, $P(B)$. Thus, for the die-toss example with event A: {Observe an even number} and event B: {Observe a number less than or equal to 3}, we find

$$P(A|B) = \frac{P(A \cap B)}{P(B)} = \frac{P(2)}{P(1) + P(2) + P(3)} = \frac{1/6}{3/6} = \frac{1}{3}$$

The formula for $P(A|B)$ is true in general:

Conditional Probability Formula

To find the *conditional probability that event A occurs given that event B occurs*, divide the probability that *both A and B* occur by the probability that *B* occurs, that is,

$$P(A|B) = \frac{P(A \cap B)}{P(B)} \qquad \text{[We assume that } P(B) \neq 0.\text{]}$$

This formula adjusts the probability of $A \cap B$ from its original value in the complete sample space **S** to a conditional probability in the reduced sample space B. If the sample points in the complete sample space are equally likely, then the formula will assign equal probabilities to the sample points in the reduced sample space, as in the die-toss experiment. If, on the other hand, the sample points have unequal probabilities, the formula will assign conditional probabilities proportional to the probabilities in the complete sample space. This is illustrated by the following examples.

EXAMPLE 3.15   APPLYING THE CONDITIONAL PROBABILITY FORMULA

Problem   Many medical researchers have conducted experiments to examine the relationship between cigarette smoking and cancer. Consider an individual randomly selected from the adult male population. Let A represent the event that the individual smokes, and let A^c denote the complement of A (the event that the individual does not smoke). Similarly, let B represent the event that the individual develops cancer, and let B^c be the complement of that event. Then the four sample points associated with the experiment are shown in Figure 3.14, and their probabilities for a certain section of the United States are given in Table 3.5. Use these sample point probabilities to examine the relationship between smoking and cancer.

Solution   One method of determining whether these probabilities indicate that smoking and cancer are related is to compare the *conditional probability* that an adult male acquires cancer given that he smokes with the conditional probability that an adult male acquires cancer given that he does not smoke [i.e., compare $P(B|A)$ to $P(B|A^c)$].

Figure 3.14
Sample Space for
Example 3.15

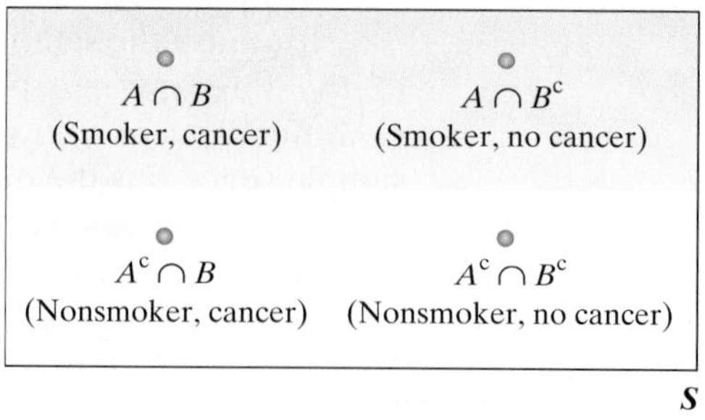

$A \cap B$ (Smoker, cancer)	$A \cap B^c$ (Smoker, no cancer)
$A^c \cap B$ (Nonsmoker, cancer)	$A^c \cap B^c$ (Nonsmoker, no cancer)

S

TABLE 3.5   Probabilities of Smoking and Developing Cancer

Smoker	Develops Cancer	
	Yes, B	No, B^c
Yes, A	.05	.20
No, A^c	.03	.72

First, we consider the reduced sample space A corresponding to adult male smokers. This reduced sample space is highlighted in Figure 3.14. The two sample points $A \cap B$ and $A \cap B^c$ are contained in this reduced sample space, and the adjusted probabilities of these two sample points are the two conditional probabilities:

$$P(B|A) = \frac{P(A \cap B)}{P(A)} \quad \text{and} \quad P(B^c|A) = \frac{P(A \cap B^c)}{P(A)}$$

The probability of event A is the sum of the probabilities of the sample points in A:

$$P(A) = P(A \cap B) + P(A \cap B^c) = .05 + .20 = .25$$

Then the values of the two conditional probabilities in the reduced sample space A are:

$$P(B|A) = \frac{.05}{.25} = .20 \quad \text{and} \quad P(B^c|A) = \frac{.20}{.25} = .80$$

These two numbers represent the probabilities that an adult male smoker develops cancer and does not develop cancer, respectively.

In a like manner, the conditional probabilities of an adult male nonsmoker developing cancer and not developing cancer are

$$P(B|A^c) = \frac{P(A^c \cap B)}{P(A^c)} = \frac{.03}{.75} = .04$$

$$P(B^c|A^c) = \frac{P(A^c \cap B^c)}{P(A^c)} = \frac{.72}{.75} = .96$$

Two of the conditional probabilities give some insight into the relationship between cancer and smoking: the probability of developing cancer given that the adult male is a smoker, and the probability of developing cancer given that the adult male is not a smoker. The conditional probability that an adult male smoker develops cancer (.20) is five times the probability that a nonsmoker develops cancer (.04). This does not imply that smoking *causes* cancer, but it does suggest a pronounced link between smoking and cancer.

Look Back Notice that the conditional probabilities $P(B^c|A) = .80$ and $P(B|A) = .20$ are in the same 4 to 1 ratio as the original unconditional probabilities, .20 and .05. The conditional probability formula simply adjusts the unconditional probabilities so that they add to 1 in the reduced sample space, A, of adult male smokers. Similarly, the conditional probabilities $P(B^c|A^c) = .96$ and $P(B|A^c) = .04$ are in the same 24 to 1 ratio as the unconditional probabilities .72 and .03.

Now Work *Exercise 3.63a-b*

■ ■ ■

EXAMPLE 3.16

FINDING A CONDITIONAL PROBABILITY

Problem The investigation of consumer product complaints by the Federal Trade Commission (FTC) has generated much interest by manufacturers in the quality of their products. A manufacturer of an electromechanical kitchen utensil conducted an analysis of a large number of consumer complaints and found that they fell into the six categories shown in Table 3.6. If a consumer complaint is received, what is the probability that the cause of the complaint was product appearance given that the complaint originated during the guarantee period?

TABLE 3.6   Distribution of Product Complaints

	Reason for Complaint			
	Electrical	Mechanical	Appearance	Totals
During Guarantee Period	18%	13%	32%	63%
After Guarantee Period	12%	22%	3%	37%
Totals	30%	35%	35%	100%

Solution Let A represent the event that the cause of a particular complaint is product appearance, and let B represent the event that the complaint occurred during the guarantee period. Checking Table 3.6, you can see that $(18 + 13 + 32)\% = 63\%$ of the complaints occur during the guarantee period. Hence, $P(B) = .63$. The percentage of complaints that were caused by appearance and occurred during the guarantee period (the event $A \cap B$) is 32%. Therefore, $P(A \cap B) = .32$.

Using these probability values, we can calculate the conditional probability $P(A|B)$ that the cause of a complaint is appearance given that the complaint occurred during the guarantee time:

$$P(A|B) = \frac{P(A \cap B)}{P(B)} = \frac{.32}{.63} = .51$$

Consequently, we can see that slightly more than half the complaints that occurred during the guarantee period were due to scratches, dents, or other imperfections in the surface of the kitchen devices.

Look Back Note that the answer $\frac{.32}{.63}$ is the proportion for the event of interest A (.32) divided by the row total proportion for the given event B (.63). That is, it is the proportion of the time that A occurs within the given event B.

Now Work   *Exercise 3.77*

■ ■ ■

3.6 The Multiplicative Rule and Independent Events

The probability of an intersection of two events can be calculated using the *multiplicative rule*, which employs the conditional probabilities we defined in the previous section. Actually, we've already developed the formula in another context. Recall that the conditional probability of B given A is

$$P(B|A) = \frac{P(A \cap B)}{P(A)}$$

Multiplying both sides of this equation by $P(A)$, we obtain a formula for the probability of the intersection of events A and B. This is often called the **multiplicative rule of probability**.

Multiplicative Rule of Probability

$P(A \cap B) = P(A)P(B|A)$ or, equivalently, $P(A \cap B) = P(B)P(A|B)$

EXAMPLE 3.17

APPLYING THE MULTIPLICATIVE RULE

Problem An agriculturist, who is interested in planting wheat next year, is concerned with the following events:

B: {The production of wheat will be profitable}
A: {A serious drought will occur}

The agriculturist believes that the probability is .01 that production of wheat will be profitable *assuming* a serious drought will occur in the same year and that the probability is .05 that a serious drought will occur. That is,

$$P(B|A) = .01 \quad \text{and} \quad P(A) = .05$$

Based on the information provided, what is the probability that a serious drought will occur *and* that a profit will be made? That is, find $P(A \cap B)$, the probability of the intersection of events A and B.

Solution We want to calculate $P(A \cap B)$. Using the formula for the multiplicative rule, we obtain

$$P(A \cap B) = P(A)P(B|A) = (.05)(.01) = .0005$$

The probability that a serious drought occurs *and* the production of wheat is profitable is only .0005.

Look Back As we might expect, this intersection is a very rare event.

Now Work   *Exercise 3.72*

━━━━━━━━ ■ ■ ■ ━━━━━━━━

Intersections often contain only a few sample points. In this case, the probability of an intersection is easy to calculate by summing the appropriate sample point probabilities. However, the formula for calculating intersection probabilities is invaluable when the intersection contains numerous sample points, as the next example illustrates.

EXAMPLE 3.18

APPLYING THE MULTIPLICATIVE RULE

Problem A county welfare agency employs 10 welfare workers who interview prospective food stamp recipients. Periodically the supervisor selects, at random, the forms completed by two workers to audit for illegal deductions. Unknown to the supervisor, three of the workers have regularly been giving illegal deductions to applicants. What is the probability that both of the two workers chosen have been giving illegal deductions?

Solution Define the following two events:

A: {First worker selected gives illegal deductions}
B: {Second worker selected gives illegal deductions}

We want to find the probability of the event that both selected workers have been giving illegal deductions. This event can be restated as: {First worker gives illegal deductions *and* second worker gives illegal deductions}. Thus, we want to find the probability of the intersection, $A \cap B$. Applying the multiplicative rule, we have

$$P(A \cap B) = P(A)P(B|A)$$

To find $P(A)$ it is helpful to consider the experiment as selecting one worker from the 10. Then the sample space for the experiment contains 10 sample points (representing the 10 welfare workers), where the three workers giving illegal deductions are denoted by the symbol I (I_1, I_2, I_3), and the seven workers not giving illegal deductions are denoted by the symbol N ($N_1, \ldots, N_7$). The resulting Venn diagram is illustrated in Figure 3.15.

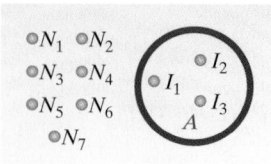

Figure 3.15

Venn Diagram for Finding $P(A)$

Since the first worker is selected at random from the 10, it is reasonable to assign equal probabilities to the 10 sample points. Thus, each sample point has a probability of $^1/_{10}$. The sample points in event A are $\{I_1, I_2, I_3\}$ — the three workers who are giving illegal deductions. Thus,

$$P(A) = P(I_1) + P(I_2) + P(I_3) = \frac{1}{10} + \frac{1}{10} + \frac{1}{10} = \frac{3}{10}$$

To find the conditional probability, $P(B|A)$, we need to alter the sample space $\mathbf{S}$. Since we know A has occurred (i.e., the first worker selected is giving illegal deductions), only two of the nine remaining workers in the sample space are giving illegal deductions. The Venn diagram for this new sample space is shown in Figure 3.16.

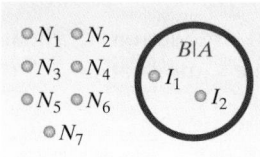

Figure 3.16

Venn Diagram for Finding $P(B|A)$

Each of these nine sample points is equally likely, so each is assigned a probability of $^1/_9$. Since the event $B|A$ contains the sample points $\{I_1, I_2\}$, we have

$$P(B|A) = P(I_1) + P(I_2) = \frac{1}{9} + \frac{1}{9} = \frac{2}{9}$$

Substituting $P(A) = ^3/_{10}$ and $P(B|A) = ^2/_9$ into the formula for the multiplicative rule, we find

$$P(A \cap B) = P(A)P(B|A) = \left(\frac{3}{10}\right)\left(\frac{2}{9}\right) = \frac{6}{90} = \frac{1}{15}$$

Thus, there is a 1 in 15 chance that both workers chosen by the supervisor have been giving illegal deductions to food stamp recipients.

Look Back The key words *both* and *and* in the statement "both A and B occur" imply an intersection of two events. This, in turn, implies that we should *multiply* probabilities to obtain the answer.

Now Work *Exercise 3.67d*

■ ■ ■

The sample space approach is only one way to solve the problem posed in Example 3.18. An alternative method employs the concept of a **tree diagram**. Tree diagrams are helpful for calculating the probability of an intersection.

To illustrate, a tree diagram for Example 3.18 is displayed in Figure 3.17. The tree begins at the far left with two branches. These branches represent the two possible outcomes N (no illegal deductions) and I (illegal deductions) for the first worker selected. The unconditional probability of each outcome is given (in parentheses) on the appropriate branch. That is, for the first worker selected, $P(N) = ^7/_{10}$

Figure 3.17

Tree Diagram for
Example 3.18

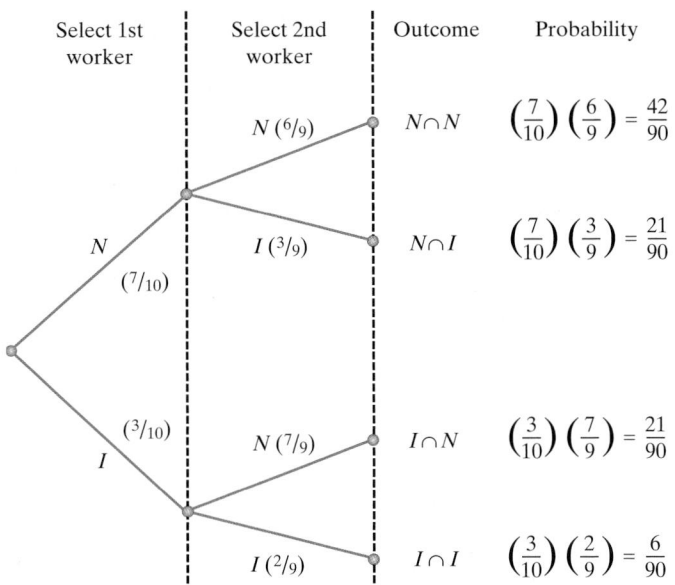

Select 1st worker	Select 2nd worker	Outcome	Probability

$N (6/9)$ $N \cap N$ $\left(\frac{7}{10}\right)\left(\frac{6}{9}\right) = \frac{42}{90}$

$I (3/9)$ $N \cap I$ $\left(\frac{7}{10}\right)\left(\frac{3}{9}\right) = \frac{21}{90}$

$N (7/9)$ $I \cap N$ $\left(\frac{3}{10}\right)\left(\frac{7}{9}\right) = \frac{21}{90}$

$I (2/9)$ $I \cap I$ $\left(\frac{3}{10}\right)\left(\frac{2}{9}\right) = \frac{6}{90}$

and $P(I) = {}^3/_{10}$. (These can be obtained by summing sample point probabilities as in Example 3.18.)

The next level of the tree diagram (moving to the right) represents the outcomes for the second worker selected. The probabilities shown here are conditional probabilities since the outcome for the first worker is assumed to be known. For example, if the first worker is giving illegal deductions (I), the probability that the second worker is also giving illegal deductions (I) is $^2/_9$ since of the nine workers left to be selected, only two remain who are giving illegal deductions. This conditional probability, $^2/_9$, is shown in parentheses on the bottom branch of Figure 3.17.

Finally, the four possible outcomes of the experiment are shown at the end of each of the four tree branches. These events are intersections of two events (outcome of first worker *and* outcome of second worker). Consequently, the multiplicative rule is applied to calculate each probability, as shown in Figure 3.17. You can see that the intersection $\{I \cap I\}$ — the event that both workers selected are giving illegal deductions — has probability $^6/_{90} = {}^1/_{15}$, which is the same value obtained in Example 3.18.

In Section 3.5 we showed that the probability of event A may be substantially altered by the knowledge that an event B has occurred. However, this will not always be the case. In some instances the assumption that event B has occurred will *not* alter the probability of event A at all. When this occurs, we say that the two events A and B are *independent events*.

DEFINITION 3.9

Events A and B are **independent events** if the occurrence of B does not alter the probability that A has occurred; that is, events A and B are independent if

$$P(A|B) = P(A)$$

When events A and B are independent, it is also true that

$$P(B|A) = P(B)$$

Events that are not independent are said to be **dependent**.

EXAMPLE 3.19   CHECKING FOR INDEPENDENCE

Problem   Consider the experiment of tossing a fair die and let

$$A = \{\text{Observe an even number}\}$$
$$B = \{\text{Observe a number less than or equal to 4}\}$$

Are events A and B independent?

Solution   The Venn diagram for this experiment is shown in Figure 3.18. We first calculate

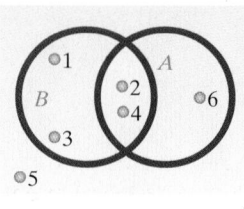

Figure 3.18
Venn Diagram for
Die-Toss Experiment

$$P(A) = P(2) + P(4) + P(6) = \frac{1}{2}$$

$$P(B) = P(1) + P(2) + P(3) + P(4) = \frac{4}{6} = \frac{2}{3}$$

$$P(A \cap B) = P(2) + P(4) = \frac{2}{6} = \frac{1}{3}$$

Now assuming B has occurred, the conditional probability of A given B is

$$P(A|B) = \frac{P(A \cap B)}{P(B)} = \frac{\frac{1}{3}}{\frac{2}{3}} = \frac{1}{2} = P(A)$$

Thus, assuming that event B occurs does not alter the probability of observing an even number — it remains $\frac{1}{2}$. Therefore, the events A and B are independent.

Look Back   Note that if we calculate the conditional probability of B given A, our conclusion is the same:

$$P(B|A) = \frac{P(A \cap B)}{P(A)} = \frac{\frac{1}{3}}{\frac{1}{2}} = \frac{2}{3} = P(B)$$

■ ■ ■

Biography

BLAISE PASCAL (1623–1662)—Solver of the Chevalier's Dilemma

As a precocious child growing up in France, Blaise Pascal showed an early inclination toward mathematics. Although his father would not permit Pascal to study mathematics before the age of 15 (removing all math texts from his house), at age 12 Blaise discovered on his own that the sum of the angles of a triangle are two right angles. Pascal went on to become a distinguished mathematician, as well as a physicist, theologian, and the inventor of the first digital calculator. Most historians attribute the beginning of the study of probability to the correspondence between Pascal and Pierre de Fermat in 1654. The two solved The Chevalier's Dilemma — a gambling problem related to Pascal by his friend and Paris gambler the Chevalier de Mere. The problem involved determining the expected number of times one could roll two dice without throwing a double 6. (Pascal proved that the "break-even" point was 25 rolls.)

EXAMPLE 3.20   CHECKING FOR INDEPENDENCE

Problem   Refer to the consumer product complaint study in Example 3.16. The percentages of complaints of various types during and after the guarantee period are shown in Table 3.6 (p. 148). Define the following events:

> *A*: {Cause of complaint is product appearance}
> *B*: {Complaint occurred during the guarantee term}

Are *A* and *B* independent events?

Solution Events *A* and *B* are independent if $P(A|B) = P(A)$. We calculated $P(A|B)$ in Example 3.16 to be .51, and from Table 3.6 we see that

$$P(A) = .32 + .03 = .35$$

Therefore, $P(A|B)$ is not equal to $P(A)$, and *A* and *B* are dependent events.

| Now Work | *Exercise 3.63c*

■ ■ ■

To gain an intuitive understanding of independence, think of situations in which the occurrence of one event does not alter the probability that a second event will occur. For example, new medical procedures are often tested on laboratory animals. The scientists conducting the tests generally try to perform the procedures on the animals so the results for one animal do not affect the results for the others. That is, the event that the procedure is successful on one animal is *independent* of the result for another. In this way, the scientists can get a more accurate idea of the efficacy of the procedure than if the results were dependent, with the success or failure for one animal affecting the results for other animals.

As a second example, consider an election poll in which 1,000 registered voters are asked their preference between two candidates. Pollsters try to use procedures for selecting a sample of voters so that the responses are independent. That is, the objective of the pollster is to select the sample so the event that one polled voter prefers candidate A does not alter the probability that a second polled voter prefers candidate A.

Now consider the world of sports. Do you think the results of a batter's successive trips to the plate in baseball, or of a basketball player's successive shots at the basket, are independent? If a basketball player makes two successive shots, is the probability of making the next shot altered from its value if the result of the first shot is not known? If a player makes two shots in a row, the probability of a third successful shot is likely to be different from what we would assign if we knew nothing about the first two shots. Why should this be so? Research has shown that many such results in sports tend to be *dependent* because players (and even teams) tend to get on "hot" and "cold" streaks, during which their probabilities of success may increase or decrease significantly.

We will make three final points about independence. The first is that the property of independence, unlike the mutually exclusive property, cannot be shown on or gleaned from a Venn diagram. This means *you can't trust your intuition*. In general, the only way to check for independence is by performing the calculations of the probabilities in the definition.

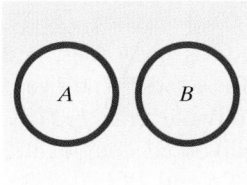

Figure 3.19
Mutually Exclusive Events are Dependent Events

The second point concerns the relationship between the mutually exclusive and independence properties. Suppose that events *A* and *B* are mutually exclusive, as shown in Figure 3.19, and both events have nonzero probabilities. Are these events independent or dependent? That is, does the assumption that *B* occurs alter the probability of the occurrence of *A*? It certainly does, because if we assume that *B* has occurred, it is impossible for *A* to have occurred simultaneously. That is, $P(A|B) = 0$. Thus, *mutually exclusive events are dependent events* since $P(A) \neq P(A|B)$.

The third point is that the probability of the intersection of independent events is very easy to calculate. Referring to the formula for calculating the probability of an intersection, we find

$$P(A \cap B) = P(A)P(B|A)$$

Thus, since $P(B|A) = P(B)$ when A and B are independent, we have the following useful rule:

> ## Probability of Intersection of Two Independent Events
>
> *If events A and B are independent,* the probability of the intersection of A and B equals the product of the probabilities of A and B; that is
>
> $$P(A \cap B) = P(A)P(B)$$
>
> The converse is also true: If $P(A \cap B) = P(A)P(B)$, then events A and B are independent.

In the die-toss experiment, we showed in Example 3.19 that the two events A A: {Observe an even number} and B B: {Observe a number less than equal to 4} are independent if the die is fair. Thus,

$$P(A \cap B) = P(A)P(B) = \left(\frac{1}{2}\right)\left(\frac{2}{3}\right) = \frac{1}{3}$$

This agrees with the result that we obtained in the example:

$$P(A \cap B) = P(2) + P(4) = \frac{2}{6} = \frac{1}{3}$$

EXAMPLE 3.21   FINDING THE PROBABILITY OF INDEPENDENT EVENTS OCCURRING SIMULTANEOUSLY

Problem   Refer to Example 3.5 (p. 127). Recall that the American Association for Marriage and Family Therapy (AAMFT) found that 25% of divorced couples are classified as "fiery foes" (i.e., they communicate through their children and are hostile toward each other).

a. What is the probability that in a sample of two divorced couples, both are classified as "fiery foes?"

b. What is the probability that in a sample of ten divorced couples, all ten are classified as "fiery foes?"

Solution   **a.** Let F_1 represent the event that divorced couple 1 is classified as "fiery foes" and F_2 represent the event that divorced couple 2 are also "fiery foes." The event that *both* couples are "fiery foes" is the intersection of the two events, $F_1 \cap F_2$. Based on the AAMFT survey that found that 25% of divorced couples are "fiery foes," we could reasonably conclude that $P(F_1) = .25$ and $P(F_2) = .25$. However, in order to compute the probability of $F_1 \cap F_2$ from the multiplicative rule, we must make the assumption that the two events are independent. Since the classification of any divorced couple is not likely to affect the classification of another divorced couple, this assumption is reasonable. Assuming independence, we have

$$P(F_1 \cap F_2) = P(F_1)P(F_2) = (.25)(.25) = .0625$$

b. To see how to compute the probability that ten of ten divorced couples will all be classified as "fiery foes," first consider the event that three of three couples are "fiery foes." If F_3 represents the event that the third divorced couple are "fiery foes," then we want to compute the probability of the intersection $F_1 \cap F_2 \cap F_3$. Again assuming independence of the classifications, we have

$$P(F_1 \cap F_2 \cap F_3) = P(F_1)P(F_2)P(F_3) = (.25)(.25)(.25) = .015625$$

Similar reasoning leads us to the conclusion that the intersection of ten such events can be calculated as follows:

$$P(F_1 \cap F_2 \cap F_3 \cap \ldots \cap F_{10}) = P(F_1)P(F_2)\cdots P(F_{10}) = (.25)^{10} = .000001$$

Thus, the probability that ten of ten sampled divorced couples are all classified as "fiery foes" is 1 in 1 million, assuming the probability of each couple being classified as "fiery foes" is .25 and the classification decisions are independent.

Look Back The very small probability in part b makes it extremely unlikely that ten of ten divorced couples are "fiery foes." If this event should actually occur, we should question the probability of .25 provided by the AAMFT and used in the calculation — it is likely to be much higher. (This conclusion is another application of the rare event approach to statistical inference.)

Now Work *Exercise 3.88a*

■ ■ ■

Statistics in Action Revisited

The Probability of Winning Cash 3 or Play 4

In addition to biweekly Lotto 6/53, the Florida Lottery runs several other games. Two popular daily games are Cash 3 and Play 4. In Cash 3, players pay $1 to select three numbers in sequential order, where each number ranges from 0 to 9. If the three numbers selected (e.g., 2-8-4) match exactly the order of the three numbers drawn, the player wins $500. Play 4 is similar to Cash 3, but players must match four numbers (each number ranging from 0 to 9). For a $1 Play 4 ticket (e.g., 3-8-3-0), the player will win $5,000 if the numbers match the order of the four numbers drawn.

During the official drawing for Cash 3, ten ping pong balls numbered 0, 1, 2, 3, 4, 5, 6, 7, 8, and 9 are placed into each of three chambers. The balls in the first chamber are colored pink, the balls in the second chamber are blue, and the balls in the third chamber are yellow. One ball of each color is randomly drawn, with the official order as pink-blue-yellow. In Play 4, a fourth chamber with orange balls is added, and the official order is pink-blue-yellow-orange. Since the draws of the colored balls are random and independent, we can apply an extension of the Probability Rule for the Intersection of Two Independent Events to find the odds of winning Cash 3 and Play 4. The probability of matching a numbered ball being drawn from a chamber is 1/10; therefore,

$$
\begin{aligned}
P(\text{Win Cash 3}) &= P(\text{match pink AND match blue} \\
&\quad \text{AND match yellow}) \\
&= P(\text{match pink}) \times P(\text{match blue}) \times \\
&\quad P(\text{match yellow}) \\
&= (1/10)(1/10)(1/10) = 1/1000 = .001 \\
P(\text{Win Play 4}) &= P(\text{match pink AND match blue AND} \\
&\quad \text{match yellow AND match orange}) \\
&= P(\text{match pink}) \times P(\text{match blue}) \times \\
&\quad P(\text{match yellow}) \times P(\text{match orange}) \\
&= (1/10)(1/10)(1/10)(1/10) \\
&= 1/10,000 = .0001
\end{aligned}
$$

Although the odds of winning one of these daily games is much better than the odds of winning Lotto 6/53, there is still only a 1 in 1,000 chance (for Cash 3) or 1 in 10,000 chance (for Play 4) of winning the daily game. And the payoffs ($500 or $5,000) are much smaller. In fact, it can be shown that you will lose an average of 50¢ every time you play either Cash 3 or Play 4!

Exercises 3.57–3.91

Understanding the Principles

3.57 Explain the difference between an unconditional probability and a conditional probability.

3.58 Give the formula for finding $P(B|A)$.

3.59 Give the Multiplicative Rule of Probability for any two events.

3.60 How are independent events different from dependent events?

3.61 Give the Multiplicative Rule of Probability for two independent events.

3.62 Defend or refute each of the following statements:
   a. Dependent events are always mutually exclusive.

b. Mutually exclusive events are always dependent.
c. Independent events are always mutually exclusive.

Learning the Mechanics

3.63 For two events, A and B, $P(A) = .4$, $P(B) = .2$, and $P(A \cap B) = .1$.
  a. Find $P(A|B)$.   **b.** Find $P(B|A)$.
  c. Are A and B independent events?

3.64 For two events, A and B, $P(A) = .4$, $P(B) = .2$, and $P(A|B) = .6$.
  a. Find $P(A \cap B)$.   **b.** Find $P(B|A)$.

3.65 For two independent events, A and B, $P(A) = .4$ and $P(B) = .2$.
  a. Find $P(A \cap B)$.   **b.** Find $P(A|B)$.
  c. Find $P(A \cup B)$.

3.66 An experiment results in one of three mutually exclusive events, A, B, or C. It is known that $P(A) = .30$, $P(B) = .55$, and $P(C) = .15$, Find each of the following probabilities:
  a. $P(A \cup B)$ **b.** $P(A \cap B)$
  c. $P(A|B)$ **d.** $P(B \cup C)$
  e. Are B and C independent events? Explain.

3.67 Consider the experiment defined by the accompany-
NW ing Venn diagram, with the sample space S containing five sample points. The sample points are assigned the following probabilities: $P(E_1) = .1$, $P(E_2) = .1$, $P(E_3) = .2$, $P(E_4) = .5$, $P(E_5) = .1$.

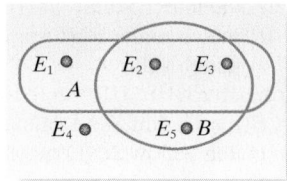

S

  a. Calculate $P(A)$, $P(B)$, and $P(A \cap B)$.
  b. Suppose we know event A has occurred, so the reduced sample space consists of the three sample points in A: E_1, E_2, and E_3. Use the formula for conditional probability to determine the probabilities of these three sample points given that A has occurred. Verify that the conditional probabilities are in the same ratio to one another as the original sample point probabilities and that they sum to 1.
  c. Calculate the conditional probability $P(B|A)$, in two ways: First, sum $P(E_2|A)$ and $P(E_3|A)$ since these sample points represent the event that B occurs given that A has occurred. Second, use the formula for conditional probability:

$$P(B|A) = \frac{P(A \cap B)}{P(A)}$$

  Verify that the two methods yield the same result.

3.68 An experiment results in one of five sample points with the following probabilities: $P(E_1) = .22$, $P(E_2) = .31$,

$P(E_3) = .15$, $P(E_4) = .22$, and $P(E_5) = .1$. The following events have been defined:

  A: $\{E_1, E_3\}$
  B: $\{E_2, E_3, E_4\}$
  C: $\{E_1, E_5\}$

Find each of the following probabilities:
  a. $P(A)$ **b.** $P(B)$ **c.** $P(A \cap B)$
  d. $P(A|B)$ **e.** $P(B \cap C)$ **f.** $P(C|B)$
  g. Consider each pair of events: A and B, A and C, and B and C. Are any of the pairs of events independent? Why?

3.69 Two fair dice are tossed, and the following events are defined:

  A: $\{$Sum of the numbers showing is odd$\}$
  B: $\{$Sum of the numbers showing is 9, 11, or 12$\}$

Are events A and B independent? Why?

3.70 A sample space contains six sample points and events A, B, and C as shown in the accompanying Venn diagram. The probabilities of the sample points are $P(1) = .20$, $P(2) = .05$, $P(3) = .30$, $P(4) = .10$, $P(6) = .25$.

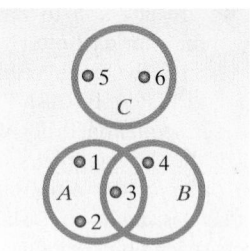

S

  a. Which pairs of events, if any, are mutually exclusive? Why?
  b. Which pairs of events, if any, are independent? Why?
  c. Find $P(A \cup B)$ by adding the probabilities of the sample points and then by using the additive rule. Verify that the answers agree. Repeat for $P(A \cup C)$.

3.71 A box contains two white, two red, and two blue poker chips. Two chips are randomly chosen without replacement, and their colors are noted. Define the following events:

  A: $\{$Both chips are of the same color$\}$
  B: $\{$Both chips are red$\}$
  C: $\{$At least one chip is red or white$\}$

Find $P(B|A)$, $P(B|A^c)$, $P(B|C)$, $P(A|C)$, and $P(C|A^c)$.

Applying the Concepts — Basic

3.72 **Treating brain cancer.** According to the Children's
NW Oncology Group Research Data Center at the University of Florida, 20% of children with neuroblastoma (a form of brain cancer) undergo surgery

rather than the traditional treatment of chemotherapy or radiation. The surgery is successful in curing the disease 95% of the time. (*Explore*, Spring 2001.) Consider a child diagnosed with neuroblastoma. What are the chances that the child undergoes surgery and is cured?

3.73 Executives who cheat at golf. To develop programs for business travelers staying at convention hotels, Hyatt Hotels Corp. commissioned a study of executives who play golf. The study revealed that 55% of the executives admitted they had cheated at golf. Also, 20% of the executives admitted they had cheated at golf and had lied in business. Given an executive had cheated at golf, what is the probability that the executive also had lied in business?

3.74 Sterile couples in Jordan. A sterile family is a couple who has no children by their deliberate choice or because they are biologically infertile. Couples who are childless by chance are not considered to be sterile. Researchers at Yarmouk University (Jordan) estimated the proportion of sterile couples in Jordan to be .06. (*Journal of Data Science*, July 2003.) Also, 64% of the sterile couples in Jordan are infertile. Find the probability that a Jordanian couple is both sterile and infertile.

3.75 Testing a psychic's ability. Consider an experiment in which 10 identical small boxes are placed side-by-side on a table. A crystal is placed, at random, inside one of the boxes. A self-professed "psychic" is asked to pick the box that contains the crystal.
  a. If the "psychic" simply guesses, what is the probability that she picks the box with the crystal?
  b. If the experiment is repeated seven times, what is the probability that the "psychic" guesses correctly at least once?
  c. A group called the Tampa Bay Skeptics recently tested a self-proclaimed "psychic" using the above test. The "psychic" failed to pick the correct box all seven times. (*Tampa Tribune*, Sept. 20, 1998.) What would you infer about this person's psychic ability?

3.76 Herbal medicines survey. Refer to *The American Association of Nurse Anesthetists Journal* (Feb. 2000) study on the use of herbal medicines before surgery, Exercise 1.20 (p. 21). Recall that 51% of surgical patients use herbal medicines against their doctor's advice prior to surgery.
  a. What is the probability that a randomly selected surgical patient will use herbal medicines against his or her doctor's advice?
  b. What is the probability that in a sample of two independently selected surgical patients, both will use herbal medicines against their doctor's advice?
  c. What is the probability that in a sample of five independently selected surgical patients, all five will use herbal medicines against their doctor's advice?
  d. Would you expect the event in part **c** to occur? Explain.

3.77 World Series winners. The New York Yankees, a member of the Eastern Division of the American League in Major League Baseball (MLB), recently won three consecutive World Series. The table summarizes the ten MLB World Series winners from 1990 to 2003 by division and league. (There was no World Series in 1994 due to a players' strike.) One of these 13 World Series winners is to be chosen at random.
  a. Given that the winner is a member of the American League, what is the probability that the winner plays in the Eastern Division?
  b. If the winner plays in the Central Division, what is the probability that the winner is a member of the National League?
  c. If the winner is a member of the National League, what is the probability that the winner plays in either the Central or Western Division?

		League	
		National	American
	Eastern	3	6
Division	Central	1	1
	Western	1	1

Source: Major League Baseball.

3.78 Study of ancient pottery. Refer to the *Chance* (Fall 2000) study of ancient pottery found at the Greek settlement of Phylakopi, Exercise 2.12 (p. 40). Of the 837 pottery pieces uncovered at the excavation site, 183 were painted. These painted pieces included 14 painted in a curvilinear decoration, 165 painted in a geometric decoration, and 4 painted in a naturalistic decoration. Suppose one of the 837 pottery pieces is selected and examined.
  a. What is the probability that the pottery piece is painted?
  b. Given that the pottery piece is painted, what is the probability that it is painted in a curvilinear decoration?

Applying the Concepts—Intermediate

NZBIRDS

3.79 Extinct New Zealand birds. Refer to the *Evolutionary Ecology Research* (July 2003) study of the patterns of extinction in the New Zealand bird population, Exercise 2.18 (p. 42). Consider the data on extinct status (extinct, absent from island, present) for the 132 bird species saved in the **NZBIRDS** file. The data are summarized in the MINITAB printout on p. 158. Suppose you randomly select 10 of the 132 bird species (without replacement) and record the extinct status of each.
  a. What is the probability that the first species you select is extinct? (*Note*: Extinct = Yes on the MINITAB printout.)
  b. Suppose the first nine species you select are all extinct. What is the probability that the tenth species you select is extinct?

Tally for Discrete Variables: Extinct

```
Extinct   Count   Percent
 Absent      16     12.12
     No      78     59.09
    Yes      38     28.79
     N=     132
```

3.80 Requirements for high school graduation. In Italy, all high school students must take a High School Diploma (HSD) exam and write a paper. In *Organizational Behavior and Human Decision Processes* (July 2000), University of Milan researcher L. Macchi provided the following information to a group of college undergraduates. *Fact 1*: In Italy, 360 out of every 1,000 students fail their HSD exam. *Fact 2*: Of those who fail the HSD, 75% also fail the written paper. *Fact 3*: Of those who pass the HSD, only 20% fail the written paper. Define events A and B as follows:

$$A = \{\text{Student fails the HSD exam}\}$$
$$B = \{\text{Student fails the written paper}\}$$

a. Write Fact 1 as a probability statement involving events A and/or B.
b. Write Fact 2 as a probability statement involving events A and/or B.
c. Write Fact 3 as a probability statement involving events A and/or B.
d. State $P(A \cap B)$ in the words of the problem.
e. Find $P(A \cap B)$.

3.81 Intrusion detection systems. A computer intrusion detection system (IDS) is designed to provide an alarm whenever an intrusion (e.g., unauthorized access) is being attempted into a computer system. A probabilistic evaluation of a system with two independently operating intrusion detection systems (a double IDS) was published in the *Journal of Research of the National Institute of Standards and Technology* (Nov.–Dec., 2003). Consider a double IDS with system A and system B. If there is an intruder, system A sounds an alarm with probability .9, and system B sounds an alarm with probability .95. If there is no intruder, the probability that system A sounds an alarm (i.e., a false alarm) is .2, and the probability that system B sounds an alarm is .1.

a. Express the four probabilities given above using symbols.
b. If there is an intruder, what is the probability that both systems sound an alarm?
c. If there is no intruder, what is the probability that both systems sound an alarm?
d. Given an intruder, what is the probability that at least one of the systems sounds an alarm?

3.82 Cigar smoking and cancer. The *Journal of the National Cancer Institute* (Feb. 16, 2000) published the results of a study that investigated the association between cigar smoking and death from tobacco-related cancers. Data were obtained for a national sample of 137,243 Ameri-

can men. The results are summarized in the table below. Each male in the study was classified according to his cigar-smoking status and whether or not he died from a tobacco-related cancer.

a. Find the probability that a randomly selected man never smoked cigars and died from cancer.
b. Find the probability that a randomly selected man was a former cigar smoker and died from cancer.
c. Find the probability that a randomly selected man was a current cigar smoker and died from cancer.
d. Given that a male was a current cigar smoker, find the probability that he died from cancer.
e. Given that a male never smoked cigars, find the probability that he died from cancer.

	Died from Cancer		
Cigars	Yes	No	Totals
Never Smoked	782	120,747	121,529
Former Smoker	91	7,757	7,848
Current Smoker	141	7,725	7,866
Totals	1,014	136,229	137,243

Source: Shapiro, J. A., Jacobs, E. J., and Thun, M. J. "Cigar smoking in men and risk of death from tobacco-related cancers." *Journal of the National Cancer Institute*, Vol. 92, No. 4, Feb. 16, 2000 (Table 2).

3.83 Stacking in the NBA. Refer to the *Sociology of Sport Journal* (Vol. 14, 1997) study of "stacking" in the National Basketball Association (NBA), Exercise 3.52 (p. 144). Reconsider the table below, which summarizes the race and positions of 368 NBA players in 1993. Suppose an NBA player is selected at random from that year's player pool.

	Position			
	Guard	Forward	Center	Totals
White	26	30	28	84
Black	128	122	34	284
Totals	154	152	62	368

a. Given the player is white, what is the probability that he is a center?
b. Given the player is African-American, what is the probability that he is a center?
c. Are the events {White player} and {Center} independent?
d. Recall that "stacking" refers to the practice of African-American players being excluded from certain positions because of race. Use your answers to parts **a–c** to make an inference about stacking in the NBA.

3.84 Refer to the *Conservation Ecology* (Dec. 2003) study of the causes of forest fragmentation, Exercise 2.148 (p. 102). Recall that the researchers used advanced high resolution satellite imagery to develop fragmentation indices for each forest. A 3×3 grid was superimposed over an aerial photo of the forest, and each square (pixel) of the grid was classified as forest (F), anthropogenic land use (A), or natural land cover (N). An example of one such grid is shown here. The edges of the grid (where an "edge" is an imaginary line that separates any two adjacent pixels) are classified as F-A, F-N, A-A, A-N, N-N, or F-F edges.

A	A	N
N	F	F
N	F	F

a. Refer to the grid shown. Note that there are 12 edges inside the grid. Classify each edge as F-A, F-N, A-A, A-N, N-N or F-F.

b. The researchers calculated the fragmentation index by considering only the F-edges in the grid. Count the number of F-edges. (These edges represent the sample space for the experiment.)

c. If an F-edge is selected at random, find the probability that it is an F-A edge. (This probability is proportional to the anthropogenic fragmentation index calculated by the researchers.)

d. If an F edge is selected at random, find the probability that it is an F-N edge. (This probability is proportional to the natural fragmentation index calculated by the researchers.)

3.85 Speeding New Jersey drivers. A certain stretch of highway on the New Jersey Turnpike has a posted speed limit of 60 mph. A survey of drivers on this portion of the turnpike revealed the following facts (reported in *The Washington Post*, Aug.16, 1998):

(1) 14% of the drivers on this stretch of highway were African Americans

(2) 98% of the drivers were exceeding the speed limit by at least 5 mph (and, thus, subject to being stopped by the state police)

(3) Of these violators, 15% of the drivers were African Americans

(4) Of the drivers stopped for speeding by the New Jersey state police, 35% were African Americans

a. Convert each of the reported percentages above into a probability statement.

b. Find the probability that a driver on this stretch of highway is African American and exceeding the speed limit.

c. Given a driver on this stretch of highway is exceeding the speed limit, what is the probability the driver is not African American?

d. Given that a driver on this stretch of highway is stopped by the New Jersey state police, what is the probability that the driver is not African American?

e. For this stretch of highway, are the events {African American driver} and {Driver exceeding the speed limit} independent (to a close approximation)? Explain.

f. Use the probabilities in *The Washington Post* report to make a statement about whether African Americans are stopped more often for speeding than expected on this stretch of turnpike.

3.86 Antibiotic for blood infections. Enterococci are bacteria that cause blood infections in hospitalized patients. One antibiotic used to battle enterococci is vancomycin. A study by the Federal Centers for Disease Control and Prevention revealed that 8% of all enterococci isolated in hospitals nationwide were resistant to vancomycin. (*New York Times*, Sept.12, 1995.) Consider a random sample of three patients with blood infections caused by the enterococci bacteria. Assume that all three patients are treated with the antibiotic vancomycin.

a. What is the probability that all three patients are successfully treated? What assumption did you make concerning the patients?

b. What is the probability that the bacteria resist the antibiotic for at least one patient?

3.87 Cornmaize seeds. The genetic origin and properties of maize (modern-day corn) was investigated in *Economic Botany* (Jan.–Mar. 1995). Seeds from maize ears carry either single spikelets or paired spikelets, but not both. Progeny tests on approximately 600 maize ears revealed the following information. Forty percent of all seeds carry single spikelets, while 60% carry paired spikelets. A seed with single spikelets will produce maize ears with single spikelets 29% of the time and paired spikelets 71% of the time. A seed with paired spikelets will produce maize ears with single spikelets 26% of the time and paired spikelets 74% of the time.

a. Find the probability that a randomly selected maize ear seed carries a single spikelet and produces ears with single spikelets.

b. Find the probability that a randomly selected maize ear seed produces ears with paired spikelets.

3.88 Lie detector test. A new type of lie detector — called the Computerized Voice Stress Analyzer (CVSA) — has been developed. The manufacturer claims that the CVSA is 98% accurate and, unlike a polygraph machine, will not be thrown off by drugs and medical factors. However, laboratory studies by the U.S. Defense Department found that the CVSA had an accuracy rate of 49.8% — slightly less than pure chance. (*Tampa Tribune*, Jan. 10, 1999.) Suppose the CVSA is used to test the veracity of four suspects. Assume the suspects' responses are independent.

NW a. If the manufacturer's claim is true, what is the probability that the CVSA will correctly determine the veracity of all four suspects?

b. If the manufacturer's claim is true, what is the probability that the CVSA will yield an incorrect result for at least one the four suspects?

c. Suppose that in a laboratory experiment conducted by the U.S. Defense Department on four suspects, the CVSA yielded incorrect results for two of the suspects. Use this result to make an inference about the true accuracy rate of the new lie detector.

Applying the Concepts—Advanced

3.89 Strategy in the game "Go." "Go" is one of the oldest and most popular strategic board games in the world, especially in Japan and Korea. The two-player game is played on a flat surface marked with 19 vertical and 19 horizontal lines. The objective is to control territory by placing pieces called "stones" on vacant points on the board. Players alternate placing their stones. The player using black stones goes first, followed by the player using white stones. [*Note:* The University of Virginia requires MBA students to learn Go to understand how the Japanese conduct business.] *Chance* (Summer 1995) published an article that investigated the advantage of playing first (i.e., using the black stones) in Go. The results of 577 games recently played by professional Go players were analyzed.

a. In the 577 games, the player with the black stones won 319 times, and the player with the white stones won 258 times. Use this information to estimate the probability of winning when you play first in Go.

b. Professional Go players are classified by level. Group C includes the top-level players, followed by Group B (middle-level) and Group A (low-level) players. The table describes the number of games won by the player with the black stones, categorized by level of the black player and level of the opponent. Estimate the probability of winning when you play first in Go for each combination of player and opponent level.

c. If the player with the black stones is ranked higher than the player with the white stones, what is the probability that black wins?

d. Given that the players are of the same level, what is the probability that the player with the black stones wins?

Black Player Level	Opponent Level	Number of Wins	Number of Games
C	A	34	34
C	B	69	79
C	C	66	118
B	A	40	54
B	B	52	95
B	C	27	79
A	A	15	28
A	B	11	51
A	C	5	39
	Totals	319	577

Source: Kim J., and Kim H. J., "The advantage of playing first in Go." *Chance,* Vol. 8, No. 3, Summer 1995, p. 26 (Table 3).

3.90 Patient medical instruction sheets. Physicians and pharmacists sometimes fail to inform patients ade-

quately about the proper application of prescription drugs and the precautions to take in order to avoid potential side effects. One method of increasing patients' awareness of the problem is for physicians to provide Patient Medication Instruction (PMI) sheets. The American Medical Association, however, has found that only 20% of the doctors who prescribe drugs frequently distribute PMI sheets to their patients. Assume that 20% of all patients receive the PMI sheet with their prescriptions and that 12% receive the PMI sheet and are hospitalized because of a drug-related problem. What is the probability that a person will be hospitalized for a drug-related problem given that the person has received the PMI sheet?

3.91 NBA draft lottery. The National Basketball Association (NBA) utilizes a lottery to determine the order in which teams draft amateur (high school and college) players. Teams with the poorest won-lost records have the best chance of obtaining the first selection (and, presumably, best player) in the draft. Under the current lottery system, 14 ping-pong balls numbered 1 through 14 are placed in a drum, and four balls are drawn. There are 1,001 possible combinations when four balls are drawn out of 14, without regard to their order of selection. Prior to the drawing, 1,000 combinations are assigned to the 13 teams with the worst records — the Lottery teams — based on their order of finish during the regular season. (The team with the worst record is assigned 250 of these combinations, the team with the second worst record is assigned 200 of these combinations, the team with the third worst record is assigned 157 of these combinations, etc. The table below gives the number combinations assigned to each Lottery team.) Once the four ping pong balls are drawn to the top of the drum to determine a four-digit combination, the team that has been assigned that combination will receive the first pick. The four balls are placed back in the drum and the process is repeated to determine the second, third, and subsequent draft picks. (*Note:* If the one unassigned combination is drawn, the balls are drawn to the top again.)

NBA Lottery Team	Number of Combinations
Worst record	250
2nd worst record	200
3rd worst record	157
4th worst record	120
5th worst record	89
6th worst record	64
7th worst record	44
8th worst record	29
9th worst record	18
10th worst record	11
11th worst record	7
12th worst record	6
13th worst record	5

Source: McCann, M. A. "Illegal defense: The irrational economics of banning high school players from the NBA draft," *Virginia Sports and Entertainment Law Journal,* Vol. 3, No. 2, Spring 2004, Table 5.

a. Demonstrate that there are 1,001 four-number combinations of the numbers 1 through 14.

b. For each of the 13 Lottery teams, find the probability that the team obtains the first pick in the draft.

c. Given that the team with the second worst record obtains the number 1 pick, find the probability that the team with the worst record obtains the second pick of the draft.

d. Calculate the conditional probability, part **c**, given that the team with the third worst record obtains the number 1 pick. Repeat these calculations for the remaining Lottery teams, assuming that Lottery team obtains the number 1 pick.

e. Find the probability that the team with the worst record obtains the second pick of the draft, assuming it did not obtain the first pick.

3.7 Random Sampling

How a sample is selected from a population is of vital importance in statistical inference because the probability of an observed sample will be used to infer the characteristics of the sampled population. To illustrate, suppose you deal yourself four cards from a deck of 52 cards and all four cards are aces. Do you conclude that your deck is an ordinary bridge deck, containing only four aces, or do you conclude that the deck is stacked with more than four aces? It depends on how the cards were drawn. If the four aces were always placed at the top of a standard bridge deck, drawing four aces is not unusual — it is certain. On the other hand, if the cards are thoroughly mixed, drawing four aces in a sample of four cards is highly improbable. The point, of course, is that in order to use the observed sample of four cards to draw inferences about the population (the deck of 52 cards), you need to know how the sample was selected from the deck.

One of the simplest and most frequently employed sampling procedures is implied in the previous examples and exercises. It produces what is known as a *random sample*. We learned in Section 1.5 (p. 14) that a random sample is likely to be *representative* of the population that it is selected from.

> **DEFINITION 3.10**
>
> If *n* elements are selected from a population in such a way that every set of *n* elements in the population has an equal probability of being selected, the *n* elements are said to be a **random sample**.*

If a population is not too large and the elements can be numbered on slips of paper, poker chips, etc., you can physically mix the slips of paper or chips and remove *n* elements from the total. The numbers that appear on the chips selected would indicate the population elements to be included in the sample. Since it is often difficult to achieve a thorough mix, such a procedure only provides an approximation to random sampling. Most researchers rely on **random number generators** to automatically generate the random sample. These random number generators are available in table form, are built into most statistical software packages, and are available on the Internet (e.g., www.random.org).

EXAMPLE 3.22   SELECTING A RANDOM SAMPLE

Problem Suppose you wish to randomly sample five households from a population of 100,000 households.

a. How many different samples can be selected?

b. Use a random number generator to select a random sample.

*Strictly speaking, this is a *simple random sample*. There are many different types of random samples. The simple random sample is the most common.

Solution

a. Since we want to select $n = 5$ objects (households) from $N = 100{,}000$, we apply the combinatorial rule, of Section 3.1:

$$\binom{N}{n} = \binom{100{,}000}{5} = \frac{100{,}000!}{5!\,99{,}995!}$$

$$= \frac{100{,}000 \cdot 99{,}999 \cdot 99{,}998 \cdot 99{,}997 \cdot 99{,}996}{5 \cdot 4 \cdot 3 \cdot 2 \cdot 1}$$

$$= 8.33 \times 10^{22}$$

Thus, there are 83.3 billion trillion different samples of five households that can be selected from 100,000.

b. To ensure that each of the possible samples has an equal chance of being selected, as required for random sampling, we can employ a **random number table**, as provided in Table I of Appendix A. Random number tables are constructed in such a way that every number occurs with (approximately) equal probability. Furthermore, the occurrence of any one number in a position is independent of any of the other numbers that appear in the table. To use a table of random numbers, number the N elements in the population from 1 to N. Then turn to Table I and select a starting number in the table. Proceeding from this number either across the row or down the column, remove and record n numbers from the table.

To illustrate, first we number the households in the population from 1 to 100,000. Then, we turn to a page of Table I, say the first page. (A partial reproduction of the first page of Table I is shown in Table 3.7.) Now, we arbitrarily select a starting number, say the random number appearing in the third row, second column. This number is 48,360. Then we proceed down the second column to obtain the remaining four random numbers. In this case we have selected five random numbers, which are highlighted in Table 3.7. Using the first five digits to represent households from 1 to 99,999 and the number 00000 to represent household 100,000, we can see that the households numbered

48,360 93,093 39,975 6,907 72,905

should be included in our sample.

Note: Use only the necessary number of digits in each random number to identify the element to be included in the sample. If, in the course of recording

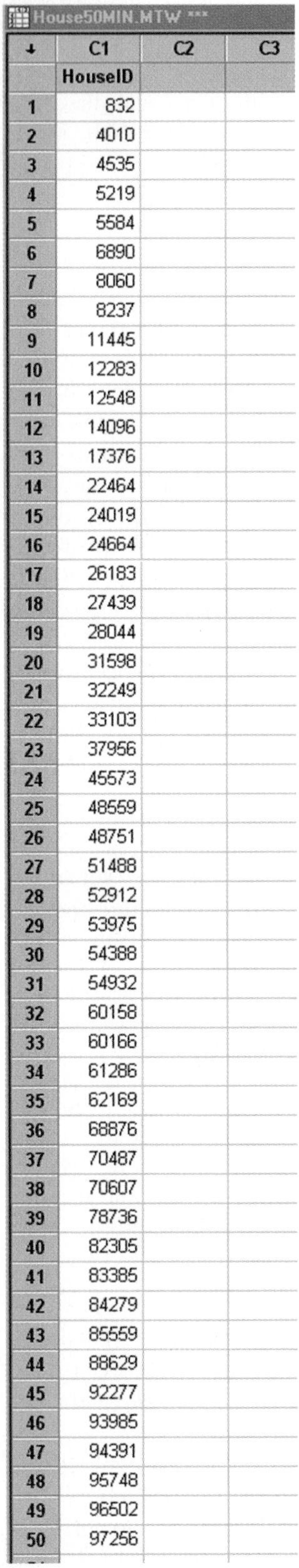

Figure 3.20

MINITAB Worksheet with Random Sample of 50

↓	C1	C2	C3
	HouseID		
1	832		
2	4010		
3	4535		
4	5219		
5	5584		
6	6890		
7	8060		
8	8237		
9	11445		
10	12283		
11	12548		
12	14096		
13	17376		
14	22464		
15	24019		
16	24664		
17	26183		
18	27439		
19	28044		
20	31598		
21	32249		
22	33103		
23	37956		
24	45573		
25	48559		
26	48751		
27	51488		
28	52912		
29	53975		
30	54388		
31	54932		
32	60158		
33	60166		
34	61286		
35	62169		
36	68876		
37	70487		
38	70607		
39	78736		
40	82305		
41	83385		
42	84279		
43	85559		
44	88629		
45	92277		
46	93985		
47	94391		
48	95748		
49	96502		
50	97256		

TABLE 3.7   Partial Reproduction of Table I in Appendix A

Row \ Column	1	2	3	4	5	6
1	10480	15011	01536	02011	81647	91646
2	22368	46573	25595	85393	30995	89198
3	24130	48360	22527	97265	76393	64809
4	42167	93093	06243	61680	07856	16376
5	37670	39975	81837	16656	06121	91782
6	77921	06907	11008	42751	27756	53498
7	99562	72905	56420	69994	98872	31016
8	96301	91977	05463	07972	18876	20922
9	89579	14342	63661	10281	17453	18103
10	85475	36857	53342	53988	53060	59533
11	28918	69578	88231	33276	70997	79936
12	63553	40961	48235	03427	49626	69445
13	09429	93969	52636	92737	88974	33488

the *n* numbers from the table, you select a number that has already been selected, simply discard the duplicate and select a replacement at the end of the sequence. Thus, you may have to record more than *n* numbers from the table to obtain a sample of *n* unique numbers.

Look Back Can we be sure that all 83.3 billion trillion samples have an equal chance of being selected? The fact is, We can't; but to the extent that the random number table contains truly random sequences of digits, the sample should be very close to random.

| Now Work | *Exercise 3.96* |

Table I in Appendix A is just one example of a random number generator. For most scientific studies that require a large random sample, computers are used to generate the random sample. The SAS, MINITAB, and SPSS statistical software packages all have easy-to-use random number generators.

For example, suppose we required a random sample of *n* = 50 households from the population of 100,000 households in Example 3.22. Here, we might employ the MINITAB random number generator. Figure 3.20 shows a MINITAB printout listing 50 random numbers (from a population of 100,000). The households with these identification numbers would be included in the random sample.

Recall our discussion of designed experiments in Section 1.5 (p. 14). The notion of random selection and randomization is key to conducting good research with a designed experiment. The next example illustrates a basic application.

ASPIRIN-STUDY.MTW

↓	C1	C2	C3
	Physician	Treatment	
1	1	3	
2	2	11	
3	3	10	
4	4	14	
5	5	2	
6	6	7	
7	7	6	
8	8	16	
9	9	17	
10	10	9	
11	11		
12	12		
13	13		
14	14		
15	15		
16	16		
17	17		
18	18		
19	19		
20	20		
21			

Figure 3.21
MINITAB Worksheet with Random Assignment of Physicians

EXAMPLE 3.23   RANDOMIZATION IN A DESIGNED EXPERIMENT

Problem A designed experiment in the medical field involving human subjects is referred to as a *clinical trial*. One recent clinical trial was designed to determine the potential of using aspirin in preventing heart attacks. Volunteer physicians were randomly divided into two groups — the *treatment* group and the *control* group. Each physician in the treatment group took one aspirin tablet a day for one year, while the physicians in the control group took an aspirin-free placebo made to look identical to an aspirin tablet. Since the physicians did not know which group, treatment or control, they were assigned to, the clinical trial is called a *blind study*. Assume 20 physicians volunteered for the study. Use a random number generator to randomly assign half of the physicians to the treatment group and half to the control group.

Solution Essentially, we want to select a random sample of 10 physicians from the 20. The first 10 selected will be assigned to the treatment group; the remaining 10 will be assigned to the control group. (Alternatively, we could randomly assign each physician, one by one, to either the treatment or control group. However, this would not guarantee exactly 10 physicians in each group.)

The MINITAB random sample procedure was employed, producing the printout shown in Figure 3.21. Numbering the physicians from 1 to 20, we see that physicians 3, 11, 10, 14, 2, 7, 6, 16, 17, and 9 are assigned to receive the aspirin (treatment). The remaining physicians are assigned the placebo (control).

Exercises 3.92–3.103

Understanding the Principles

3.92 Define a random sample.

3.93 How are random samples related to representative samples?

3.94 What is a random number generator?

3.95 Define a blind study.

Learning the Mechanics

3.96 Suppose you wish to sample $n = 2$ elements from a
NW total of $N = 10$ elements.

    a. Count the number of different samples that can be drawn, first by listing them, and then by using combinatorial mathematics.

    b. If random sampling is to be employed, what is the probability that any particular sample will be selected?

    c. Show how to use the random number table, Table I in Appendix A, to select a random sample of 2 elements from a population of 10 elements. Perform the sampling procedure 20 times. Do any two of the samples contain the same 2 elements? Given your answer to part **b**, did you expect repeated samples?

3.97 Suppose you wish to sample $n = 3$ elements from a total of $N = 600$ elements.

    a. Count the number of different samples by using combinatorial mathematics.

    b. If random sampling is to be employed, what is the probability that any particular sample will be selected?

    c. Show how to use the random number table, Table I in Appendix A, to select a random sample of three elements from a population of 600 elements. Perform the sampling procedure 20 times. Do any two of the samples contain the same three elements? Given your answer to part **b**, did you expect repeated samples?

    d. Use the computer to generate a random sample of three from the population of 600 elements.

3.98 Suppose that a population contains $N = 200,000$ elements. Use a computer or Table I of Appendix A to select a random sample of $n = 10$ elements from the population. Explain how you selected your sample.

Applying the Concepts — Basic

3.99 **Random-digit dialing.** To ascertain the effectiveness of their advertising campaigns, firms frequently conduct telephone interviews with consumers using *random-digit dialing*. With this method, a random number generator mechanically creates the sample of phone numbers to be called.

    a. Explain how the random number table (Table I of Appendix A) or a computer could be used to generate a sample of 7-digit telephone numbers.

    b. Use the procedure you described in part **a** to generate a sample of ten 7-digit telephone numbers.

    c. Use the procedure you described in part **a** to generate five 7-digit telephone numbers whose first three digits are 373.

3.100 **Census sampling.** In addition to its decennial enumeration of the population, the U.S. Bureau of the Census regularly samples the population for demographic information such as income, family size, employment, and marital status. Suppose the Bureau plans to sample 1,000 households in a city that has a total of 534,322 households. Show how the Bureau could use the random number table in Appendix A or a computer to generate the sample. Select the first 10 households to be included in the sample.

Applying the Concepts — Intermediate

3.101 **Study of the effect of songs with violent lyrics.** A series of designed experiments were conducted to examine the effects of songs with violent lyrics on aggressive thoughts and hostile feelings. (*Journal of Personality & Social Psychology*, May 2003.) In one of the experiments, college students were randomly assigned to listen to one of four types of songs: (1) humorous song with violent lyrics, (2) humorous song with nonviolent lyrics, (3) nonhumorous song with violent lyrics, and (4) nonhumorous song with nonviolent lyrics. Assuming 120 students participated in the study, use a random number generator to randomly assign 30 students to each of the four experimental conditions.

3.102 **Auditing an accounting system.** In auditing a firm's financial statements, an auditor is required to assess the operational effectiveness of the accounting system. In performing the assessment, the auditor frequently relies on a random sample of actual transactions. (Stickney and Weil, *Financial Accounting: An Introduction to Concepts, Methods, and Uses*, 1994.) A particular firm has 5,382 customer accounts that are numbered from 0001 to 5382.

    a. One account is to be selected at random for audit. What is the probability that account number 3,241 is selected?

    b. Draw a random sample of 10 accounts and explain in detail the procedure you used.

    c. Refer to part **b**. The following are two possible random samples of size 10. Is one more likely to be selected than the other? Explain.

Sample Number 1				
5011	0082	0963	0772	3415
2663	1126	0008	0026	4189

Sample Number 2				
0001	0003	0005	0007	0009
0002	0004	0006	0008	0010

Applying the Concepts—Advanced

3.103 Selecting archaeological dig sites. Archaeologists plan to perform test digs at a location they believe was inhabited several thousand years ago. The site is approximately 10,000 meters long and 5,000 meters wide. They first draw rectangular grids over the area, consisting of lines every 100 meters, creating a total of $100 \cdot 50 = 5,000$ intersections (not counting one of the outer boundaries). The plan is to randomly sample 50 intersection points and dig at the sampled intersections. Explain how you could use a random number generator to obtain a random sample of 50 intersections. Develop at least two plans: one that numbers the intersections from 1 to 5,000 prior to selection and another that selects the row and column of each sampled intersection (from the total of 100 rows and 50 columns).

3.8 Some Additional Counting Rules (Optional)

In Section 3.1 we pointed out that experiments sometimes have so many sample points that it is impractical to list them all. However, many of these experiments possess sample points with identical characteristics. If you can develop a **counting rule** to count the number of sample points for such an experiment, it can be used to aid in the solution of the problems. The Combinations Rule for selecting n elements from N elements without regard to order was presented in Section 3.1 (p. 129). In this optional section, we give three additional counting rules: the *Multiplicative Rule*, the *Permutations Rule*, and the *Partitions Rule*.

EXAMPLE 3.24   APPLYING THE MULTIPLICATIVE RULE

Problem   A product can be shipped by four airlines, and each airline can ship via three different routes. How many distinct ways exist to ship the product?

Solution   A pictorial representation of the different ways to ship the product will aid in counting them. The tree diagram (see Section 3.6) is useful for this purpose. Consider the tree diagram shown in Figure 3.22. At the starting point (stage 1), there are four choices — the different airlines — to begin the journey. Once we have chosen an airline (stage 2), there are three choices — the different routes — to complete the shipment and reach the final destination. Thus, the tree diagram clearly shows that there are $(4)(3) = 12$ distinct ways to ship the product.

Figure 3.22
Tree Diagram for Shipping Problem

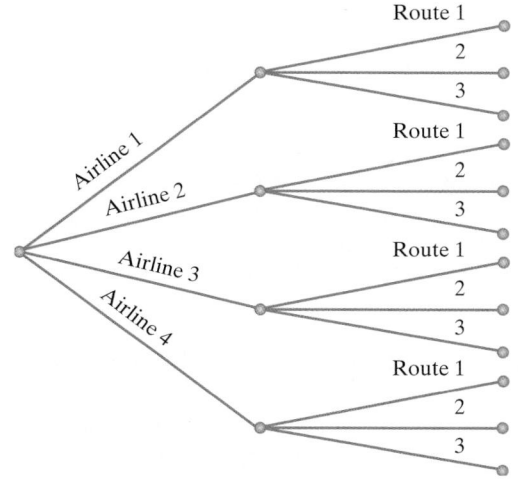

Look Back   The method of solving this example can be generalized to any number of stages with sets of different elements. This method is given by the *Multiplicative Rule*.

Now Work   *Exercise 3.114*

> ### The Multiplicative Rule
>
> You have k sets of elements, n_1 in the first set, n_2 in the second set, $\ldots$, and n_k in the kth set. Suppose you wish to form a sample of k elements by *taking one element from each* of the k sets. The number of different samples that can be formed is the product
>
> $$n_1 n_2 n_3 \cdots n_k$$

EXAMPLE 3.25   APPLYING THE MULTIPLICATIVE RULE

Problem   There are 20 candidates for three different executive positions, E_1, E_2, and E_3. How many different ways could you fill the positions?

Solution   For this example, there are $k = 3$ sets of elements corresponding to

Set 1: Candidates available to fill position E_1
Set 2: Candidates remaining (after filling E_1) that are available to fill E_2
Set 3: Candidates remaining (after filling E_1 and E_2) that are available to fill E_3

The number of elements in the sets are $n_1 = 20$, $n_2 = 19$, $n_3 = 18$. Therefore, the number of different ways of filling the three positions is given by the Multiplicative Rule as $n_1 n_2 n_3 = (20)(19)(18) = 6{,}840$.

■ ■ ■

EXAMPLE 3.26   APPLYING THE MULTIPLICATIVE RULE

Problem   Consider the experiment discussed in Example 3.12 of tossing a coin 10 times. Show how we found that there were $2^{10} = 1{,}024$ sample points for this experiment.

Solution   There are $k = 10$ sets of elements for this experiment. Each set contains two elements: a head and a tail. Thus, there are

$$(2)(2)(2)(2)(2)(2)(2)(2)(2)(2) = 2^{10} = 1{,}024$$

different outcomes (sample points) of this experiment.

■ ■ ■

EXAMPLE 3.27   APPLYING THE PERMUTATIONS RULE

Problem   Suppose there are five dangerous military missions, each requiring one soldier. In how many different ways can five soldiers from a squadron of 100 be assigned to these five missions?

Solution   We can solve this problem by using the Multiplicative Rule. The entire set of 100 soldiers is available for the first mission, and after the selection of one soldier for that mission, 99 are available for the second mission, etc. Thus, the total number of different ways of choosing five soldiers for the five missions is

$$n_1 n_2 n_3 n_4 n_5 = (100)(99)(98)(97)(96) = 9{,}034{,}502{,}400$$

Look Back   The arrangement of elements in a distinct order is called a **permutation**. Thus, there are more than 9 billion different *permutations* of five elements (soldiers) drawn from a set of 100 elements!

■ ■ ■

> **Permutations Rule**
>
> Given a *single set* of N distinctly different elements, you wish to select n elements from the N and *arrange* them within n positions. The number of different **permutations** of the N elements taken n at a time is denoted by P_n^N and is equal to
>
> $$P_n^N = N(N-1)(N-2) \cdot \cdots \cdot (N-n+1) = \frac{N!}{(N-n)!}$$
>
> where $n! = n(n-1)(n-2) \cdots (3)(2)(1)$ and is called *n factorial*. (Thus, for example, $5! = 5 \cdot 4 \cdot 3 \cdot 2 \cdot 1 = 120$.) The quantity $0!$ is defined to be 1.

EXAMPLE 3.28   APPLYING THE PERMUTATIONS RULE

Problem   You wish to drive, in sequence, from a starting point to each of five cities, and you wish to compare the distances — and ultimately the costs — of the different routes. How many different routes would you have to compare?

Solution   Denote the cities as $C_1, C_2, \ldots, C_5$. Then a route moving from the starting point to C_2 to C_1 to C_3 to C_4 to C_5 would be represented as $C_2C_1C_3C_4C_5$. The total number of routes would equal the number of ways you could rearrange the $N = 5$ cities in $n = 5$ positions. This number is

$$P_n^N = P_5^5 = \frac{5!}{(5-5)!} = \frac{5!}{0!} = \frac{5 \cdot 4 \cdot 3 \cdot 2 \cdot 1}{1} = 120$$

(Recall that $0! = 1$.)

| Now Work | *Exercise 3.113c* |

■ ■ ■

EXAMPLE 3.29   APPLYING THE PARTITIONS RULE

Problem   You are supervising four construction workers. You must assign three to job 1 and one to job 2. In how many different ways can you make this assignment?

Solution   To begin, suppose each worker is to be assigned to a distinct job. Then, using the multiplicative rule, you obtain $(4)(3)(2)(1) = 24$ ways of assigning the workers to four distinct jobs. The 24 ways are listed in four groups in Table 3.8 (where ABCD signifies that worker A was assigned the first distinct job, worker B the second, etc.).

     Now suppose the first three positions represent job 1 and the last position represents job 2. You can now see that all the listings in group 1 represent the same outcome of the experiment of interest. That is, workers A, B, and C are assigned to job 1 and worker D to job 2. Similarly, group 2 listings are equivalent, as are group 3

TABLE 3.8   Workers Assigned to Four Jobs, Example 3.29

Group 1	Group 2	Group 3	Group 4
ABCD	ABDC	ACDB	BCDA
ACBD	ADBC	ADCB	BDCA
BACD	BADC	CADB	CBDA
BCAD	BDAC	CDAB	CDBA
CABD	DABC	DACB	DBCA
CBAD	DBAC	DCAB	DCBA

TABLE 3.9 Workers Assigned to Two Jobs, Example 3.29

Job 1	Job 2
ABC	D
ABD	C
ACD	B
BCD	A

and group 4 listings. Thus, there are only four different assignments of four workers to the two jobs. These are shown in Table 3.9.

Look Back To generalize this result, we point out that the final result can be found by

$$\frac{(4)(3)(2)(1)}{(3)(2)(1)(1)} = 4$$

The $(4)(3)(2)(1)$ is the number of different ways (*permutations*) the workers could be assigned four distinct jobs. The division by $(3)(2)(1)$ is to remove the duplicated permutations resulting from the fact that three workers are assigned the same jobs. And the division by 1 is associated with the worker assigned to job 2.

Now Work *Exercise 3.113b*

■ ■ ■

Partitions Rule

There exists a *single* set of N distinctly different elements. You wish to partition them into k sets, with the first set containing n_1 elements, the second containing n_2 elements, ..., and the kth set containing n_k elements. The number of different partitions is

$$\frac{N!}{n_1! n_2! \cdots n_k!} \quad \text{where } n_1 + n_2 + n_3 + \cdots + n_k = N$$

EXAMPLE 3.30

APPLYING THE PARTITIONS RULE

Problem You have 12 construction workers. You wish to assign three to job site 1, four to job site 2, and five to job site 3. In how many different ways can you make this assignment?

Solution For this example, $k = 3$ (corresponding to the $k = 3$ different job sites), $N = 12$, and $n_1 = 3, n_2 = 4$, and $n_3 = 5$. Then the number of different ways to assign the workers to the job sites is

$$\frac{N!}{n_1! n_2! n_3!} = \frac{12!}{3! 4! 5!} = \frac{12 \cdot 11 \cdot 10 \cdot \cdots \cdot 3 \cdot 2 \cdot 1}{(3 \cdot 2 \cdot 1)(4 \cdot 3 \cdot 2 \cdot 1)(5 \cdot 4 \cdot 3 \cdot 2 \cdot 1)} = 27,720$$

■ ■ ■

EXAMPLE 3.31

APPLYING THE COMBINATIONS RULE

Problem Five soldiers from a squadron of 100 are to be chosen for a dangerous mission. In how many ways can groups of five be formed?

Solution This problem is equivalent to sampling $n = 5$ elements from a set of $N = 100$ elements. Here, we apply the Combinations Rule from Section 3.1. The number of ways is the number of possible combinations of five soldiers selected from 100, or

$$\binom{100}{5} = \frac{100!}{(5!)(95!)} = \frac{100 \cdot 99 \cdot 98 \cdot 97 \cdot 96 \cdot 95 \cdot 94 \cdot \cdots \cdot 2 \cdot 1}{(5 \cdot 4 \cdot 3 \cdot 2 \cdot 1)(95 \cdot 94 \cdot \cdots \cdot 2 \cdot 1)}$$
$$= \frac{100 \cdot 99 \cdot 98 \cdot 97 \cdot 96}{5 \cdot 4 \cdot 3 \cdot 2 \cdot 1} = 75,287,520$$

Look Back Compare this result with that of Example 3.27, where we found that the number of permutations of five elements drawn from 100 was more than 9 billion. Because the order of the elements does not affect combinations, there are *fewer* combinations than permutations.

Now Work *Exercise 3.113a*

■ ■ ■

When working a probability problem, you should carefully examine the experiment to determine whether you can use one or more of the rules we have discussed in this section. Let's see how these rules (summarized in the next box) can help solve a probability problem.

Summary of Counting Rules

1. *Multiplicative rule.* If you are drawing *one element from each of k sets of elements*, with the sizes of the sets $n_1, n_2, \ldots, n_k$, the number of different results is

$$n_1 n_2 n_3 \cdot \cdots \cdot n_k$$

2. *Permutations rule.* If you are drawing *n elements from a set of N elements and arranging the n elements in a distinct order*, the number of different results is

$$P_n^N = \frac{N!}{(N-n)!}$$

3. *Partitions rule.* If you are partitioning the *elements of a set of N elements into k groups consisting of* $n_1, n_2, \ldots, n_k$ *elements* $(n_1 + \cdots + n_k = N)$, the number of different results is

$$\frac{N!}{n_1! n_2! \cdot \cdots \cdot n_k!}$$

4. *Combinations rule.* If you are drawing *n elements from a set of N elements without regard to the order of the n elements*, the number of different results is

$$\binom{N}{n} = \frac{N!}{n!(N-n)!}$$

[*Note:* The combinations rule is a special case of the partitions rule when $k = 2$.]

EXAMPLE 3.32

APPLYING SEVERAL COUNTING RULES

Problem A consumer testing service is commissioned to rank the top three brands of laundry detergent. A total of 10 brands are to be included in the study.

a. In how many different ways can the consumer testing service arrive at the final ranking?

b. If the testing service can distinguish no difference among the brands and therefore arrives at the final ranking by random selection, what is the probability that company Z's brand is ranked first? In the top three?

Solution   **a.** Since the testing service is drawing three elements (brands) from a set of 10 elements and arranging the three elements in a distinct order, we use the permutations rule to find the number of different results:

$$P_3^{10} = \frac{10!}{(10-3)!} = 10 \cdot 9 \cdot 8 = 720$$

b. The steps for calculating the probability of interest are as follows:

Step 1 The experiment is to select and rank three brands of detergent from 10 brands.

Step 2 There are too many sample points to list. However, we know from part **a** that there are 720 different outcomes (i.e., sample points) of this experiment.

Step 3 If we assume the testing service determines the rankings at random, each of the 720 sample points should have an equal probability of occurrence. Thus,

$$P(\text{Each sample point}) = \frac{1}{720}$$

Step 4 One event of interest to company Z is that their brand receives top ranking. We will call this event A. The list of sample points that result in the occurrence of event A is long, but the *number* of sample points contained in event A is determined by breaking event A into two parts:

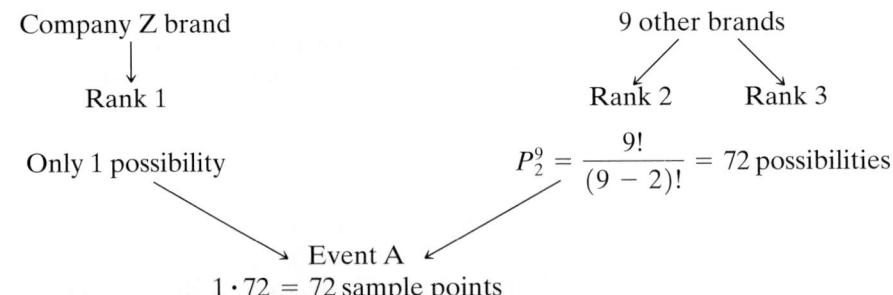

Thus, event A can occur in 72 different ways.

   Now define B as the event that company Z's brand is ranked in the top three. Since event B specifies only that brand Z appear in the top three, we repeat the calculations above, fixing brand Z in position 2 and then in position 3. We conclude that the number of sample points contained in event B is $3(72) = 216$.

Step 5 The final step is to calculate the probabilities of events A and B. Since the 720 sample points are equally likely to occur, we find

$$P(A) = \frac{\text{Number of sample points in } A}{\text{Total number of sample points}} = \frac{72}{720} = \frac{1}{10}$$

Similarly,

$$P(B) = \frac{216}{720} = \frac{3}{10}$$

■ ■ ■

EXAMPLE 3.33   APPLYING SEVERAL COUNTING RULES

Problem   Refer to Example 3.32. Suppose the consumer testing service is to choose the top three laundry detergents from the group of 10 *but is not to rank the three.*

a. In how many different ways can the testing service choose the three to be designated as top detergents?

b. Assuming that the testing service makes its choice by a random selection and that company X has two brands in the group of 10, what is the probability that exactly one of the company X brands is selected in the top three? At least one?

Solution **a.** The testing service is selecting three elements (brands) from a set of 10 elements *without regard to order*, so we can apply the combinations rule to determine the number of different results:

$$\binom{10}{3} = \frac{10!}{3!(10-3)!} = \frac{10 \cdot 9 \cdot 8}{3 \cdot 2 \cdot 1} = 120$$

b. The steps for calculating the probability of interest are as follows:

Step 1 The experiment is to select (*but not rank*) three brands from 10.
Step 2 There are 120 sample points for this experiment.
Step 3 Since the selection is made at random,

$$P(\text{Each sample point}) = \frac{1}{120}$$

Step 4 Define events A and B as follows:

A: {Exactly one company X brand is selected}
B: {At least one company X brand is selected}

Since each of the sample points is equally likely to occur, we need only know the number of sample points in A and B to determine their probabilities.

For event A to occur, exactly one company X brand must be selected, along with two of the remaining eight brands. We thus break A into two parts:

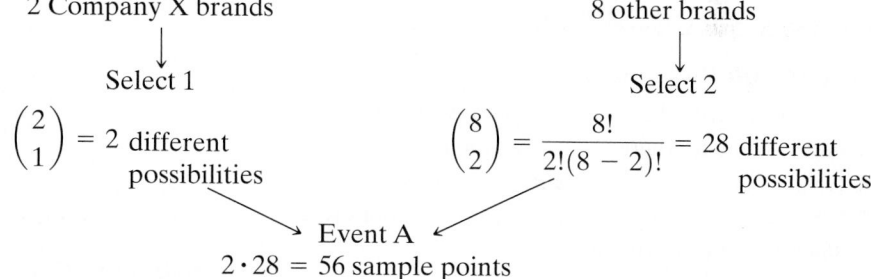

Note that the one company X brand can be selected in two ways, whereas the two other brands can be selected in 28 ways. (We use the combinations rule because the order of selection is not important.) Then, we use the multiplicative rule to combine one of the two ways to select a company X brand with one of the 28 ways to select two other brands, yielding a total of 56 sample points for event A.

To count the sample points contained in B, we first note that

$$B = A \cup C$$

where A is defined as before and

C: {Both company X brands are selected}

We have determined that A contains 56 sample points. Using the same method for event C, we find

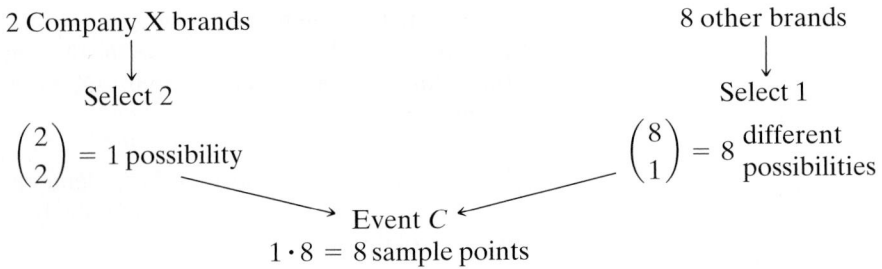

Thus, event C contains eight sample points and $P(C) = \frac{8}{120}$.

Step 5 Since all the sample points are equally likely,

$$P(A) = \frac{\text{Number of sample points in } A}{\text{Total number of sample points}} = \frac{56}{120} = \frac{7}{15}$$
$$P(B) = P(A \cup C) = P(A) + P(C) - P(A \cap C)$$

where $P(A \cap C) = 0$ because the testing service cannot select exactly one *and* exactly two of the company X brands; that is, A and C are mutually exclusive events. Thus,

$$P(B) = \frac{56}{120} + \frac{8}{120} = \frac{64}{120} = \frac{8}{15}$$

■ ■ ■

Learning how to decide whether a particular counting rule applies to an experiment takes patience and practice. If you wish to develop this skill, attempt to use the rules to solve the following exercises and especially the optional exercises (marked with an asterisk) at the end of this chapter. Proofs of the counting rules can be found in the text by W. Feller (1968, Chapter 3).

Exercises 3.104–3.125

Understanding the Principles

3.104 Give a scenario where the Multiplicative Rule applies.

3.105 Give a scenario where the Permutations Rule applies.

3.106 Give a scenario where the Partitions Rule applies.

3.107 What is the difference between the Permutations Rule and the Combinations Rule?

Learning the Mechanics

3.108 Find the numerical values of
  a. $\binom{6}{3}$ **b.** P_2^5 **c.** P_2^4 **d.** $\binom{100}{98}$
  e. $\binom{50}{50}$ **f.** $\binom{50}{0}$ **g.** P_3^5 **h.** P_0^{10}

3.109 Use the multiplicative rule to determine the number of sample points in the sample space corresponding to the experiment of tossing a coin the following number of times:
  a. 2 times   **b.** 3 times   **c.** 5 times
  d. n times

3.110 Determine the number of sample points contained in the sample space when you toss the following:
  a. 1 die   **b.** 2 dice   **c.** 4 dice
  d. n dice 6^n

3.111 An experiment consists of choosing objects without regard to order. How many sample points are there if you choose the following?
  a. 3 objects from 7   **b.** 2 objects from 6
  c. 2 objects from 30   **d.** 8 objects from 10
  e. r objects from q

Applying the Concepts—Basic

3.112 **Selecting project teams.** Suppose you are managing 10 employees, and you need to form three teams to work on different projects. Assume that each employee may serve on any team. In how many different ways can the teams be formed if the number of members on each project team are as follows:
  a. 3, 3, 4   **b.** 2, 3, 5   **c.** 1, 4, 5
  d. 2, 4, 4

3.113 **Picking a basketball team.** Suppose you are to choose
NW a basketball team (five players) from eight available athletes.
  a. How many ways can you choose a team (ignoring positions)?
  b. How many ways can you choose a team composed of two guards, two forwards, and a center?

c. How many ways can you choose a team composed of point guard, shooting guard, power forward, small forward, and center?

3.114 Air travel options. Flying into New York from a certain city, you can choose one of three airlines and can travel either first class or coach. How many travel options do you have?

3.115 Traveling between cities. A salesperson living in city A wishes to visit four cities, B, C, D, and E.

a. If the cities are all connected by airlines, how many different travel plans could be constructed to visit each city exactly once and then return home?

b. Suppose all cities are connected except that B and C are not directly connected. How many different flight plans would be available to the salesperson?

3.116 U.S. Zip codes. The 9-digit Zip code has become an integral part of the U.S. postal system.

a. How many different Zip codes are available for use by the postal service? (Assume any 9-digit number can be used as a Zip code.)

b. The first three digits of a Chicago Zip code are 606. If no other city in the United States has these first three digits as part of its Zip, how many different Zip codes may possibly exist in Chicago?

Applying the Concepts—Intermediate

3.117 Multilevel marketing schemes. Successful companies such as Amway, Herbalife, and NuSkin employ multi-level marketing (MLM) schemes to distribute their products. In MLM, each distributor recruits several additional distributors to work at the first level. Each of these first-level distributors, in turn, recruits distributors to work at the second level. Distributors at any level continue to recruit distributors, forming additional levels. A distributor at any particular level makes a 5% commission on the sales of all distributors at lower levels in his or her "group." (*Success*, Dec. 1995.)

a. Elaine is a distributor for a MLM company. Elaine recruits six distributors to work for her at the first level. Each of these distributors recruits an additional five distributors to work at the second level. How many distributors are under Elaine?

b. Refer to part **a.** Suppose each distributor at Elaine's second level recruits seven third-level distributors, and each of these third-level distributors recruits five level-four distributors. Now, how many distributors are under Elaine? (Your answer should provide insight into why some MLM distributors can earn $100,000 in commissions per month!)

3.118 Mathematical Theory of Partitions. Mathematicians at the University of Florida solved a 30-year-old math problem using the theory of partitions. (*Explore*, Fall 2000.) In math terminology, a partition is a representation of an integer as a sum of positive integers. (For example, the number 3 has three possible partitions: 3, 2 + 1 and 1 + 1 + 1.) The researchers solved the problem by using "colored partitions" of a number,

where the colors correspond to the four suits—red hearts, red diamonds, black spades, and black clubs—in a standard 52-card bridge deck. Consider forming colored partitions of an integer.

a. How many colored partitions of the number 3 are possible? [*Hint:* One partition is 3♥; another is 2♦ + 1♣.]

b. How many colored partitions of the number 5 are possible?

3.119 Matching medical students with residencies. The National Resident Matching Program (NRMP) is a service provided by the Association of American Medical Colleges to match graduating medical students with residency appointments at hospitals. After students and hospitals have evaluated each other, they submit rank-order lists of their preferences to the NRMP. Using a matching algorithm, the NRMP then generates final, nonnegotiable assignments of students to the residency programs of hospitals (*Academic Medicine*, June 1995). Assume that three graduating medical students (#1, #2, and #3) have applied for positions at three different hospitals (A, B, and C), where each hospital has one and only one resident opening.

a. How many different assignments of medical students to hospitals are possible? List them.

b. Suppose student #1 prefers hospital B. If the NRMP algorithm is entirely random, what is the probability that the student is assigned to hospital B?

3.120 Florida license plates. In the mid-1980s the state of Florida ran out of combinations of letters and numbers for its license plates. Then, each license plate contained three letters of the alphabet followed by three digits selected from the 10 digits 0, 1, 2, ..., 9.

a. How many different license plates did this system allow?

b. New Florida tags were obtained by reversing the procedure, starting with three digits followed by three letters. How many new tags did the new system provide?

c. Since the new tag numbers were added to the old numbers, what is the total number of licenses available for registration in Florida?

3.121 Selecting committee members. The Republican governor of a state with no income tax is appointing a committee of five members to consider changes in the income tax law. There are 15 state representatives available for appointment to the committee, seven Democrats and eight Republicans. Assume that the governor selects the committee of five members randomly from the 15 representatives.

a. In how many different ways can the committee members be selected?

b. What is the probability that no Democrat is appointed to the committee? If this were to occur, what would you conclude about the assumption that the appointments were made at random? Why?

c. What is the probability that the majority of the committee members are Republican? If this were to occur, would you have reason to doubt that the governor made the selections randomly? Why?

3.122 Selecting new car options. A company sells midlevel models of automobiles in five different styles. A buyer can get an automobile in one of eight colors with either standard or automatic transmission. Would it be reasonable to expect a dealer to stock at least one automobile in every style, color, and transmission combination? At a minimum, how many automobiles would the dealer have to stock?

3.123 Studying exam questions. A college professor hands out a list of 10 questions, five of which will appear on the final examination for the course. One of the students taking the course is pressed for time and can prepare for only seven of the 10 questions. Suppose the professor chooses the five questions at random from the 10.
a. What is the probability that the student will be prepared for all five questions that appear on the final examination?
b. What is the probability that the student will be prepared for fewer than three questions?
c. What is the probability that the student will be prepared for exactly four questions?

3.124 Volleyball positions. Intercollegiate volleyball rules require that after the opposing team has lost its serve, each of the six members of the serving team must rotate into new positions on the court. Hence, each player must be able to play all six different positions. How many different team combinations by player and position are possible during a volleyball game? If players are initially assigned to the positions in a random manner, find the probability that the best server on the team is in the serving position.

Applying the Concepts—Advanced

3.125 A straight flush in poker. Consider five-card poker hands dealt from a standard 52-card bridge deck. Two important events are

A: {You draw a flush}
B: {You draw a straight}

[*Note: A* flush *consists of any five cards of the same suit. A* straight *consists of any five cards with values in sequence. In a straight, the cards may be of any suit, and an ace may be considered as having a value of 1 or a value higher than a king.*]
a. How many different five-card hands can be dealt from a 52-card bridge deck?
b. Find $P(A)$.
c. Find $P(B)$.
d. The event that both A and B occur, that is $A \cap B$, is called a *straight flush*. Find $P(A \cap B)$.

3.9 Bayes's Rule (Optional)

An early attempt to employ probability in making inferences is the basis for a branch of statistical methodology known as **Bayesian statistical methods**. The logic employed by the English philosopher, Thomas Bayes, in the mid-1700s, involves converting an unknown conditional probability, say $P(B|A)$, to one involving a known conditional probability, say $P(A|B)$. The method is illustrated in the next example.

EXAMPLE 3.34   APPLYING BAYES'S LOGIC

Problem   An unmanned monitoring system uses high-tech video equipment and microprocessors to detect intruders. A prototype system has been developed and is in use outdoors at a weapons munitions plant. The system is designed to detect intruders with a probability of .90. However, the design engineers expect this probability to vary with weather condition. The system automatically records the weather condition each time an intruder is detected. Based on a series of controlled tests, in which an intruder was released at the plant under various weather conditions, the following information is available: Given the intruder was, in fact, detected by the system, the weather was clear 75% of the time, cloudy 20% of the time, and raining 5% of the time. When the system failed to detect the intruder, 60% of the days were clear, 30% cloudy, and 10% rainy. Use this information to find the probability of detecting an intruder, given rainy weather conditions. (Assume that an intruder has been released at the plant.)

Solution   Define D to be the event that the intruder is detected by the system. Then D^c is the event that the system failed to detect the intruder. Our goal is to calculate the conditional probability, $P(D|\text{Rainy})$. From the statement of the problem, the following information is available:

$$P(D) = .90 \qquad\qquad P(D^c) = .10$$
$$P(\text{Clear}|D) = .75 \qquad P(\text{Clear}|D^c) = .60$$
$$P(\text{Cloudy}|D) = .20 \qquad P(\text{Cloudy}|D^c) = .30$$
$$P(\text{Rainy}|D) = .05 \qquad P(\text{Rainy}|D^c) = .10$$

Note that $P(D|\text{Rainy})$ is not one of the conditional probabilities that is known. However, we can find

$$P(\text{Rainy} \cap D) = P(D)P(\text{Rainy}|D) = (.90)(.05) = .045$$

and

$$P(\text{Rainy} \cap D^c) = P(D^c)P(\text{Rainy}|D^c) = (.10)(.10) = .01$$

using the Multiplicative Probability Rule. Now the event Rainy is the union of two mutually exclusive events, $(\text{Rainy} \cap D)$ and $(\text{Rainy} \cap D^c)$. Thus, applying the Additive Probability Rule, we have

$$P(\text{Rainy}) = P(\text{Rainy} \cap D) + P(\text{Rainy} \cap D^c) = .045 + .01 = .055$$

We now apply the formula for conditional probability to obtain

$$P(D|\text{Rainy}) = \frac{P(\text{Rainy} \cap D)}{P(\text{Rainy})} = \frac{P(\text{Rainy} \cap D)}{P(\text{Rainy} \cap D) + P(\text{Rainy} \cap D^c)}$$
$$= .045/.055 = .818$$

Therefore, under rainy weather conditions, the prototype system can detect the intruder with a probability of .818 — a value lower than the designed probability of .90.

———— ■ ■ ■ ————

The technique utilized in Example 3.34, called **Bayes's Rule**, can be applied when an observed event A occurs with any one of several mutually exclusive and exhaustive events, $B_1, B_2, \ldots, B_k$. The formula for finding the appropriate conditional probabilities is given in the box.

Bayes's Rule

Given k mutually exclusive and exhaustive events, $B_1, B_2, \ldots, B_k$, such that $P(B_1) + P(B_2) + \cdots + P(B_k) = 1$, and an observed event A, then

$$P(B_i|A) = P(B_i \cap A)/P(A)$$

$$= \frac{P(B_i)P(A|B_i)}{P(B_1)P(A|B_1) + P(B_2)P(A|B_2) + \cdots + P(B_k)P(A|B_k)}$$

In applying Bayes's rule to Example 3.34, the observed event $A = \{\text{Rainy}\}$ and the $k = 2$ mutually exclusive and exhaustive events are the complementary events $D = \{\text{intruder detected}\}$ and $D^c = \{\text{intruder not detected}\}$. Hence, the formula

$$P(D|\text{Rainy}) = \frac{P(D)P(\text{Rainy}|D)}{P(D)P(\text{Rainy}|D) + P(D^c)P(\text{Rainy}|D^c)}$$
$$= \frac{(.90)(.05)}{(.90)(.05) + (.10)(.10)} = .818$$

Biography

**THOMAS BAYES
(1702–1761)
The Inverse Probabilist**

The Reverend Thomas Bayes was an ordained English Presbyterian minister who became a Fellow of the Royal Statistical Society without benefit of any formal training in mathematics or any published papers in science during his lifetime. His manipulation of the formula for conditional probability in 1761 is now known as Bayes's Theorem. At the time and for 200 years afterward, use of Bayes's Theorem (or inverse probability, as it was called) was controversial and, to some, considered an inappropriate practice. It was not until the 1960s that the power of the Bayesian approach to decision making began to be tapped.

Exercises 3.126–3.137

Understanding the Principles

3.126 Explain the difference between the two probabilities, $P(A|B)$ and $P(B|A)$.

3.127 Why is Bayes's Rule unnecessary for finding $P(B|A)$ if events A and B are independent?

3.128 Why is Bayes's Rule unnecessary for finding $P(B|A)$ if events A and B are mutually exclusive?

Learning the Mechanics

3.129 Suppose the events B_1 and B_2 are mutually exclusive and complementary events, such that $P(B_1) = .75$ and $P(B_2) = .25$. Consider another event A such that $P(A|B_1) = .3$ and $P(A|B_2) = .5$.
  a. Find $P(B_1 \cap A)$.   **b.** Find $P(B_2 \cap A)$.
  c. Find $P(A)$ using the results in parts a and b.
  d. Find $P(B_1|A)$.   **e.** Find $P(B_2|A)$.

3.130 Suppose the events B_1, B_2, and B_3 are mutually exclusive and complementary events, such that $P(B_1) = .2$, $P(B_2) = .15$, and $P(B_3) = .65$. Consider another event A such that $P(A|B_1) = .4$, $P(A|B_2) = .25$, and $P(A|B_3) = .6$. Use Bayes's Rule to find
  a. $P(B_1|A)$ **b.** $P(B_2|A)$
  c. $P(B_3|A)$

3.131 Suppose the events B_1, B_2, and B_3 are mutually exclusive and complementary events, such that $P(B_1) = .2$, $P(B_2) = .15$, and $P(B_3) = .65$. Consider another event A such that $P(A) = .4$. If A is independent of B_1, B_2, and B_3, use Bayes's Rule to show that $P(B_1|A) = P(B_1) = .2$.

Applying the Concepts—Basic

⚙ **DDT**
3.132 **Contaminated fish.** Refer to the U.S. Army Corps of Engineers' study on the DDT contamination of fish in the Tennessee River (Alabama), Example 1.4 (p. 00). Part of the investigation focused on how far upstream the contaminated fish have migrated. (A fish is considered to be contaminated if its measured DDT concentration is greater than 5.0 parts per million.)

  a. Considering only the contaminated fish captured from the Tennessee River, the data reveal that 52% of the fish are found between 275 and 300 miles upstream, 39% are found 305 to 325 miles upstream, and 9% are found 330 to 350 miles upstream. Use these percentages to determine the probabilities, $P(275–300)$, $P(305–325)$, and $P(330–350)$.
  b. Given a contaminated fish is found a certain distance upstream, the probability that it is a channel catfish (CC) is determined from the data as $P(CC|275–300) = .775$, $P(CC|305–325) = .77$, and $P(CC|330–350) = .86$. If a contaminated channel catfish is captured from the Tennessee River, what is the probability that it was captured 275–300 miles upstream?

3.133 **Drug testing in athletes.** Due to inaccuracies in drug testing procedures (e.g., false positives and false negatives), in the medical field the results of a drug test represent only one factor in a physician's diagnosis. Yet when Olympic athletes are tested for illegal drug use (i.e., doping), the results of a single test are used to ban the athlete from competition. In *Chance* (Spring 2004), University of Texas biostatisticians D. A. Berry and L. Chastain demonstrated the application of Bayes's Rule for making inferences about testosterone abuse among Olympic athletes. They used the following example. In a population of 1,000 athletes, suppose 100 are illegally using testosterone. Of the users, suppose 50 would test positive for testosterone. Of the nonusers, suppose 9 would test positive.
  a. Given the athlete is a user, find the probability that a drug test for testosterone will yield a positive result. (This probability represents the *sensitivity* of the drug test.)
  b. Given the athlete is a nonuser, find the probability that a drug test for testosterone will yield a negative result. (This probability represents the *specificity* of the drug test.)
  c. If an athlete tests positive for testosterone, use Bayes's Rule to find the probability that the athlete

is really doping. (This probability represents the *positive predictive value* of the drug test.)

3.134 Errors in estimating Job Costs. A construction company employs three sales engineers. Engineers 1, 2, and 3 estimate the costs of 30%, 20%, and 50%, respectively, of all jobs bid by the company. For $i = 1, 2, 3$, define E_i to be the event that a job is estimated by engineer i. The following probabilities describe the rates at which the engineers make serious errors in estimating costs:

$P(\text{error}|E_1) = .01, P(\text{error}|E_2) = .03,$ and $P(\text{error}|E_3) = .02$

a. If a particular bid results in a serious error in estimating job cost, what is the probability that the error was made by engineer 1?

b. If a particular bid results in a serious error in estimating job cost, what is the probability that the error was made by engineer 2?

c. If a particular bid results in a serious error in estimating job cost, what is the probability that the error was made by engineer 3?

d. Based on the probabilities, parts a–c, which engineer is most likely responsible for making the serious error?

Applying the Concepts—Intermediate

3.135 Intrusion detection systems. Refer to the *Journal of Research of the National Institute of Standards and Technology* (Nov.–Dec. 2003) study of a double intrusion detection system with independent systems, Exercise 3.81 (p. 158). Recall that if there is an intruder, system A sounds an alarm with probability .9 and system B sounds an alarm with probability .95. If there is no intruder, system A sounds an alarm with probability .2 and system B sounds an alarm with probability .1. Now, assume that the probability of an intruder is .4. If both systems sound an alarm, what is the probability that an intruder is detected?

3.136 Repairing a computer system. The local area network (LAN) for the College of Business computing system at a large university is temporarily shutdown for repairs. Previous shutdowns have been due to hardware failure, software failure, or power failure. Maintenance engineers have determined that the probabilities of hardware, software, and power problems are .01, .05, and .02, respectively. They have also determined that if the system experiences hardware problems, it shuts down 73% of the time. Similarly, if software problems occur, the system shuts down 12% of the time; and, if power failure occurs, the system shuts down 88% of the time. What is the probability that the current shutdown of the LAN is due to hardware failure? Software failure? Power failure?

3.137 Purchasing microchips. An important component of your desktop or laptop personal computer (PC) is a microchip. The table gives the proportions of microchips that a certain PC manufacturer purchases from seven suppliers.

Supplier	Proportion
S_1	.15
S_2	.05
S_3	.10
S_4	.20
S_5	.12
S_6	.20
S_7	.18

a. It is known that the proportions of defective microchips produced by the seven suppliers are .001, .0003, .0007, .006, .0002, .0002, and .001, respectively. If a single PC microchip failure is observed, which supplier is most likely responsible?

b. Suppose the seven suppliers produce defective microchips at the same rate, .0005. If a single PC microchip failure is observed, which supplier is most likely responsible?

Quick Review

Key Terms

[Note: Items marked with an asterisk (*)are from the optional sections in this chapter.]

Key Formulas

$$\binom{N}{n} = \frac{N!}{n!(N-n)!}$$

where $N! = N(N-1)(N-2)\cdots(2)(1)$

$P(A) + P(A^c) = 1$

$P(A \cup B) = P(A) + P(B) - P(A \cap B)$

$P(A \cap B) = 0$

$P(A \cup B) = P(A) + P(B)$

Combinations rule   129
Rule of Complements   138
Additive rule of probability   139
Mutually exclusive events   140
Additive rule of probability for mutually exclusive events   140

$$P(A|B) = \frac{P(A \cap B)}{P(B)}$$

$P(A \cap B) = P(A)P(B|A) = P(B)P(A|B)$

$P(A|B) = P(A)$

$P(A \cap B) = P(A)P(B)$

Conditional probability   146

Multiplicative rule of probability   149
Independent events   151
Multiplicative rule of probability for independent events   154

$$P(B_i|A) = \frac{P(B_i)P(A|B_i)}{P(B_1)P(A|B_1) + P(B_2)P(A|B_2) + \cdots + P(B_k)P(A|B_k)}$$

*Bayes's rule   175

Note: For a summary of counting rules, see p. 169.

Language Lab

Symbol	Pronunciation	Description
S		Sample space
$S: \{1, 2, 3, 4, 5\}$		Set of sample points, 1,2,3,4,5, in sample space
$A: \{1, 2\}$		Set of sample points, 1,2, in event A
$P(A)$	Probability of A	Probability that event A occurs
$A \cup B$	A union B	Union of events A and B (either A or B or both occur)
$A \cap B$	A intersect B	Intersection of events A and B (both A and B occur)
A^c	A complement	Complement of event A (the event that A does not occur)
$P(A|B)$	Probability of A given B	Conditional probability that event A occurs given that event B occurs
$\binom{N}{n}$	N choose n	Number of combinations of N elements taken n at a time
$N!$	N factorial	Multiply $N(N-1)(N-2)\cdots(2)(1)$

Chapter Summary

Chapter Summary Notes

- Probability rules for k **sample points**:

    (1) $0 \le P(S_i) \le 1$ and (2) $\sum_{i=1}^{k} P(S_i) = 1$

- If $A = \{S_1, S_3, S_4\}$, then
  $P(A) = P(S_1) + P(S_3) + P(S_4)$

- The number of samples of size n that can be selected from N element is $\binom{N}{n}$

- **Union**: $(A \cup B)$ implies that either A or B will occur.
- **Intersection**: $(A \cap B)$ implies that both A and B will occur.
- **Complement**: A^c is all the sample points not in A.
- **Conditional**: $(A|B)$ is the event that A occurs, given B has occurred.
- **Independent**: B occurring does not change the probability that A occurs.
- **Random sample**: All possible samples have equal probability of being selected.

Supplementary Exercises 3.138–3.179

Note: Starred (*)exercises refer to the optional sections in this chapter.

Understanding the Principles

3.138 Which of the following pairs of events are mutually exclusive?

  a. A = {San Francisco Giants win the World Series next year}
  B = {Barry Bonds, Giants outfielder, hits 75 home runs next year}

  b. A = {Psychiatric patient Tony responds to a stimulus within 5 seconds}
  B = {Psychiatric patient Tony has the fastest stimulus response time of 2.3 seconds}

  c. A = {High school graduate Cindy enrolls at the University of South Florida next year}
  B = {High school graduate Cindy does not enroll in college next year}

3.139 Convert the following statements into compound events involving events A and B using the symbols $\cap$, $\cup$, and $|$, and ^{c}, where A = {You purchase an IBM notebook computer} and B = {you vacation in Europe}.

  a. You purchase an IBM notebook computer or vacation in Europe.

  b. You will not vacation in Europe.

  c. You purchase an IBM notebook computer and vacation in Europe.

  d. Given that you vacation in Europe, you will not purchase an IBM notebook computer.

Learning the Mechanics

3.140 A sample space consists of four sample points, where $P(S_1) = .2$, $P(S_2) = .1$, $P(S_3) = .3$, and $P(S_4) = .4$.

  a. Show that the sample points obey the two probability rules for a sample space.

  b. If an event $A = \{S_1, S_4\}$, find $P(A)$.

3.141 A and B are mutually exclusive events, with $P(A) = .2$ and $P(B) = .3$

  a. Find $P(A|B)$.

  b. Are A and B independent events?

3.142 For two events A and B, suppose $P(A) = .7$, $P(B) = .5$, and $P(A \cap B) = .4$. Find $P(A \cup B)$.

3.143 Given that $P(A \cap B) = .4$ and $P(A|B) = .8$, find $P(B)$.

Learning the Mechanics

3.144 The accompanying Venn diagram illustrates a sample space containing six sample points and three events, A, B, and C. The probabilities of the sample points are $P(1) = .3$, $P(2) = .2$, $P(3) = .1$, $P(4) = .1$, $P(5) = .1$, and $P(6) = .2$.

  a. Find $P(A \cap B)$, $P(B \cap C)$, $P(A \cup C)$, $P(A \cup B \cup C)$, $P(B^c)$, $P(A^c \cap B)$, $P(B|C)$, and $P(B|A)$.

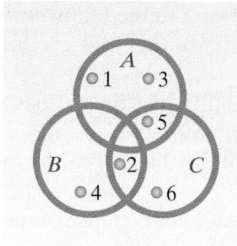

S

  b. Are A and B independent? Mutually exclusive? Why?

  c. Are B and C independent? Mutually exclusive? Why?

3.145 A fair die is tossed, and the up face is noted. If the number is even, the die is tossed again; if the number is odd, a fair coin is tossed. Define the events:

  A: {A head appears on the coin}
  B: {The die is tossed only one time}

  a. List the sample points in the sample space.

  b. Give the probability for each of the sample points.

  c. Find $P(A)$ and $P(B)$.

  d. Identify the sample points in A^c, B^c, $A \cap B$, and $A \cup B$.

  e. Find $P(A^c)$, $P(B^c)$, $P(A \cap B)$, $P(A \cup B)$, $P(A|B)$, and $P(B|A)$.

  f. Are A and B mutually exclusive events? Independent events? Why?

3.146 A balanced die is thrown once. If a 4 appears, a ball is drawn from urn 1; otherwise, a ball is drawn from urn 2. Urn 1 contains four red, three white, and three black balls. Urn 2 contains six red and four white balls.

  a. Find the probability that a red ball is drawn.

  b. Find the probability that urn 1 was used given that a red ball was drawn.

3.147 Two events, A and B, are independent, with $P(A) = .3$ and $P(B) = .1$.

  a. Are A and B mutually exclusive? Why?

  b. Find $P(A|B)$ and $P(B|A)$.

  c. Find $P(A \cup B)$.

*3.148 Find the numerical value of

  a. 6!   **b.** $\binom{10}{9}$ **c.** $\binom{10}{1}$ **d.** P_2^6

  e. $\binom{6}{3}$ **f.** 0!   **g.** P_4^{10} **h.** P_2^{50}

Applying the Concepts—Basic

⊚ CRASH

3.149 NHTSA new car crash testing. Refer to the National Highway Traffic Safety Administration (NHTSA) crash tests of new car models, Exercise 2.164 (p. 109). Recall that the NHTSA has developed a "star" scoring system, with results ranging from one star (*) to five stars (*****). The more stars in the rating, the better the level of crash protection in a head-on collision. A summary of the driver-side star ratings for 98 cars is reproduced in the accompanying MINITAB printout. Assume that one of the 98 cars is selected at random. State whether each of the following is true or false.

a. The probability that the car has a rating of two stars is 4.

b. The probability that the car has a rating of four or five stars is .7857.

c. The probability that the car has a rating of one star is 0.

d. The car has a better chance of having a 2-star rating than of having a 5-star rating.

Tally for Discrete Variables: DRIVSTAR

```
DRIVSTAR   Count   Percent
      2       4      4.08
      3      17     17.35
      4      59     60.20
      5      18     18.37
    N=       98
```

3.150 Selecting a sample. A random sample of five students is to be selected from 50 sociology majors for participation in a special program.
  a. In how many different ways can the sample be drawn?
  b. Show how the random number table, Table I of Appendix A, can be used to select the sample of students.

3.151 Forbes survey of companies. *Forbes* (July 26, 1999) conducted a survey of the 20 largest nondomestic public companies in the world. Of these 20 companies, 6 were trading companies based in Japan. A total of 11 Japanese companies were on the top 20 list. Suppose we select one of these 20 companies at random. Given the company is based in Japan, what is the probability that it is a trading company?

⊚ **EVOS**
3.152 Oilspill impact on seabirds. Refer to the *Journal of Agricultural, Biological, and Environmental Statistics* (Sept. 2000) study of the impact of the *Exxon Valdez* tanker oil spill on the seabird population in Prince William Sound, Alaska, Exercise 2.179 (p. 113). Recall that data were collected on 96 shoreline locations (called transects), and it was determined whether or not the transect was in an oiled area. The data are stored in the file called EVOS. From the data, estimate the probability that a randomly selected transect in Prince William Sound is contaminated with oil.

3.153 Education TV network. "Channel One" is an education television network for which participating secondary schools are equipped with TV sets in every classroom. According to *Educational Technology* (May–June 1995), 40% of all U.S. secondary schools subscribe to the Channel One Communications Network (CCN). Of these subscribers, 5% never use the

CCN broadcasts, while 20% use CCN more than five times per week.
  a. Find the probability that a randomly selected U.S. secondary school subscribes to CCN and never uses the CCN broadcasts.
  b. Find the probability that a randomly selected U.S. secondary school subscribes to CCN and uses the broadcasts more than five times per week.

3.154 Carbon monoxide poisoning. The *American Journal of Public Health* (July 1995) published a study on unintentional carbon monoxide (CO) poisoning of Colorado residents. A total of 981 cases of CO poisoning were reported during a six-year period. Each case was classified as fatal or nonfatal and by source of exposure. The number of cases occurring in each of the categories is shown in the accompanying table. Assume that one of the 981 cases of unintentional CO poisoning is randomly selected.

Source of Exposure	Fatal	Nonfatal	Total
Fire	63	53	116
Auto exhaust	60	178	238
Furnace	18	345	363
Kerosene or spaceheater	9	18	27
Appliance	9	63	72
Other gas-powered motor	3	73	76
Fireplace	0	16	16
Other	3	19	22
Unknown	9	42	51
Total	**174**	**807**	**981**

Source: Cook, M. C., Simon P. A., and Hoffman, R. E., "Unintentional carbon monoxide poisoning in Colorado, 1986 through 1991." *American Journal of Public Health,* Vol. 85, No. 7, July 1995, p. 989 (Table 1).

  a. List all sample points for this experiment.
  b. What is the set of all sample points called?
  c. Let A be the event that the CO poisoning is caused by fire. Find $P(A)$.
  d. Let B be the event that the CO poisoning is fatal. Find $P(B)$.
  e. Let C be the event that the CO poisoning is caused by auto exhaust. Find $P(C)$.
  f. Let D be the event that the CO poisoning is caused by auto exhaust and is fatal. Find $P(D)$.
  g. Let E be the event that the CO poisoning is caused by fire but is nonfatal. Find $P(E)$.
  h. Given that the source of the poisoning is fire, what is the probability that the case is fatal?
  i. Given that the case is nonfatal, what is the probability that it is caused by auto exhaust?
  j. If the case is fatal, what is the probability that the source is unknown?
  k. If the case is nonfatal, what is the probability that the source is not fire or a fireplace?

3.155 Media coverage of mental illness. Refer to the *Health Education Journal* (Sept. 1994) study of media coverage of stories involving mental illness in Scotland, Exercise 2.168 (p. 110). The media coverage of each of

562 stories was classified by type, with the results shown in the table. Assume that one of the 562 stories on mental illness is selected and the type of media coverage is noted.

Media Coverage	Number of Items
Violence to others	373
Sympathetic	102
Harm to self	71
Comic images	12
Criticism of definitions	4
Total	562

Source: Philo, G. et al. "The impact of the mass media on public images of mental illness: Media content and audience belief." *Health Education Journal,* Vol. 53, No. 3, Sept. 1994, p. 274 (Table 1).

a. List the sample points for this experiment.
b. Explain why the sample points, part a, are not equally likely to occur.
c. Assign reasonable probabilities to the sample points.
d. Find the probability that a randomly selected story on mental health is portrayed sympathetically or comically by the Scottish media.

3.156 Chemical insect attractant. An entomologist is studying the effect of a chemical sex attractant (pheromone) on insects. Several insects are released at a site equidistant from the pheromone under study and a control substance. If the pheromone has an effect, more insects will travel toward it rather than toward the control. Otherwise, the insects are equally likely to travel in either direction. Suppose the pheromone under study has no effect, so that it is equally likely that an insect will move toward either the pheromone or the control. Suppose five insects are released:
a. List or count the number of different way the insects can travel.
b. What is the chance that all five travel toward the pheromone?
c. What is the chance that exactly four travel toward the pheromone?
d. What inference would you make if the event in part c actually occurs? Explain.

3.157 Factors identifying urban counties. *Urban* and *rural* describe geographic areas upon which land zoning regulations, school district policy, and public service policy are often set. However, the characteristics of urban/rural areas are not clearly defined. Researchers at the University of Nevada (Reno) asked a sample of county commissioners to give their perception of the single most important factor in identifying urban counties. (*Professional Geographer,* Feb. 2000.) In all, five factors were mentioned by the commissioners: total population, agricultural change, presence of industry, growth, and population concentration. The survey results are displayed in the pie chart. Suppose one of the commissioners is selected at random, and the most important factor specified by the commissioner is recorded.

a. List the sample points for this experiment.
b. Assign reasonable probabilities to the sample points.
c. Find the probability that the most important factor specified by the commissioner is population-related.

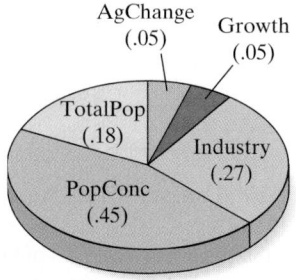

Applying the Concepts—Intermediate

3.158 Study of agressiveness and birth order. Psychologists tend to believe that there is a relationship between aggressiveness and order of birth. To test this belief, a psychologist chose 500 elementary school students at random and administered each a test designed to measure the student's aggressiveness. Each student was classified according to one of four categories. The percentages of students falling in the four categories are shown here.

	Firstborn	Not Firstborn
Aggressive	15%	15%
Not Aggressive	25%	45%

a. If one student is chosen at random from the 500, what is the probability that the student is firstborn?
b. What is the probability that the student is aggressive?
c. What is the probability that the student is aggressive, given the student was firstborn?
d. If

    A: {Student chosen is aggressive}
    B: {Student chosen is firstborn}

are A and B independent? Explain.

3.159 Elderly and their roles. Refer to the *Psychology and Aging* (Dec. 2000) study of elderly people and their roles, Exercise 2.10 (p. 40). The table summarizing the most salient roles identified by each in a national sample of 1,102 adults, 65 years or older, is reproduced on p. 182.
a. What is the probability that a randomly selected elderly person feels his/her most salient role is a spouse?
b. What is the probability that a randomly selected elderly person feels his/her most salient role is a parent or grandparent?
c. What is the probability that a randomly selected elderly person feels his/her most salient role does not involve a spouse or relative of any kind?

Most Salient Role	Number
Spouse	424
Parent	269
Grandparent	148
Other relative	59
Friend	73
Homemaker	59
Provider	34
Volunteer, club, church member	36
Total	**1,102**

Source: Krause, N., and Shaw, B. A. "Role-specific feelings of control and mortality." *Psychology and Aging,* Vol. 15, No. 4, Dec. 2000 (Table 2).

3.160 Mental health study. A Victoria University psychologist investigated whether Asian immigrants to Australia differed from Anglo-Australians in their attitudes toward mental illness. (*Community Mental Health Journal*, Feb. 1999.) Each in a sample of 139 Australian students was classified according to ethnic group and degree of contact with mentally ill people. The number in each category is displayed in the table. Suppose we randomly select one of the 139 students.
  a. Find the probability that the student is an Anglo-Australian who has had little or no contact with mentally ill people.
  b. Find the probability that the student has had close contact with mentally ill people.
  c. Find the probability that the student is not an Anglo-Australian.
  d. Find the probability that the student has had close contact with mentally ill people or is an Asian immigrant.
  e. Given the student is an Asian immigrant, what is the probability that he or she has had little or no contact with mentally ill patients?
  f. Given the student is Anglo-Australian, what is the probability that he or she has had little or no contact with mentally ill patients?
  g. Which ethnic group, Anglo-Australians or Asian immigrants, is most likely to have had little or no contact with mentally ill patients?

Mental Illness	Contact			
	Little/None	Some	Close	Totals
Anglo-Australian	30	17	16	63
Short-Term Asian	17	5	0	22
Long-Term Asian	20	3	1	24
Other	15	11	4	30
Totals	82	36	21	139

Source: Fan, C. "A comparison of attitudes towards mental illness and knowledge of mental health services between Australian and Asian students." *Community Mental Health Journal,* Vol. 35, No. 1, Feb. 1999, p. 54 (Table 2).

3.161 Shooting free throws. In college basketball games a player may be afforded the opportunity to shoot two consecutive foul shots (free throws).
  a. Suppose a player who makes (i.e., scores on) 80% of his foul shots has been awarded two free throws. If the two throws are considered independent, what is the probability that the player makes both shots? Exactly one? Neither shot?
  b. Suppose a player who makes 80% of his first attempted foul shots has been awarded two free throws, and the outcome on the second shot is dependent on the outcome of the first shot. In fact, if this player makes the first shot, he makes 90% of the second shots; and if he misses the first shot, he makes 70% of the second shots. In this case, what is the probability that the player makes both shots? Exactly one? Neither shot?
  c. In parts **a** and **b**, we considered two ways of *modeling* the probability a basketball player makes two consecutive foul shots. Which model do you think is a more realistic attempt to explain the outcome of shooting foul shots; that is, do you think two consecutive foul shots are independent or dependent? Explain.

3.162 Probability of making a sale. The probability that an Avon salesperson sells beauty products to a prospective customer on the first visit to the customer is .4. If the salesperson fails to make the sale on the first visit, the probability that the sale will be made on the second visit is .65. The salesperson never visits a prospective customer more than twice. What is the probability that the salesperson will make a sale to a particular customer?

3.163 Kidney transplant patients. Sandoz, a pharmaceutical firm, reports that kidney transplant patients who receive the drug cyclosporine have an 80% chance of surviving the first year. Suppose a hospital performs four kidney transplants, and all patients receive the drug cyclosporine. Assuming one patient's survival is independent of another's, what is the probability that
  a. All four patients are alive at the end of 1 year?
  b. None of the four patients is alive at the end of 1 year?
  c. At least one of the patients is alive at the end of 1 year?

*3.164 Producing electronic fuses. A manufacturing operation utilizes two production lines to assemble electronic fuses. Both lines produce fuses at the same rate and generally produce 2.5% defective fuses. However, production line 1 recently suffered mechanical difficulty and produced 6.0% defectives during a three-week period. This situation was not known until several lots of electronic fuses produced in this period were shipped to customers. If one of the two fuses tested by a customer was found to be defective, what is the probability that the lot from which it came was produced on malfunctioning line 1? (Assume all the fuses in the lot were produced on the same line.)

3.165 Series and parallel systems. Consider two system shown in the schematics. System A operates properly

System A: Three Components in Series

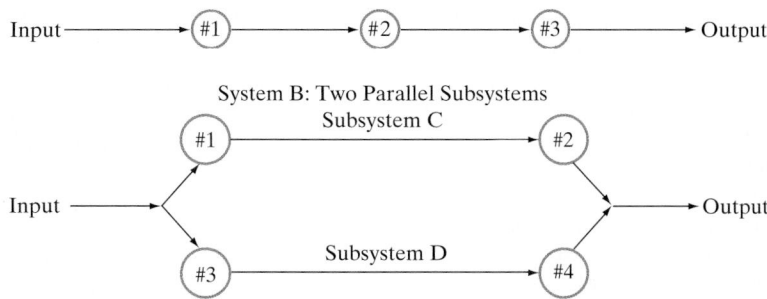

System B: Two Parallel Subsystems

only if all three components operate properly. (The three components are said to operate *in series*.) The probability of failure for system A components 1, 2, and 3 are .12, .09 and .11 respectively. Assume the components operate independently of each other.

System B comprises two subsystems said to operate *in parallel*. Each subsystem has two components that operate in series. System B will operate properly as long as at least one of the subsystems functions properly. The probability of failure for each component in the system is .1. Assume the components operate independently of each other.

a. Find the probability that System A operates properly.

b. What is the probability that at least one of the components in System A will fail and therefore that the system will fail?

c. Find the probability that System B operates properly.

d. Find the probability that exactly one subsystem in System B fails.

e. Find the probability that System B fails to operate properly.

f. How many parallel subsystems like the two shown here would be required to guarantee that the system would operate properly at least 99% of the time?

*3.166 Bottles produced at a brewery. A small brewery has two bottling machines. Machine A produces 75% of the bottles and machine B produces 25%. One out of every 20 bottles filled by A is rejected for some reason, while one out of every 30 bottles from B is rejected.

a. What proportion of bottles is rejected?

b. What is the probability that a randomly selected bottle comes from machine A, given that it is accepted?

*3.167 Allocating TVs to stores. Suppose a manufacturer of big-screen plasma television sets is prepared to send 20 new sets to three retail dealers. Dealer A is to get ten of the sets, dealer B six of the sets, and dealer C four of the sets.

a. In how many different ways can the 20 sets be divided among the dealers in the prescribed numbers?

b. Assuming the sets are randomly divided among the dealers and there are three defective sets among the 20, what is the probability that dealer A gets all three defective sets?

c. Making the assumptions given in part **b**, what is the probability that dealer A gets two defective sets and dealer B the other defective set?

*3.168 Dream experiment in psychology. A clinical psychologist is given tapes of 20 subjects discussing their recent dreams. The psychologist is told that 10 are high-anxiety individuals and 10 are low-anxiety individuals. The psychologist's task is to select the 10 high-anxiety subjects.

a. Count the number of sample points for this experiment.

b. Assuming that the psychologist is guessing, assign probabilities to each of the sample points.

c. Find the probability that the psychologist guesses all classifications correctly.

d. Find the probability that the psychologist guesses at least nine of the 10 high-anxiety subjects correctly.

Applying the Concepts—Advanced

3.169 Odds of winning a horse race. Handicappers for horse races express their belief about the probabilities of each horse winning a race in terms of **odds**. If the probability of event E is $P(E)$, then the *odds in favor of E* are $P(E)$ to $1 - P(E)$. Thus, if a handicapper assesses a probability of .25 that Smarty Jones will win the Belmont Stakes, the odds in favor of Smarty Jones are $^{25}/_{100}$ to $^{75}/_{100}$, or 1 to 3. It follows that the *odds against E* are $1 - P(E)$ to $P(E)$, or 3 to 1 against a win by Smarty Jones. In general, if the odds in favor of event E are a to b, then $P(E) = a/(a + b)$.

a. A second handicapper assesses the probability of a win by Smarty Jones to be $^1/_3$. According to the second handicapper, what are the odds in favor of a Smarty Jones win?

b. A third handicapper assesses the odds in favor of Smarty Jones to be 1 to 1. According to the third handicapper, what is the probability of a Smarty Jones win?

c. A fourth handicapper assesses the odds against Smarty Jones winning to be 3 to 2. Find this handicapper's assessment of the probability that Smarty Jones will win.

3.170 Effectiveness of an antipneumonia vaccine. Pneumovax is an antipneumonia vaccine designed especially for elderly or debilitated patients who are usually the most vulnerable to bacterial pneumonia. The vaccine is 90% effective in stimulating the production of antibodies to pneumonia-producing bacteria (i.e., it is 90% successful in preventing a person exposed to pneumonia-producing bacteria from acquiring the disease). Suppose the probability that an elderly or debilitated person is exposed to these bacteria is .40 (whether inoculated or not) and, after being exposed, the probability that the person will contract bacterial pneumonia if not inoculated with the vaccine is .95. Find the probability that an elderly or debilitated person inoculated with this new vaccine acquires pneumonia. What is the probability if this person has not been inoculated?

3.171 Finding an organ transplant match. One of the problems encountered in organ transplants is the body's rejection of the transplanted tissue. If the antigens attached to the tissue cells of the donor and receiver match, the body will accept the transplanted tissue. While the antigens in identical twins always match, the probability of a match in other siblings is .25 and that of a match in two people from the population at large is .001. Suppose you need a kidney, and you have two brothers and a sister.

a. If one of your three siblings offers a kidney, what is the probability that the antigens will match?

b. If all three siblings offer a kidney, what is the probability that all three antigens will match?

c. If all three siblings offer a kidney, what is the probability that none of the antigens will match?

d. Repeat parts **b** and **c**, this time assuming that the three donors were obtained from the population at large.

3.172 Chance of winning blackjack. Blackjack, a favorite game of gamblers, is played by a dealer and at least one opponent. At the outset of the game, two cards of a 52-card bridge deck are dealt to the player and two cards to the dealer. Drawing an ace and a face card is called *blackjack*. If the dealer does not draw a blackjack and the player does, the player wins. If both the dealer and player draw blackjack, a "push" (i.e., a tie) occurs.

a. What is the probability that the dealer will draw a blackjack?

b. What is the probability that the player wins with a blackjack?

3.173 Accuracy of pregnancy tests. Seventy-five percent of all women who submit to pregnancy tests are really pregnant. A certain pregnancy test gives a *false positive* result with probability .02 and a *valid positive result* with probability .99. If a particular woman's test is positive, what is the probability that she really is pregnant? [*Hint:* If A is the event that a woman is pregnant and B is the event that the pregnancy test is positive,

then B is the union of the two mutually exclusive events $A \cap B$ and $A^c \cap B$. Also, the probability of a false positive result may be written as $P(B|A^c) = .02$.]

3.174 Language impairment in children. Children who develop unexpected difficulties with acquisition of spoken language are often diagnosed with specific language impairment (SLI). A study published in the *Journal of Speech, Language, and Hearing Research* (Dec. 1997) investigated the incidence of SLI in kindergarten children. As an initial screen, each in a national sample of over 7,000 children was given a test for language performance. The percentages of children who passed and failed the screen were 73.8% and 26.2%, respectively. All children who failed the screen were tested clinically for SLI. About one-third of those who passed the screen were randomly selected and also tested for SLI. The percentage of children diagnosed with SLI in the "failed screen" group was 20.5%; the percentage diagnosed with SLI in the "pass screen" group was 2.8%.

a. For this problem, let "pass" represent a child who passed the language performance screen, "fail" represent a child who failed the screen, and "SLI" represent a child diagnosed with SLI. Now find each of the following probabilities: $P(\text{Pass})$, $P(\text{Fail})$, $P(\text{SLI}|\text{Pass})$, and $P(\text{SLI}|\text{Fail})$.

b. Use the probabilities, part **a**, to find $P(\text{Pass} \cap \text{SLI})$ and $P(\text{Fail} \cap \text{SLI})$. What probability law did you use to calculate these probabilities?

c. Use the probabilities, part **b**, to find $P(\text{SLI})$. What probability law did you use to calculate this probability?

*3.175 The perfect bridge hand. According to a morning news program, a very rare event recently occurred in Dubuque, Iowa. Each of four women playing bridge was astounded to note that she had been dealt a perfect bridge hand. That is, one woman was dealt all 13 spades, another all 13 hearts, another all the diamonds, and another all the clubs. What is the probability of this rare event?

3.176 Chance of winning at "craps." A version of the dice game "craps" is played in the following manner: A player starts by rolling two dice. If the result is a 7 or 11, the player wins. For most other sums appearing on the dice, the player continues to roll the dice until that sums recurs (in which case the player loses). But if on any roll the outcome is 2 or 3 (called "craps"), the game is over, and the player loses.

a. What is the probability that a player wins the game on the first roll of the dice? (Assume the dice are balanced.)

b. What is the probability that a player loses the game on the first roll of the dice?

c. If the player throws a total of 4 on the first roll, what is the probability that the game ends on the next roll?

Critical Thinking Challenges

3.177 Most likely coin flip sequence. In *Parade Magazine*'s (Nov. 26, 2000) column "Ask Marilyn," the following question was posed: "I have just tossed a [balanced] coin 10 times, and I ask you to guess which of the following three sequences was the result. One (and only one) of the sequences is genuine."

1. H H H H H H H H H H
2. H H T T H T T H H H
3. T T T T T T T T T T

Marilyn's answer to the question posed was, "Though the chances of the three specific sequences occurring randomly are equal ... it's reasonable for us to choose sequence (2) as the most likely genuine result." Do you agree?

3.178 Flawed Pentium computer chip. In October 1994, a flaw was discovered in the Pentium microchip installed in personal computers. The chip produced an incorrect result when dividing two numbers. Intel, the manufacturer of the Pentium chip, initially announced that such an error would occur once in 9 billion divides, or "once in every 27,000 years" for a typical user; consequently, it did not immediately offer to replace the chip.

Depending on the procedure, statistical software packages (e.g., SAS) may perform an extremely large number of divisions to produce the required output.

For heavy users of the software, 1 billion divisions over a short time frame is not unusual. Will the flawed chip be a problem for a heavy SAS user? [*Note:* Two months after the flaw was discovered, Intel agreed to replace all Pentium chips free of charge.]

3.179 "Let's Make a Deal." Marilyn vos Savant, who is listed in *Guinness Book of World Records Hall of Fame* for "Highest IQ," writes a weekly column in the Sunday newspaper supplement, *Parade Magazine*. Her column, "Ask Marilyn," is devoted to games of skill, puzzles, and mind-bending riddles. (See Exercise 3.177.) In one issue (*Parade Magazine*, Feb. 24, 1991), vos Savant posed the following question:

Suppose you're on a game show, and you're given a choice of three doors. Behind one door is a car; behind the others, goats. You pick a door — say, #1 — and the host, who knows what's behind the doors, opens another door — say #3 — which has a goat. He then says to you, "Do you want to pick door #2?" Is it to your advantage to switch your choice?

Marilyn's answer: "Yes, you should switch. The first door has a $1/3$ chance of winning [the car], but the second has a $2/3$ chance [of winning the car]." Predictably, vos Savant's surprising answer elicited thousands of critical letters, many of them from Ph.D. mathematicians, who disagreed with her. Who is correct, the Ph.D.'s or Marilyn?

Student Projects

Obtain a standard deck of 52 playing cards (the kind commonly used for bridge, poker, or solitaire). An experiment will consist of drawing one card at random from the deck of cards and recording which card was observed. This random drawing will be simulated by shuffling the deck thoroughly and observing the top card. Consider the following two events:

A: {Card observed is a heart}
B: {Card observed is an ace, king, queen, or jack}

a. Find $P(A)$, $P(B)$, $P(A \cap B)$, and $P(A \cup B)$.

b. Conduct the experiment 10 times, and record the observed card each time. Be sure to return the observed card each time, and thoroughly shuffle the deck before making the draw. After you've observed 10 cards, calculate the proportion of observations that satisfy event A, event B, event $A \cap B$, and event $A \cup B$. Compare the

observed proportions with the true probabilities calculated in part **a**.

c. Conduct the experiment 40 more times to obtain a total of 50 observed cards. Now calculate the proportion of observations that satisfy event A, event B, event $A \cap B$, and event $A \cup B$. Compare these proportions with those found in part **b** and the true probabilities found in part **a**.

d. Conduct the experiment 50 more times to obtain a total of 100 observations. Compare the observed proportions for the 100 trials with those found previously. What comments do you have concerning the different proportions found in parts **b**, **c**, and **d** as compared to the true probabilities found in part **a**? How do you think the observed proportions and true probabilities would compare if the experiment were conducted 1,000 times? 1 million times?

REFERENCES

Bennett, D. J. *Randomness*. Cambridge, Mass.: Harvard University Press, 1998.

Epstein, R. A. *The Theory of Gambling and Statistical Logic*, rev. ed. New York: Academic Press, 1977.

Feller, W. *An Introduction to Probability Theory and Its Applications*, 3rd ed., Vol. 1. New York: Wiley, 1968.

Lindley, D. V. *Making Decisions*, 2nd ed. London: Wiley, 1985.

Parzen, E. *Modern Probability Theory and Its Applications*. New York: Wiley, 1960.

Wackerly, D., Mendenhall, W., and Scheaffer, R. L. *Mathematical Statistics with Applications*, 6th ed. Boston: Duxbury, 2002.

Williams, B. *A Sampler on Sampling*. New York: Wiley, 1978.

Winkler, R. L. *An Introduction to Bayesian Inference and Decision*. New York: Holt, Rinehart and Winston, 1972.

Wright, G., and Ayton, P., eds. *Subjective Probability*. New York: Wiley, 1994.

Using Technology

Generating a Random Sample Using MINITAB

To obtain a random sample of observations (cases) from a data set stored in the MINITAB worksheet, click on the "Calc" button on the MINITAB menu bar, then click on "Random Data", and finally, click on "Sample from Columns", as shown in Figure 3.M.1. The resulting dialog box appears as shown in Figure 3.M.2. Specify the sample size (i.e., number of rows), the variable(s) to be sampled, and the column(s) where you want to save the sample. Click "OK" and the MINITAB worksheet will reappear with the values of the variable for the selected (sampled) cases in the column specified.

In MINITAB, you can also generate a sample of case numbers. From the MINITAB menu, click on the "Calc" button, then click on "Random Data" and finally, click on the "Uniform" option (see Figure 3.M.1). The resulting dialog box

Figure 3.M.1
MINITAB Menu Options for Sampling Your Data

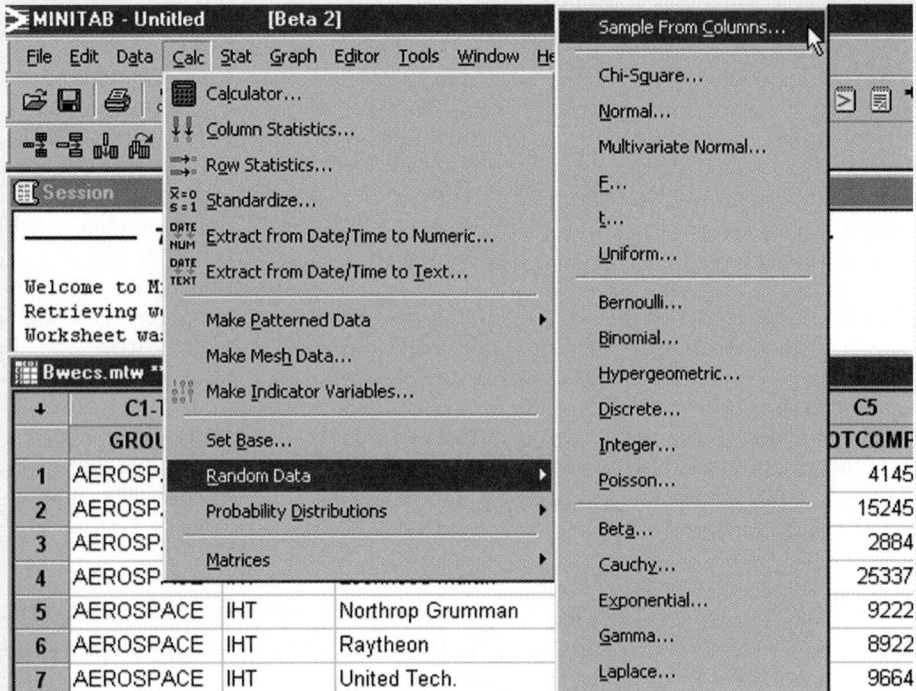

Figure 3.M.2
MINITAB Options for
Selecting a Random Sample
from Worksheet Columns

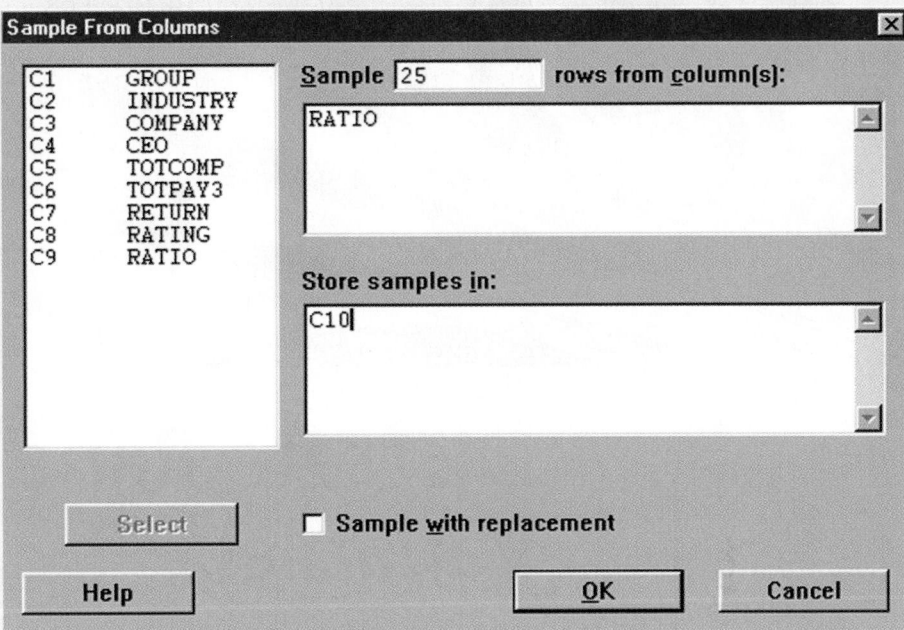

Figure 3.M.3
MINITAB Options for
Selecting a Random Sample
of Cases

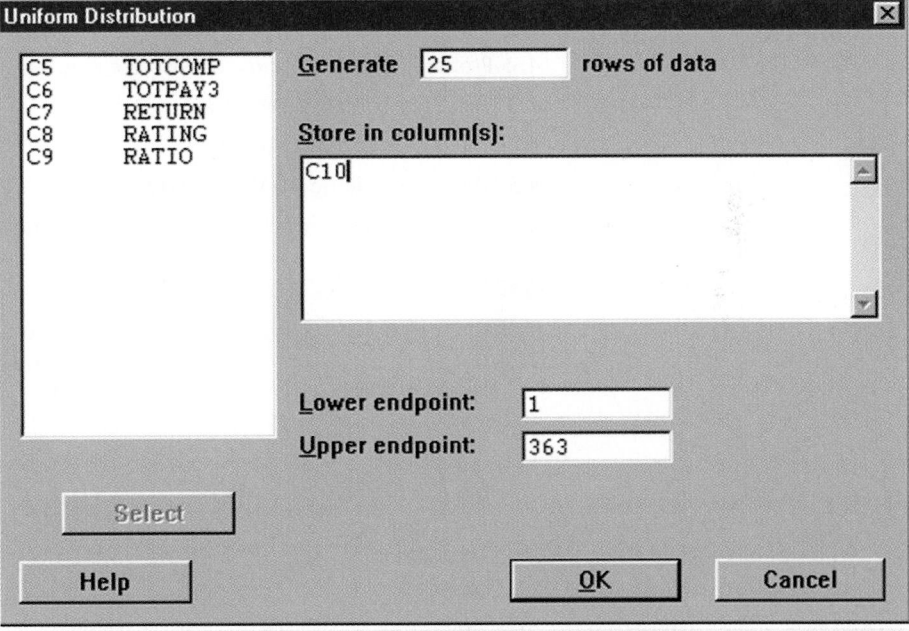

appears as shown in Figure 3.M.3. Specify the number of cases (rows) (i.e., the sample size) and the column where the case numbers selected will be stored. Click "OK" and the MINITAB worksheet will reappear with the case numbers for the selected (sampled) cases in the column specified.

[*Note:* If you want the option of generating the same (identical) sample multiple times from the data set, then first click on the "Set Base" option shown in Figure 3.M.1. Specify an integer in the resulting dialog box. If you always select the same integer, MINITAB will select the same sample when you choose the random sampling options.]

4

Discrete Random Variables

Contents

Statistics in Action

Probability in a Reverse Cocaine Sting

Using Technology

Binomial, Poisson, and Hypergeometric Probabilities Using MINITAB

Where We've Been

- Used probability to make an inference about a population from data in an observed sample
- Used probability to measure the reliability of the inference

Where We're Going

- Develop the notion of a random variable.
- Learn that many types of numerical data are observed values of discrete random variables.
- Study several important types of discrete random variables and their probability models.

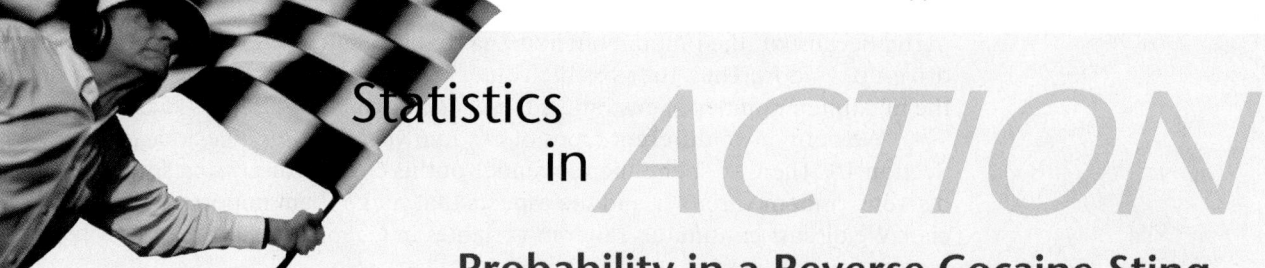

Statistics in *ACTION*

Probability in a Reverse Cocaine Sting

The American Statistician (May 1991) described an interesting application of a discrete probability distribution in a case involving illegal drugs. It all started with a "bust" in a mid-sized Florida city. During the bust, police seized approximately 500 foil packets of a white, powdery substance, presumably cocaine. Since it is not a crime to buy or sell non-narcotic cocaine look-a-likes (e.g., inert powders), detectives had to prove that the packets contained genuine cocaine in order to convict their suspects of drug trafficking. When the police laboratory randomly selected and chemically tested four of the packets, all four tested positive for cocaine. This finding led to the conviction of the traffickers.

After the conviction, the police decided to use the remaining foil packets (i.e., those not tested) in reverse sting operations. Two of these packets were randomly selected and sold by undercover officers to a buyer. Between the sale and the arrest, however, the buyer disposed of the evidence. The key question is, Beyond a reasonable doubt, did the defendant really purchase cocaine?

In court, the defendant's attorney argued that his client should not be convicted because the police could not prove that the missing foil packets contained cocaine. The police contended, however, that since four of the original packets tested positive for cocaine, the two packets sold in the reverse sting were also highly likely to contain cocaine. In this chapter, two Statistics in Action Revisited examples demonstrate how to use probability models to solve the dilemma. (The case represented Florida's first cocaine-possession conviction without the actual physical evidence.)

Statistics in Action Revisited

- Using the Binomial Model to Solve the Cocaine Sting Case (p. 215)
- Using the Hypergeometric Model to Solve the Cocaine Sting Case (p. 226)

You may have noticed that many of the examples of experiments in Chapter 3 generated quantitative (numerical) observations. The unemployment rate, the percentage of voters favoring a particular candidate, the cost of textbooks for a school term, and the amount of pesticide in the discharge waters of a chemical plant are all examples of numerical measurements of some phenomenon. Thus, most experiments have sample points that correspond to values of some numerical variable.

To illustrate, consider the coin-tossing experiment of Chapter 3. Figure 4.1 is a Venn diagram showing the sample points when two coins are tossed and the up faces (heads or tails) of the coins are observed. One possible numerical outcome is the total number of heads observed. These values (0, 1, or 2) are shown in parentheses on the Venn diagram, one numerical value associated with each sample point. In the jargon of probability, the variable "total number of heads observed in two tosses of a coin" is called a *random variable*.

HH ○ (2)	*HT* ○ (1)
TT ○ (0)	*TH* ○ (1)

S

Figure 4.1

Venn Diagram for
Coin-Tossing Experiment

DEFINITION 4.1

A **random variable** is a variable that assumes numerical values associated with the random outcomes of an experiment, where one (and only one) numerical value is assigned to each sample point.

The term *random variable* is more meaningful than the term *variable* alone because the adjective *random* indicates that the coin-tossing experiment may result in one of the several possible values of the variable—0, 1, and 2—according to the *random* outcome of the experiment, *HH, HT, TH,* and *TT*. Similarly, if the experiment is to count the number of customers who use the drive-up window of a bank each day, the random variable (the number of customers) will vary from day to day, partly because of the random phenomena that influence whether customers use the drive-up window. Thus, the possible values of this random variable range from 0 to the maximum number of customers the window could possibly serve in a day.

We define two different types of random variables, *discrete* and *continuous*, in Section 4.1. Then we spend the remainder of this chapter discussing specific types of discrete random variables and the aspects that make them important to the statistician. We discuss continuous random variables in Chapter 5.

4.1 Two Types of Random Variables

Recall that the sample point probabilities corresponding to an experiment must sum to 1. Dividing one unit of probability among the sample points in a sample space and consequently assigning probabilities to the values of a random variable is not always as easy as the examples in Chapter 3 might lead you to believe. If the number of sample points can be completely listed, the job is straightforward. But if the experiment results in an infinite number of sample points that are impossible to list, the task of assigning probabilities to the sample points is impossible without the aid of a probability model. The next three examples demonstrate the need for different probability models depending on the number of values that a random variable can assume.

EXAMPLE 4.1   VALUES OF A DISCRETE RANDOM VARIABLE

Problem   A panel of 10 experts for the *Wine Spectator* (a national publication) is asked to taste a new white wine and assign a rating of 0, 1, 2, or 3. A score is then obtained by adding together the ratings of the 10 experts. How many values can this random variable assume?

Solution   A sample point is a sequence of 10 numbers associated with the rating of each expert. For example, one sample point is

$$\{1, 0, 0, 1, 2, 0, 0, 3, 1, 0\}.$$

The random variable assigns a score to each one of these sample points by adding the 10 numbers together. Thus, the smallest score is 0 (if all 10 ratings are 0) and the largest score is 30 (if all 10 ratings are 3). Since every integer between 0 and 30 is a possible score, the random variable denoted by the symbol x can assume 31 values. Note that the value of the random variable for the sample point above is $x = 8$.*

Look Back   This is an example of a *discrete random variable*, since there is a finite number of distinct possible values. Whenever all the possible values a random variable can assume can be listed (or *counted*), the random variable is *discrete*.

■ ■ ■

EXAMPLE 4.2   VALUES OF A DISCRETE RANDOM VARIABLE

Problem   Suppose the Environmental Protection Agency (EPA) takes readings once a month on the amount of pesticide in the discharge water of a chemical company. If the amount of pesticide exceeds the maximum level set by the EPA, the company is

*The standard mathematical convention is to use a capital letter (e.g., X) to denote the theoretical random variable. The possible values (or realizations) of the random variable are typically denoted with a lowercase letter (e.g., x). Thus, in Example 4.1, the random variable X can take on the values $x = 0, 1, 2, \ldots, 30$. Since this notation can be confusing for introductory statistics students, we simplify the notation by using the lowercase x to represent the random variable throughout.

forced to take corrective action and may be subject to penalty. Consider the random variable, number, x, of months before the company's discharge exceeds the EPA's maximum level. What values can x assume?

Solution The company's discharge of pesticide may exceed the maximum allowable level on the first month of testing, the second month of testing, etc. It is possible that the company's discharge will *never* exceed the maximum level. Thus, the set of possible values for the number of months until the level is first exceeded is the set of all positive integers $1, 2, 3, 4, \ldots$.

If we can list the values of a random variable x, even though the list is never-ending, we call the list *countable* and the corresponding random variable *discrete*. Thus, the number of months until the company's discharge first exceeds the limit is a *discrete random variable*.

Now Work   *Exercise 4.7*

■ ■ ■

EXAMPLE 4.3   VALUES OF A CONTINUOUS RANDOM VARIABLE.

Problem Refer to Example 4.2. A second random variable of interest is the amount x of pesticide (in milligrams per liter) found in the monthly sample of discharge waters from the chemical company. What values can this random variable assume?

Solution Unlike the *number* of months before the company's discharge exceeds the EPA's maximum level, the set of all possible values for the *amount* of discharge *cannot* be listed (i.e., is not countable). The possible values for the amount x of pesticide would correspond to the points on the interval between 0 and the largest possible value the amount of the discharge could attain, the maximum number of milligrams that could occupy 1 liter of volume. (Practically, the interval would be much smaller, say, between 0 and 500 milligrams per liter.)

Look Back When the values of a random variable are not countable but instead correspond to the points on some interval, we call it a *continuous random variable*. Thus, the *amount* of pesticide in the chemical plant's discharge waters is a *continuous random variable*.

Now Work   *Exercise 4.8*

■ ■ ■

DEFINITION 4.2

Random variables that can assume a *countable* number of values are called **discrete**.

DEFINITION 4.3

Random variables that can assume values corresponding to any of the points contained in one or more intervals are called **continuous**.

The following are examples of discrete random variables:

1. The number of seizures an epileptic patient has in a given week: $x = 0, 1, 2, \ldots$
2. The number of voters in a sample of 500 who favor impeachment of the president: $x = 0, 1, 2, \ldots, 500$

3. The number of students applying to medical schools this year: $x = 0, 1, 2, \ldots$

4. The change received for paying a bill: $x = 1\cent, 2\cent, 3\cent, \ldots, \$1, \ldots$

5. The number of customers waiting to be served in a restaurant at a particular time: $x = 0, 1, 2, \ldots$

Note that several of the examples of discrete random variables begin with the words *The number of* .... This wording is very common, since the discrete random variables most frequently observed are counts. The following are examples of continuous random variables:

1. The length of time (in seconds) between arrivals at a hospital clinic: $0 \le x \le \infty$ (infinity)

2. The length of time (in minutes) it takes a student to complete a one-hour exam: $0 \le x \le 60$

3. The amount (in ounces) of carbonated beverage loaded into a 12-ounce can in a can-filling operation: $0 \le x \le 12$

4. The depth (in feet) at which a successful oil drilling venture first strikes oil: $0 \le x \le c$, where c is the maximum depth obtainable

5. The weight (in pounds) of a food item bought in a supermarket: $0 \le x \le 500$ [*Note:* Theoretically, there is no upper limit on x, but it is unlikely that it would exceed 500 pounds.]

Discrete random variables and their probability distributions are discussed in this chapter. Continuous random variables and their probability distributions are the topic of Chapter 5.

Exercises 4.1–4.8

Understanding the Principles

4.1 What is a random variable?

4.2 How do discrete and continuous random variables differ?

Applying the Concepts—Basic

4.3 **Type of Random Variable.** Classify the following random variables according to whether they are discrete or continuous:
 a. The number of words spelled correctly by a student on a spelling test
 b. The amount of water flowing through the Hoover Dam in a day
 c. The length of time an employee is late for work
 d. The number of bacteria in a particular cubic centimeter of drinking water
 e. The amount of carbon monoxide produced per gallon of unleaded gas
 f. Your weight

4.4 **Type of Random Variable.** Identify the following random variables as discrete or continuous:
 a. The amount of flu vaccine in a syringe
 b. The heart rate (number of beats per minute) of an American male
 c. The time it takes a student to complete an examination
 d. The barometric pressure at a given location

 e. The number of registered voters who vote in a national election
 f. Your score on the Scholastic Assessment Test (SAT)

4.5 **Type of Random Variable.** Identify the following variables as discrete or continuous:
 a. The reaction time difference to the same stimulus before and after training
 b. The number of violent crimes committed per month in your community
 c. The number of commercial aircraft near-misses per month
 d. The number of winners each week in a state lottery
 e. The number of free throws made per game by a basketball team
 f. The distance traveled by a school bus each day

4.6 **NHTSA crash tests.** The National Highway Traffic Safety Administration (NHTSA) has developed a driver-side "star" scoring system for crash-testing new cars. Each crash-tested car is given a rating ranging from one star (*) to five stars (*****); the more stars in the rating, the better the level of crash protection in a head-on collision. Suppose that a car is selected and its driver-side star rating is determined. Let x equal the number of stars in the rating. Is x a discrete or continuous random variable?

4.7 **Customers in line at a Subway shop.** The number of customers, x, waiting in line to order sandwiches at a Subway shop at noon is of interest to the store manager.

What values can x assume? Is x a discrete or continuous random variable?

⊙ **GOBIANTS**
4.8 Mongolian desert ants. Refer to the *Journal of*
NW *Biogeography* (Dec. 2003) study of ants in Mongolia,

Exercises 2.66 and 2.147 (pp. 66, 100). Two of the several variables recorded at each of 11 study sites were annual rainfall (in millimeters) and number of ant species. Identify these variables as discrete or continuous.

4.2 Probability Distributions for Discrete Random Variables

A complete description of a discrete random variable requires that we *specify the possible values the random variable can assume* and *the probability associated with each value.* To illustrate, consider Example 4.4.

EXAMPLE 4.4 — FINDING A PROBABILITY DISTRIBUTION

Problem Recall the experiment of tossing two coins (Section 4.1), and let x be the number of heads observed. Find the probability associated with each value of the random variable x, assuming the two coins are fair.

Solution The sample space and sample points for this experiment are reproduced in Figure 4.2. Note that the random variable x can assume values 0, 1, 2. Recall (from Chapter 3) that the probability associated with each of the four sample points is $\frac{1}{4}$. Then, identifying the probabilities of the sample points associated with each of these values of x, we have

$$P(x = 0) = P(TT) = \frac{1}{4}$$
$$P(x = 1) = P(TH) + P(HT) = \frac{1}{4} + \frac{1}{4} = \frac{1}{2}$$
$$P(x = 2) = P(HH) = \frac{1}{4}$$

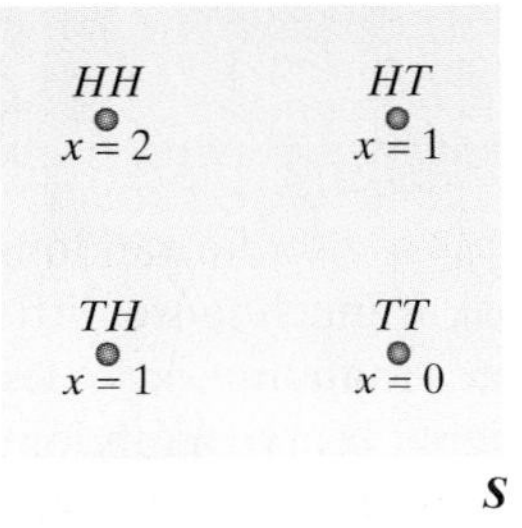

Thus, we now know the values the random variable can assume $(0, 1, 2)$ and how the probability is *distributed over* these values $\left(\frac{1}{4}, \frac{1}{2}, \frac{1}{4}\right)$. This completely describes the random variable and is referred to as the *probability distribution*, denoted by the symbol $p(x)$.* The probability distribution for the coin-toss example is shown in tabular form in Table 4.1 and in graphic form in Figure 4.3. Since the probability distribution for a discrete random variable is concentrated at specific points (values of x), the graph in Figure 4.3a represents the probabilities as the heights of vertical lines over the corresponding values of x. Although the representation of the probability distribution as a histogram, as in Figure 4.3b, is less precise (since the probability is spread over a unit interval), the histogram representation will prove useful when we approximate probabilities of certain discrete random variables in Section 4.4.

Look Back We could also present the probability distribution for x as a formula, but this would unnecessarily complicate a very simple example. We give the formulas for the probability distributions of some common discrete random variables later in this chapter.

Now Work *Exercise 4.15*

Figure 4.2
Venn Diagram for the Two-Coin-Toss Experiment

TABLE 4.1 Probability Distribution for Coin-Toss Experiment: Tabular Form

x	$p(x)$
0	$\frac{1}{4}$
1	$\frac{1}{2}$
2	$\frac{1}{4}$

*In standard mathematical notation, the probability that a random variable X takes on a value x is denoted $P(X = x) = p(x)$. Thus, $P(X = 0) = p(0)$, $P(X = 1) = p(1)$, etc. In this text, we adopt the simpler $p(x)$ notation.

Figure 4.3
Probability Distribution
for Coin-Toss Experiment:
Graphical Form

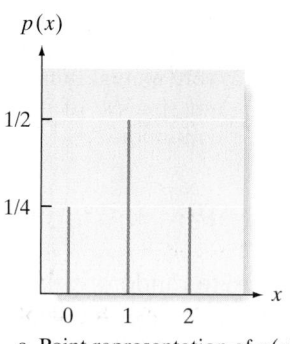

a. Point representation of $p(x)$

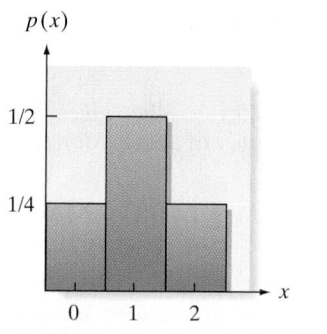

b. Histogram representation of $p(x)$

> **DEFINITION 4.4**
>
> The **probability distribution** of a discrete random variable is a graph, table, or formula that specifies the probability associated with each possible value the random variable can assume.

Two requirements must be satisfied by all probability distributions for discrete random variables.

> **Requirements for the Probability Distribution of a Discrete Random Variable x**
>
> **1.** $p(x) \geq 0$ for all values of x
>
> **2.** $\sum p(x) = 1$
>
> where the summation of $p(x)$ is over all possible values of x.*

Now Work   *Exercise 4.12*

Example 4.4 illustrates how the probability distribution for a discrete random variable can be derived, but for many practical situations the task is much more difficult. Fortunately, many experiments and associated discrete random variables observed in nature possess identical characteristics. Thus, you might observe a random variable in a psychology experiment that would possess the same probability distribution as a random variable observed in an engineering experiment or a social sample survey. We classify random variables according to type of experiment, derive the probability distribution for each of the different types, and then use the appropriate probability distribution when a particular type of random variable is observed in a practical situation. The probability distributions for most commonly occurring discrete random variables have already been derived. This fact simplifies the problem of finding the probability distributions for random variables, as the next example illustrates.

EXAMPLE 4.5

PROBABILITY DISTRIBUTION USING A FORMULA

Problem The distribution of parasites (tapeworms) found in Mediterranean brill fish was studied in the *Journal of Fish Biology* (Aug. 1990). The researchers showed that the distribution of x, the number of brill that must be sampled until a parasitic infection is found in the digestive tract, can be modeled using the formula $p(x) = (.6)(.4)^{x-1}$, $x = 1, 2, 3,$ etc. Find the probability that exactly 3 brill fish must be sampled before a tapeworm is found in the digestive tract.

*Unless otherwise indicated, summations will always be over all possible values of x.

Solution We want to find the probability that $x = 3$. Using the formula, we have

$$p(3) = (.6)(.4)^{3-1} = (.6)(.4)^2 = (.6)(.16) = .096$$

Thus, there is about a 10% chance that exactly 3 fish must be sampled before a tapeworm is found in the digestive tract.

Look Back The probability of interest can also be derived using the principles of probability developed in Chapter 3. The event of interest is $N_1 N_2 P_3$, where N_1 represents no parasite found in the first sampled fish, N_2 represents no parasite found in the second sampled fish, and P_3 represents a parasite found in the third sampled fish. The researchers discovered that the probability of finding a parasite in any sampled fish is .6 (and consequently, the probability of not finding a parasite in any sampled fish is .4). Using the Multiplicative Law of Probability for independent events, the probability of interest is $(.4)(.4)(.6) = .096$.

■ ■ ■

In Sections 4.4 and 4.5, we describe two important types of discrete random variables, give their probability distributions, and explain where and how they can be applied in practice. (Mathematical derivations of the probability distributions will be omitted, but these details can be found in the chapter references.)

But first, in Section 4.3, we discuss some descriptive measures of these sometimes complex probability distributions. Since probability distributions are analogous to the relative frequency distributions of Chapter 2, it should be no surprise that the mean and standard deviation are useful descriptive measures.

Exercises 4.9–4.26

Understanding the Principles

4.9 Give three different ways of representing the probability distribution of a discrete random variable.

Learning the Mechanics

4.10 Consider the following probability distribution:

x	−4	0	1	3
$p(x)$	.1	.2	.4	.3

a. List the values that x may assume.
b. What value of x is most probable?
c. What is the probability that x is greater than 0?
d. What is the probability that $x = -2$?

4.11 A discrete random variable x can assume five possible values: 2, 3, 5, 8, and 10. Its probability distribution is shown here:

x	2	3	5	8	10
$p(x)$	.15	.10		.25	.25

a. What is $p(5)$?
b. What is the probability that x equals 2 or 10?
c. What is $P(x \le 8)$?

4.12 Explain why each of the following is or is not a valid probability distribution for a discrete random variable x:

a.

x	0	1	2	3
$p(x)$	.2	.3	.3	.2

b.

x	−2	−1	0
$p(x)$	.25	.50	.20

c.

x	4	9	20
$p(x)$	−.3	1.0	.3

d.

x	2	3	5	6
$p(x)$	.15	.20	.40	.35

4.13 The random variable x has the following discrete probability distribution:

x	10	11	12	13	14
$p(x)$	.2	.3	.2	.1	.2

Since the values that x can assume are mutually exclusive events, the event $\{x \le 12\}$ is the union of three mutually exclusive events:

$$\{x = 10\} \cup \{x = 11\} \cup \{x = 12\}$$

a. Find $P(x \le 12)$. **b.** Find $P(x > 12)$.
c. Find $P(x \le 14)$. **d.** Find $P(x = 14)$.
e. Find $P(x \le 11 \text{ or } x > 12)$.

4.14 The random variable x has the discrete probability distribution shown here.

x	-2	-1	0	1	2
$p(x)$	.10	.15	.40	.30	.05

a. Find $P(x \le 0)$.
b. Find $P(x > -1)$.
c. Find $P(-1 \le x \le 1)$. $x \le 1$ $x \ge -1$
d. Find $P(x < 2)$.
e. Find $P(-1 < x < 2)$. $x < 2$ $x > -1$
f. Find $P(x < 1)$.

4.15 Toss three fair coins and let x equal the number of heads
NW observed.
a. Identify the sample points associated with this experiment and assign a value of x to each sample point.
b. Calculate $p(x)$ for each value of x.
c. Construct a probability histogram for $p(x)$.
d. What is $P(x = 2 \text{ or } x = 3)$?

Applying the Concepts—Basic

4.16 NHTSA crash tests. Refer to the National Highway Traffic Safety Administration (NHTSA) crash tests of new car models, Exercise 4.6 (p. 192). A summary of the driver-side star ratings for the 98 cars in the **CRASH** file is reproduced in the accompanying MINITAB printout. Assume that one of the 98 cars is selected at random, and let x equal the number of stars in the car's driver-side star rating.
a. Use the information in the printout to find the probability distribution for x.
b. Find $P(x = 5)$.
c. Find $P(x \le 2)$.

Tally for Discrete Variables: DRIVSTAR

```
DRIVSTAR   Count   Percent
    2         4      4.08
    3        17     17.35
    4        59     60.20
    5        18     18.37
   N=        98
```

4.17 Dust mite allergies. A dust mite allergen level that exceeds 2 micrograms per gram ($\mu g/g$) of dust has been associated with the development of allergies. Consider a random sample of four homes and let x be the number of homes with a dust mite level that exceeds 2 $\mu g/g$. The probability distribution for x, based on a May 2000 study

by the National Institute of Environmental Health Sciences, is shown in the following table.

x	0	1	2	3	4
$p(x)$	.09	.30	.37	.20	.04

a. Verify that the sum of the probabilities for x in the table sum to 1.
b. Find the probability that three or four of the homes in the sample have a dust mite level that exceeds 2 $\mu g/g$.
c. Find the probability that fewer than two homes in the sample have a dust mite level that exceeds 2 $\mu g/g$.

4.18 Customer arrivals at Wendy's. The probability distribution of x, the number of customer arrivals per 15-minute period, at a Wendy's restaurant in New Jersey is shown in the table at the bottom of the page.
a. Does this distribution meet the two requirements for the probability distribution of a discrete random variable? Justify your answer.
b. What is the probability that exactly 16 customers enter the restaurant in the next 15 minutes?
c. Find $P(x \le 10)$.
d. Find $P(5 \le x \le 15)$.

Applying the Concepts—Intermediate

4.19 Perfect SAT score. In Exercise 3.26 (p. 132) you learned that five in every 10,000 students who take the Scholastic Assessment Test (SAT) score a perfect 1,600. Consider a random sample of three students who take the SAT. Let x equal the number who score a perfect 1,600.
a. Find $p(x)$ for $x = 0, 1, 2, 3$.
b. Graph $p(x)$.   c. Find $P(x \le 1)$.

4.20 Contaminated gun cartridges. A weapons manufacturer uses a liquid propellant to produce gun cartridges. During the manufacturing process, the propellant can get mixed with another liquid to produce a contaminated cartridge. A University of South Florida statistician, hired by the company to investigate the level of contamination in the stored cartridges, found that 23% of the cartridges in a particular lot were contaminated. Suppose you randomly sample (without replacement) gun cartridges from this lot until you find a contaminated one. Let x be the number of cartridges sampled until a contaminated one is found. It is known that the probability distribution for x is given by the formula:

$$p(x) = (.23)(.77)^{x-1}, \quad x = 1, 2, 3, \dots$$

Table for Exercise 4.18

x	5	6	7	8	9	10	11	12	13	14	15
$p(x)$	.01	.02	.03	.05	.08	.09	.11	.13	.12	.10	.08

x	16	17	18	19	20	21
$p(x)$	.06	.05	.03	.02	.01	.01

Source: Ford, R., Roberts, D., and Saxton, P. *Queuing Models.* Graduate School of Management, Rutgers University, 1992.

a. Find $p(1)$. Interpret this result.
b. Find $p(5)$. Interpret this result.
c. Find $P(x \geq 2)$. Interpret this result.

4.21 Federal civil trial appeals. Refer to the *Journal of the American Law and Economics Association* (Vol. 3, 2001) study of appeals of federal civil trials, Exercise 3.51 (p. 144). A breakdown of the 678 civil cases that were originally tried in front of a judge (rather than a jury) and appealed by either the plaintiff or defendant is reproduced in the table. Suppose each civil case is awarded points (positive or negative) based on the outcome of the appeal for the purpose of evaluating Federal judges. If the appeal is affirmed or dismissed, +5 points are awarded. If the appeal of a plaintiff trial win is reversed, −1 point is awarded. If the appeal of a defendant trial win is reversed, −3 points are awarded. Suppose one of the 678 cases is selected at random, and the number, x, of points awarded is determined. Find and graph the probability distribution for x.

Outcome of Appeal	Number of Cases
Plaintiff trial win—reversed	71
Plaintiff trial win—affirmed/dismissed	240
Defendant trial win—reversed	68
Defendant trial win—affirmed/dismissed	299
Total	**678**

4.22 Successful NFL running plays. In his article "American Football" (*Statistics in Sport*, 1998), Iowa State University statistician Hal Stern evaluates winning strategies of teams in the National Football League (NFL). In a section on estimating the probability of winning a game, Stern used actual NFL play-by-play data to approximate the probabilities associated with certain outcomes (e.g., running plays, short pass plays, and long pass plays). The following table gives the probability distribution for the yardage gained, x, on a running play. (A negative gain represents a loss of yards on the play.)

x, Yards	Probability	x, Yards	Probability
−4	.020	6	.090
−2	.060	8	.060
−1	.070	10	.050
0	.150	15	.085
1	.130	30	.010
2	.110	50	.004
3	.090	99	.001
4	.070		

a. Find the probability of gaining 10 yards or more on a running play.
b. Find the probability of losing yardage on a running play.

4.23 Reliability of a bridge network. A team of Chinese university professors investigated the reliability of several capacitated-flow networks in the journal *Networks* (May 1995). One network examined in the article and illus-

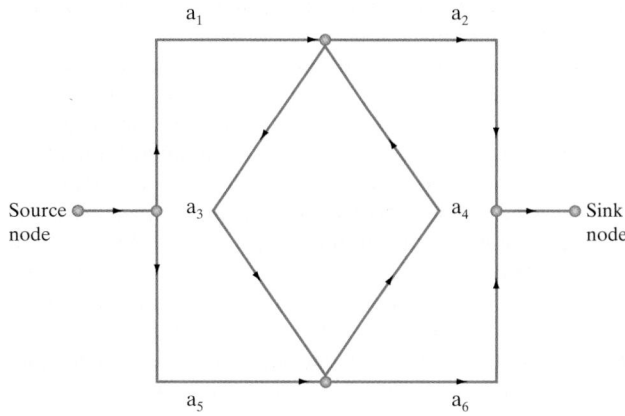

trated here is a bridge network with arcs a_1, a_2, a_3, a_4, a_5, and a_6. The probability distribution of the capacity x for each of the six arcs is provided in the table.

Arc	Capacity x	$p(x)$	Arc	Capacity x	$p(x)$
a_1	3	.60	a_4	1	.90
	2	.25		0	.10
	1	.10			
	0	.05			
a_2	2	.60	a_5	1	.90
	1	.30		0	.10
	0	.10			
a_3	1	.90	a_6	2	.70
	0	.10		1	.25
				0	.05

Source: Lin, J., et al. "On reliability evaluation of capacitated-flow network in terms of minimal pathsets." *Networks,* Vol. 25, No. 3, May 1995, p. 135 (Table 1).

a. Verify that the properties of discrete probability distributions are satisfied for each arc capacity distribution.
b. Find the probability that the capacity for arc a_1 will exceed 1.
c. Repeat part **b** for each of the remaining five arcs.

Applying the Concepts—Advanced

4.24 Bridge network reliability (cont'd). Refer to Exercise 4.23.
a. One path from the source node to the sink node is through arcs a_1 and a_2. Find the probability that the system maintains a capacity of more than 1 through the a_1–a_2 path. (Assume that the arc capacities are independent.)
b. Another path from the source node to the sink node is through a_1, a_3, and a_6. Find the probability that the system maintains a capacity of 1 through the a_1–a_3–a_6 path.

4.25 X and Y chromosomes. Every human possesses two sex chromosomes. A copy of one or the other (equally likely) is contributed to an offspring. Males have one X chromosome and one Y chromosome. Females have two X chromosomes. If a couple has three children, what is the probability that they have at least one boy?

4.26 Punnett square for earlobes. Geneticists use a grid— called a *Punnett square*—to display all possible gene combinations in genetic crosses. (The grid is named for Reginald Punnett, a British geneticist who developed the method in the early 1900s.) The accompanying figure is a Punnett square for a cross involving human earlobes. In humans, free earlobes (E) are dominant over attached earlobes (e). Consequently, the gene pairs EE and Ee will result in free earlobes, while the gene pair ee results in attached earlobes. Consider a couple with genes as shown in the Punnett square. Suppose the couple has seven children. Let x represent the number of children with attached earlobes (i.e., with the gene pair ee). Find the probability distribution of x.

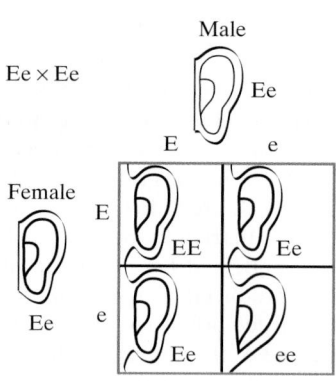

4.3 Expected Values of Discrete Random Variables

If a discrete random variable x were observed a very large number of times and the data generated were arranged in a relative frequency distribution, the relative frequency distribution would be indistinguishable from the probability distribution for the random variable. Thus, the probability distribution for a random variable is a theoretical model for the relative frequency distribution of a population. To the extent that the two distributions are equivalent (and we will assume they are), the probability distribution for x possesses a mean μ and a variance σ^2 that are identical to the corresponding descriptive measures for the population. This section explains how you can find the mean value for a random variable. We illustrate the procedure with an example.

Examine the probability distribution for x (the number of heads observed in the toss of two fair coins) in Figure 4.4. Try to locate the mean of the distribution intuitively. We may reason that the mean μ of this distribution is equal to 1 as follows: In a large number of experiments, $\frac{1}{4}$ should result in $x = 0$, $\frac{1}{2}$ in $x = 1$, and $\frac{1}{4}$ in $x = 2$ heads. Therefore, the average number of heads is

$$\mu = 0\left(\tfrac{1}{4}\right) + 1\left(\tfrac{1}{2}\right) + 2\left(\tfrac{1}{4}\right) = 0 + \tfrac{1}{2} + \tfrac{1}{2} = 1$$

Note that to get the population mean of the random variable x, we multiply each possible value of x by its probability $p(x)$, and then we sum this product over all possible values of x. The *mean of x* is also referred to as the *expected value of x*, denoted $E(x)$.

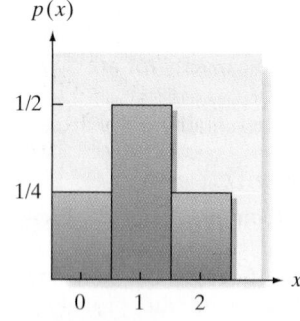

Figure 4.4

Probability Distribution for a Two-Coin Toss

> **DEFINITION 4.5**
>
> The **mean**, or **expected value**, of a discrete random variable x is
>
> $$\mu = E(x) = \Sigma x p(x)$$

The term *expected* is a mathematical term and should not be interpreted as it is typically used. Specifically, *a random variable might never be equal to its "expected value."* Rather, the expected value is the mean of the probability distribution, or a measure of its central tendency. You can think of μ as the mean value of x in a *very large* (actually, *infinite*) number of repetitions of the experiment, where the values of x occur in proportions equivalent to the probabilities of x.

EXAMPLE 4.6   FINDING AN EXPECTED VALUE

Problem   Suppose you work for an insurance company, and you sell a $10,000 1-year term insurance policy at an annual premium of $290. Actuarial tables show that the probability of death during the next year for a person of your customer's age, sex, health, etc., is .001. What is the expected gain (amount of money made by the company) for a policy of this type?

Solution The experiment is to observe whether the customer survives the upcoming year. The probabilities associated with the two sample points, Live and Die, are .999 and .001, respectively. The random variable you are interested in is the gain x, which can assume the values shown in the following table.

Gain x	Sample Point	Probability
$290	Customer lives	.999
$-$9,710	Customer dies	.001

If the customer lives, the company gains the $290 premium as profit. If the customer dies, the gain is negative because the company must pay $10,000, for a net "gain" of $(290 - 10,000) = -$9,710$. The expected gain is therefore

$$\mu = E(x) = \Sigma x p(x)$$
$$= (290)(.999) + (-9,710)(.001) = \$280$$

In other words, if the company were to sell a very large number of 1-year $10,000 policies to customers possessing the characteristics described above, it would (on the average) net $280 per sale in the next year.

Look Back Note that $E(x)$ need not equal a possible value of x. That is, the expected value is $280, but x will equal either $290 or $-$9,710 each time the experiment is performed (a policy is sold and a year elapses). The expected value is a measure of central tendency—and in this case represents the average over a very large number of 1-year policies—but is not a possible value of x.

Now Work   *Exercise 4.34*

■ ■ ■

We learned in Chapter 2 that the mean and other measures of central tendency tell only part of the story about a set of data. The same is true about probability distributions. We need to measure variability as well. Since a probability distribution can be viewed as a representation of a population, we will use the population variance to measure its variability.

The *population variance* σ^2 is defined as the average of the squared distance of x from the population mean μ. Since x is a random variable, the squared distance, $(x - \mu)^2$, is also a random variable. Using the same logic used to find the mean value of x, we find the mean value of $(x - \mu)^2$ by multiplying all possible values of $(x - \mu)^2$ by $p(x)$ and then summing over all possible x values.* This quantity

$$E[(x - \mu)^2] = \sum_{\text{all } x}(x - \mu)^2 p(x)$$

is also called the *expected value of the squared distance from the mean;* that is, $\sigma^2 = E[(x - \mu)^2]$. The standard deviation of x is defined as the square root of the variance σ^2.

---

*It can be shown that $E[(x - \mu)^2] = E(x^2) - \mu^2$, where $E(x^2) = \Sigma x^2 p(x)$. Note the similarity between this expression and the shortcut formula $\Sigma(x - \bar{x})^2 = \Sigma x^2 - (\Sigma x)^2/n$ given in Chapter 2.

DEFINITION 4.6

The **variance** of a random variable x is

$$\sigma^2 = E[(x - \mu)^2] = \Sigma(x - \mu)^2 p(x) = \Sigma x^2 p(x) - \mu^2$$

DEFINITION 4.7

The **standard deviation** of a discrete random variable is equal to the square root of the variance, i.e., $\sigma = \sqrt{\sigma^2}$.

Knowing the mean μ and standard deviation σ of the probability distribution of x, in conjunction with Chebyshev's Rule (Table 2.6) and the Empirical Rule (Table 2.7), we can make statements about the likelihood of values of x falling within the intervals $\mu \pm \sigma$, $\mu \pm 2\sigma$, and $\mu \pm 3\sigma$. These probabilities are given in the box.

Chebyshev's Rule and Empirical Rule for a Discrete Random Variable

Let x be a discrete random variable with probability distribution $p(x)$, mean μ, and standard deviation σ. Then, depending on the shape of $p(x)$, the following probability statements can be made:

	Chebyshev's Rule	Empirical Rule
	Applies to any probability distribution (see Figure 4.5a)	Applies to probability distributions that are mound-shaped and symmetric (see Figure 4.5b)
$P(\mu - \sigma < x < \mu + \sigma)$	≥ 0	$\approx .68$
$P(\mu - 2\sigma < x < \mu + 2\sigma)$	$\geq 3/4$	$\approx .95$
$P(\mu - 3\sigma < x < \mu + 3\sigma)$	$\geq 8/9$	≈ 1.00

Figure 4.5

Shapes of Two Probability Distributions for a Discrete Random Variable x

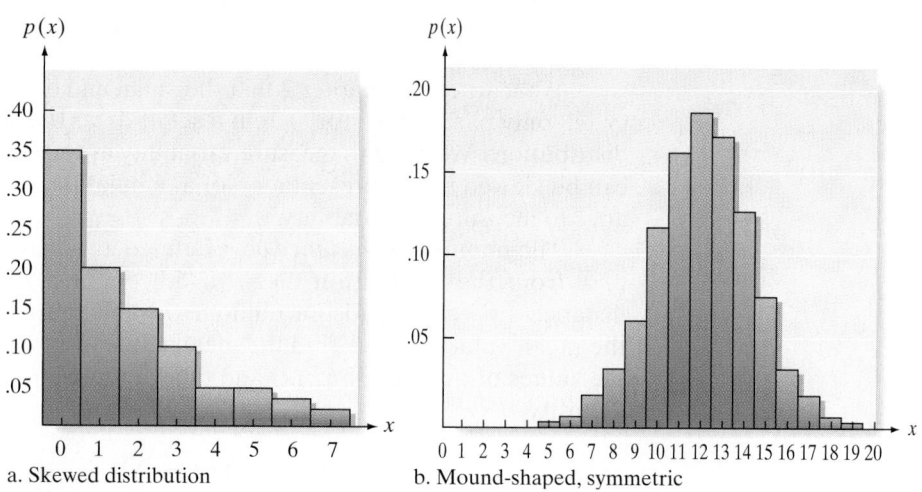

a. Skewed distribution

b. Mound-shaped, symmetric

EXAMPLE 4.7

FINDING μ AND σ

Problem Medical research has shown that a certain type of chemotherapy is successful 70% of the time when used to treat skin cancer. Suppose five skin cancer patients are treated with this type of chemotherapy and let x equal the number of successful cures out of the five. The probability distribution for the number x of successful cures out of five is given in the table:

x	0	1	2	3	4	5
$p(x)$	.002	.029	.132	.309	.360	.168

a. Find $\mu = E(x)$. Interpret the result.

b. Find $\sigma = \sqrt{E[(x - \mu)^2]}$. Interpret the result.

c. Graph $p(x)$. Locate μ and the interval $\mu \pm 2\sigma$ on the graph. Use either Chebyshev's Rule or the Empirical Rule to approximate the probability that x falls in this interval. Compare this result with the actual probability.

d. Would you expect to observe fewer than two successful cures out of five?

Solution   **a.** Applying the formula

$$\mu = E(x) = \Sigma x p(x)$$
$$= 0(.002) + 1(.029) + 2(.132) + 3(.309) + 4(.360) + 5(.168) = 3.50$$

On average, the number of successful cures out of five skin cancer patients treated with chemotherapy will equal 3.5. Remember that this expected value only has meaning when the experiment—treating five skin cancer patients with chemotherapy—is repeated a large number of times.

b. Now we calculate the variance of x:

$$\sigma^2 = E[(x - \mu)^2] = \Sigma(x - \mu)^2 p(x)$$
$$= (0 - 3.5)^2(.002) + (1 - 3.5)^2(.029) + (2 - 3.5)^2(.132)$$
$$+ (3 - 3.5)^2(.309) + (4 - 3.5)^2(.360) + (5 - 3.5)^2(.168)$$
$$= 1.05$$

Thus, the standard deviation is

$$\sigma = \sqrt{\sigma^2} = \sqrt{1.05} = 1.02$$

This value measures the spread of the probability distribution of x, the number of successful cures out of five. A more useful interpretation is obtained by answering parts **c** and **d**.

c. The graph of $p(x)$ is shown in Figure 4.6 with the mean μ and the interval $\mu \pm 2\sigma = 3.50 \pm 2(1.02) = 3.50 \pm 2.04 = (1.46, 5.54)$ shown on the graph.

Figure 4.6
Graph of $p(x)$ for
Example 4.7

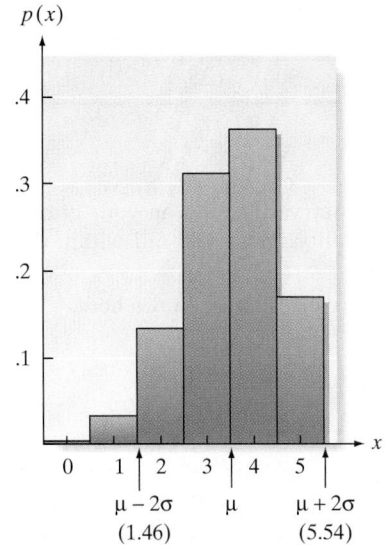

Note particularly that $\mu = 3.5$ locates the center of the probability distribution. Since this distribution is a theoretical relative frequency distribution that is moderately mound-shaped (see Figure 4.6), we expect (from Chebyshev's Rule) at least 75% and, more likely (from the Empirical Rule), approximately 95% of observed x values to fall between 1.46 and 5.54. You can see from Figure 4.6 that the actual probability that x falls in the interval $\mu \pm 2\sigma$ includes the sum of $p(x)$ for the values $x = 2$, $x = 3$, $x = 4$, and $x = 5$. This probability is $p(2) + p(3) + p(4) + p(5) = .132 + .309 + .360 + .168 = .969$. Therefore, 96.9% of the probability distribution lies within 2 standard deviations of the mean. This percentage is consistent with both Chebyshev's Rule and the Empirical Rule.

d. Fewer than two successful cures out of five implies that $x = 0$ or $x = 1$. Since both these values of x lie outside the interval $\mu \pm 2\sigma$, we know from the Empirical Rule that such a result is unlikely (approximate probability of .05). The exact probability, $P(x \le 1)$, is $p(0) + p(1) = .002 + .029 = .031$. Consequently, in a single experiment where five skin cancer patients are treated with chemotherapy, we would not expect to observe fewer than two successful cures.

Now Work *Exercise 4.32*

■ ■ ■

Exercises 4.27–4.43

Understanding the Principles

4.27 What does the expected value of a random variable represent?

4.28 Will $E(x)$ always be equal to a specific value of the random variable, x?

4.29 For a mound-shaped, symmetric distribution, what is the probability that x falls in the interval $\mu \pm 2\sigma$?

Learning the Mechanics

4.30 Consider the probability distribution for the random variable x shown here.

x	10	20	30	40	50	60
$p(x)$	.05	.20	.30	.25	.10	.10

a. Find μ, σ^2, and σ.
b. Graph $p(x)$.
c. Locate μ and the interval $\mu \pm 2\sigma$ on your graph. What is the probability that x will fall within the interval $\mu \pm 2\sigma$?

4.31 Consider the probability distribution shown here.

x	1	2	4	10
$p(x)$	.2	.4	.2	.2

a. Find $\mu = E(x)$.
b. Find $\sigma^2 = E[(x - \mu)^2]$.
c. Find σ.

d. Interpret the value you obtained for μ.
e. In this case, can the random variable x ever assume the value μ? Explain.
f. In general, can a random variable ever assume a value equal to its expected value? Explain.

4.32 Consider the probability distribution shown here.

x	-4	-3	-2	-1	0	1	2	3	4
$p(x)$	.02	.07	.10	.15	.30	.18	.10	.06	.02

a. Calculate μ, σ^2, and σ.
b. Graph $p(x)$. Locate μ, $\mu - 2\sigma$, and $\mu + 2\sigma$ on the graph.
c. What is the probability that x will fall in the interval $\mu \pm 2\sigma$?

4.33 Consider the probability distributions shown here.

x	0	1	2
$p(x)$	.3	.4	.3

y	0	1	2
$p(y)$	.1	.8	.1

a. Use your intuition to find the mean for each distribution. How did you arrive at your choice?
b. Which distribution appears to be more variable? Why?
c. Calculate μ and σ^2 for each distribution. Compare these answers to your answers in parts **a** and **b**.

Applying the Concepts—Basic

4.34 **NHTSA car crash tests.** Refer to Exercise 4.16 (p. 196), **NW** where you found the probability distribution for x, the number of stars in a randomly selected car's driver-side crash rating. Find $\mu = E(x)$ for this distribution and practically interpret the result.

4.35 Benford's Law of Numbers. Refer to the *American Scientist* (July–Aug. 1998) study of which integer is most likely to occur as the first significant digit in a randomly selected number (Benford's Law), Exercise 3.23 (p. 132). The table giving the frequency of each integer selected as the first digit in a six-digit random number is reproduced here.

First Digit	Frequency of Occurrence
1	109
2	75
3	77
4	99
5	72
6	117
7	89
8	62
9	43
Total	743

a. Construct a probability distribution for the first significant digit, x.

b. Find $E(x)$.

c. If possible, give a practical interpretation of $E(x)$.

4.36 Dust mite allergies. Exercise 4.17 (p. 196) gives the probability distribution for the number x of homes with high dust mite levels. The probability distribution is reproduced here.

Number of Homes, x	$p(x)$
0	.09
1	.30
2	.37
3	.20
4	.04

a. Find $E(x)$. Give a meaningful interpretation of the result.

b. Find σ.

c. Find the exact probability that x is in the interval $\mu \pm 2\sigma$. Compare to Chebyshev's Rule and the Empirical Rule.

4.37 Bridge-network reliabilities. A capacitated-bridge network with six arcs (*Networks*, May 1995) was presented in Exercise 4.23 (p. 197). The probability distributions for the capacities of the six arcs are reproduced here.

a. Compute the mean capacity of each arc and interpret its value.

b. Compute σ for each arc and interpret its value.

Applying the Concepts—Intermediate

4.38 Likelihood of flood damage. The National Weather Service issues precipitation forecasts that indicate the likelihood of measurable precipitation ($\geq .01$ inch) at a

Arc	Capacity x	$p(x)$	Arc	Capacity x	$p(x)$
a_1	3	.60	a_3	1	.90
	2	.25		0	.10
	1	.10			
	0	.05			
a_2	2	.60	a_4	1	.90
	1	.30		0	.10
	0	.10			
a_5	1	.90	a_6	2	.70
	0	.10		1	.25
				0	.05

Source: Lin, J., et al. "On reliability evaluation of capacitated-flow network in terms of minimal pathsets." *Networks*, Vol. 25, No. 3, May 1995, p. 135 (Table 1).

specific point—called the official rain gauge—during a given time period. Suppose that if a measurable amount of rain falls during the next 24 hours, an Ohio river will reach flood stage and a business will incur damages of $300,000. The National Weather Service has indicated that there is a 30% chance of a measurable amount of rain during the next 24 hours.

a. Construct the probability distribution that describes the potential flood damages.

b. Find the firm's expected loss due to flood damage. Interpret the result.

4.39 Expected winnings in roulette. In the popular casino game of roulette, you can bet on whether the ball will fall in an arc on the wheel colored red, black, or green. You showed (Exercise 3.49, p. 143) that the probability of a red outcome is 18/38, a black outcome is 18/38, and a green outcome is 2/38. Suppose you make a $5 bet on red. Find your expected net winnings for this single bet. Interpret the result.

4.40 Expected Lotto winnings. The chance of winning Florida's Pick-6 Lotto game is 1 in approximately 23 million. Suppose you buy a $1 Lotto ticket in anticipation of winning the $7 million grand prize. Calculate your expected net winnings for this single ticket. Interpret the result.

4.41 The Showcase Showdown. On the popular television game show, *The Price is Right*, contestants can play "The Showcase Showdown." The game involves a large wheel with twenty nickel values, 5, 10, 15, 20, ..., 95, 100, marked on it. Contestants spin the wheel once or twice, with the objective of obtaining the highest total score *without going over a dollar (100)*. [According to the *American Statistician* (Aug. 1995), the optimal strategy for the first spinner in a three-player game is to spin a second time only if the value of the initial spin is 65 or less.] Let x represent a single contestant playing "The Showcase Showdown." Assume a "fair" wheel (i.e., a wheel with equally likely outcomes). If the total of the player's spins exceeds 100, the total score is set to 0.

a. If the player is permitted only one spin of the wheel, find the probability distribution for x.

b. Refer to part **a**. Find $E(x)$ and interpret this value.

c. Refer to part **a**. Give a range of values within which x is likely to fall.

d. Suppose the player will spin the wheel twice, no matter what the outcome of the first spin. Find the probability distribution for x.

e. What assumption did you make to obtain the probability distribution, part **d**? Is it a reasonable assumption?

f. Find μ and σ for the probability distribution, part **d**, and interpret the results.

g. Refer to part **d**. What is the probability that in two spins the player's total score exceeds a dollar (i.e., is set to 0)?

h. Suppose the player obtains a 20 on the first spin and decides to spin again. Find the probability distribution for x.

i. Refer to part **h**. What is the probability that the player's total score exceeds a dollar?

j. Given the player obtains a 65 on the first spin and decides to spin again, find the probability that the player's total score exceeds a dollar.

k. Repeat part **j** for different first-spin outcomes. Use this information to suggest a strategy for the one-player game.

Applying the Concepts—Advanced

4.42 Parlay card betting. Odds makers try to predict which professional and college football teams will win and by how much (the *spread*). If the odds makers do this accurately, adding the spread to the underdog's score should make the final score a tie. Suppose a bookie will give you $6 for every $1 you risk if you pick the winners in three ballgames (adjusted by the spread) on a "parlay" card. What is the bookie's expected earnings per dollar wagered? Interpret this value.

4.43 Mastering a computer program. The number of training units that must be passed before a complex computer software program is mastered varies from one to five, depending on the student. After much experience, the software manufacturer has determined the probability distribution that describes the fraction of users mastering the software after each number of training units:

Number of Units	1	2	3	4	5
Probability of Mastery	.1	.25	.4	.15	.1

a. Calculate the mean number of training units necessary to master the program. Calculate the median. Interpret each.

b. If the firm wants to ensure that at least 75% of the students master the program, what is the minimum number of training units that must be administered? At least 90%?

c. Suppose the firm develops a new training program that increases the probability that only one unit of training is needed from .1 to .25, increases the probability that only two units are needed to .35, leaves the probability that three units are needed at .4, and completely eliminates the need for four or five units. How do your answers to parts **a** and **b** change for this new program?

4.4 The Binomial Random Variable

Many experiments result in *dichotomous* responses (i.e., responses for which there exist two possible alternatives, such as Yes–No, Pass–Fail, Defective–Nondefective, or Male–Female). A simple example of such an experiment is the coin-toss experiment. A coin is tossed a number of times, say 10. Each toss results in one of two outcomes, Head or Tail, and the probability of observing each of these two outcomes remains the same for each of the 10 tosses. Ultimately, we are interested in the probability distribution of x, the number of heads observed. Many other experiments are equivalent to tossing a coin (either balanced or unbalanced) a fixed number n of times and observing the number x of times that one of the two possible outcomes occurs. Random variables that possess these characteristics are called **binomial random variables**.

Public opinion and consumer preference polls (e.g., the CNN, Gallup, and Harris polls) frequently yield observations on binomial random variables. For example, suppose a sample of 100 students is selected from a large student body and each person is asked whether he or she favors (a Head) or opposes (a Tail) a certain campus issue. Suppose we are interested in x, the number of students in the sample who favor the issue. Sampling 100 students is analogous to tossing the coin 100 times. Thus, you can see that opinion polls that record the number of people who favor a certain issue are real-life equivalents of coin-toss experiments. We have been describing a **binomial experiment**; it is identified by the following characteristics.

Characteristics of a Binomial Random Variable

1. The experiment consists of n identical trials.

2. There are only two possible outcomes on each trial. We will denote one outcome by S (for Success) and the other by F (for Failure).

3. The probability of S remains the same from trial to trial. This probability is denoted by p, and the probability of F is denoted by q. Note that $q = 1 - p$.

4. The trials are independent.

5. The binomial random variable x is the number of S's in n trials.

Biography

JACOB BERNOULLI (1654–1705)—The Bernoulli Distribution

Son of a magistrate and spice maker in Basel, Switzerland, Jacob Bernoulli completed a degree in theology at the University of Basel. While at the university, however, he studied mathematics secretly and against the will of his father. Jacob taught mathematics to his younger brother Johan, and they both went on to become distinguished European mathematicians. At first the brothers collaborated on the problems of the time (e.g., calculus); unfortunately, they later became bitter mathematical rivals. Jacob applied his philosophical training and mathematical intuition to probability and the theory of games of chance, where he developed the law of large numbers. In his book *Ars Conjectandi*, published in 1713 (eight years after his death), the binomial distribution was first proposed. Jacob showed that the binomial distribution was a sum of independent 0-1 variables, now known as Bernoulli random variables.

EXAMPLE 4.8

ASSESSING WHETHER X IS A BINOMIAL

Problem   For the following examples, decide whether x is a binomial random variable.

a. A university scholarship committee must select two students to receive a scholarship for the next academic year. The committee receives 10 applications for the scholarships—six from male students and four from female students. Suppose the applicants are all equally qualified, so that the selections are randomly made. Let x be the number of female students who receive a scholarship.

b. Before marketing a new product on a large scale, many companies will conduct a consumer-preference survey to determine whether the product is likely to be successful. Suppose a company develops a new diet soda and then conducts a taste-preference survey in which 100 randomly chosen consumers state their preferences among the new soda and the two leading sellers. Let x be the number of the 100 who choose the new brand over the two others.

c. Some surveys are conducted by using a method of sampling other than simple random sampling (defined in Chapter 3). For example, suppose a television cable company plans to conduct a survey to determine the fraction of households in the city that would use the cable television service. The sampling method is to choose a city block at random and then survey every household on that block. This sampling technique is called *cluster sampling*. Suppose 10 blocks are so sampled, producing a total of 124 household responses. Let x be the number of the 124 households that would use the television cable service.

Solution   **a.** In checking the binomial characteristics, a problem arises with independence (characteristic 4 in the preceding box). Given that the first student selected is female, the probability that the second chosen is female is $\frac{3}{9}$. On the other hand, given that the first selection is a male student, the probability that the second is female is $^4/_9$. Thus, the conditional probability of a Success (choosing a female student to receive a scholarship) on the second trial (selection) depends on the outcome of the first trial, and the trials are therefore dependent. Since the trials are *not independent*, this is not a binomial random variable. (This variable is actually a *hypergeometric* random variable, the topic of Optional Section 4.6.)

b. Surveys that produce dichotomous responses and use random sampling techniques are classic examples of binomial experiments. In our example, each randomly selected consumer either states a preference for the new diet soda or does not. The sample of 100 consumers is a very small proportion of the totality of potential consumers, so the response of one would be, for all practical purposes, independent of another.* Thus, x is a binomial random variable.

c. This example is a survey with dichotomous responses (Yes or No to the cable service), but the sampling method is not simple random sampling. Again, the binomial characteristic of independent trials would probably not be satisfied. The responses of households within a particular block would be dependent, since households within a block tend to be similar with respect to income, level of education, and general interests. Thus, the binomial model would not be satisfactory for x if the cluster sampling technique were employed.

Look Back Nonbinomial variables with two outcomes on every trial typically occur because they do not satisfy characteristics 3 or 4 of a binomial distribution.

| Now Work | *Exercise 4.56a* |

■ ■ ■

EXAMPLE 4.9   DERIVING THE BINOMIAL PROBABILITY DISTRIBUTION

Problem   The Heart Association claims that only 10% of U.S. adults over 30 can pass the President's Physical Fitness Commission's minimum requirements. Suppose four adults are randomly selected, and each is given the fitness test.

a. Use the steps given in Chapter 3 (box on p. 127) to find the probability that none of the four adults passes the test.

b. Find the probability that three of the four adults pass the test.

c. Let x represent the number of the four adults who pass the fitness test. Explain why x is a binomial random variable.

d. Use the answers to parts **a** and **b** to derive a formula for $p(x)$, the probability distribution of the binomial random variable x.

Solution   **a.** **1.** The first step is to define the experiment. Here we are interested in observing the fitness test results of each of the four adults: pass (S) or fail (F).

2. Next, we list the sample points associated with the experiment. Each sample point consists of the test results of the four adults. For example, *SSSS* represents the sample point that all four adults pass, while *FSSS* represents the sample point that adult 1 fails, while adults 2, 3, and 4 pass the test. The 16 sample points are listed in Table 4.2.

*In most real-life applications of the binomial distribution, the population of interest has a finite number of elements (trials), denoted N. When N is large and the sample size n is small relative to N, say $n/N \leq .05$, the sampling procedure, for all practical purposes, satisfies the conditions of a binomial experiment.

TABLE 4.2 Sample Points for Fitness Test of Example 4.9

SSSS	FSSS	FFSS	SFFF	FFFF
	SFSS	FSFS	FSFF	
	SSFS	FSSF	FFSF	
	SSSF	SFFS	FFFS	
		SFSF		
		SSFF		

3. We now assign probabilities to the sample points. Note that each sample point can be viewed as the intersection of four adults' test results and, assuming the results are independent, the probability of each sample point can be obtained using the multiplicative rule, as follows:

$$P(SSSS) = P[(\text{adult 1 passes}) \cap (\text{adult 2 passes})$$
$$\cap (\text{adult 3 passes}) \cap (\text{adult 4 passes})]$$
$$= P(\text{adult 1 passes}) \times P(\text{adult 2 passes})$$
$$\times P(\text{adult 3 passes}) \times P(\text{adult 4 passes})$$
$$= (.1)(.1)(.1)(.1) = (.1)^4 = .0001$$

All other sample point probabilities are calculated using similar reasoning. For example,

$$P(FSSS) = (.9)(.1)(.1)(.1) = .0009$$

You can check that this reasoning results in sample point probabilities that add to 1 over the 16 points in the sample space.

4. Finally, we add the appropriate sample point probabilities to obtain the desired event probability. The event of interest is that all four adults fail the fitness test. In Table 4.2 we find only one sample point, *FFFF*, contained in this event. All other sample points imply that at least one adult passes. Thus,

$$P(\text{All four adults fail}) = P(FFFF) = (.9)^4 = .6561$$

b. The event that three of the four adults pass the fitness test consists of the four sample points in the second column of Table 4.2: *FSSS, SFSS, SSFS,* and *SSSF*. To obtain the event probability we add the sample point probabilities:

$$P(3 \text{ of } 4 \text{ adults pass}) = P(FSSS) + P(SFSS) + P(SSFS) + P(SSSF)$$
$$= (.1)^3(.9) + (.1)^3(.9) + (.1)^3(.9) + (.1)^3(.9)$$
$$= 4(.1)^3(.9) = .0036$$

Note that each of the four sa.mple point probabilities is the same, because each sample point consists of three *S*'s and one *F*; the order does not affect the probability because the adults' test results are (assumed) independent.

c. We can characterize the experiment as consisting of four identical trials—the four test results. There are two possible outcomes to each trial, *S* or *F*, and the probability of passing, $p = .1$, is the same for each trial. Finally, we are assuming that each adult's test result is independent of all others, so that the four trials are independent. Then it follows that *x*, the number of the four adults who pass the fitness test, is a binomial random variable.

d. The event probabilities in parts **a** and **b** provide insight into the formula for the probability distribution $p(x)$. First, consider the event that three adults pass (part **b**). We found that

$$P(x = 3) = (\text{Number of sample points for which } x = 3) \times (.1)^{\text{Number of successes}} \times (.9)^{\text{Number of failures}}$$
$$= 4(.1)^3(.9)^1$$

In general, we can use combinatorial mathematics to count the number of sample points. For example,

Number of sample points for which $x = 3$
= Number of different ways of selecting 3 successes of the 4 trials
$$= \binom{4}{3} = \frac{4!}{3!(4-3)!} = \frac{4 \cdot 3 \cdot 2 \cdot 1}{(3 \cdot 2 \cdot 1) \cdot 1} = 4$$

The formula that works for any value of x can be deduced as follows:

$$P(x = 3) = \binom{4}{3}(.1)^3(.9)^1 = \binom{4}{x}(.1)^x(.9)^{4-x}$$

The component $\binom{4}{x}$ counts the number of sample points with x successes, and the component $(.1)^x(.9)^{4-x}$ is the probability associated with each sample point having x successes.

For the general binomial experiment, with n trials and probability of Success p on each trial, the probability of x successes is

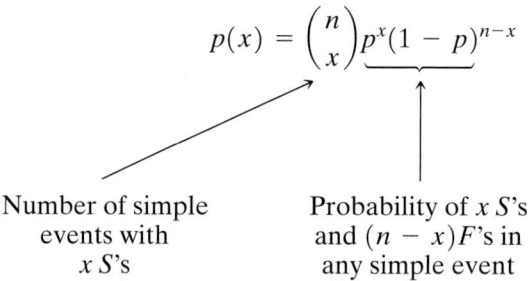

$$p(x) = \binom{n}{x}p^x(1-p)^{n-x}$$

Number of simple events with x S's

Probability of x S's and $(n-x)$ F's in any simple event

Look Back In theory, you could always resort to the principles developed in this example to calculate binomial probabilities; list the sample points and sum their probabilities. However, as the number of trials (n) increases, the number of sample points grows very rapidly (the number of sample points is 2^n). Thus, we prefer the formula for calculating binomial probabilities, since its use avoids listing sample points.

■ ■ ■

The binomial distribution* is summarized in the box.

The Binomial Probability Distribution

$$p(x) = \binom{n}{x}p^x q^{n-x} \qquad (x = 0, 1, 2, \ldots, n)$$

where

p = Probability of a success on a single trial
$q = 1 - p$
n = Number of trials
x = Number of successes in n trials
$$\binom{n}{x} = \frac{n!}{x!(n-x)!}$$

---

*The binomial distribution is so named because the probabilities, $p(x)$, $x = 0, 1, \ldots, n$, are terms of the binomial expansion, $(q + p)^n$.

As noted in Chapter 3, the symbol 5! means $5 \cdot 4 \cdot 3 \cdot 2 \cdot 1 = 120$. Similarly, $n! = n(n - 1)(n - 2) \cdots 3 \cdot 2 \cdot 1$; remember, $0! = 1$.

EXAMPLE 4.10   APPLYING THE BINOMIAL DISTRIBUTION

Problem   Refer to Example 4.9. Use the formula for a binomial random variable to find the probability distribution of x, where x is the number of adults who pass the fitness test. Graph the distribution.

Solution   For this application, we have $n = 4$ trials. Since a success S is defined as an adult who passes the test, $p = P(S) = .1$ and $q = 1 - p = .9$. Substituting $n = 4$, $p = .1$, and $q = .9$ into the formula for $p(x)$, we obtain

$$p(0) = \frac{4!}{0!(4 - 0)!}(.1)^0(.9)^{4-0} = \frac{4 \cdot 3 \cdot 2 \cdot 1}{(1)(4 \cdot 3 \cdot 2 \cdot 1)}(.1)^0(.9)^4 = 1(1)^0(.9)^4 = .6561$$

$$p(1) = \frac{4!}{1!(4 - 1)!}(.1)^1(.9)^{4-1} = \frac{4 \cdot 3 \cdot 2 \cdot 1}{(1)(3 \cdot 2 \cdot 1)}(.1)^1(.9)^3 = 4(.1)(.9)^3 = .2916$$

$$p(2) = \frac{4!}{2!(4 - 2)!}(.1)^2(.9)^{4-2} = \frac{4 \cdot 3 \cdot 2 \cdot 1}{(2 \cdot 1)(2 \cdot 1)}(.1)^2(.9)^2 = 6(.1)^2(.9)^2 = .0486$$

$$p(3) = \frac{4!}{3!(4 - 3)!}(.1)^3(.9)^{4-3} = \frac{4 \cdot 3 \cdot 2 \cdot 1}{(3 \cdot 2 \cdot 1)(1)}(.1)^3(.9)^1 = 4(.1)^3(.9) = .0036$$

$$p(4) = \frac{4!}{4!(4 - 4)!}(.1)^4(.9)^{4-4} = \frac{4 \cdot 3 \cdot 2 \cdot 1}{(4 \cdot 3 \cdot 2 \cdot 1)(1)}(.1)^4(.9)^0 = 1(.1)^4(.9) = .0001$$

Look Back   Note that these probabilities, listed in Table 4.3, sum to 1. A graph of this probability distribution is shown in Figure 4.7.

> **Now Work**   *Exercise 4.49*

■ ■ ■

TABLE 4.3   Probability Distribution for Physical Fitness Example: Tabular Form

x	$p(x)$
0	.6561
1	.2916
2	.0486
3	.0036
4	.0001

Figure 4.7

Probability Distribution for Physical Fitness Example: Graphical Form

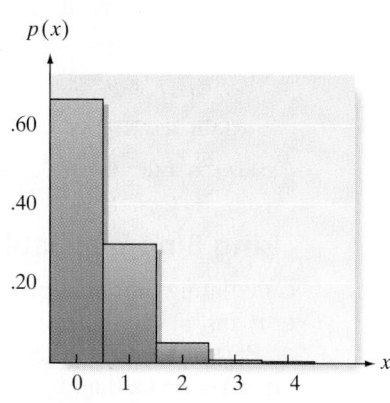

EXAMPLE 4.11   FINDING μ AND σ

Problem   Refer to Examples 4.9 and 4.10. Calculate μ and σ, the mean and standard deviation, respectively, of the number of the four adults who pass the test. Interpret the results.

Solution   From Section 4.3 we know that the mean of a discrete probability distribution is

$$\mu = \Sigma x p(x)$$

Referring to Table 4.3, the probability distribution for the number x who pass the fitness test, we find

$$\mu = 0(.6561) + 1(.2916) + 2(.0486) + 3(.0036) + 4(.0001) = .4$$
$$= 4(.1) = np$$

Thus, in the long run, the average number of adults (out of four) who pass the test is only .4. [*Note: The relationship $\mu = np$ holds in general for a binomial random variable.*]

The variance is

$$\sigma^2 = \Sigma(x - \mu)^2 p(x) = \Sigma(x - .4)^2 p(x)$$
$$= (0 - .4)^2(.6561) + (1 - .4)^2(.2916) + (2 - .4)^2(.0486)$$
$$+ (3 - .4)^2(.0036) + (4 - .4)^2(.0001)$$
$$= .104976 + .104976 + .124416 + .024336 + .001296$$
$$= .36 = 4(.1)(.9) = npq$$

[*Note: The relationship $\sigma^2 = npq$ holds in general for a binomial random variable.*]
Finally, the standard deviation of the number who pass the fitness test is

$$\sigma = \sqrt{\sigma^2} = \sqrt{.36} = .6$$

Applying the Empirical Rule, we know that approximately 95% of the x values will fall in the interval $\mu \pm 2\sigma = .4 \pm 2(.6) = (-.8, 1.6)$. Since x cannot be negative, we expect (i.e., in the long run) the number of adults out of four who pass the fitness test to be less than 1.6.

Look Back   Examining Figure 4.7, you can see that all observations equal to 0 or 1 will fall within the interval $(-.8, 1.6)$. The probabilities corresponding to these values (from Table 4.3) are .6561 and .2916, respectively. Summing them, we obtain $.6561 + .2916 = .9477 \approx .95$. This result, again, supports the Empirical Rule.

■ ■ ■

We emphasize that you need not use the expectation summation rules to calculate μ and σ^2 for a binomial random variable. You can find them easily using the formulas $\mu = np$ and $\sigma^2 = npq$.

Mean, Variance, and Standard Deviation for a Binomial Random Variable

Mean: $\mu = np$

Variance: $\sigma^2 = npq$

Standard deviation: $\sigma = \sqrt{npq}$

Using Binomial Tables

Calculating binomial probabilities becomes tedious when n is large. For some values of n and p, the binomial probabilities have been tabulated in Table II of Appendix A. Part of Table II is shown in Table 4.4; a graph of the binomial probability distribution for $n = 10$ and $p = .10$ is shown in Figure 4.8.

TABLE 4.4 Reproduction of Part of Table II of Appendix A: Binomial Probabilities for $n = 10$

k \ p	.01	.05	.10	.20	.30	.40	.50	.60	.70	.80	.90	.95	.99
0	.904	.599	.349	.107	.028	.006	.001	.000	.000	.000	.000	.000	.000
1	.996	.914	.736	.376	.149	.046	.011	.002	.000	.000	.000	.000	.000
2	1.000	.988	.930	.678	.383	.167	.055	.012	.002	.000	.000	.000	.000
3	1.000	.999	.987	.879	.650	.382	.172	.055	.011	.001	.000	.000	.000
4	1.000	1.000	.998	.967	.850	.633	.377	.166	.047	.006	.000	.000	.000
5	1.000	1.000	1.000	.994	.953	.834	.623	.367	.150	.033	.002	.000	.000
6	1.000	1.000	1.000	.999	.989	.945	.828	.618	.350	.121	.013	.001	.000
7	1.000	1.000	1.000	1.000	.998	.988	.945	.833	.617	.322	.070	.012	.000
8	1.000	1.000	1.000	1.000	1.000	.988	.989	.954	.851	.624	.264	.086	.004
9	1.000	1.000	1.000	1.000	1.000	1.000	.999	.994	.972	.893	.651	.401	.096

Table II actually contains a total of nine tables, labeled (**a**) through (**i**), one each corresponding to $n = 5, 6, 7, 8, 9, 10, 15, 20,$ and 25, respectively. In each of these tables the columns correspond to values of p, and the rows correspond to values of the random variable x. The entries in the table represent **cumulative binomial probabilities**. Thus, for example, the entry in the column corresponding to $p = .10$ and the row corresponding to $x = 2$ is .930 (highlighted), and its interpretation is

$$P(x \le 2) = P(x = 0) + P(x = 1) + P(x = 2) = .930$$

This probability is also highlighted in the graphical representation of the binomial distribution with $n = 10$ and $p = .10$ in Figure 4.8.

You can also use Table II to find the probability that x equals a specific value. For example, suppose you want to find the probability that $x = 2$ in the binomial distribution with $n = 10$ and $p = .10$. This is found by subtraction as follows:

$$P(x = 2) = [P(x = 0) + P(x = 1) + P(x = 2)] - [P(x = 0) + P(x = 1)]$$
$$= P(x \le 2) - P(x \le 1) = .930 - .736 = .194$$

The probability that a binomial random variable exceeds a specified value can be found using Table II and the notion of complementary events. For example, to find the probability that x exceeds 2 when $n = 10$ and $p = .10$, we use

$$P(x > 2) = 1 - P(x \le 2) = 1 - .930 = .070$$

Note that this probability is represented by the unhighlighted portion of the graph in Figure 4.8.

All probabilities in Table II are rounded to three decimal places. Thus, although none of the binomial probabilities in the table is exactly zero, some are small enough (less than .0005) to round to .000. For example, using the formula to find $P(x = 0)$ when $n = 10$ and $p = .6$, we obtain

$$P(x = 0) = \binom{10}{0}(.6)^0(.4)^{10-0} = .4^{10} = .00010486$$

but this is rounded to .000 in Table II of Appendix A (see Table 4.4).

Figure 4.8
Binomial Probability Distribution for $n = 10$ and $p = .10$; $P(x \le 2)$ Highlighted

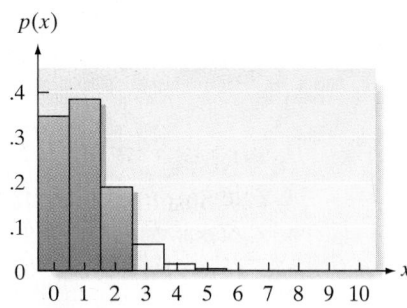

Similarly, none of the table entries is exactly 1.0, but when the cumulative probabilities exceed .9995, they are rounded to 1.000. The row corresponding to the largest possible value for x, $x = n$, is omitted, because all the cumulative probabilities in that row are equal to 1.0 (exactly). For example, in Table 4.4 with $n = 10$, $P(x \leq 10) = 1.0$, no matter what the value of p.

The following example further illustrates the use of Table II.

EXAMPLE 4.12   USING THE BINOMIAL TABLE

Problem   Suppose a poll of 20 voters is taken in a large city. The purpose is to determine x, the number who favor a certain candidate for mayor. Suppose that 60% of all the city's voters favor the candidate.

a. Find the mean and standard deviation of x.

b. Use Table II of Appendix A to find the probability that $x \leq 10$.

c. Use Table II to find the probability that $x > 12$.

d. Use Table II to find the probability that $x = 11$.

e. Graph the probability distribution of x and locate the interval $\mu \pm 2\sigma$ on the graph.

Solution   a. The number of voters polled is presumably small compared with the total number of eligible voters in the city. Thus, we may treat x, the number of the 20 who favor the mayoral candidate, as a binomial random variable. The value of p is the fraction of the total voters who favor the candidate (i.e., $p = .6$). Therefore, we calculate the mean and variance:

$$\mu = np = 20(.6) = 12$$
$$\sigma^2 = npq = 20(.6)(.4) = 4.8$$
$$\sigma = \sqrt{4.8} = 2.19$$

b. Looking in the $k = 10$ row and the $p = .6$ column of Table II (Appendix A) for $n = 20$, we find the value of .245. Thus,

$$P(x \leq 10) = .245$$

c. To find the probability

$$P(x > 12) = \sum_{x=13}^{20} p(x)$$

we use the fact that for all probability distributions,

$$\sum_{\text{all } x} p(x) = 1.$$

Therefore,

$$P(x > 12) = 1 - P(x \leq 12) = 1 - \sum_{x=0}^{12} p(x)$$

Consulting Table II, we find the entry in row $k = 12$, column $p = .6$ to be .584. Thus,

$$P(x > 12) = 1 - .584 = .416$$

Figure 4.9

The Binomial Probability Distribution for x in Example 4.12: $n = 20$ and $p = .6$

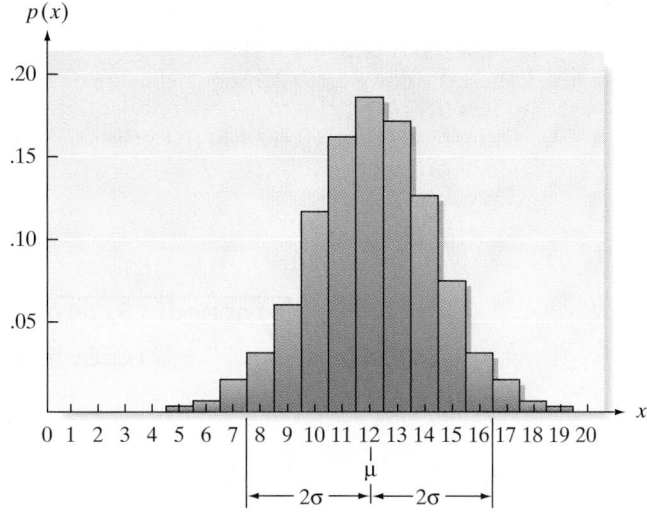

d. To find the probability that exactly 11 voters favor the candidate, recall that the entries in Table II are cumulative probabilities and use the relationship

$$P(x = 11) = [p(0) + p(1) + \cdots + p(11)] - [p(0) + p(1) + \cdots + p(10)]$$
$$= P(x \leq 11) - P(x \leq 10)$$

Then

$$P(x = 11) = .404 - .245 = .159$$

e. The probability distribution for x is shown in Figure 4.9. Note that

$$\mu - 2\sigma = 12 - 2(2.2) = 7.6 \qquad \mu + 2\sigma = 12 + 2(2.2) = 16.4$$

The interval $\mu - 2\sigma$ to $\mu + 2\sigma$ is shown in Figure 4.9. The probability that x falls in the interval $\mu + 2\sigma$ is $P(x = 8, 9, 10, \ldots, 16) = P(x \leq 16) - P(x \leq 7)$ .984 − .021 = .963. Note that this probability is very close to the .95 given by the Empirical Rule. Thus, we expect the number of voters in the sample of 20 who favor the mayoral candidate to be between 8 and 16.

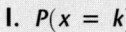 **Now Work**   *Exercise 4.52*

■ ■ ■

Binomial Probabilities

Using the TI-83 Graphing Calculator

I. $P(x = k)$

To compute the probability of k successes in n trials, where the p is probability of success for each trial, use the binompdf(command. Binompdf stands for "binomial probability density function." This command is under the DISTRibution menu and has the format binompdf(n, p, k).

Example Compute the probability of 5 successes in 8 trials where the probability of success for a single trial is 40%.

In this example, $n = 8$, $p = .4$, and $k = 5$.

Step 1   Enter the binomial parameters
Press 2nd VARS for DISTR
Press the down arrow key until 0:binompdf is highlighted
Press ENTER
After binompdf(,type 8, .4, 5) (Note: be sure to use the comma key between each parameter)
Press ENTER

You should see

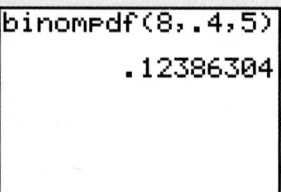

Thus, $P(x = 5)$ is about 12.4%.

II. $P(x \leq k)$

To compute the probability of k or fewer successes in n trials, where the p is probability of success for each trial, use the binomcdf(command. Binomcdf stands for "binomial *cumulative* probability density function." This command is under the DISTRibution menu and has the format binomcdf(n, p, k).

Example   Compute the probability of less than 5 successes in 8 trials where the probability of success for a single trial is 40%. In this example, $n = 8$, $p = .4$, and $k = 5$.

Step 1   Enter the binomial parameters
Press 2nd VARS for DISTR
Press down the arrow key until A:binomcdf is highlighted
Press ENTER
After binomcdf(,type 8, .4, 5)
Press ENTER

You should see

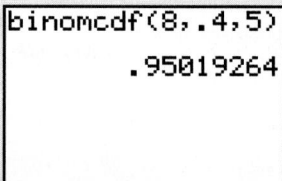

Thus, $P(x < 5)$ is about 95%.

III. $P(x < k), P(x > k), P(x \geq k)$

To find the probability of less than k successes $P(x < k)$, more than k successes $P(x > k)$, or at least k successes $P(x \geq k)$, variations of the binomcdf(command must be used as shown below.

$P(x < k)$ use **binomcdf** $(n, p, k - 1)$

$P(x > k)$ use **1–binomcdf** (n, p, k)

$P(x \geq k)$ use **1–binomcdf** $(n, p, k - 1)$

Statistics in Action Revisited
Using the Binomial Model to Solve the Cocaine Sting Case

Refer to the reverse cocaine sting case described on p. 189. During a drug bust, police seized 496 foil packets of a white, powdery substance that appeared to be cocaine. The police laboratory randomly selected four packets and found that all four tested positive for cocaine. This finding led to the conviction of the drug traffickers. Following the conviction, the police used two of the remaining 492 foil packets (i.e., those not tested) in a reverse sting operation. The two randomly selected packets were sold by undercover officers to a buyer who disposed of the evidence before being arrested. Is there evidence beyond a reasonable doubt that the two packets contained cocaine?

To solve the dilemma, we will assume that of the 496 original packets confiscated, 331 contained genuine cocaine and 165 contained an inert (legal) powder. (A statistician, hired as an expert witness on the case, showed that the chance of the defendant being found not guilty is maximized when 331 packets contain cocaine and 165 do not.) First, we'll find the probability that four packets randomly selected from the original 496 will test positive for cocaine. Then we'll find the probability that the two packets sold in the reverse sting did not contain cocaine. Finally, we'll find the probability that both events occur (i.e., that the first four packets selected test positive for cocaine but that the next two packets selected do not). In each of these probability calculations, we will apply the binomial probability distribution to approximate the probabilities.

Let x be the number of packets that contain cocaine in a sample of n selected from the 496 packets. Here, we are defining a "success" as a packet that contains cocaine. If n is small, say $n = 2$ or $n = 4$, then x has an approximate binomial distribution with probability of success $p = 331/496 \approx .67$.

The probability that the first four packets selected contain cocaine [i.e., $P(x = 4)$] is obtained using the binomial formula with $n = 4$ and $p = .67$:

$$P(x = 4) = p(4) = \binom{4}{4}p^4(1 - p)^0 = \frac{4!(.67)^4(.33)^0}{4!0!}$$
$$= (.67)^4 = .201$$

Thus, there is about a 20% chance that all four of the randomly selected packets contain cocaine.

Given that four of the original packets tested positive for cocaine, the probability that the two packets randomly selected and sold in the reverse sting *do not* contain cocaine is approximated using a binomial distribution with $n = 2$ and $p = .67$. Since a "success" is a packet with cocaine, we find $P(x = 0)$:

$$P(x = 0) = p(0) = \binom{2}{0}p^0(1 - p)^2 = \frac{2!(.67)^0(.33)^2}{0!2!}$$
$$= (.33)^2 = .109$$

Finally, to compute the probability that of the original 496 foil packets, the first four selected (at random) test positive for cocaine and the next two selected (at random) test negative for cocaine, we employ the Multiplicative Law of Probability. Let A be the event that the first four packets test positive. Let B be the event that the next two packets test negative. We want to find the probability of both events occurring [i.e., $P(A \text{ and } B) = P(A \cap B) = P(B|A)P(A)$]. Note that this probability is the product of the two previously calculated probabilities:

$$P(A \text{ and } B) = (.109)(.201) = .022$$

Consequently, there is only a .022 probability (i.e., about 2 chances in a hundred) that the first four packets will test positive for cocaine and the next two packets will test negative for cocaine. A reasonable jury would likely believe that an event with such a small probability is unlikely to occur and conclude that the two "lost" packets contained cocaine. In other words, most of us would infer that the defendant in the reverse cocaine sting was guilty of drug trafficking.

[*Epilogue*: Several of the defendant's lawyers believed that the .022 probability was too high for jurors to conclude guilt "beyond a reasonable doubt." The argument was made moot, however, when, to the surprise of the defense, the prosecution revealed that the remaining 490 packets had not been used in any other reverse sting operations and offered to test a sample of them. On the advice of the statistician, the defense requested that an additional 20 packets be tested. All 20 tested positive for cocaine! As a consequence of this new evidence, the defendant was convicted by the jury.]

Exercises 4.44–4.68

Understanding the Principles

4.44 Give the five characteristics of a binomial random variable.

4.45 Give the formula for $p(x)$ for a binomial random variable with $n = 7$ and $p = .2$.

4.46 Consider the following binomial probability distribution:

$$p(x) = \binom{5}{x}(.7)^x(.3)^{5-x} \quad (x = 0, 1, 2, \ldots, 5)$$

a. How many trials (n) are in the experiment?
b. What is the value of p, the probability of success?

Learning the Mechanics

4.47 Compute the following:

a. $\dfrac{6!}{2!(6-2)!}$ b. $\dbinom{5}{2}$ c. $\dbinom{7}{0}$

d. $\dbinom{6}{6}$ e. $\dbinom{4}{3}$

4.48 Refer to Exercise 4.46.
   a. Graph the probability distribution.
   b. Find the mean and standard deviation of x.
   c. Show the mean and the 2-standard-deviation interval on each side of the mean on the graph you drew in part **a**.

4.49 If x is a binomial random variable, compute $p(x)$ for
NW each of the following cases:
   a. $n = 5, x = 1, p = .2$
   b. $n = 4, x = 2, q = .4$
   c. $n = 3, x = 0, p = .7$
   d. $n = 5, x = 3, p = .1$
   e. $n = 4, x = 2, q = .6$
   f. $n = 3, x = 1, p = .9$

4.50 Suppose x is a binomial random variable with $n = 3$ and $p = .3$.
   a. Calculate the value of $p(x)$, $x = 0, 1, 2, 3$, using the formula for a binomial probability distribution.
   b. Using your answers to part **a**, give the probability distribution for x in tabular form.

4.51 If x is a binomial random variable, calculate μ, σ^2, and σ for each of the following:
   a. $n = 25, p = .5$
   b. $n = 80, p = .2$
   c. $n = 100, p = .6$
   d. $n = 70, p = .9$
   e. $n = 60, p = .8$
   f. $n = 1,000, p = .04$

4.52 If x is a binomial random variable, use Table II in
NW Appendix A to find the following probabilities:
   a. $P(x = 2)$ for $n = 10, p = .4$
   b. $P(x \le 5)$ for $n = 15, p = .6$
   c. $P(x > 1)$ for $n = 5, p = .1$

4.53 If x is a binomial random variable, use Table II, Appendix A to find the following probabilities:
   a. $P(x < 10)$ for $n = 25, p = .7$
   b. $P(x \ge 10)$ for $n = 15, p = .9$
   c. $P(x = 2)$ for $n = 20, p = .2$

4.54 Suppose x is a binomial random variable with $n = 5$ and $p = .5$. Compute $p(x)$ for $x = 0, 1, 2, 3, 4,$ and 5 using the following two methods:
   a. List the sample points (using S for Success and F for Failure on each trial) corresponding to each value of x, assign probabilities to each sample point, and obtain $p(x)$ by adding sample point probabilities.
   b. Use the formula for the binomial probability distribution to obtain $p(x)$.

4.55 The binomial probability distribution is a family of probability distributions with each single distribution depending on the values of n and p. Assume that x is a binomial random variable with $n = 4$.
   a. Determine a value of p such that the probability distribution of x is symmetric.
   b. Determine a value of p such that the probability distribution of x is skewed to the right.
   c. Determine a value of p such that the probability distribution of x is skewed to the left.
   d. Graph each of the binomial distributions you obtained in parts **a**, **b**, and **c**. Locate the mean for each distribution on its graph.
   e. In general, for what values of p will a binomial distribution be symmetric? Skewed to the right? Skewed to the left?

Applying the Concepts—Basic

4.56 Mobile phones with Internet access. According to a Jupiter/NPD Consumer Survey of young adults (18–24 years of age) who shop online, 20% own a mobile phone with Internet access. (*American Demographics*, May 2002.) In a random sample of 200 young adults who shop online, let x be the number who own a mobile phone with Internet access.
NW a. Explain why x is a binomial random variable (to a reasonable degree of approximation).
   b. What is the value of p? Interpret this value.
   c. What is the expected value of x? Interpret this value.

4.57 Quit smoking program. According to the University of South Florida's Tobacco Research and Intervention Program, only 5% of the nation's cigarette smokers ever enter into a treatment program to help them quit smoking. (*USF Magazine*, Spring 2000.) In a random sample of 200 smokers, let x be the number who enter into a treatment program.
   a. Explain why x is a binomial random variable (to a reasonable degree of approximation).
   b. What is the value of p? Interpret this value.
   c. What is the expected value of x? Interpret this value.

4.58 Caesarian births. The American College of Obstetricians and Gynecologists report that 22% of all births in the U.S. take place by Caesarian section each year. (*USA Today*, Sept. 19, 2000.)
   a. In a random sample of 1,000 births, how many, on average will take place by Caesarian section?
   b. What is the standard deviation of the number of Caesarian section births in a sample of 1,000 births?
   c. Use your answers to parts **a** and **b** to form an interval that is likely to contain the number of Caesarian section births in a sample of 1,000 births.

4.59 Belief in an afterlife. A national poll conducted by *The New York Times* (May 7, 2000) revealed that 80% of Americans believe that after you die, some part of you lives on, either in a next life on earth or in heaven. Consider a random sample of 10 Americans and count x, the number who believe in life after death.
   a. Find $P(x = 3)$. b. Find $P(x \le 7)$.
   c. Find $P(x > 4)$.

4.60 Parents who condone spanking. According to a nationwide survey, 60% of parents with young children condone spanking their child as a regular form of punishment. (*Tampa Tribune*, Oct. 5, 2000.) Consider a random sample of three people, each of whom is a parent with young children. Assume that x, the number in the sample that condone spanking, is a binomial random variable.

a. What is the probability that none of the three parents condones spanking as a regular form of punishment for their children?

b. What is the probability that at least one condones spanking as a regular form of punishment?

c. Give the mean and standard deviation of x. Interpret the results.

Applying the Concepts—Intermediate

4.61 Fungi in beech forest trees. Refer to the *Applied Ecology and Environmental Research* (Vol. 1, 2003) study of beech trees damaged by fungi, Exercise 3.21 (p. 131). The researchers found that 25% of the beech trees in East Central Europe have been damaged by fungi. Consider a sample of 20 beech trees from this area.

a. What is the probability that fewer than half are damaged by fungi?

b. What is the probability that more than 15 are damaged by fungi?

c. How many of the sampled trees would you expect to be damaged by fungi?

4.62 Countries that allow a free press. The degree to which democratic and nondemocratic countries attempt to control the news media was examined in the *Journal of Peace Research* (Nov. 1997). Between 1948 and 1996, 80% of all democratic regimes allowed a free press. In contrast, over the same time period, 10% of all nondemocratic regimes allowed a free press.

a. In a random sample of 50 democratic regimes, how many would you expect to allow a free press? Give a range that is highly likely to include the number of democratic regimes with a free press.

b. In a random sample of 50 nondemocratic regimes, how many would you expect to allow a free press? Give a range that is highly likely to include the number of nondemocratic regimes with a free press.

4.63 Victims of domestic abuse. According to researchers at Johns Hopkins University School of Medicine, one in every three women has been a victim of domestic abuse. (*Annals of Internal Medicine*, Nov. 1995.) This probability was obtained from a survey of nearly 2,000 adult women residing in Baltimore, Maryland. Suppose we randomly sample 15 women and find that four have been abused.

a. What is the probability of observing four or more abused women in a sample of 15 if the proportion p of women who are victims of domestic abuse is really $p = \frac{1}{3}$?

b. Many experts on domestic violence believe that the proportion of women who are domestically abused

is closer to $p = .10$. Calculate the probability of observing four or more abused women in a sample of 15 if $p = .10$.

c. Why might your answers to parts **a** and **b** lead you to believe that $p = \frac{1}{3}$?

4.64 Chickens with fecal contamination. The United States Department of Agriculture (USDA) reports that, under its standard inspection system, one in every 100 slaughtered chickens passes inspection with fecal contamination. (*Tampa Tribune*, Mar. 31, 2000.) In Exercise 3.19 (p. 141), you found the probability that a randomly selected slaughtered chicken passes inspection with fecal contamination. Now find the probability that, in a random sample of five slaughtered chickens, at least one passes inspection with fecal contamination.

4.65 Testing a psychic's ESP. Refer to Exercise 3.75 (p. 157) and the experiment conducted by the Tampa Bay Skeptics to see whether an acclaimed psychic has extrasensory perception (ESP). Recall that a crystal is placed, at random, inside one of 10 identical boxes lying side-by-side on a table. The experiment was repeated seven times, and x, the number of correct decisions, was recorded. (Assume that the seven trials are independent.)

a. If the psychic is guessing (i.e., if the psychic does *not* possess ESP) what is the value of p, the probability of a correct decision on each trial?

b. If the psychic is guessing, what is the expected number of correct decisions in seven trials?

c. If the psychic is guessing, what is the probability of no correct decisions in seven trials?

d. Now suppose the psychic has ESP and $p = .5$. What is the probability that the psychic guesses incorrectly in all seven trials?

e. Refer to part **d**. Recall that the psychic failed to select the box with the crystal on all seven trials. Is this evidence against the psychic having ESP? Explain.

Applying the Concepts—Advanced

4.66 Assigning a passing grade. A literature professor decides to give a 20-question true-false quiz to determine who has read an assigned novel. She wants to choose the passing grade such that the probability of passing a student who guesses on every question is less than .05. What score should she set as the lowest passing grade?

4.67 USGA golf ball specifications. According to the U.S. Golf Association (USGA), "The weight of the [golf] ball shall not be greater than 1.620 ounces avoirdupois (45.93 grams).... The diameter of the ball shall not be less than 1.680 inches.... The velocity of the ball shall not be greater than 250 feet per second" (USGA, 2001). The USGA periodically checks the specifications of golf balls sold in the United States by randomly sampling balls from pro shops around the country. Two dozen of each kind are sampled, and if more than three do not meet size and/or velocity requirements, that kind of ball is removed from the USGA's approved-ball list.

a. What assumptions must be made and what information must be known in order to use the binomial probability distribution to calculate the probability that the USGA will remove a particular kind of golf ball from its approved-ball list?

b. Suppose 10% of all balls produced by a particular manufacturer are less than 1.680 inches in diameter, and assume that the number of such balls, x, in a sample of two dozen balls can be adequately characterized by a binomial probability distribution. Find the mean and standard deviation of the binomial distribution.

c. Refer to part **b**. If x has a binomial distribution, then so does the number, y, of balls in the sample that meet the USGA's minimum diameter. [*Note:* $x + y = 24$.] Describe the distribution of y. In particular, what are p, q, and n? Also, find $E(y)$ and the standard deviation of y.

4.68 Premature aging gene. Ataxia-telangiectasia (A-T) is a neurological disorder that weakens immune systems and causes premature aging. According to *Science News* (June 24, 1995), when both members of a couple carry the A-T gene, their children have a one in five chance of developing the disease.

a. Consider 15 couples in which both members of each couple carry the A-T gene. What is the probability that more than 8 of 15 couples have children that develop the neurological disorder?

b. Consider 10,000 couples in which both members of each couple carry the A-T gene. Is it likely that fewer than 3,000 will have children that develop the disease? [*Hint:* Calculate the mean and standard deviation of the distribution.]

4.5 The Poisson Random Variable (Optional)

A type of probability distribution that is often useful in describing the number of events that will occur in a specific period of time or in a specific area or volume is the **Poisson distribution** (named after the 18th-century physicist and mathematician, Siméon Poisson). Typical examples of random variables for which the Poisson probability distribution provides a good model are

1. The number of traffic accidents per month at a busy intersection

2. The number of noticeable surface defects (scratches, dents, etc.) found by quality inspectors on a new automobile

3. The parts per million of some toxin found in the water or air emission from a manufacturing plant

4. The number of diseased trees per acre of a certain woodland

5. The number of death claims received per day by an insurance company

6. The number of unscheduled admissions per day to a hospital

Characteristics of a Poisson Random Variable

1. The experiment consists of counting the number of times a certain event occurs during a given unit of time or in a given area or volume (or weight, distance, or any other unit of measurement).

2. The probability that an event occurs in a given unit of time, area, or volume is the same for all the units.

3. The number of events that occur in one unit of time, area, or volume is independent of the number that occur in other units.

4. The mean (or expected) number of events in each unit is denoted by the Greek letter lambda, λ.

The characteristics of the Poisson random variable are usually difficult to verify for practical examples. The examples given satisfy them well enough that the Poisson distribution provides a good model in many instances.* As with all probability models,

*The Poisson probability distribution also provides a good approximation to a binomial probability distribution with mean $\lambda = np$ when n is large and p is small (say, $np \leq 7$).

the real test of the adequacy of the Poisson model is in whether it provides a reasonable approximation to reality—that is, whether empirical data support it.

The probability distribution, mean, and variance for a Poisson random variable are shown in the next box.

Probability Distribution, Mean, and Variance for a Poisson Random Variable

$$p(x) = \frac{\lambda^x e^{-\lambda}}{x!} \qquad (x = 0, 1, 2, \dots)$$

$$\mu = \lambda \qquad \sigma^2 = \lambda$$

where

λ = Mean number of events during given unit of time, area, volume, etc.

Biography

SIMEON D. POISSON (1781–1840)—A Lifetime Mathematician

Growing up in France during the French Revolution, Simeon-Denis Poisson was sent away by his father to become an apprentice surgeon, but he lacked the manual dexterity to perform delicate procedures required and returned home. He eventually enrolled in École Polytechnique University to study mathematics. In his final year of study, Poisson wrote a paper on the theory of equations that was of such quality that he was allowed to graduate without taking the final examination. Two years later, Poisson was named a professor at the university. During his illustrious career, Poisson published between 300 and 400 mathematics papers. He is most known for his 1837 paper where the distribution of a rare event—the Poisson distribution—first appears. (In fact, the distribution was actually described years earlier by one of the Bernoulli brothers.) Poisson dedicated his life to mathematics, once stating that "Life is good for only two things: to study mathematics and to teach it."

The calculation of Poisson probabilities is made easier by the use of Table III in Appendix A, which gives the cumulative probabilities $P(x \le k)$ for various values of λ. The use of Table III is illustrated in Example 4.13.

EXAMPLE 4.13   FINDING POISSON PROBABILITIES

Problem   Ecologists often use the number of reported sightings of a rare species of animal to estimate the remaining population size. For example, suppose the number, x, of reported sightings per week of blue whales is recorded. Assume that x has (approximately) a Poisson probability distribution. Furthermore, assume that the average number of weekly sightings is 2.6.

    a. Find the mean and standard deviation of x, the number of blue whale sightings per week. Interpret the results.

    b. Use Table III to find the probability that fewer than two sightings are made during a given week.

    c. Use Table III to find the probability that more than five sightings are made during a given week.

    d. Use Table III to find the probability that exactly five sightings are made during a given week.

Solution   **a.** The mean and variance of a Poisson random variable are both equal to λ. Thus, for this example,

$$\mu = \lambda = 2.6$$
$$\sigma^2 = \lambda = 2.6$$

Then the standard deviation of x is

$$\sigma = \sqrt{2.6} = 1.61$$

Remember that the mean measures the central tendency of the distribution and does not necessarily equal a possible value of x. In this example, the mean is 2.6 sightings, and although there cannot be 2.6 sightings during a given week, the average number of weekly sightings is 2.6. Similarly, the standard deviation of 1.61 measures the variability of the number of sightings per week. Perhaps a more helpful measure is the interval $\mu \pm 2\sigma$, which in this case stretches from $-.62$ to 5.82. We expect the number of sightings to fall in this interval most of the time—with at least 75% relative frequency (according to Chebyshev's Rule) and probably with more than 90% relative frequency (the Empirical Rule). The mean and the 2-standard-deviation interval around it are shown in Figure 4.10.

b. A partial reproduction of Table III is shown in Table 4.5. The rows of the table correspond to different values of λ, and the columns correspond to different values k of the Poisson random variable x. The entries in the table (like the binomial probabilities in Table II) give the cumulative probability $P(x \le k)$. To find the probability that fewer than two sightings are made during a given week, we first note that

$$P(x < 2) = P(x \le 1)$$

This probability is a cumulative probability and therefore is the entry in Table III in the row corresponding to $\lambda = 2.6$ and the column corresponding to $k = 1$. The entry is .267, shown highlighted in Table 4.5. This probability corresponds to the highlighted area in Figure 4.10 and may be interpreted as meaning that there is a 26.7% chance that fewer than two sightings will be made during a given week.

c. To find the probability that more than five sightings are made during a given week, we consider the complementary event

$$P(x > 5) = 1 - P(x \le 5) = 1 - .951 = .049$$

where .951 is the entry in Table III corresponding to $\lambda = 2.6$ and $k = 5$ (see Table 4.5). Note from Figure 4.10 that this is the area in the interval $\mu \pm 2\sigma$, or $-.62$ to 5.82. Then the number of sightings should exceed 5—or, equivalently, should be more than 2 standard deviations from the mean—during only about 4.9% of all weeks. Note that this percentage agrees remarkably well with that given by the Empirical Rule for mound-shaped distributions, which tells us to

Figure 4.10

Probability Distribution for Number of Blue Whale Sightings

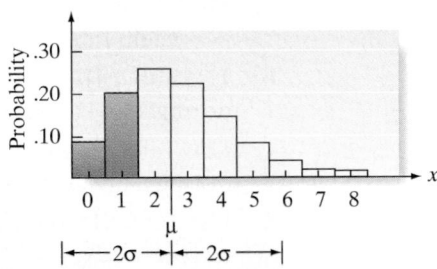

TABLE 4.5   Reproduction of Part of Table III in Appendix A

k / λ	0	1	2	3	4	5	6	7	8	9
2.2	.111	.355	.623	.819	.928	.975	.993	.998	1.000	1.000
2.4	.091	.308	.570	.779	.904	.964	.988	.997	.999	1.000
2.6	.074	.267	.518	.736	.877	.951	.983	.995	.999	1.000
2.8	.061	.231	.469	.692	.848	.935	.976	.992	.998	.999
3.0	.050	.199	.423	.647	.815	.916	.966	.988	.996	.999
3.2	.041	.171	.380	.603	.781	.895	.955	.983	.994	.998
3.4	.033	.147	.340	.558	.744	.871	.942	.977	.992	.997
3.6	.027	.126	.303	.515	.706	.844	.927	.969	.988	.996
3.8	.022	.107	.269	.473	.668	.816	.909	.960	.984	.994
4.0	.018	.092	.238	.433	.629	.785	.889	.949	.979	.992
4.2	.015	.078	.210	.395	.590	.753	.867	.936	.972	.989
4.4	.012	.066	.185	.359	.551	.720	.844	.921	.964	.985
4.6	.010	.056	.163	.326	.513	.686	.818	.905	.955	.980
4.8	.008	.048	.143	.294	.476	.651	.791	.887	.944	.975
5.0	.007	.040	.125	.265	.440	.616	.762	.867	.932	.968
5.2	.006	.034	.109	.238	.406	.581	.732	.845	.918	.960
5.4	.005	.029	.095	.213	.373	.546	.702	.822	.903	.951
5.6	.004	.024	.082	.191	.342	.512	.670	.797	.886	.941
5.8	.003	.021	.072	.170	.313	.478	.638	.771	.867	.929
6.0	.002	.017	.062	.151	.285	.446	.606	.744	.847	.916

expect approximately 5% of the measurements (values of the random variable) to lie farther than 2 standard deviations from the mean.

d. To use Table III to find the probability that *exactly* five sightings are made during a given week, we must write the probability as the difference between two cumulative probabilities:

$$P(x = 5) = P(x \le 5) - P(x \le 4) = .951 - .877 = .074$$

Now Work *Exercise 4.75*

■ ■ ■

Note that the probabilities in Table III are all rounded to three decimal places. Thus, although in theory a Poisson random variable can assume infinitely large values, the values of k in Table III are extended only until the cumulative probability is 1.000. This does not mean that x *cannot* assume larger values, but only that the likelihood is less than .001 (in fact, less than .0005) that it will do so.

Finally, you may need to calculate Poisson probabilities for values of λ not found in Table III. You may be able to obtain an adequate approximation by interpolation, but if not, consult more extensive tables for the Poisson distribution.

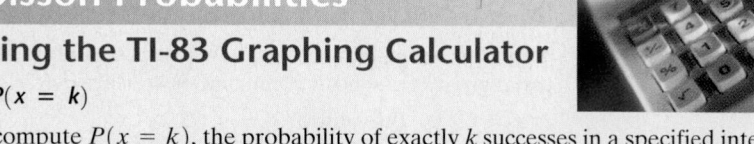

Poisson Probabilities

Using the TI-83 Graphing Calculator

I. $P(x = k)$

To compute $P(x = k)$, the probability of exactly k successes in a specified interval where λ is the mean number of successes in the interval, use the poissonpdf(command. Poissonpdf stands for "Poisson probability density function." This command is under the DISTRibution menu and has the format poissonpdf().

Example Suppose that the number, x, of reported sightings per week of blue whales is recorded. Assume that x has approximately a Poisson probability distribution, and that the average number of weekly sightings is 2.6.
Compute the probability that exactly five sightings are made during a given week. In this example, $\lambda = 2.6$ and $k = 5$.

Step 1 Enter the poisson parameters
Press 2nd VARS for DISTR
Press the down arrow key until B:poissonpdf is highlighted
Press ENTER
After poissonpdf(, type 2.6, 5) (Note: be sure to use the comma key between each parameter.)
Press ENTER

You should see

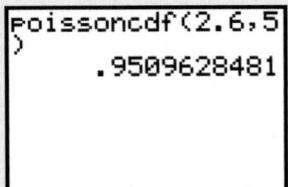

Thus, the $P(x = 5)$ is about 7.4%.

II. $P(x \leq k)$

To compute the probability of k or fewer successes in a specified interval, where λ is the mean number of successes in the interval, use the poissoncdf(command. Poissoncdf stands for "Poisson *cumulative* probability density function." This command is under the DISTRibution menu and has the format poissoncdf).

Example In the example given above, compute the probability that five or fewer sightings are made during a given week. In this example, $\lambda = 2.6$ and $k = 5$.

Step 1 Enter the poisson parameters
Press 2nd VARS for DISTR
Press the down arrow key until C:poissoncdf is highlighted
Press ENTER
After poissoncdf(, type 2.6, 5)
Press ENTER

You should see

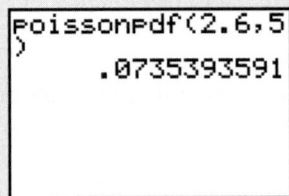

Thus, the $P(x \leq 5)$ is about 95.1%.

III. $P(x < k)$, $P(x > k)$, $P(x \geq k)$

To find the probability of less than k successes, more than k successes, or at least k successes variations of poissoncdf(command must be used as shown below.

$P(x < k)$ use **poissoncdf($\lambda, k - 1$)**

$P(x > k)$ use **$1 -$ poissoncdf(λ, k)**

$P(x \geq k)$ use **$1 -$ poissoncdf($\lambda, k - 1$)**

Exercises 4.69–4.87

Understanding the Principles

4.69 Give the four characteristics of a Poisson random variable.

4.70 Consider a Poisson random variable with probability distribution

$$p(x) = \frac{10^x e^{-10}}{x!} \qquad (x = 0, 1, 2, \dots)$$

What is the value of λ?

4.71 Consider the Poisson probability distribution shown here:

$$p(x) = \frac{3^x e^{-3}}{x!} \qquad (x = 0, 1, 2, \dots)$$

What is the value of λ?

Learning the Mechanics

4.72 Refer to Exercise 4.70.
  a. Graph the probability distribution.
  b. Find the mean and standard deviation of x.

4.73 Refer to Exercise 4.71.
  a. Graph the probability distribution.
  b. Find the mean and standard deviation of x.

4.74 Given that x is a random variable for which a Poisson probability distribution provides a good approximation, use Table III to compute the following:
  a. $P(x \le 2)$ when $\lambda = 1$
  b. $P(x \le 2)$ when $\lambda = 2$
  c. $P(x \le 2)$ when $\lambda = 3$
  d. What happens to the probability of the event $\{x \le 2\}$ as λ increases from 1 to 3? Is this intuitively reasonable?

4.75 Assume that x is a random variable having a Poisson probability distribution with a mean of 1.5. Use Table III to find the following probabilities:
  a. $P(x \le 3)$ **b.** $P(x \ge 3)$
  c. $P(x = 3)$ **d.** $P(x = 0)$
  e. $P(x > 0)$ **f.** $P(x > 6)$

4.76 Suppose x is a random variable for which a Poisson probability distribution with $\lambda = 1$ provides a good characterization.
  a. Graph $p(x)$ for $x = 0, 1, 2, \dots, 9$.
  b. Find μ and σ for x, and locate μ and the interval $\mu \pm 2\sigma$ on the graph.
  c. What is the probability that x will fall within the interval $\mu \pm 2\sigma$?

4.77 Suppose x is a random variable for which a Poisson probability distribution with $\lambda = 3$ provides a good characterization.
  a. Graph $p(x)$ for $x = 0, 1, 2, \dots, 9$.
  b. Find μ and σ for x, and locate μ and the interval $\mu \pm 2\sigma$ on the graph.
  c. What is the probability that x will fall within the interval $\mu \pm 2\sigma$?

4.78 As mentioned in Section 4.5, when n is large, p is small, and $np \le 7$, the Poisson probability distribution provides a good approximation to the binomial probability distribution. Since we provide exact binomial probabilities (Table II in Appendix A) for relatively small values of n, you can investigate the adequacy of the approximation for $n = 25$. Use Table II to find $p(0)$, $p(1)$, and $p(2)$ for $n = 25$ and $p = .05$. Calculate the corresponding Poisson approximations using $\lambda = \mu = np$. [*Note:* These approximations are reasonably good for n as small as 25, but to use the approximation in a practical situation we would prefer to have $n \ge 100$.]

Applying the Concepts—Basic

4.79 **Eye fixation experiment.** Cognitive scientists at the University of Massachusetts designed an experiment to measure x, the number of times a reader's eye fixated on a single word before moving past that word. (*Memory and Cognition*, Sept. 1997.) For this experiment, x was found to have a mean of 1. Suppose one of the readers in the experiment is randomly selected and assume that x has a Poisson distribution.
  a. Find $P(x = 0)$.
  b. Find $P(x > 1)$.
  c. Find $P(x \le 2)$.

4.80 **Rare planet transits.** The "Venus transit" describes a rare celestial event where the planet Venus appears to cross in front of the Sun as seen from Earth. The last "Venus transit" occurred more than a century ago, in 1882. (National Aeronautics and Space Administration, 2004.) When a planet crosses in front of a star, there is a noticeable dip in the star's brightness. These brightness dips allow NASA scientists to detect the presence of a planet even if the planet itself is not directly visible. In October 2007 NASA will launch its Kepler mission, measuring the brightness of 100,000 stars every 15 minutes in hopes of detecting extrasolar planet transits. The mission is expected to detect 5.6 extrasolar planets for every 10,000 stars. If the number of extrasolar planet transits for 10,000 stars follows a Poisson distribution with $\lambda = 5.6$, what is the probability that more than 10 transits will be seen?

4.81 **Airline fatalities.** U.S. airlines average about 3.8 fatalities per month. (*Statistical Abstract of the United States: 2000.*) Assume the probability distribution for x, the number of fatalities per month, can be approximated by a Poisson probability distribution.
  a. What is the probability that no fatalities will occur during any given month?
  b. What is the probability that one fatality will occur during any given month?
  c. Find $E(x)$ and the standard deviation of x.

4.82 **Deep-draft vessel casualties.** Economists at the University of New Mexico modeled the number of casualties (deaths or missing persons) experienced by a deep-draft U.S. flag vessel over a three-year period as a Poisson random variable, x. The researchers estimated $E(x)$ to be .03. (*Management Science*, Jan. 1999.)

a. Find the variance of x.

b. Discuss the conditions that would make the researchers' Poisson assumption plausible.

c. What is the probability that a deep-draft U.S. flag vessel will have no casualties in a three-year time period?

Applying the Concepts—Intermediate

4.83 Vinyl chloride emissions. The Environmental Protection Agency (EPA) limits the amount of vinyl chloride in plant air emissions to no more than 10 parts per million. Suppose the mean emission of vinyl chloride for a particular plant is 4 parts per million. Assume that the number of parts per million of vinyl chloride in air samples, x, follows a Poisson probability distribution.

a. What is the standard deviation of x for the plant?

b. Is it likely that a sample of air from the plant would yield a value of x that would exceed the EPA limit? Explain.

c. Discuss conditions that would make the Poisson assumption plausible.

4.84 Davey Crockett's use of words. Davey Crockett, a U.S. congressman during the 1830s, published *A Narrative of the Life of Davey Crockett, Written by Himself*. In a *Chance* (Spring 1999) article, researchers determined that x, the number of times (per 1,000 words) that the word *though* appears in the *Life* narrative, has a mean of .25. Assume that x has an approximate Poisson distribution.

a. Find the variance of x.

b. Calculate the interval $\mu \pm 2\sigma$. What proportion of the time will x fall in this interval?

c. In 1836, Davey Crockett died at the Alamo defending Texans from an attack by the Mexican general, Santa Anna. The following year, the narrative *Colonel Crockett's Exploits and Adventures in Texas* was published. Some historians have doubts whether Crockett was the actual author of this narrative. In the *Texas* narrative, "though" appears twice in the first 1,000 words. Use this information and your answer to part **b** to make an inference

about whether Crockett actually wrote the *Texas* narrative. (Assume the *Life* narrative was really written by Davey Crockett.)

4.85 Customer arrivals at a bakery. As part of a project targeted at improving the services of a local bakery, a management consultant (L. Lei of Rutgers University) monitored customer arrivals for several Saturdays and Sundays. Using the arrival data, she estimated the average number of customer arrivals per 10-minute period on Saturdays to be 6.2. She assumed that arrivals per 10-minute interval followed the Poisson distribution shown at the bottom of the page, some of whose values are missing.

a. Compute the missing probabilities.

b. Graph the distribution.

c. Find μ and σ and show the intervals $\mu \pm \sigma, \mu \pm 2\sigma$, and $\mu \pm 3\sigma$ on your graph of part **b**.

d. The owner of the bakery claims that more than 75 customers per hour enter the store on Saturdays. Based on the consultant's data, is this likely? Explain.

4.86 Crime Watch neighborhood. In many cities, neighborhood Crime Watch groups are formed in an attempt to reduce the amount of criminal activity. Suppose one neighborhood that has experienced an average of 10 crimes per year organizes such a group. During the first year following the creation of the group, three crimes are committed in the neighborhood.

a. Use the Poisson distribution to calculate the probability that three or fewer crimes are committed in a year assuming the average number is still 10 crimes per year.

b. Do you think this event provides some evidence that the Crime Watch group has been effective in this neighborhood?

Applying the Concepts—Advanced

4.87 Waiting for a car wash. A certain automatic car wash takes exactly 5 minutes to wash a car. On the average, 10 cars per hour arrive at the car wash. Suppose that, 30 minutes before closing time, five cars are in line. If the car wash is in continuous use until closing time, is it likely anyone will be in line at closing time?

Table for Exercise 4.85

x	0	1	2	3	4	5	6	7	8	9	10	11	12	13
$p(x)$	.002	.013	—	.081	.125	.155	—	.142	.110	.076	—	.026	.014	.007

Source: Lei, L. *Dorsi's Bakery: Modeling Service Operations.* Graduate School of Management, Rutgers University, 1993.

4.6 The Hypergeometric Random Variable (Optional)

The **hypergeometric probability distribution** provides a realistic model for some types of enumerative (countable) data. The characteristics of the hypergeometric distribution are listed in the box.

> ### Characteristics of a Hypergeometric Random Variable
>
> 1. The experiment consists of randomly drawing n elements without replacement from a set of N elements, r of which are S's (for Success) and $(N - r)$ of which are F's (for Failure).
> 2. The hypergeometric random variable x is the number of S's in the draw of n elements.

Note that both the hypergeometric and binomial characteristics stipulate that each draw, or trial, results in one of two outcomes. The basic difference between these random variables is that the hypergeometric trials are dependent, while the binomial trials are independent. The draws are dependent because the probability of drawing an S (or an F) is dependent on what occurred on preceding draws.

To illustrate the dependence between trials, we note that the probability of drawing an S on the first draw is r/N. Then the probability of drawing an S on the second draw depends on the outcome of the first. It will be either $(r - 1)/(N - 1)$, or $r/(N - 1)$, depending on whether the first draw was an S or an F. Consequently, the results of the draws represent dependent events.

For example, suppose we define x as the number of women hired in a random selection of three applicants from a total of six men and four women. This random variable satisfies the characteristics of a hypergeometric random variable with $N = 10$ and $n = 3$. The possible outcomes on each trial are either selection of a female (S) or selection of a male (F). Another example of a hypergeometric random variable is the number, x, of defective large-screen television picture tubes in a random selection of $n = 4$ from a shipment of $N = 8$ tubes. And, as a third example, suppose $n = 5$ stocks are randomly selected from a list of $N = 15$ stocks. Then, the number x of the five selected companies that pay regular dividends to stockholders is a hypergeometric random variable.

The hypergeometric probability distribution is summarized in the box.

> ### Probability Distribution, Mean, and Variance of the Hypergeometric Random Variable
>
> $$p(x) = \frac{\binom{r}{x}\binom{N - r}{n - x}}{\binom{N}{n}} \qquad [x = \text{Maximum}\,[0, n - (N - r)], \ldots, \text{Minimum}(r, n)]$$
>
> $$\mu = \frac{nr}{N} \qquad \sigma^2 = \frac{r(N - r)n(N - n)}{N^2(N - 1)}$$
>
> where
>
> N = Total number of elements
> r = Number of S's in the N elements
> n = Number of elements drawn
> x = Number of S's drawn in the n elements

EXAMPLE 4.14   APPLYING THE HYPERGEOMETRIC DISTRIBUTION

Problem   Suppose a professor randomly selects three new teaching assistants from a total of 10 applicants, six male and four female students. Let x be the number of females who are hired.

   a. Find the mean and standard deviation of x.
   b. Find the probability that no females are hired.

Solution

a. Since x is a hypergeometric random variable with $N = 10$, $n = 3$, and $r = 4$, the mean and variance are

$$\mu = \frac{nr}{N} = \frac{(3)(4)}{10} = 1.2$$

$$\sigma^2 = \frac{r(N-r)n(N-n)}{N^2(N-1)} = \frac{4(10-4)3(10-3)}{(10)^2(10-1)}$$

$$= \frac{(4)(6)(3)(7)}{(100)(9)} = .56$$

The standard deviation is

$$\sigma = \sqrt{.56} = .75$$

b. The probability that no female students are hired by the professor, assuming the selection is truly random, is

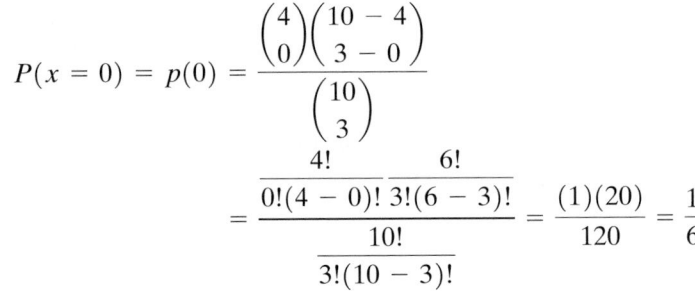

$$P(x = 0) = p(0) = \frac{\binom{4}{0}\binom{10-4}{3-0}}{\binom{10}{3}}$$

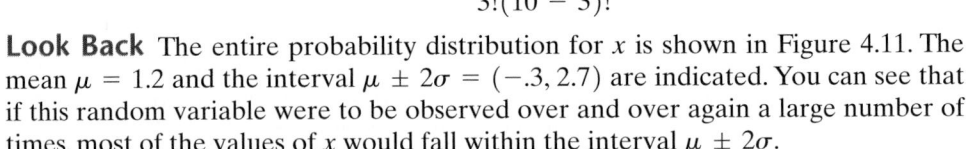

$$= \frac{\dfrac{4!}{0!(4-0)!}\dfrac{6!}{3!(6-3)!}}{\dfrac{10!}{3!(10-3)!}} = \frac{(1)(20)}{120} = \frac{1}{6}$$

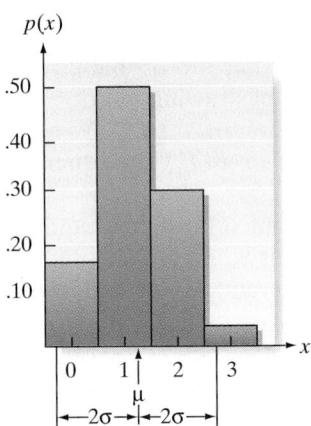

p(x)

Figure 4.11
Probability Distribution for
x in Example 4.14

Look Back The entire probability distribution for x is shown in Figure 4.11. The mean $\mu = 1.2$ and the interval $\mu \pm 2\sigma = (-.3, 2.7)$ are indicated. You can see that if this random variable were to be observed over and over again a large number of times, most of the values of x would fall within the interval $\mu \pm 2\sigma$.

Now Work **Exercise 4.103**

■ ■ ■

Statistics in Action Revisited
Using the Hypergeometric Model to Solve the Cocaine Sting Case

The reverse cocaine sting case described on p. 189 and solved in Section 4.3 can also be solved by applying the hypergeometric distribution. In fact, the probabilities obtained using the hypergeometric distribuion are *exact* probabilities, as compared to the approximate probabilities obtained using the binomial distribution.

Our objective, you will recall, is to find the probability that four packets (randomly selected from 496 packets confiscated in a drug bust) will contain cocaine and two packets (randomly selected from the remaining 492) will not contain cocaine. We assumed that of the 496 original packets, 331 contained genuine cocaine and 165 contained an inert (legal) powder. Since we are sampling *without replacement* from the 496 packets, the probability of a "Success" (i.e., the probability of a packet containing cocaine) does not remain *exactly* the same from trial to trial. For example, for the first randomly selected packet, the probability of a "Success" is

331/496 = .66734. If we find that the first three packets selected contain cocaine, the probability of "Success" for the fourth packet selected is now 328/493 = .66531. You can see that these probabilities are not exactly the same. Hence, the binomial distribution will only approximate the distribution of x, the number of packets that contain cocaine is a sample of size n.

To find the probability that four packets randomly selected from the original 496 will test positive for cocaine using the hypergeometric distribution, we first identify the parameters of the distribution:

$N = 496$ is the total number of packets in the population
$S = 331$ is the number of "successes" (cocaine packets) in the population
$n = 4$ is the sample size
$x = 4$ is the number of "successes" (cocaine packets) in the sample

Substituting into the formula for $p(x)$ (page 225), we obtain

$$P(x = 4) = p(4) = \frac{\binom{331}{4}\binom{165}{0}}{\binom{496}{4}} = \frac{\left(\frac{331!}{4!327!}\right)\left(\frac{165!}{0!165!}\right)}{\left(\frac{496!}{4!492!}\right)}$$

$$= .197$$

To find the probability that two packets randomly selected from the remaining 492 will test negative for cocaine *assuming that the first four packets tested positive*, we identify the parameters of the relevant hypergeometric distribution:

$N = 492$ is the total number of packets in the population

$S = 327$ is the number of "successes" (cocaine packets) in the population

$n = 2$ is the sample size

$x = 0$ is the number of "successes" (cocaine packets) in the sample

Again, we substitute into the formula for $p(x)$ (page 225) to obtain

$$P(x = 0) = p(0) = \frac{\binom{327}{0}\binom{165}{2}}{\binom{492}{2}} = \frac{\left(\frac{327!}{0!327!}\right)\left(\frac{165!}{2!163!}\right)}{\left(\frac{492!}{2!490!}\right)}$$

$$= .112$$

Using the Multiplicative Law of Probability, the probability that the first four packets test positive for cocaine and the next two packets test negative is the product of the two probabilities just given:

$$P(\text{first 4 positive and next 2 negative}) = (.112)(.197)$$
$$= .0221$$

Note that this exact probability is almost identical to the approximate probability computed using the binomial distribution in Section 4.3.

Exercises 4.88–4.105

Understanding the Principles

4.88 Give the characteristics of a hypergeometric distribution.

4.89 How do binomial and hypergeometric random variables differ? In what respects are they similar?

4.90 Explain the difference between sampling with replacement and sampling without replacement.

Learning the mechanics

4.91 Given that x is a hypergeometric random variable, compute $p(x)$ for each of the following cases:
  a. $N = 5, n = 3, r = 3, x = 1$
  b. $N = 9, n = 5, r = 3, x = 3$
  c. $N = 4, n = 2, r = 2, x = 2$
  d. $N = 4, n = 2, r = 2, x = 0$

4.92 Given that x is a hypergeometric random variable with $N = 8, n = 3$, and $r = 5$, compute the following:
  a. $P(x = 1)$ **b.** $P(x = 0)$
  c. $P(x = 3)$ **d.** $P(x \geq 4)$

4.93 Given that x is a hypergeometric random variable with $N = 12, n = 8$, and $r = 6$:
  a. Display the probability distribution for x in tabular form.
  b. Compute μ and σ for x.
  c. Graph $p(x)$ and locate μ and the interval $\mu \pm 2\sigma$ on the graph.
  d. What is the probability that x will fall within the interval $\mu \pm 2\sigma$?

4.94 Given that x is a hypergeometric random variable with $N = 10, n = 5$, and $r = 7$:
  a. Display the probability distribution for x in tabular form.
  b. Compute the mean and variance of x.
  c. Graph $p(x)$ and locate μ and the interval $\mu \pm 2\sigma$ on the graph.

  d. What is the probability that x will fall within the interval $\mu \pm 2\sigma$?

4.95 Suppose you plan to sample 10 items from a population of 100 items and would like to determine the probability of observing four defective items in the sample. Which probability distribution should you use to compute this probability under the following conditions? Justify your answers.
  a. The sample is drawn without replacement.
  b. The sample is drawn with replacement.

4.96 Use the results of Exercise 4.93 to find the following probabilities:
  a. $P(x = 1)$ **b.** $P(x = 4)$
  c. $P(x \leq 4)$ **d.** $P(x \geq 5)$
  e. $P(x < 3)$ **f.** $P(x \geq 8)$

Applying the Concepts—Basic

4.97 On-site disposal of hazardous waste. The Resource Conservation and Recovery Act mandates the tracking and disposal of hazardous waste produced at U.S. facilities. *Professional Geographer* (Feb. 2000) reported the hazardous waste generation and disposal characteristics of 209 facilities. Only eight of these facilities treated hazardous waste on-site.
  a. In a random sample of 10 of the 209 facilities, what is the expected number in the sample that treat hazardous waste on-site? Interpret this result.
  b. Find the probability that four of the eight selected facilities treat hazardous waste on-site.

4.98 Male nannies. According to *USA Today* (Sept. 19, 2000), there are 650 members of the International Nanny Association (INA). Of these, only three are men. In Exercise 3.18 (p. 131), you found the probability that a randomly selected member of the INA is a man. Now find the probability that in a random sample of four INA members, at least one is a man.

4.99 Contaminated gun cartridges. Refer to the investigation of contaminated gun cartridges at a weapons manufacturer, Exercise 4.20 (p. 196). In a sample of 158 cartridges from a certain lot, 36 were found to be contaminated, and 122 were "clean." If you randomly select 5 of these 158 cartridges, what is the probability that all 5 will be "clean"?

4.100 Lot inspection sampling. Imagine you are purchasing small lots of a manufactured product. If it is very costly to test a single item, it may be desirable to test a sample of items from the lot instead of testing every item in the lot. Suppose each lot contains 10 items. You decide to sample four items per lot and reject the lot if you observe one or more defectives.
  a. If the lot contains one defective item, what is the probability that you will accept the lot?
  b. What is the probability that you will accept the lot if it contains two defective items?

Applying the Concepts—Intermediate

4.101 Extinct New Zealand birds. Refer to the *Evolutionary Ecology Research* (July 2003) study of the patterns of extinction in the New Zealand bird population, Exercise 3.79 (p. 157). Of the 132 bird species saved in the **NZBIRDS** file, 38 are extinct. Suppose you randomly select 10 of the 132 bird species (without replacement) and record the extinct status of each.
  a. What is the probability that exactly five of the 10 species you select are extinct?
  b. What is the probability that at most one species is extinct?

4.102 Birds in butterfly hotspots. "Hotspots" are species-rich geographical areas. A *Nature* (Sept. 1993) study estimated the probability of a bird species in Great Britain inhabiting a butterfly hotspot at .70. Consider a random sample of four British bird species selected from a total of 10 tagged species. Assume that seven of the 10 tagged species inhabit a butterfly hotspot.
  a. What is the probability that exactly half of the four bird species sampled inhabit a butterfly hotspot?

  b. What is the probability that at least one of the four bird species sampled inhabit a butterfly hotspot?

4.103 Testing for spoiled wine. Suppose you are purchasing cases of wine (twelve bottles per case) and that, periodically, you select a test case to determine the adequacy of the bottles' seals. To do this, you randomly select and test three bottles in the case. If a case contains one spoiled bottle of wine, what is the probability that this bottle will turn up in your sample?

Applying the Concepts—Advanced

4.104 Gender discrimination suit. The *Journal of Business & Economic Statistics* (July 2000) presented a case in which a charge of gender discrimination was filed against the U.S. Postal Service. At the time, there were 302 U.S. Postal Service employees (229 men and 73 women) who applied for promotion. Of the 72 employees who were awarded promotion, 5 were female. Make an inference about whether or not females at the U.S. Postal Service were promoted fairly.

4.105 Awarding of home improvement grants. A curious event was described in the *Minneapolis Star and Tribune*. The Minneapolis Community Development Agency (MCDA) makes home improvement grants each year to homeowners in depressed city neighborhoods. Of the $708,000 granted one year, $233,000 was awarded by the city council using a "random selection" of 140 homeowners' applications from among a total of 743 applications: 601 from the north side and 142 from the south side of Minneapolis. Oddly, all 140 grants awarded were from the north side—clearly a highly improbable outcome, if, in fact, the 140 winners were randomly selected from among the 743 applicants.
  a. Suppose the 140 winning applications were randomly selected from among the total of 743, and let x equal the number in the sample from the north side. Find the mean and standard deviation of x.
  b. Use the results of part **a** to support a contention that the grant winners were not randomly selected.

Quick Review

Key Terms

Note: Starred (*) terms are from the optional sections in this chapter.

Binomial distribution   208
Binomial experiment   204
Binomial random variable   205
Continuous random variable   191
Countable   191

Cumulative binomial probabilities   211
Discrete random variable   191
Expected value   198
*Hypergeometric probability distribution   225
*Hypergeometric random variable   225
Mean of a discrete random variable   198

*Poisson distribution   219
*Poisson random variable   218
Probability distribution   194
Random variable   189
Standard deviation of a discrete random variable   200
Variance of a discrete random variable   200

Key Formulas

Note: Starred (*) formulas are from the optional sections in this chapter.

Probability Distribution	Mean (μ)	Variance (σ^2)	
$p(x)$	$\sum_{\text{all } x} xp(x)$	$\sum_{\text{all } x}(x-\mu)^2 p(x)$	General discrete random variable 198, 200
$\binom{n}{x}p^x q^{n-x}$	np	npq	Binomial random variable 208, 210
$\dfrac{\lambda^x e^{-\lambda}}{x!}$	λ	λ	Poisson random variable* 219
$\dfrac{\binom{r}{x}\binom{N-r}{n-x}}{\binom{N}{n}}$	$\dfrac{nr}{N}$	$\dfrac{r(N-r)n(N-n)}{N^2(N-1)}$	Hypergeometric random variable* 225

Language Lab

Symbol	Pronunciation	Description
$p(x)$		Probability distribution of the random variable x
S		The outcome of a binomial trial denoted a "success"
F		The outcome of a binomial trial denoted a "failure"
p		The probability of success (S) in a binomial trial
q		The probability of failure (F) in a binomial trial, where $q = 1 - p$
λ	lambda	The mean (or expected) number of events for a Poisson random variable
e		A constant used in the Poisson probability distribution, where $e = 2.71828\ldots$

Chapter Summary Notes

- Two types of random variables: **discrete** and **continuous**
- Requirements for a discrete probability distribution: $p(x) \geq 0$ and $\Sigma p(x) = 1$
- Probability models discrete random variables: **binomial, Poisson,** and **hypergeometric**
- Characteristics of a **binomial random variable**: (1) n identical trials; (2) two possible outcomes, S and F, per trial; (3) $P(S)$ and $P(F)$ remain the same from trial to trial; (4) trials are independent; (5) $x =$ number of S's in n trials

- Characteristics of a **Poisson random variable**: (1) $x =$ number of times a rare event, S, occurs in a unit of time, area, or volume; (2) $P(S)$ remains the same for all units; (3) value of x in one unit is independent of value in another unit
- Characteristics of a **hypergeometric random variable**: (1) Draw n elements without replacement from a set of N elements, r of which have outcome S and $(N - r)$ of which have outcome F; (2) $x =$ number of S's in n trials

Supplementary Exercises 4.106–4.133

Note: Starred (*) exercises refer to the optional sections in this chapter.

Understanding the Principles

4.106 Which of the following describe discrete random variables, and which describe continuous random variables?

a. The length of time that an exercise physiologist's program takes to elevate her client's heart rate to 140 beats per minute

b. The number of crimes committed on a college campus per year

c. The number of square feet of vacant office space in a large city

d. The number of voters who favor a new tax proposal

4.107 Identify the type of random variable—binomial, Poisson, or hypergeometric—described by each of the following probability distributions:

*a. $p(x) = \dfrac{.5^x e^{-.5}}{x!}$ $(x = 0, 1, 2, \dots)$

b. $p(x) = \dbinom{6}{x}(.2)^x (.8)^{6-x}$ $(x = 0, 1, 2, \dots, 6)$

c. $p(x) = \dfrac{10!}{x!(10 - x)!}(.9)^x (.1)^{10-x}$
$(x = 0, 1, 2, \dots, 10)$

4.108 For each of the following examples, decide whether x is a binomial random variable and explain your decision:

a. A manufacturer of computer chips randomly selects 100 chips from each hour's production in order to estimate the proportion defectives. Let x represent the number of defectives in the 100 sampled chips.

b. Of five applicants for a job, two will be selected. Although all applicants appear to be equally qualified, only three have the ability to fulfill the expectations of the company. Suppose that the two selections are made at random from the five applicants, and let x be the number of qualified applicants selected.

c. A software developer establishes a support hotline for customers to call in with questions regarding use of the software. Let x represent the number of calls received on the support hotline during a specified workday.

d. Florida is one of a minority of states with no state income tax. A poll of 1,000 registered voters is conducted to determine how many would favor a state income tax in light of the state's current fiscal condition. Let x be the number in the sample who would favor the tax.

Learning the Mechanics

4.109 Suppose x is a binomial random variable with $n = 20$ and $p = .7$.

a. Find $P(x = 14)$.

b. Find $P(x \le 12)$.

c. Find $P(x > 12)$.

d. Find $P(9 \le x \le 18)$.

e. Find $P(8 < x < 18)$.

f. Find μ, σ^2, and σ.

g. What is the probability that x is in the interval $\mu \pm 2\sigma$?

*4.110** Given that x is a hypergeometric random variable, compute $p(x)$ for each of the following cases:

a. $N = 8, n = 5, r = 3, x = 2$

b. $N = 6, n = 2, r = 2, x = 2$

c. $N = 5, n = 4, r = 4, x = 3$

4.111 Suppose x is a binomial random variable. Find $p(x)$ for each of the following combinations of x, n, and p.

a. $x = 1, n = 3, p = .1$

b. $x = 4, n = 20, p = .3$

c. $x = 0, n = 2, p = .4$

d. $x = 4, n = 5, p = .5$

e. $n = 15, x = 12, p = .9$

f. $n = 10, x = 8, p = .6$

4.112 Consider the discrete probability distribution shown here.

x	10	12	18	20
$p(x)$	.2	.3	.1	.4

a. Calculate μ, σ^2, and σ.

b. What is $P(x < 15)$?

c. Calculate $\mu \pm 2\sigma$.

d. What is the probability that x is in the interval $\mu \pm 2\sigma$?

*4.113** Suppose x is a Poisson random variable. Compute $p(x)$ for each of the following cases:

a. $\lambda = 2, x = 3$

b. $\lambda = 1, x = 4$

c. $\lambda = .5, x = 2$

Applying the Concepts—Basic

4.114 **Use of laughing gas.** According to the American Dental Association, 60% of all dentists use nitrous oxide ("laughing gas") in their practice. (*New York Times*, June 20, 1995.) If x equals the number of dentists in a random sample of five dentists who use laughing gas in practice, then the probability distribution of x is

x	0	1	2	3	4	5
$p(x)$	.0102	.0768	.2304	.3456	.2592	.0778

a. Find the probability that the number of dentists using laughing gas in the sample of five is less than 2.

b. Find $E(x)$ and interpret the result.

c. Show that the distribution of x is binomial with $n = 5$ and $p = .6$.

4.115 **Accuracy of checkout scanners.** A Federal Trade Commission (FTC) study of the pricing accuracy of electronic checkout scanners at stores found that one of every 30 items is priced incorrectly. (*Price Check II: A Follow-Up Report on the Accuracy of Checkout Scanner Prices*, Dec. 16, 1998.) Suppose the FTC randomly selects five items at a retail store and checks the accuracy of the scanner price of each. Let x represent the number of the five items that is priced incorrectly.

a. Show that x is (approximately) a binomial random variable.

b. Use the information in the FTC study to estimate p for the binomial experiment.

c. What is the probability that exactly one of the five items is priced incorrectly by the scanner?

d. What is the probability that at least one of the five items is priced incorrectly by the scanner?

4.116 Expected patient cost. A patient complaining of severe stomach pains checked into a local hospital. After a series of tests the doctors narrowed their diagnosis to four possible ailments. They believe there is a 40% chance that the patient has hepatitis; a 10% chance that she has cirrhosis; a 45% chance of gallstones; and a 5% chance of cancer of the pancreas. The doctors are certain that the patient has only one of the diseases, but will not know which disease until further tests are performed. The cost associated with treating each disease is given in the table:

Disease	Hepatitis	Cirrhosis	Gallstones	Pancreatic Cancer
Cost	$700	$1,100	$3,320	$16,450

a. Construct the probability distribution for the cost of treating the patient.

b. Calculate the mean of the probability distribution you constructed in part **a**. What does this number represent?

c. Further testing reveals that the patient has either hepatitis or cirrhosis. Given this information, construct the new probability distribution for the cost of treating the patient.

d. Calculate the mean of the probability distribution you constructed in part **c**. What does this number represent?

4.117 Married women study. According to *Ladies Home Journal* (June 1988), 80% of married women would marry their current husbands again, if given the chance. Consider the number x in a random sample of ten married women who would marry their husbands again. Identify the discrete probability distribution that best models the distribution of x. Explain.

*4.118 Emergency rescue vehicle use. An emergency rescue vehicle is used an average of 1.3 times daily. Use the Poisson distribution to find

a. The probability that the vehicle will be used exactly twice tomorrow

b. The probability that it will be used more than twice

c. The probability that it will be used exactly three times

4.119 New book reviews. In Exercise 2.171 (p. 108) you read about a study of book reviews in American history, geography, and area studies published in *Choice* magazine. (*Library Acquisitions: Practice and Theory*, Vol. 19, 1995.) The overall rating stated in each review was ascertained and recorded as follows: 1 = would not recommend, 2 = cautious or very little recommendation, 3 = little or no preference, 4 = favorable/recommended, 5 = outstanding/significant contribution. Based on a sample of 375 book reviews, the probability distribution of rating, x, follows.

a. Is this a valid probability distribution?

b. What is the probability that a book reviewed in *Choice* has a rating of 1?

c. What is the probability that a book reviewed in *Choice* has a rating of at least 4?

Book Rating x	$p(x)$
1	.051
2	.099
3	.093
4	.635
5	.122

d. What is the probability that a book reviewed in *Choice* has a rating of 2 or 3?

e. Find $E(x)$.

f. Interpret the result, part **e**.

*4.120 Dutch elm disease. A nursery advertises that it has 10 elm trees for sale. Unknown to the nursery, three of the trees have already been infected with Dutch elm disease and will die within a year.

a. If a buyer purchases two trees, what is the probability that both trees will be healthy?

b. Refer to part **a**. What is the probability that at least one of the trees is infected?

Applying the Concepts—Intermediate

4.121 Pesticides on food samples. The Food and Drug Administration (FDA) produces a quarterly report called *Total Diet Study*. The FDA's report covers more than 200 food items, each of which is analyzed for dangerous chemical compounds. One *Total Diet Study* reported that no pesticides at all were found on 65% of the domestically produced food samples. (*Consumer's Research*, June 1995.) Consider a random sample of 800 food items analyzed for the presence of pesticides.

a. Compute μ and σ for the random variable x, the number of food items found without any trace of pesticide.

b. Based on a sample of 800 food items, is it likely you would observe less than half without any traces of pesticide? Explain.

4.122 Factors identifying urban counties. Refer to the *Professional Geographer* (Feb. 2000) study of urban and rural counties, Exercise 3.157 (p. 181). Forty-five percent of county commissioners in Nevada feel that "population concentration" is the single most important factor in identifying urban counties. Suppose 10 county commissioners are selected at random to serve on a Nevada review board that will consider redefining urban and rural areas. What is the probability that more than half of the commissioners will specify "population concentration" as the single most important factor in identifying urban counties?

4.123 Parents behavior at a gym meet. *Pediatric Exercise Science* (Feb. 1995) published an article on the behavior of parents at competitive youth gymnastic meets. Based on a survey of the gymnasts, the researchers estimated the probability of a parent "yelling" at their child before, during, or after the meet as .05. In a random sample of 20 parents attending a gymnastic meet, find the probability that at least one parent yells at their child before, during, or after the meet.

*4.124 Accepting or rejecting a shipment of CDs. By mistake, a manufacturer of CD mini-rack systems includes three defective systems in a shipment of 10 going out to a small retailer. The retailer has decided to accept the shipment of CD systems only if none is found to be defective. Upon receipt of the shipment, the retailer examines only five of the CD systems.
a. What is the probability that the shipment will be rejected?
b. If the retailer inspects six of the CD systems, what is the probability the shipment will be accepted?

4.125 Smoking and nasal allergies. The *American Journal of Public Health* (July 1995) published a study of the relationship between passive smoking and nasal allergies in Japanese female students. The study revealed that 80% of students from heavy-smoking families showed signs of nasal allergies on physical examinations. Consider a sample of 25 Japanese female students exposed daily to heavy smoking.
a. What is the probability that fewer than 20 of the students will have nasal allergies?
b. What is the probability that more than 15 of the students will have nasal allergies?
c. On average, how many of the 25 students would you expect to show signs of nasal allergies?

*4.126 Scottish cornrakes. The cornrake is a European species of bird in danger of worldwide extinction. A census revealed that 12 cornrakes inhabit mainland Scotland. (*Journal of Applied Ecology*, 1993.) Suppose that two of these Scottish cornrakes are captured for mating purposes. Let x be the number of these captured cornrakes that are capable of mating. If exactly four of the original 12 cornrakes inhabiting Scotland are infertile and hence unable to mate, find the probability distribution for x.

4.127 Efficacy of an insecticide. The efficacy of insecticides is often measured by the dose necessary to kill a certain percentage of insects. Suppose a certain dose of a new insecticide is supposed to kill 80% of the insects that receive it. To test the claim, 25 insects are exposed to the insecticide.
a. If the insecticide really kills 80% of the exposed insects, what is the probability that fewer than 15 die?
b. If you observed such a result, what would you conclude about the new insecticide? Explain your logic.

*4.128 Arrivals at a bank counter. The number x of people who arrive at a cashier's counter in a bank during a specified period of time often exhibits (approximately) a Poisson probability distribution. If we know the mean arrival rate λ, the Poisson probability distribution can be used to aid in the design of the customer service facility. Suppose you estimate that the mean number of arrivals per minute for cashier service at a bank is one person per minute.

a. What is the probability that in a given minute the number of arrivals will equal three or more?
b. Can you tell the bank manager that the number of arrivals will rarely exceed two per minute?

4.129 Construction company profit analysis. The owner of construction company A makes bids on jobs so that if awarded the job, company A will make a $10,000 profit. The owner of construction company B makes bids on jobs so that if awarded the job, company B will make a $15,000 profit. Each company describes the probability distribution of the number of jobs x the company is awarded per year as shown in the table.

Company A		Company B	
x	$P(x)$	x	$P(x)$
2	.05	2	.15
3	.15	3	.30
4	.20	4	.30
5	.35	5	.20
6	.25	6	.05

a. Find the expected number of jobs each will be awarded in a year.
b. What is the expected profit for each company?
c. Find the variance and standard deviation of the distribution of number of jobs awarded per year for each company.
d. Graph $p(x)$ for both companies A and B. For each company, what proportion of the time will x fall in the interval $\mu \pm 2\sigma$?

Applying the Concepts—Advanced

4.130 How many questionnaires to mail? The probability that a person responds to a mailed questionnaire is .4. How many questionnaires should be mailed if you want to be reasonably certain that at least 100 will be returned?

4.131 Reliability of a space vehicle. For a space vehicle to gain reentry into the earth's atmosphere, a particular system must work properly. One component of the system operates successfully only 85% of the time. To increase the reliability of the system, four of these components are installed in such a way that the system will operate successfully if at least one component is working. What is the probability that the system will fail? Assume the components operate independently.

*4.132 Emergency room bed availability. The mean number of patients admitted per day to the emergency room of a small hospital is 2.5. If, on a given day, there are only four beds available for new patients, what is the probability the hospital will not have enough beds to accommodate its newly admitted patients?

Critical Thinking Challenge

*4.133 **Space shuttle disaster.** On January 28, 1986, the space shuttle *Challenger* exploded, killing all seven astronauts aboard. An investigation concluded that the explosion was caused by the failure of the O-ring seal in the joint between the two lower segments of the right solid rocket booster. In a report made one year prior to the catastrophe, the National Aeronautics and Space Administration (NASA) claimed that the probability of such a failure was about $1/60{,}000$, or about once in every 60,000 flights. But a risk-assessment study conducted for the Air Force at about the same time assessed the probability to be $1/35$, or about once in every 35 missions. *Note*: The shuttle had flown 24 successful missions prior to the disaster. Given the events of January 28, 1986, which risk assessment—NASA's or the Air Force's—appears to be more appropriate?

Student Projects

Consider the following random variables:

1. The number x of people who recover from a certain disease out of a sample of five patients
2. The number x of voters in a sample of five who favor a method of tax reform
3. The number x of hits a baseball player gets in five official times at bat

In each case, x is a binomial random variable (or approximately so) with $n = 5$ trials. Assume that in each case the probability of Success is .3. (What is a Success in each of the examples?) If this were true, the probability distribution for x would be the same for each of the three examples. To obtain a relative frequency histogram for x, conduct the following experiment: Place 10 poker chips (pennies, marbles, or any 10 *identical* items) in a bowl and mark three of the 10 Success—the remaining seven will represent Failure. Randomly select a chip from the 10, observing whether it was a Success or Failure. Then return the chip and randomly select a second chip from the 10 available chips, and record this outcome. Repeat this process until a total of five trials has been conducted. Count the number x of Successes observed in the five trials. Repeat the entire process 100 times to obtain 100 observed values of x.

a. Use the 100 values of x obtained from the simulation to construct a relative frequency histogram for x. Note that this histogram is an approximation to $p(x)$.

b. Calculate the exact values of $p(x)$ for $n = 5$ and $p = .3$, and compare these values with the approximations found in part **a**.

c. If you were to repeat the simulation an extremely large number of times (say 100,000), how do you think the relative frequency histogram and true probability distribution would compare?

REFERENCES

Hogg, R. V., and Craig, A. *Introduction to Mathematical Statistics*, 5th ed. Upper Saddle River, N.J.: Prentice Hall, 1995.

Mendenhall, W. *Introduction to Mathematical Statistics*, 8th ed. Boston: Duxbury, 1991.

Parzen, E. *Modern Probability Theory and Its Applications*. New York: Wiley, 1960.

Wackerly, D., Mendenhall, W., and Scheaffer, R.L. *Mathematical Statistics with Applications*, 6th ed. North Scituate, Mass.: Duxbury, 2002.

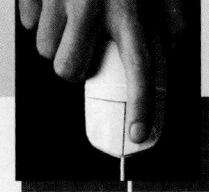

Using Technology

Binomial, Poisson, and Hypergeometric Probabilities Using MINITAB

To obtain probabilities for the discrete random variables discussed in this chapter using MINITAB, first enter the values of x that you desire probabilities for in a column (e.g., C1) on the MINITAB worksheet. Now click on the "Calc" button on the MINITAB menu bar, next click on "Probability Distributions," then finally click on the distribution of your choice (e.g., "Binomial"), as shown in Figure 4.M.1. The resulting dialog box appears as shown in Figure 4.M.2. Select either "Probabilities" or "Cumulative probabilities," specify the parameters of the distribution (e.g., sample size n and probability of success p), and enter C1 in the "Input column." When you click "OK" the binomial probabilities for the values of x (saved in C1) will appear on the MINITAB worksheet.

[*Note*: For the MINITAB Poisson option, you will need to enter the value of the mean λ in the resulting dialog box. For the hypergeometric option, you will need to enter the values of N, r, and n, in that order, in the resulting dialog box.]

Figure 4.M.1
MINITAB Menu Options for Obtaining Probabilities

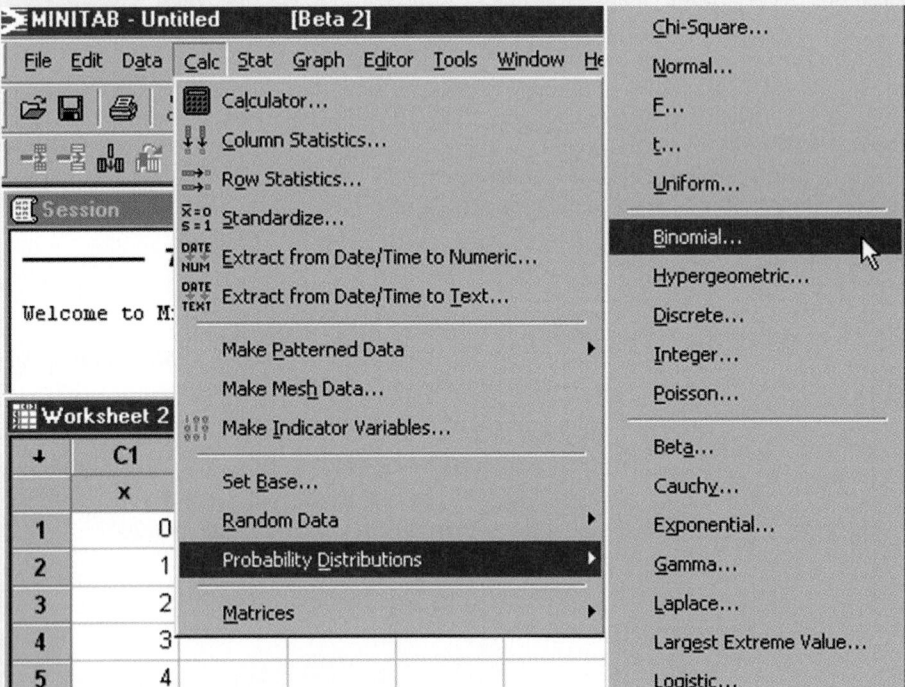

Figure 4.M.2
MINITAB Binomial
Distribution Dialog Box

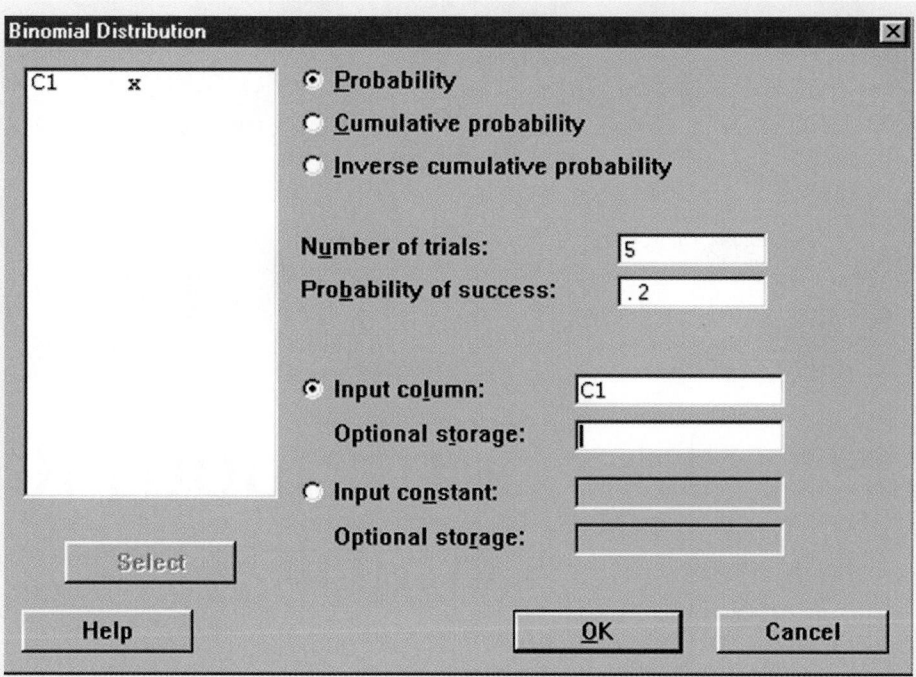

5

Continuous Random Variables

Contents

Statistics in Action

Super Weapons Development—
Optimizing the Hit Ratio

Using Technology

Cumulative Probabilities for
Continuous Random Variables
and Normal Probability Plots
Using MINITAB

Where We've Been

- Learned how to find probabilities of discrete events using the probability rules of Chapter 3
- Discussed probability models (distributions) for discrete random variables in Chapter 4

Where We're Going

- Develop the notion of a probability distribution for a continuous random variable.
- Study several important types of continuous random variables and their probability models.
- Introduce the normal probability distribution as one of the most useful distributions in statistics.

Statistics in *ACTION*

Super Weapons Development— Optimizing the Hit Ratio

The U.S. Army is working with a major defense contractor to develop a "super" weapon. The weapon is designed to fire a large number of sharp tungsten bullets—called flechettes—with a single shot that will destroy a large number of enemy soldiers. Flechettes are about the size of an average nail, with small fins at one end to stabilize them in flight. Since World War I, when France dropped them in large quantities from aircraft on masses of ground troops, munitions experts have experimented with using flechettes in a variety of guns. The problem with using flechettes as ammunition is accuracy—current weapons that fire large quantities of flechettes have unsatisfactory hit ratios when fired at long distances.

The defense contractor (not named here for both confidentiality and security reasons) has developed a prototype gun that fires 1,100 flechettes with a single round. In range tests, three 2-feet-wide targets were set up a distance of 500 meters (approximately 1,500 feet) from the weapon. Using a number line as a reference, the centers of the three targets were at 0, 5, and 10 feet, respectively, as shown in Figure SIA5.1. The prototype gun was aimed at the middle target

(center at 5 feet) and fired once. The point where each of the 1,100 flechettes landed at the 500-meter distance was measured using a horizontal and vertical grid. For the purposes of this application, only the horizontal measurements are considered. These 1,100 measurements are saved in the **MOAGUN** file. (The data are simulated for confidentiality reasons.) For example, a flechette with a value of $x = 5.5$ hit the middle target, but a flechette with a value of $x = 2.0$ did not hit any of the three targets (see Figure SIA5.1).

The defense contractor is interested in the likelihood of any one of the targets being hit by a flechette and, in particular, wants to set the gun specifications to maximize the number of target hits. The weapon is designed to have a mean horizontal value equal to the aim point (e.g., $\mu = 5$ feet when aimed at the center target). By changing specifications, the contractor can vary the standard deviation, σ. The **MOAGUN** file contains flechette measurements for three different range tests—one with a standard deviation of $\sigma = 1$ foot, one with $\sigma = 2$ feet, and one with $\sigma = 4$ feet.

In this chapter, two Statistics in Action Revisited examples demonstrate how we can use one of the probability models discussed in this chapter—the normal probability distribution—to aid the defense contractor in developing its "super" weapon.

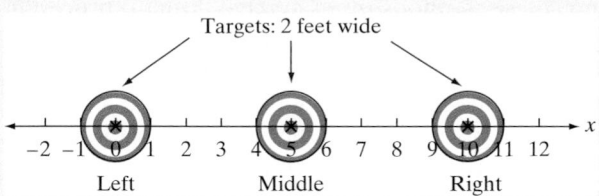

Figure SIA5.1
Target placement on gun range

Statistics in the Action Revisited

- Using the Normal Model to Maximize the Probability of a Hit with the Super Weapon (p. 256)

- Assessing whether the Normal Distribution is Appropriate for Modeling the Super Weapon Hit Data (p. 265)

In this chapter we'll consider some continuous random variables that are commonly encountered. Recall that a continuous random variable is one that can assume any value within some interval or intervals. For example, the length of time between a person's visits to a doctor, the thickness of sheets of steel produced in a rolling mill, and the yield of wheat per acre of farmland are all continuous random variables. The methodology we employ to describe continuous random variables will necessarily be somewhat different from that used to describe discrete random variables. We first discuss the general form of **continuous probability distributions**, and then we explore three specific types that are used in making statistical decisions. The normal probability distribution, which plays a basic and important role in both the theory and application of statistics, is essential to the study of most of the subsequent chapters in this book. The other types have practical applications, but a study of these topics is optional.

5.1 Continuous Probability Distributions

The graphical form of the probability distribution for a continuous random variable x is a smooth curve that might appear as shown in Figure 5.1. This curve, a function of x, is denoted by the symbol $f(x)$ and is variously called a **probability density function**, a **frequency function**, or a **probability distribution**.

The areas under a probability distribution correspond to probabilities for x. For example, the area A beneath the curve between the two points a and b, as shown in Figure 5.1, is the probability that x assumes a value between a and $b(a < x < b)$. Because there is no area over a point, say $x = a$, it follows that (according to our model) the probability associated with a particular value of x is equal to 0; that is, $P(x = a) = 0$ and hence $P(a < x < b) = P(a \le x \le b)$. In other words, the probability is the same regardless of whether or not you include the endpoints of the interval. Also, because areas over intervals represent probabilities, it follows that the total area under a probability distribution, the probability assigned to all values of x, should equal 1. Note that probability distributions for continuous random variables possess different shapes depending on the relative frequency distributions of real data that the probability distributions are supposed to model.

Figure 5.1
A Probability Distribution
$f(x)$ for a Continuous
Random Variable x

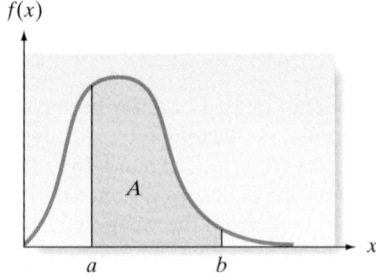

The areas under most probability distributions are obtained by using calculus or numerical methods.* Because these methods often involve difficult procedures, we will give the areas for some of the most common probability distributions in tabular form in Appendix A. Then, to find the area between two values of x, say $x = a$ and $x = b$, you simply have to consult the appropriate table.

For each of the continuous random variables presented in this chapter, we will give the formula for the probability distribution along with its mean μ and standard deviation σ. These two numbers will enable you to make some approximate probability statements about a random variable even when you do not have access to a table of areas under the probability distribution.

5.2 The Uniform Distribution

All the probability problems discussed in Chapter 3 had sample spaces that contained a finite number of sample points. In many of these problems, the sample points were assigned equal probabilities—for example, the die toss or the coin toss. For continuous random variables, there is an infinite number of values in the sample space, but in some cases the values may appear to be equally likely. For example, if a short exists in a 5-meter stretch of electrical wire, it may have an equal probability of being in any particular 1-centimeter segment along the line.

---

*Students with knowledge of calculus should note that the probability that x assumes a value in the interval $a < x < b$ is $P(a < x < b) = \int_a^b f(x)\,dx$, assuming the integral exists. Similar to the requirements for a discrete probability distribution, we require $f(x) \ge 0$ and $\int_{-\infty}^{\infty} f(x)\,dx = 1$.

Figure 5.2

The Uniform Probability Distribution

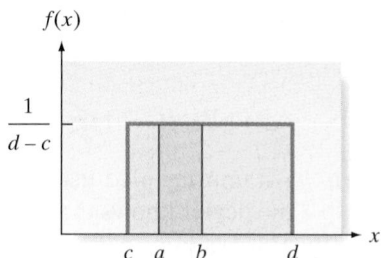

Or if a safety inspector plans to choose a time at random during the four afternoon work hours to pay a surprise visit to a certain area of a plant, then each 1-minute time interval in this 4-work-hour period will have an equally likely chance of being selected for the visit.

Continuous random variables that appear to have *equally likely outcomes over their range of possible values* possess a **uniform probability distribution**, perhaps the simplest of all continuous probability distributions. Suppose the random variable x can assume values only in an interval $c \leq x \leq d$. Then the uniform frequency function has a rectangular shape, as shown in Figure 5.2. Note that the possible values of x consist of all points in the interval between point c and point d. The height of $f(x)$ is constant in that interval and equals $1/(d - c)$. Therefore, the total area under $f(x)$ is given by

$$\text{Total area of rectangle} = (\text{Base})(\text{Height}) = (d - c)\left(\frac{1}{d - c}\right) = 1$$

The uniform probability distribution provides a model for continuous random variables that are *evenly distributed* over a certain interval. That is, a uniform random variable is one that is just as likely to assume a value in one interval as it is to assume a value in any other interval of equal size. There is no clustering of values around any value; instead, there is an even spread over the entire region of possible values.

The uniform distribution is sometimes referred to as the **randomness distribution**, since one way of generating a uniform random variable is to perform an experiment in which a point is *randomly selected* on the horizontal axis between the points c and d. If we were to repeat this experiment infinitely often, we would create a uniform probability distribution like that shown in Figure 5.2. The random selection of points in an interval can also be used to generate random numbers such as those in Table I of Appendix A. Recall that random numbers are selected in such a way that every number would have an equal probability of selection. Therefore, random numbers are realizations of a uniform random variable. (Random numbers were used to draw random samples in Section 3.7.) The formulas for the uniform probability distribution, its mean, and its standard deviation are shown in the box.

Probability Distribution for a Uniform Random Variable x

Probability density function: $f(x) = \dfrac{1}{d - c}$ $\quad (c \leq x \leq d)$

Mean: $\mu = \dfrac{c + d}{2}$ $\quad$ Standard deviation: $\sigma = \dfrac{d - c}{\sqrt{12}}$

$P(a < x < b) = (b - a)/(d - c), c \leq a < b \leq d$

Suppose the interval $a < x < b$ lies within the domain of x; that is, it falls within the larger interval $c \leq x \leq d$. Then the probability that x assumes a value within the interval $a < x < b$ is equal to the area of the rectangle over the interval, namely, $(b - a)/(d - c)$.* (See the shaded area in Figure 5.2.)

EXAMPLE 5.1   APPLYING THE UNIFORM DISTRIBUTION

Problem   An unprincipled used car dealer sells a car to an unsuspecting buyer, even though the dealer knows that the car will have a major breakdown within the next 6 months. The dealer provides a warranty of 45 days on all cars sold. Let x represent the length of time until the breakdown occurs. Assume that x is a uniform random variable with values between 0 and 6 months.

    a. Calculate and interpret the mean and standard deviation of x.

    b. Graph the probability distribution of x, and show the mean on the horizontal axis. Also show 1- and 2-standard-deviation intervals around the mean.

    c. Calculate the probability that the breakdown occurs while the car is still under warranty.

Solution   **a.** To calculate the mean and standard deviation for x, we substitute 0 and 6 months for c and d, respectively, in the formulas for uniform random variables. Thus,

$$\mu = \frac{c + d}{2} = \frac{0 + 6}{2} = 3 \text{ months}$$

$$\sigma = \frac{d - c}{\sqrt{12}} = \frac{6 - 0}{\sqrt{12}} = \frac{6}{3.464} = 1.73 \text{ months}$$

Our interpretations of μ and σ follow:

The average length of time until breakdown, x, for all similar used cars is $\mu = 3$ months. From Chebyshev's Theorem (Table 2.7, p. 74), we know that at least 75% of the values of x in the distribution will fall in the interval

$$\mu \pm 2\sigma = 3 \pm 2(1.73)$$
$$= 3 \pm 3.46$$

or, between $-.46$ and 6.46 months. Consequently, we expect the length of time until breakdown to be less than 6.46 months at least 75% of the time.

    b. The uniform probability distribution is

$$f(x) = \frac{1}{d - c} = \frac{1}{6 - 0} = \frac{1}{6} \qquad (0 \le x \le 6)$$

The graph of this function is shown in Figure 5.3. The mean and 1- and 2-standard-deviation intervals around the mean are shown on the horizontal

Figure 5.3
Distribution for x in
Example 5.1

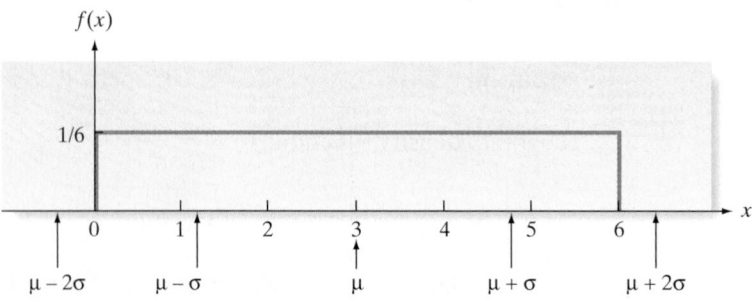

*The student with knowledge of calculus should note that

$$P(a < x < b) = \int_a^b f(x)\, d(x) = \int_a^b [1/(d - c)]\, dx = (b - a)/(d - c)$$

Figure 5.4
Probability that Car
Breaks Down Within
1.5 Months of Purchase

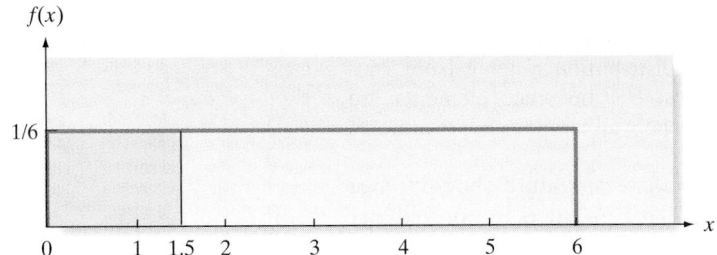

axis. Note that the entire distribution of x lies within the $\mu \pm 2\sigma$ interval. (This demonstrates, once again, the conservativeness of Chebyshev's Theorem.)

c. To find the probability that the car is still under warranty when it breaks down, we must find the probability that x is less than 45 days, or (about) 1.5 months. As indicated in Figure 5.4, we need to calculate the area under the frequency function $f(x)$ between the points $x = 0$ and $x = 1.5$. Therefore, in this case $a = 0$ and $b = 1.5$. Applying the formula in the box, we have

$$P(x < 1.5) = (1.5 - 0)/(6 - 0) = 1.5/6 = .25$$

That is, there is a 25% chance that the car will break down while under warranty.

Look Back The calculated probability in part **c** is the area of a rectangle with base $1.5 - 0 = 1.5$ and height $= 1/6$. Alternatively, the probability that the unsuspecting car buyer will be able to have the car repaired at the dealer's expense is $P(0 < x < 1.5) = (\text{Base})(\text{Height}) = (1.5)(1/6) = .25$.

> **Now Work** *Exercise 5.6*

■ ■ ■

Exercises 5.1–5.16

Understanding the Principles

5.1 Give a characteristic of a uniform random variable.

5.2 The uniform distribution is sometimes referred to as the _____ distribution. (Fill in the blank.)

Learning the Mechanics

5.3 Suppose x is a random variable best described by a uniform probability distribution with $c = 10$ and $d = 30$.
  a. Find $f(x)$.
  b. Find the mean and standard deviation of x.
  c. Graph $f(x)$ and locate μ and the interval $\mu \pm 2\sigma$ on the graph. Note that the probability that x assumes a value within the interval $\mu \pm 2\sigma$ is equal to 1.

5.4 Refer to Exercise 5.3. Find the following probabilities:
  a. $P(10 \le x \le 25)$ **b.** $P(20 < x < 30)$
  c. $P(x \ge 25)$ **d.** $P(x \le 10)$
  e. $P(x \le 25)$ **f.** $P(20.5 \le x \le 25.5)$

5.5 Suppose x is a random variable best described by a uniform probability distribution with $c = 2$ and $d = 4$.
  a. Find $f(x)$.
  b. Find the mean and standard deviation of x.
  c. Find $P(\mu - \sigma \le x \le \mu + \sigma)$.
  d. Find $P(x > 2.78)$.

  e. Find $P(2.4 \le x \le 3.7)$.
  f. Find $P(x < 2)$.

5.6 Refer to Exercise 5.5. Find the value of a that makes
  NW each of the following probability statements true.
  a. $P(x \ge a) = .5$
  b. $P(x \le a) = .2$
  c. $P(x \le a) = 0$
  d. $P(2.5 \le x \le a) = .5$

5.7 The random variable x is best described by a uniform probability distribution with $c = 100$ and $d = 200$. Find the probability that x assumes a value
  a. More than 2 standard deviations from μ
  b. Less than 3 standard deviations from μ
  c. Within 2 standard deviations of μ

5.8 The random variable x is best described by a uniform probability distribution with mean 50 and standard deviation 5. Find c, d, and $f(x)$. Graph the probability distribution.

Applying the Concepts—Basic

5.9 **Maintaining pipe wall temperature.** Maintaining a constant pipe wall temperature in some hot process applications is critical. A new technique that utilizes bolt-on trace elements to maintain temperature was presented in the *Journal of Heat Transfer* (November

2000). Without bolt-on trace elements, the pipe wall temperature of a switch condenser used to produce plastic has a uniform distribution ranging from 260° to 290°F. When several bolt-on trace elements are attached to the piping, the wall temperature is uniform from 278° to 285°F.

a. Ideally, the pipe wall temperature should range between 280° and 284°F. What is the probability that the temperature will fall in this ideal range when no bolt-on trace elements are used? When bolt-on trace elements are attached to the pipe?

b. When the temperature is 268°F or lower, the hot liquid plastic hardens (or plates), causing a buildup in the piping. What is the probability of plastic plating when no bolt-on trace elements are used? When bolt-on trace elements are attached to the pipe?

5.10 Trajectory of an electrical circuit. Researchers at the University of California–Berkeley have designed, built, and tested a switched-capacitor circuit for generating random signals. (*International Journal of Circuit Theory and Applications*, May–June 1990.) The circuit's trajectory was shown to be uniformly distributed on the interval (0, 1).

a. Give the mean and variance of the circuit's trajectory.
b. Compute the probability that the trajectory falls between .2 and .4.
c. Would you expect to observe a trajectory that exceeds .995? Explain.

5.11 Random numbers. The data set listed here was created using the MINITAB random number generator. Construct a relative frequency histogram for the data. Except for the expected variation in relative frequencies among the class intervals, does your histogram suggest that the data are observations on a uniform random variable with $c = 0$ and $d = 100$? Explain.

⚙ **RANUNI**

38.8759	98.0716	64.5788	60.8422	.8413
99.3734	31.8792	32.9847	.7434	93.3017
12.4337	11.7828	87.4506	94.1727	23.0892
47.0121	43.3629	50.7119	88.2612	69.2875
62.6626	55.6267	78.3936	28.6777	71.6829
44.0466	57.8870	71.8318	28.9622	23.0278
35.6438	38.6584	46.7404	11.2159	96.1009
95.3660	21.5478	87.7819	12.0605	75.1015

Applying the Concepts—Intermediate

5.12 Marine losses for an oil company. The frequency distribution shown in the table depicts the property and marine losses incurred by a large oil company over a 2-year period. In the insurance business, each "loss" interval is called a *layer*. *Research Review* (Summer 1998) demonstrated that analysts often treat the actual loss value within a layer as a uniform random variable.

Layer	Marine Losses ($ thousands)	Relative Frequency
1	0–10	.927
2	10–50	.053
3	50–100	.010
4	100–250	.006
5	250–500	.003
6	500–1,000	.001
7	1,000–2,500	.000

Source: Cozzolino, J. M., and Mikola, P. J. "Applications of the piecewise constant Pareto distribution." *Research Review,* Summer 1998.

a. Use the uniform distribution to find the mean loss amount in layer 2.
b. Use the uniform distribution to find the mean loss amount in layer 6.
c. If a loss occurs in layer 2, what is the probability that it exceeds $30,000?
d. If a loss occurs in layer 6, what is the probability that it is between $750,000 and $800,000?

5.13 Soft-drink dispenser. The manager of a local soft-drink bottling company believes that when a new beverage-dispensing machine is set to dispense 7 ounces, it in fact dispenses an amount x at random anywhere between 6.5 and 7.5 ounces inclusive. Suppose x has a uniform probability distribution.

a. Is the amount dispensed by the beverage machine a discrete or a continuous random variable? Explain.
b. Graph the frequency function for x, the amount of beverage the manager believes is dispensed by the new machine when it is set to dispense 7 ounces.
c. Find the mean and standard deviation for the distribution graphed in part **b**, and locate the mean and the interval $\mu \pm 2\sigma$ on the graph.
d. Find $P(x \geq 7)$.
e. Find $P(x < 6)$.
f. Find $P(6.5 \leq x \leq 7.25)$.
g. What is the probability that each of the next six bottles filled by the new machine will contain more than 7.25 ounces of beverage? Assume that the amount of beverage dispensed in one bottle is independent of the amount dispensed in another bottle.

5.14 Time delays at a bus stop. A bus is scheduled to stop at a certain bus stop every half hour on the hour and the half hour. At the end of the day, buses still stop after every 30 minutes, but because delays often occur earlier in the day, the bus is never early and likely to be late. The director of the bus line claims that the length of time a bus is late is uniformly distributed and the maximum time that a bus is late is 20 minutes.

a. If the director's claim is true, what is the expected number of minutes a bus will be late?
b. If the director's claim is true, what is the probability that the last bus on a given day will be more than 19 minutes late?

c. If you arrive at the bus stop at the end of a day at exactly half-past the hour and must wait more than 19 minutes for the bus, what would you conclude about the director's claim? Why?

Applying the Concepts—Advanced

5.15 Gouges on a spindle. A tool-and-die machine shop produces extremely high-tolerance spindles. The spindles are 18-inch slender rods used in a variety of military equipment. A piece of equipment used in the manufacture of the spindles malfunctions on occasion and places a single gouge somewhere on the spindle. However, if the spindle can be cut so that it has 14 consecutive inches without a gouge, then the spindle can be salvaged for other purposes. Assuming that the location of the gouge along the spindle is best described by a uniform distribution, what is the probability that a defective spindle can be salvaged?

5.16 Reliability of a robotic device. The *reliability* of a piece of equipment is frequently defined to be the probability, p, that the equipment performs its intended function successfully for a given period of time under specific conditions. (Render and Heizer, *Principles of Operations Management*, 1995.) Because p varies from one point in time to another, some reliability analysts treat p as if it were a random variable. Suppose an analyst characterizes the uncertainty about the reliability of a particular robotic device used in an automobile assembly line using the following distribution:

$$f(p) = \begin{cases} 1 & 0 \le p \le 1 \\ 0 & \text{otherwise} \end{cases}$$

a. Graph the analyst's probability distribution for p.
b. Find the mean and variance of p.
c. According to the analyst's probability distribution for p, what is the probability that p is greater than .95? Less than .95?
d. Suppose the analyst receives the additional information that p is definitely between .90 and .95, but that there is complete uncertainty about where it lies between these values. Describe the probability distribution the analyst should now use to describe p.

5.3 The Normal Distribution

One of the most commonly observed continuous random variables has a **bell-shaped** probability distribution (or **bell curve**) as shown in Figure 5.5. It is known as a **normal random variable** and its probability distribution is called a **normal distribution**.

The normal distribution plays a very important role in the science of statistical inference. Moreover, many phenomena generate random variables with probability distributions that are very well approximated by a normal distribution. For example, the error made in measuring a person's blood pressure may be a normal random variable, and the probability distribution for the yearly rainfall in a certain region might be approximated by a normal probability distribution. You can determine the adequacy of the normal approximation to an existing population of data by comparing the relative frequency distribution of a large sample of the data to the normal probability distribution. Methods to detect disagreement between a set of data and the assumption of normality are presented in Section 5.4.

Biography

CARL F. GAUSS (1777–1855)— The Gaussian Distribution

A well-known and respected German mathematician, physicist, and astronomer, Gauss applied the normal distribution while studying the motion of planets and stars. Gauss's prowess as a mathematician was exemplified by one of his most important discoveries. At the young age of 22, Gauss constructed a regular 17-gon by ruler and compasses—a feat that was the most major advance in mathematics since the time of the ancient Greeks. In addition to publishing close to 200 scientific papers, Gauss invented the heliograph as well as a primitive telegraph device.

The normal distribution began in the eighteenth century as a theoretical distribution for errors in disciplines where fluctuations in nature were believed to behave randomly. Although he may not have been the first to discover the formula, the normal distribution was named the Gaussian distribution after Carl Friedrich Gauss.

Figure 5.5

A Normal Probability Distribution

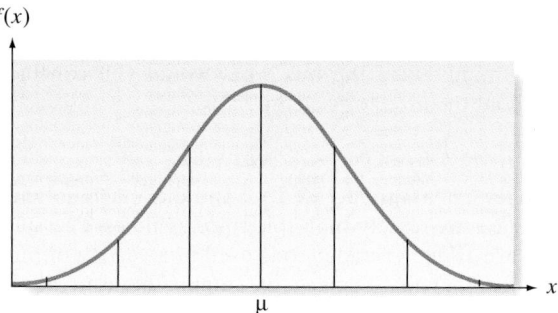

Figure 5.6

Several Normal Distributions with Different Means and Standard Deviations

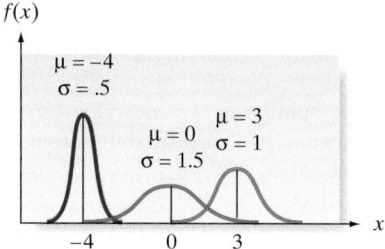

The normal distribution is perfectly symmetric about its mean μ, as can be seen in the examples in Figure 5.6. Its spread is determined by the value of its standard deviation σ.

The formula for the normal probability distribution is shown in the box. When plotted, this formula yields a curve like that shown in Figure 5.5.

Probability Distribution for a Normal Random Variable x

Probability density function: $f(x) = \dfrac{1}{\sigma\sqrt{2\pi}}e^{-(1/2)[(x-\mu)/\sigma]^2}$

where

μ = Mean of the normal random variable x
σ = Standard deviation
π = 3.1416...
e = 2.71828...

$P(x < a)$ is obtained from a table of normal probabilities.

Note that the mean μ and standard deviation σ appear in this formula, so that no separate formulas for μ and σ are necessary. To graph the normal curve we have to know the numerical values of μ and σ.

Computing the area over intervals under the normal probability distribution is a difficult task.* Consequently, we will use the computed areas listed in Table IV of Appendix A. Although there are an infinitely large number of normal curves—one for each pair of values for μ and σ—we have formed a single table that will apply to any normal curve.

Table IV is based on a normal distribution with mean $\mu = 0$ and standard deviation $\sigma = 1$, called a *standard normal distribution*. A random variable with a standard normal distribution is typically denoted by the symbol z. The formula for the probability distribution of z is given by

---

*The student with knowledge of calculus should note that there is not a closed-form expression for $P(a < x < b) = \int_a^b f(x)\,dx$ for the normal probability distribution. The value of this definite integral can be obtained to any desired degree of accuracy by numerical approximation procedures. For this reason, it is tabulated for the user.

$$f(z) = \frac{1}{\sqrt{2\pi}} e^{-(1/2)z^2}$$

Figure 5.7 shows the graph of a standard normal distribution.

Figure 5.7

Standard Normal
Distribution: $\mu = 0, \sigma = 1$

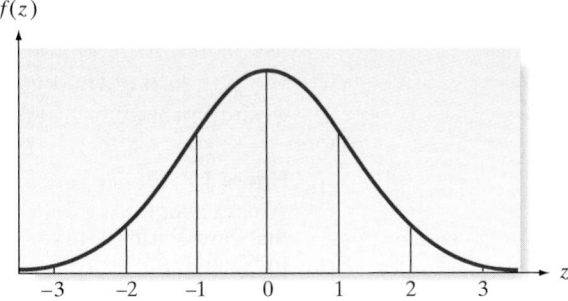

DEFINITION 5.1

The **standard normal distribution** is a normal distribution with $\mu = 0$ and $\sigma = 1$. A random variable with a standard normal distribution, denoted by the symbol z, is called a **standard normal random variable**.

Since we will ultimately convert all normal random variables to standard normal in order to use Table IV to find probabilities, it is important that you learn to use Table IV well. A partial reproduction of Table IV is shown in Table 5.1. Note that the values of the standard normal random variable z are listed in the left-hand column. The entries in the body of the table give the area (probability) between 0 and z. Examples 5.2–5.5 illustrate the use of the table.

TABLE 5.1   Reproduction of Part of Table IV in Appendix A

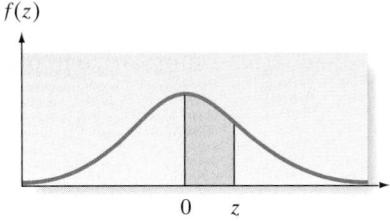

z	.00	.01	.02	.03	.04	.05	.06	.07	.08	.09
.0	.0000	.0040	.0080	.0120	.0160	.0199	.0239	.0279	.0319	.0359
.1	.0398	.0438	.0478	.0517	.0557	.0596	.0636	.0675	.0714	.0753
.2	.0793	.0832	.0871	.0910	.0948	.0987	.1026	.1064	.1103	.1141
.3	.1179	.1217	.1255	.1293	.1331	.1368	.1406	.1443	.1480	.1517
.4	.1554	.1591	.1628	.1664	.1700	.1736	.1772	.1808	.1844	.1879
.5	.1915	.1950	.1985	.2019	.2054	.2088	.2123	.2157	.2190	.2224
.6	.2257	.2291	.2324	.2357	.2389	.2422	.2454	.2486	.2517	.2549
.7	.2580	.2611	.2642	.2673	.2704	.2734	.2764	.2794	.2823	.2852
.8	.2881	.2910	.2939	.2967	.2995	.3023	.3051	.3078	.3106	.3133
.9	.3159	.3186	.3212	.3238	.3264	.3289	.3315	.3340	.3365	.3389
1.0	.3413	.3438	.3461	.3485	.3508	.3531	.3554	.3577	.3599	.3621
1.1	.3643	.3665	.3686	.3708	.3729	.3749	.3770	.3790	.3810	.3830
1.2	.3849	.3869	.3888	.3907	.3925	.3944	.3962	.3980	.3997	.4015
1.3	.4032	.4049	.4066	.4082	.4099	.4115	.4131	.4147	.4162	.4177
1.4	.4192	.4207	.4222	.4236	.4251	.4265	.4279	.4292	.4306	.4319
1.5	.4332	.4345	.4357	.4370	.4382	.4394	.4406	.4418	.4429	.4441

EXAMPLE 5.2 USING THE STANDARD NORMAL TABLE

Problem Find the probability that the standard normal random variable z falls between -1.33 and $+1.33$.

Solution The standard normal distribution is shown again in Figure 5.8. Since all probabilities associated with standard normal random variables can be depicted as areas under the standard normal curve, you should always draw the curve and then equate the desired probability to an area.

Figure 5.8
Areas under the
Standard Normal Curve
for Example 5.2

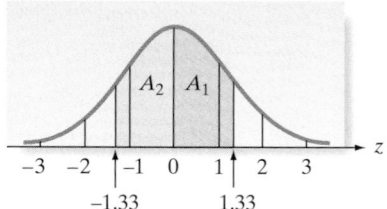

In this example we want to find the probability that z falls between -1.33 and $+1.33$, which is equivalent to the area between -1.33 and $+1.33$, shown highlighted in Figure 5.8. Table IV provides the area between $z = 0$ and any value of z, so that if we look up $z = 1.33$ (the value in the 1.3 row and .03 column, as shown in Figure 5.9), we find that the area between $z = 0$ and $z = 1.33$ is .4082. This is the area labeled A_1 in Figure 5.8. To find the area A_2 located between $z = 0$ and $z = -1.33$, we note that the symmetry of the normal distribution implies that the area between $z = 0$ and any point to the left is equal to the area between $z = 0$ and the point equidistant to the right. Thus, in this example the area between $z = 0$ and $z = -1.33$ is equal to the area between $z = 0$ and $z = +1.33$. That is,

$$A_1 = A_2 = .4082$$

The probability that z falls between -1.33 and $+1.33$ is the sum of the areas of A_1 and A_2. We summarize in probabilistic notation:

$$P(-1.33 < z < +1.33) = P(-1.33 < z < 0) + P(0 < z \leq 1.33)$$
$$= A_1 + A_2 = .4082 + .4082 = .8164$$

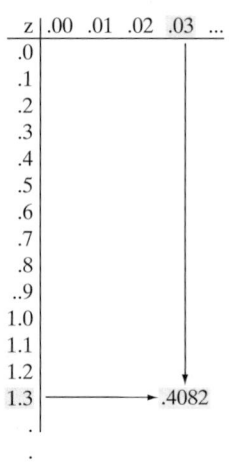

Figure 5.9
Finding $z = 1.33$ in the
Standard Normal Table,
Example 5.2

Look Back Remember that "<" and "≤" are equivalent in events involving z, because the inclusion (or exclusion) of a single point does not alter the probability of an event involving a continuous random variable.

> Now Work **Exercise 5.23ef**

■ ■ ■

EXAMPLE 5.3 USING THE STANDARD NORMAL TABLE

Problem Find the probability that a standard normal random variable exceeds 1.64; that is, find $P(z > 1.64)$.

Solution The area under the standard normal distribution to the right of 1.64 is the highlighted area labeled A_1 in Figure 5.10. This area represents the desired probability that z exceeds 1.64. However, when we look up $z = 1.64$ in Table IV, we must remember that the probability given in the table corresponds to the area between $z = 0$ and

Figure 5.10

Areas under the Standard Normal Curve for Example 5.3

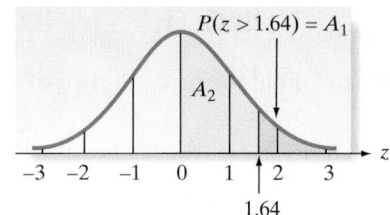

$z = 1.64$ (the area labeled A_2 in Figure 5.10). From Table IV we find that $A_2 = .4495$. To find the area A_1 to the right of 1.64, we make use of two facts:

1. The standard normal distribution is symmetric about its mean $z = 0$.

2. The total area under the standard normal probability distribution equals 1.

Taken together, these two facts imply that the areas on either side of the mean $z = 0$ equal .5; thus, the area to the right of $z = 0$ in Figure 5.10 is $A_1 + A_2 = .5$. Then

$$P(z > 1.64) = A_1 = .5 - A_2 = .5 - .4495 = .0505$$

Look Back To attach some practical significance to this probability, note that the implication is that the chance of a standard normal random variable exceeding 1.64 is only about .05.

| Now Work | *Exercise 5.22a* |

■ ■ ■

EXAMPLE 5.4   USING THE STANDARD NORMAL TABLE

Problem   Find the probability that a standard normal random variable lies to the left of .67.

Solution   The event is shown as the highlighted area in Figure 5.11. We want to find $P(z < .67)$. We divide the highlighted area into two parts: the area A_1 between $z = 0$ and $z = .67$, and the area A_2 to the left of $z = 0$. We must always make such a division when the desired area lies on both sides of the mean ($z = 0$) because Table IV contains areas between $z = 0$ and the point you look up. We look up $z = .67$ in Table IV to find that $A_1 = .2486$. The symmetry of the standard normal distribution also implies that half the distribution lies on each side of the mean, so the area A_2 to the left of $z = 0$ is .5. Then

$$P(z < .67) = A_1 + A_2 = .2486 + .5 = .7486$$

Look Back Note that this probability is approximately .75. Thus, about 75% of the time the standard normal random variable z will fall below .67. This implies that $z = .67$ represents the approximate 75th percentile for the distribution.

Figure 5.11

Areas under the Standard Normal Curve for Example 5.4

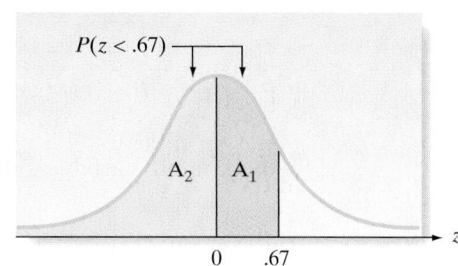

| Now Work | *Exercise 5.22h* |

■ ■ ■

EXAMPLE 5.5   USING THE STANDARD NORMAL TABLE

Problem   Find the probability that a standard normal random variable exceeds 1.96 in absolute value.

Solution   The event is shown highlighted in Figure 5.12. We want to find

$$P(|z| > 1.96) = P(z < -1.96 \text{ or } z > 1.96)$$

Figure 5.12
Areas under the
Standard Normal Curve
for Example 5.5

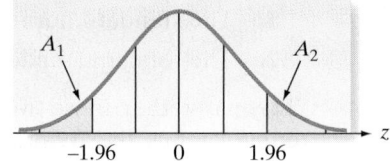

Note that the total highlighted area is the sum of two areas, A_1 and A_2—areas that are equal because of the symmetry of the normal distribution.

   We look up $z = 1.96$ and find the area between $z = 0$ and $z = 1.96$ to be .4750. Then the area to the right of 1.96, A_2, is $.5 - .4750 = .0250$ so that

$$P(|z| > 1.96) = A_1 + A_2 = .0250 + .0250 = .05$$

Look Back We emphasize, again, the importance of sketching the standard normal curve when finding normal probabilities.

━━━━━━ ■ ■ ■ ━━━━━━

   To apply Table IV to a normal random variable x with any mean μ and any standard deviation σ, we must first convert the value of x to a z-score. The population z-score for a measurement was defined (in Section 2.6) as the *distance* between the measurement and the population mean, divided by the population standard deviation. Thus, the z-score gives the distance between a measurement and the mean in units equal to the standard deviation. In symbolic form, the z-score for the measurement x is

$$z = \frac{x - \mu}{\sigma}$$

Note that when $x = \mu$, we obtain $z = 0$.

   An important property of the normal distribution is that if x is normally distributed with any mean and any standard deviation, z is *always* normally distributed with mean 0 and standard deviation 1. That is, z is a standard normal random variable.

Property of Normal Distributions

If x is a normal random variable with mean μ and standard deviation σ, then the random variable z defined by the formula

$$z = \frac{x - \mu}{\sigma}$$

has a standard normal distribution. The value z describes the number of standard deviations between x and μ.

Recall from Example 5.5 that $P(|z| > 1.96) = .05$. This probability coupled with our interpretation of z implies that any normal random variable lies more than 1.96 standard deviations from its mean only 5% of the time. Compare this to the Empirical Rule (Chapter 2), which tells us that about 5% of the measurements in mound-shaped distributions will lie beyond 2 standard deviations from the mean. The normal distribution actually provides the model on which the Empirical Rule is based, along with much "empirical" experience with real data that often approximately obey the rule, whether drawn from a normal distribution or not.

EXAMPLE 5.6

FINDING A NORMAL PROBABILITY

Problem   Assume that the length of time, x, between charges of a cellular phone is normally distributed with a mean of 10 hours and a standard deviation of 1.5 hours. Find the probability that the cell phone will last between 8 and 12 hours between charges.

Solution   The normal distribution with mean $\mu = 10$ and $\sigma = 1.5$ is shown in Figure 5.13. The desired probability that the cell phone lasts between 8 and 12 hours is highlighted. In order to find the probability, we must first convert the distribution to standard normal, which we do by calculating the z-score:

$$z = \frac{x - \mu}{\sigma}$$

The z-scores corresponding to the important values of x are shown beneath the x values on the horizontal axis in Figure 5.13. Note that $z = 0$ corresponds to the mean of $\mu = 10$ hours, whereas the x values 8 and 12 yield z-scores of -1.33 and $+1.33$, respectively. Thus, the event that the cell phone lasts between 8 and 12 hours is equivalent to the event that a standard normal random variable lies between -1.33 and $+1.33$. We found this probability in Example 5.2 (see Figure 5.8) by doubling the area corresponding to $z = 1.33$ in Table IV. That is,

$$P(8 \leq x \leq 12) = P(-1.33 \leq z \leq 1.33) = 2(.4082) = .8164$$

Figure 5.13
Areas under the Normal
Curve for Example 5.6

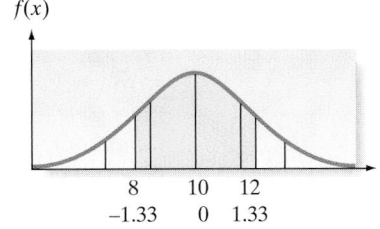

Now Work   *Exercise 5.28a*

■ ■ ■

The steps to follow when calculating a probability corresponding to a normal random variable are shown in the box.

Steps for Finding a Probability Corresponding to a Normal Random Variable

1. Sketch the normal distribution and indicate the mean of the random variable x. Then shade the area corresponding to the probability you want to find.

2. Convert the boundaries of the shaded area from x values to standard normal random variable z values using the formula

$$z = \frac{x - \mu}{\sigma}$$

Show the z values under the corresponding x values on your sketch.

3. Use Table IV in Appendix A to find the areas corresponding to the z values. If necessary, use the symmetry of the normal distribution to find areas corresponding to negative z values and the fact that the total area on each side of the mean equals .5 to convert the areas from Table IV to the probabilities of the event you have shaded.

Standard Normal Probabilities

Using the TI-83 Graphing Calculator

Graphing the Area under the Standard Normal Curve

Step 1 *Turn off all plots.*
Press **2nd PRGM** and select **1:ClrDraw**
Press **ENTER ENTER** and 'Done' will appear on the screen.
Press **2nd Y =** and select **4:PlotsOff**
Press **ENTER ENTER** and 'Done' will appear on the screen.

Step 2 *Set the viewing window. (Recall that almost all of the area under the standard normal curve falls between −5 and 5. A height of 0.5 is a good choice for Y max.)*
Set **Xmin = −5**
Xmax = 5
Xscl = 1
Ymin = 0
Ymax = .5
Yscl = 0
Xres = 1
Note: the negative sign is the gray key, not the blue key

Step 3 *View graph.*
Press **2nd VARS**
Arrow right to **DRAW**
Press **ENTER** to select **1:ShadeNorm(**
Enter your lower limit
Press **comma**
Enter your upper limit
Press) Press **ENTER**
The graph will be displayed along with the area, lower limit, and upper limit.

Example What is the probability that z is less than 1.5 under the Standard Normal curve? In this example, set the Window as shown in Step 2. For the limits in Step 3 use −5 for the lower limit and 1.5 for the upper limit.

The screens for this example are shown below.

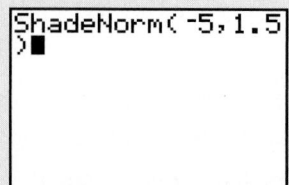
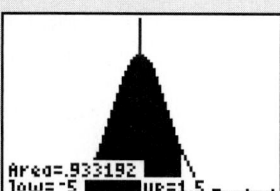

Thus, $P(z < 1.5) = .9332$.

EXAMPLE 5.7   USING NORMAL PROBABILITIES TO MAKE AN INFERENCE

Problem   Suppose an automobile manufacturer introduces a new model that has an advertised mean in-city mileage of 27 miles per gallon. Although such advertisements seldom report any measure of variability, suppose you write the manufacturer for the details of the tests, and you find that the standard deviation is 3 miles per gallon. This information leads you to formulate a probability model for the random variable x, the in-city mileage for this car model. You believe that the probability distribution of x can be approximated by a normal distribution with a mean of 27 and a standard deviation of 3.

  a. If you were to buy this model of automobile, what is the probability that you would purchase one that averages less than 20 miles per gallon for in-city driving? In other words, find $P(x < 20)$.

  b. Suppose you purchase one of these new models and it does get less than 20 miles per gallon for in-city driving. Should you conclude that your probability model is incorrect?

Solution   **a.** The probability model proposed for x, the in-city mileage, is shown in Figure 5.14. We are interested in finding the area A to the left of 20 since this area corresponds to the probability that a measurement chosen from this distribution falls below 20. In other words, if this model is correct, the area A represents the fraction of cars that can be expected to get less than 20 miles per gallon for in-city driving. To find A, we first calculate the z value corresponding to $x = 20$. That is,

$$z = \frac{x - \mu}{\sigma} = \frac{20 - 27}{3} = -\frac{7}{3} = -2.33$$

Then

$$P(x < 20) = P(z < -2.33)$$

as indicated by the highlighted area in Figure 5.14. Since Table IV gives only areas to the right of the mean (and because the normal distribution is symmetric about its mean), we look up 2.33 in Table IV and find that the corresponding area is .4901. This is equal to the area between $z = 0$ and $z = -2.33$, so we find

$$P(x < 20) = A = .5 - .4901 = .0099 \approx .01$$

According to this probability model, you should have only about a 1% chance of purchasing a car of this make with an in-city mileage under 20 miles per gallon.

Figure 5.14
Area under the Normal
Curve for Example 5.7

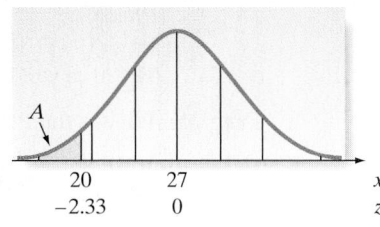

b. Now you are asked to make an inference based on a sample—the car you purchased. You are getting less than 20 miles per gallon for in-city driving. What do you infer? We think you will agree that one of two possibilities is true:

1. The probability model is correct. You simply were unfortunate to have purchased one of the cars in the 1% that get less than 20 miles per gallon in the city.
2. The probability model is incorrect. Perhaps the assumption of a normal distribution is unwarranted, or the mean of 27 is an overestimate, or the standard deviation of 3 is an underestimate, or some combination of these errors was made. At any rate, the form of the actual probability model certainly merits further investigation.

You have no way of knowing with certainty which possibility is correct, but the evidence points to the second one. We are again relying on the rare-event approach to statistical inference that we introduced earlier. The sample (one measurement in this case) was so unlikely to have been drawn from the proposed probability model that it casts serious doubt on the model. We would be inclined to believe that the model is somehow in error.

Look Back When applying the rare-event approach, the calculated probability must be small (say, less than or equal to .05) in order to infer that the observed event is, indeed, unlikely.

Now Work | *Exercise 5.40*

■ ■ ■

Nonstandard Normal Probabilities

Using the T1-83 Graphing Calculator

Finding the Area under the Normal Curve

I. Finding the area without a graph

Step 1 *Find area.*
Press 2nd DISTR and select 2:Normalcdf(
Enter the lower limit
Press comma
Enter the upper limit
Press comma
Enter the mean
Press comma
Enter the standard deviation
Press)
Press ENTER
The area will be displayed on the screen.

Example What is the P(x < 115) for a normal distribution with $\mu = 100$ and $\sigma = 10$?
In this example, the lower limit is $-\infty$, the upper limit is 115, the mean is 100, and the standard deviation is 10.

To represent $-\infty$ on the calculator, enter $(-)$ 1, press 2nd and press the comma key for EE, and then press 99. The screen is shown below.

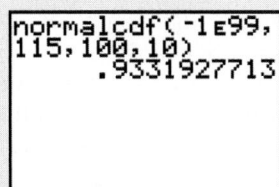 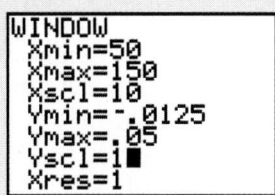

II. Finding the area with a graph

Step 1   *Turn off all plots.*
Press **2nd PRGM** and select **1:ClrDraw**
Press **ENTER ENTER** and 'Done' will appear on the screen.
Press **2nd Y =** and select **4:PlotsOff**
Press **ENTER ENTER** and 'Done' will appear on the screen.

Step 2   Set the viewing window. (These values depend on the mean and standard deviation of the data.)
Press **WINDOW**
Set **Xmin** $= \mu - 5\sigma$
Xmax $= \mu + 5\sigma$
Xscl $= \sigma$
Ymin $= -.125/\sigma$
Ymax $= .5/\sigma$
Yscl $= 1$
Xres $= 1$
Note: the negative sign is the gray key, not the blue key

Step 3   *View graph.*
Press **2nd VARS**
Arrow right to **DRAW**
Press **ENTER** to select **1:ShadeNorm(**
Enter the lower limit
Press **comma**
Enter the upper limit
Press **comma**
Enter the mean
Press **comma**
Enter the standard deviation
Press **)**
Press **ENTER**
The graph will be displayed along with the area, lower limit, and upper limit.

Example   What is the $P(x < 115)$ for a normal distribution with $\mu = 100$ and $\sigma = 10$?
In this example, the lower limit is $-\infty$, the upper limit is 115, the mean is 100, and the standard deviation is 10.
To represent $-\infty$ on the calculator, enter $(-)$ 1, press 2nd and press the comma key for EE, and then press 99. The screens are shown below.

 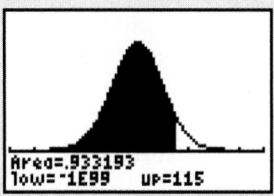

Thus, the $P(x < 115)$ is .9332.

Occasionally you will be given a probability and will want to find the values of the normal random variable that correspond to the probability. For example, suppose the scores on a college entrance examination are known to be normally distributed, and a certain prestigious university will consider for admission only those applicants whose scores exceed the 90th percentile of the test score distribution. To determine the minimum score for admission consideration, you will need to be able to use Table IV in reverse, as demonstrated in the following example.

EXAMPLE 5.8   USING THE NORMAL TABLE IN REVERSE

Problem   Find the value of z, call it z_0, in the standard normal distribution that will be exceeded only 10% of the time. That is, find z_0 such that $P(z \geq z_0) = .10$.

Solution   In this case we are given a probability, or an area, and asked to find the value of the standard normal random variable that corresponds to the area. Specifically, we want to find the value z_0 such that only 10% of the standard normal distribution exceeds z_0 (see Figure 5.15).

Figure 5.15
Areas under the Standard Normal Curve for Example 5.8

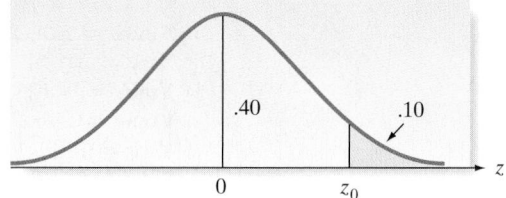

We know that the total area to the right of the mean $z = 0$ is .5, which implies that z_0 must lie to the right of (above) 0. To pinpoint the value, we use the fact that the area to the right of z_0 is .10, which implies that the area between $z = 0$ and z_0 is $.5 - .1 = .4$. But areas between $z = 0$ and some other z value are exactly the types given in Table IV. Therefore, we look up the area .4000 in the body of Table IV and find that the corresponding z value is (to the closest approximation) $z_0 = 1.28$. The implication is that the point 1.28 standard deviations above the mean is the 90th percentile of a normal distribution.

Look Back   As with earlier problems, it is critical to correctly draw the normal probability of interest on the normal curve. Placement of z_0 to the left or right of 0 is the key. Be sure to shade the probability (area) involving z_0. If it does not agree with the probability of interest (i.e., the shaded area is greater than .5 and the probability of interest is smaller than .5), then you need to place z_0 on the opposite side of 0.

■ ■ ■

EXAMPLE 5.9   USING THE NORMAL TABLE IN REVERSE

Problem   Find the value of z_0 such that 95% of the standard normal z values lie between $-z_0$ and $+z_0$—that is, $P(-z_0 \leq z \leq z_0) = .95$.

Solution   Here we wish to move an equal distance z_0 in the positive and negative directions from the mean $z = 0$ until 95% of the standard normal distribution is enclosed. This means that the area on each side of the mean will be equal to $\frac{1}{2}(.95) = .475$, as shown in Figure 5.16. Since the area between $z = 0$ and z_0 is .475, we look up .475 in

Figure 5.16
Areas under the Standard
Normal Curve for
Example 5.9

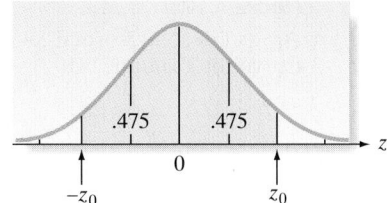

the body of Table IV to find the value $z_0 = 1.96$. Thus, as we found in the reverse order in Example 5.5, 95% of a normal distribution lies between $+1.96$ and -1.96 standard deviations of the mean.

Now Work   *Exercise 5.26a*

■ ■ ■

Now that you have learned to use Table IV to find a standard normal z value that corresponds to a specified probability, we demonstrate a practical application in Example 5.10.

EXAMPLE 5.10

APPLICATION OF THE NORMAL TABLE IN REVERSE

Problem   Suppose the scores, x, on a college entrance examination are normally distributed with a mean of 550 and a standard deviation of 100. A certain prestigious university will consider for admission only those applicants whose scores exceed the 90th percentile of the distribution. Find the minimum score an applicant must achieve in order to receive consideration for admission to the university.

Solution   In this example, we want to find a score, x_0, such that 90% of the scores (x values) in the distribution fall below x_0 and only 10% fall above x_0. That is,

$$P(x \le x_0) = .90$$

Converting x to a standard normal random variable, where $\mu = 550$ and $\sigma = 100$, we have

$$P(x \le x_0) = P\left(z \le \frac{x_0 - \mu}{\sigma}\right)$$
$$= P\left(z \le \frac{x_0 - 550}{100}\right) = .90$$

In Example 5.8 (see Figure 5.15) we found the 90th percentile of the standard normal distribution to be $z_0 = 1.28$. That is, we found $P(z \le 1.28) = .90$. Consequently, we know the minimum test score x_0 corresponds to a z-score of 1.28; that is,

$$\frac{x_0 - 550}{100} = 1.28$$

If we solve this equation for x_0, we find

$$x_0 = 550 + 1.28(100) = 550 + 128 = 678$$

This x-value is shown in Figure 5.17. Thus, the 90th percentile of the test score distribution is 678. An applicant must score at least 678 on the entrance exam to receive consideration for admission by the university.

Figure 5.17

Area under the Normal
Curve for Example 5.10

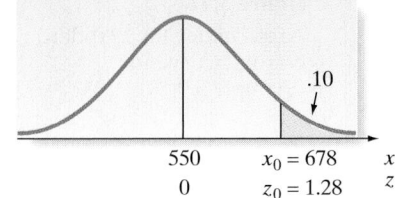

550          $x_0 = 678$ x
0            $z_0 = 1.28$ z

Look Back As the example shows, in practical applications of the normal table in
reverse, first find the value of z_0, then convert the value to the units of x using the z-
score formula in reverse.

Now Work   *Exercise 5.39e*

■ ■ ■

Statistics in Action Revisited

Using the Normal Model to Maximize the Probability of a Hit with the Super Weapon

Recall that a defense contractor has developed a prototype gun for the U.S. Army that fires 1,100 flechettes with a single round. The specifications of the weapon are set so that when the gun is aimed at a target 500 meters away, the mean horizontal grid value of the flechettes is equal to the aim point. In the range test, the weapon was aimed at the center target in Figure SIA5.1; thus, $\mu = 5$ feet. For three different tests, the standard deviation was set at $\sigma = 1$ foot, $\sigma = 2$ feet, and $\sigma = 4$ feet. From past experience, the defense contractor has found that the distribution of the horizontal flechette measurements is closely approximated by a normal distribution. Therefore, we can use the normal distribution to find the probability that a single flechette shot from the weapon will hit any one of the three targets. Recall, from Figure SIA5.1, that the three targets range from -1 to 1, 4 to 6, and 9 to 11 feet on the horizontal grid.

Consider first the middle target. Letting x represent the horizontal measurement for a flechette shot from the gun, the flechette will hit the target if $4 \le x \le 6$. The probability that this flechette will hit the target when $\mu = 5$ and $\sigma = 1$ is, using the normal probability table (Table 4, Appendix I),

$$\text{Middle:}\quad P(4 \le x \le 6) = P\left(\frac{4-5}{1} < z < \frac{6-5}{1}\right)$$
$$\sigma = 1$$
$$= P(-1 < z < 1)$$
$$= 2(.3413) = .6826$$

Similarly, we find the probability that the flechette hits the left and right targets shown in Figure SIA5.1.

$$\text{Left:}\quad P(-1 \le x \le 1) = P\left(\frac{-1-5}{1} < z < \frac{1-5}{1}\right)$$
$$\sigma = 1$$
$$= P(-6 < z < -4) \approx 0$$

$$\text{Right:}\quad P(9 \le x \le 11) = P\left(\frac{9-5}{1} < z < \frac{11-5}{1}\right)$$
$$\sigma = 1$$
$$= P(4 < z < 6) \approx 0$$

You can see that there is about a 68% chance that a flechette will hit the middle target, but virtually no chance that one will hit the left and right targets when the standard deviation is set at 1 foot.

To find these three probabilities for $\sigma = 2$ and $\sigma = 4$, we use the normal probability function in MINITAB. Figure SIA5.2 is a MINITAB worksheet giving the cumulative probabilities of a normal random variable falling below the x-values in the first column. The cumulative probabilities for $\sigma = 2$ and $\sigma = 4$ are given in the columns named "sigma2" and "sigma4," respectively.

Using the cumulative probabilities in the figure to find the three probabilities when $\sigma = 2$, we have

$$\text{Middle:}\quad P(4 \le x \le 6) = P(x \le 6) - P(x \le 4)$$
$$\sigma = 2$$
$$= .6915 - .3085 = .3830$$

$$\text{Left:}\quad P(-1 \le x \le 1) = P(x \le 1) - P(x \le -1)$$
$$\sigma = 2$$
$$= .0227 - .0013 = .0214$$

Figure SIA5.2
MINITAB Worksheet with
Cumulative Normal
Probabilities

↓	C1	C2	C3	C4	C5
	x	sigma1	sigma2	sigma4	
1	-1	0.00000	0.001350	0.066807	
2	1	0.00003	0.022750	0.158655	
3	4	0.15866	0.308538	0.401294	
4	6	0.84134	0.691462	0.598706	
5	9	0.99997	0.977250	0.841345	
6	11	1.00000	0.998650	0.933193	
7					

Right: $P(9 \leq x \leq 11) = P(x \leq 11) - P(x \leq 9)$
$\sigma = 2$
$\qquad\qquad = .9987 - .9773 = .0214$

Thus, when $\sigma = 2$, there is about a 38% chance that a flechette will hit the middle target, a 2% chance that one will hit the left target, and a 2% chance that one will hit the right target. The probability that a flechette will hit either the middle or left or right target is simply the sum of these three probabilities (an application of the Additive Rule of Probability). This sum is $.3830 + .0214 + .0214 = .4258$; consequently, there is about a 42% chance of hitting any one of the three targets when specifications are set so that $\sigma = 2$.

Now, we use the cumulative probabilities in Figure SIA5.2 to find the three hit probabilities when $\sigma = 4$:

Middle: $P(4 \leq x \leq 6) = P(x \leq 6) - P(x \leq 4)$
$\sigma = 4$
$\qquad\qquad = .5987 - .4013 = .1974$

Left: $P(-1 \leq x \leq 1) = P(x \leq 1) - P(x \leq -1)$
$\sigma = 4$
$\qquad\qquad = .1587 - .0668 = .0919$

Right: $P(9 \leq x \leq 11) = P(x \leq 11) - P(x \leq 9)$
$\sigma = 4$
$\qquad\qquad = .9332 - .8413 = .0919$

Thus, when $\sigma = 4$, there is about a 20% chance that a flechette will hit the middle target, a 9% chance that one will hit the left target, and a 9% chance that one will hit the right target. The probability that a flechette will hit any one of the three targets is $.1974 + .0919 + .0919 = .3712$.

These probability calculations reveal a few patterns. First, the probability of hitting the middle target (the target where the gun is aimed) is reduced as the standard deviation is increased. Obviously, if the U.S. Army wants to maximize the chance of hitting the target that the prototype gun is aimed at, it will want specifications set with a small value of σ. But if the Army wants to hit multiple targets with a single shot of the weapon, σ should be increased. With a larger σ, not as many of the flechettes will hit the target aimed at, but more will hit peripheral targets. Whether σ should be set at 4 or 6 (or some other value) depends on how high of a hit rate is required for the peripheral targets.

Exercises 5.17–5.48

Understanding the Principles

5.17 Describe the shape of a normal probability distribution.

5.18 What is the name given to a normal distribution when $\mu = 0$ and $\sigma = 1$?

5.19 If x has a normal distribution with mean μ and standard deviation σ, describe the distribution of $z = (x - \mu)/\sigma$.

Learning the Mechanics

5.20 Find the area under the standard normal probability distribution between the following pairs of z-scores:
  a. $z = 0$ and $z = 2.00$
  b. $z = 0$ and $z = 1.00$
  c. $z = 0$ and $z = 3$
  d. $z = 0$ and $z = .58$

5.21 Find the area under the standard normal distribution between the following pairs of z-scores:
  a. $z = -2.00$ and $z = 0$
  b. $z = -1.00$ and $z = 0$
  c. $z = -1.69$ and $z = 0$
  d. $z = -.58$ and $z = 0$

5.22 Find the following probabilities for the standard normal random variable z:
  NW **a.** $P(z > 1.46)$
  b. $P(z < -1.56)$
  c. $P(.67 \leq z \leq 2.41)$
  d. $P(-1.96 \leq z < -.33)$
  e. $P(z \geq 0)$
  f. $P(-2.33 < z < 1.50)$
  g. $P(z \geq -2.33)$
  h. $P(z < 2.33)$

5.23 Find each of the following probabilities for a standard normal random variable z:
  a. $P(z = 1)$
  b. $P(z \le 1)$
  c. $P(z < 1)$
  d. $P(z > 1)$
  NW **e.** $P(-1 \le z \le 1)$
  NW **f.** $P(-2 \le z \le 2)$
  g. $P(-2.16 \le z \le .55)$
  NW **h.** $P(-.42 < z < 1.96)$

5.24 Give the z-score for a measurement from a normal distribution for the following:
  a. 1 standard deviation above the mean
  b. 1 standard deviation below the mean
  c. Equal to the mean
  d. 2.5 standard deviations below the mean
  e. 3 standard deviations above the mean

5.25 Find a value of the standard normal random variable z, call it z_0, such that
  a. $P(z \ge z_0) = .05$
  b. $P(z \ge z_0) = .025$
  c. $P(z \le z_0) = .025$
  d. $P(z \ge z_0) = .10$
  e. $P(z > z_0) = .10$

5.26 Find a value of the standard normal random variable z, call it z_0, such that
  NW **a.** $P(z \le z_0) = .0401$
  b. $P(-z_0 \le z \le z_0) = .95$
  c. $P(-z_0 \le z \le z_0) = .90$
  d. $P(-z_0 \le z \le z_0) = .8740$
  e. $P(-z_0 \le z \le 0) = .2967$
  f. $P(-2 < z < z_0) = .9710$
  g. $P(z \ge z_0) = .5$
  h. $P(z \ge z_0) = .0057$

5.27 Suppose the random variable x is best described by a normal distribution with $\mu = 25$ and $\sigma = 5$. Find the z-score that corresponds to each of the following x values:
  a. $x = 25$ **b.** $x = 30$ **c.** $x = 37.5$
  d. $x = 10$ **e.** $x = 50$ **f.** $x = 32$

5.28 Suppose x is a normally distributed random variable with $\mu = 11$ and $\sigma = 2$. Find each of the following:
  NW **a.** $P(10 \le x \le 12)$
  b. $P(6 \le x \le 10)$
  c. $P(13 \le x \le 16)$
  d. $P(7.8 \le x \le 12.6)$
  e. $P(x \ge 13.24)$ **f.** $P(x \ge 7.62)$

5.29 The random variable x has a normal distribution with $\mu = 300$ and $\sigma = 30$.
  a. Find the probability that x assumes a value more than 2 standard deviations from its mean. More than 3 standard deviations from μ.
  b. Find the probability that x assumes a value within 1 standard deviation of its mean. Within 2 standard deviations of μ.
  c. Find the value of x that represents the 80th percentile of this distribution. The 10th percentile.

5.30 Suppose x is a normally distributed random variable with $\mu = 30$ and $\sigma = 8$. Find a value of the random variable, call it x_0, such that
  a. $P(x \ge x_0) = .5$
  b. $P(x < x_0) = .025$
  c. $P(x > x_0) = .10$
  d. $P(x > x_0) = .95$

5.31 Refer to Exercise 5.30. Find x_0 such that
  a. 10% of the values of x are less than x_0
  b. 80% of the values of x are less than x_0
  c. 1% of the values of x are greater than x_0

5.32 Suppose x is a normally distributed random variable with mean 100 and standard deviation 8. Draw a rough graph of the distribution of x. Locate μ and the interval $\mu \pm 2\sigma$ on the graph. Find the following probabilities:
  a. $P(\mu - 2\sigma \le x \le \mu + 2\sigma)$
  b. $P(x \ge \mu + 2\sigma)$
  c. $P(x \le 92)$
  d. $P(92 \le x \le 116)$
  e. $P(92 \le x \le 96)$
  f. $P(76 \le x \le 124)$

5.33 The random variable x has a normal distribution with standard deviation 25. It is known that the probability that x exceeds 150 is .90. Find the mean μ of the probability distribution.

Applying the Concepts—Basic

5.34 **Dental anxiety study.** Psychology students at Wittenberg University completed the Dental Anxiety Scale questionnaire. (*Psychological Reports*, Aug. 1997.) Scores on the scale range from 0 (no anxiety) to 20 (extreme anxiety). The mean score was 11 and the standard deviation was 3.5. Assume that the distribution of all scores on the Dental Anxiety Scale is normal with $\mu = 11$ and $\sigma = 3.5$.
  a. Suppose you score a 16 on the Dental Anxiety Scale. Find the z value for this score.
  b. Find the probability that someone scores between a 10 and a 15 on the Dental Anxiety Scale.
  c. Find the probability that someone scores above a 17 on the Dental Anxiety Scale.

5.35 **Alkalinity of river water.** The alkalinity level of water specimens collected from the Han River in Seoul, Korea, has a mean of 50 milligrams per liter and a standard deviation of 3.2 milligrams per liter. (*Environmental Science & Engineering*, Sept. 1, 2000.) Assume the distribution of alkalinity levels is approximately normal and find the probability that a water specimen collected from the river has an alkalinity level
  a. exceeding 45 milligrams per liter.
  b. below 55 milligrams per liter.
  c. between 51 and 52 milligrams per liter.

5.36 **Improving SAT scores.** Refer to the *Chance* (Winter 2001) study of students who paid a private tutor to help them improve their Standardized Admission Test (SAT) scores, Exercise 2.101 (p. 81). The table summarizing the changes in both the SAT-Mathematics and

SAT-Verbal scores for these students is reproduced here. Assume that both distributions of SAT score changes are approximately normal.

	SAT-Math	SAT-Verbal
Mean change in score	19	7
Standard deviation of score changes	65	49

a. What is the probability that a student increases his or her score on the SAT-Math test by at least 50 points?

b. What is the probability that a student increases his or her score on the SAT-Verbal test by at least 50 points?

WPOWER50

5.37 Most powerful American women. Refer to the *Fortune* (Oct. 14, 2002) list of the 50 most powerful women in America, Exercise 2.58 (p. 64). Recall that the data on age (in years) of each woman is stored in the **WPOWER50** file. The ages in the data set can be shown to be approximately normally distributed with a mean of 50 years and a standard deviation of 6 years. A powerful woman is randomly selected from the data and her age is observed.

a. Find the probability that her age will fall between 55 and 60 years.

b. Find the probability that her age will fall between 48 and 52 years.

c. Find the probability that her age will be less than 35 years.

d. Find the probability that her age will exceed 40 years.

DDT

5.38 Weights of contaminated fish. Refer to the U.S. Army Corps of Engineers data on contaminated fish in the Tennessee River, saved in the **DDT** file. Recall that one of the variables measured for each captured fish is weight (in grams). The weights in the data set can be shown to be approximately normally distributed with a mean of 1,050 grams and a standard deviation of 375 grams. An observation is randomly selected from the data and the fish weight is observed.

a. Find the probability that the weight will fall between 1,000 and 1,400 grams.

b. Find the probability that the weight will fall between 800 and 1,000 grams.

c. Find the probability that the weight will be less than 1,750 grams.

d. Find the probability that the return will exceed 500 grams.

e. Ninety-five percent of the captured fish have a weight below what value?

CRASH

5.39 NHTSA crash safety tests. Refer to the National Highway Traffic Safety Administration (NHTSA) crash test data for new cars, introduced in Exercise 2.164 (p. 107) and saved in the **CRASH** file. One of the variables measured is the severity of a driver's head injury when the car is in a head-on collision with a fixed barrier while traveling at 35 miles per hour. The more points assigned to the head injury rating, the more severe the injury. The head injury ratings can be shown to be approximately normally distributed with a mean of 605 points and a standard deviation of 185 points. One of the crash-tested cars is randomly selected from the data and the driver's head injury rating is observed.

a. Find the probability that the rating will fall between 500 and 700 points.

b. Find the probability that the rating will fall between 400 and 500 points.

c. Find the probability that the rating will be less than 850 points.

d. Find the probability that the rating will exceed 1,000 points.

NW e. For what rating will only 10% of the crash-tested cars exceed?

Applying the Concepts—Intermediate

5.40 Fitness of cardiac patients. The physical fitness of a
NW patient is often measured by the patient's maximum oxygen uptake (recorded in milliliters per kilogram, ml/kg). The mean maximum oxygen uptake for cardiac patients who regularly participate in sports or exercise programs was found to be 24.1 with a standard deviation of 6.30. (*Adapted Physical Activity Quarterly*, Oct. 1997.) Assume this distribution is approximately normal.

a. What is the probability that a cardiac patient who regularly participates in sports has a maximum oxygen uptake of at least 20 ml/kg?

b. What is the probability that a cardiac patient who regularly exercises has a maximum oxygen uptake of 10.5 ml/kg or lower?

c. Consider a cardiac patient with a maximum oxygen uptake of 10.5. Is it likely that this patient participates regularly in sports or exercise programs? Explain.

5.41 Visually impaired students. The *Journal of Visual Impairment & Blindness* (May–June 1997) published a study of the lifestyles of visually impaired students. Using diaries, the students kept track of several variables, including number of hours of sleep obtained in a typical day. These visually impaired students had a mean of 9.06 hours and a standard deviation of 2.11 hours. Assume that the distribution of the number of hours of sleep for this group of students is approximately normal.

a. Find the probability that a visually impaired student obtains less than 6 hours of sleep on a typical day.

b. Find the probability that a visually impaired student gets between 8 and 10 hours of sleep on a typical day.

c. Twenty percent of all visually impaired students obtain less than how many hours of sleep on a typical day?

5.42 Alcohol, threats, and electric shocks. A group of Florida State University psychologists examined the effects of alcohol on the reactions of people to a threat. (*Journal of Abnormal Psychology*, Vol. 107, 1998.) After obtaining a specified blood alcohol level, experimental subjects were placed in a room and threatened with electric shocks. Using sophisticated equipment to monitor the subjects' eye movements, the startle response (measured in milliseconds) was recorded for each subject. The mean and standard deviation of the startle responses were 37.9 and 12.4, respectively. Assume that the startle response x for a person with the specified blood alcohol level is approximately normally distributed.

a. Find the probability that x is between 40 and 50 milliseconds.

b. Find the probability that x is less than 30 milliseconds.

c. Give an interval for x, centered around 37.9 milliseconds, so that the probability that x falls in the interval is .95.

d. Ten percent of the experimental subjects have startle responses above what value?

5.43 Forest development following wildfires. *Ecological Applications* (May 1995) published a study on the development of forests following wildfires in the Pacific Northwest. One variable of interest to the researcher was tree diameter at breast height 110 years after the fire. The population of Douglas fir trees was shown to have an approximately normal diameter distribution, with $\mu = 50$ centimeters (cm) and $\sigma = 12$ cm. Find the diameter, d, such that 30% of the Douglas fir trees in the population have diameters that exceed d.

5.44 Gestation length of pregnant women. Based on data from the National Center for Health Statistics, N. Wetzel used the normal distribution to model the length of gestation for pregnant U.S. women (*Chance*, Spring 2001.) Gestation length has a mean of 280 days with a standard deviation of 20 days.

a. Find the probability that gestation length is between 275.5 and 276.5 days. (This estimates the probability that a woman has her baby 4 days earlier than the "average" due date.)

b. Find the probability that gestation length is between 258.5 and 259.5 days. (This estimates the probability that a woman has her baby 21 days earlier than the "average" due date.)

c. Find the probability that gestation length is between 254.5 and 255.5 days. (This estimates the probability that a woman has her baby 25 days earlier than the "average" due date.)

d. The *Chance* article referenced a newspaper story about three sisters who all gave birth on the same day (March 11, 1998). Karralee had her baby 4 days early; Marrianne had her baby 21 days early; and Jennifer had her baby 25 days early. Use the results, parts **a–c**, to estimate the probability that three women have their babies 4, 21, and 25 days early, respectively. Assume the births are independent events.

5.45 Cholesterol levels in psychiatric patients. A study of serum cholesterol levels in psychiatric patients of a maximum-security forensic hospital revealed that cholesterol level is approximately normally distributed with a mean of 208 milligrams per deciliter (mg/dL) and a standard deviation of 25 mg/dL. (*Journal of Behavioral Medicine*, Feb. 1995.) Prior research has shown that patients who exhibit violent behavior have cholesterol levels below 200 mg/dL.

a. What is the probability of observing a psychiatric patient with a cholesterol level below 200 mg/dL?

b. For three randomly selected patients, what is the probability that at least one will have a cholesterol level below 200 mg/dL?

Applying the Concepts—Advanced

5.46 Industrial filling process. The characteristics of an industrial filling process in which an expensive liquid is injected into a container were investigated in *Journal of Quality Technology* (July 1999). The quantity injected per container is approximately normally distributed with mean 10 units and standard deviation .2 units. Each unit of fill costs $20. If a container contains less than 10 units (i.e., is underfilled), it must be reprocessed at a cost of $10. A properly filled container sells for $230.

a. Find the probability that a container is underfilled.

b. A container is initially underfilled and must be reprocessed. Upon refilling it contains 10.6 units. How much profit will the company make on this container?

c. The operations manager adjusts the mean of the filling process upward to 10.5 units in order to make the probability of underfilling approximately zero. Under these conditions, what is the expected profit per container?

5.47 Box plots and the standard normal distribution. What relationship exists between the standard normal distribution and the box-plot methodology (optional Section 2.8) for describing distributions of data using quartiles? The answer depends on the true underlying probability distribution of the data. Assume for the remainder of this exercise that the distribution is normal.

a. Calculate the values of the standard normal random variable z, call them z_L and z_U, that correspond to the hinges of the box plot (i.e., the lower and upper quartiles, Q_L and Q_U) of the probability distribution.

b. Calculate the z values that correspond to the inner fences of the box plot for a normal probability distribution.

c. Calculate the z values that correspond to the outer fences of the box plot for a normal probability distribution.

d. What is the probability that an observation lies beyond the inner fences of a normal probability distribution? The outer fences?

e. Can you better understand why the inner and outer fences of a box plot are used to detect outliers in a distribution? Explain.

5.48 Dye discharged in paint. A machine used to regulate the amount of dye dispensed for mixing shades of paint can be set so that it discharges an average of μ milliliters (mL) of dye per can of paint. The amount of dye discharged is known to have a normal distribution with a standard deviation of .4 mL. If more than 6 mL of dye are discharged when making a certain shade of blue paint, the shade is unacceptable. Determine the setting for μ so that only 1% of the cans of paint will be unacceptable.

5.4 Descriptive Methods for Assessing Normality

In the chapters that follow, we learn how to make inferences about the population based on information in the sample. Several of these techniques are based on the assumption that the population is approximately normally distributed. Consequently, it will be important to determine whether the sample data come from a normal population before we can properly apply these techniques.

Several descriptive methods can be used to check for normality. In this section, we consider the three methods summarized in the box.

> ### Determining Whether the Data Are from an Approximately Normal Distribution
>
> **1.** Construct either a histogram or stem-and-leaf display for the data and note the shape of the graph. If the data are approximately normal, the shape of the histogram or stem-and-leaf display will be similar to the normal curve, Figure 5.5 (i.e., mound shaped and symmetric about the mean).
>
> **2.** Compute the intervals $\bar{x} \pm s$, $\bar{x} \pm 2s$, and $\bar{x} \pm 3s$, and determine the percentage of measurements falling in each. If the data are approximately normal, the percentages will be approximately equal to 68%, 95%, and 100%, respectively.
>
> **3.** Find the interquartile range, IQR, and standard deviation, s, for the sample, then calculate the ratio IQR/s. If the data are approximately normal, then IQR/$s \approx 1.3$.
>
> **4.** Construct a normal probability plot for the data. If the data are approximately normal, the points will fall (approximately) on a straight line.

The first two methods come directly from the properties of a normal distribution established in Section 5.3. Method 3 is based on the fact that for normal distributions, the z values corresponding to the 25th and 75th percentiles are $-.67$ and $.67$, respectively (see Example 5.4). Since $\sigma = 1$ for a standard normal distribution,

$$\frac{\text{IQR}}{\sigma} = \frac{Q_U - Q_L}{\sigma} = \frac{.67 - (-.67)}{1} = 1.34$$

The final descriptive method for checking normality is based on a *normal probability plot*. In such a plot, the observations in a data set are ordered from smallest to largest and then plotted against the expected z-scores of observations calculated under the assumption that the data come from a normal distribution. When the data are, in fact, normally distributed, a linear (straight-line) trend will result. A nonlinear trend in the plot suggests that the data are non-normal.

> ### DEFINITION 5.2
>
> A **normal probability plot** for a data set is a scatterplot with the ranked data values on one axis and their corresponding expected z-scores from a standard normal distribution on the other axis. [*Note:* Computation of the expected standard normal z-scores are beyond the scope of this text. Therefore, we will rely on available statistical software packages to generate a normal probability plot.]

EXAMPLE 5.11

CHECKING FOR NORMAL DATA

Problem   The EPA mileage ratings on 100 cars, first presented in Chapter 2 (p. 41), are reproduced in Table 5.2. Numerical and graphical descriptive measures for the data are shown on the MINITAB and SPSS printouts, Figure 5.18a–c. Determine whether the EPA mileage ratings are from an approximate normal distribution.

Solution   As a first check, we examine the MINITAB histogram of the data shown in Figure 5.18a. Clearly, the mileages fall in an approximately mound-shaped, symmetric distribution centered around the mean of approximately 37 mpg. Note that a normal curve is superimposed on the figure. Therefore, using check #1 in the box, the data appear to be approximately normal.

To apply check #2, we obtain $\bar{x} = 37$ and $s = 2.4$ from the MINITAB printout, Figure 5.18b. The intervals $\bar{x} \pm s$, $\bar{x} \pm 2s$, and $\bar{x} \pm 3s$, are shown in Table 5.3, as well as the percentage of mileage ratings that fall in each interval. (We obtained these results in Section 2.6, p. 75) These percentages agree almost exactly with those from a normal distribution.

⊚ **EPAGAS**

TABLE 5.2   EPA Gas Mileage Ratings for 100 Cars (miles per gallon)

36.3	41.0	36.9	37.1	44.9	36.8	30.0	37.2	42.1	36.7
32.7	37.3	41.2	36.6	32.9	36.5	33.2	37.4	37.5	33.6
40.5	36.5	37.6	33.9	40.2	36.4	37.7	37.7	40.0	34.2
36.2	37.9	36.0	37.9	35.9	38.2	38.3	35.7	35.6	35.1
38.5	39.0	35.5	34.8	38.6	39.4	35.3	34.4	38.8	39.7
36.3	36.8	32.5	36.4	40.5	36.6	36.1	38.2	38.4	39.3
41.0	31.8	37.3	33.1	37.0	37.6	37.0	38.7	39.0	35.8
37.0	37.2	40.7	37.4	37.1	37.8	35.9	35.6	36.7	34.5
37.1	40.3	36.7	37.0	33.9	40.1	38.0	35.2	34.8	39.5
39.9	36.9	32.9	33.8	39.8	34.0	36.8	35.0	38.1	36.9

Figure 5.18a
MINITAB Histogram for Gas Mileage Data

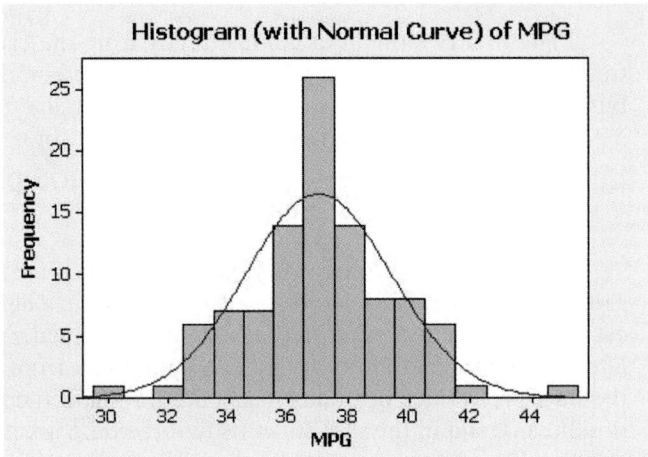

Figure 5.18b
MINITAB Descriptive Statistics for Gas Mileage Data

Descriptive Statistics: MPG

Variable	N	Mean	StDev	Minimum	Q1	Median	Q3	Maximum
MPG	100	36.994	2.418	30.000	35.625	37.000	38.375	44.900

Figure 5.18c

SPSS Normal Probability
Plot for Gas Mileage Data

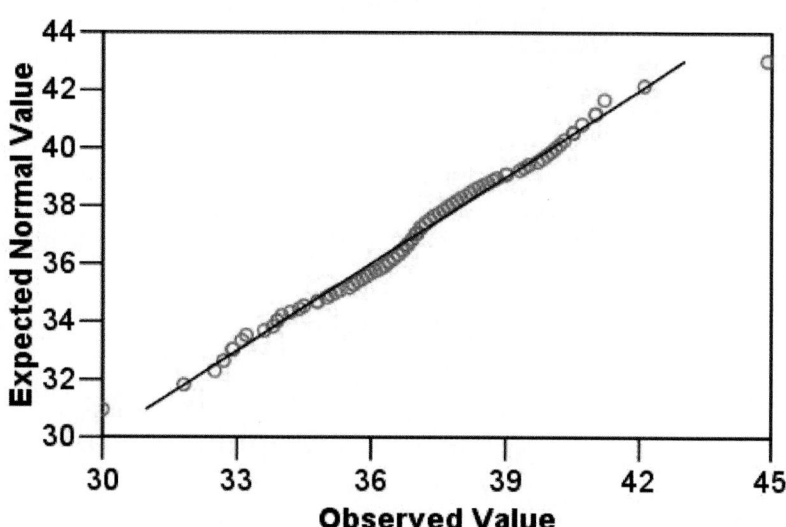

TABLE 5.3   Describing the 100 EPA Mileage Ratings

Interval	Percentage in Interval
$\bar{x} \pm s = (34.6, 39.4)$	68
$\bar{x} \pm 2s = (32.2, 41.8)$	96
$\bar{x} \pm 3s = (29.8, 44.2)$	99

Check #3 in the box requires that we find the ratio IQR/s. From Figure 5.18b, the 25th percentile (labeled Q_1 by MINITAB) is $Q_L = 35.625$ and the 75th percentile (labeled Q_3 by MINITAB) is $Q_U = 38.375$. Then IQR $= Q_U - Q_L = 2.75$ and the ratio is

$$\frac{\text{IRQ}}{s} = \frac{2.75}{2.4} = 1.15$$

Since this value is approximately equal to 1.3, we have further confirmation that the data are approximately normal.

A fourth descriptive method is to interpret a normal probability plot. An SPSS normal probability plot for the mileage data is shown in Figure 5.18c. Notice that the ordered mileage values (shown on the horizontal axis) fall reasonably close to a straight line when plotted against the expected values from a normal distribution. Thus, check #4 also suggests that the EPA mileage data are likely to be approximately normally distributed.

Look Back The checks for normality given in the box are simple, yet powerful, techniques to apply, but they are only descriptive in nature. It is possible (although unlikely) that the data are non-normal even when the checks are reasonably satisfied. Thus, we should be careful not to claim that the 100 EPA mileage ratings are, in fact, normally distributed. We can only state that it is reasonable to believe that the data are from a normal distribution.*

Now Work   *Exercise 5.56*

■ ■ ■

As we will learn in the next chapter, several inferential methods of analysis require the data to be approximately normal. If the data are clearly non-normal, inferences derived from the method may be invalid. Therefore, it is advisable to check the normality of the data prior to conducting the analysis.

Normal Probability Plot

Using the T1-83 Graphing Calculator

Step 1   *Enter the data*
press STAT and select 1:Edit
Note: If the list already contains data, clear the old data. Use the up arrow to highlight 'L1'. Press CLEAR ENTER.
Use the arrow and ENTER keys to enter the data set into L1.

Step 2   *Set up the Normal Probability Plot*
Press 2nd and press **Y =**   for STAT PLOT
Press 1 for Plot 1
Set the cursor so that ON is flashing.
For Type, use the arrow and Enter keys to highlight and select the last graph in the bottom row.
For Data List, choose the column containing the data (in most cases, L1).
(Note: Press 2nd 1 for L1)
For Data Axis, choose X.

Step 3   *View Plot*
Press ZOOM 9
Your data will be displayed against their expected *z*-scores from a normal distribution. If you see a "generally" linear relationship, your data are near normal.

Example   Using a Normal Probability Plot, test whether or not the data are normally distributed.

9.7	93.1	33.0	21.2	81.4	51.1
43.5	10.6	12.8	7.8	18.1	12.7

*Statistical tests of normality that provide a measure of reliability for the inference are available. However, these tests tend to be very sensitive to slight departures from normality (i.e., they tend to reject the hypothesis of normality for any distribution that is not perfectly symmetrical and mound shaped). Consult the references (see Ramsey & Ramsey, 1990) if you want to learn more about these tests.

The screen below shows the normal probability plot. There is a noticeable curve to the plot, indicating that the data are not normally distributed.

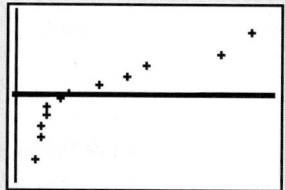

Statistics in Action Revisited

Assessing Whether the Normal Distribution Is Appropriate for Modeling the Super Weapon Hit Data

In *Statistics in Action Revisited*, Section 5.3, we used the normal distribution to find the probability that a single flechette from a super weapon that shoots 1,100 flechettes at once hits one of three targets at 500 meters. Recall that for three range tests, the weapon was always aimed at the center target (i.e., the specification mean was set at $\mu = 5$ feet), but the specification standard deviation was varied at $\sigma = 1$ foot, $\sigma = 2$ feet, and $\sigma = 4$ feet. Table SIA5.1 shows the calculated normal probabilities of hitting the

TABLE SIA5.1   **Summary of Normal Probability Calculations and Actual Range Test Results**

Target	Specification	Normal Probability	Actual Number of Hits	Hit Ratio (Hits/1,100)
LEFT (−1 to 1)	$\sigma = 1$	.0000	0	.000
	$\sigma = 2$	.0214	30	.027
	$\sigma = 4$	.0919	73	.066
MIDDLE (4 to 6)	$\sigma = 1$	.6826	764	.695
	$\sigma = 2$	.3820	409	.372
	$\sigma = 4$	.1974	242	.220
RIGHT (9 to 11)	$\sigma = 1$	.0000	0	.000
	$\sigma = 2$	.0214	23	.021
	$\sigma = 4$	.0919	93	.085

Figure SIA5.3a

MINITAB Histogram for the Horizontal Hit Measurements when $\sigma = 1$

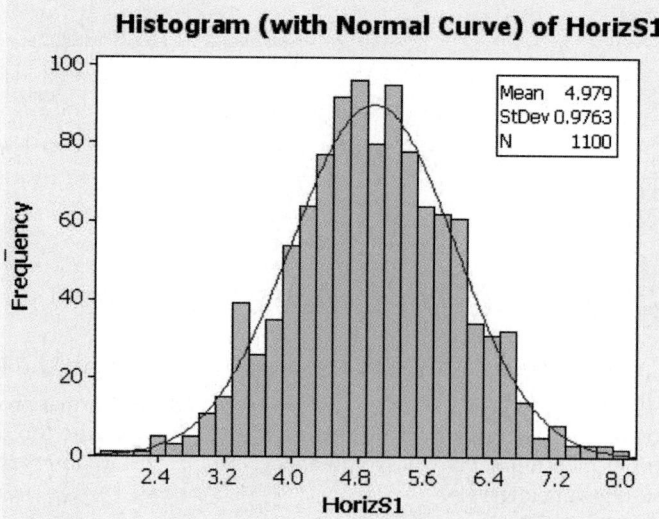

three targets for the different values of σ, as well as the actual results of the three range tests. (Recall that the actual data is saved in the **MOAGUN** file.) You can see that the proportion of the 1,100 flechettes that actually hit each target—called the hit ratio—agrees very well with the estimated probability of a hit using the normal distribution.

Consequently, it appears that our assumption that the horizontal hit measurements are approximately normally distributed is reasonably satisfied. Further evidence of this is provided by the MINITAB histograms of the horizontal hit measurements shown in Figures SIA5.3a–c. The normal curves superimposed on the histograms fit the data very well.

Figure SIA5.3b
MINITAB Histogram for the Horizontal Hit Measurements when $\sigma = 2$

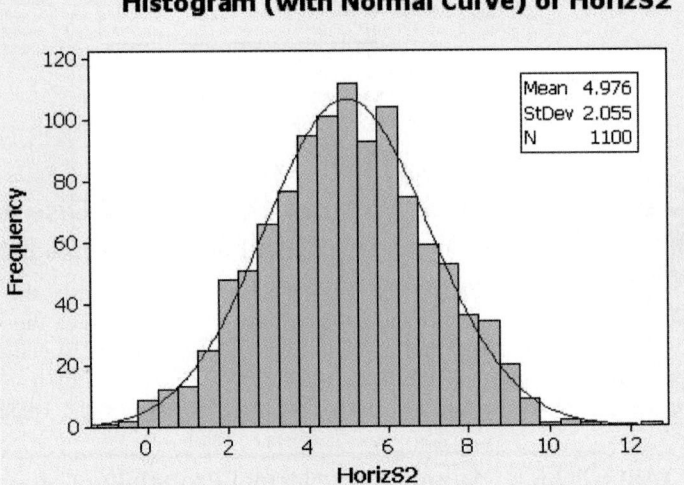

Figure SIA5.3c
MINITAB Histogram for the Horizontal Hit Measurements when $\sigma = 4$

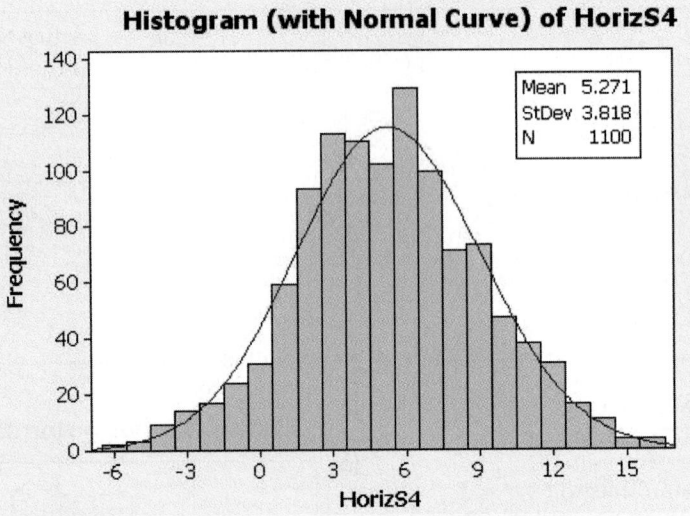

Exercises 5.49–5.65

Understanding the Principles

5.49 Why is it important to check whether the sample data come from a normal population?

5.50 Give four methods for determining whether the sample data come from a normal population.

5.51 If a population data set is normally distributed, what is the proportion of measurements you would expect to fall within the following intervals?
  a. $\mu \pm \sigma$ **b.** $\mu \pm 2\sigma$ **c.** $\mu \pm 3\sigma$

5.52 What is a normal probability plot and how is it used?

Learning the Mechanics

5.53 Normal probability plots for three data sets are shown at the top of p. 267. Which plot indicates that the data are approximately normally distributed?

5.54 Consider a sample data set with the following summary statistics: $s = 95$, $Q_L = 72$, $Q_U = 195$.

Plots for Exercise 5.53

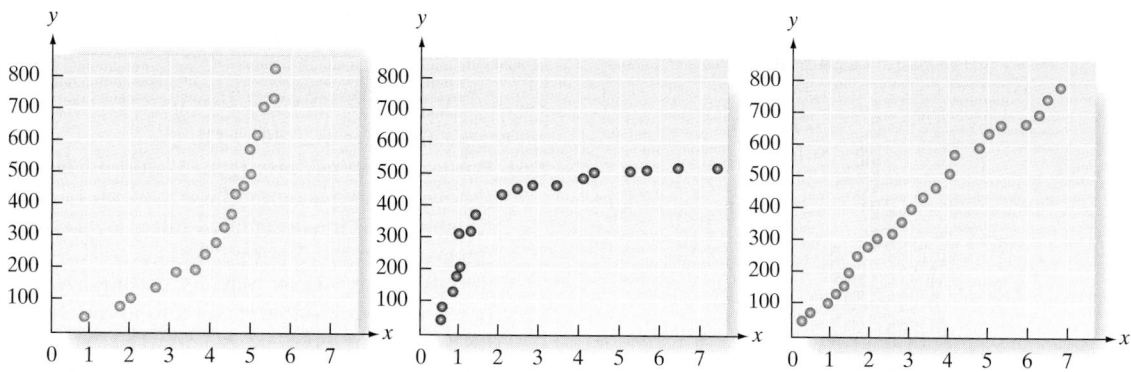

a. Calculate IQR.
b. Calculate IQR/s.
c. Is the value of IQR/s approximately equal to 1.3? What does this imply?

5.55 Examine the sample data in the table.

LM5_55

32	48	25	135	53	37	5	39	213	165
109	40	1	146	39	25	21	66	64	57
197	60	112	10	155	134	301	304	107	82
35	81	60	95	401	308	180	3	200	59

a. Construct a stem-and-leaf plot to assess whether the data are from an approximately normal distribution.
b. Find the values of Q_L, Q_U, and s for the sample data.
c. Use the results, part **b**, to assess the normality of the data.
d. Generate a normal probability plot for the data and use it to assess whether the data are approximately normal.

5.56 Examine the sample data in the table.
NW

LM5_56

5.9	5.3	1.6	7.4	8.6	3.2	2.1
4.0	7.3	8.4	5.9	6.7	4.5	6.3
6.0	9.7	3.5	3.1	4.3	3.3	8.4
4.6	8.2	6.5	1.1	5.0	9.4	6.4

a. Construct a stem-and-leaf plot to assess whether the data are from an approximately normal distribution.
b. Compute s for the sample data.
c. Find the values of Q_L and Q_U and the value of s from part **b** to assess whether the data come from an approximately normal distribution.
d. Generate a normal probability plot for the data and use it to assess whether the data are approximately normal.

Applying the Concepts—Basic

WPOWER50

5.57 Most powerful American women. Refer to the *Fortune* (Oct. 14, 2002) list of the 50 most powerful women in America. In Exercise 5.37 (p. 259), you assumed that the ages (in years) of these women are approximately normally distributed. A MINITAB printout with summary statistics for the age distribution is reproduced below.
a. Use the relevant statistics on the printout to find the interquartile range, IQR.
b. Locate the value of the standard deviation, s, on the printout.
c. Use the results, parts **a** and **b**, to demonstrate that the age distribution is approximately normal.
d. Construct a relative frequency histogram for the age data. Use this graph to support your assumption of normality.

5.58 Galaxy velocity study. Refer to the *Astronomical Journal* (July 1995) study of galaxy velocities, Exercise 2.100 (p. 80). A histogram of the velocity for 103 galaxies located in a particular cluster named A2142 is reproduced at the top of p. 268. Comment on whether or not the galaxy velocities are approximately normally distributed.

HABITAT

5.59 Habitats of endangered species. An evaluation of the habitats of endangered salmon species was performed in *Conservation Ecology* (December 2003). The researchers identified 734 sites (habitats) for Chinook, coho, or steelhead salmon species in Oregon, and assigned a habitat quality score to each. (Scores range from 0 to 36 points, with lower scores indicating poorly maintained or degraded habitats.) A MINITAB histogram for the data (saved in the **HABITAT** file) is displayed at the bottom of p. 268. Give your opinion on whether the data is normally distributed.

5.60 Dart throwing errors. How accurate are you at the game of darts? Researchers at Iowa State University

MINITAB Output for Exercise 5.57

Descriptive Statistics: AGE

| Variable | N | Mean | StDev | Minimum | Q1 | Median | Q3 | Maximum |
| AGE | 50 | 49.740 | 6.124 | 37.000 | 46.000 | 49.000 | 53.000 | 71.000 |

Figure for Exercise 5.58

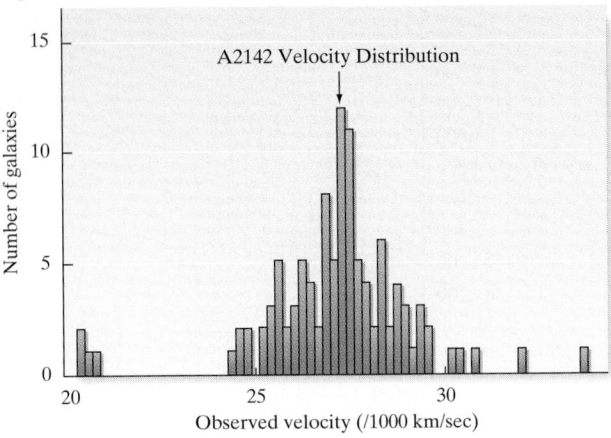

Source: Oegerle, W. R., Hill, J. M., and Fitchett, M. J. "Observations of high dispersion clusters of galaxies: Constraints on cold dark matter." *The Astronomical Journal,* Vol. 110, No. 1, July 1995, p. 37 (Figure 1).

attempted to develop a probability model for dart throws (*Chance*, Summer 1997). For each of 590 throws made at certain targets on a dart board, the distance from the dart to the target point was measured (to the nearest millimeter). The error distribution for the dart throws is described by the frequency table shown on p. 269.

a. Construct a histogram for the data. Is the error distribution for the dart throws approximately normal?

b. Descriptive statistics for the distances from the target for the 590 throws are given below. Use this information to decide whether the error distribution is approximately normal.

$$\bar{x} = 24.4 \text{ mm}$$
$$s = 12.8 \text{ mm}$$
$$Q_L = 14 \text{ mm}$$
$$Q_U = 34 \text{ mm}$$

Source: Good, T. P., Harms, T. K., and Ruckelshaus, M. H. "Misuse of checklist assessments in endangered species recovery efforts," *Conservation Ecology,* Vol. 7, No. 2, Dec. 2003 (Figure 3).

c. Construct a normal probability plot for the data and use it to assess whether the error distribution is approximately normal.

Applying the Concepts—Intermediate

MLB2003-AL & MLB2003-NL

5.61 Baseball batting averages. Major League Baseball (MLB) has two leagues: the American League (AL)—which utilizes the designated hitter (DH) to bat for the pitcher—and the National League (NL)—which does not allow the DH. A player's batting average is computed by dividing the player's total number of hits by his official number of at-bats. The batting averages for all AL and NL players with at least 100 official at-bats during the 2003 season are stored in the **MLB2003-AL** and **MLB2003-NL** files, respectively. Determine whether each batting average distribution is approximately normal.

DDT

5.62 Contaminated fish. Refer to the U.S. Army Corps of Engineers data on contaminated fish in the Tennessee River.

a. In Exercise 5.38 (p. 259), you assumed that the weight (in grams) of a captured fish is approximately normally distributed. Apply the methods of this chapter to the data saved in the **DDT** file to support this assumption.

b. Another variable saved in the **DDT** file is the amount of DDT (in parts per million) detected in each fish captured. Determine whether the DDT level is approximately normal.

CRASH

5.63 NHTSA crash tests. Refer to the National Highway Traffic Safety Administration (NHTSA) crash test data for new cars. In Exercise 5.39 (p. 259), you assumed that the driver's head injury rating is approximately normally

MINITAB Output for Exercise 5.59

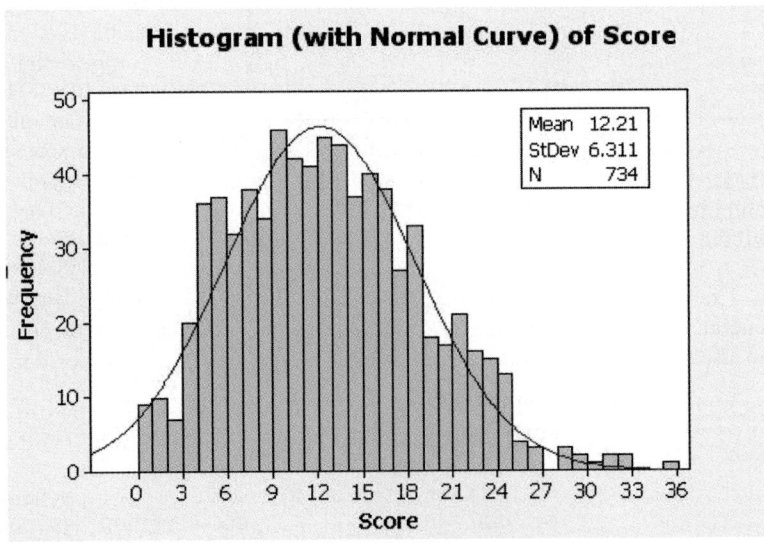

Table for Exercise 5.60

⊘ **DARTS**

Distance from Target (mm)	Frequency	Distance from Target (mm)	Frequency	Distance from Target (mm)	Frequency
0	3	21	15	41	7
2	2	22	19	42	7
3	5	23	13	43	11
4	10	24	11	44	7
5	7	25	9	45	5
6	3	26	21	46	3
7	12	27	16	47	1
8	14	28	18	48	3
9	9	29	9	49	1
10	11	30	14	50	4
11	14	31	13	51	4
12	13	32	11	52	3
13	32	33	11	53	4
14	19	34	11	54	1
15	20	35	16	55	1
16	9	36	13	56	4
17	18	37	5	57	3
18	23	38	9	58	2
19	25	39	6	61	1
20	21	40	5	62	2
				66	1

Source: Stern, H. S., and Wilcox, W. "Shooting Darts." *Chance,* Vol. 10, No. 3, Summer 1997, p. 17 (adapted from Figure 2).

distributed. Apply the methods of this chapter to the data saved in the **CRASH** file to support this assumption.

⊘ **SHIPSANIT**

5.64 Cruise ship sanitation scores. Refer to the data on the May 2004, sanitation scores for 174 cruise ships, first presented in Exercise 2.93 (p. 79). The data are saved in the **SHIPSANIT** file. Assess whether the sanitation scores are approximately normally distributed.

5.65 Language skills study. Refer to the *Applied Psycholinguistics* (June 1998) study of language skills in young children, Exercise 2.98 (p. 80). Recall that the mean sentence complexity score (measured on a 0 to 48 point scale) of low income children was 7.62 with a standard deviation of 8.91. Demonstrate why the distribution of sentence complexity scores for low income children is unlikely to be normally distributed.

5.5 Approximating a Binomial Distribution with a Normal Distribution (Optional)

When the discrete binomial random variable (Section 4.4) can assume a large number of values, the calculation of its probabilities may become very tedious. To contend with this problem, we provide tables in Appendix A to give the probabilities for some values of n and p, but these tables are by necessity incomplete. Recall that the binomial probability table (Table II) can be used only for $n = 5, 6, 7, 8, 9, 10, 15, 20,$ or 25. To deal with this limitation, we seek approximation procedures for calculating the probabilities associated with a binomial probability distribution.

When n is large, a normal probability distribution may be used to provide a good approximation to the probability distribution of a binomial random variable. To show how this approximation works, we refer to Example 4.11, in which we used the binomial distribution to model the number x of 20 eligible voters who favor a candidate. We assumed that 60% of all the eligible voters favored the candidate. The mean and standard deviation of x were found to be $\mu = 12$ and $\sigma = 2.2$. The binomial distribution for $n = 20$ and $p = .6$ is shown in Figure 5.19, and the approximating normal distribution with mean $\mu = 12$ and standard deviation $\sigma = 2.2$ is superimposed.

Figure 5.19

Binomial Distribution for $n = 20$, $p = .6$ and Normal Distribution with $\mu = 12$, $\sigma = 2.2$

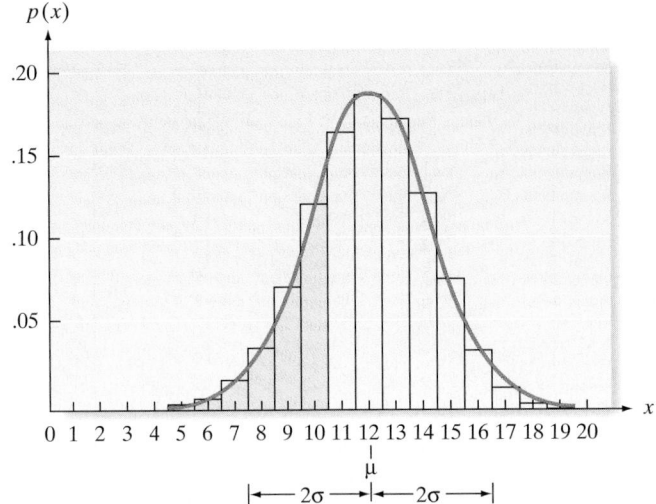

As part of Example 4.11, we used Table II to find the probability that $x \leq 10$. This probability, which is equal to the sum of the areas contained in the rectangles (shown in Figure 5.19) that correspond to $p(0)$, $p(1)$, $p(2), \ldots, p(10)$, was found to equal .245. The portion of the approximating normal curve that would be used to approximate the area $p(0) + p(1) + p(2) + \cdots + p(10)$ is highlighted in Figure 5.19. Note that this highlighted area lies to the left of 10.5 (not 10), so we may include all of the probability in the rectangle corresponding to $p(10)$. Because we are approximating a discrete distribution (the binomial) with a continuous distribution (the normal), we call the use of 10.5 (instead of 10 or 11) a **correction for continuity**. That is, we are correcting the discrete distribution so that it can be approximated by the continuous one. The use of the correction for continuity leads to the calculation of the following standard normal z-value:

$$z = \frac{x - \mu}{\sigma} = \frac{10.5 - 12}{2.2} = -.68$$

Using Table IV, we find the area between $z = 0$ and $z = .68$ to be .2517. Then the probability that x is less than or equal to 10 is approximated by the area under the normal distribution to the left of 10.5, shown highlighted in Figure 5.19. That is,

$$P(x \leq 10) \approx P(z \leq -.68) = .5 - P(-.68 < z \leq 0) = .5 - .2517 = .2483$$

The approximation differs only slightly from the exact binomial probability, .245. Of course, when tables of exact binomial probabilities are available, we will use the exact value rather than a normal approximation.

Use of the normal distribution will not always provide a good approximation for binomial probabilities. The following is a useful rule of thumb to determine when n is large enough for the approximation to be effective: *The interval $\mu \pm 3\sigma$ should lie within the range of the binomial random variable x (i.e., 0 to n) in order for the normal approximation to be adequate.* The rule works well because almost all of the normal distribution falls within 3 standard deviations of the mean, so if this interval is contained within the range of x values, there is "room" for the normal approximation to work.

As shown in Figure 5.20a for the preceding example with $n = 20$ and $p = .6$, the interval $\mu \pm 3\sigma = 12 + 3(2.19) = (5.43, 18.57)$ lies within the range 0 to 20.

Figure 5.20

Rule of Thumb for Normal
Approximation to Binomial
Probabilities

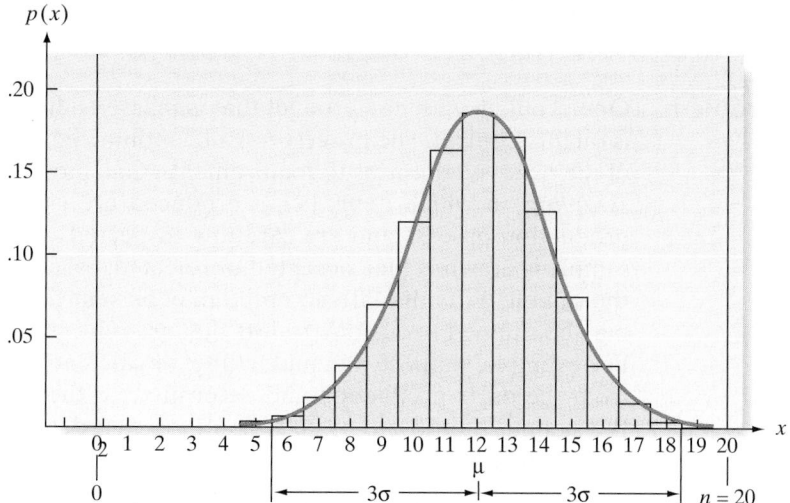

a. $n = 20$, $p = .6$: Normal approximation is good

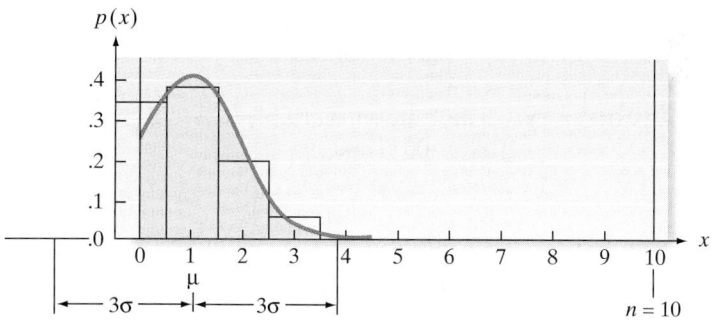

b. $n = 10$, $p = .1$: Normal approximation is poor

However, if we were to try to use the normal approximation with $n = 10$ and $p = .1$, the interval $\mu \pm 3\sigma$ is $1 \pm 3(.95)$, or $(-1.85, 3.85)$. As shown in Figure 5.20b, this interval is not contained within the range of x since $x = 0$ is the lower bound for a binomial random variable. Note in Figure 5.20b that the normal distribution will not "fit" in the range of x, and therefore it will not provide a good approximation to the binomial probabilities.

Biography

**ABRAHAM DE MOIVRE
(1667–1754)—
Advisor to Gamblers**

French-born mathematician Abraham de Moivre moved to London when he was 21 years old to escape religious persecution. In England, he earned a living first as a traveling teacher of mathematics and then as an advisor to gamblers, underwriters, and annuity brokers. De Moivre's major contributions to probability theory are contained in two of his books, *The Doctrine of Chances* (1718) and *Miscellanea Analytica* (1730). In these works he defines statistical independence, develops the formula for the normal probability distribution, and derives the normal curve as an approximation to the binomial distribution. Despite his eminence as a mathematician, de Moivre died in poverty. He is famous for predicting the day of his own death using an arithmetic progression.

EXAMPLE 5.12

APPROXIMATING A BINOMIAL PROBABILITY WITH THE NORMAL DISTRIBUTION

Problem   One problem with any product that is mass-produced (e.g., a graphing calculator) is quality control. The process must be monitored or audited to be sure the output of the process conforms to requirements. One monitoring method is *lot acceptance sampling*, in which items being produced are sampled at various stages of the production process and are carefully inspected. The lot of items from which the sample is drawn is then accepted or rejected, based on the number of defectives in the sample. Lots that are accepted may be sent forward for further processing or may be shipped to customers; lots that are rejected may be reworked or scrapped. For example, suppose a manufacturer of calculators chooses 200 stamped circuits from the day's production and determines x, the number of defective circuits in the sample. Suppose that up to a 6% rate of defectives is considered acceptable for the process.

   a. Find the mean and standard deviation of x, assuming the defective rate is 6%.

   b. Use the normal approximation to determine the probability that 20 or more defectives are observed in the sample of 200 circuits (i.e., find the approximate probability that $x \geq 20$).

Solution   **a.** The random variable x is binomial with $n = 200$ and the fraction defective $p = .06$. Thus,

$$\mu = np = 200(.06) = 12$$

$$\sigma = \sqrt{npq} = \sqrt{200(.06)(.94)} = \sqrt{11.28} = 3.36$$

We first note that

$$\mu \pm 3\sigma = 12 \pm 3(3.36) = 12 \pm 10.08 = (1.92, 22.08)$$

lies completely within the range from 0 to 200. Therefore, a normal probability distribution should provide an adequate approximation to this binomial distribution.

   b. Using the rule of complements, $P(x \geq 20) = 1 - P(x \leq 19)$. To find the approximating area corresponding to $x \leq 19$, refer to Figure 5.21. Note that we want to include all the binomial probability histograms from 0 to 19, inclusive. Since the event is of the form $x \leq a$, the proper correction for continuity is $a + .5 = 19 + .5 = 19.5$. Thus, the z value of interest is

$$z = \frac{(a + .5) - \mu}{\sigma} = \frac{19.5 - 12}{3.36} = 2.23$$

Figure 5.21

Normal Approximation to the Binomial Distribution with $n = 200$, $p = .06$

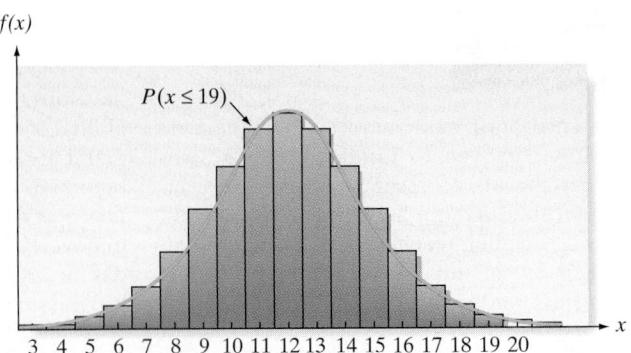

Figure 5.22
Standard Normal
Distribution

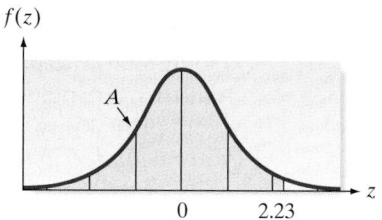

Referring to Table IV in Appendix A, we find that the area to the right of the mean 0 corresponding to $z = 2.23$ (see Figure 5.22) is .4871. So the area $A = P(z \leq 2.23)$ is

$$A = .5 + .4871 = .9871$$

Thus, the normal approximation to the binomial probability we seek is

$$P(x \geq 20) = 1 - P(x \leq 19) \approx 1 - .9871 = .0129$$

In other words, the probability is extremely small that 20 or more defectives will be observed in a sample of 200 circuits—*if in fact the true fraction of defectives is* .06.

Look Back If the manufacturer observes $x \geq 20$, the likely reason is that the process is producing more than the acceptable 6% defectives. The lot acceptance sampling procedure is another example of using the rare-event approach to make inferences.

> Now Work   *Exercise 5.69*

━━━━━━ ■ ■ ■ ━━━━━━

The steps for approximating a binomial probability by a normal probability are given in the accompanying box.

Using a Normal Distribution to Approximate Binomial Probabilities

1. After you have determined n and p for the binomial distribution, calculate the interval

$$\mu \pm 3\sigma = np \pm 3\sqrt{npq}$$

If the interval lies in the range 0 to n, the normal distribution will provide a reasonable approximation to the probabilities of most binomial events.

2. Express the binomial probability to be approximated in the form $P(x \leq a)$ or $P(x \leq b) - P(x \leq a)$. For example,

$$P(x < 3) = P(x \leq 2)$$
$$P(x \geq 5) = 1 - P(x \leq 4)$$
$$P(7 \leq x \leq 10) = P(x \leq 10) - P(x \leq 6)$$

3. For each value of interest a, the correction for continuity is $(a + .5)$, and the corresponding standard normal z value is

$$z = \frac{(a + .5) - \mu}{\sigma} \qquad \text{(see Figure 5.23)}$$

4. Sketch the approximating normal distribution and shade the area corresponding to the probability of the event of interest, as in Figure 5.23. Verify that the rectangles you have included in the shaded area correspond to the event probability you wish to approximate. Using Table IV and the z value(s) you calculated in step 3, find the shaded area. This is the approximate probability of the binomial event.

Figure 5.23
Approximating Binomial Probabilities by Normal Probabilities

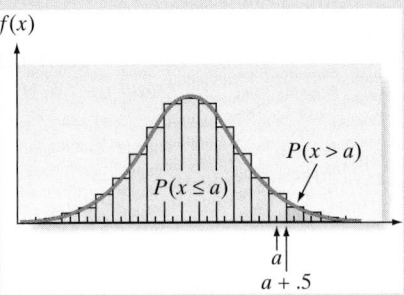

Exercises 5.66–5.84

Understanding the Principles

5.66 For large n (say, $n = 100$), why is it advantageous to use the normal distribution to approximate a binomial probability?

5.67 Why do we need a correction for continuity when approximating a binomial probability with the normal distribution?

Learning the Mechanics

5.68 Assume that x is a binomial random variable with n and p as specified in parts **a–f**. For which cases would it be appropriate to use a normal distribution to approximate the binomial distribution?
  a. $n = 100, p = .01$ **b.** $n = 20, p = .6$
  c. $n = 10, p = .4$ **d.** $n = 1,000, p = .05$
  e. $n = 100, p = .8$ **f.** $n = 35, p = .7$

5.69 Suppose x is a binomial random variable with $p = .4$
NW and $n = 25$.
  a. Would it be appropriate to approximate the probability distribution of x with a normal distribution? Explain.
  b. Assuming that a normal distribution provides an adequate approximation to the distribution of x, what are the mean and variance of the approximating normal distribution?
  c. Use Table II of Appendix A to find the exact value of $P(x \geq 9)$.
  d. Use the normal approximation to find $P(x \geq 9)$.

5.70 Assume that x is a binomial random variable with $n = 25$ and $p = .5$. Use Table II of Appendix A and the normal approximation to find the exact and approximate values, respectively, for the following probabilities:
  a. $P(x \leq 11)$
  b. $P(x \geq 16)$
  c. $P(8 \leq x \leq 16)$

5.71 Assume that x is a binomial random variable with $n = 100$ and $p = .40$. Use a normal approximation to find the following:
  a. $P(x \leq 35)$ **b.** $P(40 \leq x \leq 50)$
  c. $P(x \geq 38)$

5.72 Assume that x is a binomial random variable with $n = 1,000$ and $p = .50$. Find each of the following probabilities:
  a. $P(x > 500)$ **b.** $P(490 \leq x < 500)$
  c. $P(x > 550)$

Applying the Concepts—Basic

5.73 **Mobile phones with Internet access.** In Exercise 4.56 (p. 216), you learned that 20% of young adults who shop online own a mobile phone with Internet access. (*American Demographics*, May 2002.) In a random sample of 200 young adults who shop online, let x be the number who own a mobile phone with Internet access.
  a. Find the mean of x. (This value should agree with your answer to Exercise 4.56c.)
  b. Find the standard deviation of x.
  c. Find the z-score for the value $x = 50.5$.
  d. Find the approximate probability that the number of young adults who own a mobile phone with Internet access in a sample of 200 is less than or equal to 50.

5.74 **Melanoma deaths.** According to the *American Cancer Society*, melanoma, a form of skin cancer, kills 15% of Americans who suffer from the disease each year. Consider a sample of 10,000 melanoma patients.
  a. What are the expected value and variance of x, the number of the 10,000 melanoma patients who die of the affliction this year?
  b. Find the probability that x will exceed 1,600 patients per year.

c. Would you expect x, the number of patients dying of melanoma, to exceed 6,500 in any single year? Explain.

5.75 Quit smoking program. In Exercise 4.57 (p. 216), you learned that only 5% of the nation's cigarette smokers ever enter into a treatment program to help them quit smoking. (*USF Magazine*, Spring 2000.) In a random sample of 200 smokers, let x be the number who enter into a treatment program.

a. Find the mean of x. (This value should agree with your answer to Exercise 4.58)

b. Find the standard deviation of x.

c. Find the z-score for the value $x = 10.5$.

d. Find the approximate probability that the number of smokers in a sample of 200 who will enter into a treatment program is 11 or more.

5.76 Caesarian birth study. In Exercise 4.58 (p. 216), you learned that 22% of all births in the U.S. occur by Caesarian section each year. (*USA Today*, Sept. 19, 2000.) In a random sample of 1,000 births this year, let x be the number that occur by Caesarian section.

a. Find the mean of x. (This value should agree with your answer to Exercise 4.44**a**.)

b. Find the standard deviation of x. (This value should agree with your answer to Exercise 4.44**b**.)

c. Find the z-score for the value $x = 200.5$.

d. Find the approximate probability that the number of Caesarian sections in a sample of 1,000 births is less than or equal to 200.

Applying the Concepts—Intermediate

5.77 Defects in semiconductor wafers. The computer chips in notebook and laptop computers are produced from semiconductor wafers. Certain semiconductor wafers are exposed to an environment that generates up to 100 possible defects per wafer. The number of defects per wafer, x, was found to follow a binomial distribution if the manufacturing process is stable and generates defects that are randomly distributed on the wafers. (*IEEE Transactions on Semiconductor Manufacturing*, May 1995.) Let p represent the probability that a defect occurs at any one of the 100 points of the wafer. For each of the following cases, determine whether the normal approximation can be used to characterize x.

a. $p = .01$ **b.** $p = .50$ **c.** $p = .90$

5.78 Parents who condone spanking. In Exercise 4.60 (p. 217), you learned that 60% of parents with young children condone spanking their child as a regular form of punishment. (*Tampa Tribune*, Oct. 5, 2000.) A child psychologist with 150 parent clients claims that no more than 20 of the parents condone spanking. Do you believe this claim? Explain.

5.79 Fungi in beech forest trees. Refer to the *Applied Ecology and Environmental Research* (Vol. 1, 2003) study of beech trees damaged by fungi, Exercise 4.61 (p. 217). Recall that the researchers found that 25% of the beech trees in East Central Europe have been damaged by fungi. In an East Central European forest with 200 beech trees, how likely is it that more than half of the trees have fungi damage?

5.80 FTC price accuracy study. Refer to the FTC study of the pricing accuracy of supermarket electronic scanners, Exercise 4.115 (p. 230). Recall that the probability that a scanned item is priced incorrectly is $^1/_{30} = .033$.

a. Suppose 10,000 supermarket items are scanned. What is the approximate probability that you observe at least 100 items with incorrect prices?

b. Suppose 100 items are scanned and you are interested in the probability that fewer than five are incorrectly priced. Explain why the approximate method of part **a** may not yield an accurate estimate of the probability.

5.81 Victims of domestic abuse. In Exercise 4.63 (p. 217) you learned that some researchers believe that one in every three women are victims of domestic abuse. (*Annals of Internal Medicine*, Nov. 1995.)

a. For a random sample of 150 women, what is the approximate probability that more than half are victims of domestic abuse?

b. For a random sample of 150 women, what is the approximate probability that fewer than 50 are victims of domestic abuse?

c. Would you expect to observe fewer than 30 domestically abused women in a sample of 150? Explain.

Applying the Concepts—Advanced

5.82 Body fat in men. The percentage of fat in the bodies of American men is an approximately normal random variable with mean equal to 15% and standard deviation equal to 2%.

a. If these values were used to describe the body fat of men in the United States Army and if 20% or more body fat is characterized as obese, what is the approximate probability that a random sample of 10,000 soldiers will contain fewer than 50 who would actually be characterized as obese?

b. If the army actually were to check the percentage of body fat for a random sample of 10,000 men and if only 30 contained 20% (or higher) body fat, would you conclude that the army was successful in reducing the percentage of obese men below the percentage in the general population? Explain your reasoning.

5.83 Luggage inspection at Newark airport. According to *New Jersey Business* (Feb. 1996), Newark International Airport's new terminal handles an average of 3,000 international passengers an hour but is capable of handling twice that number. Also, 80% of arriving international passengers pass through without their luggage being inspected and the remainder are detained for inspection. The inspection facility can handle 600 passengers an hour without unreasonable delays for the travelers.

a. When international passengers arrive at the rate of 1,500 per hour, what is the expected number of passengers who will be detained for luggage inspection?

b. In the future, it is expected that as many as 4,000 international passengers will arrive per hour. When that occurs, what is the expected number of passengers who will be detained for luggage inspection?

c. Refer to part **b**. Find the approximate probability that more than 600 international passengers will be detained for luggage inspection. (This is also the probability that travelers will experience unreasonable luggage inspection delays.)

5.84 Waiting time at a doctor's office. The median time a patient waits to see a doctor in a large clinic is 20 minutes. On a day when 150 patients visit the clinic, what is the approximate probability that

a. More than half will wait more than 20 minutes?

b. More than 85 will wait more than 20 minutes?

c. More than 60 but fewer than 90 will wait more than 20 minutes?

5.6 The Exponential Distribution (Optional)

The length of time between emergency arrivals at a hospital, the length of time between breakdowns of manufacturing equipment, the length of time between catastrophic events (floods, earthquakes, etc.), and the distance traveled by a wildlife ecologist between sightings of an endangered species are all random phenomena that we might want to describe probabilistically. The amount of time or distance between occurrences of random events like these can often be described by the **exponential probability distribution**. For this reason, the exponential distribution is sometimes called the **waiting time distribution**. The formula for the exponential probability distribution is shown in the box along with its mean and standard deviation.

Probability Distribution for an Exponential Random Variable x

Probability density function: $f(x) = \dfrac{1}{\theta}e^{-x/\theta}$ $\quad (x > 0)$

Mean: $\mu = \theta$ $\quad$ Standard deviation: $\sigma = \theta$

Unlike the normal distribution, which has a shape and location determined by the values of the two quantities μ and σ, the shape of the exponential distribution is governed by a single quantity, θ. Further, it is a probability distribution with the property that its mean equals its standard deviation. Exponential distributions corresponding to $\theta = .5$, 1, and 2 are shown in Figure 5.24.

To calculate probabilities for exponential random variables, we need to be able to find areas under the exponential probability distribution. Suppose we want to find the area A to the right of some number a, as shown in Figure 5.25.

This area can be calculated by using the formula shown in the box. Use Table V in Appendix A or a pocket calculator with an exponential function to find the value of $e^{-a/\theta}$ after substituting the appropriate numerical values for θ and a.

Figure 5.24
Exponential Distributions

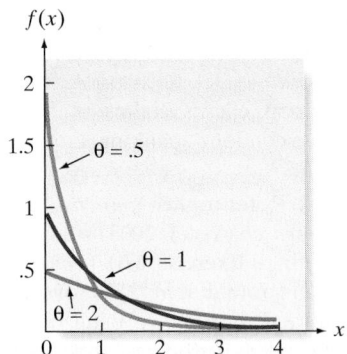

Finding the Area A to the Right of a Number a for an Exponential Distribution*

$$A = P(x \geq a) = e^{-a/\theta}$$

Figure 5.25

The Area A to the Right of a Number a for an Exponential Distribution

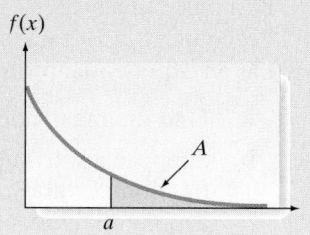

EXAMPLE 5.13   FINDING AN EXPONENTIAL PROBABILITY

Problem Suppose the length of time (in hours) between emergency arrivals at a certain hospital is modeled as an exponential distribution with $\theta = 2$. What is the probability that more than 5 hours pass without an emergency arrival?

Solution The probability we want is the area A to the right of $a = 5$ in Figure 5.26. To find this probability, use the formula given for area:

$$A = e^{-a/\theta} = e^{-(5/2)} = e^{-2.5}$$

Referring to Table V, we find

$$A = e^{-2.5} = .082085$$

Our exponential model indicates that the probability that more than 5 hours pass between emergency arrivals is about .08 for this hospital.

Look Back The value $e^{-2.5}$ can also be found using a standard hand calculator. Or, you can find the desired probability using a statistical software package.

Figure 5.26

Area to the Right of $a = 5$ for Example 5.13

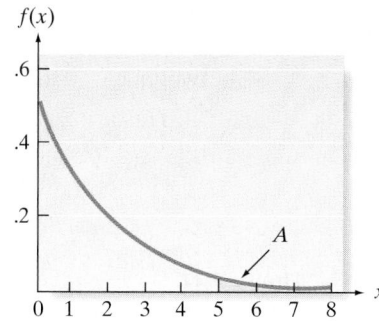

Now Work *Exercise 5.89*

■ ■ ■

---

*For students with a knowledge of calculus, the highlighted area in Figure 5.25 corresponds to the integral

$$\int_{a}^{\infty} \frac{1}{\theta} e^{-x/\theta}\, dx = -e^{-x/\theta} \Big|_{a}^{\infty} = e^{-a/\theta}.$$

EXAMPLE 5.14

THE MEAN AND VARIANCE OF AN EXPONENTIAL RANDOM VARIABLE

Problem A microwave oven manufacturer is trying to determine the length of warranty period it should attach to its magnetron tube, the most critical component in the oven. Preliminary testing has shown that the length of life (in years), x, of a magnetron tube has an exponential probability distribution with $\theta = 6.25$.

    a. Find the mean and standard deviation of x.

    b. Suppose a warranty period of 5 years is attached to the magnetron tube. What fraction of tubes must the manufacturer plan to replace, assuming that the exponential model with $\theta = 6.25$ is correct?

    c. Find the probability that the length of life of a magnetron tube will fall within the interval $\mu - 2\sigma$ to $\mu + 2\sigma$.

Solution   **a.** Since $\theta = \mu = \sigma$, both μ and σ equal 6.25.

    b. To find the fraction of tubes that will have to be replaced before the 5-year warranty period expires, we need to find the area between 0 and 5 under the distribution. This area, A, is shown in Figure 5.27. To find the required probability, we recall the formula

$$P(x > a) = e^{-a/\theta}$$

Using this formula, we can find

$$P(x > 5) = e^{-a/\theta} = e^{-5/6.25} = e^{-.80} = .449329$$

(see Table V). To find the area A, we use the complementary relationship:

$$P(x \le 5) = 1 - P(x > 5) = 1 - .449329 = .550671$$

So approximately 55% of the magnetron tubes will have to be replaced during the 5-year warranty period.

    c. We would expect the probability that the life of a magnetron tube, x, falls within the interval $\mu - 2\sigma$ to $\mu + 2\sigma$ to be quite large. A graph of the exponential distribution showing the interval $\mu - 2\sigma$ to $\mu + 2\sigma$ is given in Figure 5.28. Since the point $\mu - 2\sigma$ lies below $x = 0$, we need to find only the area between $x = 0$ and $x = \mu + 2\sigma = 6.25 + 2(6.25) = 18.75$.

This area, P, which is highlighted in Figure 5.28, is

$$P = 1 - P(x > 18.75) = 1 - e^{-18.75/\theta} = 1 - e^{-18.75/6.25} = 1 - e^{-3}$$

Figure 5.27
Area to the Left of a = 5 for Example 5.14

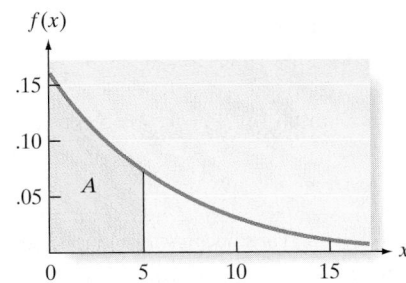

Figure 5.28
Area in the Interval $\mu \pm 2\sigma$
for Example 5.14

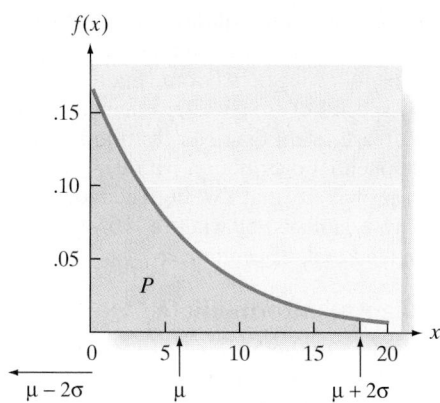

Using Table V or a calculator, we find $e^{-3} = .049787$. Therefore, the probability that the life x of a magnetron tube will fall within the interval $\mu - 2\sigma$ to $\mu + 2\sigma$ is

$$P = 1 - e^{-3} = 1 - .049787 = .950213$$

Look Back You can see that this probability agrees very well with the interpretation of a standard deviation given by the Empirical Rule (Table 2.7, p. 74) even though this probability distribution is not mound-shaped. (It is strongly skewed to the right.)

■ ■ ■

Exercises 5.85–5.101

Understanding the Principles

5.85 What are the characteristics of an exponential random variable?

5.86 The exponential distribution is often called the _____ distribution. (Fill in the blank.)

Learning the Mechanics

5.87 The random variables x and y have exponential distributions with $\theta = 3$ and $\theta = 1$, respectively. Using Table V in Appendix A, carefully plot both distributions on the same set of axes.

5.88 Use Table V in Appendix A to determine the value of $e^{-a/\theta}$ for each of the following cases.
  a. $\theta = 1, a = 1$ **b.** $\theta = 1, a = 2.5$
  c. $\theta = .4, a = 3$ **d.** $\theta = .2, a = .3$

5.89 Suppose x has an exponential distribution with $\theta = 1$.
NW Find the following probabilities:
  a. $P(x > 1)$ **b.** $P(x \le 3)$
  c. $P(x > 1.5)$ **d.** $P(x \le 5)$

5.90 Suppose x has an exponential distribution with $\theta = 2.5$. Find the following probabilities:
  a. $P(x \le 4)$ **b.** $P(x > 5)$
  c. $P(x \le 2)$ **d.** $P(x > 3)$

5.91 Suppose the random variable x has an exponential probability distribution with $\theta = 1$. Find the mean and standard deviation of x. Find the probability

that x will assume a value within the interval $\mu \pm 2\sigma$.

5.92 The random variable x can be adequately approximated by an exponential probability distribution with $\theta = 2$. Find the probability that x assumes a value
  a. More than 3 standard deviations from μ.
  b. Less than 2 standard deviations from μ.
  c. Within .5 standard deviation of μ.

Applying the Concepts—Basic

5.93 Lead in metal shredder residue. Based on data collected from metal shredders across the nation, the amount x of extractable lead in metal shredder residue has an approximate exponential distribution with mean $\theta = 2.5$ milligrams per liter (Florida Shredder's Association).
  a. Find the probability that x is greater than 2 milligrams per liter.
  b. Find the probability that x is less than 5 milligrams per liter.

5.94 Life length of bacteria. After bacteria are subjected to a certain drug, the length of time until the bacteria die follows an exponential distribution with a mean of 30 minutes.
  a. How long will it take for half the bacteria to die after the drug is administered?
  b. What proportion of bacteria will die between 12 minutes and 60 minutes?

5.95 Forest development following wildfires. Refer to the *Ecological Applications* (May 1995) study of forest development 110 years following a wildfire, Exercise 5.43 (p. 260). Another species of tree, western hemlock, was found to have a breast height diameter distribution that resembled an exponential distribution with $\theta = 30$ centimeters. Find the probability that a western hemlock tree growing in the forest damaged by wildfire 110 years ago has a diameter that exceeds 25 centimeters.

Applying the Concepts—Intermediate

5.96 NHL overtime games. In the National Hockey League (NHL), games that are tied at the end of three periods are sent to "sudden-death" overtime. In overtime, the team to score the first goal wins. An analysis of NHL overtime games showed that the length of time elapsed before the winning goal is scored has an exponential distribution with mean 9.15 minutes (*Chance*, Winter 1995.)

   a. For a randomly selected overtime NHL game, find the probability that the winning goal is scored in 3 minutes or less.

   b. In the NHL, each period (including overtime) lasts 20 minutes. If neither team scores a goal in overtime, the game is considered a tie. What is the probability of an NHL game ending in a tie?

5.97 Machine repair times. An article in *IEEE Transactions* (Mar. 1990) gave an example of a flexible manufacturing system (FMS) with four machines operating independently. The repair rates for the machines (i.e., the time, in hours, it takes to repair a failed machine) are exponentially distributed with means $\mu_1 = 1, \mu_2 = 2, \mu_3 = .5$, and $\mu_4 = .5$, respectively.

   a. Find the probability that the repair time for machine 1 exceeds 1 hour.

   b. Repeat part **a** for machine 2.

   c. Repeat part **a** for machines 3 and 4.

   d. If all four machines fail simultaneously, find the probability that the repair time for the entire system exceeds 1 hour.

FORFRAG

5.98 Forest fragmentation study. Refer to the *Conservation Ecology* (December 2003) study on the causes of fragmentation for 54 South American forests, Exercise 2.148 (p. 100). Recall that the cause is classified as either anthropogenic or natural in origin. The anthropogenic fragmentation index (saved in the **FORFRAG** file) for the South American forests has an approximate exponential distribution with a mean of 23.

   a. Find the probability that a South American forest has an anthropogenic fragmentation index between 20 and 40.

   b. Find the probability that a South American forest has an anthropogenic fragmentation index below 50.

   c. The natural fragmentation index (also saved in the **FORGRAG** file) does not have an approximate exponential distribution. Why?

5.99 Ship-to-shore transfer times. Lack of port facilities or shallow water may require cargo on a large ship to be transferred to a pier using smaller craft. This process may require the smaller craft to cycle back and forth from ship to shore many times. Researchers G. Horne (Center for Naval Analysis) and T.Irony (George Washington University) developed models of this transfer process that provide estimates of ship-to-shore transfer times. (*Naval Research Logistics*, Vol. 41, 1994.) They modeled the time between arrivals of the smaller craft at the pier using an exponential distribution.

   a. Assume the mean time between arrivals at the pier is 17 minutes. Give the value of θ for this exponential distribution. Graph the distribution.

   b. Suppose there is only one unloading zone at the pier available for the small craft to use. If the first craft docks at 10:00 am and doesn't finish unloading until 10:15 am, what is the probability that the second craft will arrive at the unloading zone and have to wait before docking?

5.100 Modeling machine downtime. The importance of modeling machine downtime correctly in simulation studies was discussed in *Industrial Engineering* (Aug. 1990). The paper presented simulation results for a single-machine-tool system with the following properties:

   1. The interarrival times of jobs are exponentially distributed with a mean of 1.25 minutes

   2. The amount of time the machine operates before breaking down is exponentially distributed with a mean of 540 minutes

   a. Find the probability that two jobs arrive for processing at most 1 minute apart.

   b. Find the probability that the machine operates for at least 720 minutes (12 hours) before breaking down.

Applying the Concepts—Advanced

5.101 Life length of a halogen bulb. For a certain type of halogen light bulb, an old bulb that has been in use for a while tends to have a longer life length than a new bulb. Let x represent the life length (in hours) of a new halogen light bulb and assume that x has an exponential distribution with mean $\theta = 250$ hours. According to *Microelectronics and Reliability* (Jan. 1986), the "life" distribution of x is considered *new better than used* (NBU) if

$$P(x > a + b) \le P(x > a)P(x > b)$$

Alternatively, a "life" distribution is considered *new worse than used* (NWU) if

$$P(x > a + b) \ge P(x > a)P(x > b)$$

   a. Show that when $a = 300$ and $b = 200$ the exponential distribution is both NBU and NWU.

   b. Choose any two positive numbers a and b, where $a > 0$ and $b > 0$, and repeat part **a**.

   c. Show that, in general, for any a and b ($a > 0$ and $b > 0$), the exponential distribution with mean θ is both NBU and NWU. Such a "life" distribution is said to be *new same as used* or *memoryless*. Explain why.

Quick Review

Key Terms

Note: Starred (*) terms are from the optional sections in this chapter.

Bell curve   243
Bell-shaped distribution   243
Continuous probability
 distribution   237
Continuous random variable   237

*Correction for continuity   270
*Exponential distribution   276
*Exponential random variable   276
Frequency function   238
Normal distribution   243
Normal probability plot   261
Normal random variable   244
Probability density function   238
Probability distribution   238

Randomness distribution   239
Standard normal distribution   245
Standard normal random
 variable   245
Uniform distribution   239
Uniform random variable   239
*Waiting time distribution   276

Key Formulas

Note: Starred (*) formulas are from the optional sections in this chapter.

Density Function	Mean	Standard Deviation	Random Variable
$f(x) = \dfrac{1}{d-c}\,(c \le x \le d)$	$\mu = \dfrac{c+d}{2}$	$\sigma = \dfrac{d-c}{\sqrt{12}}$	Uniform, x 239
$f(x) = \dfrac{1}{\sigma\sqrt{2\pi}}e^{-(1/2)[(x-\mu)/\sigma]^2}$	μ	σ	Normal, x 244
$f(z) = \dfrac{1}{\sqrt{2\pi}}e^{-(1/2)z^2}$	$\mu = 0$	$\sigma = 1$	Standard Normal, $z = \left(\dfrac{x-\mu}{\sigma}\right)$ 248
$f(x) = \dfrac{1}{\theta}e^{-x/\theta}\,(x > 0)$	$\mu = \theta$	$\sigma = \theta$	*Exponential, x 276

*Normal Approximation to Binomial: $P(x \le a) = P\left[z \le \dfrac{(a + .5) - \mu}{\sigma}\right]$ 273

Language Lab

Symbol	Pronunciation	Description
$f(x)$	f of x	Probability density function for a continuous random variable x
θ	theta	Mean of exponential random variable

Chapter Summary

- Three types of continuous random variables: **uniform**, **normal**, and **exponential**
- **Uniform probability distribution** is a model for continuous random variables that are evenly distributed over a certain interval
- **Normal probability distribution** is a model for continuous random variables that have a bell-shaped curve

- Methods for assessing normality: **histogram**, **stem-and-leaf display**, IQR/$s \approx 1.3$, and **normal probability plot**
- Normal distribution can be used to approximate a binomial probability when $\mu \pm 3\sigma$ falls within the interval $(0, n)$
- **Exponential probability distribution** is a model for continuous random variables that have a waiting time distribution

Supplementary Exercises 5.102–5.135

Note: Starred (*) exercises refer to the optional sections in this chapter.

Understanding the Principles

*5.102 Consider the following continuous random variables. Give the probability distribution (uniform, normal, or exponential) that is likely to best approximate the distribution of the random variable:
  a. Score on an IQ test
  b. Time (in minutes) waiting in line at a supermarket checkout counter
  c. Amount of liquid (in ounces) dispensed into a can of soda
  d. Difference between SAT scores for tests taken at two different times

*5.103 Identify the type of continuous random variable—uniform, normal, or exponential—described by each of the following probability density functions.
  a. $f(x) = (e^{-x/7})/7, x > 0$
  b. $f(x) = 1/20, 5 < x < 25$
  c. $f(x) = \dfrac{e^{-.5[(x-10)/5]^2}}{5\sqrt{2\pi}}$

Learning the Mechanics

5.104 Assume that x is a random variable best described by a uniform distribution with $c = 40$ and $d = 70$.
  a. Find $f(x)$.
  b. Find the mean and standard deviation of x.
  c. Graph the probability distribution for x and locate its mean and the interval $\mu \pm 2\sigma$ on the graph.
  d. Find $P(x \leq 45)$.
  e. Find $P(x \geq 58)$.
  f. Find $P(x \leq 100)$.
  g. Find $P(\mu - \sigma \leq x \leq \mu + \sigma)$.
  h. Find $P(x > 60)$.

5.105 Find the following probabilities for the standard normal random variable z:
  a. $P(z \leq 2.1)$
  b. $P(z \geq 2.1)$
  c. $P(z \geq -1.65)$
  d. $P(-2.13 \leq z \leq -.41)$
  e. $P(-1.45 \leq z \leq 2.15)$
  f. $P(z \leq -1.43)$

5.106 Find a z-score, call it z_0, such that
  a. $P(z \leq z_0) = .8708$
  b. $P(z \geq z_0) = .0526$
  c. $P(z \leq z_0) = .5$
  d. $P(-z_0 \leq z \leq z_0) = .8164$
  e. $P(z \geq z_0) = .8023$
  f. $P(z \geq z_0) = .0041$

5.107 The random variable x has a normal distribution with $\mu = 70$ and $\sigma = 10$. Find the following probabilities:
  a. $P(x \leq 75)$
  b. $P(x \geq 90)$
  c. $P(60 \leq x \leq 75)$
  d. $P(x > 75)$
  e. $P(x = 75)$
  f. $P(x \leq 95)$

5.108 The random variable x has a normal distribution with $\mu = 40$ and $\sigma^2 = 36$. Find a value of x, call it x_0, such that
  a. $P(x \geq x_0) = .5$
  b. $P(x \leq x_0) = .9911$
  c. $P(x \leq x_0) = .0028$
  d. $P(x \geq x_0) = .0228$
  e. $P(x \leq x_0) = .1003$
  f. $P(x \geq x_0) = .7995$

*5.109 Assume that x is a binomial random variable with $n = 100$ and $p = .5$. Use the normal probability distribution to approximate the following probabilities:
  a. $P(x \leq 48)$
  b. $P(50 \leq x \leq 65)$
  c. $P(x \geq 70)$
  d. $P(55 \leq x \leq 58)$
  e. $P(x = 62)$
  f. $P(x \leq 49$ or $x \geq 72)$

*5.110 Assume that x has an exponential distribution with $\theta = 3$. Find
  a. $P(x \leq 1)$
  b. $P(x > 1)$
  c. $P(x = 1)$
  d. $P(x \leq 6)$
  e. $P(2 \leq x \leq 10)$

Applying the Concepts—Basic

5.111 **Passing the FCAT math test.** All Florida high schools require their students to demonstrate competence in mathematics by scoring 70% or above on the FCAT mathematics achievement test. The FCAT math scores of those students taking the test for the first time are normally distributed with a mean of 77% and a standard deviation of 7.3%. What percentage of students who take the test for the first time will pass the test?

5.112 **No-hitters in baseball.** In baseball, a "no-hitter" is a regulation 9-inning game in which the pitcher yields no hits to the opposing batters. *Chance* (Summer 1994) reported on a study of no-hitters in Major League Baseball (MLB). The initial analysis focused on the total number of hits yielded per game per team for all 9-inning MLB games played between 1989 and 1993. The distribution of hits/9-innings is approximately normal with mean 8.72 and standard deviation 1.10.
  a. What percentage of 9-inning MLB games result in fewer than 6 hits?
  b. Demonstrate, statistically, why a no-hitter is considered an extremely rare occurrence in MLB.

*5.113 Sickle-cell anemia. Eight percent of the African-American population is known to carry the trait for sickle-cell anemia. (Sickle Cell-Information Center, Atlanta, Ga.) If 1,000 African-Americans are sampled at random, what is the approximate probability that
a. More than 175 carry the trait?
b. Fewer than 140 carry the trait?

*5.114 Laser surgery complications. According to *Time* (Oct. 11, 1999), 1% of all patients who undergo laser surgery to correct their vision have serious postlaser vision problems. In a sample of 100,000 patients, what is the approximate probability that fewer than 950 will experience serious postlaser vision problems?

5.115 Lengths of sardines. *Fisheries Science* (Feb. 1995) published a study of the length distributions of sardines inhabiting Japanese waters. At two years of age, fish have a length distribution that is approximately normal with $\mu = 20.20$ centimeters (cm) and $\sigma = .65$ cm.
a. Find the probability that a two-year-old sardine inhabiting Japanese waters is between 20 and 21 cm long.
b. Find the probability that a two-year-old sardine in Japanese waters is less than 19.84 cm long.
c. Find the probability that a two-year-old sardine in Japanese waters is greater than 22.01 cm long.

5.116 Quality of river water. The Council on Environmental Quality (CEQ) considers a river with a dissolved oxygen content of less than 5 milligrams per liter (mg/L) of water to be undesirable because it is unlikely to support aquatic life. Suppose an industrial plant discharges its waste into a river and the downstream daily oxygen content measurements are normally distributed with a mean equal to 6.3 mg/L and a standard deviation of .6 mg/L.
a. What percentage of the days would the dissolved oxygen content in the river be considered undesirable by the CEQ?
b. Within what limits would we expect the dissolved oxygen content to fall?

5.117 Tropical island temperatures. Records indicate that the daily high January temperatures on a tropical island tend to have a uniform distribution over the interval from 75°F to 90°F. A tourist arrives on the island on a randomly selected day in January.
a. What is the probability that the temperature will be above 80°F?
b. What is the probability that the temperature will be between 80°F and 85°F?
c. What is the expected temperature?

*5.118 Hospital patient interarrival times. The length of time between arrivals at a hospital clinic has an approximately exponential probability distribution. Suppose the mean time between arrivals for patients at a clinic is 4 minutes.
a. What is the probability that a particular interarrival time (the time between the arrival of two patients) is less than 1 minute?

b. What is the probability that the next four interarrival times are all less than 1 minute?
c. What is the probability that an interarrival time will exceed 10 minutes?

Applying the Concepts—Intermediate

5.119 Waiting for an elevator. The manager of a large department store with three floors reports that the time a customer on the second floor must wait for an elevator has a uniform distribution ranging from 0 to 4 minutes. If it takes the elevator 15 seconds to go from floor to floor, find the probability that a hurried customer can reach the first floor in less than 1.5 minutes after pushing the second-floor elevator button.

5.120 Comparison of exam scores: red versus blue exam. Refer to the *Teaching Psychology* (May 1998) study of how external clues influence performance, Exercise 2.116 (p. 85). Recall that two different forms of a midterm psychology examination were given—one printed on blue paper and the other printed on red paper. Grading only the difficult questions, scores on the blue exam had a distribution with a mean of 53% and a standard deviation of 15%, while scores on the red exam had a distribution with a mean of 39% and a standard deviation of 12%. Assuming that both distributions are approximately normal, on which exam form is a student more likely to score below 20% on the difficult questions, the blue or the red exam? (Compare your answer to 2.116**c**.)

5.121 Galaxy velocity study. Recall *The Astronomical Journal* (July 1995) study of galaxy velocity, Exercise 5.58 (p. 267). The observed velocity of a galaxy located in galaxy cluster A2142 was found to have a normal distribution with mean 27,117 kilometers per second (km/s) and standard deviation 1,280 km/s. A galaxy with a velocity of 24,350 km/s is observed. Comment on the likelihood of this galaxy being located in cluster A2142.

5.122 Study of children with developmental delays. It is well known that children with developmental delays (i.e., mild mental retardation) are slower, cognitively, than normally developing children. Are their social skills also lacking? A study compared the social interactions of the two groups of children in a controlled playground environment. (*American Journal on Mental Retardation*, Jan. 1992.) One variable of interest was the number of intervals of "no play" by each child. Children with developmental delays had a mean of 2.73 intervals of "no play" and a standard deviation of 2.58 intervals. Based on this information, is it possible for the variable of interest to be normally distributed? Explain.

5.123 Chemical properties of a plant extract. The extract of a plant native to Taiwan (Republic of China) is being tested as a possible treatment for leukemia. The chemical properties of the extract were examined in *Planta Medica* (Apr. 1995). A particular compound produced from the plant extract was analyzed for the protein collagen. The collagen amount x for the compound was found to be normally distributed with $\mu = 76.8$ and $\sigma = 9.2$ grams per milliliter.

a. What percentage of compounds formed from the plant extract will have a collagen amount greater than 90 grams per milliliter?

b. Give an interval around the mean μ that contains the collagen amount for 80% of the compounds formed from the plant extract.

5.124 Mile run times. A physical-fitness association is including the mile run in its secondary-school fitness test for students. The time for this event for students in secondary school is approximately normally distributed with a mean of 450 seconds and a standard deviation of 40 seconds. If the association wants to designate the fastest 10% as "excellent," what time should the association set for this criterion?

5.125 Rubidium in metamorphic rock. A chemical analysis of metamorphic rock in western Turkey was reported in *Geological Magazine* (May 1995). The trace amount (in parts per million) of the element rubidium was measured for 20 rock specimens. The data are listed here. Assess whether the sample data come from a normal population.

⊙ **RUBIDIUM**

164	286	355	308	277	330	323	370	241	402
301	200	202	341	327	285	277	247	213	424

Source: Bozkurt, E. et al. "Geochemistry and the tectonic significance of augen gneisses from the southern Menderes Massif (West Turkey)." *Geological Magazine*, Vol. 132, No. 3, May 1995, p. 291 (Table 1).

5.126 Sedimentary deposits in reservoirs. Geologists have successfully used statistical models to evaluate the nature of sedimentary deposits (called *facies*) in reservoirs. One of the model's key parameters is the proportion P of facies bodies in a reservoir. An article in *Mathematical Geology* (Apr. 1995) demonstrated that the number of facies bodies that must be sampled to satisfactorily estimate P is approximately normally distributed with $\mu = 99$ and $\sigma = 4.3$. How many facies bodies are required to satisfactorily estimate P for 99% of the reservoirs evaluated?

5.127 Normal curve approximation. A. K. Shah published a simple approximation for areas under the normal curve in *The American Statistician* (Feb. 1985). Shah showed that the area A under the standard normal curve between 0 and z is

$$A \approx \begin{cases} z(4.4 - z)/10 & \text{for } 0 \le z \le 2.2 \\ .49 & \text{for } 2.2 < z < 2.6 \\ .50 & \text{for } z \ge 2.6 \end{cases}$$

a. Use the approximation to find
  i. $P(0 < z < 1.2)$
  ii. $P(0 < z < 2.5)$
  iii. $P(z > .8)$
  iv. $P(z < 1.0)$

b. Find the exact probabilities in part **a**.

c. Shah showed that the approximation has a maximum absolute error of .0052. Verify this for the approximations in part **a**.

*5.128 Spruce budworm infestation. An infestation of a certain species of caterpillar, the spruce budworm, can cause extensive damage to the timberlands of the northern United States. It is known that an outbreak of this type of infestation occurs, on the average, every 30 years. Assuming that this phenomenon obeys an exponential probability law, what is the probability that catastrophic outbreaks of spruce budworm infestation will occur within 6 years of each other?

5.129 Maximum time to take a test. A professor believes that if a class is allowed to work on an examination as long as desired, the times spent by the students would be approximately normal with mean 40 minutes and standard deviation 6 minutes. Approximately how long should be allotted for the examination if the professor wants almost all (say, 97.5%) of the class to finish?

*5.130 Modeling an airport taxi service. In an article published in the *European Journal of Operational Research* (Vol. 21, 1985), the vehicle-dispatching decisions of an airport-based taxi service were investigated. In modeling the system, the authors assumed travel times of successive taxi trips to and from the terminal to be independent exponential random variables. Assume $\theta = 20$ minutes.

a. What is the mean trip time for the taxi service?

b. What is the probability that a particular trip will take more than 30 minutes?

c. Two taxis have just been dispatched. What is the probability that both will be gone for more than 30 minutes? That at least one of the taxis will return within 30 minutes?

*5.131 Defective CD-ROMs. A manufacturer of CD-ROMs claims that 99.4% of its CDs are defect free. A large software company that buys and uses large numbers of the CDs wants to verify this claim, so it selects 1,600 CDs to be tested. The tests reveal 12 CDs to be defective. Assuming that the manufacturer's claim is correct, what is the probability of finding 12 or more defective CDs in a sample of 1,600? Does your answer cast doubt on the manufacturer's claim? Explain.

Applying the Concepts—Advanced

*5.132 Accidents at a plant. The number of serious accidents in a manufacturing plant has (approximately) a Poisson probability distribution with a mean of two serious accidents per month. If x, the number of events per unit time, has a Poisson distribution with mean λ, then it can be shown that the time between two successive events has an exponential probability distribution with mean $\theta = 1/\lambda$.

a. If an accident occurs today, what is the probability that the next serious accident will not occur within the next month?

b. What is the probability that more than one serious accident will occur within the next month?

5.133 Water retention of soil cores. A team of soil scientists investigated the water retention properties of soil cores sampled from an uncropped field consisting of silt loam (*Soil Science*, Jan. 1995). At a pressure of .1 megapascal (MPa), the water content of the soil (measured in cubic meters of water per cubic meter of soil) was determined to be approximately normally distributed with $\mu = .27$ and $\sigma = .04$. In addition to water content readings at a pressure of .1 MPa, measurements were obtained at pressures 0, .005, .01, .03, and 1.5 MPa. Consider a soil core with a water content reading of .14. Is it likely that this reading was obtained at a pressure of .1 MPa? Explain.

Critical Thinking Challenges

*__5.134__ **Weights of corn chip bags.** The net weight per bag of a certain brand of corn chips is listed as 10 ounces. The weight of chips actually dispensed in each bag by an automated machine, when operating to specifications, is a normal random variable with mean 10.5 ounces and standard deviation .25 ounce. A quality control inspector will randomly select 1,500 bags from the automated process and measure precisely the weight of each. Ideally, all of the bags should contain at least 10 ounces of corn chips. However, if at least 97% of the bags contain 10 ounces or more, the process is considered "under control" and no adjustments to the machine are deemed necessary. Assess the likelihood that the quality control inspector will find fewer than 97% of the bags with 10 ounces or more even if the process is "under control."

5.135 IQs and "The Bell Curve". In their controversial book *The Bell Curve* (Free Press, 1994), Professors Richard J. Herrnstein (a Harvard psychologist who died while the book was in production) and Charles Murray (a political scientist at MIT) explore, as the subtitle states, "intelligence and class structure in American life." *The Bell Curve* heavily employs statistical analyses in an attempt to support the authors' positions. Since the book's publication, many expert statisticians have raised doubts about the authors' statistical methods and the inferences drawn from them. (See, for example, "Wringing *The Bell Curve*: A cautionary tale about the relationships among race, genes, and IQ," *Chance*, Summer 1995.) One of the many controversies sparked by the book is the authors' tenet that level of intelligence (or lack thereof) is a cause of a wide range of intractable social problems, including constrained economic mobility. The measure of intelligence chosen by the authors is the well known Intelligent Quotient (IQ). Numerous tests have been developed to measure IQ; Herrnstein and Murray use the Armed Forces Qualification Test (AFQT), originally designed to measure the cognitive ability of military recruits. Psychologists traditionally treat IQ as a random variable having a normal distribution with mean $\mu = 100$ and standard deviation $\sigma = 15$.

In their book, Herrnstein and Murray refer to five cognitive classes of people defined by percentiles of the normal distribution. Class I ("very bright") consists of those with IQs above the 95th percentile; Class II ("bright") are those with IQs between the 75th and 95th percentiles; Class III ("normal") includes IQs between the 25th and 75th percentiles; Class IV ("dull") are those with IQs between the 5th and 25th percentiles; and Class V ("very dull") are IQs below the 5th percentile.

a. Assuming that the distribution of IQ is accurately represented by the normal curve, determine the proportion of people with IQs in each of the five cognitive classes defined by Herrnstein and Murray.

b. Although the cognitive classes above are defined in terms of percentiles, the authors stress that IQ scores should be compared with z-scores, not percentiles. In other words, it is more informative to give the difference in z-scores for two IQ scores than it is to give the difference in percentiles. Do you agree?

c. Researchers have found that scores on many intelligence tests are decidedly nonnormal. Some distributions are skewed toward higher scores, others toward lower scores. How would the proportions in the five cognitive classes (part **a**) differ for an IQ distribution that is skewed right? Skewed left?

Student Projects

For large values of n the computational effort involved in working with the binomial probability distribution is considerable. Fortunately, in many instances the normal distribution provides a good approximation to the binomial distribution. This project was designed to enable you to demonstrate to yourself how well the normal distribution approximates the binomial distribution.

a. Let the random variable x have a binomial probability with $n = 10$ and $p = .5$. Using the binomial distribution, find the probability that x takes on a value in each of the following intervals: $\mu \pm \sigma$, $\mu \pm 2\sigma$, and $\mu \pm 3\sigma$.

b. Find the probabilities requested in part **a** by using a normal approximation to the given binomial distribution.

c. Determine the magnitude of the difference between each of the three probabilities as determined by the binomial distribution and by the normal approximation.

d. Letting x have a binomial distribution with $n = 20$ and $p = .5$, repeat parts **a**, **b**, and **c**. Notice that the probability estimates provided by the normal distribution are more accurate for $n = 20$ than for $n = 10$.

e. Letting x have a binomial distribution with $n = 20$ and $p = .01$, repeat parts **a**, **b**, and **c**. Notice that the probability estimates provided by the normal distribution are very poor in this case. Explain why this occurs.

REFERENCES

Hogg, R. V., and Craig, A. T. *Introduction to Mathematical Statistics*, 5th ed. Upper Saddle River, N.J.: Prentice Hall, 1995.

Lindgren, B. W. *Statistical Theory*, 3rd ed. New York: Macmillan, 1976.

Mood, A. M., Graybill, F. A., and Boes, D. C. *Introduction to the Theory of Statistics*, 3rd ed. New York: McGraw-Hill, 1974.

Ramsey, P. P., and Ramsey, P. H. "Simple tests of normality in small samples." *Journal of Quality Technology*, Vol. 22, 1990.

Ross, S. M. *Stochastic Processes*, 2nd ed. New York: Wiley, 1996.

Wackerly, D., Mendenhall, W., and Scheaffer, R. L. *Mathematical Statistics with Applications*, 6th ed. North Scituate, Mass.: Duxbury, 2002.

Winkler, R. L., and Hays, W. *Statistics: Probability, Inference, and Decision*, 2nd ed. New York: Holt, Rinehart and Winston, 1975.

Using Technology

Cumulative Probabilities for Continuous Random Variables and Normal Probability Plots Using MINITAB

To obtain cumulative probabilities for the continuous random variables discussed in this chapter using MINITAB, click on the "Calc" button on the MINITAB menu bar, next click on "Probability Distributions," then finally click on the distribution of your choice (e.g., "Normal"), as shown in Figure 5.M.1. The resulting dialog box appears as shown in Figure 5.M.2. Select "Cumulative probability," specify the parameters of the distribution (e.g., the mean μ and standard deviation σ), and enter the value of x in the "Input constant" box. When you click "OK," the cumulative normal probability for the value of x will appear on the MINITAB session window.

Figure 5.M.1
MINITAB Menu Options
for Obtaining Probabilities

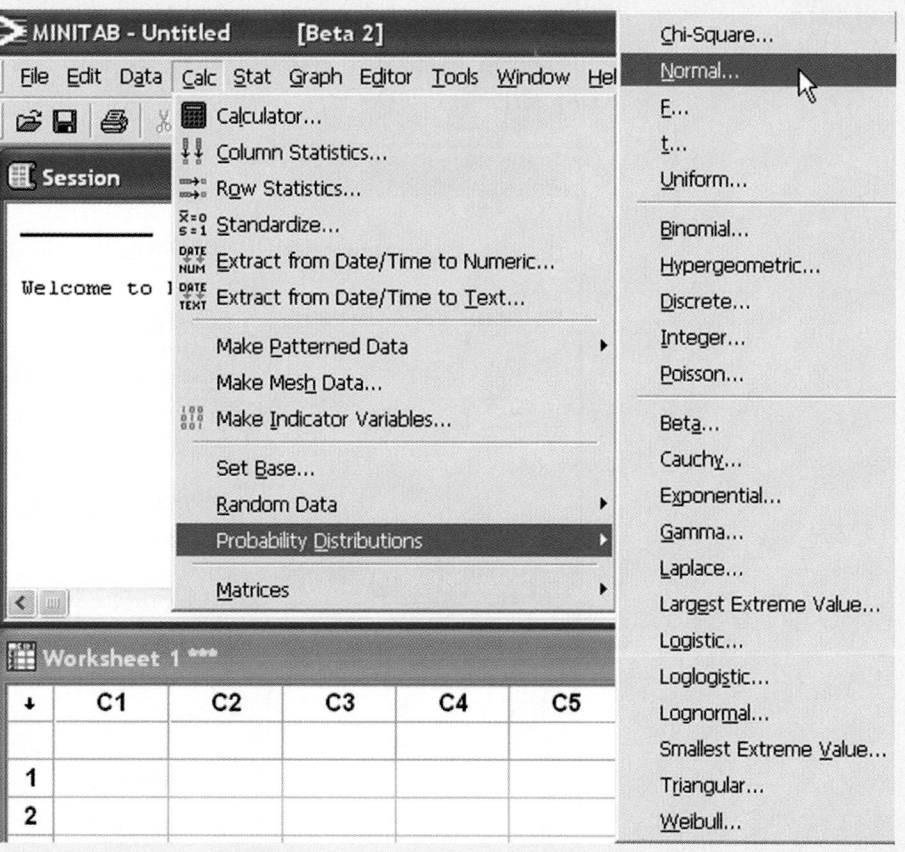

Figure 5.M.2
MINITAB Normal
Distribution Dialog Box

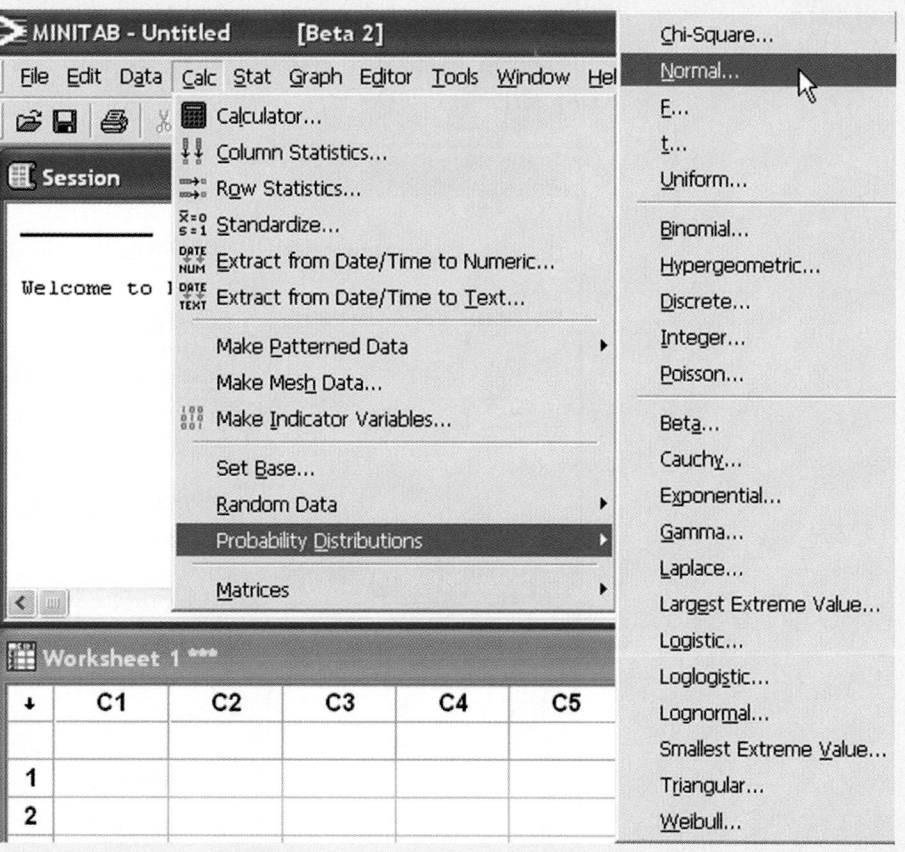

Figure 5.M.3
MINITAB Options for a
Normal Probability Plot

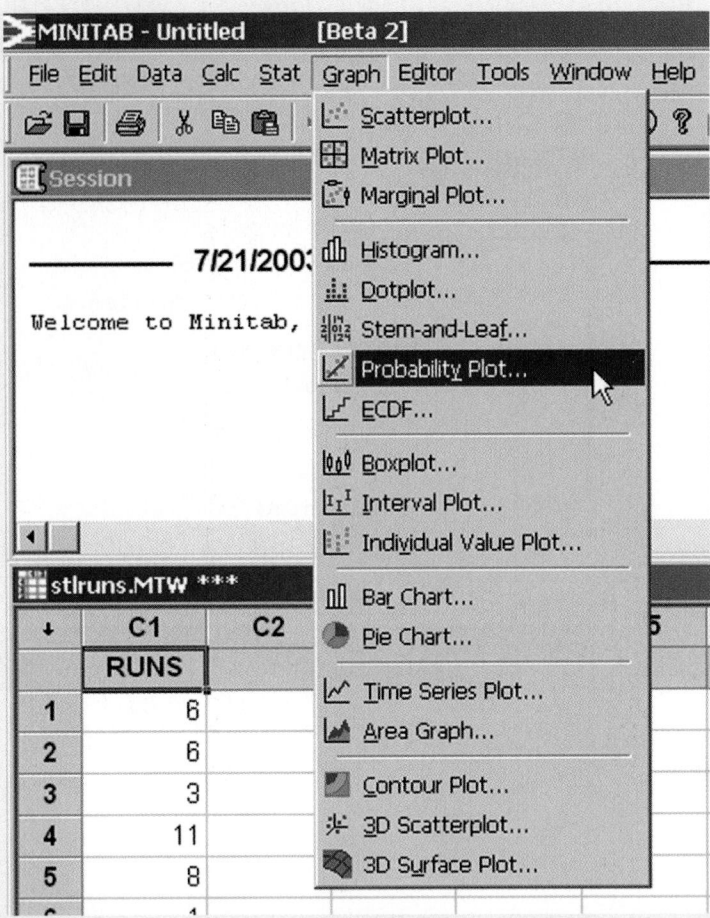

[*Note:* For the MINITAB exponential option, you will need to enter the value of the mean λ in the resulting dialog box. For the uniform option, you will need to enter the minimum and maximum values of the distribution.]

To obtain a normal probability plot using MINITAB, click on the "Graph" button on the MINITAB menu bar. Then click on "Probability Plot", as shown in Figure 5.M.3. Select "Single" (for one variable) on the next box, and the dialog box will appear as shown in Figure 5.M.4. Specify the variable of interest in the "Graph variables" box. Then click the "Distribution" button and select the "Normal" option. Click "OK" to return to the Probability Plot dialog box, and then click "OK" to generate the normal probability plot.

Figure 5.M.4
MINITAB Probability
Plot Dialog Box

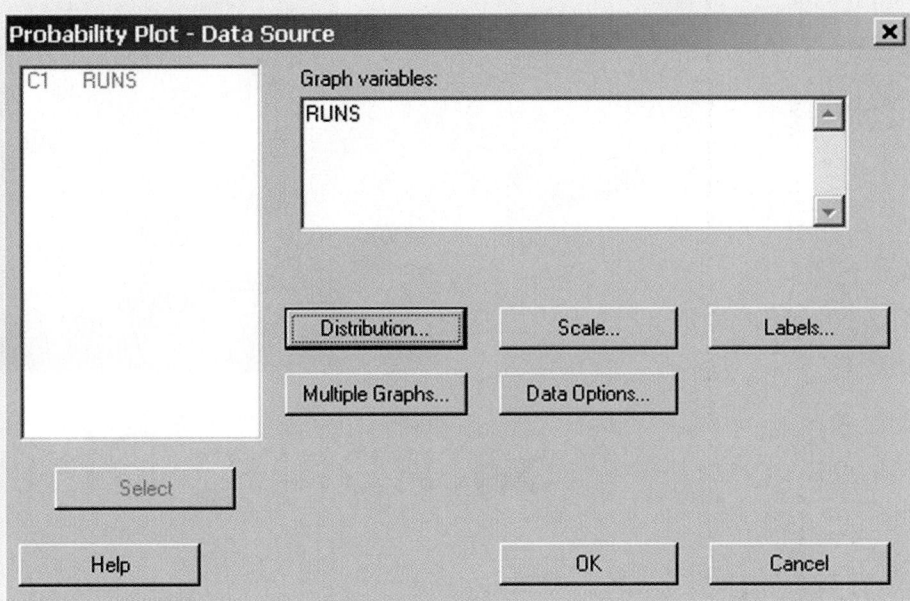

6

Sampling Distributions

Contents

Statistics in Action

The Insomnia Pill

Using Technology

Simulating a Sampling Distribution Using MINITAB

Where We've Been

- Learned that the objective of most statistical investigations is inference—that is, making decisions about a population based on a sample
- Discovered that sample statistics such as the sample mean and sample variance can be used to make the decisions
- Found that probability distributions for random variables are used to construct theoretical models of populations

Where We're Going

- Establish that a sample statistic is a random variable with a probability distribution.
- Define a *sampling distribution* as the probability distribution of a sample statistic.
- Learn that the sampling distribution of the sample mean tends to be approximately normal.

Statistics in *ACTION*

The Insomnia Pill

A research report published in the *Proceedings of the National Academy of Sciences* (March 1994) brought encouraging news to insomniacs and international business travelers who suffer from jet lag. Neuroscientists at the Massachusetts Institute of Technology (MIT) have been experimenting with melatonin—a hormone secreted by the pineal gland in the brain—as a sleep-inducing hormone. Since the hormone is naturally produced, it is nonaddictive. The researchers believe melatonin may be effective in treating jet lag—the body's response to rapid travel across many time zones so that a daylight-darkness change disrupts sleep patterns.

In the MIT study, young male volunteers were given various doses of melatonin or a placebo (a dummy medication containing no melatonin). Then they were placed in a dark room at midday and told to close their eyes for 30 minutes. The variable of interest was the time (in minutes) elapsed before each volunteer fell asleep.

According to the lead investigator, Professor Richard Wurtman, "Our volunteers fall asleep in five or six minutes on melatonin, while those on placebo take about 15 minutes." Wurtman warns, however, that uncontrolled doses of melatonin could cause mood-altering side effects. (Melatonin is sold in some health food stores. However, sales are

unregulated, and the purity and strength of the hormone are often uncertain.)

Now, consider a random sample of 40 young males, each of whom is given a dosage of the sleep-inducing hormone melatonin. The times (in minutes) to fall asleep for these 40 males are listed in Table SIA6.1 and saved in the **INSOMNIA** file. The researchers know that with the placebo (i.e., no hormone), the mean time to fall asleep is $\mu = 15$ minutes and the standard deviation is $\sigma = 10$ minutes. They want to use the data to make an inference about the true value of μ for those taking the melatonin. Specifically, the researchers want to know whether melatonin is an effective drug against insomnia.

In this chapter, a Statistics in Action Revisited example demonstrates how we can use one of the topics discussed in this chapter—the Central Limit Theorem—to make an inference about the effectiveness of melatonin as a sleep-inducing hormone.

Statistics in the Action Revisited

- Making an Inference about the Mean Sleep Time for Insomnia Pill Takers (p. 309)

☉ INSOMNIA

TABLE SIA6.1   Times (in Minutes) for 40 Male Volunteers to Fall Asleep

7.6	2.1	1.4	1.5	3.6	17.0	14.9	4.4	4.7	20.1
7.7	2.4	8.1	1.5	10.3	1.3	3.7	2.5	3.4	10.7
2.9	1.5	15.9	3.0	1.9	8.5	6.1	4.5	2.2	2.6
7.0	6.4	2.8	2.8	22.8	1.5	4.6	2.0	6.3	3.2

[Note: These data are simulated sleep times based on summary information provided in the MIT study.]

In Chapters 4 and 5 we assumed that we knew the probability distribution of a random variable, and using this knowledge we were able to compute the mean, variance, and probabilities associated with the random variable. However, in most practical applications, this information is not available. To illustrate, in Example 4.12 (p. 212) we calculated the probability that the binomial random variable x, the number of 20 polled voters who favor a certain mayoral candidate, assumed specific values. To do this, it was necessary to assume some value for p, the proportion of all voters who favor the candidate. Thus, for the purposes of illustration, we assumed $p = .6$ when, in all likelihood, the exact value of p would be unknown. In fact, the probable purpose of taking the poll is to estimate p. Similarly, when we modeled the in-city gas mileage of a certain automobile model, we used the normal probability distribution with an *assumed* mean and standard deviation of 27 and 3 miles per gallon, respectively. In most situations, the true mean and standard deviation are unknown quantities that would have to be estimated. Numerical quantities that describe probability distributions are called *parameters*. Thus, p, the probability of a success in a binomial experiment, and μ and σ, the mean and standard deviation of a normal distribution, are examples of parameters.

DEFINITION 6.1

A **parameter** is a numerical descriptive measure of a population. Because it is based on the observations in the population, its value is almost always unknown.

We have also discussed the sample mean $\bar{x}$, sample variance s^2, sample standard deviation s, etc., which are numerical descriptive measures calculated from the sample. (See Table 6.1 for a list of the statistics covered so far in this text.) We will often use the information contained in these *sample statistics* to make inferences about the parameters of a population.

DEFINITION 6.2

A **sample statistic** is a numerical descriptive measure of a sample. It is calculated from the observations in the sample.

TABLE 6.1 List of Population Parameters and Corresponding Sample Statistics

	Population Parameter	Sample Statistic
Mean:	μ	$\bar{x}$
Variance:	σ^2	s^2
Standard deviation:	σ	s
Binomial proportion:	p	$\hat{p}$

Note that the term *statistic* refers to a *sample* quantity and the term *parameter* refers to a *population* quantity.

Before we can show you how to use sample statistics to make inferences about population parameters, we need to be able to evaluate their properties. Does one sample statistic contain more information than another about a population parameter? On what basis should we choose the "best" statistic for making inferences about a parameter? The purpose of this chapter is to answer these questions.

6.1 The Concept of a Sampling Distribution

If we want to estimate a parameter of a population—say, the population mean μ—we could use a number of sample statistics for our estimate. Two possibilities are the sample mean $\bar{x}$ and the sample median M. Which of these do you think will provide a better estimate of μ?

Before answering this question, consider the following example: Toss a fair die, and let x equal the number of dots showing on the up face. Suppose the die is tossed three times, producing the sample measurements 2, 2, 6. The sample mean is $\bar{x} = 3.33$ and the sample median is $M = 2$. Since the population mean of x is $\mu = 3.5$, you can see that for this sample of three measurements, the sample mean $\bar{x}$ provides an estimate that falls closer to μ than does the sample median (see Figure 6.1a). Now suppose we toss the die three more times and obtain the sample measurements 3, 4, 6.

Figure 6.1

Comparing the Sample Mean ($\bar{x}$) and Sample Median (M) as Estimators of the Population Mean (μ)

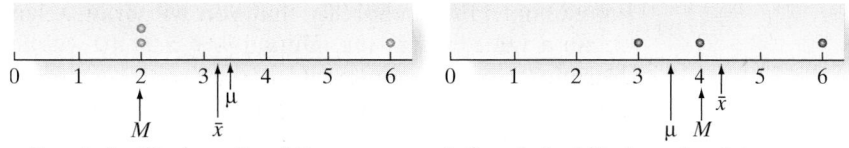

a. Sample 1: $\bar{x}$ is closer than M to μ b. Sample 2: M is closer than $\bar{x}$ to μ

The mean and median of this sample are $\bar{x} = 4.33$ and $M = 4$, respectively. This time M is closer to μ (see Figure 6.1b).

This simple example illustrates an important point: Neither the sample mean nor the sample median will *always* fall closer to the population mean. Consequently, we cannot compare these two sample statistics, or, in general, any two sample statistics, on the basis of their performance for a single sample. Instead, we need to recognize that sample statistics are themselves random variables, because different samples can lead to different values for the sample statistics. As random variables, sample statistics must be judged and compared on the basis of their probability distributions (i.e., the *collection* of values and associated probabilities of each statistic that would be obtained if the sampling experiment were repeated a *very large number of times*). We will illustrate this concept with another example.

Suppose it is known that in a certain part of Canada the daily high temperature recorded for all past months of January has a mean of $\mu = 10°F$ and a standard deviation of $\sigma = 5°F$. Consider an experiment consisting of randomly selecting 25 daily high temperatures from the records of past months of January and calculating the sample mean $\bar{x}$. If this experiment were repeated a very large number of times, the value of $\bar{x}$ would vary from sample to sample. For example, the first sample of 25 temperature measurements might have a mean $\bar{x} = 9.8$, the second sample a mean $\bar{x} = 11.4$, the third sample a mean $\bar{x} = 10.5$, etc. If the sampling experiment were repeated a very large number of times, the resulting histogram of sample means would be approximately the probability distribution of $\bar{x}$. If $\bar{x}$ is a good estimator of μ, we would expect the values of $\bar{x}$ to cluster around μ as shown in Figure 6.2. This probability distribution is called a *sampling distribution* because it is generated by repeating a sampling experiment a very large number of times.

> **DEFINITION 6.3**
>
> The **sampling distribution** of a sample statistic calculated from a sample of n measurements is the probability distribution of the statistic.

In actual practice, the sampling distribution of a statistic is obtained mathematically or (at least approximately) by simulating the sample on a computer using a procedure similar to that just described.

If $\bar{x}$ has been calculated from a sample of $n = 25$ measurements selected from a population with mean $\mu = 10$ and standard deviation $\sigma = 5$, the sampling distribution (Figure 6.2) provides information about the behavior of $\bar{x}$ in repeated sampling.

Figure 6.2

Sampling Distribution for $\bar{x}$ Based on a Sample of $n = 25$ Measurements

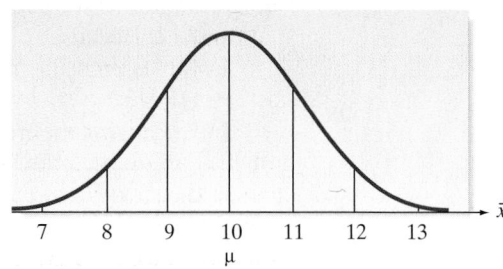

For example, the probability that you will draw a sample of 25 measurements and obtain a value of $\bar{x}$ in the interval $9 \leq \bar{x} \leq 10$ will be the area under the sampling distribution over that interval.

Since the properties of a statistic are typified by its sampling distribution, it follows that to compare two sample statistics you compare their sampling distributions. For example, if you have two statistics, A and B, for estimating the same parameter (for purposes of illustration, suppose the parameter is the population variance σ^2) and if their sampling distributions are as shown in Figure 6.3, you would choose statistic A in preference to statistic B. You would make this choice because the sampling distribution for statistic A centers over σ^2 and has less spread (variation) than the sampling distribution for statistic B. When you draw a single sample in a practical sampling situation, the probability is higher that statistic A will fall nearer σ^2.

Figure 6.3

Two Sampling Distributions for Estimating the Population Variance, σ^2

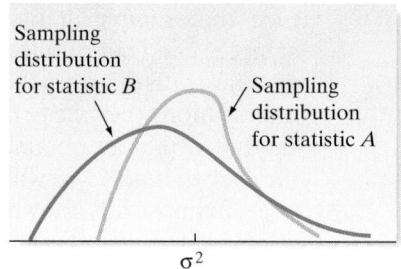

Remember that in practice we will not know the numerical value of the unknown parameter σ^2 so we will not know whether statistic A or statistic B is closer to σ^2 for a sample. We have to rely on our knowledge of the theoretical sampling distributions to choose the best sample statistic and then use it sample after sample. The procedure for finding the sampling distribution for a statistic is demonstrated in Example 6.1.

EXAMPLE 6.1

FINDING A SAMPLING DISTRIBUTION

Problem   Consider a game played with a standard 52-card bridge deck in which you can score 0, 3, or 12 points on any one hand. Suppose the population of points scored per hand is described by the probability distribution shown here. A random sample of $n = 3$ hands is selected from the population.

x	0	3	12
$p(x)$	$\frac{1}{3}$	$\frac{1}{3}$	$\frac{1}{3}$

a.   Find the sampling distribution of the sample mean $\bar{x}$.

b.   Find the sampling distribution of the sample median M.

Solution   Points for every possible sample of $n = 3$ hands is listed in Table 6.2 along with the sample mean and median. Also, because any one sample is as likely to be selected as any other (random sampling), the probability of observing any particular sample is $\frac{1}{27}$. The probability is also listed in Table 6.2.

a.   From Table 6.2 you can see that $\bar{x}$ can assume the values 0, 1, 2, 3, 4, 5, 6, 8, 9, and 12. Because $\bar{x} = 0$ occurs in only one sample, $P(\bar{x} = 0) = \frac{1}{27}$. Similarly, $\bar{x} = 1$ occurs in three samples: $(0, 0, 3)$, $(0, 3, 0)$, and $(3, 0, 0)$. Therefore, $P(\bar{x} = 1) = \frac{3}{27} = \frac{1}{9}$. Calculating the probabilities of the remaining values of $\bar{x}$ and arranging them in a table, we obtain the probability distribution shown below.

$\bar{x}$	0	1	2	3	4	5	6	8	9	12
$p(\bar{x})$	$\frac{1}{27}$	$\frac{3}{27}$	$\frac{3}{27}$	$\frac{1}{27}$	$\frac{3}{27}$	$\frac{6}{27}$	$\frac{3}{27}$	$\frac{3}{27}$	$\frac{3}{27}$	$\frac{1}{27}$

TABLE 6.2   Outcomes for Three Hands of a Card Game

Possible Samples	$\bar{x}$	M	Probability
0, 0, 0	0	0	$^1/_{27}$
0, 0, 3	1	0	$^1/_{27}$
0, 0, 12	4	0	$^1/_{27}$
0, 3, 0	1	0	$^1/_{27}$
0, 3, 3	2	3	$^1/_{27}$
0, 3, 12	5	3	$^1/_{27}$
0, 12, 0	4	0	$^1/_{27}$
0, 12, 3	5	3	$^1/_{27}$
0, 12, 12	8	12	$^1/_{27}$
3, 0, 0	1	0	$^1/_{27}$
3, 0, 3	2	3	$^1/_{27}$
3, 0, 12	5	3	$^1/_{27}$
3, 3, 0	2	3	$^1/_{27}$
3, 3, 3	3	3	$^1/_{27}$
3, 3, 12	6	3	$^1/_{27}$
3, 12, 0	5	3	$^1/_{27}$
3, 12, 3	6	3	$^1/_{27}$
3, 12, 12	9	12	$^1/_{27}$
12, 0, 0	4	0	$^1/_{27}$
12, 0, 3	5	3	$^1/_{27}$
12, 0, 12	8	12	$^1/_{27}$
12, 3, 0	5	3	$^1/_{27}$
12, 3, 3	6	3	$^1/_{27}$
12, 3, 12	9	12	$^1/_{27}$
12, 12, 0	8	12	$^1/_{27}$
12, 12, 3	9	12	$^1/_{27}$
12, 12, 12	12	12	$^1/_{27}$

This is the sampling distribution for $\bar{x}$ because it specifies the probability associated with each possible value of $\bar{x}$.

b. In Table 6.2 you can see that the median M can assume one of the three values: 0, 3, or 12. The value $M = 0$ occurs in seven different samples. Therefore, $P(M = 0) = ^7/_{27}$. Similarly, $M = 3$ occurs in 13 samples and $M = 12$ occurs in seven samples. Therefore, the probability distribution (i.e., the sampling distribution) for the median M is as shown below.

M	0	3	12
$p(M)$	$^7/_{27}$	$^{13}/_{27}$	$^7/_{27}$

Look Back The sampling distributions of parts **a** and **b** are found by first listing all the distinct values of the statistic and then calculating the probability of each value.

Now Work   *Exercise 6.3*

■ ■ ■

Example 6.1 demonstrates the procedure for finding the exact sampling distribution of a statistic when the number of different samples that could be selected from the population is relatively small. In the real world, populations often consist of a large number of different values, making samples difficult (or impossible) to enumerate. When this situation occurs, we may choose to obtain the approximate sampling distribution for a statistic by simulating the sampling over and over again and recording the proportion of times different values of the statistic occur. Example 6.2 illustrates this procedure.

EXAMPLE 6.2   SIMULATING A SAMPLING DISTRIBUTION

Problem   Suppose we perform the following experiment over and over again: Take a sample of 11 measurements from the uniform distribution shown in Figure 6.4. Calculate the two sample statistics

$$\bar{x} = \text{Sample mean} = \frac{\Sigma x}{11}$$

M = Median = Sixth sample measurement when the 11 measurements are arranged in ascending order

Obtain approximations to the sampling distributions of $\bar{x}$ and M.

Solution   We used MINITAB to generate 1,000 samples, each with $n = 11$ observations. Then we compute $\bar{x}$ and M for each sample. Our goal is to obtain approximations to the sampling distributions of $\bar{x}$ and M to find out which sample statistic ($\bar{x}$ or M) contains more information about μ. (Note that, in this particular example, we *know* the population mean for this uniform distribution is $\mu = .5$.) The first 10 of the 1,000 samples generated are presented in Table 6.3. For instance, the first computer-generated sample from the uniform distribution (arranged in ascending order) contained the following measurements: .045, .499, .860, .182, .644, .362, .502, .132, .691, .695, and .177. The sample mean $\bar{x}$ and median M computed for this sample are

$$\bar{x} = \frac{.045 + .499 + \cdots + .177}{11} = .435$$

$$M = \text{Sixth ordered measurement} = .499$$

The MINITAB relative frequency histograms for $\bar{x}$ and M for the 1,000 samples of size $n = 11$ are shown in Figure 6.5. These histograms represent approximations to the true sampling distributions of $\bar{x}$ and M.

Look Back   You can see that the values of $\bar{x}$ tend to cluster around μ to a greater extent than do the values of M. Thus, on the basis of the observed sampling distributions, we conclude that $\bar{x}$ contains more information about μ than M does—at least for samples of $n = 11$ measurements from the uniform distribution.

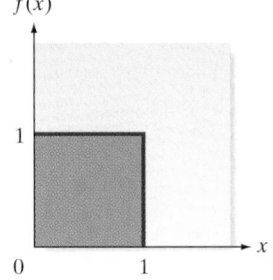

$f(x)$

Figure 6.4
Uniform Distribution from 0 to 1

TABLE 6.3   First 10 Samples of $n = 11$ Measurements from a Uniform Distribution

Sample	Measurements										
1	.045	.499	.860	.182	.644	.362	.502	.132	.691	.695	.177
2	.529	.918	.357	.156	.357	.965	.935	.515	.675	.662	.755
3	.247	.745	.085	.100	.034	.277	.002	.874	.788	.272	.156
4	.052	.454	.055	.877	.999	.467	.903	.683	.003	.845	.469
5	.082	.744	.614	.364	.882	.878	.527	.193	.186	.928	.175
6	.069	.331	.009	.083	.093	.329	.394	.732	.454	.220	.485
7	.056	.265	.367	.421	.679	.076	.548	.285	.912	.679	.862
8	.856	.528	.350	.340	.583	.785	.412	.581	.342	.022	.595
9	.689	.174	.432	.999	.186	.223	.950	.541	.053	.366	.702
10	.544	.074	.493	.326	.521	.258	.056	.087	.380	.931	.035

Figure 6.5
MINITAB Histograms for $\bar{x}$ and M, Example 6.2

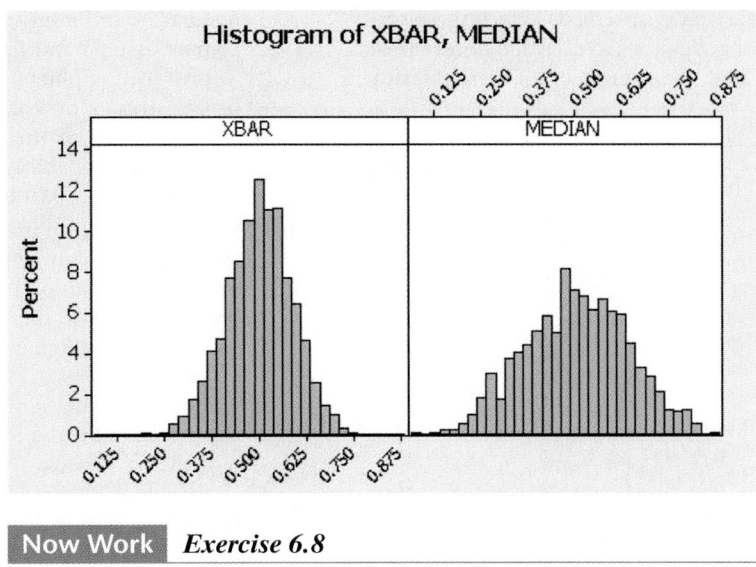

| Now Work | *Exercise 6.8* |

■ ■ ■

As noted earlier, many sampling distributions can be derived mathematically, but the theory necessary to do this is beyond the scope of this text. Consequently, when we need to know the properties of a statistic, we will present its sampling distribution and simply describe its properties. Several of the important properties we look for in sampling distributions are discussed in the next section.

Exercises 6.1–6.9

Understanding the Principles

6.1 What is the difference between a population parameter and a sample statistic?

6.2 What is a sampling distribution of a sample statistic?

Learning the Mechanics

6.3 The probability distribution shown here describes a
NW population of measurements that can assume values of 0, 2, 4, and 6, each of which occurs with the same relative frequency:

x	0	2	4	6
$p(x)$	$\frac{1}{4}$	$\frac{1}{4}$	$\frac{1}{4}$	$\frac{1}{4}$

a. List all the different samples of $n = 2$ measurements that can be selected from this population.
b. Calculate the mean of each different sample listed in part **a**.
c. If a sample of $n = 2$ measurements is randomly selected from the population, what is the probability that a specific sample will be selected?
d. Assume that a random sample of $n = 2$ measurements is selected from the population. List the different values of $\bar{x}$ found in part **b**, and find the probability of each. Then give the sampling distribution of the sample mean $\bar{x}$ in tabular form.

e. Construct a probability histogram for the sampling distribution of $\bar{x}$.

6.4 Simulate sampling from the population described in Exercise 6.3 by marking the values of x, one on each of four identical coins (or poker chips, etc.). Place the coins (marked 0, 2, 4, and 6) into a bag, randomly select one, and observe its value. Replace this coin, draw a second coin, and observe its value. Finally, calculate the mean $\bar{x}$ for this sample of $n = 2$ observations randomly selected from the population (Exercise 6.3, part **b**). Replace the coins, mix, and using the same procedure, select a sample of $n = 2$ observations from the population. Record the numbers and calculate $\bar{x}$ for this sample. Repeat this sampling process until you acquire 100 values of $\bar{x}$. Construct a relative frequency distribution for these 100 sample means. Compare this distribution to the exact sampling distribution of $\bar{x}$ found in part **e** of Exercise 6.3. [*Note:* The distribution obtained in this exercise is an approximation to the exact sampling distribution. But, if you were to repeat the sampling procedure, drawing two coins not 100 times but 10,000 times, the relative frequency distribution for the 10,000 sample means would be almost identical to the sampling distribution of $\bar{x}$ found in Exercise 6.3, part **e**.]

6.5 Consider the population described by the probability distribution shown here.

x	1	2	3	4	5
$p(x)$	.2	.3	.2	.2	.1

The random variable x is observed twice. If these observations are independent, verify that the different samples of size 2 and their probabilities are as shown below.

Sample	Probability	Sample	Probability
1, 1	.04	3, 4	.04
1, 2	.06	3, 5	.02
1, 3	.04	4, 1	.04
1, 4	.04	4, 2	.06
1, 5	.02	4, 3	.04
2, 1	.06	4, 4	.04
2, 2	.09	4, 5	.02
2, 3	.06	5, 1	.02
2, 4	.06	5, 2	.03
2, 5	.03	5, 3	.02
3, 1	.04	5, 4	.02
3, 2	.06	5, 5	.01
3, 3	.04		

a. Find the sampling distribution of the sample mean $\bar{x}$.
b. Construct a probability histogram for the sampling distribution of $\bar{x}$.
c. What is the probability that $\bar{x}$ is 4.5 or larger?
d. Would you expect to observe a value of $\bar{x}$ equal to 4.5 or larger? Explain.

6.6 Refer to Exercise 6.5 and find $E(x) = \mu$. Then use the sampling distribution of $\bar{x}$ found in Exercise 6.5 to find the expected value of $\bar{x}$. Note that $E(\bar{x}) = \mu$.

6.7 Refer to Exercise 6.5. Assume that a random sample of $n = 2$ measurements is randomly selected from the population.

a. List the different values that the sample median M may assume and find the probability of each. Then give the sampling distribution of the sample median.
b. Construct a probability histogram for the sampling distribution of the sample median and compare it with the probability histogram for the sample mean (Exercise 6.5, part **b**).

6.8 In Example 6.2 we use the computer to generate 1,000 samples, each containing $n = 11$ observations, from a uniform distribution over the interval from 0 to 1. For this exercise, use the computer to generate 500 samples, each containing $n = 15$ observations, from this population.

a. Calculate the sample mean for each sample. To approximate the sampling distribution of $\bar{x}$, construct a relative frequency histogram for the 500 values of $\bar{x}$.
b. Repeat part **a** for the sample median. Compare this approximate sampling distribution with the approximate sampling distribution of $\bar{x}$ found in part **a**.

6.9 Consider a population that contains values of x equal to 00, 01, 02, 03, ..., 96, 97, 98, 99. Assume that these values of x occur with equal probability. Use the computer to generate 500 samples, each containing $n = 25$ measurements, from this population. Calculate the sample mean $\bar{x}$ and sample variance s^2 for each of the 500 samples.

a. To approximate the sampling distribution of $\bar{x}$, construct a relative frequency histogram for the 500 values of $\bar{x}$.
b. Repeat part **a** for the 500 values of s^2.

6.2 Properties of Sampling Distributions: Unbiasedness and Minimum Variance

The simplest type of statistic used to make inferences about a population parameter is a *point estimator*. A point estimator is a rule or formula that tells us how to use the sample data to calculate a single number that is intended to estimate the value of some population parameter. For example, the sample mean $\bar{x}$ is a point estimator of the population mean μ. Similarly, the sample variance s^2 is a point estimator of the population variance σ^2.

> **DEFINITION 6.4**
>
> A **point estimator** of a population parameter is a rule or formula that tells us how to use the sample data to calculate a single number that can be used as an *estimate* of the population parameter.

Often, many different point estimators can be found to estimate the same parameter. Each will have a sampling distribution that provides information about the point estimator. By examining the sampling distribution, we can determine how large the difference between an estimate and the true value of the parameter (called the **error of estimation**) is likely to be. We can also tell whether an estimator is more likely to overestimate or to underestimate a parameter.

EXAMPLE 6.3

COMPARING TWO STATISTICS

Problem Suppose two statistics, A and B, exist to estimate the same population parameter, θ (theta). (Note that θ could be any parameter, μ, σ^2, σ, etc.) Suppose the two statistics have sampling distributions as shown in Figure 6.6. Based on these sampling distributions, which statistic is more attractive as an estimator of θ?

Figure 6.6

Sampling Distributions of Unbiased and Biased Estimators

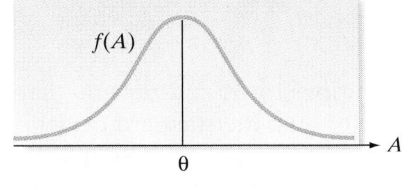

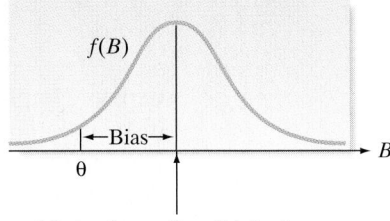

a. Unbiased sample statistic for the parameter θ

b. Biased sample statistic for the parameter θ

Solution As a first consideration, we would like the sampling distribution to center over the value of the parameter we wish to estimate. One way to characterize this property is in terms of the mean of the sampling distribution. Consequently, we say that a statistic is *unbiased* if the mean of the sampling distribution is equal to the parameter it is intended to estimate. This situation is shown in Figure 6.6a, where the mean μ_A of statistic A is equal to θ. If the mean of a sampling distribution is not equal to the parameter it is intended to estimate, the statistic is said to be *biased*. The sampling distribution for a biased statistic is shown in Figure 6.6b. The mean μ_B of the sampling distribution for statistic B is not equal to θ; in fact, it is shifted to the right of θ.

Look Back You can see that biased statistics tend either to overestimate or to underestimate a parameter. Consequently, when other properties of statistics tend to be equivalent, we will choose an unbiased statistic to estimate a parameter of interest.*

| Now Work | *Exercise 6.14* |

■ ■ ■

DEFINITION 6.5

If the sampling distribution of a sample statistic has a mean equal to the population parameter the statistic is intended to estimate, the statistic is said to be an **unbiased estimate** of the parameter.

If the mean of the sampling distribution is not equal to the parameter, the statistic is said to be a **biased estimate** of the parameter.

The standard deviation of a sampling distribution measures another important property of statistics—the spread of these estimates generated by repeated sampling. Suppose two statistics, A and B, are both unbiased estimators of the population parameter. Since the means of the two sampling distributions are the same, we turn to their standard deviations in order to decide which will provide estimates that fall closer to the unknown population parameter we are estimating. Naturally, we will choose the sample statistic that has the smaller standard deviation. Figure 6.7

*Unbiased statistics do not exist for all parameters of interest, but they do exist for the parameters considered in this text.

Figure 6.7
Sampling Distributions for
Two Unbiased Estimators

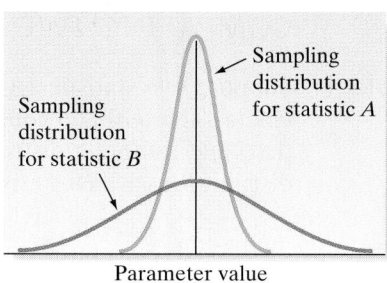

Parameter value

depicts sampling distributions for A and B. Note that the standard deviation of the distribution of A is smaller than the standard deviation for B, indicating that over a large number of samples, the values of A cluster more closely around the unknown population parameter than do the values of B. Stated differently, the probability that A is close to the parameter value is higher than the probability that B is close to the parameter value.

In summary, to make an inference about a population parameter, we use the sample statistic with a sampling distribution that is unbiased and has a small standard deviation (usually smaller than the standard deviation of other unbiased sample statistics). The derivation of this sample statistic will not concern us, because the "best" statistic for estimating specific parameters is a matter of record. We will simply present an unbiased estimator with its standard deviation for each population parameter we consider. [*Note:* The standard deviation of the sampling distribution of a statistic is also called the **standard error of the statistic**.]

EXAMPLE 6.4   BIASED AND UNBIASED ESTIMATORS

Problem   In Example 6.1, we found the sampling distributions of the sample mean $\bar{x}$ and the sample median M for random samples of $n = 3$ measurements from a population defined by the probability distribution shown below.

x	0	3	12
$p(x)$	$\frac{1}{3}$	$\frac{1}{3}$	$\frac{1}{3}$

The sampling distributions of $\bar{x}$ and M were found to be as shown here.

$\bar{x}$	0	1	2	3	4	5	6	8	9	12
$p(\bar{x})$	$\frac{1}{27}$	$\frac{3}{27}$	$\frac{3}{27}$	$\frac{1}{27}$	$\frac{3}{27}$	$\frac{6}{27}$	$\frac{3}{27}$	$\frac{3}{27}$	$\frac{3}{27}$	$\frac{1}{27}$

M	0	3	12
$p(M)$	$\frac{7}{27}$	$\frac{13}{27}$	$\frac{7}{27}$

a. Show that $\bar{x}$ is an unbiased estimator of μ in this situation.

b. Show that M is a biased estimator of μ in this situation.

Solution   **a.** The expected value of a discrete random variable x (see Section 4.3) is defined to be $E(x) = \Sigma x p(x)$, where the summation is over all values of x. Then

$$E(x) = \mu = \sum x p(x) = (0)(\tfrac{1}{3}) + (3)(\tfrac{1}{3}) + (12)(\tfrac{1}{3}) = 5$$

The expected value of the discrete random variable $\bar{x}$ is

$$E(\bar{x}) = \sum (\bar{x}) p(\bar{x})$$

summed over all values of $\bar{x}$. Or

$$E(\bar{x}) = (0)(\tfrac{1}{27}) + (1)(\tfrac{3}{27}) + 2(\tfrac{3}{27}) + \cdots + (12)(\tfrac{1}{27}) = 5$$

Since $E(\bar{x}) = \mu$, we see that $\bar{x}$ is an unbiased estimator of μ.

b. The expected value of the sample median M is

$$E(M) = \sum Mp(M) = (0)(\tfrac{7}{27}) + (3)(\tfrac{13}{27}) + (12)(\tfrac{7}{27}) = 4.56$$

Since the expected value of M is not equal to $\mu(\mu = 5)$, the sample median M is a biased estimator of μ.

Now Work   *Exercise 6.16*

■ ■ ■

EXAMPLE 6.5

VARIANCE OF ESTIMATORS

Problem   Refer to Example 6.4 and find the standard deviations of the sampling distributions of $\bar{x}$ and M. Which statistic would appear to be a better estimator of μ?

Solution   The variance of the sampling distribution of $\bar{x}$ (we denote it by the symbol $\sigma_{\bar{x}}^2$) is found to be

$$\sigma_{\bar{x}}^2 = E\{[\bar{x} - E(\bar{x})]^2\} = \sum (\bar{x} - \mu)^2 p(\bar{x})$$

where, from Example 6.4,

$$E(\bar{x}) = \mu = 5$$

Then

$$\sigma_{\bar{x}}^2 = (0 - 5)^2(\tfrac{1}{27}) + (1 - 5)^2(\tfrac{3}{27}) + (2 - 5)^2(\tfrac{3}{27})$$
$$+ \cdots + (12 - 5)^2(\tfrac{1}{27}) = 8.6667$$

and

$$\sigma_{\bar{x}} = \sqrt{8.6667} = 2.94$$

Similarly, the variance of the sampling distribution of M (we denote it by σ_M^2) is

$$\sigma_M^2 = E\{[M - E(M)]^2\}$$

where, from Example 6.4, the expected value of M is $E(M) = 4.56$. Then

$$\sigma_M^2 = E\{[M - E(M)]^2\} = \sum [M - E(M)]^2 p(M)$$
$$= (0 - 4.56)^2(\tfrac{7}{27}) + (3 - 4.56)^2(\tfrac{13}{27}) + (12 - 4.56)^2(\tfrac{7}{27}) = 20.9136$$

and

$$\sigma_M = \sqrt{20.9136} = 4.57$$

Which statistic appears to be the better estimator for the population mean μ: the sample mean $\bar{x}$ or the median M? To answer this question, we compare the sampling distributions of the two statistics. The sampling distribution of the sample median M is biased (i.e., it is located to the left of the mean μ) and its standard deviation $\sigma_M = 4.57$ is much larger than the standard deviation of the sampling distribution of $\bar{x}$, $\sigma_{\bar{x}} = 2.94$. Consequently, the sample mean $\bar{x}$ would be a better estimator of the population mean μ, for the population in question than would the sample median M.

Look Back Ideally, we desire an estimator that is unbiased *and* has the smallest variance among all unbiased estimators. We call this statistic the **minimum variance unbiased estimator (MVUE).**

| Now Work | *Exercise 6.16* |

■ ■ ■

Exercises 6.10–6.20

Understanding the Principles

6.10 What is a point estimator of a population parameter?

6.11 What is the difference between a biased and unbiased estimator?

6.12 What is the MVUE for a parameter?

6.13 What are the properties of an ideal estimator?

Learning the Mechanics

6.14 Consider the probability distribution shown here.
NW

x	0	1	4
$p(x)$	$1/3$	$1/3$	$1/3$

a. Find μ and σ^2.

b. Find the sampling distribution of the sample mean $\bar{x}$ for a random sample of $n = 2$ measurements from this distribution.

c. Show that $\bar{x}$ is an unbiased estimator for μ. [*Hint:* Show that $E(\bar{x}) = \Sigma \bar{x} p(\bar{x}) = \mu$]

d. Find the sampling distribution of the sample variance s^2 for a random sample of $n = 2$ measurements from this distribution.

e. Show that s^2 is an unbiased estimator for σ^2.

6.15 Consider the probability distribution shown here.

x	2	4	9
$p(x)$	$1/3$	$1/3$	$1/3$

a. Calculate μ for this distribution.

b. Find the sampling distribution of the sample mean $\bar{x}$ for a random sample of $n = 3$ measurements from this distribution, and show that $\bar{x}$ is an unbiased estimator of μ.

c. Find the sampling distribution of the sample median M for a random sample of $n = 3$ measurements from this distribution, and show that the median is a biased estimator of μ.

d. If you wanted to estimate μ using a sample of three measurements from this population, which estimator would you use? Why?

6.16 Consider the probability distribution shown here.
NW

x	0	1	2
$p(x)$	$1/3$	$1/3$	$1/3$

a. Find μ.

b. For a random sample of $n = 3$ observations from this distribution, find the sampling distribution of the sample mean.

c. Find the sampling distribution of the median of a sample of $n = 3$ observations from this population.

d. Refer to parts **b** and **c** and show that both the mean and median are unbiased estimators of μ for this population.

e. Find the variances of the sampling distributions of the sample mean and the sample median.

f. Which estimator would you use to estimate μ? Why?

6.17 Use the computer to generate 500 samples, each containing $n = 25$ measurements, from a population that contains values of x equal to 1, 2, ..., 48, 49, 50. Assume that these values of x are equally likely. Calculate the sample mean $\bar{x}$ and median M for each sample. Construct relative frequency histograms for the 500 values of $\bar{x}$ and the 500 values of M. Use these approximations to the sampling distributions of $\bar{x}$ and M to answer the following questions:

a. Does it appear that $\bar{x}$ and M are unbiased estimators of the population mean? [*Note:* $\mu = 25.5$.]

b. Which sampling distribution displays greater variation?

6.18 Refer to Exercise 6.5.

a. Show that $\bar{x}$ is an unbiased estimator of μ.

b. Find $\sigma_{\bar{x}}^2$

c. Find the probability that $\bar{x}$ will fall within $2\sigma_{\bar{x}}$ of μ.

6.19 Refer to Exercise 6.5.

a. Find the sampling distribution of s^2.

b. Find the population variance σ^2.

c. Show that s^2 is an unbiased estimator of σ^2.

d. Find the sampling distribution of the sample standard deviation s.

e. Show that s is a biased estimator of σ.

6.20 Refer to Exercise 6.7, where we found the sampling distribution of the sample median. Is the median an unbiased estimator of the population mean μ?

6.3 The Sampling Distribution of $\bar{x}$ and the Central Limit Theorem

Estimating the mean useful life of automobiles, the mean number of crimes per month in a large city, and the mean yield per acre of a new soybean hybrid are practical problems with something in common. In each case we are interested in making an inference about the mean μ of some population. As we mentioned in Chapter 2, the sample mean $\bar{x}$ is, in general, a good estimator of μ. We now develop pertinent information about the sampling distribution for this useful statistic. We will show that $\bar{x}$ is the minimum variance unbiased estimator (MVUE) for μ.

EXAMPLE 6.6

DESCRIBING THE SAMPLING DISTRIBUTION OF $\bar{X}$

Problem Suppose a population has the uniform probability distribution given in Figure 6.8. The mean and standard deviation of this probability distribution are $\mu = .5$ and $\sigma = .29$. (See Section 5.2 for the formulas for μ and σ.) Now suppose a sample of 11 measurements is selected from this population. Describe the sampling distribution of the sample mean $\bar{x}$ based on the 1,000 sampling experiments discussed in Example 6.2.

Figure 6.8
Sampled Uniform
Population

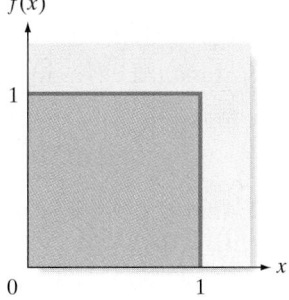

Solution You will recall that in Example 6.2 we generated 1,000 samples of $n = 11$ measurements each. The MINITAB histogram for the 1,000 sample means is shown in Figure 6.9 with a normal probability distribution superimposed. You can see that this normal probability distribution approximates the computer-generated sampling distribution very well.

Figure 6.9
MINITAB Histogram
for $\bar{x}$ in 1,000 Samples

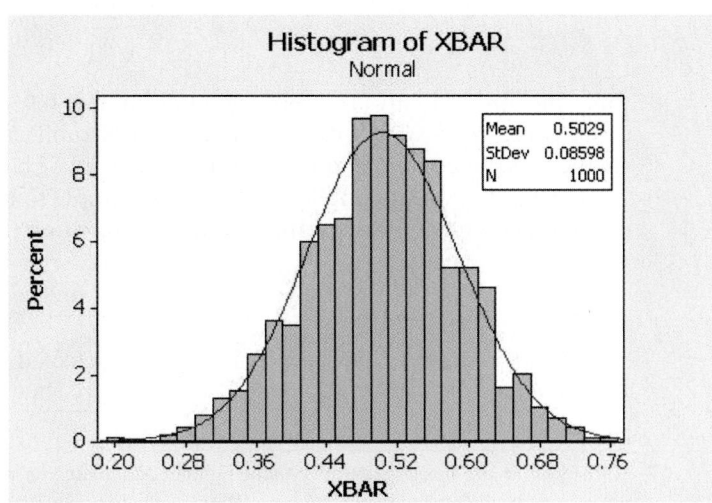

To fully describe a normal probability distribution, it is necessary to know its mean and standard deviation. MINITAB gives these statistics for the 1,000 $\bar{x}$'s in the upper right corner of the histogram, Figure 6.9. You can see that the mean is .503 and the standard deviation is .086.

To summarize our findings based on 1,000 samples, each consisting of 11 measurements from a uniform population, the sampling distribution of $\bar{x}$ appears to be approximately normal with a mean of about .5 and a standard deviation of about .09.

Look Back Note that the simulated value $\mu_{\bar{x}} = .503$ is very close to $\mu = .5$ for the uniform distribution. That is, the simulated sampling distribution of $\bar{x}$ appears to be unbiased for estimating μ.

▰ ▰ ▰

The sampling distribution of $\bar{x}$ has the properties given in the next box, assuming only that a random sample of n observations has been selected from *any* population.

Properties of the Sampling Distribution of $\bar{x}$

1. Mean of sampling distribution equals mean of sampled population. That is, $\mu_{\bar{x}} = E(\bar{x}) = \mu$.

2. Standard deviation of sampling distribution equals

$$\frac{\text{Standard deviation of sampled population}}{\text{Square root of sample size}}$$

That is, $\sigma_{\bar{x}} = \sigma/\sqrt{n}$.*
The standard deviation $\sigma_{\bar{x}}$ is often referred to as the **standard error of the mean**.

You can see that our approximation to $\mu_{\bar{x}}$ in Example 6.6 was precise, since property 1 assures us that the mean is the same as that of the sampled population: .5. Property 2 tells us how to calculate the standard deviation of the sampling distribution of $\bar{x}$. Substituting $\sigma = .29$, the standard deviation of the sampled uniform distribution, and the sample size $n = 11$ into the formula for $\sigma_{\bar{x}}$, we find

$$\sigma_{\bar{x}} = \frac{\sigma}{\sqrt{n}} = \frac{.29}{\sqrt{11}} = .09$$

Thus, the approximation we obtained in Example 6.6, $\sigma_{\bar{x}} \approx .086$, is very close to the exact value, $\sigma_{\bar{x}} = .09$. It can be shown (proof omitted) that the value of $\sigma_{\bar{x}}^2$ is the smallest variance among all unbiased estimators of μ—thus, $\bar{x}$ is the MVUE for μ.

A third property, applicable when the sample size n is large, is contained in one of the most important theoretical results in statistics: the *Central Limit Theorem*.

Central Limit Theorem

Consider a random sample of n observations selected from a population (*any* population) with mean μ and standard deviation σ. Then, when n is sufficiently large,

---

*If the sample size n is large relative to the number N of elements in the population (e.g., 5% or more), $\sigma/\sqrt{n}$ must be multiplied by a finite population correction factor, $\sqrt{(N - n)/(N - 1)}$. For most sampling situations, this correction factor will be close to 1 and can be ignored.

the sampling distribution of $\bar{x}$ will be approximately a normal distribution with mean $\mu_{\bar{x}} = \mu$ and standard deviation $\sigma_{\bar{x}} = \sigma/\sqrt{n}$. The larger the sample size, the better will be the normal approximation to the sampling distribution of $\bar{x}$.*

Biography

PIERRE-SIMON LAPLACE (1749–1827)—The CLT Originator

As a boy growing up in Normandy, France, Pierre-Simon Laplace attended a Benedictine priory school. Upon graduation, he entered Caen University to study theology. During his two years there he discovered his mathematical talents and began his career as an eminent mathematician. In fact, he considered himself the best mathematician in France. Laplace's contributions to mathe-matics ranged from introducing new methods of solving differential equations to complex analyses of motions of astronomical bodies. While studying the angles of inclination of comet orbits in 1778, Laplace showed that the sum of the angles was normally distributed. Consequently, he is considered to be the originator of the Central Limit Theorem. (A rigorous proof of the theorem, however, was not provided until the early 1930s by another French mathematician, Paul Levy.) Laplace also discovered Bayes's theorem and established Bayesian statistical analysis as a valid approach to many practical problems of his time.

Thus, for sufficiently large samples the sampling distribution of $\bar{x}$ is approximately normal. How large must the sample size n be so that the normal distribution provides a good approximation for the sampling distribution of $\bar{x}$? The answer depends on the shape of the distribution of the sampled population, as shown by Figure 6.10. Generally speaking, the greater the skewness of the sampled population distribution, the larger the sample size must be before the normal distribution is an adequate approximation for the sampling distribution of $\bar{x}$. For most sampled populations, sample sizes of $n \geq 30$ will suffice for the normal approximation to be reasonable. We will use the normal approximation for the sampling distribution of $\bar{x}$ when the sample size is at least 30.

EXAMPLE 6.7

USING THE CENTRAL LIMIT THEOREM TO FIND A PROBABILITY

Problem Suppose we have selected a random sample of $n = 36$ observations from a population with mean equal to 80 and standard deviation equal to 6. It is known that the population is not extremely skewed.

a. Sketch the relative frequency distributions for the population and for the sampling distribution of the sample mean, $\bar{x}$.

b. Find the probability that $\bar{x}$ will be larger than 82.

Solution **a.** We do not know the exact shape of the population relative frequency distribution, but we do know that it should be centered about $\mu = 80$, its spread should be measured by $\sigma = 6$, and it is not highly skewed. One possibility is shown in Figure 6.11a. From the Central Limit Theorem, we know that the sampling distribution of $\bar{x}$ will be approximately normal since the sampled

---

*Moreover, because of the Central Limit Theorem, the sum of a random sample of n observations, Σx, will possess a sampling distribution that is approximately normal for large samples. This distribution will have a mean equal to $n\mu$ and a variance equal to $n\sigma^2$. Proof of the Central Limit Theorem is beyond the scope of this book, but it can be found in many mathematical statistics texts.

Figure 6.10

Sampling Distributions of $\bar{x}$ for Different Populations and Different Sample Sizes

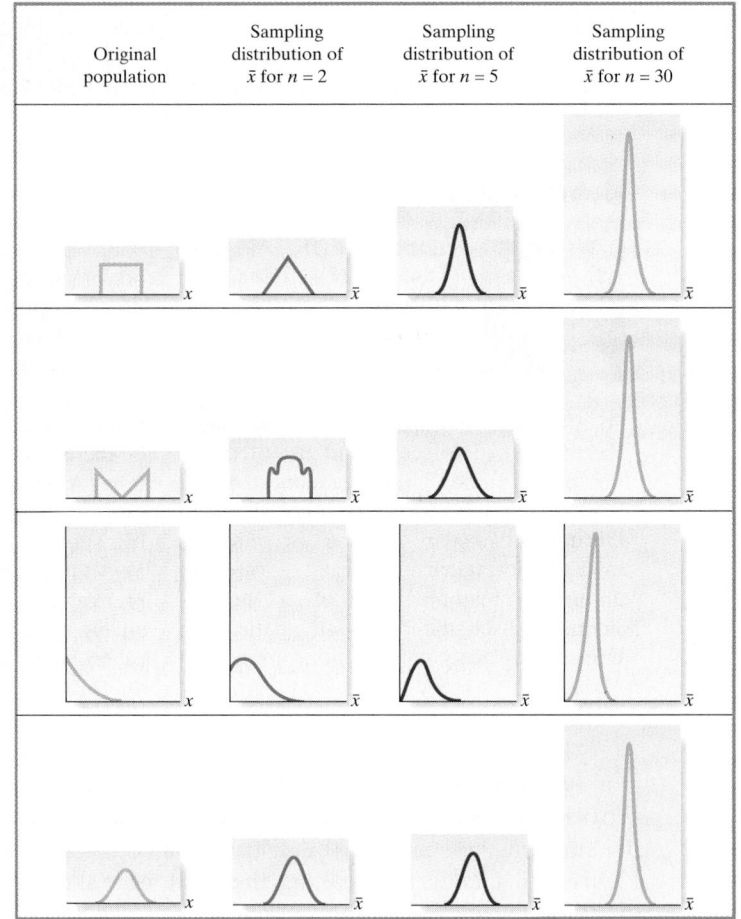

population distribution is not extremely skewed. We also know that the sampling distribution will have mean and standard deviation

$$\mu_{\bar{x}} = \mu = 80 \quad \text{and} \quad \sigma_{\bar{x}} = \frac{\sigma}{\sqrt{n}} = \frac{6}{\sqrt{36}} = 1$$

The sampling distribution of $\bar{x}$ is shown in Figure 6.11b.

b. The probability that $\bar{x}$ will exceed 82 is equal to the highlighted area in Figure 6.12. To find this area, we need to find the z value corresponding to $\bar{x} = 82$. Recall that the standard normal random variable z is the difference between any normally distributed random variable and its mean, expressed

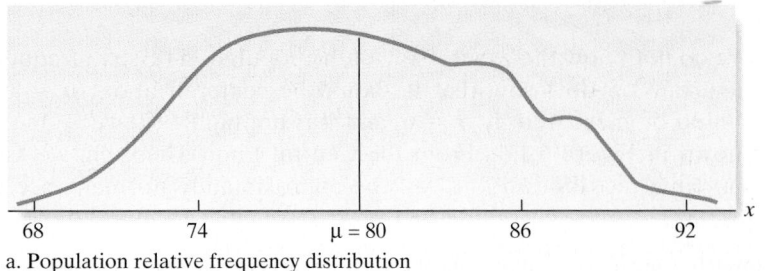

a. Population relative frequency distribution

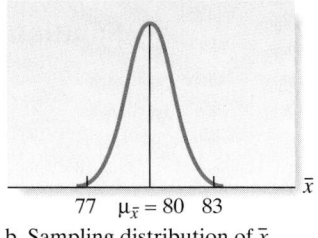

b. Sampling distribution of $\bar{x}$

Figure 6.11

A Population Relative Frequency Distribution and the Sampling Distribution for $\bar{x}$

Figure 6.12
The Sampling Distribution
of $\bar{x}$

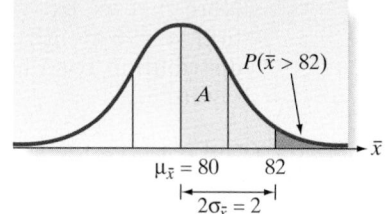

in units of its standard deviation. Since $\bar{x}$ is a normally distributed random variable with mean $\mu_{\bar{x}} = \mu$ and standard deviation $\sigma_{\bar{x}} = \sigma/\sqrt{n}$, it follows that the standard normal z value corresponding to the sample mean, $\bar{x}$, is

$$z = \frac{(\text{Normal random variable}) - (\text{Mean})}{\text{Standard Deviation}} = \frac{\bar{x} - \mu_{\bar{x}}}{\sigma_{\bar{x}}}$$

Therefore, for $\bar{x} = 82$, we have

$$z = \frac{\bar{x} - \mu_{\bar{x}}}{\sigma_{\bar{x}}} = \frac{82 - 80}{1} = 2$$

The area A in Figure 6.12 corresponding to $z = 2$ is given in the table of areas under the normal curve (see Table IV of Appendix A) as .4772. Therefore, the tail area corresponding to the probability that $\bar{x}$ exceeds 82 is

$$P(\bar{x} > 82) = P(z > 2) = .5 - .4772 = .0228$$

Look Back The key to finding the probability, part **b**, is to recognize that the distribution of $\bar{x}$ is normal with $\mu_{\bar{x}} = \mu$ and $\sigma_{\bar{x}} = \sigma/\sqrt{n}$.

Now Work   *Exercise 6.30*

■ ■ ■

EXAMPLE 6.8

A PRACTICAL APPLICATION OF THE CENTRAL LIMIT THEOREM

Problem   A manufacturer of automobile batteries claims that the distribution of the lengths of life of its best battery has a mean of 54 months and a standard deviation of 6 months. Suppose a consumer group decides to check the claim by purchasing a sample of 50 of these batteries and subjecting them to tests that determine battery life.

    a.   Assuming that the manufacturer's claim is true, describe the sampling distribution of the mean lifetime of a sample of 50 batteries.

    b.   Assuming that the manufacturer's claim is true, what is the probability the consumer group's sample has a mean life of 52 or fewer months?

Solution   **a.**   Even though we have no information about the shape of the probability distribution of the lives of the batteries, we can use the Central Limit Theorem to deduce that the sampling distribution for a sample mean lifetime of 50 batteries is approximately normally distributed. Furthermore, the mean of this sampling distribution is the same as the mean of the sampled population, which is $\mu = 54$ months according to the manufacturer's claim. Finally, the standard deviation of the sampling distribution is given by

$$\sigma_{\bar{x}} = \frac{\sigma}{\sqrt{n}} = \frac{6}{\sqrt{50}} = .85 \text{ month}$$

Note that we used the claimed standard deviation of the sampled population, $\sigma = 6$ months. Thus, if we assume that the claim is true, the sampling distribution for the mean life of the 50 batteries sampled is as shown in Figure 6.13.

Figure 6.13

Sampling Distribution of $\bar{x}$ in Example 6.8 for $n = 50$

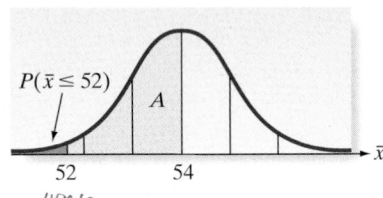

$P(\bar{x} \leq 52)$ A

52   54

4906

b. If the manufacturer's claim is true, the probability that the consumer group observes a mean battery life of 52 or fewer months for their sample of 50 batteries, $P(\bar{x} \leq 52)$, is equivalent to the highlighted area in Figure 6.13. Since the sampling distribution is approximately normal, we can find this area by computing the standard normal z value:

$$z = \frac{\bar{x} - \mu_{\bar{x}}}{\sigma_{\bar{x}}} = \frac{\bar{x} - \mu}{\sigma_{\bar{x}}} = \frac{52 - 54}{.85} = -2.35$$

where $\mu_{\bar{x}}$, the mean of the sampling distribution of $\bar{x}$, is equal to μ, the mean of the lifetimes of the sampled population, and $\sigma_{\bar{x}}$ is the standard deviation of the sampling distribution of $\bar{x}$. Note that z is the familiar standardized distance (z-score) of Section 2.6 and, since $\bar{x}$ is approximately normally distributed, it will possess the standard normal distribution of Section 5.3.

The area A shown in Figure 6.13 between $\bar{x} = 52$ and $\bar{x} = 54$ (corresponding to $z = -2.35$) is found in Table IV of Appendix A to be .4906. Therefore, the area to the left of $\bar{x} = 52$ is

$$P(\bar{x} \leq 52) = .5 - A = .5 - .4906 = .0094$$

Thus, the probability the consumer group will observe a sample mean of 52 or less is only .0094 if the manufacturer's claim is true.

Look Back If the 50 tested batteries do exhibit a mean of 52 or fewer months, the consumer group will have strong evidence that the manufacturer's claim is untrue, because such an event is very unlikely to occur if the claim is true. (This is still another application of the *rare-event approach* to statistical inference.)

Now Work   *Exercise 6.34*

■ ■ ■

We conclude this section with two final comments on the sampling distribution of $\bar{x}$. First, from the formula $\sigma_{\bar{x}} = \sigma/\sqrt{n}$, we see that the standard deviation of the sampling distribution of $\bar{x}$ gets smaller as the sample size n gets larger. For example, we computed $\sigma_{\bar{x}} = .85$ when $n = 50$ in Example 6.8. However, for $n = 100$ we obtain $\sigma_{\bar{x}} = \sigma/\sqrt{n} = 6/\sqrt{100} = .60$. This relationship will hold true for most of the sample statistics encountered in this text. That is, *The standard deviation of the sampling distribution decreases as the sample size increases.* Consequently, the larger the sample size, the more accurate the sample statistic (e.g., $\bar{x}$) is in estimating a population parameter (e.g., μ). We will use this result in Chapter 7 to help us determine the sample size needed to obtain a specified accuracy of estimation.

Our second comment concerns the Central Limit Theorem. In addition to providing a very useful approximation for the sampling distribution of a sample mean, the Central Limit Theorem offers an explanation for the fact that many relative frequency distributions of data possess mound-shaped distributions. Many of the measurements we take in various areas of research are really means or sums of a large number of small phenomena. For example, a year's growth of a pine seedling is the total of the many individual components that affect the plant's growth. Similarly, we can view the length of time a construction company takes to build a house as the total of the times taken to complete a multitude of distinct jobs, and we can regard the monthly demand for blood at a hospital as the total of the many individual patients' needs. Whether or not the observations entering into these sums satisfy the assumptions basic to the Central Limit Theorem is open to question. However, it is a fact that many distributions of data in nature are mound shaped and possess the appearance of normal distributions.

Statistics in Action Revised

Making an Inference about the Mean Sleep Time for Insomnia Pill Takers

In a Massachusetts Institute of Technology (MIT) study, each in a sample of 40 young male volunteers was given a dosage of the sleep-inducing hormone, melatonin, placed in a dark room at midday, and told to close his eyes for 30 minutes. The researchers measured the time (in minutes) elapsed before each volunteer fell asleep. Recall that the data (shown in Table SIA6.1) are saved in the **INSOMNIA** file.

Previous research established that with the placebo (i.e., no hormone), the mean time to fall asleep is $\mu = 15$ minutes and the standard deviation is $\sigma = 10$ minutes. If the true value of μ for those taking the melatonin is $\mu < 15$ (i.e., if, on average, the volunteers fall asleep faster with the drug than with the placebo), then the researchers can infer that melatonin is an effective drug against insomnia.

Descriptive statistics for the 40 sleep times are displayed in the SPSS printout, Figure SIA6.1. You can see that the mean for the sample is $\bar{x} = 5.935$ minutes. If the drug is not effective in reducing sleep times, then the distribution of sleep times will be no different than the distribution with the placebo. That is, if the drug is not effective, the mean and standard deviation of the population of sleep times are $\mu = 15$ and $\sigma = 10$. If this is true, how likely is it to observe a sample mean that is below 6 minutes?

To answer this question, we desire the probability, $P(\bar{x} < 6)$. To find this probability, we invoke the Central

Limit Theorem (CLT). According to the theorem, the sampling distribution of $\bar{x}$ has the following mean and standard deviation:

$$\mu_{\bar{x}} = \mu = 15$$
$$\sigma_{\bar{x}} = \sigma/\sqrt{n} = 10/\sqrt{40} = 1.58$$

The CLT also states that the distribution of x is approximately normally distributed. Therefore, we find the desired probability (using the standard normal table) as follows:

$$P(\bar{x} < 6) = P\left(z < \frac{6 - \mu_{\bar{x}}}{\sigma_{\bar{x}}}\right)$$
$$= P\left(z < \frac{6 - 15}{1.58}\right) = P(z < -5.70) \approx 0$$

In other words, the probability that we observe a sample mean below $\bar{x} = 6$ minutes *if the mean and standard deviation of the sleep times are $\mu = 15$ and $\sigma = 10$ (i.e., if the drug is not effective)* is almost 0. Either the drug is not effective and the researchers have observed an extremely rare event (one with almost no chance of happening) or the true value of μ for those taking the melatonin pill is much less than 15 minutes. The rare-event approach to making statistical inferences, of course, would favor the second conclusion. Melatonin appears to be an effective insomnia pill, one that lowers the average time it takes the volunteers to fall asleep.

Figure SIA6.1

SPSS Descriptive Statistics for Sleep Time Data

Descriptive Statistics

	N	Minimum	Maximum	Mean	Std. Deviation
SLEEPTIM	40	1.3	22.8	5.935	5.3917
Valid N (listwise)	40				

Exercises 6.21–6.43

Understanding the Principles

6.21 What do the symbols $\mu_{\bar{x}}$ and $\sigma_{\bar{x}}$ represent?

6.22 How does the mean of the sampling distribution of $\bar{x}$ relate to the mean of the population from which the sample is selected?

6.23 How does the standard deviation of the sampling distribution of $\bar{x}$ relate to the standard deviation of the population from which the sample is selected?

6.24 Another name given to the standard deviation of $\bar{x}$ is the _____. (Fill in the blank.)

6.25 State the Central Limit Theorem.

6.26 Will the sampling distribution of $\bar{x}$ always be approximately normally distributed? Explain.

Learning the Mechanics

6.27 Suppose a random sample of n measurements is selected from a population with mean $\mu = 100$ and variance $\sigma^2 = 100$. For each of the following values of n, give the mean and standard deviation of the sampling distribution of the sample mean $\bar{x}$.
  a. $n = 4$ **b.** $n = 25$
  c. $n = 100$ **d.** $n = 50$
  e. $n = 500$ **f.** $n = 1{,}000$

6.28 Suppose a random sample of $n = 25$ measurements is selected from a population with mean μ and standard deviation σ. For each of the following values of μ and σ, give the values of $\mu_{\bar{x}}$ and $\sigma_{\bar{x}}$.
  a. $\mu = 10, \sigma = 3$
  b. $\mu = 100, \sigma = 25$
  c. $\mu = 20, \sigma = 40$
  d. $\mu = 10, \sigma = 100$

6.29 Consider the probability distribution shown here.

x	1	2	3	8
$p(x)$	.1	.4	.4	.1

  a. Find μ, σ^2, and σ.
  b. Find the sampling distribution of $\bar{x}$ for random samples of $n = 2$ measurements from this distribution by listing all possible values of $\bar{x}$, and find the probability associated with each.
  c. Use the results of part **b** to calculate $\mu_{\bar{x}}$ and $\sigma_{\bar{x}}$. Confirm that $\mu_{\bar{x}} = \mu$ and that $\sigma_{\bar{x}} = \sigma/\sqrt{n} = \sigma/\sqrt{2}$.

6.30 A random sample of $n = 64$ observations is drawn
NW from a population with a mean equal to 20 and standard deviation equal to 16.
  a. Give the mean and standard deviation of the (repeated) sampling distribution of $\bar{x}$.
  b. Describe the shape of the sampling distribution of $\bar{x}$. Does your answer depend on the sample size?
  c. Calculate the standard normal z-score corresponding to a value of $\bar{x} = 16$.

  d. Calculate the standard normal z-score corresponding to $\bar{x} = 23$.
  e. Find $P(\bar{x} < 16)$.
  f. Find $P(\bar{x} > 23)$.
  g. Find $P(16 < \bar{x} < 23)$.

6.31 A random sample of $n = 100$ observations is selected from a population with $\mu = 30$ and $\sigma = 16$.
  a. Find $\mu_{\bar{x}}$ and $\sigma_{\bar{x}}$.
  b. Describe the shape of the sampling distribution of $\bar{x}$.
  c. Find $P(\bar{x} \geq 28)$.
  d. Find $P(22.1 \leq \bar{x} \leq 26.8)$.
  e. Find $P(\bar{x} \leq 28.2)$.
  f. Find $P(\bar{x} \geq 27.0)$.

6.32 A random sample of $n = 900$ observations is selected from a population with $\mu = 100$ and $\sigma = 10$.
  a. What are the largest and smallest values of $\bar{x}$ that you would expect to see?
  b. How far, at the most, would you expect $\bar{x}$ to deviate from μ?
  c. Did you have to know μ to answer part **b**? Explain.

6.33 Consider a population that contains values of x equal to 0, 1, 2, ..., 97, 98, 99. Assume that the values of x are equally likely. For each of the following values of n, use the computer to generate 500 random samples and calculate $\bar{x}$ for each sample. For each sample size, construct a relative frequency histogram of the 500 values of $\bar{x}$. What changes occur in the histograms as the value of n increases? What similarities exist? Use $n = 2, n = 5, n = 10, n = 30$, and $n = 50$.

Applying the Concepts—Basic

6.34 **Children's attitude toward reading.** In the journal
NW *Knowledge Quest* (January/February 2002), education professors at the University of Southern California investigated children's attitudes toward reading. One study measured third through sixth graders' attitudes toward recreational reading on a 140-point scale (where higher scores indicate a more positive attitude). The mean score for this population of children was 106 points with a standard deviation of 16.4 points. Consider a random sample of 36 children from this population and let $\bar{x}$ represent the mean recreational reading attitude score for the sample.
  a. What is $\mu_{\bar{x}}$?
  b. What is $\sigma_{\bar{x}}$?
  c. Describe the shape of the sampling distribution of $\bar{x}$.
  d. Find the z-score for the value $\bar{x} = 100$ points.
  e. Find $P(\bar{x} < 100)$.

6.35 **Dentists' use of anesthetics.** Refer to the *Current Allergy & Clinical Immunology* (March 2004) study of allergic reactions of dental patients to local anesthetics (LA), Exercise 2.95 (p. 79). The study reported that the mean number of units (ampoules) of LA used per week by dentists was 79 with a standard deviation of 23. Consider a random sample of 100 dentists and let $\bar{x}$ represent the mean number of units of LA used per week for the sample.

a. What is $\mu_{\bar{x}}$?

b. What is $\sigma_{\bar{x}}$?

c. Describe the shape of the sampling distribution of $\bar{x}$.

d. Find the z-score for the value $\bar{x} = 80$ ampoules.

e. Find $P(\bar{x} > 80)$.

6.36 Research on eating disorders. Refer to *The American Statistician* (May 2001) study of female students who suffer from bulimia, Exercise 2.38 (p. 52). Recall that each student completed a questionnaire from which a "fear of negative evaluation" (FNE) score was produced. (The higher the score, the greater the fear of negative evaluation.) Suppose the FNE scores of bulimic students have a distribution with mean $\mu = 18$ and standard deviation $\sigma = 5$. Now consider a random sample of 45 female students with bulimia.

a. What is the probability that the sample mean FNE score is greater than 17.5?

b. What is the probability that the sample mean FNE score is between 18 and 18.5?

c. What is the probability that the sample mean FNE score is less than 18.5?

6.37 Improving SAT scores. Refer to the *Chance* (Winter 2001) examination of Standardized Admission Test (SAT) scores of students who pay a private tutor to help them improve their results, Exercise 2.101 (p. 81). On the SAT-Mathematics test, these students had a mean score change of +19 points, with a standard deviation of 65 points. In a random sample of 100 students who pay a private tutor to help them improve their results, what is the likelihood that the sample mean score change is less than 10 points?

Applying the Concepts—Intermediate

6.38 Cost of unleaded fuel. According to the American Automobile Association (AAA), the average cost of a gallon unleaded fuel at gas stations in July 2004 was $1.897 (*AAA Fuel Gauge Report*). Assume that the standard deviation of such costs is $.15. Suppose that a random sample of $n = 100$ gas stations is selected from the population and the July 2004 cost per gallon of unleaded fuel is determined for each. Consider $\bar{x}$, the sample mean cost per gallon.

a. Calculate $\mu_{\bar{x}}$ and $\sigma_{\bar{x}}$.

b. What is the approximate probability that the sample has a mean fuel cost between $1.90 and $1.92?

c. What is the approximate probability that the sample has a mean fuel cost that exceeds $1.915?

d. How would the sampling distribution of $\bar{x}$ change if the sample size n were doubled from 100 to 200? How do your answers to parts **b** and **c** change when the sample size is doubled?

6.39 Violence and stress. Interpersonal violence (e.g., rape) generally leads to psychological stress for the victim.

Clinical Psychology Review (Vol. 15, 1995) reported on the results of all recently published studies of the relationship between interpersonal violence and psychological stress. The distribution of the time elapsed between the violent incident and the initial sign of stress has a mean of 5.1 years and a standard deviation of 6.1 years. Consider a random sample of $n = 150$ victims of interpersonal violence. Let $\bar{x}$ represent the mean time elapsed between the violent act and the first sign of stress for the sampled victims.

a. Give the mean and standard deviation of the sampling distribution of $\bar{x}$.

b. Will the sampling distribution of $\bar{x}$ be approximately normal? Explain.

c. Find $P(\bar{x} > 5.5)$.

d. Find $P(4 < \bar{x} < 5)$.

6.40 Susceptibility to hypnosis. The Computer-Assisted Hypnosis Scale (CAHS) is designed to measure a person's susceptibility to hypnosis. In computer-assisted hypnosis, the computer serves as a facilitator of hypnosis by using digitized speech processing coupled with interactive involvement with the hypnotic subject. CAHS scores range from 0 (no susceptibility) to 12 (extremely high susceptibility). A study in *Psychological Assessment* (March 1995) reported a mean CAHS score of 4.59 and a standard deviation of 2.95 for University of Tennessee undergraduates. Assume that $\mu = 4.29$ and $\sigma = 2.95$ for this population. Suppose a psychologist uses CAHS to test a random sample of 50 subjects.

a. Would you expect to observe a sample mean CAHS score of $\bar{x} = 6$ or higher? Explain.

b. Suppose the psychologist actually observes $\bar{x} = 6.2$. Based on your answer to part **a**, make an inference about the population from which the sample was selected.

6.41 Use of sick days. Last year a company began a program to compensate its employees for unused sick days, paying each employee a bonus of one-half the usual wage earned for each unused sick day. The question that naturally arises is, "Did this policy motivate employees to use fewer sick days?" *Before* last year, the number of sick days used by employees had a distribution with a mean of 7 days and a standard deviation of 2 days.

a. Assuming that these parameters did not change last year, find the approximate probability that the sample mean number of sick days used by 100 employees chosen at random was less than or equal to 6.4 last year.

b. How would you interpret the result if the sample mean for the 100 employees was 6.4?

Applying the Concepts—Advanced

6.42 Test of Knowledge About Epilepsy. The Test of Knowledge About Epilepsy (KAE), which is designed to measure attitudes toward persons with epilepsy, uses 20 multiple-choice items where all of the choices are incorrect. For each person, two scores (ranging from 0 to 20) are obtained, an attitude score (KAE-A) and a general knowledge score (KAE-GK). Based on a large-scale study of college students, the distribution of KAE-A scores has a mean of $\mu = 11.92$ and a standard deviation of $\sigma = 2.95$ while the distribution of KAE-GK scores has a mean of $\mu = 6.35$ and a standard deviation of $\sigma = 2.12$ (*Rehabilitative Psychology*, Spring 1995.) Consider a random sample of 100 college students, and suppose you observe a sample mean KAE score of 6.5. Is this result more likely to be the mean of the attitude scores (KAE-A) or the general knowledge scores (KAE-GK)? Explain.

6.43 Handwashing versus handrubbing. Refer to the *British Medical Journal* (August 17, 2002) study to compare the effectiveness of handwashing with soap and handrubbing with alcohol, Exercise 2.97 (p. 80). Health care workers who used handrubbing had a mean bacterial count of 35 per hand with a standard deviation of 59. Health care workers who used handwashing had a mean bacterial count of 69 per hand with a standard deviation of 106. In a random sample of 50 health care workers, all using the same method of cleaning their hands, the mean bacterial count per hand ($\bar{x}$) is less than 30. Give your opinion on whether this sample of workers used handrubbing with alcohol or handwashing with soap.

Quick Review

Key Terms

Biased estimate   299
Central Limit Theorem   304
Error of estimation   298
Minimum variance unbiased
   estimator (MVUE)   302

Parameter   292
Point estimator   298
Sample statistic   292
Sampling distribution   293
Standard error of the mean   304

Standard error of the statistic   300
Unbiased estimate   299

Key Formulas

	Mean	Standard Deviation	z-score	
Sampling distribution of $\bar{x}$	$\mu_{\bar{x}} = \mu$	$\sigma_{\bar{x}} = \dfrac{\sigma}{\sqrt{n}}$	$z = \dfrac{\bar{x} - \mu_{\bar{x}}}{\sigma_{\bar{x}}} = \dfrac{\bar{x} - \mu}{\sigma/\sqrt{n}}$	304

Language Lab

Symbol	Pronunciation	Description
θ	theta	Population parameter (general)
$\mu_{\bar{x}}$	mu of x-bar	True mean of sampling distribution of $\bar{x}$
$\sigma_{\bar{x}}$	sigma of x-bar	True standard deviation of sampling distribution of $\bar{x}$

Chapter Summary

- **Sampling distribution of a statistic** is the theoretical probability distribution of the statistic in repeated sampling.
- **Unbiased estimator** is a statistic with a sampling distribution mean equal to the population parameter being estimated

- **Central Limit Theorem**—the sampling distribution of the sample mean, $\bar{x}$, is approximately normal for large n (e.g., $n \geq 30$).
- $\bar{x}$ is the **minimum variance unbiased estimator (MVUE)** of μ.

Supplementary Exercises 6.44–6.70

Understanding the Principles

6.44 Describe how you could obtain the simulated sampling distribution of a sample statistic.

6.45 **True or False.** The sample mean, $\bar{x}$, will always be equal to $\mu_{\bar{x}}$.

6.46 **True or False.** The sampling distribution of $\bar{x}$ is normally distributed regardless of the size of the sample n.

6.47 **True or False.** The standard error of $\bar{x}$ will always be smaller than σ.

Learning the Mechanics

6.48 The standard deviation (or, as it is usually called, the *standard error*) of the sampling distribution for the sample mean, $\bar{x}$, is equal to the standard deviation of the population from which the sample was selected divided by the square root of the sample size. That is,

$$\sigma_{\bar{x}} = \frac{\sigma}{\sqrt{n}}$$

a. As the sample size is increased, what happens to the standard error of $\bar{x}$? Why is this property considered important?

b. Suppose that a sample statistic has a standard error that is not a function of the sample size. In other words, the standard error remains constant as n changes. What would this imply about the statistic as an estimator of a population parameter?

c. Suppose another unbiased estimator (call it A) of the population mean is a sample statistic with a standard error equal to

$$\sigma_A = \frac{\sigma}{\sqrt[3]{n}}$$

Which of the sample statistics, $\bar{x}$ or A, is preferable as an estimator of the population mean? Why?

d. Suppose that the population standard deviation σ is equal to 10 and that the sample size is 64. Calculate the standard errors of $\bar{x}$ and A. Assuming that the sampling distribution of A is approximately normal, interpret the standard errors. Why is the assumption of (approximate) normality unnecessary for the sampling distribution of $\bar{x}$?

6.49 Consider a sample statistic A. As with all sample statistics, A is computed by utilizing a specified function (formula) of the sample measurements. (For example, if A were the sample mean, the specified formula would be to sum the measurements and divide by the number of measurements.)

a. Describe what we mean by the phrase "the sampling distribution of the sample statistic A."

*b. Suppose A is to be used to estimate a population parameter α. What is meant by the assertion that A is an unbiased estimator of α?

*c. Consider another sample statistic, B. Assume that B is also an unbiased estimator of the population parameter α. How can we use the sampling distributions of A and B to decide which is the better estimator of α?

d. If the sample sizes on which A and B are based are large, can we apply the Central Limit Theorem and assert that the sampling distributions of A and B are approximately normal? Why or why not?

6.50 A random sample of 40 observations is to be drawn from a large population of measurements. It is known that 30% of the measurements in the population are 1's, 20% are 2's, 20% are 3's, and 30% are 4's.

a. Give the mean and standard deviation of the (repeated) sampling distribution of $\bar{x}$, the sample mean of the 40 observations.

b. Describe the shape of the sampling distribution of $\bar{x}$. Does your answer depend on the sample size?

6.51 A random sample of $n = 68$ observations is selected from a population with $\mu = 19.6$ and $\sigma = 3.2$. Approximate each of the following probabilities.

a. $P(\bar{x} \le 19.6)$
b. $P(\bar{x} \le 19)$
c. $P(\bar{x} \ge 20.1)$
d. $P(19.2 \le \bar{x} \le 20.6)$

6.52 Use a statistical software package to generate 100 random samples of size $n = 2$ from a population characterized by a normal probability distribution with mean 100 and standard deviation 10. Compute $\bar{x}$ for each sample and plot a frequency distribution for the 100 values of $\bar{x}$. Repeat this process for $n = 5, 10, 30$, and 50. How does the fact that the sampled population is normal affect the sampling distribution of $\bar{x}$?

6.53 Use a statistical software package to generate 100 random samples of size $n = 2$ from a population characterized by a uniform probability distribution with $c = 0$ and $d = 10$. Compute $\bar{x}$ for each sample, and plot a frequency distribution for the 100 $\bar{x}$ values. Repeat this process for $n = 5, 10, 30$, and 50. Explain how your plots illustrate the Central Limit Theorem.

6.54 Suppose x equals the number of heads observed when a single coin is tossed; that is, $x = 0$ or $x = 1$. The population corresponding to x is the set of 0's and 1's generated when the coin is tossed repeatedly a large number of times. Suppose we select $n = 2$ observations from this population. (That is, we toss the coin twice and observe two values of x.)

a. List the three different samples (combinations of 0's and 1's) that could be obtained.

b. Calculate the value of $\bar{x}$ for each of the samples.

c. List the values that $\bar{x}$ can assume, and find the probabilities of observing these values.

d. Construct a graph of the sampling distribution of $\bar{x}$.

6.55 A random sample of size n is to be drawn from a large population with mean 100 and standard deviation 10, and the sample mean $\bar{x}$ is to be calculated. To see the effect of different sample sizes on the standard deviation of the sampling distribution of $\bar{x}$, plot $\sigma/\sqrt{n}$ against n for $n = 1, 5, 10, 20, 30, 40,$ and 50.

Applying the Concepts—Basic

🔅 DDT

6.56 Contaminated fish. Refer to Exercise 2.33 (p. 51) and the U.S. Army Corps of Engineers data on contaminated fish saved in the **DDT** file. Recall that the length (in centimeters), weight (in grams), and DDT level (in parts per million) was measured for each of 144 fish caught from the polluted Tennessee River in Alabama.

a. In Exercise 2.33a you found that the distribution of fish lengths was skewed to the left. Assume that this is true of the population of fish lengths and that the population has mean $\mu = 43$ centimeters and $\sigma = 7$ centimeters. Use this information to describe the sampling distribution of $\bar{x}$, the mean length of a sample of $n = 40$ fish caught from the Tennessee River.

b. In Exercise 2.33b you found that the distribution of fish weights was approximately normal. Assume that this is true of the population of fish weights and that the population has mean $\mu = 1,050$ grams and $\sigma = 376$ grams. Use this information to describe the sampling distribution of $\bar{x}$, the mean weight of a sample of $n = 40$ fish caught from the Tennessee River.

c. In Exercise 2.33c you found that the distribution of fish DDT levels was highly skewed to the right. Assume that this is true of the population of fish DDT levels and that the population has mean $\mu = 24$ ppm and $\sigma = 98$ ppm. Use this information to describe the sampling distribution of $\bar{x}$, the mean DDT level of a sample of $n = 40$ fish caught from the Tennessee River.

6.57 Clam harvesting. The ocean quahog is a type of clam found in the coastal waters of the mid-Atlantic states. A federal survey of offshore ocean quahog harvesting in New Jersey revealed an average catch per unit effort (CPUE) of 89.34 clams. The CPUE standard deviation was 7.74. (*Journal of Shellfish Research*, June 1995.) Let $\bar{x}$ represent the mean CPUE for a sample of 35 attempts to catch ocean quahogs off the New Jersey shore.

a. Compute $\mu_{\bar{x}}$ and $\sigma_{\bar{x}}$. Interpret their values.

b. Sketch the sampling distribution of $\bar{x}$.

c. Find $P(\bar{x} > 88)$.

d. Find $P(\bar{x} < 87)$.

6.58 Salaries of travel professionals. According to *Business Travel News* (July 15, 2002), the average salary of a travel management professional is $74,000. Assume that the standard deviation of such salaries is $30,000. Consider a random sample of 50 travel management professionals and let $\bar{x}$ represent the mean salary for the sample.

a. What is $\mu_{\bar{x}}$?

b. What is $\sigma_{\bar{x}}$?

c. Describe the shape of the sampling distribution of $\bar{x}$.

d. Find the z-score for the value $\bar{x} = \$65,000$.

e. Find $P(\bar{x} > 65,000)$.

6.59 Violent crimes. The distribution of violent crimes per day in a certain city possesses a mean equal to 1.3 and a standard deviation equal to 1.7. A random sample of 50 days is observed, and the daily mean number of crimes for this sample, $\bar{x}$, is calculated.

a. Give the mean and standard deviation of the sampling distribution of $\bar{x}$.

b. Will the sampling distribution of $\bar{x}$ be approximately normal? Explain.

c. Find an approximate value of $P(\bar{x} < 1)$.

d. Find an approximate value of $P(\bar{x} > 1.9)$.

Applying the Concepts—Intermediate

6.60 Producing machine bearings. To determine whether a metal lathe that produces machine bearings is properly adjusted, a random sample of 25 bearings is collected and the diameter of each is measured.

a. If the standard deviation of the diameters of the bearings measured over a long period of time is .001 inch, what is the approximate probability that the mean diameter $\bar{x}$ of the sample of 25 bearings will lie within .0001 inch of the population mean diameter of the bearings?

b. If the population of diameters has an extremely skewed distribution, how will your approximation in part **a** be affected?

6.61 Quality control. Refer to Exercise 6.60. The mean diameter of the bearings produced by the machine is supposed to be .5 inch. The company decides to use the sample mean (from Exercise 6.60) to decide whether the process is in control (i.e., whether it is producing bearings with a mean diameter of .5 inch). The machine will be considered out of control if the mean of the sample of $n = 25$ diameters is less than .4994 inch or larger than .5006 inch. If the true mean diameter of the bearings produced by the machine is .501 inch, what is the approximate probability that the test will imply that the process is out of control?

6.62 Length of job tenure. Researchers at the Terry College of Business at the University of Georgia sampled 344 business students and asked them this question: "Over the course of your lifetime, what is the maximum number of years you expect to work for any one employer?" The sample resulted in $\bar{x} = 19.1$ years. Assume the sample of students was randomly selected from the 6,000 undergraduate students at the Terry College and that $\sigma = 6$ years.

a. Describe the sampling distribution of $\bar{x}$.

b. If the mean for the 6,000 undergraduate students is $\mu = 18.5$ years, find $P(\bar{x} > 19.1)$.

c. If the mean for the 6,000 undergraduate students is $\mu = 19.5$ years, find $P(\bar{x} > 19.1)$.

d. If $P(\bar{x} > 19.1) = .5$, what is μ?

e. If $P(\bar{x} > 19.1) = .2$, is μ greater than or less than 19.1 years? Explain.

6.63 New method of teaching arithmetic. This past year, an elementary school began using a new method to teach arithmetic to first graders. A standardized test, administered at the end of the year, was used to measure the effectiveness of the new method. The distribution of past scores on the standardized test produced a mean of 75 and a standard deviation of 10.

a. If the new method is no different from the old method, what is the approximate probability that the mean score $\bar{x}$ of a random sample of 36 students will be greater than 79?

b. What assumptions must be satisfied to make your answer valid?

6.64 Supercooling temperature of frogs. Many species of terrestrial tree frogs that hibernate at or near the ground surface can survive prolonged exposure to low winter temperatures. In freezing conditions, the frog's body temperature, called its *supercooling temperature*, remains relatively higher because of an accumulation of glycerol in its body fluids. Recent studies have shown that the supercooling temperature of terrestrial frogs frozen at $-6°C$ has a relative frequency distribution with a mean of $-2°C$ and a standard deviation of .3°C (The first of these studies was reported in *Science*, May 1983.) Consider the mean supercooling temperature, $\bar{x}$, of a random sample of $n = 42$ terrestrial frogs frozen at $-6°C$.

a. Find the probability that $\bar{x}$ exceeds $-2.05°C$.

b. Find the probability that $\bar{x}$ falls between $-2.20°C$ and $-2.10°C$.

6.65 Florida Employer Opinion Survey. The *Florida Employer Opinion Survey* gives the results of an extensive survey of employer opinions in Florida. Each employer was asked to rate his or her satisfaction with the preparation of employees by the public education system. Responses were 1, 1.5, or 2, representing very dissatisfied, neither satisfied nor dissatisfied, and very satisfied, respectively. A sample of 651 employers was selected. Assume that the mean for all employers in Florida is 1.50 (the "dividing line" between satisfied and dissatisfied) and the standard deviation is .45.

a. Which type of distribution describes the individual survey responses, continuous or discrete?

b. Describe the distribution that best approximates the sample mean response of 651 employers. What are the mean and standard deviation of this distribution? What assumptions, if any, are necessary to ensure the validity of your answers?

c. What is the approximate probability that the sample mean will be 1.45 or less?

d. The mean of the sample of 651 employers surveyed in a recent year was 1.36. Given this result, do you think it is likely that all Florida employers' opinions were evenly divided on the effectiveness of public education? That is, do you think the assumption that the population mean is 1.50 is correct? Why or why not?

6.66 Verbal ability of delinquents. Over the past 20 years, the *Journal of Abnormal Psychology* has published numerous studies on the relationship between juvenile delinquency and poor verbal abilities. Assume that scores on a verbal IQ test have a population mean $\mu = 107$ and a population standard deviation $\sigma = 15$.

a. What shape would you expect the sampling distribution of $\bar{x}$ for $n = 84$ juveniles to have? Does your answer depend on the shape of the distribution of verbal IQ scores for all juveniles?

b. Assuming that the population mean and standard deviation for juveniles with no record of delinquency are the same as those for all juveniles, approximate the probability that the sample mean verbal IQ for $n = 84$ juveniles will be 110 or more. State any assumptions you make.

c. In one cited study, the researchers found that a sample of $n = 84$ juveniles with no record of delinquency had a mean verbal IQ of $\bar{x} = 110$. Considering your answer to part **b**, do you think that the population mean and standard deviation for nondelinquent juveniles are the same as those for all juveniles? Explain.

Applying the Concepts—Advanced

6.67 Flaws in aluminum siding. [*Note:* This exercise refers to an optional section in Chapter 4.] A building contractor has decided to purchase a load of factory-reject aluminum siding as long as the average number of flaws per piece of siding in a sample of size 35 from the factory's reject pile is 2.1 or less. If it is known that the number of flaws per piece of siding in the factory's reject pile has a Poisson probability distribution with a mean of 2.5, find the approximate probability that the contractor will not purchase a load of siding. [*Hint:* If x is a Poisson random variable with mean λ, then $\sigma_{\bar{x}}^2$ also equals λ.]

6.68 Machine repair time. [*Note:* This exercise refers to an optional section in Chapter 5.] An article in *Industrial Engineering* (August 1990) discussed the importance of modeling machine downtime correctly in simulation studies. As an illustration, the researcher considered a single-machine-tool system with repair times (in minutes) that can be modeled by an exponential distribution

with $\theta = 60$ (see optional Section 5.6). Of interest is the mean repair time, $\bar{x}$, of a sample of 100 machine breakdowns.
   a. Find $E(\bar{x})$ and the variance of $\bar{x}$.
   b. What probability distribution provides the best model of the sampling distribution of $\bar{x}$? Why?
   c. Calculate the probability that the mean repair time, $\bar{x}$, is no longer than 30 minutes.

Critical Thinking Challenges

6.69 Soft drink bottles. A soft drink bottler purchases glass bottles from a vendor. The bottles are required to have an internal pressure of at least 150 pounds per square inch (psi). A prospective bottle vendor claims that its production process yields bottles with a mean internal pressure of 157 psi and a standard deviation of 3 psi. The bottler strikes an agreement with the vendor that permits the bottler to sample from the vendor's production process to verify the vendor's claim. The bottler randomly selects 40 bottles from the last 10,000 produced, measures the internal pressure of each, and finds the mean pressure for the sample to be 1.3 psi below the process mean cited by the vendor.
   a. Assuming the vendor's claim to be true, what is the probability of obtaining a sample mean this far or farther below the process mean? What does your answer suggest about the validity of the vendor's claim?
   b. If the process standard deviation were 3 psi as claimed by the vendor, but the mean were 156 psi, would the observed sample result be more or less likely than in part **a**? What if the mean were 158 psi?
   c. If the process mean were 157 psi as claimed, but the process standard deviation were 2 psi, would the sample result be more or less likely than in part **a**? What if instead the standard deviation were 6 psi?

6.70 Fecal pollution at Huntington Beach. The state of California mandates fecal indicator bacteria monitoring at all public beaches. When the concentration of fecal bacteria in the water exceeds a certain limit (400 colony-forming units of fecal coliform per 100 milliliters), local health officials must post a sign (called surf zone posting) warning beachgoers of potential health risks with entering the water. For fecal bacteria, the state uses a single-sample standard; that is, if the fecal limit is exceeded in a single sample of water, surf zone posting is mandatory. This single-standard policy has led to a recent rash of beach closures in California.

Joon Ha Kim and Stanley B. Grant, engineers at the University of California, Irvine, conducted a study of the surf water quality at Huntington Beach in California and reported the results in *Environmental Science & Technology* (September 2004). The researchers found that beach closings were occurring despite low pollution levels in some instances and in others, signs were not posted when the fecal limit was exceeded. They attributed these "surf zone posting errors" to the variable nature of water quality in the surf zone (for example, fecal bacteria concentration tends to be higher during ebb tide and at night) and the inherent time delay between when a water sample is collected and when a sign is posted or removed. In order to prevent posting errors, the researchers recommend using an averaging method rather than a single sample to determine unsafe water quality. (For example, one simple averaging method is to take a random sample of multiple water specimens and compare the average fecal bacteria level of the sample to the limit of 400 cfu/100 mL in order to determine whether the water is safe.)

Discuss the pros and cons of using the single-sample standard versus the averaging method. Part of your discussion should address the probability of posting a sign when, in fact, the water is safe, and the probability of posting a sign when, in fact, the water is unsafe. (Assume that the fecal bacteria concentrations of water specimens at Huntington Beach follow an approximately normal distribution.)

Student Projects

To understand the Central Limit Theorem and sampling distribution, consider the following experiment: Toss four identical coins, and record the number of heads observed. Then repeat this experiment four more times, so that you end up with a total of five observations for the random variable x, the number of heads when four coins are tossed.

Now derive and graph the probability distribution for x, assuming the coins are balanced. Note that the mean of this distribution is $\mu = 2$ and the standard deviation is $\sigma = 1$. This probability distribution represents the one from which you are drawing a random sample of five measurements.

Next, calculate the mean $\bar{x}$ of the five measurements—that is, calculate the mean number of heads you observed in five repetitions of the experiment. Although you have repeated the basic experiment five times, you have only one observed value of $\bar{x}$. To derive the probability distribution or sampling distribution of $\bar{x}$ empirically, you have to repeat the entire process (of tossing four coins five times) many times. Do it 100 times.

The approximate sampling distribution of $\bar{x}$ can be derived theoretically by making use of the Central Limit Theorem. We expect at least an approximate normal probability distribution with a mean $\mu = 2$ and a standard deviation

$$\sigma_{\bar{x}} = \frac{\sigma}{\sqrt{n}} = \frac{1}{\sqrt{5}} = .45$$

Count the number of your 100 $\bar{x}$'s that fall in each of the intervals in the accompanying figure. Use the normal probability distribution with $\mu = 2$ and $\sigma_{\bar{x}} = .45$ to calculate the expected number of the 100 $\bar{x}$'s in each of the intervals. How closely does the theory describe your experimental results?

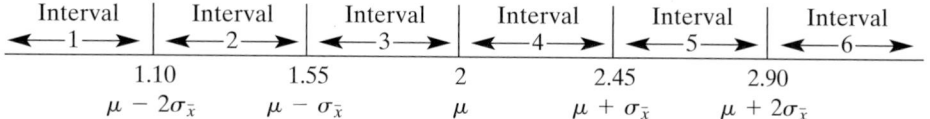

Interval ←—1—→	Interval ←—2—→	Interval ←—3—→	Interval ←—4—→	Interval ←—5—→	Interval ←—6—→
1.10 $\mu - 2\sigma_{\bar{x}}$	1.55 $\mu - \sigma_{\bar{x}}$	2 μ	2.45 $\mu + \sigma_{\bar{x}}$	2.90 $\mu + 2\sigma_{\bar{x}}$	

REFERENCES

Hogg, R. V., and Craig, A. T. *Introduction to Mathematical Statistics*, 5th ed. Upper Saddle River, N.J.: Prentice Hall, 1995.

Larsen, R. J., and Marx, M. L. *An Introduction to Mathematical Statistics and Its Applications*, 3rd ed. Upper Saddle River, N.J.: Prentice Hall, 2001.

Lindgren, B. W. *Statistical Theory*, 3rd ed. New York: Macmillan, 1976.

Wackerly, D., Mendenhall, W., and Scheaffer, R. L. *Mathematical Statistics with Applications*, 6th ed. North Scituate, Mass.: Duxbury, 2002.

Using Technology

Simulating a Sampling Distribution Using MINITAB

To generate a sampling distribution for a sample statistic using MINITAB, click on the "Calc" button on the MINITAB menu bar, next click on "Random Data," then click on the distribution of your choice (e.g., "Uniform"). A dialog box similar to the one (the Uniform Distribution) shown in Figure 6.M.1 will appear. Specify the number of samples (e.g., 1,000) to generate in the "Generate … rows of data" box and the columns where the data will be stored in the "Store in columns" box. (The number of columns will be equal to the sample size, e.g., 40.) Finally, specify the parameters of the distribution (e.g., the lower and upper range of the uniform distribution). When you click "OK," the simulated data will appear on the MINITAB worksheet.

Next, calculate the value of the sample statistic of interest for each sample. To do this, click on the "Calc" button on the MINITAB menu bar, then click on "Row Statistics," as shown in Figure 6.M.2. The resulting dialog box appears in Figure 6.M.3. Check the sample statistic (e.g., the mean) you want to calculate, and specify the "Input variables" (or columns) and the column where you want the value of the sample statistic to be saved. Click "OK" and the value of the statistic for each sample will appear on the MINITAB worksheet.

[*Note*: Use the MINITAB menu choices provided in the Chapter 2 Using Technology section to generate a histogram of the sampling distribution of the statistic or to find the mean and variance of the sampling distribution.]

Figure 6.M.1
MINITAB Dialog Box for
Simulating the Uniform
Distribution

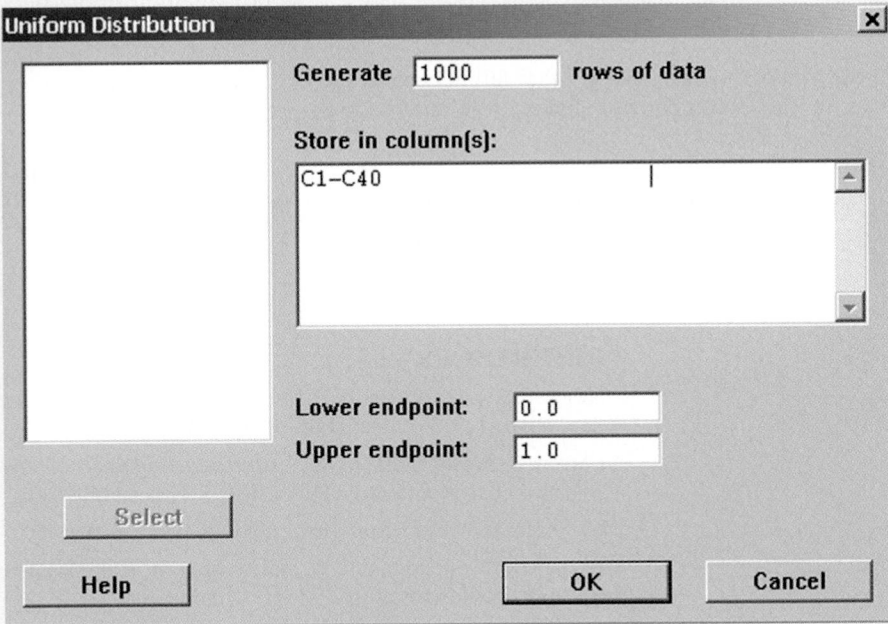

Figure 6.M.2
MINITAB Selections for
Generating Sample Statistics
for the Simulated Data

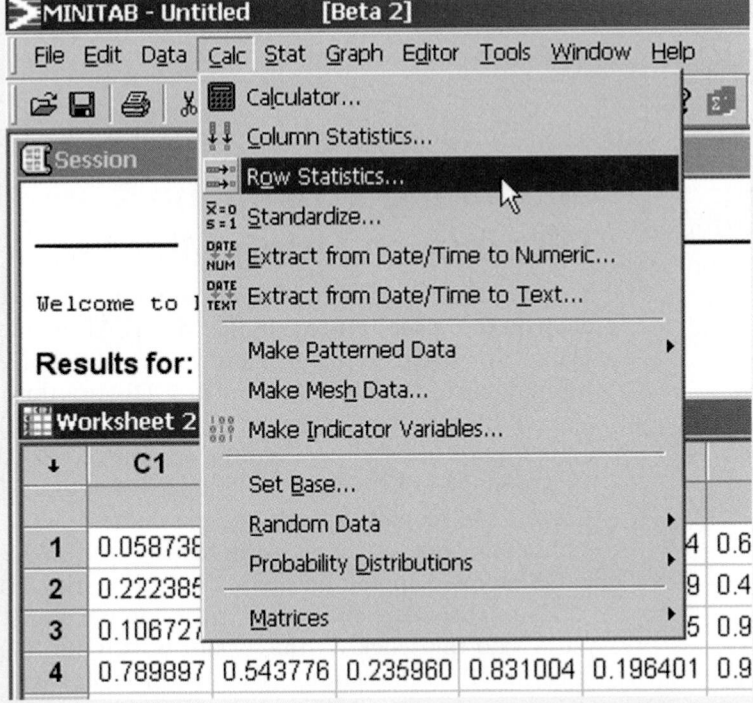

Figure 6.M.3
MINITAB Row Statistics
Dialog Box

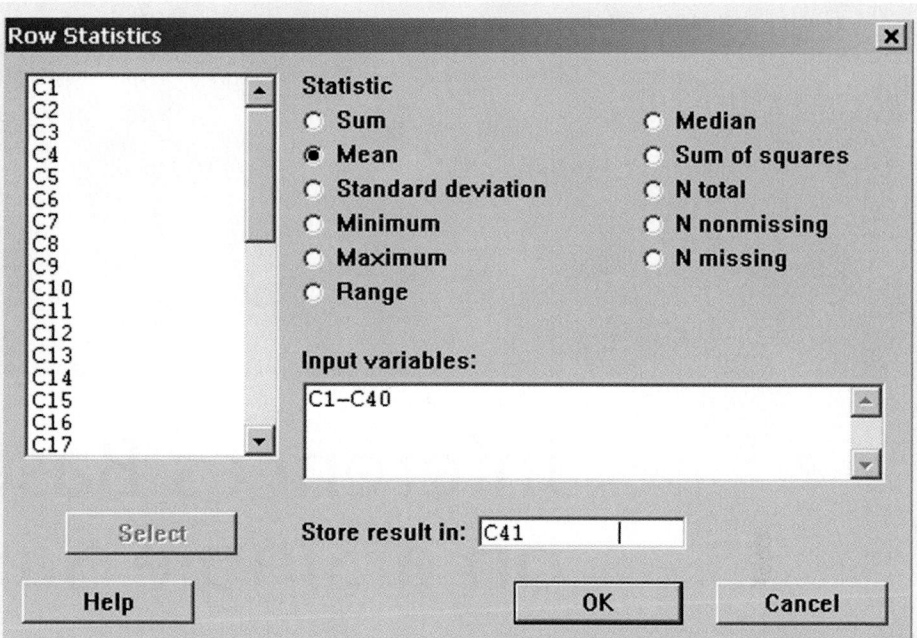

7

Inferences Based on a Single Sample

Estimation with Confidence Intervals

Contents

Statistics in Action

Speed—Improving the Sprint Times of High School Football Players

Using Technology

Confidence Intervals Using MINITAB

Where We've Been

- Learned that populations are characterized by numerical descriptive measures—called *parameters*

- Found that decisions about population parameters are based on *statistics* computed from the sample

- Discovered that *inferences* about parameters are subject to uncertainty, and that this uncertainty is reflected in the *sampling distribution* of a statistic

Where We're Going

- Estimate a population parameter (means or proportion) based on a large sample selected from the population.

- Use the sampling distribution of a statistic to form a confidence interval for the population parameter.

- Show how to select the proper sample size for estimating a population parameter.

Statistics in *ACTION*

Speed—Improving the Sprint Times of High School Football Players

The game of football requires both skill and speed. Consider that one of the key "statistics" for National Football League (NFL) coaches in evaluating college players for the NFL draft is a player's 40-yard sprint time. There are a plethora of drills available to football coaches that have been successfully used to aid a player's skill development. However, many coaches are under the impression that "you can't teach speed." Michael Gray and Jessica Sauerbeck, researchers at Northern Kentucky University, designed and tested a speed training program for junior varsity and varsity high school football players (*The Sport Journal*, Winter 2004).

The training program was carried out over a five-week period and incorporated a number of speed improvement drills, including 50-yard sprints run at varying speeds, high knee running sprints, butt kick sprints, "crazy legs" straddle runs, quick feet drills, jumping, power skipping, and all-out sprinting. A sample of 38 high school athletes participated in the study. Each participant was timed in a 40-yard sprint prior to the start of the training program and timed again after completing the program. The decrease in times (mea-

sured in seconds) was recorded for each athlete. These data, saved in the **SPRINT** file, are shown in Table SIA 7.1. [*Note:* A negative decrease implies that the athlete's time after completion of the program was higher than his time prior to training.] The goal of the research is to demonstrate that the training program is effective in improving 40-yard sprint times.

In this chapter, several Statistics in Action Revisited examples demonstrate how confidence intervals can be used to evaluate the effectiveness of the speed training program.

Statistics in Action Revisited

- Estimating the Mean Decrease in Sprint Time (p. 328)
- Estimating the Proportion of Sprinters Who Improve after Training (p. 347)
- Determining the Number of Athletes to Participate in the Training Program (p. 355)

SPRINT

TABLE SIA7.1	Decrease in 40-Yard Sprint Times for 38 Football Players								
−.01	.10	.10	.24	.25	.05	.28	.25	.20	.14
.32	.34	.30	.09	.05	0.00	.04	.17	0.00	.21
.15	.30	.02	.12	.14	.10	.08	.50	.36	.10
.01	.90	.34	.38	.44	.08	0.00	0.00		

Source: Gray, M., and Sauerbeck, J. A. "Speed training program for high school football players," *The Sport Journal,* Vol. 7, No. 1, Winter 2004 (Table 2).

7.1 Identifying the Target Parameter

In this chapter, our goal is to estimate the value of an unknown population parameter, such as a population mean or a proportion from a binomial population. For example, we might want to know the mean gas mileage for a new car model, or the average expected life of a flat screen computer monitor, or the proportion of Iraq War veterans with post-traumatic stress syndrome.

You'll see that different techniques are used for estimating a mean or proportion, depending on whether a sample contains a large or small number of measurements. Nevertheless, our objectives remain the same. We want to use the sample information to estimate the population parameter of interest (called the **target parameter**) and to assess the reliability of the estimate.

> **DEFINITION 7.1**
>
> The unknown population parameter (e.g., mean or proportion) that we are interested in estimating is called the **target parameter**.

Often, there are one or more key words in the statement of the problem that indicate the appropriate target parameter. Some key words associated with the two parameters covered in this section are listed in the accompanying box.

Determining the Target Parameter

Parameter	Key Words or Phrases	Type of Data
μ	Mean; average	Quantitative
p	Proportion; percentage; fraction; rate	Qualitative

For the examples given above, the words *mean* in "mean gas mileage" and *average* in "average life expectancy" imply that the target parameter is the population mean, μ. The word *proportion* in "proportion of Iraq War veterans with post-traumatic stress syndrome" indicated that the target parameter is the binomial proportion, p.

In addition to key words and phrases, the type of data (quantitative or qualitative) collected is indicative of the target parameter. With quantitative data, you are likely to be estimating the mean of the data. With qualitative data with two outcomes (success or failure), the binomial proportion of successes is likely to be the parameter of interest.

We consider a method of estimating a population mean using a large sample in Section 7.2 and a small sample in Section 7.3. Estimation of a population proportion is presented in Section 7.4. Finally, we show how to determine the sample sizes necessary for reliable estimates of the target parameters in Section 7.5.

7.2 Large-Sample Confidence Interval for a Population Mean

Suppose a large hospital wants to estimate the average length of time patients remain in the hospital. Hence, the hospital's target parameter is the population mean, μ. To accomplish this objective, the hospital administrators plan to randomly sample 100 of all previous patients' records and to use the sample mean, $\bar{x}$, of the lengths of stay to estimate μ, the mean of *all* patients' visits. The sample mean $\bar{x}$ represents a *point estimator* of the population mean μ (Definition 6.4). How can we assess the accuracy of this large-sample point estimator?

According to the Central Limit Theorem, the sampling distribution of the sample mean is approximately normal for large samples, as shown in Figure 7.1.

Let us calculate the interval

$$\bar{x} \pm 2\sigma_{\bar{x}} = \bar{x} \pm \frac{2\sigma}{\sqrt{n}}$$

That is, we form an interval 4 standard deviations wide—from 2 standard deviations below the sample mean to 2 standard deviations above the mean. Prior to drawing the sample, what are the chances that this interval will enclose μ, the population mean?

To answer this question, refer to Figure 7.1. If the 100 measurements yield a value of $\bar{x}$ that falls between the two lines on either side of μ (i.e., within 2 standard deviations of μ), then the interval $\bar{x} \pm 2\sigma_{\bar{x}}$ will contain μ; if $\bar{x}$ falls outside these boundaries, the interval $\bar{x} \pm 2\sigma_{\bar{x}}$ will not contain μ. Since the area under the normal curve (the sampling distribution of $\bar{x}$) between these boundaries is about .95 (more

Figure 7.1

Sampling Distribution of $\bar{x}$

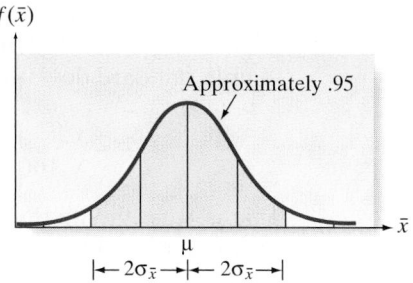

precisely, from Table IV in Appendix A the area is .9544), we know that the interval $\bar{x} \pm 2\sigma_{\bar{x}}$ will contain μ with a probability approximately equal to .95.

For instance, consider the lengths of time spent in the hospital for 100 patients shown in Table 7.1. An SAS printout of summary statistics for the sample of 100 lengths of stay is shown in Figure 7.2.

From the top of the printout, we find $\bar{x} = 4.53$ days and $s = 3.68$ days. To achieve our objective, we must construct the interval

$$\bar{x} \pm 2\sigma_{\bar{x}} = 4.53 \pm 2 \frac{\sigma}{\sqrt{100}}$$

But now we face a problem. You can see that without knowing the standard deviation σ of the original population—that is, the standard deviation of the lengths of

◉ **HOSPLOS**

TABLE 7.1 Lengths of Stay (in Days) for 100 Patients

2	3	8	6	4	4	6	4	2	5
8	10	4	4	4	2	1	3	2	10
1	3	2	3	4	3	5	2	4	1
2	9	1	7	17	9	9	9	4	4
1	1	1	3	1	6	3	3	2	5
1	3	3	14	2	3	9	6	6	3
5	1	4	6	11	22	1	9	6	5
2	2	5	4	3	6	1	5	1	6
17	1	2	4	5	4	4	3	2	3
3	5	2	3	3	2	10	2	4	2

Figure 7.2

SAS Printout with Summary Statistics and 95% Confidence Interval for Data on 100 Hospital Stays

```
Sample Statistics for LOS

     N          Mean          Std. Dev.          Std. Error
--------------------------------------------------------------
    100         4.53            3.68                0.37

Hypothesis Test

    Null hypothesis:     Mean of LOS =  0
    Alternative:         Mean of LOS ^= 0

          t Statistic          Df          Prob > t
    --------------------------------------------------
             12.318            99           <.0001


95 % Confidence Interval for the Mean

         Lower Limit:              3.80
         Upper Limit:              5.26
```

stay of *all* patients—we cannot calculate this interval. However, since we have a large sample ($n = 100$ measurements), we can approximate the interval by using the sample standard deviation s to approximate σ. Thus,

$$\bar{x} \pm 2\frac{\sigma}{\sqrt{100}} \approx \bar{x} \pm 2\frac{s}{\sqrt{100}} = 4.53 \pm 2\left(\frac{3.68}{10}\right) = 4.53 \pm .74$$

That is, we estimate the mean length of stay in the hospital for all patients to fall in the interval 3.79 to 5.27 days. (This interval is highlighted at the bottom of the SAS printout, Figure 7.2. Differences are due to rounding.)

Can we be sure that μ, the true mean, is in the interval 3.79 to 5.27? We cannot be certain, but we can be reasonably confident that it is. This confidence is derived from the knowledge that if we were to draw repeated random samples of 100 measurements from this population and form the interval $\bar{x} \pm 2\sigma_{\bar{x}}$ each time, approximately 95% of the intervals would contain μ. We have no way of knowing (without looking at all the patients' records) whether our sample interval is one of the 95% that contain μ or one of the 5% that do not, but the odds certainly favor its containing μ. Consequently, the interval 3.79 to 5.27 provides a reliable estimate of the mean length of patient stay in the hospital.

The formula that tells us how to calculate an interval estimate based on sample data is called an *interval estimator*, or *confidence interval*. The probability, .95, that measures the confidence we can place in the interval estimate is called a *confidence coefficient*. The percentage, 95%, is called the *confidence level* for the interval estimate. It is not usually possible to assess precisely the reliability of point estimators because they are single points rather than intervals. So, because we prefer to use estimators for which a measure of reliability can be calculated, we will generally use interval estimators.

> **DEFINITION 7.2**
>
> An **interval estimator** (or **confidence interval**) is a formula that tells us how to use sample data to calculate an interval that estimates a population parameter.

> **DEFINITION 7.3**
>
> The **confidence coefficient** is the probability that an interval estimator encloses the population parameter—that is, the relative frequency with which the interval estimator encloses the population parameter when the estimator is used repeatedly a very large number of times. The **confidence level** is the confidence coefficient expressed as a percentage.

Now we have seen how an interval can be used to estimate a population mean. When we use an interval estimator, we can usually calculate the probability that the estimation *process* will result in an interval that contains the true value of the population mean. That is, the probability that the interval contains the parameter in repeated usage is usually known. Figure 7.3 shows what happens when 10 different samples are drawn from a population, and a confidence interval for μ is calculated from each. The location of μ is indicated by the vertical line in the figure. Ten confidence intervals, each based on one of 10 samples, are shown as horizontal line segments. Note that the confidence intervals move from sample to sample—sometimes containing μ and other times missing μ. *If our confidence level is 95%, in the long run, 95% of our sample confidence intervals will contain μ.*

Suppose you wish to choose a confidence coefficient other than .95. Notice in Figure 7.1 that the confidence coefficient .95 is equal to the total area under the

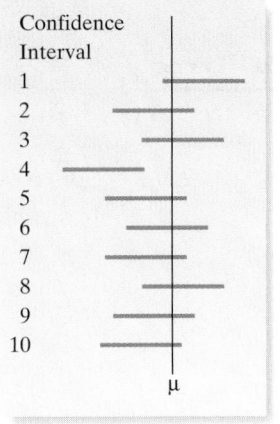

Figure 7.3

Confidence Intervals for μ: 10 Samples

Biography

JERZEY NEYMAN (1894–1981)—Speaking Statistics with a Polish Accent

Polish-born Jerzey Neyman was educated at the University of Kharkov (Russia) in elementary mathematics but taught himself graduate mathematics by studying journal articles on the subject. After receiving his doctorate in 1924 from the University of Warsaw (Poland), Neyman accepted a position at University College (London). There, he developed a friendship with Egon Pearson; Neyman and Pearson together developed the theory of hypothesis testing (Chapter 8). In 1934, in a talk to the Royal Statistical Society, Neyman first proposed the idea of interval estimation, which he called "confidence intervals." (It is interesting that Neyman rarely receives credit in textbooks as the originator of the confidence interval procedure.) In 1938, he emigrated to the United States and the University of California at Berkeley. At Berkeley, he built one of the strongest statistics departments in the country. Jerzey Neyman is considered one of the great founders of modern statistics. He was a superb teacher and innovative researcher who loved his students, always sharing his ideas with them. Neyman's influence on those he met is best expressed by a quote from prominent statistician David Salsburg: "We have all learned to speak statistics with a Polish accent."

sampling distribution, less .05 of the area, which is divided equally between the two tails. Using this idea, we can construct a confidence interval with any desired confidence coefficient by increasing or decreasing the area (call it α) assigned to the tails of the sampling distribution (see Figure 7.4). For example, if we place area $\alpha/2$ in each tail and if $z_{\alpha/2}$ is the z value such that the area $\alpha/2$ will lie to its right, then the confidence interval with confidence coefficient $(1 - \alpha)$ is

$$\bar{x} \pm z_{\alpha/2}\sigma_{\bar{x}}$$

DEFINITION 7.4

The value z_α is defined as the value of the standard normal random variable z such that the area α will lie to its right. In other words, $P(z > z_\alpha) = \alpha$.

To illustrate, for a confidence coefficient of .90 we have $(1 - \alpha) = .90$, $\alpha = .10$, and $\alpha/2 = 0.5$; $z_{.05}$ is the z value that locates area .05 in the upper tail of the sampling distribution. Recall that Table IV in Appendix A gives the areas between the mean and a specified z-value. Since the total area to the right of the mean is .5, we find that $z_{.05}$ will be the z value corresponding to an area of $.5 - .05 = .45$ to the right of the mean (see Figure 7.5). This z value is $z_{.05} = 1.645$.

Confidence coefficients used in practice usually range from .90 to .99. The most commonly used confidence coefficients with corresponding values of α and $z_{\alpha/2}$ are shown in Table 7.2.

Figure 7.4

Locating $z_{\alpha/2}$ on the Standard Normal Curve

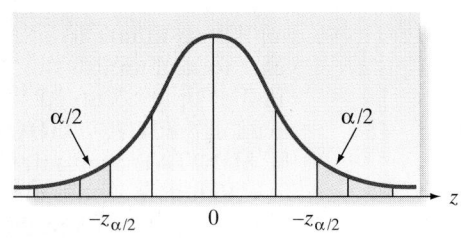

Figure 7.5

The z Value $(z_{.05})$ Corresponding to an Area Equal to .05 in the Upper Tail of the z-Distribution

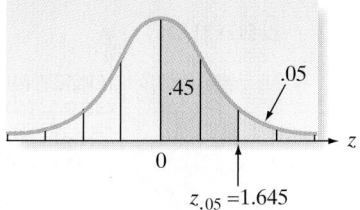

$$z_{.05} = 1.645$$

TABLE 7.2 Commonly Used Values of $z_{\alpha/2}$

Confidence Level $100(1 - \alpha)$	α	$\alpha/2$	$z_{\alpha/2}$
90%	.10	.05	1.645
95%	.05	.025	1.96
99%	.01	.005	2.575

Now Work *Exercise 7.7*

Large-Sample $100(1 - \alpha)\%$ Confidence Interval for μ

$$\bar{x} \pm z_{\alpha/2}\sigma_{\bar{x}} = \bar{x} \pm z_{\alpha/2}\frac{\sigma}{\sqrt{n}}$$

where $z_{\alpha/2}$ is the z value with an area $\alpha/2$ to its right (see Figure 7.4) and $\sigma_{\bar{x}} = \sigma/\sqrt{n}$. The parameter σ is the standard deviation of the sampled population and n is the sample size.

Note: When σ is unknown (as is almost always the case) and n is large (say, $n \geq 30$), the confidence interval is approximately equal to

$$\bar{x} \pm z_{\alpha/2}\left(\frac{s}{\sqrt{n}}\right)$$

where s is the sample standard deviation.

Conditions Required for a Valid Large-Sample Confidence Interval for μ

1. A random sample is selected from the target population.
2. The sample size n is large (i.e., $n \geq 30$). (Due to the Central Limit Theorem, this condition guarantees that the sampling distribution of $\bar{x}$ is approximately normal.)

EXAMPLE 7.1   FINDING A LARGE-SAMPLE CONFIDENCE INTERVAL FOR μ

Problem   Unoccupied seats on flights cause airlines to lose revenue. Suppose a large airline wants to estimate its average number of unoccupied seats per flight over the past year. To accomplish this, the records of 225 flights are randomly selected, and the number of unoccupied seats is noted for each of the sampled flights. (The data are saved in the **AIRNOSHOWS** file.) Descriptive statistics for the data are displayed in the MINITAB printout, Figure 7.6.

AIRNOSHOWS

Estimate μ, the mean number of unoccupied seats per flight during the past year, using a 90% confidence interval.

Figure 7.6

MINITAB Printout with
Descriptive Statistics and
90% Confidence Interval
for Example 7.1

```
Variable    N     Mean    StDev   SE Mean        90% CI
NOSHOWS   225   11.5956   4.1026   0.2735   (11.1438, 12.0473)
```

Solution   The general form of the 90% confidence interval for a population mean is

$$\bar{x} \pm z_{\alpha/2}\sigma_{\bar{x}} = \bar{x} \pm z_{.05}\sigma_{\bar{x}} = \bar{x} \pm 1.645\left(\frac{\sigma}{\sqrt{n}}\right)$$

From Figure 7.6, we find (after rounding) $\bar{x} = 11.6$. Since we do not know the value of σ (the standard deviation of the number of unoccupied seats per flight for all flights of the year), we use our best approximation—the sample standard deviation, $s = 4.1$, shown on the MINITAB printout. Then the 90% confidence interval is, approximately,

$$11.6 \pm 1.645\left(\frac{4.1}{\sqrt{225}}\right) = 11.6 \pm .45$$

or from 11.15 to 12.05. That is, at the 90% confidence level, we estimate the mean number of unoccupied seats per flight to be between 11.15 and 12.05 during the sampled year. This result is verified (except for rounding) on the right side of the MINITAB printout in Figure 7.6.

Look Back   We stress that the confidence level for this example, 90%, refers to the procedure used. If we were to apply this procedure repeatedly to different samples, approximately 90% of the intervals would contain μ. Although we do not know for sure whether this particular interval (11.15, 12.05) is one of the 90% that contain μ or one of the 10% that do not, our knowledge of probability gives us "confidence" that the interval contains μ.

> **Now Work**   *Exercise 7.11*

---   ■ ■ ■   ---

The interpretation of confidence intervals for a population mean is summarized in the next box.

Sometimes, the estimation procedure yields a confidence interval that is too wide for our purposes. In this case, we will want to reduce the width of the interval to obtain a more precise estimate of μ. One way to accomplish this is to decrease the confidence coefficient, $1 - \alpha$. For example, consider the problem of estimating the mean length of stay, μ, for hospital patients. Recall that for a sample of 100 patients, $\bar{x} = 4.53$ days and $s = 3.68$ days. A 90% confidence interval for μ is

$$\bar{x} \pm 1.645(\sigma/\sqrt{n}) \approx 4.53 \pm (1.645)(3.68/\sqrt{100}) = 4.53 \pm .61$$

or (3.92, 5.14). You can see that this interval is narrower than the previously calculated 95% confidence interval, (3.79, 5.27). Unfortunately, we also have "less confidence" in the 90% confidence interval. An alternative method used to decrease the width of an interval without sacrificing "confidence" is to increase the sample size n. We demonstrate this method in Section 7.5.

> **Interpretation of a Confidence Interval for a Population Mean**
>
> When we form a $100(1 - \alpha)\%$ confidence interval for μ, we usually express our confidence in the interval with a statement such as, "We can be $100(1 - \alpha)\%$ confident that μ lies between the lower and upper bounds of the confidence interval," where for a particular application, we substitute the appropriate numerical values

for the level of confidence and for the lower and upper bounds. *The statement reflects our confidence in the estimation process rather than in the particular interval that is calculated from the sample data.* We know that repeated application of the same procedure will result in different lower and upper bounds on the interval. Furthermore, we know that $100(1 - \alpha)\%$ of the resulting intervals will contain μ. There is (usually) no way to determine whether any particular interval is one of those that contain μ, or one that does not. However, unlike point estimators, confidence intervals have some measure of reliability, the confidence coefficient, associated with them. For that reason they are generally preferred to point estimators.

Statistics in Action Revisited

Estimating the Mean Decrease in Sprint Time

Refer to the speed training program for junior varsity and varsity high school football players described on p. 321. Recall that the 5-week training program incorporated a number of speed improvement drills for a sample of 38 high school athletes. The time in a 40-yard sprint run both before and after the training program was recorded for each athlete and the decrease in times (measured in seconds) was saved in the **SPRINT** file (Table SIA7.1). Is the training program really effective in improving 40-yard sprint times?

One way to answer this question is to form a confidence interval for the true mean decrease in sprint time for all athletes who participate in the speed training program. If the true decrease is positive (i.e., if the "before" mean time is greater than the "after" mean time), then we will conclude that the training program is effective. Since the sam-

ple size is large ($n = 38$), we can apply the large-sample methodology of this section.

The data is stored as a MINITAB worksheet and the software used to find a large-sample 95% confidence interval for the population mean decrease in sprint times. The MINITAB printout is displayed in Figure SIA7.1.

The interval, highlighted on the printout, is (.128, .247). Note that the entire interval falls above 0. Consequently, we are 95% confident that the true mean decrease in sprint times is positive. It appears that the speed training program is, in fact, effective in improving the average 40-yard sprint times of high school athletes. Note, however, that the mean decrease in time ranges from a minimum of .128 seconds to a maximum of .247 seconds. Although this decrease might be deemed critically important to world-class sprinters, it would probably be unnoticeable to high school football players.

Figure SIA7.1

MINITAB Confidence Interval for Speed Training Study

Variable	N	Mean	StDev	SE Mean	95% CI
DecrTime	38	0.187895	0.181423	0.029431	(0.128262, 0.247527)

Confidence Interval for a Population Mean (known σ or $n \geq 30$)

Using the TI-83 Graphing Calculator

Say we want to develop a **95%** confidence interval for the population mean from a sample size of **35** where we know the sample mean is **100** and the **population** deviation is **12**. For this problem we want **Z-Interval** because we are given sigma. On the calculator we will choose **"7:Zinterval"** from the **"TESTS"** menu.

Step 1   *Access the Statistical Tests Menu*
Press **STAT** (this screen will appear).
Arrow right to **"TEST"**
Arrow down to **"7:ZInterval"**
Press **ENTER**

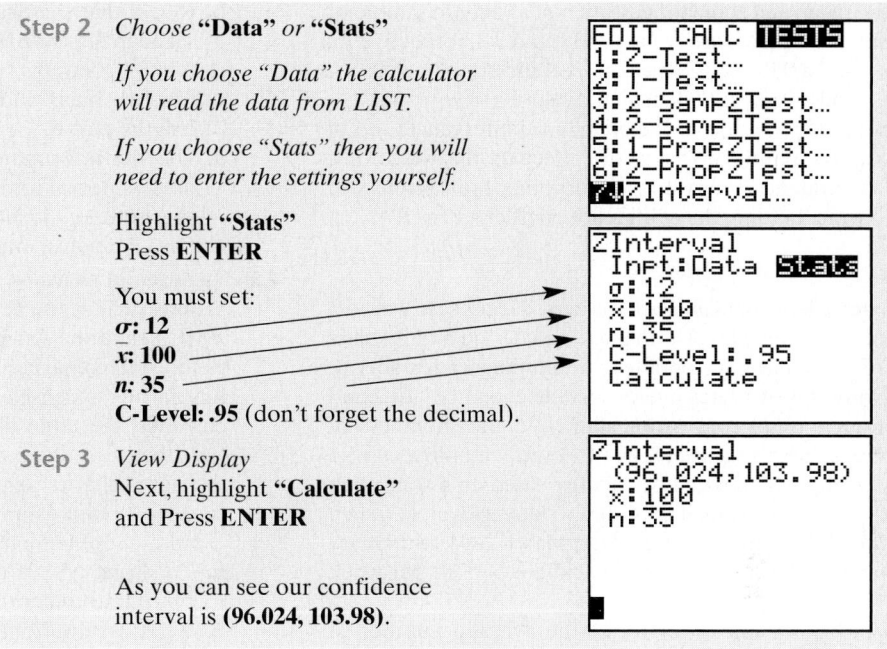

Step 2 *Choose* **"Data"** *or* **"Stats"**

If you choose "Data" the calculator will read the data from LIST.

If you choose "Stats" then you will need to enter the settings yourself.

Highlight **"Stats"**
Press **ENTER**

You must set:
σ: 12
$\bar{x}$: 100
n: 35
C-Level: .95 (don't forget the decimal).

Step 3 *View Display*
Next, highlight **"Calculate"**
and Press **ENTER**

As you can see our confidence interval is **(96.024, 103.98)**.

Exercises 7.1–7.24

Understanding the Principles

7.1 Define the target parameter.

7.2 What is the confidence coefficient in a 90% confidence interval for μ?

7.3 Explain the difference between an interval estimator and a point estimator for μ.

7.4 Explain what is meant by the statement, "We are 95% confident that an interval estimate contains μ."

7.5 Will a large-sample confidence interval be valid if the population from which the sample is taken is not normally distributed? Explain.

7.6 What conditions are required to form a valid large-sample confidence interval for μ?

Learning the Mechanics

7.7 Find $z_{\alpha/2}$ for each of the following:
NW   **a.** $\alpha = .10$ **b.** $\alpha = .01$
     c. $\alpha = .05$ **d.** $\alpha = .20$

7.8 What is the confidence level of each of the following confidence intervals for μ?

a. $\bar{x} \pm 1.96\left(\dfrac{\sigma}{\sqrt{n}}\right)$

b. $\bar{x} \pm 1.645\left(\dfrac{\sigma}{\sqrt{n}}\right)$

c. $\bar{x} \pm 2.575\left(\dfrac{\sigma}{\sqrt{n}}\right)$

d. $\bar{x} \pm 1.28\left(\dfrac{\sigma}{\sqrt{n}}\right)$

e. $\bar{x} \pm .99\left(\dfrac{\sigma}{\sqrt{n}}\right)$

7.9 A random sample of n measurements was selected from a population with unknown mean μ and standard deviation σ. Calculate a 95% confidence interval for μ for each of the following situations:
a. $n = 75$, $\bar{x} = 28$, $s^2 = 12$
b. $n = 200$, $\bar{x} = 102$, $s^2 = 22$
c. $n = 100$, $\bar{x} = 15$, $s = .3$
d. $n = 100$, $\bar{x} = 4.05$, $s = .83$
e. Is the assumption that the underlying population of measurements is normally distributed necessary to ensure the validity of the confidence intervals in parts **a–d**? Explain.

7.10 A random sample of 90 observations produced a mean $\bar{x} = 25.9$ and a standard deviation $s = 2.7$.
a. Find a 95% confidence interval for the population mean μ.
b. Find a 90% confidence interval for μ.
c. Find a 99% confidence interval for μ.

7.11 A random sample of 100 observations from a normally
NW distributed population possesses a mean equal to 83.2 and a standard deviation equal to 6.4.
a. Find a 95% confidence interval for μ.
b. What do you mean when you say that a confidence coefficient is .95?
c. Find a 99% confidence interval for μ.
d. What happens to the width of a confidence interval as the value of the confidence coefficient is increased while the sample size is held fixed?
e. Would your confidence intervals of parts **a** and **c** be valid if the distribution of the original population was not normal? Explain.

7.12 The mean and standard deviation of a random sample of n measurements are equal to 33.9 and 3.3, respectively.
   a. Find a 95% confidence interval for μ if $n = 100$.
   b. Find a 95% confidence interval for μ if $n = 400$.
   c. Find the widths of the confidence intervals found in parts **a** and **b**. What is the effect on the width of a confidence interval of quadrupling the sample size while holding the confidence coefficient fixed?

Applying the Concepts—Basic

7.13 **Latex allergy in healthcare workers.** Healthcare workers who use latex gloves with glove powder on a daily basis are particularly susceptible to developing a latex allergy. Symptoms of a latex allergy include conjunctivitis, hand eczema, nasal congestion, skin rash, and shortness of breath. Each in a sample of 46 hospital employees who were diagnosed with latex allergy based on a skin-prick test reported on their exposure to latex gloves. (*Current Allergy & Clinical Immunology,* March 2004.) Summary statistics for the number of latex gloves used per week are $\bar{x} = 19.3$, $s = 11.9$.
   a. Give a point estimate for the average number of latex gloves used per week by all healthcare workers with a latex allergy.
   b. Form a 95% confidence interval for the average number of latex gloves used per week by all healthcare workers with a latex allergy.
   c. Give a practical interpretation of the interval, part **b**.
   d. Give the conditions required for the interval, part **b**, to be valid.

7.14 **Personal networks of older adults.** In sociology, a personal network is defined as the people who you make frequent contact with. The Living Arrangements and Social Networks of Older Adults (LSN) research program used a stratified random sample of men and women born between 1908 and 1937 to gauge the personal network size of older adults. Each adult in the sample was asked to "please name the people (e.g., in your neighborhood) you have frequent contact with and who are also important to you." Based on the number of people named, the personal network size for each adult was determined. The responses for 2,819 adults in the LSN sample yielded the following statistics on network size: $\bar{x} = 14.6$, $s = 9.8$. (*Sociological Methods & Research,* August 2001.)
   a. Give a point estimate for the mean personal network size of all older adults.
   b. Form a 95% confidence interval for the mean personal network size of all older adults.
   c. Give a practical interpretation of the interval, part **b**.
   d. Give the conditions required for the interval, part **b**, to be valid.

NZBIRDS

7.15 **Extinct New Zealand birds.** Refer to the *Evolutionary Ecology Research* (July 2003) study of the patterns of extinction in the New Zealand bird population, Exercise 2.96 (p. 80). Suppose you are interested in estimating the mean egg length (in millimeters) for the New Zealand bird population.
   a. What is the target parameter?

   b. Recall that the egg lengths for 132 bird species are saved in the **NZBIRDS** file. Obtain a random sample of 50 egg lengths from the data set.
   c. Find the mean and standard deviation of the 50 egg lengths, part **b**.
   d. Use the information, part **c**, to form a 99% confidence interval for the true mean egg length of a bird species found in New Zealand.
   e. Give a practical interpretation of the interval, part **d**.

7.16 **Tax-exempt charities.** Donations to tax-exempt organizations such as the Red Cross, the Salvation Army, the YMCA, and the American Cancer Society not only go to the stated charitable purpose, but are used to cover fundraising expenses and overhead. For a sample of 30 charities, the table lists their charitable commitment (i.e., the percentage of their expenses that go toward the stated charitable purpose).
   a. Give a point estimate for the mean charitable commitment of tax-exempt organizations.
   b. Construct a 98% confidence interval for the mean charitable commitment.
   c. What assumption(s) must hold for the method of estimation used in part **b** to be appropriate?
   d. Why is the confidence interval of part **b** a better estimator of the mean charitable commitment than the point estimator of part **a**?

CHARITY

Organization	Charitable Commitment
American Cancer Society	62%
American National Red Cross	91
Big Brothers Big Sisters of America	77
Boy Scouts of America National Council	81
Boys & Girls Clubs of America	81
CARE	91
Covenant House	15
Disabled American Veterans	65
Ducks Unlimited	78
Feed The Children	90
Girl Scouts of the USA	83
Goodwill Industries International	89
Habitat for Humanity International	81
Mayo Foundation	26
Mothers Against Drunk Drivers	71
Multiple Sclerosis Association of America	56
Museum of Modern Art	79
Nature Conservancy	77
Paralyzed Veterans of America	50
Planned Parenthood Federation	81
Salvation Army	84
Shriners Hospital for Children	95
Smithsonian Institution	87
Special Olympics	72
Trust for Public Land	88
United Jewish Appeal/Federation - NY	75
United States Olympic Committee	78
United Way of New York City	85
WGBH Educational Foundation	81
YMCA of the USA	80

Source: "Look Before You Give," *Forbes,* Dec. 27, 1999, pp. 206–216.

Applying the Concepts—Intermediate

7.17 Colored string preferred by chickens. Animal behaviorists have discovered that the more domestic chickens peck at objects placed in their environment, the healthier the chickens seem to be. White string has been found to be a particularly attractive pecking stimulus. In one experiment, 72 chickens were exposed to a string stimulus. Instead of white string, blue-colored string was used. The number of pecks each chicken took at the blue string over a specified time interval was recorded. Summary statistics for the 72 chickens were: $\bar{x} = 1.13$ pecks, $s = 2.21$ pecks. (*Applied Animal Behaviour Science*, October 2000.)

a. Estimate the population mean number of pecks made by chickens pecking at blue string using a 99% confidence interval. Interpret the result.

b. Previous research has shown that $\mu = 7.5$ pecks if chickens are exposed to white string. Based on the results, part **a**, is there evidence that chickens are more apt to peck at white string than blue string? Explain.

7.18 Sentence complexity study. Refer to the *Applied Psycholinguistics* (June 1998) study of language skills of low-income children, Exercise 2.98 (p. 80). Each in a sample of 65 low-income children was administered the Communicative Development Inventory (CDI) exam. The sentence complexity scores had a mean of 7.62 and a standard deviation of 8.91.

a. Construct a 90% confidence interval for the mean sentence complexity score of all low-income children.

b. Interpret the interval, part **a**, in the words of the problem.

c. Suppose we know that the true mean sentence complexity score of middle-income children is 15.55. Is there evidence that the true mean for low-income children differs from 15.55? Explain.

7.19 Improving SAT scores. Refer to the *Chance* (Winter 2001) and National Education Longitudinal Survey (NELS) study of 265 students who paid a private tutor to help them improve their SAT scores, Exercise 2.101 (p. 81). The changes in both the SAT-Mathematics and SAT-Verbal scores for these students are reproduced in the table.

	SAT-Math	SAT-Verbal
Mean change in score	19	7
Standard deviation of score changes	65	49

a. Construct and interpret a 95% confidence interval for the population mean change in SAT-Mathematics score for students who pay a private tutor.

b. Repeat part **a** for the population mean change in SAT-Verbal score.

c. Suppose the true population mean change in score on one of the SAT tests for all students who paid a private tutor is 15. Which of the two tests, SAT-Mathematics or SAT-Verbal, is most likely to have this mean change? Explain.

7.20 Velocity of light from galaxies. Refer to *The Astronomical Journal* (July 1995) study of the velocity of light emitted from a galaxy in the universe, Exercise 2.100 (p. 80). A sample of 103 galaxies located in the galaxy cluster A2142 had a mean velocity of $\bar{x} = 27,117$ kilometers per second (km/s) and a standard deviation of $s = 1,280$ km/s. Suppose your goal is to make an inference about the population mean light velocity of galaxies in cluster A2142.

a. In part **b** of Exercise 2.100, you constructed an interval that captured approximately 95% of the galaxy velocities in the cluster. Explain why this interval is inappropriate for this inference.

b. Construct a 95% confidence interval for the mean light velocity emitted from all galaxies in cluster A2142. Interpret the result.

7.21 Psychological study of participation and satisfaction. The relationship between an employee's participation in the performance appraisal process and subsequent subordinate reactions toward the appraisal was investigated in the *Journal of Applied Psychology* (August 1998). In Chapter 11 we will discuss a quantitative measure of the relationship between two variables, called the coefficient of correlation r. The researchers obtained r for a sample of 34 studies that examined the relationship between appraisal participation and a subordinate's satisfaction with the appraisal. These correlations are listed in the table. (Values of r near $+1$ reflect a strong positive relationship between the variables.) Find a 95% confidence interval for the mean of the data and interpret it in the words of the problem.

CORR34

.50	.58	.71	.46	.63	.66	.31	.35	.51	.06	.35	.19
.40	.63	.43	.16	−.08	.51	.59	.43	.30	.69	.25	.20
.39	.20	.51	.68	.74	.65	.34	.45	.31	.27		

Source: Cawley, B. D., Keeping, L. M., and Levy, P. E. "Participation in the performance appraisal process and employee reactions: A meta-analytic review of field investigations." *Journal of Applied Psychology,* Vol. 83, No. 4, Aug. 1998, pp. 632–633 (Appendix).

7.22 Attention time given to twins. Psychologists have found that twins, in their early years, tend to have lower IQs and pick up language more slowly than nontwins. The slower intellectual growth of most twins may be caused by benign parental neglect. Suppose it is desired to estimate the mean attention time given to twins per week by their parents. A sample of 50 sets of $2\frac{1}{2}$-year-old twin boys is taken, and at the end of 1 week the attention time given to each pair is recorded. The data (in hours) are listed in the accompanying table.

ATTIMES

20.7	16.7	22.5	12.1	2.9
23.5	6.4	1.3	39.6	35.6
10.9	7.1	46.0	23.4	29.4
44.1	13.8	24.3	9.3	3.4
15.7	46.6	10.6	6.7	5.4
14.0	20.7	48.2	7.7	22.2
20.3	34.0	44.5	23.8	20.0
43.1	14.3	21.9	17.5	9.6
36.4	0.8	1.1	19.3	14.6
32.5	19.1	36.9	27.9	14.0

Find a 90% confidence interval for the mean attention time given to all twin boys by their parents. Interpret the confidence interval.

Applying the Concepts—Advanced

7.23 Job satisfaction of workers. Research reported in the *Journal of Psychology and Aging* (May 1992) studied the role that the age of workers has in determining their level of job satisfaction. The researcher hypothesized that both younger and older workers would have a higher job satisfaction rating than middle-age workers. Each of a sample of 1,686 adults was given a job satisfaction score based on answers to a series of questions. Higher job satisfaction scores indicate higher levels of job satisfaction. The data, arranged by age group, are summarized below.

	Age Group		
	Younger 18–24	Middle Age 25–44	Older 45–64
$\bar{x}$	4.17	4.04	4.31
s	.75	.81	.82
n	241	768	677

a. Construct 95% confidence intervals for the mean job satisfaction scores of each age group. Carefully interpret each interval.

b. In the construction of three 95% confidence intervals, is it more or less likely that at least one of them will *not* contain the population mean it is intended to estimate than it is for a single confidence interval to miss the population mean? [*Hint:* Assume the three intervals are independent, and calculate the probability that at least one of them will not contain the population mean it estimates. Compare this probability to the probability that a single interval fails to enclose the mean.]

c. Based on these intervals, does it appear that the researcher's hypothesis is supported? [*Caution:* We'll learn how to use sample information to compare population means in Chapter 9, and we'll return to this exercise at that time. Here, simply base your opinion on the individual confidence intervals you constructed in part **a**.]

7.24 Study of cockroach growth. According to scientists, the cockroach has had 300 million years to develop a resistance to destruction. In a study conducted by researchers for S. C. Johnson & Son, Inc. (manufacturers of Raid and Off), 5,000 roaches (the expected number in a roach-infested house) were released in the Raid test kitchen. One week later the kitchen was fumigated and 16,298 dead roaches were counted, a gain of 11,298 roaches for the 1-week period. Assume that none of the original roaches died during the 1-week period and that the standard deviation of x, the number of roaches produced per roach in a 1-week period, is 1.5. Use the number of roaches produced by the sample of 5,000 roaches to find a 95% confidence interval for the mean number of roaches produced per week for each roach in a typical roach-infested house.

7.3 Small-Sample Confidence Interval for a Population Mean

Federal legislation requires pharmaceutical companies to perform extensive tests on new drugs before they can be marketed. Initially, a new drug is tested on animals. If the drug is deemed safe after this first phase of testing, the pharmaceutical company is then permitted to begin human testing on a limited basis. During this second phase, inferences must be made about the safety of the drug based on information in very small samples.

Suppose a pharmaceutical company must estimate the average increase in blood pressure of patients who take a certain new drug. Assume that only six patients (randomly selected from the population of all patients) can be used in the initial phase of human testing. The use of a *small sample* in making an inference about μ presents two immediate problems when we attempt to use the standard normal z as a test statistic.

Problem 1 The shape of the sampling distribution of the sample mean $\bar{x}$ (and the z-statistic) now depends on the shape of the population that is sampled. We can no longer assume that the sampling distribution of $\bar{x}$ is approximately normal, because the Central Limit Theorem ensures normality only for samples that are sufficiently large.

Solution to Problem 1 The sampling distribution of $\bar{x}$ (and z) is exactly normal even for relatively small samples if the sampled population is normal. It is approximately normal if the sampled population is approximately normal.

Problem 2   The population standard deviation σ is almost always unknown. Although it is still true that $\sigma_{\bar{x}} = \sigma/\sqrt{n}$, the sample standard deviation s may provide a poor approximation for σ when the sample size is small.

Solution to Problem 2   Instead of using the standard normal statistic

$$z = \frac{\bar{x} - \mu}{\sigma_{\bar{x}}} = \frac{\bar{x} - \mu}{\sigma/\sqrt{n}}$$

which requires knowledge of or a good approximation to σ, we define and use the statistic

$$t = \frac{\bar{x} - \mu}{s/\sqrt{n}}$$

in which the sample standard deviation, s, replaces the population standard deviation, σ.

Biography

WILLIAM S. GOSSET (1876–1937)—Student's t-Distribution

At the age of 23, William Gosset earned a degree in chemistry and mathematics at prestigious Oxford University. He was immediately hired by the Guinness Brewing Company in Dublin, Ireland, for his expertise in chemistry. However, Gosset's mathematical skills allowed him to solve numerous practical problems associated with brewing beer. For example, Gosset applied the Poisson distribution to model the number of yeast cells per unit volume in the fermentation process. His most important discovery was that of the t-distribution in 1908. Since most applied researchers worked with small samples, Gosset was interested in the behavior of the mean in the small sample case. He tediously took numerous small sets of numbers, calculated the mean and standard deviation, obtained their t-ratio, and plotted the results on graph paper. The shape of the distribution was always the same—the t-distribution. Under company policy, employees were forbidden to publish their research results; so Gosset used the penname *Student* to publish a paper on the subject. Hence, the distribution has been called Student's t-distribution.

If we are sampling from a normal distribution, the **t-statistic** has a sampling distribution very much like that of the z-statistic: mound shaped, symmetric, with mean 0. The primary difference between the sampling distributions of t and z is that the t-statistic is more variable than the z, which follows intuitively when you realize that t contains two random quantities ($\bar{x}$ and s), whereas z contains only one ($\bar{x}$).

The actual amount of variability in the sampling distribution of t depends on the sample size n. A convenient way of expressing this dependence is to say that the t statistic has $(n - 1)$ **degrees of freedom (df)**.* Recall that the quantity $(n - 1)$ is the divisor that appears in the formula for s^2. This number plays a key role in the sampling distribution of s^2 and appears in discussions of other statistics in later chapters. Particularly, the smaller the number of degrees of freedom associated with the t-statistic, the more variable will be its sampling distribution.

In Figure 7.7 we show both the sampling distribution of z and the sampling distribution of a t-statistic with 4 df. You can see that the increased variability of the t-statistic means that the t-value, t_α, that locates an area α in the upper tail of the t-distribution is larger than the corresponding value z_α. For any given value of α, the t-value, t_α, increases as the number of degrees of freedom (df) decreases. Values of t that will be used in forming small-sample confidence intervals of μ are

*Since degrees of freedom are related to the sample size, n, it is helpful to think of the number of degrees of freedom as the amount of information in the sample available for estimating the target parameter.

Figure 7.7
Standard Normal (z)
Distribution and
t-Distribution with 4 df

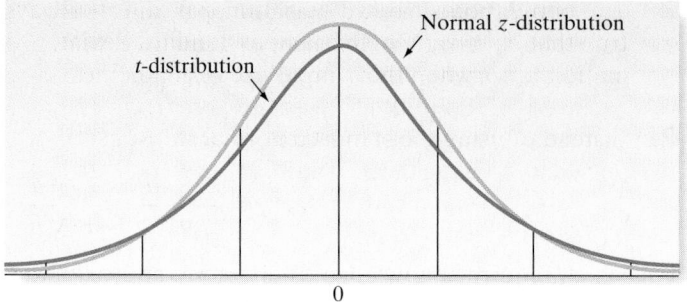

given in Table VI of Appendix A. A partial reproduction of this table is shown in Table 7.3.

Note that t_α values are listed for various degrees of freedom where α refers to the tail area under the t-distribution to the right of t_α. For example, if we want the t-value with an area of .025 to its right and 4 df, we look in the table under the column $t_{.025}$ for the entry in the row corresponding to 4 df. This entry is $t_{.025} = 2.776$, as shown in Figure 7.8. The corresponding standard normal z-score is $z_{.025} = 1.96$.

TABLE 7.3   Reproduction of Part of Table VI in Appendix A

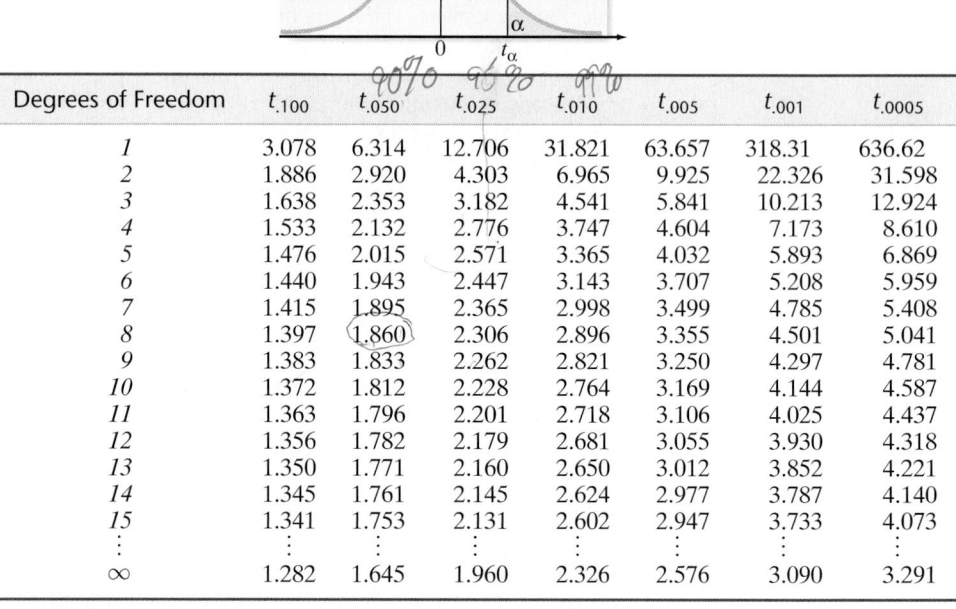

Degrees of Freedom	$t_{.100}$	$t_{.050}$	$t_{.025}$	$t_{.010}$	$t_{.005}$	$t_{.001}$	$t_{.0005}$
1	3.078	6.314	12.706	31.821	63.657	318.31	636.62
2	1.886	2.920	4.303	6.965	9.925	22.326	31.598
3	1.638	2.353	3.182	4.541	5.841	10.213	12.924
4	1.533	2.132	2.776	3.747	4.604	7.173	8.610
5	1.476	2.015	2.571	3.365	4.032	5.893	6.869
6	1.440	1.943	2.447	3.143	3.707	5.208	5.959
7	1.415	1.895	2.365	2.998	3.499	4.785	5.408
8	1.397	1.860	2.306	2.896	3.355	4.501	5.041
9	1.383	1.833	2.262	2.821	3.250	4.297	4.781
10	1.372	1.812	2.228	2.764	3.169	4.144	4.587
11	1.363	1.796	2.201	2.718	3.106	4.025	4.437
12	1.356	1.782	2.179	2.681	3.055	3.930	4.318
13	1.350	1.771	2.160	2.650	3.012	3.852	4.221
14	1.345	1.761	2.145	2.624	2.977	3.787	4.140
15	1.341	1.753	2.131	2.602	2.947	3.733	4.073
$\vdots$	$\vdots$	$\vdots$	$\vdots$	$\vdots$	$\vdots$	$\vdots$	$\vdots$
∞	1.282	1.645	1.960	2.326	2.576	3.090	3.291

Figure 7.8
The $t_{.025}$ Value in a
t-Distribution with 4 df
and the Corresponding
$z_{.025}$ Value

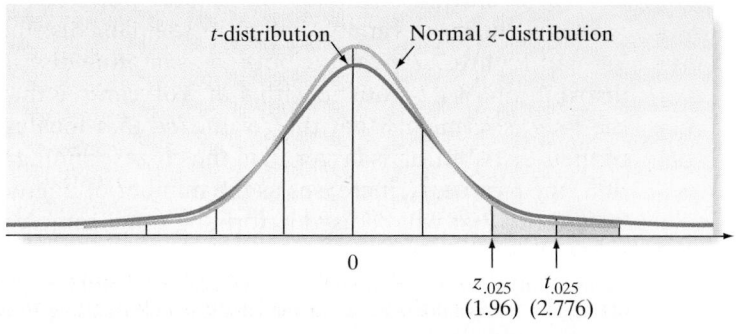

Note that the last row of Table VI, where df $= \infty$ (infinity), contains the standard normal z-values. This follows from the fact that as the sample size n grows very large, s becomes closer to σ and thus t becomes closer in distribution to z. In fact, when df $= 29$, there is little difference between corresponding tabulated values of z and t. Thus, we choose the arbitrary cutoff of $n = 30$ (df $= 29$) to distinguish between the large-sample and small-sample inferential techniques.

⊘ **BPINCR**

Returning to the example of testing a new drug, suppose that the six test patients have blood pressure increases of 1.7, 3.0, .8, 3.4, 2.7, and 2.1 points. How can we use this information to construct a 95% confidence interval for μ, the mean increase in blood pressure associated with the new drug for all patients in the population?

First, we know that we are dealing with a sample too small to assume that the sample mean $\bar{x}$ is approximately normally distributed by the Central Limit Theorem. That is, we do not get the normal distribution of $\bar{x}$ "automatically" from the Central Limit Theorem when the sample size is small. Instead, we must assume that the measured variable, in this case the increase in blood pressure, is normally distributed in order for the distribution of $\bar{x}$ to be normal.

Second, unless we are fortunate enough to know the population standard deviation σ, which in this case represents the standard deviation of *all* the patients' increases in blood pressure when they take the new drug, we cannot use the standard normal z-statistic to form our confidence interval for μ. Instead, we must use the t-distribution, with $(n - 1)$ degrees of freedom.

In this case, $n - 1 = 5$ df, and the t-value is found in Table 7.3 to be

$$t_{.025} = 2.571 \text{ with 5 df}$$

Recall that the large-sample confidence interval would have been of the form

$$\bar{x} \pm z_{\alpha/2}\sigma_{\bar{x}} = \bar{x} \pm z_{\alpha/2}\frac{\sigma}{\sqrt{n}} = \bar{x} \pm z_{.025}\frac{\sigma}{\sqrt{n}}$$

where 95% is the desired confidence level. To form the interval for a small sample from *a normal distribution, we simply substitute t for z and s for σ in the preceding formula:*

$$\bar{x} \pm t_{\alpha/2}\frac{s}{\sqrt{n}}$$

An SPSS printout showing descriptive statistics for the six blood pressure increases is displayed in Figure 7.9. Note that $\bar{x} = 2.283$ and $s = .950$. Substituting these numerical values into the confidence interval formula, we get

$$2.283 \pm (2.571)\left(\frac{.950}{\sqrt{6}}\right) = 2.283 \pm .997$$

or 1.286 to 3.280 points. Note that this interval agrees (except for rounding) with the confidence interval highlighted on the SPSS printout in Figure 7.9.

We interpret the interval as follows: We can be 95% confident that the mean increase in blood pressure associated with taking this new drug is between 1.286 and 3.28 points. As with our large-sample interval estimates, our confidence is in the process, not in this particular interval. We know that if we were to repeatedly use this estimation procedure, 95% of the confidence intervals produced would contain the true mean μ, *assuming that the probability distribution of changes in blood pressure from which our sample was selected is normal.* The latter assumption is necessary for the small-sample interval to be valid.

What price did we pay for having to utilize a small sample to make the inference? First, we had to assume the underlying population is normally distributed,

Figure 7.9

SPSS Confidence Interval
for Mean Blood Pressure
Increase

Descriptives

			Statistic	Std. Error
BPINCR	Mean		2.283	.3877
	95% Confidence Interval for Mean	Lower Bound	1.287	
		Upper Bound	3.280	
	5% Trimmed Mean		2.304	
	Median		2.400	
	Variance		.902	
	Std. Deviation		.9496	
	Minimum		.8	
	Maximum		3.4	
	Range		2.6	
	Interquartile Range		1.625	
	Skewness		-.573	.845
	Kurtosis		-.389	1.741

and if the assumption is invalid, our interval might also be invalid.* Second, we had to form the interval using a t value of 2.571 rather than a z-value of 1.96, resulting in a wider interval to achieve the same 95% level of confidence. If the interval from 1.286 to 3.28 is too wide to be of much use, we know how to remedy the situation: Increase the number of patients sampled in order to decrease the interval width (on average).

Now Work *Exercise 7.27*

The procedure for forming a small-sample confidence interval is summarized in the accompanying boxes.

Small-Sample Confidence Interval† for μ

$$\bar{x} \pm t_{\alpha/2}\left(\frac{s}{\sqrt{n}}\right)$$

where $t_{\alpha/2}$ is based on $(n-1)$ degrees of freedom

Conditions Required for a Valid Small-Sample Confidence Interval for μ

1. A random sample is selected from the target population.

2. The population has a relative frequency distribution that is approximately normal.

---

*By *invalid*, we mean that the probability that the procedure will yield an interval that contains μ is not equal to $(1 - \alpha)$. Generally, if the underlying population is approximately normal, the confidence coefficient will approximate the probability that the interval contains μ.

†The procedure given in the box assumes that the population standard deviation σ is unknown, which is almost always the case. If σ is known, we can form the small-sample confidence interval just as we would a large-sample confidence interval using a standard normal z-value instead of t. However, we must still assume that the underlying population is approximately normal.

EXAMPLE 7.2   FINDING A SMALL-SAMPLE CONFIDENCE INTERVAL FOR μ

Problem   Some quality control experiments require *destructive sampling* (i.e., the test to determine whether the item is defective destroys the item) in order to measure some particular characteristic of the product. The cost of destructive sampling often dictates small samples. Suppose a manufacturer of printers for personal computers wishes to estimate the mean number of characters printed before the printhead fails. The printer manufacturer tests $n = 15$ printheads and records the number of characters printed until failure for each. These 15 measurements (in millions of characters) are listed in Table 7.4, followed by a MINITAB summary statistics printout in Figure 7.10.

     a. Form a 99% confidence interval for the mean number of characters printed before the printhead fails. Interpret the result.

     b. What assumption is required for the interval, part **a**, to be valid? Is it reasonably satisfied?

⊚ **PRINTHEAD**

TABLE 7.4   Number of Characters (in Millions) for $n = 15$ Printhead Tests

1.13	1.55	1.43	.92	1.25	1.36	1.32	.85	1.07	1.48	1.20	1.33	1.18	1.22	1.29

Figure 7.10

MINITAB Printout with Descriptive Statistics and 90% Confidence Interval for Example 7.2

```
Variable    N     Mean    StDev   SE Mean        99% CI
NUMCHAR    15   1.23867  0.19316  0.04987   (1.09020, 1.38714)
```

Solution   **a.** For this small sample ($n = 15$), we use the t-statistic to form the confidence interval. We use a confidence coefficient of .99 and $n - 1 = 14$ degrees of freedom to find $t_{\alpha/2}$ in Table VI:

$$t_{\alpha/2} = t_{.005} = 2.977$$

[*Note:* The small sample forces us to extend the interval almost 3 standard deviations (of $\bar{x}$) on each side of the sample mean in order to form the 99% confidence interval.] From the MINITAB printout, Figure 7.10, we find $\bar{x} = 1.24$ and $s = .19$. Substituting these (rounded) values into the confidence interval formula, we obtain

$$\bar{x} \pm t_{.005}\left(\frac{s}{\sqrt{n}}\right) = 1.24 \pm 2.977\left(\frac{.19}{\sqrt{15}}\right)$$
$$= 1.24 \pm .15 \quad \text{or} \quad (1.09, 1.39)$$

This interval is highlighted on Figure 7.10.

       Our interpretation is as follows: The manufacturer can be 99% confident that the printhead has a mean life of between 1.09 and 1.39 million characters. If the manufacturer were to advertise that the mean life of its printheads is (at least) 1 million characters, the interval would support such a claim. Our confidence is derived from the fact that 99% of the intervals formed in repeated applications of this procedure will contain μ.

     b. Since n is small, we must assume that the number of characters printed before printhead failure is a random variable from a normal distribution. That is, we assume that the population from which the sample of 15 measurements is selected is distributed normally. One way to check this assumption is to graph the distribution of data in Table 7.4. If the sample data is approximately normal, then the population from which the sample is selected is very likely to be normal. A MINITAB stem-and-leaf plot for the sample data is displayed

Figure 7.11
MINITAB Stem-and-Leaf
Display of Data in Table 7.4

Stem-and-Leaf Display: NUMBER

```
Stem-and-leaf of NUMBER   N  = 15
Leaf Unit = 0.010


    1     8    5
    2     9    2
    3    10    7
    5    11    38
   (4)   12    0259
    6    13    236
    3    14    38
    1    15    5
```

in Figure 7.11. The distribution is mound shaped and nearly symmetric. Therefore, the assumption of normality appears to be reasonably satisfied.

Look Back Other checks for normality, such as a normal probability plot and the ratio IQR/s, may also be used to verify the normality condition.

| Now Work | *Exercise 7.33*

■ ■ ■

We have emphasized throughout this section that an assumption that the population is approximately normally distributed is necessary for making small-sample inferences about μ when using the t-statistic. Although many phenomena do have approximately normal distributions, it is also true that many random phenomena have distributions that are not normal or even mound shaped. Empirical evidence acquired over the years has shown that the t-distribution is rather insensitive to moderate departures from normality. That is, use of the t-statistic when sampling from slightly skewed mound-shaped populations generally produces credible results; however, for cases in which the distribution is distinctly non-normal, we must either take a large sample or use a *nonparametric method* (the topic of Chapter 14).

What Do You Do When the Population Relative Frequency Distribution Departs Greatly from Normality?

Answer: Use the nonparametric statistical methods of Chapter 14.

Confidence Interval for a Population Mean ($n < 30$)

Using the TI-83 Graphing Calculator

Step 1   *Enter the data*
Press **STAT** and select **1:Edit**
Note: If the list already contains data, clear the old data. Use the up arrow to highlight '**L1**'. Press **CLEAR ENTER**.
Use the arrow and **ENTER** keys to enter the data set into **L1**.

Step 2   *Access the Statistical Tests Menu*
Press **STAT**

Arrow right to **TESTS**
Arrow down to **8:TInterval**
Press **ENTER**

```
EDIT CALC TESTS
2↑T-Test…
3:2-SampZTest…
4:2-SampTTest…
5:1-PropZTest…
6:2-PropZTest…
7:ZInterval…
8▓TInterval…
```

Step 3   *Choose "**Data**" or "**Stats**". ("**Data**" is selected when you have entered the raw data into a List. "**Stats**" is selected when you are given only the mean, standard deviation, and sample size.)*
Press **ENTER**
If you selected "Data", set **List** to **L1**
Set **Freq** to **1**
Set **C-Level** to the confidence level
Arrow down to "**Calculate**"
Press ENTER

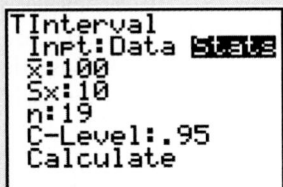

```
TInterval
 Inpt:DATA Stats
 List:L1
 Freq:1
 C-Level:.99
 Calculate
```

If you selected "Stats", enter the mean, standard deviation, and sample size.
Set **C-Level** to the confidence level
Arrow down to **"Calculate"**
Press **ENTER**
(The screen below is set up for an example with a mean of 100 and a standard deviation of 10.)

```
TInterval
 Inpt:Data Stats
 x̄:100
 Sx:10
 n:19
 C-Level:.95
 Calculate
```

The confidence interval will be displayed with the mean, standard deviation, and the sample size.

Example   Compute a 99% confidence interval for the mean using the 15 pieces of data given in Example 7.2.

1.13	1.55	1.43	0.92	1.25
1.36	1.32	0.85	1.07	1.48
1.20	1.33	1.18	1.22	1.29

As you can see from the screen, our 99% confidence interval is (1.0902, 1.3871). You will also notice it gives the mean, standard deviation, and the sample size.

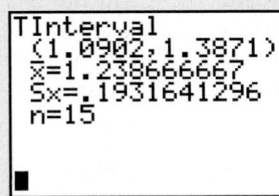

```
TInterval
 (1.0902,1.3871)
 x̄=1.238666667
 Sx=.1931641296
 n=15
▮
```

Exercises 7.25–7.41

Understanding the Principles

7.25 State the two problems (and corresponding solutions) with using a small sample to estimate μ.

7.26 Compare the shapes of the z- and t-distributions.

7.27 Explain the differences in the sampling distributions
NW of $\bar{x}$ for large and small samples under the following assumptions.
  a. The variable of interest, x, is normally distributed.
  b. Nothing is known about the distribution of the variable x.

Learning the Mechanics

7.28 Suppose you have selected a random sample of $n = 7$ measurements from a normal distribution. Compare the standard normal z-values with the corresponding t-values if you were forming the following confidence intervals.
  a. 80% confidence interval
  b. 90% confidence interval
  c. 95% confidence interval
  d. 98% confidence interval
  e. 99% confidence interval
  f. Use the table values you obtained in parts **a–e** to sketch the z- and t-distributions. What are the similarities and differences?

7.29 Let t_0 be a specific value of t. Use Table VI in Appendix A to find t_0 values such that the following statements are true.
  a. $P(t \geq t_0) = .025$ where df = 10
  b. $P(t \geq t_0) = .01$ where df = 17
  c. $P(t \leq t_0) = .005$ where df = 6
  d. $P(t \leq t_0) = .05$ where df = 13

7.30 Let t_0 be a particular value of t. Use Table VI of Appendix A to find t_0 values such that the following statements are true.
  a. $P(-t_0 < t < t_0) = .95$ where df = 16
  b. $P(t \leq -t_0 \text{ or } t \geq t_0) = .05$ where df = 16
  c. $P(t \leq t_0) = .05$ where df = 16
  d. $P(t \leq -t_0 \text{ or } t \geq t_0) = .10$ where df = 12
  e. $P(t \leq -t_0 \text{ or } t \geq t_0) = .01$ where df = 8

7.31 The following random sample was selected from a normal distribution: 4, 6, 3, 5, 9, 3.
  a. Construct a 90% confidence interval for the population mean μ.
  b. Construct a 95% confidence interval for the population mean μ.
  c. Construct a 99% confidence interval for the population mean μ.
  d. Assume that the sample mean $\bar{x}$ and sample standard deviation s remain exactly the same as those you just calculated but that they are based on a sample of $n = 25$ observations rather than $n = 6$ observations.

Repeat parts **a–c**. What is the effect of increasing the sample size on the width of the confidence intervals?

7.32 The following sample of 16 measurements was selected from a population that is approximately normally distributed:

⊙ **LM7_32**

91	80	99	110	95	106	78	121	106	100
97	82	100	83	115	104				

  a. Construct an 80% confidence interval for the population mean.
  b. Construct a 95% confidence interval for the population mean and compare the width of this interval with that of part **a**.
  c. Carefully interpret each of the confidence intervals and explain why the 80% confidence interval is narrower.

Applying the Concepts—Basic

⊙ **LICHEN**

7.33 **Radioactive lichen.** Refer to the Lichen Radionuclide
NW Baseline Research project at the University of Alaska, Exercise 2.34 (p. 51). Recall that the researchers collected 9 lichen specimens and measured the amount (in microcuries per milliliter) of the radioactive element, cesium-137, for each. (The natural logarithms of the data values are saved in the **LICHEN** file.) A MINITAB printout with summary statistics for the actual data is shown at the bottom of the page.
  a. Give a point estimate for the mean amount of cesium in lichen specimens collected in Alaska.
  b. Give the t-value used in a small-sample 95% confidence interval for the true mean amount of cesium in Alaskan lichen specimens.
  c. Use the result, part **b**, and the values of $\bar{x}$ and s shown on the MINITAB printout to form a 95% confidence interval for the true mean amount of cesium in Alaskan lichen specimens.
  d. Check your interval, part **c**, with the 95% confidence interval shown on the MINITAB printout.
  e. Give a practical interpretation for the interval, part **c**.

⊙ **GOBIANTS**

7.34 **Rainfall and desert ants.** Refer to the *Journal of Biogeography* (December 2003) study of ants and their habitat in the desert of Central Asia, Exercise 2.66 (p. 66). Recall that botanists randomly selected five sites in the Dry Steppe region and six sites in the Gobi Desert where ants were observed. One of the variables of interest is the annual rainfall (in millimeters) at each site. (The data are saved in the **GOBIANTS** file.)

MINITAB Output for Exercise 7.33

Variable	N	Mean	StDev	SE Mean	95% CI
CESIUM	9	0.009027	0.004854	0.001618	(0.005296, 0.012759)

SAS Output for Exercise 7.34

The MEANS Procedure

Analysis Variable : RAIN

REGION	N Obs	Mean	Std Dev
DryStepp	5	183.4000000	20.6470337
Gobi	6	110.0000000	15.9749804

Summary statistics for the annual rainfall at each site are provided in the above SAS printout.

a. Give a point estimate for the average annual rainfall amount at ant sites in the Dry Steppe region of Central Asia.

b. Give the *t*-value used in a small-sample 90% confidence interval for the true average annual rainfall amount at ant sites in the Dry Steppe region.

c. Use the result, part **b**, and the values of $\bar{x}$ and *s* shown on the SAS printout to form a 90% confidence interval for the target parameter.

d. Give a practical interpretation for the interval, part **c**.

e. Use the data in the **GOBIANTS** file to check the validity of the confidence interval, part **c**.

f. Repeat parts **a–e** for the Gobi Desert region of Central Asia.

7.35 Lead and copper in drinking water. Periodically, the Hillsborough County (Florida) Water Department tests the drinking water of homeowners for contaminants such as lead and copper. The lead and copper levels in water specimens collected for a sample of 10 residents of the Crystal Lake Manors subdivision are shown below.

⊙ **LEADCOPP**

Lead (μg/L)	Copper (mg/L)
1.32	.508
0	.279
13.1	.320
.919	.904
.657	.221
3.0	.283
1.32	.475
4.09	.130
4.45	.220
0	.743

Source: Hillsborough County Water Department Environmental Laboratory, Tampa, Florida.

a. Construct a 99% confidence interval for the mean lead level in water specimens from Crystal Lake Manors.

b. Construct a 99% confidence interval for the mean copper level in water specimens from Crystal Lake Manors.

c. Interpret the intervals, parts **a** and **b**, in the words of the problem.

d. Discuss the meaning of the phrase, "99% confident."

7.36 Oven cooking study. A group of Harvard University School of Public Health researchers studied the impact of cooking on the size of indoor air particles. (*Environmental Science & Technology*, September 1, 2000.) The decay rate (measured as μm/hour) for fine particles produced from oven cooking or toasting was recorded on six randomly selected days. These six measurements are

⊙ **DECAY**

.95	.83	1.20	.89	1.45	1.12

Source: Abt, E., et al. "Relative contribution of outdoor and indoor particle sources to indoor concentrations." *Environmental Science & Technology,* Vol. 34, No. 17, Sept. 1, 2000 (Table 3).

a. Find and interpret a 95% confidence interval for the true average decay rate of fine particles produced from oven cooking or toasting.

b. Explain what the phrase "95% confident" implies in the interpretation of part **a**.

c. What must be true about the distribution of the population of decay rates for the inference to be valid?

Applying the Concepts—Intermediate

7.37 DOT traffic counts. It is customary practice in the United States to base roadway design on the 30th highest hourly volume in a year. Thus, all roadway facilities are expected to operate at acceptable levels of service for all but 29 hours of the year. The Florida Department of Transportation (DOT), however, has shifted from the 30th highest hour to the 100th highest hour as the basis for level-of-service determinators. Florida Atlantic University researcher Reid Ewing investigated whether this shift was warranted in the *Journal of STAR Research* (July 1994). The table on p. 342 gives the traffic counts at the 30th highest hour and the 100th highest hour of a recent year for 20 randomly selected DOT permanent count stations.

a. Describe the population from which the sample data is selected.

b. Does the sample appear to be representative of the population? Explain.

c. Compute and interpret the 95% confidence interval for the mean traffic count at the 30th highest hour.

⊚ **TRAFFIC**

Station	Type of Route	30th Highest Hour	100th Highest Hour
0117	small city	1,890	1,736
0087	recreational	2,217	2,069
0166	small city	1,444	1,345
0013	rural	2,105	2,049
0161	urban	4,905	4,815
0096	urban	2,022	1,958
0145	rural	594	548
0149	rural	252	229
0038	urban	2,162	2,048
0118	rural	1,938	1,748
0047	rural	879	811
0066	urban	1,913	1,772
0094	rural	3,494	3,403
0105	small city	1,424	1,309
0113	small city	4,571	4,425
0151	urban	3,494	3,359
0159	rural	2,222	2,137
0160	small city	1,076	989
0164	recreational	2,167	2,039
0165	recreational	3,350	3,123

Source: Ewing, R. "Roadway levels of service in an era of growth management." *Journal of STAR Research*, Vol. 3, July 1994, p. 103 (Table 2).

d. What assumption is necessary for the confidence interval to be valid? Does it appear to be satisfied? Explain.

e. Repeat parts **c** and **d** for the 100th highest hour.

7.38 **Hearing loss study.** Patients with normal hearing in one ear and unaided sensorineural hearing loss in the other are characterized as suffering from unilateral hearing loss. In a study reported in the *American Journal of Audiology* (March 1995), 8 patients with unilateral hearing loss were fitted with a special wireless hearing aid in the "bad" ear. The absolute sound pressure level (SPL) was then measured near the eardrum of the ear when noise was produced at a frequency of 500 hertz. The SPLs of the 8 patients, recorded in decibels, are listed below.

⊚ **SOUND**

73.0	80.1	82.8	76.8	73.5	74.3	76.0	68.1

Construct and interpret a 90% confidence interval for the true mean SPL of unilateral hearing loss patients when noise is produced at a frequency of 500 hertz.

7.39 **Research on brain specimens.** In Exercise 2.37 (p. 52) you learned that postmortem interval (PMI) is the elapsed time between death and the performance of an autopsy on the cadaver. *Brain and Language* (June 1995) reported on the PMIs of 22 randomly selected human brain specimens obtained at autopsy. The data are reproduced in the following table.

⊚ **BRAINPMI**

5.5	14.5	6.0	5.5	5.3	5.8	11.0	6.4
7.0	14.5	10.4	4.6	4.3	7.2	10.5	6.5
3.3	7.0	4.1	6.2	10.4	4.9		

Source: Hayes, T. L., and Lewis, D. A. "Anatomical specialization of the anterior motor speech area: Hemispheric differences in magnopyramidal neurons." *Brain and Language*, Vol. 49, No. 3, June 1995, p. 292 (Table 1).

a. Construct a 95% confidence interval for the true mean PMI of human brain specimens obtained at autopsy.

b. Interpret the interval, part **a**.

c. What assumption is required for the interval, part **a**, to be valid? Is this assumption satisfied? Explain.

d. What is meant by the phrase "95% confidence"?

7.40 **Eating disorders in females.** The "fear of negative evaluation" (FNE) scores for 11 bulimic female students and 14 normal female students, first presented in Exercise 2.38 (p. 52) are reproduced at the bottom of the page. (Recall that the higher the score, the greater the fear of negative evaluation.)

a. Construct a 95% confidence interval for the mean FNE score of the population of bulimic female students. Interpret the result.

b. Construct a 95% confidence interval for the mean FNE score of the population of normal female students. Interpret the result.

c. What assumptions are required for the intervals of parts **a** and **b** to be statistically valid? Are these assumptions reasonably satisfied? Explain.

Applying the Concepts—Advanced

7.41 **Study on waking sleepers early.** Scientists have discovered increased levels of the hormone adrenocorticotropin in people just before they awake from sleeping (*Nature*, January 7, 1999). In the study, 15 subjects were monitored during their sleep after being told they would be woken at a particular time. One hour prior to the designated wake-up time, the adrenocorticotropin level (pg/mL) was measured in each, with the following results:

$$\bar{x} = 37.3 \qquad s = 13.9$$

a. Estimate the true mean adrenocorticotropin level of sleepers one hour prior to waking using a 95% confidence interval.

b. Interpret the interval, part **a**, in the words of the problem.

⊚ **BULIMIA**

Bulimic students	21	13	10	20	25	19	16	21	24	13	14		
Normal students	13	6	16	13	8	19	23	18	11	19	7	10	15	20

Source: Randles, R. H. "On neutral responses (zeros) in the sign test and ties in the Wilcoxon-Mann-Whitney Test." *The American Statistician*, Vol. 55, No. 2, May 2001 (Figure 3).

c. The researchers also found that if the subjects were woken three hours earlier than they anticipated, the average adrenocorticotropin level was 25.5 pg/mL. Assume that $\mu = 25.5$ for all sleepers who are woken three hours earlier than expected. Use the interval, part **a**, to make an inference about the mean adrenocorticotropin level of sleepers under two conditions: one hour before anticipated wake-up time and three hours before anticipated wake-up time.

7.4 Large-Sample Confidence Interval for a Population Proportion

The number of public opinion polls has grown at an astounding rate in recent years. Almost daily, the news media report the results of some poll. Pollsters regularly determine the percentage of people in favor of the president's welfare-reform program, the fraction of voters in favor of a certain candidate, the fraction of customers who favor a particular brand of wine, and the proportion of people who smoke cigarettes. In each case, we are interested in estimating the percentage (or proportion) of some group with a certain characteristic. In this section we consider methods for making inferences about population proportions when the sample is large.

EXAMPLE 7.3   ESTIMATING A POPULATION PROPORTION

Problem   Public opinion polls are conducted regularly to estimate the fraction of U.S. citizens who trust the president. Suppose 1,000 people are randomly chosen and 637 answer that they trust the president. How would you estimate the true fraction of *all* U.S. citizens who trust the president?

Solution   What we have really asked is how you would estimate the probability p of success in a binomial experiment, where p is the probability that a person chosen trusts the president. One logical method of estimating p for the population is to use the proportion of successes in the sample. That is, we can estimate p by calculating

$$\hat{p} = \frac{\text{Number of people sampled who trust the president}}{\text{Number of people sampled}}$$

where $\hat{p}$ is read "*p* hat." Thus, in this case,

$$\hat{p} = \frac{637}{1,000} = .637$$

Look Back   To determine the reliability of the estimator $\hat{p}$, we need to know its sampling distribution. That is, if we were to draw samples of 1,000 people over and over again, each time calculating a new estimate $\hat{p}$, what would be the frequency distribution of all the $\hat{p}$-values? The answer lies in viewing $\hat{p}$ as the average, or mean, number of successes per trial over the n trials. If each success is assigned a value equal to 1 and a failure is assigned a value of 0, then the sum of all n sample observations is x, the total number of successes, and $\hat{p} = x/n$ is the average, or mean, number of successes per trial in the n trials. The Central Limit Theorem tells us that the relative frequency distribution of the sample mean for any population is approximately normal for sufficiently large samples.

Now Work   *Exercise 7.49a*

■ ■ ■

The repeated sampling distribution of $\hat{p}$ has the characteristics listed in the next box and shown in Figure 7.12.

Figure 7.12
Sampling Distribution of $\hat{p}$

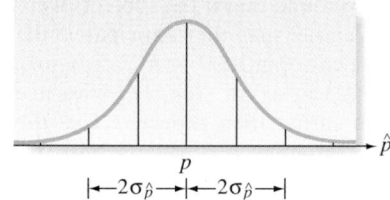

Sampling Distribution of $\hat{p}$

1. The mean of the sampling distribution of $\hat{p}$ is p; that is, $\hat{p}$ is an unbiased estimator of p.

2. The standard deviation of the sampling distribution of $\hat{p}$ is $\sqrt{pq/n}$; that is, $\sigma_p = \sqrt{pq/n}$, where $q = 1 - p$.

3. For large samples, the sampling distribution of $\hat{p}$ is approximately normal. A sample size is considered large if the interval $\hat{p} \pm 3\sigma_{\hat{p}}$ does not include 0 or 1. [*Note:* This requirement is almost equivalent to that given in Section 5.5 for approximating a binomial distribution with a normal one. The difference is that we assumed p to be known in Section 5.5; now we are trying to make inferences about an unknown p, so we use $\hat{p}$ to estimate p in checking the adequacy of the normal approximation.]

The fact that $\hat{p}$ is a "sample mean fraction of successes" allows us to form confidence intervals about p in a manner that is completely analogous to that used for large-sample estimation of μ.

Large-Sample Confidence Interval for p

$$\hat{p} \pm z_{\alpha/2}\sigma_{\hat{p}} = \hat{p} \pm z_{\alpha/2}\sqrt{\frac{pq}{n}} \approx \hat{p} \pm z_{\alpha/2}\sqrt{\frac{\hat{p}\hat{q}}{n}}$$

where $\hat{p} = \dfrac{x}{n}$ and $\hat{q} = 1 - \hat{p}$

Note: When n is large, $\hat{p}$ can approximate the value of p in the formula for $\sigma_{\hat{p}}$.

Conditions Required for a Valid Large-Sample Confidence Interval for p

1. A random sample is selected from the target population.

2. The sample size n is large. (This condition will be satisfied if $\hat{p} \pm 3\sigma_{\hat{p}}$ falls between 0 and 1.)

Thus, if 637 of 1,000 U.S. citizens say they trust the president, a 95% confidence interval for the proportion of *all* U.S. citizens who trust the president is

$$\hat{p} \pm z_{\alpha/2}\sigma_{\hat{p}} = .637 \pm 1.96\sqrt{\frac{pq}{1,000}}$$

where $q = 1 - p$. Just as we needed an approximation for σ in calculating a large-sample confidence interval for μ, we now need an approximation for p. As Table 7.5 shows, the approximation for p does not have to be especially accurate, because the value of $\sqrt{pq}$ needed for the confidence interval is relatively insensitive to changes in p. Therefore, we can use $\hat{p}$ to approximate p.

TABLE 7.5 Values of *pq* for Several Different Values of *p*

p	pq	$\sqrt{pq}$
.5	.25	.50
.6 or .4	.24	.49
.7 or .3	.21	.46
.8 or .2	.16	.40
.9 or .1	.09	.30

Keeping in mind that $\hat{q} = 1 - \hat{p}$, we substitute these values into the formula for the confidence interval:

$$\hat{p} \pm 1.96\sqrt{pq/1{,}000} \approx \hat{p} \pm 1.96\sqrt{\hat{p}\hat{q}/1{,}000}$$
$$= .637 \pm 1.96\sqrt{(.637)(.363)/1{,}000} = .637 \pm .030$$
$$= (.607, .667)$$

Then we can be 95% confident that the interval from 60.7% to 66.7% contains the true percentage of *all* U.S. citizens who trust the president. That is, in repeated construction of confidence intervals, approximately 95% of all samples would produce confidence intervals that enclose p. Note that the guidelines for interpreting a confidence interval about μ also apply to interpreting a confidence interval for p because p is the "population mean fraction of successes" in a binomial experiment.

EXAMPLE 7.4   FINDING A LARGE-SAMPLE CONFIDENCE INTERVAL FOR *p*

Problem   Many public polling agencies conduct surveys to determine the current consumer sentiment concerning the state of the economy. For example, the Bureau of Economic and Business Research (BEBR) at the University of Florida conducts quarterly surveys to gauge consumer sentiment in the Sunshine State. Suppose that BEBR randomly samples 484 consumers and finds that 257 are optimistic about the state of the economy. Use a 90% confidence interval to estimate the proportion of all consumers in Florida who are optimistic about the state of the economy. Based on the confidence interval, can BEBR infer that the majority of Florida consumers are optimistic about the economy?

Solution   The number, x, of the 484 sampled consumers who are optimistic about the Florida economy is a binomial random variable if we can assume that the sample was randomly selected from the population of Florida consumers and that the poll was conducted identically for each sampled consumer.

      The point estimate of the proportion of Florida consumers who are optimistic about the economy is

$$\hat{p} = \frac{x}{n} = \frac{257}{484} = .531$$

We first check to be sure that the sample size is sufficiently large that the normal distribution provides a reasonable approximation for the sampling distribution of $\hat{p}$. We check the 3-standard-deviation interval around $\hat{p}$:

$$\hat{p} \pm 3\sigma_{\hat{p}} \approx \hat{p} \pm 3\sqrt{\frac{\hat{p}\hat{q}}{n}}$$
$$= .531 \pm 3\sqrt{\frac{(.531)(.469)}{484}} = .531 \pm .068 = (.463, .599)$$

Since this interval is wholly contained in the interval $(0, 1)$, we may conclude that the normal approximation is reasonable.

We now proceed to form the 90% confidence interval for p, the true proportion of Florida consumers who are optimistic about the state of the economy:

$$\hat{p} \pm z_{\alpha/2}\sigma_{\hat{p}} = \hat{p} \pm z_{\alpha/2}\sqrt{\frac{pq}{n}} \approx \hat{p} \pm z_{\alpha/2}\sqrt{\frac{\hat{p}\hat{q}}{n}}$$

$$= .531 \pm 1.645\sqrt{\frac{(.531)(.469)}{484}} = .531 \pm .037 = (.494, .568)$$

Thus, we can be 90% confident that the proportion of all Florida consumers who are confident about the economy is between .494 and .568. As always, our confidence stems from the fact that 90% of all similarly formed intervals will contain the true proportion p and not from any knowledge about whether this particular interval does.

Can we conclude that the majority of Florida consumers are optimistic about the economy based on this interval? If we wished to use this interval to infer that a majority is optimistic, the interval would have to support the inference that p exceeds .5 — that is, that more than 50% of the Florida consumers are optimistic about the economy. Note that the interval contains some values below .5 (as low as .494) as well as some above .5 (as high as .568). Therefore, we cannot conclude that the true value of p exceeds .5 based on this 90% confidence interval.

Look Back If the entire confidence interval fell above .5 (e.g., an interval from .52 to .54), then we could conclude (with 90% confidence) that the true proportion of consumers who are optimistic exceeds .5.

| Now Work | *Exercise 7.49bc* |

■ ■ ■

Caution

Unless n is extremely large, the large-sample procedure presented in this section performs poorly when p is near 0 or 1. For example, suppose you want to estimate the proportion of people who die from a bee sting. This proportion is likely to be near 0 (say, $p \approx .001$). Confidence intervals for p based on a sample of size $n = 50$ will probably be misleading.

To overcome this potential problem, an *extremely* large sample size is required. Since the value of n required to satisfy "extremely large" is difficult to determine, statisticians (see Agresti & Coull, 1998) have proposed an alternative method, based on the Wilson (1927) point estimator of p. The procedure is outlined in the box. Researchers have shown that this confidence interval works well for any p even when the sample size n is very small.

Adjusted $(1 - \alpha)$ 100% Confidence Interval for a Population Proportion, p

$$\tilde{p} \pm z_{\alpha/2}\sqrt{\frac{\tilde{p}(1 - \tilde{p})}{n + 4}}$$

where $\tilde{p} = \dfrac{x + 2}{n + 4}$ is the adjusted sample proportion of observations with the characteristic of interest, x is the number of successes in the sample, and n is the sample size.

EXAMPLE 7.5

APPLYING THE ADJUSTED CONFIDENCE INTERVAL
PROCEDURE FOR *p*

Problem   According to *True Odds: How Risk Affects Your Everyday Life* (Walsh, 1997), the probability of being the victim of a violent crime is less than .01. Suppose that in a random sample of 200 Americans, 3 were victims of a violent crime. Estimate the true proportion of Americans who were victims of a violent crime using a 95% confidence interval.

Solution   Let *p* represent the true proportion of Americans who were victims of a violent crime. Since *p* is near 0, an "extremely large" sample is required to estimate its value using the usual large-sample method. Since we are unsure whether the sample size of 200 is large enough, we will apply Wilson's adjustment outlined in the box.

The number of "successes" (i.e., number of violent crime victims) in the sample is $x = 3$. Therefore, the adjusted sample proportion is

$$\widetilde{p} = \frac{x + 2}{n + 4} = \frac{3 + 2}{200 + 4} = \frac{5}{204} = .025$$

Note that this adjusted sample proportion is obtained by adding a total of four observations—two "successes" and two "failures"—to the sample data. Substituting $\widetilde{p} = .025$ into the equation for a 95% confidence interval, we obtain

$$\widetilde{p} \pm 1.96\sqrt{\frac{\widetilde{p}(1 - \widetilde{p})}{n + 4}} = .025 \pm 1.96\sqrt{\frac{(.025)(.975)}{204}}$$
$$= .025 \pm .021$$

or (.004, .046). Consequently, we are 95% confident that the true proportion of Americans who are victims of a violent crime falls between .004 and .046.

■ ■ ■

Statistics in Action Revisited

Estimating the Proportion of Sprinters Who Improve after Speed Training

In the previous Statistics in Action Revisited (p. 328), we discovered that a training program was effective in improving a high school athlete's speed in a 40-yard sprint. This inference was made by measuring the difference between an athlete's sprint time before and after the training program and then forming a confidence interval for the mean difference in times.

Another approach to the problem is to determine (based on the before and after sprint times) whether each athlete improved their speed and measure this qualitatively. Thus, each participating athlete's sprint performance will be classified as "Improved" (if the after time is less than the before time) or "Not Improved" (if the after time is greater than or the same as the before time). Now, the variable of interest is qualitative in nature with two possible outcomes and our goal is to estimate *p*, the true proportion of all high school athletes who attain improved sprint times after participating in the speed training program.

If you examine the data for the 38 sampled athletes (saved in the **SPRINT** file and listed in Table SIA7.1, p. 321), you will find that 33 of the high school athletes had an "Improved" time in the 40-yard sprint. Thus, our estimate of *p* is

$$\hat{p} = 33/38 = .868$$

A 95% confidence interval for *p* can be obtained using the confidence interval formula or with statistical software. A MINITAB printout of the analysis is displayed in Figure SIA7.2. The 95% confidence interval, highlighted on the printout, is (0.761, .976). Thus, we are 95% confident that anywhere from 76% to 97% of the athletes that participate in the speed training program will improve their 40-yard sprint time. These results present a strong case for claiming that the training program is effective in improving sprint speed. [*Note:* The MINITAB confidence interval in Figure SIA7.2 is based on a normal approximation which requires a "large" sample. A sample size of $n = 38$ athletes, although not small, may not be large

Test and CI for One Proportion: Status

```
Test of p = 0.5 vs p not = 0.5

Event = Improve


Variable   X   N   Sample p        95% CI         Z-Value   P-Value
Status    33   38  0.868421   (0.760944, 0.975898)    4.54    0.000


* NOTE * The normal approximation may be inaccurate for small samples.
```

Figure SIA7.2
MINITAB Confidence Interval for Proportion of Improved Sprint Times

enough for the sampling distribution of $\hat{p}$ to be normally distributed; thus, the interval may be inaccurate (as stated at the bottom of the MINITAB printout). To arrive at a more reliable estimate of the true proportion of improved sprint times, we may need a larger number of athletes to participate in the speed training program. In the next Statistics in Action Revisited (p. 00), we show how large of a sample needs to be selected.]

Confidence Interval for a Population Proportion

Using the TI-83 Graphing Calculator

Step 1 *Access the Statistical Tests Menu*
Press **STAT**
Arrow right to **TESTS**
Arrow down to **A:1-PropZInt**
Press **ENTER**

Step 2 *Enter the values for x, n and C-Level*
where **x** = number of successes
n = sample size
C-Level = level of confidence
Arrow down to "**Calculate**"
Press **ENTER**

Example Suppose that 1,100 U.S. citizens are randomly chosen and 532 answer that they favor a flat income tax rate. Use a 95% confidence interval to estimate the true proportion of citizens who favor a flat income tax rate.
In this example, x = 532, n = 1,100 and C-Level = .95.

The screens for this example are shown below.

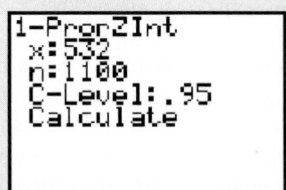 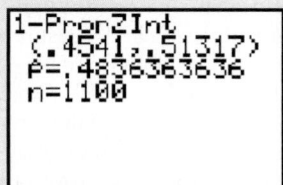

Thus, we can be 95% confident that the interval from 45.4% to 51.3% contains the true percentage of all U.S. citizens who favor a flat income tax rate.

Exercises 7.42–7.59

Understanding the Principles

7.42 Describe the sampling distribution of $\hat{p}$ based on large samples of size n. That is, give the mean, the standard deviation, and the (approximate) shape of the distribution of $\hat{p}$ when large samples of size n are (repeatedly) selected from the binomial distribution with probability of success p.

7.43 Explain the meaning of the phrase "$\hat{p}$ is an unbiased estimator of p."

7.44 If p is near 0 or 1, how large a sample is needed to employ the large-sample confidence interval procedure?

Learning the Mechanics

7.45 A random sample of size $n = 196$ yielded $\hat{p} = .64$.
  a. Is the sample size large enough to use the methods of this section to construct a confidence interval for p? Explain.
  b. Construct a 95% confidence interval for p.
  c. Interpret the 95% confidence interval.
  d. Explain what is meant by the phrase "95% confidence interval."

7.46 A random sample of size $n = 144$ yielded $\hat{p} = .76$.
  a. Is the sample size large enough to use the methods of this section to construct a confidence interval for p? Explain.
  b. Construct a 90% confidence interval for p.
  c. What assumption is necessary to ensure the validity of this confidence interval?

7.47 For the binomial sample information summarized in each part, indicate whether the sample size is large enough to use the methods of this chapter to construct a confidence interval for p.
  a. $n = 500, \hat{p} = .05$ **b.** $n = 100, \hat{p} = .05$
  c. $n = 10, \hat{p} = .5$ **d.** $n = 10, \hat{p} = .3$

7.48 A random sample of 50 consumers taste-tested a new snack food. Their responses were coded (0: do not like; 1: like; 2: indifferent) and recorded as follows:

  ⊙ **SNACK**

1	0	0	1	2	0	1	1	0	0
0	1	0	2	0	2	2	0	0	1
1	0	0	0	0	1	0	2	0	0
0	1	0	0	1	0	0	1	0	1
0	2	0	0	1	1	0	0	0	1

  a. Use an 80% confidence interval to estimate the proportion of consumers who like the snack food.
  b. Provide a statistical interpretation for the confidence interval you constructed in part **a**.

Applying the Concepts—Basic

7.49 **Cell phone use by drivers.** In a July 2001 research note, the U.S. Department of Transportation reported the results of the *National Occupant Protection Use Survey*. One focus of the survey was to determine the level of cell phone use by drivers while they are in the act of driving a motor passenger vehicle. Data collected by observers at randomly selected intersections across the country revealed that in a sample of 1,165 drivers, 35 were using their cell phone.
  a. Give a point estimate of p, the true driver cell phone use rate (i.e., the true proportion of drivers who are using a cell phone while driving).
  b. Compute a 95% confidence interval for p.
  c. Give a practical interpretation of the interval, part **b**.

7.50 **Scary movie study.** According to a University of Michigan study, many adults have experienced lingering "fright" effects from a scary movie or TV show they saw as a teenager. (*Tampa Tribune*, March 10, 1999.) In a survey of 150 college students, 39 said they still experience "residual anxiety" from a scary TV show or movie.
  a. Give a point estimate, $\hat{p}$, for the true proportion of college students who experience "residual anxiety" from a scary TV show or movie.
  b. Find a 95% confidence interval for p.
  c. Interpret the interval, part **b**.

7.51 **Ancient Greek pottery.** Refer to the *Chance* (Fall 2000) study of 837 pottery pieces found at the ancient Greek settlement at Phylakopi, Exercise 2.12 (p. 38). Of these, 183 pieces were painted with either a curvilinear, geometric, or naturalistic decoration. Find a 90% confidence interval for the population proportion of all pottery artifacts at Phylakopi that are painted. Interpret the resulting interval.

7.52 **"Made in the USA" survey.** Refer to Exercise 2.13 (p. 38) and the *Journal of Global Business* (Spring 2002) survey to determine what "Made in the USA" means to consumers. Recall that 106 shoppers at a shopping mall in Muncie, Indiana, responded to the question, " 'Made in the USA' means what percentage of U.S. labor and materials?" Sixty-four shoppers answered "100%."
  a. Define the population of interest in the survey.
  b. What is the characteristic of interest in the population?
  c. Estimate the true proportion of consumers who believe "Made in the USA" means 100% U.S. labor and materials using a 90% confidence interval.
  d. Give a practical interpretation of the interval, part **c**.
  e. Explain what the phrase "90% confidence" means for this interval.

Applying the Concepts—Intermediate

7.53 **Exercise workout dropouts.** Researchers at the University of Florida's Department of Exercise and Sport Sciences conducted a study of variety in exercise workouts. (*Journal of Sport Behavior*, 2001.) A sample of 114 men and women were randomly divided into three groups, 38 people per group. Group 1 members varied their exercise routine in workouts, group 2 members performed the same exercise at each workout, and group 3 members had no set schedule or regulations for their workouts.

a. By the end of the study, 14 people had dropped out of the first exercise group. Estimate the dropout rate for exercisers who vary their routine in workouts. Use a 90% confidence interval and interpret the result.

b. By the end of the study, 23 people had dropped out of the third exercise group. Estimate the dropout rate for exercisers who have no set schedule for their workouts. Use a 90% confidence interval and interpret the result.

7.54 Using debit cards on the Internet. The Gallup Organization surveyed 1,252 debit cardholders in the U.S. and found that 180 had used the debit card to purchase a product or service on the Internet. (*Card Fax*, November 12, 1999.)

a. Describe the population of interest to the Gallup Organization.

b. Is the sample size large enough to construct a valid confidence interval for the proportion of debit cardholders who have used their card in making purchases over the Internet?

c. Estimate the proportion referred to in part **b** using a 99% confidence interval. Interpret your result in the context of the problem.

d. If you had constructed a 90% confidence interval instead, would it be wider or narrower?

7.55 Air bags pose danger for children. By law, all new cars must be equipped with both driver-side and passenger-side safety air bags. There is concern, however, over whether air bags pose a danger for children sitting on the passenger side. In a National Highway Traffic Safety Administration (NHTSA) study of 55 people killed by the explosive force of air bags, 35 were children seated on the front-passenger side. (*Wall Street Journal*, January 22, 1997.) This study led some car owners with children to disconnect the passenger-side air bag. Consider all fatal automobile accidents in which it is determined that air bags were the cause of death. Let p represent the true proportion of these accidents involving children seated on the front-passenger side.

a. Use the data from the NHTSA study to estimate p.

b. Construct a 99% confidence interval for p.

c. Interpret the interval, part **b**, in the words of the problem.

d. NHTSA investigators determined that 24 of 35 children killed by the air bags were not wearing seat belts or were improperly restrained. How does this information impact your assessment of the risk of an air bag fatality?

DDT

7.56 Contaminated fish in the Tennessee River. Refer to the U.S. Army Corps of Engineers data on a sample of 144 contaminated fish collected from the Tennessee River, adjacent to a chemical plant in Alabama. The data are saved in the **DDT** file. Estimate the proportion of contaminated fish that are of the channel catfish species. Use a 95% confidence interval and interpret the result.

7.57 Whistling dolphins. In Exercise 2.19 (p. 40) you learned about the signature whistles of bottlenose dolphins. Marine scientists categorize signature whistles by type— type a, type b, type c, etc. For one study of a sample of 185 whistles emitted from bottlenose dolphins in captivity, 97 were categorized as type a whistles (*Ethology*, July 1995).

a. Estimate the true proportion of bottlenose dolphin signature whistles that are type a whistles. Use a 99% confidence interval.

b. Interpret the interval, part **a**.

7.58 FTC checkout scanner study. Refer to the Federal Trade Commission (FTC) "Price Check" study of electronic checkout scanners, Exercise 4.115 (p. 230). The FTC inspected 1,669 scanners at retail stores and supermarkets by scanning a sample of items at each store and determining if the scanned price was accurate. The FTC gives a store a "passing" grade if 98% or more of the items are priced accurately. Of the 1,669 stores in the study, 1,185 passed inspection.

a. Find a 90% confidence interval for the true proportion of retail stores and supermarkets with electronic scanners that pass the FTC price-check test. Interpret the result.

b. Two years prior to the study, the FTC found that 45% of the stores passed inspection. Use the interval, part **a**, to determine whether the proportion of stores that now pass inspection exceeds .45.

Applying the Concepts—Advanced

7.59 Latex allergy in healthcare workers. Refer to the *Current Allergy & Clinical Immunology* (March 2004) study of healthcare workers who use latex gloves, Exercise 7.13 (p. 330). In addition to the 46 hospital employees who were diagnosed with a latex allergy based on a skin-prick test, another 37 healthcare workers were diagnosed with the allergy using a latex-specific serum test. Of these 83 workers with confirmed latex allergy, only 36 suspected that they had the allergy when asked on a questionnaire. Make a statement about the likelihood that a healthcare worker with latex allergy suspects he or she actually has the allergy. Attach a measure of reliability to your inference.

7.5 Determining the Sample Size

Recall (Section 1.5) that one way to collect the relevant data for a study used to make inferences about the population is to implement a designed (planned) experiment. Perhaps the most important design decision faced by the analyst is to determine the size of the sample. We show in this section that the appropriate sample size for making an inference about a population mean or proportion depends on the desired reliability.

Estimating a Population Mean

Consider the example from Section 7.2 in which we estimated the mean length of stay for patients in a large hospital. A sample of 100 patients' records produced the 95% confidence interval $\bar{x} \pm 2\sigma_{\bar{x}} \approx 4.53 \pm .74$. Consequently, our estimate $\bar{x}$ was within .74 day of the true mean length of stay μ for all the hospital's patients at the 95% confidence level. That is, the 95% confidence interval for μ was $2(.74) = 1.48$ days wide when 100 accounts were sampled. This is illustrated in Figure 7.13a.

Now suppose we want to estimate μ to within .25 day with 95% confidence. That is, we want to narrow the width of the confidence interval from 1.48 days to .50 day, as shown in Figure 7.13b. How much will the sample size have to be increased to accomplish this? If we want the estimator $\bar{x}$ to be within .25 day of μ, we must have

$$2\sigma_{\bar{x}} = .25 \quad \text{or, equivalently,} \quad 2\left(\frac{\sigma}{\sqrt{n}}\right) = .25$$

Note that we are using $\sigma_{\bar{x}} = \dfrac{\sigma}{\sqrt{n}}$ in the formula since we are dealing with the sampling distribution of $\bar{x}$ (the estimator of μ).

The necessary sample size is obtained by solving this equation for n. To do this we need an approximation for σ. We have an approximation from the initial sample of 100 patients' records—namely, the sample standard deviation, $s = 3.68$. Thus,

$$2\left(\frac{\sigma}{\sqrt{n}}\right) \approx 2\left(\frac{s}{\sqrt{n}}\right) = 2\left(\frac{3.68}{\sqrt{n}}\right) = .25$$

$$\sqrt{n} = \frac{2(3.68)}{.25} = 29.44$$

$$n = (29.44)^2 = 866.71 \approx 867$$

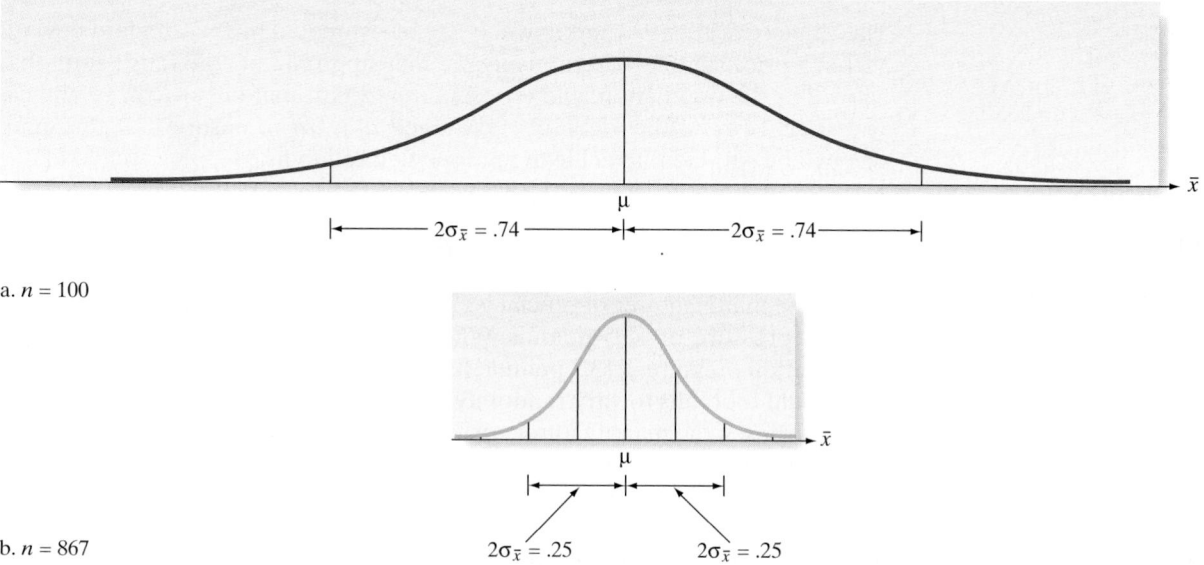

a. $n = 100$

b. $n = 867$

Figure 7.13
Relationship between Sample Size and Width of Confidence Interval: Hospital Stay Example

Approximately 867 patients' records will have to be sampled to estimate the mean length of stay μ to within .25 day with (approximately) 95% confidence. The confidence interval resulting from a sample of this size will be approximately .50 day wide (see Figure 7.13b).

In general, we express the reliability associated with a confidence interval for the population mean μ by specifying the **sampling error** within which we want to estimate μ with $100(1 - \alpha)\%$ confidence. The sampling error (denoted SE) then is equal to the half-width of the confidence interval, as shown in Figure 7.14.

Figure 7.14

Specifying the Sampling Error SE as the Half-Width of a Confidence Interval

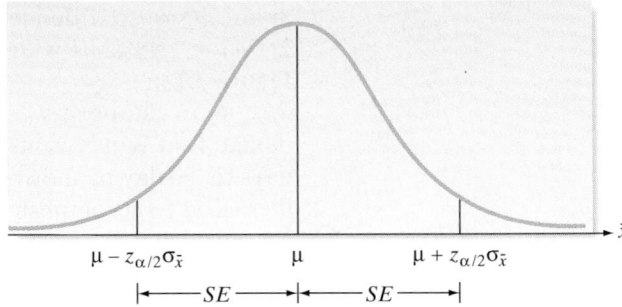

The procedure for finding the sample size necessary to estimate μ with a specific sampling error is given in the box.

Sample Size Determination for $100(1 - \alpha)\%$ Confidence Intervals for μ

In order to estimate μ with a sampling error SE and with $100(1 - \alpha)\%$ confidence, the required sample size is found as follows:

$$z_{\alpha/2}\left(\frac{\sigma}{\sqrt{n}}\right) = SE$$

The solution for n can be given by the equation

$$n = \frac{(z_{\alpha/2})^2 \sigma^2}{(SE)^2}$$

The value of σ is usually unknown. It can be estimated by the standard deviation s from a prior sample. Alternatively, we may approximate the range R of observations in the population, and (conservatively) estimate $\sigma \approx R/4$. In any case, you should round the value of n obtained *upward* to ensure that the sample size will be sufficient to achieve the specified reliability.

EXAMPLE 7.6

FINDING THE SAMPLE SIZE FOR ESTIMATING μ

Problem   Suppose the manufacturer of official NFL footballs uses a machine to inflate the new balls to a pressure of 13.5 pounds. When the machine is properly calibrated, the mean inflation pressure is 13.5 pounds, but uncontrollable factors cause the pressures of individual footballs to vary randomly from about 13.3 to 13.7 pounds. For quality control purposes, the manufacturer wishes to estimate the mean inflation pressure to within .025 pound of its true value with a 99% confidence interval. What sample size should be specified for the experiment?

Solution   We desire a 99% confidence interval that estimates μ with a sampling error of $SE = .025$ pound. For a 99% confidence interval, we have $z_{\alpha/2} = z_{.005} = 2.575$. To estimate σ, we note that the range of observations is $R = 13.7 - 13.3 = .4$

and use $\sigma \approx R/4 = .1$. Now we use the formula derived in the box to find the sample size n:

$$n = \frac{(z_{\alpha/2})^2 \sigma^2}{(SE)^2} \approx \frac{(2.575)^2 (.1)^2}{(.025)^2} = 106.09$$

We round this up to $n = 107$. Realizing that σ was approximated by $R/4$, we might even advise that the sample size be specified as $n = 110$ to be more certain of attaining the objective of a 99% confidence interval with a sampling error of .025 pound or less.

Look Back To determine the value of the sampling error SE, look for the value that follows the key words "estimate μ to within ...".

Now Work *Exercise 7.73*

■ ■ ■

Sometimes the formula will lead to a solution that indicates a small sample size is sufficient to achieve the confidence interval goal. Unfortunately, the procedures and assumptions for small samples differ from those for large samples, as we discovered in Section 7.3. Therefore, if the formulas yield a small sample size, one simple strategy is to select a sample size $n = 30$.

Estimating a Population Proportion

The method outlined above is easily applied to a population proportion p. For example, in Section 7.4 a pollster used a sample of 1,000 U.S. citizens to calculate a 95% confidence interval for the proportion who trust the president, obtaining the interval $.637 \pm .03$. Suppose the pollster wishes to estimate more precisely the proportion who trust the president, say, to within .015 with a 95% confidence interval.

The pollster wants a confidence interval for p with a sampling error of $SE = .015$. The sample size required to generate such an interval is found by solving the following equation for n:

$$z_{\alpha/2} \sigma_{\hat{p}} = SE \qquad \text{or} \qquad z_{\alpha/2} \sqrt{\frac{pq}{n}} = .015 \qquad \text{(see Figure 7.15)}$$

Since a 95% confidence interval is desired, the appropriate z-value is $z_{\alpha/2} = z_{.025} = 1.96$. We must approximate the value of the product pq before we can solve the equation for n. As shown in Table 7.5 (p. 345), the closer the values of p and q to .5, the larger the product pq. Thus, to find a conservatively large sample size that will generate a confidence interval with the specified reliability, we generally choose an approximation of p close to .5. In the case of the proportion of U.S. citizens who trust the president, however, we have an initial sample estimate of $\hat{p} = .637$.

Figure 7.15

Specifying the Sampling Error SE of a Confidence Interval for a Population Proportion p

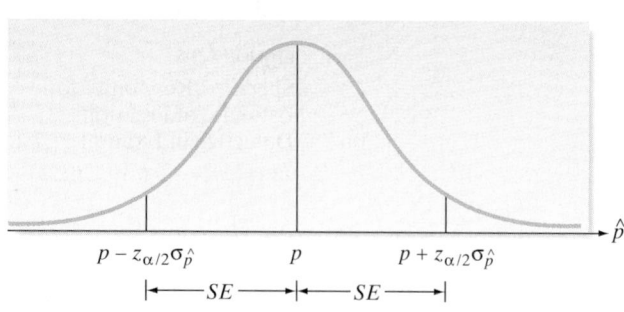

A conservatively large estimate of pq can therefore be obtained by using, say, $p = .60$. We now substitute into the equation and solve for n:

$$1.96\sqrt{\frac{(.60)(.40)}{n}} = .015$$

$$n = \frac{(1.96)^2(.60)(.40)}{(.015)^2} = 4,097.7 \approx 4,098$$

The pollster must sample about 4,098 U.S. citizens to estimate the percentage who trust the president with a confidence interval of width .03.

The procedure for finding the sample size necessary to estimate a population proportion p with a specified sampling error SE is given in the box.

Sample Size Determination for $100(1 - \alpha)\%$ Confidence Interval for p

In order to estimate a binomial probability p with sampling error SE and with $100(1 - \alpha)\%$ confidence, the required sample size is found by solving the following equation for n:

$$z_{\alpha/2}\sqrt{\frac{pq}{n}} = SE$$

The solution for n can be written as follows:

$$n = \frac{(z_{\alpha/2})^2(pq)}{(SE)^2}$$

Since the value of the product pq is unknown, it can be estimated by using the sample fraction of successes, $\hat{p}$, from a prior sample. Remember (Table 7.5) that the value of pq is at its maximum when p equals .5, so that you can obtain conservatively large values of n by approximating p by .5 or values close to .5. In any case, you should round the value of n obtained *upward* to ensure that the sample size will be sufficient to achieve the specified reliability.

EXAMPLE 7.7   FINDING THE SAMPLE SIZE FOR ESTIMATING p

Problem   Suppose a large telephone manufacturer that entered the postregulation market quickly has an initial problem with excessive customer complaints and consequent returns of the phones for repair or replacement. The manufacturer wants to estimate the magnitude of the problem in order to design a quality control program. How many telephones should be sampled and checked in order to estimate the fraction defective, p, to within .01 with 90% confidence?

Solution   In order to estimate p to within .01 of its true value, we set the half-width of the confidence interval equal to $SE = .01$, as shown in Figure 7.16.

Figure 7.16
Specified Reliability for
Estimate of Fraction
Defective in Example 7.7

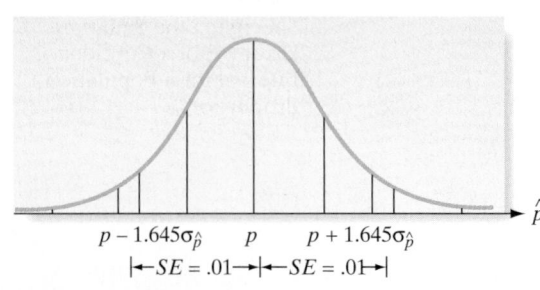

$p - 1.645\sigma_{\hat{p}}$ p $p + 1.645\sigma_{\hat{p}}$

$|\!\!\leftarrow\!\! SE = .01 \!\!\rightarrow\!\!|\!\!\leftarrow\!\! SE = .01 \!\!\rightarrow\!\!|$

The equation for the sample size n requires an estimate of the product pq. We could most conservatively estimate $pq = .25$ (i.e., use $p = .5$), but this may be overly conservative when estimating a fraction defective. A value of .1, corresponding to 10% defective, will probably be conservatively large for this application. The solution is therefore

$$n = \frac{(z_{\alpha/2})^2(pq)}{(SE)^2} = \frac{(1.645)^2(.1)(.9)}{(.01)^2} = 2,435.4 \approx 2,436$$

Thus, the manufacturer should sample 2,436 telephones in order to estimate the fraction defective, p, to within .01 with 90% confidence.

Look Back Remember that this answer depends on our approximation for pq, where we used .09. If the fraction defective is closer to .05 than .10, we can use a sample of 1,286 telephones (check this) to estimate p to within .01 with 90% confidence.

| Now Work | *Exercise 7.72* |

■ ■ ■

The cost of sampling will also play an important role in the final determination of the sample size to be selected to estimate either μ or p. Although more complex formulas can be derived to balance the reliability and cost considerations, we will solve for the necessary sample size and note that the sampling budget may be a limiting factor. Consult the references for a more complete treatment of this problem.

Statistics in Action Revisited

Determining the Number of Athletes to Participate in the Training Program

In the previous Statistics in Action applications in this chapter, we used confidence intervals to (1) estimate μ, the population mean decrease in sprint time for athletes who participate in the speed training program (p. 328), and (2) estimate p, the population proportion of athletes who improve their sprint times after participating in the speed training program (p. 347). The confidence interval for p was fairly wide due to the relatively small number of athletes ($n = 38$) who participated in the program. In order to find a valid, narrower confidence interval for the true proportion, the researchers would need to include more athletes in the study.

How many high school athletes should be sampled to estimate p to within, say, .03 with 95% confidence? Here, we have the sampling error, $SE = .03$ and $z_{.025} = 1.96$ (since, for a 95% confidence interval, $\alpha = .05$ and $\alpha/2 = .025$). From our analysis on page 00, the estimated proportion is $\hat{p} = .87$. Substituting these values into the formula given in the box (p. 354), we obtain

$$n = \frac{(z_{.025})^2(\hat{p})(1 - \hat{p})}{(SE)^2} = \frac{(1.96)^2(.87)(.13)}{(.03)^2} = 482.76$$

Consequently, the researchers would have to include 483 high school athletes in the training program in order to find a valid 95% confidence interval for p to within 3% of its true value.

Running the training program with this large number of athletes may be impractical. To reduce the size of the sample, the researchers could choose to increase the sampling error (i.e., the width) of the confidence interval for p or decrease the confidence level $(1 - \alpha)$. However, since both large- and small-sample confidence intervals are available for estimating μ, the researchers can rely on the inference made from the confidence interval for the mean decrease in sprint time (p. 328).

Exercises 7.60–7.80

Understanding the Principles

7.60 How does the sampling error SE compare to the width of a confidence interval?

7.61 **True or false.** For a specified sampling error SE, increasing the confidence level $(1 - \alpha)$ will lead to a larger n when determining sample size.

7.62 **True or false.** For a fixed confidence level $(1 - \alpha)$, increasing the sampling error SE will lead to a smaller n when determining sample size.

Learning the Mechanics

7.63 If you wish to estimate a population mean to within .2 using a 95% confidence interval and you know from prior sampling that σ^2 is approximately equal to 5.4, how many observations would you have to include in your sample?

7.64 If nothing is known about p, .5 can be substituted for p in the sample-size formula for a population proportion. But when this is done, the resulting sample size may be larger than needed. Under what circumstances will using $p = .5$ in the sample-size formula yield a sample size larger than needed to construct a confidence interval for p with a specified bound and a specified confidence level?

7.65 Suppose you wish to estimate a population mean correct to within .15 with a confidence level of .90. You do not know σ^2, but you know that the observations will range in value between 31 and 39.
   a. Find the approximate sample size that will produce the desired accuracy of the estimate. You wish to be conservative to ensure that the sample size will be ample to achieve the desired accuracy of the estimate. [*Hint:* Using your knowledge of data variation from Section 2.5, assume that the range of the observations will equal 4σ.]
   b. Calculate the approximate sample size making the less conservative assumption that the range of the observations is equal to 6σ.

7.66 In each case, find the approximate sample size required to construct a 95% confidence interval for p that has sampling error $SE = .06$.
   a. Assume p is near .3.
   b. Assume you have no prior knowledge about p, but you wish to be certain that your sample is large enough to achieve the specified accuracy for the estimate.

7.67 The following is a 90% confidence interval for p: (.26, .54). How large was the sample used to construct this interval?

7.68 It costs you $10 to draw a sample of size $n = 1$ and measure the attribute of interest. You have a budget of $1,200.
   a. Do you have sufficient funds to estimate the population mean for the attribute of interest with a 95% confidence interval 4 units in width? Assume $\sigma = 12$.

b. If a 90% confidence level were used, would your answer to part **a** change? Explain.

7.69 Suppose you wish to estimate the mean of a normal population using a 95% confidence interval, and you know from prior information that $\sigma^2 \approx 1$.
   a. To see the effect of the sample size on the width of the confidence interval, calculate the width of the confidence interval for $n = 16, 25, 49, 100$, and 400.
   b. Plot the width as a function of sample size n on graph paper. Connect the points by a smooth curve and note how the width decreases as n increases.

Applying the Concepts—Basic

7.70 **Radioactive lichen.** Refer to the Alaskan Lichen Radionuclide Baseline Research study, Exercise 7.33 (p. 340). In a sample of $n = 9$ lichen specimens, the researchers found the mean and standard deviation of the amount of the radioactive element, cesium-137, present to be .009 and .005 microcuries per milliliter, respectively. Suppose the researchers want to increase the sample size in order to estimate the mean, μ, to within .001 microcuries per milliliter of its true value using a 95% confidence interval.
   a. What is the confidence level desired by the researchers?
   b. What is the sampling error desired by the researchers?
   c. Compute the sample size necessary to obtain the desired estimate.

7.71 **"Made in the USA" survey.** Refer to Exercise 7.52 (p. 349) and the *Journal of Global Business* (Spring 2002) survey to determine what "Made in the USA" means to consumers. Recall that 64 of 106 shoppers at a shopping mall in Muncie, Indiana, believe that "Made in the USA" implies all labor and materials are produced in the United States. Suppose the researchers want to increase the sample size in order to estimate the true proportion, p, to within .05 of its true value using a 90% confidence interval.
   a. What is the confidence level desired by the researchers?
   b. What is the sampling error desired by the researchers?
   c. Compute the sample size necessary to obtain the desired estimate.

7.72 **Aluminum cans contaminated by fire.** A gigantic warehouse located in Tampa, Florida, stores approximately 60 million empty aluminum beer and soda cans. Recently, a fire occurred at the warehouse. The smoke from the fire contaminated many of the cans with blackspot, rendering them unusable. A University of South Florida statistician was hired by the insurance company to estimate p, the true proportion of cans in the warehouse that were contaminated by the fire. How many aluminum cans should be randomly sampled to estimate the true proportion to within .02 with 90% confidence?

7.73 Water pollution testing. The EPA wants to test a randomly selected sample of n water specimens and estimate μ, the mean daily rate of pollution produced by a mining operation. If the EPA wants a 95% confidence interval estimate with a sampling error of 1 milligram per liter (mg/L), how many water specimens are required in the sample? Assume that prior knowledge indicates that pollution readings in water samples taken during a day are approximately normally distributed with a standard deviation equal to 5 (mg/L).

Applying the Concepts—Intermediate

7.74 Training zoo animals. Refer to Exercise 2.63 (p. 65) and the *Teaching of Psychology* (May 1998) study in which students assisted in the training of zoo animals. A sample of 15 psychology students rated "The Training Game" as a "great" method of understanding the animal's perspective during training on a 7–1 point scale (where 1 = strongly disagree and 7 = strongly agree). The mean response was 5.87 with a standard deviation of 1.51.

a. Construct a 95% confidence interval for the true mean response of the students.

b. Suppose you want to reduce the width of the 95% confidence interval to half the size obtained in part **a**. How many students are required in the sample to obtain the desired confidence interval width?

7.75 Bacteria in bottled water. Is the bottled water you drink safe? According to an article in *U.S. News & World Report* (April 12, 1999), the Natural Resources Defense Council warns that the bottled water you are drinking may contain more bacteria and other potentially carcinogenic chemicals than allowed by state and federal regulations. Of the more than 1,000 bottles studied, nearly one-third exceeded government levels. Suppose that the Natural Resources Defense Council wants an updated estimate of the population proportion of bottled water that violates at least one government standard. Determine the sample size (number of bottles) needed to estimate this proportion to within ±0.01 with 99% confidence.

7.76 Asthma drug study. The chemical benzalkonium chloride (BAC) is an antibacterial agent that is added to some asthma medications to prevent contamination. Researchers at the University of Florida College of Pharmacy have discovered that adding BAC to asthma drugs can cause airway constriction in patients. In a sample of 18 asthmatic patients, each of whom received a heavy dose of BAC, 10 experienced a significant drop in breathing capacity (*Journal of Allergy and Clinical Immunology*, January 2001.) Based on this information, a 95% confidence interval for the true percentage of asthmatic patients who experience breathing difficulties after taking BAC is (.326, .785).

a. Why might the confidence interval lead to an erroneous inference?

b. How many asthma patients must be included in the study to estimate the true percentage who experience a significant drop in breathing capacity to within 4% using a 95% confidence interval?

7.77 Oven cooking study. Refer to Exercise 7.36 (p. 341). Suppose that we want to estimate the average decay rate of fine particles produced from oven cooking or toasting to within .04 with 95% confidence. How large a sample should be selected?

7.78 Caffeine content of coffee. According to a Food and Drug Administration (FDA) study, a cup of coffee contains an average of 115 milligrams (mg) of caffeine, with the amount per cup ranging from 60 to 180 mg. Suppose you want to repeat the FDA experiment to obtain an estimate of the mean caffeine content in a cup of coffee correct to within 5 mg with 95% confidence. How many cups of coffee would have to be included in your sample?

7.79 USGA golf ball tests. The United States Golf Association (USGA) tests all new brands of golf balls to ensure that they meet USGA specifications. One test conducted is intended to measure the average distance traveled when the ball is hit by a machine called "Iron Byron." Suppose the USGA wishes to estimate the mean distance for a new brand to within 1 yard with 90% confidence. Assume that past tests have indicated that the standard deviation of the distances Iron Byron hits golf balls is approximately 10 yards. How many golf balls should be hit by Iron Byron to achieve the desired accuracy in estimating the mean?

Applying the Concepts—Advanced

7.80 It costs more to produce defective items—since they must be scrapped or reworked—than it does to produce nondefective items. This simple fact suggests that manufacturers should ensure the quality of their products by perfecting their production processes instead of depending on inspection of finished products (Deming, 1986). In order to better understand a particular metal stamping process, a manufacturer wishes to estimate the mean length of items produced by the process during the past 24 hours.

a. How many parts should be sampled in order to estimate the population mean to within .1 millimeter (mm) with 90% confidence? Previous studies of this machine have indicated that the standard deviation of lengths produced by the stamping operation is about 2 mm.

b. Time permits the use of a sample size no larger than 100. If a 90% confidence interval for μ is constructed using $n = 100$, will it be wider or narrower than would have been obtained using the sample size determined in part **a**? Explain.

c. If management requires that μ be estimated to within .1 mm and that a sample size of no more than 100 be used, what is (approximately) the maximum confidence level that could be attained for a confidence interval that meets management's specifications?

Quick Review

Key Terms

Confidence coefficient   324
Confidence interval   324
Confidence level   324

Degrees of freedom   333
Interval estimator   324
Sampling error   352

Target parameter   322
t-statistic   333
Wilson adjustment for estimating p 346

Key Formulas

$\hat{\theta} \pm (z_{\alpha/2})\sigma_{\theta}$

Large-sample confidence interval for population parameter θ where $\hat{\theta}$ and σ_{θ} are obtained from the table below.

Parameter θ	Estimator $\hat{\theta}$	Standard Error $\sigma_{\hat{\theta}}$	
Mean, μ	$\bar{x}$	$\sigma/\sqrt{n}$	326
Proportion, p	$\hat{p}$	$\sqrt{\dfrac{pq}{n}}$	344

$\bar{x} \pm t_{\alpha/2}\left(\dfrac{s}{\sqrt{n}}\right)$
 Small-sample confidence interval for population mean μ 336

$\tilde{p} = \dfrac{x + 2}{n + 4}$
 Adjusted estimator of p 346

$n = \dfrac{(z_{\alpha/2})^2\sigma^2}{(SE)^2}$
 Determining the sample size n for estimating μ 352

$n = \dfrac{(z_{\alpha/2})^2(pq)}{(SE)^2}$
 Determining the sample size n for estimating p 354

Language Lab

Symbol	Pronunciation	Description
θ	theta	General population parameter
μ	mu	Population mean
p		Population proportion
SE		Sampling error
α	alpha	$(1 - \alpha)$ represents the confidence coefficient
$z_{\alpha/2}$	z of alpha over 2	z value used in a $100(1 - \alpha)\%$ large-sample confidence interval
$t_{\alpha/2}$	t of alpha over 2	t value used in a $100(1 - \alpha)\%$ small-sample confidence interval
$\bar{x}$	x-bar	Sample mean; point estimate of μ
$\hat{p}$	p-hat	Sample proportion; point estimate of p
σ	sigma	Population standard deviation
s		Sample standard deviation; point estimate of σ
$\sigma_{\bar{x}}$	sigma of $\bar{x}$	Standard deviation of sampling distribution of $\bar{x}$
$\sigma_{\hat{p}}$	sigma of $\hat{p}$	Standard deviation of sampling distribution of $\hat{p}$
$\tilde{p}$	p-curl	Wilson's adjusted estimate of p

Chapter Summary Notes

- **Confidence interval**—an interval which encloses an unknown population parameter with a certain level of confidence
- **Confidence coefficient**—the probability that a randomly selected confidence interval encloses the value of the population parameter
- Conditions required for a valid **large-sample confidence interval for μ**: (1) random sample, (2) large sample size ($n \geq 30$)

- Conditions required for a valid **small-sample confidence interval for μ**: (1) random sample, (2) population distribution is approximately normal
- Conditions required for a valid **large-sample confidence interval for p**: (1) random sample, (2) large sample size ($\hat{p} \pm 3\sigma_{\hat{p}}$ falls between 0 and 1)

Supplementary Exercises 7.81–7.109

Note: List the assumptions necessary for the valid implementation of the statistical procedures you use in solving all these exercises.

Understanding the Principles

7.81 For each of the following, identify the target parameter as μ or p.
  a. Average score on the SAT
  b. Mean time waiting at a supermarket checkout lane
  c. Proportion of voters in favor of legalizing marijuana
  d. Percentage of NFL players who have ever made the Pro Bowl
  e. Dropout rate of American college students

7.82 Interpret the phrase "95% confident" in the following statement: "We are 95% confident that the proportion of all PCs with a computer virus falls between .12 and .18."

7.83 In each of the following instances, determine whether you would use a z- or t-statistic (or neither) to form a 95% confidence interval; then look up the appropriate z- or t-value.
  a. Random sample of size $n = 21$ from a normal distribution with unknown mean μ and standard deviation σ
  b. Random sample of size $n = 175$ from a normal distribution with unknown mean μ and standard deviation σ
  c. Random sample of size $n = 12$ from a normal distribution with unknown mean μ and standard deviation $\sigma = 5$
  d. Random sample of size $n = 65$ from a distribution about which nothing is known
  e. Random sample of size $n = 8$ from a distribution about which nothing is known

Learning the Mechanics

7.84 Let t_0 represent a particular value of t from Table VI of Appendix A. Find the table values such that the following statements are true.

  a. $P(t \leq t_0) = .05$ where df $= 17$
  b. $P(t \geq t_0) = .005$ where df $= 14$
  c. $P(t \leq -t_0 \text{ or } t \geq t_0) = .10$ where df $= 6$
  d. $P(t \leq -t_0 \text{ or } t \geq t_0) = .01$ where df $= 22$

7.85 In a random sample of 400 measurements, 227 of the measurements possess the characteristic of interest, A.
  a. Use a 95% confidence interval to estimate the true proportion p of measurements in the population with characteristic A.
  b. How large a sample would be needed to estimate p to within .02 with 95% confidence?

7.86 A random sample of 225 measurements is selected from a population, and the sample mean and standard deviation are $\bar{x} = 32.5$ and $s = 30.0$, respectively.
  a. Use a 99% confidence interval to estimate the mean of the population, μ.
  b. How large a sample would be needed to estimate μ to within .5 with 99% confidence?
  c. What is meant by the phrase "99% confidence" as it is used in this exercise?

Applying the Concepts—Basic

7.87 **CDC health survey.** The Centers for Disease Control and Prevention (CDCP) in Atlanta, Georgia, conducts an annual survey of the general health of the U.S. population as part of its Behavioral Risk Factor Surveillance System. (*New York Times*, March 29, 1995.) Using random-digit dialing, the CDCP telephones U.S. citizens over 18 years of age and asks them the following four questions:
  (1) Is your health generally excellent, very good, good, fair, or poor?
  (2) How many days during the previous 30 days was your physical health not good because of injury or illness?
  (3) How many days during the previous 30 days was your mental health not good because of stress, depression, or emotional problems?

(4) How many days during the previous 30 days did your physical or mental health prevent you from performing your usual activities?

Identify the parameter of interest for each question.

7.88 CDC health survey. Refer to Exercise 7.87. According to the CDCP, 89,582 of 102,263 adults interviewed stated their health was good, very good, or excellent.

a. Use a 99% confidence interval to estimate the true proportion of U.S. adults who believe their health to be good to excellent. Interpret the interval.

b. Why might the estimate, part **a**, be overly optimistic (i.e., biased high)?

7.89 JAMA smoking study. The *Journal of the American Medical Association* (April 21, 1993) reported on the results of a National Health Interview Survey designed to determine the prevalence of smoking among U.S. adults. Current smokers (more than 11,000 adults in the survey) were asked, "On the average, how many cigarettes do you now smoke a day?" The results yielded a mean of 20.0 cigarettes per day with an associated 95% confidence interval of (19.7, 20.3).

a. Interpret the 95% confidence interval.

b. State any assumptions about the target population of current cigarette smokers that must be satisfied for inferences derived from the interval to be valid.

c. A tobacco industry researcher claims that the mean number of cigarettes smoked per day by regular cigarette smokers is less than 15. Comment on this claim.

7.90 JAMA smoking study. Refer to Exercise 7.89. Of the 43,732 survey respondents, 11,239 indicated that they were current smokers and 10,539 indicated they were former smokers.

a. Construct and interpret a 90% confidence interval for the percentage of U.S. adults who currently smoke cigarettes.

b. Construct and interpret a 90% confidence interval for the percentage of U.S. adults who are former cigarette smokers.

7.91 Crop weights of pigeons. The *Australian Journal of Zoology* (Vol. 43, 1995) reported on a study of the diets and water requirements of spinifex pigeons. Sixteen pigeons were captured in the desert and the crop (i.e., stomach) contents of each examined. The accompanying table reports the weight (in grams) of dry seed in the crop of each pigeon. Find a 99% confidence interval for the average weight of dry seeds in the crops of spinifex pigeons inhabiting the Western Australian desert. Interpret the result.

PIGEONS

.457	3.751	.238	2.967	2.509	1.384	1.454	.818
.335	1.436	1.603	1.309	.201	.530	2.144	.834

Source: Excerpted from Williams, J. B., Bradshaw, D., and Schmidt, L. "Field metabolism and water requirements of spinifex pigeons (*Geophaps plumifera*) in Western Australia." *Australian Journal of Zoology,* Vol. 43, No. 1, 1995, p. 7 (Table 2).

7.92 Ammonia in car exhaust. Refer to the *Environmental Science & Technology* (September 1, 2000) study on the ammonia levels near the exit ramp of a San Francisco highway tunnel, Exercise 2.59 (p. 64). The ammonia concentration (parts per million) data for eight randomly selected days during afternoon drive-time are reproduced in the table. Find a 99% confidence interval for the population mean daily ammonia level in air in the tunnel. Interpret your result.

AMMONIA

1.53	1.50	1.37	1.51	1.55	1.42	1.41	1.48

7.93 Homeless in the U.S. The *American Journal of Orthopsychiatry* (July 1995) published an article on the prevalence of homelessness in the United States. A sample of 487 U.S. adults were asked to respond to the question: "Was there ever a time in your life when you did not have a place to live?" A 95% confidence interval for the true proportion who were/are homeless, based on the survey data, was determined to be (.045, .091). Fully interpret the result.

Applying the Concepts—Intermediate

7.94 Pulse rates of joggers. Pulse rate is an important measure of the fitness of a person's cardiovascular system. The mean pulse rate for all U.S. adult males is approximately 72 heart beats per minute. A random sample of 21 U.S. adult males who jog at least 15 miles per week had a mean pulse rate of 52.6 beats per minute and a standard deviation of 3.22 beats per minute.

a. Find a 95% confidence interval for the mean pulse rate of all U.S. adult males who jog at least 15 miles per week.

b. Interpret the interval found in part **a**. Does it appear that jogging at least 15 miles per week reduces the mean pulse rate for adult males?

c. What assumptions are required for the validity of the confidence interval and the inference, part **b**?

7.95 Brown-bag lunches at work. In a study reported in *The Wall Street Journal* (April 4, 1999), the Tupperware Corporation surveyed 1,007 U.S. workers. Of the people surveyed, 665 indicated that they take their lunch to work with them. Of these 665 taking their lunch, 200 reported that they take it in brown bags.

a. Find a 95% confidence interval estimate of the population proportion of U.S. workers who take their lunch to work with them. Interpret the interval.

b. Consider the population of U.S. workers who take their lunch to work with them. Find a 95% confidence interval estimate of the population proportion who take brown-bag lunches. Interpret the interval.

7.96 Role importance for the elderly. Refer to the *Psychology and Aging* (December 2000) study of the roles that elderly people feel are the most important to them, Exercise 2.10 (p. 38). Recall that in a national sample of 1,102 adults, 65 years or older, 424 identified "spouse" as their most salient role. Use a 95% confidence interval to estimate the true percentage of adults, 65 years or older, who feel being a spouse is their most salient role. Give a practical interpretation of this interval.

7.97 Inbreeding of tropical wasps. Tropical swarm-founding wasps rely on female workers to raise their offspring. One possible explanation for this strange behavior is inbreeding, which increases relatedness among the wasps, presumably making it easier for the workers to pick out their closest relatives as propagators of their own genetic material. To test this theory, 197 swarm-founding wasps were captured in Venezuela, frozen at $-70°C$, and then subjected to a series of genetic tests (*Science*, November 1988). The data were used to generate an inbreeding coefficient, x, for each wasp specimen, with the following results: $\bar{x} = .044$ and $s = .884$.
a. Construct a 99% confidence interval for the mean inbreeding coefficient of this species of wasp.
b. A coefficient of 0 implies that the wasp has no tendency to inbreed. Use the confidence interval, part **a**, to make an inference about the tendency for this species of wasp to inbreed.

⊘ **OILSPILL**

7.98 Tanker oil spills. Refer to the *Marine Technology* (January 1995) study of the causes of fifty recent major oil spills from tankers and carriers, Exercise 2.182 (p. 112). The data are stored in the **OILSPILL** file.
a. Give a point estimate for the proportion of major oil spills that are caused by hull failure.
b. Form a 95% confidence interval for the estimate, part **a**. Interpret the result.

7.99 Time to solve a math programming problem. *IEEE Transactions* (June 1990) presented a hybrid algorithm for solving a polynomial zero-one mathematical programming problem. The algorithm incorporates a mixture of pseudo-Boolean concepts and time-proven implicit enumeration procedures. Fifty-two random problems were solved using the hybrid algorithm; the times to solution (CPU time in seconds) are listed in the table.
a. Estimate, with 95% confidence, the mean solution time for the hybrid algorithm. Interpret the result.

b. How many problems must be solved to estimate the mean, μ, to within .25 second with 95% confidence?

⊘ **MATHCPU**

.045	1.055	.136	1.894	.379	.136	.336	.258	1.070
.506	.088	.242	1.639	.912	.412	.361	8.788	.579
1.267	.567	.182	.036	.394	.209	.445	.179	.118
.333	.554	.258	.182	.070	3.985	.670	3.888	.136
.091	.600	.291	.327	.130	.145	4.170	.227	.064
.194	.209	.258	3.046	.045	.049	.079		

Source: Snyder, W. S., and Chrissis, J. W. "A hybrid algorithm for solving zero-one mathematical programming problems." *IEEE Transactions,* Vol. 22, No. 2, June 1990, p. 166 (Table 1). The Institute of Industrial Engineers, 25 Technology Park/Atlanta, Norcross, GA 30092. (770) 449-0461.

7.100 Weights of naphtha in drums. A company purchases large quantities of naphtha (a petroleum product) in 50-gallon drums. Because the purchases are ongoing, small shortages in the drums can represent a sizable loss to the company. The weights of the drums vary slightly from drum to drum, so the weight of the naphtha is determined by removing it from the drums and measuring it. Suppose the company samples the contents of 20 drums, measures the naphtha in each, and calculates $\bar{x} = 49.70$ gallons and $s = .32$ gallon. Find a 95% confidence interval for the mean number of gallons of naphtha per drum. What assumptions are necessary to assure the validity of the confidence interval?

⊘ **SUICIDE**

7.101 Jail suicide study. In Exercise 2.183 (p. 112) you considered data on suicides in correctional facilities, published in the *American Journal of Psychiatry* (July 1995). The researchers wanted to know what factors increase the risk of suicide by those incarcerated in urban jails. The data on 37 suicides that occurred in Wayne County Jail, Detroit, Michigan, are saved in the **SUICIDE** file. Answer each of the questions posed below, using a 95% confidence interval for the target parameter implied in the question. Fully interpret each confidence interval.
a. What proportion of suicides at the jail are committed by inmates charged with murder/manslaughter?
b. What proportion of suicides at the jail are committed at night?
c. What is the average length of time an inmate is in jail before committing suicide?
d. What percentage of suicides are committed by white inmates?

7.102 Recommendation letters for professors. Refer to the *American Psychologist* (July 1995) study of applications for university positions in experimental psychology, Exercise 2.62 (p. 65). One of the objectives of the analysis was to identify problems and errors (e.g., not submitting at least three letters of recommendation) in the application packages. In a sample of 148 applications for the positions, 30 applicants failed to provide *any* letters of recommendation. Estimate, with 90% confidence, the true fraction of applicants for a university position in experimental psychology that fail to provide letters of recommendation. Interpret the result.

7.103 Salmonella in ice cream bars. Recently, a case of salmonella (bacterial) poisoning was traced to a particular brand of ice cream bar, and the manufacturer removed the bars from the market. Despite this response, many consumers refused to purchase *any* brand of ice cream bars for some period of time after the event (McClave, personal consulting). One manufacturer conducted a survey of consumers 6 months after the poisoning. A sample of 244 ice cream bar consumers were contacted, and 23 of them indicated that they would not purchase ice cream bars because of the potential for food poisoning.

a. What is the point estimate of the true fraction of the entire market who refuse to purchase bars 6 months after the poisoning?

b. Is the sample size large enough to use the normal approximation for the sampling distribution of the estimator of the binomial probability? Justify your response.

c. Construct a 95% confidence interval for the true proportion of the market who still refuse to purchase ice cream bars 6 months after the event.

d. Interpret both the point estimate and confidence interval in terms of this application.

7.104 Salmonella in ice cream bars. Refer to Exercise 7.103. Suppose it is now 1 year after the poisoning was traced to ice cream bars. The manufacturer wishes to estimate the proportion who still will not purchase bars to within .02 using a 95% confidence interval. How many consumers should be sampled?

7.105 Removing a soil contaminate. A common hazardous compound found in contaminated soil is benzo(a)pyrene [B(a)p]. An experiment was conducted to determine the effectiveness of a method designed to remove B(a)p from soil. (*Journal of Hazardous Materials*, June 1995.) Three soil specimens contaminated with a known amount of B(a)p were treated with a toxin that inhibits microbial growth. After 95 days of incubation, the percentage of B(a)p removed from each soil specimen was measured. The experiment produced the following summary statistics: $\bar{x} = 49.3$ and $s = 1.5$.

a. Use a 99% confidence interval to estimate the mean percentage of B(a)p removed from a soil specimen in which the toxin was used.

b. Interpret the interval in terms of this application.

c. What assumption is necessary to ensure the validity of this confidence interval?

d. How many soil specimens must be sampled to estimate the mean percentage removed to within .5 using a 99% confidence interval?

7.106 Mail delivered on time. The accounting firm Price Waterhouse periodically monitors the U.S. Postal Service's performance. One parameter of interest is the percentage of mail delivered on time. In a sample of 332,000 mailed items, Price Waterhouse determined that 282,200 items were delivered on time. (*Tampa Tribune*, March 26, 1995.) Use this information to estimate with 99% confidence the true percentage of items delivered on time by the U.S. Postal Service. Interpret the result.

Applying the Concepts—Advanced

7.107 IMA salary survey. Each year *Management Accounting* reports the results of a salary survey of the members of the Institute of Management Accountants (IMA). One year, the 2,112 members responding had a salary distribution with a 20th percentile of $35,100; a median of $50,000; and an 80th percentile of $73,000.

a. Use this information to determine the minimum sample size that could be used in next year's survey to

estimate the mean salary of IMA members to within $2,000 with 98% confidence. [*Hint:* To estimate s, first apply Chebyshev's Theorem to find k such that at least 60% of the data fall within k standard deviations of μ. Then find $s \approx$ (80th percentile − 20th percentile)/k.]

b. Explain how you estimated the standard deviation required for the sample size calculation.

c. List any assumptions you make.

7.108 Cost of destructive sampling. Destructive sampling, in which the test to determine whether an item is defective destroys the item, is generally expensive, and the high costs involved often prohibit large sample sizes. For example, suppose the National Highway Traffic Safety Administration (NHTSA) wishes to determine the proportion of new tires that will fail when subjected to hard braking at a speed of 60 miles per hour. NHTSA can obtain the tires for $25 (wholesale price) each. Suppose the budget for the experiment is $10,000, and NHTSA wishes to estimate the percentage that will fail to within .02 with 95% confidence. Assuming that the entire $10,000 can be spent on tires (ignoring other costs) and that the true fraction that will fail is approximately .05, can NHTSA attain its goal while staying within the budget? Explain.

Critical Thinking Challenges

7.109 Scallops, sampling, and the law. MIT professor Arnold Bennett describes a recent legal case in which he served as a statistical "expert" in *Interfaces* (March–April 1995). The case involved a ship that fishes for scallops off the coast of New England. In order to protect baby scallops from being harvested, the U.S. Fisheries and Wildlife Service requires that "the average meat per scallop weigh at least $\frac{1}{36}$ of a pound." The ship was accused of violating this weight standard. Bennett lays out the scenario:

The vessel arrived at a Massachusetts port with 11,000 bags of scallops, from which the harbormaster randomly selected 18 bags for weighing. From each such bag, his agents took a large scoopful of scallops; then, to estimate the bag's average meat per scallop, they divided the total weight of meat in the scoopful by the number of scallops it contained. Based on the 18 [numbers] thus generated, the harbormaster estimated that each of the ship's scallops possessed an average $\frac{1}{39}$ of a pound of meat (that is, they were about seven percent lighter than the minimum requirement). Viewing this outcome as conclusive evidence that the weight standard had been violated, federal authorities at once confiscated 95 percent of the catch (which they then sold at auction). The fishing voyage was thus transformed into a financial catastrophe for its participants.

The actual scallop weight measurements for each of the 18 sampled bags are listed in the accompanying table. For ease of exposition, Bennett expressed each number as a multiple of $\frac{1}{36}$ of a pound, the minimum permissible average weight per scallop. Consequently, numbers below one indicate individual bags that do not meet the standard.

The ship's owner filed a lawsuit against the federal government, declaring that his vessel had fully complied with the weight standard. A Boston law firm was hired to represent the owner in legal proceedings and Bennett was retained by the firm to provide statistical litigation support and, if necessary, expert witness testimony.

a. Recall that the harbormaster sampled only 18 of the ship's 11,000 bags of scallops. One of the questions the lawyers asked Bennett was, "Can a reliable estimate of the mean weight of all the scallops be obtained from a sample of size 18?" Give your opinion on this issue.

b. As stated in the article, the government's decision rule is to confiscate a scallop catch if the sample mean weight of the scallops is less than $\frac{1}{36}$ of a pound. Do you see any flaws in this rule?

c. Develop your own procedure for determining whether a ship is in violation of the minimum weight restriction. Apply your rule to the data. Draw a conclusion about the ship in question.

SCALLOPS

.93	.88	.85	.91	.91	.84	.90	.98	.88
.89	.98	.87	.91	.92	.99	1.14	1.06	.93

Source: Bennett, A. "Misapplications review: Jail terms." *Interfaces,* Vol. 25, No. 2, Mar.–Apr. 1995, p. 20.

Student Projects

Choose a population pertinent to your major area of interest—a population that has an unknown mean or, if the population is binomial, that has an unknown probability of success. For example, a marketing major may be interested in the proportion of consumers who prefer a certain product. A sociology major may be interested in estimating the proportion of people in a certain socioeconomic group or the mean income of people living in a particular part of a city. A political science major may wish to estimate the proportion of an electorate in favor of a certain candidate, a certain amendment, or a certain presidential policy. A pre-med student might want to find the average length of time patients stay in the hospital or the average number of people treated daily in the emergency room. We could continue with examples, but the point should be clear—choose something of interest to you.

Define the parameter you want to estimate and conduct a *pilot study* to obtain an initial estimate of the parameter of interest and, more importantly, an estimate of the variability associated with the estimator. A pilot study is a small experiment (perhaps 20 to 30 observations) used to gain some information about the population of interest. The purpose of the study is to help plan more elaborate future experiments. Using the results of your pilot study, determine the sample size necessary to estimate the parameter to within a reasonable bound (of your choice) with a 95% confidence interval.

REFERENCES

Agresti, A., and Coull, B. A. "Approximate is better than 'exact' for interval estimation of binomial proportions." *The American Statistician*, Vol. 52, No. 2, May 1998, pp. 119–126.

Cochran, W. G. *Sampling Techniques*, 3rd ed. New York: Wiley, 1977.

Freedman, D., Pisani, R., and Purves, R. *Statistics*. New York: Norton, 1978.

Kish, L. *Survey Sampling*. New York: Wiley, 1965.

Mendenhall, W., Beaver, R. J., and Beaver, B. *Introduction to Probability and Statistics*, 11th ed. North Scituate, Mass.: Duxbury, 2002.

Wilson, E. G. "Probable inference, the law of succession, and statistical inference." *Journal of the American Statistical Association*, Vol. 22, 1927, pp. 209–212.

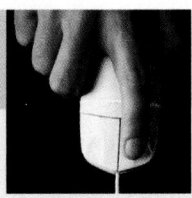

Using Technology

Confidence Intervals
Using MINITAB

MINITAB can be used to obtain one-sample confidence intervals for both a population mean and a population proportion. To generate a confidence interval for the mean using a previously created sample data set, first access the MINITAB data worksheet. Next, click on the "Stat" button on the MINITAB menu bar, then click on "Basic Statistics" and "1-sample t," as shown in Figure 7.M.1. The resulting dialog box appears as shown in Figure 7.M.2. Click on "Samples in Columns," then specify the quantitative variable of interest in the open box. Now, click on the "Options" button at the bottom of the dialog box and specify the confidence level in the resulting dialog box as shown in Figure 7.M.3. Click "OK" to return to the "1-Sample t" dialog box, then click "OK" again to produce the confidence interval.

Figure 7.M.1

MINITAB Menu Options for a Confidence Interval for μ

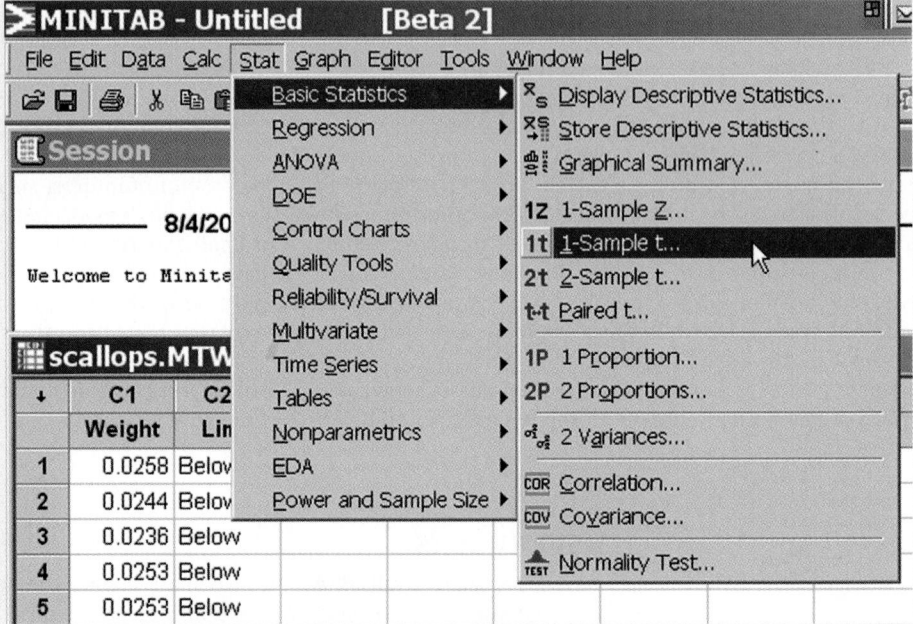

Figure 7.M.2
MINITAB 1-Sample t
Dialog Box

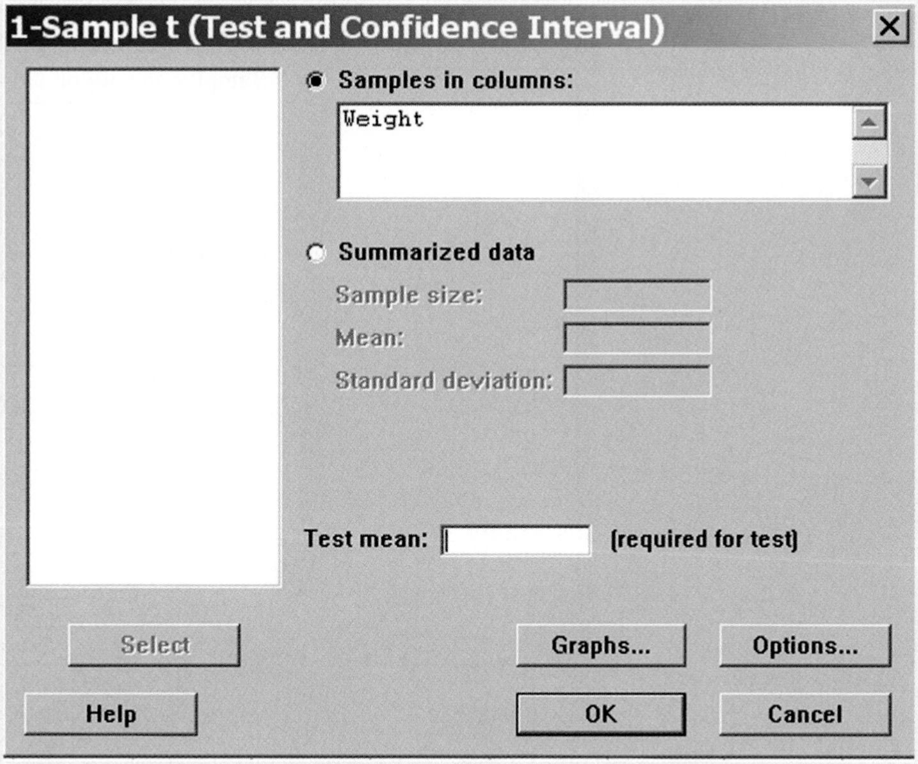

Figure 7.M.3
MINITAB 1-Sample t
Options Dialog Box

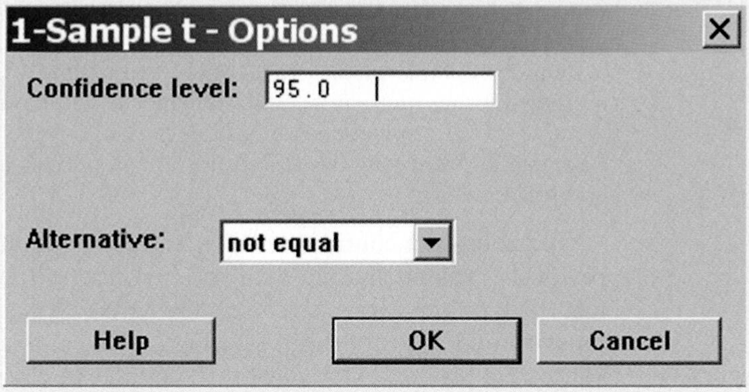

If you want to produce a confidence interval for the mean from summary information (e.g., the sample mean, sample standard deviation, and sample size), then click on "Summarized data" in the "1-Sample t" dialog box as shown in Figure 7.M.4. Enter the values of the summary statistics, then click "OK."

Figure 7.M.4
MINITAB 1-Sample t
Dialog Box with Summary
Statistics

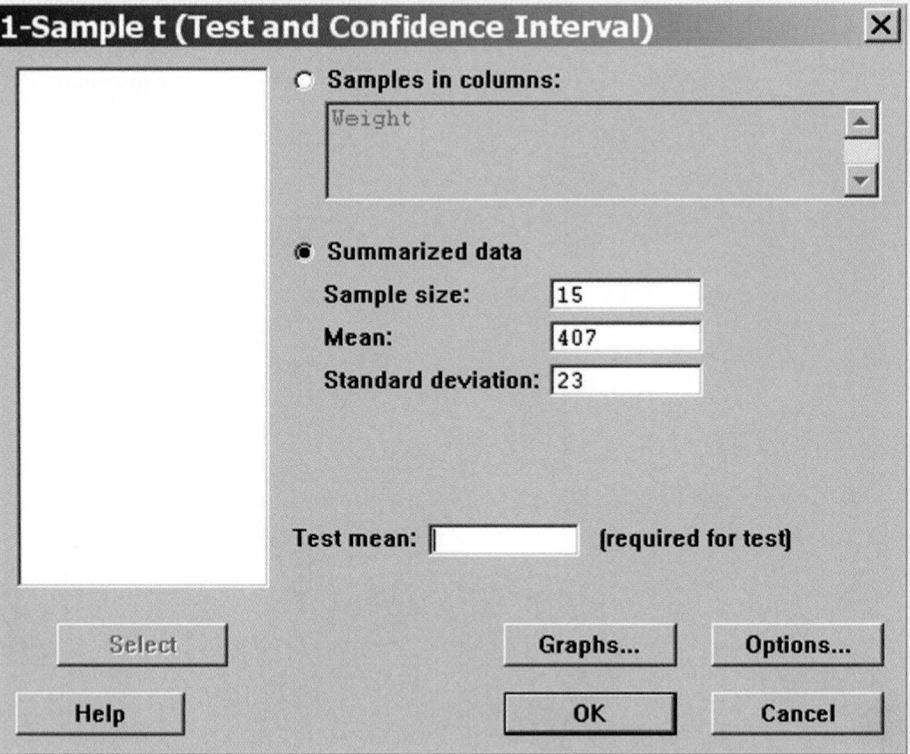

[*Important Note*: The MINITAB 1-Sample t procedure uses the *t*-statistic to generate the confidence interval. When the sample size *n* is small, this is the appropriate method. When the sample size *n* is large, the *t*-value will be approximately equal to the large-sample *z*-value and the resulting interval will still be valid. If you have a large sample and you know the value of the population standard deviation σ (which is rarely the case), then select "1-Sample Z" from the "Basic Statistics" menu options (see Figure 7.M.1) and make the appropriate selections.]

To generate a confidence interval for a population proportion mean using a previously created sample data set, first access the MINITAB data worksheet. Next, click on the "Stat" button on the MINITAB menu bar, then click on "Basic Statistics" and "1 Proportion" (see Figure 7.M.1). The resulting dialog box appears as shown in Figure 7.M.5. Click on "Samples in Columns," then specify the qualitative variable of interest in the open box. Now, click on the "Options" button at the bottom of the dialog box and specify the confidence level in the resulting dialog box, as shown in Figure 7.M.6. Also, check the "Use test and interval based on normal distribution" box at the bottom. Click "OK" to return to the "1-Proportion" dialog box, then click "OK" again to produce the confidence interval.

Figure 7.M.5
MINITAB 1-Proportion
Dialog Box

1 Proportion (Test and Confidence Interval)   ✕

| C1 | Weight |
| C2 | Limit |

⦿ **Samples in columns:**

Limit   ▲
                                                     ▼

○ **Summarized data**

Number of trials: [   ]
Number of events: [   ]

Select   Options...

Help   OK   Cancel

Figure 7.M.6
MINITAB 1-Proportion
Dialog Box

1 Proportion - Options   ✕

Confidence level: [95.0   ]

Test proportion: [0.5   ]

Alternative: [not equal   ▼]

☑ **Use test and interval based on normal distribution**

Help   OK   Cancel

If you want to produce a confidence interval for a proportion from summary information (e.g., the number of successes and the sample size), then click on "Summarized data" in the "1-Proportion" dialog box (see Figure 7.M.5). Enter the value for the number of trials (i.e., the sample size) and the number of events (i.e., the number of successes), then click "OK."

8

Inferences Based on a Single Sample

Tests of Hypothesis

Contents

Statistics in Action

Diary of a Kleenex® User

Using Technology

Tests of Hypothesis Using MINITAB

Where We've Been

- Used sample information to provide a *point estimate* of a population parameter

- Used the sampling distribution of a statistic to assess the reliability of an estimate through a *confidence interval*

Where We're Going

- Test a specific value of a population parameter (mean or proportion)—called a *test of hypothesis*.

- Provide a measure of reliability for the hypothesis test—called the *significance level* of the test.

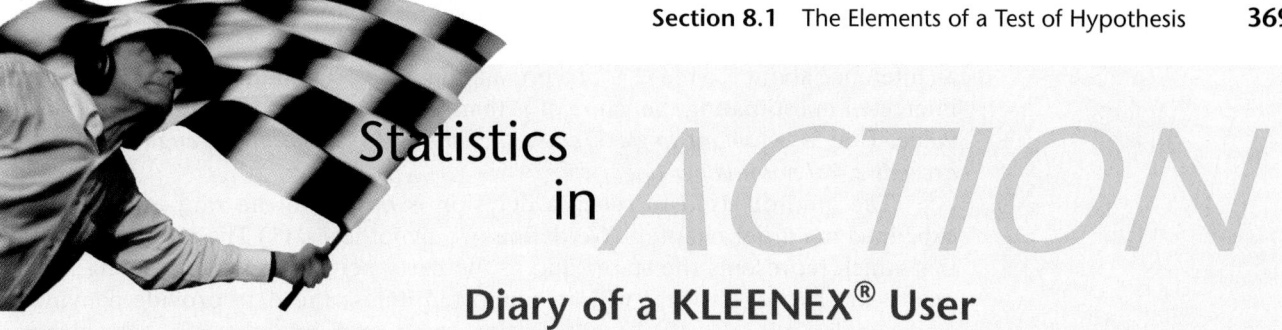

Statistics in *ACTION*

Diary of a KLEENEX® User

In 1924, Kimberly-Clark Corporation invented a facial tissue for removing cold cream and began marketing it as KLEENEX® brand tissues. Today, KLEENEX® is recognized as the top-selling brand of tissue in the world. A wide variety of KLEENEX® products is available, ranging from extra-large tissues to tissues with lotion. Over the past 80 years, Kimberly-Clark Corporation has packaged the tissues in boxes of different sizes and shapes and varied the number of tissues packaged in each box. For example, currently a family-size box contains 144 two-ply tissues, a cold-care box contains 70 tissues (coated with lotion), and a convenience pocket pack contains 15 miniature tissues.

How does Kimberly-Clark Corp. decide how many tissues to put in each box? According to the *Wall Street Journal*, marketing experts at the company use the results of a survey of KLEENEX® customers to help determine how many tissues are packed in a box. In the mid-1980s, when Kimberly-Clark Corp. developed the cold-care box designed especially for people who have a cold, the company conducted their initial survey of customers for this purpose. Hundreds of customers were asked to keep count of their KLEENEX® use in diaries. According to the *Wall Street Journal* report, the survey results left "little doubt

that the company should put 60 tissues in each box." The number 60 was "the average number of times people blow their nose during a cold." (*Note:* In 2000, the company increased the number of tissues packaged in a cold-care box to 70.)

From summary information provided in the *Wall Street Journal* (September 21, 1984) article, we constructed a data set that represents the results of a survey similar to the one just described. In the data file named **TISSUES**, we recorded the number of tissues used by each of 250 consumers during a period when they had a cold. We apply the hypothesis testing methodology presented in this chapter to this data set in several Statistics in Action Revisited examples.

Statistics in the Action Revisited

- Identifying the Key Elements of a Hypothesis Test Relevant to the KLEENEX® Survey (p. 375)
- Testing a Population Mean in the KLEENEX® Survey (p. 388)
- Testing a Population Proportion in the KLEENEX® Survey (p. 401)

Suppose you wanted to determine whether the mean level of a driver's blood alcohol exceeds the legal limit after two drinks, or whether the majority of registered voters approve of the president's performance. In both cases you are interested in making an inference about how the value of a parameter relates to a specific numerical value. Is it less than, equal to, or greater than the specified number? This type of inference, called a **test of hypothesis**, is the subject of this chapter.

We introduce the elements of a test of hypothesis in Section 8.1. We then show how to conduct a large-sample test of hypothesis about a population mean in Sections 8.2 and 8.3. In Section 8.4 we utilize small samples to conduct tests about means. Large-sample tests about binomial probabilities are the subject of Section 8.5, and some advanced methods for determining the reliability of a test are covered in optional Section 8.6. Finally, we show how to conduct a test about a population variance in optional Section 8.7.

8.1 The Elements of a Test of Hypothesis

Suppose building specifications in a certain city require that the average breaking strength of residential sewer pipe be more than 2,400 pounds per foot of length (i.e., per linear foot). Each manufacturer who wants to sell pipe in this city must demonstrate that its product meets the specification. Note that we are interested in making an inference about the mean μ of a population. However, in this example we are less interested in estimating the value of μ than we are in testing a *hypothesis* about its value. That is, *we want to decide whether the mean breaking strength of the pipe exceeds 2,400 pounds per linear foot.*

The method used to reach a decision is based on the rare-event concept explained in earlier chapters. We define two hypotheses: (1) The **null hypothesis** is that which represents the status quo to the party performing the sampling experiment—the hypothesis that will be supported unless the data provide convincing evidence that it is false. (2) The **alternative**, or **research**, **hypothesis** is that which will be accepted only if the data provide convincing evidence of its truth. From the point of view of the city conducting the tests, the null hypothesis is that the manufacturer's pipe does *not* meet specifications unless the tests provide convincing evidence otherwise. The null and alternative hypotheses are therefore

Null hypothesis (H_0): $\mu \leq 2,400$
   (i.e., the manufacturer's pipe does not meet specifications)
Alternative (research) hypothesis (H_a): $\mu > 2,400$
   (i.e., the manufacturer's pipe meets specifications)

Now Work **Exercise 8.8**

How can the city decide when enough evidence exists to conclude that the manufacturer's pipe meets specifications? Since the hypotheses concern the value of the population mean μ, it is reasonable to use the sample mean $\bar{x}$ to make the inference, just as we did when forming confidence intervals for μ in Sections 7.1 and 7.2. The city will conclude that the pipe meets specifications only when the sample mean $\bar{x}$ convincingly indicates that the population mean exceeds 2,400 pounds per linear foot.

"Convincing" evidence in favor of the alternative hypothesis will exist when the value of $\bar{x}$ exceeds 2,400 by an amount that cannot be readily attributed to sampling variability. To decide, we compute a **test statistic**, which is the z-value that measures the distance between the value of $\bar{x}$ and the value of μ specified in the null hypothesis. When the null hypothesis contains more than one value of μ, as in this case $(H_0: \mu \leq 2,400)$, we use the value of μ closest to the values specified in the alternative hypothesis. The idea is that if the hypothesis that μ *equals* 2,400 can be rejected in favor of $\mu > 2,400$, then μ *less than or equal to* 2,400 can certainly be rejected. Thus, the test statistic is

$$z = \frac{\bar{x} - 2,400}{\sigma_{\bar{x}}} = \frac{\bar{x} - 2,400}{\sigma/\sqrt{n}}$$

Note that a value of $z = 1$ means that $\bar{x}$ is 1 standard deviation above $\mu = 2,400$; a value of $z = 1.5$ means that $\bar{x}$ is 1.5 standard deviations above $\mu = 2,400$, etc. How large must z be before the city can be convinced that the null hypothesis can be rejected in favor of the alternative and conclude that the pipe meets specifications?

If you examine Figure 8.1, you will note that the chance of observing $\bar{x}$ more than 1.645 standard deviations above 2,400 is only .05—*if in fact the true mean μ is 2,400.* Thus, if the sample mean is more than 1.645 standard deviations above 2,400, either H_0 is true and a relatively rare event has occurred (.05 probability) or H_a is true and the population mean exceeds 2,400. Since we would most likely reject the notion that a rare event has occurred, we would reject the null hypothesis $(\mu \leq 2,400)$ and conclude that the alternative hypothesis $(\mu > 2,400)$ is true. What is the probability that this procedure will lead us to an incorrect decision?

Figure 8.1

The Sampling Distribution of $\bar{x}$, Assuming $\mu = 2,400$

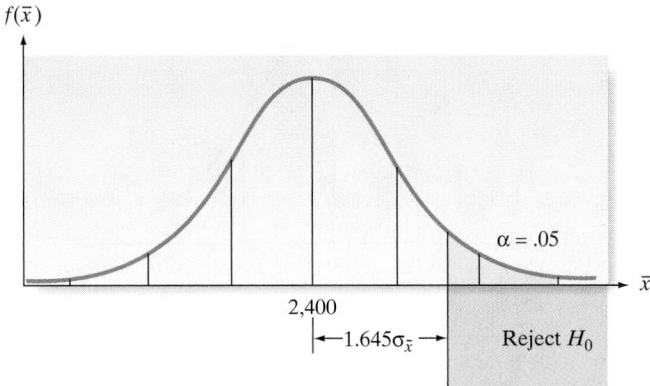

Such an incorrect decision—deciding that the null hypothesis is false when in fact it is true—is called a **Type I decision error**. As indicated in Figure 8.1, the risk of making a Type I error is denoted by the symbol α. That is,

$$\alpha = P \text{ (Type I error)}$$
$$= P \text{ (Rejecting the null hypothesis when in fact the null hypothesis is true)}$$

In our example

$$\alpha = P \text{ (}z > 1.645 \text{ when in fact } \mu = 2,400) = .05$$

We now summarize the elements of the test:

$H_0: \mu \le 2,400$
$H_a: \mu > 2,400$

Test statistic: $z = \dfrac{\bar{x} - 2,400}{\sigma_{\bar{x}}}$

Rejection region: $z > 1.645$, which corresponds to $\alpha = .05$

Note that the **rejection region** refers to the values of the test statistic for which we will *reject the null hypothesis*.

To illustrate the use of the test, suppose we test 50 sections of sewer pipe and find the mean and standard deviation for these 50 measurements to be

$$\bar{x} = 2,460 \text{ pounds per linear foot} \qquad s = 200 \text{ pounds per linear foot}$$

As in the case of estimation, we can use s to approximate σ when s is calculated from a large set of sample measurements.

The test statistic is

$$z = \frac{\bar{x} - 2,400}{\sigma_{\bar{x}}} = \frac{\bar{x} - 2,400}{\sigma/\sqrt{n}} \approx \frac{\bar{x} - 2,400}{s/\sqrt{n}}$$

Substituting $\bar{x} = 2,460$, $n = 50$, and $s = 200$, we have

$$z \approx \frac{2,460 - 2,400}{200/\sqrt{50}} = \frac{60}{28.28} = 2.12$$

Therefore, the sample mean lies $2.12\sigma_{\bar{x}}$ above the hypothesized value of $\mu = 2,400$ as shown in Figure 8.2. Since this value of z exceeds 1.645, it falls in the rejection region. That is, we reject the null hypothesis that $\mu = 2,400$ and conclude that $\mu > 2,400$. Thus, it appears that the company's pipe has a mean strength that exceeds 2,400 pounds per linear foot.

How much faith can be placed in this conclusion? What is the probability that our statistical test could lead us to reject the null hypothesis (and conclude that the company's pipe meets the city's specifications) when in fact the null hypothesis is

Figure 8.2

Location of the Test Statistic for a Test of the Hypothesis $H_0: \mu = 2{,}400$

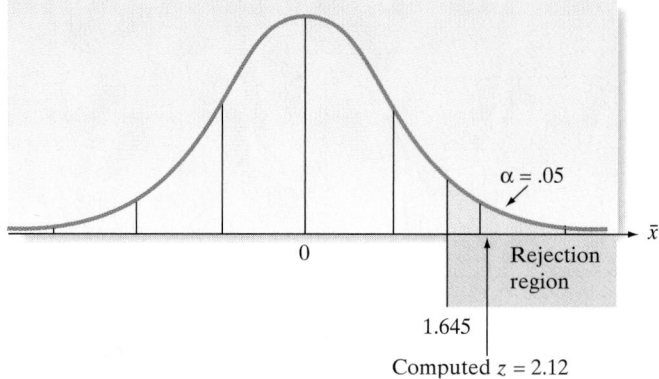

true? The answer is $\alpha = .05$. That is, we selected the level of risk, α, of making a Type I error when we constructed the test. Thus, the chance is only 1 in 20 that our test would lead us to conclude the manufacturer's pipe satisfies the city's specifications when in fact the pipe does *not* meet specifications.

Now, suppose the sample mean breaking strength for the 50 sections of sewer pipe turned out to be $\bar{x} = 2{,}430$ pounds per linear foot. Assuming that the sample standard deviation is still $s = 200$, the test statistic is

$$ z = \frac{2{,}430 - 2{,}400}{2/\sqrt{50}} = \frac{30}{28.28} = 1.06 $$

Therefore, the sample mean $\bar{x} = 2{,}430$ is only 1.06 standard deviations above the null hypothesized value of $\mu = 2{,}400$. As shown in Figure 8.3, this value does not fall into the rejection region ($z > 1.645$). Therefore, we know that we cannot reject H_0 using $\alpha = .05$. Even though the sample mean exceeds the city's specification of 2,400 by 30 pounds per linear foot, it does not exceed the specification by enough to provide *convincing* evidence that the *population mean* exceeds 2,400.

Figure 8.3

Location of the Test Statistic When $\bar{x} = 2{,}430$

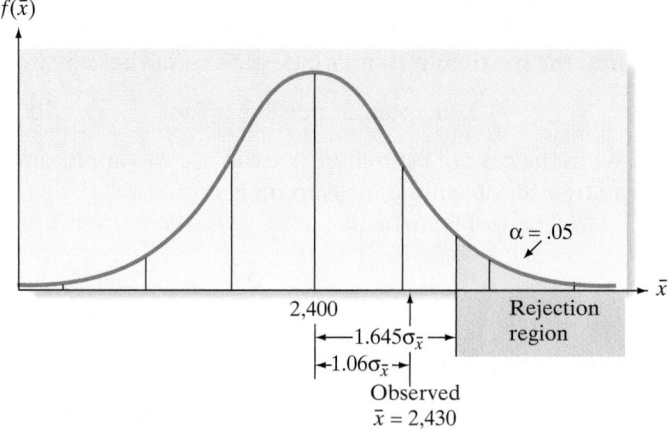

Should we accept the null hypothesis $H_0: \mu \le 2{,}400$ and conclude that the manufacturer's pipe does not meet specifications? To do so would be to risk a **Type II error**—that of concluding that the null hypothesis is true (the pipe does not meet specifications) when in fact it is false (the pipe does meet specifications). We denote the probability of committing a Type II error by β, and we show in optional Section 8.6 that β is often difficult to determine precisely. Rather than make a decision (accept H_0) for which the probability of error (β) is unknown, we avoid the potential Type II error by avoiding the conclusion that the null hypothesis is true. Instead, we will simply state that *the sample evidence is insufficient to reject H_0 at $\alpha = .05$. Since

Biography

EGON S. PEARSON (1895–1980)— The Neyman-Pearson Lemma

Egon Pearson was the only son of noteworthy British statistician Karl Pearson (see Biography, p. 786). As you might expect, Egon developed an interest in the statistical methods developed by his father and, upon completing graduate school, accepted a position to work for Karl in the Department of Applied Statistics at University College, Lon-

don. Egon is best known for his collaboration with Jerzey Neyman (see Biography, p. 325) on the development of the theory of hypothesis testing. One of the basic concepts in the Neyman-Pearson approach was that of the "null" and "alternative" hypotheses. Their famous "Neyman-Pearson" lemma was published in *Biometrika* in 1928. Egon Pearson had numerous other contributions to statistics and was known as an excellent teacher and lecturer. In his last major work, Egon fulfilled a promise made to his father by publishing an annotated version of Karl Pearson's lectures on the early history of statistics.

the null hypothesis is the "status-quo" hypothesis, the effect of not rejecting H_0 is to maintain the status quo. In our pipe-testing example, the effect of having insufficient evidence to reject the null hypothesis that the pipe does not meet specifications is probably to prohibit the utilization of the manufacturer's pipe unless and until there is sufficient evidence that the pipe does meet specifications. That is, until the data indicate convincingly that the null hypothesis is false, we usually maintain the status quo implied by its truth.

Table 8.1 summarizes the four possible outcomes of a test of hypothesis. The "true state of nature" columns in Table 8.1 refer to the fact that either the null hypothesis H_0 is true or the alternative hypothesis H_a is true. Note that the true state of nature is unknown to the researcher conducting the test. The "conclusion" rows in Table 8.1 refer to the action of the researcher, assuming that he or she will conclude either that H_0 is true or that H_a is true, based on the results of the sampling experiment. Note that a Type I error can be made *only* when the alternative hypothesis is accepted (equivalently, when the null hypothesis is rejected), and a Type II error can be made *only* when the null hypothesis is accepted. Our policy will be to make a decision only when we know the probability of making the error that corresponds to that decision. Since α is usually specified by the analyst (typically .05 in research), we will generally be able to reject H_0 (accept H_a) when the sample evidence supports that decision. However, since β is usually not specified, *we will generally avoid the decision to accept H_0, preferring instead to state that the sample evidence is insufficient to reject H_0 when the test statistic is not in the rejection region.*

 ## Caution

Be careful not to "accept H_0" when conducting a test of hypothesis since the measure of reliability, $\beta = P(\text{Type II error})$, is almost always unknown. If the test statistic does not fall into the rejection region, it is better to state the conclusion as "insufficient evidence to reject H_0."

TABLE 8.1   Conclusions and Consequences for a Test of Hypothesis

	True State of Nature	
Conclusion	H_0 True	H_a True
Accept H_0 (Assume H_0 True)	Correct decision	Type II error (probability β)
Reject H_0 (Assume H_a True)	Type I error (probability α)	Correct decision

The elements of a test of hypothesis are summarized in the following box. Note that the first four elements are all specified *before* the sampling experiment is performed. In no case will the results of the sample be used to determine the hypotheses—the data are collected to test the predetermined hypotheses, not to formulate them.

Elements of a Test of Hypothesis

1. *Null hypothesis* (H_0): A theory about the values of one or more population parameters. The theory generally represents the status quo, which we adopt until it is proven false. By convention, the theory is stated as H_0: parameter = value.

2. *Alternative (research) hypothesis* (H_a): A theory that contradicts the null hypothesis. The theory generally represents that which we will accept only when sufficient evidence exists to establish its truth.

3. *Test statistic:* A sample statistic used to decide whether to reject the null hypothesis.

4. *Rejection region:* The numerical values of the test statistic for which the null hypothesis will be rejected. The rejection region is chosen so that the probability is α that it will contain the test statistic when the null hypothesis is true, thereby leading to a Type I error. The value of α is usually chosen to be small (e.g., .01, .05, or .10), and is referred to as the **level of significance** of the test.

5. *Assumptions:* Clear statement(s) of any assumptions made about the population(s) being sampled.

6. *Experiment and calculation of test statistic:* Performance of the sampling experiment and determination of the numerical value of the test statistic.

7. *Conclusion:*
   a. If the numerical value of the test statistic falls in the rejection region, we reject the null hypothesis and conclude that the alternative hypothesis is true. We know that the hypothesis-testing process will lead to this conclusion incorrectly (Type I error) only $100\alpha\%$ of the time when H_0 is true.
   b. If the test statistic does not fall in the rejection region, we do not reject H_0. Thus, we reserve judgment about which hypothesis is true. We do not conclude that the null hypothesis is true because we do not (in general) know the probability β that our test procedure will lead to an incorrect acceptance of H_0 (Type II error).

As with confidence intervals, the methodology for testing hypotheses varies depending on the target population parameter. In this chapter, we develop methods for testing a population mean, a population proportion, and a population variance. Some key words and the type of data associated with these target parameters are listed in the accompanying box.

Determining the Target Parameter

Parameter	Key Words or Phrases	Type of Data
μ	Mean; average	Quantitative
p	Proportion; percentage; fraction; rate	Qualitative
σ^2	Variance; variability; spread	Quantitative

Statistics in Action Revisited

Identifying the Key Elements of a Hypothesis Test Relevant to the KLEENEX® Survey

In Kimberly-Clark Corporation's survey of people with colds, each of 250 customers was asked to keep count of his or her use of KLEENEX® tissues in diaries. One goal of the company was to determine how many tissues to package in a cold-care box of KLEENEX®; consequently, the total number of tissues used was recorded for each person surveyed. Since number of tissues is a quantitative variable, the parameter of interest is either μ, the mean number of tissues used by all customers with colds, or σ^2, the variance of the number of tissues used.

Now, according to a *Wall Street Journal* report, there was "little doubt that the company should put 60 tissues" in a cold-care box of KLEENEX® tissues. This statement was based on a claim made by marketing experts that 60 is the average number of times a person will blow his/her nose during a cold. The key word *average* implies that the target parameter is μ, and the marketers are claiming that $\mu = 60$. Suppose we disbelieve the claim that $\mu = 60$, believing instead that the population mean is smaller than 60 tissues. In order to test the claim against our belief, we set up the following null and alternative hypothesis:

$$H_0: \mu = 60 \qquad H_a: \mu < 60$$

We'll conduct this test in the next Statistics in Action Revisited on p. 388.

Exercises 8.1–8.17

Understanding the Principles

8.1 Which hypothesis, the null or the alternative, is the status-quo hypothesis? Which is the research hypothesis?

8.2 Which element of a test of hypothesis is used to decide whether to reject the null hypothesis in favor of the alternative hypothesis?

8.3 What is the level of significance of a test of hypothesis?

8.4 What is the difference between Type I and Type II errors in hypothesis testing? How do α and β relate to Type I and Type II errors?

8.5 List the four possible results of the combinations of decisions and true states of nature for a test of hypothesis.

8.6 We (generally) reject the null hypothesis when the test statistic falls in the rejection region, but we do not accept the null hypothesis when the test statistic does not fall in the rejection region. Why?

8.7 If you test a hypothesis and reject the null hypothesis in favor of the alternative hypothesis, does your test prove that the alternative hypothesis is correct? Explain.

Applying the Concepts—Basic

8.8 **Calories in school lunches.** A University of Florida economist conducted a study of Virginia elementary school lunch menus. During the state-mandated testing period, school lunches averaged 863 calories. (*National Bureau of Economic Research*, November 2002.) The economist claims that after the testing period ends, the average caloric content of Virginia school lunches drops significantly. Set up the null and alternative hypothesis to test the economist's claim.

8.9 **A camera that detects liars.** According to *NewScientist* (January 2, 2002), a new thermal imaging camera that detects small temperature changes is now being used as a polygraph device. The United States Department of Defense Polygraph Institute (DDPI) claims the camera can correctly detect liars 75% of the time by monitoring the temperatures of their faces. Give the null hypothesis for testing the claim made by the DDPI.

8.10 **Work travel policy.** American Express Consulting reported in *USA Today* (June 15, 2001) that 80% of U.S. companies have formal, written travel and entertainment policies for their employees. Give the null hypothesis for testing the claim made by American Express Consulting.

8.11 **Use of herbal therapy.** According to the *Journal of Advanced Nursing* (January 2001), 45% of senior women (i.e., women over the age of 65) use herbal therapies to prevent or treat health problems. Also, of senior women who use herbal therapies, they use an average of 2.5 herbal products in a year.
  a. Give the null hypothesis for testing the first claim by the journal.
  b. Give the null hypothesis for testing the second claim by the journal.

8.12 **Infants listening time.** *Science* (January 1, 1999) reported that the mean listening time of 7-month-old infants exposed to a three-syllable sentence (e.g., "ga ti ti") is 9 seconds. Set up the null and alternative hypotheses for testing the claim.

8.13 **Cocaine-exposed babies.** Infants exposed to cocaine in their mother's womb are thought to be at a high risk for major birth defects. However, according to a University of Florida study, more than 75% of all cocaine-exposed babies suffer no major problems (*Explore*, Fall 1998). Set up the null and alternative hypotheses if you want to support the findings.

Applying the Concepts—Intermediate

8.14 Susceptibility to hypnosis. The Computer-Assisted Hypnosis Scale (CAHS) is designed to measure a person's susceptibility to hypnosis. CAHS scores range from 0 (no susceptibility to hypnosis) to 12 (extremely high susceptibility to hypnosis). *Psychological Assessment* (March 1995) reported that University of Tennessee undergraduates had a mean CAHS score of $\mu = 4.6$. Suppose you want to test whether undergraduates at your college or university are more susceptible to hypnosis than University of Tennessee undergraduates.

a. Set up H_0 and H_a for the test.

b. Describe a Type I error for this test.

c. Describe a Type II error for this test.

8.15 Mercury levels in wading birds. According to a University of Florida wildlife ecology and conservation researcher, the average level of mercury uptake in wading birds in the Everglades has declined over the past several years. (*UF News*, December 15, 2000.) Five years ago, the average level was 15 parts per million.

a. Give the null and alternative hypotheses for testing whether the average level today is less than 15 ppm.

b. Describe a Type I error for this test.

c. Describe a Type II error for this test.

Applying the Concepts—Advanced

8.16 Jury trial outcomes. Sometimes, the outcome of a jury trial defies the "commonsense" expectations of the general public (e.g., the O. J. Simpson verdict in the "Trial of the Century"). Such a verdict is more acceptable if we understand that the jury trial of an accused murderer is analogous to the statistical hypothesis-testing process. The null hypothesis in a jury trial is that the accused is innocent. (The status-quo hypothesis in the U.S. system of justice is innocence, which is assumed to be true until proven *beyond a reasonable doubt*.) The alternative hypothesis is guilt, which is accepted only when sufficient evidence exists to establish its truth. If the vote of the jury is unanimous in favor of guilt, the null hypothesis of innocence is rejected and the court concludes that the accused murderer is guilty. Any vote other than a unanimous one for guilt results in a "not guilty"

verdict. The court never accepts the null hypothesis; that is, the court never declares the accused "innocent." A "not guilty" verdict (as in the O. J. Simpson case) implies that the court could not find the defendant guilty *beyond a reasonable doubt.*

a. Define Type I and Type II errors in a murder trial.

b. Which of the two errors is the more serious? Explain.

c. The court does not, in general, know the values of α and β; but ideally, both should be small. One of these probabilities is assumed to be smaller than the other in a jury trial. Which one, and why?

d. The court system relies on the belief that the value of α is made very small by requiring a unanimous vote before guilt is concluded. Explain why this is so.

e. For a jury prejudiced against a guilty verdict as the trial begins, will the value of α increase or decrease? Explain.

f. For a jury prejudiced against a guilty verdict as the trial begins, will the value of β increase or decrease? Explain.

8.17 Intrusion detection systems. Refer to the *Journal of Research of the National Institute of Standards and Technology* (November–December, 2003) study of a computer intrusion detection system (IDS), Exercise 3.81 (p. 158). Recall that an IDS is designed to provide an alarm whenever unauthorized access (e.g., an intrusion) to a computer system occurs. The probability of the system giving a false alarm (i.e., providing a warning when, in fact, no intrusion occurs) is defined by the symbol α, while the probability of a missed detection (i.e., no warning given, when, in fact, an intrusion occurs) is defined by the symbol β. These symbols are used to represent Type I and Type II error rates, respectively, in a hypothesis testing scenario.

a. What is the null hypothesis, H_0?

b. What is the alternative hypothesis, H_a?

c. According to actual data on the EMERALD system collected by the Massachusetts Institute of Technology Lincoln Laboratory, only 1 in 1,000 computer sessions with no intrusions resulted in a false alarm. For the same system, the laboratory found that only 500 of 1,000 intrusions were actually detected. Use this information to estimate the values of α and β.

8.2 Large-Sample Test of Hypothesis about a Population Mean

In Section 8.1 we learned that the null and alternative hypotheses form the basis for a test of hypothesis inference. The null and alternative hypotheses may take one of several forms. In the sewer pipe example we tested the null hypothesis that the population mean strength of the pipe is less than or equal to 2,400 pounds per linear foot against the alternative hypothesis that the mean strength exceeds 2,400. That is, we tested

$$H_0: \mu \leq 2,400$$
$$H_a: \mu > 2,400$$

This is a **one-tailed** (or **one-sided**) **statistical test** because the alternative hypothesis specifies that the population parameter (the population mean μ, in this example) is strictly greater than a specified value (2,400, in this example). If the null hypothesis had been H_0: $\mu \geq 2,400$ and the alternative hypothesis had been H_a: $\mu < 2,400$, the test would still be one-sided, because the parameter is still specified to be on "one side" of the null hypothesis value. Some statistical investigations seek to show that the population parameter is *either larger or smaller* than some specified value. Such an alternative hypothesis is called a **two-tailed** (or **two-sided**) **hypothesis**.

While alternative hypotheses are always specified as strict inequalities, such as $\mu < 2,400$, $\mu > 2,400$, or $\mu \neq 2,400$, null hypotheses are usually specified as equalities, such as $\mu = 2,400$. *Even when the null hypothesis is an inequality, such as $\mu \leq 2,400$, we specify H_0: $\mu = 2,400$, reasoning that if sufficient evidence exists to show that H_a: $\mu > 2,400$ is true when tested against H_0: $\mu = 2,400$, then surely sufficient evidence exists to reject $\mu < 2,400$ as well.* Therefore, the null hypothesis is specified as the value of μ closest to a one-sided alternative hypothesis and as the only value *not* specified in a two-tailed alternative hypothesis. The steps for selecting the null and alternative hypotheses are summarized in the box on p. 000.

The rejection region for a two-tailed test differs from that for a one-tailed test. When we are trying to detect departure from the null hypothesis in *either* direction, we must establish a rejection region in both tails of the sampling distribution of the test statistic. Figures 8.4a and 8.4b show the one-tailed rejection regions for lower- and upper-tailed tests, respectively. The two-tailed rejection region is illustrated in Figure 8.4c. Note that a rejection region is established in each tail of the sampling distribution for a two-tailed test.

The rejection regions corresponding to typical values selected for α are shown in Table 8.2 for one- and-two-tailed tests. Note that the smaller α you select, the more evidence (the larger z) you will need before you can reject H_0.

Steps for Selecting the Null and Alternative Hypotheses

1. Select the *alternative hypothesis* as that which the sampling experiment is intended to establish. The alternative hypothesis will assume one of three forms:
   a. One-tailed, upper-tailed   *Example: H_a: $\mu > 2,400$*
   b. One-tailed, lower-tailed   *Example: H_a: $\mu < 2,400$*
   c. Two-tailed   *Example: H_a: $\mu \neq 2,400$*

2. Select the *null hypothesis* as the status quo, that which will be presumed true unless the sampling experiment conclusively establishes the alternative hypothesis. The null hypothesis will be specified as that parameter value closest to the alternative in one-tailed tests, and as the complementary (or only unspecified) value in two-tailed tests.

   Example: H_0: $\mu = 2,400$

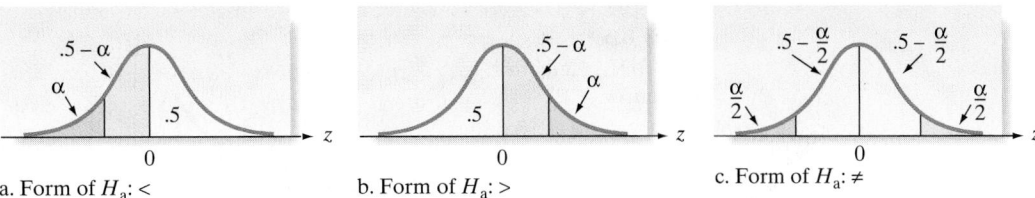

a. Form of H_a: $<$ b. Form of H_a: $>$ c. Form of H_a: $\neq$

Figure 8.4
Rejection Regions Corresponding to One- and Two-Tailed Tests

TABLE 8.2 Rejection Regions for Common Values of α

	Alternative Hypotheses		
	Lower-Tailed	Upper-Tailed	Two-Tailed
$\alpha = .10$	$z < -1.28$	$z > 1.28$	$z < -1.645$ or $z > 1.645$
$\alpha = .05$	$z < -1.645$	$z > 1.645$	$z < -1.96$ or $z > 1.96$
$\alpha = .01$	$z < -2.33$	$z > 2.33$	$z < -2.575$ or $z > 2.575$

EXAMPLE 8.1

SETTING UP A HYPOTHESIS TEST FOR μ

Problem The effect of drugs and alcohol on the nervous system has been the subject of considerable research. Suppose a research neurologist is testing the effect of a drug on response time by injecting 100 rats with a unit dose of the drug, subjecting each to a neurological stimulus, and recording its response time. The neurologist knows that the mean response time for rats not injected with the drug (the "control" mean) is 1.2 seconds. She wishes to test whether the mean response time for drug-injected rats differs from 1.2 seconds. Set up the test of hypothesis for this experiment, using $\alpha = .01$.

Solution Since the neurologist wishes to detect whether the mean response time, μ, for drug-injected rats differs from the control mean of 1.2 seconds in *either* direction—that is, $\mu < 1.2$ or $\mu > 1.2$—we conduct a two-tailed statistical test. Following the procedure for selecting the null and alternative hypotheses, we specify as the alternative hypothesis that the mean differs from 1.2 seconds, since determining whether the drug-injected mean differs from the control mean is the purpose of the experiment. The null hypothesis is the presumption that drug-injected rats have the same mean response time as control rats unless the research indicates otherwise. Thus,

$$H_0: \mu = 1.2$$
$$H_a: \mu \neq 1.2 \quad \text{(i.e., } \mu < 1.2 \text{ or } \mu > 1.2\text{)}$$

The test statistic measures the number of standard deviations between the observed value of $\bar{x}$ and the null hypothesized value $\mu = 1.2$:

$$\textit{Test statistic:} \quad z = \frac{\bar{x} - 1.2}{\sigma_{\bar{x}}}$$

The rejection region must be designated to detect a departure from $\mu = 1.2$ in *either* direction, so we will reject H_0 for values of z that are either too small (negative) or too large (positive). To determine the precise values of z that comprise the rejection region, we first select α, the probability that the test will lead to incorrect rejection of the null hypothesis. Then we divide α equally between the lower and upper tail of the distribution of z, as shown in Figure 8.5. In this example, $\alpha = .01$, so

Figure 8.5
Two-Tailed Rejection
Region: $\alpha = .01$

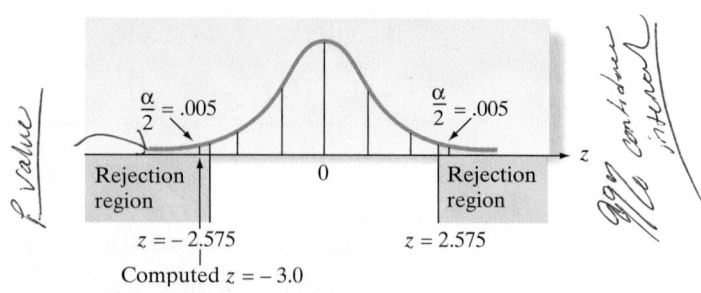

$\alpha/2 = .005$ is placed in each tail. The areas in the tails correspond to $z = -2.575$ and $z = 2.575$, respectively (from Table 8.2):

$$\text{Rejection region: } z < -2.575 \text{ or } z > 2.575 \qquad \text{(see Figure 8.5)}$$

Assumptions: Since the sample size of the experiment is large enough ($n > 30$), the Central Limit Theorem will apply, and no assumptions need be made about the population of response time measurements. The sampling distribution of the sample mean response of 100 rats will be approximately normal regardless of the distribution of the individual rats' response times.

Look Back Note that the test is set up *before* the sampling experiment is conducted. The data are not used to develop the test. Evidently, the neurologist wants to conclude that the mean response time for the drug-injected rats differs from the control mean only when the evidence is very convincing, because the value of α has been set quite low at .01. If the experiment results in the rejection of H_0, she can be 99% confident that the mean response time of the drug-injected rats differs from the control mean.

Now Work   *Exercise 8.2*

■ ■ ■

Once the test is set up, she is ready to perform the sampling experiment and conduct the test. The test is performed in Example 8.2.

EXAMPLE 8.2

CARRYING OUT A HYPOTHESIS TEST FOR μ

Problem Refer to the neurological response-time test set up in Example 8.1. Suppose the sampling experiment is conducted with the following results:

$$n = 100 \text{ response times for drug-injected rats}$$
$$\bar{x} = 1.05 \text{ seconds} \qquad s = .5 \text{ second}$$

Use the results of the sampling experiment to conduct the test of hypothesis.

Solution Since the test is completely specified in Example 8.1, we simply substitute the sample statistics into the test statistic:

$$z = \frac{\bar{x} - 1.2}{\sigma_{\bar{x}}} = \frac{\bar{x} - 1.2}{\sigma/\sqrt{n}} = \frac{1.05 - 1.2}{\sigma/\sqrt{100}}$$

$$\approx \frac{1.05 - 1.2}{s/10} = \frac{-.15}{.5/10} = -3.0$$

The implication is that the sample mean, 1.05, is (approximately) 3 standard deviations below the null hypothesized value of 1.2 in the sampling distribution of $\bar{x}$. You can see in Figure 8.5 that this value of z is in the lower-tail rejection region, which consists of all values of $z < -2.575$. This sampling experiment provides sufficient evidence to reject H_0 and conclude, at the $\alpha = .01$ level of significance, that the mean response time for drug-injected rats differs from the control mean of 1.2 seconds. It appears that the rats receiving an injection of this drug have a mean response time that is less than 1.2 seconds.

Look Back Two points about the test of hypothesis in this example apply to all statistical tests:

1. Since z is less than -2.575, it is tempting to state our conclusion at a significance level lower than $\alpha = .01$. We resist the temptation because the level of α

is determined *before* the sampling experiment is performed. If we decide that we are willing to tolerate a 1% Type I error rate, the result of the sampling experiment should have no effect on that decision. *In general, the same data should not be used both to set up and to conduct the test.*

2. When we state our conclusion at the .01 level of significance, we are referring to the failure rate of the *procedure*, not the result of this particular test. We know that the test procedure will lead to the rejection of the null hypothesis only 1% of the time when in fact $\mu = 1.2$. *Therefore, when the test statistic falls in the rejection region, we infer that the alternative $\mu \neq 1.2$ is true and express our confidence in the procedure by quoting the α level of significance, or the $100(1 - \alpha)\%$ confidence level.*

Now Work *Exercise 8.26*

■ ■ ■

The setup of a large-sample test of hypothesis about a population mean is summarized in the following box. Both the one- and two-tailed tests are shown.

Large-Sample Test of Hypothesis about μ

One-Tailed Test	**Two-Tailed Test**
$H_0: \mu = \mu_0$	$H_0: \mu = \mu_0$
$H_a: \mu < \mu_0$ (or $H_a: \mu > \mu_0$)	$H_a: \mu \neq \mu_0$
Test statistic: $z = \dfrac{\bar{x} - \mu_0}{\sigma_{\bar{x}}}$	Test statistics: $z = \dfrac{\bar{x} - \mu_0}{\sigma_{\bar{x}}}$
Rejection region: $z < -z_\alpha$ (or $z > z_\alpha$ when $H_a: \mu > \mu_0$)	Rejection region: $z < -z_{\alpha/2}$ or $z > z_{\alpha/2}$
where z_α is chosen so that $P(z > z_\alpha) = \alpha$	where $z_{\alpha/2}$ is chosen so that $P(z > z_{\alpha/2}) = \alpha/2$

Note: μ_0 is the symbol for the numerical value assigned to μ under the null hypothesis.

Conditions Required for a Valid Large-Sample Hypothesis Test for μ

1. A random sample is selected from the target population.
2. The sample size n is large (i.e., $n \geq 30$). (Due to the Central Limit Theorem, this condition guarantees that the test statistic will be approximately normal regardless of the shape of the underlying probability distribution of the population.)

Once the test has been set up, the sampling experiment is performed and the test statistic is calculated. The next box contains possible conclusions for a test of hypothesis, depending on the result of the sampling experiment.

Possible Conclusions for a Test of Hypothesis

1. If the calculated test statistic falls in the rejection region, reject H_0 and conclude that the alternative hypothesis H_a is true. State that you are rejecting H_0 at the α level of significance. Remember that the confidence is in the testing *process*, not the particular result of a single test.

> **2.** If the test statistic does not fall in the rejection region, conclude that the sampling experiment does not provide sufficient evidence to reject H_0 at the α level of significance. [Generally, we will not "accept" the null hypothesis unless the probability β of a Type II error has been calculated (see optional Section 8.6).]

Exercises 8.18–8.35

Understanding the Principles

8.18 Explain the difference between a one-tailed and a two-tailed test.

8.19 What conditions are required for a valid large-sample test for μ?

8.20 For what values of the test statistic lead you to reject H_0? Fail to reject H_0?

Learning the Mechanics

8.21 For each of the following rejection regions, sketch the
NW   sampling distribution for z and indicate the location of the rejection region.
   a. $z > 1.96$
   b. $z > 1.645$
   c. $z > 2.575$
   d. $z < -1.28$
   e. $z < -1.645$ or $z > 1.645$
   f. $z < -2.575$ or $z > 2.575$
   g. For each of the rejection regions specified in parts **a–f**, what is the probability that a Type I error will be made?

8.22 Suppose you are interested in conducting the statistical test of H_0: $\mu = 200$ against H_a: $\mu > 200$, and you have decided to use the following decision rule: Reject H_0 if the sample mean of a random sample of 100 items is more than 215. Assume that the standard deviation of the population is 80.
   a. Express the decision rule in terms of z.
   b. Find α, the probability of making a Type I error, by using this decision rule.

8.23 A random sample of 100 observations from a population with standard deviation 60 yielded a sample mean of 110.
   a. Test the null hypothesis that $\mu = 100$ against the alternative hypothesis that $\mu > 100$ using $\alpha = .05$. Interpret the results of the test.
   b. Test the null hypothesis that $\mu = 100$ against the alternative hypothesis that $\mu \neq 100$ using $\alpha = .05$. Interpret the results of the test.
   c. Compare the results of the two tests you conducted. Explain why the results differ.

8.24 A random sample of 64 observations produced the following summary statistics: $\bar{x} = .323$ and $s^2 = .034$.
   a. Test the null hypothesis that $\mu = .36$ against the alternative hypothesis that $\mu < .36$ using $\alpha = .10$.
   b. Test the null hypothesis that $\mu = .36$ against the alternative hypothesis that $\mu \neq .36$ using $\alpha = .10$. Interpret the result.

Applying the Concepts—Basic

8.25 Teacher perceptions of child behavior. *Development Psychology* (Mar. 2003) published a study on teacher perceptions of the behavior of elementary school children. Teachers rated the aggressive behavior of a sample of 11,160 New York City public school children by responding to the statement, "This child threatens or bullies others in order to get his/her own way." Responses were measured on a scale ranging from 1 (*never*) to 5 (*always*). Summary statistics for the sample of 11,160 children were reported as $\bar{x} = 2.15$ and $s = 1.05$. Let μ represent the mean response for the population of all New York City public school children. Suppose you want to test H_0: $\mu = 3$ against H_a: $\mu \neq 3$.
   a. In the words of the problem, define a Type I error and a Type II error.
   b. Use the sample information to conduct the test at a significance level of $\alpha = .05$.
   c. Conduct the test, part **b**, using $\alpha = .10$.

8.26 Latex allergy in healthcare workers. Refer to the
NW   *Current Allergy & Clinical Immunology* (March 2004) study of $n = 46$ hospital employees who were diagnosed with a latex allergy from exposure to the powder on latex gloves, Exercise 7.13 (p. 00). The number of latex gloves used per week by the sampled workers is summarized as follows: $\bar{x} = 19.3$ and $s = 11.9$. Let μ represent the mean number of latex gloves used per week by all hospital employees. Consider testing H_0: $\mu = 20$ against H_a: $\mu < 20$.
   a. Give the rejection region for the test at a significance level of $\alpha = .01$.
   b. Calculate the value of the test statistic.
   c. Use the results, parts **a** and **b**, to make the appropriate conclusion.

8.27 Alkalinity of river water. In Exercise 5.35 (p. 258) you learned that the mean alkalinity level of water specimens collected from the Han River in Seoul, Korea, is 50 milligrams per liter. (*Environmental Science & Engineering*, September 1, 2000.) Consider a random sample of 100 water specimens collected from a tributary of the Han River. Suppose the mean and standard deviation of the alkalinity levels for the sample are $\bar{x} = 67.8$ mpl and $s = 14.4$ mpl. Is there sufficient evidence (at $\alpha = .01$) to indicate that the population mean alkalinity level of water in the tributary exceeds 50 mpl?

8.28 Visual imagery of athletes. A group of St. Louis University researchers administered the Vividness of Visual Imagery Questionnaire (VVIQ) to a sample of 51 college students on varsity sports teams. (*American Statistical Association Joint Meetings,* August 1996.) The mean and standard deviation of the VVIQ scores were 65.60 and 19.38, respectively. Scores range from 16 to 80, with a lower score representing greater vividness. In the general population, the mean VVIQ score is 67. Conduct a test to determine whether the mean VVIQ score of college athletes is less than 67. Test using $\alpha = .01$.

Applying the Concepts—Intermediate

8.29 Post-traumatic stress of POWs. *Psychological Assessment* (March 1995) published the results of a study of World War II aviators who were captured by German forces after they were shot down. Having located a total of 239 World War II aviator POW survivors, the researchers asked each veteran to participate in the study; 33 responded to the letter of invitation. Each of the 33 POW survivors were administered the Minnesota Multiphasic Personality Inventory, one component of which measures level of post-traumatic stress disorder (PTSD). [*Note:* The higher the score, the higher the level of PTSD.] The aviators produced a mean PTSD score of $\bar{x} = 9.00$ and a standard deviation of $s = 9.32$.

   a. Set up the null and alternative hypotheses for determining whether the true mean PTSD score of all World War II aviator POWs is less than 16. [*Note:* The value, 16, represents the mean PTSD score established for Vietnam POWs.]

   b. Conduct the test, part **a**, using $\alpha = .10$. What are the practical implications of the test?

   c. Discuss the representativeness of the sample used in the study and its ramifications.

8.30 Point spreads of NFL games. During the National Football League (NFL) season, Las Vegas oddsmakers establish a point spread on each game for betting purposes. For example, the New England Patriots were established as 3-point favorites over the Carolina Panthers in the 2004 Super Bowl. The final scores of NFL games were compared against the final point spreads established by the oddsmakers in *Chance* (Fall 1998). The difference between the game outcome and point spread (called a point-spread error) was calculated for 240 NFL games. The mean and standard deviation of the point-spread errors are $\bar{x} = -1.6$ and $s = 13.3$. Use this information to test the hypothesis that the true mean point-spread error for all NFL games is 0. Conduct the test at $\alpha = .01$ and interpret the result.

8.31 Bone fossil study. Humerus bones from the same species of animal tend to have approximately the same length-to-width ratios. When fossils of humerus bones are discovered, archeologists can often determine the species of animal by examining the length-to-width ratios of the bones. It is known that species A exhibits a mean ratio of 8.5. Suppose 41 fossils of humerus bones were unearthed at an archeological site in East Africa, where species A is believed to have lived. (Assume that the unearthed bones were all from the same unknown species.) The length-to-width ratios of the bones were calculated and listed, as shown in the table.

BONES

10.73	8.89	9.07	9.20	10.33	9.98	9.84	9.59
8.48	8.71	9.57	9.29	9.94	8.07	8.37	6.85
8.52	8.87	6.23	9.41	6.66	9.35	8.86	9.93
8.91	11.77	10.48	10.39	9.39	9.17	9.89	8.17
8.93	8.80	10.02	8.38	11.67	8.30	9.17	12.00
9.38							

   a. Test whether the population mean ratio of all bones of this particular species differs from 8.5. Use $\alpha = .01$.

   b. What are the practical implications of the test, part **a**?

8.32 Solder joint inspections. Current technology uses X-rays and lasers for inspection of solder-joint defects on printed circuit boards (PCBs). (*Quality Congress Transactions,* 1986.) A particular manufacturer of laser-based inspection equipment claims that its product can inspect on average at least 10 solder joints per second when the joints are spaced .1 inch apart. The equipment was tested by a potential buyer on 48 different PCBs. In each case, the equipment was operated for exactly 1 second. The number of solder joints inspected on each run follows:

PCB

10	9	10	10	11	9	12	8	8	9	6	10
7	10	11	9	9	13	9	10	11	10	12	8
9	9	9	7	12	6	9	10	10	8	7	9
11	12	10	0	10	11	12	9	7	9	9	10

   a. The potential buyer wants to know whether the sample data refute the manufacturer's claim. Specify the null and alternative hypotheses that the buyer should test.

   b. In the context of this exercise, what is a Type I error? A Type II error?

   c. Conduct the hypothesis test you described in part **a**, and interpret the test's results in the context of this exercise. Use $\alpha = .05$.

8.33 Salaries of postgraduates. The *Economics of Education Review* (Vol. 21, 2002) published a paper on the relationship between education level and earnings. The data for the research were obtained from the National Adult Literacy Survey of over 25,000 respondents. The survey revealed that males with a postgraduate degree had a mean salary of $61,340 (with standard error $s_{\bar{x}} = \$2,185$), while females with a postgraduate degree had a mean of $32,227 (with standard error $s_{\bar{x}} = \$932$).

   a. The article reports that a 95% confidence interval for μ_M, the population mean salary of all males with postgraduate degrees, is ($57,050, $65,631). Based on this interval, is there evidence to say that μ_M differs from $60,000? Explain.

b. Use the summary information to test the hypothesis that the true mean salary of males with postgraduate degrees differs from $60,000. Use $\alpha = .05$. (*Note:* $s_{\bar{x}} = s/\sqrt{n}$.)

c. Explain why the inferences in parts **a** and **b** agree.

d. The article reports that a 95% confidence interval for μ_F, the population mean salary of all females with postgraduate degrees, is ($30,396, $34,058). Based on this interval, is there evidence to say that μ_F differs from $33,000? Explain.

e. Use the summary information to test the hypothesis that the true mean salary of females with postgraduate degrees differs from $33,000. Use $\alpha = .05$. (*Note:* $s_{\bar{x}} = s/\sqrt{n}$.)

f. Explain why the inferences in parts **d** and **e** agree.

8.34 Cyanide in soil. *Environmental Science & Technology* (October 1993) reported on a study of contaminated soil in The Netherlands. Seventy-two 400-gram soil specimens were sampled, dried, and analyzed for the contaminant cyanide. The cyanide concentration [in milligrams per kilogram (mg/kg) of soil] of each soil specimen was determined using an infrared microscopic method. The sample resulted in a mean cyanide level of $\bar{x} = 84$ mg/kg and a standard deviation of $s = 80$ mg/kg.

a. Test the hypothesis that the true mean cyanide level in soil in The Netherlands exceeds 100 mg/kg. Use $\alpha = .10$.

b. Would you reach the same conclusion as in part **a** using $\alpha = .05$? Using $\alpha = .01$? Why can the conclusion of a test change when the value of α is changed?

Applying the Concepts—Advanced

8.35 Social interaction of mental patients. The *Community Mental Health Journal* (Aug. 2000) presented the results of a survey of over 6,000 clients of the Department of Mental Health and Addiction Services (DMHAS) in Connecticut. One of the many variables measured for each mental health patient was frequency of social interaction (on a 5-point scale, where 1 = very infrequently, 3 = occasionally, and 5 = very frequently). The 6,681 clients who were evaluated had a mean social interaction score of 2.95 with a standard deviation of 1.10.

a. Conduct a hypothesis test (at $\alpha = .01$) to determine if the true mean social interaction score of all Connecticut mental health patients differs from 3.

b. Examine the results of the study from a practical view, then discuss why "statistical significance" does not always imply "practical significance."

c. Because the variable of interest is measured on a 5-point scale, it is unlikely that the population of ratings will be normally distributed. Consequently, some analysts may perceive the test, part **a**, to be invalid and search for alternative methods of analysis. Defend or refute this position.

8.3 Observed Significance Levels: *p*-Values

According to the statistical test procedure described in Section 8.2, the rejection region and, correspondingly, the value of α are selected prior to conducting the test, and the conclusions are stated in terms of rejecting or not rejecting the null hypothesis. A second method of presenting the results of a statistical test is one that reports the extent to which the test statistic disagrees with the null hypothesis and leaves to the reader the task of deciding whether to reject the null hypothesis. This measure of disagreement is called the *observed significance level* (or *p-value*) for the test.

> **DEFINITION 8.1**
>
> The **observed significance level**, or ***p*-value**, for a specific statistical test is the probability (assuming H_0 is true) of observing a value of the test statistic that is at least as contradictory to the null hypothesis, and supportive of the alternative hypothesis, as the actual one computed from the sample data.

For example, the value of the test statistic computed for the sample of $n = 50$ sections of sewer pipe was $z = 2.12$. Since the test is one-tailed (i.e., the alternative (research) hypothesis of interest is $H_a: \mu > 2,400$) values of the test statistic even more contradictory to H_0 than the one observed would be values larger than $z = 2.12$. Therefore, the observed significance level (*p*-value) for this test is

$$p\text{-value} = P(z \geq 2.12)$$

Figure 8.6

Finding the p-Value for an Upper-Tailed Test When $z = 2.12$

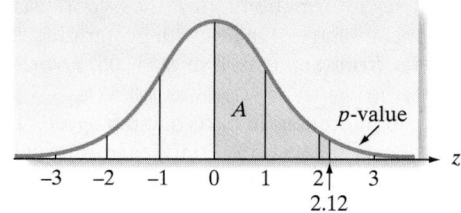

or, equivalently, the area under the standard normal curve to the right of $z = 2.12$ (see Figure 8.6).

The area A in Figure 8.6 is given in Table IV in Appendix A as .4830. Therefore, the upper-tail area corresponding to $z = 2.12$ is

$$p\text{-value} = .5 - .4830 = .0170$$

Consequently, we say that these test results are "statistically significant"; that is, they disagree (rather strongly) with the null hypothesis, $H_0: \mu = 2{,}400$, and favor $H_a: \mu > 2{,}400$. The probability of observing a z value as large as 2.12 is only .0170, if in fact the true value of μ is 2,400.

If you are inclined to select $\alpha = .05$ for this test, then you would reject the null hypothesis because the p-value for the test, .0170, is less than .05. In contrast, if you choose $\alpha = .01$, you would not reject the null hypothesis because the p-value for the test is larger than .01. Thus, the use of the observed significance level is identical to the test procedure described in the preceding sections except that the choice of α is left to you.

The steps for calculating the p-value corresponding to a test statistic for a population mean are given in the box.

Steps for Calculating the p-Value for a Test of Hypothesis

1. Determine the value of the test statistic z corresponding to the result of the sampling experiment.

2. a. If the test is one-tailed, the p-value is equal to the tail area beyond z in the same direction as the alternative hypothesis. Thus, if the alternative hypothesis is of the form $>$, the p-value is the area to the right of, or above, the observed z value. Conversely, if the alternative is of the form $<$, the p-value is the area to the left of, or below, the observed z value. (See Figure 8.7.)

b. If the test is two-tailed, the p-value is equal to twice the tail area beyond the observed z value in the direction of the sign of z. That is, if z is positive, the p-value is twice the area to the right of, or above, the observed z value. Conversely, if z is negative, the p-value is twice the area to the left of, or below, the observed z value. (See Figure 8.8.)

Figure 8.7

Finding the p-Value for a One-Tailed Test

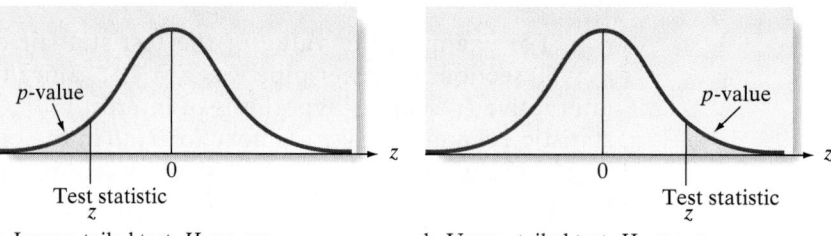

a. Lower–tailed test, $H_a: \mu < \mu_0$ b. Upper–tailed test, $H_a: \mu > \mu_0$

Figure 8.8
Finding the *p*-Value
for a Two-Tailed Test:
p-Value = $2(p/2)$

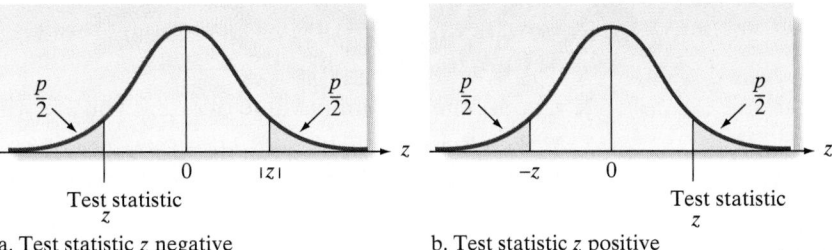

a. Test statistic *z* negative   b. Test statistic *z* positive

EXAMPLE 8.3

COMPUTING A *p*-VALUE

Problem Find the observed significance level for the test of the mean response time for drug-injected rats in Examples 8.1 and 8.2.

Solution Example 8.1 presented a two-tailed test of the hypothesis

$$H_0: \mu = 1.2 \text{ seconds}$$

against the alternative hypothesis

$$H_a: \mu \neq 1.2 \text{ seconds}$$

The observed value of the test statistic in Example 8.2 was $z = -3.0$, and any value of z less than -3.0 or greater than $+3.0$ (because this is a two-tailed test) would be even more contradictory to H_0. Therefore, the observed significance level for the test is

$$p\text{-value} = P(z < -3.0 \text{ or } z > +3.0)$$

Thus, we calculate the area below the observed z value, $z = -3.0$, and double it. Consulting Table IV in Appendix A, we find that $P(z < -3.0) = .5 - .4987 = .0013$. Therefore, the *p*-value for this two-tailed test is

$$2P(z < -3.0) = 2(.0013) = .0026$$

Look Back We can interpret this *p*-value as a strong indication that the mean reaction time of drug-injected rats differs from the control mean ($\mu \neq 1.2$), since we would observe a test statistic this extreme or more extreme only 26 in 10,000 times if the drug-injected mean were equal to the control mean ($\mu = 1.2$). The extent to which the mean differs from 1.2 could be better determined by calculating a confidence interval for μ.

Now Work *Exercise 8.42*

— ■ ■ ■ —

When publishing the results of a statistical test of hypothesis in journals, case studies, reports, etc., many researchers make use of *p*-values. Instead of selecting α beforehand and then conducting a test, as outlined in this chapter, the researcher computes (usually with the aid of a statistical software package) and reports the value of the appropriate test statistic and its associated *p*-value. It is left to the reader of the report to judge the significance of the result (i.e., the reader must determine whether to reject the null hypothesis in favor of the alternative hypothesis, based on the reported *p*-value). Usually, the null hypothesis is rejected if the observed significance level is *less than* the fixed significance level, α, chosen by the reader. The inherent advantage of reporting test results in this manner is twofold: (1) Readers are permitted to select the maximum value of α that they would be willing to tolerate if they actually carried out a standard test of hypothesis in the manner outlined in this chapter, and (2) a measure of the degree of significance of the result (i.e., the *p*-value) is provided.

Reporting Test Results as *p*-Values: How to Decide
Whether to Reject H_0

1. Choose the maximum value of α that you are willing to tolerate.

2. If the observed significance level (*p*-value) of the test is less than the chosen value of α, reject the null hypothesis. Otherwise, do not reject the null hypothesis.

EXAMPLE 8.4

AN APPLICATION USING *p*-VALUES

Problem   The lengths of stay (in days) for 100 randomly selected hospital patients, first presented in Table 7.1, are reproduced in Table 8.3. Suppose we want to test the hypothesis that the true mean length of stay (LOS) at the hospital is less than 5 days, that is,

$$H_0: \mu = 5$$
$$H_a: \mu < 5$$

Assuming $\sigma = 3.68$, use the data in the table to conduct the test at $\alpha = .05$.

⊙ **HOSPLOS**

TABLE 8.3   Lengths of Stay for 100 Hospital Patients

2	3	8	6	4	4	6	4	2	5
8	10	4	4	4	2	1	3	2	10
1	3	2	3	4	3	5	2	4	1
2	9	1	7	17	9	9	9	4	4
1	1	1	3	1	6	3	3	2	5
1	3	3	14	2	3	9	6	6	3
5	1	4	6	11	22	1	9	6	5
2	2	5	4	3	6	1	5	1	6
17	1	2	4	5	4	4	3	2	3
3	5	2	2	3	2	10	2	4	2

Solution   The data were entered into a computer and MINITAB was used to conduct the analysis. The MINITAB printout for the lower-tailed test is displayed in Figure 8.9. Both the test statistic, $z = -1.28$, and *p*-value of the test, $p = .101$, are highlighted on the MINITAB printout. Since the *p*-value exceeds our selected α value, $\alpha = .05$, we cannot reject the null hypothesis. Hence, there is insufficient evidence (at $\alpha = .05$) to conclude that the true mean LOS at the hospital is less than 5 days.

Now Work   *Exercise 8.49*

━━━━━ ■ ■ ■ ━━━━━

Note: Some statistical software packages (e.g., SPSS) will conduct only two-tailed tests of hypothesis. For these packages, you obtain the *p*-value for a one-tailed test as shown in the box on p. 388.

Figure 8.9

MINITAB Printout for the Lower-Tailed Test in Example 8.4

```
Test of mu = 5 vs < 5
The assumed standard deviation = 3.68
```

Variable	N	Mean	StDev	SE Mean	95% Upper Bound	Z	P
LOS	100	4.53000	3.67755	0.36800	5.13531	-1.28	0.101

Hypothesis Testing

Using the TI-83 Graphing Calculator

Computing the *p*-value for a *z*-Test

Step 1   *Access the Statistical Tests Menu.*
Press **STAT**
Arrow right to **TESTS**
Press **ENTER** to select **Z-Test**

Step 2   *Choose "**Data**" or "**Stats**." ("Data" is selected when you have entered the raw data into a List. "Stats" is selected when you are given only the mean, standard deviation, and sample size.)*
Press **ENTER**

If you selected "Data," enter the values for the hypothesis test where μ_0 = the value for μ in the null hypothesis, σ = assumed value of the population standard deviation.
Set **List** to **L1**
Set **Freq** to **1**
Use the arrow to highlight the appropriate alternative hypothesis.
Press **ENTER**
Arrow down to "**Calculate**"
Press **ENTER**

If you selected "Stats," enter the values for the hypothesis test where μ_0 = the value for μ in the null hypothesis, σ = assumed value of the population standard deviation.
Enter the sample mean and sample size.
Arrow down to "**Calculate**"
Press **ENTER**

The chosen test will be displayed as well as the *z*-test statistic, the *p*-value, the sample mean, and the sample size.

Example   A manufacturer claims the average life expectancy of this particular model light bulb is at least 10,000 hours with σ = 1,000 hours. A simple random sample of 40 light bulbs shows a sample mean of 9755 hours. Using α = .05, test the manufacturer's claim.

For this problem the hypotheses will be:

$$H_0: \mu \geq 10,000$$
$$H_a: \mu < 10,000$$

The screens are shown below:

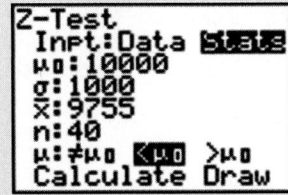
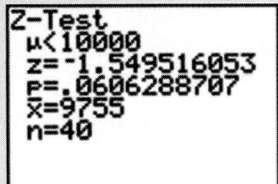

As you can see the *p*-value is 0.061. Since $p > .05$, **do not** reject H_0.

> **Converting a Two-Tailed p-Value from a Printout to a One-Tailed p-Value:**
>
> $$p = \frac{\text{Reported } p\text{-value}}{2} \quad \text{if} \begin{cases} H_a \text{ is of form} \quad > \quad \text{and } z \text{ is positive} \\ H_a \text{ is of form} \quad < \quad \text{and } z \text{ is negative} \end{cases}$$
>
> $$p = 1 - \left(\frac{\text{Reported } p\text{-value}}{2}\right) \quad \text{if} \begin{cases} H_a \text{ is of form} \quad > \quad \text{and } z \text{ is negative} \\ H_a \text{ is of form} \quad < \quad \text{and } z \text{ is positive} \end{cases}$$

Statistics in Action Revisited

Testing a Population Mean in the KLEENEX® Survey

Refer to Kimberly-Clark Corporation's survey of 250 people who kept a count of their use of KLEENEX® tissues in diaries (p. 375). We want to test the claim made by marketing experts that $\mu = 60$ is the average number of tissues used by people with colds against our belief that the population mean is smaller than 60 tissues. That is, we want to test

$$H_0: \mu = 60 \qquad H_a: \mu < 60$$

We will select $\alpha = .05$ as the level of significance for the test.

The survey results for the 250 sampled KLEENEX® users are stored in the **TISSUES** data file. A MINITAB analysis of the data yielded the printout displayed in Figure SIA8.1.

The observed significance level of the test, highlighted on the printout, is p-value $= .018$. Since this p-value is less than $\alpha = .05$, we have sufficient evidence to reject H_0; therefore, we conclude that the mean number of tissues used by a person with a cold is less than 60 tissues. [*Note:* If we conduct the same test using $\alpha = .01$ as the level of significance, we would have insufficient evidence to reject H_0 since p-value $= .018$ is greater than $\alpha = .01$. Thus, at $\alpha = .01$, there is no evidence to support our alternative that the population mean is less than 60.]

```
Test of mu = 60 vs < 60


                                              95%
                                            Upper
Variable     N      Mean      StDev   SE Mean   Bound       T       P
NUMUSED     250   56.6760   25.0343    1.5833  59.2900   -2.10   0.018
```

Figure SIA8.1
MINITAB Test of $\mu = 60$ for KLEENEX® Survey

Exercises 8.36–8.52

Understanding the Principles

8.36 How does the observed significance level (p-value) of a test differ from the value of α?

8.37 In general, do large p-values or small p-values support the alternative hypothesis, H_a?

8.38 If a hypothesis test were conducted using $\alpha = .05$, for which of the following p-values would the null hypothesis be rejected?
  a. .06   **b.** .10
  c. .01   **d.** .001
  e. .251   **f.** .042

8.39 For each α and observed significance level (p-value) pair, indicate whether the null hypothesis would be rejected.
  a. $\alpha = .05$, p-value $= .10$
  b. $\alpha = .10$, p-value $= .05$
  c. $\alpha = .01$, p-value $= .001$
  d. $\alpha = .025$, p-value $= .05$
  e. $\alpha = .10$, p-value $= .45$

Learning the Mechanics

8.40 An analyst tested the null hypothesis $\mu \geq 20$ against the alternative hypothesis $\mu < 20$. The analyst reported a *p*-value of .06. What is the smallest value of α for which the null hypothesis would be rejected?

8.41 In a test of H_0: $\mu = 100$ against H_a: $\mu > 100$, the sample data yielded the test statistic $z = 2.17$. Find the *p*-value for the test.

8.42 In a test of H_0: $\mu = 100$ against H_a: $\mu \neq 100$, the sample
[NW] data yielded the test statistic $z = 2.17$. Find the *p*-value for the test.

8.43 In a test of the hypothesis H_0: $\mu = 50$ versus H_a: $\mu > 50$, a sample of $n = 100$ observations possessed mean $\bar{x} = 49.4$ and standard deviation $s = 4.1$. Find and interpret the *p*-value for this test.

8.44 In a test of the hypothesis H_0: $\mu = 10$ versus H_a: $\mu \neq 10$, a sample of $n = 50$ observations possessed mean $\bar{x} = 10.7$ and standard deviation $s = 3.1$. Find and interpret the *p*-value for this test.

8.45 Consider a test of H_0: $\mu = 75$ performed using the computer. SPSS reports a two-tailed *p*-value of .1032. Make the appropriate conclusion for each of the following situations:
   a. H_a: $\mu < 75$, $z = -1.63$, $\alpha = .05$
   b. H_a: $\mu < 75$, $z = 1.63$, $\alpha = .10$
   c. H_a: $\mu > 75$, $z = 1.63$, $\alpha = .10$
   d. H_a: $\mu \neq 75$, $z = -1.63$, $\alpha = .01$

Applying the Concepts—Basic

8.46 **Teacher perceptions of child behavior.** Refer to the *Development Psychology* (Mar. 2003) study on the aggressive behavior of elementary school children, Exercise 8.25 (p. 381). Recall that you tested H_0: $\mu = 3$ against H_a: $\mu \neq 3$, where μ is the mean level of aggressiveness for the population of all New York City public school children as perceived by their teachers, based on the summary statistics: $n = 11,160$, $\bar{x} = 2.15$, and $s = 1.05$.
   a. Compute the *p*-value of the test.
   b. Compare the *p*-value to $\alpha = .10$ and make the appropriate conclusion.

8.47 **Latex allergy in healthcare workers.** Refer to the *Current Allergy & Clinical Immunology* (March 2004) study of latex allergy in healthcare workers, Exercise

8.26 (p. 381). Recall that you tested H_0: $\mu = 20$ against H_a: $\mu < 20$, where μ is the mean number of latex gloves used per week by all hospital employees, based on the summary statistics: $n = 46$, $\bar{x} = 19.3$ and $s = 11.9$.
   a. Compute the *p*-value of the test.
   b. Compare the *p*-value to $\alpha = .01$ and make the appropriate conclusion.

8.48 **Post-traumatic stress of POWs.** Refer to the *Psychological Assessment* study of World War II aviator POWs, Exercise 8.29 (p. 382). You tested whether the true mean post-traumatic stress disorder score of World War II aviator POWs is less than 16. Recall that $\bar{x} = 9.00$ and $s = 9.32$ for a sample of $n = 33$ POWs.
   a. Compute the *p*-value of the test.
   b. Refer to part **a**. Would the *p*-value have been larger or smaller if $\bar{x}$ had been larger? Explain.

8.49 **Bone fossil study.** In Exercise 8.31 (p. 382), you tested
[NW] H_0: $\mu = 8.5$ versus H_a: $\mu \neq 8.5$, where μ is the population mean length-to-width ratio of humerus bones of a particular species of animal. An SAS printout for the hypothesis test is provided below. Locate the *p*-value on the printout and interpret its value.

Applying the Concepts—Intermediate

8.50 **Colored string preferred by chickens.** Refer to the *Applied Animal Behaviour Science* (October 2000) study of domestic chickens exposed to a pecking stimulus, Exercise 7.17 (p. 331). Recall that the average number of pecks a chicken takes at a white string over a specified time interval is known to be $\mu = 7.5$ pecks. In an experiment where 72 chickens were exposed to blue string, the average number of pecks was $\bar{x} = 1.13$ pecks, with a standard deviation of $s = 2.21$ pecks.
   a. On average, are chickens more apt to peck at white string than at blue string? Conduct the appropriate test of hypothesis using $\alpha = .05$.
   b. Compare your answer to part **a** with your answer to Exercise 7.17**b**.
   c. Find the *p*-value for the test and interpret its value.

8.51 **Feminizing human faces.** Research published in *Nature* (August 27, 1998) revealed that people are more attracted

SAS Printout for Exercise 8.49

Sample Statistics for LWRATIO

N	Mean	Std. Dev.	Std. Error
41	9.26	1.20	0.19

Hypothesis Test

Null hypothesis: Mean of LWRATIO = 8.5
Alternative: Mean of LWRATIO ^= 8.5

With a specified known standard deviation of 1.2

Z Statistic	Prob > Z
4.042	<.0001

to "feminized" faces, regardless of gender. In one experiment, 50 human subjects viewed both a Japanese female and Caucasian male face on a computer. Using special computer graphics, each subject could morph the faces (by making them more feminine or more masculine) until they attained the "most attractive" face. The level of feminization x (measured as a percentage) was measured.

a. For the Japanese female face, $\bar{x} = 10.2\%$ and $s = 31.3\%$. The researchers used this sample information to test the null hypothesis of a mean level of feminization equal to 0%. Verify that the test statistic is equal to 2.3.

b. Refer to part **a**. The researchers reported the p-value of the test as $p \approx .02$. Verify and interpret this result.

c. For the Caucasian male face, $\bar{x} = 15.0\%$ and $s = 25.1\%$. The researchers reported the test statistic (for the test of the null hypothesis stated in part **a**) as 4.23 with an associated p-value of approximately 0. Verify and interpret these results.

8.52 Ages of cable TV shoppers. In a paper presented at the 2000 Conference of the International Association for Time Use Research, professor Margaret Sanik of Ohio State University reported the results of her study on American cable TV viewers who purchase items from one of the home shopping channels. She found that the average age of these cable TV shoppers was 51 years. Suppose you want to test the null hypothesis, $H_0: \mu = 51$, using a sample of $n = 50$ cable TV shoppers.

a. Find the p-value of a two-tailed test if $\bar{x} = 52.3$ and $s = 7.1$.

b. Find the p-value of an upper-tailed test if $\bar{x} = 52.3$ and $s = 7.1$.

c. Find the p-value of a two-tailed test if $\bar{x} = 52.3$ and $s = 10.4$.

d. For each of the tests, parts **a–c**, give a value of α that will lead to a rejection of the null hypothesis.

e. If $\bar{x} = 52.3$, give a value of s that will yield a p-value of .01 or less.

8.4 Small-Sample Test of Hypothesis about a Population Mean

Most water-treatment facilities monitor the quality of their drinking water on an hourly basis. One variable monitored is pH, which measures the degree of alkalinity or acidity in the water. A pH below 7.0 is acidic, one above 7.0 is alkaline, and a pH of 7.0 is neutral. One water-treatment plant has a target pH of 8.5 (most try to maintain a slightly alkaline level). The mean and standard deviation of 1 hour's test results, based on 17 water samples at this plant, are

$$\bar{x} = 8.42 \qquad s = .16$$

Does this sample provide sufficient evidence that the mean pH level in the water differs from 8.5?

This inference can be placed in a test of hypothesis framework. We establish the target pH as the null hypothesized value and then utilize a two-tailed alternative that the true mean pH differs from the target:

$$H_0: \mu = 8.5$$
$$H_a: \mu \neq 8.5$$

Recall from Section 7.3 that when we are faced with making inferences about a population mean using the information in a small sample, two problems emerge:

1. The normality of the sampling distribution for $\bar{x}$ does not follow from the Central Limit Theorem when the sample size is small. We must assume that the distribution of measurements from which the sample was selected is approximately normally distributed in order to ensure the approximate normality of the sampling distribution of $\bar{x}$.

2. If the population standard deviation σ is unknown, as is usually the case, then we cannot assume that s will provide a good approximation for σ when the sample size is small. Instead, we must use the t-distribution rather than the standard normal z-distribution to make inferences about the population mean μ.

Therefore, as the test statistic of a small-sample test of a population mean, we use the t-statistic:

$$Test\ statistic: t = \frac{\bar{x} - \mu_0}{s/\sqrt{n}} = \frac{\bar{x} - 8.5}{s/\sqrt{n}}$$

Figure 8.10

Two-Tailed Rejection Region
for Small-Sample t-Test

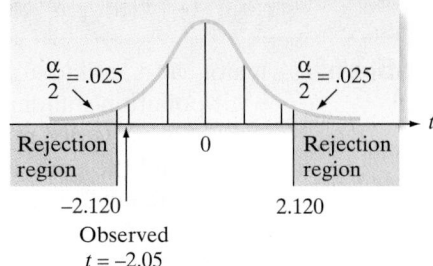

where μ_0 is the null hypothesized value of the population mean, μ. In our example, $\mu_0 = 8.5$.

To find the rejection region, we must specify the value of α, the probability that the test will lead to rejection of the null hypothesis when it is true, and then consult the t table (Table VI of Appendix A). Using $\alpha = .05$, the two-tailed rejection region is

$$\text{Rejection region: } t_{\alpha/2} = t_{.025} = 2.120 \text{ with } n - 1 = 16 \text{ degrees of freedom}$$
$$\text{Reject } H_0 \text{ if } t < -2.120 \text{ or } t > 2.120$$

The rejection region is shown in Figure 8.10.

We are now prepared to calculate the test statistic and reach a conclusion:

$$t = \frac{\bar{x} - \mu_0}{s/\sqrt{n}} = \frac{8.42 - 8.50}{.16/\sqrt{17}} = \frac{-.08}{.039} = -2.05$$

Since the calculated value of t does not fall in the rejection region (Figure 8.10), we cannot reject H_0 at the $\alpha = .05$ level of significance. Thus, the water-treatment plant should not conclude that the mean pH differs from the 8.5 target based on the sample evidence.

It is interesting to note that the calculated t value, -2.05, is *less than* the .05 level z value, -1.96. The implication is that if we had *incorrectly* used a z statistic for this test, we would have rejected the null hypothesis at the .05 level, concluding that the mean pH level differs from 8.5. The important point is that the statistical procedure to be used must always be closely scrutinized and all the assumptions understood. Many statistical lies are the result of misapplications of otherwise valid procedures.

The technique for conducting a small-sample test of hypothesis about a population mean is summarized in the following boxes.

Small-Sample Test of Hypothesis About μ

One-Tailed Test	**Two-Tailed Test**
$H_0: \mu = \mu_0$	$H_0: \mu = \mu_0$
$H_a: \mu < \mu_0$ (or $H_a: \mu > \mu_0$)	$H_a: \mu \neq \mu_0$
Test statistic: $t = \dfrac{\bar{x} - \mu_0}{s/\sqrt{n}}$	Test statistic: $t = \dfrac{\bar{x} - \mu_0}{s/\sqrt{n}}$
Rejection region: $t < -t_\alpha$ (or $t > t_\alpha$ when $H_a: \mu > \mu_0$)	Rejection region: $t < -t_{\alpha/2}$ or $t > t_{\alpha/2}$

where t_α and $t_{\alpha/2}$ are based on $(n - 1)$ degrees of freedom

---

Conditions Required for a Valid Small-Sample Hypothesis Test for μ

1. A random sample is selected from the target population.

2. The population from which the sample is selected has a distribution that is approximately normal.

EXAMPLE 8.5   CONDUCTING A SMALL-SAMPLE TEST FOR μ

Problem   A major car manufacturer wants to test a new engine to determine whether it meets new air-pollution standards. The mean emission μ of all engines of this type must be less than 20 parts per million of carbon. Ten engines are manufactured for testing purposes, and the emission level of each is determined. The data (in parts per million) are listed in Table 8.4.

⊙ **EMISSIONS**

TABLE 8.4   Emission Levels for Ten Engines

15.6	16.2	22.5	20.5	16.4	19.4	19.6	17.9	12.7	14.9

Do the data supply sufficient evidence to allow the manufacturer to conclude that this type of engine meets the pollution standard? Assume that the manufacturer is willing to risk a Type I error with probability $\alpha = .01$.

Solution   The manufacturer wants to support the research hypothesis that the mean emission level μ for all engines of this type is less than 20 parts per million. The elements of this small-sample one-tailed test are

$$H_0: \mu = 20$$
$$H_a: \mu < 20$$

Test statistic: $t = \dfrac{\bar{x} - 20}{s/\sqrt{n}}$

Rejection region: For $\alpha = .01$ and df $= n - 1 = 9$, the one-tailed rejection region (see Figure 8.11) is $t < -t_{.01} = -2.821$.

Assumption: The relative frequency distribution of the population of emission levels for all engines of this type is approximately normal. Based on the shape of the MINITAB stem-and-leaf display of the data shown in Figure 8.12, this assumption appears to be reasonably satisfied.

To calculate the test statistic, we analyzed the **EMISSIONS** data using MINITAB. The MINITAB printout is shown at the bottom of Figure 8.12. From the printout, we obtain $\bar{x} = 17.57$ and $s = 2.95$. Substituting these values into the test statistic formula, we get (rounding)

$$t = \frac{\bar{x} - 20}{s/\sqrt{n}} = \frac{17.57 - 20}{2.95/\sqrt{10}} = -2.60$$

Since the calculated t falls outside the rejection region (see Figure 8.11), the manufacturer cannot reject H_0. There is insufficient evidence to conclude that $\mu < 20$ parts per million and the new engine type meets the pollution standard.

Figure 8.11

A t-Distribution with 9 df and the Rejection Region for Example 8.5

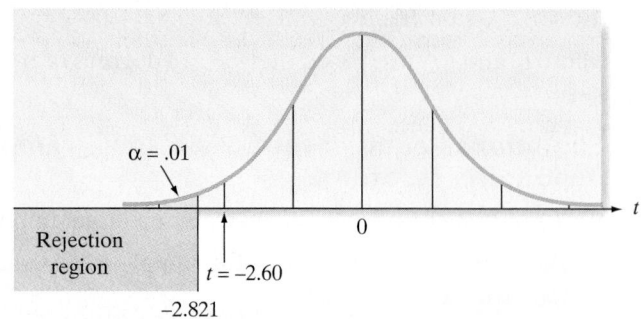

Figure 8.12

MINITAB Analysis of 10 Emission Levels, Example 8.5

Stem-and-Leaf Display: E-LEVEL

```
Stem-and-leaf of E-LEVEL   N  = 10
Leaf Unit = 1.0

    1    1   2
    3    1   45
   (3)   1   667
    4    1   99
    2    2   0
    1    2   2
```

One-Sample T: E-LEVEL

```
Test of mu = 20 vs < 20
```

| | | | | | 95% Upper | | |
Variable	N	Mean	StDev	SE Mean	Bound	T	P
E-LEVEL	10	17.5700	2.9522	0.9336	19.2814	-2.60	0.014

Look Back Are you satisfied with the reliability associated with this inference? The probability is only $\alpha = .01$ that the test would support the research hypothesis if in fact it were false.

> **Now Work** *Exercise 8.59a,b*

■ ■ ■

EXAMPLE 8.6

THE p-VALUE FOR A SMALL-SAMPLE TEST OF μ

Problem Find the observed significance level for the test described in Example 8.5. Interpret the result.

Solution The test of Example 8.5 was a lower-tailed test: $H_0: \mu = 20$ versus $H_a: \mu < 20$. Since the value of t computed from the sample data was $t = -2.60$, the observed significance level (or p-value) for the test is equal to the probability that t would assume a value less than or equal to -2.60 if in fact H_0 were true. This is equal to the area in the lower tail of the t-distribution (highlighted in Figure 8.13).

One way to find this area (i.e., the p-value for the test) is to consult the t-table (Table VI in Appendix A). Unlike the table of areas under the normal curve, Table VI gives only the t values corresponding to the areas .100, .050, .025, .010, .005, .001, and .0005. Therefore, we can only approximate the p-value for the test. Since the observed t value was based on 9 degrees of freedom, we use the df = 9 row in Table VI

Figure 8.13

The Observed Significance Level for the Test of Example 8.5

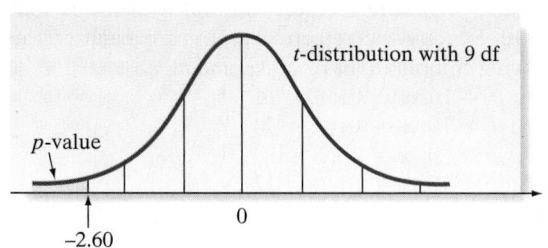

and move across the row until we reach the t values that are closest to the observed $t = -3.00$. [*Note:* We ignore the minus sign.] The t values corresponding to p-values of .010 and .005 are 2.821 and 3.250, respectively. Since the observed t value falls between $t_{.010}$ and $t_{.005}$, the p-value for the test lies between .005 and .010. In other words, $.005 < p\text{-value} < .01$. Thus, we would reject the null hypothesis, H_0: $\mu = 20$ parts per million, for any value of α larger than .01 (the upper bound of the p-value).

A second, more accurate, way to obtain the p-value is to use a statistical software package to conduct the test of hypothesis. Both the test statistic (-2.60) and p-value (.014) are highlighted on the MINITAB printout, Figure. 8.12.

You can see that the actual p-value of the test falls within the bounds obtained from Table VI. Thus, the two methods agree; we will reject H_0: $\mu = 20$ in favor of H_a: $\mu < 20$ for any α level larger than .01.

| Now Work | *Exercise 8.59c* |

■ ■ ■

Small-sample inferences typically require more assumptions and provide less information about the population parameter than do large-sample inferences. Nevertheless, the t-test is a method of testing a hypothesis about a population mean of a normal distribution when only a small number of observations is available.

What Can Be Done If the Population Relative Frequency Distribution Departs Greatly from Normal?
Answer: Use one of the nonparametric statistical methods of Chapter 14.

Exercises 8.53–8.70

Understanding the Principles

8.53 Under what circumstances should you use the t-distribution in testing a hypothesis about a population mean?

8.54 In what ways are the distributions of the z-statistic and t-statistic alike? How do they differ?

Learning the Mechanics

8.55 For each of the following rejection regions, sketch the sampling distribution of t, and indicate the location of the rejection region on your sketch:
  a. $t > 1.440$ where df $= 6$
  b. $t < -1.782$ where df $= 12$
  c. $t < -2.060$ or $t > 2.060$ where df $= 25$

8.56 For each of the rejection regions defined in Exercise 8.55, what is the probability that a Type I error will be made?

8.57 A random sample of n observations is selected from a normal population to test the null hypothesis that $\mu = 10$. Specify the rejection region for each of the following combinations of H_a, α, and n:
  a. H_a: $\mu \neq 10$; $\alpha = .05$; $n = 14$
  b. H_a: $\mu > 10$; $\alpha = .01$; $n = 24$
  c. H_a: $\mu > 10$; $\alpha = .10$; $n = 9$
  d. H_a: $\mu < 10$; $\alpha = .01$; $n = 12$
  e. H_a: $\mu \neq 10$; $\alpha = .10$; $n = 20$
  f. H_a: $\mu < 10$; $\alpha = .05$; $n = 4$

8.58 The following sample of six measurements was randomly selected from a normally distributed population: $1, 3, -1, 5, 1, 2$.
  a. Test the null hypothesis that the mean of the population is 3 against the alternative hypothesis, $\mu < 3$. Use $\alpha = .05$.
  b. Test the null hypothesis that the mean of the population is 3 against the alternative hypothesis, $\mu \neq 3$. Use $\alpha = .05$.
  c. Find the observed significance level for each test.

8.59 A sample of five measurements, randomly selected from a normally distributed population, resulted in the following summary statistics: $\bar{x} = 4.8$, $s = 1.3$.
  a. Test the null hypothesis that the mean of the population is 6 against the alternative hypothesis, $\mu < 6$. Use $\alpha = .05$.
  b. Test the null hypothesis that the mean of the population is 6 against the alternative hypothesis, $\mu \neq 6$. Use $\alpha = .05$.
  c. Find the observed significance level for each test.

8.60 Suppose you conduct a t-test for the null hypothesis H_0: $\mu = 1,000$ versus the alternative hypothesis H_a: $\mu > 1,000$ based on a sample of 17 observations. The test results are $t = 1.89$, p-value $= .038$.
  a. What assumptions are necessary for the validity of this procedure?
  b. Interpret the results of the test.

Descriptive Statistics

	N	Minimum	Maximum	Mean	Std. Deviation
BOOKS	14	16	53	31.64	10.485
Valid N (listwise)	14				

c. Suppose the alternative hypothesis had been the two-tailed H_a: $\mu \neq 1,000$. If the t-statistic were unchanged, what would the p-value be for this test? Interpret the p-value for the two-tailed test.

Applying the Concepts—Basic

8.61 A new dental bonding agent. When bonding teeth, orthodontists must maintain a dry field. A new bonding adhesive (called "Smartbond") has been developed to eliminate the necessity of a dry field. However, there is concern that the new bonding adhesive is not as strong as the current standard, a composite adhesive. (*Trends in Biomaterials & Artificial Organs*, January 2003.) Tests on a sample of 10 extracted teeth bonded with the new adhesive resulted in a mean breaking strength (after 24 hours) of $\bar{x} = 5.07$ Mpa and a standard deviation of $s = .46$ Mpa. Where Mpa = Mega pascal, a measure of force per area. Orthodontists want to know if the true mean breaking strength of the new bonding adhesive is less than 5.70 Mpa, the mean breaking strength of the composite adhesive.

a. Set up the null and alternative hypothesis for the test.

b. Find the rejection region for the test using $\alpha = .01$.

c. Compute the test statistic.

d. Give the appropriate conclusion for the test.

e. What conditions are required for the test results to be valid?

JAPANESE

8.62 Reading Japanese books. Refer to the *Reading in a Foreign Language* (April 2004) experiment to improve the Japanese reading comprehension levels

of University of Hawaii students, Exercise 2.31 (p. 51). Recall that 14 students participated in a 10-week extensive reading program in a second-semester Japanese course. The data on number of books read by each student are saved in the **JAPANESE** file. An SPSS printout giving descriptive statistics for the data is shown above.

a. Give the null and alternative hypothesis for determining whether the average number of books read by all students who participate in the extensive reading program exceeds 25.

b. Find the rejection region for the test using $\alpha = .05$.

c. Compute the test statistic.

d. Give the appropriate conclusion for the test.

e. What conditions are required for the test results to be valid?

8.63 Dental anxiety study. Refer to the *Psychological Reports* (August 1997) study of college students who completed the Dental Anxiety Scale, Exercise 5.34 (p. 258). Recall that scores range from 0 (no anxiety) to 20 (extreme anxiety). Summary statistics for the scores of the 27 students who completed the questionnaire follow: $\bar{x} = 10.7$, $s = 3.6$. Conduct a test of hypothesis to determine if the mean Dental Anxiety Scale score for the population of college students differs from $\mu = 11$. Use $\alpha = .05$.

8.64 Temperature of iron castings. The Cleveland Casting Plant produces iron automotive castings for Ford Motor Company. When the process is stable, the target pouring temperature of the molten iron is 2,550 degrees. (*Quality Engineering*, Vol. 7, 1995.) The pouring temperatures (in degrees Fahrenheit) for a random sample of 10 crankshafts produced at the plant are listed in the table. Use the SAS printout below to conduct a test to determine

```
Sample Statistics for IRONTEMP

    N         Mean        Std. Dev.      Std. Error
-----------------------------------------------------
   10        2558.70        22.75           7.19

Hypothesis Test

    Null hypothesis:    Mean of IRONTEMP  =  2550
    Alternative:        Mean of IRONTEMP ^= 2550

        t Statistic      Df        Prob > t
    ---------------------------------------------
          1.210           9         0.2573
```

whether the true mean pouring temperature differs from the target setting. Test using $\alpha = .01$.

IRONTEMP

2,543	2,541	2,544	2,620	2,560	2,559	2,562
2,553	2,552	2,553				

Source: Price, B., & Barth, B. "A structural model relating process inputs and final product characteristics." *Quality Engineering,* Vol. 7, No. 4, 1995, p. 696 (Table 2).

Applying the Concepts—Intermediate

LICHEN
8.65 Radioactive lichen. Refer to the 2003 Lichen Radionuclide Baseline Research project to monitor the level of radioactivity in lichen, Exercise 7.33 (p. 340). Recall that University of Alaska researchers collected 9 lichen specimens and measured the amount of the radioactive element cesium-137 (in microcuries per milliliter) in each specimen. (The natural logarithms of the data values are saved in the **LICHEN** file.) Assume that in previous years the mean cesium amount in lichen was $\mu = .003$ microcuries per milliliter. Is there sufficient evidence to indicate that the mean amount of cesium in lichen specimens differs from this value? Use the MINITAB printout below to conduct a complete test of hypothesis at $\alpha = .10$.

8.66 Testing a mosquito repellent. A study was conducted to evaluate the effectiveness of a new mosquito repellent designed by the U.S. Army to be applied as camouflage face paint. (*Journal of the Mosquito Control Association,* June 1995.) The repellent was applied to the forearms of 5 volunteers and then they were exposed to 15 active mosquitos for a 10-hour period. The percentage of the forearm surface area protected from bites (called percent repellency) was calculated for each of the five

volunteers. For one color of paint (loam), the following summary statistics were obtained:

$$\bar{x} = 83\% \quad s = 15\%$$

a. The new repellent is considered effective if it provides a percent repellency of at least 95. Conduct a test to determine whether the mean repellency percentage of the new mosquito repellent is less than 95. Test using $\alpha = .10$.

b. What assumptions are required for the hypothesis test in part **a** to be valid?

8.67 Cracks in highway pavement. The Mississippi Department of Transportation collected data on the number of cracks (called *crack intensity*) in an undivided two-lane highway using van-mounted state-of-the-art video technology. (*Journal of Infrastructure Systems,* March 1995.) The mean number of cracks found in a sample of eight 50-meter sections of the highway was $\bar{x} = .210$, with a variance of $s^2 = .011$. Suppose the American Association of State Highway and Transportation Officials (AASHTO) recommends a maximum mean crack intensity of .100 for safety purposes. Test the hypothesis that the true mean crack intensity of the Mississippi highway exceeds the AASHTO recommended maximum. Use $\alpha = .01$.

8.68 Mongolian desert ants. Refer to the *Journal of Biogeography* (December 2003) study of ants in Mongolia (Central Asia), Exercise 2.66 (p. 66). Recall that botanists placed seed baits at 11 study sites and observed the number of ant species attracted to each site. A portion of the data is provided in the table on p. 397. Do these data indicate that the average number of ant species at Mongolian desert sites differs from 5 species? Conduct the appropriate test at $\alpha = .05$. Are the conditions required for a valid test satisfied?

MINITAB Printout for Exercise 8.65

Stem-and-Leaf Display: CESIUM

```
Stem-and-leaf of CESIUM   N  = 9
Leaf Unit = 0.0010


    1   0  2
    2   0  4
   (3)  0  667
    4   0
    4   1  01
    2   1
    2   1  5
    1   1  6
```

One-Sample T: CESIUM

```
Test of mu = 0.003 vs not = 0.003
```

Variable	N	Mean	StDev	SE Mean	95% CI	T	P
CESIUM	9	0.009027	0.004854	0.001618	(0.005296, 0.012759)	3.72	0.006

GOBIANTS

Site	Region	Number of Ant Species
1	Dry Steppe	3
2	Dry Steppe	3
3	Dry Steppe	52
4	Dry Steppe	7
5	Dry Steppe	5
6	Gobi Desert	49
7	Gobi Desert	5
8	Gobi Desert	4
9	Gobi Desert	4
10	Gobi Desert	5
11	Gobi Desert	4

Source: Pfeiffer, M., et al. "Community organization and species richness of ants in Mongolia along an ecological gradient from steppe to Gobi desert," *Journal of Biogeography,* Vol. 30, No. 12, Dec. 2003.

8.69 Head trauma study. The SCL-90-R is a 90-item symptom inventory checklist designed to reflect the psychological status of an individual. Each symptom (e.g., obsessive-compulsive behavior) is scored on a scale of 0 (none) to 4 (extreme). The total of these scores yields an individual's Positive Symptom Total (PST). "Normal" individuals are known to have a mean PST of about 40. The *Journal of Head Trauma Rehabilitation* (April 1995) reported that a sample of 23 patients diagnosed with mild to moderate traumatic brain injury had a mean PST score of $\bar{x} = 48.43$ and a standard deviation of $s = 20.76$. Is there sufficient evidence to claim that the true mean PST score of all patients with mild to moderate traumatic brain injury exceeds the "normal" value of 40? Test using $\alpha = .05$.

Applying the Concepts—Advanced

8.70 Lengths of great white sharks. One of the most feared predators in the ocean is the great white shark. It is known that the white shark grows to a mean length of 21 feet; however, one marine biologist believes that great white sharks off the Bermuda coast grow much longer owing to unusual feeding habits. To test this claim, some full-grown great white sharks were captured off the Bermuda coast, measured, and then set free. However, because the capture of sharks is difficult, costly, and very dangerous, only three specimens were sampled. Their lengths were 24, 20, and 22 feet. Do these data support the marine biologist's claim?

8.5 Large-Sample Test of Hypothesis about a Population Proportion

Inferences about population proportions (or percentages) are often made in the context of the probability, p, of "success" for a binomial distribution. We saw how to use large samples from binomial distributions to form confidence intervals for p in Section 7.4. We now consider tests of hypotheses about p.

Consider, for example, a method currently used by doctors to screen women for possible breast cancer. The method fails to detect cancer in 20% of the women who actually have the disease. Suppose a new method has been developed that researchers hope will detect cancer more accurately. This new method was used to screen a random sample of 140 women known to have breast cancer. Of these, the new method failed to detect cancer in 12 women. Does this sample provide evidence that the failure rate of the new method differs from the one currently in use?

We first view this as a binomial experiment with 140 screened women as the trials and failure to detect breast cancer as "Success" (in binomial terminology). Let p represent the probability that the new method fails to detect the breast cancer. If the new method is no better than the current one, then the failure rate is $p = .2$. On the other hand, if the new method is either better or worse than the current method, then the failure rate is either smaller or larger than 20%; that is, $p \neq .2$.

We can now place the problem in the context of a test of hypothesis:

$$H_0: p = .2$$
$$H_a: p \neq .2$$

Recall that the sample proportion, $\hat{p}$, is really just the sample mean of the outcomes of the individual binomial trials and, as such, is approximately normally distributed (for large samples) according to the Central Limit Theorem. Thus, for large samples we can use the standard normal z as the test statistic:

$$Test\ statistic:\quad z = \frac{\text{Sample proportion } - \text{ Null hypothesized proportion}}{\text{Standard deviation of sample proportion}}$$
$$= \frac{\hat{p} - p_0}{\sigma_{\hat{p}}}$$

Figure 8.14

Rejection Region for Breast Cancer Example

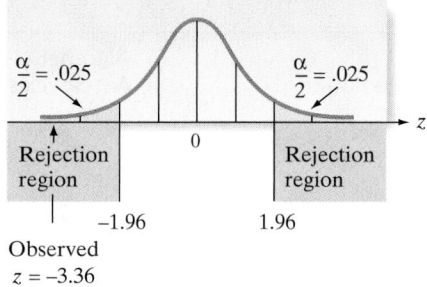

Observed
$z = -3.36$

where we use the symbol p_0 to represent the null hypothesized value of p.

Rejection region: We use the standard normal distribution to find the appropriate rejection region for the specified value of α. Using $\alpha = .05$, the two-tailed rejection region is

$$z < -z_{\alpha/2} = -z_{.025} = -1.96 \quad \text{or} \quad z > z_{\alpha/2} = z_{.025} = 1.96$$

See Figure 8.14.

We are now prepared to calculate the value of the test statistic. Before doing so, we want to be sure that the sample size is large enough to ensure that the normal approximation for the sampling distribution of $\hat{p}$ is reasonable. To check this, we calculate a 3-standard-deviation interval around the null hypothesized value, p_0, which is assumed to be the true value of p until our test procedure proves otherwise. Recall that $\sigma_{\hat{p}} = \sqrt{pq/n}$ and that we need an estimate of the product pq in order to calculate a numerical value of the test statistic z. Since the null hypothesized value is generally the accepted-until-proven-otherwise value, we use the value of p_0q_0 (where $q_0 = 1 - p_0$) to estimate pq in the calculation of z. Thus,

$$\sigma_{\hat{p}} = \sqrt{\frac{pq}{n}} \approx \sqrt{\frac{p_0q_0}{n}} = \sqrt{\frac{(.2)(.8)}{140}} = .034$$

and the 3-standard-deviation interval around p_0 is

$$p_0 \pm 3\sigma_{\hat{p}} \approx .2 \pm 3(.034) = (.098, .302)$$

As long as this interval does not contain 0 or 1 (i.e., is completely contained in the interval 0 to 1), as is the case here, the normal distribution will provide a reasonable approximation for the sampling distribution of $\hat{p}$.

Returning to the hypothesis test at hand, the proportion of the screenings that failed to detect breast cancer is

$$\hat{p} = \frac{12}{140} = .086$$

Finally, we calculate the number of standard deviations (the z value) between the sampled and hypothesized value of the binomial proportion:

$$z = \frac{\hat{p} - p_0}{\sigma_{\hat{p}}} = \frac{\hat{p} - p_0}{\sqrt{p_0q_0/n}} = \frac{.086 - .2}{.034} = \frac{-.114}{.034} = -3.36$$

The implication is that the observed sample proportion is (approximately) 3.36 standard deviations below the null hypothesized proportion .2 (Figure 8.14). Therefore, we reject the null hypothesis, concluding at the .05 level of significance that the true failure rate of the new method for detecting breast cancer differs from .20. Since $\hat{p} = .086$, it appears that the new method is better (i.e., has a smaller failure rate) than the method currently in use. (To estimate the magnitude of the failure rate for the new method, a confidence interval can be constructed.)

The test of hypothesis about a population proportion p is summarized in the next box. Note that the procedure is entirely analogous to that used for conducting large-sample tests about a population mean.

Large-Sample Test of Hypothesis about p

One-Tailed Test

$H_0: p = p_0$

$H_a: p < p_0$ (or $H_a: p > p_0$)

Test statistic: $z = \dfrac{\hat{p} - p_0}{\sigma_{\hat{p}}}$

Rejection region: $z < -z_\alpha$

(or $z > z_\alpha$ when $H_a: p > p_0$)

Two-Tailed Test

$H_0: p = p_0$

$H_a: p \neq p_0$

Test statistic: $z = \dfrac{\hat{p} - p_0}{\sigma_{\hat{p}}}$

Rejection region: $z < -z_{\alpha/2}$ or $z > z_{\alpha/2}$

where, p_0 = hypothesized value of p, $\sigma_{\hat{p}} = \sqrt{p_0 q_0 / n}$ and $q_0 = 1 - p_0$

Conditions Required for a Valid Large-Sample Hypothesis Test for p

1. A random sample is selected from a binomial population.
2. The sample size n is large. (This condition will be satisfied if $p_0 \pm 3\sigma_{\hat{p}}$ falls between 0 and 1.)

EXAMPLE 8.7

CONDUCTING A HYPOTHESIS TEST FOR p

Problem The reputations (and hence sales) of many businesses can be severely damaged by shipments of manufactured items that contain a large percentage of defectives. For example, a manufacturer of alkaline batteries may want to be reasonably certain that fewer than 5% of its batteries are defective. Suppose 300 batteries are randomly selected from a very large shipment; each is tested and 10 defective batteries are found. Does this provide sufficient evidence for the manufacturer to conclude that the fraction defective in the entire shipment is less than .05? Use $\alpha = .01$.

Solution Before conducting the test of hypothesis, we check to determine whether the sample size is large enough to use the normal approximation for the sampling distribution of $\hat{p}$. The criterion is tested by the interval

$$p_0 \pm 3\sigma_{\hat{p}} = p_0 \pm 3\sqrt{\frac{p_0 q_0}{n}} = .05 \pm 3\sqrt{\frac{(.05)(.95)}{300}}$$

$$= .05 \pm .04 \quad \text{or} \quad (.01, .09)$$

Since the interval lies within the interval $(0, 1)$, the normal approximation will be adequate.

The objective of the sampling is to determine whether there is sufficient evidence to indicate that the fraction defective, p, is less than .05. Consequently, we will test the null hypothesis that $p = .05$ against the alternative hypothesis that $p < .05$. The elements of the test are

$H_0: p = .05$

$H_a: p < .05$

Test statistic: $z = \dfrac{\hat{p} - p_0}{\sigma_{\hat{p}}}$

Rejection region: $z < -z_{.01} = -2.33$ (see Figure 8.15)

Figure 8.15
Rejection Region
for Example 8.7

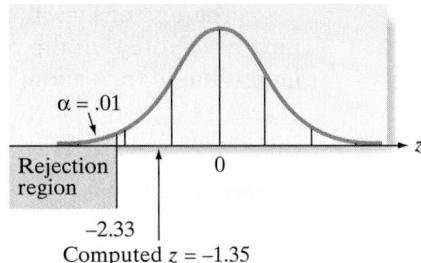

We now calculate the test statistic:

$$z = \frac{\hat{p} - .05}{\sigma_{\hat{p}}} = \frac{(10/300) - .05}{\sqrt{p_0 q_0/n}} = \frac{.033 - .05}{\sqrt{p_0 q_0/300}}$$

Notice that we use p_0 to calculate $\sigma_{\hat{p}}$ because, in contrast to calculating $\sigma_{\hat{p}}$ for a confidence interval, the test statistic is computed on the assumption that the null hypothesis is true—that is, $p = p_0$. Therefore, substituting the values for $\hat{p}$ and p_0 into the z statistic, we obtain

$$z \approx \frac{-.017}{\sqrt{(.05)(.95)/300}} = \frac{-.017}{.0126} = -1.35$$

As shown in Figure 8.15, the calculated z value does not fall in the rejection region. Therefore, there is insufficient evidence at the .01 level of significance to indicate that the shipment contains fewer than 5% defective batteries.

Now Work *Exercise 8.74ab*

■ ■ ■

EXAMPLE 8.8

FINDING THE p-VALUE FOR A TEST ABOUT p

Problem In Example 8.7 we found that we did not have sufficient evidence, at the $\alpha = .01$ level of significance, to indicate that the fraction defective p of alkaline batteries was less than $p = .05$. How strong was the weight of evidence favoring the alternative hypothesis ($H_a: p < .05$)? Find the observed significance level (p-value) for the test.

Solution The computed value of the test statistic z was $z = -1.35$. Therefore, for this lower-tailed test, the observed significance level is

$$p\text{-value} = P(z \leq -1.35)$$

This lower-tail area is shown in Figure 8.16. The area between $z = 0$ and $z = 1.35$ is given in Table IV in Appendix A as .4115. Therefore, the observed significance level is $.5 - .4115 = .0885$.

Figure 8.16
The Observed Significance
Level for Example 8.8

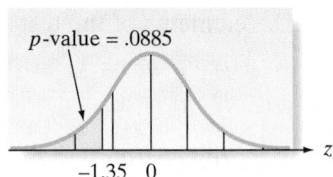

Look Back Note that this probability is quite small. Although we did not reject H_0: $p = .05$ at $\alpha = .01$, the probability of observing a z value as small as or smaller than -1.35 is only $.0885$ if in fact H_0 is true. Therefore, we would reject H_0 if we choose $\alpha = .10$ (since the observed significance level is less than .10), and we would not reject H_0 (the conclusion of Example 8.7) if we choose $\alpha = .05$ or $\alpha = .01$.

> **Now Work** *Exercise 8.74c*

■ ■ ■

Small-sample test procedures are also available for p, although most surveys use samples that are large enough to employ the large-sample tests presented in this section. A test of proportions that can be applied to small samples is discussed in Chapter 13.

Statistics in Action Revisited

Testing a Population Proportion in the KLEENEX® Survey

In the previous "Statistics in Action Revisited" (p. 388), we investigated Kimberly-Clark Corporation's assertion that the company should put 60 tissues in a cold-care box of KLEENEX® tissues. We did this by testing the claim that the mean number of tissues used by a person with a cold is $\mu = 60$ using data collected from a survey of 250 KLEENEX® users. Another approach to the problem is to consider proportion of KLEENEX® users who use fewer than 60 tissues when they have a cold. Now, the population parameter of interest is p, the proportion of all KLEENEX® users who use fewer than 60 tissues when they have a cold.

Kimberly-Clark Corporation's belief that the company should put 60 tissues in a cold-care box will be supported if half of the KLEENEX® users surveyed use less than 60 tissues and half use more than 60 tissues (i.e., if $p = .5$). Is there evidence to indicate that the population proportion differs from .5? To answer this question, we set up the following null and alternative hypothesis:

$$H_0: p = .5 \qquad H_a: p \neq .5$$

Recall that the survey results for the 250 sampled KLEENEX® users are stored in the **TISSUES** data file. In addition to the number of tissues used by each person, the file contains a qualitative variable—called USED60—representing whether the person used fewer or more than 60 tissues. (The values of USED60 in the data set are "BELOW" or "ABOVE.") A MINITAB analysis of this variable yielded the printout displayed in Figure SIA8.2.

On the MINITAB printout, x represents the number of the 250 people with colds that used less than 60 tissues. Note that $x = 143$. This value is used to compute the test statistic, $z = 2.28$, highlighted on the printout. The p-value of the test, also highlighted on the printout, is p-value $= .023$. Since this value is less than $\alpha = .05$, there is sufficient evidence (at $\alpha = .05$) to reject H_0; we conclude that the proportion of all KLEENEX® users who use fewer than 60 tissues when they have a cold differs from .5. However, if we test at $\alpha = .01$, there is insufficient evidence to reject H_0. Consequently, our choice of α (as in the previous "Statistics in the Action Revisited") is critical to our decision.

```
Test of p = 0.5 vs p not = 0.5

Event = BELOW


Variable    X    N  Sample p           95% CI        Z-Value  P-Value
USED60     143  250  0.572000  (0.510666, 0.633334)     2.28    0.023
```

Figure SIA8.2
MINITAB Test of $p = .5$ for KLEENEX® Survey

Exercises 8.71–8.89

Understanding the Principles

8.71 What type of data, quantitative or qualitative, is typically associated with making inferences about a population proportion, p?

8.72 What conditions are required for a valid large-sample test for p?

Learning the Mechanics

8.73 For the binomial sample sizes and null hypothesized values of p in each part, determine whether the sample size is large enough to use the normal approximation methodology presented in this section to conduct a test of the null hypothesis $H_0: p = p_0$.
  a. $n = 500$, $p_0 = .05$
  b. $n = 100$, $p_0 = .99$
  c. $n = 50$, $p_0 = .2$
  d. $n = 20$, $p_0 = .2$
  e. $n = 10$, $p_0 = .4$

8.74 Suppose a random sample of 100 observations from a
NW binomial population gives a value of $\hat{p} = .69$ and you wish to test the null hypothesis that the population parameter p is equal to .75 against the alternative hypothesis that p is less than .75.
  a. Noting that $\hat{p} = .69$, what does your intuition tell you? Does the value of $\hat{p}$ appear to contradict the null hypothesis?
  b. Use the large-sample z-test to test $H_0: p = .75$ against the alternative hypothesis $H_a: p < .75$. Use $\alpha = .05$. How do the test results compare with your intuitive decision from part **a**?
  c. Find and interpret the observed significance level of the test you conducted in part **b**.

8.75 Suppose the sample in Exercise 8.74 has produced $\hat{p} = .84$ and we wish to test $H_0: p = .9$ against the alternative $H_a: p < .9$.
  a. Calculate the value of the z statistic for this test.
  b. Note that the numerator of the z statistic $(\hat{p} - p_0 = .84 - .90 = -.06)$ is the same as for Exercise 8.74. Considering this, why is the absolute value of z for this exercise larger than that calculated in Exercise 8.74?
  c. Complete the test using $\alpha = .05$ and interpret the result.
  d. Find the observed significance level for the test and interpret its value.

8.76 A random sample of 100 observations is selected from a binomial population with unknown probability of success p. The computed value of $\hat{p}$ is equal to .74.
  a. Test $H_0: p = .65$ against $H_a: p > .65$. Use $\alpha = .01$.
  b. Test $H_0: p = .65$ against $H_a: p > .65$. Use $\alpha = .10$.
  c. Test $H_0: p = .90$ against $H_a: p \neq .90$. Use $\alpha = .05$.
  d. Form a 95% confidence interval for p.
  e. Form a 99% confidence interval for p.

SNACK

8.77 Refer to Exercise 7.48 (p. 349), in which 50 consumers taste-tested a new snack food.
  a. Test $H_0: p = .5$ against $H_a: p > .5$, where p is the proportion of customers who do not like the snack food. Use $\alpha = .10$.
  b. Report the observed significance level of your test.

Applying the Concepts—Basic

8.78 **"Made in the USA" survey.** Refer to the *Journal of Global Business* (Spring 2002) study of what "Made in the USA" means to consumers, Exercise 2.13 (p. 38). Recall that 64 of 106 randomly selected shoppers believed "Made in the USA" means 100% of labor and materials are from the United States. Let p represent the true proportion of consumers who believe "Made in the USA" means 100% of labor and materials are from the United States.
  a. Calculate a point estimate for p.
  b. A claim is made that $p = .70$. Set up the null and alternative hypothesis to test this claim.

  c. Calculate the test statistic for the test, part **b**.

  d. Find the rejection region for the test if $\alpha = .01$.

  e. Use the results, parts **c** and **d**, to make the appropriate conclusion.

8.79 **Unauthorized computer use.** Refer to the Computer Security Institute (CSI) survey of computer crime, Exercise 2.17 (p. 40). Recall that in 1999, 7% of reported unauthorized uses of business web sites came from inside the company. In a survey taken two years later, 7 of a sample of 163 reported unauthorized uses of business web sites came from inside the company. (*Computer Security Issues & Trends*, Vol. 7, Spring 2001.) Let p represent the true proportion of unauthorized uses of business websites in 2001 that came from inside the company.
  a. Calculate a point estimate for p.
  b. Set up the null and alternative hypothesis to test whether the value of p has changed since 1999.

  c. Calculate the test statistic for the test, part **b**.

  d. Find the rejection region for the test if $\alpha = .05$.

  e. Use the results, parts **c** and **d** to make the appropriate conclusion.
  f. Find the p-value of the test and confirm that the conclusion based on the p-value agrees with the conclusion in part **e**.

8.80 **Quality of cable-TV news.** In a survey of 500 television viewers with access to cable TV, each was asked whether they agreed with the statement, "Overall, I find the quality of news on cable networks (such as CNN, FOXNews, CNBC, and MSNBC) to be better than news on the

ABC, CBS, and NBC networks." A total of 248 viewers agreed with the statement. (*Cabletelevision Advertising Bureau*, May 2002.) *Note*: The survey respondents were contacted via e-mail on the Internet.

a. Identify the population parameter of interest in the survey.

b. Give a point estimate of the population parameter.

c. Set up H_0 and H_a for testing whether the true percentage of TV viewers who find cable news to be better quality than network news differs from 50%.

d. Conduct the test, part **c**, using $\alpha = .10$ Make the appropriate conclusion in the words of the problem.

e. What conditions are required for the inference, part **d**, to be valid? Do they appear to be satisfied?

8.81 Single-parent families. Examining data collected on 835 males from the National Youth Survey (a longitudinal survey of a random sample of U.S. households), researchers at Carnegie Mellon University found that 401 of the male youths were raised in a single-parent family. (*Sociological Methods & Research*, February 2001.) Does this information allow you to conclude that more than 45% of male youths are raised in a single-parent family? Test at $\alpha = .05$.

8.82 Cell phone use by drivers. Refer to the U.S. Department of Transportation (July 2001) study of the level of cell phone use by drivers while they are in the act of driving a motor passenger vehicle, Exercise 7.49 (p. 349). Recall that for a random sample of 1,165 drivers selected across the country, 35 were using their cell phone.

a. Conduct a test (at $\alpha = .05$) to determine if p, the true driver cell phone use rate, differs from .02.

b. Does your conclusion, part **a**, agree with the inference you derived from the 95% confidence interval for p in Exercise 7.49?

Applying the Concepts—Intermediate

8.83 Federal civil trial appeals. Refer to the *Journal of the American Law and Economics Association* (Vol. 3, 2001) study of appeals of federal civil trials, Exercise 3.51 (p. 144). A breakdown of 678 civil cases that were originally tried in front of a judge and appealed by either the plaintiff or defendant is reproduced in the table. Do the data provide sufficient evidence to indicate that the percentage of appealed civil cases that are actually reversed is less than 25%? Test using $\alpha = .01$.

Outcome of Appeal	Number of Cases
Plaintiff trial win—reversed	71
Plaintiff trial win—affirmed/dismissed	240
Defendant trial win—reversed	68
Defendant trial win—affirmed/dismissed	299
TOTAL	678

8.84 Verbs and double object datives. Any sentence that contains an animate noun as a direct object and another noun as a second object (e.g., "Sue offered Ann a cookie.") is termed a double object dative (DOD). The connection between certain verbs and a DOD was investigated in *Applied Psycholinguistics* (June 1998). The subjects were 35 native English speakers who were enrolled in an introductory English composition course at a Hawaiian community college. After viewing pictures of a family, each subject was asked to write a sentence about the family using a specified verb. Of the 35 sentences using the verb "buy," 10 had a DOD structure. Conduct a test to determine if the true fraction of sentences with the verb "buy" that are DODs is less than $1/3$. Use $\alpha = .05$.

8.85 Choosing portable grill displays. Refer to the *Journal of Consumer Research* (March 2003) experiment on influencing the choices of others by offering undesirable alternatives, Exercise 3.25 (p. 132). Recall that each of 124 college students selected three portable grills from five to display on the showroom floor. The students were instructed to include Grill #2 (a smaller-sized grill) and select the remaining two grills in the display to maximize purchases of Grill #2. If the six possible grill display combinations (1-2-3, 1-2-4, 1-2-5, 2-3-4, 2-3-5, and 2-4-5) are selected at random, then the proportion of students selecting any display will be 1/6 = .167. One theory tested by the researcher is that the students will tend to choose the three-grill display so that Grill #2 is a compromise between a more desirable and a less desirable grill. Of the 124 students, 85 students selected a three-grill display that was consistent with this theory. Use this information to test the theory proposed by the researcher at $\alpha = .05$.

8.86 Study of lunar soil. *Meteoritics* (March 1995) reported the results of a study of lunar soil evolution. Data were obtained from the Apollo 16 mission to the moon, during which a 62-cm core was extracted from the soil near the landing site. Monomineralic grains of lunar soil were separated out and examined for coating with dust and glass fragments. Each grain was then classified as coated or uncoated. Of interest is the "coat index," that is, the proportion of grains that are coated. According to soil evolution theory, the coat index will exceed .5 at the top of the core, equal .5 in the middle of the core, and fall below .5 at the bottom of the core. Use the summary data in the accompanying table to test each part of the 3-part theory. Use $\alpha = .05$ for each test.

	Location (depth)		
	Top (4.25 cm)	Middle (28.1 cm)	Bottom (54.5 cm)
Number of grains sampled	84	73	81
Number coated	64	35	29

Source: Basu, A., and McKay, D. S. "Lunar soil evolution processes and Apollo 16 core 60013/60014." *Meteoritics,* Vol. 30, No. 2, Mar. 1995, p. 166 (Table 2).

8.87 Effectiveness of skin cream. Pond's Age-Defying Complex, a cream with alpha-hydroxy acid, advertises that it can reduce wrinkles and improve the skin. In a study published in *Archives of Dermatology* (June 1996), 33 middle-aged women used a cream with alpha-hydroxy acid for twenty-two weeks. At the end of the study period, a dermatologist judged whether each woman exhibited skin improvement. The results for the 33 women (where I = improved skin and N = no improvement) are listed in the accompanying table.

   a. Do the data provide sufficient evidence to conclude that the cream will improve the skin of more than 60% of middle-aged women? Test using $\alpha = .05$.

   b. Find and interpret the *p*-value of the test.

◉ **SKINCREAM**

I	I	N	I	N	N	I	I	I	I	I	I
N	I	I	I	N	I	I	I	N	I	N	I
I	I	I	I	I	N	I	I	N			

Applying the Concepts—Advanced

8.88 Parents who condone spanking. In Exercise 4.60 (p. 217) you read about a nationwide survey that claimed that 60% of parents with young children condone spanking their child as a regular form of punishment. (*Tampa Tribune*, October 5, 2000.) In a random sample of 100 parents with young children, how many parents would need to say they condone spanking as a form of punishment in order to refute the claim?

8.89 Testing the placebo effect. The *placebo effect* describes the phenomenon of improvement in the condition of a patient taking a placebo—a pill that looks and tastes real but contains no medically active chemicals. Physicians at a clinic in La Jolla, California, gave what they thought were drugs to 7,000 asthma, ulcer, and herpes patients. Although the doctors later learned that the drugs were really placebos, 70% of the patients reported an improved condition. (*Forbes*, May 22, 1995.) Use this information to test (at $\alpha = .05$) the placebo effect at the clinic. Assume that if the placebo is ineffective, the probability of a patient's condition improving is .5.

8.6 Calculating Type II Error Probabilities: More about β (Optional)

In our introduction to hypothesis testing in Section 8.1, we showed that the probability of committing a Type I error, α, can be controlled by the selection of the rejection region for the test. Thus, when the test statistic falls in the rejection region and we make the decision to reject the null hypothesis, we do so knowing the error rate for incorrect rejections of H_0. The situation corresponding to accepting the null hypothesis, and thereby risking a Type II error, is not generally as controllable. For that reason, we adopted a policy of nonrejection of H_0 when the test statistic does not fall in the rejection region, rather than risking an error of unknown magnitude.

To see how β, the probability of a Type II error, can be calculated for a test of hypothesis, recall the example in Section 8.1 in which a city tests a manufacturer's pipe to see whether it meets the requirement that the mean strength exceed 2,400 pounds per linear foot. The setup for the test is as follows:

$$H_0: \mu = 2,400$$
$$H_a: \mu > 2,400$$

Test statistic: $z = \dfrac{\bar{x} - 2,400}{\sigma/\sqrt{n}}$

Rejection region: $z > 1.645$ for $\alpha = .05$

Figure 8.17a shows the rejection region for the **null distribution**—that is, the distribution of the test statistic assuming the null hypothesis is true. The area in the rejection region is .05, and this area represents α, the probability that the test statistic leads to rejection of H_0 when in fact H_0 is true.

The Type II error probability β is calculated assuming that the null hypothesis is false, because it is defined as the *probability of accepting H_0 when it is false*. Since H_0 is false for any value of μ exceeding 2,400, one value of β exists for each possible value of μ greater than 2,400 (an infinite number of possibilities). Figures 8.17b, 8.17c, and 8.17d show three of the possibilities, corresponding to alternative hypothesis values of μ equal to 2,425, 2,450, and 2,475, respectively. Note that β is the area in the *nonrejection* (or *acceptance*) *region* in each of these distributions and that β decreases as the true value of μ moves farther from the null hypothesized value of $\mu = 2,400$. This is sensible because

Figure 8.17

Values of α and β for Various Values of μ

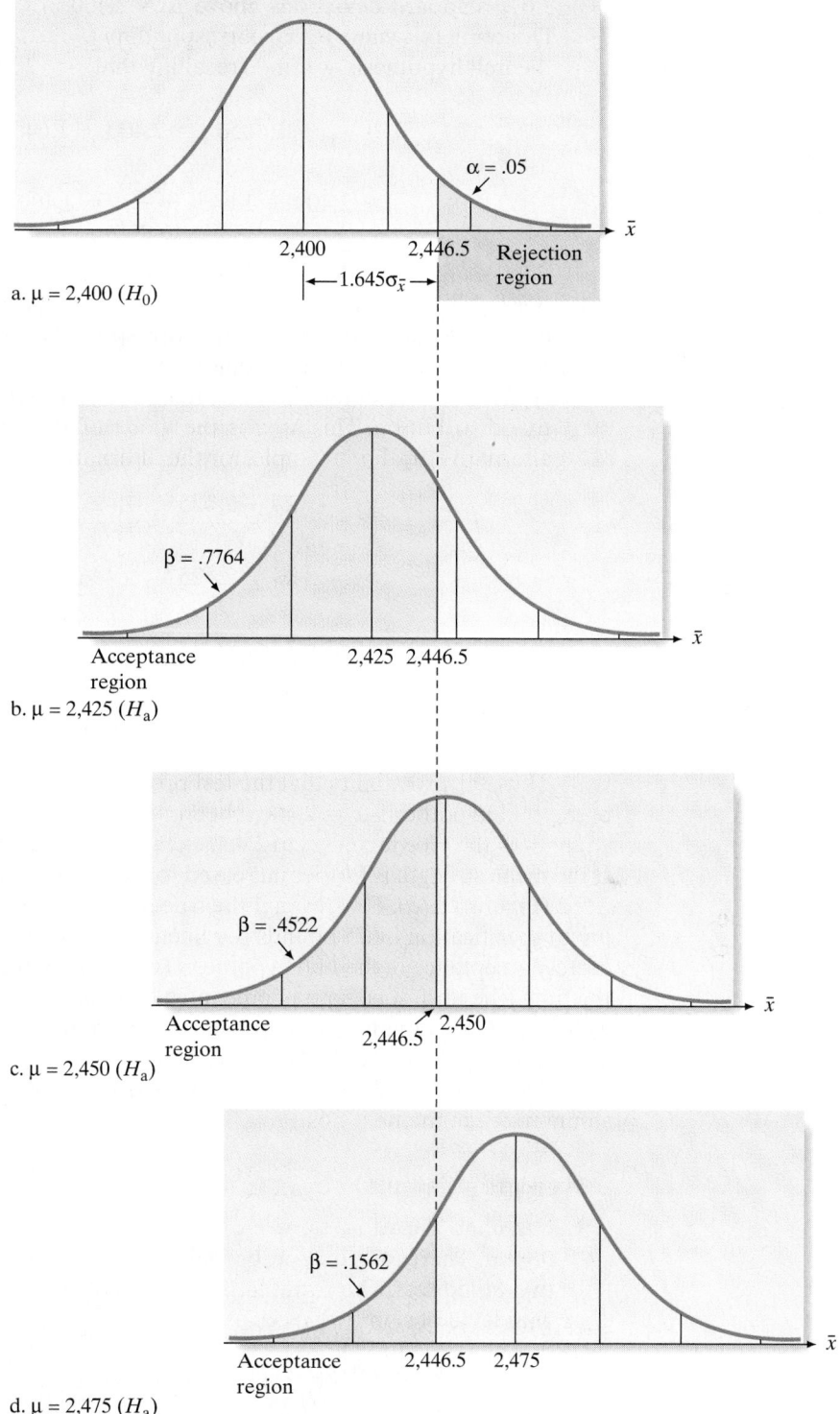

a. $\mu = 2,400$ (H_0)

b. $\mu = 2,425$ (H_a)

c. $\mu = 2,450$ (H_a)

d. $\mu = 2,475$ (H_a)

the probability of incorrectly accepting the null hypothesis should decrease as the distance between the null and alternative values of μ increases.

In order to calculate the value of β for a specific value of μ in H_a, we proceed as follows:

1. Calculate the value of $\bar{x}$ that corresponds to the border between the acceptance and rejection regions. For the sewer pipe example, this is the value of $\bar{x}$ that lies

1.645 standard deviations above $\mu = 2{,}400$ in the sampling distribution of $\bar{x}$. Denoting this value by $\bar{x}_0$, corresponding to the largest value of $\bar{x}$ that supports the null hypothesis, we find (recalling that $s = 200$ and $n = 50$)

$$\bar{x}_0 = \mu_0 + 1.645\sigma_{\bar{x}} = 2{,}400 + 1.645\left(\frac{\sigma}{\sqrt{n}}\right)$$

$$\approx 2{,}400 + 1.645\left(\frac{s}{\sqrt{n}}\right) = 2{,}400 + 1.645\left(\frac{200}{\sqrt{50}}\right)$$

$$= 2{,}400 + 1.645(28.28) = 2{,}446.5$$

2. For a particular alternative distribution corresponding to a value of μ denoted by μ_a, we calculate the z-value corresponding to $\bar{x}_0$, the border between the rejection and acceptance regions. We then use this z value and Table IV of Appendix A to determine the area in the *acceptance region* under the alternative distribution. This area is the value of β corresponding to the particular alternative μ_a. For example, for the alternative $\mu_a = 2{,}425$, we calculate

$$z = \frac{\bar{x}_0 - 2{,}425}{\sigma_{\bar{x}}} = \frac{\bar{x}_0 - 2{,}425}{\sigma/\sqrt{n}}$$

$$\approx \frac{\bar{x}_0 - 2{,}425}{s/\sqrt{n}} = \frac{2{,}446.5 - 2{,}425}{28.28} = .76$$

Note in Figure 8.17b that the area in the acceptance region is the area to the left of $z = .76$. This area is

$$\beta = .5 + .2764 = .7764$$

Thus, the probability that the test procedure will lead to an incorrect acceptance of the null hypothesis $\mu = 2{,}400$ when in fact $\mu = 2{,}425$ is about .78. As the average strength of the pipe increases to 2,450, the value of β decreases to .4522 (Figure 8.17c). If the mean strength is further increased to 2,475, the value of β is further decreased to .1562 (Figure 8.17d). Thus, even if the true mean strength of the pipe exceeds the minimum specification by 75 pounds per linear foot, the test procedure will lead to an incorrect acceptance of the null hypothesis (rejection of the pipe) approximately 16% of the time. The upshot is that the pipe must be manufactured so that the mean strength well exceeds the minimum requirement if the manufacturer wants the probability of its acceptance by the city to be large (i.e., β to be small).

The steps for calculating β for a large-sample test about a population mean are summarized in the next box.

Steps for Calculating β for a Large-Sample Test about μ

1. Calculate the value(s) of $\bar{x}$ corresponding to the border(s) of the rejection region. There will be one border value for a one-tailed test, and two for a two-tailed test. The formula is one of the following, corresponding to a test with level of significance α:

Upper-tailed test: $\bar{x}_0 = \mu_0 + z_\alpha \sigma_{\bar{x}} \approx \mu_0 + z_\alpha\left(\dfrac{s}{\sqrt{n}}\right)$

Lower-tailed test: $\bar{x}_0 = \mu_0 - z_\alpha \sigma_{\bar{x}} \approx \mu_0 - z_\alpha\left(\dfrac{s}{\sqrt{n}}\right)$

Two-tailed test: $\bar{x}_{0,\,L} = \mu_0 - z_{\alpha/2}\sigma_{\bar{x}} \approx \mu_0 - z_{\alpha/2}\left(\dfrac{s}{\sqrt{n}}\right)$

$$\bar{x}_{0,\,U} = \mu_0 + z_{\alpha/2}\sigma_{\bar{x}} \approx \mu_0 + z_{\alpha/2}\left(\frac{s}{\sqrt{n}}\right)$$

2. Specify the value of μ_a in the alternative hypothesis for which the value of β is to be calculated. Then convert the border value(s) of $\bar{x}_0$ to z value(s) using the alternative distribution with mean μ_a. The general formula for the z value is

$$z = \frac{\bar{x}_0 - \mu_a}{\sigma_{\bar{x}}}$$

Sketch the alternative distribution (centered at μ_a), and shade the area in the acceptance (nonrejection) region. Use the z-statistic(s) and Table IV of Appendix A to find the shaded area, which is β.

Following the calculation of β for a particular value of μ_a, you should interpret the value in the context of the hypothesis testing application. It is often useful to interpret the value of $1 - \beta$, which is known as the power of the test corresponding to a particular alternative, μ_a. Since β is the probability of accepting the null hypothesis when the alternative hypothesis is true with $\mu = \mu_a$, $1 - \beta$ is the probability of the complementary event, or the probability of rejecting the null hypothesis when the alternative H_a: $\mu = \mu_a$ is true. That is, the power $(1 - \beta)$ measures the likelihood that the test procedure will lead to the correct decision (reject H_0) for a particular value of the mean in the alternative hypothesis.

> **DEFINITION 8.2**
>
> The **power of a test** is the probability that the test will correctly lead to the rejection of the null hypothesis for a particular value of μ in the alternative hypothesis. The power is equal to $(1 - \beta)$ for the particular alternative considered.

For example, in the sewer pipe example we found that $\beta = .7764$ when $\mu = 2{,}425$. This is the probability that the test leads to the (incorrect) acceptance of the null hypothesis when $\mu = 2{,}425$. Or, equivalently, the power of the test is $1 - .7764 = .2236$, which means that the test will lead to the (correct) rejection of the null hypothesis only 22% of the time when the pipe exceeds specifications by 25 pounds per linear foot. When the manufacturer's pipe has a mean strength of 2,475 (that is, 75 pounds per linear foot in excess of specifications), the power of the test increases to $1 - .1562 = .8438$. That is, the test will lead to the acceptance of the manufacturer's pipe 84% of the time if $\mu = 2{,}475$.

EXAMPLE 8.9 FINDING THE POWER OF A TEST

Problem Recall the drug experiment in Examples 8.1 and 8.2, in which we tested to determine whether the mean response time for rats injected with a drug differs from the control mean response time of $\mu = 1.2$ seconds. The test setup is repeated here:

H_0: $\mu = 1.2$
H_a: $\mu \neq 1.2$ (i.e., $\mu < 1.2$ or $\mu > 1.2$)
Test statistic: $z = \dfrac{\bar{x} - 1.2}{\sigma_{\bar{x}}}$
Rejection region: $z < -1.96$ or $z > 1.96$ for $\alpha = .05$
$\quad\quad\quad\quad\quad\quad\quad z < -2.575$ or $z > 2.575$ for $\alpha = .01$

Note that two rejection regions have been specified corresponding to values of $\alpha = .05$ and $\alpha = .01$, respectively. Assume that $n = 100$ and $s = .5$.

a. Suppose drug-injected rats have a mean response time of 1.1 seconds, that is, $\mu = 1.1$. Calculate the values of β corresponding to the two rejection regions. Discuss the relationship between the values of α and β.

b. Calculate the power of the test for each of the rejection regions when $\mu = 1.1$.

Solution   **a.** We first consider the rejection region corresponding to $\alpha = .05$. The first step is to calculate the border values of $\bar{x}$ corresponding to the two-tailed rejection region, $z < -1.96$ or $z > 1.96$:

$$\bar{x}_{0,L} = \mu_0 - 1.96\sigma_{\bar{x}} \approx \mu_0 - 1.96\left(\frac{s}{\sqrt{n}}\right) = 1.2 - 1.96\left(\frac{.5}{10}\right) = 1.102$$

$$\bar{x}_{0,U} = \mu_0 + 1.96\sigma_{\bar{x}} \approx \mu_0 + 1.96\left(\frac{s}{\sqrt{n}}\right) = 1.2 + 1.96\left(\frac{.5}{10}\right) = 1.298$$

These border values are shown in Figure 8.18.

Figure 8.18
Calculation of β for
Drug-Injected Rats
(Example 8.9)

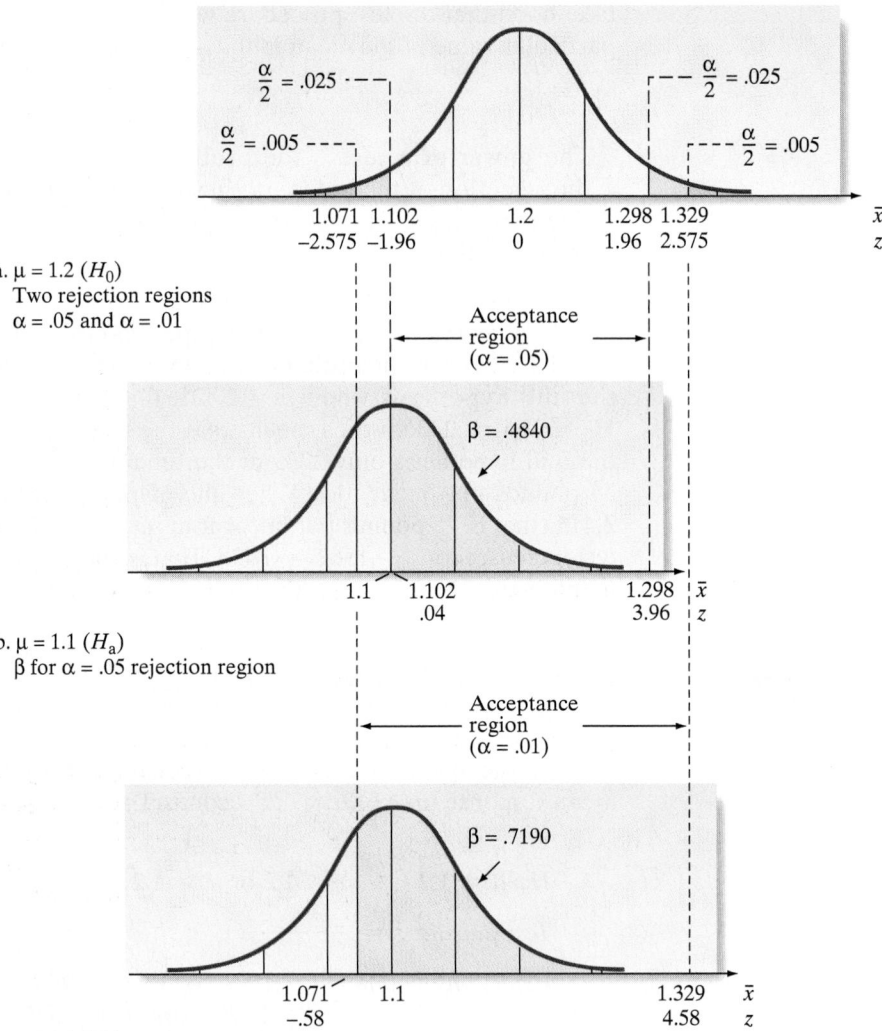

a. $\mu = 1.2$ (H_0)
   Two rejection regions
   $\alpha = .05$ and $\alpha = .01$

b. $\mu = 1.1$ (H_a)
   β for $\alpha = .05$ rejection region

c. $\mu = 1.1$ (H_a)
   β for $\alpha = .01$ rejection region

Next, we convert these values to z-values in the alternative distribution with $\mu_a = 1.1$:

$$z_L = \frac{\bar{x}_{0,\,L} - \mu_a}{\sigma_{\bar{x}}} \approx \frac{1.102 - 1.1}{.05} = .04$$

$$z_U = \frac{\bar{x}_{0,\,U} - \mu_a}{\sigma_{\bar{x}}} \approx \frac{1.298 - 1.1}{.05} = 3.96$$

These z-values are shown in Figure 8.18b. You can see that the acceptance (or nonrejection) region is the area between them. Using Table IV of Appendix A, we find that the area between $z = 0$ and $z = .04$ is .0160, and the area between $z = 0$ and $z = 3.96$ is (approximately) .5 (since $z = 3.96$ is off the scale of Table IV). Then the area between $z = .04$ and $z = 3.96$ is, approximately,

$$\beta = .5 - .0160 = .4840$$

Thus, the test with $\alpha = .05$ will lead to a Type II error about 48% of the time when the mean reaction time for drug-injected rats is .1 second less than the control mean response time.

For the rejection region corresponding to $\alpha = .01$, $z < -2.575$ or $z > 2.575$, we find

$$\bar{x}_{0,\,L} = 1.2 - 2.575\left(\frac{.5}{10}\right) = 1.0712$$

$$\bar{x}_{0,\,U} = 1.2 + 2.575\left(\frac{.5}{10}\right) = 1.3288$$

These border values of the rejection region are shown in Figure 8.18c.

Converting these to z-values in the alternative distribution with $\mu_a = 1.1$, we find $z_L = -.58$ and $z_U = 4.58$. The area between these values is, approximately,

$$\beta = .2190 + .5 = .7190$$

Thus, the chance that the test procedure with $\alpha = .01$ will lead to an incorrect acceptance of H_0 is about 72%.

Note that the value of β increases from .4840 to .7190 when we decrease the value of α from .05 to .01. This is a general property of the relationship between α and β: *as α is decreased (increased), β is increased (decreased)*.

b. The power is defined to be the probability of (correctly) rejecting the null hypothesis when the alternative is true. When $\mu = 1.1$ and $\alpha = .05$, we find

$$\text{Power} = 1 - \beta = 1 - .4840 = .5160$$

When $\mu = 1.1$ and $\alpha = .01$, we find

$$\text{Power} = 1 - \beta = 1 - .7190 = .2810$$

You can see that the power of the test is decreased as the level of α is decreased. This means that as the probability of incorrectly rejecting the null hypothesis is decreased, the probability of correctly accepting the null hypothesis for a given alternative is also decreased.

Look Back A key point in this example is that the value of α must be selected carefully, with the realization that a test is made less capable of detecting departures from the null hypothesis when the value of α is decreased.

◼ ◼ ◼

Figure 8.19

MINITAB Power Analysis for Example 8.9

Power and Sample Size

```
1-Sample Z Test

Testing mean = null (versus not = null)
Calculating power for mean = null + difference
Alpha = 0.05   Assumed standard deviation = 0.5


                Sample
Difference       Size        Power
       0.1        100     0.516005
```

Note: Most statistical software packages now have options for computing the power of standard tests of hypothesis. Usually, you will need to specify the type of test (z-test or t-test), form of H_a ($<$, $>$, or $\neq$), standard deviation, sample size, and the value of the parameter in H_a (or the difference between the value in H_0 and the value in H_a). The MINITAB power analysis for Example 8.9 when $\alpha = .05$ is displayed in Figure 8.19. The power of the test (.516) is highlighted on the printout.

We have shown that the probability of committing a Type II error, β, is inversely related to α (Example 8.9), and that the value of β decreases as the value of μ_a moves farther from the null hypothesis value (sewer pipe example). The sample size n also affects β. Remember that the standard deviation of the sampling distribution of $\bar{x}$ is inversely proportional to the square root of the sample size ($\sigma_{\bar{x}} = \sigma/\sqrt{n}$). Thus, as illustrated in Figure 8.20, the variability of both the null and alternative sampling distributions is decreased as n is increased. If the value of α is specified and remains fixed, the value of β decreases as n increases, as illustrated in Figure 8.20. Conversely, the power of the test for a given alternative hypothesis is increased as the sample size is increased.

The properties of β and power are summarized in the following box.

Properties of β and Power

1. For fixed n and α, the value of β decreases and the power increases as the distance between the specified null value μ_0 and the specified alternative value μ_a increases (see Figure 8.17).

2. For fixed n and values of μ_0 and μ_a, the value of β increases and the power decreases as the value of α is decreased (see Figure 8.18).

3. For fixed α and values of μ_0 and μ_a, the value of β decreases and the power increases as the sample size n is increased (see Figure 8.20).

Figure 8.20

Relationship between α, β, and n

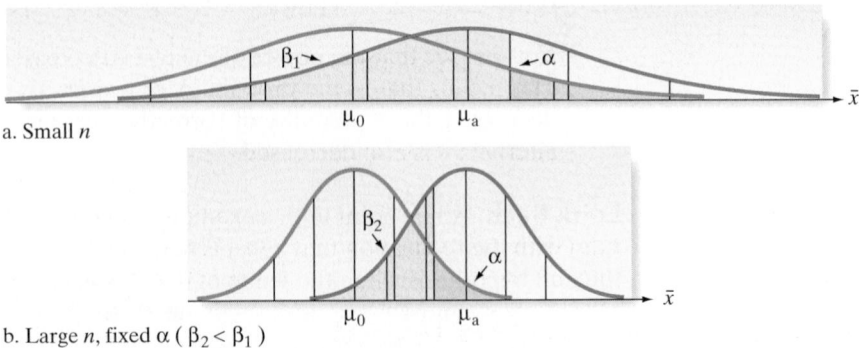

a. Small n

b. Large n, fixed α ($\beta_2 < \beta_1$)

Exercises 8.90–8.101

Understanding the Principles

8.90 Define the power of a test.

8.91 What is the relationship between β, the probability of committing a Type II error, and the power of the test?

8.92 List three factors that will increase the power of the test.

Learning the Mechanics

8.93 Suppose you want to test $H_0: \mu = 1,000$ against $H_a: \mu > 1,000$ using $\alpha = .05$. The population in question is normally distributed with standard deviation 120. A random sample of size $n = 36$ will be used.

   a. Sketch the sampling distribution of $\bar{x}$ assuming that H_0 is true.

   b. Find the value of $\bar{x}_0$, that value of $\bar{x}$ above which the null hypothesis will be rejected. Indicate the rejection region on your graph of part **a**. Shade the area above the rejection region and label it α.

   c. On your graph of part **a**, sketch the sampling distribution of $\bar{x}$ if $\mu = 1,020$. Shade the area under this distribution that corresponds to the probability that $\bar{x}$ falls in the nonrejection region when $\mu = 1,020$. Label this area β.

   d. Find β.

   e. Compute the power of this test for detecting the alternative $H_a: \mu = 1,020$.

8.94 Refer to Exercise 8.93.

   a. If $\mu = 1,040$ instead of 1,020, what is the probability that the hypothesis test will incorrectly fail to reject H_0? That is, what is β?

   b. If $\mu = 1,040$, what is the probability that the test will correctly reject the null hypothesis? That is, what is the power of the test?

   c. Compare β and the power of the test when $\mu = 1,040$ to the values you obtained in Exercise 8.93 for $\mu = 1,020$. Explain the differences.

8.95 It is desired to test $H_0: \mu = 50$ against $H_a: \mu < 50$ using $\alpha = .10$. The population in question is uniformly distributed with standard deviation 20. A random sample of size 64 will be drawn from the population.

   a. Describe the (approximate) sampling distribution of $\bar{x}$ under the assumption that H_0 is true.

   b. Describe the (approximate) sampling distribution of $\bar{x}$ under the assumption that the population mean is 45.

   c. If μ were really equal to 45, what is the probability that the hypothesis test would lead the investigator to commit a Type II error?

   d. What is the power of this test for detecting the alternative $H_a: \mu = 45$?

8.96 Refer to Exercise 8.95.

   a. Find β for each of the following values of the population mean: 49, 47, 45, 43, and 41.

   b. Plot each value of β you obtained in part **a** against its associated population mean. Show β on the vertical axis and μ on the horizontal axis. Draw a curve through the five points on your graph.

   c. Use your graph of part **b** to find the approximate probability that the hypothesis test will lead to a Type II error when $\mu = 48$.

   d. Convert each of the β values you calculated in part **a** to the power of the test at the specified value of μ. Plot the power on the vertical axis against μ on the horizontal axis. Compare the graph of part **b** to the *power curve* of this part.

   e. Examine the graphs of parts **b** and **d**. Explain what they reveal about the relationships among the distance between the true mean μ and the null hypothesized mean μ_0, the value of β, and the power.

8.97 Suppose you want to conduct the two-tailed test of $H_0: \mu = 10$ against $H_a: \mu \neq 10$ using $\alpha = .05$. A random sample of size 100 will be drawn from the population in question. Assume the population has a standard deviation equal to 1.0.

   a. Describe the sampling distribution of $\bar{x}$ under the assumption that H_0 is true.

   b. Describe the sampling distribution of $\bar{x}$ under the assumption that $\mu = 9.9$.

   c. If μ were really equal to 9.9, find the value of β associated with the test.

   d. Find the value of β for the alternative $H_a: \mu = 10.1$.

Applying the Concepts—Intermediate

8.98 **Post-traumatic stress of POWs.** Refer to Exercise 8.29 (p. 382), in which the null hypothesis that the mean PTSD score of World War II POWs is 16 is tested against the alternative hypothesis that the mean is less than 16. Recall that 33 POWs were included in the sample. Assume that the resulting standard deviation of $s = 9.32$ is a good estimate of the true standard deviation.

   a. Calculate the power of the test for the mean values of 15.5, 15.0, 14.5, 14.0, and 13.5.

   b. Plot the power of the test on the vertical axis against the mean on the horizontal axis. Draw a curve through the points.

   c. Use the power curve of part **b** to estimate the power for the mean value $\mu = 14.25$. Calculate the power for this value of μ, and compare it to your approximation.

   d. Use the power curve to approximate the power of the test when $\mu = 12.10$. If the true value of the mean PTSD score for WWII POWs is really 10, what (approximately) are the chances that the test will fail to reject the null hypothesis that the mean is 16?

8.99 **Increasing the sample size.** Refer to Exercise 8.98. Show what happens to the power curve when the sample size is increased from $n = 33$ to $n = 100$. Assume that the standard deviation is $\sigma = 9.32$.

8.100 **Gas mileage of the Honda Civic.** According to the Environmental Protection Agency (EPA) *Fuel Economy Guide*, the 2003 Honda Civic automobile obtains a mean of 38 miles per gallon (mpg) on the highway. Suppose Honda claims that the EPA has underestimated the Civic's mileage. To support its assertion, the company selects 36 model 2003 Civic cars and records the mileage obtained for each car over a driving course similar to the one used by the EPA. The following data resulted: $x = 40.3$ mpg, $s = 6.4$ mpg.

a. If Honda wishes to show that the mean mpg for 2003 Civic autos is greater than 38 mpg, what should the alternative hypothesis be? The null hypothesis?

b. Do the data provide sufficient evidence to support the auto manufacturer's claim? Test using $\alpha = .05$. List any assumptions you make in conducting the test.

c. Calculate the power of the test for the mean values of 38.5, 39.0, 39.5, 40.0, and 40.5, assuming $s = 6.4$ is a good estimate of σ.

d. Plot the power of the test on the vertical axis against the mean on the horizontal axis. Draw a curve through the points.

e. Use the power curve of part d to estimate the power for the mean value $\mu = 39.75$. Calculate the power for this value of μ, and compare it to your approximation.

f. Use the power curve to approximate the power of the test when $\mu = 43$. If the true value of the mean mpg for this model is really 43, what (approximately) are the chances that the test will fail to reject the null hypothesis that the mean is 38?

8.101 Solder joint inspections. Refer to Exercise 8.32 (p. 382), in which the performance of a particular type of laser-based inspection equipment was investigated. Assume that the standard deviation of the number of solder joints inspected on each run is 1.2. If $\alpha = .05$ is used in conducting the hypothesis test of interest using a sample of 48 circuit boards, and if the true mean number of solder joints that can be inspected is really equal to 9.5, what is the probability that the test will result in a Type II error?

8.7 Test of Hypothesis about a Population Variance (Optional)

Although many practical problems involve inferences about a population mean (or proportion), it is sometimes of interest to make an inference about a population variance, σ^2. To illustrate, a quality control supervisor in a cannery knows that the exact amount each can contains will vary since there are certain uncontrollable factors that affect the amount of fill. The mean fill per can is important, but equally important is the variation of fill. If σ^2, the variance of the fill, is large, some cans will contain too little and others too much. Suppose regulatory agencies specify that the standard deviation of the amount of fill in 16-ounce cans should be less than .1 ounce. To determine whether the process is meeting this specification, the supervisor randomly selects 10 cans and weighs the contents of each. The results are given in Table 8.5.

⊙ **FILLAMOUNTS**

TABLE 8.5   Fill Weights (ounces) of 10 Cans

16.00	15.95	16.10	16.02	15.99
16.06	16.04	16.05	16.03	16.02

Do these data provide sufficient evidence to indicate that the variability is as small as desired? To answer this question, we need a procedure for testing a hypothesis about σ^2.

Intuitively it seems that we should compare the sample variance σ^2 to the hypothesized value of σ^2 (or s to σ) in order to make a decision about the population's variability. The quantity

$$\frac{(n - 1)s^2}{\sigma^2}$$

has been shown to have a sampling distribution called a **chi-square (χ^2) distribution** when the population from which the sample is taken is *normally distributed*. Several chi-square distributions are shown in Figure 8.21.

Figure 8.21
Several χ^2 Probability Distributions

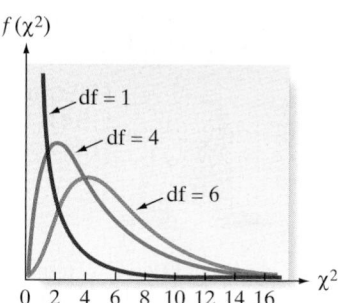

TABLE 8.6 Reproduction of Part of Table VII in Appendix A: Critical Values of Chi-Square

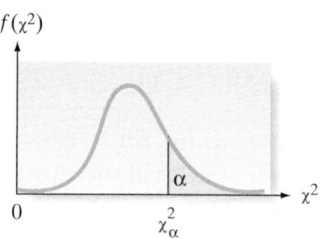

Degrees of Freedom	$\chi^2_{.100}$	$\chi^2_{.050}$	$\chi^2_{.025}$	$\chi^2_{.010}$	$\chi^2_{.005}$
1	2.70554	3.84146	5.02389	6.63490	7.87944
2	4.60517	5.99147	7.37776	9.21034	10.5966
3	6.25139	7.81473	9.34840	11.3449	12.8381
4	7.77944	9.48773	11.1433	13.2767	14.8602
5	9.23635	11.0705	12.8325	15.0863	16.7496
6	10.6446	12.5916	14.4494	16.8119	18.5476
7	12.0170	14.0671	16.0128	18.4753	20.2777
8	13.3616	15.5073	17.5346	20.0902	21.9550
9	14.6837	16.9190	19.0228	21.6660	23.5893
10	15.9871	18.3070	20.4831	23.2093	25.1882
11	17.2750	19.6751	21.9200	24.7250	26.7569
12	18.5494	21.0261	23.3367	26.2170	28.2995
13	19.8119	22.3621	24.7356	27.6883	29.8194
14	21.0642	23.6848	26.1190	29.1413	31.3193
15	22.3072	24.9958	27.4884	30.5779	32.8013
16	23.5418	26.2862	28.8454	31.9999	34.2672
17	24.7690	27.5871	30.1910	33.4087	35.7185
18	25.9894	28.8693	31.5264	34.8053	37.1564
19	27.2036	30.1435	32.8523	36.1908	38.5822

The upper-tail areas for this distribution have been tabulated and are given in Table VII of Appendix A, a portion of which is reproduced in Table 8.6. The table gives the values of χ^2, denoted as χ^2_α, that locate an area of α in the upper tail of the chi-square distribution; that is, $P(\chi^2 > \chi^2_\alpha) = \alpha$. In this case, as with the t-statistic, the shape of the chi-square distribution depends on the degrees of freedom associated with s^2, namely $(n - 1)$. Thus, for $n = 10$ and an upper-tail value $\alpha = .05$, you will have $n - 1 = 9$ df and $\chi^2_{.05} = 16.9190$ (highlighted area in Table 8.6). To further illustrate the use of Table VII, we return to the can-filling example.

Now Work *Exercise 8.106*

Biography

FRIEDRICH R. HELMERT (1843–1917)—Helmert Transformations

German Friedrich Helmert studied engineering sciences and mathematics at Dresden University, where he earned his Ph.D., then accepted a position as a professor of geodesy—the scientific study of the Earth's size and shape—at the technical school in Aachen. Helmert's mathematical solutions to geodesy problems led him to several statistics-related discoveries. His greatest statistical contribution occurred in 1876, when he was the first to prove that the sampling distribution of the sample variance, s^2, is a chi-square distribution. Helmert used a series of mathematical transformations to obtain the distribution of s^2—transformations that have since been named "Helmert transformations" in his honor. Later in life, Helmert was appointed professor of advanced geodesy at prestigious University of Berlin and director of the Prussian Geodetic Institute.

EXAMPLE 8.10

CONDUCTING A TEST FOR σ

Problem Refer to the fill weights for the sample of ten 16 ounce cans in Table 8.5. Is there sufficient evidence to conclude that the true standard deviation σ of the fill measurements of 16-ounce cans is less than .1 ounce?

Solution Here, we want to test whether $\sigma < .1$. Since the null and alternative hypotheses must be stated in terms of σ^2 (rather than σ), we want to test the null hypothesis that $\sigma^2 = (.1)^2 = .01$ against the alternative that $\sigma^2 < .01$. Therefore, the elements of the test are

$$H_0: \sigma^2 = .01$$
$$H_a: \sigma^2 < .01$$

$$\text{Test statistic: } \chi^2 = \frac{(n-1)s^2}{\sigma^2}$$

Assumption: The distribution of the amounts of fill is approximately normal.

Rejection region: The smaller the value of s^2 we observe, the stronger the evidence in favor of H_a. Thus, we reject H_0 for "small values" of the test statistic. With $\alpha = .05$ and 9 df, the χ^2 value for rejection is found in Table VII and pictured in Figure 8.22. We will reject H_0 if $\chi^2 < 3.32511$.

(Remember that the area given in Table VII is the area to the *right* of the numerical value in the table. Thus, to determine the lower-tail value, which has $\alpha = .05$ to its *left*, we used the $\chi^2_{.95}$ column in Table VII.)

Figure 8.22
Rejection Region for Example 8.10

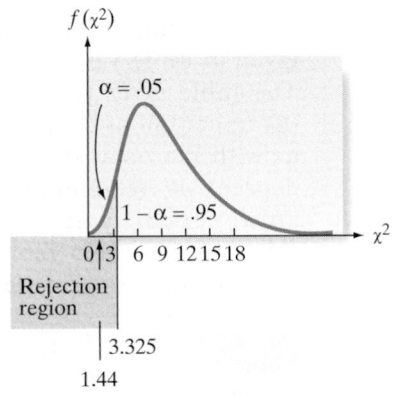

Figure 8.23

SAS Test of Fill Amount
Variance, Example 8.10

```
Sample Statistics for FILLAMT

    N         Mean        Std. Dev.        Variance
--------------------------------------------------------
    10       16.026         0.0412          0.0017


Hypothesis Test

      Null hypothesis:        Variance of FILLAMT => 0.01
      Alternative:            Variance of FILLAMT <  0.01

        Chi-square          Df            Prob
      ------------------------------------------------
          1.524              9           0.0030
```

A SAS printout of the analysis is displayed in Figure 8.23. The value of s (highlighted on the printout) is $s = .0412$. Substituting into the formula for the test statistic, we have

$$\chi^2 = \frac{(n-1)s^2}{\sigma^2} = \frac{9(.0412)^2}{.01} = 1.524$$

Since the value of the test statistic is less than 3.32511, the supervisor can conclude that the variance σ^2 of the population of all amounts of fill is less than .01 (i.e., $\sigma < .1$) with probability of a Type I error equal to $\alpha = .05$. If this procedure is repeatedly used, it will incorrectly reject H_0 only 5% of the time. Thus, the quality control supervisor is confident in the decision that the cannery is operating within the desired limits of variability.

Look Back Note that both the test statistic (1.524) and one-tailed p-value of the test (.0030) are shown on the SAS printout. Also note that the p-value is less than $\alpha = .05$, thus confirming our conclusion to reject H_0.

Now Work *Exercise 8.111*

■ ■ ■

One-tailed and two-tailed tests of hypothesis for σ^2 are given in the following box.*

Test of Hypothesis about σ^2

One-Tailed Test	**Two-Tailed Test**
$H_0: \sigma^2 = \sigma_0^2$	$H_0: \sigma^2 = \sigma_0^2$
$H_a: \sigma^2 < \sigma_0^2$ (or $H_a: \sigma^2 > \sigma_0^2$)	$H_a: \sigma^2 \neq \sigma_0^2$
Test statistic: $\chi^2 = \dfrac{(n-1)s^2}{\sigma_0^2}$	Test statistic: $\chi^2 = \dfrac{(n-1)s^2}{\sigma_0^2}$
Rejection region: $\chi^2 < \chi_{(1-\alpha)}^2$ (or $\chi^2 > \chi_\alpha^2$ when $H_a: \sigma^2 > \sigma_0^2$)	Rejection region: $\chi^2 < \chi_{(1-\alpha/2)}^2$ or $\chi^2 > \chi_{(\alpha/2)}^2$

where σ_0^2 is the hypothesized variance and the distribution of χ^2 is based on $(n-1)$ degrees of freedom.

---

*A confidence interval for σ^2 can also be formed using the χ^2-distribution with $(n-1)$ degrees of freedom. A $(1-\alpha)$ 100% confidence interval is

$$\frac{(n-1)s^2}{\chi_{\alpha/2}^2} < \sigma^2 < \frac{(n-1)s^2}{\chi_{(1-\alpha/2)}^2}$$

> **Conditions Required for a Valid Hypothesis Test for σ^2**
>
> **1.** A random sample is selected from the target population.
>
> **2.** The population from which the sample is selected has a distribution that is approximately normal.

Caution

The procedure for conducting a hypothesis test for σ^2 in the preceding examples requires an assumption regardless of whether the sample size n is large or small. We must assume that the population from which the sample is selected has an approximate normal distribution. Unlike small sample tests for μ based on the t-statistic, *slight to moderate departures from normality will render the χ^2-test invalid.*

Exercises 8.102–8.118

Understanding the Principles

8.102 What is the sampling distribution used for making inferences about σ^2?

8.103 What conditions are required for a valid test for σ^2?

8.104 *True or False.* The null hypotheses, H_0: $\sigma^2 = .25$ and H_0: $\sigma = .5$, are equivalent.

8.105 *True or False.* When the sample size n is large, no assumptions about the population are necessary to test the population variance σ^2.

Learning the Mechanics

8.106 Let χ_0^2 be a particular value of χ^2. Find the value of χ_0^2 such that
a. $P(\chi^2 > \chi_0^2) = .10$ for $n = 12$
b. $P(\chi^2 > \chi_0^2) = .05$ for $n = 9$
c. $P(\chi^2 > \chi_0^2) = .025$ for $n = 5$

8.107 A random sample of n observations is selected from a normal population to test the null hypothesis that $\sigma^2 = 25$. Specify the rejection region for each of the following combinations of H_a, α, and n:
a. H_a: $\sigma^2 \neq 25$; $\alpha = .05$; $n = 16$
b. H_a: $\sigma^2 > 25$; $\alpha = .01$; $n = 23$
c. H_a: $\sigma^2 > 25$; $\alpha = .10$; $n = 15$
d. H_a: $\sigma^2 < 25$; $\alpha = .01$; $n = 13$
e. H_a: $\sigma^2 \neq 25$; $\alpha = .10$; $n = 7$
f. H_a: $\sigma^2 < 25$; $\alpha = .05$; $n = 25$

8.108 A random sample of seven measurements gave $\bar{x} = 9.4$ and $s^2 = 1.84$.
a. What assumptions must you make concerning the population in order to test a hypothesis about σ^2?
b. Suppose the assumptions in part **a** are satisfied. Test the null hypothesis, $\sigma^2 = 1$, against the alternative hypothesis, $\sigma^2 > 1$. Use $\alpha = .05$.
c. Test the null hypothesis that $\sigma^2 = 1$ against the alternative hypothesis that $\sigma^2 \neq 1$. Use $\alpha = .05$.

8.109 Refer to Exercise 8.108. Suppose we had $n = 100$, $\bar{x} = 9.4$, and $s^2 = 1.84$.

a. Test the null hypothesis, H_0: $\sigma^2 = 1$, against the alternative hypothesis, H_a: $\sigma^2 > 1$.
b. Compare your test result with that of Exercise 8.108.

8.110 A random sample of $n = 7$ observations from a normal population produced the following measurements: 4, 0, 6, 3, 3, 5, 9. Do the data provide sufficient evidence to indicate that $\sigma^2 < 2$? Test using $\alpha = .05$.

Applying the Concepts—Basic

8.111 **Point spreads for NFL games.** Refer to the *Chance* (Fall 1998) study of point-spread errors in NFL games, Exercise 8.30 (p. 382). Recall that the difference between the actual game outcome and the point spread established by oddsmakers—the point-spread error—was calculated for 240 NFL games. The results are summarized as follows: $\bar{x} = -1.6$, $s = 13.3$. Suppose the researcher wants to know whether the true standard deviation of the point-spread errors exceeds 15.
a. Set up H_0 and H_a for the researcher.
b. Compute the value of the test statistic.
c. Conduct the test at $\alpha = .10$. Interpret the results in the words of the problem.

8.112 **A new dental bonding agent.** Refer to the *Trends in Biomaterials & Artificial Organs* (January 2003) study of a new bonding adhesive for teeth called "Smartbond," Exercise 8.61 (p. 395). Recall that tests on a sample of 10 extracted teeth bonded with the new adhesive resulted in a mean breaking strength (after 24 hours) of $\bar{x} = 5.07$ Mpa and a standard deviation of $s = .46$ Mpa. In addition to mean breaking strength, orthodontists are concerned about the variability in breaking strength of the new bonding adhesive.
a. Set up the null and alternative hypothesis for a test to determine if the breaking strength variance differs from .5 Mpa.

b. Find the rejection region for the test using $\alpha = .01$.

c. Compute the test statistic.

d. Give the appropriate conclusion for the test.

e. What conditions are required for the test results to be valid?

8.113 Birthweights of cocaine babies. A group of researchers at the University of Texas-Houston conducted a comprehensive study of pregnant cocaine-dependent women. (*Journal of Drug Issues*, Summer 1997.) All the women in the study used cocaine on a regular basis (at least three times a week) for more than a year. One of the many variables measured was birthweight (in grams) of the baby delivered. For a sample of 16 cocaine-dependent women, the mean birthweight was 2,971 grams and the standard deviation was 410 grams. Test (at $\alpha = .01$) to determine whether the variance in birthweights of babies delivered by cocaine-dependent women is less than 200,000 grams^2.

Applying the Concepts—Intermediate

⚙ **GOBIANTS**

8.114 Mongolian desert ants. Refer to the *Journal of Biogeography* (December 2003) study of ants in Mongolia (Central Asia), Exercise 8.68 (p. 396). Data on the number of ant species attracted to 11 randomly selected desert sites are saved in the **GOBIANTS** file. Do these data indicate that the standard deviation of the number of ant species at all Mongolian desert sites exceeds 15 species? Conduct the appropriate test at $\alpha = .05$. Are the conditions required for a valid test satisfied?

8.115 Extracting toxic substances. A new technique for extracting toxic organic compounds was evaluated in *Chromatographia* (March 1995). Uncontaminated fish fillets were injected with a toxic substance and the method was used to extract the toxicant. To test the precision of the new method, 7 measurements were obtained on a single fish fillet injected with a toxin. Summary statistics on percent of the toxin recovered are given as follows: $\bar{x} = 99$, $s = 9$. Determine whether the standard deviation of the percent recovery using the new method differs from 15. (Use $\alpha = .10$.)

8.116 Ball-bearing diameters. It is essential in the manufacture of machinery to utilize parts that conform to specifications. In the past, diameters of the ball-bearings produced by a certain manufacturer had a variance of .00156. To cut costs, the manufacturer instituted a less expensive production method. The variance of the diameters of 100 randomly sampled bearings produced by the new process was .00211. Do the data provide sufficient evidence to indicate that diameters of ball-bearings produced by the new process are more variable than those produced by the old process?

8.117 Laser Spectroscopy of rocks. Geologists use laser Raman microprobe (LRM) spectroscopy to analyze *fluid inclusions* (pockets of gas or liquid) in rock. A chip of natural Brazilian quartz was artificially injected with several fluid inclusions of liquid carbon dioxide (CO_2) and then subjected to LRM spectroscopy. (*Applied Spectroscopy*, February 1986.) The amount of CO_2 present in the inclusion was recorded for the same inclusion on four different days. The data (in mole percent) follow:

⚙ **CO2**

86.6	84.6	85.5	85.9

a. Do the data indicate that the variation in the CO_2 concentration measurements using the LRM method differs from 1?

b. What assumption is required for the test to be valid?

Applying the Concepts—Advanced

8.118 Homogeneity of brain signals. To improve the signal-to-noise ratio (SNR) in the electrical activity of the brain, neurologists repeatedly stimulate subjects and average the responses—a procedure that assumes that signal responses are homogeneous. A study was conducted to test the homogeneous signal theory. (*IEEE Engineering in Medicine and Biology Magazine*, March 1990.) For this study, the variance of the SNR readings of subjects will equal .54 under the homogeneous signal theory. If the SNR variance exceeds .54, the researchers will conclude that the signals are nonhomogeneous. Signal-to-noise ratios recorded for a sample of 41 normal children ranged from .03 to 3.0. Use this information to test the theory at $\alpha = .10$.

Quick Review

Key Terms

Note: Starred (*) terms are from the optional sections in this chapter.

Alternative (research) hypothesis   370
Chi-square distribution*   412
Conclusion   380
Level of significance   374
Lower-tailed test   377

Null distribution*   404
Null hypothesis   370
Observed significance level
   (*p*-value)   383
One-tailed statistical test   376
Power of a test*   407
Rejection region   371

Test of hypothesis   369
Test statistic   370
Two-tailed hypothesis   377
Two-tailed test   377
Type I error   371
Type II error   372
Upper-tailed test   377

Key Formulas

Note: Starred (*) formulas are from the optional sections in this chapter.

For testing $H_0: \theta = \theta_0$, the **large-sample test statistic** is

$$z = \frac{\hat{\theta} - \theta_0}{\sigma_{\hat{\theta}}}$$

where $\hat{\theta}$, θ_0, and $\sigma_{\hat{\theta}}$ are obtained from the table below:

Parameter, θ	Hypothesized Parameter Value, θ_0	Estimator, $\hat{\theta}$	Standard Error of Estimator, $\sigma_{\hat{\theta}}$	
μ	μ_0	$\bar{x}$	$\dfrac{\sigma}{\sqrt{n}}$	380
p	p_0	$\hat{p}$	$\sqrt{\dfrac{p_0 q_0}{n}}$	399

For testing $H_0: \mu = \mu_0$, the **small-sample test statistic** is

$$t = \frac{\bar{x} - \mu_0}{s/\sqrt{n}} \quad 391$$

*$\beta = P(\hat{\theta}$ falls in acceptance region $|\theta = \theta_a)$ 406
*Power $= 1 - \beta$ 407
**For testing $H_0: \sigma^2 = \sigma_0^2$, the test statistic is*

$$\chi^2 = \frac{(n-1)s^2}{\sigma_0^2} \quad 415$$

Language Lab

Symbol	Pronunciation	Description
H_0	*H*-oh	Null hypothesis
H_a	*H*-a	Alternative hypothesis
α	alpha	Probability of Type I error
β	beta	Probability of Type II error
χ^2	chi-square	Sampling distribution of s^2 when data are from a normal population

Chapter Summary Notes

- Elements of a **test of hypothesis: null hypothesis, alternative hypothesis, test statistic, significance level (α), rejection region, p-value,** and **conclusion.**
- Two types of errors in a hypothesis test: **Type I error** (reject H_0 when H_0 is true), **Type II error** (accept H_0 when H_0 is false).
- Probabilities of errors: $\alpha = P(\textbf{Type I error}) = P(\text{Reject } H_0 | H_0 \text{ true})$, $\beta = P(\textbf{Type II error}) = P(\text{Accept } H_0 | H_0 \text{ false})$.
- Three forms of the alternative hypothesis: **lower-tailed test** ($<$), **upper-tailed test** ($>$), **two-tailed test** ($\neq$).
- **Observed significance level (p-value)** is the smallest value of α that can be used to reject the null hypothesis.

- Conditions required for a valid **large-sample test for μ**: (1) random sample, (2) large sample size ($n \geq 30$).
- Conditions required for a valid **small-sample test for μ**: (1) random sample, (2) population distribution is approximately normal.
- Conditions required for a valid **large-sample test for p**: (1) random sample, (2) large sample size ($p_0 \pm 3\sigma$ falls between 0 and 1).
- Conditions required for a valid **test for σ^2**: (1) random sample, (2) population distribution is approximately normal, regardless of sample size.
- **Power of the test** $= 1 - \beta = P(\text{Reject } H_0 | H_0 \text{ false})$.

Supplementary Exercises 8.119–8.151

Note: List the assumptions necessary for the valid implementation of the statistical procedures you use in solving all these exercises. Starred (*) exercises refer to the optional sections in this chapter.

Understanding the Principles

8.119 *Complete the following statement:* The smaller the p-value associated with a test of hypothesis, the stronger the support for the _____ hypothesis. Explain your answer.

8.120 Specify the differences between a large-sample and small-sample test of hypothesis about a population mean μ. Focus on the assumptions and test statistics.

8.121 Which of the elements of a test of hypothesis can and should be specified *prior* to analyzing the data that are to be utilized to conduct the test?

8.122 If the rejection of the null hypothesis of a particular test would cause your firm to go out of business, would you want α to be small or large? Explain.

8.123 *Complete the following statement:* The larger the p-value associated with a test of hypothesis, the stronger the support for the_____ hypothesis. Explain your answer.

Learning the Mechanics

8.124 A random sample of 20 observations selected from a normal population produced $\bar{x} = 72.6$ and $s^2 = 19.4$.
   a. Form a 90% confidence interval for the population mean.
   b. Test H_0: $\mu = 80$ against H_a: $\mu < 80$. Use $\alpha = .05$.
   c. Test H_0: $\mu = 80$ against H_a: $\mu \neq 80$. Use $\alpha = .01$.
   d. Form a 99% confidence interval for μ.
   e. How large a sample would be required to estimate μ to within 1 unit with 95% confidence?

8.125 A random sample of $n = 200$ observations from a binomial population yields $\hat{p} = .29$.
   a. Test H_0: $p = .35$ against H_a: $p < .35$. Use $\alpha = .05$.
   b. Test H_0: $p = .35$ against H_a: $p \neq .35$. Use $\alpha = .05$.
   c. Form a 95% confidence interval for p.
   d. Form a 99% confidence interval for p.
   e. How large a sample would be required to estimate p to within .05 with 99% confidence?

8.126 A random sample of 175 measurements possessed a mean $\bar{x} = 8.2$ and a standard deviation $s = .79$.
   a. Form a 95% confidence interval for μ.
   b. Test H_0: $\mu = 8.3$ against H_a: $\mu \neq 8.3$. Use $\alpha = .05$.
   c. Test H_0: $\mu = 8.4$ against H_a: $\mu \neq 8.4$. Use $\alpha = .05$.

8.127 A random sample of 41 observations from a normal population possessed a mean $\bar{x} = 88$ and a standard deviation $s = 6.9$.
   a. Test H_0: $\sigma^2 = 30$ against H_a: $\sigma^2 > 30$. Use $\alpha = .05$.
   b. Test H_0: $\sigma^2 = 30$ against H_a: $\sigma^2 \neq 30$. Use $\alpha = .05$.

8.128 A t-test is conducted for the null hypothesis H_0: $\mu = 10$ versus the alternative H_a: $\mu > 10$ for a random sample of $n = 17$ observations. The test results are $t = 1.174$, p-value $= .1288$.
   a. Interpret the p-value.
   b. What assumptions are necessary for the validity of this test?
   c. Calculate and interpret the p-value assuming the alternative hypothesis was instead H_a: $\mu \neq 10$.

Applying the Concepts—Basic

8.129 **FDA mandatory new drug testing.** When a new drug is formulated, the pharmaceutical company must subject it to lengthy and involved testing before receiving the necessary permission from the Food and Drug Administration (FDA) to market the drug. The FDA requires the pharmaceutical company to provide substantial evidence that the new drug is safe for potential consumers.
   a. If the new drug testing were to be placed in a test of hypothesis framework, would the null hypothesis be that the drug is safe or unsafe? The alternative hypothesis?
   b. Given the choice of null and alternative hypotheses in part **a**, describe Type I and Type II errors in terms of this application. Define α and β in terms of this application.
   c. If the FDA wants to be very confident that the drug is safe before permitting it to be marketed, is it more important that α or β be small? Explain.

8.130 **Online shopping.** According to *Forbes* (December 13, 1999), Creative Good, a New York consulting firm, claims that 40% of shoppers fail in their attempts to purchase merchandise online because web sites are too complex. Another consulting firm asked a random sample of 60 online shoppers to each test a different randomly selected e-commerce web site. Only 15 reported sufficient frustration with their sites to deter making a purchase.
   a. What are the appropriate null and alternative hypotheses to test the claim made by Creative Good?
   b. Do the data provide sufficient evidence to support the claim? Use $\alpha = .01$.
   c. Calculate and interpret the p-value for the test.

8.131 **Sleep deprivation study.** In a British study, 12 healthy college students, deprived of one night's sleep, received an array of tests intended to measure thinking time, fluency, flexibility, and originality of thought. The overall test scores of the sleep-deprived students were compared to the average score expected from students who received their accustomed sleep (*Sleep*, January 1989). Suppose the overall scores of the 12 sleep-deprived students had a mean of $\bar{x} = 63$ and a standard deviation of 17.

(Lower scores are associated with a decreased ability to think creatively.)

a. Test the hypothesis that the true mean score of sleep-deprived subjects is less than 80, the mean score of subjects who received sleep prior to taking the test. Use $\alpha = .05$.

b. What assumption is required for the hypothesis test of part **a** to be valid?

8.132 The "Pepsi challenge." "Take the Pepsi Challenge" was a marketing campaign used by the Pepsi-Cola Company. Coca-Cola drinkers participated in a blind taste test in which they tasted unmarked cups of Pepsi and Coke and were asked to select their favorite. Pepsi claimed that "in recent blind taste tests, more than half the Diet Coke drinkers surveyed said they preferred the taste of Diet Pepsi." (*Consumer's Research*, May 1993.) Suppose 100 Diet Coke drinkers took the Pepsi Challenge and 56 preferred the taste of Diet Pepsi. Test the hypothesis that more than half of all Diet Coke drinkers will select Diet Pepsi in a blind taste test. Use $\alpha = .05$.

8.133 Masculinizing human faces. Refer to the *Nature* (August 27, 1998) study of facial characteristics that are deemed attractive, Exercise 8.51 (p. 389). In another experiment, 67 human subjects viewed side-by-side an image of a Caucasian male face and the same image 50% masculinized. Each subject was asked to select the facial image that they deemed more attractive. Fifty-eight of the 67 subjects felt that masculinization of face shape decreased attractiveness of the male face. The researchers used this sample information to test whether the subjects showed preference for either the unaltered or morphed male face.

a. Set up the null and alternative hypotheses for this test.

b. Compute the test statistic.

c. The researchers reported *p*-value ≈ 0 for the test. Do you agree?

d. Make the appropriate conclusion in the words of the problem. Use $\alpha = .01$.

⚙ **MATHCPU**

***8.134** Refer to the *IEEE Transactions* (June 1990) study of a new hybrid algorithm for solving polynomial 0/1 mathematical programs, Exercise 7.99 (p. 361). An SAS printout giving descriptive statistics for the sample of 52 solution times is reproduced below. Use this information to determine whether the variance of the solution times differs from 2. Use $\alpha = .05$.

8.135 Mental health of the unemployed. A study reported in the *Journal of Occupational and Organizational Psychology* (December 1992) investigated the relationship of employment status to mental health. Each of a sample of 49 unemployed men was given a mental health examination using the General Health Questionnaire (GHQ). The GHQ is a widely recognized measure of present mental health, with lower values indicating better mental health. The mean and standard deviation of the GHQ scores were $\bar{x} = 10.94$ and $s = 5.10$, respectively.

a. Specify the appropriate null and alternative hypotheses if we wish to test the research hypothesis that the mean GHQ score for all unemployed men exceeds 10. Is the test one-tailed or two-tailed? Why?

b. If we specify $\alpha = .05$, what is the appropriate rejection region for this test?

c. Conduct the test and state your conclusion clearly in the language of this exercise.

Applying the Concepts—Intermediate

8.136 Errors in medical tests. Medical tests have been developed to detect many serious diseases. A medical test is designed to minimize the probability that it will produce a "false positive" or a "false negative." A false positive is a positive test result for an individual who does not have the disease, whereas a false negative is a negative test result for an individual who does have the disease.

a. If we treat a medical test for a disease as a statistical test of hypothesis, what are the null and alternative hypotheses for the medical test?

b. What are the Type I and Type II errors for the test? Relate each to false positives and false negatives.

c. Which of these errors has graver consequences? Considering this error, is it more important to minimize α or β? Explain.

8.137 Cropweights of pigeons. In Exercise 7.91 (p. 360), you analyzed data from a study of the diets of spinifex pigeons. The data, extracted from the *Australian Journal of Zoology*, are reproduced in the table. Each measurement in the data set represents the weight (in grams) of the crop content of a spinifex pigeon. Conduct a test at $\alpha = .05$ to determine whether the mean weight in the crops of all spinifex pigeons differs from 1 gram. Test the validity of any assumptions you make.

⚙ **PIGEONS**

.457	3.751	.238	2.967	2.509	1.384	1.454	.818
.335	1.436	1.603	1.309	.201	.530	2.144	.834

SAS Output for Exercie 8.134

The MEANS Procedure

Analysis Variable : CPU

Mean	Std Dev	Variance	N	Minimum	Maximum
0.8121923	1.5047603	2.2643035	52	0.0360000	8.7880000

8.138 Production of ketchup bottles. A consumer protection group is concerned that a ketchup manufacturer is filling its 20-ounce family-size containers with less than 20 ounces of ketchup. The group purchases 10 family-size bottles of this ketchup, weighs the contents of each, and finds that the mean weight is equal to 19.86 ounces, and the standard deviation is equal to .22 ounce.

 a. Do the data provide sufficient evidence for the consumer group to conclude that the mean fill per family-size bottle is less than 20 ounces? Test using $\alpha = .05$.

 b. If the test in part **a** were conducted on a periodic basis by the company's quality control department, is the consumer group more concerned about making a Type I error or a Type II error? (The probability of making this type of error is called the *consumer's risk*.)

 c. The ketchup company is also interested in the mean amount of ketchup per bottle. It does not wish to overfill them. For the test conducted in part **a**, which type of error is more serious from the company's point of view—a Type I error or a Type II error? (The probability of making this type of error is called the *producer's risk*.)

8.139 Inbreeding of tropical wasps. Refer to the *Science* (November 1988) study of inbreeding in tropical swarm-founding wasps, Exercise 7.97 (p. 361). A sample of 197 wasps, captured, frozen, and subjected to a series of genetic tests, yielded a sample mean inbreeding coefficient of $\bar{x} = .044$ with a standard deviation of $s = .884$. Recall that if the wasp has no tendency to inbreed, the true mean inbreeding coefficient μ for the species will equal 0.

 a. Test the hypothesis that the true mean inbreeding coefficient μ for this species of wasp exceeds 0. Use $\alpha = .05$.

 b. Compare the inference, part **a**, to the inference obtained in Exercise 7.97 using a confidence interval. Do the inferences agree? Explain.

*8.140 Weights of parrotfish. A marine biologist wishes to use parrotfish for experimental purposes due to the belief that their weight is fairly stable (i.e., the variability in weights among parrotfish is small). The biologist randomly samples 10 parrotfish and finds that their mean weight is 4.3 pounds and the standard deviation is 1.4 pounds. The biologist will use the parrotfish only if there is evidence that the variance of their weights is less than 4.

 a. Is there sufficient evidence for the biologist to claim that the variability in weights among parrotfish is small enough to justify their use in the experiment? Test at $\alpha = .05$.

 b. State any assumptions that are needed for the test, part **a**, to be valid.

8.141 PCB in plant discharge. The EPA sets a limit of 5 parts per million (ppm) on PCB (a dangerous substance) in water. A major manufacturing firm producing PCB for electrical insulation discharges small amounts from the plant. The company management, attempting to control the PCB in its discharge, has given instructions to halt production if the mean amount of PCB in the effluent exceeds 3 ppm. A random sample of 50 water specimens produced the following statistics: $\bar{x} = 3.1$ ppm and $s = .5$ ppm.

 a. Do these statistics provide sufficient evidence to halt the production process? Use $\alpha = .01$.

 b. If you were the plant manager, would you want to use a large or a small value for α for the test in part **a**?

*8.142 PCB in plant discharge (cont'd). Refer to Exercise 8.141.

 a. In the context of the problem, define a Type II error.

 b. Calculate β for the test described in part **a** of Exercise 8.141 assuming that the true mean is $\mu = 3.1$ ppm.

 c. What is the power of the test to detect the effluent's departure from the standard of 3.0 ppm when the mean is 3.1 ppm?

 d. Repeat parts **b** and **c** assuming that the true mean is 3.2 ppm. What happens to the power of the test as the plant's mean PCB departs farther from the standard?

*8.143 PCB in plant discharge (cont'd). Refer to Exercises 8.141 and 8.142.

 a. Suppose an α value of .05 is used to conduct the test. Does this change favor the manufacturer? Explain.

 b. Determine the value of β and the power for the test when $\alpha = .05$ and $\mu = 3.1$.

 c. What happens to the power of the test when α is increased?

8.144 Psychological study of obese adolescents. Research reported in the *Journal of Psychology* (March 1991) studied the personality characteristics of obese individuals. One variable, the locus of control (LOC), measures the individual's degree of belief that he or she has control over situations. High scores on the LOC scale indicate less perceived control. For one sample of 19 obese adolescents, the mean LOC score was 10.89 with a standard deviation of 2.48. Suppose we wish to test whether the mean LOC score for all obese adolescents exceeds 10, the average for "normal" individuals.

 a. Specify the null and alternative hypotheses for this test.

 b. Conduct the test, part **a**, and make the proper conclusion.

8.145 Ph.D.s awarded to foreign nationals. The National Science Foundation, in a survey of 2,237 engineering graduate students who earned their Ph.D. degrees, found that 607 were U.S. citizens; the majority (1,630) of the Ph.D. degrees were awarded to foreign nationals. (*Science*, September 24, 1993.) Conduct a test to determine whether the true percentage of engineering Ph.D. degrees awarded to foreign nationals exceeds 50%. Use $\alpha = .01$.

*8.146 **Interocular eye pressure.** Ophthalmologists require an instrument that can rapidly measure interocular pressure for glaucoma patients. The device now in general use is known to yield readings of this pressure with a variance of 10.3. The variance of five pressure readings on the same eye by a newly developed instrument is equal to 9.8. Does this sample variance provide sufficient evidence to indicate that the new instrument is more reliable than the instrument currently in use? (Use $\alpha = .05$.)

Applying the Concepts—Advanced

8.147 **Polygraph test error rates.** A group of physicians subjected the *polygraph* (or *lie detector*) to the same careful testing given to medical diagnostic tests. They found that if 1,000 people were subjected to the polygraph and 500 told the truth and 500 lied, the polygraph would indicate that approximately 185 of the truth-tellers were liars and that approximately 120 of the liars were truth-tellers (*Discover*, 1986).

   a. In the application of a polygraph test, an individual is presumed to be a truth-teller (H_0) until "proven" a liar (H_a). In this context, what is a Type I error? A Type II error?

   b. According to the study, what is the probability (approximately) that a polygraph test will result in a Type I error? A Type II error?

8.148 **NCAA March Madness.** For three weeks each March, the National Collegiate Athletic Association (NCAA) holds its annual men's basketball championship tournament. The 64 best college basketball teams in the nation play a single-elimination tournament—a total of 63 games—to determine the NCAA champion. Tournament followers, from hardcore gamblers to the casual fan who enters the office betting pool, have a strong interest in handicapping the games. To provide insight into this phenomenon, statisticians Hal Stern and Barbara Mock analyzed data from 13 previous NCAA tournaments and published their results in *Chance* (Winter 1998). The re-

sults of first-round games are summarized in the table at the bottom of the page.

   a. A common perception among fans, media, and gamblers is that the higher seeded team has a better than 50-50 chance of winning a first-round game. Is there evidence to support this perception? Conduct the appropriate test for each matchup. What trends do you observe?

   b. Is there evidence to support the claim that a 1-, 2-, 3-, or 4-seeded team will win by an average of more than 10 points in first-round games? Conduct the appropriate test for each matchup.

   c. Is there evidence to support the claim that a 5-, 6-, 7-, or 8-seeded team will win by an average of less than 5 points in first-round games? Conduct the appropriate test for each matchup.

* **d.** For each matchup, test the null hypothesis that the standard deviation of the victory margin is 11 points.

   e. The researchers also calculated the difference between the game outcome (victory margin, in points) and point spread established by Las Vegas oddsmakers for a sample of 360 recent NCAA tournament games. The mean difference is .7 and the standard deviation of the difference is 11.3. If the true mean difference is 0, then the point spread can be considered a good predictor of the game outcome. Use this sample information to test the hypothesis that the point spread, on average, is a good predictor of the victory margin in NCAA tournament games.

8.149 **Homes sold in California.** In 1999, the average size of single-family homes built in the U.S. was 2,230 square feet, an increase of over 200 square feet from a decade earlier. (*Wall Street Journal Interactive Edition*, January 7, 2000.) A random sample of 100 new homes sold in California in late 1999 yielded the following size information: $\bar{x} = 2,347$ square feet and $s = 257$ square feet.

   a. Assume the average size of U.S. homes is known with certainty. Do the sample data provide sufficient

Summary of First-Round NCAA Tournament Games, 1985–1997

Matchup (Seeds)	Number of Games	Number Won by Favorite (Higher Seed)	Margin of Victory (Points)	
			Mean	Standard Deviation
1 vs 16	52	52	22.9	12.4
2 vs 15	52	49	17.2	11.4
3 vs 14	52	41	10.6	12.0
4 vs 13	52	42	10.0	12.5
5 vs 12	52	37	5.3	10.4
6 vs 11	52	36	4.3	10.7
7 vs 10	52	35	3.2	10.5
8 vs 9	52	22	−2.1	11.0

Source: Stern, H. S., and Mock, B. "College basketball upsets: Will a 16-seed ever beat a 1-seed?" *Chance,* Vol. 11, No. 1, Winter 1998, p. 29 (Table 3).

evidence to conclude that the mean size of all California homes built in late 1999 exceeds the national average? Test using $\alpha = .01$.

b. What is the probability that the test will fail to reject the null hypothesis when in fact $\mu = 2,330$ square feet? What is the name given to this type of error?

c. What is the probability that the test will reject the null hypothesis when in fact $\mu = 2,330$ square feet? What is the name given to this type of error?

***d.** Suppose the actual mean size of new California homes was $\mu = 2,330$ square feet. What is the power of the test in part **a** to detect this 100-square-feet difference?

***e.** If the California mean were actually $\mu = 2,280$ square feet, what is the power of the test in part **a** to detect this 50-square-feet difference?

Critical Thinking Challenges

8.150 The Hot Tamale caper. "Hot Tamales" are chewy, cinnamon-flavored candies. A bulk vending machine is known to dispense, on average, 15 Hot Tamales per bag. *Chance* (Fall 2000) published an article on a classroom project in which students were required to purchase bags of Hot Tamales from the machine and count the number of candies per bag. One student group claimed they purchased five bags that had the following candy counts: 25, 23, 21, 21, and 20. There was some question as to whether the students had fabricated the data. Use a hypothesis test to gain insight into whether or not the data collected by the students are fabricated. Use a level of significance that gives the benefit of the doubt to the students.

8.151 Verifying voter petitions. To get their names on the ballot of a local election, political candidates often must obtain petitions bearing the signatures of a minimum number of registered voters. In Pinellas County, Florida, a certain political candidate obtained petitions with 18,200 signatures. (*St. Petersburg Times*, April 7, 1992.) To verify that the names on the petitions were signed by actual registered voters, election officials randomly sampled 100 of the names and checked each for authenticity. Only two were invalid signatures.

a. Is 98 out of 100 verified signatures sufficient to believe that more than 17,000 of the total 18,200 signatures are valid?

b. Repeat part **a** if only 16,000 valid signatures are required.

Student Projects

The "efficient market" theory postulates that the best predictor of a stock's price at some point in the future is the current price of the stock (with some adjustments for inflation and transaction costs, which we shall assume to be negligible for the purpose of this exercise). To test this theory, select a random sample of 25 stocks on the New York Stock Exchange and record the closing prices on the last days of two recent consecutive months. Calculate the increase or decrease in the stock price over the 1-month period.

a. Define μ as the mean change in price of all stocks over a 1-month period. Set up the appropriate null and alternative hypotheses in terms of μ.

b. Use the sample of the 25 stock price differences to conduct the test of hypothesis established in part **a**. Use $\alpha = .05$.

c. Tabulate the number of rejections and nonrejections of the null hypothesis in your class. If the null hypothesis were true, how many rejections of the null hypothesis would you expect among those in your class? How does this expectation compare with the actual number of nonrejections? What does the result of this exercise indicate about the efficient market theory?

REFERENCES

Snedecor, G. W., and Cochran, W. G. *Statistical Methods*, 7th ed. Ames: Iowa State University Press, 1980.

Wackerly, D., Mendenhall, W., and Scheaffer, R. *Mathematical Statistics with Applications*, 6th ed. North Scituate, Mass.: Duxbury, 2002.

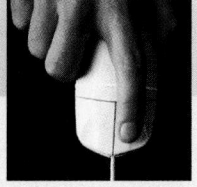

Using Technology

Tests of Hypothesis
Using MINITAB

MINITAB can be used to obtain one-sample tests for both a population mean and a population proportion but cannot currently produce a test for a population variance. To generate a hypothesis test for the mean using a previously created sample data set, first access the MINITAB data worksheet. Next, click on the "Stat" button on the MINITAB menu bar, then click on "Basic Statistics" and "1-Sample t," as shown in Figure 8.M.1. The resulting dialog box appears as shown in Figure 8.M.2. Click on "Samples in Columns," then specify the quantitative variable of interest in the open box. Specify the value of μ_0 for the null hypothesis in the "Test mean" box. Now, click on the "Options" button at the bottom of the dialog box and specify the form of the alternative hypothesis as shown in Figure 8.M.3. Click "OK" to return to the "1-Sample t" dialog box, then click "OK" again to produce the hypothesis test.

Figure 8.M.1
MINITAB Menu Options for a Test about μ

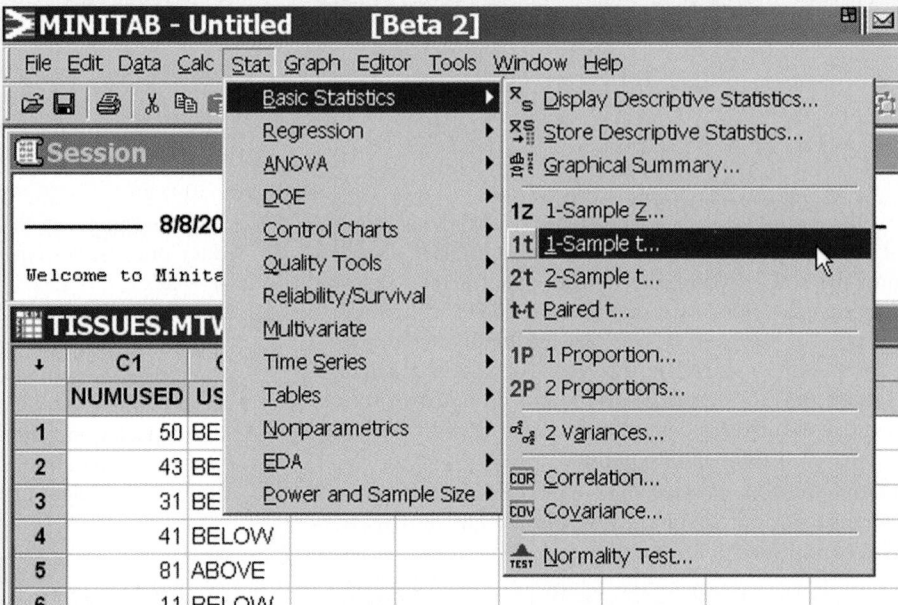

Figure 8.M.2
MINITAB 1-Sample t
Dialog Box

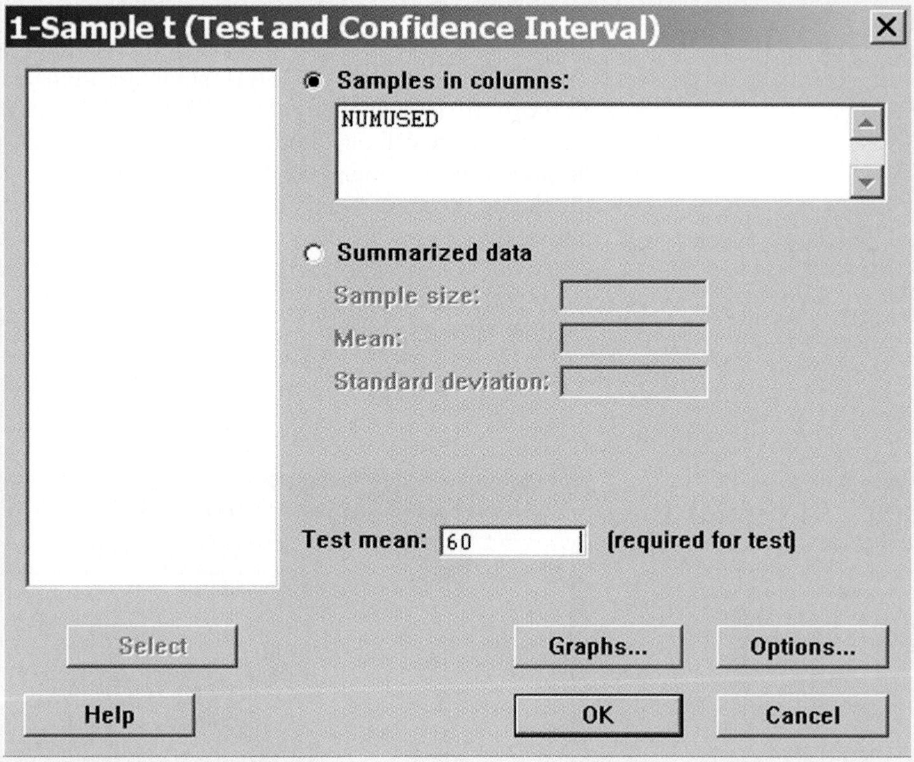

If you want to produce a test for the mean from summary information (e.g., the sample mean, sample standard deviation, and sample size), then click on "Summarized data" in the "1-Sample t" dialog box, enter the values of the summary statistics and μ_0, then click "OK."

[*Important Note*: The MINITAB one-sample *t*-procedure uses the *t*-statistic to generate the hypothesis test. When the sample size *n* is small, this is the appropriate method. When the sample size *n* is large, the *t*-value will be approximately equal to the large-sample *z*-value and the resulting test will still be valid. If you have a large sample and you know the value of the population standard deviation σ (which is rarely the case), then select "1-sample Z" from the "Basic Statistics" menu options (see Figure 8.M.1) and make the appropriate selections.]

Figure 8.M.3
MINITAB 1-Sample t
Options Dialog Box

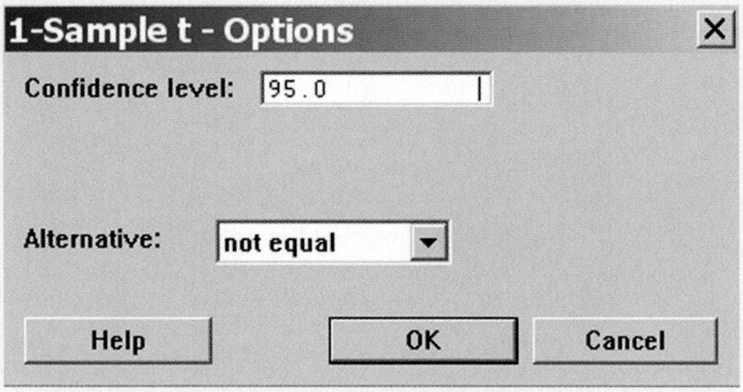

To generate a test for a population proportion using a previously created sample data set, first access the MINITAB data worksheet. Next, click on the "Stat" button on the MINITAB menu bar, then click on "Basic Statistics" and "1 Proportion" (see Figure 8.M.1). The resulting dialog box appears as shown in Figure 8.M.4. Click on "Samples in Columns," then specify the qualitative variable of interest in the open box. Now, click on the "Options" button at the bottom of the dialog box and specify the null hypothesis value p_0 and the form of the alternative hypothesis in the resulting dialog box, as shown in Figure 8.M.5. Also, check the "Use test and interval based on normal distribution" box at the bottom. Click "OK" to return to the "1-Proportion" dialog box, then click "OK" again to produce the test results.

Figure 8.M.4
MINITAB 1-Proportion
Dialog Box

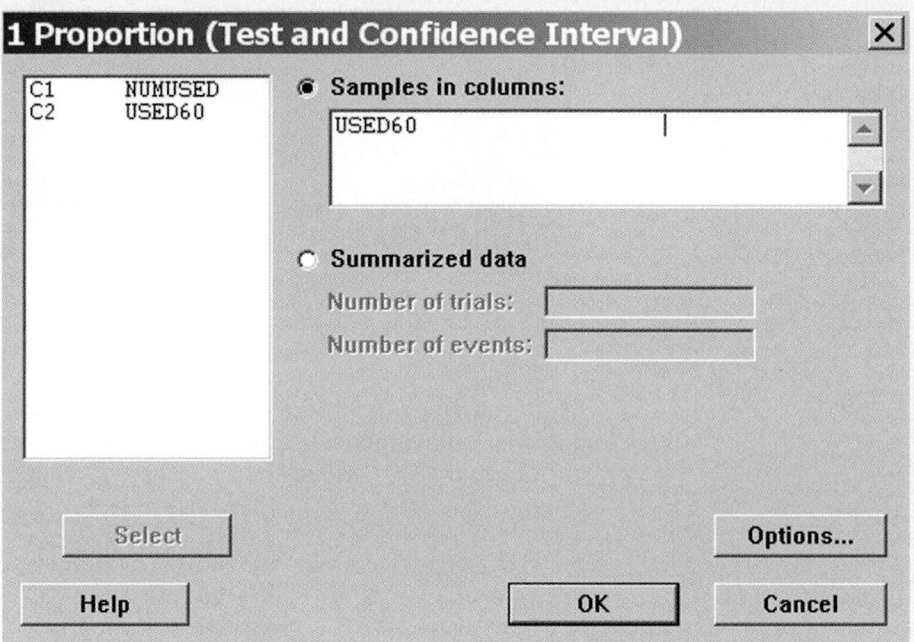

Figure 8.M.5
MINITAB 1-Proportion
Options

If you want to produce a confidence interval for a proportion from summary information (e.g., the number of successes and the sample size), then click on "Summarized data" in the "1-Proportion" dialog box (see Figure 8.M.4). Enter the value for the number of trials (i.e., the sample size) and the number of events (i.e., the number of successes), then click "OK."

9 Inferences Based on Two Samples

Confidence Intervals and Tests of Hypotheses

Contents

Statistics in Action

Detection of Rigged Milk Prices

Using Technology

Two-Sample Inferences Using MINITAB

Where We've Been

- Explored two methods for making statistical inferences: *confidence intervals* and *tests of hypotheses*

- Studied confidence intervals and tests for a single population mean μ, a single population proportion p, and a single population variance σ^2

- Learned how to select the sample size necessary to estimate a population parameter with a specified margin of error

Where We're Going

- Learn how to compare two populations using confidence intervals and tests of hypotheses.

- Apply these inferential methods to problems where we want to compare two population means, two population proportions, or two population variances.

- Determine the sizes of the samples necessary to estimate the difference between two population parameters with a specified margin of error.

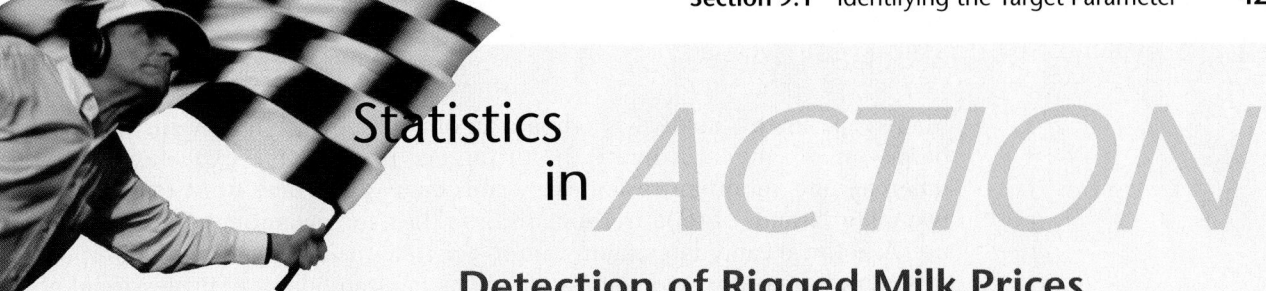

Statistics in ACTION

Detection of Rigged Milk Prices

Many products and services are purchased by governments, cities, states, and businesses on the basis of sealed bids, and contracts are awarded to the lowest bidders. This process works extremely well in competitive markets, but it has the potential to increase the cost of purchasing if the markets are noncompetitive or if bidders are engaged in collusive practices. An investigation that began with a statistical analysis of bids in the Florida school milk market in the mid-1980s led to the recovery of more than $33 million from dairies who had conspired to rig the bids there. The investigation spread quickly to other states, and settlements and fines from dairies exceed $100 million for school milk bid-rigging in twenty other states. This case concerns a school milk bid-rigging investigation in Kentucky.

Each year, the state of Kentucky invites bids from dairies to supply half-pint containers of fluid milk products for its school districts. The products include whole white milk, low-fat white milk, and low-fat chocolate milk. In several school districts in northern Kentucky, the suppliers (dairies) were accused of "price fixing"—that is, conspiring to allocate the districts so that the "winner" was predetermined and the price per pint was set above the fair, or competitive, price. A market analysis disclosed that two dairies—Meyer Dairy and Trauth Dairy—were the only two bidders on the milk contracts in the school districts in the tricounty market of Boone, Campbell, and Kenton counties between 1983 and 1991. Consequently, these two companies were awarded all the milk contracts in the market. (In contrast, a large number of different dairies won the milk contracts for school districts in the remainder of the northern Kentucky market—called the surrounding market.) Speculation is that Meyer and Trauth conspired to rig their bids in the tricounty market.

The Commonwealth of Kentucky maintains a database on all bids received from the dairies competing for the milk contracts. With permission, some of these data have been made available for analysis to determine whether there is empirical evidence of bid collusion in the tricounty market. The data, saved in the **MILK** file, are described in detail in Table SIA9.1.

In this chapter, several Statistics in Action Revisited examples demonstrate how we can use the methodology of this chapter to investigate the possibility of bid-rigging in the northern Kentucky milk market.

Statistics in the Action Revisited

- Comparing the Mean Winning Bid Prices of Two Kentucky Milk Markets (p. 446)
- Comparing the Incumbency Rates of Two Kentucky Milk Markets (p. 476)
- Comparing the Dispersion of Bid Prices in Two Kentucky Milk Markets (p. 490)

⊙ MILK

TABLE SIA9.1   Description of Data Saved in the MILK file

Number of observations: 392

Variable	Type	Description
YEAR	QN	Year in which milk contract awarded
MARKET	QL	Northern Kentucky market (TRI-COUNTY or SURROUND)
WINNER	QL	Name of winning dairy
WWBID	QN	Winning bid price of whole white milk (dollars per half-pint)
LFWBID	QN	Winning bid price of low-fat white milk (dollars per half-pint)
LFCBID	QN	Winning bid price of low-fat chocolate milk (dollars per half-pint)
DISTRICT	QL	School district number

9.1 Identifying the Target Parameter

Many experiments involve a comparison of two populations. For instance, a sociologist may want to estimate the difference in mean life expectancy between innercity and suburban residents. A consumer group may want to test whether two major brands of food freezers differ in the average amount of electricity they use. A political candidate might want to estimate the difference in the proportions of voters in two districts who favor his or her candidacy. A professional golfer might be interested in comparing the variability in the distance that two competing brands of golf balls travel when struck with the same club. In this chapter, we consider techniques for using two samples to compare the populations from which they were selected.

The same procedures that are used to estimate and test hypotheses about a single population can be modified to make inferences about two populations. As in Chapters 7 and 8, the methodology used will depend on the sizes of the samples and the parameter of interest (i.e., the *target parameter*). Some key words and the type of data associated with the parameters covered in this chapter are listed in the box.

Determining the Target Parameter

Parameter	Key Words or Phrases	Type of Data
$\mu_1 - \mu_2$	Mean difference; difference in averages	Quantitative
$p_1 - p_2$	Difference between proportions, percentages, fractions, or rates; compare proportions	Qualitative
$(\sigma_1)^2/(\sigma_2)^2$	Ratio of variances; difference in variability or spread; compare variation	Quantitative

You can see that the key words *difference* and *compare* help identify the fact that two populations are to be compared. For the examples given above, the words *mean* in *mean life expectancy* and *average* in *average amount of electricity* imply that the target parameter is the difference in population means, $\mu_1 - \mu_2$. The word *proportions* in *proportions of voters in two districts* indicates that the target parameter is the difference in proportions, $p_1 - p_2$. Finally, the key word *variability* in *variability in the distance* identifies the ratio of population variances, $(\sigma_1)^2/(\sigma_2)^2$, as the target parameter.

As with inferences about a single population, the type of data (quantitative or qualitative) collected on the two samples is also indicative of the target parameter. With quantitative data, you are likely to be interested in comparing the means or variances of the data. With qualitative data with two outcomes (success or failure), a comparison of the proportions of successes is likely to be of interest.

We consider methods for comparing two population means in Sections 9.2 and 9.3. A comparison of population proportions is presented in Section 9.4 and population variances in optional Section 9.6. We show how to determine the sample sizes necessary for reliable estimates of the target parameters in Section 9.5.

9.2 Comparing Two Population Means: Independent Sampling

In this section we develop both large-sample and small-sample methodologies for comparing two population means. In the large-sample case we use the z-statistic, while in the small-sample case we use the t-statistic.

Large Samples

EXAMPLE 9.1   A LARGE-SAMPLE CONFIDENCE INTERVAL FOR $\mu_1 - \mu_2$

Problem   A dietitian has developed a diet that is low in fats, carbohydrates, and cholesterol. Although the diet was initially intended to be used by people with heart disease, the dietitian wishes to examine the effect this diet has on the weights of obese people. Two random samples of 100 obese people each are selected, and one group of 100 is placed on the low-fat diet. The other 100 are placed on a diet that contains approximately the same quantity of food but is not as low in fats, carbohydrates, and cholesterol. For each person, the amount of weight lost (or gained) in a 3-week period is recorded. The data, saved in the **DIETSTUDY** file, are listed in Table 9.1. Form a 95% confidence interval for the difference between the population mean weight losses for the two diets. Interpret the result.

Solution   Recall that the general form of a large-sample confidence interval for a single mean μ is $\bar{x} \pm z_{\alpha/2}\sigma_{\bar{x}}$. That is, we add and subtract $z_{\alpha/2}$ standard deviations of the sample estimate, $\bar{x}$, to the value of the estimate. We employ a similar procedure to form the confidence interval for the difference between two population means.

Let μ_1 represent the mean of the conceptual population of weight losses for all obese people who could be placed on the low-fat diet. Let μ_2 be similarly defined for the other diet. We wish to form a confidence interval for $(\mu_1 - \mu_2)$. An intuitively appealing estimator for $(\mu_1 - \mu_2)$ is the difference between the sample means, $(\bar{x}_1 - \bar{x}_2)$. Thus, we will form the confidence interval of interest by

$$(\bar{x}_1 - \bar{x}_2) \pm z_{\alpha/2}\sigma_{(\bar{x}_1 - \bar{x}_2)}$$

◉ **DIETSTUDY**

TABLE 9.1   Diet Study Data, Example 9.1

Weight Losses for Lowfat Diet									
8	10	10	12	9	3	11	7	9	2
21	8	9	2	2	20	14	11	15	6
13	8	10	12	1	7	10	13	14	4
8	12	8	10	11	19	0	9	10	4
11	7	14	12	11	12	4	12	9	2
4	3	3	5	9	9	4	3	5	12
3	12	7	13	11	11	13	12	18	9
6	14	14	18	10	11	7	9	7	2
16	16	11	11	3	15	9	5	2	6
5	11	14	11	6	9	4	17	20	10

Weight Losses for Regular Diet									
6	6	5	5	2	6	10	3	9	11
14	4	10	13	3	8	8	13	9	3
4	12	6	11	12	9	8	5	8	7
6	2	6	8	5	7	16	18	6	8
13	1	9	8	12	10	6	1	0	13
11	2	8	16	14	4	6	5	12	9
11	6	3	9	9	14	2	10	4	13
8	1	1	4	9	4	1	1	5	6
14	0	7	12	9	5	9	12	7	9
8	9	8	10	5	8	0	3	4	8

Figure 9.1

SPSS Analysis of Diet Study Data

Group Statistics

	DIET	N	Mean	Std. Deviation	Std. Error Mean
WTLOSS	LOWFAT	100	9.31	4.668	.467
	REGULAR	100	7.40	4.035	.404

Independent Samples Test

		Levene's Test for Equality of Variances		t-test for Equality of Means						
									95% Confidence Interval of the Difference	
		F	Sig.	t	df	Sig. (2-tailed)	Mean Difference	Std. Error Difference	Lower	Upper
WTLOSS	Equal variances assumed	1.367	.244	3.095	198	.002	1.910	.617	.693	3.127
	Equal variances not assumed			3.095	193.940	.002	1.910	.617	.693	3.127

Assuming the two samples are independent, the standard deviation of the difference between the sample means is

$$\sigma_{(\bar{x}_1 - \bar{x}_2)} = \sqrt{\frac{\sigma_1^2}{n_1} + \frac{\sigma_2^2}{n_2}} \approx \sqrt{\frac{s_1^2}{n_1} + \frac{s_2^2}{n_2}}$$

Summary statistics for the diet data are displayed at the top of the SPSS printout, Figure 9.1. Note that $\bar{x}_1 = 9.31$, $\bar{x}_2 = 7.40$, $s_1 = 4.67$, and $s_2 = 4.04$. Using these values and noting that $\alpha = .05$ and $z_{.025} = 1.96$, we find that the 95% confidence interval is, approximately,

$$(9.31 - 7.40) \pm 1.96\sqrt{\frac{(4.67)^2}{100} + \frac{(4.04)^2}{100}} = 1.91 \pm (1.96)(.62) = 1.91 \pm 1.22$$

or $(.69, 3.13)$. This interval (rounded) is also given at the bottom of Figure 9.1.

Using this estimation procedure over and over again for different samples, we know that approximately 95% of the confidence intervals formed in this manner will enclose the difference in population means $(\mu_1 - \mu_2)$. Therefore, we are highly confident that the mean weight loss for the low-fat diet is between .69 and 3.13 pounds more than the mean weight loss for the other diet. With this information, the dietitian better understands the potential of the low-fat diet as a weight-reducing diet.

Look Back If the confidence interval for $(\mu_1 - \mu_2)$ contains 0 [e.g., $(-2.5, 1.3)$], then it is possible for the difference between the population means to be 0 (i.e., $\mu_1 - \mu_2 = 0$). In this case we could not conclude that a significant difference exits between the mean weight losses for the two diets.

Now Work *Exercise 9.6a*

■ ■ ■

Figure 9.2

Sampling Distribution of $(\bar{x}_1 - \bar{x}_2)$

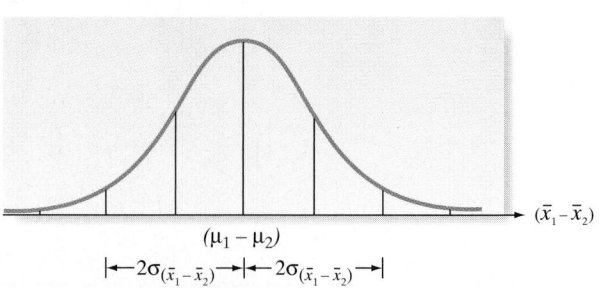

The justification for the procedure used in Example 9.1 to estimate $(\mu_1 - \mu_2)$ relies on the properties of the sampling distribution of $(\bar{x}_1 - \bar{x}_2)$. The performance of the estimator in repeated sampling is pictured in Figure 9.2, and its properties are summarized in the next box.

Properties of the Sampling Distribution of $(\bar{x}_1 - \bar{x}_2)$

1. The mean of the sampling distribution of $(\bar{x}_1 - \bar{x}_2)$ is $(\mu_1 - \mu_2)$.

2. If the two samples are independent, the standard deviation of the sampling distribution is

$$\sigma_{(\bar{x}_1 - \bar{x}_2)} = \sqrt{\frac{\sigma_1^2}{n_1} + \frac{\sigma_2^2}{n_2}}$$

where σ_1^2 and σ_2^2 are the variances of the two populations being sampled and n_1 and n_2 are the respective sample sizes. We also refer to $\sigma_{(\bar{x}_1 - \bar{x}_2)}$ as the **standard error** of the statistic $(\bar{x}_1 - \bar{x}_2)$.

3. The sampling distribution of $(\bar{x}_1 - \bar{x}_2)$ is approximately normal for *large samples* by the Central Limit Theorem.

In Example 9.1, we noted the similarity in the procedures for forming a large-sample confidence interval for one population mean and a large-sample confidence interval for the difference between two population means. When we are testing hypotheses, the procedures are again very similar. The general large-sample procedures for forming confidence intervals and testing hypotheses about $(\mu_1 - \mu_2)$ are summarized in the following boxes.

Large Sample Confidence Interval for $(\mu_1 - \mu_2)$

$$(\bar{x}_1 - \bar{x}_2) \pm z_{\alpha/2}\sigma_{(\bar{x}_1 - \bar{x}_2)} = (\bar{x}_1 - \bar{x}_2) \pm z_{\alpha/2}\sqrt{\frac{\sigma_1^2}{n_1} + \frac{\sigma_2^2}{n_2}}$$

$$\approx (\bar{x}_1 - \bar{x}_2) \pm z_{\alpha/2}\sqrt{\frac{s_1^2}{n_1} + \frac{s_2^2}{n_2}}$$

Large-Sample Test of Hypothesis for $(\mu_1 - \mu_2)$

One-Tailed Test	**Two-Tailed Test**
$H_0: (\mu_1 - \mu_2) = D_0$	$H_0: (\mu_1 - \mu_2) = D_0$
$H_a: (\mu_1 - \mu_2) < D_0$	$H_a: (\mu_1 - \mu_2) \neq D_0$
[or $H_a: (\mu_1 - \mu_2) > D_0$]	

where D_0 = Hypothesized difference between the means (this difference is often hypothesized to be equal to 0)

Test statistic:

$$z = \frac{(\bar{x}_1 - \bar{x}_2) - D_0}{\sigma_{(\bar{x}_1 - \bar{x}_2)}} \quad \text{where} \quad \sigma_{(\bar{x}_1 - \bar{x}_2)} = \sqrt{\frac{\sigma_1^2}{n_1} + \frac{\sigma_2^2}{n_2}} \approx \sqrt{\frac{s_1^2}{n_1} + \frac{s_2^2}{n_2}}$$

Rejection region: $z < -z_\alpha$ *Rejection region:* $|z| > z_{\alpha/2}$
  [or $z > z_\alpha$ when
  $H_a: (\mu_1 - \mu_2) > D_0$]

> **Conditions Required for Valid Large-Sample Inferences about $\mu_1 - \mu_2$**
>
> 1. The two samples are randomly selected in an independent manner from the two target populations.
> 2. The sample sizes, n_1 and n_2, are both large (i.e., $n_1 \geq 30$ and $n_2 \geq 30$). (Due to the Central Limit Theorem, this condition guarantees that the sampling distribution of $(\bar{x}_1 - \bar{x}_2)$ will be approximately normal regardless of the shapes of the underlying probability distributions of the populations. Also, s_1^2 and s_2^2 will provide good approximations to σ_1^2 and σ_2^2 when both the samples are large.)

EXAMPLE 9.2

A LARGE SAMPLE TEST FOR $\mu_1 - \mu_2$

Problem Refer to the study of obese people on a low-fat diet and a regular diet, Example 9.1. Another way to compare the mean weight losses for the two different diets is to conduct a test of hypothesis. Use the information on the SPSS printout, Figure 9.1, to conduct the test. Use $\alpha = .05$.

Solution Again, we let μ_1 and μ_2 represent the population mean weight losses of obese people on the low-fat diet and regular diet, respectively. If one diet is more effective in reducing the weights of obese people, then either $\mu_1 < \mu_2$ or $\mu_2 < \mu_1$; that is, $\mu_1 \neq \mu_2$.

Thus, the elements of the test are as follows:

H_0: $(\mu_1 - \mu_2) = 0$ (i.e., $\mu_1 = \mu_2$; note that $D_0 = 0$ for this hypothesis test)
H_a: $(\mu_1 - \mu_2) \neq 0$ (i.e., $\mu_1 \neq \mu_2$)

Test statistic: $z = \dfrac{(\bar{x}_1 - \bar{x}_2) - D_0}{\sigma_{(\bar{x}_1 - \bar{x}_2)}} = \dfrac{\bar{x}_1 - \bar{x}_2 - 0}{\sigma_{(\bar{x}_1 - \bar{x}_2)}}$

Rejection region: $z < -z_{\alpha/2} = -1.96$ or $z > z_{\alpha/2} = 1.96$ (see Figure 9.3)

Substituting the summary statistics given in Figure 9.1 into the test statistic, we obtain

$$z = \frac{(\bar{x}_1 - \bar{x}_2) - 0}{\sigma_{(\bar{x}_1 - \bar{x}_2)}} = \frac{9.31 - 7.40}{\sqrt{\dfrac{\sigma_1^2}{n_1} + \dfrac{\sigma_2^2}{n_2}}}$$

$$\approx \frac{1.91}{\sqrt{\dfrac{s_1^2}{n_1} + \dfrac{s_2^2}{n_2}}} = \frac{1.91}{\sqrt{\dfrac{(4.67)^2}{100} + \dfrac{(4.04)^2}{100}}} = \frac{1.91}{.617} = 3.09$$

[*Note:* The value of the test statistic is also shown (highlighted) at the bottom of the SPSS printout, Figure 9.1.]

As you can see in Figure 9.3, the calculated z-value clearly falls in the rejection region. Therefore, the samples provide sufficient evidence, at $\alpha = .05$, for the dietitian to conclude that the mean weight losses for the two diets differ.

Figure 9.3
Rejection Region for Example 9.2

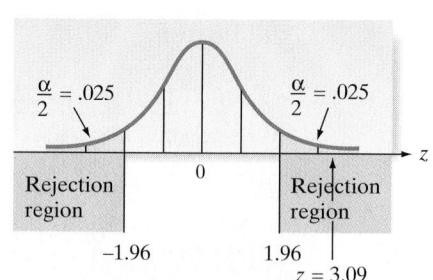

Look Back This conclusion agrees with the inference drawn from the 95% confidence interval in Example 9.1. However, the confidence interval provides more information on the mean weight losses. From the hypothesis test, we know only that the two means differ; that is, $\mu_1 \neq \mu_2$. From the confidence interval in Example 9.1, we found that the mean weight loss μ_1 of the low-fat diet was between .69 and 3.13 pounds more than the mean weight loss μ_2 of the regular diet. In other words, the test tells us that the means differ, but the confidence interval tells us how large the difference is. Both inferences are made with the same degree of reliability—namely, with 95% confidence (or, at $\alpha = .05$).

■ ■ ■

EXAMPLE 9.3

THE p-VALUE FOR A TEST OF $\mu_1 - \mu_2$

Problem Find the observed significance level for the test in Example 9.2. Interpret the result.

Solution The alternative hypothesis in Example 9.2, $H_a: \mu_1 - \mu_2 \neq 0$, required a two-tailed test using

$$z = \frac{\overline{x}_1 - \overline{x}_2}{\sigma_{(\overline{x}_1 - \overline{x}_2)}}$$

as a test statistic. Since the z-value calculated from the sample data was 3.09, the observed significance level (p-value) for the two-tailed test is the probability of observing a value of z at least as contradictory to the null hypothesis as $z = 3.09$; that is,

$$p\text{-value} = 2 \cdot P(z \geq 3.09)$$

This probability is computed assuming H_0 is true and is equal to the highlighted area shown in Figure 9.4.

The tabulated area corresponding to $z = 3.09$ in Table IV of Appendix A is .4990. Therefore,

$$P(z \geq 3.09) = .5 - .4990 = .0010$$

and the observed significance level for the test is

$$p\text{-value} = 2(.001) = .002$$

Since our selected α value, .05, exceeds this p-value, we have sufficient evidence to reject $H_0: \mu_1 - \mu_2 = 0$.

Figure 9.4
The Observed Significance
Level for Example 9.2

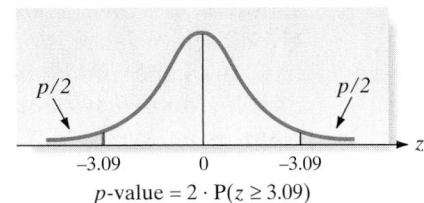

$p/2$ $p/2$

-3.09 0 -3.09
$p\text{-value} = 2 \cdot P(z \geq 3.09)$

Look Back The p-value of the test is more easily obtained from a statistical software package. The p-value is highlighted at the bottom of the SPSS printout, Figure 9.1. This value agrees with our calculated p-value.

Now Work *Exercise 9.6b*

■ ■ ■

Small Samples

When comparing two population means with small samples (say, $n_1 < 30$ and $n_2 < 30$), the methodology of the previous three examples is invalid. The reason? When the sample sizes are small, estimates of σ_1^2 and σ_2^2 are unreliable and the Central Limit Theorem (which guarantees that the z statistic is normal) can no longer be applied. But as in the case of a single mean (Section 8.4), we use the familiar Student's t-distribution described in Chapter 7.

To use the t-distribution, both sampled populations must be approximately normally distributed with equal population variances, and the random samples must be selected independently of each other. The normality and equal variances assumptions imply relative frequency distributions for the populations that would appear as shown in Figure 9.5.

Since we assume the two populations have equal variances ($\sigma_1^2 = \sigma_2^2 = \sigma^2$), it is reasonable to use the information contained in both samples to construct a **pooled sample estimator σ^2** for use in confidence intervals and test statistics. Thus, if s_1^2 and s_2^2 are the two sample variances (both estimating the variance σ^2 common to both populations), the pooled estimator of σ^2, denoted as s_p^2, is

$$s_p^2 = \frac{(n_1 - 1)s_1^2 + (n_2 - 1)s_2^2}{(n_1 - 1) + (n_2 - 1)} = \frac{(n_1 - 1)s_1^2 + (n_2 - 1)s_2^2}{n_1 + n_2 - 2}$$

or

$$s_p^2 = \frac{\overbrace{\sum(x_1 - \bar{x}_1)^2}^{\text{From sample 1}} + \overbrace{\sum(x_2 - \bar{x}_2)^2}^{\text{From sample 2}}}{n_1 + n_2 - 2}$$

where x_1 represents a measurement from sample 1 and x_2 represents a measurement from sample 2. Recall that the term *degrees of freedom* was defined in Section 7.2 as 1 less than the sample size. Thus, in this case, we have $(n_1 - 1)$ degrees of freedom for sample 1 and $(n_2 - 1)$ degrees of freedom for sample 2. Since we are pooling the information on σ^2 obtained from both samples, the degrees of freedom associated with the pooled variance s_p^2 is equal to the sum of the degrees of freedom for the two samples, namely, the denominator of s_p^2; that is, $(n_1 - 1) + (n_2 - 1) = n_1 + n_2 - 2$.

Note that the second formula given for s_p^2 shows that the pooled variance is simply a *weighted average* of the two sample variances s_1^2 and s_2^2. The weight given each variance is proportional to its degrees of freedom. If the two variances have the same number of degrees of freedom (i.e., *if the sample sizes are equal*), *then the pooled variance is a simple average of the two sample variances*. The result is an average or "pooled" variance that is a better estimate of σ^2 than either s_1^2 or s_2^2 alone.

Figure 9.5

Assumptions for the Two-Sample t: (1) Normal Populations, (2) Equal Variances

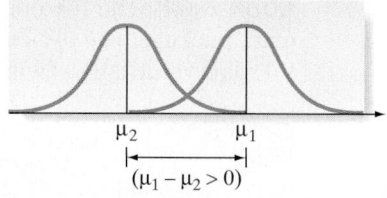

Biography

Bradley Efron (1938–present)—The Bootstrap Method

Bradley Efron was raised in St. Paul, Minnesota, the son of a truck driver who was the amateur statistician for his bowling and baseball leagues. Efron received a B.S. in mathematics from California Institute of Technology in 1960, but, by his own admission, had no talent for modern abstract math. His interest in the science of statistics developed after reading a book, cover-to-cover, by Harold Cramer. Efron went to the University of Stanford to study statistics and earned his Ph.D there in 1964. He has been a faculty member in Stanford's Department of Statistics since 1966. Over his career, Efron has received numerous awards and prizes for his contributions to modern statistics, including the MacArthur Prize Fellow (1983), the American Statistical Association Wilks Medal (1990), and the Parzen Prize for Statistical Innovation (1998). In 1979, Efron invented a method—called the *bootstrap*—of estimating and testing population parameters in situations where either the sampling distribution is unknown or the assumptions are violated. The method involves repeatedly taking samples of size n (with replacement) from the original sample and calculating the value of the point estimate. Efron showed that the sampling distribution of the estimator is simply the frequency distribution of the bootstrap estimates.

Both the confidence interval and the test of hypothesis procedures for comparing two population means with small samples are summarized in the accompanying boxes.

Small-Sample Confidence Interval for $(\mu_1 - \mu_2)$: Independent Samples

$$(\bar{x}_1 - \bar{x}_2) \pm t_{\alpha/2}\sqrt{s_p^2\left(\frac{1}{n_1} + \frac{1}{n_2}\right)}$$

where $s_p^2 = \dfrac{(n_1 - 1)s_1^2 + (n_2 - 1)s_2^2}{n_1 + n_2 - 2}$

and $t_{\alpha/2}$ is based on $(n_1 + n_2 - 2)$ degrees of freedom.

Small-Sample Test of Hypothesis for $(\mu_1 - \mu_2)$: Independent Samples

One-Tailed Test

$H_0: (\mu_1 - \mu_2) = D_0$

$H_a: (\mu_1 - \mu_2) < D_0$
   [or $H_a: (\mu_1 - \mu_2) > D_0$]

Two-Tailed Test

$H_0: (\mu_1 - \mu_2) = D_0$

$H_a: (\mu_1 - \mu_2) \neq D_0$

Test statistic:

$$t = \frac{(\bar{x}_1 - \bar{x}_2) - D_0}{\sqrt{s_p^2\left(\frac{1}{n_1} + \frac{1}{n_2}\right)}}$$

Rejection region: $t < -t_\alpha$
   [or $t > t_\alpha$ when
   $H_a: (\mu_1 - \mu_2) > D_0$]

Rejection region: $|t| > t_{\alpha/2}$

where t_α and $t_{\alpha/2}$ are based on $(n_1 + n_2 - 2)$ degrees of freedom.

> **Conditions Required for Valid Small-Sample Inferences about $(\mu_1 - \mu_2)$**
>
> 1. The two samples are randomly selected in an independent manner from the two target populations.
> 2. Both sampled populations have distributions that are approximately normal.
> 3. The population variances are equal (i.e., $\sigma_1^2 = \sigma_2^2$).

EXAMPLE 9.4

A SMALL-SAMPLE CONFIDENCE INTERVAL FOR $\mu_1 - \mu_2$

Problem   Suppose you wish to compare a new method of teaching reading to "slow learners" to the current standard method. You decide to base this comparison on the results of a reading test given at the end of a learning period of 6 months. Of a random sample of 22 slow learners, 10 are taught by the new method and 12 are taught by the standard method. All 22 children are taught by qualified instructors under similar conditions for a 6-month period. The results of the reading test at the end of this period are given in Table 9.2.

⊚ **READING**

TABLE 9.2   Reading Test Scores for Slow Learners

New Method				Standard Method			
80	80	79	81	79	62	70	68
76	66	71	76	73	76	86	73
70	85			72	68	75	66

a. Use the data in the table to estimate the true mean difference between the test scores for the new method and the standard method. Use a 95% confidence interval.

b. Interpret the interval, part **a**.

c. What assumptions must be made in order that the estimate be valid? Are they reasonably satisfied?

Solution   **a.** For this experiment, let μ_1 and μ_2 represent the mean reading test scores of slow learners taught with the new and standard methods, respectively. Then the objective is to obtain a 95% confidence interval for $(\mu_1 - \mu_2)$.

The first step in constructing the confidence interval is to obtain summary statistics (e.g., $\bar{x}$ and s) on reading test scores for each method. The data of Table 9.2 were entered into a computer, and SAS was used to obtain these descriptive statistics. The SAS printout appears in Figure 9.6. Note that $\bar{x}_1 = 76.4$, $s_1 = 5.8348$, $\bar{x}_2 = 72.333$, and $s_2 = 6.3437$

Next, we calculate the pooled estimate of variance:

$$s_p^2 = \frac{(n_1 - 1)s_1^2 + (n_2 - 1)s_2^2}{n_1 + n_2 - 2}$$

$$= \frac{(10 - 1)(5.8348)^2 + (12 - 1)(6.3437)^2}{10 + 12 - 2} = 37.45$$

where s_p^2 is based on $(n_1 + n_2 - 2) = (10 + 12 - 2) = 20$ degrees of freedom. Also, we find $t_{\alpha/2} = t_{.025} = 2.086$ (based on 20 degrees of freedom) from Table VI of Appendix A.

Figure 9.6
SAS Printout for Example 9.4

```
              Two Sample t-test for the Means of SCORE within METHOD

Sample Statistics

     Group          N        Mean      Std. Dev.    Std. Error
     --------------------------------------------------------------
     NEW           10        76.4        5.8348       1.8451
     STD           12     72.33333       6.3437       1.8313

Hypothesis Test

     Null hypothesis:      Mean 1 - Mean 2 =  0
     Alternative:          Mean 1 - Mean 2 ^= 0

     If Variances Are      t statistic       Df        Pr > t
     --------------------------------------------------------------
     Equal                    1.552          20        0.1364
     Not Equal                1.564        19.77        0.1336

95% Confidence Interval for the Difference between Two Means

               Lower Limit      Upper Limit
               -----------      -----------
                  -1.40             9.53
```

Finally, the 95% confidence interval for $(\mu_1 - \mu_2)$, the difference between mean test scores for the two methods, is

$$(\bar{x}_1 - \bar{x}_2) \pm t_{\alpha/2}\sqrt{s_p^2\left(\frac{1}{n_1} + \frac{1}{n_2}\right)} = (76.4 - 72.33) \pm t_{.025}\sqrt{37.45\left(\frac{1}{10} + \frac{1}{12}\right)}$$

$$= 4.07 \pm (2.086)(2.62)$$
$$= 4.07 \pm 5.47$$

or $(-1.4, 9.54)$. This interval agrees (except for rounding) with the one shown at the bottom of the SAS printout, Figure 9.6.

b. The interval can be interpreted as follows. With a confidence coefficient equal to .95, we estimate the difference in mean test scores between using the new method of teaching and using the standard method to fall in the interval −1.4 to 9.54. In other words, we estimate (with 95% confidence) the mean test score for the new method to be anywhere from 1.4 points less than to 9.54 points more than the mean test score for the standard method. Although the sample means seem to suggest that the new method is associated with a higher mean test score, there is insufficient evidence to indicate that $(\mu_1 - \mu_2)$ differs from 0 because the interval includes 0 as a possible value for $(\mu_1 - \mu_2)$. To show a difference in mean test scores (if it exists), you could increase the sample size and thereby narrow the width of the confidence interval for $(\mu_1 - \mu_2)$. An alternative is to design the experiment differently. This possibility is discussed in the next section.

c. To properly use the small-sample confidence interval, the following assumptions must be satisfied:

(1) The samples are randomly and independently selected from the populations of slow learners taught by the new method and the standard method.

(2) The test scores are normally distributed for both teaching methods.

(3) The variance of the test scores are the same for the two populations, that is, $\sigma_1^2 = \sigma_2^2$.

Figure 9.7
MINITAB Normal
Probability Plots for
Example 9.4

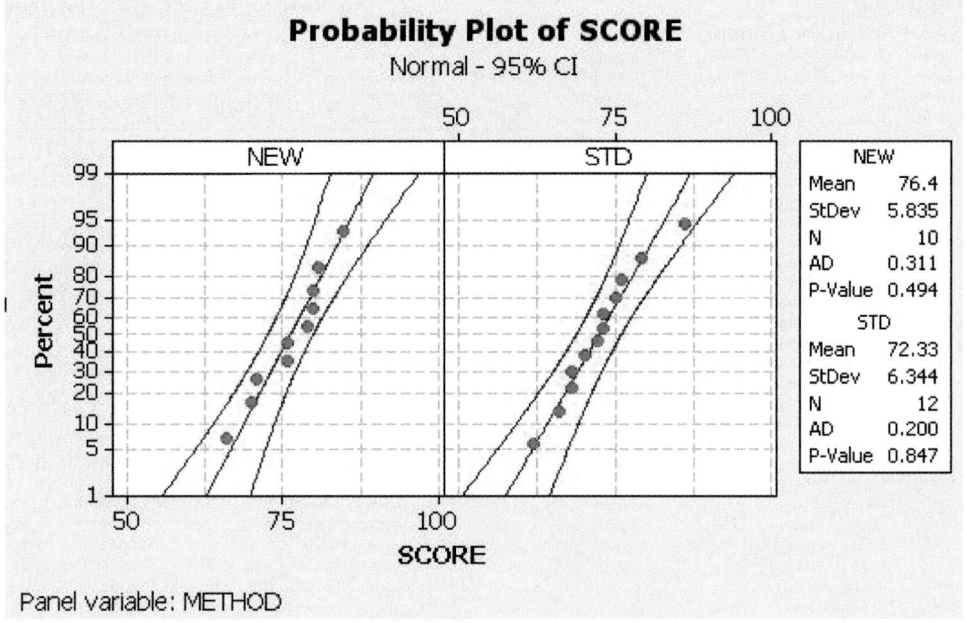

The first assumption is satisfied, based on the information provided about the sampling procedure in the problem description. To check the plausibility of the remaining two assumptions, we resort to graphical methods. Figure 9.7 is a MINITAB printout that gives normal probability plots for the test scores of the two samples of slow learners. The near straight-line trends on both plots indicate that the score distributions are approximately mound-shaped and symmetric. Consequently, each sample data set appears to come from a population that is approximately normal.

One way to check assumption #3 is to test the null hypothesis $H_0: \sigma_1^2 = \sigma_2^2$. This test is covered in Section 9.6. Another approach is to examine box plots for the sample data. Figure 9.8 is a MINITAB printout

Figure 9.8
MINITAB Box Plots
for Example 9.4

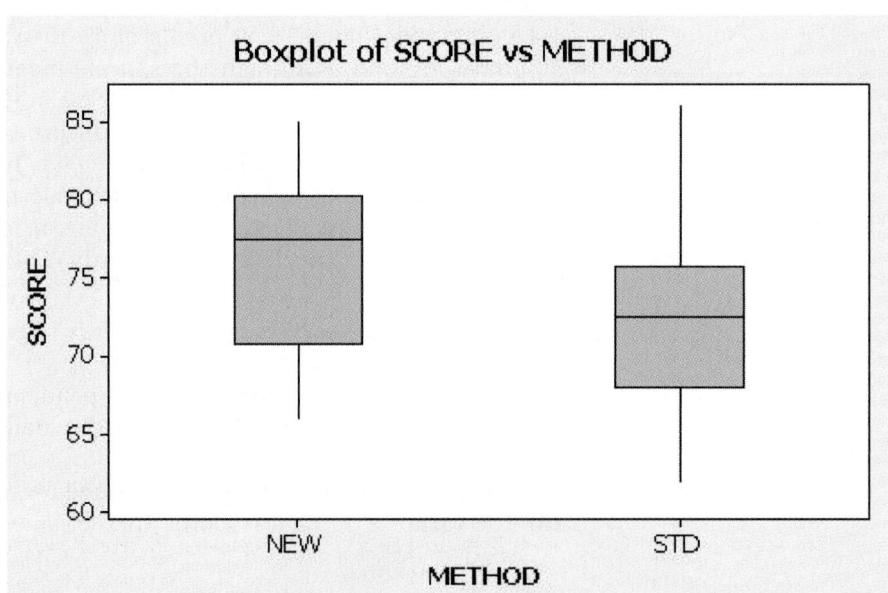

that shows side-by-side vertical box plots for the test scores in the two samples. Recall, from Section 2.9, that the box plot represents the "spread" of a data set. The two box plots appear to have about the same spread; thus, the samples appear to come from populations with approximately the same variance.

Look Back All three assumptions, then, appear to be reasonably satisfied for this application of the small-sample confidence interval.

| Now Work | *Exercise 9.9* |

━━━━━━━ ■ ■ ■ ━━━━━━━

The two-sample *t*-statistic is a powerful tool for comparing population means when the assumptions are satisfied. It has also been shown to retain its usefulness when the sampled populations are only approximately normally distributed. And when the sample sizes are equal, the assumption of equal population variances can be relaxed. That is, if $n_1 = n_2$, then σ_1^2 and σ_2^2 can be quite different and the test statistic will still possess, approximately, a Student's *t*-distribution. In the case where $\sigma_1^2 \neq \sigma_2^2$ and $n_1 \neq n_2$, an approximate small-sample confidence interval or test can be obtained by modifying the degrees of freedom associated with the *t*-distribution.

Confidence Interval for $\mu_1 - \mu_2$

Using The TI-83 Graphing Calculator

For this calculation, you have the option of entering the raw data or using the summary statistics

Step 1   *If you are using the raw data, enter the data into L1 and L2.*

Press **STAT** and select **1:Edit**
Note: If a list already contains data, clear the old data. Use the up arrow to highlight the name at the top of the list. Press **CLEAR ENTER**.
Use the **Down Arrow** or **ENTER** key to enter the data sets into **L1** and **L2**.

Step 2   *Access the Statistical Tests Menu*

Press **STAT**
Arrow right to **TESTS**
Arrow down to **0: 2-Samp TInt**
Press **ENTER**

Step 3   *Choose "**Data**" or "**Stats**". ("Data" is selected when you have entered the raw data into L1 and L2. "Stats" is selected when you are using the summary statistics.)*

Press **ENTER**

If you selected "Data":
Set **List1** to **L1.** Note: If List1 is not already set to L1, arrow down to List1: and hit **2nd** and '**1**'.
Set **List2** to **L2.** Note: If List2 is not already set to L2, arrow down to List2: and hit **2nd** and '**2**'.
Set **Freq1** to **1.** Note: If Freq1 is not already set to 1, arrow down to Freq1 and hit **Alpha** and '**1**'.

Set **Freq2** to **1.** Note: If Freq2 is not already set to 1, arrow down to Freq2 and hit **Alpha** and '**1**'.
Arrow down to **C-Level** and enter the confidence level.
Arrow down to **Pooled**: highlight **No** or **Yes** and Press **Enter**. Note: Select **Yes** if you have made the assumption of equal population variances, otherwise select **No.**
Arrow down to **"Calculate"**
Press **ENTER**

If you selected "Stats", enter the mean, standard deviation and sample size for each of the two datasets.
Arrow down to **C-Level** and enter the confidence level.
Arrow down to **Pooled**: highlight **No** or **Yes** and Press **Enter**. Note: Select **Yes** if you have made the assumption of equal population variances, otherwise select **No.**
Arrow down to **"Calculate"**
Press **ENTER**

Step 4   *View the Output*

The confidence interval will be displayed along with the calculated degrees of freedom, the means, the standard deviations, the pooled standard deviation (if **Yes** was highlighting for **Pooled**) and the sample sizes.

Example   *Consider Example 9.4 with data in Table 9.2 on pg. 438. For this example, the raw data will be used, so begin with Step 1 of the instructions.*

Step 1   *Enter the data into L1 and L2.*

Press **STAT** and select **1:Edit**
Note: If a list already contains data, clear the old data. Use the up arrow to highlight the name at the top of the list. Press **CLEAR ENTER**.
Use the **Down Arrow** or **ENTER** key to enter the data sets into **L1 and L2.**

Step 2   *Access the Statistical Tests Menu*

Press **STAT**
Arrow right to **TESTS**
Arrow down to **0: 2-Samp TInt**
Press **ENTER**

Step 3   *Choose "Data"*

Set **List1** to **L1.**
Set **List2** to **L2.**

Set **Freq1** to **1.**
Set **Freq2** to **1.**
Arrow down to **C-Level** and enter .95 for the confidence level.
Arrow down to **Pooled**: highlight **Yes** and Press **Enter**. Note: You have made
the assumption of equal population variances.

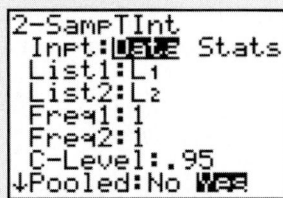

Arrow down to **"Calculate"**
Press **ENTER**

Step 4   *View the Output*

The confidence interval will be displayed with the sample information.

```
2-SampTInt
 (-1.399,9.5327)
 df=20
 x̄₁=76.4
 x̄₂=72.33333333
 Sx₁=5.83476173
↓Sx₂=6.34369169
█
```

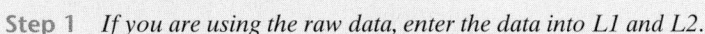

Hypothesis Test
for $\mu_1 - \mu_2$

Using the TI-83 Graphing Calculator

**For this calculation, you have the option
of entering the raw data or using the
summary statistics**

Step 1   *If you are using the raw data, enter the data into L1 and L2.*

Press **STAT** and select **1:Edit**
Note: If a list already contains data, clear the old data. Use the up arrow to
highlight the name at the top of the list. Press **CLEAR ENTER**.
Use the **Down Arrow** or **ENTER** key to enter the data sets into **L1** and **L2**.

Step 2   *Access the Statistical Tests Menu*

Press **STAT**
Arrow right to **TESTS**
Arrow down to **4 : 2-Samp TTest**
Press **ENTER**

Step 3   *Choose "**Data**" or "**Stats**". ("Data" is selected when you have entered
the raw data into L1 and L2. "Stats" is selected when you are using the
summary statistics.)*

Press **ENTER**

If you selected "Data":

Set **List1** to **L1**. Note: If List1 is not already set to L1, arrow down to List1: and hit **2nd** and **'1'**.

Set **List2** to **L2**. Note: If List2 is not already set to L2, arrow down to List2: and hit **2nd** and **'2'**.

Set **Freq1** to **1**. Note: If Freq1 is not already set to 1, arrow down to Freq1 and hit **Alpha** and **'1'**.

Set **Freq2** to **1**. Note: If Freq2 is not already set to 1, arrow down to Freq2 and hit **Alpha** and **'1'**.

Arrow down to the next line and select the appropriate alternative hypothesis by using the right arrow to highlight $\neq$**u2**, $<$**u2**, or $>$**u2** and press **ENTER**.

Arrow down to **Pooled**: highlight **No** or **Yes** and Press **Enter**. Note: Select **Yes** if you have made the assumption of equal population variances, otherwise select **No**.

Arrow down to **Calculate Draw**. You may choose either one of these for your output.

Press **ENTER**

If you selected "Stats", enter the mean, standard deviation and sample size for each of the two datasets. Arrow down to the next line and select the appropriate alternative hypothesis by using the right arrow to highlight $\neq$**u2**, $<$**u2**, or $>$**u2** and press **ENTER**.

Arrow down to **Pooled**: highlight **No** or **Yes** and Press **Enter**. Note: Select **Yes** if you have made the assumption of equal population variances, otherwise select **No**.

Arrow down to **Calculate Draw**. You may choose either one of these for your output.

Press **ENTER**

Step 4 *View the Output*

If you chose **Calculate**, the output will display the alternative hypothesis, the t-value, the p-value, the calculated degrees of freedom, the means, the standard deviations, the pooled standard deviation (if **Yes** was highlighting for **Pooled**) and the sample sizes.

If you chose **Draw**, a graph of the normal curve will be displayed along with the t-value and the p-value. The area under the curve that represents the p-value is shaded. Note: In order to use the **Draw** feature, you must first clear all equations from the **Y-registers**. Hit '**Y =**' and delete all equations.

Example *Consider Exercise 9.12 part a on pg. 449. For this example, the summary statistics will be used so begin with Step2 of the instructions.*

Step 2 *Access the Statistical Tests Menu*

Press **STAT**
Arrow right to **TESTS**
Arrow down to **4:2-SampTTest**
Press **ENTER**

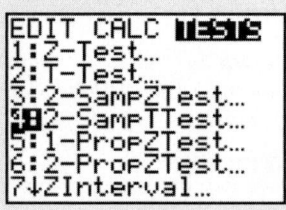

Step 3 *Choose "Stats"*

Enter the standard deviations, means, and sample sizes.

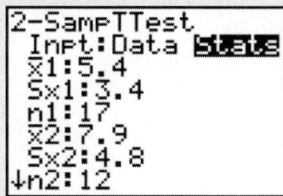

Arrow down to the next line and select ≠**u2** and press **ENTER**
Arrow down to **Pooled**: highlight **No** and Press **Enter**. Note: You are not
making the assumption of equal population variances
Arrow down to **Calculate Draw**. You may choose either one of these for your
output.

Press **ENTER**

Step 4 *View the Output*

If you chose **Calculate**, the output will display the alternative hypothesis, the
t-value, the p-value, the calculated degrees of freedom, the means, the standard
deviations, and the sample sizes.

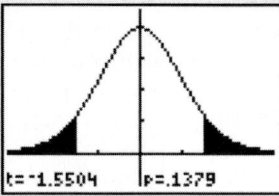

If you chose **Draw**, a graph of the normal curve will be displayed along with
the t-value and the p-value. The area under the curve that represents the
p-value is shaded. Note: In order to use the **Draw** feature, you must first clear
all equations from the **Y-registers**. Hit 'Y=' and delete all equations.

The next box gives the approximate small-sample procedures to use when the
assumption of equal variances is violated. The test for the "unequal sample sizes"
case is based on Satterthwaite's (1946) approximation.

Approximate Small-Sample Procedures when $\sigma_1^2 \neq \sigma_2^2$

Equal Sample Sizes ($n_1 = n_2 = n$)

Confidence interval: $$(\bar{x}_1 - \bar{x}_2) \pm t_{\alpha/2}\sqrt{(s_1^2 + s_2^2)/n}$$

Test statistic for H_0: $(\mu_1 - \mu_2) = 0$: $\quad t = (\bar{x}_1 - \bar{x}_2)/\sqrt{(s_1^2 + s_2^2)/n}$
where t is based on $\nu = n_1 + n_2 - 2 = 2(n - 1)$ degrees of freedom.

Unequal Sample Sizes ($n_1 \neq n_2$)

Confidence interval: $(\bar{x}_1 - \bar{x}_2) \pm t_{\alpha/2} \sqrt{(s_1^2/n_1) + (s_2^2/n_2)}$

Test statistic for H_0: $(\mu_1 - \mu_2) = 0$: $\qquad t = (\bar{x}_1 - \bar{x}_2)/\sqrt{(s_1^2/n_1) + (s_2^2/n_2)}$
where t is based on degrees of freedom equal to

$$\nu = \frac{(s_1^2/n_1 + s_2^2/n_2)^2}{\dfrac{(s_1^2/n_1)^2}{n_1 - 1} + \dfrac{(s_2^2/n_2)^2}{n_2 - 1}}$$

Note: The value of ν will generally not be an integer. Round ν down to the nearest integer to use the t-table.

When the assumptions are not clearly satisfied, you can select larger samples from the populations or you can use other available statistical tests (**nonparametric statistical tests**, which are described in Chapter 14).

What Should You Do If the Assumptions Are Not Satisfied?

Answer: If you are concerned that the assumptions are not satisfied, use the Wilcoxon rank sum test for independent samples to test for a shift in population distributions. See Chapter 14.

Statistics in Action Revisited

Comparing the Mean Winning Bid Prices of Two Kentucky Milk Markets

In the Statistics in Action study for this chapter (see p. 00), we want to determine whether there is evidence of collusive bidding in the northern Kentucky school milk market. Numerous economic and statistical methods exist for detecting the possibility of collusive practices among bidders. These typically involve (1) market-shares analyses, (2) an analysis of incumbency rates (i.e., the percentage of markets that are won by the same vendor who won the previous year), (3) an examination of bid levels and their dispersion, (4) price-distance analyses, (5) bid-sequence analyses, (6) comparison of mean winning prices, and (7) econometric modeling of the winning price. Here, we focus on a comparison of the mean winning bid prices.

Consider two similar markets, one in which the bids are rigged and the other in which bids are made competitively. In theory, the mean winning price in the "rigged" market will be higher than the mean winning price in the competitive market. Recall that in this case, the speculation is that from 1983 to 1991 the tricounty milk market (where Meyer and Trauth dairies supplied almost all of the milk) was rigged while the school districts in the surrounding market were competitive. Consequently, we want to compare the mean winning price in the tricounty market with that of the surrounding market

for each milk product sold (whole white, low-fat white, and low-fat chocolate).

If we let μ_1 represent the population mean winning price in the surrounding market and μ_2 represent the population mean winning price in the tricounty market, then the speculation is that $\mu_1 < \mu_2$ (i.e., the mean winning price in the "competitive" market is less than the mean winning price in the "rigged" market). Of interest, then, is the one-tailed test, H_0: $(\mu_1 - \mu_2) = 0$ versus H_a: $(\mu_1 - \mu_2) < 0$, for each of the three milk price variables. An examination of the data saved in the **MILK** file reveals that both samples are large ($n_1 \geq 200$ and $n_2 \geq 70$); thus, the large-sample z-test is appropriate and no assumptions about the price distributions are required. We conducted these tests using MINITAB. The output is shown in Figure SIA9.1.

The one-tailed p-values of the three tests are highlighted on Figure SIA9.1. All three milk price variables (WWBID, LFWBID, and LFCBID) show a significant difference at $\alpha = .05$ (p-values all $= .0000$). Therefore, there is sufficient evidence to conclude that the mean winning bid price for the surrounding ("competitive") market is significantly less than the mean winning bid price for the tricounty ("rigged") market. Consequently, there is support for the claim that the dairies in the tricounty market participated in collusive practices.

Two-Sample T-Test and CI: WWBID, Market

```
Two-sample T for WWBID

Market        N     Mean    StDev    SE Mean
SURROUND    254   0.1331   0.0158    0.00099
TRI-COUNTY  100   0.1431   0.0133    0.0013


Difference = mu (SURROUND) - mu (TRI-COUNTY)
Estimate for difference:  -0.009970
95% upper bound for difference:  -0.007232
T-Test of difference = 0 (vs <): T-Value = -6.02  P-Value = 0.000  DF = 213
```

Two-Sample T-Test and CI: LFWBID, Market

```
Two-sample T for LFWBID

Market        N     Mean    StDev    SE Mean
SURROUND    258   0.1242   0.0166    0.0010
TRI-COUNTY  114   0.1366   0.0136    0.0013


Difference = mu (SURROUND) - mu (TRI-COUNTY)
Estimate for difference:  -0.012445
95% upper bound for difference:  -0.009744
T-Test of difference = 0 (vs <): T-Value = -7.61  P-Value = 0.000  DF = 261
```

Two-Sample T-Test and CI: LFCBID, Market

```
Two-sample T for LFCBID

Market        N     Mean    StDev    SE Mean
SURROUND    259   0.1251   0.0172    0.0011
TRI-COUNTY   70   0.1401   0.0149    0.0018


Difference = mu (SURROUND) - mu (TRI-COUNTY)
Estimate for difference:  -0.015091
95% upper bound for difference:  -0.011642
T-Test of difference = 0 (vs <): T-Value = -7.25  P-Value = 0.000  DF = 122
```

Figure SIA9.1
MINITAB Comparison of Means for Two Kentucky Milk Markets

Exercises 9.1–9.29

Understanding the Principles

9.1 Describe the sampling distribution of $(\bar{x}_1 - \bar{x}_2)$ when the samples are large.

9.2 To use the t-statistic to test for a difference between the means of two populations, what assumptions must be made about the two populations? About the two samples?

9.3 Two populations are described in each of the following cases. In which cases would it be appropriate to apply the small-sample t-test to investigate the difference between the population means?

a. Population 1: Normal distribution with variance σ_1^2
Population 2: Skewed to the right with variance $\sigma_2^2 = \sigma_1^2$

b. Population 1: Normal distribution with variance σ_1^2
Population 2: Normal distribution with variance $\sigma_2^2 \neq \sigma_1^2$

c. Population 1: Skewed to the left with variance σ_1^2 Population 2: Skewed to the left with variance $\sigma_2^2 = \sigma_1^2$

d. Population 1: Normal distribution with variance σ_1^2
Population 2: Normal distribution with variance $\sigma_2^2 = \sigma_1^2$

e. Population 1: Uniform distribution with variance σ_1^2
Population 2: Uniform distribution with variance $\sigma_2^2 = \sigma_1^2$

9.4 A confidence interval for $(\mu_1 - \mu_2)$ is $(-10, 4)$. Which of the following inferences is correct?

a. $\mu_1 > \mu_2$
b. $\mu_1 < \mu_2$
c. $\mu_1 = \mu_2$
d. no significant difference between means

9.5 A confidence interval for $(\mu_1 - \mu_2)$ is $(-10, -4)$. Which of the following inferences is correct?

a. $\mu_1 > \mu_2$
b. $\mu_1 < \mu_2$
c. $\mu_1 = \mu_2$
d. no significant difference between means

Learning the Mechanics

9.6 In order to compare the means of two populations, independent random samples of 400 observations are selected from each population, with the following results:

Sample 1	Sample 2
$\bar{x}_1 = 5{,}275$	$\bar{x}_2 = 5{,}240$
$s_1 = 150$	$s_2 = 200$

a. Use a 95% confidence interval to estimate the difference between the population means $(\mu_1 - \mu_2)$. Interpret the confidence interval.
b. Test the null hypothesis $H_0: (\mu_1 - \mu_2) = 0$ versus the alternative hypothesis $H_a: (\mu_1 - \mu_2) \neq 0$. Give the significance level of the test, and interpret the result.
c. Suppose the test in part **b** was conducted with the alternative hypothesis $H_a: (\mu_1 - \mu_2) > 0$. How would your answer to part **b** change?
d. Test the null hypothesis $H_0: (\mu_1 - \mu_2) = 25$ versus the alternative $H_a: (\mu_1 - \mu_2) \neq 25$. Give the significance level, and interpret the result. Compare your answer to the test conducted in part **b**.
e. What assumptions are necessary to ensure the validity of the inferential procedures applied in parts **a–d**?

9.7 Independent random samples of 100 observations each are chosen from two normal populations with the following means and standard deviations:

Population 1	Population 2
$\mu_1 = 14$	$\mu_2 = 10$
$\sigma_1 = 4$	$\sigma_2 = 3$

Let $\bar{x}_1$ and $\bar{x}_2$ denote the two sample means.

a. Give the mean and standard deviation of the sampling distribution of $\bar{x}_1$.
b. Give the mean and standard deviation of the sampling distribution of $\bar{x}_2$.
c. Suppose you were to calculate the difference $(\bar{x}_1 - \bar{x}_2)$ between the sample means. Find the mean and standard deviation of the sampling distribution of $(\bar{x}_1 - \bar{x}_2)$.

d. Will the statistic $(\bar{x}_1 - \bar{x}_2)$ be normally distributed? Explain.

9.8 Assume that $\sigma_1^2 = \sigma_2^2 = \sigma^2$. Calculate the pooled estimator of σ^2 for each of the following cases:

a. $s_1^2 = 200$, $s_2^2 = 180$, $n_1 = n_2 = 25$
b. $s_1^2 = 25$, $s_2^2 = 40$, $n_1 = 20$, $n_2 = 10$
c. $s_1^2 = .20$, $s_2^2 = .30$, $n_1 = 8$, $n_2 = 12$
d. $s_1^2 = 2{,}500$, $s_2^2 = 1{,}800$, $n_1 = 16$, $n_2 = 17$

e. Note that the pooled estimate is a weighted average of the sample variances. To which of the variances does the pooled estimate fall nearer in each of the above cases?

9.9 Independent random samples from normal populations
NW produced the results shown below:

⊛ **LM9_9**

Sample 1	Sample 2
1.2	4.2
3.1	2.7
1.7	3.6
2.8	3.9
3.0	

a. Calculate the pooled estimate of σ^2.
b. Do the data provide sufficient evidence to indicate that $\mu_2 > \mu_1$? Test using $\alpha = .10$.
c. Find a 90% confidence interval for $(\mu_1 - \mu_2)$.
d. Which of the two inferential procedures, the test of hypothesis in part **b** or the confidence interval in part **c**, provides more information about $(\mu_1 - \mu_2)$?

9.10 Two independent random samples have been selected, 100 observations from population 1 and 100 from population 2. Sample means $\bar{x}_1 = 70$ and $\bar{x}_2 = 50$ were obtained. From previous experience with these populations, it is known that the variances are $\sigma_1^2 = 100$ and $\sigma_2^2 = 64$.

a. Find $\sigma_{(\bar{x}_1 - \bar{x}_2)}$.
b. Sketch the approximate sampling distribution $(\bar{x}_1 - \bar{x}_2)$ assuming $(\mu_1 - \mu_2) = 5$.
c. Locate the observed value of $(\bar{x}_1 - \bar{x}_2)$ on the graph you drew in part **b**. Does it appear that this value contradicts the null hypothesis $H_0: (\mu_1 - \mu_2) = 5$?
d. Use the z-table to determine the rejection region for the test of $H_0: (\mu_1 - \mu_2) = 5$ against $H_a: (\mu_1 - \mu_2) \neq 5$. Use $\alpha = .05$.

e. Conduct the hypothesis test of part **d** and interpret your result.
f. Construct a 95% confidence interval for $(\mu_1 - \mu_2)$. Interpret the interval.
g. Which inference provides more information about the value of $(\mu_1 - \mu_2)$—the test of hypothesis in part **e** or the confidence interval in part **f**?

9.11 Independent random samples are selected from two populations and used to test the hypothesis $H_0: (\mu_1 - \mu_2) = 0$ against the alternative $H_a: (\mu_1 - \mu_2) \neq 0$. An analysis of 233 observations

from population 1 and 312 from population 2 yielded a p-value of .115.
a. Interpret the results of the computer analysis.
b. If the alternative hypothesis had been H_a: $(\mu_1 - \mu_2) < 0$, how would the p-value change? Interpret the p-value for this one-tailed test.

9.12 Independent random samples selected from two normal populations produced the sample means and standard deviations shown below:

Sample 1	Sample 2
$n_1 = 17$	$n_2 = 12$
$\bar{x}_1 = 5.4$	$\bar{x}_2 = 7.9$
$s_1 = 3.4$	$s_2 = 4.8$

a. Conduct the test $H_0: (\mu_1 - \mu_2) = 0$ against H_a: $(\mu_1 - \mu_2) \neq 0$.
b. Find and interpret the 95% confidence interval for $(\mu_1 - \mu_2)$.

Applying the Concepts—Basic

9.13 Reading Japanese books. Refer to the *Reading in a Foreign Language* (Apr. 2004) experiment to improve the Japanese reading comprehension levels of University of Hawaii students, Exercise 2.31 (p. 51). Recall that 14 students participated in a ten-week extensive reading program in a second semester Japanese course. The numbers of books read by each student and the student's course grade are repeated in the table.

JAPANESE

Number of Books	Course Grade	Number of Books	Course Grade
53	A	30	A
42	A	28	B
40	A	24	A
40	B	22	C
39	A	21	B
34	A	20	B
34	A	16	B

Source: Hitosugi, C. I., and Day, R. R. "Extensive Reading in Japanese," *Reading in a Foreign Language*, Vol. 16, No. 1, Apr. 2004 (Table 4).

a. Consider two populations of students who participate in the reading program prior to taking a second semester Japanese course—those who earn an A grade and those who earn a B or "C" grade. Of interest is the difference in the mean number of books read by the two populations of students. Identify the parameter of interest in words and symbols.
b. Form a 95% confidence interval for the target parameter identified in part **a**.
c. Give a practical interpretation of the confidence interval, part **b**.
d. Compare the inference in part **c** with the inference you derived from stem-and-leaf plots in Exercise 2.31b.

9.14 Rating service at five-star hotels. A study published in *The Journal of American Academy of Business, Cambridge* (March 2002) examined whether the perception of service quality at five-star hotels in Jamaica differed by gender. Hotel guests were randomly selected from the lobby and restaurant areas and asked to rate ten service-related items (e.g., "The personal attention you received from our employees"). Each item was rated on a 5-point scale (1 = "much worse than I expected," 5 = "much better than I expected") and the sum of the items for each guest was determined. A summary of the guest scores are provided in the table.

Gender	Sample Size	Mean Score	Standard Deviation
Males	127	39.08	6.73
Females	114	38.79	6.94

a. Construct a 90% confidence interval for the difference between the population mean service-rating scores given by male and female guests at Jamaican 5-star hotels.
b. Use the interval, part **a**, to make an inference about whether the perception of service quality at five-star hotels in Jamaica differs by gender.

9.15 Heights of grade school repeaters. Are children who repeat a grade in elementary school shorter on average, than their peers? To answer this question, researchers compared the heights of Australian school children who repeated a grade to those who did not. (*The Archives of Disease in Childhood*, Apr. 2000.) All height measurements were standardized using z-scores. A summary of the results, by gender, is shown in the table.

	Never Repeated	Repeated a Grade
Boys	$n = 1,349$ $\bar{x} = .30$ $s = .97$	$n = 86$ $\bar{x} = -.04$ $s = 1.17$
Girls	$n = 1,366$ $\bar{x} = .22$ $s = 1.04$	$n = 43$ $\bar{x} = .26$ $s = .94$

Source: Wake, M., Coghlan, D., and Hesketh, K. "Does height influence progression through primary school grades?" *The Archives of Disease in Childhood*, Vol. 82, Apr. 2000 (Table 3).

a. Set up the null and alternative hypothesis for determining whether the average height of Australian boys who repeated a grade is less than the average height of boys who never repeated.
b. Conduct the test, part **a**, using $\alpha = .05$.
c. Repeat parts **a** and **b** for Australian girls.

9.16 Short-term memory study. A group of University of Florida psychologists investigated the effects of age and gender on the short-term memory of adults. (*Cognitive Aging Conference*, Apr. 1996.) Each in a sample of 152 adults was asked to place 20 common household items (e.g., eyeglasses, keys, hat, hammer) into the rooms of

BULIMIA

Bulimic students	21	13	10	20	25	19	16	21	24	13	14			
Normal students	13	6	16	13	8	19	23	18	11	19	7	10	15	20

Source: Randles, R. H. "On neutral responses (zeros) in the sign test and ties in the Wilcoxon-Mann-Whitney test." *The American Statistician,* Vol. 55, No. 2, May 2001 (Figure 3).

a computer-image house. After performing some unrelated activities, each subject was asked to recall the locations of the objects they had placed. The number of correct responses (out of 20) was recorded.

a. The researchers theorized that women will have a higher mean recall score than men. Set up the null and alternative hypotheses to test this theory.

b. Refer to part **a.** The 43 men in the study had a mean recall score of 13.5, while the 109 women had a mean recall score of 14.4. The observed significance level for comparing these two means was found to be $p = .0001$. Interpret this value.

c. The researchers also hypothesized that younger adults would have a higher mean recall score than older adults. Set up H_0 and H_a to test this theory.

d. The observed significance level for the test of part **c** was reported as $p = .0001$. Interpret this result.

9.17 Watching cable TV shopping networks. Refer to the International Association for Time Use Research study on cable TV viewers who purchase items from one of the home shopping channels, Exercise 8.52 (p. 390). The 1,600 sampled viewers described their motivation for watching cable TV shopping networks by giving their level of agreement (on a 5-point scale, where 1 = strongly disagree and 5 = strongly agree) with the statement "I have nothing else to do." The researcher wanted to compare the mean responses of viewers who watch the shopping network at noon with those viewers who do not watch at noon.

a. Give the null and alternative hypotheses for determining whether the mean response of noontime watchers differs from the mean response of non-noontime watchers.

b. The researcher found the *p*-value for the test, part **a**, to be .02. Interpret this result, assuming $\alpha = .05$.

c. Interpret the result, part **b**, assuming $\alpha = .01$.

d. The sample means for noontime watchers and non-watchers were found to be 3.3 and 3.4, respectively. Comment on the practical significance of this result.

9.18 Impact of exam question order. Do the early questions in an exam affect your performance on later questions? Research reported in the *Journal of Psychology* (Jan. 1991) investigated this relationship by conducting the following experiment: 140 college students were randomly assigned to two groups. Group A took an exam that contained 15 difficult questions followed by five moderate questions. Group B took an exam that contained 15 easy questions followed by the same five moderate questions. One of the goals of the study was to compare the average score on the moderate questions for the two groups.

a. The researchers hypothesized that the students who had the easy questions would score better than the students who had the difficult questions. Set up the appropriate null and alternative hypotheses to test their research hypothesis, defining any symbols you use.

b. Suppose the *p*-value for this test was reported as .3248. What conclusion would you reach based on this *p*-value?

c. What assumptions are necessary to ensure the validity of this conclusion?

9.19 Bulimia study. The "fear of negative evaluation" (FNE) scores for 11 female students known to suffer from the eating disorder bulimia and 14 female students with normal eating habits, first presented in Exercise 2.38 (p. 52), are reproduced above. (Recall that the higher the score, the greater the fear of negative evaluation.)

a. Find a 95% confidence interval for the difference between the population means of the FNE scores for bulimic and normal female students. Interpret the result.

b. What assumptions are required for the interval of part **a** to be statistically valid? Are these assumptions reasonably satisfied? Explain.

Applying the Concepts—Intermediate

9.20 Patent infringement case. *Chance* (Fall 2002) described a lawsuit where Intel Corp. was charged with infringing on a patent for an invention used in the automatic manufacture of computer chips. In response, Intel accused the inventor of adding material to his patent notebook after the patent was witnessed and granted. The case rested on whether a patent witness' signature was written on top of key text in the notebook, or under the key text. Intel hired a physicist who used an X-ray beam to measure the relative concentration of certain elements (e.g., nickel, zinc, potassium) at several spots on the notebook page. The zinc measurements for three notebook locations—on a text line, on a witness line, and on the intersection of the witness and text line—are provided in the table.

PATENT

Text line:	.335	.374	.440			
Witness line:	.210	.262	.188	.329	.439	.397
Intersection:	.393	.353	.285	.295	.319	

a. Use a test or a confidence interval (at $\alpha = .05$) to compare the mean zinc measurement for the text line with the mean for the intersection.

b. Use a test or a confidence interval (at $\alpha = .05$) to compare the mean zinc measurement for the witness line with the mean for the intersection.

c. From the results, parts **a** and **b**, what can you infer about the mean zinc measurements at the three notebook locations?

d. What assumptions are required for the inferences to be valid? Are they reasonably satisfied?

9.21 Family involvement in homework. Teachers Involve Parents in Schoolwork (TIPS) is an interactive homework process designed to improve the quality of homework assignments for elementary, middle, and high school students. TIPS homework assignments require students to conduct interactions with family partners (parents, guardians, etc.) while completing the homework. Frances Van Voorhis (Johns Hopkins University) conducted a study to investigate the effects of TIPS in science, mathematics, and language arts homework assignments (April 2001). Each in a sample of 128 middle school students was assigned to complete TIPS homework assignments, while 98 students in a second sample were assigned traditional, noninteractive, homework assignments (called ATIPS). At the end of the study, all students reported on the level of family involvement in their homework on a 4-point scale (0 = never, 1 = rarely, 2 = sometimes, 3 = frequently, 4 = always). Three scores were recorded for each student: one for science homework, one for math homework, and one for language arts homework. The data for the study are saved in the **HWSTUDY** file. (The first 5 and last 5 observations in the data set are listed in the table.)

HWSTUDY (First and last 5 observations)

Homework Condition	Science	Math	Language
ATIPS	1	0	0
ATIPS	0	1	1
ATIPS	0	1	0
ATIPS	1	2	0
ATIPS	1	1	2
TIPS	2	3	2
TIPS	1	4	2
TIPS	2	4	2
TIPS	4	0	3
TIPS	2	0	1

Source: Van Voorhis, F. L., "Teachers' use of interactive homework and its effects on family involvement and science achievement of middle grade students." Paper presented at the annual meeting of the American Educational Research Association, Seattle, April 2001.

a. Conduct an analysis to compare the mean level of family involvement in science homework assignments of TIPS and ATIPS students. Use $\alpha = .05$. Make a practical conclusion.

b. Repeat part **a** for mathematics homework assignments.

c. Repeat part **a** for language arts homework assignments.

d. What assumptions are necessary for the inferences of parts **a–c** to be valid? Are they reasonably satisfied?

9.22 Pig castration study. Two methods of castrating male piglets were investigated in *Applied Animal Behaviour Science* (Nov. 1, 2000). Method 1 involved an incision of the spermatic cords while Method 2 involved pulling and severing the cords. Forty-nine male piglets were randomly allocated to one of the two methods. During castration, the researchers measured the number of high frequency vocal responses (squeals) per second over a 5-second period. The data are summarized in the table. Conduct a test of hypothesis to determine whether the population mean number of high frequency vocal responses differs for piglets castrated by the two methods. Use $\alpha = .05$.

	Method 1	Method 2
Sample size	24	25
Mean number of squeals	.74	.70
Standard deviation	.09	.09

Source: Taylor, A. A., and Weary, D. M. "Vocal responses of piglets to castration: Identifying procedural sources of pain." *Applied Animal Behaviour Science*, Vol. 70, No. 1, November 1, 2000.

9.23 Mongolian desert ants. Refer to the *Journal of Biogeography* (Dec. 2003) study of ants in Mongolia (Central Asia), Exercise 2.66 (p. 66). Recall that botanists placed seed baits at 5 sites in the Dry Steppe region and 6 sites in the Gobi Desert and observed the number of ant species attracted to each site. These data are listed in the accompanying table. Is there evidence to conclude that a difference exists between the average number of ant species found at sites in the two regions of Mongolia? Draw the appropriate conclusion using $\alpha = .05$.

GOBIANTS

Site	Region	Number of Ant Species
1	Dry Steppe	3
2	Dry Steppe	3
3	Dry Steppe	52
4	Dry Steppe	7
5	Dry Steppe	5
6	Gobi Desert	49
7	Gobi Desert	5
8	Gobi Desert	4
9	Gobi Desert	4
10	Gobi Desert	5
11	Gobi Desert	4

Source: Pfeiffer, M., et al. "Community organization and species richness of ants in Mongolia along an ecological gradient from steppe to Gobi desert," *Journal of Biogeography*, Vol. 30, No. 12, Dec. 2003.

9.24 Accuracy of mental maps. To help students organize global information about people, places, and environments, geographers encourage them to develop "mental maps" of the world. A series of lessons was designed to aid students in the development of mental maps. (*Journal of Geography*, May/June 1997.) In one experiment, a class of 24 seventh-grade geography students was given mental map lessons while a second class of 20 students received traditional instruction. All students were asked to sketch a map of the world, and each portion of the map was evaluated for accuracy using a 5-point scale (1 = low accuracy, 5 = high accuracy).

a. The mean accuracy scores of the two groups of seventh-graders were compared using a test of hypothesis. State H_0 and H_a for a test to determine whether the mental map lessons improve a student's ability to sketch a world map.

b. What assumptions (if any) are required for the test to be statistically valid?

c. The observed significance level of the test for comparing the mean accuracy scores for continents drawn is .0507. Interpret this result.

d. The observed significance level of the test for comparing the mean accuracy scores for labeling oceans is .7371. Interpret the result.

e. The observed significance level of the test for comparing the mean accuracy scores for the entire map is .0024. Interpret the result.

9.25 Masculinity and crime. The *Journal of Sociology* (July 2003) published a study on the link between the level of masculinity and criminal behavior in men. Using a sample of newly incarcerated men in Nebraska, the researcher identified 1,171 violent events and 532 avoided-violent events that the men were involved in. (A violent event involved use of a weapon, throwing of objects, punching, choking, or kicking. An avoided-violent event included pushing, shoving, grabbing, or threats of violence that did not escalate into a violent event.) Each of the sampled men took the Masculinity-Femininity Scale (MFS) test to determine his level of masculinity based on common male stereotyped traits. MFS scores ranged from 0 to 56 points, with lower scores indicating a more masculine orientation. One goal of the research was to compare the mean MFS scores for two groups of men—those involved in violent events and those involved in avoided-violent events.

a. Identify the target parameter for this study.

b. The sample mean MFS score for the violent event group was 44.50 while the sample mean MFS score for the avoided-violent event group was 45.06. Is this sufficient information to make the comparison desired by the researcher? Explain.

c. In a large-sample test of hypothesis to compare the two means, the test statistic was computed to be $z = 1.21$. Compute the two-tailed p-value of the test.

d. Make the appropriate conclusion using $\alpha = .10$.

9.26 Mating habits of snails. Hermaphrodites are animals that possess the reproductive organs of both sexes. *Genetical Research* (June 1995) published a study of the mating systems of hermaphroditic snail species. The mating habits of the snails were classified into two groups: (1) self-fertilizing (selfing) snails that mate with snails of the same sex and (2) cross-fertilizing (outcrossing) snails that mate with snails of the opposite sex. One variable of interest in the study was the effective population size of the snail species. The means and standard deviations of the effective population size for independent random samples of 17 outcrossing snail species and 5 selfing snail species are given

in the table. Compare the mean effective population sizes of the two types of snail species with a 90% confidence interval. Interpret the result.

Snail Mating System	Sample Size	Effective Population Size	
		Mean	Standard Deviation
Outcrossing	17	4,894	1,932
Selfing	5	4,133	1,890

Source: Jarne, P. "Mating system, bottlenecks, and genetic polymorphism in hermaphroditic animals." *Genetical Research,* Vol. 65, No. 3, June 1995, p. 197 (Table 4).

9.27 Children's use of pronouns. Refer to the *Journal of Communication Disorders* (Mar. 1995) study of specifically language-impaired (SLI) children, Exercise 2.65 (p. 65). The data on deviation intelligence quotient (DIQ) for 10 SLI children and 10 younger, normally developing (YND) children are reproduced in the table. Use the methodology of this section to compare the mean DIQ of the two groups of children. (Use $\alpha = .10$.) What do you conclude?

💿 **SLI**

SLI Children			YND Children		
86	87	84	110	90	105
94	86	107	92	92	96
89	98	95	86	100	92
110			90		

9.28 Personalities of cocaine users. Do cocaine abusers have radically different personalities than nonabusing college students? This was one of the questions researched in *Psychological Assessment* (June 1995). A personality questionnaire (ZKPQ) was administered to a sample of 450 cocaine abusers and a sample of 589 college students. The ZKPQ yields scores (measured on a 20-point scale) on each of five dimensions: impulsive–sensation seeking, sociability, neuroticism–anxiety, aggression–hostility, and activity. The results are summarized in the table. Compare the mean ZKPQ scores of the two groups on each dimension using a statistical test of hypothesis. Interpret the results at $\alpha = .01$.

ZKPQ Dimension	Cocaine Abusers ($n = 450$)		College Students ($n = 589$)	
	Mean	Std. Dev.	Mean	Std. Dev.
Impulsive–sensation seeking	9.4	4.4	9.5	4.4
Sociability	10.4	4.3	12.5	4.0
Neuroticism–anxiety	8.6	5.1	9.1	4.6
Aggression–hostility	8.6	3.9	7.3	4.1
Activity	11.1	3.4	8.0	4.1

Source: Ball, S. A. "The validity of an alternative five-factor measure of personality in cocaine abusers." *Psychological Assessment,* Vol. 7, No. 2, June 1995, p. 150 (Table 1).

Applying the Concepts—Advanced

9.29 Ethnicity and pain perception. An investigation of ethnic differences in reports of pain perception was presented at the annual meeting of the American Psychosomatic Society (March 2001). A sample of 55 blacks and 159 whites participated in the study. Subjects rated (on a 13-point scale) the intensity and unpleasantness of pain felt when a bag of ice was placed on their foreheads for two minutes. (Higher ratings correspond to higher pain intensity.) A summary of the results is provided in the accompanying table.

	Blacks	Whites
Sample size	55	159
Mean pain intensity	8.2	6.9

a. Why is it dangerous to draw a statistical inference from the summarized data? Explain.

b. Give values of the missing sample standard deviations that would lead you to conclude (at $\alpha = .05$) that blacks, on average, have a higher pain intensity rating than whites.

c. Give values of the missing sample standard deviations that would lead you to an inconclusive decision (at $\alpha = .05$) regarding whether blacks or whites have a higher mean intensity rating.

9.3 Comparing Two Population Means: Paired Difference Experiments

In Example 9.4, we compared two methods of teaching reading to slow learners by means of a 95% confidence interval. Suppose it is possible to measure the slow learners' "reading IQs" *before* they are subjected to a teaching method. Eight pairs of slow learners with similar reading IQs are found, and one member of each pair is randomly assigned to the standard teaching method while the other is assigned to the new method. The data are given in Table 9.3. Do the data support the hypothesis that the population mean reading test score for slow learners taught by the new method is greater than the mean reading test score for those taught by the standard method?

We want to test

$$H_0: (\mu_1 - \mu_2) = 0$$
$$H_a: (\mu_1 - \mu_2) > 0$$

One way to conduct this test is to use the t statistic for two independent samples (Section 9.2). The analysis is shown on the MINITAB printout, Figure 9.9. The test statistic, $t = 1.26$, is highlighted on the printout as well as the p-value of the test, $p = .115$. At $\alpha = .10$, the p-value exceeds α. Thus, from *this* analysis we might conclude that we do not have sufficient evidence to infer a difference in the mean test scores for the two methods.

If you carefully examine the data in Table 9.3, however, you will find this result difficult to accept. The test score of the new method is larger than the corresponding test score for the standard method *for every one of the eight pairs of slow learners.* This, in itself, seems to provide strong evidence to indicate that μ_1 exceeds μ_2. Why,

⊙ **PAIREDSCORES**

TABLE 9.3 Reading Test Scores for Eight Pairs of Slow Learners

Pair	New Method (1)	Standard Method (2)
1	77	72
2	74	68
3	82	76
4	73	68
5	87	84
6	69	68
7	66	61
8	80	76

Figure 9.9

MINITAB Analysis
of Reading Test Scores
in Table 9.3

```
Two-sample T for NEW vs STANDARD

            N    Mean   StDev   SE Mean
NEW         8   76.00    6.93      2.4
STANDARD    8   71.63    7.01      2.5


Difference = mu (NEW) - mu (STANDARD)
Estimate for difference:   4.37500
95% lower bound for difference:   -1.76200
T-Test of difference = 0 (vs >): T-Value = 1.26   P-Value = 0.115   DF = 14
Both use Pooled StDev = 6.9687
```

then, did the *t*-test fail to detect this difference? The answer is; *The independent samples t-test is not a valid procedure to use with this set of data.*

The *t*-test is inappropriate because the assumption of independent samples is invalid. We have randomly chosen *pairs of test scores,* and thus, once we have chosen the sample for the new method, we have *not* independently chosen the sample for the standard method. The dependence between observations within pairs can be seen by examining the pairs of test scores, which tend to rise and fall together as we go from pair to pair. This pattern provides strong visual evidence of a violation of the assumption of independence required for the two-sample *t*-test of Section 9.2. Note, also, that

$$s_p^2 = \frac{(n_1 - 1)s_1^2 + (n_2 - 1)s_2^2}{n_1 + n_2 - 2} = \frac{(8 - 1)(6.93)^2 + (8 - 1)(7.01)^2}{8 + 8 - 2} = 48.55$$

Thus, there is a *large variation within samples* (reflected by the large value of s_p^2) in comparison to the relatively *small difference between the sample means.* Because s_p^2 is so large, the *t*-test of Section 9.2 is unable to detect a difference between μ_1 and μ_2.

We now consider a valid method of analyzing the data of Table 9.3. In Table 9.4 we add the column of differences between the test scores of the pairs of slow learners. We can regard these differences in test scores as a random sample of differences for all pairs (matched on reading IQ) of slow learners, past and present. Then we can use this sample to make inferences about the mean of the population of differences, μ_D, which is equal to the difference ($\mu_1 - \mu_2$). That is, the mean of the population (and sample) of differences equals the difference between the population (and sample) means. Thus, our test becomes

$$H_0: \mu_D = 0 \quad (\mu_1 - \mu_2 = 0)$$
$$H_a: \mu_D > 0 \quad (\mu_1 - \mu_2 > 0)$$

TABLE 9.4   Differences in Reading Test Scores

Pair	New Method	Standard Method	Difference (New Method − Standard Method)
1	77	72	5
2	74	68	6
3	82	76	6
4	73	68	5
5	87	84	3
6	69	68	1
7	66	61	5
8	80	76	4

The test statistic is a one-sample t (Section 8.4), since we are now analyzing a single sample of differences for small n:

$$Test\ statistic:\quad t = \frac{\bar{x}_D - 0}{s_D/\sqrt{n_D}}$$

where

$$\bar{x}_D = \text{Sample mean difference}$$
$$s_D = \text{Sample standard deviation of differences}$$
$$n_D = \text{Number of differences} = \text{Number of pairs}$$

Assumptions: The population of differences in test scores is approximately normally distributed. The sample differences are randomly selected from the population differences. [*Note:* We do not need to make the assumption that $\sigma_1^2 = \sigma_2^2$.]

Rejection region: At significance level $\alpha = .05$, we will reject H_0: if $t > t_{.05}$, where $t_{.05}$ is based on $(n_D - 1)$ degrees of freedom.

Referring to Table VI in Appendix A, we find the t-value corresponding to $\alpha = .05$ and $n_D - 1 = 8 - 1 = 7$ df to be $t_{.05} = 1.895$. Then we will reject the null hypothesis if $t > 1.895$ (see Figure 9.10). Note that the number of degrees of freedom decreases from $n_1 + n_2 - 2 = 14$ to 7 when we use the paired difference experiment rather than the two independent random samples design.

Figure 9.10

Rejection Region for Example 9.4

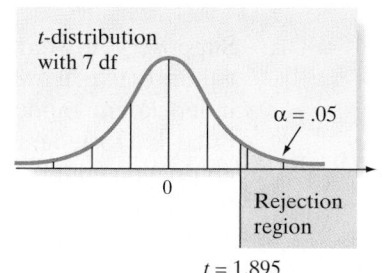

$t = 1.895$

Summary statistics for the $n = 8$ differences are shown in the MINITAB printout, Figure 9.11. Note that $\bar{x}_D = 4.375$ and $s_D = 1.685$. Substituting these values into the formula for the test statistic, we have

$$t = \frac{\bar{x}_D - 0}{s_D/\sqrt{n_D}} = \frac{4.375}{1.685/\sqrt{8}} = 7.34$$

Because this value of t falls in the rejection region, we conclude (at $\alpha = .05$) that the population mean test score for slow learners taught by the new method exceeds the population mean score for those taught by the standard method. We can reach the same conclusion by noting that the p-value of the test, highlighted in Figure 9.11, is much smaller than $\alpha = .05$.

Figure 9.11

MINITAB Paired Difference Analysis of Reading Test Scores

```
Paired T for NEW - STANDARD

              N      Mean    StDev   SE Mean
NEW           8   76.0000   6.9282    2.4495
STANDARD      8   71.6250   7.0089    2.4780
Difference    8    4.37500  1.68502   0.59574


95% lower bound for mean difference: 3.24632
T-Test of mean difference = 0 (vs > 0): T-Value = 7.34   P-Value = 0.000
```

Now Work   *Exercise 9.35ab*

This kind of experiment, in which observations are paired and the differences are analyzed, is called a **paired difference experiment**. In many cases, a paired difference experiment can provide more information about the difference between population means than an independent samples experiment. The idea is to compare population means by comparing the differences between pairs of experimental units (objects, people, etc.) that were very similar prior to the experiment. The differencing removes sources of variation that tend to inflate σ^2. For example, when two children are taught to read by two different methods, the observed difference in achievement may be due to a difference in the effectiveness of the two teaching methods *or* it may be due to differences in the initial reading levels and IQs of the two children (random error). To reduce the effect of differences in the children on the observed differences in reading achievement, the two methods of reading are imposed on two children who are more likely to possess similar intellectual capacity, namely children with nearly equal IQs. The effect of this pairing is to remove the larger source of variation that would be present if children with different abilities were randomly assigned to the two samples. Making comparisons within groups of similar experimental units is called **blocking**, and the paired difference experiment is a simple example of a **randomized block experiment**. In our example, pairs of children with matching IQ scores represent the blocks.

Some other examples for which the paired difference experiment might be appropriate are the following:

1. Suppose you want to estimate the difference $(\mu_1 - \mu_2)$ in mean price per gallon between two major brands of premium gasoline. If you choose two independent random samples of stations for each brand, the variability in price due to geographic location may be large. To eliminate this source of variability you could choose pairs of stations of similar size, one station for each brand, in close geographic proximity and use the sample of differences between the prices of the brands to make an inference about $(\mu_1 - \mu_2)$.

2. Suppose a college placement center wants to estimate the difference $(\mu_1 - \mu_2)$ in mean starting salaries for men and women graduates who seek jobs through the center. If it independently samples men and women, the starting salaries may vary because of their different college majors and differences in grade point averages. To eliminate these sources of variability, the placement center could match male and female job-seekers according to their majors and grade point averages. Then the differences between the starting salaries of each pair in the sample could be used to make an inference about $(\mu_1 - \mu_2)$.

3. Suppose you wish to estimate the difference $(\mu_1 - \mu_2)$ in mean absorption rate into the bloodstream for two drugs that relieve pain. If you independently sample people, the absorption rates might vary because of age, weight, sex, blood pressure, etc. In fact, there are many possible sources of nuisance variability, and pairing individuals who are similar in all the possible sources would be quite difficult. However, it may be possible to obtain two measurements *on the same person*. First, we administer one of the two drugs and record the time until absorption. After a sufficient amount of time, the other drug is administered and a second measurement on absorption time is obtained. The differences between the measurements for each person in the sample could then be used to estimate $(\mu_1 - \mu_2)$. This procedure would be advisable only if the amount of time allotted between drugs is sufficient to guarantee little or no carryover effect. Otherwise, it would be better to use different people matched as closely as possible on the factors thought to be most important.

Now Work   *Exercise 9.33*

The hypothesis-testing procedures and the method of forming confidence intervals for the difference between two means using a paired difference experiment are summarized in the following boxes for both large and small n.

Paired Difference Confidence Interval for $\mu_D = \mu_1 - \mu_2$

Large Sample

$$\bar{x}_D \pm z_{\alpha/2} \frac{\sigma_D}{\sqrt{n_D}} \approx \bar{x}_D \pm z_{\alpha/2} \frac{s_D}{\sqrt{n_D}}$$

Small Sample

$$\bar{x}_D \pm t_{\alpha/2} \frac{s_D}{\sqrt{n_D}}$$

where $t_{\alpha/2}$ is based on $(n_D - 1)$ degrees of freedom

---

Paired Difference Test of Hypothesis for $\mu_D = \mu_1 - \mu_2$

One-Tailed Test

H_0: $\mu_D = D_0$
H_a: $\mu_D < D_0$
  [or H_a: $\mu_D > D_0$]

Two-Tailed Test

H_0: $\mu_D = D_0$
H_a: $\mu_D \neq D_0$

Large Sample

Test statistic: $z = \dfrac{\bar{x}_D - D_0}{\sigma_D/\sqrt{n_D}} \approx \dfrac{\bar{x}_D - D_0}{s_D/\sqrt{n_D}}$

Rejection region: $z < -z_\alpha$
  [or $z > z_\alpha$ when H_a: $\mu_D > D_0$]

Rejection region: $|z| > z_{\alpha/2}$

Small Sample

Test statistic: $t = \dfrac{\bar{x}_D - D_0}{s_D/\sqrt{n_D}}$

Rejection region: $t < -t_\alpha$
  [or $t > t_\alpha$ when H_a: $\mu_D > D_0$]

Rejection region: $|t| > t_{\alpha/2}$

where t_α and $t_{\alpha/2}$ are based on $(n_D - 1)$ degrees of freedom

---

Conditions Required for Valid Large-Sample Inferences about μ_D

1. A random sample of differences is selected from the target population of differences.

2. The sample size n_D is large (i.e., $n_D \geq 30$). (Due to the Central Limit Theorem, this condition guarantees that the test statistic will be approximately normal regardless of the shape of the underlying probability distribution of the population.)

Conditions Required for Valid Small-Sample Inferences about μ_D

1. A random sample of differences is selected from the target population of differences.

2. The population of differences has a distribution that is approximately normal.

EXAMPLE 9.5 A CONFIDENCE INTERVAL FOR μ_D

Problem An experiment is conducted to compare the starting salaries of male and female college graduates who find jobs. Pairs are formed by choosing a male and a female with the same major and similar grade point averages (GPA). Suppose a random sample of 10 pairs is formed in this manner and the starting annual salary of each person is recorded. The results are shown in Table 9.5. Compare the mean starting salary, μ_1, for males to the mean starting salary, μ_2, for females using a 95% confidence interval. Interpret the results.

⊚ GRADPAIRS

TABLE 9.5 Data on Annual Salaries for Matched Pairs of College Graduates

Pair	Male	Female	Difference Male − Female	Pair	Male	Female	Difference Male − Female
1	$29,300	$28,800	$ 500	6	$37,800	$38,000	$−200
2	41,500	41,600	−100	7	69,500	69,200	300
3	40,400	39,800	600	8	41,200	40,100	1,100
4	38,500	38,500	0	9	38,400	38,200	200
5	43,500	42,600	900	10	59,200	58,500	700

Solution Since the data on annual salary are collected in pairs of males and females matched on GPA and major, a paired difference experiment is performed. To conduct the analysis, we first compute the differences between the salaries, as shown in Table 9.5. Summary statistics for these $n = 10$ differences are displayed at the top of the SAS printout, Figure 9.12.

Figure 9.12
SAS Analysis of Salary
Differences

The MEANS Procedure

Analysis Variable : DIFF

Mean	Std Dev	N	Minimum	Maximum
400.0000000	434.6134937	10	−200.0000000	1100.00

Two Sample Paired t-test for the Means of MALE and FEMALE

Sample Statistics

Group	N	Mean	Std. Dev.	Std. Error
MALE	10	43930	11665	3688.8
FEMALE	10	43530	11617	3673.6

Hypothesis Test

Null hypothesis: Mean of (MALE - FEMALE) = 0
Alternative: Mean of (MALE - FEMALE) ^= 0

t Statistic	Df	Prob > t
2.910	9	0.0173

95% Confidence Interval for the Difference between Two Paired Means

Lower Limit	Upper Limit
89.10	710.90

The 95% confidence interval for $\mu_D = (\mu_1 - \mu_2)$ for this small sample is

$$\bar{x}_D \pm t_{\alpha/2} \frac{s_D}{\sqrt{n_D}}$$

where $t_{\alpha/2} = t_{.025} = 2.262$ (obtained from Table VI, Appendix A) is based on $n - 1 = 9$ degrees of freedom. Substituting the values of $\bar{x}_D$ and s_D shown on the printout, we obtain

$$\bar{x}_D \pm 2.262 \frac{s_D}{\sqrt{n_D}} = 400 \pm 2.262 \left(\frac{434.613}{\sqrt{10}} \right)$$
$$= 400 \pm 310.88 \approx 400 \pm 311 = (\$89, \$711)$$

[*Note:* This interval is also shown highlighted at the bottom of the SAS printout, Figure 9.12.] Our interpretation is that the true mean difference between the starting salaries of males and females falls between $89 and $711, with 95% confidence. Since the interval falls above 0, we infer that $\mu_1 - \mu_2 > 0$; that is, the mean salary for males exceeds the mean salary for females.

Look Back Remember that $\mu_D = \mu_1 - \mu_2$. So if $\mu_D > 0$, then $\mu_1 > \mu_2$. Alternatively, if $\mu_D < 0$, then $\mu_1 < \mu_2$.

Now Work   *Exercise 9.40*

■ ■ ■

To measure the amount of information about $(\mu_1 - \mu_2)$ gained by using a paired difference experiment in Example 9.5 rather than an independent samples experiment, we can compare the relative widths of the confidence intervals obtained by the two methods. A 95% confidence interval for $(\mu_1 - \mu_2)$ using the paired difference experiment is, from Example 9.5, ($89, $711). If we analyzed the same data as though this were an independent samples experiment,* we would first obtain the descriptive statistics shown in the SAS printout, Figure 9.13. Then we substitute the sample means and standard deviations shown on the printout into the formula for a 95% confidence interval for $(\mu_1 - \mu_2)$ using independent samples:

$$(\bar{x}_1 - \bar{x}_2) \pm t_{.025} \sqrt{s_p^2 \left(\frac{1}{n_1} + \frac{1}{n_2} \right)}$$

Figure 9.13

SPSS Analysis of Salaries, assuming Independent Samples

Group Statistics

	GENDER	N	Mean	Std. Deviation	Std. Error Mean
SALARY	M	10	43930.00	11665.148	3688.844
	F	10	43530.00	11616.946	3673.601

Independent Samples Test

		Levene's Test for Equality of Variances		t-test for Equality of Means					95% Confidence Interval of the Difference	
		F	Sig.	t	df	Sig. (2-tailed)	Mean Difference	Std. Error Difference	Lower	Upper
SALARY	Equal variances assumed	.000	.991	.077	18	.940	400.00	5206.046	-10537.5	11337.50
	Equal variances not assumed			.077	18.000	.940	400.00	5206.046	-10537.5	11337.51

*This is done only to provide a measure of the increase in the amount of information obtained by a paired design in comparison to an unpaired design. Actually, if an experiment is designed using pairing, an unpaired analysis would be invalid because the assumption of independent samples would not be satisfied.

where

$$s_p^2 = \frac{(n_1 - 1)s_1^2 + (n_2 - 1)s_2^2}{n_1 + n_2 - 2}$$

SPSS performed these calculations and obtained the interval ($-10,537.50, $11,337.50). This interval is highlighted on Figure 9.13.

Notice that the independent samples interval includes 0. Consequently, if we were to use this interval to make an inference about $(\mu_1 - \mu_2)$, we would incorrectly conclude that the mean starting salaries of males and females do not differ! You can see that the confidence interval for the independent sampling experiment is about five times wider than for the corresponding paired difference confidence interval. Blocking out the variability due to differences in majors and grade point averages significantly increases the information about the difference in male and female mean starting salaries by providing a much more accurate (smaller confidence interval for the same confidence coefficient) estimate of $(\mu_1 - \mu_2)$.

You may wonder whether conducting a paired difference experiment is always superior to an independent samples experiment. The answer is, Most of the time, but not always. We sacrifice half the degrees of freedom in the t-statistic when a paired difference design is used instead of an independent samples design. This is a loss of information, and unless this loss is more than compensated for by the reduction in variability obtained by blocking (pairing), the paired difference experiment will result in a net loss of information about $(\mu_1 - \mu_2)$. Thus, we should be convinced that the pairing will significantly reduce variability before performing the paired difference experiment. Most of the time this will happen.

One final note: The pairing of the observations is determined *before* the experiment is performed (that is, by the *design* of the experiment). A paired difference experiment is *never* obtained by pairing the sample observations after the measurements have been acquired.

What Do You Do when the Assumption of a Normal Distribution for the Population of Differences Is Not Satisfied?

Answer: Use the Wilcoxon signed rank test for the paired difference design (Chapter 14).

Confidence Interval for a Paired Difference

Using The TI-83 Graphing Calculator

For this calculation, you have the option of entering the raw data or using the summary statistics

Step 1   *If you are using the raw data, enter the data into L1 and L2 and create a column of differences in L3.*

Press **STAT** and select **1:Edit**
Note: If a list already contains data, clear the old data. Use the up arrow to highlight the name at the top of the list. Press **CLEAR ENTER**.
Use the **Down Arrow** or **ENTER** key to enter the data sets into **L1** and **L2**.
Move the cursor so that it is flashing on '**L3**' at the top of the third column and press **ENTER**.
The cursor will be positioned at the bottom of the screen to the right of
'**L3 = .**' Type in 'L1-L2' by hitting **2nd** and '**1**' - **2nd** and '**2.**' Press **ENTER**.

Step 2 *Access the Statistical Tests Menu*

Press **STAT**
Arrow right to **TESTS**
Arrow down to **8: TInterval**
Press **ENTER**

Step 3 *Choose "**Data**" or "**Stats**". ("Data" is selected when you have entered the raw data into L1 and L2. "Stats" is selected when you are using the summary statistics.)*

Press **ENTER**

If you selected "Data":
Set **List** to **L3**. Note: Arrow down to List: and hit **2nd** and **'3'**.
Set **Freq** to **1**. Note: If Freq is not already set to 1, arrow down to Freq and hit **Alpha** and **'1'**.
Arrow down to **C-Level** and enter the confidence level.
Arrow down to **"Calculate"**
Press **ENTER**

If you selected "Stats", enter the mean of the paired differences, the standard deviation of the paired differences and sample size of the paired differences.
Arrow down to **C-Level** and enter the confidence level.
Arrow down to **"Calculate"**
Press **ENTER**

Step 4 *View the Output*

The confidence interval will be displayed with the mean and standard deviation of the paired differences and the sample size.

Example *Consider Example 9.5 with data in Table 9.5 on pg. 458. For this example, the raw data will be used, so begin with Step1 of the instructions.*

Step 1 *Enter the salaries for the Males into L1 and the salaries for the Females into L2 and create a column of differences in L3.*

Press **STAT** and select **1:Edit**
Note: If a list already contains data, clear the old data. Use the up arrow to highlight the name at the top of the list. Press **CLEAR ENTER**.
Use the **Down Arrow** or **Enter** key to enter the data sets into **L1** and **L2**.
Move the cursor so that it is flashing on **'L3'** at the top of the third column and press **ENTER**.
The cursor will be positioned at the bottom of the screen to the right of **'L3 = .'** Type in 'L1-L2' by hitting **2nd** and **'1'** - **2nd** and **'2.'** Press **ENTER**.

L1	L2	L3 3
29300	28800	500
41500	41600	-100
40400	39800	600
38500	38500	0
43500	42600	900
37800	38000	-200
69500	69200	300

L3(1)=500

Step 2 *Access the Statistical Tests Menu*

Press **STAT**
Arrow right to **TESTS**
Arrow down to **8: TInterval**
Press **ENTER**

Step 3 *Choose "**Data**"*

Press **ENTER**

Set **List** to **L3**. Note: Arrow down to List: and hit **2nd** and **'3'**.
Set **Freq** to **1**. Note: If Freq is not already set to 1, arrow down to Freq and hit **Alpha** and **'1'**.

Arrow down to **C-Level** and enter the confidence level.
Arrow down to **"Calculate"**
Press **ENTER**

```
TInterval
 Inpt:DATA Stats
 List:L₃
 Freq:1
 C-Level:.95
 Calculate
```

Step 4   *View the Output*

The confidence interval will be displayed with the mean and standard deviation of the paired differences and the sample size.

```
TInterval
 (89.096,710.9)
 x̄=400
 Sx=434.6134937
 n=10
```

Hypothesis Test for a Paired Difference

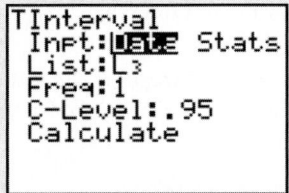

Using The TI-83 Graphing Calculator

For this calculation, you have the option of entering the raw data or using the summary statistics

Step 1   *If you are using the raw data, enter the data into L1 and L2 and create a column of differences in L3.*

Press **STAT** and select **1:Edit**
Note: If a list already contains data, clear the old data. Use the up arrow to highlight the name at the top of the list. Press **CLEAR ENTER**.
Use the **Down Arrow** or **ENTER** key to enter the data sets into **L1** and **L2**.
Move the cursor so that it is flashing on **'L3'** at the top of the third column and press **ENTER**.
The cursor will be positioned at the bottom of the screen to the right of
'L3 = .' Type in 'L1-L2' by hitting **2nd** and **'1' - 2nd** and **'2.'** Press **ENTER**.

Step 2   *Access the Statistical Tests Menu*

Press **STAT**
Arrow right to **TESTS**
Arrow down to **2 : TTest**
Press **ENTER**

Step 3   *Choose "**Data**" or "**Stats**". ("Data" is selected when you have entered the raw data into L1 and L2. "Stats" is selected when you are using the summary statistics.)*

Press **ENTER**

If you selected "Data":
Arrow down to μ_0 and enter the value for D_0 from the null hypothesis.
Note: In most cases, the value for D_0 is 0.
Set **List** to **L3**. Note: Arrow down to List: and hit **2nd** and **'3'**.
Set **Freq** to **1**. Note: If Freq is not already set to 1, arrow down to Freq and hit **Alpha** and **'1'**.

Arrow down to the next line and select the appropriate alternative hypothesis by using the right arrow to highlight $\neq u_0$, $<u_0$, or $>u_0$ and press **ENTER**.
Arrow down to **Calculate Draw**. You may choose either one of these for your output.
Press **ENTER**

If you selected "Stats":
Arrow down to μ_0 and enter the value for D_0 from the null hypothesis.
Note: In most cases, the value for D_0 is 0.
Enter the mean of the paired differences, the standard deviation of the paired differences and sample size of the paired differences.
Arrow down to the next line and select the appropriate alternative hypothesis by using the right arrow to highlight $\neq u_0$, $<u_0$, or $>u_0$ and press **ENTER**.
Arrow down to **Calculate Draw**. You may choose either one of these for your output.
Press **ENTER**

Step 4 *View the Output*

If you chose **Calculate,** the output will display the alternative hypothesis, the t-value, the p-value, the mean and standard deviation of the paired differences and the sample size.
If you chose **Draw,** a graph of the normal curve will be displayed along with the t-value and the p-value.
The area under the curve that represents the p-value is shaded. Note: In order to use the **Draw** feature, you must first clear all equations from the **Y-registers**.
Hit '**Y =** ' and delete all equations.

Example *Consider Exercise 9.42 on pg. 466. For this example, the raw data will be used, so begin with Step 1 of the instructions.*

Step 1 *Enter the 'Attitude toward Father' into L1 and the 'Attitude toward Mother' into L2 and create a column of differences in L3.*

Press **STAT** and select **1:Edit**
Note: If a list already contains data, clear the old data. Use the up arrow to highlight the name at the top of the list. Press **CLEAR ENTER**.
Use the **Down Arrow** or **ENTER** key to enter the data sets into **L1** and **L2**.
Move the cursor so that it is flashing on '**L3**' at the top of the third column and press **ENTER**.
The cursor will be positioned at the bottom of the screen to the right of '**L3 = .**' Type in 'L1-L2' by hitting **2nd** and '**1**' - **2nd** and '**2.**' Press **ENTER**.

Step 2 *Access the Statistical Tests Menu*

Press **STAT**
Arrow right to **TESTS**
Arrow down to **2 : TTest**
Press **ENTER**

Step 3 *Choose "Data"*

Press **ENTER**
Arrow down to μ_0 and enter 0, the value for D_0 from the null hypothesis.
Set **List** to **L3**. Note: Arrow down to List: and hit **2nd** and '**3**'.
Set **Freq** to **1**. Note: If Freq is not already set to 1, arrow down to Freq and hit **Alpha** and '**1**'.

Arrow down to the next line and select $\neq u_0$ and press **ENTER**.
Arrow down to **Calculate Draw**. You may choose either one of these for your
output.
Press **ENTER**

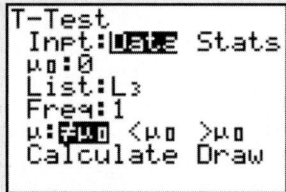

Step 4 *View the Output*

If you chose **Calculate,** the output will display the alternative hypothesis, the t-
value, the p-value, the mean and standard deviation of the paired differences
and the sample size.

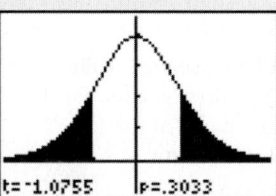

If you chose **Draw,** a graph of the normal curve will be displayed along with
the t-value and the p-value. The area under the curve that represents the
p-value is shaded. Note: In order to use the **Draw** feature, you must first
clear all equations from the **Y-registers**. Hit 'Y= ' and delete all equations.

Exercises 9.30–9.48

Understanding the Principles

9.30 What are the advantages of using a paired difference
experiment over an independent samples design?

9.31 In a paired difference experiment, when should the
observations be paired—before or after collecting the
data?

9.32 What conditions are required for valid large-sample
inferences about μ_D? Small-sample inferences?

Learning the Mechanics

9.33 A paired difference experiment yielded n_D pairs of
NW observations. In each case, what is the rejection region
for testing $H_0: \mu_D = 2$ against $H_a: \mu_D > 2$?
a. $n_D = 10, \alpha = .05$
b. $n_D = 20, \alpha = .10$
c. $n_D = 5, \alpha = .025$
d. $n_D = 9, \alpha = .01$

9.34 A paired difference experiment produced the following data:

$$n_D = 16 \quad \bar{x}_1 = 143 \quad \bar{x}_2 = 150 \quad \bar{x}_D = -7 \quad s_D^2 = 64$$

a. Determine the values of t for which the null
hypothesis, $\mu_1 - \mu_2 = 0$, would be rejected in favor
of the alternative hypothesis, $\mu_1 - \mu_2 < 0$. Use
$\alpha = .10$.
b. Conduct the paired difference test described in part
a. Draw the appropriate conclusions.
c. What assumptions are necessary so that the paired
difference test will be valid?
d. Find a 90% confidence interval for the mean difference μ_D.
e. Which of the two inferential procedures, the confidence interval of part **d** or the test of hypothesis of
part **b**, provides more information about the difference between the population means?

9.35 The data for a random sample of six paired observations are shown in the table.

NW

⊚ LM9_35

Pair	Sample from Population 1	Sample from Population 2
1	7	4
2	3	1
3	9	7
4	6	2
5	4	4
6	8	7

a. Calculate the difference between each pair of observations by subtracting observation 2 from observation 1. Use the differences to calculate $\bar{x}_D$ and s_D^2.

b. If μ_1 and μ_2 are the means of populations 1 and 2, respectively, express μ_D in terms of μ_1 and μ_2.

c. Form a 95% confidence interval for μ_D.

d. Test the null hypothesis $H_0: \mu_D = 0$ against the alternative hypothesis $H_a: \mu_D \neq 0$. Use $\alpha = .05$.

9.36 The data for a random sample of 10 paired observations are shown in the next table.

⊚ LM9_36

Pair	Population 1	Population 2
1	19	24
2	25	27
3	31	36
4	52	53
5	49	55
6	34	34
7	59	66
8	47	51
9	17	20
10	51	55

a. If you wish to test whether these data are sufficient to indicate that the mean for population 2 is larger than that for population 1, what are the appropriate null and alternative hypotheses? Define any symbols you use.

b. Conduct the test, part a, using $\alpha = .10$. What is your decision?

c. Find a 90% confidence interval for μ_D. Interpret this interval.

d. What assumptions are necessary to ensure the validity of this analysis?

9.37 A paired difference experiment yielded the following results:

$$n_D = 40$$
$$\bar{x}_D = 11.7$$
$$s_D = 6.$$

a. Test $H_0: \mu_D = 10$ against $H_a: \mu_D \neq 10$, where $\mu_D = (\mu_1 - \mu_2)$. Use $\alpha = .05$.

b. Report the p-value for the test you conducted in part a. Interpret the p-value.

Applying the Concepts—Basic

9.38 Life expectantcy of oscar winners. Does winning an Academy of Motion Picture Arts and Sciences award lead to long-term mortality for movie actors? In an article in the *Annals of Internal Medicine* (May 15, 2001), researchers sampled 762 Academy Award winners and matched each one with another actor of the same sex who was in the same winning film and was born in the same era. The life expectancy (age) of each pair of actors was compared.

a. Explain why the data should be analyzed as a paired difference experiment.

b. Set up the null hypothesis for a test to compare the mean life expectancies of Academy Award winners and nonwinners.

c. The sample mean life expectancies of Academy Award winners and nonwinners were reported as 79.7 years and 75.8 years, respectively. The p-value for comparing the two population means was reported as $p = .003$. Interpret this value in the context of the problem.

9.39 The placebo effect and pain. According to research published in *Science* (Feb. 20, 2004), the mere belief that you are receiving an effective, treatment for pain can reduce the pain you actually feel. Researchers from the University of Michigan and Princeton University tested this placebo effect on 24 volunteers as follows. Each volunteer was put inside a magnetic resonance imaging machine (MRI) for two consecutive sessions. During the first session, electric shocks were applied to their arms and the blood oxygen level-dependent (BOLD) signal (a measure related to neural activity in the brain) was recorded during pain. The second session was identical to the first, except that prior to applying the electric shocks the researchers smeared a cream on the volunteer's arms. The volunteers were informed that the cream would block the pain, when, in fact, it was just a regular skin lotion (i.e., a placebo). If the placebo is effective in reducing the pain experience, the BOLD measurements should be higher, on average, in the first MRI session than in the second MRI session.

a. Identify the target parameter for this study.

b. What type of design was used to collect the data?

c. Give the null and alternative hypothesis for testing the placebo effect theory.

d. The differences between the BOLD measurements in the first and second sessions were computed and summarized in the study as follows: $n_D = 24$, $\bar{x}_D = .21$, $s_D = .47$. Use this information to calculate the test statistic.

e. The p-value of the test was reported as p-value $= .02$. Make the appropriate conclusion at $\alpha = .05$.

⊚ CRASH

9.40 NHTSA new car crash tests. Refer to the National

NW Highway Traffic Safety, Administration (NHTSA) crash test data for new cars saved in the **CRASH** file.

Crash test dummies were placed in the driver's seat and front passenger's seat of a new car model, and the car was steered by remote control into a head-on collision with a fixed barrier while traveling at 35 miles per hour. Two of the variables measured for each of the 98 new cars in the data set are (1) the severity of the driver's chest injury and (2) the severity of the passenger's chest injury. (The more points assigned to the chest injury rating, the more severe the injury.) Suppose the NHTSA wants to determine whether the true mean driver chest injury rating exceeds the true mean passenger chest injury rating, and if so, by how much.

a. State the parameter of interest to the NHTSA.
b. Explain why the data should be analyzed as matched pairs.
c. Find a 99% confidence interval for the true difference between the mean chest injury ratings of drivers and front-seat passengers.
d. Interpret the interval, part c. Does the true mean driver chest injury rating exceed the true mean passenger chest injury rating? If so, by how much?
e. What conditions are required for the analysis to be valid? Do these conditions hold for these data?

9.41 Depo-Provera drug study. Hypersexual behavior caused by traumatic brain injury (TBI) is often treated with the drug Depo-Provera. In one clinical study, eight young male TBI patients who exhibited hypersexual behavior were treated weekly with 400 milligrams of Depo-Provera for six months. (*Journal of Head Trauma Rehabilitation,* June 1995.) The testosterone levels (in nanograms per deciliter) of the patients both prior to treatment and at the end of the 6-month treatment period are given in the table below.

DEPOPROV

Patient	Pretreatment	After 6 months on Depo-Provera
1	849	96
2	903	41
3	890	31
4	1,092	124
5	362	46
6	900	53
7	1,006	113
8	672	174

Source: Emory, L. E., Cole, C. M., and Meyer, W. J. "Use of Depo-Provera to control sexual aggression in persons with traumatic brain injury." *Journal of Head Trauma Rehabilitation,* Vol. 10, No. 3, June 1995, p. 52 (Table 2).

a. Construct a 99% confidence interval for the true mean difference between the pretreatment and after-treatment testosterone levels of young male TBI patients.
b. Use the interval, part a, to make an inference about the effectiveness of Depo-Provera in reducing testosterone levels of young male TBI patients.

9.42 Students attitudes toward parents. Researchers at the University of South Alabama compared the attitudes of male college students toward their fathers with their attitudes toward their mothers. (*Journal of Genetic Psychology*, March 1998.) Each of a sample of 13 males was asked to complete the following statement about each of their parents: My relationship with my father (mother) can best be described as (1) Awful, (2) Poor, (3) Average, (4) Good, or (5) Great. The following data were obtained:

FMATTITUDES

Student	Attitude toward Father	Attitude toward Mother
1	2	3
2	5	5
3	4	3
4	4	5
5	3	4
6	5	4
7	4	5
8	2	4
9	4	5
10	5	4
11	4	5
12	5	4
13	3	3

Source: Adapted from Vitulli, W. F., and Richardson, D. K. "College student's attitudes toward relationships with parents: A five-year comparative analysis." *Journal of Genetic Psychology*, Vol. 159, No. 1, (March 1998), pp. 45–52.

a. Specify the appropriate hypotheses for testing whether male students' attitudes toward their fathers differ from their attitudes toward their mothers, on average.
b. Conduct the test of part a at $\alpha = .05$. Interpret the results in the context of the problem.

Applying the Concepts—Intermediate

9.43 Reading tongue-twisters. According to *Webster's New World Dictionary*, a tongue-twister is "a phrase that is hard to speak rapidly." Do tongue-twisters have an effect on the length of time it takes to read silently? To answer this question, 42 undergraduate psychology students participated in a reading experiment. (*Memory & Cognition*, Sept. 1997.) Two lists, each comprised of 600 words, were constructed. One list contained a series of tongue-twisters and the other list (called the *control*) did not contain any tongue-twisters. Each student read both lists and the length of time (in minutes) required to complete the lists was recorded. The researchers used a test of hypothesis to compare the mean reading response times for the tongue-twister and control lists.

a. Set up the null hypothesis for the test.
b. Use the information in the table at the top of p. 467 to find the test statistic and *p*-value of the test.
c. Give the appropriate conclusion. Use $\alpha = .05$.

List Type	Response Time (minutes)	
	Mean	Standard Deviation
Tongue-twister	6.59	1.94
Control	6.34	1.92
Difference	.25	.78

Source: Robinson, D. H., and Katayama, A. D. "At-lexical, articulatory interference in silent reading: The 'upstream' tongue-twister effect." *Memory & Cognition,* Vol. 25, No. 5, Sept. 1997, p. 663.

9.44 Visual search and memory study. When searching for an item (e.g., a roadside traffic sign, a lost earring, or a tumor in a mammogram), common sense dictates that you will not re-examine items previously rejected. However, researchers at Harvard Medical School found that a visual search has no memory. (*Nature,* Aug. 6, 1998.) In their experiment, nine subjects searched for the letter "T" mixed among several letters "L." Each subject conducted the search under two conditions: random and static. In the random condition, the location of the letters were changed every 111 milliseconds; in the static condition, the location of the letters remained unchanged. In each trial, the reaction time (i.e., the amount of time it took the subject to locate the target letter) was recorded in milliseconds.

a. One goal of the research is to compare the mean reaction times of subjects in the two experimental conditions. Explain why the data should be analyzed as a paired-difference experiment.

b. If a visual search has no memory, then the main reaction times in the two conditions will not differ. Specify H_0 and H_a for testing the "no memory" theory.

c. The test statistic was calculated as $t = 1.52$ with p-value $= .15$. Make the appropriate conclusion.

9.45 Linking dementia and leisure activities. Does participation in leisure activities in your youth reduce the risk of Alzheimer's disease and other forms of dementia? To answer this question, a group of university researchers studied a sample of 107 same-sex Swedish twin pairs. (*Journal of Gerontology: Psychological Sciences and Social Sciences*, Sept. 2003.) Each twin pair was discordant for dementia—that is, one member of each pair was diagnosed with Alzheimer's disease while the other member (the control) was nondemented for at least five years after the sibling's onset of dementia. The level of overall leisure activity (measured on an 80-point scale, where higher values indicate higher levels of leisure activity) of each twin pair member 20 years prior to onset of dementia was obtained from the Swedish Twin Registry database. The leisure activity scores (simulated, based on summary information presented in the journal article) are saved in the **DEMENTIA** file. The first five and last five observations are shown in the table.

DEMENTIA (first and last 5 observations)

Pair	Control	Demented
1	27	13
2	57	57
3	23	31
4	39	46
5	37	37
.	.	.
.	.	.
103	22	14
104	32	23
105	33	29
106	36	37
107	24	1

a. Explain why the data should be analyzed as a paired difference experiment.

b. Conduct the appropriate analysis using $\alpha = .05$. Make an inference about which group, the demented or control (nondemented) twin members, had the largest average level of leisure activity.

9.46 Swim maze study. Merck Research Labs conducted an experiment to evaluate the effect of a new drug using the single-T swim maze. Nineteen impregnated dam rats were allocated a dosage of 12.5 milligrams of the drug. One male and one female rat pup were randomly selected from each resulting litter to perform in the swim maze. Each pup was placed in the water at one end of the maze and allowed to swim until it escaped at the opposite end. If the pup failed to escape after a certain period of time, it was placed at the beginning of the maze and given another chance. The experiment was repeated until each pup accomplished three successful escapes. The accompanying table reports the number of swims required by each pup to perform three successful escapes. Is there sufficient evidence of a difference between the mean number of swims required by male and female pups? Conduct the test (at $\alpha = .10$). Comment on the assumptions required for the test to be valid.

RATPUPS

Litter	Male	Female	Litter	Male	Female
1	8	5	11	6	5
2	8	4	12	6	3
3	6	7	13	12	5
4	6	3	14	3	8
5	6	5	15	3	4
6	6	3	16	8	12
7	3	8	17	3	6
8	5	10	18	6	4
9	4	4	19	9	5
10	4	4			

Source: Thomas E. Bradstreet, Merck Research Labs, BL 3–2, West Point, PA 19486.

Applying the Concepts—Advanced

9.47 Homophone confusion in Alzheimer's patients. A *homophone* is a word whose pronunciation is the same as that of another word having a different meaning and spelling (e.g., *nun* and *none, doe* and *dough,* etc.). *Brain and Language* (Apr. 1995) reported on a study of homophone spelling in patients with Alzheimer's disease. Twenty Alzheimer's patients were asked to spell 24 homophone pairs given in random order, then the number of homophone confusions (e.g., spelling *doe* given the context, *bake bread dough*) was recorded for each patient. One year later, the same test was given to the same patients. The data for the study are provided in the table. The researchers posed the following question: "Do Alzheimer's patients show a significant increase in mean homophone confusion errors over time?" Perform an analysis of the data to answer the researchers' question. What assumptions are necessary for the procedure used to be valid? Are they satisfied?

HOMOPHONE

Patient	Time 1	Time 2
1	5	5
2	1	3
3	0	0
4	1	1
5	0	1
6	2	1
7	5	6
8	1	2
9	0	9
10	5	8
11	7	10
12	0	3
13	3	9
14	5	8
15	7	12
16	10	16
17	5	5
18	6	3
19	9	6
20	11	8

Source: Neils, J., Roeltgen, D. P., and Constantinidou, F. "Decline in homophone spelling associated with loss of semantic influence on spelling in Alzheimer's disease." *Brain and Language,* Vol. 49, No. 1, Apr. 1995, p. 36 (Table 3).

9.48 Alcoholic fermentation in wines. Determining alcoholic fermentation in wine is critical to the wine-making process. Must/wine density is a good indicator of the fermentation point since the density value decreases as sugars are converted into alcohol. For decades, winemakers have measured must/wine density with a hydrometer. Although accurate, the hydrometer employs a manual process that is very time consuming. Consequently, large wineries are searching for more rapid measures of density measurement. An alternative method utilizes the hydrostatic balance instrument (similar to the hydrometer, but digital). A winery in Portugal collected the must/wine density measurements for white wine samples randomly selected from the fermentation process for a recent harvest. For each sample, the density of the wine at 20°C was measured with both the hydrometer and the hydrostat balance. The densities for 40 wine samples are saved in the **WINE40** file. The first five and last five observations are shown in the table. The winery will use the alternative method of measuring wine density only if it can be demonstrated that the mean difference between the density measurements of the two methods does not exceed .002. Perform the analysis for the winery. Provide the winery with a written report of your conclusions.

WINE40 (first and last 5 observations)

Sample	Hydrometer	Hydrostatic
1	1.08655	1.09103
2	1.00270	1.00272
3	1.01393	1.01274
4	1.09467	1.09634
5	1.10263	1.10518
.	.	.
.	.	.
36	1.08084	1.08097
37	1.09452	1.09431
38	0.99479	0.99498
39	1.00968	1.01063
40	1.00684	1.00526

Source: Cooperative Cellar of Borba *(Adega Cooperative de Borba),* Portugal.

9.4 Comparing Two Population Proportions: Independent Sampling

Suppose a presidential candidate wants to compare the preference of registered voters in the northeastern United States (NE) to those in the southeastern United States (SE). Such a comparison would help determine where to concentrate campaign efforts. The candidate hires a professional pollster to randomly choose 1,000 registered voters in the northeast and 1,000 in the southeast and interview each to learn her or his voting preference. The objective is to use this sample information to make an inference about the difference $(p_1 - p_2)$ between the proportion p_1 of *all* registered voters in the northeast and the proportion p_2 of *all* registered voters in the southeast who plan to vote for the presidential candidate.

The two samples represent independent binomial experiments. (See Section 4.4 for the characteristics of binomial experiments.) The binomial random variables are the numbers x_1 and x_2 of the 1,000 sampled voters in each area who indicate they will vote for the candidate. The results are summarized in Table 9.6

We can now calculate the sample proportions $\hat{p}_1$ and $\hat{p}_2$ of the voters in favor of the candidate in the northeast and southeast, respectively:

$$\hat{p} = \frac{x_1}{n_1} = \frac{546}{1,000} = .546 \quad \hat{p}_2 = \frac{x_2}{n_2} = \frac{475}{1,000} = .475$$

The difference between the sample proportions $(\hat{p}_1 - \hat{p}_2)$ makes an intuitively appealing point estimator of the difference between the population $(p_1 - p_2)$. For our example, the estimate is

$$(\hat{p}_1 - \hat{p}_2) = .546 - .475 = .071$$

To judge the reliability of the estimator $(\hat{p}_1 - \hat{p}_2)$, we must observe its performance in repeated sampling from the two populations. That is, we need to know the sampling distribution of $(\hat{p}_1 - \hat{p}_2)$. The properties of the sampling distribution are given in the following box. Remember that $\hat{p}_1$ and $\hat{p}_2$ can be viewed as means of the number of successes per trial in the respective samples, so the Central Limit Theorem applies when the sample sizes are large.

> **Properties of the Sampling Distribution of $(\hat{p}_1 - \hat{p}_2)$**
>
> **1.** The mean of the sampling distribution of $(\hat{p}_1 - \hat{p}_2)$ is $(p_1 - p_2)$ that is,
>
> $$E(\hat{p}_1 - \hat{p}_2) = p_1 - p_2$$
>
> Thus, $(\hat{p}_1 - \hat{p}_2)$ is an unbiased estimator of $(p_1 - p_2)$.
>
> **2.** The standard deviation of the sampling distribution of $(\hat{p}_1 - \hat{p}_2)$ is
>
> $$\sigma_{(\hat{p}_1 - \hat{p}_2)} = \sqrt{\frac{p_1 q_1}{n_1} + \frac{p_2 q_2}{n_2}}$$
>
> **3.** If the sample sizes n_1 and n_2 are large (see Section 7.3 for a guideline), the sampling distribution of $(\hat{p}_1 - \hat{p}_2)$ is approximately normal.

Since the distribution of $(\hat{p}_1 - \hat{p}_2)$ in repeated sampling is approximately normal, we can use the z statistic to derive confidence intervals for $(p_1 - p_2)$ or to test a hypothesis about $(p_1 - p_2)$.

For the voter example, a 95% confidence interval for the difference $(p_1 - p_2)$ is

$$(\hat{p}_1 - \hat{p}_2) \pm 1.96\sigma_{(\hat{p}_1 - \hat{p}_2)} \text{ or } (\hat{p}_1 - \hat{p}_2) \pm 1.96\sqrt{\frac{p_1 q_1}{n_1} + \frac{p_2 q_2}{n_2}}$$

The quantities $p_1 q_1$ and $p_2 q_2$ must be estimated in order to complete the calculation of the standard deviation, $\sigma_{(\hat{p}_1 - \hat{p}_2)}$, and hence the calculation of the confidence interval. In Section 7.3 we showed that the value of pq is relatively insensitive to the value chosen to approximate p. Therefore, $\hat{p}_1 \hat{q}_1$ and $\hat{p}_2 \hat{q}_2$ will provide satisfactory estimates to approximate $p_1 q_1$ and $p_2 q_2$, respectively. Then

$$\sqrt{\frac{p_1 q_1}{n_1} + \frac{p_2 q_2}{n_2}} \approx \sqrt{\frac{\hat{p}_1 \hat{q}_1}{n_1} + \frac{\hat{p}_2 \hat{q}_2}{n_2}}$$

and we will approximate the 95% confidence interval by

$$(\hat{p}_1 - \hat{p}_2) \pm 1.96\sqrt{\frac{\hat{p}_1\hat{q}_1}{n_1} + \frac{\hat{p}_2\hat{q}_2}{n_2}}$$

Substituting the sample quantities yields

$$(.546 - .475) \pm 1.96\sqrt{\frac{(.546)(.454)}{1,000} + \frac{(.475)(.525)}{1,000}}$$

or $.071 \pm .044$. Thus, we are 95% confident that the interval from .027 to .115 contains $(p_1 - p_2)$.

We infer that there are between 2.7% and 11.5% more registered voters in the northeast than in the southeast who plan to vote for the presidential candidate. It seems that the candidate should direct a greater campaign effort in the southeast compared to the northeast.

Now Work *Exercise 9.59*

The general form of a confidence interval for the difference $(p_1 - p_2)$ between population proportions is given in the following box.

Large-Sample $100(1 - \alpha)\%$ Confidence Interval for $(p_1 - p_2)$

$$(\hat{p}_1 - \hat{p}_2) \pm z_{\alpha/2}\sigma_{(\hat{p}_1-\hat{p}_2)} = (\hat{p}_1 - \hat{p}_2) \pm z_{\alpha/2}\sqrt{\frac{p_1 q_1}{n_1} + \frac{p_2 q_2}{n_2}}$$

$$\approx (\hat{p}_1 - \hat{p}_2) \pm z_{\alpha/2}\sqrt{\frac{\hat{p}_1\hat{q}_1}{n_1} + \frac{\hat{p}_2\hat{q}_2}{n_2}}$$

The z-statistic,

$$z = \frac{(\hat{p}_1 - \hat{p}_2) - (p_1 - p_2)}{\sigma_{(\hat{p}_1-\hat{p}_2)}}$$

is used to test the null hypothesis that $(p_1 - p_2)$ equals some specified difference, say D_0. For the special case where $D_0 = 0$, that is, where we want to test the null hypothesis H_0: $(p_1 - p_2) = 0$ (or, equivalently, H_0: $p_1 = p_2$), the best estimate of $p_1 = p_2 = p$ is obtained by dividing the total number of successes $(x_1 + x_2)$ for the two samples by the total number of observations $(n_1 + n_2)$; that is,

$$\hat{p} = \frac{x_1 + x_2}{n_1 + n_2} \quad \text{or} \quad \hat{p} = \frac{n_1\hat{p}_1 + n_2\hat{p}_2}{n_1 + n_2}$$

The second equation shows that $\hat{p}$ is a weighted average of $\hat{p}_1$ and $\hat{p}_2$, with the larger sample receiving more weight. If the sample sizes are equal, then $\hat{p}$ is a simple average of the two sample proportions of successes.

We now substitute the weighted average $\hat{p}$ for both p_1 and p_2 in the formula for the standard deviation of $(\hat{p}_1 - \hat{p}_2)$:

$$\sigma_{(\hat{p}_1-\hat{p}_2)} = \sqrt{\frac{p_1 q_1}{n_1} + \frac{p_2 q_2}{n_2}} \approx \sqrt{\frac{\hat{p}\hat{q}}{n_1} + \frac{\hat{p}\hat{q}}{n_2}} = \sqrt{\hat{p}\hat{q}\left(\frac{1}{n_1} + \frac{1}{n_2}\right)}$$

The test is summarized in the next box.

Large-Sample Test of Hypothesis About $(p_1 - p_2)$

One-Tailed Test	**Two-Tailed Test**

One-Tailed Test

$H_0: (p_1 - p_2) = 0*$
$H_a: (p_1 - p_2) < 0$
$\quad$ [or $H_a: (p_1 - p_2) > 0$]

Test statistic:

Two-Tailed Test

$H_0: (p_1 - p_2) = 0$
$H_a: (p_1 - p_2) \neq 0$

$$z = \frac{(\hat{p}_1 - \hat{p}_2)}{\sigma_{(\hat{p}_1 - \hat{p}_2)}}$$

Rejection region: $z < -z_\alpha$
$\quad$ [or $z > z_\alpha$ when $H_a: (p_1 - p_2) > 0$]

Rejection region: $|z| > z_{\alpha/2}$

Note: $\sigma_{(\hat{p}_1 - \hat{p}_2)} = \sqrt{\dfrac{p_1 q_1}{n_1} + \dfrac{p_2 q_2}{n_2}} \approx \sqrt{\hat{p}\hat{q}\left(\dfrac{1}{n_1} + \dfrac{1}{n_2}\right)}$ $\quad$ where $\quad \hat{p} = \dfrac{x_1 + x_2}{n_1 + n_2}$

Conditions Required for Valid Large-Sample Inferences about $(p_1 - p_2)$

1. The two samples are randomly selected in an independent manner from the two target populations.
2. The sample sizes, n_1 and n_2, are both large so that the sampling distribution of $(\hat{p}_1 - \hat{p}_2)$ will be approximately normal. (This condition will be satisfied if both intervals, $\hat{p}_1 \pm 3\sigma_{\hat{p}_1}$ and $\hat{p}_2 \pm 3\sigma_{\hat{p}_2}$, fall between 0 and 1.)

EXAMPLE 9.6

LARGE-SAMPLE TEST ABOUT $p_1 - p_2$

Problem In the past decade intensive antismoking campaigns have been sponsored by both federal and private agencies. Suppose the American Cancer Society randomly sampled 1,500 adults in 1995 and then sampled 1,750 adults in 2005 to determine whether there was evidence that the percentage of smokers had decreased. The results of the two sample surveys are shown in Table 9.7, where x_1 and x_2 represent the numbers of smokers in the 1995 and 2005 samples, respectively. Do these data indicate that the fraction of smokers decreased over this 10-year period? Use $\alpha = .05$.

Solution If we define p_1 and p_2 as the true proportions of adult smokers in 1995 and 2005, the elements of our test are

TABLE 9.7 Results of Smoking Survey

1995	2005
$n_1 = 1,500$	$n_2 = 1,750$
$x_1 = 555$	$x_2 = 578$

$$H_0: (p_1 - p_2) = 0$$
$$H_a: (p_1 - p_2) > 0$$

(The test is one-tailed since we are interested only in determining whether the proportion of smokers *decreased*.)

$$\textit{Test statistic: } z = \frac{(\hat{p}_1 - \hat{p}_2) - 0}{\sigma_{(\hat{p}_1 - \hat{p}_2)}}$$

Rejection region using $\alpha = .05$:
$$z > z_\alpha = z_{.05} = 1.645 \qquad \text{(see Figure 9.14)}$$

---

*The test can be adapted to test for a difference $D_0 \neq 0$. Because most applications call for a comparison of p_1 and p_2, implying $D_0 = 0$. we will confine our attention to this case.

Figure 9.14
Rejection Region
for Example 9.6

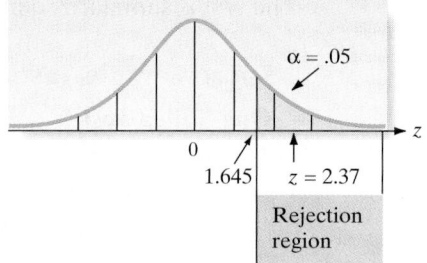

We now calculate the sample proportions of smokers

$$\hat{p}_1 = \frac{555}{1{,}500} = .37 \qquad \hat{p}_2 = \frac{578}{1{,}750} = .33$$

Then

$$z = \frac{(\hat{p}_1 - \hat{p}_2) - 0}{\sigma_{(\hat{p}_1 - \hat{p}_2)}} \approx \frac{(\hat{p}_1 - \hat{p}_2)}{\sqrt{\hat{p}\hat{q}\left(\dfrac{1}{n_1} + \dfrac{1}{n_2}\right)}}$$

where

$$\hat{p} = \frac{x_1 + x_2}{n_1 + n_2} = \frac{555 + 578}{1{,}500 + 1{,}750} = .349$$

Note that $\hat{p}$ is a weighted average of $\hat{p}_1$ and $\hat{p}_2$, with more weight given to the larger (2005) sample.

Thus, the computed value of the test statistic is

$$z = \frac{.37 - .33}{\sqrt{(.349)(.651)\left(\dfrac{1}{1{,}500} + \dfrac{1}{1{,}750}\right)}} = \frac{.040}{.0168} = 2.37$$

There is sufficient evidence at the $\alpha = .05$ level to conclude that the proportion of adults who smoke has decreased over the 1995–2005 period.

Look Back We could place a confidence interval on $(p_1 - p_2)$ if we were interested in estimating the extent of the decrease.

Now Work   *Exercise 9.60*

■ ■ ■

EXAMPLE 9.7   FINDING THE OBSERVED SIGNIFICANCE LEVEL
OF A TEST FOR $p_1 - p_2$

Problem   Use a statistical software package to conduct the test in Example 9.6. Find and interpret the *p*-value of the test.

Solution   We entered the sample sizes (n_1 and n_2) and numbers of successes (x_1 and x_2) into MINITAB and obtained the printout shown in Figure 9.15. The test statistic for this one-tailed test, $z = 2.37$, is shaded on the printout, as well as the *p*-value of the test. Note that *p*-value $= .009$ is smaller than $\alpha = .05$. Consequently, we have strong evidence to reject H_0 and conclude that p_1 exceeds p_2.

Figure 9.15

MINITAB Output for Test of Two Proportions

Test and CI for Two Proportions

```
Sample    X     N   Sample p
1        555  1500   0.370000
2        578  1750   0.330286


Difference = p (1) - p (2)
Estimate for difference:  0.0397143
95% lower bound for difference:  0.0121024
Test for difference = 0 (vs > 0):   Z = 2.37   P-Value = 0.009
```

■ ■ ■

Confidence Interval for $p_1 - p_2$

Using the TI-83 Graphing Calculator

Step 1 *Access the Statistical Tests Menu*

Press **STAT**
Arrow right to **TESTS**
Arrow down to **B: 2-PropZint**
Press **ENTER**

Step 2 *Enter the summary statistics.*

For x_1, enter the number of successes in the first data set.
For n_1, enter the sample size for the first data set.
For x_2, enter the number of successes in the second data set.
For n_2, enter the sample size for the second data set.
For **C-Level**, enter the confidence level.
Arrow down to "**Calculate**"
Press **ENTER**

Step 3 *View the Output*

The confidence interval will be displayed along with the two sample proportions ($\hat{p}_1$ and $\hat{p}_2$) and the sample sizes.

Example *Consider the data in Table 9.6 on pg. ###*

Step 1 *Access the Statistical Tests Menu*

Press **STAT**
Arrow right to **TESTS**
Arrow down to **B: 2-PropZint**
Press **ENTER**

```
EDIT CALC TESTS
6↑2-PropZTest…
7:ZInterval…
8:TInterval…
9:2-SampZInt…
0:2-SampTInt…
A:1-PropZInt…
B:2-PropZInt…
```

Step 2 *Enter the summary statistics.*

For x_1, enter the number of successes in the first data set.
For n_1, enter the sample size for the first data set.

For x_2, enter the number of successes in the second data set.
For n_2, enter the sample size for the second data set.
For **C-Level**, enter the confidence level.
Arrow down to "**Calculate**"
Press **ENTER**

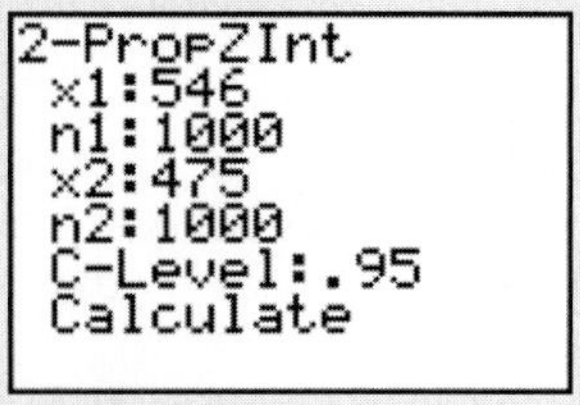

```
2-PropZInt
 x1:546
 n1:1000
 x2:475
 n2:1000
 C-Level:.95
 Calculate
```

Step 3   *View the Output*

The confidence interval will be displayed along with the two sample
proportions ($\hat{p}_1$ and $\hat{p}_2$) and the sample sizes.

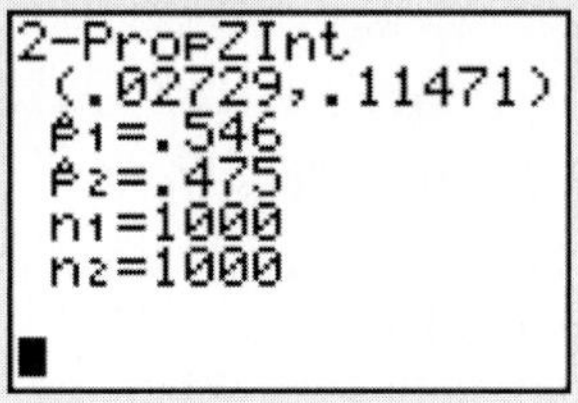

```
2-PropZInt
 (.02729,.11471)
 p̂1=.546
 p̂2=.475
 n1=1000
 n2=1000

■
```

Hypothesis Test for $p_1 - p_2$

Using the TI-83 Graphing Calculator

Step 1   *Access the Statistical Tests Menu*

Press **STAT**
Arrow right to **TESTS**
Arrow down to **6: 2-PropZTest**
Press **ENTER**

Step 2   *Enter the summary statistics.*

For x_1, enter the number of successes in the first data set.
For n_1, enter the sample size for the first data set.
For x_2, enter the number of successes in the second data set.
For n_2, enter the sample size for the second data set.
Arrow down to the next line and select the appropriate alternative hypothesis
by using the right arrow to highlight $\neq$**p2, <p2, or >p2** and press **ENTER**
Arrow down to **Calculate Draw**. You may choose either one of these for your
output.
Press **ENTER**

Step 3   *View the Output*

If you chose **Calculate**, the output will display the alternative hypothesis, the
z-value and the p-value, along with the two sample proportions ($\hat{p}_1$ and $\hat{p}_2$), and
$\hat{p}$, the weighted average of the two sample proportions and the sample sizes.

If you chose **Draw**, a graph of the normal curve will be displayed along
with the z-value and the p-value. The area under the curve that represents

the p-value is shaded. Note: In order to use the **Draw** feature, you must first clear all equations from the **Y - registers**. Hit "**Y =** " and delete all equations.

Example *Consider Example 9.6 with data in Table 9.7 on pg. 471*

Step 1 *Access the Statistical Tests Menu*

Press **STAT**
Arrow right to **TESTS**
Arrow down to **6: 2-PropZTest**
Press **ENTER**

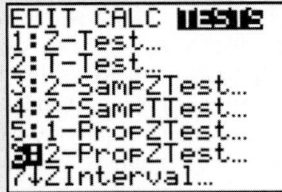

```
EDIT CALC TESTS
1:Z-Test…
2:T-Test…
3:2-SampZTest…
4:2-SampTTest…
5:1-PropZTest…
6▋2-PropZTest…
7↓ZInterval…
```

Step 2 *Enter the summary statistics.*

For **x₁**, enter the number of successes in the first data set.
For **n₁**, enter the sample size for the first data set.
For **x₂**, enter the number of successes in the second data set.
For **n₂**, enter the sample size for the second data set.
Arrow down to the next line and select **>p2** and press **ENTER**.
Arrow down to **Calculate Draw**. You may choose either one of these for your output.
Press **ENTER**

```
2-PropZTest
 x1:555
 n1:1500
 x2:578
 n2:1750
 p1:≠p2 <p2 >p2
Calculate Draw
```

Step 3 *View the Output*

If you chose **Calculate**, the output will display the alternative hypothesis, the z-value and the p-value, along with the two sample proportions ($\hat{p}_1$ and $\hat{p}_2$), and $\hat{p}$, the weighted average of the two sample proportions and the sample sizes.

```
2-PropZTest
 p1>p2
 z=2.368523545
 p=.0089296078
 p̂1=.37
 p̂2=.3302857143
↓p̂=.3486153846
█
```

If you chose **Draw**, a graph of the normal curve will be displayed along with the z-value and the p-value. The are under the curve that represents the p-value is shaded. Note: In order to use the **Draw** feature, you must first clear all equations from the **Y-registers**. Hit '**Y =** ' and delete all equations.

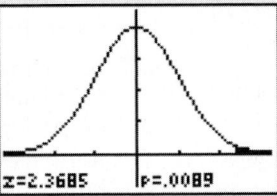

```
z=2.3685    p=.0089
```

Statistics in Action Revisited

Comparing the Incumbency Rates of Two Kentucky Milk Markets

We now return to the Statistics in Action investigation of collusive bidding in the northern Kentucky school milk market (p. 429). Market allocation is a common form of collusive behavior in bid-rigging conspiracies. Under collusion, the same dairy usually controls the same school districts year after year. The *incumbency rate* for a market is defined as the proportion of school districts that are won by the same vendor who won the previous year. Past experience with milk bids in a competitive environment reveals that a typical incumbency rate is .7. That is, 70% of the school districts are expected to purchase their milk from the same dairy who won the previous year. Incumbency rates of .9 or higher are strong indicators of collusive bidding.

Another method of detecting bid collusion is to compare the incumbency rates in the two milk markets— the tricounty ("rigged") and the surrounding ("competitive") markets. Since an incumbency rate is a population proportion, the target parameter is $(p_1 - p_2)$, where p_1 is the incumbency rate for the surrounding market and p_2 is the incumbency rate for the tricounty market. If, in fact, the two dairies bidding in the tricounty market were rigging the bids, then $p_1 < p_2$. Thus, we are interested in the one-tailed test, $H_0: (p_1 - p_2) = 0$ versus $H_a: (p_1 - p_2) < 0$.

Over the years 1985–1988, when bid collusion was alleged to have occurred in northern Kentucky, there were 51 potential vendor transitions (i.e., changes in milk supplier from one year to the next in a district) in the tricounty market and 134 potential vendor transitions in the surrounding market. These values represent the sample sizes ($n_1 = 134$ and $n_2 = 51$) for calculating incumbency rates. Examining the data saved in the **MILK** file, we find that in 50 of the 51 potential vendor transitions for the tricounty market, the winning dairy from the previous year won the bid the next year; similarly, we find 91 of the 134 potential vendor transitions for the surrounding area had the same dairy winning the bid the next year. Consequently, the estimated incumbency rates are

$$p_1 = 91/134 = .679 \quad \text{and} \quad p_2 = 50/51 = .980$$

The MINITAB large-sample z-test for comparing the two proportions is shown in Figure SIA9.2. The one-tailed p-value of the test (highlighted at the bottom of the printout) is .000. Therefore, there is sufficient evidence to conclude (at $\alpha = .01$) that the true incumbency rate for the surrounding ("competitive") market is significantly less than the true incumbency rate for the tricounty ("rigged") market. This result, as well as the fact that the sample incumbency rate for the tricounty market ($\hat{p}_2 = .980$) is much higher than the typical rate of about .7, lends further support for the bid collusion theory.

Figure SIA9.2

MINITAB Comparison of Incumbency Rate for Two Kentucky Milk Markets

Test and CI for Two Proportions

```
Sample    X    N   Sample p
1        91   134  0.679104
2        50    51  0.980392


Difference = p (1) - p (2)
Estimate for difference:  -0.301288
95% upper bound for difference:  -0.227669
Test for difference = 0 (vs < 0):   Z = -4.30   P-Value = 0.000
```

Exercises 9.49–9.69

Understanding the Principles

9.49 Consider making an inference about $p_1 - p_2$, where there are x_1 successes in n_1 binomial trials and x_2 successes in n_2 binomial trials.
   a. Describe the distributions of x_1 and x_2.
   b. For large samples, describe the sampling distribution of $(\hat{p}_1 - \hat{p}_2)$.

9.50 What is the problem with using the z-statistic to make inferences about $p_1 - p_2$ when the sample sizes are both small?

9.51 What conditions are required for valid large-sample inferences about $p_1 - p_2$?

Learning the Mechanics

9.52 In each case, determine whether the sample sizes are large enough to conclude that the sampling distribution of $(\hat{p}_1 - \hat{p}_2)$ is approximately normal.
   a. $n_1 = 10, n_2 = 12, \hat{p}_1 = .50, \hat{p}_2 = .50$
   b. $n_1 = 10, n_2 = 12, \hat{p}_1 = .10, \hat{p}_2 = .08$
   c. $n_1 = n_2 = 30, \hat{p}_1 = .20, \hat{p}_2 = .30$

d. $n_1 = 100, n_2 = 200, \hat{p}_1 = .05, \hat{p}_2 = .09$
e. $n_1 = 100, n_2 = 200, \hat{p}_1 = .95, \hat{p}_2 = .91$

9.53 For each of the following values of α, find the values of z for which $H_0: (p_1 - p_2) = 0$ would be rejected in favor of $H_a: (p_1 - p_2) < 0$.
a. $\alpha = .01$ **b.** $\alpha = .025$
c. $\alpha = .05$ **d.** $\alpha = .10$

9.54 Independent random samples, each containing 800 observations, were selected from two binomial populations. The samples from populations 1 and 2 produced 320 and 400 successes, respectively.
a. Test $H_0: (p_1 - p_2) = 0$ against $H_a: (p_1 - p_2) \neq 0$. Use $\alpha = .05$.
b. Test $H_0: (p_1 - p_2) = 0$ against $H_a: (p_1 - p_2) \neq 0$. Use $\alpha = .01$.
c. Test $H_0: (p_1 - p_2) = 0$ against $H_a: (p_1 - p_2) < 0$. Use $\alpha = .01$.
d. Form a 90% confidence interval for $(p_1 - p_2)$.

9.55 Construct a 95% confidence interval for $(p_1 - p_2)$ in each of the following situations:
a. $n_1 = 400, \hat{p}_1 = .65; n_2 = 400, \hat{p}_2 = .58$
b. $n_1 = 180, \hat{p}_1 = .31; n_2 = 250, \hat{p}_2 = .25$
c. $n_1 = 100, \hat{p}_1 = .46; n_2 = 120, \hat{p}_2 = .61$

9.56 Sketch the sampling distribution of $(\hat{p}_1 - \hat{p}_2)$ based on independent random samples of $n_1 = 100$ and $n_2 = 200$ observations from two binomial populations with success probabilities $p_1 = .1$ and $p_2 = .5$, respectively.

9.57 Random samples of size $n_1 = 50$ and $n_2 = 60$ were drawn from populations 1 and 2, respectively. The samples yielded $\hat{p}_1 = .4$ and $\hat{p}_2 = .2$. Test $H_0: (p_1 - p_2) = .1$ against $H_a: (p_1 - p_2) > .1$ using $\alpha = .05$.

Applying the Concepts—Basic

9.58 Executive workout dropouts. Refer to the *Journal of Sport Behavior* (2001) study of variety in exercise workouts, Exercise 7.53 (p. 349). One group of 38 people varied their exercise routine in workouts while a second group of 38 exercisers had no set schedule or regulations for their workouts. By the end of the study, 14 people had dropped out of the first exercise group and 23 had dropped out of the second group.
a. Find the dropout rates (i.e., the percentage of exercisers who had dropped out of the exercise group) for each of the two groups of exercisers.
b. Find a 90% confidence interval for the difference between the dropout rates of the two groups of exercisers.
c. Give a practical interpretation of the confidence interval, part **c**.

9.59 The "winner's curse" in auction bidding. In auction bidding, the "winner's curse" is the phenomenon of the winning (or highest) bid price being above the expected value of the item being auctioned. *The Review of Economics and Statistics* (Aug. 2001) published a study on whether bid experience impacts the likelihood of the winner's curse occurring. Two groups of bidders in a sealed-bid auction were compared: (1) super-experienced bidders

and (2) less-experienced bidders. In the super-experienced group, 29 of 189 winning bids were above the item's expected value; in the less-experienced group, 32 of 149 winning bids were above the item's expected value.
a. Find an estimate of p_1, the true proportion of super-experienced bidders who fall prey to the winner's curse.
b. Find an estimate of p_2, the true proportion of less-experienced bidders who fall prey to the winner's curse.
c. Construct a 90% confidence interval for $p_1 - p_2$.
d. Give a practical interpretation of the confidence interval, part **c**. Make a statement about whether bid experience impacts the likelihood of the winner's curse occurring.

9.60 Planning habits survey. *American Demographics* (Jan. 2002) reported the results of a survey on the planning habits of men and women. In response to the question, "What is your preferred method of planning and keeping track of meetings, appointments and deadlines?", 56% of the men and 46% of the women answered "keep them in my head." A nationally representative sample of 1,000 adults participated in the survey; therefore, assume that 500 were men and 500 were women.
a. Set up the null and alternative hypotheses for testing whether the percentage of men who prefer keeping track of appointments in their head is larger than the corresponding percentage of women.
b. Compute the test statistic for the test.
c. Give the rejection region for the test using $\alpha = .01$.
d. Find the p-value for the test.
e. Make the appropriate conclusion.

9.61 Treating depression with St. John's wort. The *Journal of the American Medical Association* (April 18, 2001) published a study of the effectiveness of using extracts of the herbal medicine, St. John's wort, in treating major depression. In an 8-week randomized, controlled trial, 200 patients diagnosed with major depression were divided into two groups: one group ($n_1 = 98$) received St. John's wort extract while the other group ($n_2 = 102$) received a placebo (no drug). At the end of the study period, 14 of the St. John's wort patients were in remission compared to 5 of the placebo patients.
a. Compute the proportion of the St. John's wort patients who were in remission.
b. Compute the proportion of the placebo patients who were in remission.
c. If St. John's wort is effective in treating major depression, then the proportion of St. John's wort patients in remission will exceed the proportion of placebo patients in remission. At $\alpha = .01$, is St. John's wort effective in treating major depression?
d. Repeat part **c**, but use $\alpha = .10$.
e. Explain why the choice of α is critical for this study.

Applying the Concepts—Intermediate

9.62 Effectiveness of drug tests of Olympic athletes. Erythropoietin (EPO) is a banned drug used by athletes to increase the oxygen carrying capacity of their blood. New tests for EPO were first introduced prior to the 2000 Olympic Games held in Sydney, Australia. *Chance* (Spring 2004) reported that of a sample of 830 world-class athletes, 159 did not compete in the 1999 World Championships (a year prior to the new EPO test). Similarly, 133 of 825 potential athletes did not compete in the 2000 Olympic games. Was the new test effective in deterring an athlete's participation in the 2000 Olympics? If so, then the proportion of non-participating athletes in 2000 will be more than the proportion of nonparticipating athletes in 1999. Conduct the analysis (at $\alpha = .10$) and make the proper conclusion.

9.63 Killing moths with carbon dioxide. A University of South Florida biologist conducted an experiment to determine whether increased levels of carbon dioxide kill leaf-eating moths. (*USF Magazine*, Winter 1999.) Moth larvae were placed in open containers filled with oak leaves. Half the containers had normal carbon dioxide levels while the other half had double the normal level of carbon dioxide. Ten percent of the larvae in the containers with high carbon dioxide levels died, compared to 5 percent in the containers with normal levels. Assume that 80 moth larvae were placed, at random, in each of the two types of containers. Do the experimental results demonstrate that an increased level of carbon dioxide is effective in killing a higher percentage of leaf-eating moth larvae? Test using $\alpha = .01$.

9.64 Switching majors in college. When female undergraduates switch from science, mathematics, and engineering (SME) majors into disciplines that are not science-based, are their reasons different from those of their male counterparts? This question was investigated in *Science Education* (July 1995). A sample of 335 junior/senior undergraduates—172 females and 163 males—at two large research universities were identified as "switchers," that is, they left a declared SME major for a non-SME major. Each student listed one or more factors that contributed to the switching decision.

a. Of the 172 females in the sample, 74 listed lack or loss of interest in SME (i.e., "turned off" by science) as a major factor, compared to 72 of the 163 males. Conduct a test (at $\alpha = .10$) to determine whether the proportion of female switchers who give "lack of interest in SME" as a major reason for switching differs from the corresponding proportion of males.

b. Thirty-three of the 172 females in the sample indicated that they were discouraged or lost confidence because of low grades in SME during their early years, compared to 44 of 163 males. Construct a 90% confidence interval for the difference between the proportions of female and male switchers who lost confidence due to low grades in SME. Interpret the result.

9.65 "Tip-of-the-tongue" study. Trying to think of a word you know, but can't instantly retrieve, is called the "tip of the tongue" phenomenon. *Psychology and Aging* (Sept. 2001) published a study of this phenomenon in senior citizens. The researchers compared 40 people between 60 and 72 years of age with 40 between 73 and 83 years of age. When primed with the initial syllable of a missing word (e.g., seeing the word *include* to help recall the word *incisor*), the younger seniors had a higher recall rate. Suppose 31 of the 40 seniors in the younger group could recall the word when primed with the initial syllable, while only 22 of the 40 seniors could recall the word. Compare the recall rates of the two groups using $\alpha = .05$. Does one group of elderly people have a significantly higher recall rate than the other?

9.66 Gambling in public high schools. With the rapid growth in legalized gambling in the United States, there is concern that the involvement of youth in gambling activities is also increasing. University of Minnesota professor Randy Stinchfield compared the rates of gambling among Minnesota public school students between 1992 and 1998. (*Journal of Gambling Studies*, Winter 2001.) Based on survey data, the table shows the percentages of ninth-grade boys who gambled weekly or daily on any game (e.g., cards, sports betting, lotteries) for the two years.

	1992	1998
Number of ninth-grade boys in survey	21,484	23,199
Number who gambled weekly/daily	4,684	5,313

a. Are the percentages of ninth-grade boys who gambled weekly or daily on any game in 1992 and 1998 significantly different? (Use $\alpha = .01$.)

b. Professor Stinchfield states that "because of the large sample sizes, even small differences may achieve statistical significance, so interpretations of the differences should include a judgement regarding the magnitude of the difference and its public health significance." Do you agree with this statement? If not, why not? If so, obtain a measure of the magnitude of the difference between 1992 and 1998 and attach a measure of reliability to the difference.

NZBIRDS

9.67 Extinct New Zealand birds. Refer to the *Evolutionary Ecology Research* (July 2003) study of the patterns of extinction in the New Zealand bird population, Exercise 2.18 (p. 40). Recall that the **NZBIRDS** file contains data on flight capability and extinct status for 132 bird species at the time of the Maori colonization of New Zealand. Ecologists want to compare the proportions of flightless birds for two New Zealand bird populations—those that are extinct and those that are not extinct. The sample information shown on p. 479 is obtained from the **NZBIRDS** file. The ecologists are investigating the theory that the proportion of flightless birds will be greater for extinct species than for nonextinct species. Test the theory using $\alpha = .01$.

Bird Population	Number of Species Sampled	Number of Flightless Species
Extinct	38	21
Nonextinct	78	7

9.68 Reducing the risk of heart attacks. Can stress management help heart patients reduce their risk of heart attacks or heart surgery? In a study published in the *Archives of Internal Medicine* (Oct. 27, 1997), 107 heart patients were randomly divided into three groups. One group participated in a stress management program, the second group underwent an exercise program, and the third group received usual heart care from their doctors. After three years, the number in each group suffering "cardiac events" (e.g., heart attack, bypass surgery, or angioplasty) was recorded. The results are shown in the table.

	Management	Exercise	Usual Care
Number of heart patients	33	34	40
Number who suffer a "cardiac event"	3	7	12

a. Use a 99% confidence interval to compare the proportion of stress management patients who suffer a cardiac event to the proportion of exercise patients who suffer a cardiac event. What inference can you make?

b. Repeat part **a,** but compare the stress management group to the usual care group.

c. Repeat part **a,** but compare the exercise group to the usual care group.

Applying the Concepts—Advanced

9.69 Food craving study. Do you have an insatiable craving for chocolate or some other food? Since many people apparently do, psychologists are designing scientific studies to examine the phenomenon. According to the *New York Times* (Feb. 22, 1995), one of the largest studies of food cravings involved a survey of 1,000 McMaster University (Canada) students. The survey revealed that 97% of the women in the study acknowledged specific food cravings while only 67% of the men did.

a. How large do n_1 and n_2 have to be to conclude that the true proportion of women who acknowledge having food cravings exceeds the corresponding proportion of men? Assume $\alpha = .01$.

b. Why is it dangerous to conclude from the study that women have a higher incidence of food cravings than men?

9.5 Determining the Sample Size

You can find the appropriate sample size to estimate the difference between a pair of parameters with a specified sampling error (SE) and degree of reliability by using the method described in Section 7.4. That is, to estimate the difference between a pair of parameters correct to within *SE* units with confidence level $(1 - \alpha)$, let $z_{\alpha/2}$ standard deviations of the sampling distribution of the estimator equal *SE*. Then solve for the sample size. To do this, you have to solve the problem for a specific ratio between n_1 and n_2. Most often, you will want to have equal sample sizes, that is, $n_1 = n_2 = n$. We will illustrate the procedure with two examples.

EXAMPLE 9.8   FINDING THE SAMPLE SIZES FOR ESTIMATING $\mu_1 - \mu_2$

Problem   New fertilizer compounds are often advertised with the promise of increased crop yields. Suppose we want to compare the mean yield μ_1 of wheat when a new fertilizer is used to the mean yield μ_2 with a fertilizer in common use. The estimate of the difference in mean yield per acre is to be correct to within .25 bushel with a confidence coefficient of .95. If the sample sizes are to be equal, find $n_1 = n_2 = n$, the number of 1-acre plots of wheat assigned to each fertilizer.

Solution   To solve the problem, you need to know something about the variation in the bushels of yield per acre. Suppose from past records you know the yields of wheat possess a range of approximately 10 bushels per acre. You could then approximate $\sigma_1 = \sigma_2 = \sigma$ by letting the range equal 4σ. Thus,

$$4\sigma \approx 10 \text{ bushels}$$
$$\sigma \approx 2.5 \text{ bushels}$$

The next step is to solve the equation

$$z_{\alpha/2}\sigma_{(\bar{x}_1-\bar{x}_2)} = SE \quad \text{or} \quad z_{\alpha/2}\sqrt{\frac{\sigma_1^2}{n_1} + \frac{\sigma_2^2}{n_2}} = SE$$

for n, where $n = n_1 = n_2$. Since we want the estimate to lie within $SE = .25$ of $(\mu_1 - \mu_2)$ with confidence coefficient equal to .95, we have $z_{\alpha/2} = z_{.025} = 1.96$. Then, letting $\sigma_1 = \sigma_2 = 2.5$ and solving for n, we have

$$1.96\sqrt{\frac{(2.5)^2}{n} + \frac{(2.5)^2}{n}} = .25$$

$$1.96\sqrt{\frac{2(2.5)^2}{n}} = .25$$

$$n = 768.32 \approx 769 \text{ (rounding up)}$$

Consequently, you will have to sample 769 acres of wheat for each fertilizer to estimate the difference in mean yield per acre to within .25 bushel.

Look Back Since $n = 769$ would necessitate extensive and costly experimentation, you might decide to allow a larger sampling error (say, $SE = .50$ or $SE = 1$) in order to reduce the sample size, or you might decrease the confidence coefficient. The point is that we can obtain an idea of the experimental effort necessary to achieve a specified precision in our final estimate by determining the approximate sample size *before* the experiment is begun.

Now Work **Exercise 9.74**

■ ■ ■

EXAMPLE 9.9

FINDING THE SAMPLE SIZES FOR ESTIMATING $p_1 - p_2$

Problem A production supervisor suspects that a difference exists between the proportions p_1 and p_2 of defective items produced by two different machines. Experience has shown that the proportion defective for each of the two machines is in the neighborhood of .03. If the supervisor wants to estimate the difference in the proportions to within .005 using a 95% confidence interval, how many items must be randomly sampled from the production of each machine? (Assume that the supervisor wants $n_1 = n_2 = n$.)

Solution In this sampling problem, the sampling error $SE = .005$, and for the specified level of reliability, $z_{\alpha/2} = z_{.025} = 1.96$. Then, letting $p_1 = p_2 = .03$ and $n_1 = n_2 = n$, we find the required sample size per machine by solving the following equation for n:

$$z_{\alpha/2}\sigma_{(\hat{p}_1-\hat{p}_2)} = SE$$

or

$$z_{\alpha/2}\sqrt{\frac{p_1 q_1}{n_1} + \frac{p_2 q_2}{n_2}} = SE$$

$$1.96\sqrt{\frac{(.03)(.97)}{n} + \frac{(.03)(.97)}{n}} = .005$$

$$1.96\sqrt{\frac{2(.03)(.97)}{n}} = .005$$

$$n = 8{,}943.2$$

Look Back This large n will likely result in a tedious sampling procedure. If the supervisor insists on estimating $(p_1 - p_2)$ correct to within .005 with 95% confidence, approximately 9,000 items will have to be inspected for each machine.

Now Work   *Exercise 9.75a*

━━━━━ ■ ■ ■ ━━━━━

You can see from the calculations in Example 9.9 that $\sigma_{(\hat{p}_1-\hat{p}_2)}$ (and hence the solution, $n_1 = n_2 = n$) depends on the actual (but unknown) values of p_1 and p_2. In fact, the required sample size $n_1 = n_2 = n$ is largest when $p_1 = p_2 = .5$. Therefore, if you have no prior information on the approximate values of p_1 and p_2, use $p_1 = p_2 = .5$ in the formula for $\sigma_{(\hat{p}_1-\hat{p}_2)}$. If p_1 and p_2 are in fact close to .5, then the values of n_1 and n_2 that you have calculated will be correct. If p_1 and p_2 differ substantially from .5, then your solutions for n_1 and n_2 will be larger than needed. Consequently, using $p_1 = p_2 = .5$ when solving for n_1 and n_2 is a conservative procedure because the sample sizes n_1 and n_2 will be at least as large as (and probably larger than) needed.

The procedures for determining sample sizes necessary for estimating $(\mu_1 - \mu_2)$ or $(p_1 - p_2)$ for the case $n_1 = n_2$ are given in the following boxes.

Determination of Sample Size for Estimating $\mu_1 - \mu_2$

To estimate $(\mu_1 - \mu_2)$ to within a given sampling error SE and with confidence level $(1 - \alpha)$, use the following formula to solve for equal sample sizes that will achieve the desired reliability:

$$n_1 = n_2 = \frac{(z_{\alpha/2})^2(\sigma_1^2 + \sigma_2^2)}{(SE)^2}$$

You will need to substitute estimates for the values of σ_1^2 and σ_2^2 before solving for the sample size. These estimates might be sample variances s_1^2 and s_2^2 from prior sampling (e.g., a pilot study), or from an educated (and conservatively large) guess based on the range—that is, $s \approx R/4$.

Determination of Sample Size for Estimating $p_1 - p_2$

To estimate $(p_1 - p_2)$ to within a given sampling error SE and with confidence level $(1 - \alpha)$, use the following formula to solve for equal sample sizes that will achieve the desired reliability:

$$n_1 = n_2 = \frac{(z_{\alpha/2})^2(p_1q_1 + p_2q_2)}{(SE)^2}$$

You will need to substitute estimates for the values of p_1 and p_2 before solving for the sample size. These estimates might be based on prior samples, obtained from educated guesses or, most conservatively, specified as $p_1 = p_2 = .5$.

Exercises 9.70–9.83

Understanding the Principles

9.70 When determining the sample sizes for estimating $\mu_1 - \mu_2$, how do you obtain estimates of the population variances $(\sigma_1)^2$ and $(\sigma_2)^2$ used in the calculations?

9.71 When determining the sample sizes for estimating $p_1 - p_2$, how do you obtain estimates of the binomial proportions p_1 and p_2 used in the calculations?

9.72 If the sample size calculation yields a value of n that is too large to be practical, how should you proceed?

Learning the Mechanics

9.73 Suppose you want to estimate the difference between two population means correct to within 2.2 with probability .95. If prior information suggests that the population variances are approximately equal to $\sigma_1^2 = \sigma_2^2 = 15$ and you want to select independent random samples of equal size from the populations, how large should the sample sizes, n_1 and n_2, be?

9.74 Find the appropriate values of n_1 and n_2 (assume NW $n_1 = n_2$) needed to estimate $(\mu_1 - \mu_2)$ with
  a. A sampling error equal to 3.2 with 95% confidence. From prior experience it is known that $\sigma_1 \approx 15$ and $\sigma_2 \approx 17$.
  b. A sampling error equal to 8 with 99% confidence. The range of each population is 60.
  c. A 90% confidence interval of width 1.0. Assume that $\sigma_1^2 \approx 5.8$ and $\sigma_2^2 \approx 7.5$.

9.75 Assuming that $n_1 = n_2$, find the sample sizes needed to NW estimate $(p_1 - p_2)$ for each of the following situations:
  a. $SE = .01$ with 99% confidence. Assume that $p_1 \approx .4$ and $p_2 \approx .7$.
  b. A 90% confidence interval of width .05. Assume there is no prior information available to obtain approximate values of p_1 and p_2.
  c. $SE = .03$ with 90% confidence. Assume that $p_1 \approx .2$ and $p_2 \approx .3$.

9.76 Enough money has been budgeted to collect independent random samples of size $n_1 = n_2 = 100$ from populations 1 and 2 in order to estimate $(p_1 - p_2)$. Prior information indicates that $p_1 = p_2 \approx .6$. Have sufficient funds been allocated to construct a 90% confidence interval for $(p_1 - p_2)$ of width .1 or less? Justify your answer.

Applying the Concepts—Basic

9.77 **Bulimia study.** Refer to *The American Statistician* (May 2001) study comparing the "fear of negative evaluation" (FNE) scores for bulimic and normal female students, Exercise 9.19 (p. 450). Suppose you want to estimate $(\mu_B - \mu_N)$, the difference between the population means of the FNE scores for bulimic and normal female students, using a 95% confidence interval with a sampling error of two points. Find the sample sizes required to obtain such an estimate. Assume equal sample sizes of $\sigma_B^2 = \sigma_N^2 = 25$.

9.78 **Students attitudes towards parents.** Refer to the *Journal of Genetic Psychology* (Mar. 1998) study comparing male students' attitudes toward their fathers and mothers, Exercise 9.42 (p. 466). Suppose you want to estimate $\mu_D = (\mu_F - \mu_M)$, the difference between the population mean attitudes toward fathers and mothers, using a 90% confidence interval with a sampling error of .3. Find the number of male students required to obtain such an estimate. Assume the variance of the paired differences is $\sigma_D^2 = 1$.

9.79 **Executive workout dropouts.** Refer to the *Journal of Sport Behavior* (2001) study comparing the dropout rates of two groups of exercisers, Exercise 9.58 (p. 477). Suppose you want to reduce the sampling error for estimating $(p_1 - p_2)$ with a 90% confidence interval to .1.

Determine the number of exercisers to be sampled from each group in order to obtain such an estimate. Assume equal sample sizes, $p_1 \approx .4$, and $p_2 \approx .6$.

Applying the Concepts—Intermediate

9.80 **Cable-TV home shoppers.** All cable television companies carry at least one home shopping channel. Who uses these home shopping services? Are the shoppers primarily men or women? Suppose you want to estimate the difference in the proportions of men and women who say they have used or expect to use televised home shopping using an 80% confidence interval of width .06 or less.
  a. Approximately how many people should be included in your samples?
  b. Suppose you want to obtain individual estimates for the two proportions of interest. Will the sample size found in part **a** be large enough to provide estimates of each proportion correct to within .02 with probability equal to .90? Justify your response.

9.81 **Health hazards of housework.** Is housework hazardous to your health? A study in the *Public Health Reports* (July–Aug. 1992) compares the life expectancies of 25-year-old white women in the labor force to those who are housewives. How large a sample would have to be taken from each group in order to be 95% confident that the estimate of difference in mean life expectancies for the two groups is within 1 year of the true difference? Assume that equal sample sizes will be selected from the two groups, and that the standard deviation for both groups is approximately 15 years.

9.82 **Rat damage in sugarcane.** Poisons are used to prevent rat damage in sugarcane fields. The U.S. Department of Agriculture is investigating whether the rat poison should be located in the middle of the field or on the outer perimeter. One way to answer this question is to determine where the greater amount of damage occurs. If damage is measured by the proportion of cane stalks that have been damaged by rats, how many stalks from each section of the field should be sampled in order to estimate the true difference between proportions of stalks damaged in the two sections to within .02 with 95% confidence?

9.83 **Scouting an NFL free agent.** In seeking a free agent NFL running back, a general manager is looking for a player with high mean yards gained per carry and a small standard deviation. Suppose the GM wishes to compare the mean yards gained per carry for two free agents based on independent random samples of their yards gained per carry. Data from last year's pro football season indicate that $\sigma_1 = \sigma_2 \approx 5$ yards. If the GM wants to estimate the difference in means correct to within 1 yard with a confidence level of .90, how many runs would have to be observed for each player? (Assume equal sample sizes.)

9.6 Comparing Two Population Variances: Independent Sampling (Optional)

Many times it is of practical interest to use the techniques developed in this chapter to compare the means or proportions of two populations. However, there are also important instances when we wish to compare two population variances. For example, when two devices are available for producing precision measurements (scales, calipers, thermometers, etc.), we might want to compare the variability of the measurements of the devices before deciding which one to purchase. Or when two standardized tests can be used to rate job applicants, the variability of the scores for both tests should be taken into consideration before deciding which test to use.

For problems like these we need to develop a statistical procedure to compare population variances. The common statistical procedure for comparing population variances, σ_1^2 and σ_2^2, makes an inference about the ratio σ_1^2/σ_2^2. In this section, we will show how to test the null hypothesis that the ratio σ_1^2/σ_2^2 equals 1 (the variances are equal) against the alternative hypothesis that the ratio differs from 1 (the variances differ):

$$H_0: \frac{\sigma_1^2}{\sigma_2^2} = 1 \quad (\sigma_1^2 = \sigma_2^2)$$

$$H_a: \frac{\sigma_1^2}{\sigma_2^2} \neq 1 \quad (\sigma_1^2 \neq \sigma_2^2)$$

To make an inference about the ratio σ_1^2/σ_2^2, it seems reasonable to collect sample data and use the ratio of the sample variances, s_1^2/s_2^2. We will use the test statistic

$$F = \frac{s_1^2}{s_2^2}$$

To establish a rejection region for the test statistic, we need to know the sampling distribution of s_1^2/s_2^2. As you will subsequently see, the sampling distribution of s_1^2/s_2^2 is based on two of the assumptions already required for the t-test:

1. The two sampled populations are normally distributed.
2. The samples are randomly and independently selected from their respective populations.

When these assumptions are satisfied and when the null hypothesis is true (that is, $\sigma_1^2 = \sigma_2^2$), the sampling distribution of $F = s_1^2/s_2^2$ is the **F-distribution** with $(n_1 - 1)$ numerator degrees of freedom and $(n_2 - 1)$ denominator degrees of freedom, respectively. The shape of the F-distribution depends on the degrees of freedom associated with s_1^2 and s_2^2—that is, on $(n_1 - 1)$ and $(n_2 - 1)$. An F-distribution with 7 and 9 df is shown in Figure 9.16. As you can see, the distribution is skewed to the right, since s_1^2/s_2^2 cannot be less than 0 but can increase without bound.

Figure 9.16

An F-Distribution with 7 Numerator and 9 Denominator Degrees of Freedom

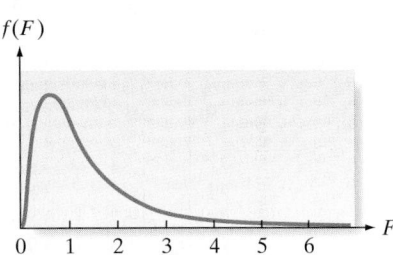

Biography

GEORGE W. SNEDECOR (1882–1974)—Snedecor's F-test

George W. Snedecor's education began at the University of Alabama, where he obtained his bachelor's degree in mathematics and physics. He went on to the University of Michigan for his master's degree in physics and finally earned his Ph.D in mathematics at the University of Kentucky. Snedecor learned of an opening for an assistant professor of mathematics at the University of Iowa, packed his belongings in his car, and began driving to apply for the position. By mistake, he ended up in Ames, Iowa, home of Iowa State University—then an agricultural school that had no need for a mathematics teacher. Nevertheless, Snedecor stayed and founded a statistics laboratory, eventually teaching the first course in statistics at Iowa State in 1915. In 1933, Snedecor turned the statistics laboratory into the first ever Department of Statistics in the United States. During his tenure as chair of the department, Snedecor published his landmark textbook, *Statistical Methods* (1937). The text contained the first published reference for a test of hypothesis to compare two variances. Although Snedecor named it the *F*-test in honor of statistician R. A. Fisher (who had developed the *F*-distribution a few years earlier—see Biography, p. 516), many researchers still refer to it as Snedecor's *F*-test. Now in its ninth edition, *Statistical Methods* (with William Cochran as a coauthor) continues to be one of the most frequently cited texts in the statistics field.

We need to be able to find *F*-values corresponding to the tail areas of this distribution in order to establish the rejection region for our test of hypothesis because we expect the ratio *F* of the sample variances to be either very large or very small when the population variances are unequal. The upper-tail *F* values for $\alpha = .10, .05, .025$, and $.01$ can be found in Tables VIII, IX, X, and XI of Appendix A. Table IX is partially reproduced in Table 9.8. It gives *F* values that correspond to $\alpha = .05$ upper-tail areas for

TABLE 9.8 **Reproduction of Part of Table IX in Appendix A: Percentage Points of the F-distribution, $\alpha = .05$**

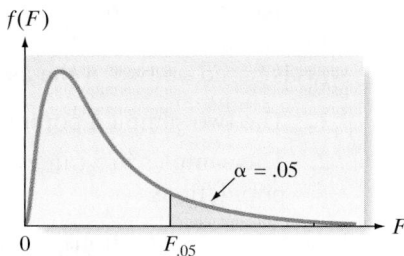

ν_1	Numerator Degrees of Freedom								
ν_2	1	2	3	4	5	6	7	8	9
1	161.4	199.5	215.7	224.6	230.2	234.0	236.8	238.9	240.5
2	18.51	19.00	19.16	19.25	19.30	19.33	19.35	19.37	19.38
3	10.13	9.55	9.28	9.12	9.01	8.94	8.89	8.85	8.81
4	7.71	6.94	6.59	6.39	6.26	6.16	6.09	6.04	6.00
5	6.61	5.79	5.41	5.19	5.05	4.95	4.88	4.82	4.77
6	5.99	5.14	4.76	4.53	4.39	4.28	4.21	4.15	4.10
7	5.59	4.74	4.35	4.12	3.97	3.87	3.79	3.73	3.68
8	5.32	4.46	4.07	3.84	3.69	3.58	3.50	3.44	3.39
9	5.12	4.26	3.86	3.63	3.48	3.37	3.29	3.23	3.18
10	4.96	4.10	3.71	3.48	3.33	3.22	3.14	3.07	3.02
11	4.84	3.98	3.59	3.36	3.20	3.09	3.01	2.95	2.90
12	4.75	3.89	3.49	3.25	3.11	3.00	2.91	2.85	2.80
13	4.67	3.81	3.41	3.18	3.03	2.92	2.83	2.77	2.71
14	4.60	3.74	3.34	3.11	2.96	2.85	2.76	2.70	2.65

Denominator Degrees of Freedom

Figure 9.17
An F-Distribution for $\nu_1 = 7$
and $\nu_2 = 9$ df; $\alpha = .05$

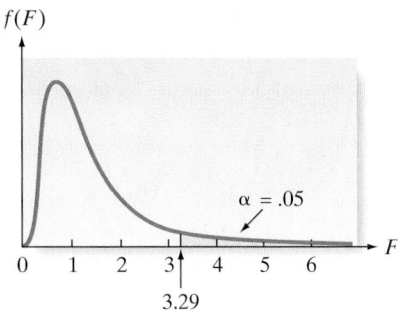

different degrees of freedom; the columns correspond to the degrees of freedom ν_1 for the numerator sample variance, s_1^2, whereas the rows correspond to the degrees of freedom ν_2 for the denominator sample variance, s_2^2. Thus, if the numerator degrees of freedom is $\nu_1 = 7$ and the denominator degrees of freedom is $\nu_2 = 9$, we look in the seventh column and ninth row to find $F_{.05} = 3.29$. As shown in Figure 9.17, $\alpha = .05$ is the tail area to the right of 3.29 in the F-distribution with 7 and 9 df. That is, if $\sigma_1^2 = \sigma_2^2$, then the probability that the F-statistic will exceed 3.29 is $\alpha = .05$.

Now Work *Exercise 9.88*

EXAMPLE 9.10   CONDUCTING AN F-TEST

Problem   An experimenter wants to compare the metabolic rates of white mice subjected to different drugs. The weights of the mice may affect their metabolic rates, and thus the experimenter wishes to obtain mice that are relatively homogeneous with respect to weight. Five hundred mice will be needed to complete the study. Currently, 13 mice from supplier 1 and another 18 mice from supplier 2 are available for comparison. The experimenter weighs these mice and obtains the data shown in Table 9.9. Do these data provide sufficient evidence to indicate a difference in the variability of weights of mice obtained from the two suppliers? (Use $\alpha = .10$.) Using the results of this analysis, what would you suggest to the experimenter?

⬡ MICEWTS

TABLE 9.9   Weights (in ounces) of Experimental Mice

Supplier 1					
4.23	4.35	4.05	3.75	4.41	4.37
4.01	4.06	4.15	4.19	4.52	4.21
4.29					

Supplier 2					
4.14	4.26	4.05	4.11	4.31	4.12
4.17	4.35	4.25	4.21	4.05	4.28
4.15	4.20	4.32	4.25	4.02	4.14

Solution   Let

$$\sigma_1^2 = \text{Population variance of weights of white mice from supplier 1}$$
$$\sigma_2^2 = \text{Population variance of weights of white mice from supplier 2}$$

The hypotheses of interest are then

$$H_0: \frac{\sigma_1^2}{\sigma_2^2} = 1 \quad (\sigma_1^2 = \sigma_2^2)$$

$$H_a: \frac{\sigma_1^2}{\sigma_2^2} \neq 1 \quad (\sigma_1^2 \neq \sigma_2^2)$$

The nature of the F-tables given in Appendix A affects the form of the test statistic. To form the rejection region for a two-tailed F-test, we want to make certain that the upper tail is used, because only the upper-tail values of F are shown in Tables VIII, IX, X, and XI. To accomplish this, *we will always place the larger sample variance in the numerator of the F-test statistic.* This has the effect of doubling the tabulated value for α, since we double the probability that the F-ratio will fall in the upper tail by always placing the larger sample variance in the numerator. That is, we establish a one-tailed rejection region by putting the larger variance in the numerator rather than establishing rejection regions in both tails.

To calculate the value of the test statistic, we require the sample variances. These are shown on the MINITAB printout, Figure 9.19. The sample variances are $s_1^2 = .0409$ and $s_2^2 = .00964$. Therefore,

$$F = \frac{\text{Larger sample variance}}{\text{Smaller sample variance}} = \frac{s_1^2}{s_2^2} = \frac{(.0409)}{(.00964)} = 4.24$$

For our example, we have a numerator s_1^2 with df $= \nu_1 = n_1 - 1 = 12$ and a denominator s_2^2 with df $= \nu_2 = n_2 - 1 = 17$. Therefore, we will reject $H_0: \sigma_1^2 = \sigma_2^2$ for $\alpha = .10$ when the calculated value of F exceeds the tabulated value:

$$F_{\alpha/2} = F_{.05} = 2.38 \quad \text{(see Figure 9.18)}$$

Note that $F = 4.24$ falls in the rejection region. Therefore, the data provide sufficient evidence to indicate that the population variances differ. It appears that the weights of mice obtained from supplier 2 tend to be more homogeneous (less variable) than the weights of mice obtained from supplier 1. On the basis of this evidence, we would advise the experimenter to purchase the mice from supplier 2.

Look Back What would you have concluded if the value of F calculated from the samples had not fallen in the rejection region? Would you conclude that the null hypothesis of equal variances is true? No, because then you risk the possibility of a Type II error (accepting H_0 if H_a is true) without knowing the value of β, the

Figure 9.18
Rejection Region
for Example 9.10

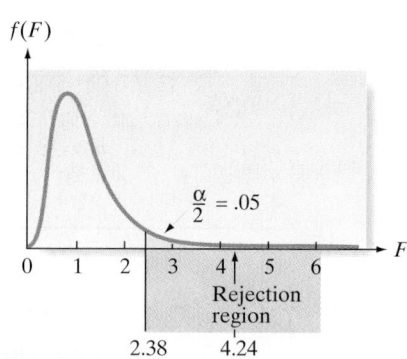

Figure 9.19

MINITAB Summary
Statistics and *F*-Test
for Mice Weights

Descriptive Statistics: WEIGHT

```
Variable  SUPPLIER   N    Mean    StDev   Variance  Minimum   Maximum
WEIGHT       1       13  4.1992  0.2021    0.0409    3.7500    4.5200
             2       18  4.1878  0.0982    0.00964   4.0200    4.3500
```

Test for Equal Variances: WEIGHT versus SUPPLIER

```
95% Bonferroni confidence intervals for standard deviations

SUPPLIER   N     Lower      StDev      Upper
      1   13   0.138579   0.202133   0.361467
      2   18   0.070860   0.098193   0.156836


F-Test (normal distribution)
Test statistic = 4.24,  p-value = 0.007


Levene's Test (any continuous distribution)
Test statistic = 4.31,  p-value = 0.047
```

probability of accepting H_0: $\sigma_1^2 = \sigma_2^2$ if in fact it is false. Since we will not consider the calculation of β for specific alternatives in this text, when the F-statistic does not fall in the rejection region we simply conclude that insufficient sample evidence exists to refute the null hypothesis that $\sigma_1^2 = \sigma_2^2$.

Now Work | *Exercise 9.93a*

■ ■ ■

EXAMPLE 9.11

THE *p*-VALUE OF AN *F*-TEST

Problem Find the *p*-value for the test in Example 9.10 using the *F*-tables in Appendix A. Compare this to the exact *p*-value obtained from a computer printout.

Solution Since the observed value of the *F*-statistic in Example 9.10 was 4.24, the observed significance level of the test would equal the probability of observing a value of F at least as contradictory to H_0: $\sigma_1^2 = \sigma_2^2$ as $F = 4.24$, if in fact H_0 is true. Since we give the *F*-tables in Appendix A only for values of α equal to .10, .05, .025, and .01, we can only approximate the observed significance level. Checking Table XI, we find $F_{.01} = 3.46$. Since the observed value of F is greater than $F_{.01}$, the observed significance level for the test is less than $2(.01) = .02$. (Note that we double the α value shown in Table XI because this is a two-tailed test.) The exact *p*-value, $p = .007$, is highlighted at the bottom of the MINITAB printout, Figure 9.19.

Look Back We double the α value in Table XI because this is a two-tailed test.

Now Work | *Exercise 9.93b*

■ ■ ■

In addition to a hypothesis test for σ_1^2/σ_2^2, we can use the F-statistic to estimate the ratio with a confidence interval. The following boxes summarize the confidence interval and testing procedures.

A $(1 - \alpha) \times 100\%$ Confidence Interval for $(\sigma_1)^2/(\sigma_2)^2$

$$\left(\frac{s_1^2}{s_2^2}\right)\left(\frac{1}{F_{L,\alpha/2}}\right) < \left(\frac{\sigma_1^2}{\sigma_2^2}\right) < \left(\frac{s_1^2}{s_2^2}\right)F_{U,\alpha/2}$$

where $F_{L,\alpha/2}$ is the value of F that places an area $\alpha/2$ in the upper tail of an F-distribution with $v_1 = (n_1 - 1)$ numerator and $v_2 = (n_2 - 1)$ denominator degrees of freedom, and $F_{U,\alpha/2}$ is the value of F that places an area $\alpha/2$ in the upper tail of an F-distribution with $v_1 = (n_2 - 1)$ numerator and $v_2 = (n_1 - 1)$ denominator degrees of freedom.

F-Test for Equal Population Variances*

One-Tailed Test

$H_0: \sigma_1^2 = \sigma_2^2$
$H_a: \sigma_1^2 < \sigma_2^2$ (or $H_a: \sigma_1^2 > \sigma_2^2$)

Test statistic:

$$F = \frac{s_2^2}{s_1^2}$$

$$\left(\text{or } F = \frac{s_1^2}{s_2^2} \text{ when } H_a: \sigma_1^2 > \sigma_2^2\right)$$

Rejection region:
$F > F_\alpha$

Two-Tailed Test

$H_0: \sigma_1^2 = \sigma_2^2$
$H_a: \sigma_1^2 \neq \sigma_2^2$

Test statistic:

$$F = \frac{\text{Larger sample variance}}{\text{Smaller sample variance}}$$

$$= \frac{s_1^2}{s_2^2} \text{ when } s_1^2 > s_2^2$$

$$\left(\text{or } \frac{s_2^2}{s_1^2} \text{ when } s_2^2 > s_1^2\right)$$

Rejection region:
$F > F_{\alpha/2}$

where F_α and $F_{\alpha/2}$ are based on $v_1 =$ numerator degrees of freedom and $v_2 =$ denominator degrees of freedom; v_1 and v_2 are the degrees of freedom for the numerator and denominator sample variances, respectively.

Conditions Required for Valid Inferences about $(\sigma_1)^2/(\sigma_2)^2$

1. The samples are random and independent.
2. Both populations are normally distributed.

To conclude this section, we consider the comparison of population variances as a check of the assumption $\sigma_1^2 = \sigma_2^2$ needed for the two-sample t-test. Rejection of the null hypothesis $\sigma_1^2 = \sigma_2^2$ would indicate that the assumption is invalid. [*Note:* Nonrejection of the null hypothesis does *not* imply that the assumption is valid.]

---

*Although a test of a hypothesis of equality of variances is the most common application of the F-test, it can also be used to test a hypothesis that the ratio between the population variances is equal to some specified value, $H_0: \sigma_1^2/\sigma_2^2 = k$. The test is conducted in exactly the same way as specified in the box, except that we use the test statistic

$$F = \left(\frac{s_1^2}{s_2^2}\right)\left(\frac{1}{k}\right)$$

EXAMPLE 9.12

CHECKING THE ASSUMPTION OF EQUAL VARIANCES

Problem   In Example 9.4 (Section 9.2) we used the two-sample t-statistic to compare the mean reading scores of two groups of slow learners who had been taught to read using two different methods. The data are repeated in Table 9.10 for convenience. The use of the t-statistic was based on the assumption that the population variances of the test scores were equal for the two methods. Conduct a test of hypothesis to check this assumption at $\alpha = .10$.

💿 **READING**

TABLE 9.10   Reading Test Scores for Slow Learners

New Method				Standard Method			
80	80	79	81	79	62	70	68
76	66	71	76	73	76	86	73
70	85			72	68	75	66

Solution   We want to test

$$H_0: \sigma_1^2/\sigma_2^2 = 1 \quad (\text{i.e., } \sigma_1^2 = \sigma_2^2)$$
$$H_a: \sigma_1^2/\sigma_2^2 \neq 1 \quad (\text{i.e., } \sigma_1^2 \neq \sigma_2^2)$$

The data were entered into SAS, and the SAS printout shown in Figure 9.20 was obtained. Both the test statistic, $F = .85$, and two-tailed p-value, .8148, are highlighted on the printout. Since $\alpha = .10$ is less than the p-value, we do not reject the null hypothesis that the population variances of the reading test scores are equal. It is here that the temptation to misuse the F-test is strongest. *We cannot conclude that the data justify the use of the t-statistic.* This is equivalent to accepting H_0, and we have repeatedly warned against this conclusion because the probability of a Type II error, β, is unknown. The α level of .10 protects us only against rejecting H_0 if it is true. This use of the F-test may prevent us from abusing the t procedure when we obtain a value of F that leads to a rejection of

Figure 9.20
SAS F-Test for Testing
Assumption of Equal
Variances

```
                    Two Sample Test for Variances of SCORE within METHOD

        Sample Statistics

           METHOD
           Group        N        Mean     Std. Dev.    Variance
           ------------------------------------------------------
           NEW          10       76.4      5.8348      34.04444
           STD          12    72.33333     6.3437      40.24242


        Hypothesis Test

           Null hypothesis:      Variance 1 / Variance 2 =  1
           Alternative:          Variance 1 / Variance 2 ^= 1

                          - Degrees of Freedom -
                F            Numer.    Denom.              Pr > F
           ------------------------------------------------------
              0.85             9         11              0.8148


        95% Confidence Interval of the Ratio of Two Variances

              Lower Limit      Upper Limit
              -----------      -----------
                0.2358           3.3096
```

the assumption that $\sigma_1^2 = \sigma_2^2$. But when the F-statistic does not fall in the rejection region, we know little more about the validity of the assumption than before we conducted the test.

Look Back A 95% confidence interval for the ratio $(\sigma_1)^2/(\sigma_2)^2$ is shown at the bottom of the SAS printout, Figure 9.20. Note that the interval, (.2358, 3.3096), includes 1; hence, we cannot conclude that the ratio differs significantly from 1. Thus the confidence interval leads to the same conclusion as the two-tailed test—there is insufficient evidence of a difference between the population variances.

> Now Work   *Exercise 9.97*

■ ■ ■

What Do You Do If the Assumption of Normal Population Distributions Is Not Satisfied?

Answer: The F-test is much less robust (i.e., much more sensitive) to departures from normality than the t-test for comparing the population means (Section 9.2). If you have doubts about the normality of the population frequency distributions, use a *nonparametric method* for comparing the two population variances. A method can be found in the nonparametric statistics texts listed in the references.

Statistics in Action Revisited

Comparing the Dispersion of Bid Prices in Two Kentucky Milk Markets

We return, once more, to the Statistics in Action investigation of collusive bidding in the northern Kentucky school milk market (p. 429). In competitive sealed bid markets, vendors do not share information about their bids. Consequently, more dispersion or variability among the bids is typically observed than in collusive markets, where vendors communicate about their bids and have a tendency to submit bids in close proximity to one another in an attempt to make the bidding appear competitive.

If collusion exists in the tricounty milk market, the variation in winning bid prices in the surrounding ("competitive") market will be significantly larger than the corresponding variation in the tricounty ("rigged") market. Let $(\sigma_1)^2$ and $(\sigma_2)^2$ represent the population milk price variances for the surrounding market and the tricounty market, respectively. To test this theory, we test the null hypothesis, $H_0: (\sigma_1)^2 = (\sigma_2)^2$, against the one-sided alternative hypothesis, $H_a: (\sigma_1)^2 > (\sigma_2)^2$, for each of the three milk products (whole white, low-fat white, and low-fat chocolate).

A MINITAB analysis of the the data in the **MILK** file yielded the printouts in Figure SIA9.3. The two-tailed p-values of the F-tests are highlighted on the figure. To

Figure SIA9.3
MINITAB Comparison of Bid Price Variances for Two Kentucky Milk Markets

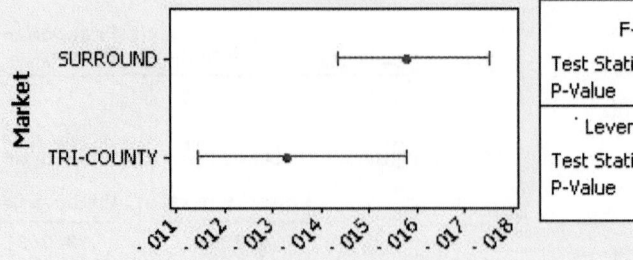

Figure SIA9.3
Continued

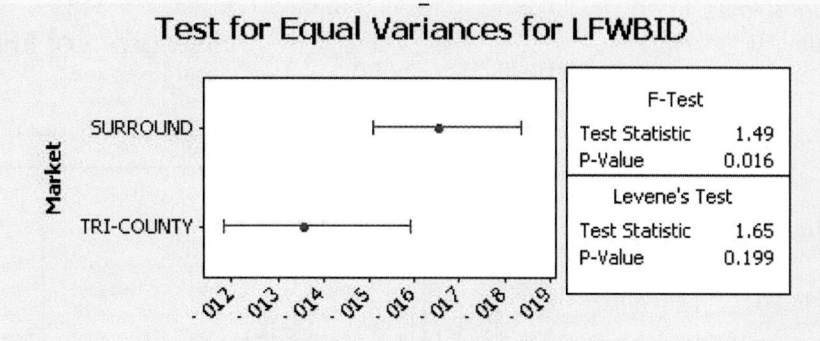

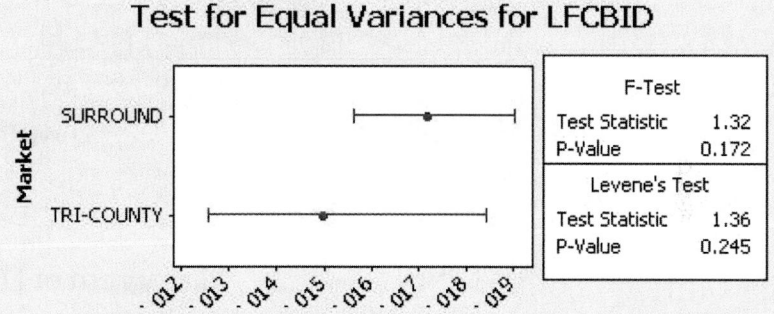

obtain the one-tailed *p*-values, we divide these values in half. Therefore, the one-tailed *p*-values are .048/2 = .024 (for whole white milk), .016/2 = .008 (for low-fat white milk), and .172/2 = .086 (for low-fat chocolate milk). At $\alpha = .05$, the tests provide evidence that the bid price variance for the surrounding market exceeds the bid price variance for the tricounty market for two of the three milk products—whole milk and low-fat white milk. Only for

low-fat chocolate is there insufficient evidence to support the bid collusion theory.

[*Note:* For the inferences derived from these *F*-tests to be statistically valid, however, we require the bid price distributions for each market to be approximately normal. (Remember, the normality assumption is required regardless of the sizes of the samples.) The MINITAB printouts in Figure SIA9.4 show histograms (with the

Figure SIA9.4
MINITAB Histograms of Bid Prices for Two Kentucky Milk Markets

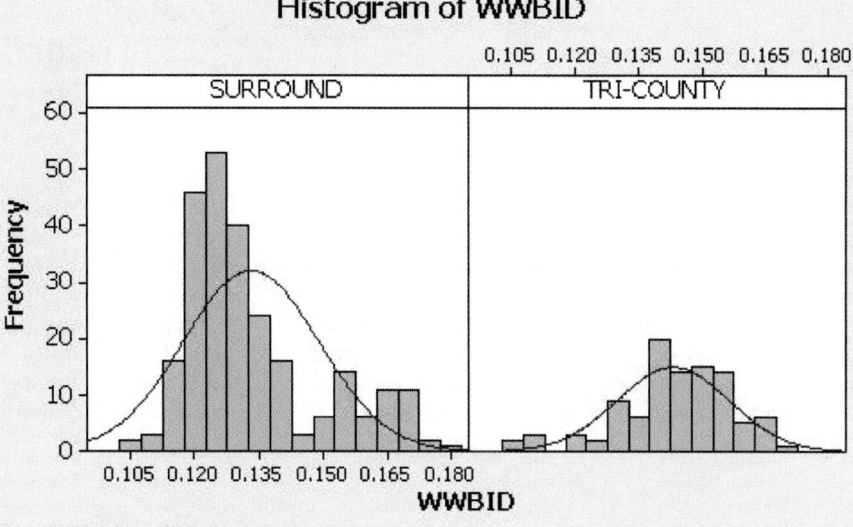

Figure SIA9.4
Continued

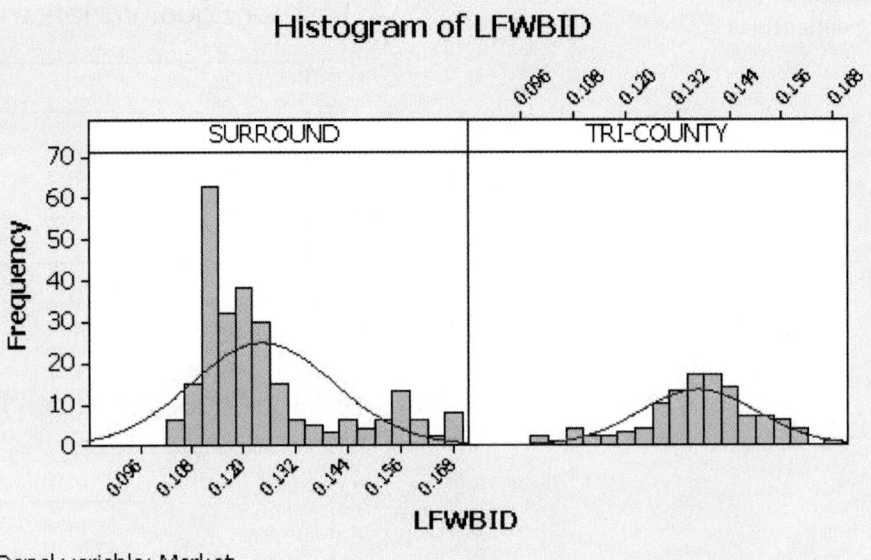

Panel variable: Market

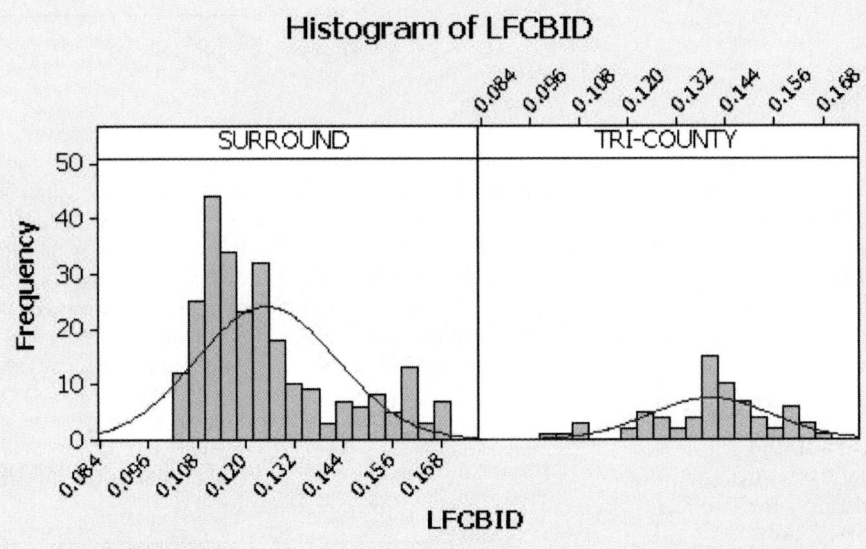

Panel variable: Market

normal distribution superimposed) of the milk prices for each of the two markets. The data for the tricounty market appear to be approximately normal; but the data for the surrounding market have a tendency to be rightward skewed. Since the *F*-test is not robust for nonnormal data, the test results may not be valid. A more robust test for comparing variances is the Levene test. (See box, p. 00) The *p*-values for the Levene tests are shown below the *F*-tests on the MINITAB printout, Figure SIA9.3. When converted to one-tailed *p*-values, these values all exceed $\alpha = .10$. Thus, the more conservative Levene test does not find significant differences between the bid price variances for any of the three milk products. Consequently, the state of Kentucky may choose to omit the bid dispersion theory in its argument that bid collusion existed in the tricounty market.]

Hypothesis Test for Two Population Variances

Using the TI-83 Graphing Calculator

For this calculation, you have the option of entering the raw data or using the summary statistics

Step 1 *If you are using the raw data, enter the data into L1 and L2.*

Press **STAT** and select **1:Edit**
Note: If a list already contains data, clear the old data. Use the up arrow to highlight the name at the top of the list. Press **CLEAR ENTER**.
Use the **Down Arrow** or **ENTER** key to enter the data sets into **L1** and **L2**.

Step 2 *Access the Statistical Tests Menu*

Press **STAT**
Arrow right to **TESTS**
Arrow down to **D : 2-SampFTest**
Press **ENTER**

Step 3 *Choose "**Data**" or "**Stats**".* (*"Data" is selected when you have entered the raw data into L1 and L2. "Stats" is selected when you are using the summary statistics.*)

Press **ENTER**

If you selected "Data":
Set **List1** to **L1**. Note: If List1 is not already set to L1, arrow down to List1: and hit **2nd** and '**1**'.
Set **List2** to **L2**. Note: If List2 is not already set to L2, arrow down to List2: and hit **2nd** and '**2**'.
Set **Freq1** to **1**. Note: If Freq1 is not already set to 1, arrow down to Freq1 and hit **Alpha** and '**1**'.
Set **Freq2** to **1**. Note: If Freq2 is not already set to 1, arrow down to Freq2 and hit **Alpha** and '**1**'.
Arrow down to the next line and select the appropriate alternative hypothesis by using the right arrow to highlight $\neq$ $\sigma 2$, $<\sigma 2$, or $>\sigma 2$ and press **ENTER**.
Arrow down to **Calculate Draw**. You may choose either one of these for your output.
Press **ENTER**

If you selected "Stats", enter the sample standard deviations and sample sizes.
Arrow down to the next line and select the appropriate alternative hypothesis by using the right arrow to highlight $\neq$ $\sigma 2$, $<\sigma 2$, or $>\sigma 2$ and press **ENTER**.
Arrow down to **Calculate Draw**. You may choose either one of these for your output.
Press **ENTER**.

Step 4 *View the Output*

If you chose **Calculate**, the output will display the alternative hypothesis, the F-value, the p-value and the summary statistics (sample standard deviations and sample sizes.)

If you chose **Draw**, a graph of the F-distribution will be displayed along with the F-value and the p-value. The area under the curve that represents the p-value is shaded. Note: In order to use the **Draw** feature, you must first clear all equations from the **Y-registers**. Hit '**Y =**' and delete all equations.

Example   *Consider Example 9.10 with data in Table 9.9 on pg. 485. For this example, the raw data will be used, so begin with Step1 of the instructions.*

Step 1   *Enter the data for Supplier 1 into L1 and the data for Supplier 2 into L2.*

Press **STAT** and select **1:Edit**
Note: If a list already contains data, clear the old data. Use the up arrow to highlight the name at the top of the list. Press **CLEAR ENTER**.
Use the **Down Arrow** or **ENTER** key to enter the data sets into **L1 and L2.**

L1	L2	L3	2
4.23	4.14	------	
4.35	4.26		
4.05	4.05		
3.75	4.11		
4.41	4.31		
4.37	4.12		
4.01	4.17		

L2(1)=4.14

Step 2   *Access the Statistical Tests Menu*

Press **STAT**
Arrow right to **TESTS**
Arrow down to **D : 2-SampFTest**
Press **ENTER**

EDIT CALC **TESTS**
8↑TInterval…
9:2-SampZInt…
0:2-SampTInt…
A:1-PropZInt…
B:2-PropZInt…
C:χ²-Test…
D:2-SampFTest…

Step 3   *Choose "Data"*

Set **List1** to **L1**. Note: If List1 is not already set to L1, arrow down to List1: and hit **2nd** and **'1'**.
Set **List2** to **L2.** Note: If List2 is not already set to L2, arrow down to List2: and hit **2nd** and **'2'**.
Set **Freq1** to **1.** Note: If Freq1 is not already set to 1, arrow down to Freq1 and hit **Alpha** and **'1'**.
Set **Freq2** to **1.** Note: If Freq2 is not already set to 1, arrow down to Freq2 and hit **Alpha** and **'1'**.
Arrow down to the next line and select $\neq \sigma 2,$ and press **ENTER**
Arrow down to **Calculate Draw.** You may choose either one of these for your output.
Press **ENTER**

2-SampFTest
 Inpt:**Data** Stats
 List1:L₁
 List2:L₂
 Freq1:1
 Freq2:1
 σ1:**≠σ2** <σ2 >σ2
 Calculate Draw

Step 4   *View the Output*

If you chose **Calculate**, the output will display the alternative hypothesis, the F-value, the p-value and the summary statistics (sample standard deviations and sample sizes.)

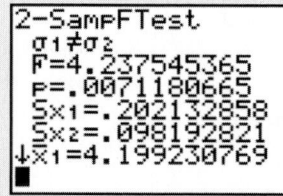

If you chose **Draw**, a graph of the F-distribution will be displayed along with the F-value and the p-value. The area under the curve that represents the p-value is shaded. Note: In order to use the **Draw** feature, you must first clear all equations from the **Y-registers**. Hit 'Y =' and delete all equations.

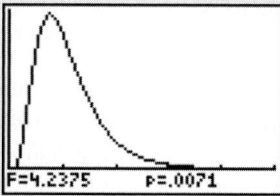

Note: The shaded area under the curve is not noticeable because the p-value is so small.

Exercises 9.84–9.102

Understanding the Principles

9.84 Describe the sampling distribution of $(s_1)^2/(s_2)^2$ for normal data.

9.85 What conditions are required for valid inferences about $(\sigma_1)^2/(\sigma_2)^2$?

9.86 *True or false.* The F-statistic used for testing $H_0: \sigma_1^2 = \sigma_2^2$ against $H_a: \sigma_1^2 < \sigma_2^2$ is $F = (s_1)^2/(s_2)^2$.

9.87 *True or false.* $H_0: \sigma_1^2 = \sigma_2^2$ is equivalent to $H_0: \sigma_1^2/\sigma_2^2 = 0$.

Learning the Mechanics

9.88 Use Tables VIII, IX, X, and XI of Appendix A to find
NW each of the following F-values:
    a. $F_{.05}$ where $\nu_1 = 8$ and $\nu_2 = 5$
    b. $F_{.01}$ where $\nu_1 = 20$ and $\nu_2 = 14$
    c. $F_{.025}$ where $\nu_1 = 10$ and $\nu_2 = 5$
    d. $F_{.10}$ where $\nu_1 = 20$ and $\nu_2 = 5$

9.89 Given ν_1 and ν_2, find the following probabilities:
    a. $\nu_1 = 2, \nu_2 = 30, P(F \geq 4.18)$
    b. $\nu_1 = 24, \nu_2 = 14, P(F < 1.94)$
    c. $\nu_1 = 9, \nu_2 = 1, P(F \leq 6,022.0)$
    d. $\nu_1 = 30, \nu_2 = 30, P(F > 1.84)$

9.90 For each of the following cases, identify the rejection region that should be used to test $H_0: \sigma_1^2 = \sigma_2^2$ against $H_a: \sigma_1^2 > \sigma_2^2$. Assume $\nu_1 = 20$ and $\nu_2 = 30$.
    a. $\alpha = .10$ **b.** $\alpha = .05$
    c. $\alpha = .025$ **d.** $\alpha = .01$

9.91 For each of the following cases, identify the rejection region that should be used to test $H_0: \sigma_1^2 = \sigma_2^2$ against $H_a: \sigma_1^2 \neq \sigma_2^2$. Assume $\nu_1 = 8$ and $\nu_2 = 40$.
    a. $\alpha = .20$

    b. $\alpha = .10$
    c. $\alpha = .05$
    d. $\alpha = .02$

9.92 Specify the appropriate rejection region for testing against $H_0: \sigma_1^2 = \sigma_2^2$ in each of the following situations:
    a. $H_a: \sigma_1^2 > \sigma_2^2; \alpha = .05, n_1 = 25, n_2 = 20$

    b. $H_a: \sigma_1^2 < \sigma_2^2; \alpha = .05, n_1 = 10, n_2 = 15$

    c. $H_a: \sigma_1^2 \neq \sigma_2^2; \alpha = .10, n_1 = 21, n_2 = 31$

    d. $H_a: \sigma_1^2 < \sigma_2^2; \alpha = .01, n_1 = 31, n_2 = 41$

    e. $H_a: \sigma_1^2 \neq \sigma_2^2; \alpha = .05, n_1 = 7, n_2 = 16$

9.93 Independent random samples were selected from each
NW of two normally distributed populations, $n_1 = 16$ from population 1 and $n_2 = 25$ from population 2. The means and variances for the two samples are shown in the table.

Sample 1	Sample 2
$n_1 = 16$	$n_2 = 25$
$\bar{x}_1 = 22.5$	$\bar{x}_2 = 28.2$
$s_1^2 = 2.87$	$s_2^2 = 9.85$

    a. Test the null hypothesis $H_0: \sigma_1^2 = \sigma_2^2$ against the alternative hypothesis $H_a: \sigma_1^2 \neq \sigma_2^2$. Use $\alpha = .05$.
    b. Find and interpret the p-value of the test.

9.94 Independent random samples were selected from each of two normally distributed populations, $n_1 = 6$ from population 1 and $n_2 = 4$ from population 2. The data are shown in the table.

⊘ **LM9_94**

Sample 1	Sample 2
3.1	2.3
4.3	1.4
1.2	3.7
1.7	8.9
.6	
3.4	

a. Test $H_0: \sigma_1^2 = \sigma_2^2$ against $H_a: \sigma_1^2 < \sigma_2^2$. Use $\alpha = .01$.
b. Test $H_0: \sigma_1^2 = \sigma_2^2$ against $H_a: \sigma_1^2 \neq \sigma_2^2$. Use $\alpha = .10$.

Applying the Concepts—Basic

9.95 Rating service at five-star hotels. Refer to *The Journal of American Academy of Business, Cambridge* (March 2002) study of how guests perceive the service quality at five-star hotels in Jamaica, Exercise 9.14 (p. 449). A summary of the guest perception scores, by gender, are reproduced in the table. Let σ_M^2 and σ_F^2 represent the true variances in scores for male and female guests, respectively.

Gender	Sample size	Mean score	Standard deviation
Males	127	39.08	6.73
Females	114	38.79	6.94

a. Set up H_0 and H_a for determining whether σ_M^2 is less than σ_F^2.
b. Find the test statistic for the test.
c. Give the rejection region for the test if $\alpha = .10$.
d. Find the approximate *p*-value of the test.
e. Make the appropriate conclusion in the words of the problem.
f. What conditions are required for the test results to be valid?

⊘ **GOBIANTS**

9.96 Mongolian desert ants. Refer to the *Journal of Biogeography* (Dec. 2003) study of ants in Mongolia (Central Asia), Exercise 9.23 (p. 451), where you compared the mean number of ants at two desert sites. Since the sample sizes were small, the variances of the populations at the two sites must be equal in order for the inference to be valid.
a. Set up H_0 and H_a for determining whether the variances are the same.
b. Use the data in the **GOBIANTS** file to find the test statistic for the test.

c. Give the rejection region for the test if $\alpha = .05$.
d. Find the approximate *p*-value of the test.
e. Make the appropriate conclusion in the words of the problem.
f. What conditions are required for the test results to be valid?

9.97 Bulimia study. Refer to Exercise 9.19 (p. 450). The "fear of negative evaluation" (FNE) scores for the 11 bulimic females and 14 females with normal eating habits are reproduced in the table at the bottom of the page. The confidence interval you constructed in Exercise 9.19 requires that the variance of the FNE scores of bulimic females is equal to the variance of the FNE scores of normal females. Conduct a test (at $\alpha = .05$) to determine the validity of this assumption.

9.98 Gender differences in math test scores. The *American Educational Research Journal* (Fall 1998) published a study to compare the mathematics achievement test scores of male and female students. The researchers hypothesized that the distribution of test scores for males is more variable than the corresponding distribution for females. Use the summary information reproduced in the table to test this claim at $\alpha = .01$.

	Males	Females
Sample size	1,764	1,739
Mean	48.9	48.4
Standard deviation	12.96	11.85

Source: Bielinski, J., and Davison, M. L. "Gender differences by item difficulty interactions in multiple-choice mathematics items." *American Educational Research Journal,* Vol. 35, No. 3, Fall 1998, p. 464 (Table 1).

Applying the Concepts—Intermediate

9.99 Patent infringement case. Refer to the *Chance* (Fall 2002) description of a patent infringement case against Intel Corp., Exercise 9.20 (p. 450). The zinc measurements for three locations on the original inventor's notebook—on a text line, on a witness line, and on the intersection of the witness and text line—are reproduced in the table.

⊘ **PATENT**

Text line:	.335	.374	.440			
Witness line:	.210	.262	.188	.329	.439	.397
Intersection:	.393	.353	.285	.295	.319	

a. Use a test (at $\alpha = .05$) to compare the variation in zinc measurements for the text line with the corresponding variation for the intersection.

BULIMIA

Bulimic students	21	13	10	20	25	19	16	21	24	13	14			
Normal students	13	6	16	13	8	19	23	18	11	19	7	10	15	20

Source: Randles, R. H. "On neutral responses (zeros) in the sign test and ties in the Wilcoxon-Mann-Whitney test." *The American Statistician.* Vol. 55, No. 2, May 2001 (Figure 3).

b. Use a test (at $\alpha = .05$) to compare the variation in zinc measurements for the witness line with the corresponding variation for the intersection.

c. From the results, parts **a** and **b**, what can you infer about the variation in zinc measurements at the three notebook locations?

d. What assumptions are required for the inferences to be valid? Are they reasonably satisfied? (You checked these assumptions when answering Exercise 9.20d.)

9.100 Human inspection errors. Tests of product quality using human inspectors can lead to serious inspection error problems (*Journal of Quality Technology,* Apr. 1986.) To evaluate the performance of inspectors in a new company, a quality manager had a sample of 12 novice inspectors evaluate 200 finished products. The same 200 items were evaluated by 12 experienced inspectors. The quality of each item—whether defective or nondefective—was known to the manager. The following table lists the number of inspection errors (classifying a defective item as nondefective or vice versa) made by each inspector.

INSPECT

Novice Inspectors				Experienced Inspectors			
30	35	26	40	31	15	25	19
36	20	45	31	28	17	19	18
33	29	21	48	24	10	20	21

a. Prior to conducting this experiment, the manager believed the variance in inspection errors was lower for experienced inspectors than for novice inspectors. Do the sample data support her belief? Test using $\alpha = .05$.

b. What is the appropriate p-value of the test you conducted in part **a**?

9.101 Mating habits of snails. Refer to the *Genetical Research* (June 1995) study of the mating habits of hermaphroditic snails, Exercise 9.26 (p. 452). Recall that the snails were identified as either selfers or outcrossers and that the mean effective population sizes of the two groups were compared. The data for the study are reproduced in the table. Geneticists are often more interested in comparing the variation in population size of the two types of mating systems.

Conduct this analysis for the researcher. Interpret the result.

Snail Mating System	Sample Size	Effective Population Size	
		Mean	Standard Deviation
Outcrossing	17	4,894	1,932
Selfing	5	4,133	1,890

Source: Jarne, P. "Mating system, bottlenecks, and genetic polymorphism in hermaphroditic animals." *Genetical Research,* Vol. 65, No. 3, June 1995, p. 197 (Table 4).

9.102 Following the initial Persian Gulf War, the Pentagon changed its logistics processes to be more corporate-like. The extravagant "just-in-case" mentality was replaced with "just-in-time" systems. Emulating Federal Express and United Parcel Service, deliveries from factories to foxholes are now expedited using bar codes, laser cards, radio tags, and databases to track supplies. The table contains order-to-delivery times (in days) for a sample of shipments from the U.S. to the Persian Gulf in 1991 and a sample of shipments to Bosnia in 1995.

ORDTIMES

Persian Gulf	Bosnia
28.0	15.1
20.0	6.4
26.5	5.0
10.6	11.4
9.1	6.5
35.2	6.5
29.1	3.0
41.2	7.0
27.5	5.5

Source: Adapted from Crock, S. "The Pentagon goes to B-school." *Business Week,* Dec. 11, 1995, p. 98.

a. Determine whether the variances in order-to-delivery times for Persian Gulf and Bosnia shipments are equal. Use $\alpha = .05$.

b. Given your answer to part **a**, is it appropriate to construct a confidence interval for the difference between the mean order-to-delivery times? Explain.

Quick Review

Key Terms

Note: Starred (*) terms are from the optional section in this chapter.

Key Formulas

Note: Starred (*) formulas are from the optional section in this chapter.

$(1 - \alpha)100\%$ confidence interval for θ: (see table)

$$\text{Large samples:} \quad \hat{\theta} \pm z_{\alpha/2}\sigma_{\hat{\theta}}$$

$$\text{Small samples:} \quad \hat{\theta} \pm t_{\alpha/2}\sigma_{\hat{\theta}}$$

For testing $H_0: \theta = D_0$: (see table)

$$\text{Large samples:} \quad z = \frac{\hat{\theta} - D_0}{\sigma_{\hat{\theta}}}$$

$$\text{Small samples:} \quad t = \frac{\hat{\theta} - D_0}{\sigma_{\hat{\theta}}}$$

Parameter, θ	Estimator, $\hat{\theta}$	Standard Error of Estimator, $\sigma_{\hat{\theta}}$	Estimated Standard Error	
$\mu_1 - \mu_2$ (independent samples)	$\bar{x}_1 - \bar{x}_2$	$\sqrt{\dfrac{\sigma_1^2}{n_1} + \dfrac{\sigma_2^2}{n_2}}$	Large n: $\sqrt{\dfrac{s_1^2}{n_1} + \dfrac{s_2^2}{n_2}}$	433
			Small n: $\sqrt{s_p^2\left(\dfrac{1}{n_1} + \dfrac{1}{n_2}\right)}$	437
μ_D (paired sample)	$\bar{x}_D$	$\dfrac{\sigma_D}{\sqrt{n_D}}$	$\dfrac{s_D}{\sqrt{n_D}}$	457
$p_1 - p_2$ (large, independent samples)	$\hat{p}_1 - \hat{p}_2$	$\sqrt{\dfrac{p_1 q_1}{n_1} + \dfrac{p_2 q_2}{n_2}}$	Confidence intervals: $\sqrt{\dfrac{\hat{p}_1\hat{q}_1}{n_1} + \dfrac{\hat{p}_2\hat{q}_2}{n_2}}$	469, 470
			Hypothesis tests: $\sqrt{\hat{p}\hat{q}\left(\dfrac{1}{n_1} + \dfrac{1}{n_2}\right)}$	471

Pooled sample variance:	$s_p^2 = \dfrac{(n_1 - 1)s_1^2 + (n_2 - 1)s_2^2}{n_1 + n_2 - 2}$	437
Pooled sample proportion:	$\hat{p} = \dfrac{x_1 + x_2}{n_1 + n_2}, \quad \hat{q} = 1 - \hat{p}$	471
Determining the sample size for estimating $\mu_1 - \mu_2$:	$n_1 = n_2 = \dfrac{(z_{\alpha/2})^2(\sigma_1^2 + \sigma_2^2)}{(\text{SE})^2}$	481
Determining the sample size for estimating $p_1 - p_2$:	$n_1 = n_2 = \dfrac{(z_{\alpha/2})^2(p_1 q_1 + p_2 q_2)}{(\text{SE})^2}$	481
*Test statistic for testing $H_0: \dfrac{\sigma_1^2}{\sigma_2^2}$	$F = \dfrac{\text{larger } s^2}{\text{smaller } s^2}$ if $H_a: \dfrac{\sigma_1^2}{\sigma_2^2} \neq 1$	488
	$F = \dfrac{s_1^2}{s_2^2}$ if $H_a: \dfrac{\sigma_1^2}{\sigma_2^2} > 1$	

Language Lab

Symbol	Pronunciation	Description
$\mu_1 - \mu_2$	mu-1 minus mu-2	Difference between population means
$\bar{x}_1 - \bar{x}_2$	x-bar-1 minus x-bar-2	Difference between sample means
$\sigma_{(\bar{x}_1 - \bar{x}_2)}$	sigma of x-bar-1 minus x-bar-2	Standard deviation of the sampling distribution of $(\bar{x}_1 - \bar{x}_2)$
s_p^2	s-p squared	Pooled sample variance
D_0	D naught	Hypothesized value of difference

μ_D	mu D	Difference between population means, paired data
$\bar{x}_D$	x-bar D	Mean of sample differences
s_D	s-D	Standard deviation of sample differences
n_D	n-D	Number of differences in sample
$p_1 - p_2$	p-1 minus p-2	Difference between population proportions
$\hat{p}_1 - \hat{p}_2$	p-1 hat minus p-2 hat	Difference between sample proportions
$\sigma_{(\hat{p}_1 - \hat{p}_2)}$	sigma of p-1 hat minus p-2 hat	Standard deviation of the sampling distribution of $(\hat{p}_1 - \hat{p}_2)$
F_α	F-alpha	Critical value of F associated with tail area α
ν_1	nu-1	Numerator degrees of freedom for F statistic
ν_2	nu-2	Denominator degrees of freedom for F statistic
$\dfrac{\sigma_1^2}{\sigma_2^2}$	sigma-1 squared over sigma-2 squared	Ratio of two population variances

Chapter Summary Notes

- Key words/phrases for identifying $\mu_1 - \mu_2$ as the parameter of interest: *difference between means or averages, compare two means using independent samples*
- Key words/phrases for identifying μ_D as the parameter of interest: *mean or average of paired differences, compare two means using matched pairs*
- Key words/phrases for identifying $p_1 - p_2$ as the parameter of interest: *difference between proportions or percentages, compare two proportions using independent samples*
- Key words/phrases for identifying σ_1^2/σ_2^2 as the parameter of interest: *difference between variances, compare variation in two populations using independent samples*
- Conditions required for a valid **large-sample inferences for $\mu_1 - \mu_2$:** (1) independent random samples, (2) large sample sizes ($n_1 \geq 30$ and $n_2 \geq 30$)

- Conditions required for a valid **small-sample inferences for $\mu_1 - \mu_2$:** (1) independent random samples, (2) both populations are approximately normal, (3) $\sigma_1^2 = \sigma_2^2$
- Conditions required for a valid **large-sample inferences for μ_D** (1) random sample of paired differences, (2) large sample size ($n_d \geq 30$)
- Conditions required for a valid **small-sample inferences for μ_D** (1) random sample of paired differences, (2) population of differences are approximately normal
- Conditions required for a valid **large-sample inferences for $p_1 - p_2$:** (1) independent random samples, (2) large sample sizes (both $p_1 \pm 3\sigma$ and $p_2 \pm 3\sigma$ fall between 0 and 1)
- Conditions required for a valid **inferences for σ_1^2/σ_2^2:** (1) independent random samples, (2) both populations are approximately normal

Supplementary Exercises 9.103–9.136

Note: Starred (*) exercises refer to the optional section in this chapter.

Understanding the Principles

9.103 For each of the following, identify the target parameter as $\mu_1 - \mu_2$, $p_1 - p_2$, or σ_1^2/σ_2^2.

  a. Compare average SAT scores of males and females.

  b. Difference between mean waiting times at two supermarket checkout lanes

  c. Compare proportions of Democrats and Republicans who favor legalization of marijuana.

  *d. Compare variation in salaries of NBA players picked in the first round and the second round.

  e. Difference in dropout rates of college student athletes and regular students

9.104 List the assumptions necessary for each of the following inferential techniques:

  a. Large-sample inferences about the difference $(\mu_1 - \mu_2)$ between population means using a two-sample z-statistic

  b. Small-sample inferences about $(\mu_1 - \mu_2)$ using an independent samples design and a two-sample t-statistic

  c. Small-sample inferences about $(\mu_1 - \mu_2)$ using a paired difference design and a single-sample t-statistic to analyze the differences

  d. Large-sample inferences about the differences $(p_1 - p_2)$ between binomial proportions using a two-sample z-statistic

  *e. Inferences about the ratio σ_1^2/σ_2^2 of two population variances using an F-test

Learning the Mechanics

9.105 Independent random samples were selected from two normally distributed populations with means μ_1 and μ_2, respectively. The sample sizes, means, and variances are shown in the following table.

Sample 1	Sample 2
$n_1 = 12$	$n_2 = 14$
$\bar{x}_1 = 17.8$	$\bar{x}_2 = 15.3$
$s_1^2 = 74.2$	$s_2^2 = 60.5$

a. Test $H_0: (\mu_1 - \mu_2) = 0$ against $H_a: (\mu_1 - \mu_2) > 0$. Use $\alpha = .05$.

b. Form a 99% confidence interval for $(\mu_1 - \mu_2)$.

c. How large must n_1 and n_2 be if you wish to estimate $(\mu_1 - \mu_2)$ to within 2 units with 99% confidence? Assume that $n_1 = n_2$.

9.106 Two independent random samples were selected from normally distributed populations with means and variances (μ_1, σ_1^2) and (μ_2, σ_2^2), respectively. The sample sizes, means, and variances are shown in the table.

Sample 1	Sample 2
$n_1 = 20$	$n_2 = 15$
$\bar{x}_1 = 123$	$\bar{x}_2 = 116$
$s_1^2 = 31.3$	$s_2^2 = 120.1$

*a. Test $H_0: \sigma_1^2 = \sigma_2^2$ against $H_a: \sigma_1^2 \neq \sigma_2^2$. Use $\alpha = .05$.

b. Would you be willing to use a t-test to test the null hypothesis $H_0: (\mu_1 - \mu_2) = 0$ against the alternative hypothesis $H_a: (\mu_1 - \mu_2) \neq 0$? Why?

9.107 Two independent random samples are taken from two populations. The results of these samples are summarized in the following table.

Sample 1	Sample 2
$n_1 = 135$	$n_2 = 148$
$\bar{x}_1 = 12.2$	$\bar{x}_2 = 8.3$
$s_1^2 = 2.1$	$s_2^2 = 3.0$

a. Form a 90% confidence interval for $(\mu_1 - \mu_2)$.

b. Test $H_0: (\mu_1 - \mu_2) = 0$ against $H_a: (\mu_1 - \mu_2) \neq 0$. Use $\alpha = .01$.

c. What sample size would be required if you wish to estimate $(\mu_1 - \mu_2)$ to within .2 with 90% confidence? Assume that $n_1 = n_2$.

9.108 Independent random samples were selected from two binomial populations. The size and number of observed successes for each sample are shown in the table.

Sample 1	Sample 2
$n_1 = 200$	$n_2 = 200$
$x_1 = 110$	$x_2 = 130$

a. Test $H_0: (p_1 - p_2) = 0$ against $H_a: (p_1 - p_2) < 0$. Use $\alpha = .10$.

b. Form a 95% confidence interval for $(p_1 - p_2)$.

c. What sample sizes would be required if we wish to use a 95% confidence interval of width .01 to estimate $(p_1 - p_2)$?

9.109 A random sample of five pairs of observations were selected, one observation from a population with mean μ_1, the other from a population with mean μ_2. The data are shown in the accompanying table.

LM9_109

Pair	Value from Population 1	Value from Population 2
1	28	22
2	31	27
3	24	20
4	30	27
5	22	20

a. Test the null hypothesis $H_0: \mu_D = 0$ against $H_a: \mu_D \neq 0$, where $\mu_D = \mu_1 - \mu_2$. Use $\alpha = .05$.

b. Form a 95% confidence interval for μ_D.

c. When are the procedures you used in parts **a** and **b** valid?

Applying the Concepts—Basic

OILSPILL

9.110 **Hull failure of oil tankers.** Refer to the *Marine Technology* (Jan. 1995) study of major oil spills from tankers and carriers, Exercise 2.182 (p. 112). The data for the 50 spills are saved in the **OILSPILL** file.

a. Construct a 90% confidence interval for the difference between the mean spillage amount of accidents caused by collision and the mean spillage amount of accidents caused by fire/explosion. Interpret the result.

b. Conduct a test of hypothesis to compare the mean spillage amount of accidents caused by grounding to the corresponding mean of accidents caused by hull failure. Use $\alpha = .05$.

c. Refer to parts **a** and **b**. State any assumptions required for the inferences derived from the analyses to be valid. Are these assumptions reasonably satisfied?

*d. Conduct a test of hypothesis to compare the variation in spillage amounts for accidents caused by collision and accidents caused by grounding. Use $\alpha = .02$.

EVOS

9.111 **Oil spill impact on seabirds.** Refer to the *Journal of Agricultural, Biological, and Environmental Statistics* (Sept. 2000) impact study of a tanker oil spill on the seabird population in Alaska, Exercise 2.179 (p. 111). Recall that for each of 96 shoreline locations (called transects), the number of seabirds found, the length (in kilometers) of the transect, and whether or not the transect was in an oiled area were recorded. (The data are saved in the **EVOS** file.) Observed seabird density is defined as the observed count divided by the length of

MINITAB Output for Exercise 9.111

Two-Sample T-Test and CI: Density, Oil

```
Two-sample T for Density

Oil   N  Mean  StDev  SE Mean
no   36  3.27   6.70      1.1
yes  60  3.50   5.97     0.77


Difference = mu (no ) - mu (yes)
Estimate for difference:  -0.221165
95% CI for difference:  (-2.927767, 2.485436)
T-Test of difference = 0 (vs not =): T-Value = -0.16  P-Value = 0.871  DF = 67
```

the transect. A comparison of the mean densities of oiled and unoiled transects is displayed in the MINITAB printout above. Use this information to make an inference about the difference in the population mean seabird densities of oiled and unoiled transects.

9.112 **Tapeworms in brill fish.** The *Journal of Fish Biology* (Aug. 1990) reported on a study to compare the incidence of parasites (tapeworms) in species of Mediterranean and Atlantic fish. In the Mediterranean Sea, 588 brill were captured and dissected, and 211 were found to be infected by the parasite. In the Atlantic Ocean, 123 brill were captured and dissected, and 26 were found to be infected. Compare the proportions of infected brill at the two capture sites using a 90% confidence interval. Interpret the interval.

9.113 **Effect of altitude on climbers.** Dr. Philip Lieberman, a neuroscientist at Brown University, recently conducted a field experiment to gauge the effect of high altitude on a person's ability to think critically. (*New York Times*, Aug. 23, 1995.) The subjects of the experiment were five male members of an American expedition climbing Mount Everest. At the base camp, Lieberman read sentences to the climbers while they looked at simple pictures in a book. The length of time (in seconds) it took for each climber to match the picture with a sentence was recorded. Using a radio, Lieberman repeated the task when the climbers reached a camp 5 miles above sea level. At this altitude, he noted that the climbers took 50% longer to complete the task.
a. What is the variable measured for this experiment?
b. What are the experimental units?
c. Discuss how the data should be analyzed.

*9.114 **Bear vs. pig bile study.** Bear gallbladder is used in Chinese medicine to treat inflamation. A study in the *Journal of Ethnopharmacology* (June 1995) examined the easier to obtain pig gallbladder as an effective substitute for bear gallbladder. Twenty male mice were divided randomly into two groups: 10 were given a dosage of bear bile and 10 were given a dosage of pig bile. All the mice then received an injection of croton oil in the left ear lobe to induce inflammation. Four hours later, both the left and right ear lobes were weighed, with the difference (in milligrams) representing the degree of swelling. Summary statistics on the degree of swelling are provided in the following table.

Bear Bile	Pig Bile
$n_1 = 10$	$n_2 = 10$
$\bar{x}_1 = 9.19$	$\bar{x}_2 = 9.71$
$s_1 = 4.17$	$s_2 = 3.33$

a. Use a hypothesis test (at $\alpha = .05$) to compare the variation in degree of swelling for mice treated with bear and pig bile.
b. What assumptions are necessary for the inference, part **a**, to be valid?

9.115 **Self-concepts of female students.** A study reported in the *Journal of Psychology* (Mar. 1991) measures the change in female students' self-concepts as they move from high school to college. A sample of 133 Boston College female students were selected for the study. Each was asked to evaluate several aspects of her life at two points in time: at the end of her senior year of high school, and during her sophomore year of college. Each student was asked to evaluate where she believed she stood on a scale that ranged from top 10% of class (1) to lowest 10% of class (5). The results for three of the traits evaluated are reported in the table below.

Trait	n	Senior Year of High School $\bar{x}$	Sophomore Year of College $\bar{x}$
Leadership	133	2.09	2.33
Popularity	133	2.48	2.69
Intellectual self-confidence	133	2.29	2.55

a. What null and alternative hypotheses would you test to determine whether the mean self-concept of females decreases between the senior year of high school and the sophomore year of college as measured by each of these three traits?
b. Are these traits more appropriately analyzed using an independent samples test or a paired difference test? Explain.
c. Noting the size of the sample, what assumptions are necessary to ensure the validity of the tests?
d. The article reports that the leadership test results in a *p*-value greater than .05, while the tests for popularity and intellectual self-confidence result in *p*-values less than .05. Interpret these results.

9.116 Environmental impact study. Some power plants are located near rivers or oceans so that the available water can be used for cooling the condensers. Suppose that, as part of an environmental impact study, a power company wants to estimate the difference in mean water temperature between the discharge of its plant and the offshore waters. How many sample measurements must be taken at each site in order to estimate the true difference between means to within .2°C with 95% confidence? Assume that the range in readings will be about 4°C at each site and the same number of readings will be taken at each site.

9.117 Hatching tree frog eggs. Scientists have linked a catastrophic decline in the number of frogs inhabiting the world to ultraviolet radiation. (*Tampa Tribune*, Mar. 1, 1994.) The Pacific tree frog, however, is not believed to be in decline, possibly because it manufactures an enzyme that protects its eggs from ultraviolet radiation. Researchers at Oregon State University compared the hatching rates of two groups of Pacific tree frog eggs. One group of eggs was shielded with ultraviolet-blocking sun shades, while the second group was not. The number of eggs successfully hatched in each group is provided in the following table.

	Sun-Shaded Eggs	Unshaded Eggs
Total number	70	80
Number hatched	34	31

a. Set up H_0 and H_a for testing whether the hatching rates of the two groups of Pacific tree frog eggs differ.
b. Conduct the test at $\alpha = .01$. What do you conclude?

9.118 Pulling strength of firefighters. Consider a study conducted to compare the abilities of men and women to perform the strenuous tasks required of a shipboard firefighter. (*Human Factors*, Vol. 24, 1982.) The researchers measured the pulling force (in newtons) that a firefighter was able to exert in pulling the starter cord of a P-250 water pump. Firefighters were matched in pairs according to weight, thus producing data for a matched pairs (or paired difference) experiment, as shown below.

⬙ **PULLFORCE**

Pair	Female	Male
1	40.03	84.51
2	75.62	80.06
3	53.38	102.30
4	62.27	88.96

Source: Phillips, M. D., and Pepper, R. L., "Shipboard firefighting performance of females and males." *Human Factors*, Vol. 24, No. 3, 1982.

a. Do the data provide sufficient evidence to indicate a difference in mean pulling force between female and male firefighters? Test using $\alpha = .05$.
b. Find a 95% confidence interval for the difference in mean pulling force between female and male firefighters. Interpret the interval.

9.119 Job satisfaction study. Does the time of day during which one works affect job satisfaction? A study in *The Journal of Occupational Psychology* (Sept. 1991) examined differences in job satisfaction between day-shift and night-shift nurses. Nurses' satisfaction with their hours of work, free time away from work, and breaks during work were measured. The table shows the mean scores for each measure of job satisfaction (higher scores indicate greater satisfaction), along with the observed significance level comparing the means for the day-shift and night-shift samples:

	Day Shift	Night Shift	p-Value
Mean satisfaction with:			
Hours of work	3.91	3.56	.813
Free time	2.55	1.72	.047
Breaks	2.53	3.75	.0073

a. Specify the null and alternative hypotheses if we wish to test whether a difference in job satisfaction exists between day-shift and night-shift nurses on each of the three measures. Define any symbols you use.
b. Interpret the p-value for each of the tests. (Each of the p-values in the table is two-tailed.)
c. Assume that each of the tests is based on small samples of nurses from each group. What assumptions are necessary for the tests to be valid?

Applying the Concepts—Intermediate

9.120 Consumer reaction to silverware patterns. A pupillometer is a device used to observe changes in an individual's pupil dilations as he or she is exposed to different visual stimuli. The Design and Market Research Laboratories of the Container Corporation of America used a pupillometer to evaluate consumer reaction to different silverware patterns for one of its clients. Suppose 15 consumers were chosen at random, and each was shown two different silverware patterns. The pupillometer readings (in millimeters) for each consumer are shown below.

⬙ **PUPILL**

Consumer	Pattern 1	Pattern 2
1	1.00	.80
2	.97	.66
3	1.45	1.22
4	1.21	1.00
5	.77	.81
6	1.32	1.11
7	1.81	1.30
8	.91	.32
9	.98	.91
10	1.46	1.10
11	1.85	1.60
12	.33	.21
13	1.77	1.50
14	.85	.65
15	.15	.05

a. Which type of experiment does this represent—independent samples or paired difference? Explain.

b. Use a 90% confidence interval to estimate the difference in mean pupil dilation per consumer for silverware patterns 1 and 2. Interpret the confidence interval, assuming that the pupillometer indeed measures consumer interest.

c. Test the hypothesis that the mean dilation differs for the two patterns. Use $\alpha = .10$. Does the test conclusion support your interpretation of the confidence interval in part **b**?

d. What assumptions are necessary to ensure the validity of the inferences in parts **b** and **c**?

9.121 Teaching nursing skills. A traditional approach to teaching basic nursing skills was compared with an innovative approach in the *Journal of Nursing Education* (Jan. 1992). The innovative approach utilizes Vee heuristics and concept maps to link theoretical concepts with practical skills. Forty-two students enrolled in an upper-division nursing course participated in the study. Half (21) were randomly assigned to labs that utilized the innovative approach. After completing the course, all students were given short-answer questions about scientific principles underlying each of 10 nursing skills. The objective of the research is to compare the mean scores of the two groups of students.

a. What is the appropriate test to use to compare the two groups?

b. Are any assumptions required for the test?

c. One question dealt with the use of clean/sterile gloves. The mean scores for this question were 3.28 (traditional) and 3.40 (innovative). Is there sufficient information to perform the test?

d. Refer to part **c**. The *p*-value for the test was reported as $p = .79$. Interpret this result.

e. Another question concerned the choice of a stethoscope. The mean scores of the two groups were 2.55 (traditional) and 3.60 (innovative) with an associated *p*-value of .02. Interpret these results.

9.122 Identical twins reared apart. Twins, because they share an identical genotype, make ideal subjects for investigating the degree to which various environmental conditions impact personality. The classical method of studying this phenomenon and the subject of an interesting book by Susan Farber (*Identical Twins Reared Apart*, New York: Basic Books, 1981) is the study of identical twins separated early in life and reared apart. Much of Farber's discussion focuses on a comparison of IQ scores. The data for this analysis appear in the table. One member (A) of each of the $n = 32$ pairs of twins was reared by a natural parent; the other member (B) was reared by a relative or some other person. Is there a significant difference between the average IQ scores of identical twins, where one member of the pair is reared by the natural parents and the other member of the pair is not? Use $\alpha = .05$ to make your conclusion.

⊘ **TWINSIQ**

Pair ID	Twin A	Twin B	Pair ID	Twin A	Twin B
112	113	109	228	100	88
114	94	100	232	100	104
126	99	86	236	93	84
132	77	80	306	99	95
136	81	95	308	109	98
148	91	106	312	95	100
170	111	117	314	75	86
172	104	107	324	104	103
174	85	85	328	73	78
180	66	84	330	88	99
184	111	125	338	92	111
186	51	66	342	108	110
202	109	108	344	88	83
216	122	121	350	90	82
218	97	98	352	79	76
220	82	94	416	97	98

Source: Adapted from *Identical Twins Reared Apart*, by Susan L. Farber 1981 by Basic Books, Inc.

9.123 Treatments for panic disorder. Inositol is a complex cyclic alcohol found to be effective against clinical depression. Medical researchers believe inositol may also be used to treat panic disorder. To test this theory, a double-blind, placebo-controlled study of 21 patients diagnosed with panic disorder was conducted. (*American Journal of Psychiatry*, July 1995.) Patients completed daily panic diaries recording the occurrence of their panic attacks. The data for a week in which patients received a glucose placebo and for a week when they were treated with inositol are provided in the table. [*Note:* Neither the patients nor the treating physicians knew which week the placebo was given.] Analyze the data and interpret the results. Comment on the validity of the assumptions.

⊘ **INOSITOL**

Patient	Placebo	Inositol	Patient	Placebo	Inositol
1	0	0	12	0	2
2	2	1	13	3	1
3	0	3	14	3	1
4	1	2	15	3	4
5	0	0	16	4	2
6	10	5	17	6	4
7	2	0	18	15	21
8	6	4	19	28	8
9	1	1	20	30	0
10	1	0	21	13	0
11	1	3			

Source: Benjamin, J., et al. "Double-blind, placebo-controlled, crossover trial of inositol treatment for panic disorder." *American Journal of Psychiatry,* Vol. 152, No. 7, July 1995, p. 1085 (Table 1).

9.124 Aerobic exercise study. Excess postexercise oxygen consumption (EPOC) describes energy expended during the body's recovery period immediately following aerobic exercise. *The Journal of Sports Medicine and Physical Fitness* (Dec. 1994) published a study designed to investigate the effect of fitness level on the magnitude and duration of EPOC. Ten healthy young adult males volunteered for the study. Five of these were endurance-trained and comprised the fit group; the other five were not engaged in any systematic training and comprised the sedentary group. Each volunteer engaged in a weight-supported exercise on a cycle ergometer until 300 kilocalories were expended. The magnitude (in kilocalories) and duration (in minutes) of the EPOC of each exerciser were measured. The study results are summarized in the table below.

Variable		Fit (n = 5)	Sedentary (n = 5)	p-value
Magnitude	Mean	12.2	12.2	.998
(kcal)	Std. dev.	3.1	4.3	
Duration	Mean	16.6	20.4	.344
(min)	Std. dev.	3.1	7.8	

Source: Sedlock, D. A. "Fitness levels and postexercise energy expenditure." *The Journal of Sports Medicine and Physical Fitness*, Vol. 34, No. 4, Dec. 1994, p. 339 (Table III).

a. Conduct a test of hypothesis to determine whether the true mean magnitude of EPOC differs for fit and sedentary young adult males. Use $\alpha = .10$.

b. The p-value for the test, part **a**, is given in the table. Interpret this value.

c. Conduct a test of hypothesis to determine whether the true mean duration of EPOC differs for fit and sedentary young adult males. Use $\alpha = .10$.

d. The p-value for the test, part **c**, is given in the table. Interpret this value.

9.125 Inositol-fed infants. Inositol—a complex alcohol nutrient found in breast milk—has been found to reduce the risk of lung and eye damage in premature infants. The study, published in the *New England Journal of Medicine* (May 7, 1992), involved 220 premature infants. The infants were randomly divided into two groups; half (110) of the infants received an intravenous feeding of inositol, while the other half received a standard diet. The researchers found that 14 of the inositol-fed infants suffered retinopathy of prematurity, an eye injury that results from the high oxygen levels needed to compensate for poorly developed lungs. In contrast, 29 of the infants on the standard diet developed this disease. Analyze the results with a test of hypothesis. Use $\alpha = .01$.

9.126 Radiation exposure of sailors. Operation Crossroads was a 1946 military exercise in which atomic bombs were detonated over empty target ships in the Pacific Ocean. The Navy assigned sailors to wash down the test ships immediately after the atomic blasts. The National Academy of Science reported "that the overall death rate among Operation Crossroads sailors was 4.6% higher than among a comparable group of sailors.... However, this increase was not statistically significant." (*Tampa Tribune*, Oct. 30, 1996.)

a. Describe the parameter of interest in the National Academy of Science study.

b. Interpret the statement: "This increase was not statistically significant."

9.127 Rating music teachers. Students enrolled in music classes at the University of Texas (Austin) participated in a study to compare the observations and teacher evaluations of music education majors and nonmusic majors. (*Journal of Research in Music Education*, Winter 1991.) Independent random samples of 100 music majors and 100 nonmajors rated the overall performance of their teacher using a 6-point scale, where 1 = lowest rating and 6 = highest rating. Use the information in the table to compare the mean teacher ratings of the two groups of music students with a 95% confidence interval. Interpret the result.

	Music Majors	Nonmusic Majors
Sample size	100	100
Mean "overall" rating	4.26	4.59
Standard deviation	.81	.78

Source: Duke, R. A., and Blackman, M. D. "The relationship between observers' recorded teacher behavior and evaluation of music instruction." *Journal of Research in Music Education*, Vol. 39, No. 4, Winter 1991 (Table 2).

9.128 Identifying the target parameter. For each of the following studies, give the parameter of interest and state any assumptions that are necessary for the inferences to be invalid.

a. To investigate a possible link between jet lag and memory impairment, a University of Bristol (England) neurologist recruited 20 female flight attendants who worked flights across several time zones. Half of the attendants had only a short recovery time between flights and half had a long recovery time between flights. The average size of the right temporal lobe of the brain for the short-recovery group was significantly smaller than the average size of the right temporal lobe of the brain for the long-recovery group. (*Tampa Tribune*, May 23, 2001.)

b. In a study presented at the March, 2001, meeting of the Association for the Advancement of Applied Sport Psychology, researchers revealed that the proportion of athletes who have a good self-image of their body is 20% higher than the corresponding proportion of nonathletes.

c. A University of Florida animal sciences professor has discovered that feeding chickens corn oil causes them to produce larger eggs. (*UF News*, April 11, 2001.) The weight of eggs produced by each of a sample of chickens on a regular feed diet was recorded. Then, the same chickens were fed a diet supplemented by corn oil, and the weight of eggs produced by each was recorded. The mean weight of the eggs produced with corn oil was 3 grams heavier than the mean weight produced with the regular diet.

*9.129 **Instrument precision.** The quality control department of a paper company measures the brightness (a measure of reflectance) of finished paper on a periodic basis throughout the day. Two instruments that are available to measure the paper specimens are subject to error, but they can be adjusted so that the mean readings for a control paper specimen are the same for both instruments. Suppose you are concerned about the precision of the two instruments and want to compare the variability in the readings of instrument 1 to those of instrument 2. Five brightness measurements were made on a single paper specimen using each of the two instruments. The data are shown in the table.

○ **BRIGHT**

Instrument 1	Instrument 2
29	26
28	34
30	30
28	32
30	28

a. Is the variance of the measurements obtained by instrument 1 significantly different from the variance of the measurements obtained using instrument 2?

b. What assumptions must be satisfied for the test in part **a** to be valid?

9.130 **Genetic Study of cell growth.** Geneticists at Duke University Medical Center have identified the E2F1 transcription factor as an important component of cell proliferation control. (*Nature*, Sept. 23, 1993.) The researchers induced DNA synthesis in two batches of serum-starved cells. Each cell in one batch was microinjected with the E2F1 gene, while the cells in the second batch (the controls) were not exposed to E2F1. After 30 hours, the number of cells in each batch that exhibited altered growth was determined. The results of the experiment are summarized in the table.

	Control	E2F1-Treated Cells
Total number of cells	158	92
Number of growth-altered cells	15	41

Source: Johnson, D. G., et al. "Expression of transcription factor E2F1 induces quiescent cells to enter S phase." *Nature,* Vol. 365, No. 6444, Sept, 23, 1993, p. 351 (Table 1).

a. Compare the percentages of cells exhibiting altered growth in the two batches with a 90% confidence interval.

b. Use the interval, part **a**, to make an inference about the ability of the E2F1 transcription factor to induce cell growth.

9.131 **Kicking the cigarette habit.** Can taking an antidepressant drug help cigarette smokers kick their habit? The *New England Journal of Medicine* (Oct. 23, 1997) published a study in which 615 smokers (all who wanted to give up smoking) were randomly assigned to receive either Zyban (an antidepressant) or a placebo (a dummy pill) for six weeks. Of the 309 patients who received Zyban, 71 were not smoking one year later. Of the 306 patients who received a placebo, 37 were not smoking one year later. Conduct a test of hypothesis (at $\alpha = .05$) to answer the research question posed above.

9.132 **Muscle properties of crayfish.** Marine biochemists at the University of Tokyo studied the properties of crustacean striated muscles. (*The Journal of Experimental Zoology*, Sept. 1993.) The main purpose of the experiment was to compare the biochemical properties of fast and slow muscles of crayfish. Using crayfish obtained from a local supplier, the researchers excised twelve fast-muscle fiber bundles and tested each fiber bundle for uptake of calcium. Twelve slow-muscle fiber bundles were excised from a second sample of crayfish, and calcium uptake was measured. The results of the experiment are summarized here. (All calcium measurements are in moles per milligram.) Analyze the data using a 95% confidence interval. Make an inference about the difference between the calcium uptake means of fast and slow muscles.

Fast Muscle	Slow Muscle
$n_1 = 12$	$n_2 = 12$
$\bar{x}_1 = .57$	$\bar{x}_2 = .37$
$s_1 = .104$	$s_2 = .035$

Source: Ushio, H., and Watabe, S. "Ultrastructural and biochemical analysis of the sarcoplasmic reticulum from crayfish fast and slow striated muscles." *The Journal of Experimental Zoology*, Vol. 267, Sept. 1993, p. 16 (Table 1). Copyright © 1993 John Wiley & Sons, Inc. Reprinted with permission.

Applying the Concepts—Advanced

9.133 **Isolation timeouts for students.** Researchers at Rochester Institute of Technology investigated the use of isolation timeout as a behavioral management technique (*Exceptional Children*, Feb. 1995.) Subjects for the study were 155 emotionally disturbed students enrolled in a special education facility. The students were randomly assigned to one of two types of classrooms—Option II classrooms (one teacher, one paraprofessional, and a maximum of 12 students) and Option III classrooms (one teacher, one paraprofessional, and a maximum of 6 students). Over the academic year the number of behavioral incidents resulting in an isolation timeout was recorded for each student. Summary statistics for the two groups of students are shown in the following table.

	Option II	Option III
Number of students	100	55
Mean number of timeout incidents	78.67	102.87
Standard deviation	59.08	69.33

Source: Costenbader, V., and Reading-Brown, M. "Isolation timeout used with students with emotional disturbance." *Exceptional Children,* Vol. 61, No. 4, Feb. 1995, p. 359 (Table 3).

Do you agree with the following statement: "On average, students in Option III classrooms had significantly more timeout incidents than students in Option II classrooms?"

9.134 Feeding habits of sea urchins. The *Florida Scientist* (Summer/Autumn 1991) reported on a study of the feeding habits of sea urchins. A sample of 20 urchins were captured from Biscayne Bay (Miami), placed in marine aquaria, then starved for 48 hours. Each sea urchin was then fed a 5-cm blade of turtle grass. Ten of the urchins received only green blades, while the other half received only decayed blades. (Assume that the two samples of sea urchins—10 urchins per sample—were randomly and independently selected.) The ingestion time, measured from the time the blade first made contact with the urchin's mouth to the time the urchin had finished ingesting the blade, was recorded. A summary of the results is provided in the next table.

	Green Blades	Decayed Blades
Number of sea urchins	10	10
Mean ingestion time (hours)	3.35	2.36
Standard deviation (hours)	.79	.47

Source: Montague, J. R., et al. "Laboratory measurement of ingestion rate for the sea urchin *Lytechinus variegatus.*" *Florida Scientist,* Vol. 54, Nos. 3/4, Summer/Autumn, 1991 (Table 1).

According to the researchers, "the difference in rates at which the urchins ingested the blades suggest that green, unblemished turtle grass may not be a particularly palatable food compared with decayed turtle grass. If so, urchins in the field may find it more profitable to selectively graze on decayed portions of the leaves." Do the results support this conclusion?

Critical Thinking Challenges

9.135 Self-managed work teams and family life. To improve quality, productivity, and timeliness, more and more American industries are utilizing self-managed work teams (SMWTs). A team typically consists of 5 to 15 workers who are collectively responsible for making decisions and performing all tasks related to a particular project. Researchers L. Stanley-Stevens (Tarleton State University), D. E. Yeatts, and R. R. Seward (both University of North Texas) investigated the connection between SMWT, work characteristics and workers' perceptions of positive spillover into family life (*Quality Management Journal*, Summer 1995.) Survey data were collected from 114 AT&T employees who work in one of fifteen SMWTs at an AT&T technical division. The workers were divided into two groups: (1) those who reported positive spillover of work skills to family life and (2) those who did not report positive work spillover. The two groups were compared on a variety of job and demographic characteristics, several of which are shown below. All but the demographic characteristics were measured on a 7-point scale, ranging from 1 = "strongly disagree" to 7 = "strongly agree"; thus, the larger the number, the more of the characteristic indicated.

SPILLOVER

Variables Measured in the SMWT Survey

Characteristic	Variable
Information Flow	Use of creative ideas (7-point scale)
Information Flow	Utilization of information (7-point scale)
Decision-Making	Participation in decisions regarding personnel matters (7-pt. scale)
Job	Good use of skills (7-point scale)
Job	Task identity (7-point scale)
Demographic	Age (years)
Demographic	Education (years)
Demographic	Gender (male or female)
Comparison	Group (positive spillover or no spillover)

The file named **SPILLOVER** includes the values of the variables listed above for each of the 114 survey participants. The researchers' objectives were to compare the two groups of workers on each characteristic. In particular, they want to know which job-related characteristics are most highly associated with positive work spillover. Conduct a complete analysis of the data for the researchers.

9.136 MS and exercise study. A study published in *Clinical Kinesiology* (Spring 1995) was designed to examine the metabolic and cardiopulmonary responses during exercise of persons diagnosed with multiple sclerosis (MS). Leg cycling and arm cranking exercises were performed by 10 MS patients and 10 healthy (non-MS) control subjects. Each member of the control group was selected based on gender, age, height, and weight to match (as closely as possible) with one member of the MS group. Consequently, the researchers compared the MS and non-MS groups by matched-pairs *t*-tests on such outcome variables as oxygen uptake, carbon dioxide output, and peak aerobic power. The data on the matching variables used in the experiment are shown in the table on p. 507. Have the researchers successfully matched the MS and non-MS subjects?

MSSTUDY

Matched Pair	MS Subjects				Non-MS Subjects			
	Gender	Age (years)	Height (cm)	Weight (kg)	Gender	Age (years)	Height (cm)	Weight (kg)
1	M	48	171.0	80.8	M	45	173.0	76.3
2	F	34	158.5	75.0	F	34	158.0	75.6
3	F	34	167.6	55.5	F	34	164.5	57.7
4	M	38	167.0	71.3	M	34	161.3	70.0
5	M	45	182.5	90.9	M	39	179.0	96.0
6	F	42	166.0	72.4	F	42	167.0	77.8
7	M	32	172.0	70.5	M	34	165.8	74.7
8	F	35	166.5	55.3	F	43	165.1	71.4
9	F	33	166.5	57.9	F	31	170.1	60.4
10	F	46	175.0	79.9	F	43	175.0	77.9

Source: Ponichtera-Mulcare, J. A., et al. "Maximal aerobic exercise of individuals with multiple sclerosis using three modes of ergometry." *Clinical Kinesiology,* Vol. 49, No. 1, Spring 1995, p. 7 (Table 1).

Student Projects

We have now discussed two methods of collecting data to compare two population means. In many experimental situations a decision must be made either to collect two independent samples or to conduct a paired difference experiment. The importance of this decision cannot be overemphasized, since the amount of information obtained and the cost of the experiment are both directly related to the method of experimentation that is chosen.

Choose two populations (pertinent to your major area) that have unknown means and for which you could both collect two independent samples and collect paired observations. Before conducting the experiment, state which method of sampling you think will provide more information (and why). Compare the two methods, first performing the independent sampling procedure by collecting 10 observations from each population (a total of 20 measurements), then performing the paired difference experiment by collecting 10 pairs of observations.

Construct two 95% confidence intervals, one for each experiment you conduct. Which method provides the narrower confidence interval and hence more information on this performance of the experiment? Does this result agree with your preliminary expectations?

REFERENCES

Freedman, D., Pisani, R., and Purves, R. *Statistics*. New York: W. W. Norton and Co., 1978.

Gibbons, J. D. *Nonparametric Statistical Inference*, 2nd ed. New York: McGraw-Hill, 1985.

Hollander, M., and Wolfe, D. A. *Nonparametric Statistical Methods*. New York: Wiley, 1973.

Mendenhall, W., Beaver, R. J., and Beaver, B. *Introduction to Probability and Statistics*, 11th ed. North Scituate, Mass.: Duxbury, 2002.

Satterthwaite, F. W. "An approximate distribution of estimates of variance components." *Biometrics Bulletin*, Vol. 2, 1946, pp. 110–114.

Snedecor, G. W., and Cochran, W. *Statistical Methods*, 7th ed. Ames: Iowa State University Press, 1980.

Steel, R. G. D., and Torrie, J. H. *Principles and Procedures of Statistics*, 2nd ed. New York: McGraw-Hill, 1980.

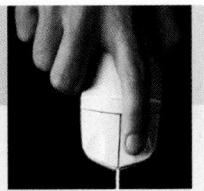

Using Technology

Two Sample Inferences Using MINITAB

MINITAB can be used to make two-sample inferences about $\mu_1 - \mu_2$ for independent samples, μ_D for paired samples, $p_1 - p_2$ and σ_1^2/σ_2^2.

To carry out an analysis for $\mu_1 - \mu_2$, first access the MINITAB worksheet that contains the sample data. Next, click on the "Basic Statistics" button on the MINITAB menu bar, then click on "2-Sample t", as shown in Figure 9.M.1. The resulting dialog box appears as shown in Figure 9.M.2.

If the worksheet contains data for one quantitative variable (which the means will be computed on) and one qualitative variable (which represents the two groups or populations), select "Samples in one column", then specify the quantitative variable in the "Samples" area and the qualitative variable in the "Subscripts" area. (See Figure 9.M.2.)

If the worksheet contains the data for the first sample in one column and the data for the second sample in another column, select "Samples in different columns", then specify the "First" and "Second" variables. Alternatively, if you have only summarized data (i.e., sample sizes, sample means, and sample standard deviations), select "Summarized data" and enter these summarized values in the appropriate boxes.

Once you have made the appropriate menu selection, click the "Options" button on the MINITAB "2-Sample T" dialog box. Specify the confidence level for a confidence interval, the null hypothesized value of the difference, $\mu_1 - \mu_2$, and the form of the alternative hypothesis (lower tailed, two tailed, or upper tailed) in the resulting dialog box, as shown in Figure 9.M.3. Click "OK" to return to the "2-Sample T" dialog box, then click "OK" again to generate the MINITAB printout.

Figure 9.M.1

MINITAB Menu Options for Inferences about $\mu_1 - \mu_2$

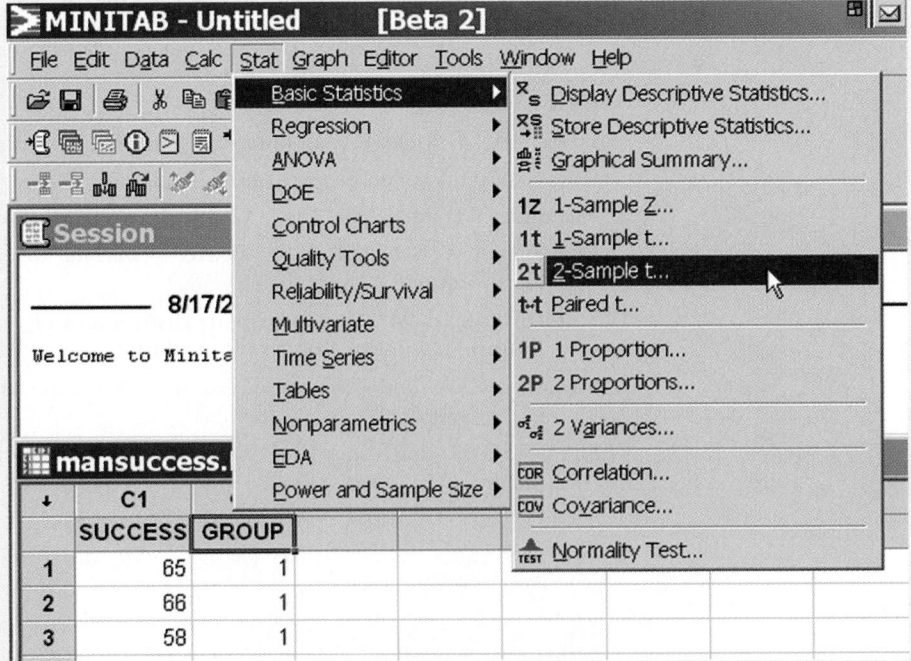

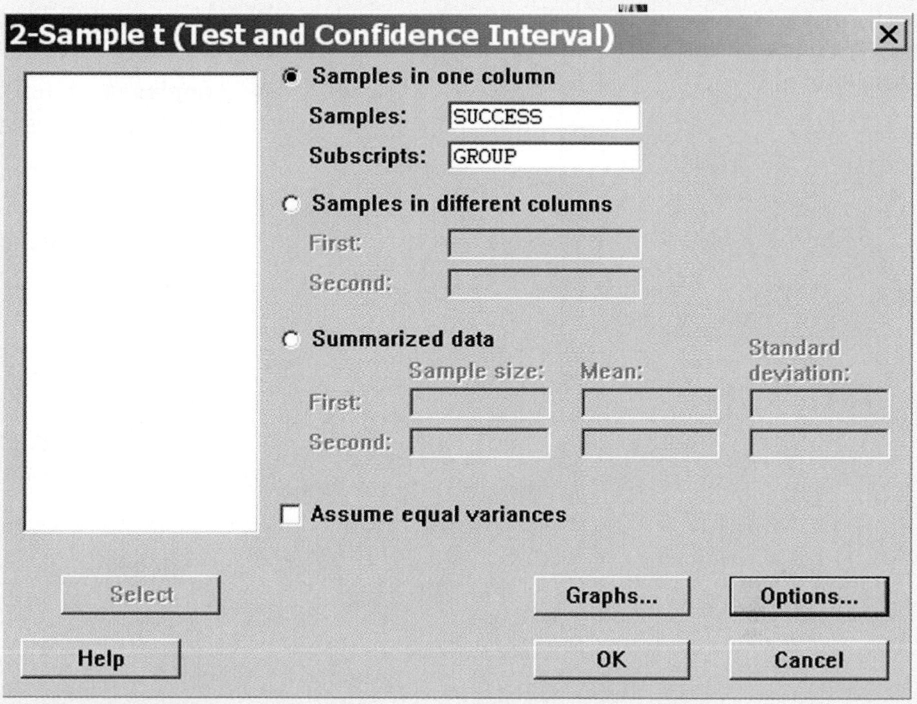

Figure 9.M.2
MINITAB 2-Sample
T Dialog Box

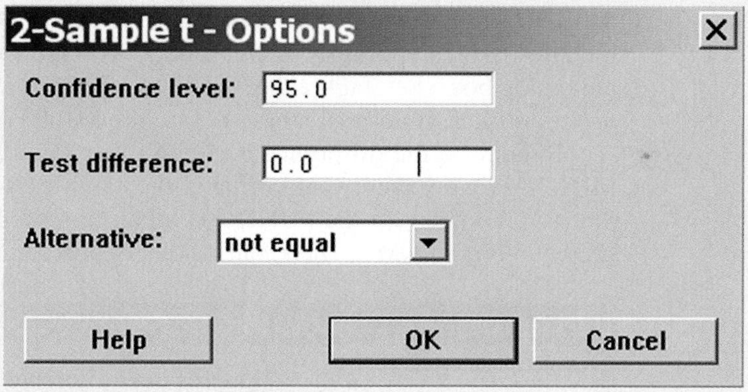

Figure 9.M.3
MINITAB Options
Dialog Box

[*Important Note:* The MINITAB two-sample *t*-procedure uses the *t*-statistic to conduct the test of hypothesis. When the sample sizes are small, this is the appropriate method. When the sample sizes are large, the *t*-value will be approximately equal to the large-sample *z*-value and the resulting test will still be valid.]

To carry out an analysis of μ_d for matched pairs, first access the MINITAB worksheet that contains the sample data. The data file should contain two quantitative variables—one with the data values for the first group (or, population) and one with the data values for the second group. (*Note:* The sample size should be the same for each group.) Next, click on the "Basic Statistics" button on the MINITAB menu bar, then click on "Paired t" (see Figure 9.M.1). The resulting dialog box appears as shown in Figure 9.M.4.

Select the "Samples in columns" option and specify the two quantitative variables of interest in the "First sample" and "Second sample" boxes, as shown in Figure 9.M.4. [Alternatively, if you have only summarized data of the paired differences, select the "Summarized data (differences)" option and enter the sample size, sample mean, and sample standard deviation in the appropriate boxes.]

Figure 9.M.4
MINITAB Paired-Samples
Dialog Box

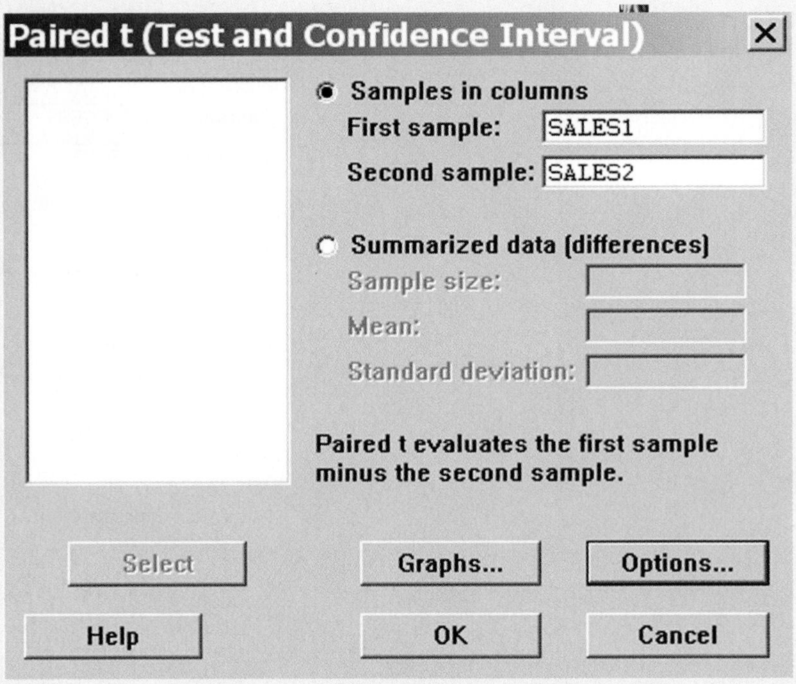

Paired t (Test and Confidence Interval) ☒

⦿ **Samples in columns**

First sample: |SALES1|

Second sample: |SALES2|

○ **Summarized data (differences)**

Sample size: | |

Mean: | |

Standard deviation: | |

Paired t evaluates the first sample minus the second sample.

| Select | | Graphs... | | Options... |

| Help | | OK | | Cancel |

Next, click the "Options" button and specify the confidence level for a confidence interval, the null hypothesized value of the difference, μ_D, and the form of the alternative hypothesis (lower tailed, two tailed, or upper tailed) in the resulting dialog box. (See Figure 9.M.3.) Click "OK" to return to the "Paired t" dialog box, then click "OK" again to generate the MINITAB printout.

To analyze the difference between two proportions, $p_1 - p_2$, first access the MINITAB worksheet that contains the sample data. Next, click on the "Basic Statistics" button on the MINITAB menu bar, then click on "2 Proportions", as shown in Figure 9.M.1. The resulting dialog box appears as shown in Figure 9.M.5.

Figure 9.M.5
MINITAB 2 Proportions
Dialog Box

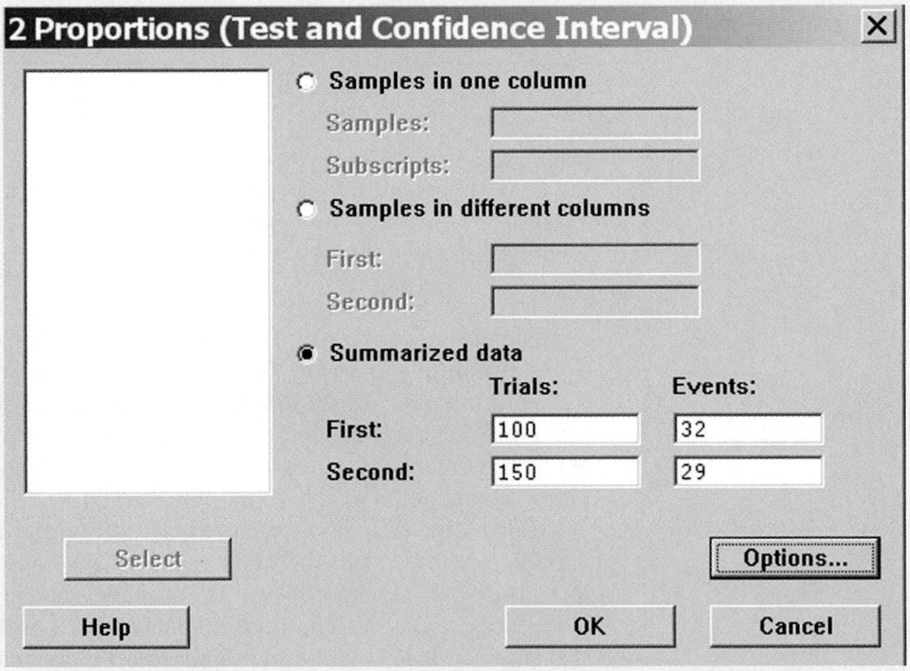

2 Proportions (Test and Confidence Interval) ☒

○ **Samples in one column**

Samples: | |

Subscripts: | |

○ **Samples in different columns**

First: | |

Second: | |

⦿ **Summarized data**

	Trials:	**Events:**
First:	|100|	|32|
Second:	|150|	|29|

| Select | | Options... |

| Help | | OK | | Cancel |

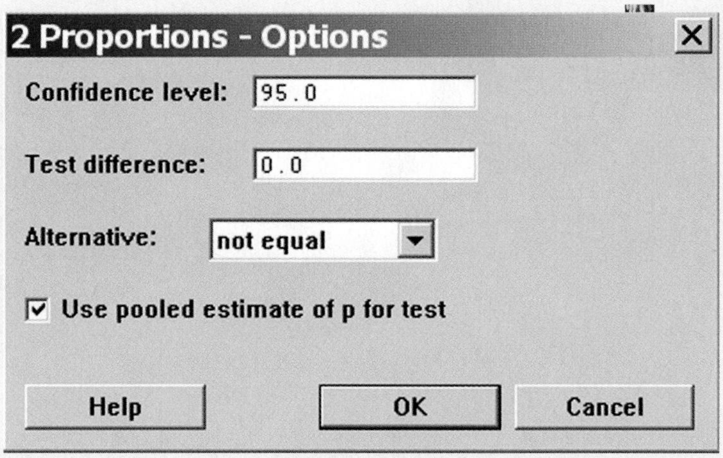

Figure 9.M.6
MINITAB 2 Proportions
Options Box

Figure 9.M.6
MINITAB 2 Proportions
Options Box

Select the data option ("Samples in one column" or "Samples in different columns" or "Summarized data") and make the appropriate menu choices. (Figure 9.M.5 shows the menu options when you select "Summarized data".)

Next, click the "Options" button and specify the confidence level for a confidence interval, the null hypothesized value of the difference, and the form of the alternative hypothesis (lower tailed, two tailed, or upper tailed) in the resulting dialog box, as shown in Figure 9.M.6. (If you desire a pooled estimate of p for the test, be sure to check the appropriate box.) Click "OK" to return to the "2 Proportions" dialog box, then click "OK" again to generate the MINITAB printout.

To carry out an analysis for the ratio of two variances, σ_1^2/σ_2^2, first access the MINITAB worksheet that contains the sample data. Next, click on the "Basic Statistics" button on the MINITAB menu bar, then click on "2 Variances", as shown in Figure 9.M.1. The resulting dialog box appears as shown in Figure 9.M.7. The menu selections and options are similar to those for the two-sample t-test. Once the selections are made, click "OK" to produce the MINITAB F-test printout.

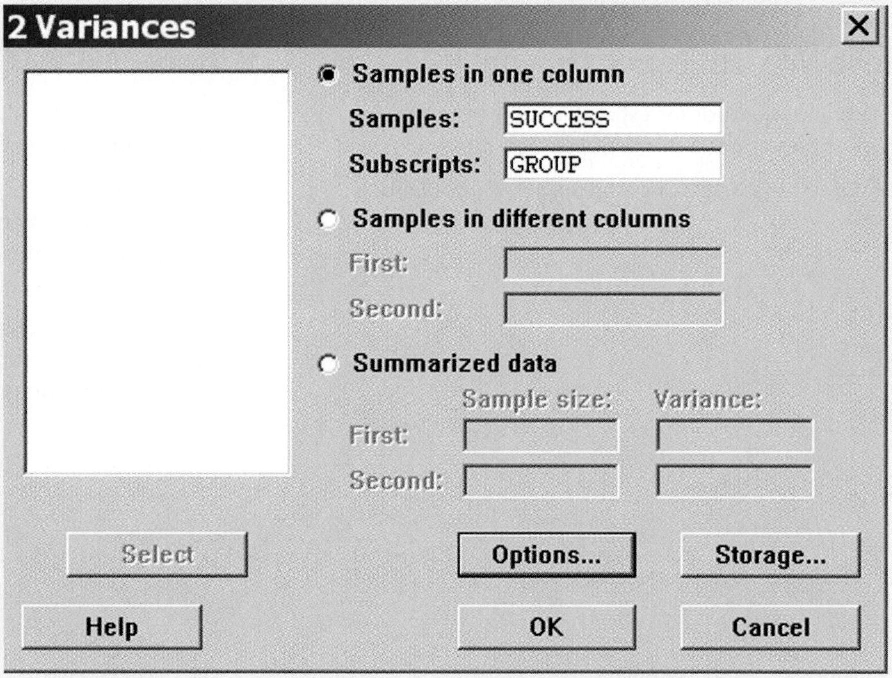

Figure 9.M.7
MINITAB 2 Variances
Dialog Box

10

Analysis of Variance
Comparing More Than Two Means

Contents

Statistics in Action
On the Trail of the Cockroach

Using Technology
Analysis of Variance Using MINITAB

Where We've Been

- Presented methods for estimating and testing hypotheses about a single population mean
- Presented methods for comparing two population means

Where We're Going

- Discuss the critical elements in the *design* of a sampling experiment.
- Learn how to set up three of the more popular experimental designs for comparing more than two population means: *completely randomized*, *randomized block*, and *factorial designs*.
- Show how to analyze data collected from a designed experiment using a technique called an *analysis of variance*.

Statistics in *ACTION*

On the Trail of the Cockroach

Entomologists have long established that insects such as ants, bees, caterpillars, and termites use chemical or "odor" trails for navigation. These trails are used as highways between sources of food and the insect nest. Until recently, however, "bug" researchers believed that the navigational behavior of cockroaches scavenging for food was random and not linked to a chemical trail.

One of the first researchers to challenge the "random-walk" theory for cockroaches was professor and entomologist Dini Miller of Virginia Tech University. According to Miller, "The idea that roaches forage randomly means that they would have to come out of their hiding places every night and bump into food and water by accident. But roaches never seem to go hungry." Since cockroaches had never before been evaluated for trail following behavior, Miller designed an experiment to test a cockroach's ability to follow a trail of their fecal material (*Explore*, Research at the University of Florida, Fall 1998).

First, Dr. Miller developed a methanol extract from roach feces—called a pheromone. She theorized that "pheromones are communication devices between cockroaches. If you have an infestation and have a lot of fecal material around, it advertises, 'Hey, this is a good cockroach place.'" Then she created a chemical trail with the pheromone on a strip of white chromatography paper and placed the paper at the bottom of a plastic, V-shaped container, 122 square centimeters in size. German cockroaches were released into the container at the beginning of the trail, one at a time, and a video surveillance camera was used to monitor the roach's movements.

In addition to the trail containing the fecal extract (the treatment), a trail using methanol only was created. This second trail served as a control to compare back against the treated trail. Because Dr. Miller also wanted to determine if trail-following ability differed among cockroaches of different age, sex, and reproductive status, four roach groups were utilized in the experiment: adult males, adult females, gravid (pregnant) females, and nymphs (immatures). Twenty roaches of each type were randomly assigned to the treatment trail and ten of each type were randomly assigned to the control trail. Thus, a total of 120 roaches were used in the experiment.

The movement pattern of each cockroach tested was translated into *xy* coordinates every one-tenth of a second by the Dynamic Animal Movement Analyzer (DAMA) program. Miller measured the perpendicular distance of each *xy*-coordinate from the trail and then averaged these distances, or deviations, for each cockroach. The average trail deviations (measured in pixels, where 1 pixel equals approximately 2 centimeters) for each of the 120 cockroaches in the study are stored in the data file named **ROACH**.

We apply the statistical methodology presented in this chapter to the cockroach data in the following Statistics in the Action Revisited sections.

Statistics in the Action Revisited

- A One-Way Analysis of the Cockroach Data (p. 531)
- Ranking the Means of the Cockroach Groups (p. 541)
- A Two-Way Analysis of the Cockroach Data (p. 571)

Most of the data analyzed in previous chapters were collected in *observational* sampling experiments rather than *designed* sampling experiments. In *observational experiments* the analyst has little or no control over the variables under study and merely observes their values. In contrast, *designed experiments* are those in which the analyst attempts to control the levels of one or more variables to determine their effect on a variable of interest. Although many practical situations do not present the opportunity for such control, it is instructive, even for observational experiments, to have a working knowledge of the analysis and interpretation of data that result from designed experiments and to know the basics of how to design experiments when the opportunity arises.

We first present the basic elements of an experimental design in Section 10.1. We then discuss two of the simpler, and more popular, experimental designs in Sections 10.2 and 10.4. Slightly more complex experiments are discussed in Section 10.5.

10.1 Elements of a Designed Experiment

Certain elements are common to almost all designed experiments, regardless of the specific area of application. For example, the *response* is the variable of interest in the experiment. The response might be the SAT scores of a high school senior, the total sales of a firm last year, or the total income of a particular household this year. We will also refer to the response as the *dependent variable*.

> **DEFINITION 10.1**
>
> The **response variable** is the variable of interest to be measured in the experiment. We also refer to the response as the **dependent variable**.

The intent of most statistical experiments is to determine the effect of one or more variables on the response. These variables are usually referred to as the *factors* in a designed experiment. Factors are either *quantitative* or *qualitative*, depending on whether the variable is measured on a numerical scale or not. For example, we might want to explore the effect of the qualitative factor Gender on the response SAT score. In other words, we want to compare the SAT scores of male and female high school seniors. Or, we might wish to determine the effect of the quantitative factor Number of salespeople on the response Total sales for retail firms. Often two or more factors are of interest. For example, we might want to determine the effect of the quantitative factor Number of wage earners and the qualitative factor Location on the response Household income.

> **DEFINITION 10.2**
>
> **Factors** are those variables whose effect on the response is of interest to the experimenter. **Quantitative factors** are measured on a numerical scale, whereas **qualitative factors** are those that are not (naturally) measured on a numerical scale.

Levels are the values of the factors that are utilized in the experiment. The levels of qualitative factors are usually nonnumerical. For example, the levels of Gender are Male and Female, and the levels of Location might be North, East, South, and West.* The levels of quantitative factors are the numerical values of the variable utilized in the experiment. The Number of salespeople for each of a set of companies, the Number of wage earners in each of a set of households, and the GPAs for a set of high school seniors all represent levels of the respective quantitative factors.

> **DEFINITION 10.3**
>
> **Factor levels** are the values of the factor utilized in the experiment.

When a *single factor* is employed in an experiment, the *treatments* of the experiment are the levels of the factor. For example, if the effect of the factor Gender on the response SAT score is being investigated, the treatments of the experiment are the two levels of Gender—Female and Male. Or, if the effect of the Number of wage earners on Household income is the subject of the experiment, the numerical values assumed by the quantitative factor Number of wage earners are the treatments. If *two or more factors* are utilized in an experiment, the treatments are the

---

*The levels of a qualitative variable may bear numerical labels. For example, the Locations could be numbered 1, 2, 3, and 4. However, in such cases the numerical labels for a qualitative variable will usually be codes representing non-numerical levels.

factor–level combinations used. For example, if the effects of the factors Gender and Socioeconomic status (SES) on the response SAT score are being investigated, the treatments are the combinations of the levels of Gender and SES used; thus (Female, high SES) and (Male, low SES) would be treatments.

The **treatments** of an experiment are the factor–level combinations utilized.

The objects on which the response variable and factors are observed are the *experimental units*. For example, SAT score, High school GPA, and Gender are all variables that can be observed on the same experimental unit—a high school senior. Or, the Total sales, the Earnings per share, and the Number of salespeople can be measured on a particular firm in a particular year, and the firm–year combination is the experimental unit. The Total income, the Number of female wage earners, and the Location can be observed for a household at a particular point in time, and the household–time combination is the experimental unit. Every experiment, whether observational or designed, has experimental units on which the variables are observed. However, the identification of the experimental units is more important in designed experiments, when the experimenter must actually sample the experimental units and measure the variables.

An **experimental unit** is the object on which the response and factors are observed or measured.*

When the specification of the treatments and the method of assigning the experimental units to each of the treatments is controlled by the analyst, the experiment is said to be *designed*. In contrast, if the analyst is just an observer of the treatments on a sample of experimental units, the experiment is *observational*. For example, if you specify the number of female and male high school students within each GPA range to be randomly selected in order to evaluate the effect of gender and GPA on SAT scores, you are designing the experiment. If, on the other hand, you simply observe the SAT scores, gender, and GPA for all students who took the SAT test last month at a particular high school, the experiment is observational.

A **designed experiment** is one for which the analyst controls the specification of the treatments and the method of assigning the experimental units to each treatment. An **observational experiment** is one for which the analyst simply observes the treatments and the response on a sample of experimental units.

The diagram in Figure 10.1 provides an overview of the experimental process and a summary of the terminology introduced in this section. Note that the experimental unit is at the core of the process. The method by which the sample of experimental units is selected from the population determines the type of experiment. The level of every factor (the treatment) and the response are all variables that are observed or measured on each experimental unit.

*Recall (Chapter 1) that the set of all experimental units is the population.

Biography

SIR RONALD A. FISHER (1890–1962)—The Founder of Modern Statistics

At a young age, Ronald Fisher demonstrated special abilities in mathematics, astronomy, and biology. (Fisher's biology teacher once divided all his students for "sheer brilliance" into two groups—Fisher and the rest.) Fisher graduated from prestigious Cambridge University in London in 1912 with a B.A. degree in astronomy, and, after several years teaching mathematics, he found work at the Rothamsted Agricultural Experiment station. There, Fisher began his extraordinary career as a statistician. Many consider Fisher to be the leading founder of modern statistics. His contributions to the field include the notion of unbiased statistics, the develop-

ment of p-values for hypothesis tests, the invention of analysis of variance for designed experiments, the maximum likelihood estimation theory, and the mathematical distributions of several well-known statistics. Fisher's book, *Statistical Methods for Research Workers* (written in 1925), revolutionized applied statistics, demonstrating how to analyze data and interpret the results with very readable and practical examples. In 1935, Fisher wrote *The Design of Experiments*, where he first described his famous experiment on the "lady tasting tea." (Fisher showed, through a designed experiment, that the lady really could determine whether tea poured into milk tastes better than milk poured into tea.) Before his death, Fisher was elected a Fellow of the Royal Statistical Society, was awarded numerous medals, and was knighted by the Queen of England.

Figure 10.1
Sampling Experiment:
Process and Terminology

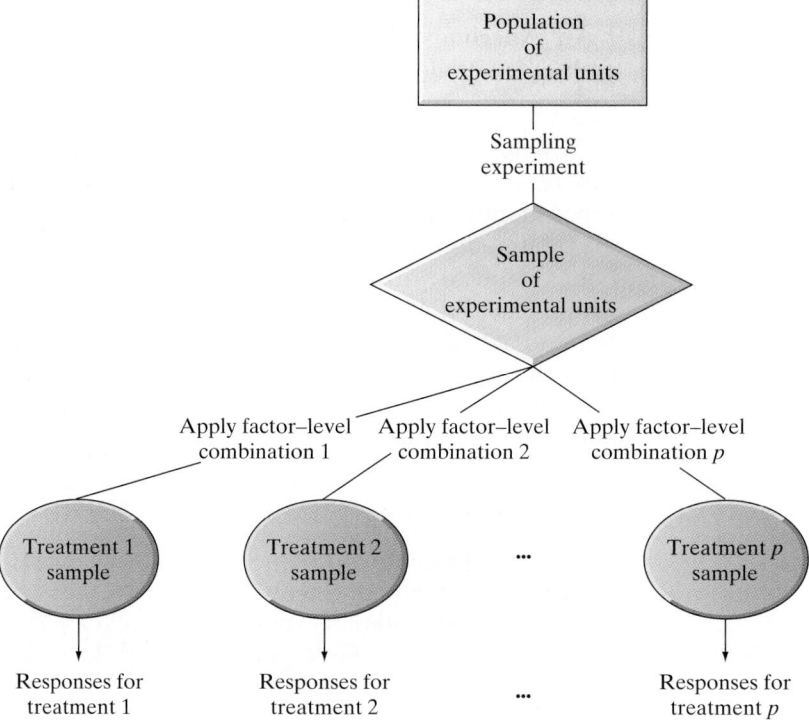

EXAMPLE 10.1   THE KEY ELEMENTS OF A DESIGNED EXPERIMENT

Problem   The USGA (United States Golf Association) regularly tests golf equipment to ensure that it conforms to USGA standards. Suppose it wishes to compare the mean distance traveled by four different brands of golf balls when struck by a driver (the club used to maximize distance). The following experiment is conducted: 10 balls of each brand are randomly selected. Each is struck by "Iron Byron" (the USGA's golf robot named for the famous golfer, Byron Nelson) using a driver, and the distance traveled is recorded. Identify each of the following elements in this experiment: response, factors, factor types, levels, treatments, and experimental units.

Solution The response is the variable of interest, Distance traveled. The only factor being investigated is Brand of golf ball, and it is nonnumerical and therefore qualitative. The four brands (say A, B, C, and D) represent the levels of this factor. Since only one factor is utilized, the treatments are the four levels of this factor—that is, the four brands. The experimental unit is a golf ball; more specifically, it is a golf ball at a particular position in the striking sequence, since the distance traveled can be recorded only when the ball is struck, and we would expect the distance to be different (due to random factors such as wind resistance, landing place, and so forth) if the same ball is struck a second time. Note that 10 experimental units are sampled for each treatment, generating a total of 40 observations.

Look Back This experiment, like many real applications, is a blend of designed and observational: The analyst cannot control the assignment of the brand to each golf ball (observational), but he or she can control the assignment of each ball to the position in the striking sequence (designed).

Now Work *Exercise 10.5*

■ ■ ■

EXAMPLE 10.2   A TWO-FACTOR EXPERIMENT

Problem Suppose the USGA is also interested in comparing the mean distances the four brands of golf balls travel when struck by a five-iron and by a driver. Ten balls of each brand are randomly selected, five to be struck by the driver, and five by the five-iron. Identify the elements of the experiment, and construct a schematic diagram similar to Figure 10.1 to provide an overview of this experiment.

Solution The response is the same as in Example 10.1—Distance traveled. The experiment now has two factors, Brand of golf ball and Club utilized. There are four levels of Brand (A, B, C, and D) and two of Club (driver and five-iron, or 1 and 5). Treatments are factor–level combinations, so there are $4 \times 2 = 8$ treatments in this experiment: (A, 1), (A, 5), (B, 1), (B, 5), (C, 1), (C, 5), (D, 1), and (D, 5). The experimental units are still the combinations of golf ball and hitting position. Note that five experimental units are sampled per treatment, generating 40 observations. The experiment is summarized in Figure 10.2.

Look Back Whenever there are two or more factors in an experiment, remember to combine the levels of the factors—one from each factor—to obtain the treatments.

Now Work *Exercise 10.10*

■ ■ ■

Our objective in designing an experiment is usually to maximize the amount of information obtained about the relationship between the treatments and the response. Of course, we are almost always subject to constraints on budget, time, and even the availability of experimental units. Nevertheless, designed experiments are generally preferred to observational experiments. Not only do we have better control of the amount and quality of the information collected, but we also avoid the biases

Figure 10.2
Two-Factor Golf Experiment
Summary: Example 10.2

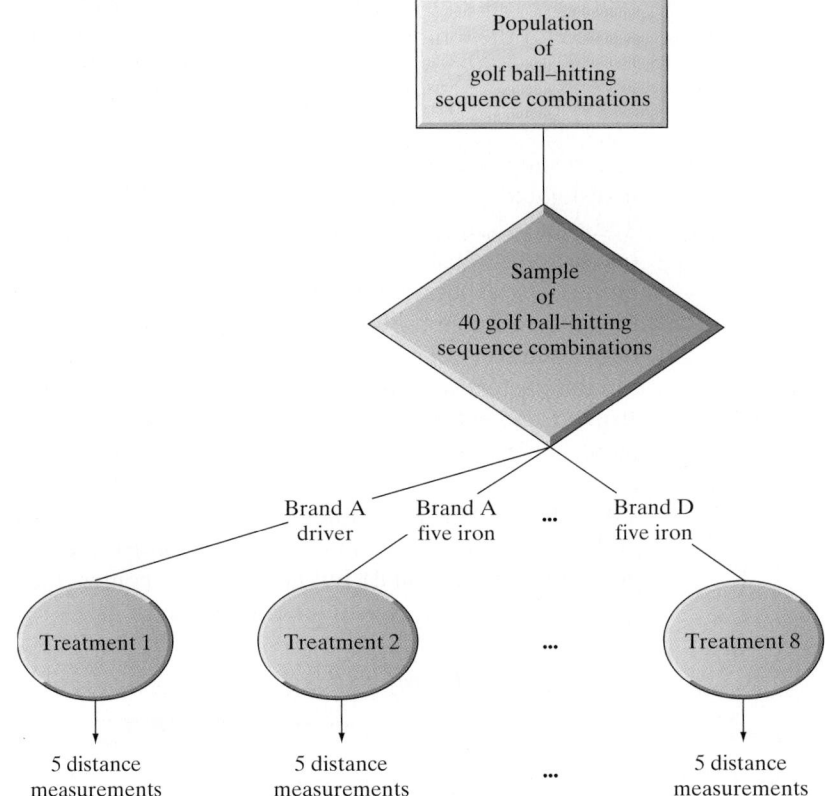

inherent in observational experiments in the selection of the experimental units representing each treatment. Inferences based on observational experiments always
carry the implicit assumption that the sample has no hidden bias that was not considered in the statistical analysis. Better understanding of the potential problems
with observational experiments is a by-product of our study of experimental design
in the remainder of this chapter.

Exercises 10.1–10.12

Understanding the Principles

10.1 What are the treatments for a designed experiment that
utilizes one qualitative factor with four levels—A, B, C,
and D?

10.2 What are the treatments for a designed experiment
with two factors, one qualitative with two levels (A
and B) and one quantitative with five levels (50, 60, 70,
80, and 90)?

10.3 What is the difference between an observational and a
designed experiment?

Applying the Concepts—Basic

10.4 What are the experimental units on which each of the
following responses are observed?
  a. GPA of a college student
  b. Household income
  c. Your time in running the 100-yard dash
  d. A patient's reaction to a new drug

10.5 Brief descriptions of a number of experiments are given
next. Determine whether each is observational or
designed, and explain your reasoning.
  a. An economist obtains the unemployment rate and
  gross state product for a sample of states over the past
  10 years, with the objective of examining the relationship between the unemployment rate and the gross
  state product by census region.
  b. A psychologist tests the effects of three different
  feedback programs by randomly assigning five rats
  to each program and recording their response times
  at specified intervals during the program.
  c. A marketer of computers runs ads in each of four
  national publications for one quarter, and keeps
  track of the number of sales that are attributable to
  each publication's ad.
  d. An electric utility engages a consultant to monitor
  the discharge from its smokestack on a monthly basis
  over a 1-year period in order to relate the level of

sulfur dioxide in the discharge to the load on the facility's generators.

e. Intrastate trucking rates are compared before and after governmental deregulation of prices charged, with the comparison also taking into account distance of haul, goods hauled, and the price of diesel fuel.

f. An agriculture student compares the amount of rainfall in four different states over the past 5 years.

10.6 Extinct New Zealand birds. Refer to the *Evolutionary Ecology Research* (July 2003) study of extinction in the New Zealand bird population, Exercise 1.18 (p. 20). Recall that biologists measured the body mass (in grams) and habitat type (aquatic, ground terrestrial, or aerial terrestrial) for each bird species. One objective is to compare the body mass means of birds with the three different habitat types.

a. Identify the response variable of the study.
b. Identify the experimental units of the study.
c. Identify the factor(s) in the study.
d. Identify the treatments in the study

10.7 CT scanning for lung cancer. Refer to Exercise 1.23 (p. 21) and the University of South Florida clinical trial of 50,000 smokers to compare the effectiveness of CT scans with X-rays for detecting lung cancer. (*Today's Tomorrows*, Fall 2002.) Recall that each participating smoker will be randomly assigned to one of two screening methods, CT or chest X-ray, and the age (in years) at which the scanning method first detects a tumor will be determined. One goal of the study is to compare the mean ages when cancer is first detected of the two screening methods.

a. Identify the response variable of the study.
b. Identify the experimental units of the study.
c. Identify the factor(s) in the study.
d. Identify the treatments in the study.

Applying the Concepts—Intermediate

10.8 Treatment for tendon pain. Chronic Achilles tendon pain (i.e., tendinosis) is common among middle-aged recreational athletes. A group of Swedish physicians investigated the use of heavy load eccentric calf muscle training to treat Achilles tendinosis. (*British Journal of Sports Medicine*, Feb. 1, 2004.) A sample of 25 patients with chronic Achilles tendinosis undertook the treatment. Data on tendon thickness (measured in millimeters) were collected using ultrasonography both before and following treatment of each patient. The researchers want to compare the mean tendon thickness before treatment to the mean tendon thickness after treatment.

a. Is this a designed experiment or an observational study? Explain.
b. What is the experimental unit for this study?
c. What is the response variable for this study?

d. What are the treatments for this experiment?

e. After reading Section 10.3, you will learn that patients represent a blocking factor in this study. How many levels are in the blocking factor?

10.9 Taste preferences of cockatiels. *Applied Animal Behaviour Science* (Oct. 2000) published a study of the taste preferences of caged cockatiels. A sample of birds bred at the University of California, Davis, were randomly divided into three experimental groups. Group 1 was fed purified water in bottles on both sides of the cage. Group 2 was fed water on one side and a liquid sucrose (sweet) mixture on the opposite side of the cage. Group 3 was fed water on one side and a liquid sodium chloride (salty) mixture on the opposite side of the cage. One variable of interest to the researchers was total consumption of liquid by each cockatiel.

a. What is the experimental unit for this study?
b. Is the study a designed experiment? Why?
c. What are the factors in the study?
d. Give the levels of each factor.
e. How many treatments are in the study? Identify them.
f. What is the response variable?

10.10 Exam performance study. In *Teaching of Psychology* [NW] (Aug. 1998), a study investigated whether final exam performance is affected by whether or not students take a practice test. Students in an introductory psychology class at Pennsylvania State University were initially divided into three groups based on their class standing: Low, Medium, or High. Within each group, the students were randomly assigned to either attend a review session or take a practice test prior to the final exam. Thus, six groups were formed: (Low, Review), (Low, Practice exam), (Medium, Review), (Medium, Practice exam), (High, Review), and (High, Practice exam). One goal of the study was to compare the mean final exam scores of the six groups of students.

a. What is the experimental unit for this study?
b. Is the study a designed experiment? Why?
c. What are the factors in the study?
d. Give the levels of each factor.
e. How many treatments are in the study? Identify them.
f. What is the response variable?

Applying the Concepts—Advanced

10.11 Ethics of salespeople. Within marketing, the area of personal sales has long suffered from a poor ethical image, particularly in the eyes of college students. An article in *Journal of Business Ethics* (Vol. 15, 1996) investigated whether such opinions by college students are a function of the type of sales jobs (high tech versus low tech) and/or the sales task (new account development versus account maintenance). Four different samples of college students were confronted with the four different situations (new account development in a high-tech sales task; new account development in a low-tech sales task; account maintenance in a high-tech sales task; and account maintenance in a low-tech sales task), and were asked to evaluate the ethical behavior of the salesperson on a 7-point scale ranging from 1 (not a serious ethical violation) to 7 (a very serious ethical violation). Identify each of the following elements of the experiment:

a. Response
b. Factor(s) and factor level(s)
c. Treatments
d. Experimental units

10.12 Steel ingot experiment. A quality control supervisor measures the quality of a steel ingot on a scale from 0 to 10. He designs an experiment in which three different temperatures (ranging from 1,100 to 1,200°F) and five different pressures (ranging from 500 to 600 psi) are utilized, with 20 ingots produced at each Temperature–Pressure combination. Identify the following elements of the experiment:

a. Response
b. Factor(s) and factor type(s)
c. Treatments
d. Experimental units

10.2 The Completely Randomized Design

The simplest experimental design, a *completely randomized design*, consists of the *independent random selection* of experimental units representing each treatment. For example, we could independently select random samples of 20 female and 15 male high school seniors to compare their mean SAT scores. Or, we could randomly assign cancer patients to receive one of three experimental treatments, then compare the mean pain levels of patients in the treatment groups. In both examples our objective is to compare treatment means by selecting random, independent samples for each treatment.

> ### DEFINITION 10.7
>
> A **completely randomized design** is a design for which treatments are randomly assigned to the experimental units, or in which independent random samples of experimental units are selected for each treatment.*

EXAMPLE 10.3

ASSIGNING TREATMENTS IN A COMPLETELY RANDOMIZED DESIGN

Problem   Suppose we want to compare the taste preferences of consumers for three different brands of bottled water (say, Brands A, B, and C) using a random sample of 15 bottled water consumers. Set up a completely randomized design for this purpose. That is, assign the treatments to the experimental units for this design.

Solution   In this study, the experimental units are the 15 consumers and the treatments are the three brands of bottled water. One way to set up the completely randomized design is to assign randomly one of the three brands to each consumer to taste. Then we could measure (say, on a 1- to 10-point scale) the taste preference of each consumer. A good practice is to assign the same number of consumers to each brand—in this case, five consumers to each of the three brands. (When an equal number of experimental units is assigned to each treatment, we call the design a *balanced design*.)

A random number table (Table 1, Appendix A) or computer software can be used to make the random assignments. Figure 10.3 is a MINITAB worksheet showing the random assignments made with the MINITAB "Random Data" function. You can see that MINITAB randomly assigned consumers numbered 2, 11, 1, 13, and 3 to taste Brand A; consumers numbered 15, 14, 7, 10, and 8 to taste Brand B; and consumers numbered 6, 5, 12, 9, and 4 to taste Brand C.

*We use *completely randomized design* to refer to both designed and observational experiments. Thus, the only requirement is that the experimental units to which treatments are applied (designed) or on which treatments are observed (observational) are independently selected for each treatment.

Figure 10.3

MINITAB Random
Assignments of Consumers
to Brands

CRD3brands.MTW ***				
↓	C1	C2	C3	C4	C5
	Consumer	BrandA	BrandB	BrandC	
1	1	2	15	6	
2	2	11	14	5	
3	3	1	7	12	
4	4	13	10	9	
5	5	3	8	4	
6	6				
7	7				
8	8				
9	9				
10	10				
11	11				
12	12				
13	13				
14	14				
15	15				
16					

Look Back In some experiments, it will not be possible to randomly assign treatments to the experimental units—the units will already be associated with one of the treatments. (For example, if the treatments are "Male" and "Female," you cannot change a person's gender.) In this case, a completely randomized design is one where you select independent random samples of experimental units from each treatment.

■ ■ ■

The objective of a completely randomized design is usually to compare the treatment means. If we denote the true, or population, means of the k treatments as $\mu_1, \mu_2, \ldots, \mu_k$, then we will test the null hypothesis that the treatment means are all equal against the alternative that at least two of the treatment means differ:

$$H_0: \mu_1 = \mu_2 = \cdots = \mu_k$$
$$H_a: \text{At least two of the } k \text{ treatment means differ}$$

The μ's might represent the means of *all* female and male high school seniors' SAT scores or the means of *all* households' income in each of four census regions.

To conduct a statistical test of these hypotheses, we will use the means of the independent random samples selected from the treatment populations using the completely randomized design. That is, we compare the k sample means $\bar{x}_1, \bar{x}_2, \ldots, \bar{x}_k$.

For example, suppose you select independent random samples of five female and five male high school seniors and obtain sample mean SAT scores of 550 and 590, respectively. Can we conclude that males score 40 points higher, on average, than females? To answer this question, we must consider the amount of sampling variability among the experimental units (students). If the scores are as depicted in the dot plot shown in Figure 10.4, then the difference between the means is small relative to the sampling variability of the scores within the treatments, Female and Male. We would be inclined not to reject the null hypothesis of equal population means in this case.

Figure 10.4

Dot Plot of SAT Scores: Difference between Means Dominated by Sampling Variability

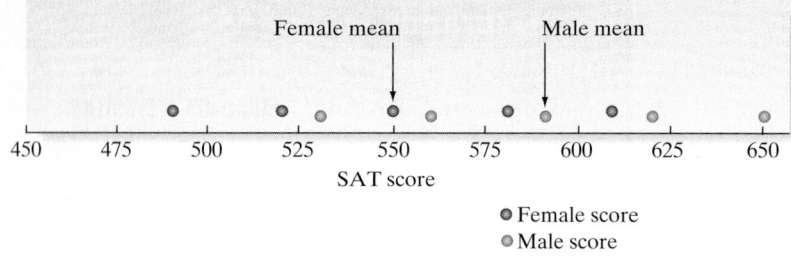

Figure 10.5

Dot Plot of SAT Scores: Difference between Means Large Relative to Sampling Variability

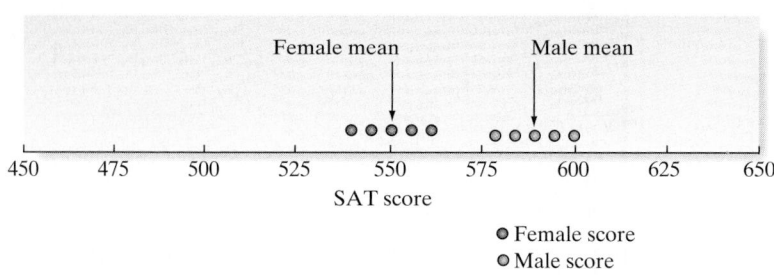

In contrast, if the data are as depicted in the dot plot of Figure 10.5, then the sampling variability is small relative to the difference between the two means. We would be inclined to favor the alternative hypothesis that the population means differ in this case.

You can see that the key is to compare the difference between the treatment means to the amount of sampling variability. To conduct a formal statistical test of the hypothesis requires numerical measures of the difference between the treatment means and the sampling variability within each treatment. The variation between the treatment means is measured by the **Sum of Squares for Treatments** (SST), which is calculated by squaring the distance between each treatment mean and the overall mean of *all* sample measurements, multiplying each squared distance by the number of sample measurements for the treatment, and adding the results over all treatments:

$$\text{SST} = \sum_{i=1}^{k} n_i(\bar{x}_i - \bar{x})^2 = 5(550 - 570)^2 + 5(590 - 570)^2 = 4{,}000$$

where we use $\bar{x}$ to represent the overall mean response of all sample measurements, that is, the mean of the combined samples. The symbol n_i is used to denote the sample size for the ith treatment. You can see that the value of SST is 4,000 for the two samples of five female and five male SAT scores depicted in Figures 10.4 and 10.5.

Next, we must measure the sampling variability within the treatments. We call this the **Sum of Squares for Error** (SSE) because it measures the variability around the treatment means that is attributed to sampling error. Suppose the 10 measurements in the first dot plot (Figure 10.4) are 490, 520, 550, 580, and 610 for females and 530, 560, 590, 620, and 650 for males. Then the value of SSE is computed by summing the squared distance between each response measurement and the corresponding treatment mean, and then adding the squared differences over all measurements in the entire sample:

$$\text{SSE} = \sum_{j=1}^{n_1}(x_{1j} - \bar{x}_1)^2 + \sum_{j=1}^{n_2}(x_{2j} - \bar{x}_2)^2 + \dots \sum_{j=1}^{n_k}(x_{kj} - \bar{x}_k)^2$$

where the symbol x_{1j} is the jth measurement in sample 1, x_{2j} is the jth measurement in sample 2, and so on. This rather complex-looking formula can be simplified by recalling the formula for the sample variance, s^2, given in Chapter 2:

$$s^2 = \sum_{i=1}^{n} \frac{(x_i - \bar{x})^2}{n - 1}$$

Note that each sum in SSE is simply the numerator of s^2 for that particular treatment. Consequently, we can rewrite SSE as

$$\text{SSE} = (n_1 - 1)s_1^2 + (n_2 - 1)s_2^2 + \cdots + (n_k - 1)s_k^2$$

where $s_1^2, s_2^2, \ldots, s_k^2$ are the sample variances for the k treatments. For our samples of SAT scores, we find $s_1^2 = 2{,}250$ (for females) and $s_2^2 = 2{,}250$ (for males); then we have

$$\text{SSE} = (5 - 1)(2{,}250) + (5 - 1)(2{,}250) = 18{,}000$$

To make the two measurements of variability comparable, we divide each by the degrees of freedom to convert the sums of squares to mean squares. First, the **Mean Square for Treatments** (MST), which measures the variability *among* the treatment means, is equal to

$$\text{MST} = \frac{\text{SST}}{k - 1} = \frac{4{,}000}{2 - 1} = 4{,}000$$

where the number of degrees of freedom for the k treatments is $(k - 1)$. Next, the **Mean Square for Error** (MSE), which measures the sampling variability *within* the treatments, is

$$\text{MSE} = \frac{\text{SSE}}{n - k} = \frac{18{,}000}{10 - 2} = 2{,}250$$

Finally, we calculate the ratio of MST to MSE—an ***F*-statistic**:

$$F = \frac{\text{MST}}{\text{MSE}} = \frac{4{,}000}{2{,}250} = 1.78$$

Values of the F-statistic near 1 indicate that the two sources of variation, between treatment means and within treatments, are approximately equal. In this case, the difference between the treatment means may well be attributable to sampling error, which provides little support for the alternative hypothesis that the population treatment means differ. Values of F well in excess of 1 indicate that the variation among treatment means well exceeds that within means and therefore support the alternative hypothesis that the population treatment means differ.

When does F exceed 1 by enough to reject the null hypothesis that the means are equal? This depends on the degrees of freedom for treatments and for error, and on the value of α selected for the test. We compare the calculated F-value to a table F-value (Tables VIII–XI of Appendix A) with $v_1 = (k - 1)$ degrees of freedom in the numerator and $v_2 = (n - k)$ degrees of freedom in the denominator and corresponding to a Type I error probability of α. For the SAT score example, the F-statistic has $v_1 = (2 - 1)$ numerator degree of freedom and $v_2 = (10 - 2) = 8$ denominator degrees of freedom. Thus, for $\alpha = .05$ we find (Table IX of Appendix A)

$$F_{.05} = 5.32$$

The implication is that MST would have to be 5.32 times greater than MSE before we could conclude at the .05 level of significance that the two population treatment means differ. Since the data yielded $F = 1.78$, our initial impressions for the dot plot in Figure 10.4 are confirmed—there is insufficient information to conclude that the mean SAT scores differ for the populations of female and male high school seniors. The rejection region and the calculated F value are shown in Figure 10.6.

In contrast, consider the dot plot in Figure 10.5. Since the means are the same as in the first example, 550 and 590, respectively, the variation between the means is the same, MST = 4,000. But the variation within the two treatments appears to be considerably smaller. The observed SAT scores are 540, 545, 550, 555, and 560 for

Figure 10.6

Rejection Region and Calculated F-Values for SAT Score Samples

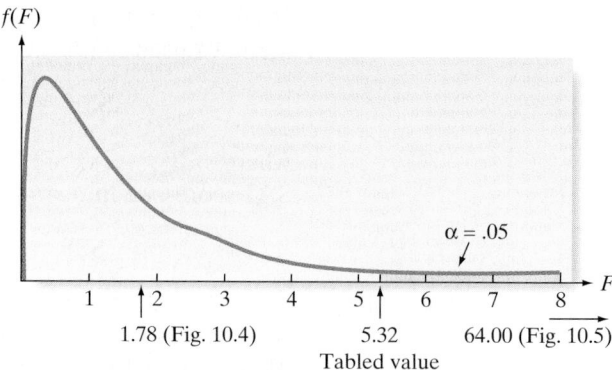

females and 580, 585, 590, 595, and 600 for males. These values yield $s_1^2 = 62.5$ and $s_2^2 = 62.5$. Thus, the variation within the treatments is measured by

$$SSE = (5 - 1)(62.5) + (5 - 1)(62.5) = 500$$
$$MSE = \frac{SSE}{n - k} = \frac{500}{8} = 62.5$$

Then the F-ratio is

$$F = \frac{MST}{MSE} = \frac{4,000}{62.5} = 64.0$$

Again, our visual analysis of the dot plot is confirmed statistically: $F = 64.0$ well exceeds the tabled F value, 5.32, corresponding to the .05 level of significance. We would therefore reject the null hypothesis at that level and conclude that the SAT mean score of males differs from that of females.

Now Work   *Exercise 10.20*

Recall that we performed a hypothesis test for the difference between two means in Section 9.2 using a two-sample t-statistic for two independent samples. When two independent samples are being compared, the t- and F-tests are equivalent. To see this, recall the formula

$$t = \frac{\overline{x}_1 - \overline{x}_2}{\sqrt{s_p^2 \left(\frac{1}{n_1} + \frac{1}{n_2} \right)}} = \frac{590 - 550}{\sqrt{(62.5) \left(\frac{1}{5} + \frac{1}{5} \right)}} = \frac{40}{5} = 8$$

where we used the fact that $s_p^2 = MSE$, which you can verify by comparing the formulas. Note that the calculated F for these samples ($F = 64$) equals the square of the calculated t for the same samples ($t = 8$). Likewise, the tabled F-value (5.32) equals the square of the tabled t-value at the two-sided .05 level of significance ($t_{.025} = 2.306$ with 8 df). Since both the rejection region and the calculated values are related in the same way, the tests are equivalent. Moreover, the assumptions that must be met to ensure the validity of the t- and F-tests are the same:

1. The probability distributions of the populations of responses associated with each treatment must all be normal.

2. The probability distributions of the populations of responses associated with each treatment must have equal variances.

3. The samples of experimental units selected for the treatments must be random and independent.

In fact, the only real difference between the tests is that the F-test can be used to compare *more than two* treatment means, whereas the t-test is applicable to two samples only. The **F-test** is summarized in the accompanying box.

ANOVA F-Test to Compare k Treatment Means for a Completely Randomized Design

H_0: $\mu_1 = \mu_2 = \cdots = \mu_k$

H_a: At least two treatment means differ

Test statistic: $F = \dfrac{\text{MST}}{\text{MSE}}$

Rejection region: $F > F_\alpha$, where F_α is based on $v_1 = (k - 1)$ numerator degrees of freedom (associated with MST) and $v_2 = (n - k)$ denominator degrees of freedom (associated with MSE).

Conditions Required for a Valid ANOVA F-Test: Completely Randomized Design

1. The samples are randomly selected in an independent manner from the k treatment populations. (This can be accomplished by randomly assigning the experimental units to the treatments.)

2. All k sampled populations have distributions that are approximately normal.

3. The k population variances are equal (i.e., $\sigma_1^2 = \sigma_2^2 = \sigma_3^2 = \cdots = \sigma_k^2$).

Computational formulas for MST and MSE are given in Appendix B. We will rely on statistical software to compute the F statistic, concentrating on the interpretation of the results rather than their calculations.

EXAMPLE 10.4   CONDUCTING AN ANOVA F-TEST

Problem   Suppose the USGA wants to compare the mean distances associated with four different brands of golf balls when struck with a driver. A completely randomized design is employed, with Iron Byron, the USGA's robotic golfer, using a driver to hit a random sample of 10 balls of each brand in a random sequence. The distance is recorded for each hit, and the results are shown in *Table 10.1*, organized by brand.

⊚ **GOLFCRD**

TABLE 10.1   Results of Completely Randomized Design: Iron Byron Driver

	Brand A	Brand B	Brand C	Brand D
	251.2	263.2	269.7	251.6
	245.1	262.9	263.2	248.6
	248.0	265.0	277.5	249.4
	251.1	254.5	267.4	242.0
	260.5	264.3	270.5	246.5
	250.0	257.0	265.5	251.3
	253.9	262.8	270.7	261.8
	244.6	264.4	272.9	249.0
	254.6	260.6	275.6	247.1
	248.8	255.9	266.5	245.9
Sample means	250.8	261.1	270.0	249.3

a. Set up the test to compare the mean distances for the four brands. Use $\alpha = .10$.

b. Use Statistical Software to obtain the test statistic and p-value. Interpret the results.

Solution **a.** To compare the mean distances of the $k = 4$ brands, we first specify the hypotheses to be tested. Denoting the population mean of the ith brand by μ_i, we test

$$H_0: \mu_1 = \mu_2 = \mu_3 = \mu_4$$
$$H_a: \text{The mean distances differ for at least two of the brands.}$$

The test statistic compares the variation among the four treatment (Brand) means to the sampling variability within each of the treatments.

$$\textit{Test statistic:} \quad F = \frac{\text{MST}}{\text{MSE}}$$
$$\textit{Rejection region:} \quad F > F_\alpha = F_{.10} \text{ with } v_1 = (k - 1) = 3 \text{ df}$$
$$\text{and } v_2 = (n - k) = 36 \text{ df}$$

From Table VIII of Appendix A, we find $F_{.10} \approx 2.25$ for 3 and 36 df. Thus, we will reject H_0 if $F > 2.25$. (See Figure 10.7.)

Figure 10.7
F-Test for Completely Randomized Design: Golf Ball Experiment

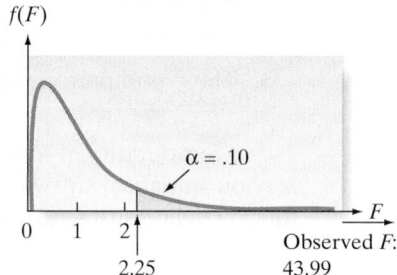

The assumptions necessary to ensure the validity of the test are as follows:

1. The samples of 10 golf balls for each brand are selected randomly and independently.

2. The probability distributions of the distances for each brand are normal.

3. The variances of the distance probability distributions for each brand are equal.

b. The MINITAB printout for the data in Table 10.1 resulting from this completely randomized design is given in Figure 10.8. The Total Sum of Squares is designated "Total," and it is partitioned into the "Brand" (i.e., treatments) and "Error" Sum of Squares (SS).

The values of the mean squares, MST and MSE (highlighted on the printout), are 931.5 and 21.2, respectively. The F-ratio, 43.99, also highlighted on the printout, exceeds the tabled value of 2.25. We therefore reject the null hypothesis at the .10 level of significance, concluding that at least two of the brands differ with respect to mean distance traveled when struck by the driver.

Look Back We can also arrive at the appropriate conclusion by noting that the observed significance level of the F-test (highlighted on the printout) is p-value $= .0001$. This implies that we would reject the null hypothesis that the means are equal at any α level.

Figure 10.8

MINITAB ANOVA for Completely Randomized Design

One-way ANOVA: DISTANCE versus BRAND

```
Source   DF       SS      MS      F       P
BRAND     3   2794.4   931.5   43.99   0.000
Error    36    762.3    21.2
Total    39   3556.7

S = 4.602      R-Sq = 78.57%    R-Sq(adj) = 76.78%


                                   Individual 95% CIs For Mean Based on
                                   Pooled StDev
Level    N    Mean   StDev    --------+---------+---------+---------+-
BrandA   10  250.78    4.74    (---*---)
BrandB   10  261.06    3.87                   (---*---)
BrandC   10  269.95    4.50                                 (----*---)
BrandD   10  249.32    5.20    (---*---)
                              --------+---------+---------+---------+-
                                  252.0     259.0     266.0     273.0


Pooled StDev = 4.60
```

Now Work   *Exercise 10.25*

◼ ◼ ◼

The results of an **analysis of variance (ANOVA)** can be summarized in a simple tabular format similar to that obtained from the MINITAB program in Example 10.4. The general form of the table is shown in Table 10.2, where the symbols df, SS, and MS stand for degrees of freedom, Sum of Squares, and Mean Square, respectively. Note that the two sources of variation, Treatments and Error, add to the Total Sum of Squares, SS(Total). The ANOVA summary table for Example 10.4 is given in Table 10.3, and the partitioning of the Total Sum of Squares into its two components is illustrated in Figure 10.9.

Suppose the *F*-test results in a rejection of the null hypothesis that the treatment means are equal. Is the analysis complete? Usually, the conclusion that at least

TABLE 10.2   General ANOVA Summary Table for a Completely Randomized Design

Source	df	SS	MS	F
Treatments	$k - 1$	SST	$MST = \dfrac{SST}{k - 1}$	$\dfrac{MST}{MSE}$
Error	$n - k$	SSE	$MSE = \dfrac{SSE}{n - k}$	
Total	$n - 1$	SS(Total)		

TABLE 10.3   ANOVA Summary Table for Example 10.4

Source	df	SS	MS	F	*p*-Value
Brands	3	2,794.39	931.46	43.99	.0001
Error	36	762.30	21.18		
Total	39	3,556.69			

Figure 10.9

Partitioning of the Total Sum of Squares for the Completely Randomized Design

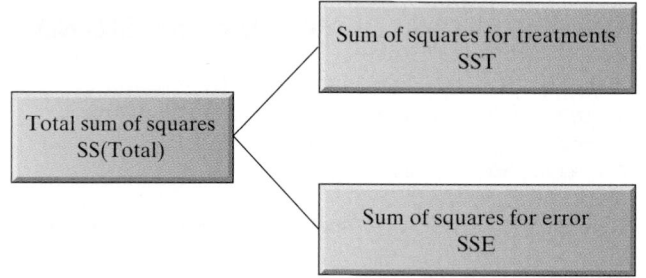

two of the treatment means differ leads to other questions. Which of the means differ, and by how much? For example, the F-test in Example 10.3 leads to the conclusion that at least two of the brands of golf balls have different mean distances traveled when struck with a driver. Now the question is, which of the brands differ? How are the brands ranked with respect to mean distance?

One way to obtain this information is to construct a confidence interval for the difference between the means of any pair of treatments using the method of Section 9.2. For example, if a 95% confidence interval for $\mu_A - \mu_C$ in Example 10.4 is found to be $(-24, -13)$, we are confident that the mean distance for Brand C exceeds the mean for Brand A (since all differences in the interval are negative). Constructing these confidence intervals for all possible brand pairs allows you to rank the brand means. A method for conducting these *multiple comparisons*—one that controls for Type I errors—is presented in Section 10.3.

Analysis of Variance

Using the TI-83 Graphing Calculator

Computing a One-Way ANOVA

Step 1   *Enter each data set into its own list (i.e., sample 1 into L1, sample 2 into L2, sample 3 into L3, etc.).*

Step 2   *Access the Statistical Test Menu*
Press **STAT**
Arrow right to TESTS
Arrow up to F: ANOVA(
Press **ENTER**

Step 3   *Enter each list, separated by commas, for which you want to perform the analysis (e.g., L1, L2, L3, L4)*
Press **ENTER**

Step 4   *View Display*
The calculator will display the F-test statistic, as well as the p-value, the Factor degrees of freedom, sum of squares, mean square, and by arrowing Down, the Error degrees of freedom, sum of squares, mean square, and the pooled standard deviation.

Example   Below are four different samples. At the $\alpha = .05$ level of significance test whether the four population means are equal. The null hypothesis will be $H_0: \mu_1 = \mu_2 = \mu_3 = \mu_4$. The alternative hypothesis is H_a: At least one mean is different.

SAMPLE 1	SAMPLE 2	SAMPLE 3	SAMPLE 4
60	59	55	58
61	52	55	58
56	51	52	55

The screens for this example are shown below.

```
ANOVA(L₁,L₂,L₃,L
4)■
```

```
One-way ANOVA
 F=2.25
 p=.1597672711
 Factor
  df=3
  SS=54
↓ MS=18
■
```

```
One-way ANOVA
↑ MS=18
 Error
  df=8
  SS=64
  MS=8
 Sxp=2.82842712
■
```

As you can see from the screen, the p-value is 0.1598 which is *not less than* 0.05 therefore we *should not* reject H_0. The differences are not significant.

EXAMPLE 10.5   CHECKING THE ANOVA ASSUMPTIONS

Problem   Refer to the completely randomized design ANOVA conducted in Example 10.4. Are the assumptions required for the test approximately satisfied?

Solution   The assumptions for the test are repeated below.

1. The samples of golf balls for each brand are selected randomly and independently.
2. The probability distributions of the distances for each brand are normal.
3. The variances of the distance probability distributions for each brand are equal.

Since the sample consisted of 10 randomly selected balls of each brand and the robotic golfer Iron Byron was used to drive all the balls, the first assumption of independent random samples is satisfied. To check the next two assumptions, we will employ two graphical methods presented in Chapter 2: histograms and box plots. A MINITAB histogram of driving distances for each brand of golf ball is shown in Figure 10.10, and SAS box plots in Figure 10.11.

The normality assumption can be checked by examining the histograms in Figure 10.10. With only 10 sample measurements for each brand, however, the displays are not very informative. More data would need to be collected for each brand before we could assess whether the distances come from normal distributions. Fortunately, analysis of variance has been shown to be a very **robust method** when the assumption of normality is not satisfied exactly: That is, *moderate departures from normality do not have much effect on the significance level of the ANOVA F-test or on confidence coefficients.* Rather than spend the time, energy, or money to collect additional data for this experiment in order to verify the normality assumption, we will rely on the robustness of the ANOVA methodology.

Box plots are a convenient way to obtain a rough check on the assumption of equal variances. With the exception of a possible outlier for Brand D, the box plots in Figure 10.11 show that the spread of the distance measurements is about the same for each brand. Since the sample variances appear to be the same, the assumption of equal population variances for the brands is probably satisfied. Although robust with respect to the normality assumption, ANOVA is *not robust* with respect to the equal variances assumption. Departures from the assumption of equal population variances can affect the associated measures of reliability (e.g., p-values and confidence levels). Fortunately, the effect is slight when the sample sizes are equal, as in this experiment.

Figure 10.10

MINITAB Histograms for Golf Ball Distances

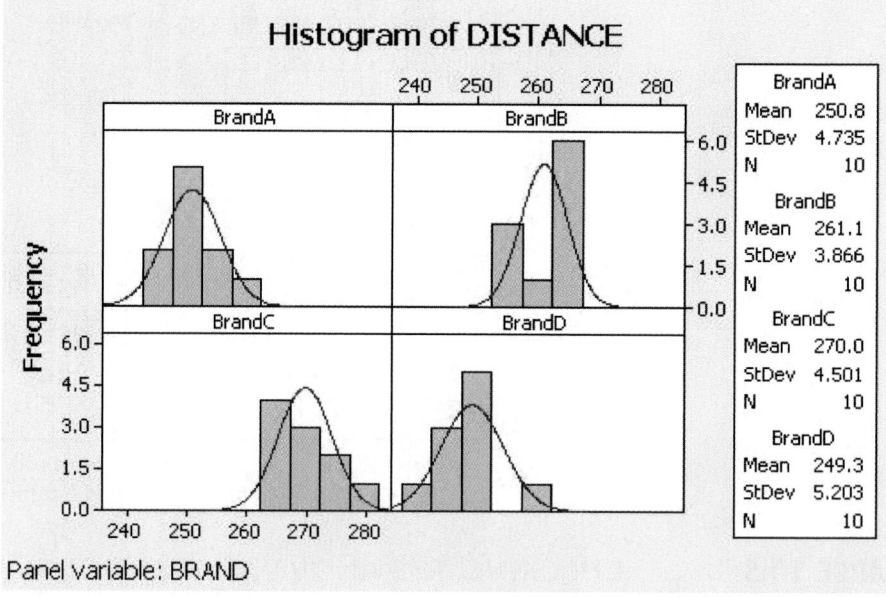

Figure 10.11

SAS Box Plots for Golf Ball Distances

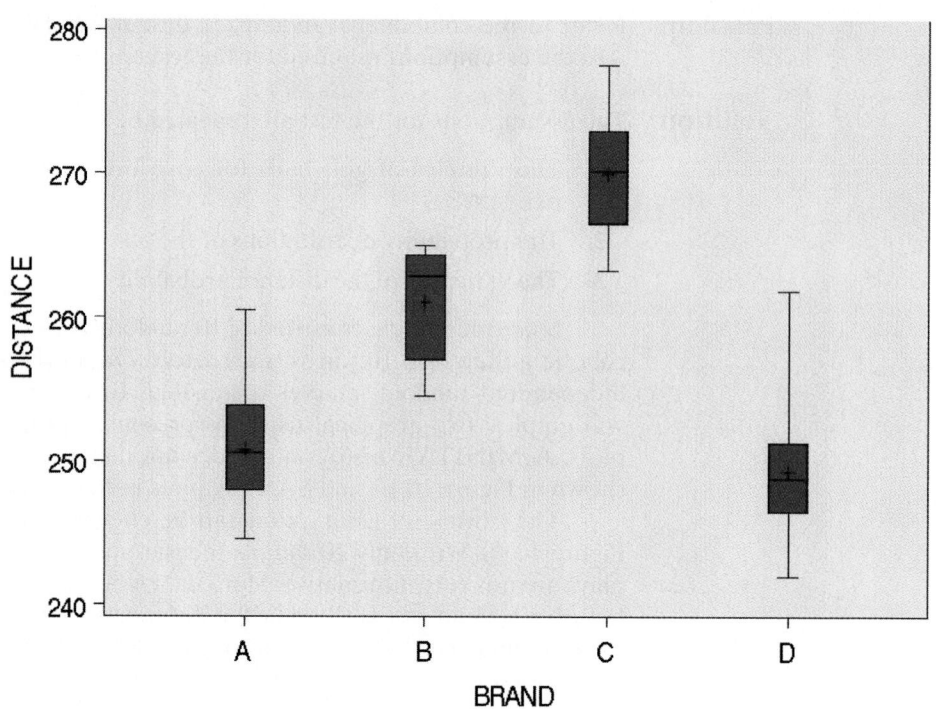

Now Work *Exercise 10.30*

■ ■ ■

Although graphs can be used to check the ANOVA assumptions, as in Example 10.5, no measures of reliability can be attached to these graphs. When you have a plot that is unclear as to whether or not an assumption is satisfied, you can use formal statistical tests that are beyond the scope of this text. Consult the references at the end of the chapter for information on these tests. When the validity of the ANOVA assumptions is in doubt, nonparametric statistical methods are useful.

What Do You Do When the Assumptions Are Not Satisfied for the Analysis of Variance for a Completely Randomized Design?

Answer: Use a nonparametric statistical method such as the Kruskal-Wallis *H*-Test of Section 14.4.

The procedure for conducting an analysis of variance for a completely randomized design is summarized in the following box. Remember that the hallmark of this design is independent random samples of experimental units associated with each treatment. We discuss a design with dependent samples in Section 10.4.

Steps for Conducting an ANOVA for a Completely Randomized Design

1. Be sure the design is truly completely randomized, with independent random samples for each treatment.

2. Check the assumptions of normality and equal variances.

3. Create an ANOVA summary table that specifies the variability attributable to treatments and error, making sure that it leads to the calculation of the *F*-statistic for testing the null hypothesis that the treatment means are equal in the population. Use a statistical software program to obtain the numerical results. If no such package is available, use the calculation formulas in Appendix C.

4. If the *F*-test leads to the conclusion that the means differ:
   a. Conduct a multiple comparisons procedure for as many of the pairs of means as you wish to compare (see Section 10.3). Use the results to summarize the statistically significant differences among the treatment means.
   b. If desired, form confidence intervals for one or more individual treatment means.

5. If the *F*-test leads to the nonrejection of the null hypothesis that the treatment means are equal, consider the following possibilities:
   a. The treatment means are equal—that is, the null hypothesis is true.
   b. The treatment means really differ, but other important factors affecting the response are not accounted for by the completely randomized design. These factors inflate the sampling variability, as measured by MSE, resulting in smaller values of the *F*-statistic. Either increase the sample size for each treatment or use a different experimental design (as in Section 10.4) that accounts for the other factors affecting the response.

[*Note:* Be careful not to automatically conclude that the treatment means are equal since the possibility of a Type II error must be considered if you accept H_0.]

Statistics in Action Revisited

A One-Way Analysis of the Cockroach Data

Consider the experiment designed to investigate the trail-following ability of German cockroaches (p. 513). Recall that an entomologist created a chemical trail with either a methanol extract from roach feces or just methanol (control). Cockroaches were then released into a container at the beginning of the trail, one at a time, and a video surveillance camera was used to monitor the roach's movements. The movement pattern of each cockroach was measured by its average trail deviation (in pixels) and the data stored in the **ROACH** file.

For this application, consider only the cockroaches assigned to the fecal extract trail. Four roach groups were utilized in the experiment—adult males, adult females, gravid females, and nymphs—with 20 roaches of each type independently and randomly selected. Is there sufficient evidence to say that the ability to follow the extract trail differs among cockroaches of different age, sex, and reproductive status? In other words, is there evidence to suggest that the mean trail deviation μ differs for the four roach groups?

To answer this question, we conduct a one-way analysis of variance on the **ROACH** data. The dependent (response) variable of interest is extract trail deviation, while the treatments are the four different roach groups. Thus, we want to test the null hypothesis:

$$H_0: \mu_{\text{Male}} = \mu_{\text{Female}} = \mu_{\text{Gravid}} = \mu_{\text{Nymph}}$$

A MINITAB printout of the ANOVA is displayed in Figure SIA10.1. The p-value of the test (highlighted on the printout) is 0. Since this value is less than, say, $\alpha = .05$, we

reject the null hypothesis and conclude (at the .05 level of significance) that the mean extract trail deviation differs among the populations of adult male, adult female, gravid, and nymph cockroaches.

The sample means for the four cockroach groups are also highlighted in Figure SIA10.1. Note that adult males have the smallest sample mean deviation (7.38) while gravids have the largest sample mean deviation (44.03). In the next Statistics in Action Revisited application (p. 00), we demonstrate how to rank, statistically, the four population means based on their respective sample means.

Results for Trail = Extract

One-way ANOVA: Deviate versus Group

```
Source   DF     SS     MS      F      P
Group     3  14164   4721  11.61  0.000
Error    76  30918    407
Total    79  45083

S = 20.17    R-Sq = 31.42%    R-Sq(adj) = 28.71%


                              Individual 95% CIs For Mean Based on
                              Pooled StDev
Level    N    Mean   StDev  -+---------+---------+---------+--------
Female  20   21.07   26.13                (-----*-----)
Gravid  20   44.03   24.84                           (-----*-----)
Male    20    7.38    8.61  (-----*-----)
Nymph   20   18.73   15.92             (-----*-----)
                           -+---------+---------+---------+--------
                            0        15        30        45

Pooled StDev = 20.17
```

Figure SIA10.1
MINITAB One-Way ANOVA for Extract Trail Deviation

Exercises 10.13–10.36

Understanding the Principles

10.13 Explain how to collect the data for a completely randomized design.

10.14 Explain the concept of a balanced design.

10.15 What conditions are required for a valid ANOVA F-test in a completely randomized design?

10.16 True or False. The ANOVA method is robust when the assumption of normality is not exactly satisfied in a completely randomized design.

Learning the Mechanics

10.17 Use Tables VIII, IX, X, and XI of Appendix A to find each of the following F values:
 a. $F_{.05}, v_1 = 3, v_2 = 4$
 b. $F_{.01}, v_1 = 3, v_2 = 4$

 c. $F_{.10}, v_1 = 20, v_2 = 40$
 d. $F_{.025}, v_1 = 12, v_2 = 9$

10.18 Find the following probabilities:
 a. $P(F \leq 3.48)$ for $v_1 = 5, v_2 = 9$
 b. $P(F > 3.09)$ for $v_1 = 15, v_2 = 20$
 c. $P(F > 2.40)$ for $v_1 = 15, v_2 = 15$
 d. $P(F \leq 1.83)$ for $v_1 = 8, v_2 = 40$

10.19 Consider dot plots a. and b. shown on p. 533. In which
 NW   dot plot is the difference between the sample means small relative to the variability within the sample observations? Justify your answer.

10.20 Refer to Exercise 10.19. Assume that the two samples
 NW   represent independent, random samples corresponding to two treatments in a completely randomized design.

Plots for Exercise 10.19

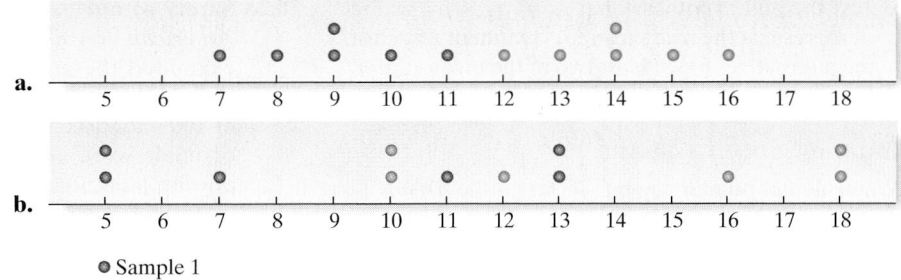

● Sample 1
● Sample 2

a. Calculate the treatment means, i.e., the means of samples 1 and 2, for both dot plots.
b. Use the means to calculate the Sum of Squares for Treatments (SST) for each dot plot.
c. Calculate the sample variance for each sample and use these values to obtain the Sum of Squares for Error (SSE) for each dot plot.
d. Calculate the Total Sum of Squares [SS(Total)] for the two dot plots by adding the Sums of Squares for Treatment and Error. What percentage of SS(Total) is accounted for by the treatments—that is, what percentage of the Total Sum of Squares is the Sum of Squares for Treatment—in each case?
e. Convert the Sum of Squares for Treatment and Error to mean squares by dividing each by the appropriate number of degrees of freedom. Calculate the F-ratio of the Mean Square for Treatment (MST) to the Mean Square for Error (MSE) for each dot plot.
f. Use the F-ratios to test the null hypothesis that the two samples are drawn from populations with equal means. Use $\alpha = .05$.
g. What assumptions must be made about the probability distributions corresponding to the responses for each treatment in order to ensure the validity of the F-tests conducted in part **f**?

10.21 Refer to Exercises 10.19 and 10.20. Conduct a two-sample t-test (Section 9.2) of the null hypothesis that the two treatment means are equal for each dot plot. Use $\alpha = .05$ and two-tailed tests. In the course of the test, compare each of the following with the F-tests in Exercise 10.20.
a. The pooled variances and the MSEs
b. The t- and the F-test statistics
c. The tabled values of t and F that determine the rejection regions
d. The conclusions of the t- and F-tests
e. The assumptions that must be made in order to ensure the validity of the t- and F-tests

10.22 Refer to Exercises 10.19 and 10.20. Complete the following ANOVA table for each of the two dot plots:

Source	df	SS	MS	F
Treatments				
Error				
Total				

10.23 Suppose the Total Sum of Squares for a completely randomized design with $p = 5$ treatments and $n = 30$ total measurements (six per treatment) is equal to 500. In each of the following cases, conduct an F-test of the null hypothesis that the mean responses for the five treatments are the same. Use $\alpha = .10$.
a. Sum of Squares for Treatment (SST) is 20% of SS(Total)
b. SST is 50% of SS(Total)
c. SST is 80% of SS(Total)
d. What happens to the F-ratio as the percentage of the Total Sum of Squares attributable to treatments is increased?

10.24 A partially completed ANOVA table for a completely randomized design is shown here:

Source	df	SS	MS	F
Treatments	6	18.4		
Error				
Total	41	45.2		

a. Complete the ANOVA table.
b. How many treatments are involved in the experiment?
c. Do the data provide sufficient evidence to indicate a difference among the population means? Test using $\alpha = .10$.
d. Find the approximate observed significance level for the test in part **c**, and interpret it.

10.25 The data in the table below resulted from an experiment that utilized a completely randomized design.
NW

⚙ **LM10_25**

Treatment 1	Treatment 2	Treatment 3
3.9	5.4	1.3
1.4	2.0	.7
4.1	4.8	2.2
5.5	3.8	
2.3	3.5	

a. Use statistical software (or the formulas in Appendix C) to complete the following ANOVA table:

Source	df	SS	MS	F
Treatments				
Error				
Total				

b. Test the null hypothesis that $\mu_1 = \mu_2 = \mu_3$, where μ_i represents the true mean for treatment i, against the alternative that at least two of the means differ. Use $\alpha = .01$.

Applying the Concepts—Basic

10.26 A new dental bonding agent. Refer to the *Trends in Biomaterials & Artificial Organs* (Jan. 2003) study of a new bonding adhesive for teeth, Exercise 8.61 (p. 395). Recall that the new adhesive (called "Smartbond") has been developed to eliminate the necessity of a dry field. In one portion of the study, 30 extracted teeth were bonded with Smartbond and each was randomly assigned one of three different bonding times: 1 hour, 24 hours, or 48 hours. At the end of the bonding period, the breaking strength (in Mpa) of each tooth was determined. The data were analyzed using analysis of variance in order to determine if true mean breaking strength of the new adhesive differs depending on the length of bonding time.

a. Identify the experimental units, treatments, and response variable for this completely randomized design.

b. Set up the null and alternative hypothesis for the ANOVA.

c. Find the rejection region for the test using $\alpha = .01$.

d. The test results were $F = 61.62$ and p-value ≈ 0. Give the appropriate conclusion for the test.

e. What conditions are required for the test results to be valid?

10.27 Robots trained to behave like ants. Robotics researchers investigated whether robots could be trained to behave like ants in an ant colony (*Nature*, Aug. 2000). Robots were trained and randomly assigned to "colonies" (i.e., groups) consisting of 3, 6, 9, or 12 robots. The robots were assigned the task of foraging for "food" and to recruit another robot when they identified a resource-rich area. One goal of the experiment was to compare the mean energy expended (per robot) of the four different colony sizes.

a. What type of experimental design was employed?

b. Identify the treatments and the dependent variable.

c. Set up the null and alternative hypotheses of the test.

d. The following ANOVA results were reported: $F = 7.70$, numerator df = 3, denominator df = 56, p-value < .001. Conduct the test at a significance level of $\alpha = .05$ and interpret the result.

10.28 Safety of nuclear power plants. An article in the *American Journal of Political Science* (Jan. 1998) examined the attitudes of three groups of professionals that influence U.S. policy. Random samples of 100 scientists, 100 journalists, and 100 government officials were asked about the safety of nuclear power plants. Responses were made on a seven-point scale, where 1 = very unsafe and 7 = very safe. The mean safety scores for the groups are: scientists, 4.1; journalists, 3.7; government officials, 4.2.

a. Identify the response variable for this study.

b. How many treatments are included in this study? Describe them.

c. Specify the null and alternative hypotheses that should be used to investigate whether there are differences in the attitudes of scientists, journalists, and government officials regarding the safety of nuclear power plants.

d. The MSE for the sample data is 2.355. At least how large must MST be in order to reject the null hypothesis of the test of part **a** using $\alpha = .05$?

e. If the MST = 11.280, what is the approximate p-value of the test of part **a**?

10.29 Heights of grade school repeaters. Refer to *The Archives of Disease in Childhood* (Apr. 2000) study of whether height influences a child's progression through elementary school, Exercise 9.15 (p. 449). Within each grade, Australian school children were divided into equal thirds (tertiles) based on age (youngest third, middle third, and oldest third). The researchers compared the average heights of the three groups using an analysis of variance. (All height measurements were standardized using z-scores.) A summary of the results for all grades combined, by gender, is shown in the table below.

a. What is the null hypothesis for the ANOVA of the boys' data?

b. Interpret the results of the test, part **a**. Use $\alpha = .05$.

c. Repeat parts **a** and **b** for the girls' data.

d. Summarize the results of the hypothesis tests in the words of the problem.

10.30 Most powerful American women. Refer to *Fortune* (Oct. 14, 2002) magazine's study of the most powerful women in America, Exercise 2.58 (p. 64). Recall that the data on age (in years) and title of each of the 50 women in the survey are stored in the **WPOWER50** file. (Some of the data are listed in the table on p. 535.) Suppose you want to compare the average ages of all powerful American women in three groups based on their position (title) within the firm: Group 1 (CEO, CFO, COO, or CRO);

	Sample Size	Youngest Tertile Mean Height	Middle Tertile Mean Height	Oldest Tertile Mean Height	F-Value	p-Value
Boys	1439	0.33	0.33	0.16	4.57	0.01
Girls	1409	0.27	0.18	0.21	0.85	0.43

Source: Wake, M., Coghlan, D., and Hesketh, K. "Does height influence progression through primary school grades?" *The Archives of Disease in Childhood*, Vol. 82, Apr. 2000 (Table 2).

MINITAB Output for Exercise 10.30
One-way ANOVA: AGE versus GROUP

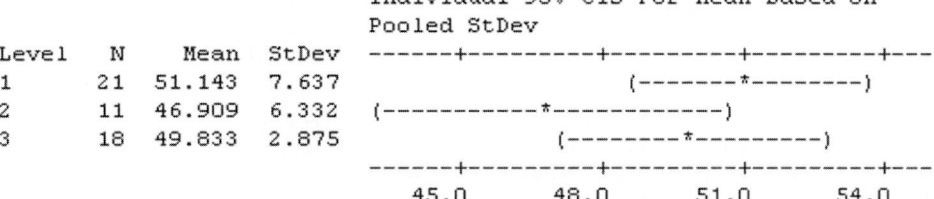

```
Source    DF      SS     MS     F      P
GROUP      2   129.6   64.8   1.78  0.179
Error     47  1708.0   36.3
Total     49  1837.6


S = 6.028    R-Sq = 7.05%    R-Sq(adj) = 3.10%



                              Individual 95% CIs For Mean Based on
                              Pooled StDev
Level   N    Mean    StDev   ------+---------+---------+---------+---
1      21  51.143   7.637                      (-------*--------)
2      11  46.909   6.332    (-----------*-------------)
3      18  49.833   2.875                (--------*---------)
                            ------+---------+---------+---------+---
                              45.0      48.0      51.0      54.0
```

Group 2 (Chairman or President); and, Group 3 (EVP, SVP, and Vice Chair).

a. Give the null and alternative hypothesis to be tested.

b. A MINITAB analysis of variance printout for the test, part **a**, is shown above. The sample means for the three groups are shown at the bottom of the printout. Why is it insufficient to make a decision about the null hypothesis based solely on these sample means?

c. Locate the test statistic and p-value on the printout. Use this information to make the appropriate conclusion at $\alpha = .10$.

d. Use the data in the **WPOWER50** file to determine whether the ANOVA assumptions are reasonably satisfied.

WPOWER50 (First 5 and last 2 observations)

Rank	Name	Age	Company	Title
1	Carly Fiorina	48	Hewlet-Packard	CEO
2	Betsy Holden	46	Kraft Foods	CEO
3	Meg Whitman	46	eBay	CEO
4	Indra Nooyi	46	PepsiCo	CFO
5	Andrea Jung	44	Avon Products	CEO
.	.	.		
.	.	.		
49	Fran Keeth	56	Royal Dutch Petrol.	CEO
50	Heidi Miller	49	Bank One	EVP

Source: Fortune, Oct. 14, 2002.

Applying the Concepts—Intermediate

10.31 Income and road rage. The phenomenon of road rage has received much media attention in recent years. Is a driver's propensity to engage in road rage related to his/her income? Researchers at Mississippi State University attempted to answer this question by conducting a survey of a representative sample of over 1,000 U.S. adult drivers. (*Accident Analysis and Prevention*, Vol. 34, 2002.) Based on how often each driver engaged in certain road rage behaviors (e.g., making obscene gestures

at, tailgating, and thinking about physically hurting another driver), a road rage score was assigned. (Higher scores indicate a greater pattern of road rage behavior.) The drivers were also grouped by annual income: under $30,000, between $30,000 and $60,000, and over $60,000. The data were subjected to an analysis of variance, with the results summarized in the table. Fully interpret the results. Is a driver's propensity to engage in road rage related to his/her income?

Income Group	Sample Size	Mean Road Rage Score
Under $30,000	379	4.60
$30,000 to $60,000	392	5.08
Over $60,000	267	5.15
ANOVA results:	F-value = 3.90	p-value < .01

10.32 Hair color and pain. Studies conducted at the University of Melbourne (Australia) indicate that there may be a difference between the pain thresholds of blondes and brunettes. Men and women of various ages were divided into four categories according to hair color: light blond, dark blond, light brunette, and dark brunette. The purpose of the experiment was to determine whether hair color is related to the amount of pain evoked by common types of mishaps and assorted types of trauma. Each person in the experiment was given a pain threshold score based on his or her performance in a pain sensitivity test (the higher the score, the higher the person's pain tolerance). The data are listed in the table.

HAIRPAIN

Light Blond	Dark Blond	Light Brunette	Dark Brunette
62	63	42	32
60	57	50	39
71	52	41	51
55	41	37	30
48	43		35

a. What type of experimental design appears to have been employed?

b. Conduct a test to determine whether the mean pain thresholds differ among people possessing the four types of hair color. Use $\alpha = .05$.

c. What is the observed significance level for the test in part **b**? Interpret it.

d. What assumptions must be met in order to ensure the validity of the inferences you made in part **b**?

10.33 Effect of scopolamine on memory. The drug scopolamine is often used as a sedative to induce sleep in patients. In *Behavioral Neuroscience* (Feb., 2004), medical researchers examined scopolamine's effects on memory for word-pair associates. A total of 28 human subjects, recruited from a university community, were given a list of related word pairs to memorize. For every word pair in the list (e.g., robber-jail), there was an associated word pair with the same first word but a different second word (e.g., robber-police). The subjects were then randomly divided into three treatment groups. Group 1 subjects were administered an injection of scopolamine, group 2 subjects were given an injection of glycopyrrolate (an active placebo), and group 3 subjects were not given any drug. Four hours later, subjects were shown 12 word pairs from the list and tested on how many of the associated word pairs they could recall. The data on number of pairs recalled (simulated based on summary information provided in the research article) are listed below. Prior to the analysis, the researchers theorized that the mean number of word pairs recalled for the scopolamine subjects (group 1) would be less than the corresponding means for the other two groups.

a. Explain why this is a completely randomized design.

b. Identify the treatments and response variable.

c. Find the sample means for the three groups. Is this sufficient information to support the researchers' theory? Explain.

d. Conduct the ANOVA *F*-test on the data. Is there sufficient evidence (at $\alpha = .05$) to conclude that the mean number of word pairs recalled differs among the three treatment groups?

⊚ SCOPOLAMINE

Group 1 (Scopolamine):	5 8 8 6 6 6 6 8 6 4 5 6
Group 2 (Placebo):	8 10 12 10 9 7 9 10
Group 3 (No drug):	8 9 11 12 11 10 12 12

10.34 The "name game." Psychologists at Lancaster University (United Kingdom) evaluated three methods of name retrieval in a controlled setting. (*Journal of Experimental Psychology-Applied*, June 2000.) A sample of 139 students was randomly divided into three groups, and each group of students used a different method to learn the names of the other students in the group. Group 1 used the "simple name game," where the first student states his/her full name, the second student announces his/her name and the name of the first student, the third student says his/her name and the names of the first two students, etc. Group 2 used the "elaborate name game,"

a modification of the simple name game where the students not only state their names but also their favorite activity (e.g., sports). Group 3 used "pairwise introductions," where students are divided into pairs and each student must introduce the other member of the pair. One year later, all subjects were sent pictures of the students in their group and asked to state the full name of each. The researchers measured the percentage of names recalled for each student respondent. The data (simulated based on summary statistics provided in the research article) are shown in the table. Conduct an analysis of variance to determine whether the mean percentages of names recalled differ for the three name retrieval methods. Use $\alpha = .05$.

⊚ NAMEGAME

Simple Name Game

24	43	38	65	35	15	44	44	18	27	0	38	50	31
7	46	33	31	0	29	0	0	52	0	29	42	39	26
51	0	42	20	37	51	0	30	43	30	99	39	35	19
24	34	3	60	0	29	40	40						

Elaborate Name Game

39	71	9	86	26	45	0	38	5	53	29	0	62	0
1	35	10	6	33	48	9	26	83	33	12	5	0	0
25	36	39	1	37	2	13	26	7	35	3	8	55	50

Pairwise Introductions

5	21	22	3	32	29	32	0	4	41	0	27	5	9
66	54	1	15	0	26	1	30	2	13	0	2	17	14
5	29	0	45	35	7	11	4	9	23	4	0	8	2
18	0	5	21	14									

Source: Morris, P. E., and Fritz, C. O. "The name game: Using retrieval practice to improve the learning of names." *Journal of Experimental Psychology—Applied*, Vol. 6, No. 2, June 2000 (data simulated from Figure 1).

10.35 Facial expression study. What do people infer from facial expressions of emotion? This was the research question of interest in an article published in the *Journal of Nonverbal Behavior* (Fall 1996). A sample of 36 introductory psychology students was randomly divided into six groups. Each group was assigned to view one of six slides showing a person making a facial expression.* The six expressions were (1) angry, (2) disgusted, (3) fearful, (4) happy, (5) sad, and (6) neutral faces. After viewing the slides, the students rated the degree of dominance they inferred from the facial expression (on a scale ranging from -15 to $+15$). The data (simulated from summary information provided in the article) are listed in the table on p. 537. Conduct an analysis of variance to determine whether the mean dominance ratings differ among the six facial expressions. Use $\alpha = .10$.

*In the actual experiment, each group viewed all six facial expression slides and the design employed was a Latin Square (beyond the scope of this text).

FACES

Angry	Disgusted	Fearful	Happy	Sad	Neutral
2.10	.40	.82	1.71	.74	1.69
.64	.73	−2.93	−.04	−1.26	−.60
.47	−.07	−.74	1.04	−2.27	−.55
.37	−.25	.79	1.44	−.39	.27
1.62	.89	−.77	1.37	−2.65	−.57
−.08	1.93	−1.60	.59	−.44	−2.16

Applying the Concepts—Advanced

10.36 Therapy for binge eaters. Do you experience episodes of excessive eating accompanied by being overweight? If so, you may suffer from binge eating disorder. Cognitive-behavioral therapy (CBT), in which patients are taught how to make changes in specific behavior patterns (e.g., exercise, eat only low-fat foods), can be effective in treating the disorder. A group of Stanford University researchers investigated the effectiveness of interpersonal therapy (IPT) as a second level of treatment for binge eaters. (*Journal of Consulting and Clinical Psychology*, June 1995.) The researchers employed a design that randomly assigned a sample of 41 overweight individuals diagnosed with binge eating disorder to either a treatment group (30 subjects) or a control group (11 subjects). Subjects in the treatment group received 12 weeks of cognitive-behavior therapy, then were subdivided into two groups. Those who responded successfully to CBT (17 subjects) were assigned to a weight loss therapy (WLT) program for the next 12 weeks. Those CBT subjects who did not respond to treatment (13 subjects) received 12 weeks of IPT. The subjects in the control group received no therapy of any type. Thus, the study ultimately consisted of three groups of overweight binge eaters: the CBT-WLT group, the CBT-IPT group, and the control group. One outcome (response) variable measured for each subject was the number of binge eating episodes per week, x. Summary statistics for each of the three groups at the end of the 24-week period are shown in the table. The data were analyzed as a completely randomized design with three treatments (CBT-WLT, CBT-IPT, and Control). Although the ANOVA tables were not provided in the article, sufficient information is provided in the table to reconstruct them.

	CBT-WLT	CBT-IPT	Control
Sample size	17	13	11
Mean number of binges per week	0.2	1.9	2.9
Standard deviation	0.4	1.7	2.0

Source: Agras, W. S., et al. "Does interpersonal therapy help patients with binge eating disorder who fail to respond to cognitive-behavioral therapy?" *Journal of Consulting and Clinical Psychology*, Vol. 63, No. 3, June 1995, p. 358 (Table 1).

a. Compute SST for the ANOVA using the formula on p. 000:

$$\text{SST} = \sum_{i=1}^{3} n_i(\bar{x}_i - \bar{x})^2$$

where $\bar{x}$ is the overall mean number of binges per week of all 41 subjects. [*Hint:* $\bar{x} = \left(\sum_{i=1}^{3} n_i\bar{x}_i\right)/41$.]

b. Recall that SSE for the ANOVA can be written as

$$\text{SSE} = (n_1 - 1)s_1^2 + (n_2 - 1)s_2^2 + (n_3 - 1)s_3^2$$

where s_1^2, s_2^2, and s_3^2 are the sample variances associated with the three treatments. Compute SSE for the ANOVA.

c. Use the results, parts **a** and **b**, to construct the ANOVA table.

d. Is there sufficient evidence (at $\alpha = .01$) of differences among the mean number of binges experienced per week by the subjects in the three groups?

e. Comment on the validity of the ANOVA assumptions. How might this impact the results of the study?

f. Comment on the use of a completely randomized design for this study. Have subjects been randomly and independently assigned to each group? How might this impact the results of the study?

10.3 Multiple Comparisons of Means

Consider a completely randomized design with three treatments, A, B, and C. Suppose we determine that the treatment means are statistically different via the ANOVA F-test of Section 10.2. To complete the analysis, we want to rank the three treatment means. As mentioned in Section 10.2, we start by placing confidence intervals on the difference between various pairs of treatment means in the experiment. In the three-treatment experiment, for example, we would construct confidence intervals for the following differences: $\mu_A - \mu_B$, $\mu_A - \mu_C$, and $\mu_B - \mu_C$.

> ### Determining the Number of Pairwise Comparisons of Treatment Means
>
> In general, if there are k treatment means, there are
>
> $$c = k(k - 1)/2$$
>
> pairs of means that can be compared.

If we want to have $100(1 - \alpha)\%$ confidence that each of the c confidence intervals contains the true difference it is intended to estimate, we must use a smaller value of α for each individual confidence interval than we would use for a single interval. For example, suppose we want to rank the means of the three treatments, A, B, and C, with 95% confidence that all three confidence intervals comparing the means contain the true differences between the treatment means. Then each individual confidence interval will need to be constructed using a level of significance smaller than $\alpha = .05$ in order to have 95% confidence that the three intervals collectively include the true differences.*

| Now Work | *Exercise 10.41* |

To make **multiple comparisons** of a set of treatment means we can use a number of procedures that, under various assumptions, ensure that the overall confidence level associated with all the comparisons remains at or above the specified $100(1 - \alpha)\%$ level. Three widely used techniques are the Bonferroni, Scheffé, and Tukey methods. For each of these procedures, the risk of making a Type I error applies to the comparisons of the treatment means in the experiment; thus, the value of α selected is called an **experimentwise error rate** (in contrast to a **comparisonwise error rate**).

The choice of a multiple comparisons method in ANOVA will depend on the type of experimental design used and the comparisons of interest to the analyst. For example, **Tukey** (1949) developed his procedure specifically for pairwise comparisons when the sample sizes of the treatments are equal. The **Bonferroni** method (see Miller, 1981), like the Tukey procedure, can be applied when pairwise comparisons are of interest; however, Bonferroni's method does not require equal sample sizes. **Scheffé** (1953) developed a more general procedure for comparing all possible linear combinations of treatment means (called *contrasts*). Consequently, when making pairwise comparisons, the confidence intervals produced by Scheffé's method will generally be wider than the Tukey or Bonferroni confidence intervals.

Biography

Carlo E. Bonferroni (1892–1960)—
Bonferroni Inequalities

During his childhood years in Turin, Italy, Carlo Bonferroni developed an aptitude for mathematics while studying music. He went on to obtain a degree in mathematics at the University of Turin. Bonferroni's first appointment as a professor of mathematics was at the University of Bari in 1923. Ten years later, he became chair of financial mathematics at the University of Florence, where he remained until his death. Bonferroni was a prolific writer, authoring over 65 research papers and books. His interest in statistics included various methods of calculating a mean and a correlation coefficient. Among statisticians, however, Bonferroni is most well known for developing his Bonferroni inequalities in probability theory in 1935. Later, other statisticians proposed using these inequalities for finding simultaneous confidence intervals, which led to the development of the Bonferroni multiple comparisons method in ANOVA. Bonferroni balanced these scientific accomplishments with his music, becoming an excellent pianist and composer.

---

*The reason each interval must be formed at a higher confidence level than that specified for the collection of intervals can be demonstrated as follows:

$$P \{\text{At least one of } c \text{ intervals fails to contain the true difference}\}$$
$$= 1 - P \{\text{All } c \text{ intervals contain the true differences}\}$$
$$= 1 - (1 - \alpha)^c \geq \alpha$$

Thus, to make this probability of at least one failure equal to α, we must specify the individual levels of significance to be less than α.

The formulas for constructing confidence intervals for differences between treatment means using the Tukey, Bonferroni, or Scheffé method are provided in Appendix B. However, since these procedures (and many others) are available in the ANOVA programs of most statistical software packages, we will use the computer to conduct the analysis. The programs generate a confidence interval for the difference between two treatment means for all possible pairs of treatments, based on the experimentwise error rate (α) selected by the analyst.

EXAMPLE 10.6   RANKING TREATMENT MEANS

Problem   Refer to the completely randomized design of Example 10.4, in which we concluded that at least two of the four brands of golf balls are associated with different mean distances traveled when struck with a driver.

    a. Use Tukey's multiple comparisons procedure to rank the treatment means with an overall confidence level of 95%.

    b. Estimate the mean distance traveled for balls manufactured by the brand with the highest rank.

Solution   **a.** To rank the treatment means with an overall confidence level of .95, we require the experimentwise error rate of $\alpha = .05$. The confidence intervals generated by Tukey's method appear at the bottom of the SAS ANOVA printout, Figure 10.12. Note that for any pair of means, μ_i and μ_j, SAS computes two confidence intervals—one for $(\mu_i - \mu_j)$ and one for $(\mu_j - \mu_i)$. Only one of these intervals is necessary to decide whether the means differ significantly.

     In this example, we have $k = 4$ brand means to compare. Consequently, the number of relevant pairwise comparisons—that is, the number of nonredundant confidence intervals—is $c = 4(3)/2 = 6$. These six intervals, highlighted in Figure 10.12, are given in Table 10.4.

     We are 95% confident that the intervals *collectively* contain all the differences between the true brand mean distances. Note that intervals that contain 0, such as the (Brand A–Brand D) interval from −4.08 to 7.00, do not support

Figure 10.12
SAS Multiple Comparisons
Printout for Example 10.6

```
                        The ANOVA Procedure

              Tukey's Studentized Range (HSD) Test for DISTANCE

      NOTE: This test controls the Type I experimentwise error rate.


              Alpha                                  0.05
              Error Degrees of Freedom                 36
              Error Mean Square                  21.17503
              Critical Value of Studentized Range 3.80880
              Minimum Significant Difference       5.5424


         Comparisons significant at the 0.05 level are indicated by ***.


                            Difference
              BRAND          Between        Simultaneous 95%
            Comparison        Means        Confidence Limits

            C  - B           8.890          3.348   14.432  ***
            C  - A          19.170         13.628   24.712  ***
            C  - D          20.630         15.088   26.172  ***
            B  - C          -8.890        -14.432   -3.348  ***
            B  - A          10.280          4.738   15.822  ***
            B  - D          11.740          6.198   17.282  ***
            A  - C         -19.170        -24.712  -13.628  ***
            A  - B         -10.280        -15.822   -4.738  ***
            A  - D           1.460         -4.082    7.002
            D  - C         -20.630        -26.172  -15.088  ***
            D  - B         -11.740        -17.282   -6.198  ***
            D  - A          -1.460         -7.002    4.082
```

TABLE 10.4   Pairwise Comparisons for Example 10.6

Brand Comparison	Confidence Interval
$(\mu_A - \mu_B)$	$(-15.82, -4.74)$
$(\mu_A - \mu_C)$	$(-24.71, -13.63)$
$(\mu_A - \mu_D)$	$(-4.08, 7.00)$
$(\mu_B - \mu_C)$	$(-14.43, -3.35)$
$(\mu_B - \mu_D)$	$(6.20, 17.28)$
$(\mu_C - \mu_D)$	$(15.09, 26.17)$

a conclusion that the true brand mean distances differ. If both endpoints of the interval are positive, as with the (Brand B–Brand D) interval from 6.20 to 17.28, the implication is that the first brand (B) mean distance exceeds the second (D). Conversely, if both endpoints of the interval are negative, as with the (Brand A–Brand C) interval from −24.71 to −13.63, the implication is that the second brand (C) mean distance exceeds the first brand (A) mean distance.

A convenient summary of the results of the Tukey multiple comparisons is a listing of the brand means from highest to lowest, with a solid line connecting those that are *not* significantly different. This summary is shown in Figure 10.13. The interpretation is that brand C's mean distance exceeds all others; brand B's mean exceeds that of brands A and D; and the means of brands A and D do not differ significantly. All these inferences are made with 95% confidence, the overall confidence level of the Tukey multiple comparisons.

Mean: $\overline{249.3 \quad 250.8}$ 261.1   270.0
Brand: D   A   B   C

Figure 10.13

Summary of Tukey Multiple Comparisons

b. Brand C is ranked highest; thus, we want a confidence interval for μ_C. Since the samples were selected independently in a completely randomized design, a confidence interval for an individual treatment mean is obtained with the one-sample t confidence interval of Section 7.2, using the standard deviation, $s = \sqrt{MSE}$, as the measure of sampling variability for the experiment. A 95% confidence interval on the mean distance traveled by brand C (apparently the "longest ball" of those tested), is

$$\overline{x}_C \pm t_{.025}s\sqrt{1/n}$$

where $n = 10, t_{.025} \approx 2$ (based on 36 degrees of freedom), and $s = 4.602$ (obtained from the MINITAB printout, Figure 10.7). Substituting, we obtain

$$270.0 \pm (2)(4.60)(\sqrt{.1})$$
$$270.0 \pm 2.9 \quad \text{or} \quad (267.1, 272.9)$$

Thus, we are 95% confident that the true mean distance traveled for brand C is between 267.1 and 272.9 yards, when hit with a driver by Iron Byron.

Look Back The easiest way to create a summary table like Figure 10.13 is to first list the treatment means in rank order. Begin with the largest mean and compare it to (in order), the second largest mean, the third largest mean, etc., by examining the appropriate confidence intervals shown on the computer printout. If a confidence interval contains 0, then connect the two means with a line. (These two means are not significantly different.) Continue in this manner by comparing the second largest mean with the third largest, fourth largest, etc., until all possible $c = (k)(k + 1)/2$ comparisons are made.

Now Work *Exercise 10.43*

■ ■ ■

Many of the available statistical software packages that have multiple comparisons routines will also produce the rankings shown in Figure 10.13. For example, the

Figure 10.14

Alternative SAS Multiple
Comparisons Output for
Example 10.6

```
                    The ANOVA Procedure

        Tukey's Studentized Range (HSD) Test for DISTANCE

NOTE: This test controls the Type I experimentwise error rate, but it generally has a higher
                    Type II error rate than REGWQ.


        Alpha                                    0.05
        Error Degrees of Freedom                   36
        Error Mean Square                    21.17503
        Critical Value of Studentized Range   3.80880
        Minimum Significant Difference         5.5424


    Means with the same letter are not significantly different.

        Tukey Grouping        Mean      N    BRAND

                     A      269.950     10    C

                     B      261.060     10    B

                     C      250.780     10    A
                     C
                     C      249.320     10    D
```

SAS printout in Figure 10.14 displays the ranking of the mean distances for the four golf ball brands using Tukey's method.

The experimentwise error rate (.05), MSE value (21.17503), and minimum significant difference (5.5424) used in the analysis are highlighted on the printout. The Tukey rankings of the means are displayed at the bottom of Figure 10.14. Instead of a solid line, note that SAS uses a "Tukey Grouping" letter to connect means that are not significantly different. You can see that the mean for brand C is ranked highest, followed by the mean for brand B. These two brand means are significantly different since they are associated with a different "Tukey Grouping" letter. Brands A and D are ranked lowest and their means are not significantly different (since they have the same "Tukey Grouping" letter).

Remember that the Tukey method—designed for comparing treatments with equal sample sizes—is just one of numerous multiple comparisons procedures available. Another technique may be more appropriate for the experimental design you employ. Consult the references for details on these other methods and when they should be applied. Guidelines for using the Tukey, Bonferroni, and Scheffé methods are given in the box.

Guidelines for Selecting a Multiple Comparisons Method in ANOVA

Method	Treatment Sample Sizes	Types of Comparisons
Tukey	Equal	Pairwise
Bonferroni	Equal or unequal	Pairwise
Scheffé	Equal or unequal	General contrasts

Note: For equal sample sizes and pairwise comparisons, Tukey's method will yield simultaneous confidence intervals with the smallest width and the Bonferroni intervals will have smaller widths than the Scheffé intervals.

Statistics in Action Revisited

Ranking the Means of the Cockroach Groups

Refer to the experiment designed to investigate the trail-following ability of German cockroaches. In the previous Statistics in the Action Revisited (p. 531), we applied a one-way ANOVA to the **ROACH** data and discovered statistically significant differences among the mean extract trail deviations for the four groups of cockroaches, adult males, adult females, gravid females, and nymphs. In order to determine which group has the highest degree of trail-following ability, we want to rank the population means from largest to smallest. That is, we want to follow up the ANOVA by conducting multiple comparisons of the four treatment means.

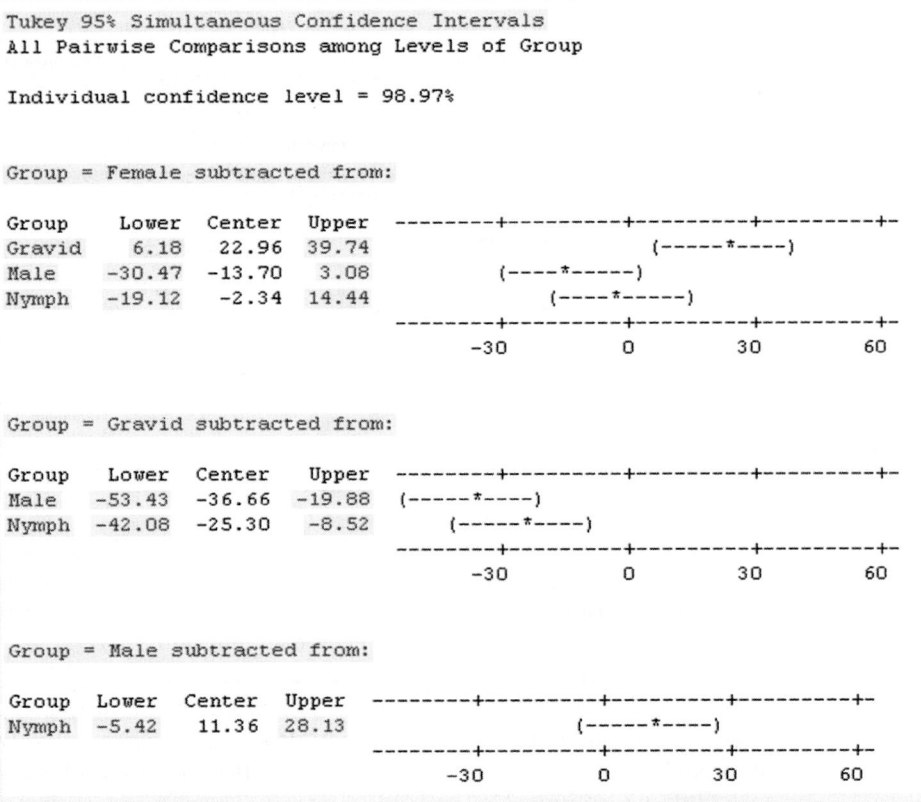

```
Tukey 95% Simultaneous Confidence Intervals
All Pairwise Comparisons among Levels of Group

Individual confidence level = 98.97%


Group = Female subtracted from:

Group     Lower   Center   Upper   --------+---------+---------+---------+-
Gravid     6.18    22.96   39.74                        (-----*----)
Male     -30.47   -13.70    3.08        (----*-----)
Nymph    -19.12    -2.34   14.44            (----*-----)
                                   --------+---------+---------+---------+-
                                         -30        0        30       60


Group = Gravid subtracted from:

Group     Lower   Center   Upper   --------+---------+---------+---------+-
Male     -53.43   -36.66  -19.88   (-----*----)
Nymph    -42.08   -25.30   -8.52       (-----*----)
                                   --------+---------+---------+---------+-
                                         -30        0        30       60


Group = Male subtracted from:

Group     Lower   Center   Upper   --------+---------+---------+---------+-
Nymph     -5.42    11.36   28.13                  (-----*----)
                                   --------+---------+---------+---------+-
                                         -30        0        30       60
```

Figure SIA10.2

MINITAB Multiple Comparisons of Extract Trail Deviation Means

Figure SIA10.2 is a MINITAB printout of the ANOVA and the multiple comparisons results. MINITAB uses Tukey's method to compare the means. The experimentwise confidence level (95%) is highlighted on the printout; also highlighted are the Tukey confidence intervals for all possible pairs of means. The information contained in these confidence intervals will enable us to rank the treatment (population) means.

For example, the confidence interval for $(\mu_{Gravid} - \mu_{Female})$ is (6.18, 39.74). Since the endpoints of the interval are both positive, the difference between the means is positive. This implies that the population mean trail deviation for gravid cockroaches is greater than the population mean for adult females (i.e., $\mu_{Gravid} > \mu_{Female}$).

Now consider the confidence interval for $(\mu_{Male} - \mu_{Female})$. The interval shown on the printout is $(-30.47, 3.08)$. Since the value 0 is included in the interval,

there is no evidence of a significant difference between the two treatment means. Similar interpretations are made for the confidence intervals for $(\mu_{Nymph} - \mu_{Female})$ and $(\mu_{Nymph} - \mu_{Male})$ since both these intervals contain 0.

Finally, note that the confidence intervals for $(\mu_{Male} - \mu_{Gravid})$ and $(\mu_{Nymph} - \mu_{Gravid})$ both have negative endpoints, implying that the differences between the means is negative. Thus, the population mean trail deviation for gravids is greater than either the population mean for adult males ($\mu_{Gravid} > \mu_{Male}$) or the population mean for nymphs ($\mu_{Gravid} > \mu_{Nymph}$).

The results of these multiple comparisons are summarized in Table SIA10.1. With an overall confidence level of .95, we conclude that gravid cockroaches have a mean extract trail deviation larger than any of the other three groups; there are no significant differences among adult males, adult females, or nymphs.

TABLE SIA10.1 Ranking of the Cockroach Group Means

Treatment Mean:	7.38	18.73	21.07	44.03
Cockroach Group:	Male	Nymph	Female	Gravid

Exercises 10.37–10.53

Understanding the Principles

10.37 Define an experimentwise error rate.

10.38 Define a comparisonwise error rate.

10.39 Confidence intervals for the difference between two means, $(\mu_1 - \mu_2)$, are shown below. For each part, which mean is significantly larger?
a. $(-10, 5)$ **b.** $(-10, -5)$ **c.** $(5, 10)$

10.40 Give a situation when it is most appropriate to apply Tukey's multiple comparisons of means method.

Learning the Mechanics

10.41 Consider a completely randomized design with p treatments. Assume all pairwise comparisons of treatment means are to be made using a multiple comparisons procedure. Determine the total number of treatment means to be compared for the following values of k.
a. $k = 3$ **b.** $k = 5$
c. $k = 4$ **d.** $k = 10$

10.42 Consider a completely randomized design with 5 treatments, A, B, C, D, and E. The ANOVA F-test revealed significant differences among the means. A multiple comparisons procedure was used to compare all possible pairs of treatment means at $\alpha = .05$. The ranking of the 5 treatment means is summarized below. Identify which pairs of means are significantly different.
a. $\overline{A \quad C} \quad \overline{E \quad B \quad D}$ **b.** $\overline{A \quad C \quad E} \quad B \quad D$
c. A $\overline{C \quad E \quad B} \quad D$ **d.** A $\overline{C \quad E \quad B \quad D}$

10.43 A multiple comparisons procedure for comparing four treatment means produced the confidence intervals shown below. Rank the means from smallest to largest. Which means are significantly different?
$(\mu_1 - \mu_2): (2, 15)$
$(\mu_1 - \mu_3): (4, 7)$
$(\mu_1 - \mu_4): (-10, 3)$
$(\mu_2 - \mu_3): (-5, 11)$
$(\mu_2 - \mu_4): (-12, -6)$
$(\mu_3 - \mu_4): (-8, -5)$

Applying the Concepts—Basic

10.44 A new dental bonding agent. Refer to the *Trends in Biomaterials & Artificial Organs* (Jan. 2003) study of a new bonding adhesive for teeth, Exercise 10.26 (p. 534). A completely randomized design was used to compare the mean breaking strengths of teeth bonded for three different bonding times: 1 hour, 24 hours, or 48 hours. The sample mean breaking strengths were $\overline{x}_{1 \text{ hour}} = 3.32$ Mpa, $\overline{x}_{24 \text{ hours}} = 5.07$ Mpa, and $\overline{x}_{48 \text{ hours}} = 5.03$ Mpa. Using an experimentwise error rate of .05, Tukey's method detected no significant difference between the means at 24 and 48 hours; however, the mean at 1 hour was found to be significantly smaller than the other two means.

a. Illustrate the results of the multiple comparisons of means by ordering the sample means and connecting means that are not significantly different.
b. What practical conclusions can you draw from the analysis?
c. Give a measure of reliability (i.e., overall confidence level) for the inferences drawn in part **b**.

10.45 Robots trained to behave like ants. Refer to the *Nature* (Aug. 2000) study of robots trained to behave like ants, Exercise 10.27 (p. 534). Multiple comparisons of mean energy expended for the four colony sizes were conducted using an experimentwise error rate of .05. The results are summarized below.

Sample mean:	.97	.95	.93	.80
Group size:	3	6	9	12

a. How many pairwise comparisons are conducted in this analysis?
b. Interpret the results shown in the table.

10.46 Safety of nuclear power plants. Refer to the *American Journal of Political Science* (Jan. 1998) study of the attitudes of three groups of professionals (scientists, journalists, and federal government policy-makers) regarding the safety of nuclear power plants, Exercise 10.28 (p. 534). The mean safety scores for the groups were

Government officials	4.2
Scientists	4.1
Journalists	3.7

a. Determine the number of pairwise comparisons of treatment means that can be made in this study.
b. Using an experimentwise error rate of $\alpha = .05$, Tukey's minimum significant difference for comparing means is .23. Use this information to conduct a multiple comparisons of the safety score means. Fully interpret the results.

10.47 Heights of grade school repeaters. Refer to *The Archives of Disease in Childhood* (Apr. 2000) comparison of the average heights of the three groups of Australian school children based on age, Exercise 10.29 (p. 534). The three height means for boys were ranked using the Bonferroni method at $\alpha = .05$. The results are summarized below. (Recall that all height measurements were standardized using z-scores.)

Sample mean:	0.16	0.33	0.33
Age group:	Oldest	Youngest	Middle

a. Is there a significant difference between the standardized height means for the oldest and youngest boys?
b. Is there a significant difference between the standardized height means for the oldest and middle-aged boys?

c. Is there a significant difference between the standardized height means for the youngest and middle-aged boys?

d. What is the experimentwise error rate for the inferences made in parts **a–c**? Interpret this value.

e. The researchers did not perform a Bonferroni analysis of the height means for the three groups of girls. Explain why not.

DDT

10.48 **Contaminated fish in the Tennessee River.** Refer to the U.S. Army Corps of Engineers data on contaminated fish saved in the **DDT** file. The results of an ANOVA and Tukey multiple comparisons to compare the three fish species (channel catfish, largemouth bass, and smallmouth buffalo) on the dependent variable length (in centimeters) are shown in the MINITAB printout below.

a. Is there sufficient evidence (at $\alpha = .01$) to conclude that the mean lengths differ among the three fish species?

b. What is the experimentwise error rate for the Tukey multiple comparisons?

c. Locate the confidence interval for the difference, $(\mu_{bass} - \mu_{catfish})$. Are the two means significantly different? If so, which mean is significantly larger?

d. Locate the confidence interval for the difference, $(\mu_{buffalo} - \mu_{catfish})$. Are the two means significantly different? If so, which mean is significantly larger?

e. Locate the confidence interval for the difference, $(\mu_{buffalo} - \mu_{bass})$. Are the two means significantly different? If so, which mean is significantly larger?

Applying the Concepts—Intermediate

10.49 **Dental fear study.** Does recalling a traumatic dental experience increase your level of anxiety at the dentist's office? In a study published in *Psychological Reports* (Aug. 1997), researchers at Wittenberg University randomly assigned 74 undergraduate psychology students to one of three experimental conditions. Subjects in the "Slide" condition viewed ten slides of scenes from a dental office. Subjects in the "Questionnaire" condition completed a full dental history questionnaire; one of the questions asked them to describe their worst dental experience. Subjects in the "Control" condition received no actual treatment. All students then completed the Dental Fear Scale, with scores ranging from 27 (no fear) to 135 (extreme fear). The sample dental fear means for the Slide,

MINITAB Output for Exercise 10.48

One-way ANOVA: LENGTH versus SPECIES

```
Source    DF      SS      MS      F       P
SPECIES    2   3533.1  1766.5  76.88   0.000
Error    141   3239.9    23.0
Total    143   6772.9

S = 4.794    R-Sq = 52.16%    R-Sq(adj) = 51.49%


Level              N     Mean    StDev
CHANNELCATFISH    96   44.729   4.581
LARGEMOUTHBASS    12   26.542   4.480
SMALLMOUTHBUFF    36   43.125   5.413

Pooled StDev = 4.794


Tukey 95% Simultaneous Confidence Intervals
All Pairwise Comparisons among Levels of SPECIES

Individual confidence level = 98.08%

SPECIES = CHANNELCATFISH subtracted from:

SPECIES            Lower    Center    Upper
LARGEMOUTHBASS   -21.664   -18.188  -14.711
SMALLMOUTHBUFF    -3.823    -1.604    0.615

SPECIES = LARGEMOUTHBASS subtracted from:

SPECIES          Lower   Center   Upper
SMALLMOUTHBUFF  12.798   16.583  20.368
```

Questionnaire, and Control groups were reported as 43.1, 53.8, and 41.8, respectively.

a. A completely randomized design ANOVA was carried out on the data, with the following results: $F = 4.43$, p-value $< .05$. Interpret these results.

b. According to the article, a Bonferroni ranking of the three dental fear means (at $\alpha = .05$) "indicated a significant difference between the mean scores on the Dental Fear Scale for the Control and Questionnaire groups, but not for the means between the Control and Slide groups." Summarize these results in a chart similar to Figure 10.13 (p. 540).

10.50 Income and road rage. Refer to the *Accident Analysis and Prevention* (Vol. 34, 2002) study of road rage, Exercise 10.31 (p. 535). Recall that the mean road rage scores of drivers in the three income groups, Under $30 thousand, Between $30 and $60 thousand, and Over $60 thousand, were 4.60, 5.08, and 5.15, respectively.

a. An experimentwise error rate of .01 was used to rank the three means. Give a practical interpretation of this error rate.

b. How many pairwise comparisons are necessary to compare the three means? List them.

c. A multiple comparisons procedure revealed that the means for the two income groups, Between $30 and $60 thousand and Over $60 thousand, were not significantly different. All other pairs of means were found to be significantly different. Summarize these results in table form.

d. Which of the comparisons of part **b** will yield a confidence interval that does not contain 0?

⊙ SCOPOLAMINE

10.51 Effect of scopolamine on memory. Refer to the *Behavioral Neuroscience* (Feb. 2004) study of the drug scopolamine's effects on memory for word-pair associates, Exercise 10.33 (p. 536). Recall that the researchers theorized that the mean number of word pairs recalled for the scopolamine subjects (group 1) would be less than the corresponding means for the placebo subjects (group 2) and the no drug subjects (group 3). Conduct multiple comparisons of the three means (using an experimentwise error rate of .05). Do the results support the researchers' theory? Explain.

10.52 Refer to the *Journal of Nonverbal Behavior* study of students' evaluations of facial expressions, Exercise 10.35 (p. 536). Use a statistical software package with Tukey's method to rank the dominance rating means of the six facial expressions. (Use an experimentwise error rate of $\alpha = .10$.)

Applying the Concepts—Advanced

10.53 Therapy for binge eaters. Refer to the *Journal of Consulting and Clinical Psychology* study of binge eaters, Exercise 10.36 (p. 537). You found evidence of a difference among the treatment means for the three treatments, CBT-WT, CBT-IPT, and Control. Conduct a Bonferroni analysis to rank the three treatment means. Use an experimentwise error rate of $\alpha = .03$. Interpret the results for the researchers. [*Hint:* As shown in Appendix B, the Bonferroni formula for a confidence interval for the difference $(\mu_i - \mu_j)$ is

$$(\bar{x}_i - \bar{x}_j) \pm t_{\alpha*/2}(s)\sqrt{(1/n_i) + (1/n_j)}$$

where $\alpha* = 2\alpha/[(k)(k-1)]$, α is the experimentwise error rate, and k is the total number of treatment means compared.]

10.4 The Randomized Block Design

If the completely randomized design results in nonrejection of the null hypothesis that the treatment means differ because the sampling variability (as measured by MSE) is large, we may want to consider an experimental design that better controls the variability. In contrast to the selection of independent samples of experimental units specified by the completely randomized design, the *randomized block design* utilizes experimental units that are *matched sets*, assigning one from each set to each treatment. The matched sets of experimental units are called *blocks*. The theory behind the randomized block design is that the sampling variability of the experimental units in each block will be reduced, in turn reducing the measure of error, MSE.

> **DEFINITION 10.8**
>
> The **randomized block design** consists of a two-step procedure:
>
> **1.** Matched sets of experimental units, called **blocks**, are formed, each block consisting of k experimental units (where k is the number of treatments). The b blocks should consist of experimental units that are as similar as possible.
>
> **2.** One experimental unit from each block is randomly assigned to each treatment, resulting in a total of $n = bk$ responses.

TABLE 10.5   Randomized Block Design: SAT Score Comparison

Block	Female SAT Score	Male SAT Score	Block Mean
1 (School A, 2.75 GPA)	540	530	535
2 (School B, 3.00 GPA)	570	550	560
3 (School C, 3.25 GPA)	590	580	585
4 (School D, 3.50 GPA)	640	620	630
5 (School E, 3.75 GPA)	690	690	690
Treatment mean	606	594	

For example, if we wish to compare SAT scores of female and male high school seniors, we could select independent random samples of five females and five males, and analyze the results of the completely randomized design as outlined in Section 10.2. Or, we could select matched pairs of females and males according to their scholastic records, and analyze the SAT scores of the pairs. For instance, we could select pairs of students with approximately the same GPAs from the same high school. Five such pairs (blocks) are depicted in Table 10.5. Note that this is just a *paired difference experiment*, first discussed in Section 9.3.

As before, the variation between the treatment means is measured by squaring the distance between each treatment mean and the overall mean, multiplying each squared distance by the number of measurements for the treatment, and summing over treatments:

$$\text{SST} = \sum_{i=1}^{k} b(\overline{x}_{T_i} - \overline{x})^2$$
$$= 5(606 - 600)^2 + 5(594 - 600)^2 = 360$$

where $\overline{x}_{T_i}$ represents the sample mean for the ith treatment, b (the number of blocks) is the number of measurements for each treatment, and k is the number of treatments.

The blocks also account for some of the variation among the different responses. That is, just as SST measures the variation between the female and male means, we can calculate a measure of variation among the five block means representing different schools and scholastic abilities. Analogous to the computation of SST, we sum the squares of the differences between each block mean and the overall mean, multiplying each squared difference by the number of measurements for each block, and sum over blocks to calculate the **Sum of Squares for Blocks** (SSB):

$$\text{SSB} = \sum_{i=1}^{b} k(\overline{x}_{B_i} - \overline{x})^2$$
$$= 2(535 - 600)^2 + 2(560 - 600)^2 + 2(585 - 600)^2 +$$
$$2(630 - 600)^2 + 2(690 - 600)^2$$
$$= 30{,}100$$

where $\overline{x}_{B_i}$ represents the sample mean for the ith block and k (the number of treatments) is the number of measurements in each block. As we expect, the variation in SAT scores attributable to Schools and Levels of scholastic achievement is apparently large.

As before, we want to compare the variability attributed to treatments with that which is attributed to sampling variability. In a randomized block design, the sampling variability is measured by subtracting that portion attributed to treatments and blocks from the Total Sum of Squares, SS(Total). The total variation is the sum of squared differences of each measurement from the overall mean:

$$
\begin{aligned}
\text{SS(Total)} &= \sum_{i=1}^{n}(x_i - \bar{x})^2 \\
&= (540 - 600)^2 + (530 - 600)^2 + (570 - 600)^2 + (550 - 600)^2 + \\
&\quad \cdots + (690 - 600)^2 \\
&= 30{,}600
\end{aligned}
$$

Then the variation attributable to sampling error is found by subtraction:

$$
\text{SSE} = \text{SS(Total)} - \text{SST} - \text{SSB} = 30{,}600 - 360 - 30{,}100 = 140
$$

In summary, the Total Sum of Squares, 30,600, is divided into three components: 360 attributed to treatments (Gender), 30,100 attributed to blocks (Scholastic ability), and 140 attributed to sampling error.

The mean squares associated with each source of variability are obtained by dividing the sum of squares by the appropriate number of degrees of freedom. The partitioning of the Total Sum of Squares and the total degrees of freedom for a randomized block experiment is summarized in Figure 10.15.

To determine whether we can reject the null hypothesis that the treatment means are equal in favor of the alternative that at least two of them differ, we calculate

$$
\text{MST} = \frac{\text{SST}}{k-1} = \frac{360}{2-1} = 360
$$

$$
\text{MSE} = \frac{\text{SSE}}{n-b-k+1} = \frac{140}{10-5-2+1} = 35
$$

The F-ratio that is used to test the hypothesis is

$$
F = \frac{360}{35} = 10.29
$$

Figure 10.15

Partitioning of the Total Sum of Squares for the Randomized Block Design

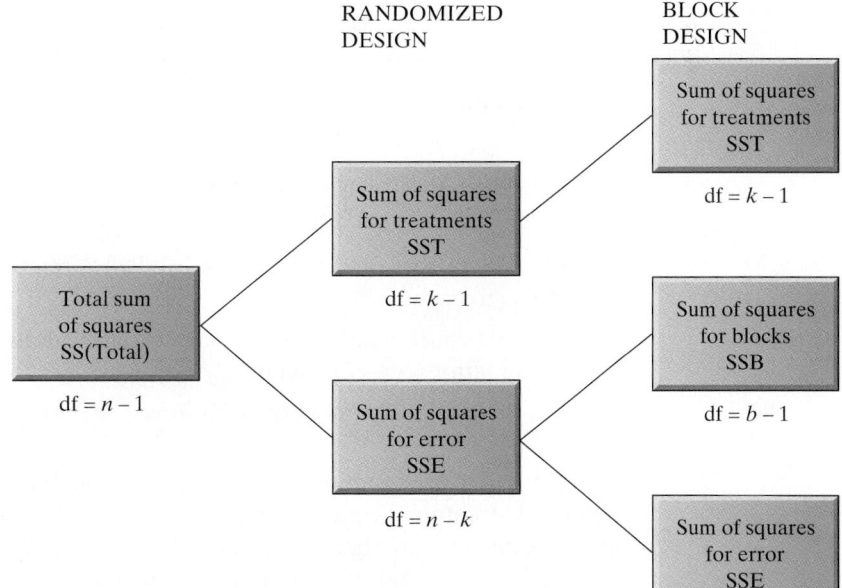

Comparing this ratio to the tabled F-value corresponding to $\alpha = .05$, $v_1 = (k - 1) = 1$ degree of freedom in the numerator, and $v_2 = (n - b - k + 1) = 4$ degrees of freedom in the denominator, we find that

$$F = 10.29 > F_{.05} = 7.71$$

which indicates that we should reject the null hypothesis and conclude that the mean SAT scores differ for females and males.

If you review Section 9.3, you will find that the analysis of a paired difference experiment results in a one-sample t-test on the differences between the treatment responses within each block. Applying the procedure to the differences between female and male scores in Table 10.5, we find

$$t = \frac{\bar{x}_D}{s_D / \sqrt{n_D}} = \frac{12}{\sqrt{70} / \sqrt{5}} = 3.207$$

At the .05 level of significance with $(n_D - 1) = 4$ degrees of freedom,

$$t = 3.207 > t_{.025} = 2.776$$

Since $t^2 = (3.207)^2 = 10.29$ and $t_{.025}^2 = (2.776)^2 = 7.71$, we find that the paired difference t-test and the ANOVA F-test are equivalent, with both the calculated test statistics and the rejection region related by the formula $F = t^2$. The difference between the tests is that the paired difference t-test can be used to compare only two treatments in a randomized block design, whereas the F-test can be applied to *two or more* treatments in a randomized block design. The F-test is summarized in the following box.

ANOVA F-Test to Compare k Treatment Means: Randomized Block Design

H_0: $\mu_1 = \mu_2 = \cdots = \mu_p$

H_a: At least two treatment means differ

Test statistic: $F = \dfrac{\text{MST}}{\text{MSE}}$

Rejection region: $F > F_\alpha$, where F_α is based on $(k - 1)$ numerator degrees of freedom and $(n - b - k + 1)$ denominator degrees of freedom.

Conditions Required for a Valid ANOVA F-Test: Randomized Block Design

1. The b blocks are randomly selected and all k treatments are applied (in random order) to each block.

2. The distributions of observations corresponding to all bk block–treatment combinations are approximately normal.

3. The bk block–treatment distributions have equal variances.

Note that the assumptions concern the probability distributions associated with each block–treatment combination. The experimental unit selected for each combination is assumed to have been randomly selected from all possible experimental units for that combination, and the response is assumed to be normally distributed with the same variance for each of the block–treatment combinations. For example, the F-test comparing female and male SAT score means requires the scores for each combination of gender and scholastic ability (e.g., females with 3.25 GPA) to be normally distributed with the same variance as the other combinations employed in the experiment.

The calculation formulas for randomized block designs are given in Appendix C. We will rely on statistical software packages to analyze randomized block designs and to obtain the necessary ingredients for testing the null hypothesis that the treatment means are equal.

EXAMPLE 10.7

EXPERIMENTAL DESIGN PRINCIPLES

Problem   Refer to Examples 10.4–10.6. Suppose the USGA wants to compare the mean distances associated with the four brands of golf balls when struck by a driver, but wishes to employ human golfers rather than the robot, Iron Byron. Assume that 10 balls of each brand are to be utilized in the experiment.

  a. Explain how a completely randomized design could be employed.

  b. Explain how a randomized block design could be employed.

  c. Which design is likely to provide more information about the differences among the brand mean distances?

Solution   **a.** Since the completely randomized design calls for independent samples, we can employ such a design by randomly selecting 40 golfers and then randomly assigning 10 golfers to each of the four brands. Finally, each golfer will strike the ball of the assigned brand, and the distance will be recorded. The design is illustrated in Figure 10.16a.

  b. The randomized block design employs blocks of relatively homogeneous experimental units. For example, we could randomly select 10 golfers, and permit each golfer to hit four balls, one of each brand, in a random sequence.

Figure 10.16

Illustration of Completely Randomized Design and Randomized Block Design: Comparison of Four Golf Ball Brands

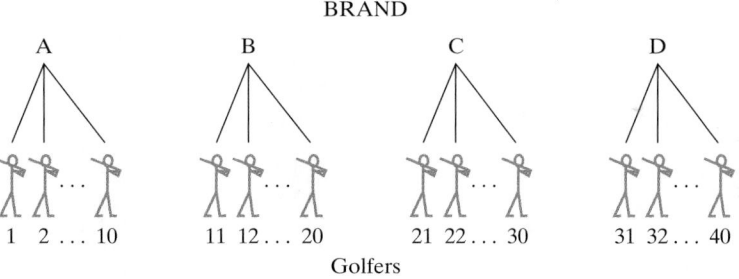

a. Completely randomized design

b. Randomized block design

Then each golfer is a block, with each treatment (brand) assigned to each block (golfer). The design is summarized in Figure 10.16b.

c. Because we expect much more variability among distances generated by "real" golfers than by Iron Byron, we would expect the randomized block design to control the variability better than the completely randomized design. That is, with 40 different golfers, we would expect the sampling variability among the measured distances within each brand to be greater than that among the four distances generated by each of 10 golfers hitting one ball of each brand.

Now Work   *Exercise 10.63a,b*

■ ■ ■

EXAMPLE 10.8

AN ANOVA FOR A RANDOMIZED BLOCK DESIGN

Problem   Refer to Example 10.7. Suppose the randomized block design of part **b** is employed, utilizing a random sample of 10 golfers with each golfer using a driver to hit four balls, one of each brand, in a random sequence.

a. Set up a test of the research hypothesis that the brand mean distances differ. Use $\alpha = .05$.

b. The data for the experiment are given in Table 10.6. Use statistical software to analyze the data, and conduct the test set up in part **a**.

⊚ **GOLFRBD**

TABLE 10.6   Distance Data for Randomized Block Design

Golfer (Block)	Brand A	Brand B	Brand C	Brand D
1	202.4	203.2	223.7	203.6
2	242.0	248.7	259.8	240.7
3	220.4	227.3	240.0	207.4
4	230.0	243.1	247.7	226.9
5	191.6	211.4	218.7	200.1
6	247.7	253.0	268.1	244.0
7	214.8	214.8	233.9	195.8
8	245.4	243.6	257.8	227.9
9	224.0	231.5	238.2	215.7
10	252.2	255.2	265.4	245.2
Sample means	227.0	233.2	245.3	220.7

Solution   a. We want to test whether the data in Table 10.6 provide sufficient evidence to conclude that the brand mean distances differ. Denoting the population mean of the ith brand by μ_i, we test

$$H_0: \mu_1 = \mu_2 = \mu_3 = \mu_4$$
$$H_a: \text{The mean distances differ for at least two of the brands}$$

The test statistic compares the variation among the four treatment (brand) means to the sampling variability within each of the treatments.

$$\text{Test statistic:} \quad F = \frac{\text{MST}}{\text{MSE}}$$

Rejection region: $F > F_\alpha = F_{.05}$, with $v_1 = (k - 1) = 3$ numerator degrees of freedom and $v_2 = (n - k - b + 1) = 27$ denominator degrees of freedom.

Figure 10.17

SPSS Printout for
Randomized Block
Design ANOVA of
Data in Table 10.6

Tests of Between-Subjects Effects

Dependent Variable: DISTANCE

Source	Type III Sum of Squares	df	Mean Square	F	Sig.
Corrected Model	15372.539ª	12	1281.045	63.276	.000
Intercept	2145032.91	1	2145032.910	105952.6	.000
BRAND	3298.657	3	1099.552	54.312	.000
GOLFER	12073.882	9	1341.542	66.265	.000
Error	546.621	27	20.245		
Total	2160952.07	40			
Corrected Total	15919.160	39			

a. R Squared = .966 (Adjusted R Squared = .950)

From Table IX of Appendix A, we find $F_{.05} = 2.96$. Thus, we will reject H_0 if $F > 2.96$.

The assumptions necessary to ensure the validity of the test are as follows: (1) The probability distributions of the distances for each brand–golfer combination are normal. (2) The variances of the distance probability distributions for each brand–golfer combination are equal.

b. SPSS was used to analyze the data in Table 10.6, and the result is shown in Figure 10.17. The values of MST and MSE (highlighted on the printout) are 1,099.552 and 20.245, respectively. The F-ratio for Brand (also highlighted on the printout) is $F = 54.312$, which exceeds the tabled value of 2.96. We therefore reject the null hypothesis at the $\alpha = .05$ level of significance, concluding that at least two of the brands differ with respect to mean distance traveled when struck by the driver.

Look Back The result of part **b** is confirmed by noting that the observed significance level of the test, highlighted on the printout, is $p \approx 0$.

Now Work *Exercise 10.63c*

■ ■ ■

The results of an ANOVA can be summarized in a simple tabular format, similar to that utilized for the completely randomized design in Section 10.2. The general form of the table is shown in Table 10.7, and that for Example 10.8 is given in Table 10.8. Note that the randomized block design is characterized by three sources of variation—Treatments, Blocks, and Error—which sum to the Total Sum of Squares. We hope that employing blocks of experimental units will reduce the error variability, thereby making the test for comparing treatment means more powerful.

TABLE 10.7   General ANOVA Summary Table for a Randomized Block Design

Source	df	SS	MS	F
Treatment	$k - 1$	SST	MST	MST/MSE
Block	$b - 1$	SSB	MSB	
Error	$n - k - b + 1$	SSE	MSE	
Total	$n - 1$	SS(Total)		

TABLE 10.8   ANOVA Table for Example 10.8

Source	df	SS	MS	F	p
Treatment (Brand)	3	3,298.7	1,099.6	54.31	.000
Block (Golfer)	9	12,073.9	1,341.5		
Error	27	546.6	20.2		
Total	39	15,919.2			

When the F-test results in the rejection of the null hypothesis that the treatment means are equal, we will usually want to compare the various pairs of treatment means to determine which specific pairs differ. We can employ a multiple comparisons procedure as in Section 10.3. The number of pairs of means to be compared will again be $c = k(k - 1)/2$, where p is the number of treatment means. In Example 10.8, $c = 4(3)/2 = 6$; that is, there are six pairs of golf ball brand means to be compared.

EXAMPLE 10.9   RANKING MEANS IN A RANDOMIZED BLOCK DESIGN

Problem   Bonferroni's procedure is used to compare the mean distances of the four golf ball brands in Example 10.8. The resulting confidence intervals, with an experimentwise error rate of $\alpha = .05$, are shown in the SPSS printout, Figure 10.18. Interpret the results.

Solution   Note that 12 confidence intervals are shown in Figure 10.18 rather than 6. SPSS (like SAS) computes intervals for both $\mu_i - \mu_j$ and $\mu_j - \mu_i, i \neq j$. Only half of these are necessary to conduct the analysis, and these are highlighted on the printout. The intervals (rounded) are summarized below:

$$
\begin{array}{ll}
(\mu_A - \mu_B): & (-11.9, -.4) \\
(\mu_A - \mu_C): & (-24.0, -12.6) \\
(\mu_A - \mu_D): & (.6, 12.0) \\
(\mu_B - \mu_C): & (-17.9, -6.4) \\
(\mu_B - \mu_D): & (6.7, 18.2) \\
(\mu_C - \mu_D): & (18.9, 30.3)
\end{array}
$$

Figure 10.18
SPSS Printout of Bonferroni Confidence Intervals for the Randomized Block Design

Multiple Comparisons

Dependent Variable: DISTANCE
Bonferroni

(I) BRAND	(J) BRAND	Mean Difference (I-J)	Std. Error	Sig.	95% Confidence Interval Lower Bound	95% Confidence Interval Upper Bound
A	B	-6.1300*	2.01222	.031	-11.8586	-.4014
	C	-18.2800*	2.01222	.000	-24.0086	-12.5514
	D	6.3200*	2.01222	.024	.5914	12.0486
B	A	6.1300*	2.01222	.031	.4014	11.8586
	C	-12.1500*	2.01222	.000	-17.8786	-6.4214
	D	12.4500*	2.01222	.000	6.7214	18.1786
C	A	18.2800*	2.01222	.000	12.5514	24.0086
	B	12.1500*	2.01222	.000	6.4214	17.8786
	D	24.6000*	2.01222	.000	18.8714	30.3286
D	A	-6.3200*	2.01222	.024	-12.0486	-.5914
	B	-12.4500*	2.01222	.000	-18.1786	-6.7214
	C	-24.6000*	2.01222	.000	-30.3286	-18.8714

Based on observed means.
*. The mean difference is significant at the .05 level.

Figure 10.19

Listing of Brand Means for Randomized Block Design [*Note:* All differences are statistically significant.]

Mean:	220.7	227.0	233.2	245.3
Brand:	D	A	B	C

Note that we are 95% confident that all the brand means differ because none of the intervals contains 0. The listing of the brand means in Figure 10.19 has no lines connecting them because there are no nonsignificant differences at the .05 level.

Now Work *Exercise 10.63d*

■ ■ ■

Unlike the completely randomized design, the randomized block design cannot, in general, be used to estimate individual treatment means. Whereas the completely randomized design employs a random sample for each treatment, the randomized block design does not necessarily employ a random sample of experimental units for each treatment. The experimental units within the blocks are assumed to be randomly selected, but the blocks themselves may not be randomly selected.

We can, however, test the hypothesis that the block means are significantly different. We simply compare the variability attributable to differences among the block means to that associated with sampling variability. The ratio of MSB to MSE is an *F*-ratio similar to that formed in testing treatment means. The *F*-statistic is compared to a tabled value for a specific value of α, with numerator degrees of freedom $(b - 1)$ and denominator degrees of freedom $(n - k - b + 1)$. The test is usually given on the same printout as the test for treatment means. Refer to the SPSS printout in Figure 10.17, and note that the test statistic for comparing the block means is

$$F = \frac{\text{MSB}}{\text{MSE}} = \frac{\text{MS(Golfers)}}{\text{MS(Error)}} = \frac{1{,}341.5}{20.2} = 66.26$$

with a *p*-value of .000. Since $\alpha = .05$ exceeds this *p*-value, we conclude that the block means are different. The results of the test are summarized in Table 10.9.

In the golf example, the test for block means confirms our suspicion that the golfers vary significantly; therefore, the use of the block design was a good decision. However, be careful not to conclude that the block design was a mistake if the *F*-test for blocks does not result in rejection of the null hypothesis that the block means are the same. Remember that the possibility of a Type II error exists, and we are not controlling its probability as we are the probability α of a Type I error. If the experimenter believes that the experimental units are more homogeneous within blocks than between blocks, then he or she should use the

TABLE 10.9 ANOVA Table for Randomized Block Design: Test for Blocks Included

Source	df	SS	MS	F	p
Treatments (Brands)	3	3,298.7	1,099.6	54.31	.000
Blocks (Golfers)	9	12,073.9	1,341.5	66.26	.000
Error	27	546.6	20.2		
Total	39	15,919.2			

randomized block design regardless of the results of a single test comparing the block means.

The procedure for conducting an analysis of variance for a randomized block design is summarized in the next box. Remember that the hallmark of this design is the utilization of blocks of homogeneous experimental units in which each treatment is represented.

Steps for Conducting an ANOVA for a Randomized Block Design

1. Be sure the design consists of blocks (preferably, blocks of homogeneous experimental units) and that each treatment is randomly assigned to one experimental unit in each block.

2. If possible, check the assumptions of normality and equal variances for all block–treatment combinations. [*Note:* This may be difficult to do since the design will likely have only one observation for each block–treatment combination.]

3. Create an ANOVA summary table that specifies the variability attributable to Treatments, Blocks, and Error, and that leads to the calculation of the *F*-statistic to test the null hypothesis that the treatment means are equal in the population. Use a statistical software package or the calculation formulas in Appendix C to obtain the necessary numerical ingredients.

4. If the *F*-test leads to the conclusion that the means differ, use the Bonferroni, Tukey, or similar procedure to conduct multiple comparisons of as many of the pairs of means as you wish. Use the results to summarize the statistically significant differences among the treatment means. Remember that, in general, the randomized block design cannot be used to form confidence intervals for individual treatment means.

5. If the *F*-test leads to the nonrejection of the null hypothesis that the treatment means are equal, several possibilities exist:
   a. The treatment means are equal—that is, the null hypothesis is true.
   b. The treatment means really differ, but other important factors affecting the response are not accounted for by the randomized block design. These factors inflate the sampling variability, as measured by MSE, resulting in smaller values of the *F*-statistic. Either increase the sample size for each treatment, or conduct an experiment that accounts for the other factors affecting the response (as in Section 10.5). Do not automatically reach the former conclusion, since the possibility of a Type II error must be considered if you accept H_0.

6. If desired, conduct the *F*-test of the null hypothesis that the block means are equal. Rejection of this hypothesis lends statistical support to the utilization of the randomized block design.

Note: It is often difficult to check whether the assumptions for a randomized block design are satisfied. When you feel these assumptions are likely to be violated, a nonparametric procedure is advisable.

What Do You Do When the Assumptions Are Not Satisfied for the Analysis of Variance for a Completely Randomized Design?

Answer: Use a nonparametric statistical method such as the Friedman F_r test of Section 14.5.

Exercises 10.54–10.69

Understanding the Principles

10.54 When is it advantageous to use a randomized block design over a completely randomized design?

10.55 Explain the difference between a randomized block design and a paired difference experiment.

10.56 What conditions are required for valid inferences from a randomized block design ANOVA?

Learning the Mechanics

10.57 A randomized block design yielded the ANOVA table shown here.

Source	df	SS	MS	F
Treatments	4	501	125.25	9.109
Blocks	2	225	112.50	8.182
Error	8	110	13.75	
Total	14	836		

a. How many blocks and treatments were used in the experiment?

b. How many observations were collected in the experiment?

c. Specify the null and alternative hypotheses you would use to compare the treatment means.

d. What test statistic should be used to conduct the hypothesis test of part **c**?

e. Specify the rejection region for the test of parts **c** and **d**. Use $\alpha = .01$.

f. Conduct the test of parts **c–e**, and state the proper conclusion.

g. What assumptions are necessary to ensure the validity of the test you conducted in part **f**?

10.58 An experiment was conducted using a randomized block design. The data from the experiment are displayed in the following table.

⊙ **LM10_58**

		Block	
Treatment	1	2	3
1	2	3	5
2	8	6	7
3	7	6	5

a. Fill in the missing entries in the ANOVA table.

Source	df	SS	MS	F
Treatments		21.5555		
Blocks				
Error				
Total		30.2222		

b. Specify the null and alternative hypotheses you would use to investigate whether a difference exists among the treatment means.

c. What test statistic should be used in conducting the test of part **b**?

d. Describe the Type I and Type II errors associated with the hypothesis test of part **b**.

e. Conduct the hypothesis test of part **b** using $\alpha = .05$.

10.59 A randomized block design was used to compare the mean responses for three treatments. Four blocks of three homogeneous experimental units were selected, and each treatment was randomly assigned to one experimental unit within each block. The data are shown in the table; a MINITAB ANOVA printout for this experiment is displayed on p. 556.

⊙ **LM10_59**

		Block		
Treatment	1	2	3	4
A	3.4	5.5	7.9	1.3
B	4.4	5.8	9.6	2.8
C	2.2	3.4	6.9	.3

a. Use the printout to fill in the entries in the following ANOVA table.

Source	df	SS	MS	F
Treatments				
Blocks				
Error				
Total				

b. Do the data provide sufficient evidence to indicate that the treatment means differ? Use $\alpha = .05$.

c. Do the data provide sufficient evidence to indicate that blocking was effective in reducing the experimental error? Use $\alpha = .05$.

d. Use the printout to rank the treatment means at $\alpha = .05$.

e. What assumptions are necessary to ensure the validity of the inferences made in parts **b, c**, and **d**?

10.60 Suppose an experiment utilizing a randomized block design has four treatments and nine blocks, for a total of $4 \times 9 = 36$ observations. Assume that the Total Sum of Squares for the response is $SS(Total) = 500$. For each of the following partitions of $SS(Total)$, test the null hypothesis that the treatment means are equal, and the null hypothesis that the block means are equal. Use $\alpha = .05$ for each test.

a. The Sum of Squares for Treatments (SST) is 20% of $SS(Total)$, and the Sum of Squares for Blocks (SSB) is 30% of $SS(Total)$.

b. SST is 50% of $SS(Total)$, and SSB is 20% of $SS(Total)$.

c. SST is 20% of $SS(Total)$, and SSB is 50% of $SS(Total)$.

MINITAB Output for Exercise 10.59

Two-way ANOVA: X versus Treatment, Block

```
Source      DF      SS       MS       F      P
Treatment    2  12.0317   6.0158   50.96  0.000
Block        3  71.7492  23.9164  202.59  0.000
Error        6   0.7083   0.1181
Total       11  84.4892


S = 0.3436   R-Sq = 99.16%   R-Sq(adj) = 98.46%


Tukey 95.0% Simultaneous Confidence Intervals
Response Variable X
All Pairwise Comparisons among Levels of Treatment
Treatment = A  subtracted from:


Treatment   Lower  Center   Upper   -+---------+---------+---------+-----
B           0.379   1.125  1.8706                                (----*---)
C          -2.071  -1.325 -0.5794                   (----*----)
                                    -+---------+---------+---------+-----
                                  -3.0      -1.5       0.0       1.5




Treatment = B  subtracted from:


Treatment   Lower  Center   Upper   -+---------+---------+---------+-----
C          -3.196  -2.450 -1.704    (----*----)
                                    -+---------+---------+---------+-----
                                  -3.0      -1.5       0.0       1.5
```

d. SST is 40% of SS(Total), and SSB is 40% of SS(Total).

e. SST is 20% of SS(Total), and SSB is 20% of SS(Total).

Applying the Concepts—Basic

10.61 Rotary oil rigs. An economist wants to compare the average monthly number of rotary oil rigs running in three states—California, Utah, and Alaska. In order to account for month-to-month variation, three months were randomly selected over a two-year period and the number of oil rigs running in each state in each month was obtained from data provided from *World Oil* (Jan. 2002) magazine. The data, reproduced in the accompanying table, were analyzed using a randomized block design. The MINITAB printout is shown below.

OILRIGS

Month/Year	California	Utah	Alaska
Nov. 2000	27	17	11
Oct. 2001	34	20	14
Nov. 2001	36	15	14

Two-way ANOVA: NumRigs versus State, Month/Year

```
Source       DF      SS       MS      F      P
State         2  617.556  308.778  38.07  0.002
Month/Year    2   30.889   15.444   1.90  0.262
Error         4   32.444    8.111
Total         8  680.889


S = 2.848   R-Sq = 95.23%   R-Sq(adj) = 90.47%




                    Individual 95% CIs For Mean Based on
                    Pooled StDev
State    Mean    ---------+---------+---------+---------+
AL     13.0000   (----*-----)
CAL    32.3333                          (----*-----)
UT     17.3333        (-----*----)
                 ---------+---------+---------+---------+
                       16.0      24.0      32.0      40.0
```

NUMRIGS

Tukey HSD[a,b]

STATE	N	Subset 1	Subset 2
AL	3	13.00	
UT	3	17.33	
CAL	3		32.33
Sig.		.262	1.000

Means for groups in homogeneous subsets are displayed.
Based on Type III Sum of Squares
The error term is Mean Square(Error) = 8.111.

a. Uses Harmonic Mean Sample Size = 3.000.

b. Alpha = .05.

a. Why is a randomized block design preferred over a completely randomized design for comparing the mean number of oil rigs running monthly in California, Utah, and Alaska?

b. Identify the treatments for the experiment.

c. Identify the blocks for the experiment.

d. State the null hypothesis for the ANOVA F-test.

e. Locate the test statistic and p-value on the MINITAB printout. Interpret the results.

f. A Tukey multiple comparison of means (at $\alpha = .05$) is summarized in the SPSS printout shown above. Which state(s) have the significantly largest mean number of oil rigs running monthly?

10.62 Treatment for tendon pain. Refer to the *British Journal of Sports Medicine* (Feb. 1, 2004) study of Chronic Achilles tendon pain, Exercise 10.8 (p. 519). Recall that each in a sample of 25 patients with chronic Achilles tendinosis was treated with heavy load eccentric calf muscle training. Tendon thickness (in millimeters) was measured both before and following treatment of each patient. These experimental data are shown in the table.

a. What experimental design was employed in this study?

b. How many treatments are used in this study? How many blocks?

c. Give the null and alternative hypothesis for determining whether the mean tendon thickness before treatment differs from the mean after treatment.

d. Compute the test statistic for the test, part **c**, using the paired difference t formula of Section 9.3.

e. Use statistical software (or the formulas in Appendix C) to compute the ANOVA F-statistic for the test, part **c**.

f. Compare the test statistics, parts **d** and **e**. You should find that $F = t^2$.

g. Show that the two tests, parts **d** and **e**, yield identical p-values.

h. Give the appropriate conclusion using $\alpha = .05$.

TENDON

Patient	Before Thickness (millimeters)	After Thickness (millimeters)
1	11.0	11.5
2	4.0	6.4
3	6.3	6.1
4	12.0	10.0
5	18.2	14.7
6	9.2	7.3
7	7.5	6.1
8	7.1	6.4
9	7.2	5.7
10	6.7	6.5
11	14.2	13.2
12	7.3	7.5
13	9.7	7.4
14	9.5	7.2
15	5.6	6.3
16	8.7	6.0
17	6.7	7.3
18	10.2	7.0
19	6.6	5.3
20	11.2	9.0
21	8.6	6.6
22	6.1	6.3
23	10.3	7.2
24	7.0	7.2
25	12.0	8.0

Source: Ohberg, L., et al. "Eccentric training in patients with chronic Achilles tendinosis: normalized tendon structure and decreased thickness at follow up," *British Journal of Sports Medicine,* Vol. 38, No. 1, Feb. 1, 2004 (Table 2).

10.63 Endangered dwarf shrubs. Rugel's pawpaw (yellow NW squirrel banana) is an endangered species of a dwarf shrub. Biologists from Stetson University conducted an experiment to determine the effects of fire on the shrub's growth. (*Florida Scientist*, Spring 1997.) Twelve experimental plots of land were selected in a pasture where the shrub is abundant. Within each plot, three pawpaws were randomly selected and treated as follows: one shrub was subjected to fire, another to clipping, and the third was left unmanipulated (a control). After five months, the number of flowers produced by each of the 36 shrubs was determined. The objective of the study was to compare the mean number of flowers produced by pawpaws for the three treatments (fire, clipping, and control).

a. Identify the type of experimental design employed, including the treatments, response variable, and experimental units.

b. Illustrate the layout of the design using a graphic similar to Figure 10.16.

c. The ANOVA of the data resulted in a test statistic of $F = 5.42$ for treatments with p-value = .009. Interpret this result.

d. The three treatment means were compared using Tukey's method at $\alpha = .05$. Interpret the results shown below.

Mean number of flowers:	1.17	10.58	17.08
Treatment:	Control	Clipping	Burning

10.64 Effectiveness of geese decoys. Using decoys is a common method of hunting waterfowl. A study in the *Journal of Wildlife Management* (July 1995) compared the effectiveness of three different decoy types—taxidermy-mounted decoys, plastic shell decoys, and full-bodied plastic decoys—in attracting Canada geese to sunken pit blinds. In order to account for an extraneous source of variation, three pit blinds were used as blocks in the experiment. Thus, a randomized block design with three treatments (decoy types) and three blocks (pit blinds) was employed. The response variable was the percentage of a goose flock to approach within 46 meters of the pit blind on a given day. The data are given in the table,* followed by a MINITAB printout of the analysis (shown below).

*The actual design employed in the study was more complex than the randomized block design shown here. In the actual study, each number in the table represents the mean daily percentage of goose flocks attracted to the blind, averaged over 13–17 days.

⚙ **DECOY**

Blind	Shell	Full-Bodied	Taxidermy-Mounted
1	7.3	13.6	17.8
2	12.6	10.4	17.0
3	16.4	23.4	13.6

Source: Harrey, W. F., Hindman, L. J., and Rhodes, W. E. "Vulnerability of Canada geese to taxidermy-mounted decoys." *Journal of Wildlife Management*, Vol. 59, No. 3, July 1995, p. 475 (Table 1).

a. Find and interpret the *F*-statistic for comparing the response means of the three decoy types.

b. What assumptions are necessary for the validity of the inference made in part **a**?

c. Why is it not necessary to conduct multiple comparisons of the response means for the three decoy types?

Applying the Concepts—Intermediate

10.65 Prompting walkers to walk. A study was conducted to investigate the effect of prompting in a walking program. (*Health Psychology*, Mar. 1995.) Five groups of walkers—27 in each group—agreed to participate by walking for 20 minutes at least one day per week over a 24-week period. The participants were prompted to walk each week via telephone calls, but different prompting schemes were used for each group. Walkers in the control group received no prompting phone calls; walkers in the "frequent/low" group received a call once a week with low structure (i.e., "just touching base"); walkers in the "frequent/high" group received a call once a week with high structure (i.e., goals are set); walkers in the "infrequent/low" group received a call once every 3 weeks with low structure; and walkers in the "infrequent/high" group received a call once every 3 weeks with high structure. The table on p. 559 lists the number of participants in each group who actually walked the minimum requirement each week for weeks 1, 4, 8, 12, 16, and 24. The data were subjected to an analysis of variance for a randomized block design, with the five walker groups representing the treatments and the six time periods (weeks) representing the blocks.

a. What is the purpose of blocking on weeks in this study?

b. Construct an ANOVA summary table using statistical software (or the formulas in Appendix C).

Two-way ANOVA: PERCENT versus DECOY, BLIND

```
Source   DF       SS        MS       F      P
DECOY     2   30.069   15.0344   0.61   0.589
BLIND     2   44.149   22.0744   0.89   0.479
Error     4   99.338   24.8344
Total     8  173.556

S = 4.983    R-Sq = 42.76%    R-Sq(adj) = 0.00%
```

WALKERS

Week	Control	Frequent/Low	Frequent/High	Infrequent/Low	Infrequent/High
1	7	23	25	21	19
4	2	19	25	10	12
8	2	18	19	9	9
12	2	7	20	8	2
16	2	18	18	8	7
24	1	17	17	7	6

Source: Lombard, D. N., et al. "Walking to meet health guidelines: The effect of prompting frequency and prompt structure." *Health Psychology*, Vol. 14, No. 2, Mar. 1995, p. 167 (Table 2).

c. Is there sufficient evidence of a difference in the mean number of walkers per week among the five walker groups? Use $\alpha = .05$.

d. Use Tukey's technique to compare all pairs of treatment means with an experimentwise error rate of $\alpha = .05$. Interpret these results.

e. What assumptions must hold to ensure the validity of the inferences in parts **c** and **d**?

10.66 Effect of massage on boxers. Eight amateur boxers participated in an experiment to investigate the effect of massage on boxing performance. (*British Journal of Sports Medicine*, Apr. 2000.) The punching power of each boxer (measured in Newtons) was recorded in the round following each of four different interventions: (M1) in round 1 following a pre-bout sports massage, (R1) in round 1 following a pre-bout period of rest, (M5) in round 5 following a sports massage between rounds, and (R5) in round 5 following a period of rest between rounds. Based on information provided in the article, the data in the table below were obtained. The main goal of the experiment is to compare the punching power means of the four interventions.

a. Set up H_0 and H_a for this analysis.

b. Identify the treatments and blocks in this experiment.

c. Conduct the test, part **a**. What conclusions can you draw regarding the effect of massage on punching power?

BOXING

Boxer	Intervention			
	M1	R1	M5	R5
1	1243	1244	1291	1262
2	1147	1053	1169	1177
3	1247	1375	1309	1321
4	1274	1235	1290	1285
5	1177	1139	1233	1238
6	1336	1313	1366	1362
7	1238	1279	1275	1261
8	1261	1152	1289	1266

Source: Hemmings, B., Smith, M., Graydon, J., and Dyson, R. "Effects of massage on physiological restoration, perceived recovery, and repeated sports performance." *British Journal of Sports Medicine*, Vol. 34, No. 2, Apr. 2000 (adapted from Table 3).

10.67 Plants and stress reduction. Plant therapists believe that plants can reduce the stress levels of humans. A Kansas State University study was conducted to investigate this phenomenon. Two weeks prior to final exams, 10 undergraduate students took part in an experiment to determine what effect the presence of a live plant, a photo of a plant, or absence of a plant has on the student's ability to relax while isolated in a dimly lit room. Each student participated in three sessions—one with a live plant, one with a plant photo, and one with no plant (control).* During each session, finger temperature was measured at one-minute intervals for 20 minutes. Since increasing finger temperature indicates an increased level of relaxation, the maximum temperature (in degrees) was used as the response variable. The data for the experiment are provided in the table. Conduct an ANOVA and make the proper inferences at $\alpha = .10$.

PLANTS

Student	Live Plant	Plant Photo	No Plant (control)
1	91.4	93.5	96.6
2	94.9	96.6	90.5
3	97.0	95.8	95.4
4	93.7	96.2	96.7
5	96.0	96.6	93.5
6	96.7	95.5	94.8
7	95.2	94.6	95.7
8	96.0	97.2	96.2
9	95.6	94.8	96.0
10	95.6	92.6	96.6

Source: Elizabeth Schreiber, Department of Statistics, Kansas State University, Manhattan, Kansas.

10.68 Absentee rates at a jeans plant. A plant that manufactures denim jeans in the United Kingdom recently introduced a computerized automated handling system. The new system delivers garments to the assembly line operators by means of an overhead conveyor. While the automated system minimizes operator handling time, it inhibits operators from working ahead and taking breaks from their machine. A study in *New Technology, Work, and Employment* (July 2001) investigated the impact of the new handling system on

*The experiment is simplified for this exercise. The actual experiment involved 30 students who participated in 12 sessions.

worker absentee rates at the jeans plant. One theory is that the mean absentee rate will vary by day of the week, as operators decide to indulge in one-day absences to relieve work pressure. Nine weeks were randomly selected and the absentee rate (percentage of workers absent) determined for each day (Monday through Friday) of the work week. The data are listed in the table. Conduct a complete analysis of the data to determine whether the mean absentee rate differs across the five days of the work week.

JEANS

Week	Monday	Tuesday	Wednesday	Thursday	Friday
1	5.3	0.6	1.9	1.3	1.6
2	12.9	9.4	2.6	0.4	0.5
3	0.8	0.8	5.7	0.4	1.4
4	2.6	0.0	4.5	10.2	4.5
5	23.5	9.6	11.3	13.6	14.1
6	9.1	4.5	7.5	2.1	9.3
7	11.1	4.2	4.1	4.2	4.1
8	9.5	7.1	4.5	9.1	12.9
9	4.8	5.2	10.0	6.9	9.0

Source: Boggis, J. J. "The eradication of leisure." *New Technology, Work, and Employment*, Volume 16, Number 2, July 2001 (Table 3).

Applying the Concepts—Advanced

10.69 Listening ability of infants. *Science* (Jan. 1, 1999) reported on the ability of 7-month-old infants to learn an unfamiliar language. In one experiment, 16 infants were trained in an artificial language. Then, each infant was presented with two 3-word sentences that consisted entirely of new words (e.g., "wo fe wo"). One sentence was consistent (i.e., constructed from the same grammar as in the training session) and one sentence was inconsistent (i.e., constructed from grammar in which the infant was not trained). The variable measured in each trial was the time (in seconds) the infant spent listening to the speaker, with the goal to compare the mean listening times of consistent and inconsistent sentences.

a. The data were analyzed as a randomized block design with the 16 infants representing the blocks and the two sentence types (consistent and inconsistent) representing the treatments. Do you agree with this data analysis method? Explain.

b. Refer to part **a**. The test statistic for testing treatments was $F = 25.7$ with an associated observed significance level of $p < .001$. Interpret this result.

c. Explain why the data could also be analyzed as a paired difference experiment, with a test statistic of $t = 5.07$.

d. The mean listening times and standard deviations for the two treatments are given here. Use this information to calculate the F-statistic for comparing the treatment means in an ANOVA for a completely randomized design. Explain why this test statistic provides weaker evidence of a difference between treatment means than the test in part **b**.

	Consistent Sentences	Inconsistent Sentences
Mean	6.3	9.0
Standard deviation	2.6	2.16

e. Explain why there is no need to control the experimentwise error rate when ranking the treatment means for this experiment.

10.5 Factorial Experiments

All the experiments discussed in Sections 10.2 and 10.4 were **single-factor experiments**. The treatments were levels of a single factor, with the sampling of experimental units performed using either a completely randomized or randomized block design. However, most responses are affected by more than one factor, and we will therefore often wish to design experiments involving more than one factor.

Consider an experiment in which the effects of two factors on the response are being investigated. Assume that factor A is to be investigated at a levels, and factor B at b levels. Recalling that treatments are factor–level combinations, you can see that the experiment has, potentially, ab treatments that could be included in the experiment. A *complete factorial experiment* is one in which all possible ab treatments are utilized.

> **DEFINITION 10.9**
>
> A **complete factorial experiment** is one in which every factor–level combination is utilized. That is, the number of treatments in the experiment equals the total number of factor–level combinations.

For example, suppose the USGA wants to determine not only the relationship between distance and brand of golf ball, but also between distance and the

TABLE 10.10 Schematic Layout of Two-Factor Factorial Experiment

	Level	Factor B at b Levels				
		1	2	3	$\cdots$	b
Factor A	1	Trt. 1	Trt. 2	Trt. 3	$\ldots$	Trt. b
at a levels	2	Trt. $b + 1$	Trt. $b + 2$	Trt. $b + 3$	$\ldots$	Trt. $2b$
	3	Trt. $2b + 1$	Trt. $2b + 2$	Trt. $2b + 3$	$\ldots$	Trt. $3b$
	$\vdots$	$\vdots$	$\vdots$	$\vdots$	$\ldots$	$\vdots$
	a	Trt. $(a-1)b + 1$	Trt. $(a-1)b + 2$	Trt. $(a-1)b + 3$	$\ldots$	Trt. ab

club used to hit the ball. If they decide to use four brands and two clubs (say, driver and five-iron) in the experiment, then a complete factorial would call for utilizing all $4 \times 2 = 8$ Brand–Club combinations. This experiment is referred to more specifically as a *complete 4×2 factorial.* A layout for a two-factor factorial experiment (we are henceforth referring to a *complete factorial* when we use the term *factorial*) is given in Table 10.10. The factorial experiment is also referred to as a **two-way classification** because it can be arranged in the row–column format exhibited in Table 10.10.

In order to complete the specification of the experimental design, the treatments must be assigned to the experimental units. If the assignment of the *ab* treatments in the factorial experiment is random and independent, the design is completely randomized. For example, if the machine Iron Byron is used to hit 80 golf balls, 10 for each of the eight Brand–Club combinations, in a random sequence, the design would be completely randomized. On the other hand, if the assignment is made within homogeneous blocks of experimental units, then the design is a randomized block. For example, if ten golfers are employed to hit each of the eight golf balls, and each golfer hits all eight Brand–Club combinations in a random sequence, then the design is randomized block, with the golfers serving as blocks. In the remainder of this section, we confine our attention to factorial experiments utilizing completely randomized designs.

If we utilize a completely randomized design to conduct a factorial experiment with *ab* treatments, we can proceed with the analysis in exactly the same way as we did in Section 10.2. That is, we calculate (or let the computer calculate) the measure of treatment mean variability (MST) and the measure of sampling variability (MSE) and use the F-ratio of these two quantities to test the null hypothesis that the treatment means are equal. However, if this hypothesis is rejected, so that we conclude some differences exist among the treatment means, important questions remain. Are both factors affecting the response, or only one? If both, do they affect the response independently, or do they interact to affect the response?

For example, suppose the distance data indicate that at least two of the eight treatment (Brand–Club combinations) means differ in the golf experiment. Does the brand of ball (factor A) or the club utilized (factor B) affect mean distance, or do both affect it? Several possibilities are shown in Figure 10.20. In Figure 10.20a, the brand means are equal (only three are shown for the purpose of illustration), but the distances differ for the two levels of factor B (Club). Thus, there is no effect of Brand on distance, but a Club main effect is present. In Figure 10.20b, the Brand means differ, but the Club means are equal for each Brand. Here a Brand main effect is present, but no effect of Club is present.

Figures 10.20c and 10.20d illustrate cases in which both factors affect the response. In Figure 10.20c, the mean distance between clubs does not change for the three Brands, so that the effect of Brand on distance is independent of Club. That is, the two factors Brand and Club *do not interact.* In contrast, Figure 10.20d shows that the difference between mean distances between clubs varies with

Figure 10.20

Illustration of Possible Treatment Effects: Factorial Experiment

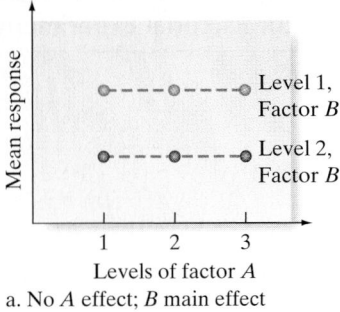

a. No A effect; B main effect

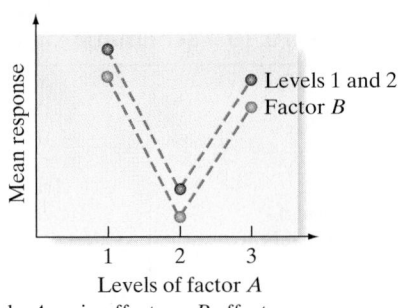

b. A main effect; no B effect

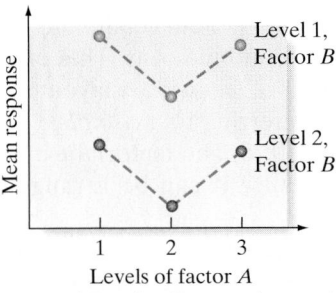

c. A and B main effects; no interaction

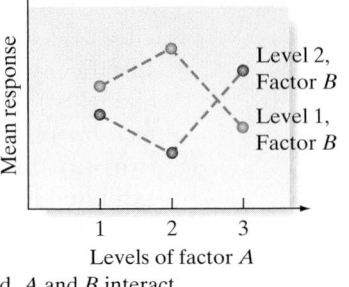

d. A and B interact

Brand. Thus, the effect of Brand on distance depends on Club; therefore, the two factors *do interact*.

In order to determine the nature of the treatment effect, if any, on the response in a factorial experiment, we need to break the treatment variability into three components: Interaction Between Factors A and B, Main Effect of Factor A, and Main Effect of Factor B. The **Factor Interaction** component is used to test whether the factors combine to affect the response, while the **Factor Main Effect** components are used to determine whether the factors separately affect the response.

> **Now Work** *Exercise 10.83*

The partitioning of the Total Sum of Squares into its various components is illustrated in Figure 10.21. Notice that at stage 1 the components are identical to those in the one-factor, completely randomized designs of Section 10.2; the Sums of Squares for Treatment and Error sum to the Total Sum of Squares. The degrees of freedom for treatments is equal to $(ab - 1)$, one less than the number of treatments. The degrees of freedom for error is equal to $(n - ab)$, the total sample size minus the number of treatments. Only at stage 2 of the partitioning does the factorial experiment differ from those previously discussed. Here we divide the Treatment Sum of Squares into its three components: Interaction and the two Main Effects. These components can then be used to test the nature of the differences, if any, among the treatment means.

There are a number of ways to proceed in the testing and estimation of factors in a factorial experiment. We present one approach in the following box.

> **Procedure for Analysis of Two-Factor Factorial Experiment**
>
> **1.** Partition the Total Sum of Squares into the Treatment and Error components (stage 1 of Figure 10.21). Use either a statistical software package or the calculation formulas in Appendix C to accomplish the partitioning.

Figure 10.21

Partitioning the Total Sum of Squares for a Two-Factor Factorial

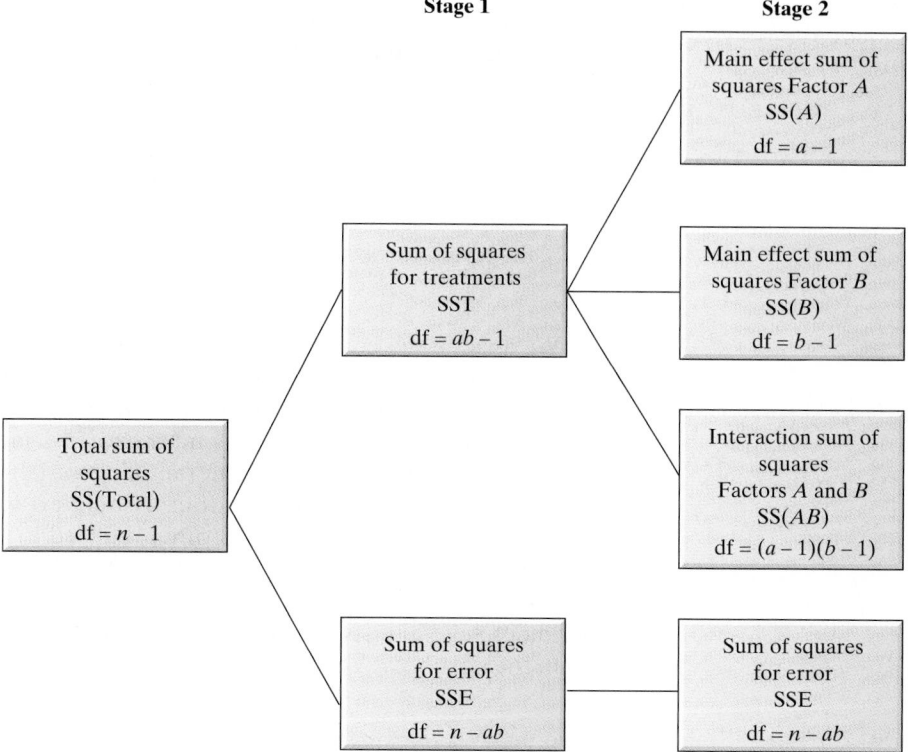

2. Use the F-ratio of Mean Square for Treatments to Mean Square for Error to test the null hypothesis that the treatment means are equal.*

   a. If the test results in nonrejection of the null hypothesis, consider refining the experiment by increasing the number of replications or introducing other factors. Also consider the possibility that the response is unrelated to the two factors.

   b. If the test results in rejection of the null hypothesis, then proceed to step 3.

3. Partition the Treatment Sum of Squares into the Main Effect and Interaction Sum of Squares (stage 2 of Figure 10.21). Use either a statistical software package or the calculation formulas in Appendix C to accomplish the partitioning.

4. Test the null hypothesis that factors A and B do not interact to affect the response by computing the F-ratio of the Mean Square for Interaction to the Mean Square for Error.

   a. If the test results in nonrejection of the null hypothesis, proceed to step 5.

   b. If the test results in rejection of the null hypothesis, conclude that the two factors interact to affect the mean response. Then proceed to step 6a.

5. Conduct tests of two null hypotheses that the mean response is the same at each level of factor A and factor B. Compute two F-ratios by comparing the Mean Square for each Factor Main Effect to the Mean Square for Error.

   a. If one or both tests result in rejection of the null hypothesis, conclude that the factor affects the mean response. Proceed to step 6b.

   b. If both tests result in nonrejection, an apparent contradiction has occurred. Although the treatment means apparently differ (step 2 test), the interaction

*Some analysts prefer to proceed directly to test the interaction and main effect components, skipping the test of treatment means. We begin with this test to be consistent with our approach in the one-factor completely randomized design.

(step 4) and main effect (step 5) tests have not supported that result. Further experimentation is advised.

6. Compare the means:
   a. If the test for interaction (step 4) is significant, use a multiple comparisons procedure to compare any or all pairs of the treatment means.
   b. If the test for one or both main effects (step 5) is significant, use a multiple comparisons procedure to compare the pairs of means corresponding to the levels of the significant factor(s).

We assume the completely randomized design is a **balanced design**, meaning that the same number of observations are made for each treatment. That is, we assume that r experimental units are randomly and independently selected for each treatment. The numerical value of r must exceed 1 in order to have any degrees of freedom with which to measure the sampling variability. [Note that if $r = 1$, then $n = ab$, and the degrees of freedom associated with Error (Figure 10.21) is df $= n - ab = 0$.] The value of r is often referred to as the number of **replicates** of the factorial experiment since we assume that all ab treatments are repeated, or replicated, r times. Whatever approach is adopted in the analysis of a factorial experiment, several tests of hypotheses are usually conducted. The tests are summarized in the next box.

Tests Conducted in Analyses of Factorial Experiments: Completely Randomized Design, r Replicates per Treatment

Test for Treatment Means

H_0: No difference among the ab treatment means

H_a: At least two treatment means differ

Test statistic: $F = \dfrac{\text{MST}}{\text{MSE}}$

Rejection region: $F \geq F_\alpha$, based on $(ab - 1)$ numerator and $(n - ab)$ denominator degrees of freedom [*Note: $n = abr$.*]

Test for Factor Interaction

H_0: Factors A and B do not interact to affect the response mean

H_a: Factors A and B do interact to affect the response mean

Test statistic: $F = \dfrac{\text{MS}(AB)}{\text{MSE}}$

Rejection region: $F \geq F_\alpha$, based on $(a - 1)(b - 1)$ numerator and $(n - ab)$ denominator degrees of freedom

Test for Main Effect of Factor A

H_0: No difference among the a mean levels of factor A

H_a: At least two factor A mean levels differ

Test statistic: $F = \dfrac{\text{MS}(A)}{\text{MSE}}$

Rejection region: $F \geq F_\alpha$, based on $(a - 1)$ numerator and $(n - ab)$ denominator degrees of freedom

Test for Main Effect of Factor B

H_0: No difference among the b mean levels of factor B

H_a: At least two factor B mean levels differ

Test statistic: $F = \dfrac{MS(B)}{MSE}$

Rejection region: $F \geq F_\alpha$, based on $(b - 1)$ numerator and $(n - ab)$ denominator degrees of freedom

---

Conditions Required for Valid F-Tests in Factorial Experiments

1. The response distribution for each factor–level combination (treatment) is normal.

2. The response variance is constant for all treatments.

3. Random and independent samples of experimental units are associated with each treatment.

EXAMPLE 10.10 CONDUCTING A FACTORIAL ANOVA

Problem Suppose the USGA tests four different brands (A, B, C, D) of golf balls and two different clubs (driver, five-iron) in a 4 × 2 factorial design. Each of the eight Brand–Club combinations (treatments) is randomly and independently assigned to four experimental units, each experimental unit consisting of a specific position in the sequence of hits by Iron Byron. The distance response is recorded for each of the 32 hits, and the results are shown in Table 10.11.

⊚ **GOLFFAC1**

TABLE 10.11 Distance Data for 4 × 2 Factorial Golf Experiment

Club	Brand			
	A	B	C	D
Driver	226.4	238.3	240.5	219.8
	232.6	231.7	246.9	228.7
	234.0	227.7	240.3	232.9
	220.7	237.2	244.7	237.6
Five-iron	163.8	184.4	179.0	157.8
	179.4	180.6	168.0	161.8
	168.6	179.5	165.2	162.1
	173.4	186.2	156.5	160.3

a. Use statistical software to partition the Total Sum of Squares into the components necessary to analyze this 4 × 2 factorial experiment.

b. Conduct the appropriate ANOVA tests and interpret the results of your analysis. Use $\alpha = .10$ for the tests you conduct.

c. If appropriate, conduct multiple comparisons of the treatment means. Use an experimentwise error rate of .10. Illustrate the comparisons with a graph.

Solution a. The SAS printout that partitions the Total Sum of Squares for this factorial experiment is given in Figure 10.22. The partitioning takes place in two stages: First, the Total Sum of Squares is partitioned into the Model (Treatment) and Error Sums of Squares at the top of the printout. Note that SST is 33,659.8 with 7 degrees of freedom, and SSE is 822.2 with 24 degrees of

Figure 10.22

SAS Printout for Factorial
ANOVA of Golf Data

Dependent Variable: DISTANCE

Source	DF	Sum of Squares	Mean Square	F Value	Pr > F
Model	7	33659.80875	4808.54411	140.35	<.0001
Error	24	822.24000	34.26000		
Corrected Total	31	34482.04875			

R-Square	Coeff Var	Root MSE	DISTANCE Mean
0.976155	2.896461	5.853204	202.0813

Source	DF	Type III SS	Mean Square	F Value	Pr > F
CLUB	1	32093.11125	32093.11125	936.75	<.0001
BRAND	3	800.73625	266.91208	7.79	0.0008
CLUB*BRAND	3	765.96125	255.32042	7.45	0.0011

freedom, adding to 34,482.0 and 31 degrees of freedom. In the second stage of partitioning, the Treatment Sum of Squares is further divided into the Main Effect and Interaction Sums of Squares. At the bottom of the printout we see that SS(Club) is 32,093.1 with 1 degree of freedom, SS(Brand) is 800.7 with 3 degrees of freedom, and SS(Club $\times$ Brand) is 766.0 with 3 degrees of freedom, adding to 33,659.8 and 7 degrees of freedom.

b. Once partitioning is accomplished, our first test is

H_0: The eight treatment means are equal

H_a: At least two of the eight means differ

$Test\ statistic: F = \dfrac{\text{MST}}{\text{MSE}} = 140.35$ (top of printout)

$Observed\ significance\ level: p < .0001$ (top of printout)

Since $\alpha = .10$ exceeds p, we reject this null hypothesis and conclude that at least two of the Brand–Club combinations differ in mean distance.

After accepting the hypothesis that the treatment means differ, and therefore that the factors Brand and/or Club somehow affect the mean distance, we want to determine how the factors affect the mean response. We begin with a test of interaction between Brand and Club:

H_0: The factors Brand and Club do not interact to affect the mean response

H_a: Brand and Club interact to affect mean response

$Test\ statistic: F = \dfrac{\text{MS}(AB)}{\text{MSE}} = \dfrac{\text{MS(Brand} \times \text{Club)}}{\text{MSE}}$

$= \dfrac{255.32}{34.26} = 7.45$ (bottom of printout)

$Observed\ significance\ level: p = .0011$ (bottom of printout)

Since $\alpha = .10$ exceeds the p-value, we conclude that the factors Brand and Club interact to affect mean distance.

Because the factors interact, we do not test the main effects for Brand and Club. Instead, we compare the treatment means in an attempt to learn the nature of the interaction.

c. Rather than compare all $8(7)/2 = 28$ pairs of treatment means, we test for differences only between pairs of brands within each club. That differences exist *between* clubs can be assumed. Therefore, only $4(3)/2 = 6$ pairs of means need to be compared for each club, or a total of 12 comparisons for the two clubs. The results of these comparisons using Tukey's method with an experimentwise error rate of $\alpha = .10$ for each club are displayed in the SAS printout,

Figure 10.23

SAS Printout of Tukey Rankings of Golf Ball Brand Means for Each Club

```
---------------------------------------- CLUB=5IRON ----------------------------------------
                          The ANOVA Procedure

                Tukey's Studentized Range (HSD) Test for DISTANCE

NOTE: This test controls the Type I experimentwise error rate, but it generally has a higher
                      Type II error rate than REGWQ.


                Alpha                                   0.1
                Error Degrees of Freedom                 12
                Error Mean Square                  36.10792
                Critical Value of Studentized Range 3.62071
                Minimum Significant Difference       10.878


           Means with the same letter are not significantly different.


         Tukey Grouping            Mean      N    BRAND

                   A             182.675      4    B

                   B             171.300      4    A
                   B
                   B             167.175      4    C
                   B
                   B             160.500      4    D


---------------------------------------- CLUB=DRIVER ----------------------------------------
                          The ANOVA Procedure

                Tukey's Studentized Range (HSD) Test for DISTANCE

NOTE: This test controls the Type I experimentwise error rate, but it generally has a higher
                      Type II error rate than REGWQ.


                Alpha                                   0.1
                Error Degrees of Freedom                 12
                Error Mean Square                  32.41208
                Critical Value of Studentized Range 3.62071
                Minimum Significant Difference       10.307


           Means with the same letter are not significantly different.


         Tukey Grouping            Mean      N    BRAND

                       A         243.100      4    C
                       A
                   B   A         233.725      4    B
                   B
                   B             229.750      4    D
                   B
                   B             228.425      4    A
```

Figure 10.23. For each club, the brand means are listed in descending order in Figure 10.23, and those not significantly different are connected by the same letter in the "Tukey Grouping" column.

As shown in Figure 10.23, the picture is unclear with respect to Brand means. For the five-iron (top of Figure 10.23), the brand B mean significantly exceeds all other brands. However, when hit with a driver (bottom of Fig. 10.23), brand B's mean is not significantly different from any of the other brands.

The Club × Brand interaction can be seen in the plot of means in Figure 10.24. Note that the difference between the mean distances of the two clubs (driver and five-iron) varies depending on brand. The biggest difference appears for Brand C while the smallest difference is for Brand B.

Look Back Note the nontransitive nature of the multiple comparisons. For example, for the driver the brand C mean can be "the same" as the brand B mean, and the

Figure 10.24

Sample Mean Plot for Factorial Golf Experiment

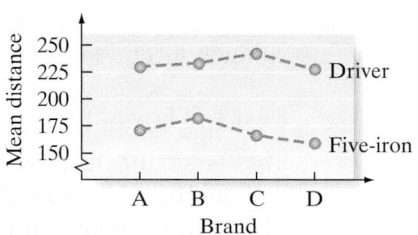

brand B mean can be "the same" as the brand D mean, and yet the brand C mean can significantly exceed the brand D mean. The reason lies in the definition of "the same"—we must be careful not to conclude two means are equal simply because they are connected by a line or letter. The line indicates only that *the connected means are not significantly different*. You should conclude (at the overall α level of significance) only that means *not* connected are different, while withholding judgment on those that are connected. The picture of which means differ and by how much will become clearer as we increase the number of replicates of the factorial experiment.

Now Work | *Exercise 10.80*

As with completely randomized and randomized block designs, the results of a factorial ANOVA are typically presented in an ANOVA summary table. Table 10.12 gives the general form of the ANOVA table, while Table 10.13 gives the ANOVA table for the golf ball data analyzed in Example 10.10. A two-factor factorial is characterized by four sources of variation—Factor A, Factor B, A × B interaction, and Error—which sum to the total sum of squares.

TABLE 10.12 General ANOVA Summary Table for a Two-Factor Factorial Experiment with *r* Replicates, Where Factor A Has *a* Levels and Factor B Has *b* Levels

Source	df	SS	MS	F
A	$a - 1$	SSA	MSA	MSA/MSE
B	$b - 1$	SSB	MSB	MSB/MSE
AB	$(a - 1)(b - 1)$	SSAB	MSAB	MSAB/MSE
Error	$ab(r - 1)$	SSE	MSE	
Total	$n - 1$	SS(Total)		

TABLE 10.13 ANOVA Summary Table for Example 10.10

Source	df	SS	MS	F
Brand	1	32,093.11	32,093.11	936.75
Club	3	800.74	266.91	7.79
Interaction	3	765.96	255.32	7.45
Error	24	822.24	34.26	
Total	31	34,482.05		

EXAMPLE 10.11 MORE PRACTICE ON CONDUCTING A FACTORIAL ANOVA

Problem Refer to Example 10.10. Suppose the same factorial experiment is performed on four other brands (E, F, G, and H), and the results are as shown in Table 10.14. Repeat the factorial analysis and interpret the results.

Solution An SPSS printout for the second factorial experiment is shown in Figure 10.25. We conduct several tests, as outlined in the box.

Test for Equality of Treatment Means:

The *F*-ratio for Treatments is $F = 290.1$ (highlighted at the top of the printout), which exceeds the tabled value of $F_{.10} = 1.98$ for 7 numerator and 24 denominator degrees of freedom. (Note that the same rejection regions will apply in this example as in

⊚ **GOLFFAC2**

TABLE 10.14   Distance Data for Second Factorial Golf Experiment

Club	Brand			
	E	F	G	H
Driver	238.6	261.4	264.7	235.4
	241.9	261.3	262.9	239.8
	236.6	254.0	253.5	236.2
	244.9	259.9	255.6	237.5
Five-iron	165.2	179.2	189.0	171.4
	156.9	171.0	191.2	159.3
	172.2	178.0	191.3	156.6
	163.2	182.7	180.5	157.4

Example 10.10 since the factors, treatments, and replicates are the same.) We conclude that at least two of the Brand–Club combinations have different mean distances.

Test for Interaction:

We next test for interaction between Brand and Club. The F-value (highlighted on the SPSS printout) is

$$F = \frac{\text{MS}(\text{Brand} \times \text{Club})}{\text{MSE}} = 1.425$$

Since this F-ratio does *not* exceed the tabled value of $F_{.10} = 2.33$ with 3 and 24 df, we cannot conclude at the .10 level of significance that the factors interact. In fact, note that the observed significance level (highlighted on the SPSS printout) for the test of interaction is .260. Thus, at any level of significance lower than $\alpha = .26$, we could not conclude that the factors interact. We therefore test the main effects for Brand and Club.

Test for Brand Main Effect:

We first test the Brand main effect:

H_0: No difference exists among the true Brand mean distances
H_a: At least two Brand mean distances differ
Test statistic:

$$F = \frac{\text{MS}(\text{Brand})}{\text{MSE}} = \frac{1,136.772}{24.600} = 46.210 \quad \text{(highlighted on printout)}$$

Figure 10.25
SPSS Printout of Second
Factorial ANOVA

Tests of Between-Subjects Effects

Dependent Variable: DISTANCE

Source	Type III Sum of Squares	df	Mean Square	F	Sig.
Corrected Model	49959.375[a]	7	7137.054	290.120	.000
Intercept	1423532.83	1	1423532.828	57866.45	.000
BRAND	3410.316	3	1136.772	46.210	.000
CLUB	46443.900	1	46443.900	1887.939	.000
BRAND * CLUB	105.158	3	35.053	1.425	.260
Error	590.407	24	24.600		
Total	1474082.61	32			
Corrected Total	50549.782	31			

a. R Squared = .988 (Adjusted R Squared = .985)

Observed significance level: p = .000 (highlighted on printout)

Since α = .10 exceeds the *p*-value, we conclude that at least two of the brand means differ. We will subsequently determine which brand means differ using Tukey's multiple comparisons procedure. But first, we want to test the Club main effect:

Test for Club Main Effect:

H_0: No differences exist between the Club mean distances
H_a: The Club mean distances differ
Test statistic:

$$F = \frac{\text{MS(Club)}}{\text{MSE}} = \frac{46{,}443.900}{24.600} = 1{,}887.939 \quad \text{(highlighted on printout)}$$

Observed significance level: p = .000 (highlighted on printout)

Since α = .10 exceeds the *p*-value, we conclude that the two clubs are associated with different mean distances. Since only two levels of Club were utilized in the experiment, this *F*-test leads to the inference that the mean distance differs for the two clubs. It is no surprise (to golfers) that the mean distance for balls hit with the driver is significantly greater than the mean distance for those hit with the five-iron.

Ranking of Means:

To determine which of the Brands' mean distances differ, we wish to compare the four Brand means using Tukey's method at α = .10. The results of these multiple comparisons are displayed in the SPSS printout, Figure 10.26. The Brand means are shown, grouped by subset, in Figure 10.26, with means that are not significantly different in the same subset. Brands G and F (subset 2) are associated with significantly greater mean distances than brands E and H (subset 1). However, since G and F are in the same Tukey subset and E and H are in the same subset, we cannot distinguish between brands G and F or between brands E and H using these data.

Look Back Since the interaction between Brand and Club was not significant, we conclude that this difference among brands applies to both clubs. The sample means for all Club–Brand combinations are shown in Figure 10.27 and appear to support the conclusions of the tests and comparisons. Note that the Brand means maintain their relative positions for each Club—brands F and G dominate brands E and H for both the driver and five-iron.

Figure 10.26
SPSS Printout of Tukey
Rankings of Brand Means

DISTANCE

Tukey HSD[a,b]

BRAND	N	Subset 1	Subset 2
H	8	199.2000	
E	8	202.4375	
F	8		218.4375
G	8		223.5875
Sig.		.568	.189

Means for groups in homogeneous subsets are displayed.
Based on Type III Sum of Squares
The error term is Mean Square(Error) = 24.600.
  a. Uses Harmonic Mean Sample Size = 8.000.
  b. Alpha = .05.

Figure 10.27
Sample Mean Plot for
Second Factorial Golf
Experiment

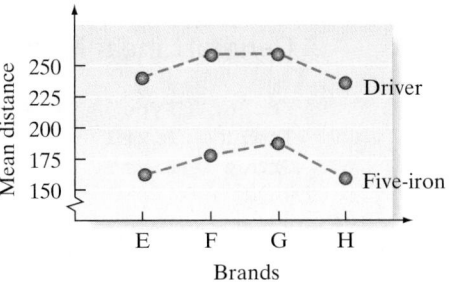

Now Work	*Exercise 10.81*

■ ■ ■

The analysis of factorial experiments can become complex if the number of factors is increased. Even the two-factor experiment becomes more difficult to analyze if some factor combinations have different numbers of observations than others. We have provided an introduction to these important experiments using two-factor factorials with equal numbers of observations for each treatment. Although similar principles apply to most factorial experiments, you should consult the references at the end of this chapter if you need to design and analyze more complex factorials.

Statistics in Action Revisited

A Two-Way Analysis of the Cockroach Data

We now return to the study of the trail-following ability of German cockroaches (p. 513). Recall that an entomologist created a chemical trail with either a methanol extract from roach feces or just methanol (control). Cockroaches from one of four age-sex groups were released into a container at the beginning of the trail, one at a time, and the movement pattern of each cockroach—the trail deviation (in pixels)—was measured. The layout for the experimental design is shown in Figure SIA10.3. You can see that 20 roaches of each type were randomly assigned to the treatment trail and 10 of each type were randomly assigned to the control trail. Thus, a total of 120 roaches were used in the experiment. The design is a factorial with two factors—Trail (extract or control) and Group (adult males, adult females, gravid females, or nymphs).

The entomologist wants to determine whether cockroaches in different age-sex groups differ in their ability to follow either the extract trail or the control trail. In other words, how do the two factors, age-sex group and trail type, impact the mean trail deviation of cockroaches? To answer this question, we conduct a two-way factorial analysis of variance on the data saved in the **ROACH** file. A MINITAB printout of the ANOVA is displayed in Figure SIA10.4.

First, note that the p-value for the test for factor interaction (highlighted on the printout) is .386. Thus, there is insufficient evidence (at $\alpha = .05$) of interaction between the two factors. This implies that the impact of one factor (say, age-sex group) on mean trail deviation does not depend on the level of the other factor (trail type). With no evidence of interaction, it is appropriate to conduct main effect tests on the two factors.

The p-values for the main effects of Trail type and age-sex Group (both highlighted on the printout) are .000 and .001, respectively. Since both p-values are less than $\alpha = .05$, there is sufficient evidence of (1) a difference between the mean deviations of cockroaches following the fecal extract trail and the control, and (2) differences in mean deviations among the four age-sex groups. To determine which main effect means are largest, we perform multiple comparisons of the means for both main effects.

Figure SIA10.5 is a SAS printout showing (at the top) the Bonferroni rankings of the age-sex Group means and

		Roach Type			
		Adult Male	Adult Female	Gravid Female	Nymph
Trail	**Extract**	$n = 20$	$n = 20$	$n = 20$	$n = 20$
	Control	$n = 10$	$n = 10$	$n = 10$	$n = 10$

Figure SIA10.3 Layout of Experimental Design for Cockroach Study

Figure SIA10.4
MINITAB Two-Way
ANOVA for Extract Trail
Deviation

```
General Linear Model: Deviate versus Trail, Group

Factor   Type    Levels   Values
Trail    fixed        2   Control, Extract
Group    fixed        4   Female, Gravid, Male, Nymph


Analysis of Variance for Deviate, using Adjusted SS for Tests

Source          DF     Seq SS     Adj SS     Adj MS       F       P
Trail            1    46445.5    46445.5    46445.5   63.34   0.000
Group            3    16271.2    13000.6     4333.5    5.91   0.001
Trail*Group      3     2245.2     2245.2      748.4    1.02   0.386
Error          112    82131.7    82131.7      733.3
Total          119   147093.6


S = 27.0799    R-Sq = 44.16%    R-Sq(adj) = 40.67%
```

(at the bottom) the Bonferroni rankings of the Trail type means. Both comparisons are made using an experiment-wise error rate of .05. Based on the "Bon Grouping" letters shown at the top of the printout, the only significant difference between age-sex group means is for the adult male and gravid cockroaches. Adult males have a significantly smaller mean trail deviation than gravids. No other pair of age-sex group means are significantly different. The bottom of

Figure SIA 10.5 shows that the mean deviation for cockroaches following the fecal extract trail is significantly smaller than the mean for cockroaches following the control.

The final conclusions of the entomologist were as follows. There is evidence that cockroaches exhibit more of an ability to follow a fecal extract trail than a control (methanol) trail. Also, adult males appear to have more of an ability to follow a trail than gravid (pregnant female) cockroaches.

Figure SIA10.5
SAS Bonferroni Rankings
of Group Means and Trail
Means

```
                Bonferroni (Dunn) t Tests for DEVIATE

NOTE: This test controls the Type I experimentwise error rate, but it generally has a higher
                   Type II error rate than REGWQ.


              Alpha                            0.05
              Error Degrees of Freedom          112
              Error Mean Square            733.3187
              Critical Value of t          2.68593
              Minimum Significant Difference  18.78


        Means with the same letter are not significantly different.


            Bon Grouping          Mean     N    GROUP
                       A        53.747     30    Gravid
                       A
                   B   A        37.077     30    Nymph
                   B
                   B   A        35.110     30    Female
                   B
                   B            20.917     30    Male


                Bonferroni (Dunn) t Tests for DEVIATE

NOTE: This test controls the Type I experimentwise error rate, but it generally has a higher
                   Type II error rate than REGWQ.


              Alpha                            0.05
              Error Degrees of Freedom          112
              Error Mean Square            733.3187
              Critical Value of t          1.98137
              Minimum Significant Difference  10.39
              Harmonic Mean of Cell Sizes  53.33333

                 NOTE: Cell sizes are not equal.


        Means with the same letter are not significantly different.


            Bon Grouping          Mean     N    TRAIL
                       A        64.535     40    Control

                       B        22.801     80    Extract
```

Exercises 10.70–10.89

Understanding the Principles

10.70 Describe how the treatments are formed in a complete factorial experiment.

10.71 What conditions are required for valid inferences from a factorial ANOVA?

10.72 Describe what is meant by factor interaction.

10.73 Suppose you conduct a 3×5 factorial experiment.
- **a.** How many factors are used in the experiment?
- **b.** Can you determine the factor type(s)—qualitative or quantitative—from the information given? Explain.
- **c.** Can you determine the number of levels used for each factor? Explain.
- **d.** Describe a treatment for this experiment, and determine the number of treatments used.
- **e.** What problem is caused by using a single replicate of this experiment? How is the problem solved?

Learning the Mechanics

10.74 The partially complete ANOVA table given here is for a two-factor factorial experiment.

Source	df	SS	MS	F
A	3		.75	
B	1	.95		
AB			.30	
Error				
Total	23	6.5		

- **a.** Give the number of levels for each factor.
- **b.** How many observations were collected for each factor–level combination?
- **c.** Complete the ANOVA table.
- **d.** Test to determine whether the treatment means differ. Use $\alpha = .10$.
- **e.** Conduct the tests of factor interaction and mean effects, each at the $\alpha = .10$ level of significance. Which of the tests are warranted as part of the factorial analysis? Explain.

10.75 The partially completed ANOVA table for a 3×4 factorial experiment with two replications is shown in the next column.

Source	df	SS	MS	F
A		.8		
B		5.3		
AB		9.6		
Error				
Total		18.1		

- **a.** Complete the ANOVA table.
- **b.** Which sums of squares are combined to find the Sum of Squares for Treatment? Do the data provide sufficient evidence to indicate that the treatment means differ? Use $\alpha = .05$.
- **c.** Does the result of the test in part **b** warrant further testing? Explain.
- **d.** What is meant by factor interaction, and what is the practical implication if it exists?
- **e.** Test to determine whether these factors interact to affect the response mean. Use $\alpha = .05$, and interpret the result.
- **f.** Does the result of the interaction test warrant further testing? Explain.

10.76 The following two-way table gives data for a 2×3 factorial experiment with two observations for each factor–level combination.

⊙ **LM10_76**

	Level	Factor B 1	2	3
Factor A	1	3.1, 4.0	4.6, 4.2	6.4, 7.1
	2	5.9, 5.3	2.9, 2.2	3.3, 2.5

- **a.** Identify the treatments for this experiment. Calculate and plot the treatment means, using the response variable as the y-axis and the levels of factor B as the x-axis. Use the levels of factor A as plotting symbols. Do the treatment means appear to differ? Do the factors appear to interact?
- **b.** The MINITAB ANOVA printout for this experiment is shown below. Test to determine whether the treatment means differ at the $\alpha = .05$ level of significance.

MINITAB Output for Exercise 10.76

Two-way ANOVA: RESPONSE versus A, B

```
Source        DF      SS       MS       F      P
A              1   4.4408  4.44083  18.06  0.005
B              2   4.1267  2.06333   8.39  0.018
Interaction    2  18.0067  9.00333  36.62  0.000
Error          6   1.4750  0.24583
Total         11  28.0492

S = 0.4958   R-Sq = 94.74%   R-Sq(adj) = 90.36%
```

Does the test support your visual interpretation from part **a**?

c. Does the result of the test in part **b** warrant a test for interaction between the two factors? If so, perform it using $\alpha = .05$.

d. Do the results of the previous tests warrant tests of the two factor main effects? If so, perform them using $\alpha = .05$.

e. Interpret the results of the tests. Do they support your visual interpretation from part **a**?

10.77 Suppose a 3×3 factorial experiment is conducted with three replications. Assume that SS(Total) = 1,000. For each of the following scenarios, form an ANOVA table, conduct the appropriate tests, and interpret the results.

a. The Sum of Squares of factor A main effect [SS(A)] is 20% of SS(Total), the Sum of Squares for factor B main effect [SS(B)] is 10% of SS(Total), and the Sum of Squares for interaction [SS(AB)] is 10% of SS(Total).

b. SS(A) is 10%, SS(B) is 10%, and SS(AB) is 50% of SS(Total).

c. SS(A) is 40%, SS(B) is 10%, and SS(AB) is 20% of SS(Total).

d. SS(A) is 40%, SS(B) is 40%, and SS(AB) is 10% of SS(Total).

10.78 The two-way table below gives data for a 2×2 factorial experiment with two observations per factor–level combination.

🔘 **LM10_78**

		Factor B	
	Level	1	2
Factor A	1	29.6, 35.2	47.3, 42.1
	2	12.9, 17.6	28.4, 22.7

a. Identify the treatments for this experiment. Calculate and plot the treatment means, using the response variable as the y-axis and the levels of factor B as the x-axis. Use the levels of factor A as plotting symbols. Do the treatment means appear to differ? Do the factors appear to interact?

b. Construct an ANOVA table for this experiment.

c. Test to determine whether the treatment means differ at the $\alpha = .05$ level of significance. Does the test support your visual interpretation from part **a**?

d. Does the result of the test in part **b** warrant a test for interaction between the two factors? If so, perform it using $\alpha = .05$.

e. Do the results of the previous tests warrant tests of the two factor main effects? If so, perform them using $\alpha = .05$.

f. Interpret the results of the tests. Do they support your visual interpretation from part **a**?

g. Given the results of your tests, which pairs of means, if any, should be compared?

Applying the Concepts—Basic

10.79 Removing bacteria from water. A coagulation–microfiltration process for removing bacteria from water was investigated in *Environmental Science & Engineering* (Sept. 1, 2000). Chemical engineers at Seoul National University performed a designed experiment to estimate the effect of both the level of the coagulant and acidity (pH) level on the coagulation efficiency of the process. Six levels of coagulant (5, 10, 20, 50, 100, and 200 milligrams per liter) and six pH levels (4.0, 5.0, 6.0, 7.0, 8.0, and 9.0) were employed. Water specimens collected from the Han River in Seoul, Korea, were placed in jars, and each jar randomly assigned to receive one of the $6 \times 6 = 36$ combinations of coagulant level and pH level.

a. What type of experimental design was applied in this study?

b. Give the factors, factor levels, and treatments for the study.

10.80 Baker's versus brewer's yeast. The *Electronic Journal of Biotechnology* (Dec. 15, 2003) published an article on a comparison of two yeast extracts, baker's yeast and brewer's yeast. Brewer's yeast is a surplus by-product obtained from a brewery, hence it is less expensive than primary-grown baker's yeast. Samples of both yeast extracts were prepared at four different temperatures (45, 48, 51, and 54°C); thus, a 2×4 factorial design with yeast extract at two levels and temperature at four levels was employed. The response variable was the autolysis yield (recorded as a percentage).

a. How many treatments are included in the experiment?

b. An ANOVA found sufficient evidence of factor interaction at $\alpha = .05$. Interpret this result practically.

c. Give the null and alternative hypotheses for testing the main effects of yeast extract and temperature.

d. Explain why the tests, part **c**, should not be conducted.

e. Multiple comparisons of the four temperature means were conducted for each of the two yeast extracts. Interpret the results shown below.

Baker's yeast:	Mean yield (%):	41.1	47.5	48.6	50.3
	Temperature (°C):	54	45	48	51
Brewer's yeast:	Mean yield (%):	39.4	47.3	49.2	49.6
	Temperature (°C):	54	51	48	45

10.81 Mussel settlement patterns on algae. Mussel larvae are in great abundance in the drift material that washes up on Ninety Mile Beach in New Zealand. These larvae tend to settle on algae. Environmentalists at the University of Auckland investigated the impact of algae type on the abundance of mussel larvae in drift material. (*Malacologia*, Feb. 8, 2002.) Drift material from three different wash-up events on Ninety Mile Beach were collected; for each wash-up, the algae was separated into four strata—coarse-branching, medium-branching, fine-branching, and hydroid algae. Two samples were randomly selected for each of the $3 \times 4 = 12$

event/strata combinations, and the mussel density (percent per square centimeter) was measured for each. The data were analyzed as a complete 3×4 factorial design. The ANOVA summary table is shown below.

Source	df	F	p-Value
Event	2	.35	>.05
Strata	3	217.33	<.05
Interaction	6	1.91	>.05
Error	12		
Total	23		

a. Identify the factors (and levels) in this experiment.
b. How many treatments are included in the experiment?
c. How many replications are included in the experiment?
d. What is the total sample size for the experiment?
e. What is the response variable measured?

f. Which ANOVA F-test should be conducted first? Conduct this test (at $\alpha = .05$) and interpret the results.
g. If appropriate, conduct the F-tests (at $\alpha = .05$) for the main effects. Interpret the results.

h. Tukey multiple comparisons of the four algae strata means (at $\alpha = .05$) are summarized below. Which means are significantly different?

Mean abundance ($\%/cm^2$):	9	10	27	55
Algae stratum	Coarse	Medium	Fine	Hydroid

10.82 Are you lucky? Parapsychologists define "lucky" people as individuals who report that seemingly chance events consistently tend to work out in their favor. A team of British psychologists designed a study to examine the effects of luckiness and competition on performance in a guessing task. (*The Journal of Parapsychology*, Mar. 1997.) Each in a sample of 56 college students was classified as lucky, unlucky, or uncertain based on their responses to a Luckiness Questionnaire. In addition, the participants were randomly assigned to either a competitive or noncompetitive condition. All students were then asked to guess the outcomes of 50 flips of a coin. The response variable measured was percentage of coin-flips correctly guessed.
a. An ANOVA for a 2×3 factorial design was conducted on the data. Identify the factors and their levels for this design.
b. The results of the ANOVA are summarized in the table. Fully interpret the results.

Source	df	F	p-Value
Luckiness (L)	2	1.39	.26
Competition (C)	1	2.84	.10
L × C	2	0.72	.72
Error	50		
Total	55		

10.83 Impact of paper color on exam scores. A study published in *Teaching Psychology* (May, 1998) examined how external clues influence student performance. Introductory psychology students were randomly assigned to one of four different midterm examinations. Form 1 was printed on blue paper and contained difficult questions, while form 2 was also printed on blue paper but contained simple questions. Form 3 was printed on red paper, with difficult questions; form 4 was printed on red paper with simple questions. The researchers were interested in the impact that Color (red or blue) and Question (simple or difficult) had on mean exam score.
a. What experimental design was employed in this study? Identify the factors and treatments.
b. The researchers conducted an ANOVA and found a significant interaction between Color and Question (p-value <.03). Interpret this result.
c. The sample mean scores (percentage correct) for the four exam forms are listed below. Plot the four means on a graph to illustrate the Color × Question interaction.

Form	Color	Question	Mean Score
1	Blue	Difficult	53.3
2	Blue	Simple	80.0
3	Red	Difficult	39.3
4	Red	Simple	73.6

10.84 Student use of the computer lab. A computer lab at the University of Oklahoma is open 24 hours a day, seven days a week. In *Production and Inventory Management Journal* (3rd Quarter, 1999), S. Barman investigated whether computer usage differed significantly (1) among the seven days of the week and (2) among the 24 hours of the day. Using student log-on records, data on hourly student loads (number of users per hour) were collected during a 7-week period. A factorial ANOVA was employed with the results presented in the table on p. 576.
a. Is this an observational or a designed experiment? Explain.
b. What are the two factors of the experiment and how many levels of each factor are used?
c. This is an $a \times b$ factorial experiment. What are a and b?
d. Specify the null and alternative hypotheses for testing the equality of all treatment means.
e. Conduct the test, part d, using $\alpha = .05$. Interpret your result in the context of the problem.

f. Specify the null and alternative hypotheses that should be used to test for an interaction effect between the two factors of the study.
g. Conduct the test of part d using $\alpha = .05$. Interpret your result in the context of the problem.
h. If appropriate, conduct main effects tests for both day and time. Use $\alpha = .05$. Interpret your results in the context of the problem.

Table for Exercise 10.84

Source	df	Sum of Squares	Mean Square		
Model	167	191047.21	1143.99		
Error	1004	45829.97	45.65		
Total	1171	236877.18			

$R = .8065$ F value $= 25.06$ $\text{Pr} > F = .001$

Source	df	Sum of Squares	Mean Square	F Value	Pr > F
Day	6	18732.13	3122.02	68.39	.0001
Time	23	164629.86	7157.82	156.80	.0001
Day × Time	138	7685.22	55.69	1.22	.0527

Source: Barman, S. "A statistical analysis of the attendance pattern of a computer laboratory." *Production and Inventory Management Journal*, 3rd Quarter, 1999, pp. 26–30.

Applying the Concepts—Intermediate

10.85 The thrill of a close game. Do women enjoy the thrill of a close basketball game as much as men? To answer this question, male and female undergraduate students were recruited to participate in an experiment. (*Journal of Sport & Social Issues*, Feb. 1997.) The students watched one of eight live televised games of a recent NCAA basketball tournament. (None of the games involved a home team to which the students could be considered emotionally committed.) The "suspense" of each game was classified into one of four categories according to the closeness of scores at the game's conclusions: minimal (15 point or greater differential), moderate (10–14 point differential), substantial (5–9 point differential), and extreme (1–4 point differential). After the game, each student rated his or her enjoyment on an 11-point scale ranging from 0 (not at all) to 10 (extremely). The enjoyment rating data were analyzed as a 4 × 2 factorial design, with suspense (four levels) and gender (two levels) as the two factors. The 4 × 2 = 8 treatment means are shown in the accompanying table.

Suspense	Gender	
	Male	Female
Minimal	1.77	2.73
Moderate	5.38	4.34
Substantial	7.16	7.52
Extreme	7.59	4.92

Source: Gan, Su-lin, et al. "The thrill of a close game: Who enjoys it and who doesn't?" *Journal of Sport & Social Issues*, Vol. 21, No. 1, Feb. 1997, pp. 59–60.

a. Plot the treatment means in a graph similar to Figure 10.24. Does the pattern of means suggest interaction between suspense and gender? Explain.

b. The ANOVA F-test for interaction yielded the following results: numerator df = 3, denominator df = 68, $F = 4.42$, p-value = .007. What can you infer from these results?

c. Based on the test, part **b**, is the difference between the mean enjoyment levels of males and females the same, regardless of the suspense level of the game?

10.86 Alcohol and marriage study. An experiment was conducted to examine the effects of alcohol on the marital interactions of husbands and wives (*Journal of Abnormal Psychology*, Nov. 1998.) A total of 135 couples participated in the experiment. The husband in each couple was classified as aggressive (60 husbands) or nonaggressive (75 husbands), based on an interview and his response to a questionnaire. Before the marital interactions of the couples were observed, each husband was randomly assigned to three groups: receive no alcohol, receive several alcoholic mixed drinks, or receive placebos (nonalcoholic drinks disguised as mixed drinks). Consequently, a 2 × 3 factorial design was employed, with husband's aggressiveness at 2 levels (aggressive or nonaggressive) and husband's alcohol condition at 3 levels (no alcohol, alcohol, and placebo). The response variable observed during the marital interaction was severity of conflict (measured on a 100-point scale).

a. A partial ANOVA table is shown below. Fill in the missing degrees of freedom.

Source	df	F
Aggressiveness (A)	–	16.43 ($p < .001$)
Alcohol Condition (C)	–	6.00 ($p < .01$)
A × C	–	–
Error	129	
Total	–	

b. Interpret the p-value of the F-test for Aggressiveness.

c. Interpret the p-value of the F-test for Alcohol Condition.

d. The F-test for interaction was omitted from the article. Discuss the dangers of making inferences based on the tests, parts **a** and **b**, without knowing the result of the interaction test.

10.87 Impact of vitamin-B supplement. In the *Journal of Nutrition* (July 1995), University of Georgia researchers examined the impact of a vitamin-B supplement (nicotinamide) on the kidney. The experimental "subjects" were 28 Zucker rats—a species that tends to develop kidney problems. Half of the rats were classified as obese and half as lean. Within each group, half were randomly assigned to receive a vitamin-B supplemented diet and half were not. Thus, a 2×2 factorial experiment was conducted with seven rats assigned to each of the four combinations of size (lean or obese) and diet (supplemental or not). One of the response variables measured was weight (in grams) of the kidney at the end of a 20-week feeding period. The data (simulated from summary information provided in the journal article) are shown in the table.

⊙ VITAMINB

		Diet			
		Regular		Vitamin-B Supplement	
Rat Size	**Lean**	1.62	1.47	1.51	1.63
		1.80	1.37	1.65	1.35
		1.71	1.71	1.45	1.66
		1.81		1.44	
	Obese	2.35	2.84	2.93	2.63
		2.97	2.05	2.72	2.61
		2.54	2.82	2.99	2.64
		2.93		2.19	

a. Conduct an analysis of variance on the data. Summarize the results in an ANOVA table.

b. Conduct the appropriate ANOVA F-tests at $\alpha = .01$. Interpret the results.

10.88 Violent lyrics and aggressiveness. In the *Journal of Personality and Social Psychology* (May 2003), psychologists investigated the potentially harmful effects of violent music lyrics. The researchers theorized that listening to a song with violent lyrics will lead to more violent thoughts and actions. A total of 60 undergraduate college students participated in one experiment designed by the researchers. Half of the students were volunteers and half were required to participate as part of their introductory psychology class. Each student listened to a song by the group "Tool"—half the students were randomly assigned a song with violent lyrics and half assigned a song with nonviolent lyrics. Consequently, the experiment used a 2×2 factorial design with the factors Song (violent, nonviolent) and Pool (volunteer, psychology class). After listening to the song, each student was given a list of word pairs and asked to rate the similarity of each word in the pair on a 7-point scale. One word in each pair was aggressive in meaning (e.g., *choke*) and the other was ambiguous (e.g., *night*). An aggressive cognition score was assigned based on the average word-pair scores. (The higher the score, the more the subject associated an ambiguous word with a violent word.) The data (simulated) are shown in the next table. Conduct a complete analysis of variance on the data.

Applying the Concepts—Advanced

10.89 Comparing male and female firefighters. How do women compare with men in their ability to perform laborious tasks that require strength? Some information on this question is provided in a study of the

⊙ LYRICS

	Volunteer					Psychology Class				
Violent Song	4.1	3.5	3.4	4.1	3.7	3.4	3.9	4.2	3.2	4.3
	2.8	3.4	4.0	2.5	3.0	3.3	3.1	3.2	3.8	3.1
	3.4	3.5	3.2	3.1	3.6	3.8	4.1	3.3	3.8	4.5
Non-Violent Song	2.4	2.4	2.5	2.6	3.6	2.5	2.9	2.9	3.0	2.6
	4.0	3.3	3.7	2.8	2.9	2.4	3.5	3.3	3.7	3.3
	3.2	2.5	2.9	3.0	2.4	2.8	2.5	2.8	2.0	3.1

firefighting ability of men and women. (*Human Factors*, 1982.) The researchers conducted a 2×2 factorial experiment to investigate the effect of the factor Sex (male or female) and the factor Weight (light or heavy) on the length of time required for a person to perform a particular firefighting task. Eight persons were selected for each of the $2 \times 2 = 4$ Sex–Weight categories of the 2×2 factorial experiment, and the length of time needed to complete the task was recorded for each of the 32 persons. The means and standard deviations of the four samples are shown in the following table.

	Light		Heavy	
	Mean	Standard Deviation	Mean	Standard Deviation
Female	18.30	6.81	14.50	2.93
Male	13.00	5.04	12.25	5.70

Source: M. D. Phillips and R. L. Pepper, "Shipboard firefighting performance of females and males." Reprinted with permission from *Human Factors*, Vol. 24, No. 3, 1982. Copyright 1982 by the Human Factors and Ergonomics Society. All rights reserved.

a. Calculate the total of the $n = 8$ time measurements for each of the four categories of the 2×2 factorial experiment.

b. Calculate the correction for mean, CM. (See Appendix C for computational formulas.)

c. Use the results of parts **a** and **b** to calculate the sums of squares for Sex, Weight, and for the Sex $\times$ Weight interaction.

d. Calculate each sample variance. Then calculate the sum of squares of deviations *within* each sample for each of the four samples.

e. Calculate SSE. [*Hint:* SSE is the pooled sum of squares for the deviations calculated in part **d**.]

f. Now that you know SS(Sex), SS(Weight), SS(Sex $\times$ Weight), and SSE, find SS(Total).

g. Summarize the calculations in an ANOVA table.

h. Conduct a complete analysis of these data. Use $\alpha = .05$ for any inferential techniques you employ. Interpret your conclusions graphically.

i. What assumptions are necessary to ensure the validity of the inferential techniques you utilized? State them in terms of this experiment.

Quick Review

Key Terms

Analysis of variance (ANOVA) 527
Balanced design 564
Blocks 545
Bonferroni multiple comparisons
  procedure 538
Comparisonwise error rate 538
Complete factorial experiment 560
Completely randomized design 520
Dependent variable 514
Designed experiment 515
Experimental unit 515
Experimentwise error rate 538
Factorial experiment 560

Factor interaction 562
Factor levels 514
Factor main effect 562
Factors 514
F-statistic 523
Mean square for error 523
Mean square for treatments 523
Multiple comparisons of means 538
Observational experiment 515
Qualitative factors 514
Quantitative factors 514
Randomized block design 545
Replicates of the experiment 564

Response variable 514
Robust method 529
Scheffé multiple comparisons
  procedure 538
Single-factor experiment 560
Sum of squares for blocks 546
Sum of squares for error 522
Sum of squares for treatment 522
Treatments 515
Tukey multiple comparisons
  procedure 538
Two-factor experiment 562
Two-way classification 561

Key Formulas

Note: Computing formulas for sums of squares (SS) and mean squares (MS) in ANOVA are provided in Appendix C.

Completely randomized design:

$$F = \frac{MST}{MSE}$$ Testing treatments 525

$$c = k(k - 1)/2$$ Number of pairwise comparisons for p treatment means 537

Randomized block design:

$$F = \frac{MST}{MSE}$$ Testing treatments 548

$$F = \frac{MSB}{MSE}$$ Testing blocks 554

Factorial design with 2 factors:

$$F = \frac{MS(A)}{MSE}$$ Testing main effect A 564

$$F = \frac{MS(B)}{MSE}$$ Testing main effect B 565

$$F = \frac{MS(AB)}{MSE}$$ Testing $A \times B$ interaction 564

Language Lab

Symbol	Description
ANOVA	Analysis of variance
SST	Sum of Squares for Treatments (i.e., the variation among treatment means)
SSE	Sum of Squares for Error (i.e., the variability around the treatment means due to sampling error)
MST	Mean Square for Treatments
MSE	Mean Square for Error (an estimate of σ^2)
SSB	Sum of Squares for Blocks
MSB	Mean Square for Blocks

$a \times b$ factorial	Two-factor factorial experiment with one factor at a levels and the other at b levels (thus, there are $a \times b$ treatments in the experiment)
SS(A)	Sum of Squares for Factor A
MS(A)	Mean Square for Factor A
SS(B)	Sum of Squares for Factor B
MS(B)	Mean Square for Factor B
SS(AB)	Sum of Square for $A \times B$ interaction
MS(AB)	Mean Square for $A \times B$ interaction

Chapter Summary Notes

- Key elements of a **designed experiment**: *response variable* (quantitative), *factors* (quantitative or qualitative), *factor levels* (values of each factor selected by the experimenter), *treatments* (combinations of factor-levels), *experimental units*.

- Characteristics of a **completely randomized design**: a single factor; levels of the factor are the treatments; independent random samples selected for each treatment, or, experimental units randomly assigned to a treatment.

- Characteristics of a **randomized block design**: a single factor and a set of matched experimental units (blocks); levels of the factor are the treatments; one experimental unit from each block is randomly assigned to each treatment.

- Characteristics of a **complete factorial design**: two factors; combinations of all possible factor levels are the treatments; independent random samples selected for each treatment, or experimental units randomly assigned to a treatment.

- A **balanced design** is one where the sample sizes for each treatment are equal.

- Conditions required for a valid **ANOVA F-test in a completely randomized design**: (1) all k treatment populations are approximately normal, (2) $\sigma_1^2 = \sigma_2^2 = \cdots = \sigma_k^2$.

- Conditions required for valid **ANOVA F-tests in a randomized block design**: (1) all treatment–block populations are approximately normal, (2) all treatment–block populations have the same variance.

- Conditions required for valid **ANOVA F-tests in a factorial design**: (1) all treatment populations are approximately normal, (2) all treatment populations have the same variance.

- ANOVA is a **robust method**—slight to moderate departures from normality do not have an impact on the validity of the results.

- The **experimentwise error rate** is the risk of making at least one Type I error when making multiple comparisons in an ANOVA.

- **Multiple comparisons methods** for controlling the experimentwise error rate: *Tukey, Bonferroni*, and *Scheffé*.

- **Tukey's method** is appropriate when (1) the treatment sample sizes are equal and (2) pairwise comparisons of treatment means are desired.

- **Bonferroni's method** is appropriate when (1) the treatment sample sizes are equal or unequal and (2) pairwise comparisons of treatment means are desired.

- **Scheffé's method** is appropriate when (1) the treatment sample sizes are equal or unequal and (2) general contrasts involving treatment means are desired.

- **Tests for main effects** in a factorial design are only appropriate if the **test for interaction is nonsignificant**.

Supplementary Exercises 10.90–10.114

Understanding the Principles

10.90 What is the difference between a one-way ANOVA and a two-way ANOVA?

10.91 Explain the difference between an experiment that utilizes a completely randomized design and one that utilizes a randomized block design.

10.92 What are the treatments in a two-factor experiment, with factor A at three levels and factor B at two levels?

10.93 Why does the experimentwise error rate of a multiple comparisons procedure differ from the significance level for each comparison (assuming the experiment has more than two treatments)?

Learning the Mechanics

10.94 A completely randomized design is utilized to compare four treatment means. The data are shown in the table on p. 580.

a. Given that SST = 36.95 and SS(Total) = 62.55, complete an ANOVA table for this experiment.

b. Is there evidence that the treatment means differ? Use $\alpha = .10$.

LM10_94

Treatment 1	Treatment 2	Treatment 3	Treatment 4
8	6	9	12
10	9	10	13
9	8	8	10
10	8	11	11
11	7	12	11

10.95 An experiment utilizing a randomized block design was conducted to compare the mean responses for four treatments, A, B, C, and D. The treatments were randomly assigned to the four experimental units in each of five blocks. The data are shown in the following table.

EX10_95

	Block				
Treatment	1	2	3	4	5
A	8.6	7.5	8.7	9.8	7.4
B	7.3	6.3	7.3	8.4	6.3
C	9.1	8.3	9.0	9.9	8.2
D	9.3	8.2	9.2	10.0	8.4

a. Given that SS(Total) = 22.31 and SS(Block) = 10.688 and SSE = .288, complete an ANOVA table for the experiment.
b. Do the data provide sufficient evidence to indicate a difference among treatment means? Test using $\alpha = .05$
c. Does the result of the test in part **b** warrant further comparison of the treatment means? If so, how many pairwise comparisons need to be made?
d. Is there evidence that the block means differ? Use $\alpha = .05$.

10.96 The table shows a partially completed ANOVA table for a two-factor factorial experiment.

Source	df	SS	MS	F
A	3	2.6		
B	5	9.2		
A × B				3.1
Error		18.7		
Total	47			

a. Complete the ANOVA table.
b. How many levels were used for each factor? How many treatments were used? How many replications were performed?
c. Find the value of the Sum of Squares for Treatments. Test to determine whether the data provide evidence that the treatment means differ. Use $\alpha = .05$.
d. Is further testing of the nature of factor effects warranted? If so, test to determine whether the factors interact. Use $\alpha = .05$. Interpret the result.

Applying the Concepts—Basic

10.97 Strength of fiberboard boxes. The *Journal of Testing and Evaluation* (July 1992) published an investigation of the mean compression strength of corrugated fiberboard shipping containers. Comparisons were made for boxes of five different sizes: A, B, C, D, and E. Twenty identical boxes of each size were tested and the peak compression strength (pounds) was recorded for each box. The figure below shows the sample means for the five box types as well as the variation around each sample mean.

a. Explain why the data are collected as a completely randomized design.
b. Refer to box types B and D. Based on the graph, does it appear that the mean compressive strengths of these two box types are significantly different? Explain.
c. Based on the graph, does it appear that the mean compressive strengths of all five box types are significantly different? Explain.

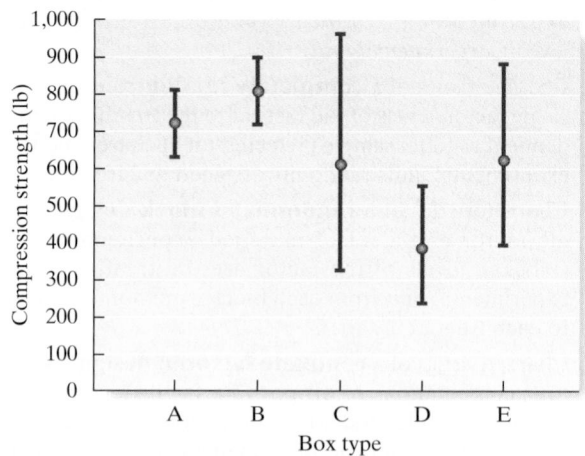

Source: Singh, S. P., et al. "Compression of single-wall corrugated shipping containers using fixed and floating test platens." *Journal of Testing and Evaluation*, Vol. 20, No. 4, July 1992, p. 319 (Figure 3). Copyright American Society for Testing and Materials.

10.98 Shoulder rotation experiment. *Physical Therapy* (Aug. 1986) reported on a study of the medial rotation of the shoulder. Ten college students, who exhibited no evidence of disease or limitation of movement in their shoulders, served as the subjects for the study. For each subject, the medial rotation was measured (in degrees) at each of three positions: (1) beginning position, (2) point at which rotation changed directions, and (3) ending position. The goal of the analysis is to compare the mean medial rotation measurements of the three positions.

a. Explain why a randomized block design is appropriate in this experiment.
b. Identify the treatments in this experiment.
c. Identify the blocks in this experiment.
d. Identify the response variable.

10.99 Children's use of pronouns. Refer to the *Journal of Communication Disorders* (Mar. 1995) study of pronoun misusage by specifically language-impaired (SLI) children, Exercise 2.65 (p. 65). Recall that percentage of pronoun errors was recorded for three groups of children: ten 5-year-old SLI children, ten younger (3-year-old) normally developing (YND) children, and ten older (5-year-old) normally developing (OND) children. A completely randomized design ANOVA with multiple comparisons of means was conducted on the data, producing the results shown here. Interpret these results.

Source	df	SS	MS	F	p-Value
Groups	2	11,292.45	5,646.22	8.295	.0016
Error	27	18,377.65	680.65		
Total	29	29,670.10			

Mean percentage of errors:	0.00	30.17	46.88
Group:	OND	SLI	YND

Source: Moore, M. E. "Error analysis of pronouns by normal and language-impaired children." *Journal of Communication Disorders*, Vol. 28, No. 1, Mar. 1995, p. 67 (Table 6).

10.100 Retelling a story problem. In a study published in the *American Journal of Psychology* (Winter 1991), 14-year-old students from New York City were divided into four groups: (1) learning-disabled, low socioeconomic status; (2) nondisabled, low socioeconomic status; (3) learning-disabled, high socioeconomic status; and (4) nondisabled, high socioeconomic status. Each student was asked to read and retell a "story problem"; however, the way in which the problem was presented was varied. Some students read a "no-priority" problem (i.e., a problem with no clear goals and/or objectives), while others read a "priority" problem (i.e., a problem with a clear statement of the character's priority). The experiment was designed as a 4 × 2 factorial, with Group at 4 levels and Problem type (priority or no-priority) at 2 levels. One of the dependent variables measured was proportion of ideas recalled correctly.

a. The test for Group by Problem type interaction was nonsignificant at $\alpha = .01$. Interpret this result.

b. The test for Problem type main effect was nonsignificant at $\alpha = .01$. Interpret this result.

c. The test for Group main effect was statistically significant at $\alpha = .01$. Interpret this result.

10.101 Detecting distorted responses. The Minnesota Multiphasic Personality Inventory (MMPI) is a questionnaire used to gauge personality type. *Psychological Assessment* (Mar. 1995) published a study that investigated the effectiveness of the MMPI scales in detecting deliberately distorted responses. A completely randomized design with four treatments was employed. The treatments consisted of independent random samples of females in the following four groups: nonforensic psychiatric patients ($n_1 = 65$), forensic psychiatric patients ($n_2 = 28$), college students who were requested to respond honestly ($n_3 = 140$), and college students who were instructed to provide "fake bad" responses ($n_4 = 45$). All 278 participants were given the MMPI and the score on a scale designed to assess response distortion was recorded for each. An analysis of variance was conducted on the data.

a. The ANOVA yielded $F = 39.1$. Do the mean scores of the four groups completing the MMPI differ significantly? Use $\alpha = .05$.

b. If the MMPI is effective in detecting distorted responses, then the mean score for the "fake bad" treatment group will be largest. Based on the information provided, can the researchers make an inference about the effectiveness of the MMPI? Explain.

c. The Bonferroni method was used to rank the means of the four groups using an experimentwise error rate of $\alpha = .05$. The results are shown in the table. Interpret the results.

Mean Score	12.0	12.1	20.8	21.0
Group	FP	NFP	CSH	CSFB

Source: Bagby, R. M., Burs, T., and Nicholson, R. A. "Relative effectiveness of the standard validity scales in detecting fake-bad and fake-good responding: Replication and extension." *Psychological Assessment*, Vol. 7, No. 1, Mar. 1995, p. 86 (Table 1).

Applying the Concepts—Intermediate

10.102 Hazardous organic solvents. The *Journal of Hazardous Materials* (July 1995) published the results of a study of the chemical properties of three different types of hazardous organic solvents used to clean metal parts: aromatics, chloroalkanes, and esters. One variable studied was sorption rate, measured as mole percentage. Independent samples of solvents from each type were tested and their sorption rates were recorded, as shown in the table.

⊙ SORPRATE

Aromatics		Chloroalkanes		Esters		
1.06	.95	1.58	1.12	.29	.43	.06
.79	.65	1.45	.91	.06	.51	.09
.82	1.15	.57	.83	.44	.10	.17
.89	1.12	1.16	.43	.61	.34	.60
1.05				.55	.53	.17

Source: Reprinted from *Journal of Hazardous Materials*, Vol. 42, No. 2, J. D. Ortego et al., "A review of polymeric geosynthetics used in hazardous waste facilities," p. 142 (Table 9), July 1995, Elsevier Science-NL, Sara Burgerhartstraat 25, 1055 KV Amsterdam. The Netherlands.

a. Construct an ANOVA table for the data.

b. Is there evidence of differences among the mean sorption rates of the three organic solvent types? Test using $\alpha = .10$.

c. If appropriate, conduct multiple comparisons using $\alpha = .10$.

⊙ OILSPILL

10.103 Hull failures of oil tankers Refer to the *Marine Technology* (Jan. 1995) study of major ocean oil spills by tanker vessels, Exercise 2.182 (p. 112). The spillage amounts (thousands of metric tons) and cause of accident for 48 tankers are saved in the OILSPILL file. (*Note:* Delete the two tankers with oil spills of unknown causes.)

a. Conduct an analysis of variance (at $\alpha = .01$) to compare the mean spillage amounts for the four accident types: (1) collision, (2) grounding, (3) fire/explosion, and (4) hull failure. Interpret your results.

b. If appropriate, use Bonferroni's method to rank the spillage means of the four accident types. Use $\alpha = .01$.

10.104 Who makes the decisions? An article in the *Journal of Family Violence* (Mar. 1992) investigated the differences in the opinions of adolescents and their parents when determining who should make decisions about a variety of issues. Thirty-five adolescents and their parents were independently given a list of 44 issues common to adolescent children (e.g., curfew, homework, chores, etc.). Each was asked to state how many of the 44 issues the adolescent should make decisions about on his or her own. The results are presented below.

	Mother	Father	Adolescent
Mean	7.57	7.51	15.94
Std. dev.	6.07	5.93	9.11

a. State the hypotheses necessary to determine whether parents and adolescents disagree on who should make decisions about issues common to adolescent children.

b. The article reported an F-statistic of $F = 15.92$. Conduct the test using $\alpha = .05$. [*Hint:* A total of 105 people were sampled.]

c. The Bonferroni technique was used to compare the pairs of means at an overall significance level of $\alpha = .06$. The resulting confidence intervals are shown here. Interpret the intervals.

Comparison	Confidence Interval
(Mother – Father)	$.06 \pm 4.06$
(Mother – Adolescent)	-8.37 ± 4.06
(Father – Adolescent)	-8.43 ± 4.06

10.105 Availability of food for marsh ducks. Ducks inhabiting the Great Salt Lake marshes feed on a variety of animals, including water boatmen, brine shrimp, beetles, and snails. The changes in the availability of these animal species for ducks during the summer was investigated (*Wetlands*, March 1995). The goal was to compare the mean amount (measured as biomass) of a particular duck food species across four different summer time periods: (1) July 9–23, (2) July 24–Aug. 8, (3) Aug. 9–23, and (4) Aug. 24–31. Ten stations in the marshes were randomly selected, and the biomass density in a water specimen collected from each was measured. Biomass measurements (milligrams per square meter) were collected during each of the four summer time periods at each station, with stations treated as a blocking factor. Thus, the data were analyzed as a randomized block design.

a. Fill in the missing degrees of freedom in the randomized block ANOVA table shown here.

Source	df	F	p-Value
Time Period	—	11.25	.0001
Station	—	—	—
Error	—		
Total	39		

b. The F-value (and corresponding p-value) shown in the ANOVA table, part **a**, were computed from an analysis of biomass of water boatmen nymphs (a common duck food). Interpret these results.

c. A multiple comparisons of time period means was conducted using an experimentwise error rate of .05. The results are summarized at the bottom of the page. Identify the time period(s) with the largest and smallest mean biomass.

10.106 Testing the ability to perform left-handed tasks. Most people are right-handed due to the propensity of the left hemisphere of the brain to control sequential movement. Similarly, the fact that some tasks are performed better with the left hand is likely due to the superiority of the right hemisphere of the brain to process the necessary information. Does such cerebral specialization in spatial processing occur in adults with Down syndrome? A 2×2 factorial experiment was conducted to answer this question. (*American Journal on Mental Retardation*, May 1995.) A sample of adults with Down syndrome was compared to a control group of normal individuals of a similar age. Thus, one factor was Group at two levels (Down syndrome and control) and the second factor was the Handedness (left or right) of the subject. All the subjects performed a task that typically yields a left-hand advantage. The response variable was "laterality index," measured on a -100- to 100-point scale. (A large positive index indicates a right-hand advantage, while a large negative index indicates a left-hand advantage.)

a. Identify the treatments in this experiment.

b. Construct a graph that would support a finding of no interaction between the two factors.

c. Construct a graph that would support a finding of interaction between the two factors.

d. The F-test for factor interaction yielded an observed significance level of $p < .05$. Interpret this result.

e. Multiple comparisons of all pairs of treatment means yielded the rankings shown at the top of p. 583. Interpret the results.

Multiple Comparisons for Exercise 10.105

Mean biomass (mg/m²):	19	54.5	90	148
Time period:	8/24–8/31	8/9–8/23	7/24–8/8	7/9–7/23

| Mean laterality index: | −30 | −4 | −.5 | +.5 |
| Group/Handed: | Down/Left | Control/Right | Control/Left | Down/Right |

f. The experimentwise error rate for part **e** was .05. Interpret this value.

10.107 Tests of bonding agents. An evaluation of diffusion bonding of zircaloy components is performed. The main objective is to determine which of three elements— nickel, iron, or copper—is the best bonding agent. A series of zircaloy components are bonded with each of the possible bonding agents. Since there is a great deal of variation in components machined from different ingots, a randomized block design is used, blocking on the ingots. A pair of components from each ingot are bonded together using each of the three agents, and the pressure (in units of 1,000 pounds per square inch) required to separate the bonded components is measured. The data in the table are obtained.

⊙ **INGOT**

| | Bonding Agent | | |
Ingot	Nickel	Iron	Copper
1	67.0	71.9	72.2
2	67.5	68.8	66.4
3	76.0	82.6	74.5
4	72.7	78.1	67.3
5	73.1	74.2	73.2
6	65.8	70.8	68.7
7	75.6	84.9	69.0

a. Identify the following elements of the experiment: experimental units, blocks, response, factor(s) and factor type(s), and treatments.

b. Is the experiment designed or observational? Explain.

c. Construct an ANOVA summary table for the experiment.

d. Is there sufficient evidence that the mean pressure required to separate the components differs for the three bonding agents? Use $\alpha = .05$.

e. If warranted, perform multiple comparisons of bonding agents means using $\alpha = .05$.

f. What assumptions must hold to ensure the validity of the inferences in parts **d** and **e**?

10.108 Modifying the work environment to improve attitude. Sixteen workers were randomly selected to participate in an experiment to determine the effects of work scheduling and method of payment on attitude toward the job. Two types of scheduling were employed, the standard 8-to-5 workday and a modification in which the worker was permitted to start the day at either 7 or 8 a.m. and to vary the starting time as desired; in addition, the worker was allowed to choose, on a daily basis, either a $\frac{1}{2}$-hour or 1-hour lunch period. The two methods of payment were a standard hourly rate and a reduced hourly rate with an added piece rate based on the worker's production. Four workers were randomly assigned to each of the four Scheduling–Payment combinations, and each completed an attitude test after 1 month on the job. The test scores are shown in the table.

⊙ **WORKSCHD**

Scheduling	Hourly Rate	Hourly and Piece Rate
8–5	54, 68, 55, 63	89, 75, 71, 83
Worker modified	79, 65, 62, 74	83, 94, 91, 86

a. What type of experiment was performed? Identify the response, factor(s), factor type(s), treatments, and experimental units.

b. Perform an ANOVA to compare the treatment means. Use $\alpha = .05$.

c. If the test in part **b** warrants further analysis, conduct the appropriate tests of interaction and main effects. Interpret your results.

d. What assumptions are necessary to ensure the validity of the inferences? State the assumptions in terms of this experiment.

10.109 Price & display marketing study. Complete factorial designs are commonly employed in marketing research to evaluate the effectiveness of sales strategies. At one supermarket, two of the factors were Price level (regular, reduced price, cost to supermarket) and Display level (normal display space, normal display space plus end-of-aisle display, twice the normal display space). Each of the $3 \times 3 = 9$ treatments was applied three times to a particular product for a full week. The dependent variable of interest was unit sales for the week. To minimize treatment carryover effects, each treatment was preceded and followed by a week in which the product was priced at its regular price and was displayed in its normal manner. The accompanying table reports the data collected.

a. Form an ANOVA table for the study.

b. Do the data indicate that the mean sales differ among the nine treatments? Test using $\alpha = .10$.

⊙ **SUPMARK**

Display	Regular			Reduced			Cost to Supermarket		
Normal	989	1,025	1,030	1,211	1,215	1,182	1,577	1,559	1,598
Normal Plus	1,191	1,233	1,221	1,860	1,910	1,926	2,492	2,527	2,511
Twice Normal	1,226	1,202	1,180	1,516	1,501	1,498	1,801	1,833	1,852

c. Is the test of interaction between the factors Price and Display warranted as a result of the test in part **b**? If so, conduct the test using $\alpha = .10$.

d. Are the tests of the main effects for Price and Display warranted as a result of the previous tests? If so, conduct them using $\alpha = .10$.

e. Which pairs of treatment means should be compared as a result of the tests in parts **b–d**? Perform the comparisons.

10.110 Mosquito insecticide study. A species of Caribbean mosquito is known to be resistant against certain insecticides. The effectiveness of five different types of insecticides—temephos, malathion, fenitrothion, fenthion, and chlorpyrifos—in controlling this mosquito species was investigated in the *Journal of the American Mosquito Control Association* (March 1995). Mosquito larvae were collected from each of seven Caribbean locations. In a laboratory, the larvae from each location were divided into five batches and each batch exposed to one of the five insecticides. The dosage of insecticide required to kill 50% of the larvae was recorded and divided by the known dosage for a susceptible mosquito strain. The resulting value is called the resistance ratio. (The higher the ratio, the more resistant the mosquito species is to the insecticide relative to the susceptible mosquito strain.) The resistance ratios for the study are listed in the table at the bottom of the page. The researchers want to compare the mean resistance ratios of the five insecticides.

a. Explain why the experimental design is a randomized block design. Identify the treatments and the blocks.

b. Conduct a complete analysis of the data. Are any of the insecticides more effective than any of the others?

10.111 Short-day traits of lemmings. Many temperate-zone animal species exhibit physiological and morphological changes when the hours of daylight begin to decrease during autumn months. A study was conducted to investigate the "short-day" traits of collared lemmings. (*The Journal of Experimental Zoology*, Sept. 1993.) A total of 124 lemmings were bred in a colony maintained with a photoperiod of 22 hours of light per day.

At weaning (19 days of age), the lemmings were weighed and randomly assigned to live under one of two photoperiods: 16 hours or less of light per day and more than 16 hours of light per day. (Each group was assigned the same number of males and females.) After 10 weeks, the lemmings were weighed again. The response variable of interest was the gain in body weight (measured in grams) over the 10-week experimental period. The researchers analyzed the data using an ANOVA for a 2×2 factorial design, where the two factors are Photoperiod (at two levels) and Gender (at two levels).

a. Construct an ANOVA table for the experiment, listing the sources of variation and associated degrees of freedom.

b. The *F*-test for interaction was not significant. Interpret this result practically.

c. The *p*-values for testing for Photoperiod and Gender main effects were both smaller than .001. Interpret these results practically.

10.112 Ranging behavior of Spanish cattle. The cattle inhabiting the Biological Reserve of Doñana (Spain), live under free-range conditions, with virtually no human interference. The cattle population is organized into four herds (LGN, MTZ, PLC, and QMD). The *Journal of Zoology* (July 1995) investigated the ranging behavior of the four herds across the four seasons. Thus, a 4×4 factorial experiment was employed, with Herd and Season representing the two factors. Three animals from each herd during each season were sampled and the home range of each individual was measured (in square kilometers). The data were subjected to an ANOVA, with the results shown in the table.

Source	df	F	p-Value
Herd (H)	3	17.2	$p < .001$
Season (S)	3	3.0	$p < .05$
H × S	9	1.2	$p > .05$
Error	32		
Total	47		

a. Conduct the appropriate ANOVA *F*-tests and interpret the results.

MOSQUITO

	Insecticide				
Location	Temephos	Malathion	Fenitrothion	Fenthion	Chlorpyrifos
Anguilla	4.6	1.2	1.5	1.8	1.5
Antigua	9.2	2.9	2.0	7.0	2.0
Dominica	7.8	1.4	2.4	4.2	4.1
Guyana	1.7	1.9	2.2	1.5	1.8
Jamaica	3.4	3.7	2.0	1.5	7.1
St. Lucia	6.7	2.7	2.7	4.8	8.7
Suriname	1.4	1.9	2.0	2.1	1.7

Source: Rawlins, S. C., and Oh Hing Wan, J. "Resistance in some Caribbean population of *Aedes aegypti* to several insecticides." *Journal of the American Mosquito Control Association,* Vol. 11, No. 1, Mar. 1995 (Table 1).

b. The researcher ranked the four herd means independently of season. Do you agree with this strategy? Explain.

c. Refer to part **b**. The Bonferroni rankings of the four herd means (at $\alpha = .05$) are shown below. Interpret the results.

Mean (km)^2:	.75	1.0	2.7	3.8
Herd:	PLC	LGN	QMD	MTZ

Applying the Concepts—Advanced

10.113 Testing a new insecticide. Traditionally, people protect themselves from mosquito bites by applying insect repellent to their skin and clothing. Recent research suggests that peremethrin, an insecticide with low toxicity to humans, can provide protection from mosquitoes. A study in the *Journal of the American Mosquito Control Association* (Mar. 1995) investigated whether a tent sprayed with a commercially available 1% peremethrin formulation would protect people, both inside and outside the tent, against biting mosquitoes. Two canvas tents—one treated with peremethrin, the other untreated—were positioned 25 meters apart on flat dry ground in an area infested with mosquitoes. Eight people participated in the experiment, with four randomly assigned to each tent. Of the four stationed at each tent, two were randomly assigned to stay inside the tent (at opposite corners) and two to stay outside the tent (at opposite corners). During a specified 20-minute period during the night, each person kept count of the number of mosquito bites received. The goal of the study was to determine the effect of both Tent type (treated or untreated) and Location (inside or outside the tent) on the mean mosquito bite count.

a. What type of design was employed in the study?

b. Identify the factors and treatments.

c. Identify the response variable.

d. The study found statistical evidence of interaction between Tent type and Location. Give a practical interpretation of this result.

Critical Thinking Challenge

10.114 Exam performance study. Refer to the *Teaching of Psychology* (Aug. 1998) study of whether a practice test helps students prepare for a final exam, Exercise 10.10 (p. 519). Recall that students in an introductory psychology class were grouped according to their class standing and whether they attended a review session or took a practice test prior to the final exam. The experimental design was a 3×2 factorial design, with Class Standing at 3 levels (low, medium, or high) and Exam Preparation at 2 levels (practice exam or review session). There were 22 students in each of the $3 \times 2 = 6$ treatment groups. After completing the final exam, each student rated her or his exam preparation on an 11-point scale ranging from 0 (not helpful at all) to 10 (extremely helpful). The data for this experiment (simulated from summary statistics provided in the article) are saved in the **PRACEXAM** file. The first five and last five observations in the data set are listed below. Conduct a complete analysis of variance of the helpfulness ratings data, including (if warranted) multiple comparisons of means. Do your findings support the research conclusion that "students at all levels of academic ability benefit from a... practice exam"?

🌐 PRACEXAM

Exam Preparation	Class Standing	Helpfulness Rating
PRACTICE	LOW	6
PRACTICE	LOW	7
PRACTICE	LOW	7
PRACTICE	LOW	5
PRACTICE	LOW	3
⋮	⋮	⋮
REVIEW	HI	5
REVIEW	HI	2
REVIEW	HI	5
REVIEW	HI	4
REVIEW	HI	3

Source: Balch, W. R. "Practice versus review exams and final exam performance." *Teaching of Psychology*, Vol. 25, No. 3, Aug. 1998 (adapted from Table 1).

Student Projects

Due to ever-increasing food costs, consumers are becoming more discerning in their choice of supermarkets. It is usually more convenient to shop at just one market, as opposed to buying different items at different markets. Thus, it would be useful to compare the mean food expenditure for a market basket of food items from store to store. Since there is a great deal of variability in the prices of products sold at any supermarket, we will consider an experiment that blocks on products.

Choose three (or more) supermarkets in your area that you want to compare. Then choose approximately 10 (or more) food products you typically purchase. For each food item, record the price each store charges in the following manner:

Food Item 1	Food Item 2	· · ·	Food Item 10
Price store 1	Price store 1	· · ·	Price store 1
Price store 2	Price store 2	· · ·	Price store 2
Price store 3	Price store 3	· · ·	Price store 3

Use the data you obtain to test

H_0: Mean expenditures at the stores are the same

H_a: Mean expenditures for at least two of the stores are different

Also, test to determine whether blocking on food items is advisable in this kind of experiment. Fully interpret the results of your analysis.

REFERENCES

Cochran, W. G., and Cox, G. M. *Experimental Designs*, 2nd ed. New York: Wiley, 1957.

Hsu, J. C. *Multiple Comparisons: Theory and Methods*. London: Chapman & Hall, 1996.

Kramer, C. Y. "Extension of multiple range tests to group means with unequal number of replications." *Biometrics*, Vol. 12, 1956, pp. 307–310.

Mason, R. L., Gunst, R. F., and Hess, J. L. *Statistical Design and Analysis of Experiments*. New York: Wiley, 1989.

Mendenhall, W. *Introduction to Linear Models and the Design and Analysis of Experiments*. Belmont, Calif.: Wadsworth, 1968.

Miller, R. G., Jr. *Simultaneous Statistical Inference*. New York: Springer-Verlag, 1981.

Neter, J., Kutner, M., Nachtsheim, C., and Wasserman, W. *Applied Linear Statistical Models*, 4th ed. Homewood, Ill.: Richard D. Irwin, 1996.

Scheffé, H. "A method for judging all contrasts in the Analysis of Variance," *Biometrica*, Vol. 40, 1953, pp. 87–104.

Scheffé, H. *The Analysis of Variance*. New York: Wiley, 1959.

Snedecor, G. W., and Cochran, W. G. *Statistical Methods*, 7th ed. Ames: Iowa State University Press, 1980.

Steele, R. G. D., and Torrie, J. H. *Principles and Procedures of Statistics: A Biometrical Approach*, 2d ed. New York: McGraw-Hill, 1980.

Tukey, J. "Comparing individual means in the Analysis of Variance," *Biometrics*, Vol. 5, 1949, pp. 99–114.

Winer, B. J. *Statistical Principles in Experimental Design*, 2d ed. New York: McGraw-Hill, 1971.

Using Technology

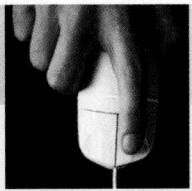

Analysis of Variance Using MINITAB

MINITAB can conduct ANOVAs for all three types of experimental designs discussed in this chapter: completely randomized, randomized block, and factorial designs.

Completely Randomized Design

To conduct a completely randomized design ANOVA, first access the MINITAB worksheet file that contains the sample data. The data file should contain one quantitative variable (the response, or dependent, variable) and one qualitative factor variable with at least two levels. Next, click on the "Stat" button on the MINITAB menu bar, then click on "ANOVA" and "One-Way," as shown in Figure 10.M.1.

The resulting dialog box appears as shown in Figure 10.M.2. Specify the response variable in the "Response" box and the factor variable in the "Factor" box. Click the "Comparisons" button, and select a multiple comparisons method

Figure 10.M.1
MINITAB Menu Options
for One-Way ANOVA

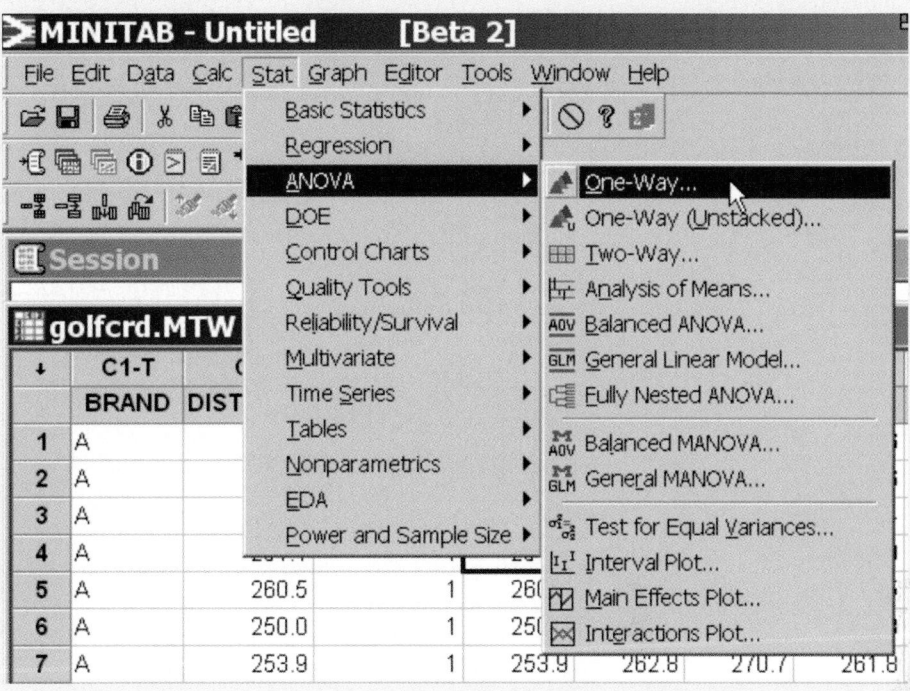

Figure 10.M.2
MINITAB One-Way
ANOVA Dialog Box

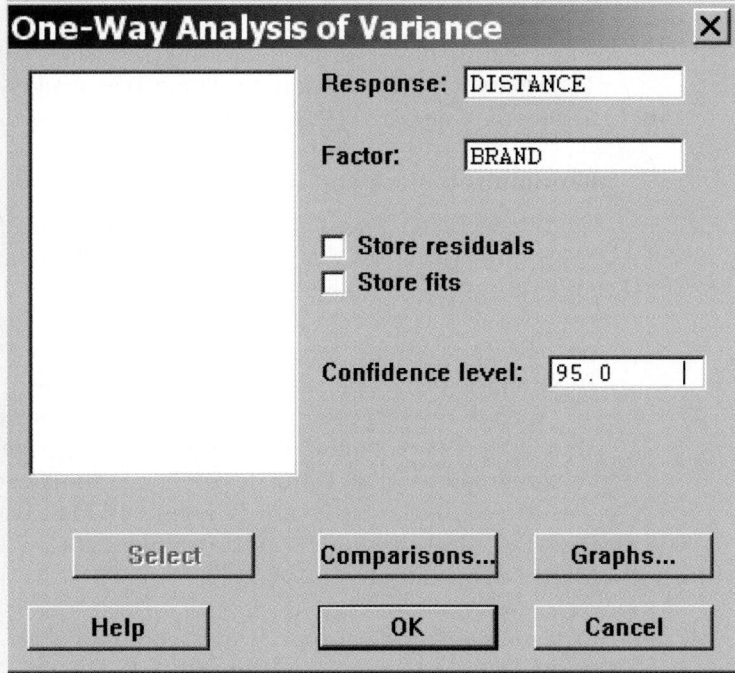

Figure 10.M.3
MINITAB Multiple
Comparisons Dialog Box

and experimentwise error rate in the resulting dialog box (see Figure 10.M.3). Then click "OK" to return to the "One-Way ANOVA" dialog screen. Click "OK" to generate the MINITAB printout.

Randomized Block and Factorial Designs

To conduct either a randomized block or factorial design ANOVA, first access the SPSS spreadsheet file that contains the sample data. The data file should contain one quantitative variable (the response, or dependent, variable) and two other variables that represent the factors and/or blocks. Next, click on the "Stat" button on the MINITAB menu bar, then click on "ANOVA" and "Two-Way" (see Figure 10.M.1). The resulting dialog box appears as shown in Figure 10.M.4.

Specify the response variable in the "Response" box, the first factor variable in the "Row factor" box, and the second factor or block variable in the "Column factor" box. If the design is a randomized block, select the "Fit additive model" option as shown in Figure 10.M.4. If the design is factorial, leave the "Fit additive model" option unselected. Click "OK" to generate the MINITAB printout.

[*Note:* Multiple comparisons of treatment means are unavailable in MINITAB for randomized block and factorial designs.]

Figure 10.M.4
MINITAB Two-Way
ANOVA Dialog Box

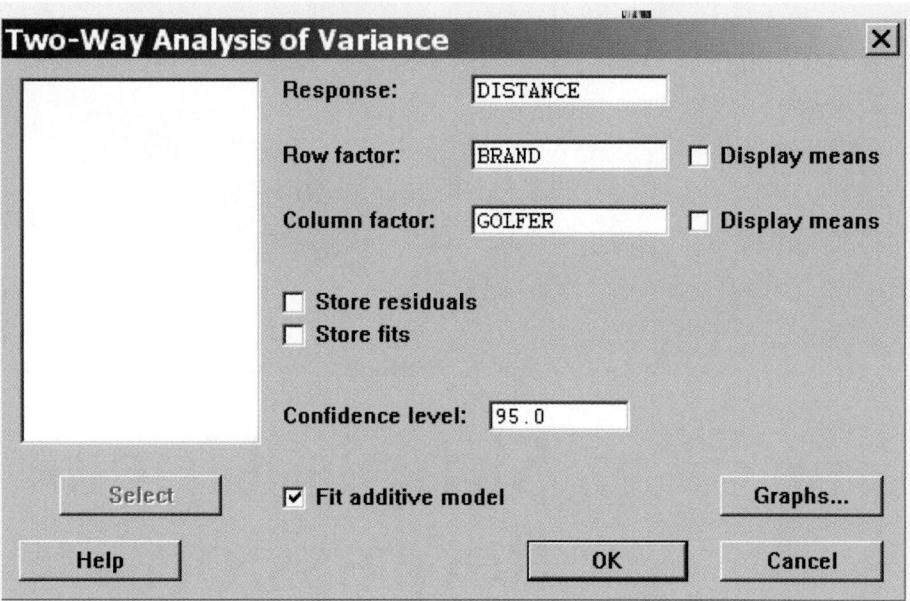

11

Simple Linear Regression

Contents

Where We've Been

- Presented methods for estimating and testing population parameters (e.g., the mean, proportion, and variance) for a single sample
- Extended these methods to allow for a comparison of population parameters for multiple samples

Where We're Going

- Introduce the straight-line (*simple linear regression*) model as a means of relating one quantitative variable to another quantitative variable.
- Introduce the *correlation coefficient* as a means of relating one quantitative variable to another quantitative variable.
- Assess how well the simple linear regression model fits the sample data.
- Utilize the simple linear regression model for predicting the value of one variable from a specified value of another variable.

Statistics in *ACTION*

Can Dowsers Really Detect Water?

The act of searching for and finding underground supplies of water using nothing more than a divining rod is commonly known as "dowsing." Although widely regarded among scientists as no more than a superstitious relic from medieval times, dowsing remains popular in folklore and, to this day, there are individuals who claim to have this mysterious skill.

Many dowsers in Germany claim that they respond to "earthrays" that emanate from the water source. These earthrays, say the dowsers, are a subtle form of radiation potentially hazardous to human health. As a result of these claims, the German government in the mid-1980s conducted a 2-year experiment to investigate the possibility that dowsing is a genuine skill. If such a skill could be demonstrated, reasoned government officials, then dangerous levels of radiation in Germany could be detected, avoided, and disposed of.

A group of university physicists in Munich, Germany, were provided a grant of 400,000 marks ($\approx$\$250,000) to conduct the study. Approximately 500 candidate dowsers were recruited to participate in preliminary tests of their skill. To avoid fraudulent claims, the 43 individuals who seemed to be the most successful in the preliminary tests were selected for the final, carefully controlled, experiment.

The researchers set up a 10-meter-long line on the ground floor of a vacant barn, along which a small wagon could be moved. Attached to the wagon was a short length of pipe, perpendicular to the test line, that was connected by hoses to a pump with running water. The location of the pipe along the line for each trial of the experiment was assigned using a computer-generated random number. On the upper floor of the barn, directly above the experimental line, a 10-meter test line was painted. In each trial, a dowser was admitted to this upper level and required, with his or her rod, stick, or other tool of choice, to ascertain where the pipe with running water on the ground floor was located.

Each dowser participated in at least one test series, that is, a sequence of from 5 to 15 trials (typically 10), with the pipe randomly repositioned after each trial. (Some

dowsers undertook only one test series, selected others underwent more than 10 test series.) Over the 2-year experimental period, the 43 dowsers participated in a total of 843 tests. The experiment was "double blind" in that neither the observer (researcher) on the top floor nor the dowser knew the pipe's location, even after a guess was made. [*Note:* Before the experiment began, a professional magician inspected the entire arrangement for potential deception or cheating by the dowsers.]

For each trial, two variables were recorded: the actual pipe location (in decimeters from the beginning of the line) and the dowser's guess (also measured in decimeters). Based on an examination of these data, the German physicists concluded in their final report that although most dowsers did not do particularly well in the experiments, "some few dowsers, in particular tests, showed an extraordinarily high rate of success, which can scarcely if at all be explained as due to chance . . . a real core of dowser-phenomena can be regarded as empirically proven . . ." (Wagner, Betz, and König, 1990).

This conclusion was critically assessed by Professor J. T. Enright of the University of California–San Diego. (*Skeptical Inquirer*, Jan./Feb. 1999.) In the Statistics in Action Revisited sections of this chapter, we demonstrate how Enright concluded the exact opposite of the German physicists.

Statistics in Action Revisited

- Estimating a Straight-Line Regression Model for the Dowsing Data (p. 603)
- Assessing How Well the Straight-Line Model Fits the Dowsing Data (p. 621)
- Using the Correlation Coefficient to Assess the Dowsing Data (p. 628)
- Predicting Pipe Location for the Dowsing Data Using the Straight-Line Model (p. 640)

In Chapters 7–10 we described methods for making inferences about population means. The mean of a population has been treated as a *constant*, and we have shown how to use sample data to estimate or to test hypotheses about this constant mean. In many applications, the mean of a population is not viewed as a constant, but rather as a variable. For example, the mean sale price of residences in a large city might be treated as a variable that depends on the square feet of living space in the residence. The relationship might be

$$\text{Mean sale price} = \$30,000 + \$60 \text{ (Square feet)}$$

This formula implies that the mean sale price of 1,000-square-foot homes is $90,000, the mean sale price of 2,000-square-foot homes is $150,000, and the mean sale price of 3,000-square-foot homes is $210,000.

In this chapter we discuss situations in which the mean of the population is treated as a variable, dependent on the value of another variable. The dependence of residential sale price on the square feet of living space is one illustration. Other examples include the dependence of mean reaction time on the amount of a drug in the bloodstream, the dependence of mean starting salary of a college graduate on the student's GPA, and the dependence of mean number of years to which a criminal is sentenced on the number of previous convictions.

Here, we present the simplest of all models relating a population mean to another variable, the *straight-line model*. We show how to use the sample data to estimate the straight-line relationship between the mean value of one variable, y, as it relates to a second variable, x. The methodology of estimating and using a straight-line relationship is referred to as *simple linear regression analysis*.

11.1 Probabilistic Models

An important consideration when taking a drug is how it may affect one's perception or general awareness. Suppose you want to model the length of time it takes to respond to a stimulus (a measure of awareness) as a function of the percentage of a certain drug in the bloodstream. The first question to be answered is this: "Do you think an exact relationship exists between these two variables?" That is, do you think it is possible to state the exact length of time it takes an individual (subject) to respond if the amount of the drug in the bloodstream is known? We think you will agree with us that this is *not* possible for several reasons. The reaction time depends on many variables other than the percentage of the drug in the bloodstream—for example, the time of day, the amount of sleep the subject had the night before, the subject's visual acuity, the subject's general reaction time without the drug, and the subject's age would all probably affect reaction time. Even if many variables are included in a model (the topic of Chapter 12), it is still unlikely that we would be able to predict *exactly* the subject's reaction time. There will almost certainly be some variation in response times due strictly to *random phenomena* that cannot be modeled or explained.

If we were to construct a model that hypothesized an exact relationship between variables, it would be called a **deterministic model**. For example, if we believe that y, the reaction time (in seconds), will be exactly one and one-half times x, the amount of drug in the blood, we write

$$y = 1.5x$$

This represents a **deterministic relationship** between the variables y and x. It implies that y can always be determined exactly when the value of x is known. *There is no allowance for error in this prediction*.

If, on the other hand, we believe there will be unexplained variation in reaction times—perhaps caused by important but unincluded variables or by random phenomena—we discard the deterministic model and use a model that accounts for this **random error**. This **probabilistic model** includes both a deterministic component and a random error component. For example, if we hypothesize that the response time y is related to the percentage of drug x by

$$y = 1.5x + \text{Random error}$$

we are hypothesizing a **probabilistic relationship** between y and x. Note that the deterministic component of this probabilistic model is $1.5x$.

Figure 11.1

Possible Reaction Times, *y*, for Five Different Drug Percentages, *x*

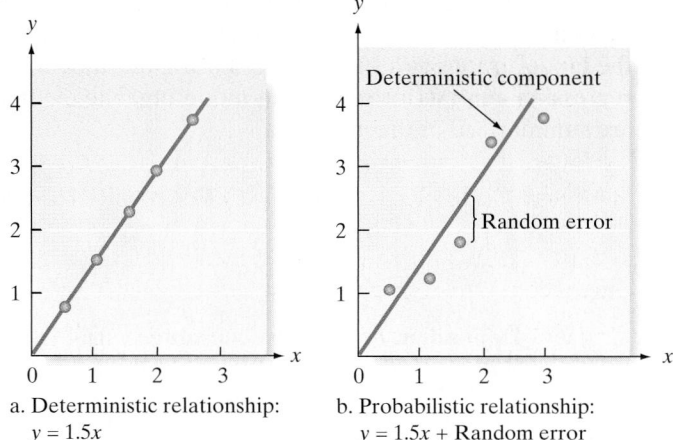

a. Deterministic relationship:
 $y = 1.5x$

b. Probabilistic relationship:
 $y = 1.5x + \text{Random error}$

Figure 11.1a shows the possible responses for five different values of *x*, the percentage of drug in the blood, when the model is deterministic. All the responses must fall exactly on the line because a deterministic model leaves no room for error.

Figure 11.1b shows a possible set of responses for the same values of *x* when we are using a probabilistic model. Note that the deterministic part of the model (the straight line itself) is the same. Now, however, the inclusion of a random error component allows the response times to vary from this line. Since we know that the response time does vary randomly for a given value of *x*, the probabilistic model provides a more realistic model for *y* than does the deterministic model.

General Form of Probabilistic Models

$$y = \text{Deterministic component} + \text{Random error}$$

where *y* is the variable of interest. We always assume that the mean value of the random error equals 0. This is equivalent to assuming that the mean value of *y*, $E(y)$, equals the deterministic component of the model; that is,

$$E(y) = \text{Deterministic component}$$

Biography

FRANCIS GALTON (1822–1911)—The Law of Universal Regression

Francis Galton was the youngest of seven children born to a middle-class English family of Quaker faith. A cousin of Charles Darwin, Galton attended Trinity College (Cambridge, England) to study medicine. Due to the death of his father, Galton was unable to obtain his degree. His competence in both medicine and mathematics, however, led Galton to pursue a career as a scientist. Galton made major contributions to the fields of genetics, psychology, meteorology, and anthropology. Some consider Galton to be the first social scientist for his applications of the novel statistical concepts of the time—in particular, regression and correlation. While studying natural inheritance in 1886, Galton collected data on heights of parents and adult children. He noticed the tendency for tall (or short) parents to have tall (or short) children, but that the children were not as tall (or short), on average as their parents. Galton called this phenomenon the "law of universal regression," for the average heights of adult children tended to "regress" to the mean of the population. Galton, with the help of his friend and disciple, Karl Pearson, applied the straight-line model to the height data, and the term *regression model* was coined.

In this chapter we present the simplest of probabilistic models—the **straight-line model**—which derives its name from the fact that the deterministic portion of the model graphs as a straight line. Fitting this model to a set of data is an example of **regression analysis**, or **regression modeling**. The elements of the straight-line model are summarized in the next box.

A First-Order (Straight-Line) Probabilistic Model

$$y = \beta_0 + \beta_1 x + \varepsilon$$

where

$y = $ **Dependent** *or* **response variable** (variable to be modeled)
$x = $ **Independent** *or* **predictor variable** (variable used as a predictor of y)*

$E(y) = \beta_0 + \beta_1 x = $ Deterministic component
ε (epsilon) $ = $ Random error component
β_0 (beta zero) $ = $ **y-intercept of the line**, that is, the point at which the line intersects or cuts through the y-axis (see Figure 11.2)
β_1 (beta one) $ = $ **Slope of the line**, that is, the amount of increase (or decrease) in the deterministic component of y for every 1-unit increase in x.

[*Note:* A *positive* slope implies that $E(y)$ *increases* by the amount β_1 (see Figure 11.2). A *negative* slope implies that $E(y)$ *decreases* by the amount β_1.]

Figure 11.2
The Straight-Line Model

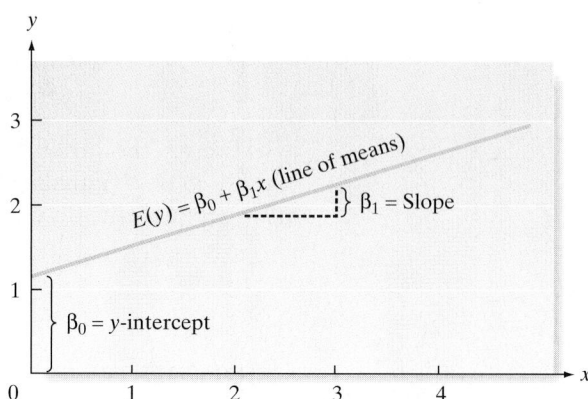

In the probabilistic model, the deterministic component is referred to as the **line of means**, because the mean of y, $E(y)$, is equal to the straight-line component of the model. That is,

$$E(y) = \beta_0 + \beta_1 x$$

Note that the Greek symbols β_0 and β_1, respectively, represent the y-intercept and slope of the model. They are population parameters that will be known only if we have access to the entire population of (x, y) measurements. Together with a specific value of the independent variable x, they determine the mean value of y, which is just a specific point on the line of means (Figure 11.2).

The values of β_0 and β_1 will be unknown in almost all practical applications of regression analysis. The process of developing a model, estimating the unknown parameters, and using the model can be viewed as the five-step procedure shown in the next box.

---

*The word *independent* should not be interpreted in a probabilistic sense, as defined in Chapter 3. The phrase *independent variable* is used in regression analysis to refer to a predictor variable for the response y.

> **Step 1** Hypothesize the deterministic component of the model that relates the mean, $E(y)$, to the independent variable x (Section 11.1).
>
> **Step 2** Use the sample data to estimate unknown parameters in the model (Section 11.2).
>
> **Step 3** Specify the probability distribution of the random error term and estimate the standard deviation of this distribution (Sections 11.3 and 11.4).
>
> **Step 4** Statistically evaluate the usefulness of the model (Sections 11.5, 11.6, and 11.7).
>
> **Step 5** When satisfied that the model is useful, use it for prediction, estimation, and other purposes (Section 11.8).

Exercises 11.1–11.10

Understanding the Principles

11.1 Why do we generally prefer a probabilistic model to a deterministic model? Give examples for which the two types of models might be appropriate.

11.2 What is the difference between a dependent variable and an independent variable in a probabilistic model?

11.3 What is the line of means?

11.4 If a straight-line probabilistic relationship relates the mean $E(y)$ to an independent variable x, does it imply that every value of the variable y will always fall exactly on the line of means? Why or why not?

Learning the Mechanics

11.5 In each case, graph the line that passes through the given points.
  a. $(1, 1)$ and $(5, 5)$
  b. $(0, 3)$ and $(3, 0)$
  c. $(-1, 1)$ and $(4, 2)$
  d. $(-6, -3)$ and $(2, 6)$

11.6 Give the slope and y-intercept for each of the lines graphed in Exercise 11.5.

11.7 The equation for a straight line (deterministic) is

$$y = \beta_0 + \beta_1 x$$

If the line passes through the point $(-2, 4)$, then $x = -2$, $y = 4$ must satisfy the equation; that is,

$$4 = \beta_0 + \beta_1(-2)$$

Similarly, if the line passes through the point $(4, 6)$, then $x = 4$, $y = 6$ must satisfy the equation; that is,

$$6 = \beta_0 + \beta_1(4)$$

Use these two equations to solve for β_0 and β_1; then find the equation of the line that passes through the points $(-2, 4)$ and $(4, 6)$.

11.8 Refer to Exercise 11.7. Find the equations of the lines that pass through the points listed in Exercise 11.5.

11.9 Plot the following lines:
  a. $y = 4 + x$ **b.** $y = 5 - 2x$
  c. $y = -4 + 3x$ **d.** $y = -2x$
  e. $y = x$ **f.** $y = .50 + 1.5x$

11.10 Give the slope and y-intercept for each of the lines defined in Exercise 11.9.

11.2 Fitting the Model: The Least Squares Approach

After the straight-line model has been hypothesized to relate the mean $E(y)$ to the independent variable x, the next step is to collect data and to estimate the (unknown) population parameters, the y-intercept β_0 and the slope β_1.

To begin with a simple example, suppose an experiment involving five subjects is conducted to determine the relationship between the percentage of a certain drug in the bloodstream and the length of time it takes to react to a stimulus. The results are shown in Table 11.1. (The number of measurements and the measurements themselves are unrealistically simple in order to avoid arithmetic confusion in this introductory example.) This set of data will be used to demonstrate the five-step procedure of regression modeling given in Section 11.1. In this section we hypothesize the deterministic component of the model and estimate its unknown parameters (steps 1 and 2). The model assumptions and the random error component (step 3) are the subjects of

⊙ STIMULUS

TABLE 11.1   Reaction Time versus Drug Percentage

Subject	Amount of Drug x (%)	Reaction Time y (seconds)
1	1	1
2	2	1
3	3	2
4	4	2
5	5	4

Sections 11.3 and 11.4, whereas Sections 11.5–11.7 assess the utility of the model (step 4). Finally, we use the model for prediction and estimation (step 5) in Section 11.8.

Step 1 *Hypothesize the deterministic component of the probabilistic model.* As stated before, we will consider only straight-line models in this chapter. Thus the complete model to relate mean response time $E(y)$ to drug percentage x is given by

$$E(y) = \beta_0 + \beta_1 x$$

Step 2 *Use sample data to estimate unknown parameters in the model.* This step is the subject of this section — namely, how can we best use the information in the sample of five observations in Table 11.1 to estimate the unknown y-intercept β_0 and slope β_1?

To determine whether a linear relationship between y and x is plausible, it is helpful to plot the sample data in a **scattergram**. Recall (Section 2.9) that a scattergram locates each of the five data points on a graph, as shown in Figure 11.3. Note that the scattergram suggests a general tendency for y to increase as x increases. If you place a ruler on the scattergram, you will see that a line may be drawn through three of the five points, as shown in Figure 11.4. To obtain the equation of this visually fitted line, note that the line intersects the y-axis at $y = -1$, so the y-intercept is -1. Also, y increases exactly 1 unit for every 1-unit increase in x, indicating that the slope is $+1$. Therefore, the equation is

$$\tilde{y} = -1 + 1(x) = -1 + x$$

where $\tilde{y}$ is used to denote the predicted y from the visual model.

Figure 11.3
Scattergram for Data
in Table 11.1

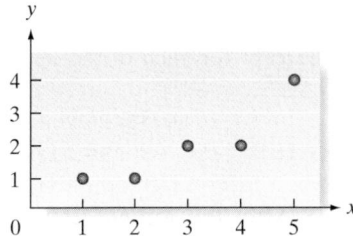

Figure 11.4
Visual Straight Line Fitted
to the Data in Figure 11.3

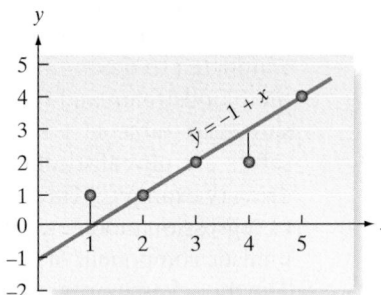

TABLE 11.2   Comparing Observed and Predicted Values for the Visual Model

x	y	$\widetilde{y} = -1 + x$	$(y - \widetilde{y})$	$(y - \widetilde{y})^2$
1	1	0	$(1 - 0) = 1$	1
2	1	1	$(1 - 1) = 0$	0
3	2	2	$(2 - 2) = 0$	0
4	2	3	$(2 - 3) = -1$	1
5	4	4	$(4 - 4) = 0$	0
			Sum of errors = 0	Sum of squared errors (SSE) = 2

One way to decide quantitatively how well a straight line fits a set of data is to note the extent to which the data points deviate from the line. For example, to evaluate the model in Figure 11.4, we calculate the magnitude of the *deviations* (i.e., the differences between the observed and the predicted values of y). These deviations, or **errors of prediction**, are the vertical distances between observed and predicted values (see Figure 11.4). The observed and predicted values of y, their differences, and their squared differences are shown in Table 11.2. Note that the *sum of errors* equals 0 and the *sum of squares of the errors* (SSE), which gives greater emphasis to large deviations of the points from the line, is equal to 2.

You can see by shifting the ruler around the graph that it is possible to find many lines for which the sum of errors is equal to 0, but it can be shown that there is one (and only one) line for which the SSE is a *minimum*. This line is called the **least squares line**, the **regression line**, or the **least squares prediction equation**. The methodology used to obtain this line is called the **method of least squares**.

> **Now Work**   *Exercise 11.16a-d*

■ ■ ■

To find the least squares prediction equation for a set of data, assume that we have a sample of n data points consisting of pairs of values of x and y, say $(x_1, y_1), (x_2, y_2), \ldots, (x_n, y_n)$. For example, the $n = 5$ data points shown in Table 11.2 are $(1, 1), (2, 1), (3, 2), (4, 2)$, and $(5, 4)$. The fitted line, which we will calculate based on the five data points, is written as

$$\hat{y} = \hat{\beta}_0 + \hat{\beta}_1 x$$

The "hats" indicate that the symbols below them are estimates: $\hat{y}$ (y-hat) is an estimator of the mean value of y, $E(y)$, and a predictor of some future value of y; and $\hat{\beta}_0$ and $\hat{\beta}_1$ are estimators of β_0 and β_1, respectively.

For a given data point, say the point (x_i, y_i), the observed value of y is y_i and the predicted value of y would be obtained by substituting x_i into the prediction equation:

$$\hat{y}_i = \hat{\beta}_0 + \hat{\beta}_1 x_i$$

And the deviation of the ith value of y from its predicted value is

$$(y_i - \hat{y}_i) = [y_i - (\hat{\beta}_0 + \hat{\beta}_1 x_i)]$$

Then the sum of squares of the deviations of the y-values about their predicted values for all the n data points is

$$\text{SSE} = \sum [y_i - (\hat{\beta}_0 + \hat{\beta}_1 x_i)]^2$$

The quantities $\hat{\beta}_0$ and $\hat{\beta}_1$ that make the SSE a minimum are called the **least squares estimates** of the population parameters β_0 and β_1, and the prediction equation $\hat{y} = \hat{\beta}_0 + \hat{\beta}_1 x$ is called the *least squares line*.

> **DEFINITION 11.1**
>
> The **least squares line** is one that has the following two properties:
>
> **1.** the sum of the errors (SE) equals 0
> **2.** the sum of squared errors (SSE) is smaller than that for any other straight-line model

The values of $\hat{\beta}_0$ and $\hat{\beta}_1$ that minimize the SSE are (proof omitted) given by the formulas in the next box.*

> **Formulas for the Least Squares Estimates**
>
> *Slope:* $\hat{\beta}_1 = \dfrac{SS_{xy}}{SS_{xx}}$
>
> *y-intercept:* $\hat{\beta}_0 = \bar{y} - \hat{\beta}_1 \bar{x}$
>
> where
>
> $$SS_{xy} = \sum (x_i - \bar{x})(y_i - \bar{y}) = \sum x_i y_i - \frac{\left(\sum x_i\right)\left(\sum y_i\right)}{n}$$
>
> $$SS_{xx} = \sum (x_i - \bar{x})^2 = \sum x_i^2 - \frac{\left(\sum x_i\right)^2}{n}$$
>
> n = Sample size

EXAMPLE 11.1   APPLYING THE METHOD OF LEAST SQUARES

Problem Refer to the reaction data presented in Table 11.1. Consider the straight-line model, $E(y) = \beta_0 + \beta_1 x$, where y = reaction time (in seconds) and x = amount of drug (percentage).

a. Use the method of least squares to estimate the values of β_0 and β_1.
b. Predict the reaction time when $x = 2\%$.
c. Find SSE for the analysis.
d. Give practical interpretations to $\hat{\beta}_0$ and $\hat{\beta}_1$.

Solution **a.** Preliminary computations for finding the least squares line for the drug reaction example are presented in Table 11.3. We can now calculate

$$SS_{xy} = \sum x_i y_i - \frac{\left(\sum x_i\right)\left(\sum y_i\right)}{5} = 37 - \frac{(15)(10)}{5} = 37 - 30 = 7$$

*Students who are familiar with calculus should note that the values of β_0 and β_1 that minimize SSE $= \Sigma(y_i - \hat{y}_i)^2$ are obtained by setting the two partial derivatives $\partial SSE/\partial \beta_0$ and $\partial SSE/\partial \beta_1$ equal to 0. The solutions to these two equations yield the formulas shown in the box. Furthermore, we denote the *sample* solutions to the equations by $\hat{\beta}_0$ and $\hat{\beta}_1$, where the "hat" denotes that these are sample estimates of the true population intercept β_0 and slope β_1.

TABLE 11.3   Preliminary Computations for the Drug Reaction Example

	x_i	y_i	x_i^2	$x_i y_i$
	1	1	1	1
	2	1	4	2
	3	2	9	6
	4	2	16	8
	5	4	25	20
Totals	$\sum x_i = 15$	$\sum y_i = 10$	$\sum x_i^2 = 55$	$\sum x_i y_i = 37$

$$\text{SS}_{xx} = \sum x_i^2 - \frac{\left(\sum x_i\right)^2}{5} = 55 - \frac{(15)^2}{5} = 55 - 45 = 10$$

Then the slope of the least squares line is

$$\hat{\beta}_1 = \frac{\text{SS}_{xy}}{\text{SS}_{xx}} = \frac{7}{10} = .7$$

and the y-intercept is

$$\hat{\beta}_0 = \bar{y} - \hat{\beta}_1 \bar{x} = \frac{\sum y_i}{5} - \hat{\beta}_1 \frac{\sum x_i}{5}$$

$$= \frac{10}{5} - (.7)\left(\frac{15}{5}\right) = 2 - (.7)(3) = 2 - 2.1 = -.1$$

The least squares line is thus

$$\hat{y} = \hat{\beta}_0 + \hat{\beta}_1 x = -.1 + .7x$$

The graph of this line is shown in Figure 11.5.

b. The predicted value of y for a given value of x can be obtained by substituting into the formula for the least squares line. Thus, when $x = 2$ we predict y to be

$$\hat{y} = -.1 + .7x = -.1 + .7(2) = 1.3$$

We show how to find a prediction interval for y in Section 11.8.

c. The observed and predicted values of y, the deviations of the y values about their predicted values, and the squares of these deviations are shown in Table 11.4. Note that the sum of squares of the deviations, SSE, is 1.10, and (as we would expect) this is less than the SSE = 2.0 obtained in Table 11.2 for the visually fitted line.

d. The estimated y-intercept, $\hat{\beta}_0 = -.1$, appears to imply that the estimated mean reaction time is equal to $-.1$ second when the amount of drug, x, is equal to 0%. Since negative reaction times are not possible, this seems

Figure 11.5
The line $\hat{y} = -.1 + .7x$
Fitted to the Data

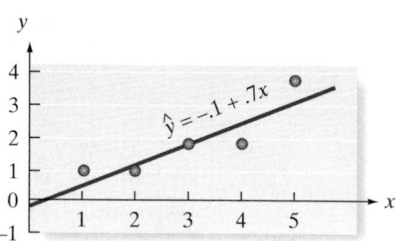

TABLE 11.4 Comparing Observed and Predicted Values for the Least Squares Prediction Equation

x	y	$\hat{y} = -.1 + .7x$	$(y - \hat{y})$	$(y - \hat{y})^2$
1	1	.6	$(1 - .6) =$ .4	.16
2	1	1.3	$(1 - 1.3) = -.3$	.09
3	2	2.0	$(2 - 2.0) =$ 0	.00
4	2	2.7	$(2 - 2.7) = -.7$	.49
5	4	3.4	$(4 - 3.4) =$ .6	.36
			Sum of errors = 0	SSE = 1.10

to make the model nonsensical. However, *the model parameters should be interpreted only within the sampled range of the independent variable*—in this case, for amounts of drug in the bloodstream between 1% and 5%. Thus, the y-intercept—which is, by definition, at $x = 0$ (0% drug)—is not within the range of the sampled values of x and is not subject to meaningful interpretation.

The slope of the least squares line, $\hat{\beta} = .7$, implies that for every unit increase of x, the mean value of y is estimated to increase by .7 unit. In terms of this example, for every 1% increase in the amount of drug in the bloodstream, the mean reaction time is estimated to increase by .7 second *over the sampled range of drug amounts from 1% to 5%*. Thus, the model does not imply that increasing the drug amount from 5% to 10% will result in an increase in mean reaction time of 3.5 seconds, because the range of x in the sample does not extend to 10% ($x = 10$). In fact, 10% might be such a high concentration that the drug would kill the subject! Be careful to interpret the estimated parameters only within the sampled range of x.

Look Back The calculations required to obtain $\hat{\beta}_0$, $\hat{\beta}_1$, and SSE in simple linear regression, although straightforward, can become rather tedious. Even with the use of a pocket calculator, the process is laborious and susceptible to error, especially when the sample size is large. Fortunately, the use of statistical computer software can significantly reduce the labor involved in regression calculations. The SAS, SPSS, and MINITAB, outputs for the simple linear regression of the data in Table 11.1 are displayed in Figure 11.6a–c. The values of $\hat{\beta}_0$ and $\hat{\beta}_1$ are highlighted

Figure 11.6a

SAS Printout for the Time-Drug Regression

```
                              Dependent Variable: TIME_Y

                    Number of Observations Read        5
                    Number of Observations Used        5


                              Analysis of Variance

                                   Sum of          Mean
    Source              DF        Squares        Square    F Value    Pr > F

    Model                1        4.90000       4.90000      13.36    0.0354
    Error                3        1.10000       0.36667
    Corrected Total      4        6.00000


              Root MSE              0.60553    R-Square     0.8167
              Dependent Mean       2.00000    Adj R-Sq     0.7556
              Coeff Var           30.27650


                              Parameter Estimates

                          Parameter     Standard
    Variable     DF        Estimate        Error    t Value    Pr > |t|

    Intercept     1        -0.10000      0.63509      -0.16      0.8849
    DRUG_X        1         0.70000      0.19149       3.66      0.0354
```

Figure 11.6b
SPSS Printout for the
Time-Drug Regression

Model Summary

Model	R	R Square	Adjusted R Square	Std. Error of the Estimate
1	.904[a]	.817	.756	.606

a. Predictors: (Constant), DRUG_X

ANOVA[b]

Model		Sum of Squares	df	Mean Square	F	Sig.
1	Regression	4.900	1	4.900	13.364	.035[a]
	Residual	1.100	3	.367		
	Total	6.000	4			

a. Predictors: (Constant), DRUG_X

b. Dependent Variable: TIME_Y

Coefficients[a]

Model		Unstandardized Coefficients		Standardized Coefficients	t	Sig.
		B	Std. Error	Beta		
1	(Constant)	-.100	.635		-.157	.885
	DRUG_X	.700	.191	.904	3.656	.035

a. Dependent Variable: TIME_Y

Figure 11.6c
MINITAB Printout for
the Time-Drug Regression

Regression Analysis: TIME_Y versus DRUG_X

```
The regression equation is
TIME_Y = - 0.100 + 0.700 DRUG_X


Predictor      Coef   SE Coef      T       P
Constant    -0.1000    0.6351   -0.16   0.885
DRUG_X       0.7000    0.1915    3.66   0.035


S = 0.605530    R-Sq = 81.7%    R-Sq(adj) = 75.6%


Analysis of Variance

Source           DF       SS       MS       F       P
Regression        1   4.9000   4.9000   13.36   0.035
Residual Error    3   1.1000   0.3667
Total             4   6.0000
```

on the printouts. These values, $\hat{\beta}_0 = -.1$ and $\hat{\beta}_1 = .7$, agree exactly with our hand-calculated values. The value of SSE = 1.10 is also highlighted on the printouts.

Now Work *Exercise 11.18*

■ ■ ■

Simple Linear Regression

Using the TI-83 Graphing Calculator

I. Finding the least squares regression equation

Step 1   *Enter the data*
Press **STAT** and select **1:Edit**
Note: If a list already contains data, clear the old data. Use the up arrow to highlight the list name, '**L1**' or '**L2**.'

Press **CLEAR ENTER**.
Enter your x-data in **L1** and your y-data in **L2**.

Step 2   *Find the equation*
Press **STAT** and highlight **CALC**
Press 4 for **LinReg(ax + b)**
Press **ENTER**
The screen will show the values for *a* and *b* in the equation **y = ax + b.**

Example   The figures below show a table of data entered on the TI-83 and the regression equation obtained using the steps given above.

```
L1      L2      L3     2

2       67      ------
1.9     68
2.3     70
3.9     74
6.4     78
7.4     81
6.8     82

L2(7) =82
```

```
LinReg
 y=ax+b
 a=2.495967742
 b=63.3391129
```

II. Graphing the least squares line with the scatterplot

Step 1   *Enter the data as shown in part I above*

Step 2   *Set up the data plot*
Press **2nd Y =** for **STAT PLOT**
Press **1** for **Plot1**
Set the cursor so that **ON** is flashing.
For **Type**, use the arrow and Enter keys to highlight and select the scatterplot (first icon in the first row).
For **Xlist**, choose the column containing the x-data.
For **Freq**, choose the column containing the y-data.

Step 3   *Find the regression equation and store the equation in Y1*
Press **STAT** and highlight **CALC**

Press **4** for **LinReg(ax + b)** (Note: Don't press ENTER here because you want to store the regression equation in Y1)
Press **VARS**
Use the right arrow to highlight **Y-VARS**
Press **ENTER** to select **1:Function**
Press **ENTER** to select **1:Y1**
Press **ENTER**

Step 4 *View the scatterplot and regression line*
Press **ZOOM** and then press 9 to select **9:ZoomStat**
You should see the data graphed along with the regression line.

Example The figure below shows a graph of the scatterplot and least squares line obtained using the steps given above.

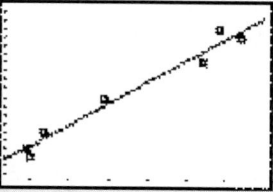

Even when the interpretations of the estimated parameters in a simple linear regression are meaningful, we need to remember that they are only estimates based on the sample. As such, their values will typically change in repeated sampling. How much confidence do we have that the estimated slope, $\hat{\beta}_1$, accurately approximates the true slope, β_1? This requires statistical inference, in the form of confidence intervals and tests of hypotheses, which we address in Section 11.5.

To summarize, we defined the best-fitting straight line to be the one that minimizes the sum of squared errors around the line, and we called it the least squares line. We should interpret the least squares line only within the sampled range of the independent variable. In subsequent sections we show how to make statistical inferences about the model.

Statistics in Action Revisited

Estimating a Straight-Line Regression Model for the Dowsing Data

After conducting a series of experiments in a Munich barn, a group of German physicists concluded that dowsing (i.e., the ability to find underground water with a divining rod) "can be regarded as empirically proven." This observation was based on the data collected for three (of the participating 500) dowsers who had particularly impressive results. All three of these "best" dowsers (numbered 99, 18, and 108) performed the experiment multiple times and the best test series (sequence of trials) for each of these three dowsers was identified. These data, saved in the **DOWSING** file, are listed in Table SIA11.1.

Recall (p. 00) that for various hidden pipe locations, each dowser provided a guess of where the pipe with running water was located. Let x = dowser's guess (in meters) and y = pipe location (in meters) for each trial. One way to determine whether the "best" dowsers are effective is to fit the straight-line model, $E(y) = \beta_0 + \beta_1 x$, to the data in Table SIA11.1.

DOWSING

TABLE SIA11.1   Dowsing Trial Results: Best Series for the Three Best Dowsers

Trial	Dowser Number	Pipe Location	Dowser's Guess
1	99	4	4
2	99	5	87
3	99	30	95
4	99	35	74
5	99	36	78
6	99	58	65
7	99	40	39
8	99	70	75
9	99	74	32
10	99	98	100
11	18	7	10
12	18	38	40
13	18	40	30
14	18	49	47
15	18	75	9
16	18	82	95
17	108	5	52
18	108	18	16
19	108	33	37
20	108	45	40
21	108	38	66
22	108	50	58
23	108	52	74
24	108	63	65
25	108	72	60
26	108	95	49

Source: Enright, J. T. "Testing dowsing: The failure of the Munich experiments." *Skeptical Inquirer*, Jan./Feb. 1999, p.45 (Figure 6a).

A MINITAB scatterplot of the data is shown in Figure SIA11.1. The least squares line, obtained from the MINITAB regression printout shown in Figure SIA11.2, is also displayed on the scatterplot. Although the least squares line has a slight upward trend, the variation of the data points around the line is large. It does not appear that a dowser's guess (x) will be a very good predictor of actual pipe location (y). In fact, the estimated slope

Figure SIA11.1
MINITAB Scatterplot of Dowsing Data

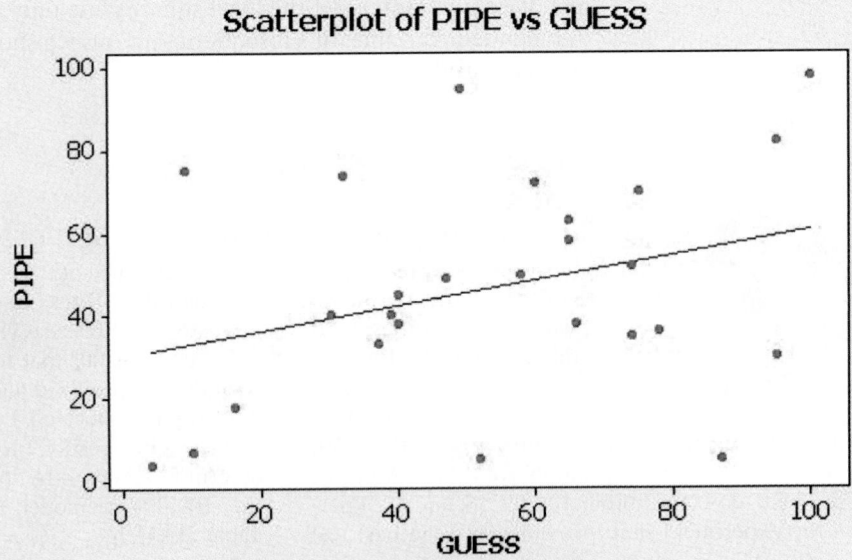

Figure SIA11.2
MINITAB Simple Linear
Regression for Dowsing
Data

Regression Analysis: PIPE versus GUESS

```
The regression equation is
PIPE = 30.1 + 0.308 GUESS


Predictor      Coef   SE Coef      T       P
Constant      30.07     11.41   2.63   0.015
GUESS        0.3079    0.1900   1.62   0.118


S = 26.0298    R-Sq = 9.9%    R-Sq(adj) = 6.1%


Analysis of Variance

Source             DF      SS      MS      F       P
Regression          1   1778.9  1778.9   2.63   0.118
Residual Error     24  16261.2   677.6
Total              25  18040.2
```

(obtained from Figure SIA11.2) is $\hat{\beta}_1 = .31$. Thus, for every 1-meter increase in a dowser's guess, we estimate that the actual pipe location will increase only .31 meter.

In the Statistics in Action Revisited sections that follow, we will provide a measure of reliability to this inference and investigate the phenomenon of dowsing further.

Exercises 11.11–11.29

Understanding the Principles

11.11 In regression, what is an error of prediction?

11.12 Give two properties of the line estimated using the method of least squares.

11.13 *True or False*. The estimates of β_0 and β_1 should be interpreted only within the sampled range of the independent variable, x.

Learning the Mechanics

11.14 The following table is similar to Table 11.3. It is used for making the preliminary computations for finding the least squares line for the given pairs of x and y values.
   a. Complete the table.   **b.** Find SS_{xy}.
   c. Find SS_{xx}.   **d.** Find $\hat{\beta}_1$.
   e. Find $\bar{x}$ and $\bar{y}$.   **f.** Find $\hat{\beta}_0$.
   g. Find the least squares line.

x_i	y_i	x_i^2	$x_i y_i$
7	2	—	—
4	4	—	—
6	2	—	—
2	5	—	—
1	7	—	—
1	6	—	—
3	5	—	—
Totals $\sum x_i =$	$\sum y_i =$	$\sum x_i^2 =$	$\sum x_i y_i =$

11.15 Refer to Exercise 11.14. After the least squares line has been obtained, the table below (which is similar to Table 11.4) can be used for (1) comparing the observed and the predicted values of y, and (2) computing SSE.

x	y	$\hat{y}$	$(y - \hat{y})$	$(y - \hat{y})^2$
7	2	—	—	—
4	4	—	—	—
6	2	—	—	—
2	5	—	—	—
1	7	—	—	—
1	6	—	—	—
3	5	—	—	—
			$\sum (y - \hat{y}) =$	SSE $= \sum (y - \hat{y})^2 =$

   a. Complete the table.
   b. Plot the least squares line on a scattergram of the data. Plot the following line on the same graph:
$$\hat{y} = 14 - 2.5x$$
   c. Show that SSE is larger for the line in part **b** than it is for the least squares line.

11.16 Construct a scattergram for the data in the following
   NW table.

x	.5	1	1.5
y	2	1	3

   a. Plot the following two lines on your scattergram:
$$y = 3 - x \quad \text{and} \quad y = 1 + x$$

b. Which of these lines would you choose to characterize the relationship between x and y? Explain.

c. Show that the sum of errors for both of these lines equals 0.

d. Which of these lines has the smaller SSE?

e. Find the least squares line for the data and compare it to the two lines described in part **a**.

11.17 Consider the following pairs of measurements:

x	5	3	−1	2	7	6	4
y	4	3	0	1	8	5	3

a. Construct a scattergram for these data.

b. What does the scattergram suggest about the relationship between x and y?

c. Given that $SS_{xx} = 43.4286$, $SS_{xy} = 39.8571$, $\bar{y} = 3.4286$, and $\bar{x} = 3.7143$, calculate the least squares estimates of β_0 and β_1.

d. Plot the least squares line on your scattergram. Does the line appear to fit the data well? Explain.

e. Interpret the y-intercept and slope of the least squares line. Over what range of x are these interpretations meaningful?

Applying the Concepts—Basic

11.18 Predicting sale prices of homes. Real estate investors, home buyers, and home owners often use the appraised value of a property as a basis for predicting sale price. Data on sale prices and total appraised values of 92 residential properties sold in 1999 in an upscale Tampa, Florida, neighborhood named Tampa Palms are saved in the **TAMPALMS** file. The first five and last five observations of the data set are listed in the table on p. 607.

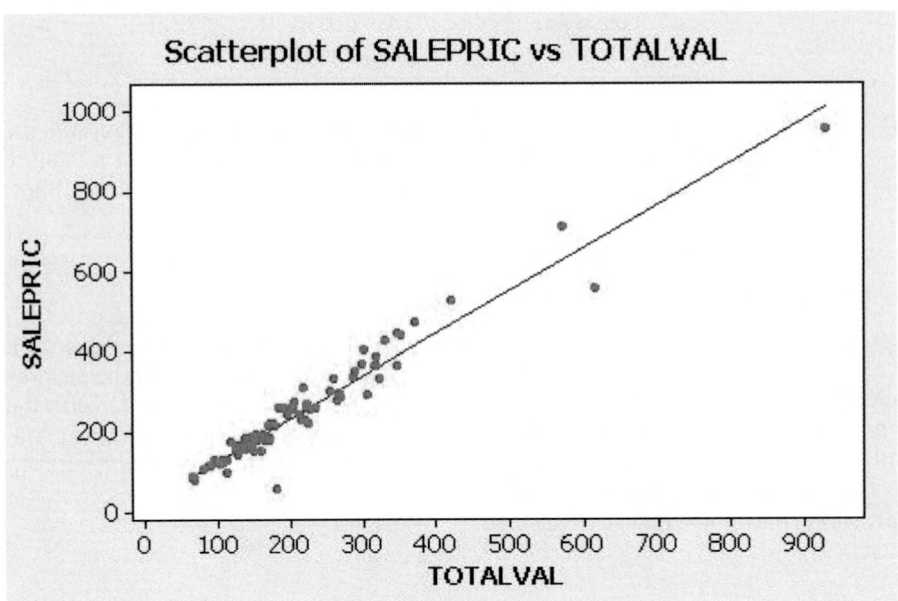

The regression equation is
SALEPRIC = 20.9 + 1.07 TOTALVAL

```
Predictor       Coef   SE Coef       T       P
Constant      20.942     6.446    3.25   0.002
TOTALVAL     1.06873   0.02709   39.45   0.000


S = 32.7865    R-Sq = 94.5%    R-Sq(adj) = 94.5%


Analysis of Variance

Source          DF        SS        MS       F       P
Regression       1   1673142   1673142  1556.48   0.000
Residual Error  90     96746      1075
Total           91   1769888
```

⊚ **TAMPALMS**

Property	Appraised Value	Sale Price
1	$ 170,432	$ 180,000
2	212,827	245,100
3	68,130	85,400
4	65,505	87,900
5	68,655	84,200
⋮	⋮	⋮
88	195,862	244,000
89	176,850	219,000
90	95,718	132,000
91	137,108	156,900
92	183,704	263,000

Source: Hillsborough County (Florida) Property Appraiser's Office.

a. Propose a straight-line model to relate the appraised property value x to the sale price y for residential properties in this neighborhood.

b. A MINITAB scatterplot of the data is shown on p. 606. [*Note:* Both sale price and total appraised value are shown in thousands of dollars.] Does it appear that a straight-line model will be an appropriate fit to the data?

c. A MINITAB simple linear regression printout is also shown on p. 606. Find the equation of the best-fitting line through the data on the printout.

d. Interpret the y-intercept of the least squares line. Does it have a practical meaning for this application? Explain.

e. Interpret the slope of the least squares line. Over what range of x is the interpretation meaningful?

f. Use the least squares model to estimate the mean sale price of a property appraised at $300,000.

11.19 College protests of labor exploitation. Refer to the *Journal of World-Systems Research* (Winter 2004) study of student "sit-ins" for a "sweat free campus" at universities, Exercise 2.145 (p. 100). Recall that the **SITIN** file contains data on the duration (in days) of each sit-in as well as the number of student arrests. The data for 5 sit-ins where there was at least one arrest are shown below. Let y = number of arrests and x = duration.

a. Give the equation of a straight-line model relating y to x.

b. Fit the model to the data for the 5 sit-ins using the method of least squares. Give the equation of the least squares prediction equation.

c. Interpret the estimates of β_0 and β_1 in the context of the problem.

⊚ **SITIN2** (Selected observations)

Sit-In	University	Duration (days)	Number of Arrests
12	Wisconsin	4	54
14	SUNY Albany	1	11
15	Oregon	3	14
17	Iowa	4	16
18	Kentucky	1	12

Source: Ross, R. J. S. "From antisweatshop to global justice to antiwar: How the new new left is the same and different from the old new left," *Journal of Word-Systems Research*, Vol. X, No. 1, Winter 2004 (Tables 1 and 3).

11.20 Winning marathon times. In *Chance* (Winter, 2000), statistician Howard Wainer and two students compared men's and women's winning times in the Boston Marathon. One of the graphs used to illustrate gender differences is reproduced below. The scattergram plots the winning times (in minutes) against year in which

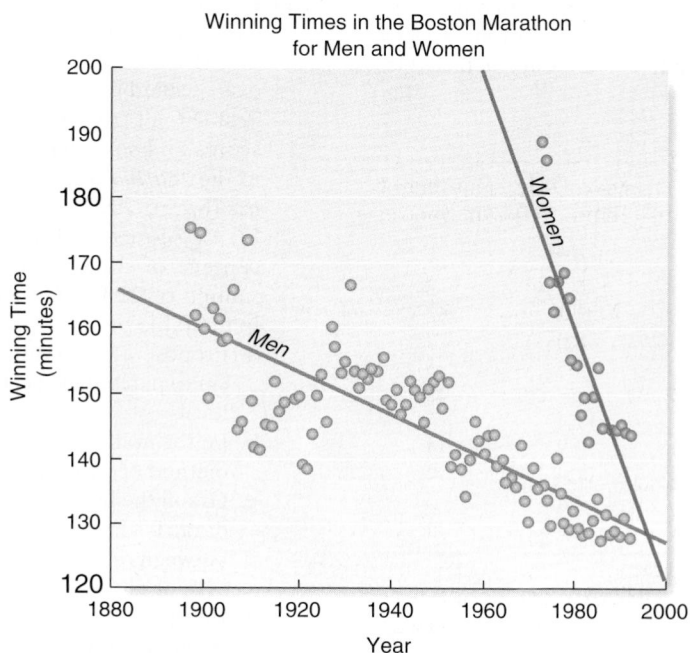

the race was run. Men's times are represented by solid dots and women's times by open circles.

a. Consider only the winning times for men. Is there evidence of a linear trend? If so, propose a straight-line model for predicting winning time (y) based on year (x). Would you expect the slope of this line to be positive or negative?

b. Repeat part **b** for women's times.

c. Which slope, men's or women's, will be greater in absolute value?

d. Would you recommend using the straight-line models to predict the winning time in the 2020 Boston Marathon? Why or why not?

11.21 Redshifts of quasi stellar objects. Astronomers call a shift in the spectrum of galaxies a "redshift." A correlation between redshift level and apparent magnitude (i.e., brightness on a logarithmic scale) of a Quasi Stellar Object (QSO) was discovered and reported in the *Journal of Astrophysics & Astronomy* (Mar./Jun., 2003). Physicist D. Basu (Carleton University, Ottawa) applied simple linear regression to data collected for a sample of over 6,000 QSOs with confirmed redshift. The analysis yielded the following results for a specific magnitude range: $\hat{y} = 18.13 + 6.21x$, where y = magnitude and x = redshift level.

a. Graph the least squares line. Is the slope of the line positive or negative?

b. Interpret the estimate of the y-intercept in the words of the problem.

c. Interpret the estimate of the slope in the words of the problem.

⊚ **GOBIANTS**
11.22 Mongolian desert ants. Refer to the *Journal of Biogeography* (Dec. 2003) study of ants in Mongolia, Exercise 2.147 (p. 100). Data on annual rainfall, maximum daily temperature, and number of ant species recorded at each of 11 study sites are listed in the table.

⊚ **GOBIANTS**

Site	Region	Annual Rainfall (mm)	Max. Daily Temp. (°C)	Number of Ant Species
1	Dry Steppe	196	5.7	3
2	Dry Steppe	196	5.7	3
3	Dry Steppe	179	7.0	52
4	Dry Steppe	197	8.0	7
5	Dry Steppe	149	8.5	5
6	Gobi Desert	112	10.7	49
7	Gobi Desert	125	11.4	5
8	Gobi Desert	99	10.9	4
9	Gobi Desert	125	11.4	4
10	Gobi Desert	84	11.4	5
11	Gobi Desert	115	11.4	4

Source: Pfeiffer, M., et al. "Community organization and species richness of ants in Mongolia along an ecological gradient from steppe to Gobi desert," *Journal of Biogeography*, Vol. 30, No. 12, Dec. 2003 (Tables 1 and 2).

a. Consider a straight-line model relating annual rainfall (y) and maximum daily temperature (x). Fit the model to the data in the **GOBIANTS** file using the method of least squares. Give the least squares prediction equation.

b. Construct a scatterplot for the analysis, part **a**. Include the least square line on the plot. Does the line appear to be a good predictor of annual rainfall?

c. Now consider a straight-line model relating number of ant species (y) to annual rainfall (x). Repeat parts **a** and **b**.

Applying the Concepts—Intermediate

11.23 Arsenic in soil. In Denver, Colorado, environmentalists have discovered a link between high arsenic levels in soil and a crabgrass killer used in the 1950s and 1960s. (*Environmental Science & Technology*, Sept. 1, 2000.) The recent discovery was based, in part, on the scattergrams shown on p. 609. The graphs plot the level of the metals cadmium and arsenic, respectively, against the distance from a former smelter plant for samples of soil taken from Denver residential properties.

a. Normally, the metal level in soil decreases as distance from the source (e.g., a smelter plant) increases. Propose a straight-line model relating metal level y to distance from the plant x. Based on the theory, would you expect the slope of the line to be positive or negative?

b. Examine the scatterplot for cadmium. Does the plot support the theory, part **a**?

c. Examine the scatterplot for arsenic. Does the plot support the theory, part **a**? (*Note:* This finding led investigators to discover the link between high arsenic levels and the use of the crabgrass killer.)

11.24 FCAT scores and poverty. In the state of Florida, elementary school performance is based on the average score obtained by students on a standardized exam, called the Florida Comprehensive Assessment Test (FCAT). An analysis of the link between FCAT scores and sociodemographic factors was published in the *Journal of Educational and Behavioral Statistics* (Spring 2004). Data on average math and reading FCAT scores of third graders, as well as the percentage of students below the poverty level, for a sample of 22 Florida elementary schools are listed in the table on p. 609.

a. Propose a straight-line model relating math score (y) to percentage (x) of students below the poverty level.

b. Fit the model to the data in the FCAT file using the method of least squares.

c. Graph the least squares line on a scattergram of the data. Is there visual evidence of a relationship between the two variables? Is the relationship positive or negative?

d. Interpret the estimates of the y-intercept and slope in the words of the problem.

Scattergrams for Exercise 11.23:

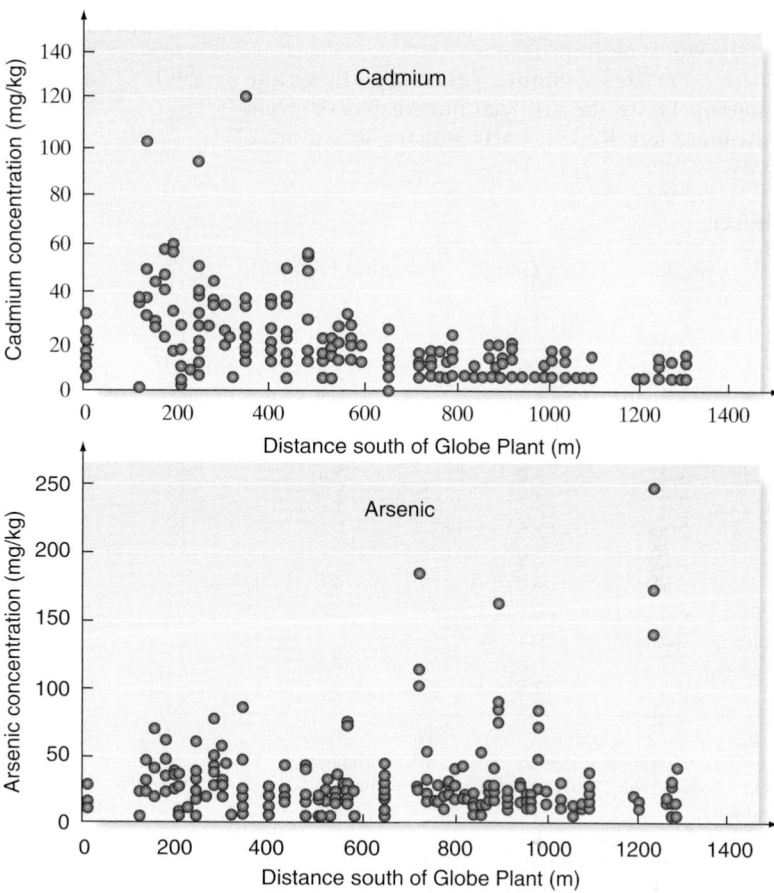

FCAT

Elementary School	FCAT—Math	FCAT—Reading	% Below Poverty
1	166.4	165.0	91.7
2	159.6	157.2	90.2
3	159.1	164.4	86.0
4	155.5	162.4	83.9
5	164.3	162.5	80.4
6	169.8	164.9	76.5
7	155.7	162.0	76.0
8	165.2	165.0	75.8
9	175.4	173.7	75.6
10	178.1	171.0	75.0
11	167.1	169.4	74.7
12	177.0	172.9	63.2
13	174.2	172.7	52.9
14	175.6	174.9	48.5
15	170.8	174.8	39.1
16	175.1	170.1	38.4
17	182.8	181.4	34.3
18	180.3	180.6	30.3
19	178.8	178.0	30.3
20	181.4	175.9	29.6
21	182.8	181.6	26.5
22	186.1	183.8	13.8

Source: Tekwe, C. D., et al. "An empirical comparison of statistical models for value-added assessment of school performance," *Journal of Educational and Behavioral Statistics*, Vol. 29, No. 1, Spring 2004 (Table 2).

e. Now consider a model relating reading score (y) to percentage (x) of students below the poverty level. Repeat parts **a–d** for this model.

11.25 New method of estimating rainfall. Accurate measurements of rainfall are critical for many hydrological and meteorological projects. Two standard methods of monitoring rainfall use rain gauges and weather radar. Both, however, can be contaminated by human and environmental interference. In the *Journal of Data Science* (Apr. 2004), researchers employed artificial neural networks (i.e., computer-based mathematical models) to estimate rainfall at a meteorological station in Montreal. Rainfall estimates were made every five minutes over a 70-minute period by each of the three methods. The data (in millimeters) are listed in the table on p. 610.

a. Propose a straight-line model relating rain gauge amount (y) to weather radar rain estimate (x).

b. Fit the model to the data in the **RAINFALL** file using the method of least squares.

c. Graph the least squares line on a scattergram of the data. Is there visual evidence of a relationship between the two variables? Is the relationship positive or negative?

d. Interpret the estimates of the y-intercept and slope in the words of the problem.

e. Now consider a model relating rain gauge amount (y) to the artificial neural network rain estimate (x). Repeat parts a–d for this model.

⊚ RAINFALL

Time	Radar	Rain Gauge	Neural Network
8:00 A.M.	3.6	0	1.8
8:05	2.0	1.2	1.8
8:10	1.1	1.2	1.4
8:15	1.3	1.3	1.9
8:20	1.8	1.4	1.7
8:25	2.1	1.4	1.5
8:30	3.2	2.0	2.1
8:35	2.7	2.1	1.0
8:40	2.5	2.5	2.6
8:45	3.5	2.9	2.6
8:50	3.9	4.0	4.0
8:55	3.5	4.9	3.4
9:00 A.M.	6.5	6.2	6.2
9:05	7.3	6.6	7.5
9:10	6.4	7.8	7.2

Source: Hessami, M. et al. "Selection of an artificial neural network model for the post-calibration of weather radar rainfall estimation," *Journal of Data Science*, Vol. 2, No. 2, Apr. 2004. (Adapted from Figures 2 and 4.)

11.26 **Sweetness of orange juice** The quality of the orange juice produced by a manufacturer (e.g., Minute Maid, Tropicana) is constantly monitored. There are numerous sensory and chemical components that combine to make the best tasting orange juice. For example, one manufacturer has developed a quantitative index of the "sweetness" of orange juice. (The higher the index, the sweeter the juice.) Is there a relationship between the sweetness index and a chemical measure such as the amount of water soluble pectin (parts per million) in the orange juice? Data collected on these two variables for 24 production runs at a juice manufacturing plant are shown in the next table. Suppose a manufacturer wants to use simple linear regression to predict the sweetness (y) from the amount of pectin (x).
a. Find the least squares line for the data.
b. Interpret $\hat{\beta}_0$ and $\hat{\beta}_1$ in the words of the problem.
c. Predict the sweetness index if amount of pectin in the orange juice is 300 ppm. [*Note:* A measure of reliability of such a prediction is discussed in Section 11.8.]

11.27 **Conversing with the hearing impaired.** A study was conducted to investigate how people with a hearing impairment communicate with their conversational partners. (*Journal of the Academy of Rehabilitative Audiology*, Vol. 27, 1994.) Each of thirteen hearing-impaired subjects, all fitted with a cochlear implant, participated in a structured communication interaction with a familiar conversational partner (a family member) and with an unfamiliar conversational partner (who was instructed not to take the initiative to repair breakdowns in

⊚ OJUICE

Run	Sweetness Index	Pectin (ppm)
1	5.2	220
2	5.5	227
3	6.0	259
4	5.9	210
5	5.8	224
6	6.0	215
7	5.8	231
8	5.6	268
9	5.6	239
10	5.9	212
11	5.4	410
12	5.6	256
13	5.8	306
14	5.5	259
15	5.3	284
16	5.3	383
17	5.7	271
18	5.5	264
19	5.7	227
20	5.3	263
21	5.9	232
22	5.8	220
23	5.8	246
24	5.9	241

Note: The data in the table are authentic. For confidentiality reasons, the manufacturer cannot be disclosed.

communication). The total number of words used by the subject in each of the two conversations is given in the table below.
a. Plot the data in a scattergram. Is there visual evidence of a linear relationship between x and y? If so, is it positive or negative?
b. Propose a straight-line model relating y to x.
c. Use the method of least squares to find the estimates of β_0 and β_1.
d. Interpret the values of $\hat{\beta}_0$ and $\hat{\beta}_1$.

⊚ HEARAID

Subject	Words with Familiar Partner x	Words with Unfamiliar Partner y
1	65	47
2	160	78
3	55	90
4	83	75
5	0	6
6	140	101
7	49	40
8	164	215
9	62	29
10	56	75
11	207	121
12	207	139
13	93	83

Source: Tye-Murray, N., et al., "Communication breakdowns: Partner contingencies and partner reactions." *Journal of the Academy of Rehabilitative Audiology*, Vol. 27, 1994, pp. 116–117 (Tables 6 & 7).

11.28 The "name game." Refer to the *Journal of Experimental Psychology—Applied* (June 2000) study in which the "name game" was used to help groups of students learn the names of other students in the group, Exercise 10.34 (p. 536). Recall that the "name game" requires the first student in the group to state his/her full name, the second student to say his/her name and the name of the first student, the third student to say his/her name and the names of the first two students, etc. After making their introductions, the students listened to a seminar speaker for 30 minutes. At the end of the seminar, all students were asked to remember the full name of each of the other students in their group and the researchers measured the proportion of names recalled for each. One goal of the study was to investigate the linear trend between y = recall proportion and x = position (order) of the student during the game. The data (simulated based on summary statistics provided in the research article) for 144 students in the first eight positions are saved in the **NAMEGAME2** file. The first five and last five observations in the data set are listed below. [*Note:* Since the student in position 1 actually must recall the names of all the other students, he or she is assigned the position number 9 in the data set.] Use the method of least squares to estimate the line, $E(y) = \beta_0 + \beta_1 x$. Interpret the β estimates in the words of the problem.

NAMEGAME2

Position	Recall
2	0.04
2	0.37
2	1.00
2	0.99
2	0.79
⋮	⋮
9	0.72
9	0.88
9	0.46
9	0.54
9	0.99

Source: Morris, P.E., and Fritz, C.O. "The name game: Using retrieval practice to improve the learning of names." *Journal of Experimental Psychology—Applied*, Vol. 6, No. 2, June 2000 (data simulated from Figure 2).

Applying the Concepts—Advanced

11.29 Long jump "takeoff error." The long jump is a track and field event in which a competitor attempts to jump a maximum distance into a sand pit after a running start. At the edge of the sand pit is a takeoff board. Jumpers usually try to plant their toes at the front edge of this board to maximize jumping distance. The absolute distance between the front edge of the takeoff board and the spot where the toe actually lands on the board prior to jumping is called "takeoff error." Is takeoff error in the long jump linearly related to best jumping distance? To answer this question, kinesiology researchers videotaped the performances of 18 novice long jumpers at a high school track meet. (*Journal of Applied Biomechanics*, May 1995.) The average takeoff error, x, and best jumping distance (out of three jumps), y, for each jumper are recorded in the table. If a jumper can reduce his/her average takeoff error by .1 meter, how much would you estimate his/her best jumping distance to change? Based on your answer, comment on the usefulness of the model for predicting best jumping distance.

LONGJUMP

Jumper	Best Jumping Distance y (meters)	Average Takeoff Error x (meters)
1	5.30	.09
2	5.55	.17
3	5.47	.19
4	5.45	.24
5	5.07	.16
6	5.32	.22
7	6.15	.09
8	4.70	.12
9	5.22	.09
10	5.77	.09
11	5.12	.13
12	5.77	.16
13	6.22	.03
14	5.82	.50
15	5.15	.13
16	4.92	.04
17	5.20	.07
18	5.42	.04

Source: Berg, W. P., and Greer, N. L. "A kinematic profile of the approach run of novice long jumpers." *Journal of Applied Biomechanics*, Vol. 11, No. 2, May 1995, p. 147 (Table 1).

11.3 Model Assumptions

In Section 11.2 we assumed that the probabilistic model relating drug reaction time y to the percentage of drug x in the bloodstream is

$$y = \beta_0 + \beta_1 x + \varepsilon$$

We also recall that the least squares estimate of the deterministic component of the model, $\beta_0 + \beta_1 x$, is

$$\hat{y} = \hat{\beta}_0 + \hat{\beta}_1 x = -.1 + .7x$$

Now we turn our attention to the random component ε of the probabilistic model and its relation to the errors in estimating β_0 and β_1. We will use a probability distribution to characterize the behavior of ε. We will see how the probability distribution of ε determines how well the model describes the relationship between the dependent variable y and the independent variable x.

Step 3 in a regression analysis requires us to specify the probability distribution of the random error ε. We will make four basic assumptions about the general form of this probability distribution:

Assumption 1 The mean of the probability distribution of ε is 0. That is, the average of the values of ε over an infinitely long series of experiments is 0 for each setting of the independent variable x. This assumption implies that the mean value of y, $E(y)$, for a given value of x is $E(y) = \beta_0 + \beta_1 x$.

Assumption 2 The variance of the probability distribution of ε is constant for all settings of the independent variable x. For our straight-line model, this assumption means that the variance of ε is equal to a constant, say σ^2, for all values of x.

Assumption 3 The probability distribution of ε is normal.

Assumption 4 The values of ε associated with any two observed values of y are independent. That is, the value of ε associated with one value of y has no effect on the values of ε associated with other y values.

The implications of the first three assumptions can be seen in Figure 11.7, which shows distributions of errors for three values of x, namely, 5, 10, and 15. Note that the relative frequency distributions of the errors are normal with a mean of 0 and a constant variance σ^2. (All the distributions shown have the same amount of spread or variability.) The straight line shown in Figure 11.7 is the line of means. It indicates the mean value of y for a given value of x. We denote this mean value as $E(y)$. Then, the line of means is given by the equation

$$E(y) = \beta_0 + \beta_1 x$$

These assumptions make it possible for us to develop measures of reliability for the least squares estimators and to develop hypothesis tests for examining the usefulness of the least squares line. We have various techniques for checking the validity of these assumptions, and we have remedies to apply when they appear to be invalid. Several of these remedies are discussed in Chapter 12. Fortunately, the assumptions need not hold exactly in order for least squares estimators to be useful. The assumptions will be satisfied adequately for many applications encountered in practice.

Figure 11.7

The Probability Distribution of ε

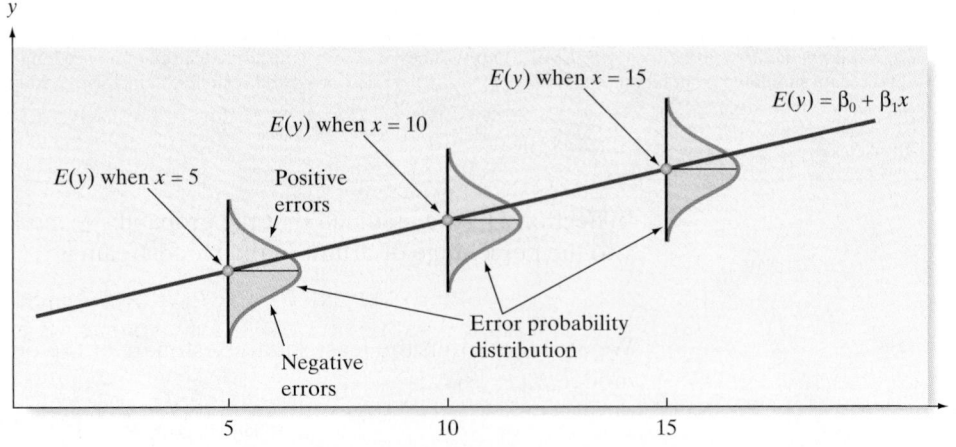

11.4 An Estimator of σ^2

It seems reasonable to assume that the greater the variability of the random error ε (which is measured by its variance σ^2), the greater will be the errors in the estimation of the model parameters β_0 and β_1 and in the error of prediction when $\hat{y}$ is used to predict y for some value of x. Consequently, you should not be surprised, as we proceed through this chapter, to find that σ^2 appears in the formulas for all confidence intervals and test statistics that we will be using.

In most practical situations, σ^2 is unknown and we must use our data to estimate its value. The best estimate of σ^2, denoted by s^2, is obtained by dividing the sum of squares of the deviations of the y values from the prediction line,

$$\text{SSE} = \sum(y_i - \hat{y}_i)^2$$

by the number of degrees of freedom associated with this quantity. We use 2 df to estimate the two parameters β_0 and β_1 in the straight-line model, leaving $(n - 2)$ df for the error variance estimation.

Estimation of σ^2 for a (First-Order) Straight-Line Model

$$s^2 = \frac{\text{SSE}}{\text{Degrees of freedom for error}} = \frac{\text{SSE}}{n - 2}$$

where $\text{SSE} = \sum(y_i - \hat{y}_i)^2 = \text{SS}_{yy} - \hat{\beta}_1\text{SS}_{xy}$

$$\text{SS}_{yy} = \sum(y_i - \bar{y})^2 = \sum y_i^2 - \frac{\left(\sum y_i\right)^2}{n}$$

To estimate the standard deviation σ of ε, we calculate

$$s = \sqrt{s^2} = \sqrt{\frac{\text{SSE}}{n - 2}}$$

We will refer to s as the **estimated standard error of the regression model**.

Warning

When performing these calculations, you may be tempted to round the calculated values of SS_{yy}, $\hat{\beta}_1$, and SS_{xy}. Be certain to carry at least six significant figures for each of these quantities to avoid substantial errors in calculation of the SSE.

EXAMPLE 11.2   ESTIMATING σ

Problem   Refer to Example 11.1 and the simple linear regression of the drug reaction data in Table 11.3.

   a.   Compute an estimate of σ.

   b.   Give a practical interpretation of the estimate.

Solution   **a.**   We previously calculated SSE = 1.10 for the least squares line $\hat{y} = -.1 + .7x$. Recalling that there were $n = 5$ data points, we have $n - 2 = 5 - 2 = 3$ df for estimating σ^2. Thus,

$$s^2 = \frac{\text{SSE}}{n - 2} = \frac{1.10}{3} = .367$$

is the estimated variance, and

$$s = \sqrt{.367} = .61$$

is the standard error of the regression model.

b. You may be able to grasp s intuitively by recalling the interpretation of a standard deviation given in Chapter 2 and remembering that the least squares line estimates the mean value of y for a given value of x. Since s measures the spread of the distribution of y values about the least squares line and these errors of prediction are assumed to be normally distributed, we should not be surprised to find that most (about 95%) of the observations lie within $2s$, or $2(.61) = 1.22$, of the least squares line. For this simple example (only five data points), all five data points fall within $2s$ of the least squares line. In Section 11.8, we use s to evaluate the error of prediction when the least squares line is used to predict a value of y to be observed for a given value of x.

Look Back The values of s^2 and s can also be obtained from a simple linear regression printout. The SAS printout for the drug reaction example is reproduced in Figure 11.8. The value of s^2 is highlighted on the printout (in the **Mean Square** column in the row labeled **Error**). The value, $s^2 = .36667$, rounded to three decimal places, agrees with the one calculated by hand. The value of s is also highlighted in Figure 11.8 (next to the heading **Root MSE**). This value, $s = .60553$, agrees (except for rounding) with our hand-calculated value.

Figure 11.8
SAS Printout for the
Time-Drug Regression

Dependent Variable: TIME_Y

| | | | | Number of Observations Read | 5 | |
| | | | | Number of Observations Used | 5 | |

Analysis of Variance

Source	DF	Sum of Squares	Mean Square	F Value	Pr > F
Model	1	4.90000	4.90000	13.36	0.0354
Error	3	1.10000	0.36667		
Corrected Total	4	6.00000			

Root MSE	0.60553	R-Square	0.8167
Dependent Mean	2.00000	Adj R-Sq	0.7556
Coeff Var	30.27650		

Parameter Estimates

Variable	DF	Parameter Estimate	Standard Error	t Value	Pr > |t|
Intercept	1	-0.10000	0.63509	-0.16	0.8849
DRUG_X	1	0.70000	0.19149	3.66	0.0354

Now Work   *Exercise 11.34ab*

■ ■ ■

Interpretation of s, the Estimated Standard Deviation of ε

We expect most ($\approx$95%) of the observed y values to lie within $2s$ of their respective least squares predicted values, $\hat{y}$.

Exercises 11.30–11.42

Understanding the Principles

11.30 What are the four assumptions made about the probability distribution of ε in regression?

11.31 Illustrate the assumptions, Exercise 11.30, with a graph.

11.32 Visually compare the scattergrams shown below. If a least squares line were determined for each data set, which do you think would have the smallest variance, s^2? Explain.

a.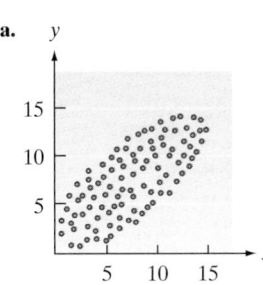

b.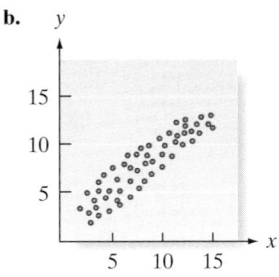

c.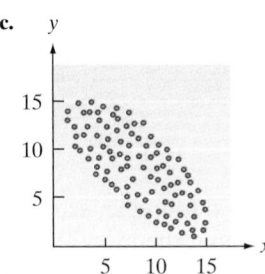

Learning the Mechanics

11.33 Calculate SSE and s^2 for each of the following cases:

a. $n = 20$, $SS_{yy} = 95$, $SS_{xy} = 50$, $\hat{\beta}_1 = .75$

b. $n = 40$, $\sum y^2 = 860$, $\sum y = 50$, $SS_{xy} = 2,700$, $\hat{\beta}_1 = .2$

c. $n = 10$, $\sum (y_i - \bar{y})^2 = 58$, $SS_{xy} = 91$, $SS_{xx} = 170$

11.34 Suppose you fit a least squares line to 12 data points
NW and the calculated value of SSE is .429.
a. Find s^2, the estimator of σ^2 (the variance of the random error term ε).
b. Find s, the estimate of σ.
c. What is the largest deviation that you might expect between any one of the 12 points and the least squares line?

11.35 Refer to Exercises 11.14 and 11.17 (p. 605-6). Calculate SSE, s^2, and s for the least squares lines obtained in those exercises. Interpret the standard error of the regression model, s, for each.

Applying the Concepts—Basic

11.36 Predicting sale prices of homes. Refer to the simple linear regression relating sale price y to appraised value x for residential properties, Exercise 11.18 (p. 606).
a. Find SSE, s^2, and s on the MINITAB printout (p. 606).
b. Interpret the value of s.

SITIN

11.37 College protests of labor exploitation. Refer to the *Journal of World-Systems Research* (Winter 2004) study of college student "sit-ins," Exercise 11.19 (p. 607). You applied the method of least squares to the data in the SITIN file to estimate the straight-line model relating number of arrests (y) and duration of sit-in (x).
a. Give the values of SSE, s^2, and s.

b. Give a practical interpretation of the value of s.

GOBIANTS

11.38 Mongolian desert ants. Refer to the *Journal of Biogeography* (Dec. 2003) study of ant sites in Mongolia, Exercise 11.22 (p. 608). You applied the method of least squares to the data in the **GOBIANTS** file to estimate the straight-line model relating annual rainfall (y) and maximum daily temperature (x).
a. Give the values of SSE, s^2, and s.

b. Give a practical interpretation of the value of s.

Applying the Concepts—Intermediate

FCAT

11.39 FCAT scores and poverty. Refer to the *Journal of Educational and Behavioral Statistics* (Spring 2004) study of scores on the Florida Comprehensive Assessment Test (FCAT), Exercise 11.24 (p. 608).
a. Consider the simple linear regression relating math score (y) to percentage (x) of students below the poverty level. Find and interpret the value of s for this regression.
b. Consider the simple linear regression relating reading score (y) to percentage (x) of students below the poverty level. Find and interpret the value of s for this regression.
c. Which dependent variable, math score or reading score, can be more accurately predicted by percentage (x) of students below the poverty level? Explain.

RAINFALL

11.40 New method of estimating rainfall. Refer to the *Journal of Data Science* (Apr. 2004) comparison of methods for estimating rainfall, Exercise 11.25 (p. 609).
a. Consider the simple linear regression relating rain gauge amount (y) to weather radar rain estimate (x). Find and interpret the value of s for this regression.
b. Consider the simple linear regression relating rain gauge amount (y) to the artificial neural network

rain estimate (x). Find and interpret the value of s for this regression.

c. Which independent variable, radar estimate or neural network estimate, is a more accurate predictor of rain gauge amount (y)? Explain.

11.41 Conversing with the hearing impaired. Refer to the simple linear regression relating total number of words in a conversation with an unfamiliar partner y to total number of words with a familiar partner x for hearing-impaired subjects, Exercise 11.27 (p. 610).

a. Find the estimated standard error of the regression, s.

b. Within how many words can you expect to predict y using this model?

Applying the Concepts—Advanced

11.42 Life tests of cutting tools. To improve the quality of the output of any production process, it is necessary first to understand the capabilities of the process (*Out of the Crisis*, Deming, 1982). In a particular manufacturing process, the useful life of a cutting tool is linearly related to the speed at which the tool is operated. The data in the accompanying table were derived from life tests for the two different brands of cutting tools

currently used in the production process. For which brand would you feel more confident in using the least squares line to predict useful life for a given cutting speed? Explain.

CUTTOOL

Cutting Speed (meters per minute)	Useful Life (hours)	
	Brand A	Brand B
30	4.5	6.0
30	3.5	6.5
30	5.2	5.0
40	5.2	6.0
40	4.0	4.5
40	2.5	5.0
50	4.4	4.5
50	2.8	4.0
50	1.0	3.7
60	4.0	3.8
60	2.0	3.0
60	1.1	2.4
70	1.1	1.5
70	.5	2.0
70	3.0	1.0

11.5 Assessing the Utility of the Model: Making Inferences About the Slope β_1

Now that we have specified the probability distribution of ε and found an estimate of the variance σ^2, we are ready to make statistical inferences about the model's usefulness for predicting the response y. This is step 4 in our regression modeling procedure.

Refer again to the data of Table 11.1 and suppose the reaction times are *completely unrelated* to the percentage of drug in the bloodstream. What could be said about the values of β_0 and β_1 in the hypothesized probabilistic model

$$y = \beta_0 + \beta_1 x + \varepsilon$$

if x contributes no information for the prediction of y? The implication is that the mean of y—that is, the deterministic part of the model $E(y) = \beta_0 + \beta_1 x$—does not change as x changes. In the straight-line model, this means that the true slope, β_1, is equal to 0 (see Figure 11.9). Therefore, to test the null hypothesis that the linear model contributes no information for the prediction of y against the alternative hypothesis that the linear model is useful for predicting y, we test

$$H_0: \beta_1 = 0$$
$$H_a: \beta_1 \neq 0$$

Figure 11.9

Graphing the Model with $\beta_1 = 0$: $y = \beta_0 + \varepsilon$

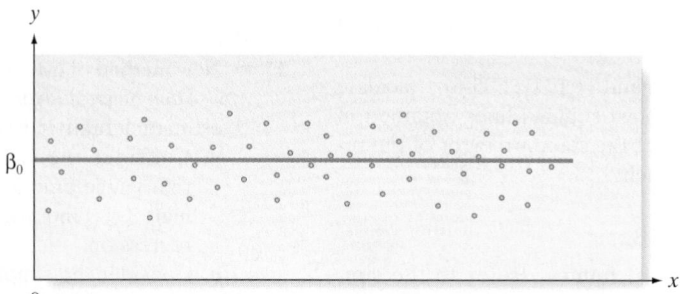

If the data support the alternative hypothesis, we will conclude that x does contribute information for the prediction of y using the straight-line model [although the true relationship between $E(y)$ and x could be more complex than a straight line]. Thus, in effect, this is a test of the usefulness of the hypothesized model.

The appropriate test statistic is found by considering the sampling distribution of $\hat{\beta}_1$, the least squares estimator of the slope β_1, as shown in the following box.

Sampling Distribution of $\hat{\beta}_1$

If we make the four assumptions about ε (see Section 11.3), the sampling distribution of the least squares estimator $\hat{\beta}_1$ of the slope will be normal with mean β_1 (the true slope) and standard deviation

$$\sigma_{\hat{\beta}_1} = \frac{\sigma}{\sqrt{SS_{xx}}} \quad \text{(see Figure 11.10)}$$

We estimate $\sigma_{\hat{\beta}_1}$ by $s_{\hat{\beta}_1} = \dfrac{s}{\sqrt{SS_{xx}}}$ and refer to this quantity as the **estimated standard error of the least squares slope $\hat{\beta}_1$.**

Figure 11.10
Sampling Distribution of $\hat{\beta}_1$

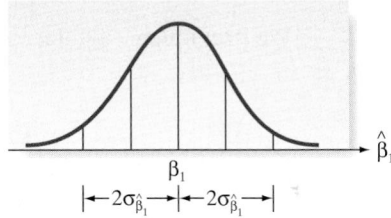

Since σ is usually unknown, the appropriate test statistic is a t-statistic, formed as follows:

$$t = \frac{\hat{\beta}_1 - \text{Hypothesized value of } \beta_1}{s_{\hat{\beta}_1}} \quad \text{where} \quad s_{\hat{\beta}_1} = \frac{s}{\sqrt{SS_{xx}}}$$

Thus,

$$t = \frac{\hat{\beta}_1 - 0}{s/\sqrt{SS_{xx}}}$$

Note that we have substituted the estimator s for σ and then formed the estimated standard error $s_{\hat{\beta}_1}$ by dividing s by $\sqrt{SS_{xx}}$. The number of degrees of freedom associated with this t statistic is the same as the number of degrees of freedom associated with s. Recall that this number is $(n - 2)$ df when the hypothesized model is a straight line (see Section 11.4). The setup of our test of the usefulness of the straight-line model is summarized in the boxes on p. 618.

EXAMPLE 11.3   TESTING THE SLOPE, β_1

Problem Refer to the simple linear regression analysis of the drug reaction data, Examples 11.1 and 11.2. Conduct a test (at $\alpha = .05$) to determine if reaction time (y) is linearly related to amount of drug (x).

Solution For the drug reaction example, $n = 5$. Thus, t will be based on $n - 2 = 3$ df and the rejection region t (at $\alpha = .05$) will be

$$|t| > t_{.025} = 3.182$$

A Test of Model Utility: Simple Linear Regression

One-Tailed Test

$H_0: \beta_1 = 0$

$H_a: \beta_1 < 0 \quad (\text{or } H_a: \beta_1 > 0)$

Two-Tailed Test

$H_0: \beta_1 = 0$

$H_a: \beta_1 \neq 0$

Test statistic: $\quad t = \dfrac{\hat{\beta}_1}{s_{\hat{\beta}_1}} = \dfrac{\hat{\beta}_1}{s/\sqrt{SS_{xx}}}$

Rejection region: $\quad t < -t_\alpha$

(or $t > t_\alpha$ when $H_a: \beta_1 > 0$)

Rejection region: $\quad |t| > t_{\alpha/2}$

where t_α and $t_{\alpha/2}$ are based on $(n - 2)$ degrees of freedom

Conditions Required for a Valid Test: Simple Linear Regression

The four assumptions about ε listed in Section 11.3.

We previously calculated $\hat{\beta}_1 = .7$, $s = .61$, and $SS_{xx} = 10$. Thus,

$$t = \frac{\hat{\beta}_1}{s/\sqrt{SS_{xx}}} = \frac{.7}{.61\sqrt{10}} = \frac{.7}{.19} = 3.7$$

Since this calculated t-value falls in the upper-tail rejection region (see Figure 11.11), we reject the null hypothesis and conclude that the slope β_1 is not 0. The sample evidence indicates that amount of drug x in the blood stream contributes information for the prediction of reaction time y when a linear model is used.

[*Note:* We can reach the same conclusion by using the observed significance level (*p*-value) of the test from a computer printout. The MINITAB printout for the drug reaction example is reproduced in Figure 11.12. The test statistic and two-tailed *p*-value are highlighted on the printout. Since the *p*-value = .035 is smaller than $\alpha = .05$, we will reject H_0.]

Look Back What conclusion can be drawn if the calculated t-value does not fall in the rejection region or if the observed significance level of the test exceeds α?

Figure 11.11

Rejection Region and Calculated t Value for Testing $H_0: \beta_1 = 0$ Versus $H_a: \beta_1 \neq 0$

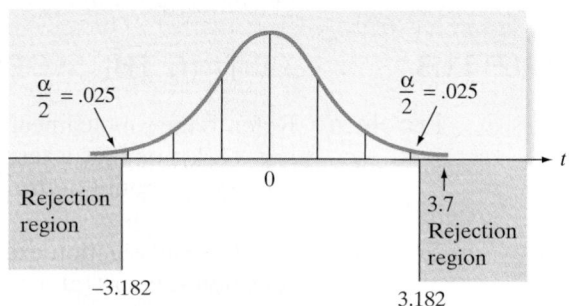

Figure 11.12
MINITAB Printout for the
Time-Drug Regression

Regression Analysis: TIME_Y versus DRUG_X

```
The regression equation is
TIME_Y = - 0.100 + 0.700 DRUG_X


Predictor      Coef   SE Coef       T      P
Constant    -0.1000    0.6351   -0.16  0.885
DRUG_X       0.7000    0.1915    3.66  0.035



S = 0.605530    R-Sq = 81.7%    R-Sq(adj)  = 75.6%



Analysis of Variance

Source           DF       SS       MS       F      P
Regression        1   4.9000   4.9000   13.36  0.035
Residual Error    3   1.1000   0.3667
Total             4   6.0000
```

We know from previous discussions of the philosophy of hypothesis testing that such a t-value does *not* lead us to accept the null hypothesis. That is, we do not conclude that $\beta_1 = 0$. Additional data might indicate that β_1 differs from 0, or a more complex relationship may exist between x and y, requiring the fitting of a model other than the straight-line model. We discuss several such models in Chapter 12.

Now Work   *Exercise 11.50ab*

■ ■ ■

Interpreting *p*-Values for β Coefficients in Regression

Almost all statistical computer software packages report a *two-tailed p*-value for each of the β parameters in the regression model. For example, in simple linear regression, the *p*-value for the two-tailed test H_0: $\beta_1 = 0$ versus H_a: $\beta_1 \neq 0$ is given on the printout. If you want to conduct a *one-tailed* test of hypothesis, you will need to adjust the *p*-value reported on the printout as follows:

Upper-tailed test $(H_a: \beta_1 > 0)$: p-value $= \begin{cases} p/2 & \text{if } t > 0 \\ 1 - p/2 & \text{if } t < 0 \end{cases}$

Lower-tailed test $(H_a: \beta_1 < 0)$: p-value $= \begin{cases} p/2 & \text{if } t < 0 \\ 1 - p/2 & \text{if } t > 0 \end{cases}$

where p is the *p*-value reported on the printout and t is the value of the test statistic.

Another way to make inferences about the slope β_1 is to estimate it using a confidence interval. This interval is formed as shown in the next box.

A 100(1 − α)% Confidence Interval for the Simple Linear Regression Slope β_1

$$\hat{\beta}_1 \pm t_{\alpha/2} s_{\hat{\beta}_1}$$

where the estimated standard error of $\hat{\beta}_1$ is calculated by

$$s_{\hat{\beta}_1} = \frac{s}{\sqrt{SS_{xx}}}$$

and $t_{\alpha/2}$ is based on $(n - 2)$ degrees of freedom.

Conditions Required for a Valid Confidence Interval: Simple Linear Regression

The four assumptions about ε listed in Section 11.3.

For the drug reaction simple linear regression (Examples 11.1–11.3), $t_{\alpha/2}$ is based on $(n - 2) = 3$ degrees of freedom. Therefore, a 95% confidence interval for the slope β_1, the expected change in reaction time for a 1% increase in the amount of drug in the bloodstream, is

$$\hat{\beta}_1 \pm t_{.025} s_{\hat{\beta}_1} = .7 \pm 3.182\left(\frac{s}{\sqrt{SS_{xx}}}\right) = .7 \pm 3.182\left(\frac{.61}{\sqrt{10}}\right) = .7 \pm .61$$

Thus, the interval estimate of the slope parameter β_1 is .09 to 1.31. [*Note:* This interval can also be obtained using statistical software and is highlighted on the SPSS printout, Figure 11.13.] In terms of this example, the implication is that we can be 95% confident that the *true* mean increase in reaction time per additional 1% of the drug is between .09 and 1.31 seconds. This inference is meaningful only over the sampled range of x—that is, from 1% to 5% of the drug in the bloodstream.

Now Work *Exercise 11.50c*

Since all the values in this interval are positive, it appears that β_1 is positive and that the mean of y, $E(y)$, increases as x increases. However, the rather large width of the confidence interval reflects the small number of data points (and, consequently, a lack of information) in the experiment. We would expect a narrower interval if the sample size were increased.

Figure 11.13

SPSS Printout with 95% Confidence Intervals for the Time-Drug Regression Betas

Coefficients^a

Model		Unstandardized Coefficients B	Std. Error	Standardized Coefficients Beta	t	Sig.	95% Confidence Interval for B Lower Bound	Upper Bound
1	(Constant)	-.100	.635		-.157	.885	-2.121	1.921
	TIME_X	.700	.191	.904	3.656	.035	.091	1.309

a. Dependent Variable: DRUG_Y

Statistics in Action Revisited

Assessing How Well the Straight-Line Model Fits the Dowsing Data

In the previous Statistics in Action Revisited, we fit the straight-line model, $E(y) = \beta_0 + \beta_1 x$, where x = dowser's guess (in meters) and y = pipe location (in meters) for each trial. The MINITAB regression printout is reproduced in Figure SIA11.3. The two-tailed p-value for testing the null hypothesis, H_0: $\beta_1 = 0$, (highlighted on the printout) is p-value = .118. Even for an α-level as high as $\alpha = .10$, there is insufficient evidence to reject H_0. Consequently, the dowsing data in Table SIA11.1 provide no statistical support for the German researchers' claim that the three best dowsers have an ability to find underground water with a divining rod.

This lack of support for the dowsing theory is made clearer with a confidence interval for the slope of the line. When $n = 26$, df = $(n - 2) = 24$, and $t_{.025} = 2.064$.

Substituting this value and the relevant values shown on the MINITAB printout, a 95% confidence interval for β_1 is

$$\hat{\beta}_1 \pm t_{.025}(S_{\hat{\beta}_1}) = .31 \pm (2.064)(.19)$$
$$= 31 \pm .39, \text{ or } (-.08, .70)$$

Thus, for every 1-meter increase in a dowser's guess, we estimate (with 95% confidence) that the change in the actual pipe location will range anywhere from a decrease of .08 meter to an increase of .70 meter. In other words, we're not sure whether the pipe location will increase or decrease along the 10-meter pipeline! Keep in mind, also, that the data in Table SIA11.1 represent the "best" performances of the three dowsers (i.e., the outcome of the dowsing experiment in its most favorable light). When the data for all trials are considered and plotted, there is not even a hint of a trend.

Figure SIA11.3
MINITAB Simple Linear Regression for Dowsing Data

Regression Analysis: PIPE versus GUESS

```
The regression equation is
PIPE = 30.1 + 0.308 GUESS


Predictor      Coef   SE Coef     T      P
Constant      30.07     11.41   2.63  0.015
GUESS        0.3079    0.1900   1.62  0.118


S = 26.0298    R-Sq = 9.9%    R-Sq(adj) = 6.1%


Analysis of Variance

Source            DF       SS       MS      F      P
Regression         1   1778.9   1778.9   2.63  0.118
Residual Error    24  16261.2    677.6
Total             25  18040.2
```

Exercises 11.43–11.60

Understanding the Principles

11.43 In the equation $E(y) = \beta_0 + \beta_1 x$, what is the value of β_1 if x has no linear relationship with y?

11.44 What conditions are required for valid inferences about the β's in simple linear regression?

11.45 How do you adjust the p-value obtained from a computer printout when performing a one-tailed test of β_1 in simple linear regression?

11.46 For each of the following 95% confidence intervals for β_1 in simple linear regression, decide whether there is evidence of a positive or negative linear relationship between y and x.
a. $(22, 58)$ **b.** $(-30, 111)$ **c.** $(-45, -7)$

Learning the Mechanics

11.47 Construct both a 95% and a 90% confidence interval for β_1 for each of the following cases:
a. $\hat{\beta}_1 = 31$, $s = 3$, $SS_{xx} = 35$, $n = 12$
b. $\hat{\beta}_1 = 64$, SSE $= 1,960$, $SS_{xx} = 30$, $n = 18$
c. $\hat{\beta}_1 = -8.4$, SSE $= 146$, $SS_{xx} = 64$, $n = 24$

11.48 Consider the following pairs of observations:

x	1	5	3	2	6	6	0
y	1	3	3	1	4	5	1

a. Construct a scattergram for the data.
b. Use the method of least squares to fit a straight line to the seven data points in the table.
c. Plot the least squares line on your scattergram of part **a**.
d. Specify the null and alternative hypotheses you would use to test whether the data provide sufficient evidence to indicate that x contributes information for the (linear) prediction of y.
e. What is the test statistic that should be used in conducting the hypothesis test of part **d**? Specify the degrees of freedom associated with the test statistic.
f. Conduct the hypothesis test of part **d** using $\alpha = .05$.
g. Construct 95% confidence interval for β_1.

11.49 Consider the following pairs of observations:

y	4	2	5	3	2	4
x	1	4	5	3	2	4

a. Construct a scattergram for the data.
b. Use the method of least squares to fit a straight line to the six data points.
c. Plot the least squares line on the scattergram of part **a**.
d. Compute the test statistic for determining whether x and y are linearly related.
e. Carry out the test, part **d**, using $\alpha = .01$.

f. Find a 99% confidence interval for β_1.

Applying the Concepts—Basic

⊚ TAMPALMS

11.50 Predicting sale prices of homes. Refer to the data on sale prices and total appraised values of 92 residential properties in an upscale Tampa, Florida, neighborhood, Exercise 11.18 (p. 606). A SAS simple linear regression printout for the analysis is given below.

a. Specify H_0 and H_a for testing whether there is a positive linear relationship between appraised property value x and sale price y for residential properties sold in this neighborhood.
b. Locate the p-value of the test on the printout. Interpret the result.
c. Find a 95% confidence interval for the slope, β_1, on the printout. Interpret the result practically.
d. What can be done to obtain a narrower confidence interval in part **c**?

⊚ FCAT

11.51 FCAT scores and poverty. Refer to the *Journal of Educational and Behavioral Statistics* (Spring 2004) study of scores on the Florida Comprehensive Assessment Test (FCAT), Exercises 11.24 and 11.39 (p. 608). Consider the simple linear regression relating math score (y) to percentage (x) of students below the poverty level.

a. Test whether y is negatively related to x. Use $\alpha = .01$.
b. Construct a 99% confidence interval for β_1. Practically interpret the result.

⊚ RAINFALL

11.52 New method of estimating rainfall. Refer to the *Journal of Data Science* (Apr. 2004) comparison of methods for estimating rainfall, Exercises 11.25 and 11.40 (p. 609). Consider the simple linear regression relating rain gauge amount (y) to the artificial neural network rain estimate (x).

a. Test whether y is positively related to x. Use $\alpha = .10$.
b. Construct a 90% confidence interval for β_1. Practically interpret the result.

⊚ OJUICE

11.53 Sweetness of orange juice. Refer to Exercise 11.26 (p. 610) and the simple linear regression relating the sweetness index (y) of an orange juice sample to the amount of water soluble pectin (x) in the juice. Find a 90% confidence interval for the true slope of the line. Interpret the result.

11.54 Australian seagrass study. The abundance of tropical seagrasses in Australia was the subject of research

---

Dependent Variable: SALEPRIC

Analysis of Variance

Source	DF	Sum of Squares	Mean Square	F Value	Pr > F
Model	1	1673142	1673142	1556.48	<.0001
Error	90	96746	1074.95355		
Corrected Total	91	1769888			

Root MSE	32.78648	R-Square	0.9453	
Dependent Mean	236.55761	Adj R-Sq	0.9447	
Coeff Var	13.85983			

Parameter Estimates

Variable	DF	Parameter Estimate	Standard Error	t Value	Pr > |t|	95% Confidence Limits	
Intercept	1	20.94193	6.44617	3.25	0.0016	8.13550	33.74837
TOTALVAL	1	1.06873	0.02709	39.45	<.0001	1.01491	1.12254

published in *Aquatic Botany* (Mar. 1995). Simple linear regression was used to relate the standing crop (*y*) of a seagrass species (measured in grams per meter squared) to the percentage (*x*) of land covered by the plants. Data collected for *n* = 12 sites at Rowes Bay, Australia, yielded the following results:

$$\hat{y} = .031 + .089x \qquad t = 2.34 \qquad p\text{-value} = .042$$

a. Give the practical interpretation, if possible, of the estimated *y*-intercept of the line.
b. Give the practical interpretation, if possible, of the estimated slope of the line.
c. At what α-level is there sufficient evidence of a linear relationship between standing crop and percentage cover of a seagrass species? Explain.

Applying the Concepts—Intermediate

11.55 Forest fragmentation study. Refer to the *Conservation Ecology* (Dec. 2003) study on the causes of fragmentation for 54 South American forests, Exercise 2.148 (p. 100). Recall that researchers developed two fragmentation indices for each forest—one index for anthropogenic (human development activities) fragmentation and one for fragmentation from natural causes. The data for five of the 54 forests saved in the **FORFRAG** file are listed below.

FORFRAG (First 5 observations listed)

Ecoregion (forest)	Anthropogenic Index, y	Natural Origin Index, x
Araucaria moist forests	34.09	30.08
Atlantic Coast restingas	40.87	27.60
Bahia coastal forests	44.75	28.16
Bahia interior forests	37.58	27.44
Bolivian Yungas	12.40	16.75

Source: Wade, T. G., et al. "Distribution and causes of global forest fragmentation," *Conservation Ecology*, Vol. 72, No. 2, Dec. 2003 (Table 6).

a. Ecologists theorize that a linear relationship exists between the two fragmentation indices. Write the model relating *y* to *x*.
b. Fit the model to the data in the **FORFRAG** file using the method of least squares. Give the equation of the least squares prediction equation.
c. Interpret the estimates of β_0 and β_1 in the context of the problem.
d. Is there sufficient evidence to indicate that natural origin index (*x*) and anthropogenic index (*y*) are positively linearly related? Test using $\alpha = .05$.
e. Find and interpret a 95% confidence interval for the change in the anthropogenic index (*y*) for every 1-point increase in the natural origin index (*x*).

11.56 Effect of massage on boxers. Refer to the *British Journal of Sports Medicine* (Apr. 2000) study of the effect of massage on boxing performance, Exercise 10.66 (p. 559). Two other variables measured on the boxers were blood

lactate concentration (mM) and the boxer's perceived recovery (28-point scale). Based on information provided in the article, the following data were obtained for 16 five-round boxing performances, where a massage was given to the boxer between rounds. Conduct a test to determine whether blood lactate level (*y*) is linearly related to perceived recovery (*x*). Use $\alpha = .10$.

BOXING2

Blood Lactate Level	Perceived Recovery
3.8	7
4.2	7
4.8	11
4.1	12
5.0	12
5.3	12
4.2	13
2.4	17
3.7	17
5.3	17
5.8	18
6.0	18
5.9	21
6.3	21
5.5	20
6.5	24

Source: Hemmings, B., Smith, M., Graydon, J., and Dyson, R. "Effects of massage on physiological restoration, perceived recovery, and repeated sports performance." *British Journal of Sports Medicine*, Vol. 34, No. 2, Apr. 2000 (data adapted from Figure 3).

11.57 Relation of eye and head movements. How do eye and head movements relate to body movements when reacting to a visual stimulus? Scientists at the California Institute of Technology designed an experiment to answer this question and reported their results in *Nature* (Aug. 1998). Adult male rhesus monkeys were exposed to a visual stimulus (i.e., a panel of light-emitting diodes) and their eye, head, and body movements were electronically recorded. In one variation of the experiment, two variables were measured: active head movement (*x*, percent per degree) and body plus head rotation (*y*, percent per degree). The data for *n* = 39 trials were subjected to a simple linear regression analysis, with the following results: $\hat{\beta}_1 = .88$, $s_{\hat{\beta}_1} = .14$

a. Conduct a test to determine whether the two variables, active head movement *x* and body plus head rotation *y* are positively linearly related. Use $\alpha = .05$.
b. Construct and interpret a 90% confidence interval for β_1.
c. The scientists want to know if the true slope of the line differs significantly from 1. Based on your answer to part **b**, make the appropriate inference.

11.58 Pain empathy and brain activity. *Empathy* refers to being able to understand and vicariously feel what others actually feel. Neuroscientists at University

College of London investigated the relationship between brain activity and pain-related empathy in persons who watch others in pain. (*Science*, Feb. 20, 2004.) Sixteen couples participated in the experiment. The female partner watched while painful stimulation was applied to the finger of her male partner. Two variables were measured for each female: y = pain-related brain activity (measured on a scale ranging from -2 to 2) and x = score on the Empathic Concern Scale (0 to 25 points). The data are listed in the accompanying table. The research question of interest was, "Do people scoring higher in empathy show higher pain-related brain activity?" Use simple linear regression analysis to answer the research question.

BRAINPAIN

Couple	Brain Activity (y)	Empathic Concern (x)
1	.05	12
2	−.03	13
3	.12	14
4	.20	16
5	.35	16
6	0	17
7	.26	17
8	.50	18
9	.20	18
10	.21	18
11	.45	19
12	.30	20
13	.20	21
14	.22	22
15	.76	23
16	.35	24

Source: Singer, T. et al. "Empathy for pain involves the affective but not sensory components of pain," *Science*, Vol. 303, Feb. 20, 2004. (Adapted from Figure 4.)

NAMEGAME2

11.59 The "name game." Refer to the *Journal of Experimental Psychology-Applied* (June 2000) name retrieval study, Exercise 11.28 (p. 611). Recall that the goal of the study was to investigate the linear trend between proportion of names recalled (y) and position (order) of the student (x) during the "name game." Is there sufficient evidence (at $\alpha = .01$) of a linear trend? Answer the question by analyzing the data for 144 students saved in the **NAMEGAME2** file.

Applying the Concepts—Advanced

11.60 Foreign investment risk. One of the most difficult tasks of developing and managing a global portfolio is assessing the risks of potential foreign investments. Duke University researcher C. R. Henry collaborated with two First Chicago Investment Management Company directors to examine the use of country credit ratings as a means of evaluating foreign investments. (*Journal of Portfolio Management*, Winter 1995.) To be effective, such a measure should help explain and predict the volatility of the foreign market in question. The researchers analyzed data on annualized risk (y) and average credit rating (x) for 40 countries. The data are saved in the **GLOBRISK** file. (The first and last 5 countries are displayed in the table.)

GLOBRISK (First and last 5 observations listed)

Country	Annualized Risk (%)	Average Credit Rating
Argentina	87.0	31.8
Australia	26.9	78.2
Austria	26.3	83.8
Belgium	22.0	78.4
Brazil	64.8	36.2
⋮	⋮	⋮
Turkey	74.1	32.6
United Kingdom	21.8	87.6
United States	15.4	93.4
Venezuela	46.0	45.0
Zimbabwe	35.6	24.5

Source: Erb, C. B., Harvey, C. R., and Viskanta, T. E. "Country risk and global equity selection." *Journal of Portfolio Management*, Vol. 21, No. 2, Winter 1995, p. 76.

a. Do the data provide sufficient evidence to conclude that credit rating (x) contributes information for the prediction of annualized risk (y)?

b. Use the plot, part **a**, to locate any unusual data points (outliers).

c. Eliminate the outlier(s), part **b**, from the data set and rerun the simple linear regression analysis. Note any dramatic changes in the results.

11.6 The Coefficient of Correlation

Recall (from optional Section 2.9) that a **bivariate relationship** describes a relationship—or correlation—between two variables, x and y. Scattergrams are used to graphically describe a bivariate relationship. In this section we will discuss the concept of **correlation** and how it can be used to measure the linear relationship between two variables, x and y. A numerical descriptive measure of correlation is provided by the *Pearson product moment coefficient of correlation, r*.

> **DEFINITION 11.2**
>
> The **coefficient of correlation**,* r, is a measure of the strength of the *linear* relationship between two variables x and y. It is computed (for a sample of n measurements on x and y) as follows:
>
> $$r = \frac{SS_{xy}}{\sqrt{SS_{xx}\,SS_{yy}}}$$

Note that the computational formula for the correlation coefficient r given in Definition 11.2 involves the same quantities that were used in computing the least squares prediction equation. In fact, since the numerators of the expressions for $\hat{\beta}_1$ and r are identical, you can see that $r = 0$ when $\hat{\beta}_1 = 0$ (the case where x contributes no information for the prediction of y) and that r is positive when the slope is positive and negative when the slope is negative. Unlike $\hat{\beta}_1$, the correlation coefficient r is *scaleless* and assumes a value between -1 and $+1$, regardless of the units of x and y.

A value of r near or equal to 0 implies little or no linear relationship between y and x. In contrast, the closer r comes to 1 or -1, the stronger the linear relationship between y and x. And if $r = 1$ or $r = -1$, all the sample points fall exactly on the least squares line. Positive values of r imply a positive linear relationship between y and x; that is, y increases as x increases. Negative values of r imply a negative linear relationship between y and x; that is, y decreases as x increases. Each of these situations is portrayed in Figure 11.14.

Now Work | *Exercise 11.65*

We demonstrate how to calculate the coefficient of correlation r using the data in Table 11.1 for the drug reaction example. The quantities needed to calculate r are SS_{xy}, SS_{xx}, and SS_{yy}. The first two quantities have been calculated previously and are repeated here for convenience:

$$SS_{xy} = 7, \quad SS_{xx} = 10, \quad SS_{yy} = \sum y^2 - \frac{\left(\sum y\right)^2}{n}$$

$$= 26 - \frac{(10)^2}{5} = 26 - 20 = 6$$

We now find the coefficient of correlation:

$$r = \frac{SS_{xy}}{\sqrt{SS_{xx}\,SS_{yy}}} = \frac{7}{\sqrt{(10)(6)}} = \frac{7}{\sqrt{60}} = .904$$

The fact that r is positive and near 1 in value indicates that the reaction time tends to increase as the amount of drug in the bloodstream increases—*for this sample of five subjects*. This is the same conclusion we reached when we found the calculated value of the least squares slope to be positive.

---

*The value of r is often called the *Pearson correlation coefficient* to honor its developer, Karl Pearson. (See Biography, p. 786.)

Figure 11.14

Values of r and Their
Implications

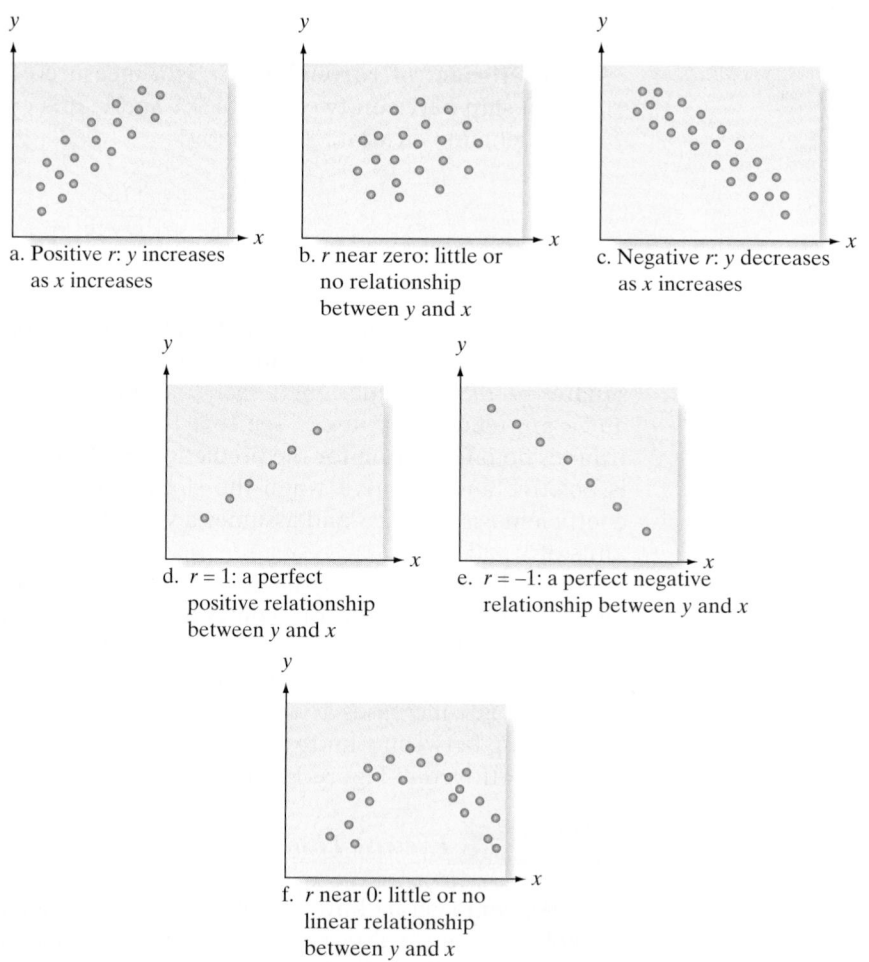

a. Positive r: y increases
as x increases

b. r near zero: little or
no relationship
between y and x

c. Negative r: y decreases
as x increases

d. $r = 1$: a perfect
positive relationship
between y and x

e. $r = -1$: a perfect negative
relationship between y and x

f. r near 0: little or no
linear relationship
between y and x

EXAMPLE 11.4   FINDING THE CORRELATION COEFFICIENT

Problem Legalized gambling is available on several riverboat casinos operated by a city in
Mississippi. The mayor of the city wants to know the correlation between the number
of casino employees and yearly crime rate. The records for the past 10 years are
examined, and the results listed in Table 11.5 are obtained. Calculate the coefficient of
correlation r for the data. Interpret the result.

☉ CASINO

TABLE 11.5   Data on Casino Employees and Crime Rate, Example 11.1

Year	Number of Casino Employees x (thousands)	Crime Rate y (number of crimes per 1,000 population)
1996	15	1.35
1997	18	1.63
1998	24	2.33
1999	22	2.41
2000	25	2.63
2001	29	2.93
2002	30	3.41
2003	32	3.26
2004	35	3.63
2005	38	4.15

Figure 11.15

MINITAB Correlation Printout and Scattergram for Example 12.4

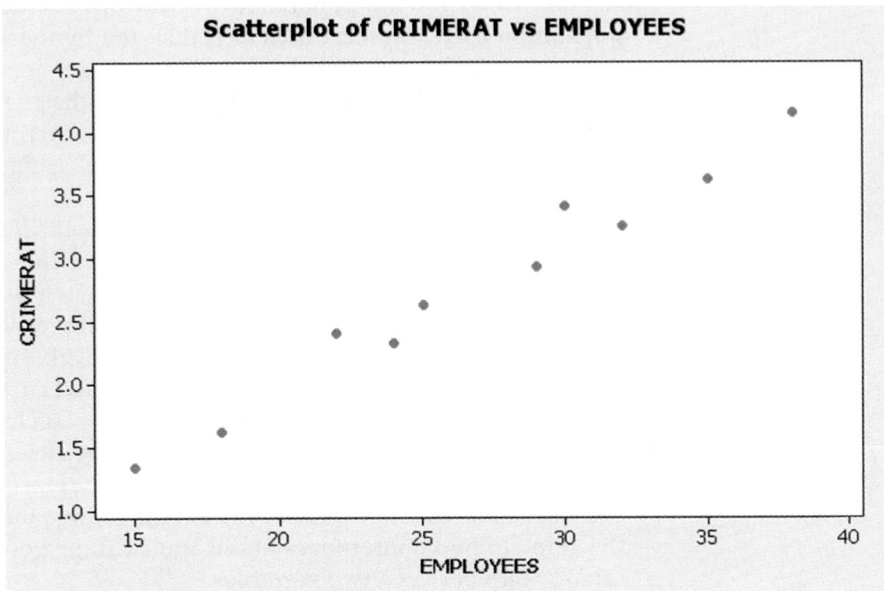

Correlations: EMPLOYEES, CRIMERAT

```
Pearson correlation of EMPLOYEES and CRIMERAT = 0.987
P-Value = 0.000
```

Solution Rather than use the computing formula given in Definition 11.2, we resort to a statistical software package. The data of Table 11.5 were entered into a computer and MINITAB was used to compute r. The MINITAB printout is shown in Figure 11.15.

The coefficient of correlation, highlighted at the top on the printout, is $r = .987$. Thus, the size of the casino workforce and crime rate in this city are very highly correlated—at least over the past 10 years. The implication is that a strong positive linear relationship exists between these variables (see Figure 11.15). We must be careful, however, not to jump to any unwarranted conclusions. For instance, the mayor may be tempted to conclude that hiring more casino workers next year will increase the crime rate—that is, that there is a *causal relationship* between the two variables. However, high correlation does not imply causality. The fact is, many things have probably contributed both to the increase in the casino workforce and to the increase in crime rate. The city's tourist trade has undoubtedly grown since legalizing riverboat casinos and it is likely that the casinos have expanded both in services offered and in number. *We cannot infer a causal relationship on the basis of high sample correlation. When a high correlation is observed in the sample data, the only safe conclusion is that a linear trend may exist between x and y.*

Look Back Another variable, such as the increase in tourism, may be the underlying cause of the high correlation between x and y.

Now Work *Exercise 11.70a*

■ ■ ■

Warning

When using the sample correlation coefficient, r, to infer the nature of the relationship between x and y, two caveats exist: (1) A *high correlation* does not necessarily imply that a causal relationship exists between x and y—only that a linear trend may exist;

(2) a *low correlation* does not necessarily imply that x and y are unrelated—only that x and y are not strongly linearly related.

Keep in mind that the correlation coefficient r measures the linear correlation between x values and y values in the sample, and a similar linear coefficient of correlation exists for the population from which the data points were selected. The **population correlation coefficient** is denoted by the symbol ρ (rho). As you might expect, ρ is estimated by the corresponding sample statistic, r. Or, instead of estimating ρ, we might want to test the null hypothesis H_0: $\rho = 0$ against H_a: $\rho \neq 0$—that is, we can test the hypothesis that x contributes no information for the prediction of y by using the straight-line model against the alternative that the two variables are at least linearly related.

However, we already performed this *identical* test in Section 11.5 when we tested H_0: $\beta_1 = 0$ against H_a: $\beta_1 \neq 0$. That is, the null hypothesis H_0: $\rho = 0$ is equivalent to the hypothesis H_0: $\beta_1 = 0$.* When we tested the null hypothesis H_0: $\beta_1 = 0$ in connection with the drug reaction example, the data led to a rejection of the null hypothesis at the $\alpha = .05$ level. This rejection implies that the null hypothesis of a 0 linear correlation between the two variables (drug and reaction time) can also be rejected at the $\alpha = .05$ level. The only real difference between the least squares slope $\hat{\beta}_1$ and the coefficient of correlation r is the measurement scale. Therefore, the information they provide about the usefulness of the least squares model is to some extent redundant. For this reason, we will use the slope to make inferences about the existence of a positive or negative linear relationship between two variables.

Statistics in Action Revisited

Using the Correlation Coefficient to Assess the Dowsing Data

In the previous Statistics in Action Revisited, we discovered that using a dowser's guess (x) in a straight-line model was not statistically useful for predicting actual pipe location (y). The coefficient of correlation, shown on the MINITAB printout in Figure SIA11.4, also supports this conclusion. The value, $r = .314$, indicates a fairly weak positive linear relationship between the variables. This value, however, is not statistically significant (p-value $= .118$). In other words, there is no evidence to indicate that the population correlation coefficient is different from 0.

Figure SIA11.4

MINITAB Correlation Printout for Dowsing Data

Correlations: PIPE, GUESS

```
Pearson correlation of PIPE and GUESS = 0.314
P-Value = 0.118
```

11.7 The Coefficient of Determination

Another way to measure the usefulness of the model is to measure the contribution of x in predicting y. To accomplish this, we calculate how much the errors of prediction of y were reduced by using the information provided by x. To illustrate, consider the sample shown in the scattergram of Figure 11.16a. If we assume that x contributes no information for the prediction of y, the best prediction for a value of y is the sample mean $\bar{y}$, which is shown as the horizontal line in Figure 11.16b. The vertical line segments in Figure 11.16b are the deviations of the points

*The correlation test statistic that is equivalent to $t = \hat{\beta}_1/s_{\hat{\beta}_1}$, is $t = \dfrac{r}{\sqrt{(1 - r^2)/(n - 2)}}$

Figure 11.16

A Comparison of the Sum of Squares of Deviations for Two Models

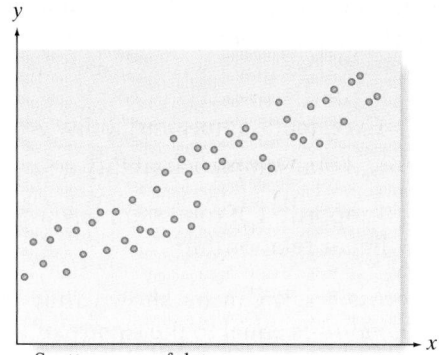

a. Scattergram of data

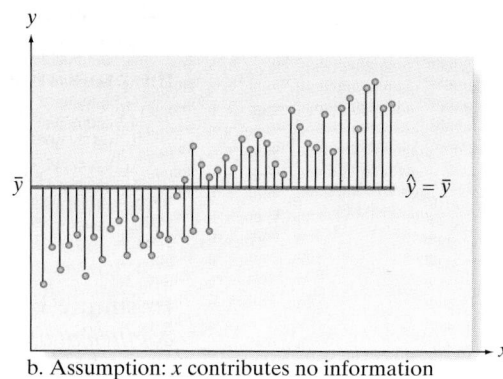

b. Assumption: x contributes no information for predicting y, $\hat{y} = \bar{y}$

$$\hat{y} = \hat{\beta}_0 + \hat{\beta}_1 x$$

c. Assumption: x contributes information for predicting y, $\hat{y} = \hat{\beta}_0 + \hat{\beta}_1 x$

about the mean $\bar{y}$. Note that the sum of squares of deviations for the prediction equation $\hat{y} = \bar{y}$ is

$$SS_{yy} = \sum (y_i - \bar{y})^2$$

Now suppose you fit a least squares line to the same set of data and locate the deviations of the points about the line as shown in Figure 11.16c. Compare the deviations about the prediction lines in Figures 11.16b and 11.16c. You can see that

1. If x contributes little or no information for the prediction of y, the sums of squares of deviations for the two lines,

$$SS_{yy} = \sum (y_i - \bar{y})^2 \quad \text{and} \quad SSE = \sum (y_i - \hat{y}_i)^2$$

will be nearly equal.

2. If x does contribute information for the prediction of y, the SSE will be smaller than SS_{yy}. In fact, if all the points fall on the least squares line, then SSE = 0.

Then, the reduction in the sum of squares of deviations that can be attributed to x, expressed as a proportion of SS_{yy}, is

$$\frac{SS_{yy} - SSE}{SS_{yy}}$$

Note that SS_{yy} is the "total sample variation" of the observations around the mean $\bar{y}$ and that SSE is the remaining "unexplained sample variability" after fitting the line $\hat{y}$. Thus, the difference $(SS_{yy} - SSE)$ is the "explained sample variability" attributable to the linear relationship with x. Then a verbal description of the proportion is

$$\frac{SS_{yy} - SSE}{SS_{yy}} = \frac{\text{Explained sample variability}}{\text{Total sample variability}}$$

$$= \text{Proportion of total sample variability explained by the linear relationship}$$

In simple linear regression, it can be shown that this proportion—called the *coefficient of determination*—is equal to the square of the simple linear coefficient of correlation r

---

DEFINITION 11.3

The **coefficient of determination** is

$$r^2 = \frac{SS_{yy} - SSE}{SS_{yy}} = 1 - \frac{SSE}{SS_{yy}}$$

It represents the proportion of the total sample variability around $\bar{y}$ that is explained by the linear relationship between y and x. (In simple linear regression, it may also be computed as the square of the coefficient of correlation r.)

Note that r^2 is always between 0 and 1, because r is between -1 and $+1$. Thus, an r^2 of .60 means that the sum of squares of deviations of the y values about their predicted values has been reduced 60% by the use of the least squares equation $\hat{y}$, instead of $\bar{y}$, to predict y.

EXAMPLE 11.5   FINDING THE VALUE OF r^2

Problem   Calculate the coefficient of determination for the drug reaction example. The data are repeated in Table 11.6 for convenience. Interpret the result.

STIMULUS

TABLE 11.6

Amount of Drug x (%)	Reaction Time y (seconds)
1	1
2	1
3	2
4	2
5	4

Solution   From previous calculations,

$$SS_{yy} = 6 \quad \text{and} \quad SSE = \sum(y - \hat{y})^2 = 1.10$$

Then, from Definition 11.3, the coefficient of determination is given by

$$r^2 = \frac{SS_{yy} - SSE}{SS_{yy}} = \frac{6.0 - 1.1}{6.0} = \frac{4.9}{6.0} = .817$$

Figure 11.17

Portion of SPSS Printout for Time-Drug Regression

Model Summary

Model	R	R Square	Adjusted R Square	Std. Error of the Estimate
1	.904[a]	.817	.756	.606

a. Predictors: (Constant), DRUG_X

Another way to compute r^2 is to recall (Section 11.6) that $r = .904$. Then we have $r^2 = (.904)^2 = .817$. A third way to obtain r^2 is from a computer printout. This value is highlighted on the SPSS printout in Figure 11.17. Our interpretation is as follows: We know that using the amount of drug in the blood, x, to predict y with the least squares line

$$\hat{y} = -.1 + .7x$$

accounts for nearly 82% of the total sum of squares of deviations of the five sample y values about their mean. Or, stated another way, 82% of the sample variation in reaction time (y) can be "explained" by using amount (x) of drug in a straight-line model.

Now Work   *Exercise 11.70b*

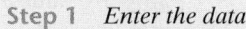

Practical Interpretation of the Coefficient of Determination, r^2

$100(r^2)\%$ of the sample variation in y (measured by the total sum of squares of deviations of the sample y values about their mean $\bar{y}$) can be explained by (or attributed to) using x to predict y in the straight-line model.

Simple Linear Regression

Using the TI-83 Graphing Calculator

Finding r and r^2

Step 1   *Enter the data*
Press **STAT** and select **1:Edit**
Note: If a list already contains data, clear the old data. Use the up arrow to highlight the list name, **L.1** or **L.2'**

Press **CLEAR ENTER.**
Enter your x-data in **L1** and your y-data in **L2.**

Step 2   *Turn the diagnostics feature on*
Press **2nd 0** for **CATALOG**
Press the key for **D**
Press the down arrow key until **Diagnostics On** is highlighted
Press **ENTER** twice

Step 3   *Find the equation*
Press **STAT** and highlight **CALC**
Press 4 for LinReg (**ax + b**)
Press **ENTER**
The screen will show the values for a and b in the equation y = ax + b.

The values for r and r^2 will appear on the screen as well.

Example The figure below shows the output with the **DiagnosticsOn.**

```
LinReg
y=ax+b
a=2.495967742
b=63.3391129
r²=.9620199454
r=.9808261545
```

Exercises 11.61–11.80

Understanding the Principles

11.61 Explain the difference between r and r^2.

11.62 Describe the slope of the least squares line if
   a. $r = .7$ **b.** $r = -.7$ **c.** $r = 0$ **d.** $r^2 = .64$

11.63 *True or False.* The correlation coefficient is a measure of the strength of the linear relationship between x and y.

11.64 *True or False.* A value of the correlation coefficient near 1 or near -1 implies a causal relationship between x and y.

11.65 Explain what each of the following sample correlation
NW coefficients tells you about the relationship between the x and y values in the sample:
   a. $r = 1$
   b. $r = -1$
   c. $r = 0$
   d. $r = .90$
   e. $r = .10$
   f. $r = -.88$

Learning the Mechanics

11.66 Calculate r^2 for the least squares line in Exercise 11.14 (p. 605).

11.67 Calculate r^2 for the least squares line in Exercise 11.17 (p. 606).

11.68 Construct a scattergram for each data set. Then calculate r and r^2 for each data set. Interpret their values.

a. x	-2	-1	0	1	2
y	-2	1	2	5	6

b. x	-2	-1	0	1	2
y	6	5	3	2	0

c. x	1	2	2	3	3	3	4
y	2	1	3	1	2	3	2

d. x	0	1	3	5	6
y	0	1	2	1	0

Applying the Concepts—Basic

11.69 **Sports news on local TV broadcasts.** *The Sports Journal* (Winter 2004) published the results of a study conducted to assess the factors that impact the time allotted to sports news on local television news broadcasts. Information on total time (in minutes) allotted to sports and audience ratings of the TV news broadcast (measured on a 100-point scale) was obtained from a national sample of 163 news directors. A correlation analysis on the data yielded $r = .43$.
   a. Interpret the value of the correlation coefficient, r.
   b. Find and interpret the value of the coefficient of determination, r^2.

⊙ TAMPALMS

11.70 **Predicting sale prices of homes.** Refer to the data on
NW sale prices and total appraised values of 92 residential properties recently sold in an upscale Tampa, Florida, neighborhood, Exercise 11.18 (p. 606). The MINITAB simple linear regression printout relating sale price (y) to appraised property value (x) is reproduced on p. 633, followed by a MINITAB correlation printout.
   a. Find the coefficient of correlation between appraised property value and sale price on the printout. Interpret this value.
   b. Find the coefficient of determination between appraised property value and sale price on the printout. Interpret this value.

11.71 **Redshifts of Quasi Stellar Objects.** Refer to the *Journal of Astrophysics & Astronomy* (Mar./Jun., 2003) study of redshifts in Quasi Stellar Objects (QSOs), Exercise 11.21 (p. 608). Recall that simple linear regression was used to model the magnitude (y) of a QSO as a function of redshift level (x). In addition to the least squares line, $\hat{y} = 18.13 + 6.21x$, the coefficient of correlation was determined as $r = .84$.
   a. Interpret the value of r in the words of the problem.
   b. What is the relationship between r and the estimated slope of the line?
   c. Find the value of r^2 and interpret its value.

MINITAB Output for Exercise 11.70

```
The regression equation is
SALEPRIC = 20.9 + 1.07 TOTALVAL


Predictor      Coef  SE Coef       T      P
Constant     20.942    6.446    3.25  0.002
TOTALVAL    1.06873  0.02709   39.45  0.000


S = 32.7865    R-Sq = 94.5%   R-Sq(adj) = 94.5%


Analysis of Variance

Source           DF       SS      MS       F      P
Regression        1  1673142  1673142  1556.48  0.000
Residual Error   90    96746     1075
Total            91  1769888
```

Correlations: SALEPRIC, TOTALVAL

```
Pearson correlation of SALEPRIC and TOTALVAL = 0.972
P-Value = 0.000
```

11.72 Removing metal from water. In the *Electronic Journal of Biotechnology* (Apr. 15, 2004), Egyptian scientists studied a new method for removing heavy metals from water. Metal solutions were prepared in glass vessels, then biosorption was used to remove the metal ions. Two variables were measured for each test vessel: y = the metal uptake (milligrams of metal per gram of biosorbent) and x = final concentration of metal in the solution (milligrams per liter).

a. Write a simple linear regression model relating y to x.

b. For one metal, simple linear regression analysis yielded $r^2 = .92$. Interpret this result.

11.73 "Metaskills" and career management. In today's business environment, effective management of one's own career requires a skill set that includes adaptability, tolerance for ambiguity, self-awareness, and identity change. Management professors at Pace University (New York) used correlation coefficients to investigate the relationship between these "metaskills" and effective career management. (*International Journal of Manpower,* Aug. 2000.) Data were collected for 446 business graduates who all completed a management "metaskills" course. Two of the many variables measured were self-knowledge skill level (x) and goal-setting ability (y). The correlation coefficient for these two variables was $r = .70$.

a. Give a practical interpretation of the value of r.

b. The p-value for a test of no correlation between the two variables was reported as p-value $= .001$. Interpret this result.

c. Find the coefficient of determination, r^2, and interpret the result.

Applying the Concepts—Intermediate

11.74 Salary linked to height. Are short people short-changed when it comes to salary? According to business professors T. A. Judge (University of Florida) and D. M. Cable (University of North Carolina), tall people tend to earn more money over their career than short people. (*Journal of Applied Psychology,* June 2004.) Using data collected from participants in the National Longitudinal Surveys begun in 1979, the researchers computed the correlation between average earnings from 1985 to 2000 (in dollars) and height (in inches) for several occupations. The results are given in the table.

Occupation	Correlation, r	Sample Size, n
Sales	.41	117
Managers	.35	455
Blue Collar	.32	349
Service Workers	.31	265
Professional/Technical	.30	453
Clerical	.25	358
Crafts/Forepersons	.24	250

Source: Judge, T. A., & Cable, D. M. "The effect of physical height on workplace success and income: Preliminary test of a theoretical model," *Journal of Applied Psychology,* Vol. 89, No. 3, June 2004 (Table 5).

a. Interpret the value of r for people in sales occupations.

b. Compute r^2 for people in sales occupations. Interpret the result.

c. Give H_0 and H_a for testing whether average earnings and height are positively correlated.

d. The test statistic for testing H_0 and H_a in part **c** is $t = \dfrac{r\sqrt{n-2}}{\sqrt{1-r^2}}$. Compute this value for people in sales occupations.

e. Use the result, part **d**, to conduct the test at $\alpha = .01$. Give the appropriate conclusion.

f. Select another occupation and repeat parts **a–e**.

11.75 View of rotated objects. *Perception & Psychophysics* (July 1998) reported on a study of how people view the 3-dimensional objects projected onto a rotating 2-dimensional image. Each in a sample of 25 university students viewed various depth-rotated objects (e.g., hairbrush, duck, shoe) until they recognized the object. The recognition exposure time—that is, the minimum time (in milliseconds) required for the subject to recognize the object—was recorded for each. In addition, each subject rated the "goodness of view" of the object on a numerical scale, where lower scale values correspond to better views. The table gives the correlation coefficient, r, between recognition exposure time and goodness of view for several different rotated objects.

Object	r	t
Piano	.447	2.40
Bench	−.057	.27
Motorbike	.619	3.78
Armchair	.294	1.47
Teapot	.949	14.50

a. Interpret the value of r for each object.

b. Calculate and interpret the value of r^2 for each object.

c. The table also includes the t-value for testing the null hypothesis of no correlation (i.e., for testing $H_0: \beta_1 = 0$). Interpret these results.

11.76 Snow geese feeding trial. Botanists at the University of Toronto conducted a series of experiments to investigate the feeding habits of baby snow geese. (*Journal of Applied Ecology,* Vol. 32, 1995.) Goslings were deprived of food until their guts were empty, then were allowed to feed for 6 hours on a diet of plants or Purina Duck Chow. For each feeding trial, the change in the weight of the gosling after 2.5 hours was recorded as a percentage of initial weight. Two other variables recorded were digestion efficiency (measured as a percentage) and amount of acid-detergent fibre in the digestive tract (also measured as a percentage). The data for 42 feeding trials saved in the **SNOWGEESE** file are listed below.

a. The botanists were interested in the correlation between weight change (y) and digestion efficiency (x). Plot the data for these two variables in a scattergram. Do you observe a trend?

b. Find the coefficient of correlation relating weight change y to digestion efficiency x. Interpret this value.

c. Conduct a test to determine whether weight change y is correlated with a digestion efficiency x. Use $\alpha = .01$.

d. Repeat parts **b** and **c**, but exclude the data for trials that used duck chow from the analysis. What do you conclude?

e. The botanists were also interested in the correlation between digestion efficiency y and acid-detergent fibre x. Repeat parts **a–d** for these two variables.

11.77 Dance/movement therapy. In cotherapy two or more therapists lead a group. An article in the *American Journal of Dance Therapy* (Spring/Summer 1995) examined the use of cotherapy in dance/movement therapy. Two of several variables measured on each of a sample of 136 professional dance/movement therapists were years of formal training x and reported success rate y (measured as a percentage) of coleading dance/movement therapy groups.

a. Propose a linear model relating y to x.

b. The researcher hypothesized that dance/movement therapists with more years in formal dance training

⊙ **SNOWGEESE** (First and last 5 observations listed)

Feeding Trial	Diet	Weight Change (%)	Digestion Efficiency (%)	Acid-Detergent Fibre (%)
1	Plants	−6	0	28.5
2	Plants	−5	2.5	27.5
3	Plants	−4.5	5	27.5
4	Plants	0	0	32.5
5	Plants	2	0	32
⋮	⋮	⋮	⋮	⋮
38	Duck Chow	9	59	8.5
39	Duck Chow	12	52.5	8
40	Duck Chow	8.5	75	6
41	Duck Chow	10.5	72.5	6.5
42	Duck Chow	14	69	7

Source: Gadallah, F. L., and Jefferies, R. L. "Forage quality in brood rearing areas of the lesser snow goose and the growth of captive goslings." *Journal of Applied Biology,* Vol. 32, No. 2, 1995, pp. 281–282 (adapted from Figures 2 and 3).

will report higher perceived success rates in cotherapy relationships. State the hypothesis in terms of the parameter of the model, part **a**.

c. The correlation coefficient for the sample data was reported as $r = -.26$. Interpret this result.

d. Does the value of r in part **c** support the hypothesis in part **b**? Test using $\alpha = .05$. [*Hint:* See the last paragraph and accompanying footnote of Section 11.6]

NAMEGAME2

11.78 The "name game." Refer to the *Journal of Experimental Psychology-Applied* (June 2000) name retrieval study, Exercises 11.28 (p. 611) and 11.59 (p. 624). Find and interpret the values of r and r^2 for the simple linear regression relating the proportion of names recalled (y) and position (order) of the student (x) during the "name game."

BOXING2

11.79 Effect of massage on boxing. Refer to the *British Journal of Sports Medicine* (April 2000) study of the effect of massage on boxing performance, Exercise 11.56 (p. 623). Find and interpret the values of r and r^2 for the simple linear regression relating the blood lac-

tate concentration and the boxer's perceived recovery.

Applying the Concepts—Advanced

11.80 Pain tolerance study. A study published in *Psychosomatic Medicine* (Mar./Apr. 2001) explored the relationship between reported severity of pain and actual pain tolerance in 337 patients who suffer from chronic pain. Each patient reported his/her severity of chronic pain on a 7-point scale (1 = no pain, 7 = extreme pain). To obtain a pain tolerance level, a tourniquet was applied to the arm of each patient and twisted. The maximum pain level tolerated was measured on a quantitative scale.

a. According to the researchers, "correlational analysis revealed a small but significant inverse relationship between [actual] pain tolerance and the reported severity of chronic pain." Based on this statement, is the value of r for the 337 patients positive or negative?

b. Suppose that the result reported in part **a** is significant at $\alpha = .05$. Find the approximate value of r for the sample of 337 patients. [*Hint:* Use the formula $t = r\sqrt{(n-2)}/\sqrt{(1-r^2)}$.]

11.8 Using the Model for Estimation and Prediction

If we are satisfied that a useful model has been found to describe the relationship between reaction time and amount of drug in the bloodstream, we are ready for step 5 in our regression modeling procedure: using the model for estimation and prediction.

> *The most common uses of a probabilistic model for making inferences can be divided into two categories. The first is the use of the model for estimating the mean value of y, E(y), for a specific value of x.*

For our drug reaction example, we may want to estimate the mean response time for all people whose blood contains 4% of the drug.

> *The second use of the model entails predicting a new individual y value for a given x.*

That is, we may want to predict the reaction time for a specific person who possesses 4% of the drug in the bloodstream.

In the first case, we are attempting to estimate the mean value of y for a very large number of experiments at the given x value. In the second case, we are trying to predict the outcome of a single experiment at the given x value. Which of these model uses—estimating the mean value of y or predicting an individual new value of y (for the same value of x)—can be accomplished with the greater accuracy?

Before answering this question, we first consider the problem of choosing an estimator (or predictor) of the mean (or a new individual) y value. We will use the least squares prediction equation

$$\hat{y} = \hat{\beta}_0 + \hat{\beta}_1 x$$

both to estimate the mean value of y and to predict a specific new value of y for a given value of x. For our example, we found

$$\hat{y} = -.1 + .7x$$

Figure 11.18

Estimated Mean Value and Predicted Individual Value of Reaction Time y for $x = 4$

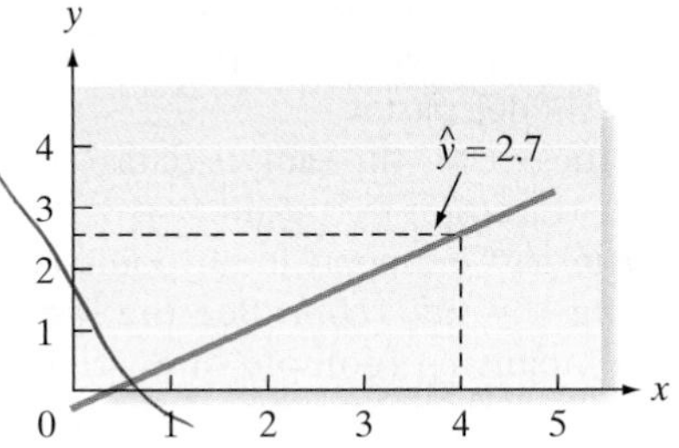

so that the estimated mean reaction time for all people when $x = 4$ (drug is 4% of blood content) is

$$\hat{y} = -.1 + .7(4) = 2.7 \text{ seconds}$$

The same value is used to predict a new y value when $x = 4$. That is, both the estimated mean and the predicted value of y are $\hat{y} = 2.7$ when $x = 4$, as shown in Figure 11.18.

The difference between these two model uses lies in the relative accuracy of the estimate and the prediction. These accuracies are best measured by using the sampling errors of the least squares line when it is used as an estimator and as a predictor, respectively. These errors are reflected in the standard deviations given in the next box.

Sampling Errors for the Estimator of the Mean of y and the Predictor of an Individual New Value of y

1. The standard deviation of the sampling distribution of the estimator $\hat{y}$ of the mean value of y at a specific value of x, say x_p, is

$$\sigma_{\hat{y}} = \sigma \sqrt{\frac{1}{n} + \frac{(x_p - \bar{x})^2}{SS_{xx}}}$$

where σ is the standard deviation of the random error ε. We refer to $\sigma_{\hat{y}}$ as the standard error of $\hat{y}$.

2. The standard deviation of the prediction error for the predictor $\hat{y}$ of an individual new y value at a specific value of x is

$$\sigma_{(y-\hat{y})} = \sigma \sqrt{1 + \frac{1}{n} + \frac{(x_p - \bar{x})^2}{SS_{xx}}}$$

where σ is the standard deviation of the random error ε. We refer to $\sigma_{(y-\hat{y})}$ as the standard error of prediction.

The true value of σ is rarely known, so we estimate σ by s and calculate the estimation and prediction intervals as shown in the next two boxes.

A $100(1 - \alpha)\%$ Confidence Interval for the Mean Value of y at $x = x_p$

$$\hat{y} + t_{\alpha/2}(\text{Estimated standard error of } \hat{y})$$

or

$$\hat{y} \pm t_{\alpha/2}s \sqrt{\frac{1}{n} + \frac{(x_p - \bar{x})^2}{SS_{xx}}}$$

where $t_{\alpha/2}$ is based on $(n - 2)$ degrees of freedom.

> **A 100(1 − α)% Prediction Interval* for an Individual New Value of y at x = x_p**
>
> $$\hat{y} \pm t_{\alpha/2}(\text{Estimated standard error of prediction})$$
>
> or
>
> $$\hat{y} \pm t_{\alpha/2}s\sqrt{1 + \frac{1}{n} + \frac{(x_p - \bar{x})^2}{SS_{xx}}}$$
>
> where $t_{\alpha/2}$ is based on $(n - 2)$ degrees of freedom.

EXAMPLE 11.6

ESTIMATING THE MEAN OF y

Problem Refer to the drug reaction simple linear regression. Find a 95% confidence interval for the mean reaction time when the concentration of the drug in the bloodstream is 4%.

Solution For a 4% concentration, $x = 4$ and the confidence interval for the mean value of y is

$$\hat{y} \pm t_{\alpha/2}s\sqrt{\frac{1}{n} + \frac{(x_p - \bar{x})^2}{SS_{xx}}} = \hat{y} \pm t_{.025}s\sqrt{\frac{1}{5} + \frac{(4 - \bar{x})^2}{SS_{xx}}}$$

where $t_{.025}$ is based on $n - 2 = 5 - 2 = 3$ degrees of freedom. Recall that $\hat{y} = 2.7$, $s = .61$, $\bar{x} = 3$, and $SS_{xx} = 10$. From Table VI in Appendix A, $t_{.025} = 3.182$. Thus, we have

$$2.7 \pm (3.182)(.61)\sqrt{\frac{1}{5} + \frac{(4 - 3)^2}{10}} = 2.7 \pm (3.182)(.61)(.55)$$

$$= 2.7 \pm (3.182)(.34)$$

$$= 2.7 \pm 1.1$$

Therefore, when the percentage of drug in the bloodstream is 4%, we can be 95% confident that the mean reaction time for all possible subjects is 1.6 to 3.8 seconds.

Look Back Note that we used a small amount of data (small sample size) for purposes of illustration in fitting the least squares line. The interval would probably be narrower if more information had been obtained from a larger sample.

Now Work *Exercise 11.86a-d*

■ ■ ■

EXAMPLE 11.7

PREDICTING AN INDIVIDUAL VALUE OF y

Problem Refer, again, to the drug reaction regression. Predict the reaction time for the next performance of the experiment for a subject with a drug concentration of 4%. Use a 95% prediction interval.

---

*The term *prediction interval* is used when the interval formed is intended to enclose the value of a random variable. The term *confidence interval* is reserved for estimation of population parameters (such as the mean).

Solution To predict the response time for an individual new subject for whom $x = 4$, we calculate the 95% prediction interval as

$$\hat{y} \pm t_{\alpha/2} s \sqrt{1 + \frac{1}{n} + \frac{(x_p - \bar{x})^2}{SS_{xx}}} = 2.7 \pm (3.182)(.61)\sqrt{1 + \frac{1}{5} + \frac{(4-3)^2}{10}}$$

$$= 2.7 \pm (3.182)(.61)(1.14)$$
$$= 2.7 \pm (3.182)(.70)$$
$$= 2.7 \pm 2.2$$

Therefore, when the drug concentration for an individual is 4%, we predict with 95% confidence that the reaction time for this new individual will fall in the interval from .5 to 4.9 seconds.

Look Back Like the confidence interval for the mean value of y, the prediction interval for y is quite large. This is because we have chosen a simple example (only five data points) to fit the least squares line. The width of the prediction interval could be reduced by using a larger number of data points.

Now Work   *Exercise 11.86e*

■ ■ ■

Both the confidence interval for $E(y)$ and the prediction interval for y can be obtained using a statistical software package. Figure 11.19 is a MINITAB printout showing the confidence interval and prediction interval, respectively, for the data in the drug example.

The 95% confidence interval for $E(y)$ when $x = 4$, highlighted under "95% CI" in Figure 11.19, is (1.645, 3.755). The 95% prediction interval for y when $x = 4$, highlighted in Figure 11.19 under "95% PI", is (.503, 4.897). These agree with the ones computed in Examples 11.6–7.

Note that the prediction interval for an individual new value of y is *always* wider than the corresponding confidence interval for the mean value of y. Will this always be true? The answer is "yes." The error in estimating the mean value of y, $E(y)$, for a given value of x, say x_p, is the distance between the least squares line and the true line of means, $E(y) = \beta_0 + \beta_1 x$. This error, $[\hat{y} - E(y)]$, is shown in Figure 11.20. In contrast, *the error $(y_p - \hat{y})$ in predicting some future value of y is the sum of two errors*—the error of estimating the mean of y, $E(y)$, shown in Figure 11.20, plus the random error that is a component of the value of y to be

Figure 11.19

MINITAB Printout Giving 95% Confidence Interval for E(y) and 95% Prediction Interval for *y*

```
Predicted Values for New Observations

New
Obs    Fit   SE Fit      95% CI            95% PI
  1  2.700   0.332   (1.645, 3.755)   (0.503, 4.897)


Values of Predictors for New Observations

New
Obs   DRUG_X
  1    4.00
```

Figure 11.20

Error of Estimating the
Mean Value of y for a Given
Value of x

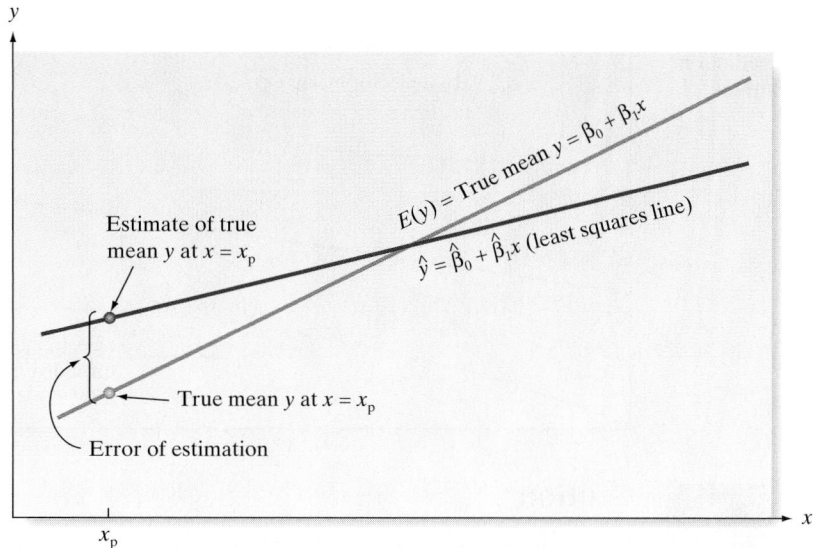

predicted (see Figure 11.21). Consequently, the error of predicting a particular value of y will be larger than the error of estimating the mean value of y for a particular value of x. Note from their formulas that both the error of estimation and the error of prediction take their smallest values when $x_p = \bar{x}$. The farther x_p lies from $\bar{x}$, the larger will be the errors of estimation and prediction. You can see why this is true by noting the deviations for different values of x_p between the line of means $E(y) = \beta_0 + \beta_1 x$ and the predicted line of means $\hat{y} = \hat{\beta}_0 + \hat{\beta}_1 x$ shown in Figure 11.21. The deviation is larger at the extremes of the interval where the largest and smallest values of x in the data set occur.

Both the confidence intervals for mean values and the prediction intervals for new values are depicted over the entire range of the regression line in Figure 11.22. You can see that the confidence interval is always narrower than the prediction interval, and that they are both narrowest at the mean $\bar{x}$, increasing steadily as the distance $|x - \bar{x}|$ increases. In fact, when x is selected far enough away from $\bar{x}$ so that it falls outside the range of the sample data, it is dangerous to make any inferences about $E(y)$ or y.

Figure 11.21

Error of Predicting a
Future Value of y for
a Given Value of x

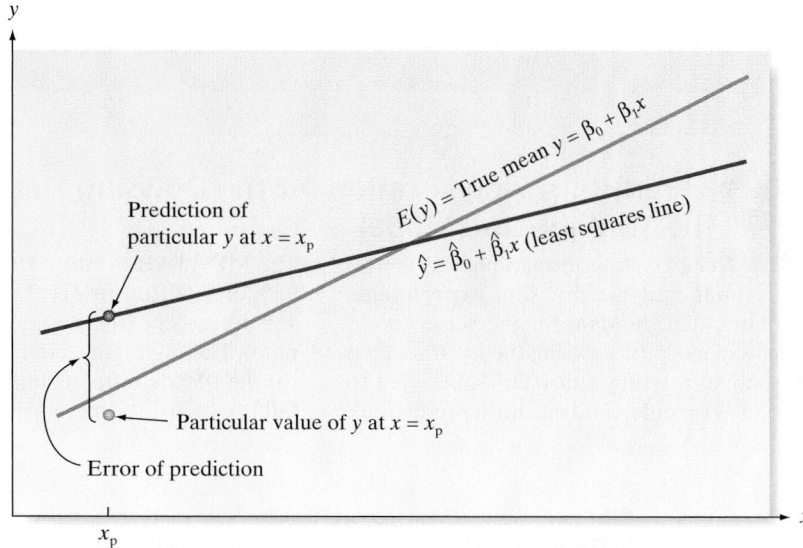

Figure 11.22

Confidence Intervals for
Mean Values and Prediction
Intervals for New Values

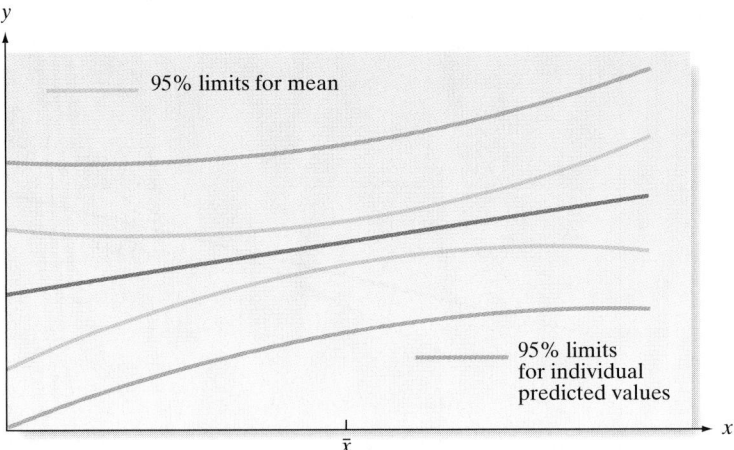

 **Caution**

Using the least squares prediction equation to estimate the mean value of y or to predict a particular value of y for values of x that fall outside the range of the values of x contained in your sample data may lead to errors of estimation or prediction that are much larger than expected. Although the least squares model may provide a very good fit to the data over the range of x values contained in the sample, it could give a poor representation of the true model for values of x outside this region.

The confidence interval width grows smaller as n is increased; thus, in theory, you can obtain as precise an estimate of the mean value of y as desired (at any given x) by selecting a large enough sample. The prediction interval for a new value of y also grows smaller as n increases, but there is a lower limit on its width. If you examine the formula for the prediction interval, you will see that the interval can get no smaller than $\hat{y} \pm z_{\alpha/2}\sigma$.* Thus, the only way to obtain more accurate predictions for new values of y is to reduce the standard deviation of the regression model, σ. This can be accomplished only by improving the model, either by using a curvilinear (rather than linear) relationship with x or by adding new independent variables to the model, or both. Methods of improving the model are discussed in Chapter 12.

Now Work *Exercise 11.86f*

Statistics in Action Revisited

Predicting Pipe Location for the Dowsing Data Using the Straight-Line Model

The group of German physicists who conducted the dowsing experiments stated that the data for the three "best" dowsers empirically support the dowsing theory. If so, then the straight-line model relating a dowser's guess (x) to actual pipe location (y) should yield accurate predictions.

The MINITAB printout shown in Figure SIA11.5 gives a 95% prediction interval for y when a dowser guesses $x = 50$ meters (the middle of the 100-meter-long waterpipe). The highlighted interval is $(-9.3, 100.23)$. Thus, we can be 95% confident that the actual pipe location will fall between -9.3 meters and 100.23 meters for this guess.

*The result follows from the facts that, for large n, $t_{\alpha/2} \approx z_{\alpha/2}$, $s \approx \sigma$, and the last two terms under the radical in the standard error of the predictor are approximately 0.

Since the pipe is only 100 meters long, in effect the interval ranges from 0 to 100 meters— the entire length of the pipe! This result, of course, is due to the fact that the straight-line model is not a statistically useful predictor of pipe location—a fact we discovered in the previous Statistics in Action Revisited sections.

Figure SIA11.5
MINITAB Prediction
Interval for Dowsing Data

```
Predicted Values for New Observations

New
Obs     Fit   SE Fit       95% CI           95% PI
 1    45.47    5.15   (34.83, 56.10)   (-9.30, 100.23)


Values of Predictors for New Observations

New
Obs   GUESS
 1    50.0
```

Exercises 11.81–11.95

Understanding the Principles

11.81 Explain the difference between y and $E(y)$ for a given x.

11.82 *True or False.* For a given x, a confidence interval for $E(y)$ will always be wider than a prediction interval for y.

11.83 *True or False.* The greater the deviation between x and $\bar{x}$, the wider the prediction interval for y will be.

11.84 For each of the following, decide whether the proper inference is a prediction interval for y or a confidence interval for $E(y)$.
   a. A jeweler wants to predict the selling price of a diamond stone based on its size (number of carats).
   b. A psychologist wants to estimate the average IQ of all patients that have a certain income level.

Learning the Mechanics

11.85 In fitting a least squares line to $n = 10$ data points, the following quantities were computed:
$$SS_{xx} = 32, \bar{x} = 3, SS_{yy} = 26, \bar{y} = 4, SS_{xy} = 28$$
   a. Find the least squares line.
   b. Graph the least squares line.
   c. Calculate SSE.
   d. Calculate s^2.
   e. Find a 95% confidence interval for the mean value of y when $x_p = 2.5$.
   f. Find a 95% prediction interval for y when $x_p = 4$.

11.86 Consider the following pairs of measurements:
NW

⊚ **LM11_86**

x	1	2	3	4	5	6	7
y	3	5	4	6	7	7	10

   a. Construct a scattergram for these data.
   b. Find the least squares line, and plot it on your scattergram.
   c. Find s^2.
   d. Find a 90% confidence interval for the mean value of y when $x = 4$. Plot the upper and lower bounds of the confidence interval on your scattergram.
   e. Find a 90% prediction interval for a new value of y when $x = 4$. Plot the upper and lower bounds of the prediction interval on your scattergram.
   f. Compare the widths of the intervals you constructed in parts **d** and **e**. Which is wider and why?

11.87 Consider the pairs of measurements shown in the following table.

⊚ **LM11_87**

x	4	6	0	5	2	3	2	6	2	1
y	3	5	-1	4	3	2	0	4	1	1

For these data, $SS_{xx} = 38.900$, $SS_{yy} = 33.600$, $SS_{xy} = 32.8$, and $\hat{y} = -4.14 + .843x$.
   a. Construct a scattergram for these data.
   b. Plot the least squares line on your scattergram.
   c. Use a 95% confidence interval to estimate the mean value of y when $x_p = 6$. Plot the upper and lower bounds of the interval on your scattergram.
   d. Repeat part **c** for $x_p = 3.2$ and $x_p = 0$.
   e. Compare the widths of the three confidence intervals you constructed in parts **c** and **d** and explain why they differ.

11.88 Refer to Exercise 11.87.
   a. Using no information about x, estimate and calculate a 95% confidence interval for the mean value of y. [*Hint:* Use the one-sample t methodology of Section 7.3.]

b. Plot the estimated mean value and the confidence interval as horizontal lines on your scattergram.

c. Compare the confidence intervals you calculated in parts **c** and **d** of Exercise 11.87 with the one you calculated in part **a** of this exercise. Does x appear to contribute information about the mean value of y?

d. Check the answer you gave in part **c** with a statistical test of the null hypothesis $H_0: \beta_1 = 0$ against $H_a: \beta_1 \neq 0$. Use $\alpha = .05$.

Applying the Concepts—Basic

11.89 Predicting the sale prices of homes. Refer to the data on sale prices and total appraised values of 92 residential properties in an upscale Tampa, Florida, neighborhood, Exercises 11.18 and 11.50 (pp. 607, 622).

a. In Exercise 11.50, you determined that appraised property value x and sale price y are positively related for homes sold in the Tampa Palms subdivision. Does this result guarantee that appraised value will yield accurate predictions of sale price? Explain.

b. MINITAB was used to predict the sale price of a residential property in the subdivision with a total appraised value of $300,000. Locate a 95% prediction interval for the sale price of this property on the printout below and interpret the result.

c. Locate a 95% confidence interval for $E(y)$ on the printout and interpret the result.

11.90 Sweetness of orange juice. Refer to the simple linear regression of sweetness index y and amount of pectin x for $n = 24$ orange juice samples, Exercise 11.26 (p. 610). The SPSS printout of the analysis is shown at the top of p. 643. A 90% confidence interval for the mean sweetness index, $E(y)$, for each value of x is shown on the SPSS spreadsheet. Select an observation and interpret this interval.

GOBIANTS

11.91 Mongolian desert ants. Refer to the *Journal of Biogeography* (Dec. 2003) study of ant sites in Mongolia, Exercises 11.22 and 11.38 (p. 608, 615). You applied the method of least squares to the data in the **GOBIANTS** file to estimate the straight-line model relating annual rainfall (y) and maximum daily temperature (x). A SAS

printout giving 95% prediction intervals for the amount of rainfall at each of the 11 sites is shown on p. 643. Select the interval associated with site (observation) 7 and interpret it practically.

Applying the Concepts—Intermediate

NAMEGAME2

11.92 The "name game." Refer to the *Journal of Experimental Psychology-Applied* (June 2000) name retrieval study, Exercises 11.28, 11.59, and 11.78 (pp. 611, 624, 635).

a. Find a 99% confidence interval for the mean recall proportion for students in the fifth position during the "name game." Interpret the result.

b. Find a 99% prediction interval for the recall proportion of a particular student in the fifth position during the "name game." Interpret the result.

c. Compare the two intervals, parts **a** and **b**. Which interval is wider? Will this always be the case? Explain.

11.93 Sports participation survey. The Sasakawa Sports Foundation conducted a national survey to assess the physical activity patterns of Japanese adults. The table below lists the frequency (average number of days in the past year) and duration of time (average number of minutes per single activity) Japanese adults spent participating in a sample of 11 sports activities.

JAPANSPORTS

Activity	Frequency x (days/year)	Duration y (minutes)
Jogging	135	43
Cycling	68	99
Aerobics	44	61
Swimming	39	60
Volleyball	30	80
Tennis	21	100
Softball	16	91
Baseball	19	127
Skating	7	115
Skiing	10	249
Golf	5	262

Source: J. Bennett, ed. *Statistics in Sport.* London: Arnold, 1998 (adapted from Figure 11.6).

MINITAB Output for Exercise 11.89

```
Predicted Values for New Observations

New
Obs      Fit   SE Fit       95% CI             95% PI
  1   341.56     4.33   (332.95, 350.17)   (275.86, 407.26)



Values of Predictors for New Observations

New
Obs   TOTALVAL
  1        300
```

SPSS Output for Exercise 11.90

	run	sweet	pectin	lower90m	upper90m
1	1	5.2	220	5.64898	5.83848
2	2	5.5	227	5.63898	5.81613
3	3	6.0	259	5.57819	5.72904
4	4	5.9	210	5.66194	5.87173
5	5	5.8	224	5.64337	5.82560
6	6	6.0	215	5.65564	5.85493
7	7	5.8	231	5.63284	5.80379
8	8	5.6	268	5.55553	5.71011
9	9	5.6	239	5.61947	5.78019
10	10	5.9	212	5.65946	5.86497
11	11	5.4	410	5.05526	5.55416
12	12	5.6	256	5.58517	5.73592
13	13	5.8	306	5.43785	5.65219
14	14	5.5	259	5.57819	5.72904
15	15	5.3	284	5.50957	5.68213
16	16	5.3	383	5.15725	5.57694
17	17	5.7	271	5.54743	5.70434
18	18	5.5	264	5.56591	5.71821
19	19	5.7	227	5.63898	5.81613
20	20	5.3	263	5.56843	5.72031
21	21	5.9	232	5.63125	5.80075
22	22	5.8	220	5.64898	5.83848
23	23	5.8	246	5.60640	5.76091
24	24	5.9	241	5.61587	5.77454

SAS Output for Exercise 11.91   Predictions

Obs	Region	Rain	Temp	Predicted Rain	Lower prediction limit of Rain	Upper prediction limit of Rain
1	DryStepp	196	5.7	201.977	179.525	224.430
2	DryStepp	196	5.7	201.977	179.525	224.430
3	DryStepp	179	7.0	180.704	163.694	197.714
4	DryStepp	197	8.0	164.340	150.594	178.085
5	DryStepp	149	8.5	156.157	143.513	168.802
6	GobiDese	112	10.7	120.156	106.038	134.274
7	GobiDese	125	11.4	108.701	92.298	125.104
8	GobiDese	99	10.9	116.883	102.172	131.595
9	GobiDese	125	11.4	108.701	92.298	125.104
10	GobiDese	84	11.4	108.701	92.298	125.104
11	GobiDese	115	11.4	108.701	92.298	125.104

a. Write the equation of a straight-line model relating duration (y) to frequency (x).

b. Find the least squares prediction equation.

c. Is there evidence of a linear relationship between y and x? Test using $\alpha = .05$.

d. Use the least squares line to predict the duration of time Japanese adults participate in a sport that they play 25 times a year. Form a 95% confidence interval around the prediction and interpret the result.

⊙ **SNOWGEESE**

11.94 Feeding habits of snow geese. Refer to the *Journal of Applied Ecology* feeding study of the relation-ship between weight change y of baby snow geese and digestion efficiency x, Exercise 11.76 (p. 634).

a. Fit the simple linear regression model to the data.

b. Do you recommend using the model to predict weight change y? Explain.

c. Use the model to form a 95% confidence interval for the mean weight change of all baby snow geese with a digestion efficiency of $x = 15\%$. Interpret the interval.

Applying the Concepts—Advanced

CUTTOOL

11.95 Life tests of cutting tools. Refer to the data saved in the **CUTTOOL** file, Exercise 11.42 (p. 616).

a. Use a 90% confidence interval to estimate the mean useful life of a brand A cutting tool when the cutting speed is 45 meters per minute. Repeat for brand B. Compare the widths of the two intervals and comment on the reasons for any difference.

b. Use a 90% prediction interval to predict the useful life of a brand A cutting tool when the cutting speed is 45 meters per minute. Repeat for brand B. Compare the widths of the two intervals to each other and to the two intervals you calculated in part a. Comment on the reasons for any differences.

c. Note that the estimation and prediction you performed in parts **a** and **b** were for a value of x that

was not included in the original sample. That is, the value $x = 45$ was not part of the sample. However, the value is within the range of x values in the sample, so that the regression model spans the x value for which the estimation and prediction were made. In such situations, estimation and prediction represent **interpolations**.

Suppose you were asked to predict the useful life of a brand A cutting tool for a cutting speed of $x = 100$ meters per minute. Since the given value of x is outside the range of the sample x values, the prediction is an example of **extrapolation**. Predict the useful life of a brand A cutting tool that is operated at 100 meters per minute, and construct a 95% confidence interval for the actual useful life of the tool. What additional assumption do you have to make in order to ensure the validity of an extrapolation?

11.9 A Complete Example

In the preceding sections we presented the basic elements necessary to fit and use a straight-line regression model. In this section we will assemble these elements by applying them in an example with the aid of computer software.

Suppose a fire insurance company wants to relate the amount of fire damage in major residential fires to the distance between the burning house and the nearest fire station. The study is to be conducted in a large suburb of a major city; a sample of 15 recent fires in this suburb is selected. The amount of damage, y, and the distance between the fire and the nearest fire station, x, are recorded for each fire. The results are given in Table 11.7.

Step 1 First, we hypothesize a model to relate fire damage, y, to the distance from the nearest fire station, x. We hypothesize a straight-line probabilistic model:

$$y = \beta_0 + \beta_1 x + \varepsilon$$

Step 2 Next, we enter the data of Table 11.7 into a computer and use statistical software to estimate the unknown parameters in the deterministic component

FIREDAM

TABLE 11.7 Fire Damage Data

Distance from Fire Station x (miles)	Fire Damage y (thousands of dollars)
3.4	26.2
1.8	17.8
4.6	31.3
2.3	23.1
3.1	27.5
5.5	36.0
.7	14.1
3.0	22.3
2.6	19.6
4.3	31.3
2.1	24.0
1.1	17.3
6.1	43.2
4.8	36.4
3.8	26.1

Figure 11.23

SAS Printout for Fire
Damage Regression

Dependent Variable: DAMAGE

Analysis of Variance

Source	DF	Sum of Squares	Mean Square	F Value	Pr > F
Model	1	841.76636	841.76636	156.89	<.0001
Error	13	69.75098	5.36546		
Corrected Total	14	911.51733			

Root MSE	2.31635	R-Square	0.9235	
Dependent Mean	26.41333	Adj R-Sq	0.9176	
Coeff Var	8.76961			

Parameter Estimates

Variable	DF	Parameter Estimate	Standard Error	t Value	Pr > |t|	95% Confidence Limits	
Intercept	1	10.27793	1.42028	7.24	<.0001	7.20960	13.34625
DISTANCE	1	4.91933	0.39275	12.53	<.0001	4.07085	5.76781

Output Statistics

Obs	DISTANCE	Dep Var DAMAGE	Predicted Value	Std Error Mean Predict	95% CL Predict		Residual
1	3.4	26.2000	27.0037	0.5999	21.8344	32.1729	-0.8037
2	1.8	17.8000	19.1327	0.8340	13.8141	24.4514	-1.3327
3	4.6	31.3000	32.9068	0.7915	27.6186	38.1951	-1.6068
4	2.3	23.1000	21.5924	0.7112	16.3577	26.8271	1.5076
5	3.1	27.5000	25.5279	0.6022	20.3573	30.6984	1.9721
6	5.5	36.0000	37.3342	1.0573	31.8334	42.8351	-1.3342
7	0.7	14.1000	13.7215	1.1766	8.1087	19.3342	0.3785
8	3	22.3000	25.0359	0.6081	19.8622	30.2097	-2.7359
9	2.6	19.6000	23.0682	0.6550	17.8678	28.2686	-3.4682
10	4.3	31.3000	31.4311	0.7198	26.1908	36.6713	-0.1311
11	2.1	24.0000	20.6085	0.7566	15.3442	25.8729	3.3915
12	1.1	17.3000	15.6892	1.0444	10.1999	21.1785	1.6108
13	6.1	43.2000	40.2858	1.2587	34.5906	45.9811	2.9142
14	4.8	36.4000	33.8907	0.8450	28.5640	39.2175	2.5093
15	3.8	26.1000	28.9714	0.6320	23.7843	34.1585	-2.8714
16	3.5	.	27.4956	0.6043	22.3239	32.6672	.

of the hypothesized model. The SAS printout for the simple linear regression analysis is shown in Figure 11.23. The least squares estimate of the slope β_1 and intercept β_0, highlighted on the printout, are

$$\hat{\beta}_1 = 4.91933$$
$$\hat{\beta}_0 = 10.27793$$

and the least squares equation is (rounded)

$$\hat{y} = 10.28 + 4.92x$$

This prediction equation is graphed using MINITAB in Figure 11.24 along with a plot of the data points.

The least squares estimate of the slope, $\hat{\beta}_1 = 4.92$, implies that the estimated mean damage increases by \$4,920 for each additional mile from the fire station. This interpretation is valid over the range of x, or from .7 to 6.1 miles from the station. The estimated y-intercept, $\hat{\beta}_0 = 10.28$, has the interpretation that a fire 0 miles from the fire station has an estimated mean damage of \$10,280. Although this would seem to apply to the fire station itself, remember that the y-intercept is meaningfully interpretable only if $x = 0$ is within the sampled range of the independent variable. Since $x = 0$ is outside the range, $\hat{\beta}_0$ has no practical interpretation.

Step 3 Now we specify the probability distribution of the random error component ε. The assumptions about the distribution are identical to those listed in Section 11.3. Although we know that these assumptions are not completely satisfied (they rarely are for practical problems), we are willing to assume they are approximately satisfied for this example. The estimate of the standard deviation σ of ε, highlighted on the printout, is

$$s = 2.31635$$

Figure 11.24

MINITAB Scatterplot with Least Squares Line for Fire Damage Regression Analysis

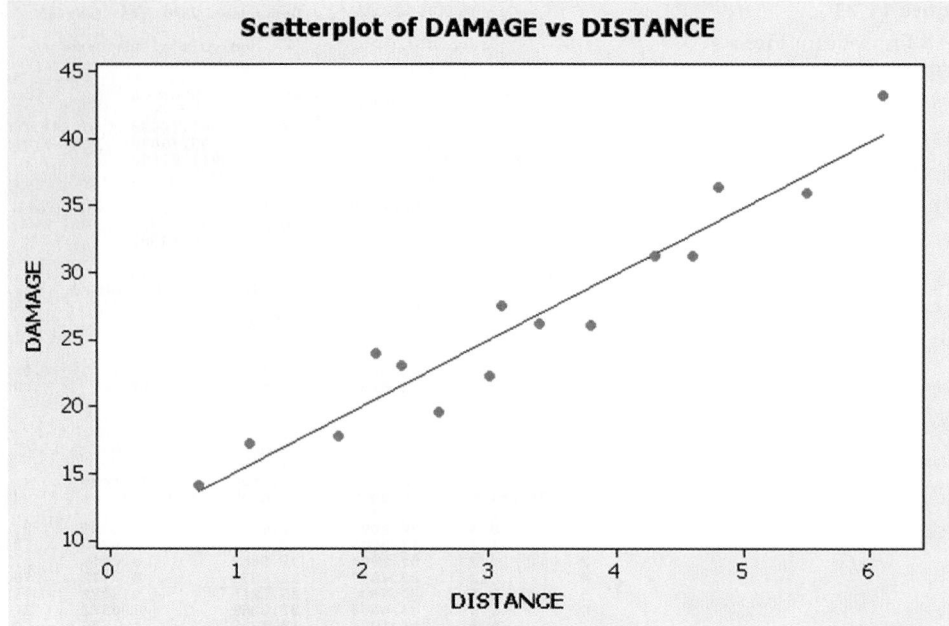

This implies that most of the observed fire damage (y) values will fall within approximately $2s = 4.64$ thousand dollars of their respective predicted values when using the least squares line.

Step 4 We can now check the usefulness of the hypothesized model—that is, whether x really contributes information for the prediction of y using the straight-line model. First, test the null hypothesis that the slope β_1 is 0, that is, that there is no linear relationship between fire damage and the distance from the nearest fire station, against the alternative hypothesis that fire damage increases as the distance increases. We test

$$H_0: \beta_1 = 0$$
$$H_a: \beta_1 > 0$$

The two-tailed observed significance level for testing $H_a: \beta_1 \neq 0$, highlighted on the printout, is less than .0001. Thus, the p-value for our one-tailed test is less than half of this value (.00005). This small p-value leaves little doubt that mean fire damage and distance between the fire and station are at least linearly related, with mean fire damage increasing as the distance increases.

We gain additional information about the relationship by forming a confidence interval for the slope β_1. A 95% confidence interval, highlighted on the SAS printout, is (4.071, 5.768). With 95% confidence, we estimate that the interval from \$4,071 to \$5,768 encloses the mean increase (β_1) in fire damage per additional mile distance from the fire station.

Another measure of the utility of the model is the coefficient of determination, r^2. The value (also highlighted on the printout) is $r^2 = .9235$, which implies that about 92% of the sample variation in fire damage (y) is explained by the distance (x) between the fire and the fire station.

The coefficient of correlation, r, that measures the strength of the linear relationship between y and x is not shown on the SAS printout and must be calculated. Using the facts that $r = \sqrt{r^2}$ in simple linear regression and that r and $\hat{\beta}_1$ have the same sign, we find

$$r = +\sqrt{r^2} = \sqrt{.9235} = .96$$

The high correlation confirms our conclusion that β_1 is greater than 0; it appears that fire damage and distance from the fire station are positively correlated. All signs point to a strong linear relationship between y and x.

Step 5 We are now prepared to use the least squares model. Suppose the insurance company wants to predict the fire damage if a major residential fire were to occur 3.5 miles from the nearest fire station. The predicted value (highlighted at the bottom of the printout) is $\hat{y} = 27.496$, while the 95% prediction interval (also highlighted) is (22.324, 32.667). Therefore, with 95% confidence we predict fire damage in a major residential fire 3.5 miles from the nearest station to be between $22,324 and $32,667.

Caution

We would not use this prediction model to make predictions for homes less than .7 mile or more than 6.1 miles from the nearest fire station. A look at the data in Table 11.7 reveals that all the x-values fall between .7 and 6.1. It is dangerous to use the model to make predictions outside the region in which the sample data fall. A straight line might not provide a good model for the relationship between the mean value of y and the value of x when stretched over a wider range of x-values.

Quick Review

Key Terms

Bivariate relationship   624
Coefficient of correlation   625
Coefficient of determination   630
Confidence interval for mean of y 636
Dependent variable   594
Deterministic model   592
Deterministic relationship   592
Errors of prediction   597
Independent variable   594
Least squares estimates   598

Least squares line (or prediction equation)   598
Line of means   594
Method of least squares   597
Population correlation coefficient   628
Prediction interval for y 637
Predictor variable   594
Probabilistic model   592
Probabilistic relationship   592
Random error   592

Regression analysis   594
Response variable   594
Scattergram   596
Slope   594
Straight-line (first-order) model   594
Standard error of the regression model   613
Standard error of the estimated slope   617
y-intercept   594

Key Formulas

$$\hat{\beta}_1 = \frac{SS_{xy}}{SS_{xx}}, \qquad \hat{\beta}_0 = \bar{y} - \hat{\beta}_1 \bar{x}$$
Least squares estimates of β's   598

where $SS_{xy} = \sum xy - \dfrac{(\Sigma x)(\Sigma y)}{n}$

$$SS_{xx} = \sum x^2 - \frac{(\Sigma x)^2}{n}$$

$\hat{y} = \hat{\beta}_0 + \hat{\beta}_1 x$
Least squares line   598

$SSE = \sum (y_i - \hat{y}_i)^2 = SS_{yy} - \hat{\beta}_1 SS_{xy}$
Sum of squared errors   613

where $SS_{yy} = \sum y^2 - \dfrac{(\Sigma y)^2}{n}$

$s^2 = \dfrac{SSE}{n - 2}$
Estimated variance of σ^2 of ε 613

$s_{\hat{\beta}_1} = \dfrac{s}{\sqrt{SS_{xx}}}$
Estimated standard error of $\hat{\beta}_1$ 617

$$t = \frac{\hat{\beta}_1}{s_{\hat{\beta}_1}}$$

Test statistic for H_0: $\beta_1 = 0$ 618

$$\hat{\beta}_1 \pm (t_{\alpha/2})s_{\hat{\beta}_1}$$

$(1 - \alpha)100\%$ confidence interval for β_1 620

$$r = \frac{SS_{xy}}{\sqrt{SS_{xx}\,SS_{yy}}} = \pm\sqrt{r^2}\ (\text{same sign as}\ \hat{\beta}_1)$$

Coefficient of correlation   625

$$r^2 = \frac{SS_{yy} - SSE}{SS_{yy}}$$

Coefficient of determination   630

$$\hat{y} \pm (t_{\alpha/2})s\sqrt{\frac{1}{n} + \frac{(x_p - \bar{x})^2}{SS_{xx}}}$$

$(1 - \alpha)100\%$ confidence interval for $E(y)$ when $x = x_p$ 636

$$\hat{y} \pm (t_{\alpha/2})s\sqrt{1 + \frac{1}{n} + \frac{(x_p - \bar{x})^2}{SS_{xx}}}$$

$(1 - \alpha)100\%$ prediction interval for y when $x = x_p$ 637

Language Lab

Symbol	Pronunciation	Description
y		Dependent variable (variable to be predicted or modeled)
x		Independent (predictor) variable
$E(y)$		Expected (mean) value of y
β_0	beta-zero	y-intercept of true line
β_1	beta-one	Slope of true line
$\hat{\beta}_0$	beta-zero hat	Least squares estimate of y-intercept
$\hat{\beta}_1$	beta-one hat	Least squares estimate of slope
ε	epsilon	Random error
$\hat{y}$	y-hat	Predicted value of y
$(y - \hat{y})$		Error of prediction
SE		Sum of errors (will equal zero with least squares line)
SSE		Sum of squared errors (will be smallest for least squares line)
SS_{xx}		Sum of squares of x-values
SS_{yy}		Sum of squares of y-values
SS_{xy}		Sum of squares of cross-products, $x \cdot y$
r		Coefficient of correlation
r^2	R-squared	Coefficient of determination
x_p		Value of x used to predict y

Chapter Summary Notes

- Two quantitative variables in *simple linear regression*: **y = dependent** variable (i.e., the variable to be predicted) and **x = independent** (i.e., predictor) variable.
- General form of a **probabilistic model** for y: $y = E(y) + \varepsilon$
- **First-order (straight-line) model**: $y = \beta_0 + \beta_1 x + \varepsilon$
- **Slope (β_1)** represents the change in y for every 1-unit increase in x.

- **y-intercept (β_0)** represents the value where the line intercepts the y-axis.
- **Steps** in simple linear regression: (1) Hypothesize the model, (2) use the method of least squares to estimate the unknown β's, (3) make assumptions on the random error (ε), (4) statistically evaluate the adequacy of the model, and (5) if deemed useful, use the model for estimation and prediction.

- Properties of **method of least squares**: (1) sum of errors of prediction is 0, (2) sum of squared errors of prediction is minimized.

- Estimates of slope and y-intercept should only be interpreted *over the range of x-values in the sample.*

- **Four assumptions for ε**: (1) mean of ε is 0, (2) variance of ε is constant for all x-values, (3) distribution of ε is normal, (4) values of ε are independent.

- *Interpretation of estimated standard deviation of ε*: About 95% of the observed y-values will lie within $2s$ of the respective predicted values.

- *Statistics used to assess the adequacy of the model:* (1) test of hypothesis for β_1, (2) confidence interval for β_1, (3) coefficient of correlation r, (4) coefficient of determination, r^2.

- Range of correlation coefficient: $-1 \leq r \leq 1$.

- Range of coefficient of determination: $0 \leq r^2 \leq 1$.

- **Correlation coefficient** measures the strength of the linear relationship between x and y.

- **Coefficient of determination** gives the proportion of the sample variation in y that can be explained by the straight-line model.

- Do not assume that a high correlation implies that x causes y.

- For a given x-value, a confidence interval for $E(y)$ will be narrower than a prediction interval for y.

Supplementary Exercises 11.96–11.119

Understanding the Principles

11.96 Explain the difference between a probabilistic model and a deterministic model.

11.97 Give the general form of a straight-line model for $E(y)$.

11.98 Outline the five steps in a simple linear regression analysis.

11.99 *True or False.* In simple linear regression, about 95% of the y-values in the sample will fall within $2s$ of their respective predicted values.

Learning the Mechanics

11.100 In fitting a least squares line to $n = 15$ data points, the following quantities were computed: $SS_{xx} = 55$, $SS_{yy} = 198$, $SS_{xy} = -88$, $\bar{x} = 1.3$, and $\bar{y} = 35$.
  a. Find the least squares line.
  b. Graph the least squares line.
  c. Calculate SSE.
  d. Calculate s^2.
  e. Find a 90% confidence interval for β_1. Interpret this estimate.
  f. Find a 90% confidence interval for the mean value of y when $x = 15$.
  g. Find a 90% prediction interval for y when $x = 15$.

11.101 Consider the following sample data:

y	5	1	3
x	5	1	3

  a. Construct a scattergram for the data.
  b. It is possible to find many lines for which $\Sigma(y - \hat{y}) = 0$. For this reason, the criterion $\Sigma(y - \hat{y}) = 0$ is not used for identifying the "best-fitting" straight line. Find two lines that have $\Sigma(y - \hat{y}) = 0$.
  c. Find the least squares line.
  d. Compare the value of SSE for the least squares line to that of the two lines you found in part **b**. What principle of least squares is demonstrated by this comparison?

11.102 Consider the following 10 data points:

⊙ **LM11_102**

x	3	5	6	4	3	7	6	5	4	7
y	4	3	2	1	2	3	3	5	4	2

  a. Plot the data on a scattergram.
  b. Calculate the values of r and r^2.
  c. Is there sufficient evidence to indicate that x and y are linearly correlated? Test at the $\alpha = .10$ level of significance.

Applying the Concepts—Basic

11.103 Baseball batting averages versus wins. Is the number of games won by a major league baseball team in a season related to the team's batting average? In Exercise 2.141 (p. 99), you examined data from the *Baseball Almanac* (2003) on the number of games won and the batting averages for the 14 teams in the American League for the 2002 Major League Baseball season. The data are repeated in the table.

⊙ **ALWINS**

Team	Games Won	Batting Ave.
New York	103	.275
Toronto	78	.261
Baltimore	67	.246
Boston	93	.277
Tampa Bay	55	.253
Cleveland	74	.249
Detroit	55	.248
Chicago	81	.268
Kansas City	62	.256
Minnesota	94	.272
Anaheim	99	.282
Texas	72	.269
Seattle	93	.275
Oakland	103	.261

Source: Baseball Almanac, 2003.

```
The regression equation is
Wins = - 214 + 1118 BatAve


Predictor      Coef  SE Coef      T      P
Constant    -214.30    68.33  -3.14  0.009
BatAve       1118.4    258.9   4.32  0.001


S = 11.1244    R-Sq = 60.9%    R-Sq(adj) = 57.6%


Analysis of Variance

Source            DF      SS      MS      F      P
Regression         1  2310.2  2310.2  18.67  0.001
Residual Error    12  1485.0   123.8
Total             13  3795.2


Predicted Values for New Observations

New
Obs    Fit  SE Fit       95% CI              95% PI
  1  93.27    4.17  (84.18, 102.35)  (67.38, 119.15)


Values of Predictors for New Observations

New
Obs   BatAve
  1    0.275
```

a. If you were to model the relationship between the mean (or expected) number of games won by a major league team and the team's batting average x, using a straight line, would you expect the slope of the line to be positive or negative? Explain.

b. Construct a scattergram of the data. Does the pattern revealed by the scattergram agree with your answer to part a?

c. A MINITAB printout of the simple linear regression is provided above. Find the estimates of the β's on the printout and write the equation of the least squares line.

d. Graph the least squares line on your scattergram. Does your least squares line seem to fit the points on your scattergram?

e. Interpret the estimates of β_0 and β_1 in the words of the problem.

f. Conduct a test (at $\alpha = .05$) to determine if the mean (or expected) number of games won by a major league baseball team is positively linearly related to the team's batting average.

g. Find the coefficient of determination, r^2, and interpret its value.

h. Predict the number of games won by team with a .275 batting average.

i. Find a 95% prediction interval for the number of games won by team with a .275 batting average. Interpret the interval.

11.104 Life span of a horse. A breeder of thoroughbred horses wishes to model the relationship between the gestation period and the life span of a horse. The breeder believes that the two variables may follow a linear trend. The information in the table was supplied to the breeder from various thoroughbred stables across the state. (Note that the horse has the greatest variation of gestation period of any species owing to seasonal and nutritional factors.) An SPSS printout of the simple linear regression analysis is shown on p. 651.

⊙ **HORSES**

Horse	Gestation Period x (days)	Life Span y (years)
1	416	24
2	279	25.5
3	298	20
4	307	21.5
5	356	22
6	403	23.5
7	265	21

Coefficients[a]

Model		Unstandardized Coefficients		Standardized Coefficients	t	Sig.	95% Confidence Interval for B	
		B	Std. Error	Beta			Lower Bound	Upper Bound
1	(Constant)	18.890	4.499		4.198	.009	7.324	30.457
	GESTATE	.011	.013	.342	.813	.453	-.023	.045

a. Dependent Variable: LIFESPAN

a. Do the data provide sufficient evidence to support the breeder's hypothesis? Test using $\alpha = .05$.

b. Find a 95% confidence interval for β_1. Interpret this interval. Does the interval support part **a**?

11.105 Feeding habits of fish. Refer to the *Brain and Behavior Evolution* (Apr. 2000) study of the feeding behavior of blackbream fish, Exercise 2.142 (p. 99). Recall that the zoologists recorded the number of aggressive strikes of two blackbream fish feeding at the bottom of an aquarium in the 10-minute period following the addition of food. The table listing the weekly number of strikes and age of the fish (in days) is reproduced here.

BLACKBREAM

Week	Number of Strikes	Age of Fish (days)
1	85	120
2	63	136
3	34	150
4	39	155
5	58	162
6	35	169
7	57	178
8	12	184
9	15	190

Source: Shand, J., et al. "Variability in the location of the retinal ganglion cell area centralis is correlated with ontogenetic changes in feeding behavior in the Blackbream, Acanthopagrus 'butcher'." *Brain and Behavior*, Vol. 55, No. 4, Apr. 2000 (Figure H).

a. Write the equation of a straight-line model relating number of strikes (y) to age of fish (x).

b. Fit the model to the data using the method of least squares and give the least squares prediction equation.

c. Give a practical interpretation of the value of $\hat{\beta}_0$, if possible.

d. Give a practical interpretation of the value of $\hat{\beta}_1$, if possible.

e. Test $H_0: \beta_1 = 0$ versus $H_a: \beta_1 < 0$ using $\alpha = .10$. Interpret the result.

11.106 Math anxiety study. Many high school students experience "math anxiety." Does such an attitude carry over to learning computer skills? A researcher at Duquesne University investigated this question and published her results in *Educational Technology* (May–June 1995). A sample of high school students— 902 boys and 828 girls—from public schools in Pittsburgh, Pennsylvania, participated in the study. Using 5-point Likert scales, where 1 = "strongly disagree" and 5 = "strongly agree," the researcher measured the students' interest and confidence in both mathematics and computers.

a. For boys, math confidence and computer interest were correlated at $r = .14$. Fully interpret this result.

b. For girls, math confidence and computer interest were correlated at $r = .33$. Fully interpret this result.

11.107 Leisure research. Is there a relationship between leisure activities and high school performance? This question was investigated in the *Journal of Leisure Research* (Vol. 24, 1992). A list of 43 leisure activities was presented to 159 high school students and the students were asked to identify the activities they participated in each week. The article reports that the correlation between high school GPA and the number of leisure activities is $r = .13$, which has a two-tailed p-value of .0512.

a. What are the appropriate null and alternative hypotheses to test whether the number of leisure activities and GPA are linearly related?

b. Interpret the p-value in terms of this test.

11.108 Walking study. Refer to the *American Scientist* (July–Aug. 1998) study of the relationship between self-avoiding and unrooted walks, Exercise 2.143 (p. 99). Recall that in a self-avoiding walk you never retrace or cross your own path; while an unrooted walk is a path in which the starting and ending points are impossible to distinguish. The possible number of walks of each type of various lengths are reproduced in the table. Consider the straight-line model $y = \beta_0 + \beta_1 x + \varepsilon$, where x is walk length (number of steps).

WALK

Walk Length (number of steps)	Unrooted Walks	Self-Avoiding Walks
1	1	4
2	2	12
3	4	36
4	9	100
5	22	284
6	56	780
7	147	2,172
8	388	5,916

Source: Hayes, B. "How to avoid yourself." *American Scientist*, Vol. 86, No. 4, July–Aug. 1988, p. 317 (Figure 5).

a. Use the method of least squares to fit the model to the data if y is the possible number of unrooted walks.

b. Interpret $\hat{\beta}_0$ and $\hat{\beta}_1$ in the estimated model, part **a**.

c. Repeat parts **a** and **b** if y is the possible number of self-avoiding walks.

d. Find a 99% confidence interval for the number of unrooted walks possible when walk length is four steps.

BEANIE (First and last 5 observations shown.)

Name	Age (months) as of Sept. 1998	Retired (R)/ Current (C)	Value ($)
1. Ally the Alligator	52	R	55.00
2. Batty the Bat	12	C	12.00
3. Bongo the Brown Monkey	28	R	40.00
4. Blackie the Bear	52	C	10.00
5. Bucky the Beaver	40	R	45.00
⋮	⋮	⋮	⋮
46. Stripes the Tiger (Gold/Black)	40	R	400.00
47. Teddy the 1997 Holiday Bear	12	R	50.00
48. Tuffy the Terrier	17	C	10.00
49. Tracker the Basset Hound	5	C	15.00
50. Zip the Black Cat	28	R	40.00

Source: Beanie World Magazine, Sept. 1998.

e. Would you recommend using simple linear regression to predict the number of walks possible when walk length is 15 steps? Explain.

Applying the Concepts—Intermediate

11.109 Beanie Babies. Refer to Exercise 2.167 (p. 107) and the data on 50 Beanie Babies collector's items, published in *Beanie World Magazine*. Can age of a Beanie Baby be used to accurately predict its market value? Answer this question by conducting a complete simple linear regression analysis on the data saved in the **BEANIE** file. (The first and last 5 are shown in the table above.)

11.110 Organic chemistry experiment. Chemists at Kyushu University (Japan) examined the linear relationship between the maximum absorption rate y (in nanomoles) and the Hammett substituent constant x for metacyclophane compounds (*Journal of Organic Chemistry*, July 1995.) The data for variants of two compounds are given in the table. The variants of compound 1 are labeled 1a, 1b, 1d, 1e, 1f, 1g, and 1h; the variants of compound 2 are 2a, 2b, 2c, and 2d.

a. Plot the data in a scattergram. Use two different plotting symbols for the two compounds. What do you observe?

ORGCHEM

Compound	Maximum Absorption y	Hammett Constant x
1a	298	0.00
1b	346	.75
1d	303	.06
1e	314	−.26
1f	302	.18
1g	332	.42
1h	302	−.19
2a	343	.52
2b	367	1.01
2c	325	.37
2d	331	.53

Source: Adapted from Tsuge, A., et al. "Preparation and spectral properties of disubstituted [2-2] metacyclophanes." Journal of Organic Chemistry, Vol. 60, No. 15, July 1995, pp. 4390–4391 (Table 1 and Figure 1).

b. Using only the data for compound 1 fit the model $E(y) = \beta_0 + \beta_1 x$.

c. Assess the adequacy of the model, part **b**. Using $\alpha = .01$.

d. Repeat parts **b** and **c** using only the data for compound 2.

11.111 Mortality of predatory birds. Two species of predatory birds, collard flycatchers and tits, compete for nest holes during breeding season on the island of Gotland, Sweden. Frequently, dead flycatchers are found in nest boxes occupied by tits. A field study examined whether the risk of mortality to flycatchers is related to the degree of competition between the two bird species for nest sites. (*The Condor*, May 1995.) The table gives data on the number y of flycatchers killed at each of 14 discrete locations (plots) on the island as well as the nest box tit occupancy x (that is, the percentage of nest boxes occupied by tits) at each plot. Consider the simple linear regression model, $E(y) = \beta_0 + \beta_1 x$.

CONDOR2

Plot	Number of Flycatchers Killed y	Nest Box Tit Occupancy x (%)
1	0	24
2	0	33
3	0	34
4	0	43
5	0	50
6	1	35
7	1	35
8	1	38
9	1	40
10	2	31
11	2	43
12	3	55
13	4	57
14	5	64

Source: Merila, J., and Wiggins, D. A. "Interspecific competition for nest holes causes adult mortality in the collard flycatcher." The Condor, Vol. 97, No. 2, May 1995, p. 449 (Figure 2), Cooper Ornithological Society.

a. Plot the data in a scattergram. Does the frequency of flycatcher casualties per plot appear to increase linearly with increasing proportion of nest boxes occupied by tits?

b. Use the method of least squares to find the estimates of β_0 and β_1. Interpret their values.

c. Test the utility of the model using $\alpha = .05$.

d. Find r and r^2 and interpret their values.

e. Find s and interpret the result.

f. Do you recommend using the model to predict number of flycatchers killed? Explain.

11.112 Snowmelt runoff erosion. The U.S. Department of Agriculture has developed and adopted the Universal Soil Loss Equation (USLE) for predicting water erosion of soils. In geographic areas where runoff from melting snow is common, calculating the USLE requires an accurate estimate of snowmelt runoff erosion. An article in the *Journal of Soil and Water Conservation* (Mar.–Apr. 1995) used simple linear regression to develop a snowmelt erosion index. Data for 54 climatological stations in Canada were used to model the McCool winter-adjusted rainfall erosivity index, y, as a straight-line function of the once-in-5-year snowmelt runoff amount, x (measured in millimeters).

a. The data points are plotted in the scattergram shown below. Is there visual evidence of a linear trend?

b. The data for seven stations were removed from the analysis due to lack of snowfall during the study period. Why is this strategy advisable?

c. The simple linear regression on the remaining $n = 47$ data points yielded the following results:

$\hat{y} = -6.72 + 1.39x$, $s_{\hat{\beta}_1} = .06$. Use this information to construct a 90% confidence interval for β_1.

d. Interpret the interval, part **c**.

11.113 Quantum tunneling. At temperatures approaching absolute zero ($-273°C$), helium exhibits traits that seem to defy many laws of Newtonian physics. An experiment has been conducted with helium in solid form at various temperatures near absolute zero. The solid helium is placed in a dilution refrigerator along with a solid impure substance, and the fraction (in weight) of the impurity passing through the solid helium is recorded. (This phenomenon of solids passing directly through solids is known as *quantum tunneling*.) The data are given in the table.

HELIUM

Temperature $x(°C)$	Proportion of Impurity
−262.0	.315
−265.0	.202
−256.0	.204
−267.0	.620
−270.0	.715
−272.0	.935
−272.4	.957
−272.7	.906
−272.8	.985
−272.9	.987

a. Find the least squares estimates of the intercept and slope. Interpret them.

b. Use a 95% confidence interval to estimate the slope β_1. Interpret the interval in terms of this application. Does the interval support the hypothesis that temperature contributes information

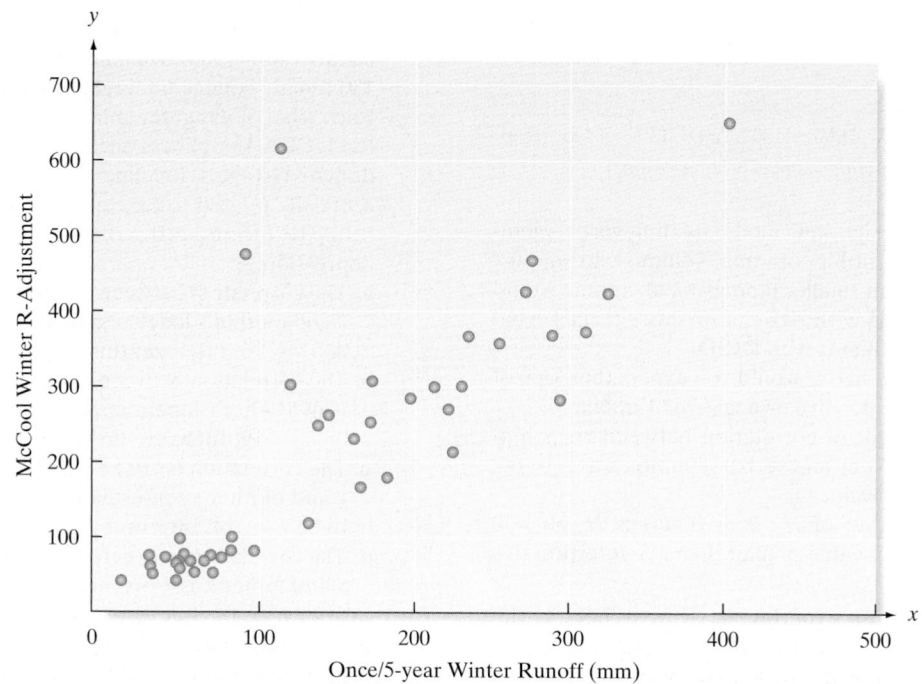

McCool Winter R-Adjustment (y-axis)

Once/5-year Winter Runoff (mm) (x-axis)

about the proportion of impurity passing through helium?

c. Interpret the coefficient of determination for this model.

d. Find a 95% prediction interval for the percentage of impurity passing through solid helium at −273°C. Interpret the result.

e. Note that the value of x in part **d** is outside the experimental region. Why might this lead to an unreliable prediction?

11.114 Short-term memory study. Neurologists have found that the hippocampus, a structure found in the brain, plays an important role in short-term memory. Refer to the *American Journal of Psychiatry* (July 1995) study of the relationship between hippocampal volume and short-term verbal memory of 21 Vietnam vets with combat related posttraumatic stress disorder (PTSD), Exercise 2.146 (p. 100). Recall that magnetic resonance imaging was used to measure the volume, x, of the right hippocampus (in cubic millimeters) of each subject while verbal memory retention, y, of each subject was measured by the percent retention subscale of the Wechsler Memory Scale. The scattergram for the data is reproduced here.

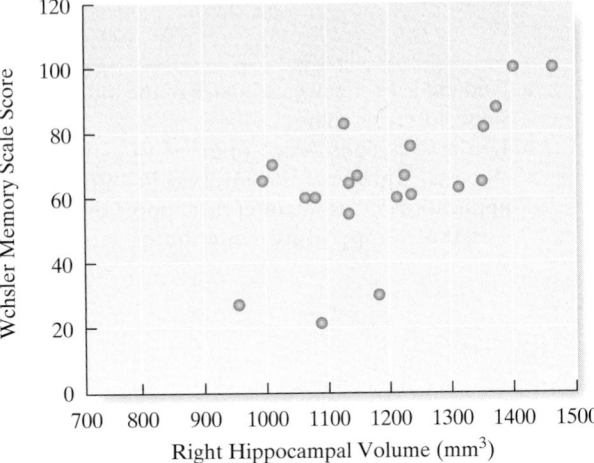

a. Propose a straight-line model relating verbal memory y with right hippocampal volume x to support the theory that smaller hippocampal volume would be associated with deficits in short-term verbal memory in patients with PTSD.

b. Based on the theory, would you expect the slope of the line to be positive or negative? Explain.

c. The coefficient of correlation between right hippocampal volume and verbal retention was $r = .64$. Interpret this value.

d. A statistical test of $H_0: \beta_1 = 0$ versus $H_a: \beta_1 > 0$ resulted in a p-value smaller than .05. Interpret this result.

11.115 Seed germination experiment. Researchers at the University of North Carolina-Greensboro investigated a model for the rate of seed germination. (*Journal*

of Experimental Botany, Jan. 1993.) In one experiment, alfalfa seeds were placed in a specially constructed germination chamber. Eleven hours later, the seeds were examined and the change in free energy (a measure of germination rate) was recorded. The results for seeds germinated at seven different temperatures are given in the table. The data were used to fit a simple linear regression model, with y = change in free energy and x = temperature.

⊚ SEEDGERM

Change in Free Energy (kJ/mol) y	Temperature (K) x
7	295
6.2	297.5
9	291
9.5	289
8.5	301
7.8	293
11.2	286.5

Source: Hageseth, G. T., and Cody, A. L. "Energy-level model for isothermal seed germination." *Journal of Experimental Botany*, Vol. 44, No. 258, Jan. 1993, p. 123 (Figure 9).

a. Plot the points in a scattergram.

b. Find the least squares prediction equation.

c. Plot the least squares line, part **b** on the scattergram of part **a**.

d. Conduct a test of model adequacy. Use $\alpha = .01$.

e. Use the plot, part **c**, to locate any unusual data points (outliers).

f. Eliminate the outlier, part **e**, from the data set and repeat parts **a–d**.

11.116 Loneliness in families. Is there a link between the loneliness of parents and that of their offspring? This question was examined in the *Journal of Marriage and Family* (Aug. 1986). The participants in the study were 130 female college undergraduates and their parents. Each triad of daughter, mother, and father completed the UCLA Loneliness Scale, a 20-item questionnaire designed to assess loneliness and several variables theoretically related to loneliness, such as social accessibility to others, difficulty in making friends, and depression.

a. The correlation between daughter's loneliness score y and mother's loneliness score x was determined to be $r = .26$. Interpret this value.

b. The correlation between daughter's loneliness score y and father's loneliness score x was determined to be $r = .19$. Interpret this value.

c. The correlation between daughter's loneliness score y and mother's self-esteem score x was determined to be $r = .14$. Interpret this value.

d. The correlation between daughter's loneliness score y and father's assertiveness score x was determined to be $r = .01$. Interpret this value.

e. Calculate the coefficient of determination r^2, for parts **a–d**. Interpret the results.

Applying the Concepts—Advanced

11.117 Regression through the origin. Sometimes it is known from theoretical considerations that the straight-line relationship between two variables, x and y, passes through the origin of the xy-plane. Consider the relationship between the total weight of a shipment of 50-pound bags of flour, y, and the number of bags in the shipment, x. Since a shipment containing $x = 0$ bags (i.e., no shipment at all) has a total weight of $y = 0$, a straight-line model of the relationship between x and y should pass through the point $x = 0$, $y = 0$. In such a case you could assume $\beta_0 = 0$ and characterize the relationship between x and y with the following model:

$$y = \beta_1 x + \varepsilon$$

The least squares estimate of β_1 for this model is

$$\hat{\beta}_1 = \frac{\sum x_i y_i}{\sum x_i^2}$$

From the records of past flour shipments, 15 shipments were randomly chosen and the data shown in the table were recorded.

FLOUR

Weight of Shipment	Number of 50-Pound Bags in Shipment
5,050	100
10,249	205
20,000	450
7,420	150
24,685	500
10,206	200
7,325	150
4,958	100
7,162	150
24,000	500
4,900	100
14,501	300
28,000	600
17,002	400
16,100	400

a. Find the least squares line for the given data under the assumption that $\beta_0 = 0$. Plot the least squares line on a scattergram of the data.

b. Find the least squares line for the given data using the model

$$y = \beta_0 + \beta_1 x + \varepsilon$$

(i.e., do not restrict β_0 to equal 0). Plot this line on the same scatterplot you constructed in part **a**.

c. Refer to part **b**. Why might $\hat{\beta}_0$ be different from 0 even though the true value of β_0 is known to be 0?

d. The estimated standard error of $\hat{\beta}_0$ is equal to

$$s\sqrt{\frac{1}{n} + \frac{\bar{x}^2}{SS_{xx}}}$$

Use the t-statistic

$$t = \frac{\hat{\beta}_0 - 0}{s\sqrt{(1/n) + (\bar{x}^2/SS_{xx})}}$$

to test the null hypothesis H_0: $\beta_0 = 0$ against the alternative H_a: $\beta_0 \neq 0$. Use $\alpha = .10$. Should you include β_0 in your model?

Critical Thinking Challenge

11.118 Study of fertility rates. The fertility rate of a country is defined as the number of children a woman citizen bears, on average, in her lifetime. *Scientific American* (Dec. 1993) reported on the declining fertility rate in developing countries. The researchers found that family planning can have a great effect on fertility rate. The table below gives the fertility rate, y, and contraceptive prevalence, x, (measured as the percentage of married women who use contraception) for each of 27 developing countries.

a. According to the researchers, "the data reveal that differences in contraceptive prevalence explain about 90% of the variation in fertility rates." Do you concur?

FERTRATE

Country	Contraceptive Prevalence x	Fertility Rate y
Mauritius	76	2.2
Thailand	69	2.3
Colombia	66	2.9
Costa Rica	71	3.5
Sri Lanka	63	2.7
Turkey	62	3.4
Peru	60	3.5
Mexico	55	4.0
Jamaica	55	2.9
Indonesia	50	3.1
Tunisia	51	4.3
El Salvador	48	4.5
Morocco	42	4.0
Zimbabwe	46	5.4
Egypt	40	4.5
Bangladesh	40	5.5
Botswana	35	4.8
Jordan	35	5.5
Kenya	28	6.5
Guatemala	24	5.5
Cameroon	16	5.8
Ghana	14	6.0
Pakistan	13	5.0
Senegal	13	6.5
Sudan	10	4.8
Yemen	9	7.0
Nigeria	7	5.7

Source: Robey, B., et al. "The fertility decline in developing countries." *Scientific American*, Dec. 1993, p. 62. [*Note:* The data values are estimated from a scatterplot.]

b. The researchers also concluded that "if contraceptive use increases by 18 percent, women bear, on average, one fewer child." Is this statement supported by the data? Explain.

11.119 Spall damage in bricks. A recent civil suit revolved around a 5-building brick apartment complex located in the Bronx, New York, which began to suffer *spalling* damage (i.e., a separation of some portion of the face of a brick from its body). The owner of the complex alleged that the bricks were defectively manufactured. The brick manufacturer countered that poor design and shoddy management led to the damage. To settle the suit, an estimate of the rate of damage per 1,000 bricks, called the spall rate, was required. (*Chance*, Summer 1994.) The owner estimated the spall rate using several *scaffold-drop* surveys. (With this method, an engineer lowers a scaffold down at selected places on building walls and counts the number of visible spalls for every 1,000 bricks in the observation area.) The brick manufacturer conducted its own survey by dividing the walls of the complex into 83 wall segments and taking a photograph of each wall segment. (The number of spalled bricks that could be made out from each photo was recorded and the sum over all 83 wall segments used as an estimate of total spall damage.) In this court case, the jury was faced with the following dilemma: The scaffold-drop survey provided the most accurate estimate of spall rates in a given wall segment. Unfortunately, the drop areas were not selected at random from the entire complex; rather, drops were made at areas with high spall concentrations, leading to an overestimate of the total damage. On the other hand, the photo survey was complete in that all 83 wall segments in the complex

were checked for spall damage. But the spall rate estimated by the photos, at least in areas of high spall concentration, was biased low (spalling damage cannot always be seen from a photo), leading to an underestimate of the total damage.

The data in the table are the spall rates obtained using the two methods at 11 drop locations. Use the data, as did expert statisticians who testified in the case, to help the jury estimate the true spall rate at a given wall segment. Then explain how this information, coupled with the data (not given here) on all 83 wall segments, can provide a reasonable estimate of the total spall damage (i.e., total number of damaged bricks).

BRICKS

Drop Location	Drop Spall Rate (per 1,000 bricks)	Photo Spall Rate (per 1,000 bricks)
1	0	0
2	5.1	0
3	6.6	0
4	1.1	.8
5	1.8	1.0
6	3.9	1.0
7	11.5	1.9
8	22.1	7.7
9	39.3	14.9
10	39.9	13.9
11	43.0	11.8

Source: Fairley, W. B., *et al.* "Bricks, buildings, and the Bronx: Estimating masonry deterioration." *Chance*, Vol. 7. No. 3, Summer 1994, p. 36 (Figure 3). [*Note:* The data points are estimated from the points shown on a scatterplot.]

Student Projects

Many dependent variables in all areas of research serve as the subjects of regression modeling efforts. We list five such variables here:

1. Crime rate in various communities
2. Daily maximum temperature in your town
3. Grade point average of students who have completed one academic year at your college
4. Gross Domestic Product of the United States
5. Points scored by your favorite football team in a single game

Choose one of these dependent variables or choose some other dependent variable for which you want to construct a prediction model. There may be a large number of independent variables that should be included in a prediction equation for the dependent variable you choose. List three potentially important independent variables, x_1, x_2, and x_3, that you think might be (individually) strongly related to your dependent variable. Next, obtain 10 data values, each of

which consists of a measure of your dependent variable y and the corresponding values of x_1, x_2, and x_3.

a. Use the least squares formulas given in this chapter to fit three straight-line models—one for each independent variable—for predicting y.

b. Interpret the sign of the estimated slope coefficient $\hat{\beta}_1$ in each case, and test the utility of each model by testing $H_0: \beta_1 = 0$ against $H_a: \beta_1 \neq 0$. What assumptions must be satisfied to ensure the validity of these tests?

c. Calculate the coefficient of determination r^2 for each model. Which of the independent variables predicts y best for the 10 sampled sets of data? Is this variable necessarily best in general (that is, for the entire population)? Explain.

Be sure to keep the data and the results of your calculations, since you will need them for the Student Projects section in Chapter 12.

REFERENCES

Chatterjee, S., and Price, B. *Regression Analysis by Example*, 2nd ed. New York: Wiley, 1991.

Draper, N., and Smith, H. *Applied Regression Analysis*, 3rd ed. New York: Wiley, 1987.

Graybill, F. *Theory and Application of the Linear Model*. North Scituate, Mass.: Duxbury, 1976.

Kleinbaum, D., and Kupper, L. *Applied Regression Analysis and Other Multivariable Methods*. 2nd ed. North Scituate, Mass.: Duxbury, 1997.

Mendenhall, W. *Introduction to Linear Models and the Design and Analysis of Experiments*, Belmont, Calif.: Wadsworth, 1968.

Mendenhall, W., and Sincich, T. *A Second Course in Statistics: Regression Analysis*, 6th ed. Upper Saddle River, N.J.: Prentice Hall, 2003.

Montgomery, D., Peck, E., and Vining, G. *Introduction to Linear Regression Analysis*, 3rd ed. New York: Wiley, 2001.

Mosteller, F., and Tukey, J. W. *Data Analysis and Regression: A Second Course in Statistics*. Reading, Mass.: Addison-Wesley, 1977.

Neter, J., Kutner, M., Nachtsheim, C., and Wasserman, W. *Applied Linear Statistical Models*, 4th ed. Homewood, III.: Richard Irwin, 1996.

Rousseeuw, P. J., and Leroy, A. M. *Robust Regression and Outlier Detection*. New York: Wiley, 1987.

Weisburg, S. *Applied Linear Regression*, 2nd ed. New York: Wiley, 1985.

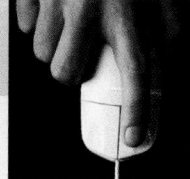

Using Technology

Simple Linear Regression Using MINITAB

To conduct a simple linear regression analysis, first access the MINITAB worksheet file that contains the two quantitative variables (dependent and independent variables). Next, click on the "Stat" button on the MINITAB menu bar, then click on "Regression" and "Regression" again, as shown in Figure 11.M.1.

Figure 11.M.1
MINITAB Menu Options for Regression

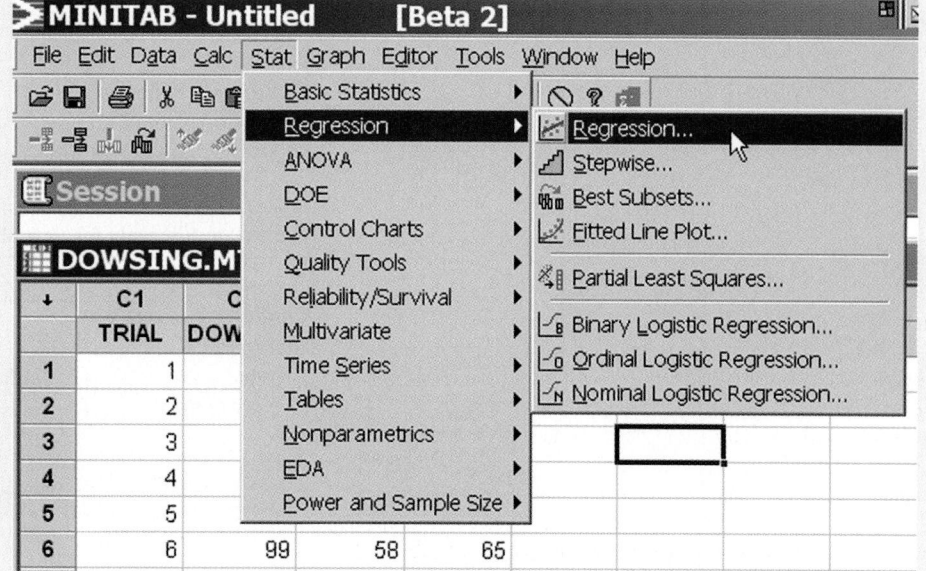

Figure 11.M.2
MINITAB Regression
Dialog Box

Figure 11.M.2
MINITAB Regression
Dialog Box

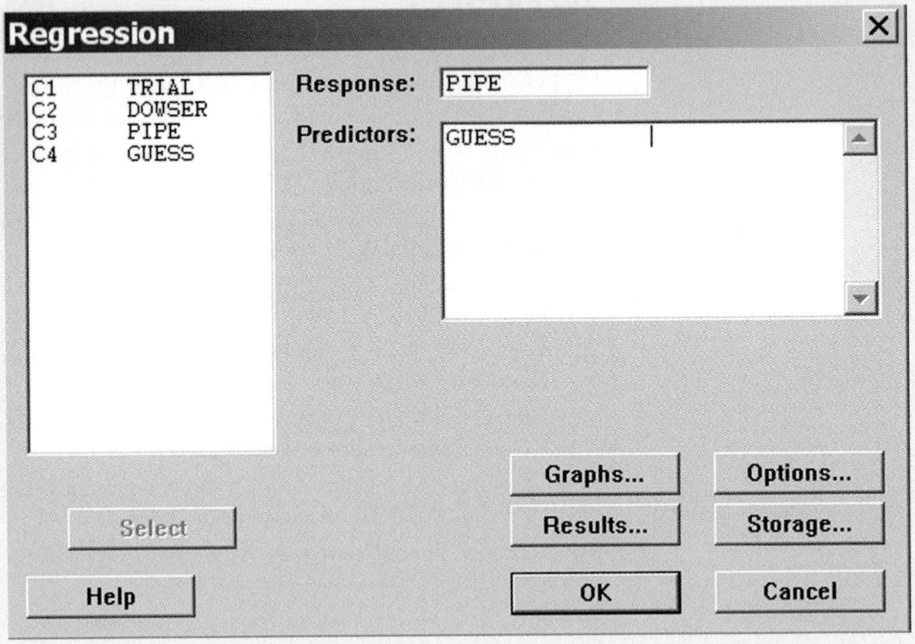

The resulting dialog box appears as shown in Figure 11.M.2. Specify the dependent variable in the "Response" box and the independent variable in the "Predictors" box.

Optionally, you can get MINITAB to produce prediction intervals for y and confidence intervals for $E(y)$ by clicking the "Options" button. The resulting dialog box is shown in Figure 11.M.3. Check "Confidence limits" and/or "Prediction limits," specify the "Confidence level", and enter the value of x in the "Prediction intervals for new observations" box. Click "OK" to return to the main Regression dialog box, then click "OK" again to produce the MINITAB simple linear regression printout.

Figure 11.M.3
MINITAB Regression
Options

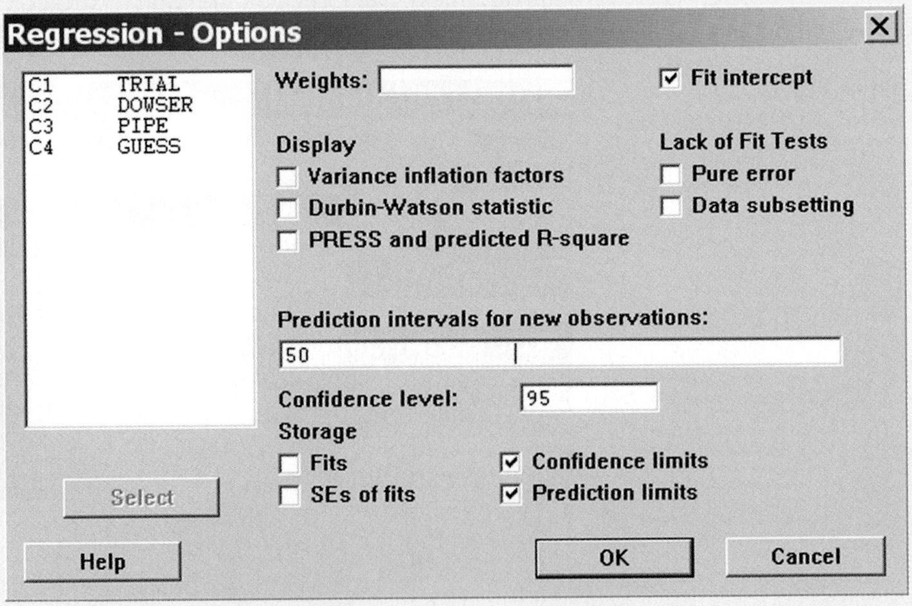

Figure 11.M.4
MINITAB Menu Options
for Correlation

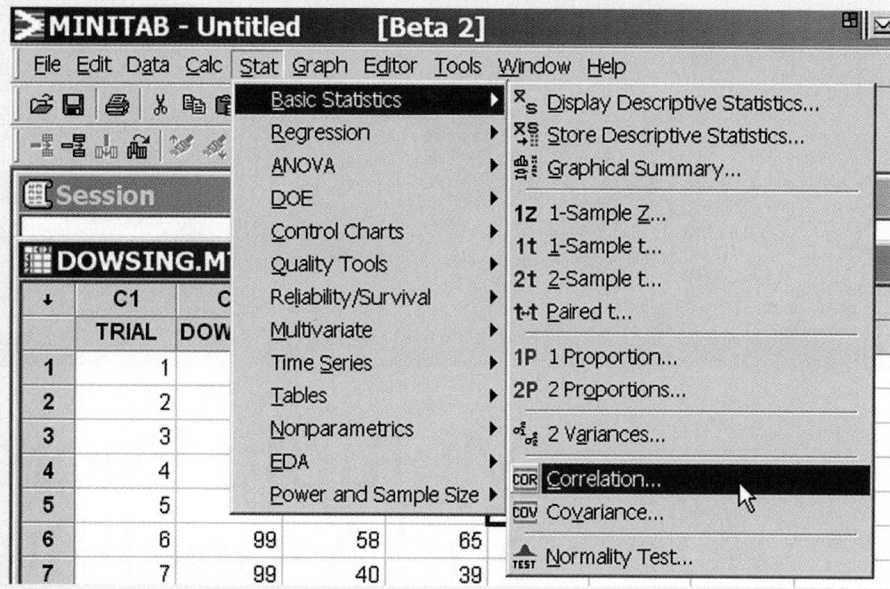

To obtain the correlation coefficient for the two quantitative variables, click on the "Stat" button on the MINITAB main menu bar, then click on "Basic Statistics", then click on "Correlation," as shown in Figure 11.M.4. The resulting dialog box appears in Figure 11.M.5. Enter the two variables of interest in the "Variables" box, then click "OK" to obtain a printout of the correlation.

Figure 11.M.5
MINITAB Correlation
Dialog Box

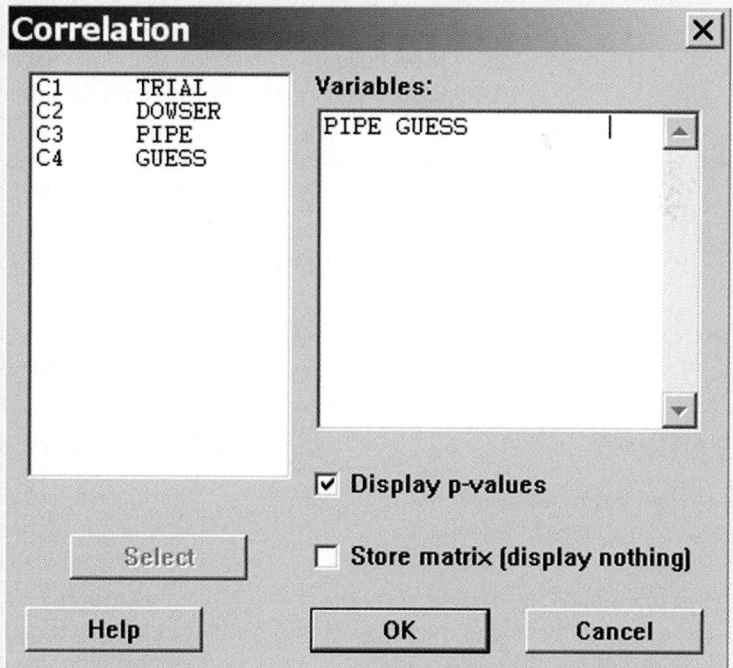

12

Multiple Regression and Model Building

Contents

Statistics in Action
Modeling Condominium Sale Prices

Using Technology
Multiple Regression Using MINITAB

Where We've Been

- Introduced the straight-line model relating a dependent variable y to a single independent variable x
- Demonstrated how to estimate the parameters of the straight-line model using the method of least squares
- Showed how to statistically assess the adequacy of the model
- Showed how to use the model to estimate $E(y)$ and predict y for a given value of x

Where We're Going

- Introduce a *multiple regression* model as a means of relating a dependent variable y to two or more independent variables.
- Present several different multiple regression models involving both quantitative and qualitative independent variables.
- Assess how well the multiple regression model fits the sample data.
- Show how an analysis of the model's *residuals* can aid in detecting violations of model assumptions and in identifying model modifications.

Statistics in *ACTION*

Modeling Condominium Sale Prices

This application involves an investigation of the factors that affect the sale price of oceanside condominium units. It represents an extension of an analysis of the same data by Herman Kelting (1979). Although condo sale prices have increased dramatically over the past 20 years, the relationship between these factors and sale price remain about the same. Consequently, the data provide valuable insight into today's condominium sales market.

The sales data were obtained for a newly built oceanside condominium complex consisting of two adjacent and connected eight-floor buildings. The complex contains 209 units of equal size (approximately 500 square feet each). The locations of the buildings relative to the ocean, the swimming pool, the parking lot, etc., are shown in Figure SIA 12.1.

There are several features of the complex that you should note:

1. The units facing south, called *oceanview*, face the beach and ocean. In addition, units in building 1 have a good view of the pool. Units to the rear of the building, called *bayview*, face the parking lot and an area of land that,

ultimately, borders a bay. The view from the upper floors of these units is primarily of wooded, sandy terrain. The bay is very distant and barely visible.

2. The only elevator in the complex is located at the east end of building 1, next to the office and the game room. People moving to or from the higher floor units in building 2 would probably use the elevator, then move through the passages to their units. Thus, units on the higher floors and at a greater distance from the elevator would be less convenient for their occupants; they would expend greater effort in moving baggage, groceries, etc., and would be farther away from the game room, the office, and the swimming pool. These units also possess an advantage: Because traffic through the hallways in the area would be minimal, these units would be the most private.

3. Lower-floor oceanside units are most suited to active people; they open onto the beach, ocean, and pool. They are within easy reach of the game room and they are easily reached from the parking area.

(continues on next page)

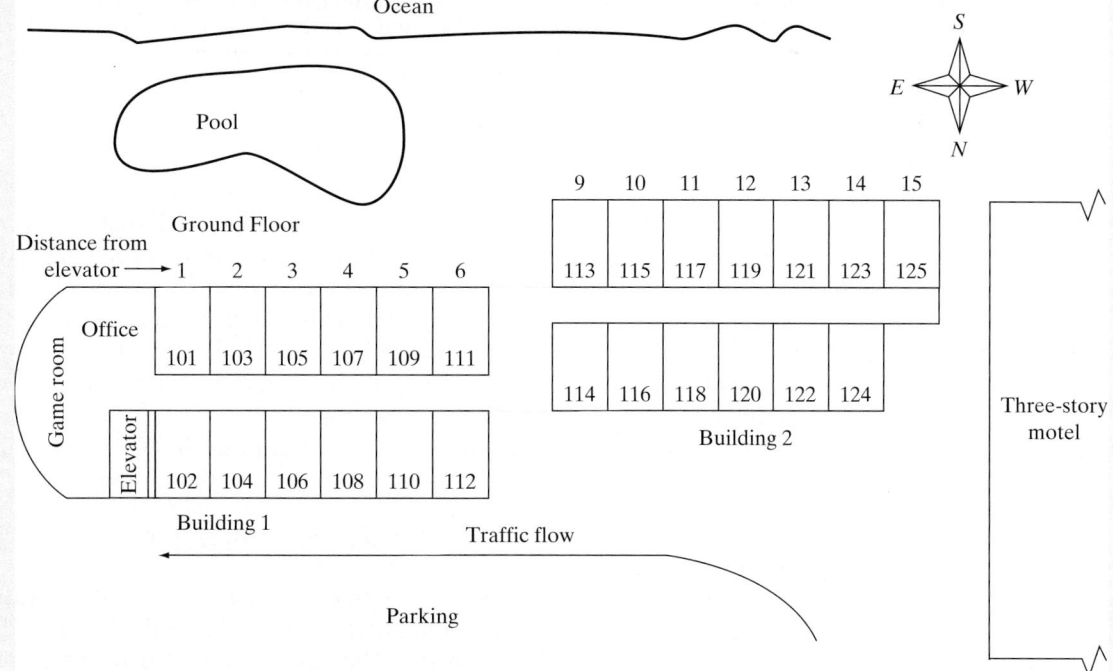

Figure SIA12.1
Layout of Condominium Complex

4. Checking Figure SIA 12.1, you will see that the views in some of the units at the center of the complex, units ending in numbers 11 and 14, are partially blocked.

5. The condominium complex was completed at the time of the 1975 recession; sales were slow and the developer was forced to sell about half of the units at auction approximately 18 months after opening. Many unsold units were furnished by the developer and rented prior to the auction.

This condominium complex is particularly suited to our study. Because the single elevator is located at one end of the complex, it is the source of a remarkably high level of both inconvenience and privacy for the people occupying units on the top floors in building 2. Consequently, the data provide a good opportunity to investigate the relationship that might exist between sale price, height of the unit (floor number), distance of the unit from the elevator, and presence or absence of an ocean view. In addition, the presence or absence of furniture in each of the units permits an investigation of the effect of the availability of furniture on

sale price. Finally, the units sold at auction are completely buyer-specified and hence consumer-oriented, in contrast to most other real estate sales data that are, to a high degree, seller-oriented and broker-specified.

The **CONDO** file contains data for each of the 209 units sold—106 sold at a public auction and 103 sold at the developer's fixed price. The variables measured for each condominium unit are listed in Table SIA 12.1. We want to build a model for the auction price, and, ultimately, to use the model to predict the sales prices of future units.

In several Statistics in Action Revisited sections (see below), we show how to analyze the data using a multiple regression analysis.

Statistics in Action Revisited

- A First-Order Model for Condominium Sales Prices (p. 687)

- Building a Model for Condominium Sales Prices (p. 730)

- A Residual Analysis for the Condo Sales Model (p. 756)

CONDO

TABLE SIA12.1   Variables in the CONDO Data File

Variable Name	Type	Description
PRICE100	Quantitative	Sales price (hundreds of dollars)
FLOOR	Quantitative	Floor height $(1, 2, 3, \ldots, 8)$
DIST	Quantitative	Distance, in units, from the elevator $(1, 2, 3, \ldots, 15)$
VIEW	Qualitative	View $(1 = \text{ocean view}, 0 = \text{non–ocean view})$
END	Qualitative	Location of unit $(1 = \text{end of complex}, 0 = \text{not an end unit})$
FURNISH	Qualitative	Furniture status $(1 = \text{furnished}, 0 = \text{nonfurnished})$
AUCTION	Qualitative	Method of sale $(1 = \text{public auction}, 0 = \text{fixed price})$

12.1 Multiple Regression Models

Most practical applications of regression analysis utilize models that are more complex than the simple straight-line model. For example, a realistic probabilistic model for reaction time would include more than just the amount of a particular drug in the bloodstream. Factors such as age, a measure of visual perception, and sex of the subject are a few of the many variables that might be related to reaction time. Thus, we would want to incorporate these and other potentially important independent variables into the model in order to make accurate predictions.

Probabilistic models that include more than one independent variable are called **multiple regression models**. The general form of these models is

$$y = \beta_0 + \beta_1 x_1 + \beta_2 x_2 + \cdots + \beta_k x_k + \varepsilon$$

The dependent variable y is now written as a function of k independent variables, $x_1, x_2, \ldots, x_k$. The random error term is added to make the model probabilistic rather than deterministic. The value of the coefficient β_i determines the contribution of the independent variable x_i, and β_0 is the y-intercept. The coefficients $\beta_0, \beta_1, \ldots, \beta_k$ are usually unknown because they represent population parameters.

At first glance it might appear that the regression model shown above would not allow for anything other than straight-line relationships between y and the independent variables, but this is not true. Actually, $x_1, x_2, \ldots, x_k$ can be functions of variables as long as the functions do not contain unknown parameters. For example, the reaction time, y, of a subject to a visual stimulus could be a function of the independent variables

$$x_1 = \text{Age of the subject}$$
$$x_2 = (\text{Age})^2 = x_1^2$$
$$x_3 = 1 \text{ if male subject, 0 if female subject}$$

The x_2-term is called a **higher-order term**, since it is the value of a quantitative variable (x_1) squared (i.e., raised to the second power). The x_3-term is a **coded variable** representing a qualitative variable (gender). The multiple regression model is quite versatile and can be made to model many different types of response variables.

The General Multiple Regression Model

$$y = \beta_0 + \beta_1 x_1 + \beta_2 x_2 + \cdots + \beta_k x_k + \varepsilon$$

where

y is the dependent variable

$x_1, x_2, \ldots, x_k$ are the independent variables

$E(y) = \beta_0 + \beta_1 x_1 + \beta_2 x_2 + \cdots + \beta_k x_k$ is the deterministic portion of the model

β_i determines the contribution of the independent variable x_i

Note: The symbols $x_1, x_2, \ldots, x_k$ may represent higher-order terms for quantitative predictors or terms that represent qualitative predictors.

As shown in the next box, the steps used to develop the multiple regression model are similar to those used for the simple regression model.

Analyzing a Multiple Regression Model

Step 1 Hypothesize the deterministic component of the model. This component relates the mean, $E(y)$, to the independent variables $x_1, x_2, \ldots, x_k$. This involves the choice of the independent variables to be included in the model (Sections 12.2, 12.5–12.10).

Step 2 Use the sample data to estimate the unknown model parameters $\beta_0, \beta_1, \beta_2, \ldots, \beta_k$ in the model (Section 12.2).

Step 3 Specify the probability distribution of the random error term, ε, and estimate the standard deviation of this distribution, σ (Section 12.3).

Step 4 Check that the assumptions on ε are satisfied, and make model modifications if necessary (Section 12.11).

Step 5 Statistically evaluate the usefulness of the model (Section 12.3).

Step 6 When satisfied that the model is useful, use it for prediction, estimation, and other purposes (Section 12.4).

The assumptions we make about the random error ε of the multiple regression model are also similar to those in a simple linear regression. These are summarized as follows.

> ### Assumptions for Random Error ε
>
> For any given set of values of $x_1, x_2, \ldots, x_k$, the random error ε has a probability distribution with the following properties:
>
> **1.** Mean equal to 0
> **2.** Variance equal to σ^2
> **3.** Normal distribution
> **4.** Random errors are independent (in a probabilistic sense)

Throughout this chapter, we introduce several different types of models that form the foundation of **model building** (or useful model construction). In the next several sections, we consider the most basic multiple regression model, called the *first-order model*.

12.2 The First-Order Model: Estimating and Interpreting the β Parameters

A model that includes only terms for *quantitative* independent variables, called a **first-order model**, is described in the box. Note that the first-order model does not include any higher-order terms (such as x_1^2).

> ### A First-Order Model in Five Quantitative Independent Variables*
>
> $$E(y) = \beta_0 + \beta_1 x_1 + \beta_2 x_2 + \beta_3 x_3 + \beta_4 x_4 + \beta_5 x_5$$
>
> where $x_1, x_2, \ldots, x_5$ are all quantitative variables that *are not* functions of other independent variables.
>
> *Note:* β_i represents the slope of the line relating y to x_i when all the other x's are held fixed.

The method of fitting first-order models—and multiple regression models in general—is identical to that of fitting the simple straight-line model: the method of least squares. That is, we choose the estimated model

$$\hat{y} = \hat{\beta}_0 + \hat{\beta}_1 x_1 + \cdots + \hat{\beta}_k x_k$$

that minimizes

$$\text{SSE} = \sum (y - \hat{y})^2$$

As in the case of the simple linear model, the sample estimates $\hat{\beta}_0, \hat{\beta}_1, \ldots, \hat{\beta}_k$ are obtained as a solution to a set of simultaneous linear equations.†

The primary difference between fitting the simple and multiple regression models is computational difficulty. The $(k + 1)$ simultaneous linear equations that must be solved to find the $(k + 1)$ estimated coefficients $\hat{\beta}_0, \hat{\beta}_1, \ldots, \hat{\beta}_k$ are difficult (sometimes nearly impossible) to solve with a calculator. Consequently, we resort to the use of computers. Instead of presenting the tedious hand calculations required to fit the models, we present output from SAS, SPSS, and MINITAB.

---

*The terminology "first-order" is derived from the fact that each x in the model is raised to the first power.
†Students who are familiar with calculus should note that $\hat{\beta}_0, \hat{\beta}_1, \ldots, \hat{\beta}_k$ are the solutions to the set of equations $\partial\text{SSE}/\partial\hat{\beta}_0 = 0, \partial\text{SSE}/\partial\hat{\beta}_1 = 0, \ldots, \partial\text{SSE}/\partial\hat{\beta}_k = 0$. The solution is usually given in matrix form, but we do not present the details here. See the references for details.

Biography

GEORGE U. YULE
(1871–1951)—
Yule Processes

Born on a small farm in Scotland, George Yule received an extensive childhood education. After graduating from University College (London), where he studied civil engineering, Yule spent a year employed in engineering workshops. However, he made a career change in 1893, accepting a teaching position back at University College under the guidance of statistician Karl Pearson (see p. 786). Inspired by Pearson's work, Yule produced a series of important articles on the statistics of regression and correlation. Yule is considered the first to apply the method of least squares in regression analysis and he developed the theory of multiple regression. He eventually was appointed a lecturer in statistics at Cambridge University and later became the president of the prestigious Royal Statistical Society. Yule made many other contributions to the field, including the invention of time series analysis and the development of "Yule" processes and the "Yule" distribution.

EXAMPLE 12.1

FITTING A FIRST-ORDER MULTIPLE REGRESSION MODEL

Problem   Suppose a property appraiser wants to model the relationship between the sale price of a residential property in a midsize city and the following three independent variables: (1) appraised land value of the property, (2) appraised value of improvements (i.e., home value) on the property, and (3) area of living space on the property (i.e., home size). Consider the first-order model

$$y = \beta_0 + \beta_1 x_1 + \beta_2 x_2 + \beta_3 x_3 + \varepsilon$$

where

$$
\begin{aligned}
y &= \text{Sale price (dollars)} \\
x_1 &= \text{Appraised land value (dollars)} \\
x_2 &= \text{Appraised improvements (dollars)} \\
x_3 &= \text{Area (square feet)}
\end{aligned}
$$

To fit the model, the appraiser selected a random sample of $n = 20$ properties from the thousands of properties that were sold in a particular year. The resulting data are given in Table 12.1.

a.   Use scattergrams to plot the sample data. Interpret the plots.

b.   Use the method of least squares to estimate the unknown parameters β_0, β_1, β_2, and β_3 in the model.

c.   Find the value of SSE that is minimized by the least squares method.

d.   Estimate σ_1, the standard deviation of the model, and interpret the result.

Solution   **a.**   MINITAB side-by-side scatterplots for examining the bivariate relationships between y and x_1, y and x_2, and y and x_3 are shown in Figure 12.1. Of the three variables, appraised improvements (x_2) appears to have the strongest linear relationship with sale price (y).

b.   The model hypothesized above is fit to the data of Table 12.1 using MINITAB. A portion of the printout is reproduced in Figure 12.2. The least squares estimates of the β parameters (highlighted) are $\hat{\beta}_0 = 1{,}470$. $\hat{\beta}_1 = .8145$, $\hat{\beta}_2 = .8204$, and $\hat{\beta}_3 = 13.529$. Therefore, the equation that minimizes SSE for this data set (i.e., the **least squares prediction equation**) is

$$\hat{y} = 1{,}470 + .8145 x_1 + .8204 x_2 + 13.53 x_3$$

⊘ **REALESTATE**

TABLE 12.1   Real Estate Appraisal Data for 20 Properties

Property # (Obs.)	Sale Price y	Land Value x_1	Improvements Value x_2	Area x_3
1	68,900	5,960	44,967	1,873
2	48,500	9,000	27,860	928
3	55,500	9,500	31,439	1,126
4	62,000	10,000	39,592	1,265
5	116,500	18,000	72,827	2,214
6	45,000	8,500	27,317	912
7	38,000	8,000	29,856	899
8	83,000	23,000	47,752	1,803
9	59,000	8,100	39,117	1,204
10	47,500	9,000	29,349	1,725
11	40,500	7,300	40,166	1,080
12	40,000	8,000	31,679	1,529
13	97,000	20,000	58,510	2,455
14	45,500	8,000	23,454	1,151
15	40,900	8,000	20,897	1,173
16	80,000	10,500	56,248	1,960
17	56,000	4,000	20,859	1,344
18	37,000	4,500	22,610	988
19	50,000	3,400	35,948	1,076
20	22,400	1,500	5,779	962

Source: Alachua County (Florida) Property Appraisers Office.

c. The minimum value of the sum of squared errors, highlighted in Figure 12.2, is SSE = 1,003,491,259.

d. Recall that the estimator of σ^2 for the straight-line model is $s^2 = \text{SSE}/(n-2)$ and note that the denominator is (n − Number of estimated β parameters), which is ($n-2$) in the straight-line model. Since we must estimate four parameters, $\beta_0, \beta_1, \beta_2,$ and β_3 for the first-order model, the estimator of σ^2 is

$$s^2 = \frac{\text{SSE}}{n-4} = \frac{\text{SSE}}{20-4} = \frac{1,003,491,259}{16} = 62,718,204$$

Figure 12.1
MINITAB Side-by-Side
Scatterplots for the Data
of Table 12.1

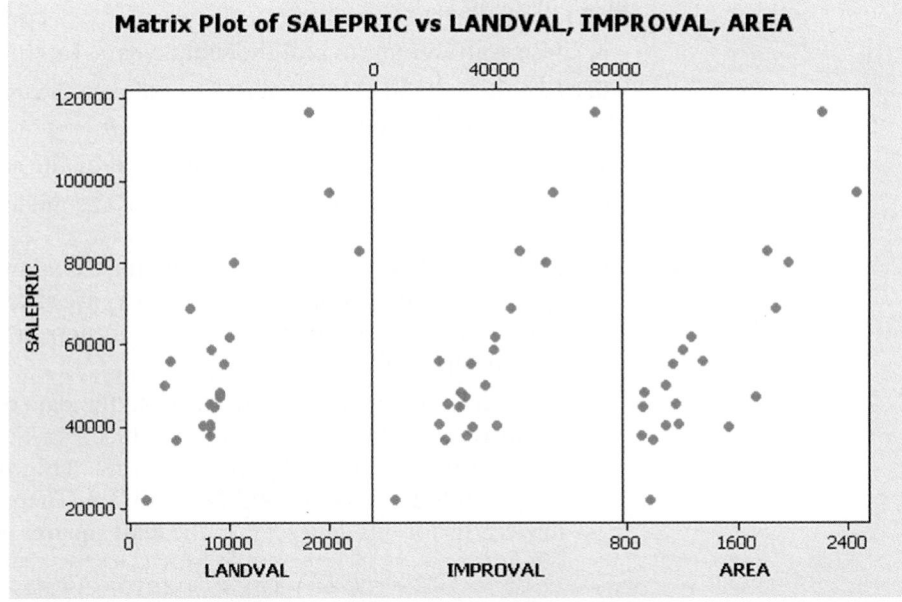

Figure 12.2
MINITAB Analysis
of Sale Price Model

Regression Analysis: SALEPRIC versus LANDVAL, IMPROVAL, AREA

```
The regression equation is
SALEPRIC = 1470 + 0.814 LANDVAL + 0.820 IMPROVAL + 13.5 AREA


Predictor    Coef   SE Coef     T      P
Constant     1470      5746   0.26  0.801
LANDVAL    0.8145    0.5122   1.59  0.131
IMPROVAL   0.8204    0.2112   3.88  0.001
AREA       13.529     6.586   2.05  0.057


S = 7919.48    R-Sq = 89.7%    R-Sq(adj) = 87.8%


Analysis of Variance

Source          DF          SS           MS       F      P
Regression       3  8779676741   2926558914   46.66  0.000
Residual Error  16  1003491259     62718204
Total           19  9783168000
```

This value, often called the **mean square for error (MSE)** is also highlighted at the bottom of the MINITAB printout in Figure 12.2.

The estimate of σ, then, is

$$s = \sqrt{62{,}718{,}204} = 7{,}919.5$$

which is highlighted in the middle of the printout in Figure 12.2. One useful interpretation of the estimated standard deviation s is that the interval $\pm 2s$ will provide a rough approximation to the accuracy with which the model will predict future values of y for given values of x. Thus, we expect the model to provide predictions of sale price to within about $\pm 2s = \pm 2(7{,}919.5) = \pm 15{,}839$ dollars.*

Look Back As with simple linear regression, we will use the estimator of σ^2 both to check the utility of the model (Section 12.3) and to provide a measure of reliability of predictions and estimates when the model is used for those purposes (Section 12.4). Thus, you can see that the estimation of σ^2 plays an important part in the development of a regression model.

Now Work   *Exercise 12.6a–c*

■ ■ ■

Estimator of σ^2 for a Multiple Regression Model with k Independent Variables

$$s^2 = \frac{\text{SSE}}{n - \text{Number of estimated } \beta \text{ parameters}} = \frac{\text{SSE}}{n - (k + 1)}$$

After obtaining the least squares prediction equation, the analyst will usually want to make meaningful interpretations of the β estimates. Recall that in the straight-line model (Chapter 11)

$$y = \beta_0 + \beta_1 x + \varepsilon$$

*The $\pm 2s$ approximation will improve as the sample size is increased. We will provide more precise methodology for the construction of prediction intervals in Section 12.4.

β_0 represents the y-intercept of the line and β_1 represents the slope of the line. From our discussion in Chapter 11, β_1 has a practical interpretation—it represents the mean change in y for every 1-unit increase in x. When the independent variables are quantitative, the β parameters in the first-order model specified in Example 12.1 have similar interpretations. The difference is that when we interpret the β that multiplies one of the variables (e.g., x_1), we must be certain to hold the values of the remaining independent variables (e.g., x_2, x_3) fixed.

To see this, suppose that the mean $E(y)$ of a response y is related to two quantitative independent variables, x_1 and x_2, by the first-order model

$$E(y) = 1 + 2x_1 + x_2$$

In other words, $\beta_0 = 1$, $\beta_1 = 2$, and $\beta_2 = 1$.

Now, when $x_2 = 0$, the relationship between $E(y)$ and x_1 is given by

$$E(y) = 1 + 2x_1 + (0) = 1 + 2x_1$$

A graph of this relationship (a straight line) is shown in Figure 12.3. Similar graphs of the relationship between $E(y)$ and x_1 for $x_2 = 1$,

$$E(y) = 1 + 2x_1 + (1) = 2 + 2x_1$$

and for $x_2 = 2$,

$$E(y) = 1 + 2x_1 + (2) = 3 + 2x_1$$

also are shown in Figure 12.3. Note that the slopes of the three lines are all equal to $\beta_1 = 2$, the coefficient that multiplies x_1.

Figure 12.3 exhibits a characteristic of all first-order models: If you graph $E(y)$ versus any one variable—say, x_1—for fixed values of the other variables, the result will always be a *straight line* with slope equal to β_1. If you repeat the process for other values of the fixed independent variables, you will obtain a set of *parallel* straight lines. This indicates that the effect of the independent variable x_i on $E(y)$ is independent of all the other independent variables in the model, and this effect is measured by the slope β_i (see the box on p. 664).

A three-dimensional graph of the model $E(y) = 1 + 2x_1 + x_2$ is shown in Figure 12.4. Note that the model graphs as a plane. If you slice the plane at a particular value of x_2 (say, $x_2 = 0$), you obtain a straight line relating $E(y)$ to x_1 (e.g., $E(y) = 1 + 2x_1$). Similarly, if you slice the plane at a particular value of x_1, you

Figure 12.3
Graphs of
$E(y) = 1 + 2x_1 + x_2$
for $x_2 = 0, 1, 2$

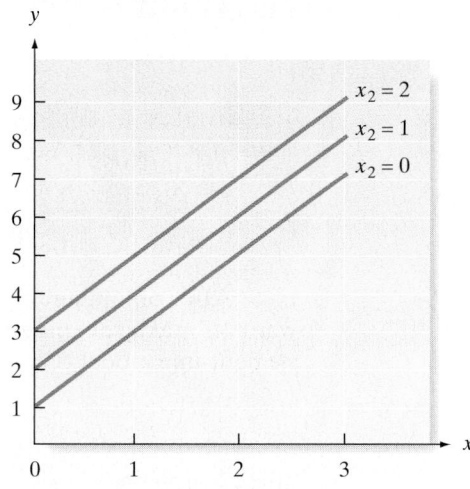

Figure 12.4
The Plane
$E(y) = 1 + 2x_1 + x_2$

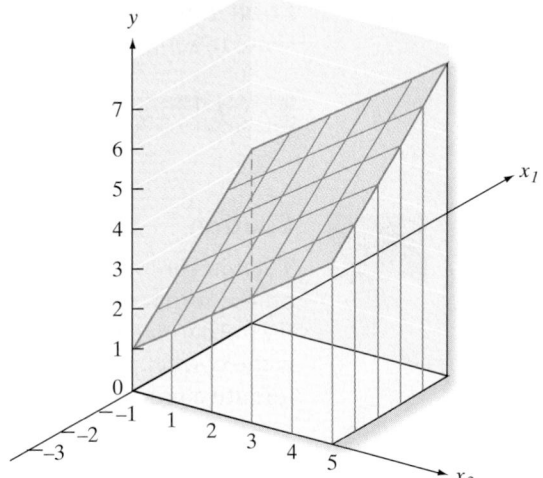

obtain a straight line relating $E(y)$ to x_2. Since it is more difficult to visualize three-dimensional and, in general, k-dimensional surfaces, we will graph all the models presented in this chapter in two dimensions. The key to obtaining these graphs is to hold fixed all but one of the independent variables in the model.

EXAMPLE 12.2

INTERPRETING THE β ESTIMATES

Problem   Refer to the first-order model for sale price y considered in Example 12.1. Interpret the estimates of the β parameters in the model.

Solution   The least squares prediction equation, as given in Example 12.1, is $\hat{y} = 1{,}470 + .8145x_1 + .8204x_2 + 13.53x_3$. We know that with first-order models β_1 represents the slope of the $y - x_1$ line for fixed x_2 and x_3. That is, β_1 measures the change in $E(y)$ for every 1-unit increase in x_1 when all other independent variables in the model are held fixed. Similar statements can be made about β_2 and β_3; e.g., β_2 measures the change in $E(y)$ for every 1-unit increase in x_2 when all other x's in the model are held fixed. Consequently, we obtain the following interpretations:

$\hat{\beta}_1 = .8145$: We estimate the mean sale price of a property, $E(y)$, to increase .8145 dollar for every \$1 increase in appraised land value (x_1) when both appraised improvements (x_2) and area (x_3) are held fixed.

$\hat{\beta}_2 = .8204$: We estimate the mean sale price of a property, $E(y)$, to increase .8204 dollar for every \$1 increase in appraised improvements (x_2) when both appraised land value (x_1) and area (x_3) are held fixed.

$\hat{\beta}_3 = 13.53$: We estimate the mean sale price of a property, $E(y)$, to increase \$13.53 for each additional square foot of living area (x_3) when both appraised land value (x_1) and appraised improvements (x_2) are held fixed.

The value $\hat{\beta}_0 = 1{,}470$ does not have a meaningful interpretation in this example. To see this, note that $\hat{y} = \hat{\beta}_0$ when $x_1 = x_2 = x_3 = 0$. Thus, $\hat{\beta}_0 = 1{,}470$ represents the estimated mean sale price when the values of all the independent variables are set equal to 0. Since a residential property with these characteristics—appraised land value of \$0, appraised improvements of \$0, and 0 square feet of living area—is not practical, the value of $\hat{\beta}_0$ has no meaningful interpretation. In general, $\hat{\beta}_0$ will not have a practical interpretation unless it makes sense to set the values of the x's simultaneously equal to 0.

Look Back In general, $\hat{\beta}_0$ will not have a practical interpretation unless it makes sense to set the values of the x's simultaneously equal to 0.

Now Work   *Exercise 12.11a–b*

■ ■ ■

Caution

The interpretation of the β parameters in a multiple regression model will depend on the terms specified in the model. The interpretations in Example 12.2 are for a first-order linear model only. In practice, you should be sure that a first-order model is the correct model for $E(y)$ before making these β interpretations. [We discuss alternative models for $E(y)$ in Sections 12.5–12.8.]

12.3 Inferences About the β Parameters and the Overall Model Utility

Inferences about the individual β parameters in a model are obtained using either a confidence interval or a test of hypothesis, as outlined in the following boxes.*

Test of an Individual Parameter Coefficient in the Multiple Regression Model

One-Tailed Test   **Two-Tailed Test**

$H_0: \beta_i = 0$ $H_0: \beta_i = 0$

$H_a: \beta_i < 0$ [or $H_a: \beta_i > 0$] $H_a: \beta_i \neq 0$

$$\text{Test statistic: } t = \frac{\hat{\beta}_i}{s_{\hat{\beta}_i}}$$

Rejection region: $t < -t_\alpha$ Rejection region: $|t| > t_{\alpha/2}$
[or $t > t_\alpha$ when $H_a: \beta_i > 0$]
where t_α and $t_{\alpha/2}$ are based on $n - (k + 1)$ degrees of freedom and

$$n = \text{Number of observations}$$
$$k + 1 = \text{Number of } \beta \text{ parameters in the model}$$

A $100(1 - \alpha)\%$ Confidence Interval for a β Parameter

$$\hat{\beta}_i \pm t_{\alpha/2}s_{\hat{\beta}_i}$$

where $t_{\alpha/2}$ is based on $n - (k + 1)$ degrees of freedom and

$$n = \text{Number of observations}$$
$$k + 1 = \text{Number of } \beta \text{ parameters in the model}$$

Conditions Required for Valid Inferences About Individual β Parameters

The four assumptions about the probability distribution for the random error, ε (p. 664).

---

*The formulas for computing $\hat{\beta}_i$ and its standard error are so complex, the only reasonable way to present them is by using matrix algebra. We do not assume a prerequisite of matrix algebra for this text and, in any case, we think the formulas can be omitted in an introductory course without serious loss. They are programmed into almost all statistical software packages with multiple regression routines and are presented in some of the texts listed in the references.

We illustrate these methods with another example.

EXAMPLE 12.3 INFERENCES ABOUT THE β PARAMETERS

Problem A collector of antique grandfather clocks knows that the price received for the clocks increases linearly with the age of the clocks. Moreover, the collector hypothesizes that the auction price of the clocks will increase linearly as the number of bidders increases. Thus, the following first-order model is hypothesized:

$$y = \beta_0 + \beta_1 x_1 + \beta_2 x_2 + \varepsilon$$

where

$$y = \text{Auction price}$$
$$x_1 = \text{Age of clock (years)}$$
$$x_2 = \text{Number of bidders}$$

A sample of 32 auction prices of grandfather clocks, along with their age and the number of bidders, is given in Table 12.2. The model $y = \beta_0 + \beta_1 x_1 + \beta_2 x_2 + \varepsilon$ is fit to the data, and a portion of the SAS printout is shown in Figure 12.5.

a. Test the hypothesis that the mean auction price of a clock increases as the number of bidders increases when age is held constant, that is, $\beta_2 > 0$. Use $\alpha = .05$.

b. Find a 90% confidence interval for β_1 and interpret the result.

Solution **a.** The hypotheses of interest concern the parameter β_2. Specifically,

$$H_0: \beta_2 = 0$$
$$H_a: \beta_2 > 0$$

The test statistic is a t statistic formed by dividing the sample estimate $\hat{\beta}_2$ of the parameter β_2 by the estimated standard error of $\hat{\beta}_2$ (denoted $s_{\hat{\beta}_2}$). These estimates, $\hat{\beta}_2 = 85.953$ and $s_{\hat{\beta}_2} = 8.729$, as well as the calculated t value, are highlighted on the SAS printout, Figure 12.5.

$$\textit{Test statistic: } t = \frac{\hat{\beta}_2}{s_{\hat{\beta}_2}} = \frac{85.953}{8.729} = 9.85$$

⊚ **GFCLOCKS**

TABLE 12.2 Auction Price Data

Age x_1	Number of Bidders x_2	Auction Price y	Age x_1	Number of Bidders x_2	Auction Price y
127	13	$1,235	170	14	$2,131
115	12	1,080	182	8	1,550
127	7	845	162	11	1,884
150	9	1,522	184	10	2,041
156	6	1,047	143	6	845
182	11	1,979	159	9	1,483
156	12	1,822	108	14	1,055
132	10	1,253	175	8	1,545
137	9	1,297	108	6	729
113	9	946	179	9	1,792
137	15	1,713	111	15	1,175
117	11	1,024	187	8	1,593
137	8	1,147	111	7	785
153	6	1,092	115	7	744
117	13	1,152	194	5	1,356
126	10	1,336	168	7	1,262

Figure 12.5

SAS Regression Printout
for First-Order Grandfather
Clock Model

Dependent Variable: PRICE

Number of Observations Read	32
Number of Observations Used	32

Analysis of Variance

Source	DF	Sum of Squares	Mean Square	F Value	Pr > F
Model	2	4283063	2141531	120.19	<.0001
Error	29	516727	17818		
Corrected Total	31	4799790			

Root MSE	133.48467	R-Square	0.8923
Dependent Mean	1326.87500	Adj R-Sq	0.8849
Coeff Var	10.06008		

Parameter Estimates

Variable	DF	Parameter Estimate	Standard Error	t Value	Pr > |t|	90% Confidence Limits	
Intercept	1	-1338.95134	173.80947	-7.70	<.0001	-1634.27571	-1043.62697
AGE	1	12.74057	0.90474	14.08	<.0001	11.20331	14.27784
NUMBIDS	1	85.95298	8.72852	9.85	<.0001	71.12211	100.78385

The rejection region for the test is found in exactly the same way as the rejection regions for the t-tests in previous chapters. That is, we consult Table VI in Appendix A to obtain an upper-tail value of t. This is a value t_α such that $P(t > t_\alpha) = \alpha$. We can then use this value to construct rejection regions for either one-tailed or two-tailed tests.

For $\alpha = .05$ and $n - (k + 1) = 32 - (2 + 1) = 29$ df, the critical t value obtained from Table VI is $t_{.05} = 1.699$. Therefore,

$$\text{Rejection region: } t > 1.699 \quad \text{(see Figure 12.6)}$$

Since the test statistic value, $t = 9.85$, falls in the rejection region, we have sufficient evidence to reject H_0. Thus, the collector can conclude that the mean auction price of a clock increases as the number of bidders increases, when age is held constant. Note that the two-tailed observed significance level of the test (highlighted on the printout) is less than .0001. Since the one-tailed p-value (half this value) is less than .00005, any nonzero α will lead us to reject H_0.

b. A 90% confidence interval for β_1 is (from the box):

$$\hat{\beta}_1 \pm t_{\alpha/2}s_{\hat{\beta}_1} = \hat{\beta}_1 \pm t_{.05}s_{\hat{\beta}_1}$$

Substituting $\hat{\beta}_1 = 12.74$, $s_{\hat{\beta}_i} = .905$ (both obtained from the SAS printout, Figure 12.5) and $t_{.05} = 1.699$ (from part **a**) into the equation, we obtain

$$12.74 \pm 1.699(.905) = 12.74 \pm 1.54$$

or (11.20,14.28). (This interval is also highlighted on Figure 12.5.) Thus, we are 90% confident that β_1 falls between 11.20 and 14.28. Since β_1 is the slope of the line relating auction price (y) to age of the clock (x_1), we conclude that price increases between \$11.20 and \$14.28 for every 1-year increase in age, holding number of bidders (x_2) constant.

Figure 12.6

Rejection Region for
$H_0: \beta_2 = 0 \ vs \ H_a: \beta_2 > 0$

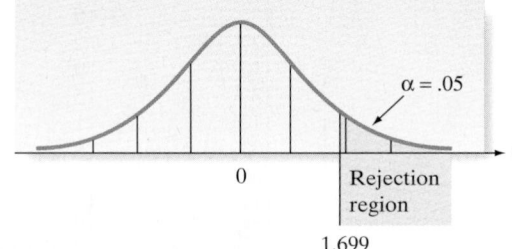

Look Back When interpreting the β multiplied by one x, be sure to hold fixed the values of the other x's in the model.

> **Now Work**   *Exercise 12.13c–d*

■ ■ ■

There are caveats with conducting t-tests on individual β parameters in a model for the purposes of determining which x's are useful for predicting y. Several of these are listed in the next box.

Caution

It is dangerous to conduct t-tests on the individual β parameters in a *first-order linear model* for the purpose of determining which independent variables are useful for predicting y and which are not. If you fail to reject H_0: $\beta_i = 0$, several conclusions are possible:

1. There is no relationship between y and x_i.
2. A straight-line relationship between y and x_i exists (holding the other x's in the model fixed), but a Type II error occurred.
3. A relationship between y and x_i (holding the other x's in the model fixed) exists, but is more complex than a straight-line relationship (e.g., a curvilinear relationship may be appropriate). The most you can say about a β parameter test is that there is either sufficient (if you reject H_0: $\beta_i = 0$) or insufficient (if you do not reject H_0: $\beta_i = 0$) evidence of a *linear (straight-line)* relationship between y and x_i.

In addition, conducting t-tests on each β parameter in a model is *not* the best way to determine whether the *overall* model is contributing information for the prediction of y. If we were to conduct a series of t-tests to determine whether the independent variables are contributing to the predictive relationship, we would be very likely to make one or more errors in deciding which terms to retain in the model and which to exclude.

For example, suppose you fit a first-order model in 10 quantitative x variables and decide to conduct t-tests on all 10 of the individual β's in the model, each at $\alpha = .05$. Even if all the β parameters (except β_0) are equal to 0, approximately 40% of the time you will incorrectly reject the null hypothesis at least once and conclude that some β parameter differs from 0.* Thus, in multiple regression models for which a large number of independent variables are being considered, conducting a series of t-tests may include a large number of insignificant variables and exclude some useful ones. If we want to test the utility of a multiple regression model, we will need a *global test* (one that encompasses all the β parameters). We would also like to find some statistical quantity that measures how well the model fits the data.

We commence with the easier problem—finding a measure of how well a linear model fits a set of data. For this we use the multiple regression equivalent of r^2, the coefficient of determination for the straight-line model (Chapter 11), as shown in the next box.

---

*The proof of this result (assuming independence of tests) proceeds as follows:

$P(\text{Reject } H_0 \text{ at least once} \mid \beta_1 = \beta_2 = \cdots = \beta_{10} = 0)$
$= 1 - P(\text{Reject } H_0 \text{ no times} \mid \beta_1 = \beta_2 = \cdots = \beta_{10} = 0)$
$\leq 1 - [P(\text{Accept } H_0: \beta_1 = 0 \mid \beta_1 = 0) \cdot P(\text{Accept } H_0: \beta_2 = 0 \mid \beta_2 = 0)| \cdots \cdot P(\text{Accept } H_0: \beta_{10} = 0 \mid \beta_{10} = 0)]$
$= 1 - [(1 - \alpha)^{10}] = 1 - (.95)^{10} = .401.$

For dependent tests, the Bonferroni inequality states that

$$P(\text{Reject } H_0 \text{ at least once} \mid \beta_1 = \beta_2 = \cdots = \beta_{10} = 0) \leq 10(\alpha) = 10(.05) = .50$$

> **DEFINITION 12.1**
>
> The **multiple coefficient of determination, R^2,** is defined as
>
> $$R^2 = 1 - \frac{\text{SSE}}{\text{SS}_{yy}} = \frac{\text{SS}_{yy} - \text{SSE}}{\text{SS}_{yy}} = \frac{\text{Explained variablilty}}{\text{Total variability}}$$

Just as for the simple linear model, R^2 represents the fraction of the sample variation of the y-values (measured by SS_{yy}) that is explained by the least squares prediction equation. Thus, $R^2 = 0$ implies a complete lack of fit of the model to the data and $R^2 = 1$ implies a perfect fit with the model passing through every data point. In general, the larger the value of R^2, the better the model fits the data.

To illustrate, the value $R^2 = .8923$ for the grandfather clock model of Example 12.3 is highlighted in Figure 12.5. This high value of R^2 implies that using the independent variables age and number of bidders in a first-order model explains 89.2% of the total *sample variation* (measured by SS_{yy}) of auction price y. Thus, R^2 is a sample statistic that tells how well the model fits the data and thereby represents a measure of the usefulness of the entire model.

A large value of R^2 computed from the *sample* data does not necessarily mean that the model provides a good fit to all of the data points in the *population*. For example, a first-order linear model that contains three parameters will provide a perfect fit to a sample of three data points and R^2 will equal 1. Likewise, you will always obtain a perfect fit ($R^2 = 1$) to a set of n data points if the model contains exactly n parameters. Consequently, if you want to use the value of R^2 as a measure of how useful the model will be for predicting y, it should be based on a sample that contains substantially more data points than the number of parameters in the model.

Caution

In a multiple regression analysis, use the value of R^2 as a measure of how useful a linear model will be for predicting y only if the sample contains substantially more data points than the number of β parameters in the model.

As an alternative to using R^2 as a measure of model adequacy, the *adjusted multiple coefficient of determination*, denoted R_a^2, is often reported. The formula for R_a^2 is shown in the box.

> **DEFINITION 12.2**
>
> The **adjusted multiple coefficient of determination** is given by
>
> $$R_a^2 = 1 - \left[\frac{(n-1)}{n-(k+1)}\right]\left(\frac{\text{SSE}}{\text{SS}_{yy}}\right)$$
>
> $$= 1 - \left[\frac{(n-1)}{n-(k+1)}\right](1 - R^2)$$
>
> *Note:* $R_a^2 \le R^2$

R^2 and R_a^2 have similar interpretations. However, unlike R^2, R_a^2 takes into account ("adjusts" for) both the sample size n and the number of β parameters in the model. R_a^2 will always be smaller than R^2, and more importantly, cannot be "forced" to 1 by simply adding more and more independent variables to the model. Consequently, analysts prefer the more conservative R_a^2 when choosing a measure of model adequacy. The value of R_a^2 is also highlighted in Figure 12.5. Note that $R_a^2 = .8849$, a value only slightly smaller than R^2.

Despite their utility, R^2 and R_a^2 are only sample statistics. Therefore, it is dangerous to judge the global usefulness of the model based solely on these values. A better method is to conduct a test of hypothesis involving *all* the β parameters (except β_0) in a model. In particular, for the general multiple regression model $E(y) = \beta_0 + \beta_1 x_1 + \beta_2 x_2 + \cdots + \beta_k x_k$, we would test

$$H_0: \beta_1 = \beta_2 = \cdots = \beta_k = 0$$
$$H_a: \text{At least one of the coefficients is nonzero}$$

The test statistic used to test this hypothesis is an F-statistic, and several equivalent versions of the formula can be used (although we will usually rely on the computer to calculate the F-statistic):

$$\textit{Test statistic: } F = \frac{(\text{SS}_{yy} - \text{SSE})/k}{\text{SSE}/[n - (k + 1)]} = \frac{\text{Mean square (Model)}}{\text{Mean square (Error)}}$$

$$= \frac{R^2/k}{(1 - R^2)/[n - (k + 1)]}$$

Both these formulas indicate that the F-statistic is the ratio of the *explained* variability divided by the model degrees of freedom to the *unexplained* variability divided by the error degrees of freedom. (For this reason, the test is often called the "analysis of variance" F-Test.) Thus, the larger the proportion of the total variability accounted for by the model, the larger the F-statistic.

To determine when the ratio becomes large enough that we can confidently reject the null hypothesis and conclude that the model is more useful than no model at all for predicting y, we compare the calculated F-statistic to a tabulated F-value with k df in the numerator and $[n - (k + 1)]$ df in the denominator. Recall that tabulations of the F-distribution for various values of α are given in Tables VIII, IX, X, and XI of Appendix A.

Rejection region: $F > F_\alpha$, where F is based on k numerator and $n - (k + 1)$ denominator degrees of freedom

The analysis of variance F-test for testing the overall utility of the multiple regression model is summarized in the next box.

Testing Global Usefulness of the Model: The Analysis of Variance F-Test

$H_0: \beta_1 = \beta_2 = \cdots = \beta_k = 0$ (All model terms are unimportant for predicting y)

$H_a:$ At least one $\beta_i \neq 0$ (At least one model term is useful for predicting y)

$$\textit{Test statistic: } F = \frac{(\text{SS}_{yy} - \text{SSE})/k}{\text{SSE}/[n - (k + 1)]} = \frac{R^2/k}{(1 - R^2)/[n - (k + 1)]}$$

$$= \frac{\text{Mean square (Model)}}{\text{Mean square (Error)}}$$

where n is the sample size and k is the number of terms in the model.

Rejection region: $F > F_\alpha$, with k numerator degrees of freedom and $[n - (k + 1)]$ denominator degrees of freedom.

Conditions Required for the Global F-Test to be Valid

The standard regression assumptions about the random error component (Section 12.1).

Caution

A rejection of the null hypothesis H_0: $\beta_1 = \beta_2 = \cdots = \beta_k = 0$ in the global F-test leads to the conclusion [with $100(1 - \alpha)\%$ confidence] that the model is statistically useful. However, statistically "useful" does not necessarily mean "best." Another model may prove even more useful in terms of providing more reliable estimates and predictions. This global F-test is usually regarded as a test that the model *must* pass to merit further consideration.

EXAMPLE 12.4   ASSESSING THE OVERALL ADEQUACY OF THE MODEL
_____

Problem   Refer to Example 12.3, in which an antique collector modeled the auction price y of grandfather clocks as a function of the age of the clock, x_1, and the number of bidders, x_2. Recall that the hypothesized first-order model is

$$y = \beta_0 + \beta_1 x_1 + \beta_2 x_2 + \varepsilon$$

a. Find and interpret the adjusted coefficient of determination R_a^2 for this example.

b. Conduct the global F-test of model usefulness at the $\alpha = .05$ level of significance.

Solution   **a.** The R_a^2 value (highlighted in the SAS printout, Figure 12.5) is .8849. This implies that the least squares model has explained about 88.5% of the total sample variation in y values (auction prices), after adjusting for sample size and number of independent variables in the model.

b. The elements of the global test of the model follow:

H_0: $\beta_1 = \beta_2 = 0$ [*Note:* $k = 2$]
H_a: At least one of the two model coefficients is nonzero

$$Test\ statistic:\ F = \frac{MS(\text{Model})}{MSE} = \frac{2{,}141{,}531}{17{,}818} = 120.19\ (\text{see Figure 12.5})$$

p-value: less than .0001

Conclusion: Since $\alpha = .05$ exceeds the observed significance level, $p < .0001$, the data provide strong evidence that at least one of the model coefficients is nonzero. The overall model appears to be statistically useful for predicting auction prices.

Look Back Can we be sure that the best prediction model has been found if the global F-test indicates that a model is useful? Unfortunately, we cannot. The addition of other independent variables may improve the usefulness of the model. (See the box, above) We consider more complex multiple regression models in Sections 12.5–12.8.

Now Work   *Exercise 12.18b,d*

———————   ■ ■ ■   ———————

In this section we discussed several different statistics for assessing the utility of a multiple regression model: t-tests on the individual β parameters, R^2, R_{adj}^2, and the global F-test. Both R^2 and R_a^2 are indicators of how well the prediction equation fits the data. Intuitive evaluations of the contribution of the model based on R^2 must be examined with care. Unlike R_a^2, the value of R^2 increases as more and more variables are added to the model. Consequently, you could force R^2 to take a value very close to 1 even though the model contributes no information for the prediction of y. In fact, R^2 equals 1 when the number of terms in the model (including β_0) equals the number of data points. Therefore, you should not rely solely on the value of R^2 (or even R_a^2) to tell you whether the model is useful for predicting y.

Conducting t-tests on all the β parameters is also not the best method of testing the global utility of the model, since these multiple tests result in a high probability of making at least one Type I error. Use the F-test for testing the global utility of the model.

After we have determined that the overall model is useful for predicting y using the F-test, we may elect to conduct one or more t-tests on the individual β parameters. However, the test (or tests) to be conducted should be decided *a priori*, that is, prior to fitting the model. Also, we should limit the number of t-tests conducted to avoid the potential problem of making too many Type I errors. Generally, the regression analyst will conduct t-tests only on the "most important" β's. We provide insight in identifying the most important β's in a linear model in Sections 12.5–12.8.

> ### Recommendation for Checking the Utility of a Multiple Regression Model
>
> **1.** First, conduct a test of overall model adequacy using the F-test, that is, test
> $$H_0: \beta_1 = \beta_2 = \cdots = \beta_k = 0$$
> If the model is deemed adequate (that is, if you reject H_0), then proceed to step 2. Otherwise, you should hypothesize and fit another model. The new model may include more independent variables or higher-order terms.
>
> **2.** Conduct t-tests on those β parameters in which you are particularly interested (that is, the "most important" β's). These usually involve only the β's associated with higher-order terms (x_2, x_1x_2, etc.). However, it is a safe practice to limit the number of β's that are tested. Conducting a series of t-tests leads to a high overall Type I error rate α.
>
> **3.** Examine the values of R_a^2 and $2s$ to evaluate how well, numerically, the model fits the data.

Exercises 12.1–12.27

Understanding the Principles

12.1 Write a first-order model relating $E(y)$ to
  a. two quantitative independent variables
  b. four quantitative independent variables
  c. five quantitative independent variables

12.2 List the four assumptions about the random error ε required for a multiple regression analysis.

12.3 Outline the six steps in a multiple regression analysis.

12.4 What are the caveats to conducting t-tests on all of the individual β parameters in a multiple regression model?

12.5 How should you test the overall adequacy of a multiple regression model?

Learning the Mechanics

12.6 MINITAB was used to fit the model $y = \beta_0 + \beta_1x_1 + \beta_2x_2 + \varepsilon$ to $n = 20$ data points and the printout shown on p. 678 was obtained.
  a. What are the sample estimates of β_0, β_1, and β_2?
  b. What is the least squares prediction equation?
  c. Find SSE, MSE, and s. Interpret the standard deviation in the context of the problem.
  d. Test $H_0: \beta_1 = 0$ against $H_a: \beta_1 \neq 0$. Use $\alpha = .05$.
  e. Use a 95% confidence interval to estimate β_2.
  f. Find R^2 and R_a^2 and interpret these values.

  g. Calculate the test statistic for the null hypothesis $H_0: \beta_1 = \beta_2 = 0$ using the two formulas given in this section. Compare your results to the test statistic shown on the printout.
  h. Find the observed significance level of the test, part **g.** Interpret the value.

12.7 Suppose you fit the model
$$y = \beta_0 + \beta_1x_1 + \beta_2x_2 + \beta_3x_3 + \varepsilon$$
to $n = 30$ data points and obtain the following result:
$$\hat{y} = 3.4 - 4.6x_1 + 2.7x_2 + .93x_3$$
The estimated standard errors of $\hat{\beta}_2$ and $\hat{\beta}_3$ are 1.86 and .29, respectively.
  a. Test the null hypothesis $H_0: \beta_2 = 0$ against the alternative hypothesis $H_a: \beta_2 \neq 0$. Use $\alpha = .05$.
  b. Test the null hypothesis $H_0: \beta_3 = 0$ against the alternative hypothesis $H_a: \beta_3 \neq 0$. Use $\alpha = .05$.
  c. The null hypothesis $H_0: \beta_2 = 0$ is not rejected. In contrast, the null hypothesis $H_0: \beta_3 = 0$ is rejected. Explain how this can happen even though $\hat{\beta}_2 > \hat{\beta}_3$.

12.8 Suppose you fit the first-order multiple regression model
$$y = \beta_0 + \beta_1x_1 + \beta_2x_2 + \varepsilon$$

MINITAB Output for
Exercise 12.6

```
The regression equation is
Y = 506.35 - 941.9 X1 - 429.1 X2


Predictor     Coef   SE Coef       T       P
Constant   506.346     45.17   11.21   0.000
X1        -941.900    275.08   -3.42   0.003
X2        -429.060    379.83   -1.13   0.274


S = 94.251     R-Sq = 45.9%     R-Sq(adj) = 39.6%


Analysis of Variance

Source           DF       SS      MS      F       P
Regression        2   128329   64165   7.22   0.005
Residual Error   17   151016    8883
Total            19   279345
```

to $n = 25$ data points and obtain the prediction equation

$$\hat{y} = 6.4 + 3.1x_1 + .92x_2$$

The estimated standard deviations of the sampling distributions of $\hat{\beta}_1$ and $\hat{\beta}_2$ are 2.3 and .27, respectively.

a. Test H_0: $\beta_1 = 0$ against H_a: $\beta_1 > 0$. Use $\alpha = .05$.
b. Test H_0: $\beta_2 = 0$ against H_a: $\beta_2 \neq 0$. Use $\alpha = .05$.
c. Find a 90% confidence interval for β_1. Interpret the interval.
d. Find a 99% confidence interval for β_2. Interpret the interval.

12.9 How is the number of degrees of freedom available for estimating σ^2 (the variance of ε) related to the number of independent variables in a regression model?

12.10 Consider the first-order model equation in three quantitative independent variables

$$E(y) = 1 + 2x_1 + x_2 - 3x_3$$

a. Graph the relationship between y and x_1 for $x_2 = 1$ and $x_3 = 3$.
b. Repeat part **a** for $x_2 = -1$ and $x_3 = 1$.
c. How do the graphed lines in parts **a** and **b** relate to each other? What is the slope of each line?
d. If a linear model is first-order in three independent variables, what type of geometric relationship will you obtain when $E(y)$ is graphed as a function of one of the independent variables for various combinations of values of the other independent variables?

12.11 Suppose you fit the first-order model

$$y = \beta_0 + \beta_1 x_1 + \beta_2 x_2 + \beta_3 x_3 + \beta_4 x_4 + \beta_5 x_5 + \varepsilon$$

to $n = 30$ data points and obtain

$$\text{SSE} = .33 \quad R^2 = .92$$

a. Do the values of SSE and R^2 suggest that the model provides a good fit to the data? Explain.
b. Is the model of any use in predicting y? Test the null hypothesis H_0: $\beta_1 = \beta_2 = \cdots = \beta_5 = 0$ against the alternative hypothesis H_a: At least one

of the parameters $\beta_1, \beta_2, \ldots, \beta_5$ is nonzero. Use $\alpha = .05$.

12.12 If the analysis of variance F-test leads to the conclusion that at least one of the model parameters is nonzero, can you conclude that the model is the best predictor for the dependent variable y? Can you conclude that all of the terms in the model are important for predicting y? What is the appropriate conclusion?

Applying the Concepts—Basic

12.13 Predicting runs scored in baseball. In *Chance* (Fall NW 2000), statistician Scott Berry built a multiple regression model for predicting total number of runs scored by a Major League Baseball team during a season. Using data on all teams from 1990–1998 (a sample of $n = 234$), the results in the next table were obtained.

a. Write the least squares prediction equation for y = total number of runs scored by a team in a season.
b. Give practical interpretations of the β estimates.
c. Conduct a test of H_0: $\beta_7 = 0$ against H_a: $\beta_7 < 0$ at $\alpha = .05$. Interpret the results.
d. Form a 95% confidence interval for β_5. Interpret the results.

Independent Variable	β Estimate	Standard Error
Intercept	3.70	15.00
Walks (x_1)	.34	.02
Singles (x_2)	.49	.03
Doubles (x_3)	.72	.05
Triples (x_4)	1.14	.19
Home Runs (x_5)	1.51	.05
Stolen Bases (x_6)	.26	.05
Caught Stealing (x_7)	−.14	.14
Strikeouts (x_8)	−.10	.01
Outs (x_9)	−.10	.01

Source: Berry, S. M. "A statistician reads the sports pages: Modeling offensive ability in baseball." *Chance*, Vol. 13, No. 4, Fall 2000 (Table 2).

e. Predict the number of runs scored by your favorite Major League Baseball team last year. How close is the predicted value to the actual number of runs scored by your team? (*Note*: You can find data on your favorite team on the Internet at www.majorleaguebaseball.com.)

12.14 Growth of Japanese beetles. In the *Journal of Insect Behavior* (Nov. 2001), biologists at Eastern Illinois University published the results of their study on Japanese beetles. The biologists collected beetles over a period of $n = 13$ summer days in a soybean field. For one portion of the study, the biologists modeled y, the average size (in millimeters) of female beetles as a function of the average daily temperature x_1 (degrees) and Julian date x_2.

a. Write a first-order model for $E(y)$ as a function of x_1 and x_2.

b. The model was fit to the data, with the following results. Interpret the estimate of β_1.

Variable	Parameter Estimate	t-value	p-value
Intercept	6.51	26.0	<.0001
Temperature (x_1)	−.002	−0.72	.49
Date (x_2)	−.010	−3.30	.008

c. Conduct a test to determine whether the average size of female Japanese beetles decreases linearly as temperature increases. Use $\alpha = .05$.

12.15 Study of adolescents with ADHD. Children with attention-deficit/hyperactivity disorder (ADHD) were monitored to evaluate their risk for substance (e.g., alcohol, tobacco, illegal drug) use. (*Journal of Abnormal Psychology*, Aug. 2003.) The following data were collected on 142 adolescents diagnosed with ADHD:

y = frequency of marijuana use the past six months
x_1 = severity of inattention (5-point scale)
x_2 = severity of impulsivity-hyperactivity (5-point scale)
x_3 = level of oppositional-defiant and conduct disorder (5-point scale)

a. Write the equation of a first-order model for $E(y)$.

b. The coefficient of determination for the model is $R^2 = .08$. Interpret this value.

c. The global F-test for the model yielded a p-value less than .01. Interpret this result.

d. The t-test for $H_0: \beta_1 = 0$ resulted in a p-value less than .01. Interpret this result.

e. The t-test for $H_0: \beta_2 = 0$ resulted in a p-value greater than .05. Interpret this result.

f. The t-test for $H_0: \beta_3 = 0$ resulted in a p-value greater than .05. Interpret this result.

12.16 Global warming and foreign investments. Scientists believe that a major cause of global warming is higher levels of carbon dioxide (CO_2) in the atmosphere. In the *Journal of World-Systems Research* (Summer 2003), sociologists examined the impact of foreign investment dependence on CO_2 emissions in $n = 66$ developing countries. In particular, the researchers modeled the level of CO_2 emissions in 1996 based on foreign investments made 16 years earlier and several other independent variables. The variables and the model results are listed in the accompanying table.

y = ln(level of CO_2 emissions in 1996)	β Estimate	t-value	p-value
x_1 = ln(foreign investments in 1980)	.79	2.52	<.05
x_2 = gross domestic investment in 1980)	.01	.13	>.10
x_3 = trade exports in 1980	−.02	−1.66	>.10
x_4 = ln(GNP in 1980)	−.44	−.97	>.10
x_5 = agricultural production in 1980	−.03	−.66	>.10
x_6 = 1 if African country, 0 if not	−1.19	−1.52	>.10
x_7 = ln(level of CO_2 emissions in 1980)	.56	3.35	<.001
$R^2 = .31$			

Source: Grimes, P., & Kentor, J. "Exporting the greenhouse: Foreign capital penetration and CO_2 emissions 1980–1996," *Journal of World-Systems Research*, Vol. IX, No. 2, Summer 2003 (Table 1).

a. Interpret the value of R^2.

b. Use the value of R^2 to test the null hypothesis, $H_0: \beta_1 = \beta_2 = \cdots = \beta_7 = 0$, at $\alpha = .01$. Give the appropriate conclusion.

c. What null hypothesis would you test to determine if foreign investments in 1980 is a statistically useful predictor of CO_2 emissions in 1996?

d. Conduct the test, part **c**, at $\alpha = .05$. Give the appropriate conclusion.

12.17 Novelty of a vacation destination. Many tourists choose a vacation destination based on the newness or uniqueness (i.e., the novelty) of the itinerary. Texas A&M University professor J. Petrick investigated the relationship between novelty and vacationing golfers' demographics. (*Annals of Tourism Research*, Vol. 29, 2002.) Data were obtained from a mail survey of 393 golf vacationers to a large coastal resort in southeastern United States. Several measures of novelty level (on a numerical scale) were obtained for each vacationer, including "change from routine," "thrill," "boredom-alleviation," and "surprise." The researcher employed four independent variables in a regression model to predict each of the novelty measures. The independent variables were x_1 = number of rounds of golf per year, x_2 = total number of golf vacations taken, x_3 = number of years played golf, and x_4 = average golf score.

a. Give the hypothesized equation of a first-order model for y = change from routine.

b. A test of $H_0: \beta_3 = 0$ versus $H_a: \beta_3 < 0$ yielded a p-value of .005. Interpret this result if $\alpha = .01$.

c. The estimate of β_3 was found to be negative. Based on this result (and the result of part **b**), the researcher concluded that "those who have played golf for more years are less apt to seek change from their normal routine in their golf vacations." Do you agree with this statement? Explain.

d. The regression results for the three other dependent novelty measures are summarized in the table. Give the null hypothesis for testing the overall adequacy of each first-order regression model.

Dependent Variable	F-value	p-value	R^2
Thrill	5.56	<.001	.055
Change from routine	3.02	.018	.030
Surprise	3.33	.011	.023

Source: Petrick, J. F. "An examination of golf vacationers' novelty," *Annals of Tourism Research*, Vol. 29, 2002.

e. Give the rejection region for the test, part **d**, using $\alpha = .01$.

f. Use the test statistics reported in the table and the rejection region from part **e** to conduct the test for each of the dependent measures of novelty.

g. Verify that the p-values in the table support your conclusions in part **f**.

h. Interpret the values of R^2 reported in the table.

12.18 Factors identifying urban counties. Refer to the
NW *Professional Geographer* (Feb. 2000) study of urban and rural counties in the western United States, Exercise 3.157 (p. 181). University of Nevada (Reno) researchers asked a sample of 256 county commissioners to rate their "home" county on a scale of 1 (most rural) to 10 (most urban). The urban/rural rating (y) was used as the dependent variable in a first-order multiple regression model with six independent variables: total county population (x_1), population density (x_2), population concentration (x_3), population growth (x_4), proportion of county land in farms (x_5), and 5-year change in agricultural land base (x_6). Some of the regression results are shown in the table.

Independent Variable	β Estimate	p-value
x_1: Total population	0.110	0.045
x_2: Population density	0.065	0.230
x_3: Population concentration	0.540	0.000
x_4: Population growth	−0.009	0.860
x_5: Farm land	−0.150	0.003
x_6: Agricultural change	−0.027	0.58

Overall model: $R^2 = .44$ $R_a^2 = .43$ $F = 32.47$ p-value <.001

Source: Berry, K. A., et al. "Interpreting what is rural and urban for western U.S. counties." *Professional Geographer*, Vol. 52, No. 1, Feb. 2000 (Table 2).

a. Write the least squares prediction equation for y.

b. Give the null hypothesis for testing overall model adequacy.

c. Conduct the test, part **b**, at $\alpha = .01$ and give the appropriate conclusion.

d. Interpret the values of R^2 and R_a^2.

e. Give the null hypothesis for testing the contribution of population growth (x_4) to the model.

f. Conduct the test, part **e**, at $\alpha = .01$ and give the appropriate conclusion.

Applying the Concepts—Intermediate

12.19 Incomes of Mexican street vendors. Detailed interviews were conducted with over 1,000 street vendors in the city of Puebla, Mexico, in order to study the factors influencing vendors' incomes. (*World Development*, Feb. 1998.) Vendors were defined as individuals working in the street, and included vendors with carts and stands on wheels and excluded beggars, drug dealers, and prostitutes. The researchers collected data on gender, age, hours worked per day, annual earnings, and education level. A subset of these data appears in the table.

💿 **STREETVEN**

Vendor Number	Annual Earnings y	Age x_1	Hours Worked per Day x_2
21	$2841	29	12
53	1876	21	8
60	2934	62	10
184	1552	18	10
263	3065	40	11
281	3670	50	11
354	2005	65	5
401	3215	44	8
515	1930	17	8
633	2010	70	6
677	3111	20	9
710	2882	29	9
800	1683	15	5
914	1817	14	7
997	4066	33	12

Source: Adapted from Smith, P. A., and Metzger, M. R. "The return to education: Street vendors in Mexico." *World Development*, Vol. 26, No. 2, Feb. 1998, pp. 289–296.

a. Write a first-order model for mean annual earnings, $E(y)$, as a function of age (x_1) and hours worked (x_2).

b. Fit the model to the data and give the least squares prediction equation.

c. Interpret the estimated β coefficients in your model.

d. Is age (x_1) a statistically useful predictor of annual earnings? Test using $\alpha = .01$.

e. Find a 95% confidence interval for β_2. Interpret the interval in the words of the problem.

f. Find and interpret the value of R^2.

g. Find and interpret the value of R_a^2. Explain the relationship between R^2 and R_a^2.

h. Conduct a test of the global utility of the model (at $\alpha = .01$). Interpret the results.

12.20 Study of contaminated fish. Refer to the U.S. Army Corps of Engineers data on fish contaminated from the toxic discharges of a chemical plant located on the banks of the Tennessee River in Alabama. Recall that the engineers measured the length (in centimeter),

⊙ DDT (First and last 5 observations listed.)

River	Mile	Species	Length	Weight	DDT
FC	5	CHANNELCATFISH	42.5	732	10.00
FC	5	CHANNELCATFISH	44.0	795	16.00
FC	5	CHANNELCATFISH	41.5	547	23.00
FC	5	CHANNELCATFISH	39.0	465	21.00
FC	5	CHANNELCATFISH	50.5	1252	50.00
⋮	⋮	⋮	⋮	⋮	⋮
TR	345	LARGEMOUTHBASS	23.5	358	2.00
TR	345	LARGEMOUTHBASS	30.0	856	2.20
TR	345	LARGEMOUTHBASS	29.0	793	7.40
TR	345	LARGEMOUTHBASS	17.5	173	0.35
TR	345	LARGEMOUTHBASS	36.0	1433	1.90

weight (in grams), and DDT level (in parts per million) for 144 captured fish. In addition, the number of miles upstream from the river was recorded. The data are saved in the DDT file. (The first and last five observations are shown in the table above.)

a. Fit the first-order model, $E(y) = \beta_0 + \beta_1 x_1 + \beta_2 x_2 + \beta_3 x_3$, to the data, where y = DDT level, x_1 = mile, x_2 = length, and x_3 = weight. Report the least squares prediction equation.

b. Find the estimate of the standard deviation of ε for the model and give a practical interpretation of its value.

c. Do the data provide sufficient evidence to conclude that DDT level increases as length increases? Report the observed significance level of the test and reach a conclusion using $\alpha = .05$.

d. Find and interpret a 95% confidence interval for β_3.

e. Test the overall adequacy of the model using $\alpha = .05$.

12.21 Snow geese feeding trial. Refer to the *Journal of Applied Ecology* (Vol. 32, 1995) study of the feeding habits of baby snow geese, Exercise 11.76 (p. 634). The data on gosling weight change, digestion efficiency, acid-detergent fibre (all measured as percentages) and diet (plants or duck chow) for 42 feeding trials are saved in the **SNOWGEESE** file. Selected observations are shown in the table below. The botanists were interested in predicting weight change (y) as a function of the other variables. Consider the first-order model $E(y) = \beta_0 + \beta_1 x_1 + \beta_2 x_2$, where x_1 is digestion efficiency and x_2 is acid-detergent fibre.

a. Find the least squares prediction equation for weight change, y.

b. Interpret the β-estimates in the equation, part **a**.

c. Conduct a test to determine if digestion efficiency, x_1, is a useful linear predictor of weight change. Use $\alpha = .01$.

d. Form a 99% confidence interval for β_2. Interpret the result.

e. Find and interpret R^2 and R_a^2. Which statistic is the preferred measure of model fit? Explain.

f. Is the overall model statistically useful for predicting weight change? Test using $\alpha = .05$.

12.22 Deep space survey of quasars. A quasar is a distant celestial object (at least four billion light-years away) that provides a powerful source of radio energy. The *Astronomical Journal* (July 1995) reported on a study of 90 quasars detected by a deep space survey. The survey enabled astronomers to measure several different

⊙ SNOWGEESE (First and last 5 trials shown)

Feeding Trial	Diet	Weight Change (%)	Digestion Efficiency (%)	Acid-Detergent Fibre (%)
1	Plants	−6	0	28.5
2	Plants	−5	2.5	27.5
3	Plants	−4.5	5	27.5
4	Plants	0	0	32.5
5	Plants	2	0	32
⋮	⋮	⋮	⋮	⋮
38	Duck Chow	9	59	8.5
39	Duck Chow	12	52.5	8
40	Duck Chow	8.5	75	6
41	Duck Chow	10.5	72.5	6.5
42	Duck Chow	14	69	7

Source: Gadallah, F. L., and Jefferies, R. L. "Forage quality in brood rearing areas of the lesser snow goose and the growth of captive goslings." *Journal of Applied Biology*, Vol. 32, No. 2, 1995, pp. 281–282 (adapted from Figures 2 and 3).

QUASAR (First 5 quasars shown.)

Quasar	Redshift (x_1)	Line Flux (x_2)	Line Luminosity (x_3)	AB_{1450} x_4	Absolute Magnitude (x_5)	Rest Frame Equivalent Width (y)
1	2.81	−13.48	45.29	19.50	−26.27	117
2	3.07	−13.73	45.13	19.65	−26.26	82
3	3.45	−13.87	45.11	18.93	−27.17	33
4	3.19	−13.27	45.63	18.59	−27.39	92
5	3.07	−13.56	45.30	19.59	−26.32	114

Source: Schmidt, M., Schneider, D. P., and Gunn, J. E. "Spectroscopic CCD surveys for quasars at large redshift." *The Astronomical Journal,* Vol. 110, No. 1, July 1995, p. 70 (Table 1).

quantitative characteristics of each quasar, including redshift range, line flux (erg/cm$^2 \cdot$s), line luminosity (erg/s), AB_{1450} magnitude, absolute magnitude, and rest frame equivalent width. The data for a sample of 25 large (redshift) quasars are saved in the **QUASAR** file. (Several quasars are listed in the table above.)

a. Hypothesize a first-order model for equivalent width, y, as a function of the first four variables in the table.

b. Fit the first-order model to the data. Give the least squares prediction equation.

c. Interpret the β estimates in the model.

d. Test the overall adequacy of the model using $\alpha = .05$.

e. Test to determine whether redshift (x_1) is a useful linear predictor of equivalent width (y), using $\alpha = .05$.

12.23 Extracting water from oil. In the oil industry, water that mixes with crude oil during production and transportation must be removed. Chemists have found that the oil

can be extracted from the water/oil mix electrically. Researchers at the University of Bergen (Norway) conducted a series of experiments to study the factors that influence the voltage (y) required to separate the water from the oil. (*Journal of Colloid and Interface Science,* Aug.1995.) The seven independent variables investigated in the study are listed in the table. (Each variable was measured at two levels—a "low" level and a "high" level.) Sixteen water/oil mixtures were prepared using different combinations of the independent variables; then each emulsion was exposed to a high electric field. In addition, three mixtures were tested when all independent variables were set to 0. The data for all 19 experiments are also given in the table below.

a. Propose a first-order model for y as a function of all seven independent variables.

b. Use a statistical software package to fit the model to the data in the table.

c. Fully interpret the β estimates.

WATEROIL

Experiment Number	Voltage y (kw/cm)	Disperse Phase Volume x_1 (%)	Salinity x_2 (%)	Temperature x_3 (°C)	Time Delay x_4 (hours)	Surfactant Concentration x_5 (%)	Span: Triton x_6	Solid Particles x_7 (%)
1	.64	40	1	4	.25	2	.25	.5
2	.80	80	1	4	.25	4	.25	2
3	3.20	40	4	4	.25	4	.75	.5
4	.48	80	4	4	.25	2	.75	2
5	1.72	40	1	23	.25	4	.75	2
6	.32	80	1	23	.25	2	.75	.5
7	.64	40	4	23	.25	2	.25	2
8	.68	80	4	23	.25	4	.25	.5
9	.12	40	1	4	24	2	.75	2
10	.88	80	1	4	24	4	.75	.5
11	2.32	40	4	4	24	4	.25	2
12	.40	80	4	4	24	2	.25	.5
13	1.04	40	1	23	24	4	.25	.5
14	.12	80	1	23	24	2	.25	2
15	1.28	40	4	23	24	2	.75	.5
16	.72	80	4	23	24	4	.75	2
17	1.08	0	0	0	0	0	0	0
18	1.08	0	0	0	0	0	0	0
19	1.04	0	0	0	0	0	0	0

Source: Førdedal, H., et al. "A multivariate analysis of W/O emulsions in high external electric fields as studied by means of dielectric time domain spectroscopy." *Journal of Colloid and Interface Science,* Vol. 173, No. 2, Aug. 1995, p. 398 (Table 2).

12.24 Students ability in science. An article published in the *American Educational Research Journal* (Fall 1998) used multiple regression to model the students' perceptions of their ability in science classes. The sample consisted of 165 Grade 5–Grade 8 students in six performance-based science classrooms, all of which use hands-on activities as the main teaching tool. The dependent variable of interest, the student's ability perception (y), was measured on a 4-point scale (where $1 =$ little or no ability and $4 =$ high ability). The independent variables included in the model are listed in the table at the bottom of the page, as well as the results of the multiple regression.

a. Which independent variables appear to contribute to the prediction of a student's perception of his/her ability in science? Explain.

b. Construct a 95% confidence interval for the β coefficient of the variable active-leading behavior. Interpret the interval.

c. The following statistics for evaluating the overall prediction power of the model were reported: $R^2 = .48$, $F = 12.84$, $p < .001$. Interpret the results.

12.25 Occupational safety study. An important goal in occupational safety is "active caring." Employees demonstrate active caring (AC) about the safety of their co-workers when they identify environmental hazards and unsafe work practices and then implement appropriate corrective actions for these unsafe conditions or behaviors. Three factors hypothesized to increase the propensity for an employee to actively care for safety are (1) high self-esteem, (2) optimism, and (3) group cohesiveness. *Applied & Preventive Psychology* (Winter 1995) attempted to establish empirical support for the AC hypothesis by fitting the model $E(y) = \beta_0 + \beta_1 x_1 + \beta_2 x_2 + \beta_3 x_3$, where

$$
\begin{aligned}
y &= \text{AC score (measuring active caring} \\
&\quad \text{on a 15-point scale)} \\
x_1 &= \text{Self-esteem score} \\
x_2 &= \text{Optimism score} \\
x_3 &= \text{Group cohesion score}
\end{aligned}
$$

The regression analysis, based on data collected for $n = 31$ hourly workers at a large fiber-manufacturing plant, yielded a multiple coefficient of determination of $R^2 = .362$.

a. Interpret the value of R^2.

b. Use the R^2 value to test the global utility of the model. Use $\alpha = .05$.

12.26 R^2 and model fit. Because the coefficient of determination R^2 always increases when a new independent variable is added to the model, it is tempting to include many variables in a model to force R^2 to be near 1. However, doing so reduces the degrees of freedom available for estimating σ^2, which adversely affects our ability to make reliable inferences. Suppose you want to use 18 psychological and sociological factors to predict a student's Scholastic Assessment Test (SAT) score. You fit the model

$$
y = \beta_0 + \beta_1 x_1 + \beta_2 x_2 + \cdots + \beta_{17} x_{17} + \beta_{18} x_{18} + \varepsilon
$$

where $y =$ SAT score and $x_1, x_2, \ldots, x_{18}$ are the psychological and sociological factors. Only 20 years of data ($n = 20$) are used to fit the model, and you obtain $R^2 = .95$. Test to see whether this impressive-looking R^2 is large enough for you to infer that the model is useful, that is, that at least one term in the model is important for predicting SAT scores. Use $\alpha = .05$.

Applying the Concepts—Advanced

12.27 The vineyards in the Bordeaux region of France are known for producing excellent red wines. However, the uncertainty of the weather during the growing season, the phenomenon that wine tastes better with age, and the fact that some Bordeaux vineyards produce better wines than others, encourages speculation concerning the value of a case of wine produced by a certain vineyard during a certain year (or vintage). As a result, many wine experts attempt to predict the auction price of a case of Bordeaux wine. The publishers of a newsletter titled *Liquid Assets: The International Guide to Fine Wine* discussed a multiple regression approach to predicting the London auction price of red Bordeaux wine

Results for Exercise 12.24

Independent Variable	β Estimate	Standard Error	t	p-Value
Prior science attitude (4-point scale)	.60	.07	8.13	$p < .001$
Science ability (standardized test score)	.003	.005	.57	NS
Gender (1 if boy, 0 if girl)	.22	.07	2.98	$p < .01$
Class 1 (1 if classroom 1, 0 if not)	.15	.153	0.98	NS
Class 3 (1 if classroom 3, 0 if not)	−.16	.13	−1.17	NS
Class 4 (1 if classroom 4, 0 if not)	.00	.13	0.00	NS
Class 5 (1 if classroom 5, 0 if not)	.08	.13	0.64	NS
Class 6 (1 if classroom 6, 0 if not)	−.15	.14	−1.12	NS
Active-leading behavior (0 to 1 scale)	.88	.34	2.62	$p < .01$
Passive-assisting behavior (0 to 1 scale)	.43	.51	0.85	NS
Active-manipulating behavior (0 to 1 scale)	.77	.61	1.25	NS

Note: Base level for classroom dummy variables is classroom 2.
NS = *not significant*
Source: Jovanovic, J., and King, S. S. "Boys and girls in the performance-based science classroom: Who's doing the performing?" *American Educational Research Journal*, Vol. 35, No. 3, Fall 1998, p. 489 (Table 8).

in *Chance* (Fall 1995). The natural logarithm of the price y (in dollars) of a case containing a dozen bottles of red wine was modeled as a function of weather during growing season and age of vintage using data collected for the vintages of 1952–1980. Three models were fit to the data. The results of the regressions are summarized in the table below.

a. For each model, conduct a t-test for each of the β parameters in the model. Interpret the results.

b. When the natural log of y is used as a dependent variable, the antilogarithm of a β coefficient minus 1, that is $e^{\beta_i} - 1$, represents the percentage change in y for every 1-unit increase in the associated x value.* Use this information to interpret the β estimates of each model.

c. Based on the values of R^2 and s shown, which of the three models would you use to predict red Bordeaux wine prices? Explain.

	Beta Estimates (Standard Errors)		
Independent Variables	Model 1	Model 2	Model 3
x_1 = Vintage year	.0354 (.0137)	.0238 (.00717)	.0240 (.00747)
x_2 = Average growing season temperature (°C)	(not included)	.616 (.0952)	.608 (.116)
x_3 = Sept./Aug. rainfall (cm)	(not included)	−.00386 (.00081)	−.00380 (.00095)
x_4 = Rainfall in months preceding vintage (cm)	(not included)	.0001173 (.000482)	.00115 (.000505)
x_5 = Average Sept. temperature (°C)	(not included)	(not included)	.00765 (.565)
	$R^2 = .212, s = .575$	$R^2 = .828, s = .287$	$R^2 = .828, s = .293$

Source: Ashenfelter, O., Ashmore, D., and LaLonde, R. "Bordeaux wine vintage quality and weather." *Chance*, Vol. 8, No. 4, Fall 1995, p. 116 (Table 2).

12.4 Using the Model for Estimation and Prediction

In Section 11.8 we discussed the use of the least squares line for estimating the mean value of y, $E(y)$, for some particular value of x, say $x = x_p$. We also showed how to use the same fitted model to predict, when $x = x_p$, some new value of y to be observed in the future. Recall that the least squares line yielded the same value for both the estimate of $E(y)$ and the prediction of some future value of y. That is, both are the result of substituting x_p into the prediction equation $\hat{y} = \hat{\beta}_0 + \hat{\beta}_1 x$ and calculating $\hat{y}_p$. There the equivalence ends. The confidence interval for the mean $E(y)$ is narrower than the prediction interval for y because of the additional uncertainty attributable to the random error ε when predicting some future value of y.

These same concepts carry over to the multiple regression model. Consider, again, the first-order model in Section 12.2 relating sale price of a residential property to land value (x_1), improvements (x_2), and home size (x_3). Suppose we want to estimate the mean sale price for a given property with $x_1 = \$15,000$, $x_2 = \$50,000$, and $x_3 = 1,800$ square feet. Assuming that the first-order model represents the true relationship between sale price and the three independent variables, we want to estimate

$$E(y) = \beta_0 + \beta_1 x_1 + \beta_2 x_2 + \beta_3 x_3$$
$$= \beta_0 + \beta_1(15,000) + \beta_2(50,000) + \beta_3(1,800)$$

Substituting into the least squares prediction equation, we find the estimate of $E(y)$ to be

$$\hat{y} = \hat{\beta}_0 + \hat{\beta}_1(15,000) + \hat{\beta}_2(50,000) + \hat{\beta}_3(1,800)$$
$$= 1,470.27 + .814(50,000) + .820(50,000) + 13.529(1,800) = 79,061.4$$

*The result is derived by expressing the percentage change in price y as $(y_1 - y_0)/y_0$, where $y_1 =$ the value of y when, say, $x = 1$, and $y_0 =$ the value of y when $x = 0$. Now let $y^* = \ln(y)$ and assume the model is $y^* = \beta_0 + \beta_1 x$. Then

$$y = e^{y^*} = e^{\beta_0}e^{\beta_1 x} = \begin{cases} e^{\beta_0} & \text{when } x = 0 \\ e^{\beta_0}e^{\beta_1} & \text{when } x = 1 \end{cases}$$

Substituting, we have

$$\frac{y_1 - y_0}{y_0} = \frac{e^{\beta_0}e^{\beta_1} - e^{\beta_0}}{e^{\beta_0}} = e^{\beta_1} - 1$$

Figure 12.7

SPSS Spreadsheet with
95% Confidence Intervals
for Sale Price Model

	PROPERTY	LANDVAL	IMPROVAL	AREA	PREDICT	LLMean95	ULMean95	LLPred95	ULPred95
1	1	5960	44967	1873	68557	59404	77709	49435	87678
2	2	9000	27860	928	44213	37904	50522	26278	62148
3	3	9500	31439	1126	50235	45338	55132	32747	67723
4	4	10000	39592	1265	59212	54285	64139	41715	76709
5	5	18000	72827	2214	105834	95659	116009	86203	125465
6	6	8500	27317	912	43144	36932	49355	25243	61044
7	7	8000	29856	899	44644	38243	51044	26677	62611
8	8	23000	47752	1803	83774	71575	95972	63021	104526
9	9	8100	39117	1204	58449	50998	61901	38798	74101
10	10	9000	29349	1725	56217	48550	63884	37760	74673
11	11	7300	40166	1080	54981	47863	62099	36746	73216
12	12	8000	31679	1529	54662	49555	59770	37114	72211
13	13	20000	58510	2455	98977	88619	109336	79250	118704
14	14	8000	23454	1151	42800	37649	47951	25239	60361
15	15	8000	20897	1173	41000	35078	46922	23198	58802
16	16	10500	56248	1960	82687	74643	90731	64071	101303
17	17	4000	20859	1344	40024	32814	47235	21753	58296
18	18	4500	22610	988	37052	31827	42278	19469	54635
19	19	3400	35948	1076	48290	40504	56076	29784	66796
20	20	1500	5779	962	20448	11231	29664	1296	39600
21	21	15000	50000	1800	79061	73381	84742	61338	96785

To form a confidence interval for the mean, we need to know the standard deviation of the sampling distribution for the estimator $\hat{y}$. For multiple regression models, the form of this standard deviation is rather complex. However, the regression routines of statistical software packages allow us to obtain the confidence intervals for mean values of y for any given combination of values of the independent variables. A portion of the SPSS output for the sale price example is shown in Figure 12.7.

The estimated mean value (i.e., predicted value) for the selected x values is highlighted on the SPSS printout under "**PREDICT**." We observe that $\hat{y} = 79,061$, which agrees (except for rounding) with our calculation. The corresponding 95% confidence interval for the true mean of y (highlighted under "**LLMean95**" and "**ULMean95**") is (73,381, 84,742). Thus, with 95% confidence, we conclude that the mean sale price for all properties with $x_1 = \$15,000$, $x_2 = \$50,000$, and $x_3 = 1,800$ square feet will fall between $73,381 and $84,742.

If we were interested in predicting the sale price for a particular (single) property with $x_1 = \$15,000$, $x_2 = \$50,000$, and $x_3 = 1,800$ square feet, we would use $\hat{y} = \$79,061$ as the predicted value. However, the prediction interval for a new value of y is wider than the confidence interval for the mean value. This is reflected by the SPSS 95% prediction interval, (61,338, 96,785) highlighted under "**LLPred95**" and "**ULPred95**." Thus, with 95% confidence, we conclude that the sale price for an individual property with the characteristics $x_1 = \$15,000$, $x_2 = \$50,000$, and $x_3 = 1,800$ square feet will fall between $61,338 and $96,785.

Now Work *Exercise 12.30*

EXAMPLE 12.5

Problem

ESTIMATING $E(y)$ AND PREDICTING y

Refer to Examples 12.3–12.4 and the first-order model, $E(y) = \beta_0 + \beta_1 x_1 + \beta_2 x_2$, where y = auction price of a grandfather clock, x_1 = age of the clock, and x_2 = number of bidders.

a. Estimate the average auction price for all 150-year-old clocks sold at an auction with 10 bidders using a 95% confidence interval. Interpret the result.

b. Suppose you want to predict the auction price for one clock that is 50 years old and has 2 bidders. How should you proceed?

Figure 12.8

MINITAB Printout with
95% Confidence Intervals
for Grandfather Clock
Model

Regression Analysis: PRICE versus AGE, NUMBIDS

```
The regression equation is
PRICE = - 1339 + 12.7 AGE + 86.0 NUMBIDS


Predictor        Coef   SE Coef       T       P
Constant      -1339.0     173.8   -7.70   0.000
AGE           12.7406    0.9047   14.08   0.000
NUMBIDS        85.953     8.729    9.85   0.000


S = 133.485    R-Sq = 89.2%    R-Sq(adj) = 88.5%


Analysis of Variance

Source            DF        SS        MS        F       P
Regression         2   4283063   2141531   120.19   0.000
Residual Error    29    516727     17818
Total             31   4799790


Predicted Values for New Observations


New
Obs      Fit   SE Fit        95% CI                95% PI
  1   1431.7     24.6   (1381.4, 1481.9)   (1154.1, 1709.3)


Values of Predictors for New Observations


New
Obs   AGE   NUMBIDS
  1   150      10.0
```

Solution

a. Here, the key words *average* and *for all* imply that we want to estimate the mean of y, $E(y)$. We want a 95% confidence interval for $E(y)$ when $x_1 = 150$ years and $x_2 = 10$ bidders. A MINITAB printout for this analysis is shown in Figure 12.8. The confidence interval (highlighted under "**95% CI**") is (1381.4, 1481.9). Thus, we are 95% confident that the mean auction price for all 150-year-old clocks sold at an auction with 10 bidders lies between $1,381.4 and $1,481.9.

b. Now, we want to predict the auction price, y, for a single ("one") grandfather clock when $x_1 = 50$ years and $x_2 = 2$ bidders. Consequently, we desire a 95% prediction interval for y. However, before we form this prediction interval, we should check to make sure that the selected values of the independent variables, $x_1 = 50$ and $x_2 = 2$, are both reasonable and within their respective sample ranges. If you examine the sample data shown in Table 12.2 (p. 671), you will see that the range for age is $108 \le x_1 \le 194$ and the range for number of bidders is $5 \le x_2 \le 15$. Thus, both selected values fall well *outside* their respective ranges. Recall the *Caution* box in Section 11.8 (p. 640) warning about the dangers of using the model to predict y for a value of an independent variable that is not within the range of the sample data. This may lead to an unreliable prediction. If want to make such a prediction, we would need to collect additional data on clocks with these characteristics (i.e., $x_1 = 50$ years and $x_2 = 2$ bidders) and then refit the model.

■ ■ ■

Statistics in Action Revisited

A First-Order Model for Condominium Sale Price

The developer of a Florida condominium complex wants to build a model for the sales price (y) of a condo unit (recorded in hundreds of dollars) and to use the model to predict the prices of future units sold. In addition to sale price, the **CONDO** file contains data on six potential predictor variables for a sample of 209 units sold. (See Table SIA12.1 on p. 661) These independent variables are defined as follows:

x_1 = floor location (i.e., floor height) of the unit $(1, 2, 3, \ldots,$ or 8)

x_2 = distance, in units, from the elevator $(1, 2, 3 \ldots,$ or 15)

x_3 = 1 if ocean view, 0 if non–ocean view

x_4 = 1 if end unit, 0 if not an end unit

x_5 = 1 if furnished unit, 0 if unfurnished

x_6 = 1 if sold at public auction, 0 if sold at developer's price

As our initial model, we consider the first-order regression model

$$E(y) = \beta_0 + \beta_1 x_1 + \beta_2 x_2 + \beta_3 x_3 + \beta_4 x_4 + \beta_5 x_5 + \beta_6 x_6$$

The MINITAB printout for the regression analysis is shown in Figure SIA12.2. The global F-statistic ($F = 49.99$) and

associated p-value (.000) shown on the printout indicate that the overall model is statistically useful for predicting auction price. The value of adjusted R^2, however, indicates that the model explains only about 59% of the sample variation in price and the standard deviation of the model ($s = 21.8$) implies that the model can predict price to within about $2s = 43.6$ hundred dollars (i.e., \$4,360). While the model appears to be "statistically" useful for predicting auction price, the moderate adjusted-R^2 value and relatively large $2s$ value indicate that the model may not yield accurate predictions.

[*Note:* Not all of the independent variables have statistically significant t-values. However, we caution against dropping the insignificant variables from the model at this stage. One reason (discussed in Section 12.3) is that performing a large number of t-tests will yield an inflated probability of at least one Type I error. In later sections of this chapter, we develop other reasons for why the multiple t-test approach is not a good strategy for determining which independent variables to keep in the model.]

The MINITAB printout shown in Figure SIA12.3 gives a 95% prediction interval for auction price and a 95% confidence interval for the mean price for the following x-values:

$x_1 = 5$ (i.e., unit on the 5th floor)

$x_2 = 9$ (i.e., distance of 9 units from the elevator)

$x_3 = 1$ (i.e., ocean view)

$x_4 = 0$ (i.e., not an end unit)

Figure SIA12.2

MINITAB Regression Output for First-Order Model of Condominium Unit Sale Price

```
The regression equation is
PRICE100 = 187 - 1.16 FLOOR + 0.917 DISTANCE + 48.3 VIEW - 23.6 END
           + 6.95 FURNISH - 28.0 AUCTION


Predictor      Coef   SE Coef       T       P
Constant    186.990     4.693   39.84   0.000
FLOOR       -1.1599    0.7636   -1.52   0.130
DISTANCE     0.9168    0.3544    2.59   0.010
VIEW         48.250     3.281   14.70   0.000
END         -23.633     8.661   -2.73   0.007
FURNISH       6.945     3.243    2.14   0.033
AUCTION     -27.970     3.886   -7.20   0.000


S = 21.8163    R-Sq = 59.8%    R-Sq(adj) = 58.6%


Analysis of Variance

Source            DF      SS      MS      F       P
Regression         6  142750   23792  49.99   0.000
Residual Error   202   96142     476
Total            208  238893
```

$x_5 = 0$ (i.e., an unfurnished unit)
$x_6 = 1$ (i.e., a unit sold at public auction)

The 95% confidence interval of $(204.05, 215.39)$ implies that for all condo units with these x-values, the mean auction price falls between 204.05 and 215.39 hundred dollars, with 95% confidence. The 95% prediction interval of $(166.33, 253.11)$ implies that for an individual condo unit with these x-values, the auction price falls between 166.33 and 253.11 hundred dollars, with 95% confidence. Note the wide range of the prediction interval. This is due to the large magnitude of the model standard deviation, $s = 21.8$ hundred dollars. Again, although the model is deemed statistically useful for predicting auction price, it may not be "practically" useful. To reduce the magnitude of s, we will need to improve the model's predictive ability. (We consider such a model in the next Statistics in Action Revisited section.)

Figure SIA12.3
MINITAB Printout with 95% Confidence and Prediction Intervals

```
Predicted Values for New Observations

New
Obs     Fit  SE Fit         95% CI              95% PI
  1  209.72    2.87  (204.05, 215.39)  (166.33, 253.11)


Values of Predictors for New Observations

New
Obs  FLOOR  DISTANCE  VIEW      END   FURNISH  AUCTION
  1   5.00      9.00  1.00  0.000000  0.000000     1.00
```

Exercises 12.28–12.36

Understanding the Principles

12.28 Explain why we use $\hat{y}$ as an estimate of $E(y)$ and to predict y.

12.29 Which interval will be narrower, a 95% confidence interval for $E(y)$ or a 95% prediction interval for y? (Assume the values of the x's are the same for both intervals.)

Applying the Concepts—Basic

STREETVEN

12.30 Incomes of Mexican street vendors. Refer to the *World Development* (Feb. 1998) study of street vendors' earnings (y), Exercise 12.19 (p. 680). The MINITAB printout in the next column shows both a 95% prediction interval for y and a 95% confidence interval for $E(y)$ for a 45-year-old vendor who works 10 hours a day (i.e., for $x_1 = 45$ and $x_2 = 10$).

a. Interpret the 95% prediction interval for y in the words of the problem.

b. Interpret the 95% confidence interval for $E(y)$ in the words of the problem.

c. Note that the interval of part **a** is wider than the interval of part **b**. Will this always be true? Explain.

```
Predicted Values for New Observations

New
Obs   Fit  SE Fit       95% CI           95% PI
  1  3018     182  (2620, 3415)  (1760, 4275)


Values of Predictors for New Observations

New
Obs   Age  Hours
  1  45.0   10.0
```

SNOWGEESE

12.31 Snow geese feeding trial. Refer to the *Journal of Applied Ecology* study of the feeding habits of baby snow geese, Exercise 12.21 (p. 681). Recall that a first-order model was used to relate gosling weight change (y) to digestion efficiency (x_1) and acid-detergent fibre (x_2). The SAS printout below shows a confidence interval for $E(y)$ for the first 5 feeding trials.

a. Interpret the confidence interval for $E(y)$ for trial #1.

b. Interpret the confidence interval for $E(y)$ for trial #2.

```
               Dependent Variable: WTCHANGE

                    Output Statistics

                              Dependent  Predicted      Std Error
Obs  DIET   DIGEFF  ADFIBER    Variable      Value  Mean Predict      95% CL Mean
  1  Plants       0     28.5     -6.0000    -0.8678        1.1392    -3.1722    1.4365
  2  Plants     2.5     27.5     -5.0000    -0.4763        1.1038    -2.7090    1.7563
  3  Plants       5     27.5     -4.5000    -0.5427        0.9964    -2.5580    1.4726
  4  Plants       0     32.5           0    -2.6992        0.9482    -4.6170   -0.7813
  5  Plants       0       32      2.0000    -2.4703        0.9592    -4.4104   -0.5301
```

	REDSHIFT	LINEFLUX	LINELUM	AB1450	RFEWIDTH	LCL_95	UCL_95
1	2.81	-13.48	45.29	19.50	117	101.29	172.22
2	3.07	-13.73	45.13	19.65	82	59.62	125.89
3	3.45	-13.87	45.11	18.93	33	-35.09	37.04
4	3.19	-13.27	45.63	18.59	92	63.76	139.25
5	3.07	-13.56	45.30	19.59	114	90.69	158.57

12.32 Deep space survey of quasars. Refer to *The Astronomical Journal* study of quasars, Exercise 12.22 (p. 681). Recall that a first-order model was used to relate a quasar's equivalent width (y) to redshift (x_1), line flux (x_2), line luminosity (x_3), and $AB_{1450}(x_4)$. A portion of the SPSS spreadsheet showing 95% prediction intervals for y for the first five observations in the data set is reproduced above. Interpret the interval corresponding to the fifth observation.

Applying the Concepts—Intermediate

⊚ DDT
12.33 Study of contaminated fish. Refer to Exercise 12.20 (p. 680) and the U.S. Army Corps of Engineers data on contaminated fish. You fit the first-order model, $E(y) = \beta_0 + \beta_1 x_1 + \beta_2 x_2 + \beta_3 x_3$, to the data, where

y = DDT level (parts per million), x_1 = number of miles upstream, x_2 = length (centimeters), and x_3 = weight (in grams). Predict, with 95% confidence, the DDT level of a fish caught 100 miles upstream with a length of 40 centimeters and a weight of 800 grams. Interpret the result.

12.34 California rain levels. An article published in *Geography* (July 1980) used multiple regression to predict annual rainfall levels in California. Data on the average annual precipitation (y), altitude (x_1), latitude (x_2), and distance from the Pacific coast (x_3) for 30 meteorological stations scattered throughout California are listed in the table. Consider the first-order model $y = \beta_0 + \beta_1 x_1 + \beta_2 x_2 + \beta_3 x_3 + \varepsilon$.
a. Fit the model to the data and give the least squares prediction equation.

⊚ CALIRAIN

Station	Avg. Annual Precipitation y (inches)	Altitude x_1 (feet)	Latitude x_2 (degrees)	Distance from Coast x_3 (miles)
1. Eureka	39.57	43	40.8	1
2. Red Bluff	23.27	341	40.2	97
3. Thermal	18.20	4152	33.8	70
4. Fort Bragg	37.48	74	39.4	1
5. Soda Springs	49.26	6752	39.3	150
6. San Francisco	21.82	52	37.8	5
7. Sacramento	18.07	25	38.5	80
8. San Jose	14.17	95	37.4	28
9. Giant Forest	42.63	6360	36.6	145
10. Salinas	13.85	74	36.7	12
11. Fresno	9.44	331	36.7	114
12. Pt. Piedras	19.33	57	35.7	1
13. Pasa Robles	15.67	740	35.7	31
14. Bakersfield	6.00	489	35.4	75
15. Bishop	5.73	4108	37.3	198
16. Mineral	47.82	4850	40.4	142
17. Santa Barbara	17.95	120	34.4	1
18. Susanville	18.20	4152	40.3	198
19. Tule Lake	10.03	4036	41.9	140
20. Needles	4.63	913	34.8	192
21. Burbank	14.74	699	34.2	47
22. Los Angeles	15.02	312	34.1	16
23. Long Beach	12.36	50	33.8	12
24. Los Banos	8.26	125	37.8	74
25. Blythe	4.05	268	33.6	155
26. San Diego	9.94	19	32.7	5
27. Daggett	4.25	2105	34.1	85
28. Death Valley	1.66	−178	36.5	194
29. Crescent City	74.87	35	41.7	1
30. Colusa	15.95	60	39.2	91

Source: Taylor, P. J. "A pedagogic application of multiple regression analysis." *Geography*, July 1980, Vol. 65, pp. 203–212.

b. Is there evidence that the first-order model is useful for predicting annual precipitation y? Test using $\alpha = .05$.

c. Find a 95% prediction interval for y for the Giant Forest meteorological station (station #9). Interpret the interval.

⊙ WATEROIL

12.35 Removing water from oil. Refer to Exercise 12.23 (p. 682). The researchers concluded that "in order to break a water-oil mixture with the lowest possible voltage, the volume fraction of the disperse phase (x_1) should be high, while the salinity (x_2) and the amount of surfactant (x_5) should be low." Use this information and the first-order model of Exercise 12.23 to find a 95% prediction interval for this "low" voltage y. Interpret the interval.

12.36 Boiler drum production. In a production facility, an accurate estimate of hours needed to complete a task is crucial to management in making such decisions as the proper number of workers to hire, an accurate deadline to quote a client, or cost-analysis decisions regarding

budgets. A manufacturer of boiler drums wants to use regression to predict the number of hours needed to erect the drums in future projects. To accomplish this, data for 35 boilers were collected. In addition to hours (y), the variables measured were boiler capacity $(x_1 = \text{lb/hr})$, boiler design pressure $(x_2 = \text{pounds per square inch or psi})$, boiler type $(x_3 = 1$ if industry field erected, 0 if utility field erected), and drum type $(x_4 = 1$ if steam, 0 if mud). The data are saved in the **BOILERS** file. (Selected observations are shown in the table.)

a. Fit the model $E(y) = \beta_0 + \beta_1 x_1 + \beta_2 x_2 + \beta_3 x_3 + \beta_4 x_4$ to the data and give the prediction equation.

b. Conduct a test for the global utility of the model. Use $\alpha = .01$.

c. Find a 95% confidence interval for $E(y)$ when $x_1 = 150,000$, $x_2 = 500$, $x_3 = 1$, and $x_4 = 0$. Interpret the result.

d. What type of interval, would you use if you want to estimate the average hours required to erect all industrial, mud boilers with a capacity of 150,000 lb/hr and a design pressure of 500 psi?

⊙ BOILERS (First and last 5 observations shown)

Hours y	Boiler Capacity x_1	Design Pressure x_2	Boiler Type x_3	Drum Type x_4
3,137	120,000	375	1	1
3,590	65,000	750	1	1
4,526	150,000	500	1	1
10,825	1,073,877	2,170	0	1
4,023	150,000	325	1	1
⋮	⋮	⋮	⋮	⋮
4,206	441,000	410	1	0
4,006	441,000	410	1	0
3,728	627,000	1,525	0	0
3,211	610,000	1,500	0	0
1,200	30,000	325	1	0

Source: Dr. Kelly Uscategui, University of Connecticut.

12.5 Model Building: Interaction Models

In Section 12.2, we demonstrated the relationship between $E(y)$ and the independent variables in a first-order model. When $E(y)$ is graphed against any one variable (say, x_1) for fixed values of the other variables, the result is a set of *parallel* straight lines (see Figure 12.3, p. 668). When this situation occurs (as it always does for a first-order model), we say that the relationship between $E(y)$ and any one independent variable *does not depend* on the values of the other independent variables in the model.

However, if the relationship between $E(y)$ and x_1 does, in fact, depend on the values of the remaining x's held fixed, then the first-order model is not appropriate for predicting y. In this case, we need another model that will take into account this dependence. Such a model includes the *cross products* of two or more x's.

For example, suppose that the mean value $E(y)$ of a response y is related to two quantitative independent variables, x_1 and x_2, by the model

$$E(y) = 1 + 2x_1 - x_2 + x_1 x_2$$

A graph of the relationship between $E(y)$ and x_1 for $x_2 = 0, 1,$ and 2 is displayed in Figure 12.9.

Figure 12.9
Graphs of
$1 + 2x_1 - x_2 + x_1x_2$
for $x_2 = 0, 1, 2$

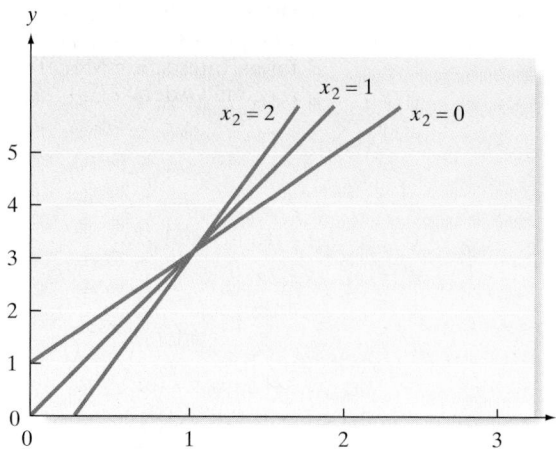

Note that the graph shows three nonparallel straight lines. You can verify that the slopes of the lines differ by substituting each of the values $x_2 = 0$, 1, and 2 into the equation. For $x_2 = 0$:

$$E(y) = 1 + 2x_1 - (0) + x_1(0) = 1 + 2x_1 \qquad \text{(slope} = 2)$$

For $x_2 = 1$:

$$E(y) = 1 + 2x_1 - (1) + x_1(1) = 3x_1 \qquad \text{(slope} = 3)$$

For $x_2 = 2$:

$$E(y) = 1 + 2x_1 - (2) + x_1(2) = -1 + 4x_1 \qquad \text{(slope} = 4)$$

Note that the slope of each line is represented by $\beta_1 + \beta_3x_2 = 2 + x_2$. Thus, the effect on $E(y)$ of a change in x_1 (i.e., the slope) now *depends* on the value of x_2. When this situation occurs, we say that x_1 and x_2 **interact**. The cross-product term, x_1x_2, is called an **interaction term**, and the model $E(y) = \beta_0 + \beta_1x_1 + \beta_2x_2 + \beta_3x_1x_2$ is called an **interaction model** with two quantitative variables.

> **An Interaction Model Relating $E(y)$ to Two Quantitative Independent Variables**
>
> $$E(y) = \beta_0 + \beta_1x_1 + \beta_2x_2 + \beta_3x_1x_2$$
>
> where
>
> $(\beta_1 + \beta_3x_2)$ represents the change in $E(y)$ for every 1-unit increase in x_1, holding x_2 fixed
>
> $(\beta_2 + \beta_3x_1)$ represents the change in $E(y)$ for every 1-unit increase in x_2, holding x_1 fixed

A three-dimensional graph (generated by computer) of an interaction model in two quantitative x's is shown in Figure 12.10. Unlike the flat planar surface displayed in Figure 12.4 (p. 669), the interaction model traces a ruled surface (twisted plane) in three-dimensional space. If we slice the twisted plane at a fixed value of x_2, we obtain a straight line relating $E(y)$ to x_1; however, the slope of the line will change as we change the value of x_2.

EXAMPLE 12.6   EVALUATING AN INTERACTION MODEL

Problem   Refer to Examples 12.3 and 12.4. Suppose the collector of grandfather clocks, having observed many auctions, believes that the *rate of increase* of the auction price with age will be driven upward by a large number of bidders. Thus, instead of a relationship like that shown in Figure 12.11a, in which the rate of increase in price with age is

Figure 12.10

Computer-Generated Graph for an Interaction Model (Second Order)

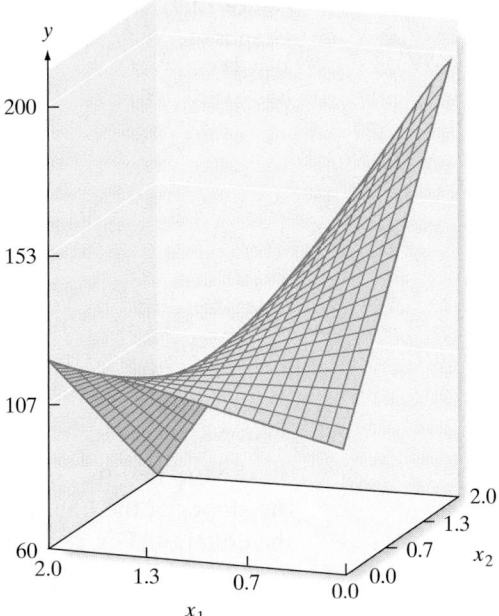

the same for any number of bidders, the collector believes the relationship is like that shown in Figure 12.11b. Note that as the number of bidders increases from 5 to 15, the slope of the price versus age line increases.

Consequently, the interaction model is proposed:

$$y = \beta_0 + \beta_1 x_1 + \beta_2 x_2 + \beta_3 x_1 x_2 + \varepsilon$$

The 32 data points listed in Table 12.2 were used to fit the model with interaction. A portion of the MINITAB printout is shown in Figure 12.12.

a. Test the overall utility of the model using the global F-test at $\alpha = .05$.

b. Test the hypothesis (at $\alpha = .05$) that the price-age slope increases as the number of bidders increases—that is, that age and number of bidders, x_2, interact positively.

c. Estimate the change in auction price of a 150-year-old grandfather clock, y, for each additional bidder.

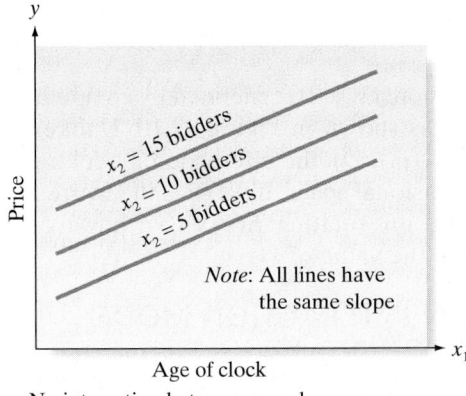

a. No interaction between x_1 and x_2

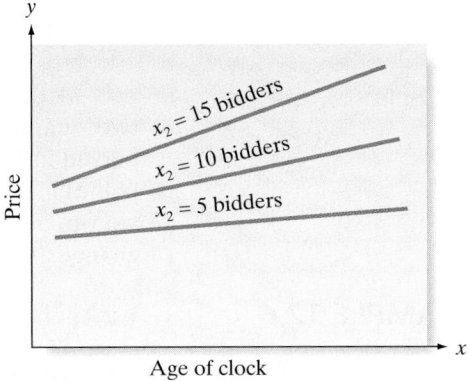

b. Interaction between x_1 and x_2

Figure 12.11

Examples of No-Interaction and Interaction Models

Figure 12.12

MINITAB Printout for
Interaction Model of
Auction Price

Regression Analysis: PRICE versus AGE, NUMBIDS, AGEBID

```
The regression equation is
PRICE = 320 + 0.88 AGE - 93.3 NUMBIDS + 1.30 AGEBID


Predictor      Coef   SE Coef      T      P
Constant      320.5     295.1   1.09  0.287
AGE           0.878     2.032   0.43  0.669
NUMBIDS      -93.26     29.89  -3.12  0.004
AGEBID       1.2978    0.2123   6.11  0.000


S = 88.9145    R-Sq = 95.4%    R-Sq(adj) = 94.9%


Analysis of Variance

Source           DF        SS       MS       F      P
Regression        3   4578427  1526142  193.04  0.000
Residual Error   28    221362     7906
Total            31   4799790
```

Solution   **a.** The global F-test is used to test the null hypothesis

$$H_0: \beta_1 = \beta_2 = \beta_3 = 0$$

The test statistic and p-value of the test (highlighted on the MINITAB printout)
are $F = 193.04$ and $p = 0$, respectively. Since $\alpha = .05$ exceeds the p-value,
there is sufficient evidence to conclude that the model fit is a statistically useful
predictor of auction price, y.

b. The hypotheses of interest to the collector concern the interaction parameter
β_3. Specifically,

$$H_0: \beta_3 = 0$$
$$H_a: \beta_3 > 0$$

Since we are testing an individual β parameter, a t-test is required. The test statistic and two-tailed p-value (highlighted on the printout) are $t = 6.11$ and $p = 0$,
respectively. The upper-tailed p-value, obtained by dividing the two-tailed p-value
in half, is $0/2 = 0$. Since $\alpha = .05$ exceeds the p-value, the collector can reject H_0
and conclude that the rate of change of the mean price of the clocks with age
increases as the number of bidders increases; that is, x_1 and x_2 interact positively.
Thus, it appears that the interaction term should be included in the model.

c. To estimate the change in auction price, y, for every 1-unit increase in number
of bidders, x_2, we need to estimate the slope of the line relating y to x_2 when
the age of the clock, x_1, is 150 years old. An analyst who is not careful may
estimate this slope as $\hat{\beta}_2 = -93.26$. Although the coefficient of x_2 is negative,
this does *not* imply that auction price decreases as the number of bidders
increases. Since interaction is present, the rate of change (slope) of mean
auction price with the number of bidders *depends* on x_1, the age of the
clock. Thus, the estimated rate of change of y for a unit increase in x_2 (one new
bidder) for a 150-year-old clock is

$$\text{Estimated } x_2 \text{ slope} = \hat{\beta}_2 + \hat{\beta}_3 x_1 = -93.26 + 1.30(150) = 101.74$$

In other words, we estimate that the auction price of a 150-year-old clock will
increase by about \$101.74 for every additional bidder.

Look Back Although the rate of increase will vary as x_1 is changed, it will remain positive for the range of values of x_1 included in the sample. Extreme care is needed in interpreting the signs and sizes of coefficients in a multiple regression model.

Now Work   *Exercise 12.42*

■ ■ ■

Example 12.6 illustrates an important point about conducting t-tests on the β parameters in the interaction model. The "most important" β parameter in this model is the interaction β, β_3. [Note that this β is also the one associated with the highest-order term in the model, $x_1x_2.*$] Consequently, we will want to test $H_0\colon \beta_3 = 0$ after we have determined that the overall model is useful for predicting y. Once interaction is detected (as in Example 12.6), however, tests on the first-order terms x_1 and x_2 should *not* be conducted since they are meaningless tests; the presence of interaction implies that both x's are important.

Caution

Once interaction has been deemed important in the model $E(y) = \beta_0 + \beta_1x_1 + \beta_2x_2 + \beta_3x_1x_2$, do not conduct t-tests on the β coefficients of the first-order terms x_1 and x_2. These terms should be kept in the model regardless of the magnitude of their associated p-values shown on the printout.

We close this section with a comment. You will probably never know *a priori* whether interaction exists between two independent variables. Consequently, you will need to fit and test the interaction term to determine its importance.

Exercises 12.37–12.48

Understanding the Principles

12.37 Write an interaction model relating the mean value of y, $E(y)$ to
  a. two quantitative independent variables
  b. three quantitative independent variables [*Hint:* Include all possible two-way cross-product terms.]

12.38 If two variables, x_1 and x_2, do not interact, how would you describe their effect on the mean response, $E(y)$?

Learning the Mechanics

12.39 Suppose the true relationship between $E(y)$ and the quantitative independent variables x_1 and x_2 is

$$E(y) = 3 + x_1 + 2x_2 - x_1x_2$$

  a. Describe the corresponding three-dimensional response surface.
  b. Plot the linear relationship between y and x_2 for $x_2 = 0, 1, 2$, where $0 \le x_2 \le 5$.
  c. Explain why the lines you plotted in part **b** are not parallel.
  d. Use the lines you plotted in part **b** to explain how changes in the settings of x_1 and x_2 affect $E(y)$.

  e. Use your graph from part **b** to determine how much $E(y)$ changes when x_1 is changed from 2 to 0 and x_2 is simultaneously changed from 4 to 5.

12.40 Suppose you fit the interaction model

$$y = \beta_0 + \beta_1x_1 + \beta_2x_2 + \beta_3x_1x_2 + \varepsilon$$

to $n = 32$ data points and obtain the following results:

$$SS_{yy} = 479 \qquad SSE = 21 \qquad \hat{\beta}_3 = 10 \qquad s_{\hat{\beta}_3} = 4$$

  a. Find R^2 and interpret its value.
  b. Is the model adequate for predicting y? Test at $\alpha = .05$.
  c. Use a graph to explain the contribution of the x_1x_2 term to the model.
  d. Is there evidence that x_1 and x_2 interact? Test at $\alpha = .05$.

12.41 MINITAB was used to fit the model

$$y = \beta_0 + \beta_1x_1 + \beta_2x_2 + \beta_3x_1x_2 + \varepsilon$$

to $n = 15$ data points. The printout is shown on p. 695.
  a. What is the prediction equation for the response surface?

---

*The order of a term is equal to the sum of the exponents of the quantitative variables included in the term. Thus, when x_1 and x_2 are both quantitative variables, the cross product, x_1x_2, is a second-order term.

MINITAB Output
for Exercise 12.41

```
The regression equation is
Y = -2.55 + 3.82 X1 + 2.63 X2 -1.29 X1X2


Predictor      Coef   SE Coef      T       P
Constant    -2.550     1.142   -2.23   0.043
X1           3.815     0.529    7.22   0.000
X2           2.630     0.344    7.64   0.000
X1X2        -1.285     0.159   -8.06   0.000


S = 0.713      R-Sq = 85.6%    R-Sq(adj) = 81.6%


Analysis of Variance

Source            DF       SS       MS       F       P
Regression         3   33.149   11.050   21.75   0.000
Residual Error    11    5.587    0.508
Total             14   38.736
```

b. Describe the geometric form of the response surface of part **a**.
c. Plot the prediction equation for the case when $x_2 = 1$. Do this twice more on the same graph for the cases when $x_2 = 3$ and $x_2 = 5$.
d. Explain what it means to say that x_1 and x_2 interact. Explain why your graph of part **c** suggests that x_1 and x_2 interact.
e. Specify the null and alternative hypotheses you would use to test whether x_1 and x_2 interact.
f. Conduct the hypothesis test of part **e** using $\alpha = .01$.

Applying the Concepts—Basic

12.42 Incomes of Mexican street vendors. Refer to the
NW *World Development* (Feb. 1998) study of street vendors in the city of Puebla, Mexico, Exercise 12.19 (p. 680). Recall that the vendors' mean annual earnings, $E(y)$, was modeled as a first-order function of age (x_1) and hours worked (x_2). Now, consider the interaction model $E(y) = \beta_0 + \beta_1 x_1 + \beta_2 x_2 + \beta_3 x_1 x_2$.
a. Find the least squares prediction equation.
b. What is the estimated slope relating annual earnings (y) to age (x_1) when number of hours worked (x_2) is 10? Interpret the result.
c. What is the estimated slope relating annual earnings (y) to hours worked (x_2) when age (x_1) is 40? Interpret the result.
d. Give the null hypothesis for testing whether age (x_1) and hours worked (x_2) interact.
e. Find the p-value of the test, part **d**.
f. Refer to part **e**. Give the appropriate conclusion in the words of the problem.

12.43 Frequency of drinking alcohol. To what degree do the attitudes of your peers influence your behavior? A study presented in *Social Psychology Quarterly* (Vol. 50, 1987) included a sample of $n = 143$ adult drinkers in an urban setting characterized by high physical availability of alcoholic beverages. The goal of the study was to build a model relating frequency of drinking alcoholic beverages, y, to attitude toward drinking (x_1) and social support (x_2). Consider the interaction model

$$E(y) = \beta_0 + \beta_1 x_1 + \beta_2 x_2 + \beta_3 x_1 x_2$$

a. Interpret the phrase "x_1 and x_2 interact" in terms of the problem.
b. Write the null and alternative hypotheses for determining whether attitude (x_1) and social support (x_2) interact.
c. The reported p-value for the test, part **b**, was $p < .001$. Interpret this result.

12.44 Psychology of waiting in line. While waiting in a long line for service (e.g., to use an ATM, or, at the post office), at some point you may decide to leave the queue. The *Journal of Consumer of Research* (Nov. 2003) published a study of consumer behavior while waiting in a queue. College students (sample size $n = 148$) were asked to imagine that they were waiting in line at a post office to mail a package and that the estimated waiting time is 10 minutes or less. After a 10-minute wait, students were asked about their level of negative feelings (annoyed, anxious) on a scale of 1 (strongly disagree) to 9 (strongly agree). Before answering, however, the students were informed about how many people were ahead of them and behind them in the line. The researchers used regression to relate negative feelings score (y) to number ahead in line (x_1) and number behind in line (x_2).
a. The researchers fit an interaction model to the data. Write the hypothesized equation of this model.
b. In the words of the problem, explain what it means to say that "x_1 and x_2 interact to effect y."
c. A t-test for the interaction β in the model resulted in a p-value greater than .25. Interpret this result.

d. From their analysis, the researchers concluded that "the greater the number of people ahead, the higher the negative feeling score" and "the greater the number of people behind, the lower the negative feeling score." Use this information to determine the signs of $\hat{\beta}_1$ and $\hat{\beta}_2$ in the model.

Applying the Concepts—Intermediate

12.45 Child abuse report. Licensed therapists are mandated, by law, to report child abuse by their clients. This requires the therapist to breach confidentiality and possibly lose the client's trust. A national survey of licensed psychotherapists was conducted to investigate clients' reactions to legally mandated child-abuse reports. (*American Journal of Orthopsychiatry*, Jan. 1997.) The sample consisted of 303 therapists who had filed a child-abuse report against one of their clients. The researchers were interested in finding the best predictors of a client's reaction (y) to the report, where y is measured on a 30-point scale. (The higher the value, the more favorable the client's response to the report.) The independent variables found to have the most predictive power are listed here.

x_1: Therapist's age (years)
x_2: Therapist's gender (1 if male, 0 if female)
x_3: Degree of therapist's role strain (25-point scale)
x_4: Strength of client-therapist relationship (40-point scale)
x_5: Type of case (1 if family, 0 if not)
x_1x_2: Age × Gender interaction

a. Hypothesize a first-order model relating y to each of the five independent variables.
b. Give the null hypothesis for testing the contribution of x_4, strength of client-therapist relationship, to the model.
c. The test statistic for the test, part **b**, was $t = 4.408$ with an associated p-value of .001. Interpret this result.
d. The estimated β coefficient for the x_1x_2 interaction term was positive and highly significant ($p < .001$). According to the researchers, "this interaction suggests that . . . as the age of the therapist increased, . . . male therapists were less likely to get negative client reactions than were female therapists." Do you agree?
e. For this model, $R^2 = .2946$. Interpret this value.

12.46 Treatment for a parasitic worm. *Trichuristrichiura*, a parasitic worm, affects millions of school-age children each year, especially children from developing countries. A study was conducted to determine the effects of treatment of the parasite on school achievement in 407 school-age Jamaican children infected with the disease. (*Journal of Nutrition*, July 1995.) About half the children in the sample received the treatment, while the others received a placebo. Multiple regression was used to model spelling test score y, measured as number correct, as a function of the following independent variables:

$$\text{Treatment (T): } x_1 = \begin{cases} 1 & \text{if treatment} \\ 0 & \text{if placebo} \end{cases}$$

$$\text{Disease intensity (I): } x_2 = \begin{cases} 1 & \text{if more than 7,000 eggs} \\ & \text{per gram of stool} \\ 0 & \text{if not} \end{cases}$$

a. Propose a model for $E(y)$ that includes interaction between treatment and disease intensity.
b. The estimates of the β's in the model, part **a**, and the respective p-values for t-tests on the β's are given in the table. Is there sufficient evidence to indicate that the effect of the treatment on spelling score depends on disease intensity? Test using $\alpha = .05$.

Variable	β Estimate	p-Value
Treatment (x_1)	−.1	.62
Intensity (x_2)	−.3	.57
T × I(x_1x_2)	1.6	.02

c. Based on the result, part **b**, explain why the analyst should avoid conducting t-tests for the treatment (x_1) and intensity (x_2) β's or interpreting these β's individually.

12.47 Factors that impact an auditor's judgement. A study was conducted to determine the effects of linguistic delivery style and client credibility on auditors' judgments. (*Advances in Accounting and Behavioral Research*, 2003.) Two hundred auditors from Big 5 accounting firms were asked to assume that he or she was an audit team supervisor of a new manufacturing client and was performing an analytical review of the client's financial statement. The researchers gave the auditors different information on the client's credibility and linguistic delivery style of the client's explanation. Each auditor then provided an assessment of the likelihood that the client-provided explanation accounts for the fluctuation in the financial statement. The three variables of interest, credibility (x_1), linguistic delivery style (x_2), and likelihood (y), were all measured on a numerical scale. Regression analysis was used to fit the interaction model, $y = \beta_0 + \beta_1x_1 + \beta_2x_2 + \beta_3x_1x_2 + \varepsilon$. The results are summarized in the table on p. 697.

a. Interpret the phrase "client credibility and linguistic delivery style interact" in the words of the problem.
b. Give the null and alternative hypotheses for testing the overall adequacy of the model.
c. Conduct the test, part **b**, using the information in the table.
d. Give the null and alternative hypotheses for testing whether client credibility and linguistic delivery style interact.

Results for Exercise 12.47

	Beta Estimate	Std Error	t-statistic	p-value
Constant	15.865	10.980	1.445	0.150
Client credibility (x_1)	0.037	0.339	0.110	0.913
Linguistic delivery Style (x_2)	−0.678	0.328	−2.064	0.040
Interaction (x_1x_2)	0.036	0.009	4.008	<0.005

F-statistic = 55.35 ($p < 0.0005$); Adjusted $R^2 = .450$

e. Conduct the test, part **d**, using the information in the table.

f. The researchers estimated the slope of the Likelihood–Linguistic delivery style line at a low level of client credibility ($x_1 = 22$). Obtain this estimate and interpret it in the words of the problem.

g. The researchers also estimated the slope of the Likelihood–Linguistic delivery style line at a high level of client credibility ($x_1 = 46$). Obtain this estimate and interpret it in the words of the problem.

12.48 Media coverage of a war. Does extensive media coverage of a military crisis influence public opinion on how to respond to the crisis? Political scientists at UCLA researched this question and reported their results in *Communication Research* (June 1993). The military crisis of interest was the 1990 Persian Gulf War, precipitated by Iraqi leader Saddam Hussein's invasion of Kuwait. The researchers used multiple regression analysis to model the level y of support Americans had for a military (rather than a diplomatic) response to the crisis. Values of y ranged from 0 (preference for a diplomatic response) to 4 (preference for a military response). The following independent variables were used in the model:

x_1 = Level of TV news exposure in a selected week (number of days)

x_2 = Knowledge of seven political figures (1 point for each correct answer)

x_3 = Gender (1 if male, 0 if female)

x_4 = Race (1 if nonwhite, 0 if white)

x_5 = Partisanship (0–6 scale, where 0 = strong Democrat and 6 = strong Republican)

x_6 = Defense spending attitude (1–7 scale, where 1 = greatly decrease spending and 7 = greatly increase spending)

x_7 = Education level (1–7 scale, where 1 = less than eight grades and 7 = college)

Data from a survey of 1,763 U.S. citizens were used to fit the model

$$E(y) = \beta_0 + \beta_1x_1 + \beta_2x_2 + \beta_3x_3 + \beta_4x_4 + \beta_5x_5 + \beta_6x_6 + \beta_7x_7 + \beta_8x_2x_3 + \beta_9x_2x_4$$

The regression results are shown in the table at the bottom of the page.

a. Interpret the β estimate for the variable x_1, TV news exposure.

b. Conduct a test to determine whether an increase in TV news exposure is associated with an increase in support for a military resolution of the crisis. Use $\alpha = .05$.

c. Is there sufficient evidence to indicate that the relationship between support for a military resolution (y) and gender (x_1) depends on political knowledge (x_2)? Test using $\alpha = .05$.

d. Is there sufficient evidence to indicate that the relationship between support for a military resolution (y) and race (x_4) depends on political knowledge (x_2)? Test using $\alpha = .05$.

e. The coefficient of determination for the model was $R^2 = .194$. Interpret this value.

f. Use the value of R^2, part **e**, to conduct a global test for model utility. Use $\alpha = .05$.

Variable	β Estimate	Standard Error	Two-Tailed p-Value
TV news exposure (x_1)	.02	.01	.03
Political knowledge (x_2)	.07	.03	.03
Gender (x_3)	.67	.11	<.001
Race (x_4)	−.76	.13	<.001
Partisanship (x_5)	.07	.01	<.001
Defense spending (x_6)	.20	.02	<.001
Education (x_7)	.07	.02	<.001
Knowledge × Gender (x_2x_3)	−.09	.04	.02
Knowledge × Race (x_2x_4)	.10	.06	.08

Source: Iyengar, S., and Simon, A. "News coverage of the Gulf Crisis and public opinion." *Communication Research*, Vol. 20, No. 3, June 1993, p. 380 (Table 2).

12.6 Model Building: Quadratic and Other Higher-Order Models

All of the models discussed in the previous sections proposed straight-line relationships between $E(y)$ and each of the independent variables in the model. In this section, we consider models that allow for curvature in the relationships. Each of these models is a **second-order model** because it will include an x^2-term.

First, we consider a model that includes only one independent variable x. The form of this model, called the **quadratic model**, is

$$y = \beta_0 + \beta_1 x + \beta_2 x^2 + \varepsilon$$

The term involving x^2, called a **quadratic term** (or **second-order term**), enables us to hypothesize curvature in the graph of the response model relating y to x. Graphs of the quadratic model for two different values of β_2 are shown in Figure 12.13. When the curve opens upward, the sign of β_2 is positive (see Figure 12.13a); when the curve opens downward, the sign of β_2 is negative (see Figure 12.13b).

Figure 12.13
Graphs for Two Quadratic Models

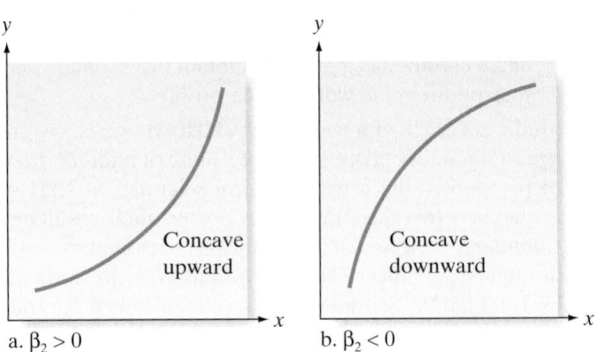

a. $\beta_2 > 0$ b. $\beta_2 < 0$

> **A Quadratic (Second-Order) Model in a Single Quantitative Independent Variable**
>
> $$E(y) = \beta_0 + \beta_1 x + \beta_2 x^2$$
>
> where
>
> β_0 is the y-intercept of the curve
> β_1 is a shift parameter
> β_2 is the rate of curvature

EXAMPLE 12.7   ANALYZING A QUADRATIC MODEL

Problem   In all-electric homes, the amount of electricity expended is of interest to consumers, builders, and groups involved with energy conservation. Suppose we wish to investigate the monthly electrical usage, y, in all-electric homes and its relationship to the size, x, of the home. Moreover, suppose we think that monthly electrical usage in all-electric homes is related to the size of the home by the quadratic model

$$y = \beta_0 + \beta_1 x + \beta_2 x^2 + \varepsilon$$

To fit the model, the values of y and x are collected for 10 homes during a particular month. The data are shown in Table 12.3.

a.   Construct a scatterplot for the data. Is there evidence to support the use of a quadratic model?

b.   Use the method of least squares to estimate the unknown parameters β_0, β_1, and β_2 in the quadratic model.

⊙ **ELECTRIC**

TABLE 12.3   Home Size–Electrical Usage Data

Size of Home x (sq. ft)	Monthly Usage y (kilowatt-hours)
1,290	1,182
1,350	1,172
1,470	1,264
1,600	1,493
1,710	1,571
1,840	1,711
1,980	1,804
2,230	1,840
2,400	1,956
2,930	1,954

c. Graph the prediction equation and assess how well the model fits the data, both visually and numerically.

d. Interpret the β estimates.

e. Is the overall model useful (at $\alpha = .01$) for predicting electrical usage y?

f. Is there sufficient evidence of concave downward curvature in the electrical usage–home size relationship? Test using $\alpha = .01$.

Solution   **a.** A scattergram for the data of Table 12.3, produced using MINITAB, is shown in Figure 12.14. The figure illustrates that the electrical usage appears to increase in a curvilinear manner with the size of the home. This provides some support for the inclusion of the quadratic term x^2 in the model.

b. We used SAS to fit the model to the data in Table 12.3. Part of the SAS regression output is displayed in Figure 12.15. The least squares estimates of the β parameters (highlighted) are $\hat{\beta}_0 = -1,216.14$, $\hat{\beta}_1 = 2.39893$, and $\hat{\beta}_2 = -.00045$. Therefore, the equation that minimizes the SSE for the data is

$$\hat{y} = -1,216.14 + 2.3989x - .00045x^2$$

Figure 12.14

MINITAB Scatterplot for Electrical Usage Data

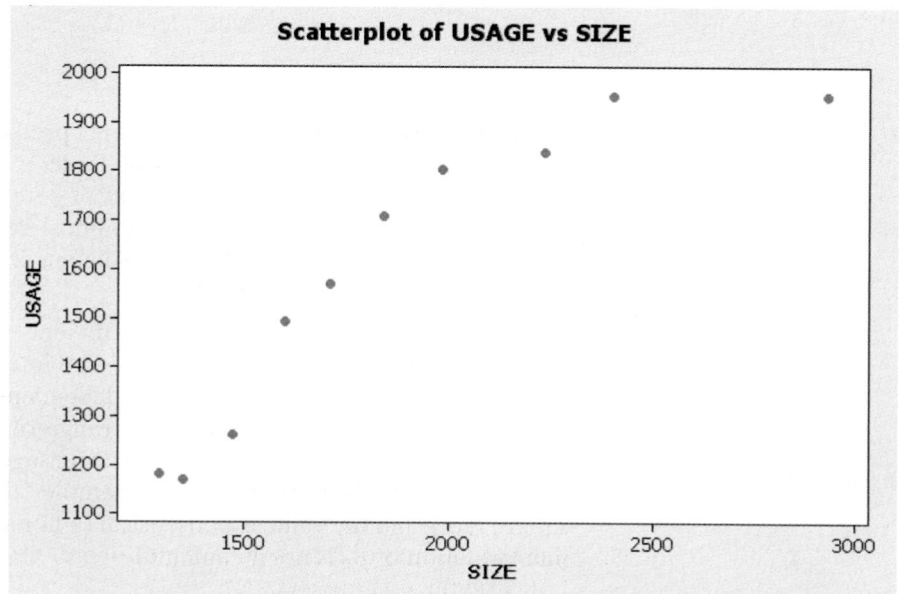

Figure 12.15

SAS Regression Output for Electrical Usage Model

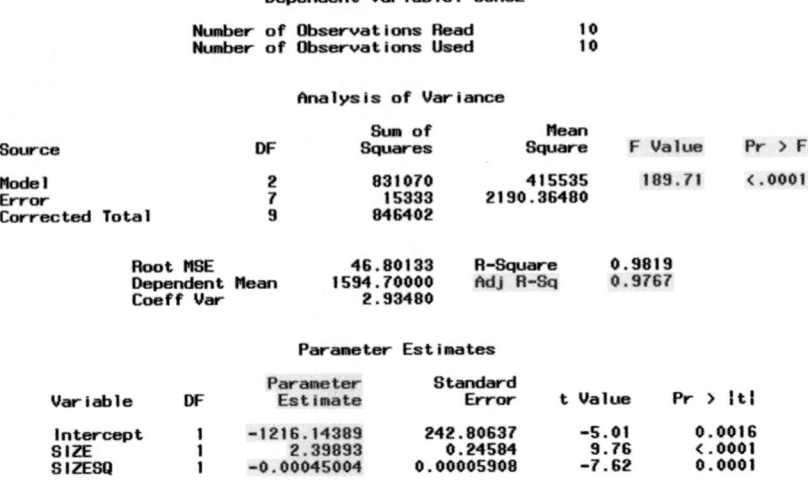

Dependent Variable: USAGE

Number of Observations Read   10
Number of Observations Used   10

Analysis of Variance

Source	DF	Sum of Squares	Mean Square	F Value	Pr > F
Model	2	831070	415535	189.71	<.0001
Error	7	15333	2190.36480		
Corrected Total	9	846402			

Root MSE	46.80133	R-Square	0.9819
Dependent Mean	1594.70000	Adj R-Sq	0.9767
Coeff Var	2.93480		

Parameter Estimates

Variable	DF	Parameter Estimate	Standard Error	t Value	Pr > |t|
Intercept	1	-1216.14389	242.80637	-5.01	0.0016
SIZE	1	2.39893	0.24584	9.76	<.0001
SIZESQ	1	-0.00045004	0.00005908	-7.62	0.0001

Figure 12.16

MINITAB Plot of Least Squares Model for Electrical Usage

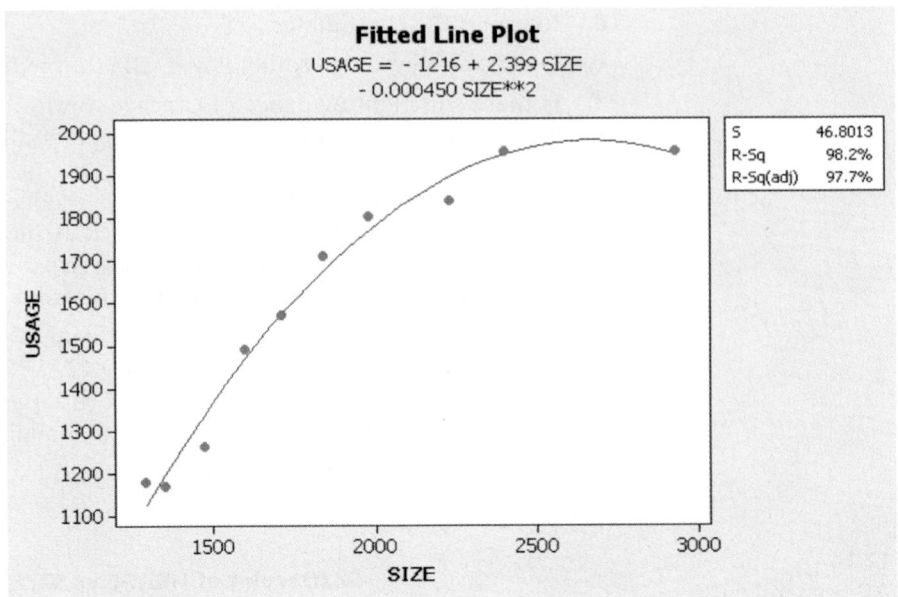

Fitted Line Plot
USAGE = - 1216 + 2.399 SIZE
- 0.000450 SIZE**2

S	46.8013
R-Sq	98.2%
R-Sq(adj)	97.7%

c. Figure 12.16 is a graph of the least squares prediction equation. Note that the graph provides a good fit to the data of Table 12.3. A numerical measure of fit is obtained with the adjusted coefficient of determination, R_a^2. From the SAS printout, $R_a^2 = .9767$. This implies that almost 98% of the sample variation in electrical usage (y) can be explained by the quadratic model (after adjusting for sample size and degrees of freedom).

d. The interpretation of the estimated coefficients in a quadratic model must be undertaken cautiously. First, the estimated y-intercept, $\hat{\beta}_0$, can be meaningfully interpreted only if the range of the independent variable includes zero—that is, if $x = 0$ is included in the sampled range of x. Although $\hat{\beta}_0 = -1,216.14$ seems to imply that the estimated electrical usage is negative when $x = 0$, this zero point is not in the range of the sample (the lowest value of x is 1,290 square feet), and the value is nonsensical (a home with 0 square feet); thus the interpretation of $\hat{\beta}_0$ is not meaningful.

Quadratic Regression

Using the TI-83 Graphing Calculator

I. Finding the quadratic regression equation

Step 1 *Enter the data*
Press **STAT** and select **1:Edit**
Note: If the list already contains data, clear the old data. Use the up arrow to highlight '**L1**' or '**L2**'.
Press **CLEAR ENTER**.
Use the arrow and **ENTER** keys to enter the data set into **L1** and **L2**.

Step 2 *Find the quadratic regression equation*
Press **STAT** and highlight **CALC**
Press **5** for **QuadReg**
Press **ENTER**
The screen will show the values for a, b, and c in the equation
If the diagnostics are on, the screen will also give the value for r^2
To turn the diagnostics feature on:
  1. Press **2nd 0** for **CATALOG**
  2. Press the down arrow key until **DiagnosticsOn** is highlighted
  3. Press **ENTER** twice

II. Graphing the quadratic curve with the scatterplot

Step 1 *Enter the data as shown in part I above*

Step 2 *Graph the regression curve with the scatterplot*
Press **STAT** and highlight **CALC**
Press **5** for **QuadReg** (Note: Don't press ENTER here because you want to store the regression equation in Y1)
Press **VARS**
Use the right arrow to highlight **Y-VARS**
Press **ENTER** to select **1:Function**
Press **ENTER** to select **1:Y1**
Press **ENTER**

Step 4 *View the scatterplot and regression line*

Press **ZOOM** and then press **9** to select **9:ZoomStat**

Example The figures below show a table of data entered on the TI-83, the quadratic regression equation and the graph obtained using the steps given above.

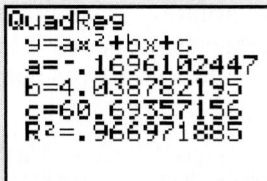
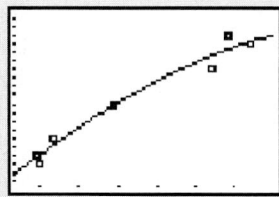

Figure 12.17

Potential Misuse of
Quadratic Model

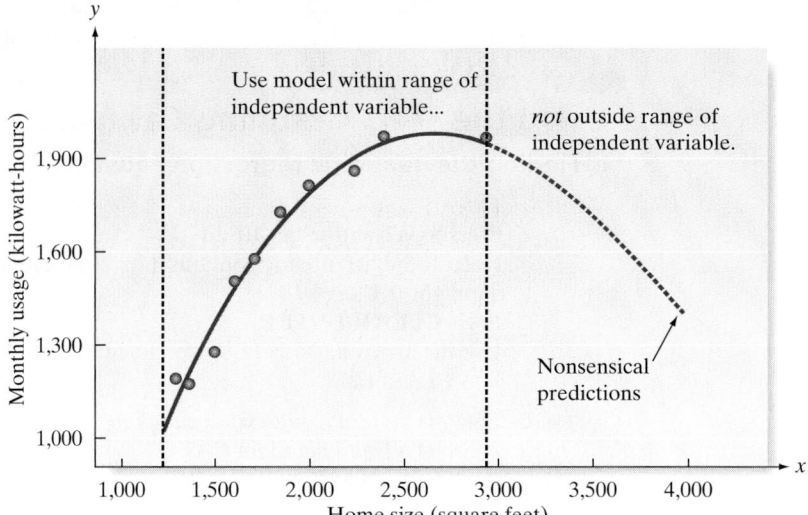

The estimated coefficient of x is $\hat{\beta}_1 = 2.3989$, but it no longer represents a slope in the presence of the quadratic term x^2.* The estimated coefficient of the first-order term x will not, in general, have a meaningful interpretation in the quadratic model.

The sign of the coefficient, $\hat{\beta}_2 = -.00045$, of the quadratic term, x^2, is the indicator of whether the curve is concave downward (mound shaped) or concave upward (bowlshaped). A negative $\hat{\beta}_2$ implies downward concavity, as in this example (Figure 12.16), and a positive $\hat{\beta}_2$ implies upward concavity. Rather than interpreting the numerical value of $\hat{\beta}_2$ itself, we utilize a graphical representation of the model, as in Figure 12.16, to describe the model.

Note that Figure 12.16 implies that the estimated electrical usage is leveling off as the home sizes increase beyond 2,500 square feet. In fact, the concavity of the model would lead to decreasing usage estimates if we were to display the model out to 4,000 square feet and beyond (see Figure 12.17). However, model interpretations are not meaningful outside the range of the independent variable, which has a maximum value of 2,930 square feet in this example. Thus, although the model appears to support the hypothesis that the *rate of increase* per square foot *decreases* for the home sizes near the high end of the sampled values, the conclusion that usage will actually begin to decrease for very large homes would be a *misuse* of the model, since no homes of 3,000 square feet or more were included in the sample.

e. To test whether the quadratic model is statistically useful, we conduct the global F-test:

$$H_0: \beta_1 = \beta_2 = 0$$
$$H_a: \text{At least one of the above coefficients is nonzero}$$

From the SAS printout, Figure 12.15, the test statistic is $F = 189.71$ with an associated p-value near 0. For any reasonable α, we reject H_0 and conclude that the overall model is a useful predictor of electrical usage, y.

*For students with knowledge of calculus, note that the slope of the quadratic model is the first derivative $\partial y / \partial x = \beta_1 + 2\beta_2 x$. Thus, the slope varies as a function of x, rather than the constant slope associated with the straight-line model.

f. Figure 12.16 shows concave downward curvature in the relationship between size of a home and electrical usage in the sample of 10 data points. To determine if this type of curvature exists in the population, we want to test

$$H_0: \beta_2 = 0 \text{ (no curvature in the response curve)}$$
$$H_a: \beta_2 < 0 \text{ (downward concavity exists in the response curve)}$$

The test statistic for testing β_2, highlighted on the printout, is $t = -7.62$ and the associated two-tailed p-value is .0001. Since this is a one-tailed test, the appropriate p-value is $.0001/2 = .00005$. Now $\alpha = .01$ exceeds this p-value. Thus, there is very strong evidence of downward curvature in the population; that is, electrical usage increases more slowly per square foot for large homes than for small homes.

Look Back Note that the SAS printout in Figure 12.15 also provides the t-test statistic and corresponding two-tailed p-values for the tests of $H_0: \beta_0 = 0$ and $H_0: \beta_1 = 0$. Since the interpretation of these parameters is not meaningful for this model, the tests are not of interest.

| Now Work | *Exercise 12.58* |

■ ■ ■

When two or more quantitative independent variables are included in a second-order model, we can incorporate squared terms for each x in the model, as well as the interaction between the two independent variables. A model that includes all possible second-order terms in two independent variables—called a **complete second-order model**—is given in the box.

Complete Second-Order Model with Two Quantitative Independent Variables

$$E(y) = \beta_0 + \beta_1 x_1 + \beta_2 x_2 + \beta_3 x_1 x_2 + \beta_4 x_1^2 + \beta_5 x_2^2$$

Comments on the Parameters

β_0: y-intercept, the value of $E(y)$ when $x_1 = x_2 = 0$

β_1, β_2: Changing β_1 and β_2 causes the surface to shift along the x_1- and x_2-axes

β_3: Controls the rotation of the surface

β_4, β_5: Signs and values of these parameters control the type of surface and the rates of curvature.

Three types of surfaces are produced by a second-order model:* a **paraboloid** that opens upward (Figure 12.18a), a paraboloid that opens downward (Figure 12.18b), and a **saddle-shaped surface** (Figure 12.18c).

A complete second-order model is the three-dimensional equivalent of a quadratic model in a single quantitative variable. Instead of tracing parabolas, it traces paraboloids and saddle-shaped surfaces. Since only a portion of the complete surface is used to fit the data, this model provides a very large variety of gently curving surfaces that can be used to fit data. It is a good choice for a model if you expect curvature in the response surface relating $E(y)$ to x_1 and x_2.

---

*The saddle-shaped surface (Figure 12.18c) is produced when $\beta_3^2 > 4\beta_4\beta_5$; the paraboloid opens upward (Figure 12.18a) when $\beta_4 + \beta_5 > 0$ and opens downward (Figure 12.18b) when $\beta_4 + \beta_5 < 0$.

Figure 12.18

Graphs for Three
Second-Order Surfaces

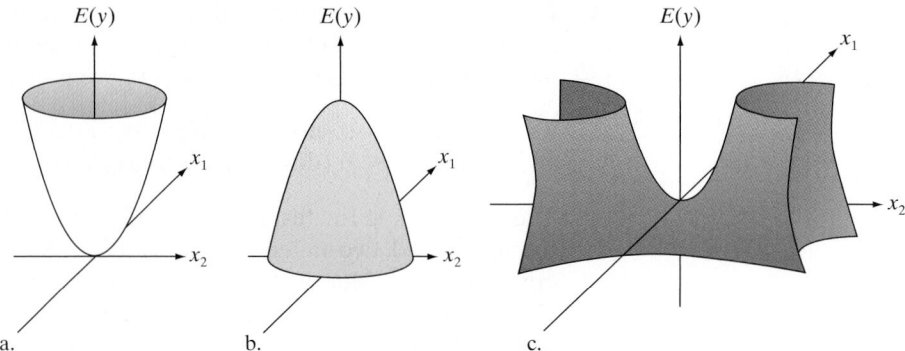

EXAMPLE 12.8   A SECOND-ORDER MODEL IN TWO x'S

Problem   A social scientist would like to relate the number of hours worked per week (outside
the home) by a married woman to the number of years of formal education she has
completed and the number of children in her family.

   a. Identify the dependent variable and the independent variables.

   b. Write the first-order model for this example.

   c. Modify the model in part b to include an interaction term.

   d. Write a complete second-order model for $E(y)$.

Solution   **a.** The dependent variable is

$$y = \text{Number of hours worked per week by a married woman}$$

The two independent variables, both quantitative in nature, are

$$x_1 = \text{Number of years of formal education completed by the woman}$$
$$x_2 = \text{Number of children in the family}$$

   b. The first-order model is

$$E(y) = \beta_0 + \beta_1 x_1 + \beta_2 x_2$$

This model would probably not be appropriate in this situation because x_1
and x_2 may interact and/or curvature terms corresponding to x_1^2 and x_2^2 may
be needed to obtain a good model for $E(y)$.

   c. Adding the interaction term, we obtain

$$E(y) = \beta_0 + \beta_1 x_1 + \beta_2 x_2 + \beta_3 x_1 x_2$$

This model should be better than the model in part **b**, since we have now
allowed for interaction between x_1 and x_2.

   d. The complete second-order model is

$$E(y) = \beta_0 + \beta_1 x_1 + \beta_2 x_2 + \beta_3 x_1 x_2 + \beta_4 x_1^2 + \beta_5 x_2^2$$

Since it would not be surprising to find curvature in the response surface,
the complete second-order model would be preferred to the models in
parts **b** and **c**. How can we tell whether the complete second-order model
really does provide better predictions of hours worked than the models in
parts **b** and **c**? The answers to these and similar questions are examined in
Section 12.9.

Most relationships between $E(y)$ and two or more quantitative independent variables are second-order and require the use of either the interactive or the complete second-order model to obtain a good fit to a data set. As in the case of a single quantitative independent variable, however, the curvature in the response surface may be very slight over the range of values of the variables in the data set. When this happens, a first-order model may provide a good fit to the data.

Exercises 12.49–12.63

Understanding the Principles

12.49 Write a second-order model relating the mean of y, $E(y)$, to
  a. one quantitative independent variable
  b. two quantitative independent variables
  c. three quantitative independent variables [*Hint:* Include all possible two-way cross-product terms and squared terms.]

12.50 In the model $E(y) = \beta_0 + \beta_1 x + \beta_2 x^2$
  a. Which β represents the y-intercept?
  b. Which β represents the shift?
  c. Which β represents the rate of curvature?

Learning the Mechanics

12.51 Suppose you fit the second-order model

$$y = \beta_0 + \beta_1 x + \beta_2 x^2 + \varepsilon$$

to $n = 25$ data points. Your estimate of β_2 is $\hat{\beta}_2 = .47$, and the estimated standard error of the estimate is .15.
  a. Test H_0: $\beta_2 = 0$ against H_a: $\beta_2 \neq 0$. Use $\alpha = .05$.
  b. Suppose you want to determine only whether the quadratic curve opens upward; that is, as x increases, the slope of the curve increases. Give the test statistic and the rejection region for the test for $\alpha = .05$. Do the data support the theory that the slope of the curve increases as x increases? Explain.

12.52 Suppose you fit the quadratic model

$$E(y) = \beta_0 + \beta_1 x + \beta_2 x^2$$

to a set of $n = 20$ data points and found $R^2 = .91$, $SS_{yy} = 29.94$, and SSE = 2.63.
  a. Is there sufficient evidence to indicate that the model contributes information for predicting y? Test using $\alpha = .05$.
  b. What null and alternative hypotheses would you test to determine whether upward curvature exists?
  c. What null and alternative hypotheses would you test to determine whether downward curvature exists?

12.53 Consider the following quadratic models:

$$
\begin{aligned}
(1)\ &y = 1 - 2x + x^2 \\
(2)\ &y = 1 + 2x + x^2 \\
(3)\ &y = 1 + x^2 \\
(4)\ &y = 1 - x^2 \\
(5)\ &y = 1 + 3x^2
\end{aligned}
$$

  a. Graph each of the quadratic models, side by side, on the same sheet of graph paper.
  b. What effect does the first-order term $(2x)$ have on the graph of the curve?
  c. What effect does the second-order term (x^2) have on the graph of the curve?

12.54 MINITAB (shown below) was used to fit the complete second-order model

$$E(y) = \beta_0 + \beta_1 x_1 + \beta_2 x_2 + \beta_3 x_1 x_2 + \beta_4 x_1^2 + \beta_5 x_2^2$$

```
The regression equation is
Y = -24.56 + 1.12 X1 + 27.99 X2 - 0.54 X1X2 - 0.004 X1SQ + 0.002 X2SQ


Predictor        Coef      SE Coef       T       P
Constant       -24.563       6.531    -3.76   0.001
X1              1.19848      0.1103    10.86   0.000
X2             27.988       79.489     0.35   0.727
X1X2           -0.5397       1.0338   -0.52   0.605
X1SQ           -0.0043       0.0004  -10.74   0.000
X2SQ            0.0020       0.0033    0.60   0.550


S = 2.762      R-Sq = 79.7%    R-Sq(adj) = 76.6%

Analysis of Variance

Source          DF      SS       MS       F       P
Regression       5   989.30   197.86   25.93   0.000
Residual Error  33   251.81     7.63
Total           38  1241.11
```

to $n = 39$ data points.

a. Is there sufficient evidence to indicate that at least one of the parameters, β_1, β_2, β_3, β_4, and β_5 is nonzero? Test using $\alpha = .05$.

b. Test H_0: $\beta_4 = 0$ against H_a: $\beta_4 \neq 0$. Use $\alpha = .01$.

c. Test H_0: $\beta_5 = 0$ against H_a: $\beta_5 \neq 0$. Use $\alpha = .01$.

d. Use graphs to explain the consequences of the tests in parts **b** and **c**.

Applying the Concepts—Basic

12.55 Testing tires for wear. Underinflated or overinflated tires can increase tire wear. A new tire was tested for wear at different pressures with the results shown in the table.

☉ TIRES

Pressure x (pounds per square inch)	Mileage y (thousands)
30	29
31	32
32	36
33	38
34	37
35	33
36	26

a. Plot the data on a scattergram.

b. If you were given only the information for $x = 30$, 31, 32, 33, what kind of model would you suggest? For $x = 33, 34, 35, 36$? For all the data?

12.56 Catalytic converters in cars. A quadratic model was applied to motor vehicle toxic emissions data collected between 1984 and 1999 in Mexico City. (*Environmental Science & Engineering*, Sept. 1, 2000.) The following equation was used to predict the percentage (y) of motor vehicles without catalytic converters in the Mexico City fleet for a given year (x): $\hat{y} = 325{,}790 - 321.67x + 0.794x^2$.

a. Explain why the value $\hat{\beta}_0 = 325{,}790$ has no practical interpretation.

b. Explain why the value $\hat{\beta}_1 = -321.67$ should not be interpreted as a slope.

c. Examine the value of $\hat{\beta}_2$ to determine the nature of the curvature (upward or downward) in the sample data.

d. The researchers used the model to estimate "that just after the year 2021 the fleet of cars with catalytic converters will completely disappear." Comment on the danger of using the model to predict y in the year 2021.

12.57 Oil evaporation study. The *Journal of Hazardous Materials* (July 1995) presented a literature review of models designed to predict oil spill evaporation. One model discussed in the article used boiling point (x_1) and API specific gravity (x_2) to predict the molecular weight (y) of the oil that is spilled. A complete second-order model for y was proposed.

a. Write the equation of the model.

b. Identify the terms in the model that allow for curvilinear relationships.

12.58 Violent behavior in children. Refer to the *Development Psychology* (Mar. 2003) study on the behavior of elementary school children, Exercise 8.25 (p. 381). The researchers used a quadratic equation to model the level (y) of aggressive fantasies experienced by a child as a function of the child's age (x). [*Note:* Level y was measured as an average of responses to six questions (e.g., "Do you sometimes have daydreams about hitting or hurting someone you don't like?"). Responses were measured on a scale ranging from 1 (*never*) to 5 (*always*).]

a. Write the equation of the hypothesized model for $E(y)$.

b. Research psychologists theorized that a child's aggressive fantasies will increase with age, but at a slower rate of acceleration for older children. Sketch the curve hypothesized by the researchers.

c. Set up H_0 and H_a for testing the researcher's theory.

d. The model was fit to data collected for over 11,000 elementary school children, with the following results: $\hat{y} = 1.926 + .097x - .003x^2$, standard error of $\hat{\beta}_2 = .001$. Compute the test statistic for the test, part **c**.

e. Use the result, part **d**, to make the appropriate conclusion using $\alpha = .05$.

Applying the Concepts—Intermediate

12.59 University of Florida fitness study. Does exercise improve the human immune system? An experiment was conducted by a physiologist at the University of Florida to determine whether such a relationship exists. Thirty subjects volunteered to participate in the study. The amount of immunoglobulin known as IgG (an indicator of long-term immunity) and the maximal oxygen uptake (a measure of aerobic fitness level) were recorded for each subject. The resulting data are given in the table on p. 707.

a. Construct a scattergram for the IgG–maximal oxygen uptake data.

b. Hypothesize a probabilistic model relating IgG to maximal oxygen uptake.

c. Fit the model to the data. Is there sufficient evidence to indicate that the model provides information for the prediction of IgG, y? Test using $\alpha = .05$.

d. Does the second-order term contribute information for the prediction of y? Test using $\alpha = .05$.

12.60 Revenues of popular movies. The *Internet Movie Database* (www.imdb.com) monitors the gross revenues for all major motion pictures. The 2nd table on p. 707 gives both the domestic (U.S. and Canada) and international gross revenues for a sample of 15 popular movies.

Table for Exercise 12.59

⊙ AEROBIC

Subject	IgG y	Maximum Oxygen Uptake x	Subject	IgG y	Maximum Oxygen Uptake x
1	881	34.6	16	1,660	52.5
2	1,290	45.0	17	2,121	69.9
3	2,147	62.3	18	1,382	38.8
4	1,909	58.9	19	1,714	50.6
5	1,282	42.5	20	1,959	69.4
6	1,530	44.3	21	1,158	37.4
7	2,067	67.9	22	965	35.1
8	1,982	58.5	23	1,456	43.0
9	1,019	35.6	24	1,273	44.1
10	1,651	49.6	25	1,418	49.8
11	752	33.0	26	1,743	54.4
12	1,687	52.0	27	1,997	68.5
13	1,782	61.4	28	2,177	69.5
14	1,529	50.2	29	1,965	63.0
15	969	34.1	30	1,264	43.2

a. Write a first-order model for international gross revenues, y, as a function of domestic gross revenues, x.

b. Write a second-order model for international gross revenues, y, as a function of domestic gross revenues, x.

c. Construct a scatterplot for these data. Which of the models appears to be a better choice for explaining variation in international gross revenues?

d. Fit the model of part **b** to the data and investigate its usefulness. Is there evidence of a curvilinear relationship between international and domestic gross revenues? Test using $\alpha = .05$.

e. Based on your analysis in part **d**, which of the two models better explains the variation in international gross revenues?

12.61 Genetics of a brain disease. Spinocerebellar ataxia type 1 (SCA1) is an inherited neurodegenerative disorder characterized by dysfunction of the brain. From a DNA analysis of SCA1 chromosomes, researchers discovered the presence of repeat gene sequences. (*Cell Biology*, Feb. 1995.) In general, the more repeat sequences observed, the earlier the onset of the disease (in years of age). The scatterplot on p. 708 shows this relationship for data collected on 113 individuals diagnosed with SCA1.

a. Suppose you want to model the age y of onset of the disease as a function of number x of repeat gene sequences in SCA1 chromosomes. Propose a quadratic model for y.

b. Will the sign of β_2 in the model, part **a**, be positive or negative? Base your decision on the results shown in the scatterplot.

⊙ IMDB

Movie Title (year)	Domestic Gross ($ millions)	International Gross ($ millions)
Titanic (1997)	600.7	1,234.6
E.T. (1982)	439.9	321.8
Jurassic Park (1993)	356.8	563.0
Lion King (1994)	328.4	455.0
Harry Potter and the Sorcerer's Stone (2001)	317.6	651.1
Sixth Sense (1999)	293.5	368.0
Jaws (1975)	260.0	210.6
Ghost (1990)	217.6	300.0
Saving Private Ryan (1998)	216.1	263.2
Gladiator (2000)	187.7	268.6
Dances with Wolves (1990)	184.2	240.0
The Exorcist (1973)	204.6	153.0
My Big Fat Greek Wedding (2002)	241.4	115.1
Rocky IV (1985)	127.9	172.6
Star Wars: The Phantom Menace (1999)	431.1	491.3

Source: The *Internet Movie Database* (www.imdb.com).

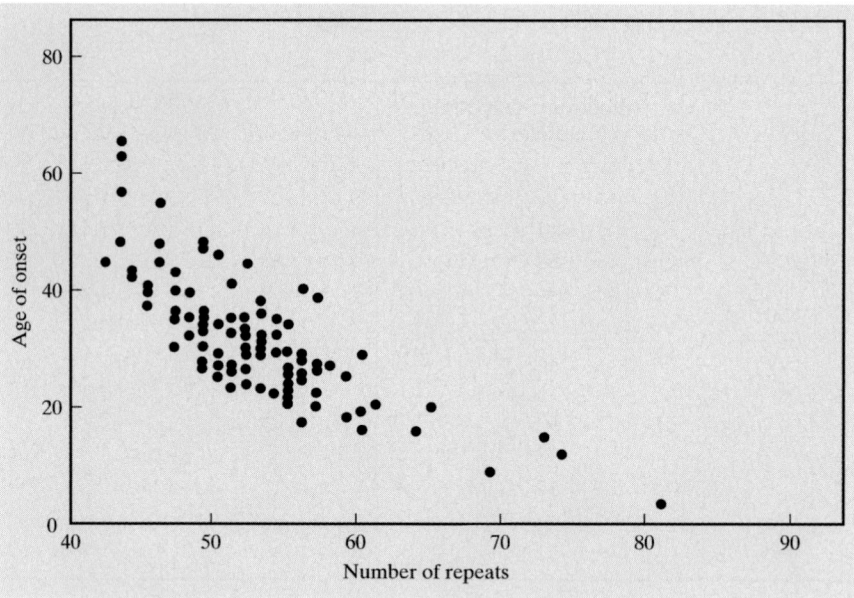

c. The researchers reported a correlation of $r = -.815$ between age and number of repeats. Since $r^2 = (-.815)^2 = .664$, they concluded that about "66% of the variability in the age of onset can be accounted for by the number of repeats." Does this statement apply to the quadratic model $E(y) = \beta_0 + \beta_1 x + \beta_2 x^2$? If not, give the equation of the model for which it does apply.

12.62 Walking study. Refer to the *American Scientist* (July–Aug. 1998) study of the relationship between self-avoiding and unrooted walks, Exercise 11.108 (p. 651). Recall that in a self-avoiding walk you never retrace or cross your own path, while an unrooted walk is a path in which the starting and ending points are impossible to distinguish. The possible number of walks of each type of various lengths are reproduced in the table. In Exercise 11.108 you analyzed the straight-line model relating number of unrooted walks (*y*) to walk length *(x)*. Now consider the quadratic model

⊚ **WALK**

Walk Length (Number of steps)	Unrooted Walks	Self-Avoiding Walks
1	1	4
2	2	12
3	4	36
4	9	100
5	22	284
6	56	780
7	147	2,172
8	388	5,916

Source: Hayes, B. "How to avoid yourself." *American Scientist*, Vol. 86, No. 4, July–Aug. 1998, p. 317 (Figure 5).

$E(y) = \beta_0 + \beta_1 x + \beta_2 x^2$. Is there sufficient evidence of an upward concave curvilinear relationship between *y* and *x*? Test at $\alpha = .10$.

Applying the Concepts—Advanced

12.63 Tree frog study. The optomotor responses of tree frogs were studied in the *Journal of Experimental Zoology* (Sept. 1993). Microspectrophotometry was used to measure the threshold quantal flux (the light intensity at which the optomotor response was first observed) of tree frogs tested at different spectral wavelengths. The data revealed the relationship between the log of quantal flux (*y*) and wavelength (*x*) shown in the accompanying graph.

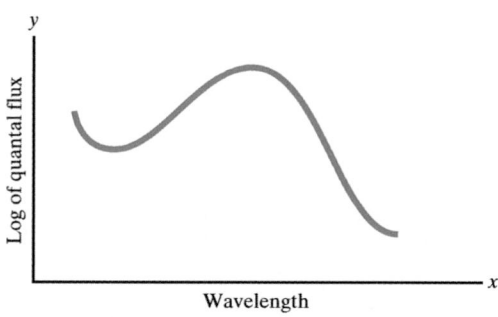

a. Explain why a first-order model would not be appropriate for modeling $E(y)$.
b. Explain why a second-order model would not be appropriate for modeling $E(y)$.
c. Demonstrate that the third-order model $E(y) = \beta_0 + \beta_1 x + \beta_2 x^2 + \beta_3 x^3$ may be the most appropriate model for $E(y)$.

12.7 Model Building: Qualitative (Dummy) Variable Models

Multiple regression models can also be written to include **qualitative** (or **categorical**) independent variables. Qualitative variables, unlike quantitative variables, cannot be measured on a numerical scale. Therefore, we must code the values of the qualitative variable (called **levels**) as numbers before we can fit the model. These coded qualitative variables are called **dummy** (or **indicator**) **variables** since the numbers assigned to the various levels are arbitrarily selected.

To illustrate, suppose a female executive at a certain company claims that male executives earn higher salaries, on average, than female executives with the same education, experience, and responsibilities. To support her claim, she wants to model the salary y of an executive using a qualitative independent variable representing the gender of an executive (male or female).

A convenient method of coding the values of a qualitative variable at two levels involves assigning a value of 1 to one of the levels and a value of 0 to the other. For example, the dummy variable used to describe gender could be coded as follows:

$$x = \begin{cases} 1 & \text{if male} \\ 0 & \text{if female} \end{cases}$$

The choice of which level is assigned to 1 and which is assigned to 0 is arbitrary. The model then takes the following form:

$$E(y) = \beta_0 + \beta_1 x$$

The advantage of using a 0–1 coding scheme is that the β coefficients are easily interpreted. The model above allows us to compare the mean executive salary $E(y)$ for males with the corresponding mean for females.

Males $(x = 1)$: $E(y) = \beta_0 + \beta_1(1) = \beta_0 + \beta_1$
Females $(x = 0)$: $E(y) = \beta_0 + \beta_1(0) = \beta_0$

These two means are illustrated in the bar graph in Figure 12.19.

First note that β_0 represents the mean salary for females (say, μ_F). When a 0–1 coding convention is used, β_0 will always represent the mean response associated with the level of the qualitative variable assigned the value 0 (called the **base level**). The difference between the mean salary for males and the mean salary for females, $\mu_M - \mu_F$, is represented by β_1—that is,

$$\mu_M - \mu_F = (\beta_0 + \beta_1) - (\beta_0) = \beta_1$$

This difference is shown in Figure 12.19.* With a 0–1 coding convention, β_1 will always represent the difference between the mean response for the level assigned

Figure 12.19
Bar Chart Comparing $E(y)$ for Males and Females

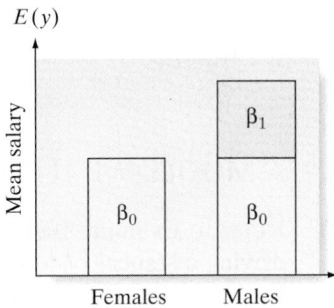

*Note that β_1 could be negative. If β_1 were negative, the height of the bar corresponding to males would be *reduced* (rather than increased) from the height of the bar for females by the amount β_1. Figure 12.19 is constructed assuming that β_1 is a positive quantity.

the value 1 and the mean for the base level. Thus, for the executive salary model, we have

$$\beta_0 = \mu_F$$
$$\beta_1 = \mu_M - \mu_F$$

Now carefully examine the model with a single qualitative independent variable at two levels, because we will use exactly the same pattern for any number of levels. Moreover, the interpretation of the parameters will always be the same.

One level (say, level A) is selected as the *base level*. Then for the 0–1 coding* for the dummy variables,

$$\mu_A = \beta_0$$

The coding for all dummy variables is as follows: To represent the mean value of y for a particular level, let that dummy variable equal 1; otherwise, the dummy variable is set equal to 0. Using this system of coding,

$$\mu_B = \beta_0 + \beta_1$$
$$\mu_C = \beta_0 + \beta_2$$

and so on. Because $\mu_A = \beta_0$, any other model parameter will represent the difference between means for that level and the base level:

$$\beta_1 = \mu_B - \mu_A$$
$$\beta_2 = \mu_C - \mu_A$$

and so on.

Procedure for Writing a Model with One Qualitative Independent Variable with k Levels

Always use a number of dummy variables that is one less than the number of levels of the qualitative variable. Thus, for a qualitative variable with k levels, use $k - 1$ dummy variables:

$$y = \beta_0 + \beta_1 x_1 + \beta_2 x_2 + \cdots + \beta_{k-1} x_{k-1} + \varepsilon$$

where x_i is the dummy variable for level $i + 1$ and

$$x_i = \begin{cases} 1 & \text{if } y \text{ is observed at level } i + 1 \\ 0 & \text{otherwise} \end{cases}$$

Then, for this system of coding

$$\mu_A = \beta_0 \qquad\qquad \text{and} \qquad\qquad \beta_1 = \mu_B - \mu_A$$
$$\mu_B = \beta_0 + \beta_1 \qquad\qquad\qquad\qquad \beta_2 = \mu_C - \mu_A$$
$$\mu_C = \beta_0 + \beta_2 \qquad\qquad\qquad\qquad \beta_3 = \mu_D - \mu_A$$
$$\mu_D = \beta_0 + \beta_3 \qquad\qquad\qquad\qquad \vdots$$
$$\vdots$$

EXAMPLE 12.9   A MODEL WITH ONE QUALITATIVE INDEPENDENT VARIABLE

Problem   Refer to Example 10.4 (p. 525). Recall that the USGA wants to compare the mean driving distances of four different golf ball brands (A, B, C, and D). Iron Byron, the USGA's robotic golfer, is used to hit a sample of 10 balls of each brand. The distance data are reproduced in Table 12.4.

*You do not have to use a 0–1 system of coding for the dummy variables. Any two-value system will work, but the interpretation given to the model parameters will depend on the code. Using the 0–1 system makes the model parameters easy to interpret.

⊙ **GOLFCRD**

TABLE 12.4   Driving Distances (in feet) for Four Golf Ball Brands

Brand A	Brand B	Brand C	Brand D
251.2	263.2	269.7	251.6
245.1	262.9	263.2	248.6
248.0	265.0	277.5	249.4
251.1	254.5	267.4	242.0
260.5	264.3	270.5	246.5
250.0	257.0	265.5	251.3
253.9	262.8	270.7	261.8
244.6	264.4	272.9	249.0
254.6	260.6	275.6	247.1
248.8	255.9	266.5	245.9

a. Hypothesize a regression model for driving distance y using Brand as an independent variable.

b. Interpret the β's in the model.

c. Use the model to determine if the mean driving distances for the four brands are significantly different at $\alpha = .05$.

Solution   **a.** Note that golf ball brand (A, B, C, and D) is a qualitative variable (measured on a nominal scale). According to the box, for a four-level qualitative variable we require three dummy variables in the regression model. The model relating $E(y)$, where y is the distance the ball is driven by Iron Byron, to this single qualitative variable, golf ball brand, is

$$E(y) = \beta_0 + \beta_1 x_1 + \beta_2 x_2 + \beta_3 x_3$$

where

$$x_1 = \begin{cases} 1 & \text{if Brand B} \\ 0 & \text{if not} \end{cases} \qquad x_2 = \begin{cases} 1 & \text{if Brand C} \\ 0 & \text{if not} \end{cases} \qquad x_3 = \begin{cases} 1 & \text{if Brand D} \\ 0 & \text{if not} \end{cases}$$

b. Since Brand A is the base level, β_0 represents the mean driving distance for Brand A (i.e., $\beta_0 = \mu_A$). The other β's are differences in means:

$$\beta_1 = \mu_B - \mu_A$$
$$\beta_2 = \mu_C - \mu_A$$
$$\beta_3 = \mu_D - \mu_A$$

where μ_A, μ_B, μ_C, and μ_D are the mean distances for Brands A, B, C, and D, respectively.

c. Testing the null hypothesis that the means for the four brands are equal, that is, $\mu_A = \mu_B = \mu_C = \mu_D$, is equivalent to testing

$$H_0: \beta_1 = \beta_2 = \beta_3 = 0$$

You can see this by observing that if $\beta_1 = \mu_B - \mu_A = 0$, then $\mu_A = \mu_B$. Similarly, $\beta_2 = \mu_C - \mu_A = 0$ implies that $\mu_A = \mu_C$, and $\beta_3 = \mu_D - \mu_A = 0$ implies that $\mu_A = \mu_D$. The alternative hypothesis is

$$H_a: \text{At least one of the parameters, } \beta_1, \beta_2 \text{ and } \beta_3, \text{ differs from } 0$$

which implies that at least two of the four means (μ_A, μ_B, μ_C, and μ_D) differ.

Figure 12.20

SPSS Regression Printout
for Dummy Variable Model

Model Summary

Model	R	R Square	Adjusted R Square	Std. Error of the Estimate
1	.886[a]	.786	.768	4.60163

a. Predictors: (Constant), X3, X2, X1

ANOVA[b]

Model		Sum of Squares	df	Mean Square	F	Sig.
1	Regression	2794.389	3	931.463	43.989	.000[a]
	Residual	762.301	36	21.175		
	Total	3556.690	39			

a. Predictors: (Constant), X3, X2, X1

b. Dependent Variable: DISTANCE

Coefficients[a]

Model		Unstandardized Coefficients		Standardized Coefficients	t	Sig.
		B	Std. Error	Beta		
1	(Constant)	250.780	1.455		172.338	.000
	X1	10.280	2.058	.472	4.995	.000
	X2	19.170	2.058	.880	9.315	.000
	X3	-1.460	2.058	-.067	-.709	.483

a. Dependent Variable: DISTANCE

To test this hypothesis, we conduct the global F-test for the model. The SPSS printout for fitting the model

$$E(y) = \beta_0 + \beta_1 x_1 + \beta_2 x_2 + \beta_3 x_3$$

is shown in Figure 12.20. The value of the F statistic for testing the adequacy of the model, $F = 43.99$, and the observed significance level of the test, $p \approx .000$, are both highlighted. Since $\alpha = .05$ exceeds the p-value, we reject H_0 and conclude that at least one of the parameters differs from 0. Or, equivalently, we conclude that the data provide sufficient evidence to indicate that the mean driving distance does vary from one golf ball brand to another.

Look Back This global F-test is equivalent to the analysis of variance F-test for a completely randomized design of Chapter 10.

Now Work *Exercise 12.67*

▬▬▬ ■ ■ ■ ▬▬▬

Caution

A common mistake by regression analysts is to use a single dummy variable x for a qualitative variable at k levels, where $x = 1, 2, 3, \ldots, k$. Such a regression model will have unestimable β's and β's that are difficult to interpret. Remember, when modeling $E(y)$ with a single qualitative independent variable, the number of 0–1 dummy variables to include in the model will always be one less than the number of levels of the qualitative variable.

Exercises 12.64–12.76

Understanding the Principles

12.64 Write a regression model relating the mean value of y to a qualitative independent variable that can assume two levels. Interpret all the terms in the model.

12.65 Write a regression model relating $E(y)$ to a qualitative independent variable that can assume three levels. Interpret all the terms in the model.

Learning the Mechanics

12.66 The following model was used to relate $E(y)$ to a single qualitative variable with four levels: $E(y) = \beta_0 + \beta_1 x_1 + \beta_2 x_2 + \beta_3 x_3$, where

$$x_1 = \begin{cases} 1 & \text{if level 2} \\ 0 & \text{if not} \end{cases}$$

$$x_2 = \begin{cases} 1 & \text{if level 3} \\ 0 & \text{if not} \end{cases}$$

$$x_3 = \begin{cases} 1 & \text{if level 4} \\ 0 & \text{if not} \end{cases}$$

This model was fit to $n = 30$ data points and the following result was obtained:

$$\hat{y} = 10.2 - 4x_1 + 12x_2 + 2x_3$$

a. Use the least squares prediction equation to find the estimate of $E(y)$ for each level of the qualitative independent variable.

b. Specify the null and alternative hypotheses you would use to test whether $E(y)$ is the same for all four levels of the independent variable.

12.67 MINITAB (printout shown below) was used to fit the following model to $n = 15$ data points:

NW

$$y = \beta_0 + \beta_1 x_1 + \beta_2 x_2 + \varepsilon$$

where

$$x_1 = \begin{cases} 1 & \text{if level 2} \\ 0 & \text{if not} \end{cases}$$

$$x_2 = \begin{cases} 1 & \text{if level 3} \\ 0 & \text{if not} \end{cases}$$

a. Report the least squares prediction equation.

b. Interpret the values of β_1 and β_2.

c. Interpret the following hypotheses in terms of μ_1, μ_2, and μ_3:

$$H_0: \beta_1 = \beta_2 = 0$$
$$H_a: \text{At least one of the parameters } \beta_1 \text{ and } \beta_2 \text{ differs from 0}$$

d. Conduct the hypothesis test of part **c**.

Applying the Concepts—Basic

12.68 Improving SAT scores. Refer to the *Chance* (Winter 2001) study of students who paid a private tutor (or coach) to help them improve their Scholastic Assessment Test (SAT) scores, Exercise 2.101 (p. 81). Multiple regression was used to estimate the effect of coaching on SAT-Mathematics scores. Data on 3,492 students (573 of whom were coached) were used to fit the model $E(y) = \beta_0 + \beta_1 x_1 + \beta_2 x_2$, where $y = $ SAT-Math score, $x_1 = $ score on PSAT, and $x_2 = \{1$ if student was coached, 0 if not$\}$.

a. The fitted model had an adjusted R^2 value of .76. Interpret this result.

b. The estimate of β_2 in the model was 19, with a standard error of 3. Use this information to form a 95% confidence interval for β_2. Interpret the interval.

c. Based on the interval, part **b**, what can you say about the effect of coaching on SAT-Math scores?

```
The regression equation is
Y = 80.0 + 16.8 X1 + 40.4 X2


Predictor    Coef   SE Coef      T       P
Constant   80.000     4.082   19.60   0.000
X1         16.800     5.774    2.91   0.013
X2         40.400     5.774    7.00   0.000


S = 9.129      R-Sq = 80.5%   R-Sq(adj) = 77.2%


Analysis of Variance

Source          DF      SS       MS      F      P
Regression       2   4118.9   2059.5   24.72  0.000
Residual Error  12   1000.0     83.3
Total           14   5118.9
```

12.69 Assassination risk. *New Scientist* (Apr. 3, 1993) published an article on strategies for foiling assassination attempts on politicians. The strategies are based on the findings of researchers at Middlesex University (United Kingdom), who used a multiple regression model for predicting the level y of assassination risk. One of the variables used in the model was political status of a country (communist, democratic, or dictatorship).

a. Propose a model for $E(y)$ as a function of political status.

b. Interpret the β's in the model, part **a**.

12.70 Comparing mosquito repellents. Which insect repellents protect best against mosquitoes? *Consumer Reports* (June 2000) tested 14 products that all claim to be an effective mosquito repellent. Each product was classified as either lotion/cream or aerosol/spray. The cost of the product (in dollars) was divided by the amount of the repellent needed to cover exposed areas of the skin (about 1/3 ounce) to obtain a cost-per-use value. Effectiveness was measured as the maximum number of hours of protection (in half-hour increments) provided when human testers exposed their arms to 200 mosquitoes. The data from the report are listed in the table below.

a. Suppose you want to use repellent type to model the cost per use (y). Create the appropriate number of dummy variables for repellent type and write the model.

b. Fit the model, part **a**, to the data.

c. Give the null hypothesis for testing whether repellent type is a useful predictor of cost per use (y).

d. Conduct the test, part **c**, and give the appropriate conclusion. Use $\alpha = .10$.

e. Repeat parts **a–d** if the dependent variable is maximum number of hours of protection (y).

NZBIRDS

12.71 Extinct New Zealand birds. Refer to the *Evolutionary Ecology Research* (July 2003) study of the patterns of extinction in the New Zealand bird population, Exercise 2.18 (p. 40). Recall that the **NZBIRDS** file contains qualitative data on flight capability (volant or flightless), habitat (aquatic, ground terrestrial, or aerial terrestrial), nesting site (ground, cavity within ground, tree, cavity above ground), nest density (high or low), diet (fish, vertebrates, vegetables, or invertebrates), and extinct status (extinct, absent from island, present), and quantitative data on body mass (grams) and egg length (millimeters) for 132 bird species at the time of the Maori colonization of New Zealand.

a. Write a model for mean body mass as a function of flight capability.

b. Write a model for mean body mass as a function of diet.

c. Write a model for mean egg length as a function of nesting site.

d. Fit the model, part **a**, to the data and interpret the estimates of the β's.

e. Conduct a test to determine if the model, part **a**, is statistically useful (at $\alpha = .01$) for estimating mean body mass.

f. Fit the model, part **b**, to the data and interpret the estimates of the β's.

g. Conduct a test to determine if the model, part **b**, is statistically useful (at $\alpha = .01$) for estimating mean body mass.

h. Fit the model, part **c**, to the data and interpret the estimates of the β's.

i. Conduct a test to determine if the model, part **c**, is statistically useful (at $\alpha = .01$) for estimating mean egg length.

Applying the Concepts—Intermediate

12.72 Listen and look study. Where do you look when you are listening to someone speak? Researchers have discovered that listeners tend to gaze at the eyes or mouth of the speaker. In a study published in *Perception & Psychophysics* (Aug. 1998), subjects watched a videotape of a speaker giving a series of

REPELLENT

Insect Repellent	Type	Cost/Use	Maximum Protection
Amway HourGuard 12	Lotion/Cream	$2.08	13.5 hours
Avon Skin-So-Soft	Aerosol/Spray	0.67	0.5
Avon BugGuard Plus	Lotion/Cream	1.00	2.0
Ben's Backyard Formula	Lotion/Cream	0.75	7.0
Bite Blocker	Lotion/Cream	0.46	3.0
BugOut	Aerosol/Spray	0.11	6.0
Cutter Skinsations	Aerosol/Spray	0.22	3.0
Cutter Unscented	Aerosol/Spray	0.19	5.5
Musko11 Ultra6Hours	Aerosol/Spray	0.24	6.5
Natrapel	Aerosol/Spray	0.27	1.0
Off! Deep Woods	Aerosol/Spray	1.77	14.0
Off! Skintastic	Lotion/Cream	0.67	3.0
Sawyer Deet Formula	Lotion/Cream	0.36	7.0
Repel Permanone	Aerosol/Spray	2.75	24.0

Source: "Buzz off." *Consumer Reports*, June 2000.

short monologues at a social gathering (e.g., a party). The level of background noise (multilingual voices and music) was varied during the listening sessions. Each subject wore a pair of clear plastic goggles on which an infrared corneal detection system was mounted, enabling the researchers to monitor the subject's eye movements. One response variable of interest was the proportion y of times the subject's eyes fixated on the speaker's mouth.

a. The researchers wanted to estimate $E(y)$ for four different noise levels: none, low, medium, and high. Hypothesize a model that will allow the researchers to obtain these estimates.

b. Interpret the β's in the model, part **a**.

c. Explain how to test the hypothesis of no differences in the mean proportions of mouth fixations for the four background noise levels.

12.73 Density of mosquito larvae. A field experiment was conducted to assess the effect of organic enrichment on the mean density of mosquito larvae. (*Journal of the American Mosquito Control Association*, June 1995.) Larval specimens were collected from a pond three days after the pond was flooded with canal water. A second sample of specimens was collected three weeks after flooding and enriching the pond with rabbit pellets. All specimens were returned to the laboratory and the number y of mosquito larvae counted in each specimen.

a. Write a model that will allow you to compare the mean number of mosquito larvae found in the enriched pond to the corresponding mean for the natural pond.

b. Interpret the β coefficients in the model, part **a**.

c. Set up the null and alternative hypotheses for testing whether the mean larval density for the enriched pond exceeds the mean for the natural pond.

d. The p-value associated with the global F-test for the model, part **a**, was determined to be .004. Interpret this result.

12.74 Yields of pea varieties. Five varieties of peas (A, B, C, D, and E) are currently being tested by a large agribusiness cooperative to determine which is best suited for production. A field was divided into 20 plots, with each variety of peas planted in four plots. The yields (in bushels of peas) produced from each plot are shown in the table.

⊛ **PEAS**

A	B	C	D	E
26.2	29.2	29.1	21.3	20.1
24.3	28.1	30.8	22.4	19.3
21.8	27.3	33.9	24.3	19.9
28.1	31.2	32.8	21.8	22.1

a. Use the data to fit the model

$$y = \beta_0 + \beta_1 x_1 + \beta_2 x_2 + \beta_3 x_3 + \beta_4 x_4 + \varepsilon$$

where $x_1 = 1$ for variety A, $x_2 = 1$ for variety B, $x_3 = 1$ for variety C, and $x_4 = 1$ for variety D.

b. Interpret all the estimated parameters in the model.

c. What null and alternative hypotheses are tested by the global F-test for this model? Interpret the hypotheses both in terms of the β parameters and the mean yields for the five varieties of peas.

d. Test the hypotheses of part **c** using $\alpha = .05$.

e. Place a 95% confidence interval on the differences in the mean yields of varieties D and E.

f. Perform an analysis of variance for a completely randomized design on the data. Show that the ANOVA results are equivalent to the regression results.

12.75 Habitats of grizzly bears. Do grizzly bears segregate on the basis of sex? One hypothesis is that female grizzlies avoid male-occupied habitats because of competition for food and cannibalism. A competing theory is that females do not avoid males, but simply have different habitats available to them. These hypotheses were investigated in the *Journal of Wildlife Management* (July 1995). Grizzly bears were trapped, radio-collared, and released in the Highwood trapping zone (HTZ) in Alberta, Canada. The percentage of time, y, each bear used the HTZ as a habitat over a 4-year period was recorded. The researchers modeled $E(y)$ as a function of reproductive class at five levels: estrous adult females, adult females with offspring, independent subadult females, adult males, and independent subadult males. One goal was to compare the mean percent use of HTZ for the five classes of grizzly bears.

Grizzly Class	n	Mean Percent Use
Estrous adult females	5	38
Adult females with offspring	7	16
Independent subadult females	2	89
Adult males	7	43
Independent subadult males	8	58

Source: Wielgus, R. B., and Bunnell, F. L. "Tests of hypotheses for sexual segregation in grizzly bears." *Journal of Wildlife Management*, Vol. 59, No. 3, July 1995, p. 555 (Table 1).

a. Write a model for $E(y)$ that will enable the researchers to carry out the comparison.

b. The sample sizes and sample means for the five classes are shown in the table above. Use this information to find estimates of the β's in the model, part **a**.

c. Give the null hypothesis for a test to determine whether the mean percent use of HTZ differs among the grizzly bear classes.

d. The p-value for the test, part **c**, was reported as .15. Interpret this result.

Applying the Concepts—Advanced

12.76 Heights of grade school repeaters. Refer to *The Archives of Disease in Childhood* (Apr. 2000) study of whether height influences a child's progression through elementary school, Exercise 10.29 (p. 534). Recall that Australian school children were divided into equal thirds (tertiles) based on age (youngest third, middle third, and oldest third). The average heights of the three groups (where all height measurements were standardized using z-scores), by gender, are shown in the table below.

a. Propose a regression model that will enable you to compare the average heights of the three age groups for boys.
b. Find the estimates of the β's in the model, part **a**.
c. Repeat parts **a** and **b** for girls.

	Sample Size	Youngest Tertile Mean Height	Middle Tertile Mean Height	Oldest Tertile Mean Height
Boys	1439	0.33	0.33	0.16
Girls	1409	0.27	0.18	0.21

Source: Wake, M., Coghlan, D., and Hesketh, K. "Does height influence progression through primary school grades?" *The Archives of Disease in Childhood*, Vol. 82, Apr. 2000 (Table 2).

12.8 Model Building: Models with Both Quantitative and Qualitative Variables (Optional)

Suppose you want to relate the mean monthly sales $E(y)$ of a company to monthly advertising expenditure x for three different advertising media (say newspaper, radio, and television) and you wish to use first-order (straight-line) models to model the responses for all three media. Graphs of these three relationships might appear as shown in Figure 12.21.

Since the lines in Figure 12.21 are hypothetical, a number of practical questions arise. Is one advertising medium as effective as any other? That is, do the three mean sales lines differ for the three advertising media? Do the increases in mean sales per dollar input in advertising differ for the three advertising media? That is, do the slopes of the three lines differ? Note that the two practical questions have been rephrased into questions about the parameters that define the three lines of Figure 12.21. To answer them, we must write a single regression model that will characterize the three lines of Figure 12.21 and that, by testing hypotheses about the lines, will answer the questions.

Figure 12.21
Graphs of the Relationship Between Mean Sales $E(y)$ and Advertising Expenditure x

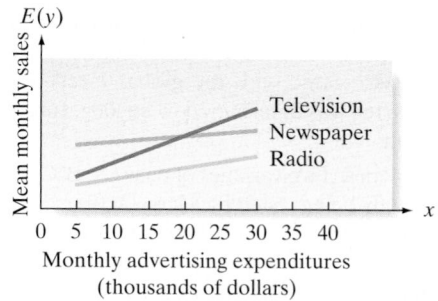

The response described previously, monthly sales, is a function of *two* independent variables, one quantitative (advertising expenditure x_1) and one qualitative (type of medium). We will proceed, in stages, to build a model relating $E(y)$ to these variables and will show graphically the interpretation we would give to the model at each stage. This will help you to see the contributions of the various terms in the model.

1. The straight-line relationship between mean sales $E(y)$ and advertising expenditure is the same for all three media—that is, a single line will describe the relationship between $E(y)$ and advertising expenditure x_1 for all the media (see Figure 12.22).

$$E(y) = \beta_0 + \beta_1 x_1 \quad \text{where } x_1 = \text{Advertising expenditure}$$

Figure 12.22

The Relationship between $E(y)$ and x_1 Is the Same for All Media

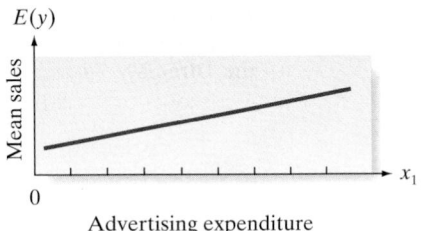

2. The straight lines relating mean sales $E(y)$ to advertising expenditure x_1 differ from one medium to another, but the rate of increase in mean sales per increase in dollar advertising expenditure x_1 is the same for all media—that is, the lines are parallel but possess different y-intercepts (see Figure 12.23).

$$E(y) = \beta_0 + \beta_1 x_1 + \beta_2 x_2 + \beta_3 x_3$$

where

$$x_1 = \text{Advertising expenditure}$$
$$x_2 = \begin{cases} 1 & \text{if radio medium} \\ 0 & \text{if not} \end{cases}$$
$$x_3 = \begin{cases} 1 & \text{if television medium} \\ 0 & \text{if not} \end{cases}$$

Figure 12.23

Parallel Response Lines for the Three Media

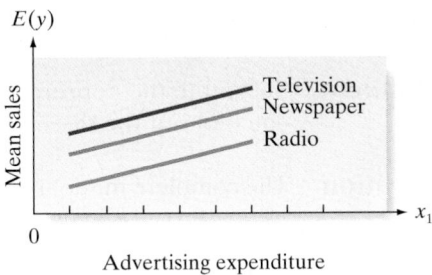

Notice that this model is essentially a combination of a first-order model with a single quantitative variable and the model with a single qualitative variable:

First-order model with a single quantitative variable: $\qquad E(y) = \beta_0 + \beta_1 x_1$

Model with single qualitative variable at three levels: $\qquad E(y) = \beta_0 + \beta_2 x_2 + \beta_3 x_3$

where x_1, x_2, and x_3 are as just defined. The model described here implies no interaction between the two independent variables, which are advertising expenditure x_1 and the qualitative variable (type of advertising medium). The change in $E(y)$ for a 1-unit increase in x_1 is identical (the slopes of the lines are equal) for all three advertising media. The terms corresponding to each of the independent variables are called **main effect terms** because they imply no interaction.

3. The straight lines relating mean sales $E(y)$ to advertising expenditure x_1 differ for the three advertising media—that is, both the line intercepts and the slopes differ (see Figure 12.24).

Figure 12.24
Different Response Lines
for the Three Media

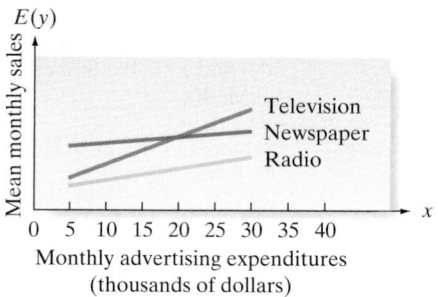

As you will see, this interaction model is obtained by adding terms involving the cross-product terms, one each from each of the two independent variables:

$$E(y) = \beta_0 + \underbrace{\beta_1 x_1}_{\substack{\text{Main effect,} \\ \text{advertising} \\ \text{expenditure}}} + \underbrace{\beta_2 x_2 + \beta_3 x_3}_{\substack{\text{Main effect,} \\ \text{type of} \\ \text{medium}}} + \underbrace{\beta_4 x_1 x_2 + \beta_5 x_1 x_3}_{\text{Interaction}}$$

Note that each of the preceding models is obtained by adding terms to model 1, the single first-order model used to model the responses for all three media. Model 2 is obtained by adding the main effect terms for type of medium, the qualitative variable. Model 3 is obtained by adding the interaction terms to model 2.

EXAMPLE 12.10 INTERPRETING THE β's IN A MODEL WITH MIXED VARIABLES

Problem Substitute the appropriate values of the dummy variables in model 3 to obtain the equations of the three response lines in Figure 12.24.

Solution The complete model that characterizes the three lines in Figure 12.24 is

$$E(y) = \beta_0 + \beta_1 x_1 + \beta_2 x_2 + \beta_3 x_3 + \beta_4 x_1 x_2 + \beta_5 x_1 x_3$$

where

$$x_1 = \text{Advertising expenditure}$$

$$x_2 = \begin{cases} 1 & \text{if radio medium} \\ 0 & \text{if not} \end{cases}$$

$$x_3 = \begin{cases} 1 & \text{if television medium} \\ 0 & \text{if not} \end{cases}$$

Examining the coding, you can see that $x_2 = x_3 = 0$ when the advertising medium is newspaper. Substituting these values into the expression for $E(y)$, we obtain the newspaper medium line:

$$E(y) = \beta_0 + \beta_1 x_1 + \beta_2(0) + \beta_3(0) + \beta_4 x_1(0) + \beta_5 x_1(0) = \beta_0 + \beta_1 x_1$$

Similarly, we substitute the appropriate values of x_2 and x_3 into the expression for $E(y)$ to obtain the radio medium line ($x_2 = 1, x_3 = 0$):

$$E(y) = \beta_0 + \beta_1 x_1 + \beta_2(1) + \beta_3(0) + \beta_4 x_1(1) + \beta_5 x_1(0)$$

$$= \underbrace{(\beta_0 + \beta_2)}_{y\text{-intercept}} + \underbrace{(\beta_1 + \beta_4)}_{\text{Slope}} x_1$$

and the television medium line ($x_2 = 0, x_3 = 1$)

$$E(y) = \beta_0 + \beta_1 x_1 + \beta_2(0) + \beta_3(1) + \beta_4 x_1(0) + \beta_5 x_1(1)$$

$$= \underbrace{(\beta_0 + \beta_3)}_{y\text{-intercept}} + \underbrace{(\beta_1 + \beta_5)}_{\text{Slope}} x_1$$

Look Back If you were to fit model 3, obtain estimates of $\beta_0, \beta_1, \beta_2, \ldots, \beta_5$, and substitute them into the equations for the three media lines, you would obtain exactly the same prediction equations as you would obtain if you were to fit three separate straight lines, one to each of the three sets of media data. You may ask why we would not fit the three lines separately. Why bother fitting a model that combines all three lines (model 3) into the same equation? The answer is that you need to use this procedure if you wish to use statistical tests to compare the three media lines. We need to be able to express a practical question about the lines in terms of a hypothesis that a set of parameters in the model equals 0. (We demonstrate this procedure in the next section.) You could not do this if you were to perform three separate regression analyses and fit a line to each set of media data.

Now Work   *Exercise 12.80*

■ ■ ■

EXAMPLE 12.11

TESTING FOR TWO DIFFERENT SLOPES

Problem An industrial psychologist conducted an experiment to investigate the relationship between worker productivity and a measure of salary incentive for two manufacturing plants; one plant had union representation and the other plant had nonunion representation. The productivity y per worker was measured by recording the number of machined castings that a worker could produce in a 4-week period of 40 hours per week. The incentive was the amount x_1 of bonus (in cents per casting) paid for all castings produced in excess of 1,000 per worker for the 4-week period. Nine workers were selected from each plant, and three from each group of nine were assigned to receive a 20¢ bonus per casting, three a 30¢ bonus, and three a 40¢ bonus. The productivity data for the 18 workers, three for each plant type and incentive combination, are shown in Table 12.5.

a. Write a model for mean productivity, $E(y)$, assuming that the relationship between $E(y)$ and incentive, x_1, is first order.

b. Fit the model and graph the prediction equations for the union and nonunion plants.

c. Do the data provide sufficient evidence to indicate that the rate of increase of worker productivity is different for union and nonunion plants? Test at $\alpha = .10$.

🔵 CASTING

TABLE 12.5 Productivity Data (Number of Castings) for Example 12.11

Type of Plant	Incentive								
	20¢/casting			30¢/casting			40¢/casting		
Union	1,435	1,512	1,491	1,583	1,529	1,610	1,601	1,574	1,636
Nonunion	1,575	1,512	1,488	1,635	1,589	1,661	1,645	1,616	1,689

Figure 12.25

MINITAB Printout of the
Complete Model for the
Casting Data

```
The regression equation is
CASTINGS = 1366 + 6.22 INCENTIVE + 47.8 PDUMMY + 0.03 INC_PDUM


Predictor      Coef  SE Coef      T      P
Constant    1365.83    51.84  26.35  0.000
INCENTIVE     6.217    1.667   3.73  0.002
PDUMMY        47.78    73.31   0.65  0.525
INC_PDUM      0.033    2.358   0.01  0.989


S = 40.8387    R-Sq = 71.1%    R-Sq(adj) = 64.9%


Analysis of Variance

Source           DF     SS     MS      F      P
Regression        3  57332  19111  11.46  0.000
Residual Error   14  23349   1668
Total            17  80682
```

Solution

a. If we assume that a first-order model* is adequate to detect a change in mean productivity as a function of incentive x_1, then the model that produces two straight lines, one for each plant, is

$$E(y) = \beta_0 + \beta_1 x_1 + \beta_2 x_2 + \beta_3 x_1 x_2$$

where

$$x_1 = \text{Incentive} \qquad x_2 = \begin{cases} 1 & \text{if nonunion plant} \\ 0 & \text{if union plant} \end{cases}$$

b. The MINITAB printout for the regression analysis is shown in Figure 12.25. Reading the parameter estimates highlighted on the printout, you can see that

$$\hat{y} = 1{,}365.83 + 6.217 x_1 + 47.78 x_2 + .033 x_1 x_2$$

The prediction equation for the union plant can be obtained (see the coding) by substituting $x_2 = 0$ into the general prediction equation. Then

$$\hat{y} = \hat{\beta}_0 + \hat{\beta}_1 x_1 + \hat{\beta}_2(0) + \hat{\beta}_3 x_1(0) = \hat{\beta}_0 + \hat{\beta}_1 x_1$$
$$= 1{,}365.83 + 6.217 x_1$$

Similarly, the prediction equation for the nonunion plant can be obtained by substituting $x_2 = 1$ into the general prediction equation. Then

$$\hat{y} = \hat{\beta}_0 + \hat{\beta}_1 x_1 + \hat{\beta}_2 x_2 + \hat{\beta}_3 x_1 x_2$$
$$= \hat{\beta}_0 + \hat{\beta}_1 x_1 + \hat{\beta}_2(1) + \hat{\beta}_3 x_1(1)$$
$$= \underbrace{(\hat{\beta}_0 + \hat{\beta}_2)}_{y\text{-intercept}} + \underbrace{(\hat{\beta}_1 + \hat{\beta}_3)}_{\text{Slope}} x_1$$
$$= (1{,}365.83 + 47.78) + (6.217 + .033) x_1$$
$$= 1{,}413.61 + 6.250 x_1$$

A MINITAB graph of these prediction equations is shown in Figure 12.26. Note that the slopes of the two lines are nearly identical (6.217 for union and 6.250 for nonunion).

---

*Although the model contains a term involving $x_1 x_2$, it is first-order (graphs as a straight line) in the quantitative variable x_1. The variable x_2 is a dummy variable that introduces or deletes terms in the model. The order of a model is determined only by the quantitative variables that appear in the model.

Figure 12.26

MINITAB Plot of Prediction Equations for the Two Productivity Lines

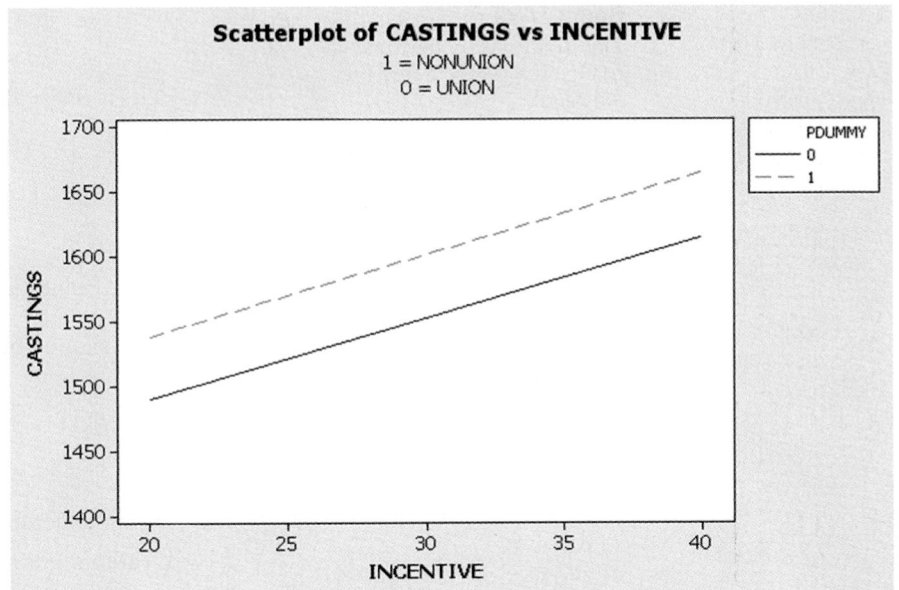

c. If the rate of increase of productivity with incentive (i.e, the slope) for nonunion plants is different than the corresponding slope for union plants, then the interaction β (i.e., β_3) will differ from 0. Consequently, we want to test

$$H_0: \beta_3 = 0$$
$$H_a: \beta_3 \neq 0$$

This test is conducted using the t-test of Section 12.3. From the MINITAB printout, the test statistic and corresponding p-value (highlighted) are

$$t = .014 \qquad p\text{-value} = .989$$

Since $\alpha = .10$ is less than the p-value, we fail to reject H_0. There is insufficient evidence to conclude that the union and nonunion shapes differ. Thus, the test supports our observation of two nearly identical slopes in part b.

Look Back Since interaction is not significant, we will drop the x_1x_2 term from the model and use the simpler model, $E(y) = \beta_0 + \beta_1x_1 + \beta_2x_2$, to predict productivity.

> **Now Work** *Exercise 12.84*

━━━━━━━━ ■ ■ ■ ━━━━━━━━

Models with both quantitative and qualitative x's may also include higher-order (e.g., second-order) terms. In the problem of relating mean monthly sales $E(y)$ of a company to monthly advertising expenditure x_1 and type of medium, suppose we think that the relationship between $E(y)$ and x_1 is curvilinear. We will construct the model, stage by stage, to enable you to compare the procedure with the stage-by-stage construction of the first-order model in the beginning of this section. The graphical interpretations will help you understand the contributions of the model terms.

1. The mean sales curves are identical for all three advertising media, that is, a single second-order curve will suffice to describe the relationship between $E(y)$ and x_1 for all the media (see Figure 12.27):

$$E(y) = \beta_0 + \beta_1x_1 + \beta_2x_1^2$$

where x_1 = Advertising expenditure

Figure 12.27

The Relationship between $E(y)$ and x_1 Is the Same for All Media

2. The response curves possess the same shapes but different y-intercepts (see Figure 12.28):

$$E(y) = \beta_0 + \beta_1 x_1 + \beta_2 x_1^2 + \beta_3 x_2 + \beta_4 x_3$$

where

$$x_1 = \text{Advertising expenditure}$$

$$x_2 = \begin{cases} 1 & \text{if radio medium} \\ 0 & \text{if not} \end{cases}$$

$$x_3 = \begin{cases} 1 & \text{if television medium} \\ 0 & \text{if not} \end{cases}$$

Figure 12.28

The Response Curves Have the Same Shapes But Different y-Intercepts

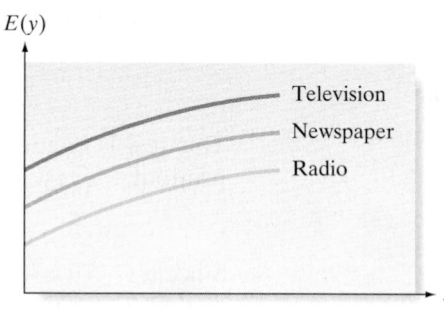

3. The response curves for the three advertising media are different (i.e., Advertising expenditure and Type of medium interact), as shown in Figure 12.29:

$$E(y) = \beta_0 + \beta_1 x_1 + \beta_2 x_1^2 + \beta_3 x_2 + \beta_4 x_3 + \beta_5 x_1 x_2 + \beta_6 x_1 x_3 + \beta_7 x_1^2 x_2 + \beta_8 x_1^2 x_3$$

Now that you know how to write a model with two independent variables—one qualitative and one quantitative—we ask a question. Why do it? Why not write a separate second-order model for each type of medium where $E(y)$ is a function of only advertising expenditure? As stated earlier, one reason we wrote the single model representing all three response curves is so that we can test to determine whether the curves are different. We illustrate this procedure in optional Section 12.9. A second reason for writing a single model is that we obtain a pooled estimate of σ^2, the

Figure 12.29

The Response Curves for the Three Media Differ

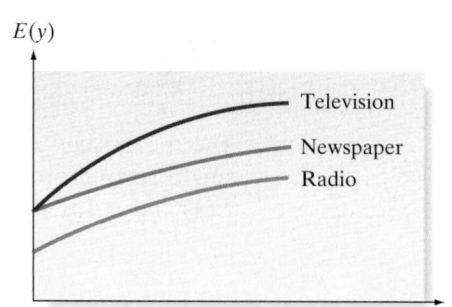

variance of the random error component ε. If the variance of ε is truly the same for each type of medium, the pooled estimate is superior to three separate estimates calculated by fitting a separate model for each type of medium.

Exercises 12.77–12.88

Understanding the Principles

12.77 Consider a multiple regression model for a response y, with one quantitative independent variable x_1, and one qualitative variable at three levels.

   a. Write a first-order model that relates the mean response $E(y)$ to the quantitative independent variable.

   b. Add the main effect terms for the qualitative independent variable to the model of part **a**. Specify the coding scheme you use.

   c. Add terms to the model of part **b** to allow for interaction between the quantitative and qualitative independent variables.

   d. Under what circumstances will the response lines of the model in part **c** be parallel?

   e. Under what circumstances will the model in part **c** have only one response line?

12.78 Refer to Exercise 12.77.

   a. Write a complete second-order model that relates $E(y)$ to the quantitative variable.

   b. Add the main effect terms for the qualitative variable (at three levels) to the model of part **a**.

   c. Add terms to the model of part **b** to allow for interaction between the quantitative and qualitative independent variables.

   d. Under what circumstances will the response curves of the model have the same shape but different y-intercepts?

   e. Under what circumstances will the response curves of the model be parallel lines?

   f. Under what circumstances will the response curves of the model be identical?

12.79 Write a model that relates $E(y)$ to two independent variables—one quantitative and one qualitative at four levels. Construct a model that allows the associated response curves to be second order but does not allow for interaction between the two independent variables.

Learning the Mechanics

12.80 Consider the model
NW

$$y = \beta_0 + \beta_1 x_1 + \beta_2 x_2 + \beta_3 x_3 + \varepsilon$$

where x_1 is a quantitative variable and x_2 and x_3 are dummy variables describing a qualitative variable at three levels using the coding scheme

$$x_2 = \begin{cases} 1 & \text{if level 2} \\ 0 & \text{otherwise} \end{cases} \qquad x_3 = \begin{cases} 1 & \text{if level 3} \\ 0 & \text{otherwise} \end{cases}$$

The resulting least squares prediction equation is

$$\hat{y} = 44.8 + 2.2x_1 + 9.4x_2 + 15.6x_3$$

   a. What is the response line (equation) for $E(y)$ when $x_2 = x_3 = 0$? When $x_2 = 1$ and $x_3 = 0$? When $x_2 = 0$ and $x_3 = 1$?

   b. What is the least squares prediction equation associated with level 1? Level 2? Level 3? Plot these on the same graph.

12.81 Consider the model

$$y = \beta_0 + \beta_1 x_1 + \beta_2 x_1^2 + \beta_3 x_2 + \beta_4 x_3 + \beta_5 x_1 x_2 + \beta_6 x_1 x_3 + \beta_7 x_1^2 x_2 + \beta_8 x_1^2 x_3 + \varepsilon$$

where x_1 is a quantitative variable and

$$x_2 = \begin{cases} 1 & \text{if level 2} \\ 0 & \text{otherwise} \end{cases} \qquad x_3 = \begin{cases} 1 & \text{if level 3} \\ 0 & \text{otherwise} \end{cases}$$

The resulting least squares prediction equation is

$$\hat{y} = 48.8 - 3.4x_1 + .07x_1^2$$
$$- 2.4x_2 - 7.5x_3 + 3.7x_1 x_2$$
$$+ 2.7x_1 x_3 - .02x_1^2 x_2$$
$$- .04x_1^2 x_3$$

   a. What is the equation of the response curve for $E(y)$ when $x_2 = 0$ and $x_3 = 0$? When $x_2 = 1$ and $x_3 = 0$? When $x_2 = 0$ and $x_3 = 1$?

   b. On the same graph, plot the least squares prediction equation associated with level 1, with level 2, and with level 3.

Applying the Concepts—Basic

12.82 **Winning marathon times.** Refer to the *Chance* (Winter 2000) study of men's and women's winning times in the Boston Marathon, Exercise 11.20 (p. 607). Suppose the researchers want to build a model for predicting winning time (y) of the marathon as a function of year (x_1) in which race is run and gender of winning runner (x_2).

   a. Set up the appropriate dummy variables (if necessary) for x_1 and x_2.

   b. Write the equation of a model that proposes parallel straight-line relationships between winning time (y) and year (x_1), one line for each gender.

   c. Write the equation of a model that proposes nonparallel straight-line relationships between winning time (y) and year (x_1), one line for each gender.

   d. Which of the models do you think will provide the best predictions of winning time (y)? Base your answer on the graph displayed in Exercise 11.20.

⊙ **DDT**

12.83 **Study of contaminated fish.** Refer to Exercise 12.20 (p. 681) and the model relating the mean DDT level

$E(y)$ of contaminated fish to x_1 = miles captured upstream, x_2 = length, and x_3 = weight. Now consider a model for $E(y)$ as a function of both weight and species (Channel catfish, Largemouth bass, and Small-mouth buffalo).

a. Set up the appropriate dummy variables for species.

b. Write the equation of a model that proposes parallel straight-line relationships between mean DDT level $E(y)$ and weight, one line for each species.

c. Write the equation of a model that proposes non-parallel straight-line relationships between mean DDT level $E(y)$ and weight, one line for each species.

d. Fit the model, part **b**, to the data saved in the **DDT** file. Give the least squares prediction equation.

e. Refer to part **d**. Interpret the value of the least squares estimate of the beta coefficient multiplied by weight.

f. Fit the model, part **c**, to the data saved in the **DDT** file. Give the least squares prediction equation.

g. Refer to part **f**. Find the estimated slope of the line relating DDT level (y) to weight for the Channel catfish species.

12.84 Smoking and resting energy. The influence of cigarette smoking on resting energy expenditure (REE) in normal-weight and obese smokers was recently investigated. (*Health Psychology*, Mar. 1995.) The researchers hypothesized that the relationship between a smoker's REE and length of time since smoking differs for normal-weight and obese smokers. Consequently, the interaction model was examined:

$$E(y) = \beta_0 + \beta_1 x_1 + \beta_2 x_2 + \beta_3 x_1 x_2$$

where

y = REE, measured in kilocalories per day
x_1 = Time, in minutes after smoking, of metabolic energy reading (levels = 10, 20, and 30 minutes)
$x_2 = \begin{cases} 1 & \text{if normal weight} \\ 0 & \text{if obese} \end{cases}$

a. Give the equation of the hypothesized line relating mean REE to time after smoking for obese smokers. What is the slope of the line?

b. Repeat part **a** for normal-weight smokers.

c. A test for interaction resulted in an observed significance level of .044. Interpret this value.

Applying the Concepts—Intermediate

SNOWGEESE

12.85 Snowgeese feeding trial. Refer to the *Journal of Applied Ecology* study of feeding habits of baby snow geese, Exercise 12.21 (p. 681).

a. Write a first-order model relating gosling weight change (y) to digestion efficiency (x_1) and diet (plants or duck chow) that allows for different slopes for each diet.

b. Fit the model, part **a**, to the data saved in the **SNOWGEESE** trial. Give the least squares prediction equation.

c. Find the estimated slope of the line for goslings fed a diet of plants. Interpret its value.

d. Find the estimated slope of the line for goslings fed a diet of duck chow. Interpret its value.

e. Conduct a test to determine whether the slopes associated with the two diets are significantly different. Use $\alpha = .05$.

12.86 Lead levels in mountain moss. A study of the atmospheric pollution on the slopes of the Blue Ridge Mountains (Tennessee) was conducted. The file **LEADMOSS** contains the levels of lead found in 70 fern moss specimens (in micrograms of lead per gram of moss tissue) collected from the mountain slopes, as well as the elevation of the moss specimen (in feet) and the direction (1 if east, 0 if west) of the slope face. The first five and last five observations of the data set are listed in the table.

LEADMOSS (First and last 5 specimens shown.)

Specimen	Lead Level	Elevation	Slope Face
1	3.475	2000	0
2	3.359	2000	0
3	3.877	2000	0
4	4.000	2500	0
5	3.618	2500	0
⋮	⋮	⋮	⋮
66	5.413	2500	1
67	7.181	2500	1
68	6.589	2500	1
69	6.182	2000	1
70	3.706	2000	1

Source: Schilling, J. "Bioindication of atmospheric heavy metal deposition in the Blue Ridge using the moss, *Thuidium delicatulum*." Master of Science Thesis, Spring 2000.

a. Write the equation of a first-order model relating mean lead level, $E(y)$, to elevation (x_1) and slope face (x_2). Include interaction between elevation and slope face in the model.

b. Graph the relationship between mean lead level and elevation for the different slope faces that is hypothesized by the model, part **a**.

c. In terms of the β's of the model, part **a**, give the change in lead level for every one foot increase in elevation for moss specimens on the east slope.

d. Fit the model, part **a**, to the data using an available statistical software package. Is the overall model statistically useful for predicting lead level? Test using $\alpha = .10$.

12.87 "Sun safety" study. Excessive exposure to solar radiation is known to increase the risk of developing skin cancer, yet many people do not practice "sun safety." A group of University of Arizona researchers examined the feasibility of educating preschool (4- to 5-year-old) children about sun safety. (*American Journal of Public Health*, July 1995.) A sample of 122 preschool children was divided into two groups, the control group and the intervention group. Children in the intervention group received a *Be Sun Safe* curriculum in preschool, while the control group did not. All children were tested for their knowledge, comprehension, and application of sun safety at two points in time: prior to the sun safety curriculum (pretest, x_1) and seven weeks following the curriculum (posttest, y).

a. Write a first-order model for mean posttest score, $E(y)$, as a function of pretest score, x_1 and group. Assume that no interaction exists between pretest score and group.

b. For the model, part **a**, show that the slope of the line relating posttest score to pretest score is the same for both groups of children.

c. Repeat part **a**, but assume that pretest score and group interact.

d. For the model, part **c**, show that the slope of the line relating posttest score to pretest score differs for the two groups of children.

Applying the Concepts—Advanced

12.88 Iron supplement for anemia. Many women suffer from anemia. A female physician, who is also an avid jogger, wanted to know if women who exercise regularly have a different mean red blood cell count than women who do not. She also wanted to know if the amount of a particular iron supplement a woman takes has any effect and whether the effect is the same for both groups. Write a model that will reflect the relationship between red blood cell count and the two independent variables described previously, assuming that

a. the effect of the iron supplement on mean blood cell count is the same regardless of whether a woman exercises regularly.

b. the effect of the iron supplement on mean blood cell count depends on whether a woman exercises regularly.

12.9 Model Building: Comparing Nested Models (Optional)

To be successful model builders, we require a statistical method that will allow us to determine (with a high degree of confidence) which one among a set of candidate models best fits the data. In this section, we present such a technique for *nested models*.

> **DEFINITION 12.3**
>
> Two models are **nested** if one model contains all the terms of the second model and at least one additional term. The more complex of the two models is called the **complete** model, and the simpler of the two is called the **reduced** model.

To illustrate the concept of nested models, consider the straight-line interaction model for the mean auction price $E(y)$ of a grandfather clock as a function of two quantitative variables: age of the clock (x_1) and number of bidders (x_2). The interaction model, fit in Example 12.6, is

$$E(y) = \beta_0 + \beta_1 x_1 + \beta_2 x_2 + \beta_3 x_1 x_2$$

If we assume that the relationship between auction price (y), age (x_1), and bidders (x_2) is curvilinear, then the complete second-order model is more appropriate:

$$E(y) = \overbrace{\beta_0 + \beta_1 x_1 + \beta_2 x_2 + \beta_3 x_1 x_2}^{\text{Terms in interaction model}} + \overbrace{\beta_4 x_1^2 + \beta_5 x_2^2}^{\text{Quadratic terms}}$$

Note that the curvilinear model contains quadratic terms for x_1 and x_2, as well as the terms in the interaction model. Therefore, the models are nested models. In this case, the interaction model is nested within the more complex curvilinear model. Thus, the curvilinear model is the *complete* model and the interaction model is the *reduced* model.

Suppose we want to know whether the more complex curvilinear model contributes more information for the prediction of y than the straight-line interaction model. This is equivalent to determining whether the quadratic betas β_4 and β_5 should be retained in the model. To test whether these terms should be retained, we set up the null and alternative hypotheses as follows:

H_0: $\beta_4 = \beta_5 = 0$ (i.e., quadratic terms are not important for predicting y).

H_a: At least one of the parameters β_4 and β_5 is nonzero (i.e., at least one of the quadratic terms is useful for predicting y).

Note that the terms being tested are those additional terms in the complete (curvilinear) model that are not in the reduced (straight-line interaction) model.

In Section 12.3, we presented the t-test for a single β coefficient and the global F-test for *all* the β parameters (except β_0) in the model. We now need a test for a *subset* of the β parameters in the complete model. The test procedure is intuitive. First, we use the method of least squares to fit the reduced model and calculate the corresponding sum of squares for error, SSE_R (the sum of squares of the deviations between observed and predicted y-values). Next, we fit the complete model and calculate its sum of squares for error, SSE_C. Then, we compare SSE_R to SSE_C by calculating the difference, $SSE_R - SSE_C$. If the additional terms in the complete model are significant, then SSE_C should be much smaller than SSE_R, and the difference $SSE_R - SSE_C$ will be large.

Since SSE will always decrease when new terms are added to the model, the question is whether the difference $SSE_R - SSE_C$ is large enough to conclude that it is due to more than just an increase in the number of model terms and to chance. The formal statistical test utilizes an F-statistic, as shown in the box.

F-Test for Comparing Nested Models

Reduced model: $E(y) = \beta_0 + \beta_1 x_1 + \cdots + \beta_g x_g$

Complete model: $E(y) = \beta_0 + \beta_1 x_1 + \cdots + \beta_g x_g + \beta_{g+1} x_{g+1} + \cdots + \beta_k x_k$

H_0: $\beta_{g+1} = \beta_{g+2} = \cdots = \beta_k = 0$

H_a: At least one of the β parameters specified in H_0 is nonzero.

Test statistic: $F = \dfrac{(SSE_R - SSE_C)/(k - g)}{SSE_C/[n - (k + 1)]}$

$\qquad\qquad = \dfrac{(SSE_R - SSE_C)/\#\beta\text{'s tested in } H_0}{MSE_C}$

where

$\quad SSE_R$ = Sum of squared errors for the reduced model
$\quad SSE_C$ = Sum of squared errors for the complete model
$\quad MSE_C$ = Mean square error (s^2) for the complete model
$\quad k - g$ = Number of β parameters specified in H_0 (i.e., number of β's tested)
$\quad k + 1$ = Number of β parameters in the complete model (including β_0)
$\qquad n$ = Total sample size

Rejection region: $F > F_\alpha$

where F is based on $\nu_1 = k - g$ numerator degrees of freedom and $\nu_2 = n - (k + 1)$ denominator degrees of freedom

---

When the assumptions listed in Section 12.1 about the random error term are satisfied, this F-statistic has an F-distribution with ν_1 and ν_2 df. Note that ν_1 is the number of β parameters being tested and ν_2 is the number of degrees of freedom associated with s^2 in the complete model.

EXAMPLE 12.12   ANALYZING A COMPLETE SECOND-ORDER MODEL

Problem   A botanist conducted an experiment to study the growth of carnations as a function of the temperature x_1 (°F) in a greenhouse and the amount of fertilizer x_2 [kilograms (kg) per plot] applied to the soil. A total of 27 plots of equal size are treated with fertilizer in amounts varying between 50 and 60 kg per plot, and these plots are

○ **CARNATIONS**

TABLE 12.6   Temperature (x_1), Amount of Fertilizer (x_2), and Height of Carnations (y)

x_1	x_2	y	x_1	x_2	y	x_1	x_2	y
80	50	50.8	90	50	63.4	100	50	46.6
80	50	50.7	90	50	61.6	100	50	49.1
80	50	49.4	90	50	63.4	100	50	46.4
80	55	93.7	90	55	93.8	100	55	69.8
80	55	90.9	90	55	92.1	100	55	72.5
80	55	90.9	90	55	97.4	100	55	73.2
80	60	74.5	90	60	70.9	100	60	38.7
80	60	73.0	90	60	68.8	100	60	42.5
80	60	71.2	90	60	71.3	100	60	41.4

mechanically kept at constant temperatures between 80 and 100°F. Small carnation plants [approximately 15 centimeters (cm) in height] are planted in each plot, and their height y (cm) is measured after a 6-week growing period. The resulting data are shown in Table 12.6.

a. Fit a complete second-order model to the data.

b. Sketch the fitted model in three dimensions.

c. Do the data provide sufficient evidence to indicate that the second-order terms, β_3, β_4, and β_5, contribute information for the prediction of y?

Solution   **a.** The complete second-order model is

$$E(y) = \beta_0 + \beta_1 x_1 + \beta_2 x_2 + \beta_3 x_1 x_2 + \beta_4 x_1^2 + \beta_5 x_2^2$$

The data in Table 12.6 were used to fit this model, and a portion of the SAS output is shown in Figure 12.30.

Figure 12.30

SAS Printout of Complete Second-Order Model for Height

```
                          Dependent Variable: HEIGHT

                  Number of Observations Read          27
                  Number of Observations Used          27

                           Analysis of Variance

                                  Sum of          Mean
   Source              DF        Squares        Square    F Value    Pr > F
   Model                5     8402.26454    1680.45291     596.32    <.0001
   Error               21       59.17843       2.81802
   Corrected Total     26     8461.44296

               Root MSE              1.67870    R-Square     0.9930
               Dependent Mean       66.96296    Adj R-Sq     0.9913
               Coeff Var             2.50690

                           Parameter Estimates

                             Parameter      Standard
   Variable      DF          Estimate         Error    t Value    Pr > |t|
   Intercept      1      -5127.89907      110.29601     -46.49     <.0001
   TEMP           1         31.09639        1.34441      23.13     <.0001
   FERT           1        139.74722        3.14005      44.50     <.0001
   TEM_FERT       1         -0.14550        0.00969     -15.01     <.0001
   TEMPSQ         1         -0.13339        0.00685     -19.46     <.0001
   FERTSQ         1         -1.14422        0.02741     -41.74     <.0001


          Test HIGHORD Results for Dependent Variable HEIGHT

                                    Mean
         Source            DF      Square    F Value    Pr > F
         Numerator          3  2204.11003     782.15    <.0001
         Denominator       21     2.81802
```

Figure 12.31

Plot of Second-Order Least Squares Model for Example 12.12

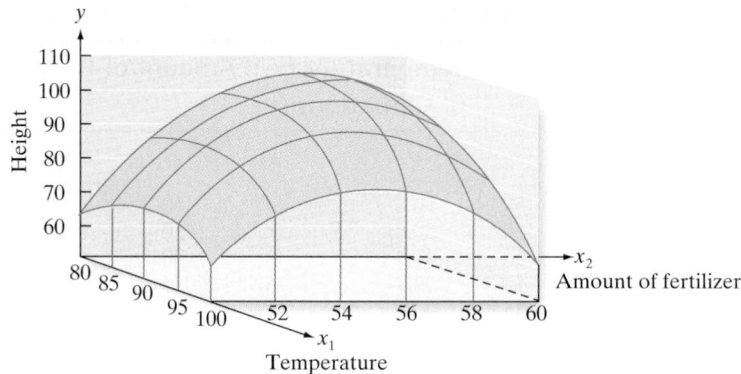

The least squares prediction equation (rounded) is

$$\hat{y} = -5{,}127.90 + 31.10x_1 + 139.75x_2 - .146x_1x_2 - .133x_1^2 - 1.14x_2^2$$

b. A three-dimensional graph of this prediction model, called a **response surface**, is shown in Figure 12.31. Note that the height seems to be greatest for temperatures of about 85–90°F and for applications of about 55–57 kg of fertilizer per plot.* Further experimentation in these ranges might lead to a more precise determination of the optimal temperature–fertilizer combination.

c. To determine whether the data provide sufficient information to indicate that the second-order terms contribute information for the prediction of y, we wish to test

$$H_0: \beta_3 = \beta_4 = \beta_5 = 0$$

against the alternative hypothesis

H_a: At least one of the parameters, β_3, β_4, β_5, differs from 0.

The first step in conducting the test is to drop the second-order terms out of the complete (second-order) model and fit the reduced model

$$E(y) = \beta_0 + \beta_1x_1 + \beta_2x_2$$

to the data. The SAS printout for this model is shown in Figure 12.32.

Figure 12.32

SAS Printout of 1st-Order Model for Height

Dependent Variable: HEIGHT

Number of Observations Read	27
Number of Observations Used	27

Analysis of Variance

Source	DF	Sum of Squares	Mean Square	F Value	Pr > F
Model	2	1789.93444	894.96722	3.22	0.0577
Error	24	6671.50852	277.97952		
Corrected Total	26	8461.44296			

Root MSE	16.67272	R-Square	0.2115
Dependent Mean	66.96296	Adj R-Sq	0.1458
Coeff Var	24.89842		

Parameter Estimates

Variable	DF	Parameter Estimate	Standard Error	t Value	Pr > |t|
Intercept	1	106.08519	55.94500	1.90	0.0700
TEMP	1	-0.91611	0.39298	-2.33	0.0285
FERT	1	0.78778	0.78596	1.00	0.3262

*Students with knowledge of calculus should note that we can solve for the exact temperature and amount of fertilizer that maximize height in the least squares model by solving $\partial\hat{y}/\partial x_1 = 0$ and $\partial\hat{y}/\partial x_2 = 0$ for x_1 and x_2. Sample estimates of these estimated optimal values are $x_1 = 86.25°F$ and $x_2 = 55.58$ kg per plot.

The sums of squares for error, highlighted in Figures 12.30 and 12.32 for the complete and reduced models, respectively, are

$$SSE_C = 59.17843$$
$$SSE_R = 6{,}671.50852$$

and that s^2 for the complete model (highlighted on Figure 12.30) is

$$s^2 = MSE_C = 2.81802$$

Recall that $n = 27$, $k = 5$, and $g = 2$. Therefore, the calculated value of the F statistic, based on $\nu_1 = (k - g) = 3$ numerator df and $\nu_2 = [n - (k + 1)] = 21$ denominator df, is

$$F = \frac{(SSE_R - SSE_C)/(k - g)}{SSE_C/[n - (k + 1)]} = \frac{(SSE_R - SSE_C)/(k - g)}{MSE_C}$$

where $\nu_1 = (k - g)$ is equal to the number of parameters involved in H_0. Therefore,

$$Test\ statistic: \quad F = \frac{(6{,}671.50852 - 59.17843)/3}{2.81802} = 782.15$$

The final step in the test is to compare this computed value of F with the tabulated value based on $\nu_1 = 3$ and $\nu_2 = 21$ df. If we choose $\alpha = .05$, then $F_{.05} = 3.07$ and the rejection region is

$$Rejection\ region: \quad F > 3.07.$$

Since the computed value of F falls in the rejection region (i.e., it exceeds $F_{.05} = 3.07$), we reject H_0 and conclude that at least one of the second-order terms contributes information for the prediction of y. The second-order model appears to provide better predictions of y than a first-order model.

Look Back Using special commands, you can get SAS to perform the desired nested model F-test. The test statistic and p-value for the test above are highlighted at the bottom of the SAS printout, Figure 12.30.

| Now Work | *Exercise 12.92* |

■ ■ ■

The nested model F-test can be used to determine whether *any* subset of terms should be included in a complete model by testing the null hypothesis that a particular set of β parameters simultaneously equals 0. For example, we may want to test to determine whether a set of interaction terms for quantitative variables or a set of main effect terms for a qualitative variable should be included in a model. If we reject H_0, the complete model is the better of the two nested models.

Suppose the F-test in Example 12.12 yielded a test statistic that did not fall in the rejection region. Although we must be cautious about accepting H_0, most practitioners of regression analysis adopt the principle of *parsimony*. That is, in situations where two competing models are found to have essentially the same predictive power (as in this case), the model with the fewer number of β's (i.e., the more parsimonious model) is selected. Based on this principle, we would drop the three second-order terms and select the straight-line (reduced) model over the second-order (complete) model.

> **DEFINITION 12.4**
>
> A **parsimonious model** is a general linear model with a small number of β parameters. In situations where two competing models have essentially the same predictive power (as determined by an F-test), choose the more parsimonious of the two.

When the candidate models in model building are nested models, the F-test developed in this section is the appropriate procedure to apply to compare the models. However, if the models are not nested, this F-test is not applicable. In this situation, the analyst must base the choice of the best model on statistics such as R_a^2 and s. It is important to remember that decisions based on these and other numerical descriptive measures of model adequacy cannot be supported with a measure of reliability and are often very subjective in nature.

Statistics in Action Revisited
Building a Model for Condominium Sale Price

In the previous Statistics in Action Revisited section (p. 687), we fit a first-order model for the auction price (y) of a condominium unit using the six independent variables listed in Table SIA12.1. Although the model was deemed statistically useful for predicting y, the standard deviation of the model ($s = 21.8$ hundred dollars) was probably too large for the model to be "practically" useful. A more complex model—one involving higher-order terms (interactions and squared terms)—needs to be considered. We start with a second-order model involving only the two quantitative independent variables, FLOOR (x_1) and DISTANCE (x_2). The model is given by the equation

$$E(y) = \beta_0 + \beta_1 x_1 + \beta_2 x_2 + \beta_3 x_1 x_2 + \beta_4 (x_1)^2 + \beta_5 (x_2)^2$$

The SAS printout for this model is shown in Figure SIA12.4. Note that the global F-test for the model is statistically significant (p-value $<.0001$).

Are the higher (second)-order terms in the model, $\beta_3 x_1 x_2$, $\beta_4 (x_1)^2$, and $\beta_5 (x_2)^2$ necessary? If not, we can simplify the model by dropping these curvature terms. The hypothesis of interest is $H_0: \beta_3 = \beta_4 = \beta_5 = 0$. To test this subset of β's, we compare the second-order model to a model without the interaction and curvilinear terms. The reduced model takes the form

$$E(y) = \beta_0 + \beta_1 x_1 + \beta_2 x_2$$

The results of this nested model (or partial) F-test are shown at the bottom of the SAS printout, Figure SIA12.4. The p-value of the test (highlighted) is less than .0001. Since this p-value is smaller than $\alpha = .01$, there is sufficient evidence to reject H_0. That is, there is evidence to indicate that at least one of the three higher-order terms is a useful predictor of auction price.

To improve the model, we now add terms for the qualitative variables VIEW (x_3), END (x_4), FURNISH (x_5), and AUCTION (x_6). The developer theorizes that the impact of floor height and distance from elevator on price will vary depending on the unit's view. Consequently, we also add interaction between floor and view and between distance and view. The complete model takes the form

$$\begin{aligned} E(y) = {} & \beta_0 + \beta_1 x_1 + \beta_2 x_2 + \beta_3 x_1 x_2 + \beta_4 (x_1)^2 \\ & + \beta_5 (x_2)^2 + \beta_6 x_3 + \beta_7 x_3 x_1 + \beta_8 x_3 x_2 \\ & + \beta_9 x_3 x_1 x_2 + \beta_{10} x_3 (x_1)^2 + \beta_{11} x_3 (x_2)^2 \\ & + \beta_{12} x_4 + \beta_{13} x_5 + \beta_{14} x_6 \end{aligned}$$

The SAS printout for this complete model is shown in Figure SIA12.5. The overall model is statistically useful (p-value $<.0001$ for global F-test), explaining about 68% (adjusted $R^2 = .6815$) of the sample variation in auction prices. The model standard deviation, $s = 19$, implies that we can predict price to within about 38 hundred dollars. Both the adjusted R^2 and $2s$ values are improvements over the corresponding values for the first-order model of the previous Statistics in Action Revisited (p. 687).

To test the developer's theory of how view impacts the sales price relationship, we conduct a nested model F-test of all the VIEW (x_3) interaction terms. The null hypothesis of interest is $H_0: \beta_7 = \beta_8 = \beta_9 = \beta_{10} = \beta_{11} = 0$ and the reduced model takes the form

$$\begin{aligned} E(y) = {} & \beta_0 + \beta_1 x_1 + \beta_2 x_2 + \beta_3 x_1 x_2 \\ & + \beta_4 (x_1)^2 + \beta_5 (x_2)^2 + \beta_6 x_3 + \beta_{12} x_4 \\ & + \beta_{13} \beta_5 + \beta_{14} x_6 \end{aligned}$$

The p-value of the test (highlighted at the bottom of the SAS printout, Figure SIA12.5) is less than .0001. Since this p-value is smaller than $\alpha = .01$, there is sufficient evidence to conclude that at least one of the view interaction terms is useful for predicting auction price. This implies, as theorized by the developer, that the price–floor and price–distance relationships depend on the unit's view (ocean view or not).

Figure SIA12.4

SAS Printout of the Second-Order Model for Condo Sale Price—Quantitative Variables Only

Dependent Variable: PRICE100

Number of Observations Read   209
Number of Observations Used   209

Analysis of Variance

Source	DF	Sum of Squares	Mean Square	F Value	Pr > F
Model	5	60858	12172	13.88	<.0001
Error	203	178035	877.01813		
Corrected Total	208	238893			

Root MSE	29.61449	R-Square	0.2548	
Dependent Mean	201.28708	Adj R-Sq	0.2364	
Coeff Var	14.71256			

Parameter Estimates

Variable	DF	Parameter Estimate	Standard Error	t Value	Pr > |t|
Intercept	1	229.80506	13.17675	17.44	<.0001
FLOOR	1	-8.76315	4.49100	-1.95	0.0524
DISTANCE	1	-7.33316	2.31456	-3.17	0.0018
FLR_DIST	1	-0.17739	0.20153	-0.88	0.3798
FLOORSQ	1	0.76065	0.44691	1.70	0.0903
DISTSQ	1	0.66948	0.13221	5.06	<.0001

Test HIORDER Results for Dependent Variable PRICE100

Source	DF	Mean Square	F Value	Pr > F
Numerator	3	8500.32594	9.69	<.0001
Denominator	203	877.01813		

Figure SIA12.5

SAS Regression Printout for the Complete Second-Order Model of Condo Sale Price—Qualitative Variables Added

Dependent Variable: PRICE100

Number of Observations Read   209
Number of Observations Used   209

Analysis of Variance

Source	DF	Sum of Squares	Mean Square	F Value	Pr > F
Model	14	167924	11995	32.79	<.0001
Error	194	70968	365.81593		
Corrected Total	208	238893			

Root MSE	19.12632	R-Square	0.7029	
Dependent Mean	201.28708	Adj R-Sq	0.6815	
Coeff Var	9.50201			

Parameter Estimates

Variable	DF	Parameter Estimate	Standard Error	t Value	Pr > |t|
Intercept	1	188.72646	13.15224	14.35	<.0001
FLOOR	1	-4.61416	4.49901	-1.03	0.3064
DISTANCE	1	-2.35297	2.43410	-0.97	0.3349
FLR_DIST	1	-0.33458	0.20177	-1.66	0.0989
FLOORSQ	1	1.01858	0.42699	2.39	0.0180
DISTSQ	1	0.29095	0.14111	2.06	0.0406
VIEW	1	74.33636	17.73275	4.19	<.0001
VU_FLOOR	1	-4.77034	5.90393	-0.81	0.4201
VU_DIST	1	-1.62826	3.17204	-0.51	0.6083
VU_FLR_DIST	1	0.07496	0.27422	0.27	0.7849
VU_FLRSQ	1	-0.26670	0.57992	-0.46	0.6461
VU_DISTSQ	1	0.13340	0.18392	0.73	0.4691
END	1	-16.27750	7.90038	-2.06	0.0407
FURNISH	1	7.96051	2.99043	2.66	0.0084
AUCTION	1	-25.59314	3.66294	-6.99	<.0001

Test VUINT Results for Dependent Variable PRICE100

Source	DF	Mean Square	F Value	Pr > F
Numerator	5	2565.21064	7.01	<.0001
Denominator	194	365.81593		

Exercises 12.89–12.101

Understanding the Principles

12.89 Determine which pairs of the following models are "nested" models. For each pair of nested models, identify the complete and reduced model.

a. $E(y) = \beta_0 + \beta_1 x_1 + \beta_2 x_2$
b. $E(y) = \beta_0 + \beta_1 x_1$
c. $E(y) = \beta_0 + \beta_1 x_1 + \beta_2 x_1^2$
d. $E(y) = \beta_0 + \beta_1 x_1 + \beta_2 x_2 + \beta_3 x_1 x_2$
e. $E(y) = \beta_0 + \beta_1 x_1 + \beta_2 x_2 + \beta_3 x_1 x_2 + \beta_4 x_1^2 + \beta_5 x_2^2$

12.90 Explain why the F-test used to compare complete and reduced models is a one-tailed, upper-tailed test.

12.91 What is a parsimonious model?

Learning the Mechanics

12.92 Suppose you fit the regression model

NW
$$y = \beta_0 + \beta_1 x_1 + \beta_2 x_2 + \beta_3 x_1 x_2 + \beta_4 x_1^2 + \beta_5 x_2^2 + \varepsilon$$

to $n = 30$ data points and you wish to test

$$H_0: \beta_3 = \beta_4 = \beta_5 = 0$$

a. State the alternative hypothesis H_a.
b. Give the reduced model appropriate for conducting the test.
c. What are the numerator and denominator degrees of freedom associated with the F statistic?
d. Suppose the SSE's for the complete and reduced models are $SSE_R = 1,250.2$ and $SSE_C = 1,125.2$. Conduct the hypothesis test and interpret the results of your test. Test using $\alpha = .05$.

12.93 The complete model

$$y = \beta_0 + \beta_1 x_1 + \beta_2 x_2 + \beta_3 x_3 + \beta_4 x_4 + \varepsilon$$

was fit to $n = 20$ data points, with SSE $= 152.66$. The independent variables x_3 and x_4 were dropped from the preceding model, yielding SSE $= 160.44$.

a. How many β parameters are in the complete model? The reduced model?
b. Specify the null and alternative hypotheses you would use to investigate whether the complete model contributes more information for the prediction of y than the reduced model.
c. Conduct the hypothesis test of part **b**. Use $\alpha = .05$.

Applying the Concepts—Basic

12.94 **Mental health of a community.** An article in the *Community Mental Health Journal* (Aug. 2000) used multiple regression analysis to model the level of community adjustment of clients of the Department of Mental Health and Addiction Services in Connecticut. The dependent variable, community adjustment (y), was measured quantitatively based on staff ratings of the clients. (Lower scores indicate better adjustment.) The complete model was a first-order model with 21 independent variables. The independent variables were categorized as Demographic (four variables), Diagnostic (seven variables), Treatment (four variables), and Community (six variables).

a. Write the equation of $E(y)$ for the complete model.
b. Give the null hypothesis for testing whether the seven Diagnostic variables contribute information for the prediction of y.
c. Give the equation of the reduced model appropriate for the test, part **b**.
d. The test, part **b**, resulted in a test statistic of $F = 59.3$ and p-value $<.0001$. Interpret this result in the words of the problem.

12.95 **Students' ability in science.** Refer to the *American Educational Research Journal* (Fall 1998) study of students' perceptions of their science ability in hands-on classrooms, Exercise 12.24 (p. 683). Recall that the first-order, main effects model that was used to predict ability perception (y) included the following independent variables:

Control Variables

$x_1 =$ Prior science attitude score
$x_2 =$ Science ability test score
$x_3 = 1$ if boy, 0 if girl
$x_4 = 1$ if classroom 1 student, 0 if not
$x_5 = 1$ if classroom 3 student, 0 if not
$x_6 = 1$ if classroom 4 student, 0 if not
$x_7 = 1$ if classroom 5 student, 0 if not
$x_8 = 1$ if classroom 6 student, 0 if not

Performance Behaviors

$x_9 =$ Active-leading behavior score
$x_{10} =$ Passive-assisting behavior score
$x_{11} =$ Active-manipulating behavior score

a. Hypothesize the equation of the first-order, main effects model for $E(y)$.
b. The researchers also considered a model that included all possible interactions between the control variables and the performance behavior variables. Write the equation for this model for $E(y)$.
c. The researchers determined that the interaction terms in the model, part **b**, were not significant, and therefore used the model, part **a**, to make inferences. Explain the best way to conduct this test for interaction. Give the null hypothesis of the test.

BEANIE

12.96 **Beanie babies.** Refer to Exercise 11.109 (p. 652) and the data on the values of 50 beanie babies collector's items, published in *Beanie World Magazine*. Suppose we want to predict the market value of a beanie baby using age (in months since Sept. 1998) and whether the beanie baby has been retired or is current (i.e., still in production).

a. Write a complete second-order model for market value as a function of age and current/retired status.

SPSS Output for Exercise 12.96

ANOVA[b]

Model		Sum of Squares	df	Mean Square	F	Sig.
1	Regression	1186549	5	237309.713	2.885	.024[a]
	Residual	3618994	44	82249.862		
	Total	4805543	49			

a. Predictors: (Constant), AGESQ_STATUS, STATDUM, AGE, AGESQ, AGE_STATUS

b. Dependent Variable: VALUE

ANOVA[b]

Model		Sum of Squares	df	Mean Square	F	Sig.
1	Regression	1116017	3	372005.590	4.638	.006[a]
	Residual	3689526	46	80207.081		
	Total	4805543	49			

a. Predictors: (Constant), AGE_STATUS, AGE, STATDUM

b. Dependent Variable: VALUE

ANOVA[b]

Model		Sum of Squares	df	Mean Square	F	Sig.
1	Regression	1082210	3	360736.761	4.457	.008[a]
	Residual	3723332	46	80942.005		
	Total	4805543	49			

a. Predictors: (Constant), AGESQ, STATDUM, AGE

b. Dependent Variable: VALUE

b. Specify the null hypothesis for testing whether the quadratic terms in the model, part **a**, are important for predicting market value.

c. Specify the null hypothesis for testing whether the interaction terms in the model, part **a**, are important for predicting market value.

d. Three models were fit to the data using SPSS. The relevant SPSS printouts are shown above. Use this information to conduct the tests specified in parts **b** and **c**. Interpret the results.

Applying the Concepts—Intermediate

WATEROIL

12.97 Extracting water from oil. Refer to the *Journal of Colloid and Interface Science* study of water/oil mixtures, Exercise 12.23 (p. 682). Recall that three of the seven variables used to predict voltage (y) were volume (x_1), salinity (x_2), and surfactant concentration (x_5). The model the researchers fit is

$$E(y) = \beta_0 + \beta_1 x_1 + \beta_2 x_2 + \beta_3 x_5 + \beta_4 x_1 x_2 + \beta_5 x_1 x_5$$

a. Note that the model includes interaction between disperse phase volume (x_1) and salinity (x_2) as well as interaction between disperse phase volume (x_1) and surfactant concentration (x_5). Discuss how

these interaction terms affect the hypothetical relationship between y and x_1. Draw a sketch to support your answer.

b. Fit the interaction model to the data. Does this model appear to fit the data better than the first-order model in Exercise 12.23? Explain.

c. Interpret the β estimates of the interaction model.

d. Conduct a test to determine whether the interaction terms contribute significantly to the prediction of voltage(y). Use $\alpha = .05$.

12.98 "Sun Safety" study. Refer to the *American Journal of Public Health* study of preschool children's awareness of sun safety, Exercise 12.87 (p. 724). Consider the first-order, interaction model

$$E(y) = \beta_0 + \beta_1 x_1 + \beta_2 x_2 + \beta_3 x_1 x_2$$

where

y = sun safety posttest score

x_1 = sun safety pretest score

$x_2 = \begin{cases} 1 & \text{if in the } \textit{Be Sun Safe} \text{ intervention group} \\ 0 & \text{if in the control group} \end{cases}$

a. Assuming interaction exists, give the reduced model for testing whether the mean posttest scores differ for the intervention and control groups.

b. When sun safety knowledge was used as the dependent variable, the test of part **a** resulted in a p-value of .03. Interpret this result.

c. When sun safety comprehension was used as the dependent variable, the test of part **a** resulted in a p-value of .033. Interpret this result.

d. When sun safety application was used as the dependent variable, the test of part **a** resulted in a p-value of .322. Interpret this result.

12.99 Improving SAT scores. Refer to the *Chance* (Winter 2001) study of students who paid a private tutor (or coach) to help them improve their Scholastic Assessment Test (SAT) scores, Exercise 12.68 (p. 713). Recall that the baseline model, $E(y) = \beta_0 + \beta_1 x_1 + \beta_2 x_2$, where y = SAT-Math score, x_1 = score on PSAT, and x_2 = {1 if student was coached, 0 if not}, had the following results: $R_a^2 = .76$, $\hat{\beta}_2 = 19$, and $s_{\hat{\beta}_2} = 3$. As an alternative model, the researcher added several "control" variables, including dummy variables for student ethnicity (x_3, x_4, and x_5), a socioeconomic status index variable (x_6), two variables that measured high school performance (x_7 and x_8), the number of math courses taken in high school (x_9), and the overall GPA for the math courses (x_{10}).

a. Write the hypothesized equation for $E(y)$ for the alternative model.

b. Give the null hypothesis for a nested model F-test comparing the initial and alternative models.

c. The nested model F-test, part **b**, was statistically significant at $\alpha = .05$. Practically interpret this result.

d. The alternative model, part **a**, resulted in $R_a^2 = .79$, $\hat{\beta}_2 = 14$, and $s_{\hat{\beta}_2} = 3$. Interpret the value of R_a^2.

e. Refer to part **d**. Find and interpret a 95% confidence interval for β_2.

f. The researcher concluded that "the estimated effect of SAT coaching decreases from the baseline model when control variables are added to the model." Do you agree? Justify your answer.

g. As a modification to the model of part **a**, the researcher added all possible interactions between the coaching variable (x_2) and the other independent variables in the model. Write the equation for $E(y)$ for this modified model.

h. Give the null hypothesis for comparing the models, parts **a** and **g**. How would you perform this test?

12.100 Glass as a waste encapsulant. Since glass is not subject to radiation damage, encapsulation of waste in glass is considered to be one of the most promising solutions to the problem of low-level nuclear waste in the environment. However, chemical reactions may weaken the glass. This concern led to a study undertaken jointly by the Department of Materials Science and Engineering at the University of Florida and the U.S. Department of Energy to assess the utility of glass as a waste encapsulant.* Corrosive chemical solutions (called corrosion baths) were prepared and applied directly to glass samples containing one of three types of waste (TDS-3A, FE, and AL); the chemical reactions were observed over time. A few of the key variables measured were

y = Amount of silicon (in parts per million) found in solution at end of experiment. (This is both a measure of the degree of breakdown in the glass and a proxy for the amount of radioactive species released into the environment.)

x_1 = Temperature (°C) of the corrosion bath

x_2 = 1 if waste type TDS-3A, 0 if not

x_3 = 1 if waste type FE, 0 if not

(Waste type AL is the base level.) Suppose we want to model amount y of silicon as a function of temperature (x_1) and type of waste (x_2, x_3).

a. Write a model that proposes parallel straight-line relationships between amount of silicon and temperature, one line for each of the three waste types.

b. Add terms for the interaction between temperature and waste type to the model of part **a**.

c. Refer to the model of part **b**. For each waste type, give the slope of the line relating amount of silicon to temperature.

d. Explain how you could test for the presence of temperature–waste type interaction.

Applying the Concepts—Advanced

12.101 Emotional distress in firefighters. The *Journal of Human Stress* (Summer 1987) reported on a study of "psychological response of firefighters to chemical fire." It is thought that the following complete second-order model will be adequate to describe the relationship between emotional distress and years of experience for two groups of firefighters—those exposed to a chemical fire and those unexposed.

$$E(y) = \beta_0 + \beta_1 x_1 + \beta_2 x_1^2 + \beta_3 x_2 + \beta_4 x_1 x_2 + \beta_5 x_1^2 x_2$$

where

y = Emotional distress

x_1 = Experience (years)

x_2 = 1 if exposed to chemical fire, 0 if not

a. How would you determine whether the *rate* of increase of emotional distress with experience is different for the two groups of firefighters?

b. How would you determine whether there are differences in mean emotional distress levels that are attributable to exposure group?

---

*The background information for this exercise was provided by Dr. David Clark, Department of Materials Science and Engineering, University of Florida.

12.10 Model Building: Stepwise Regression (Optional)

Consider the problem of predicting the salary y of an executive. Perhaps the biggest problem in building a model to describe executive salaries is choosing the important independent variables to be included in the model. The list of potentially important independent variables is extremely long (e.g., age, experience, tenure, education level, etc.), and we need some objective method of screening out those that are not important.

The problem of deciding which of a large set of independent variables to include in a model is a common one. Trying to determine which variables influence the profit of a firm, affect blood pressure of humans, or are related to a student's performance in college are only a few examples.

A systematic approach to building a model with a large number of independent variables is difficult because the interpretation of multivariable interactions and higher-order terms is tedious. We therefore turn to a screening procedure, available in most statistical software packages, known as *stepwise regression*.

The most commonly used **stepwise regression** procedure works as follows. The user first identifies the response, y, and the set of potentially important independent variables, $x_1, x_2, \ldots, x_k$, where k is generally large. [*Note:* This set of variables could include both first-order and higher-order terms. However, we may often include only the main effects of both quantitative variables (first-order terms) and qualitative variables (dummy variables), since the inclusion of second-order terms greatly increases the number of independent variables.] The response and independent variables are then entered into the computer software, and the stepwise procedure begins.

Step 1 The software program fits all possible one-variable models of the form

$$E(y) = \beta_0 + \beta_1 x_i$$

to the data, where x_i is the ith independent variable, $i = 1, 2, \ldots, k$. For each model, the test of the null hypothesis

$$H_0: \beta_1 = 0$$

against the alternative hypothesis

$$H_a: \beta_1 \neq 0$$

is conducted using the t-test (or the equivalent F-test) for a single β parameter. The independent variable that produces the largest (absolute) t value is declared the best one-variable predictor of y.* Call this independent variable x_1.

Step 2 The stepwise program now begins to search through the remaining $(k - 1)$ independent variables for the best two-variable model of the form

$$E(y) = \beta_0 + \beta_1 x_1 + \beta_2 x_i$$

This is done by fitting all two-variable models containing x_1 and each of the other $(k - 1)$ options for the second variable x_i. The t-values for the test $H_0: \beta_2 = 0$ are computed for each of the $(k - 1)$ models (corresponding to the remaining independent variables, $x_i, i = 2, 3, \ldots, k$), and the variable having the largest t is retained. Call this variable x_2. At this point, some software packages diverge in methodology. The better packages now go back and check the t-value of $\hat{\beta}_1$ after $\hat{\beta}_2 x_2$ has been added to the model. If the t-value

*Note that the variable with the largest t value is also the one with the largest (absolute) Pearson product moment correlation, r (Section 11.6), with y.

has become nonsignificant at some specified α level (say $\alpha = .05$), the variable x_1 is removed and a search is made for the independent variable with a β parameter that will yield the most significant t-value in the presence of $\hat{\beta}_2 x_2$. Other packages do not recheck the significance of $\hat{\beta}_1$ but proceed directly to step 3.

The reason the t-value for x_1 may change from step 1 to step 2 is that the meaning of the coefficient $\hat{\beta}_1$ changes. In step 2, we are approximating a complex response surface in two variables with a plane. The best-fitting plane may yield a different value for $\hat{\beta}_1$ than that obtained in step 1. Thus, both the value of $\hat{\beta}_1$ and its significance usually changes from step 1 to step 2. For this reason, the software packages that recheck the t-values at each step are preferred.

Step 3 The stepwise procedure now checks for a third independent variable to include in the model with x_1 and x_2. That is, we seek the best model of the form

$$E(y) = \beta_0 + \beta_1 x_1 + \beta_2 x_2 + \beta_3 x_i$$

To do this, we fit all the $(k - 2)$ models using x_1, x_2, and each of the $(k - 2)$ remaining variables, x_i, as a possible x_3. The criterion is again to include the independent variable with the largest t-value. Call this best third variable x_3. The better programs now recheck the t-values corresponding to the x_1- and x_2-coefficients, replacing the variables that yield nonsignificant t-values. This procedure is continued until no further independent variables can be found that yield significant t-values (at the specified α level) in the presence of the variables already in the model.

The result of the stepwise procedure is a model containing only those terms with t-values that are significant at the specified α level. Thus, in most practical situations only several of the large number of independent variables remain. However, it is very important *not* to jump to the conclusion that all the independent variables important for predicting y have been identified or that the unimportant independent variables have been eliminated. Remember, the stepwise procedure is using only *sample estimates* of the true model coefficients (β's) to select the important variables. An extremely large number of single β parameter t-tests have been conducted, and the probability is very high that one or more errors have been made in including or excluding variables. That is, we have very probably included some unimportant independent variables in the model (Type I errors) and eliminated some important ones (Type II errors).

There is a second reason why we might not have arrived at a good model. When we choose the variables to be included in the stepwise regression, we may often omit high-order terms (to keep the number of variables manageable). Consequently, we may have initially omitted several important terms from the model. Thus, we should recognize stepwise regression for what it is: an objective *variable screening* procedure.

Successful model builders will now consider second-order terms (for quantitative variables) and other interactions among variables screened by the stepwise procedure. It would be best to develop this response surface model with a second set of data independent of that used for the screening, so the results of the stepwise procedure can be partially verified with new data. This is not always possible, however, because in many modeling situations only a small amount of data is available.

Do not be deceived by the impressive-looking t values that result from the stepwise procedure—it has retained only the independent variables with the largest t values. Also, be certain to consider second-order terms in systematically developing the prediction model. Finally, if you have used a first-order model for your stepwise procedure, remember that it may be greatly improved by the addition of higher-order terms.

Caution

Be wary of using the results of stepwise regression to make inferences about the relationship between $E(y)$ and the independent variables in the resulting first-order model. First, an extremely large number of t-tests have been conducted, leading to a high probability of making one or more Type I or Type II errors. Second, the stepwise model does not include any higher-order or interaction terms. Stepwise regression should be used only when necessary, that is, when you want to determine which of a large number of potentially important independent variables should be used in the model-building process.

EXAMPLE 12.13   RUNNING A STEPWISE REGRESSION

Problem   An international management consulting company develops multiple regression models for executive salaries of its client firms. The consulting company has found that models that use the natural logarithm of salary as the dependent variable have better predictive power than those using salary as the dependent variable.* A preliminary step in the construction of these models is the determination of the most important independent variables. For one firm, 10 potential independent variables (seven quantitative and three qualitative) were measured in a sample of 100 executives. The data, described in Table 12.7, are saved in the **EXECSAL** file. Since it would be very difficult to construct a complete second-order model with all of the 10 independent variables, use stepwise regression to decide which of the 10 variables should be included in the building of the final model for the log of executive salaries.

⊚ **EXECSAL**

TABLE 12.7   Independent Variables in the Executive Salary Example

Independent Variable	Description
x_1	Experience (years)—quantitative
x_2	Education (years)—quantitative
x_3	Bonus eligibility (1 if yes, 0 if no)—qualitative
x_4	Number of employees supervised—quantitative
x_5	Corporate assets (millions of dollars)—quantitative
x_6	Board member (1 if yes, 0 if no)—qualitative
x_7	Age (years)—quantitative
x_8	Company profits (past 12 months, millions of dollars)—quantitative
x_9	Has international responsibility (1 if yes, 0 if no)—qualitative
x_{10}	Company's total sales (past 12 months, millions of dollars)—quantitative

Solution   We will use stepwise regression with the main effects of the 10 independent variables to identify the most important variables. The dependent variable y is the natural logarithm of the executive salaries. The MINITAB stepwise regression printout is shown in Figure 12.33.

   Note that the first variable included in the model is x_1, years of experience. At the second step, x_3, a dummy variable for the qualitative variable, bonus eligibility, is entered into the model. In steps 3, 4, and 5, the variables x_4 (number of employees supervised), x_2 (years of education), and x_5 (corporate assets), respectively, are selected for model inclusion. MINITAB stops after five steps because no other independent variables met the criterion for admission into the model. As

---

*This is probably because salaries tend to be incremented in *percentages* rather than dollar values. When a response variable undergoes percentage changes as the independent variables are varied, the logarithm of the response variable will be more suitable as a dependent variable.

Figure 12.33

MINITAB Stepwise
Regression Printout for
Executive Salary Data

```
           Alpha-to-Enter: 0.15   Alpha-to-Remove: 0.15


Response is Y on 10 predictors, with N = 100


Step               1        2        3        4        5
Constant      11.091   10.968   10.783   10.278    9.962

X1            0.0278   0.0273   0.0273   0.0273   0.0273
T-Value        12.62    15.13    18.80    24.68    26.50
P-Value        0.000    0.000    0.000    0.000    0.000

X3                     0.197    0.233    0.232    0.225
T-Value                 7.10    10.17    13.30    13.74
P-Value                 0.000    0.000    0.000    0.000

X4                             0.00048  0.00055  0.00052
T-Value                          7.32    10.92    11.06
P-Value                          0.000    0.000    0.000

X2                                      0.0300   0.0291
T-Value                                   8.38     8.72
P-Value                                   0.000    0.000

X5                                               0.00196
T-Value                                             3.95
P-Value                                            0.000

S             0.161    0.131    0.106   0.0807   0.0751
R-Sq          61.90    74.92    83.91    90.75    92.06
R-Sq(adj)     61.51    74.40    83.41    90.36    91.64
Mallows C-p   343.9    195.5     93.8     16.8      3.6
```

a default, MINITAB uses $\alpha = .15$ in the t-tests conducted. In other words, if the p-value associated with a β coefficient exceeds $\alpha = .15$, the variable is *not* included in the model.

The results of the stepwise regression suggest that we should concentrate on these five independent variables. Models with second-order terms and interactions should be proposed and evaluated to determine the best model for predicting executive salaries.

Now Work   *Exercise 12.104*

■ ■ ■

We conclude this section with some advice on the use of stepwise regression.

Recommendation

Do *not* use the stepwise regression model as the final model for predicting y. Recall that the stepwise procedure tends to perform a large number of t-tests, inflating the overall probability of a Type I error, and does not automatically include higher-order terms (e.g., interactions and squared terms) in the final model. Use stepwise regression as a variable screening tool when there exists a large number of potentially important independent variables. Then begin building models for y using the variables identified by stepwise regression.

Exercises 12.102–12.107

Understanding the Principles

12.102 Explain the difference between a stepwise model and a standard regression model.

12.103 Give two coveats with using the stepwise regression results as the final model for predicting y.

Learning the Mechanics

12.104 There are six independent variables, x_1, x_2, x_3, x_4, x_5, and x_6, that might be useful in predicting a response y. A total of $n = 50$ observations are available, and it is decided to employ stepwise regression to help in selecting the independent variables that appear to be useful. The computer fits all possible one-variable models of the form

$$E(y) = \beta_0 + \beta_1 x_i$$

where x_i is the ith independent variable, $i = 1, 2, \ldots, 6$. The information in the table is provided from the computer printout.

Independent Variable	$\hat{\beta}_i$	$s_{\hat{\beta}_i}$
x_1	1.6	.42
x_2	−.9	.01
x_3	3.4	1.14
x_4	2.5	2.06
x_5	−4.4	.73
x_6	.3	.35

a. Which independent variable is declared the best one-variable predictor of y? Explain.
b. Would this variable be included in the model at this stage? Explain.
c. Describe the next phase that a stepwise procedure would execute.

Applying the Concepts—Basic

12.105 Entry level job preferences. *Benefits Quarterly* (First Quarter, 1995) published a study of entry level job preferences. A number of independent variables were used to model the job preferences (measured on a 10-point scale) of 164 business school graduates. Suppose stepwise regression is used to build a model for job preference score (y) as a function of the following independent variables:

$x_1 = \begin{cases} 1 & \text{if flextime position} \\ 0 & \text{if not} \end{cases}$

$x_2 = \begin{cases} 1 & \text{if day care support required} \\ 0 & \text{if not} \end{cases}$

$x_3 = \begin{cases} 1 & \text{if spousal transfer support required} \\ 0 & \text{if not} \end{cases}$

$x_4 =$ Number of sick days allowed

$x_5 = \begin{cases} 1 & \text{if applicant married} \\ 0 & \text{if not} \end{cases}$

$x_6 =$ Number of children of applicant

$x_7 = \begin{cases} 1 & \text{if male applicant} \\ 0 & \text{if female applicant} \end{cases}$

a. How many models are fit to the data in step 1? Give the general form of these models.
b. How many models are fit to the data in step 2? Give the general form of these models.
c. How many models are fit to the data in step 3? Give the general form of these models.
d. Explain how the procedure determines when to stop adding independent variables to the model.
e. Describe two major drawbacks to using the final stepwise model as the "best" model for job preference score (y).

Applying the Concepts—Intermediate

12.106 A marine biologist was hired by the EPA to determine whether the hot-water runoff from a particular power plant located near a large gulf is having an adverse effect on the marine life in the area. The biologist's goal is to acquire a prediction equation for the number of marine animals located at certain designated areas, or stations, in the gulf. Based on past experience, the EPA considered the following environmental factors as predictors for the number of animals at a particular station:

$x_1 =$ Temperature of water (TEMP)
$x_2 =$ Salinity of water (SAL)
$x_3 =$ Dissolved oxygen content of water (DO)
$x_4 =$ Turbidity index, a measure of the turbidity of the water (TI)
$x_5 =$ Depth of the water at the station (ST DEPTH)
$x_6 =$ Total weight of sea grasses in sampled area (TGRSWT)

As a preliminary step in the construction of this model, the biologist used a stepwise regression procedure to identify the most important of these six variables. A total of 716 samples were taken at different stations in the gulf, producing the SPSS printout shown on p. 740. (The response measured was y, the logarithm of the number of marine animals found in the sampled area.)

a. According to the SPSS printout, which of the six independent variables should be used in the model?
b. Are we able to assume that the marine biologist has identified all the important independent variables for the prediction of y? Why?
c. Using the variables identified in part **a**, write the first-order model with interaction that may be used to predict y.
d. How would the marine biologist determine whether the model specified in part **c** is better than the first-order model?
e. Note the small value of R_2. What action might the biologist take to improve the model?

12.107 Bus Rapid Transit study. Bus Rapid Transit (BRT) is a rapidly growing trend in the provision of public

SPSS Output for Exercise 12.106

Variables Entered/Removed[a]

Model	Variables Entered	Variables Removed	Method
1	ST_DEPTH	.	Stepwise (Criteria: Probability-of-F-to-enter <= .050, Probability-of-F-to-remove >= .100).
2	TGRSWT	.	Stepwise (Criteria: Probability-of-F-to-enter <= .050, Probability-of-F-to-remove >= .100).
3	TI	.	Stepwise (Criteria: Probability-of-F-to-enter <= .050, Probability-of-F-to-remove >= .100).

a. Dependent Variable: LOGNUM

Model Summary

Model	R	R Square	Adjusted R Square	Std. Error of the Estimate
1	.329[a]	.122	.121	.7615773
2	.427[b]	.182	.180	.7348470
3	.432[c]	.187	.184	.7348469

a. Predictors: (Constant), ST_DEPTH

b. Predictors: (Constant), ST_DEPTH, TGRSWT

c. Predictors: (Constant), ST_DEPTH, TGRSWT, TI

transportation in America. The Center for Urban Transportation Research (CUTR) at the University of South Florida conducted a survey of BRT customers in Miami. (*Transportation Research Board* Annual Meeting, Jan. 2003.) Data on the following variables (all measured on a 5-point scale, where $1 =$ "very unsatisfied" and $5 =$ "very satisfied") were collected for a sample of over 500 bus riders: overall satisfaction with BRT (y), safety on bus (x_1), seat availability (x_2), dependability (x_3), travel time (x_4), cost (x_5), information/maps (x_6), convenience of routes (x_7), traffic signals (x_8), safety at bus stops (x_9), hours of service (x_{10}), and frequency of service (x_{11}). CUTR analysts used stepwise regression to model overall satisfaction (y).

a. How many models are fit at Step 1 of the stepwise regression?

b. How many models are fit at Step 2 of the stepwise regression?

c. How many models are fit at Step 11 of the stepwise regression?

d. The stepwise regression selected the following eight variables to include in the model (in order of selection): $x_{11}, x_4, x_2, x_7, x_{10}, x_1, x_9,$ and x_3. Write the equation for $E(y)$ that results from stepwise.

e. The model, part **d**, resulted in $R^2 = .677$. Interpret this value.

f. Explain why the CUTR analysts should be cautious in concluding that the "best" model for $E(y)$ has been found.

12.11 Residual Analysis: Checking the Regression Assumptions

When we apply regression analysis to a set of data, we never know for certain whether the assumptions of Section 12.1 are satisfied. How far can we deviate from the assumptions and still expect regression analysis to yield results that will have the reliability stated in this chapter? How can we detect departures (if they exist) from the assumptions and what can we do about them? We provide some answers to these questions in this section.

Recall from Section 12.1 that for any given set of values of $x_1, x_2, \ldots, x_k$ we assume that the random error term ε has a normal probability distribution with mean equal to 0 and variance equal to σ^2. Also, we assume that the random errors are probabilistically independent. It is unlikely that these assumptions are ever satisfied exactly in a practical application of regression analysis. Fortunately, experience has shown that least squares regression analysis produces reliable statistical tests, confidence intervals, and prediction intervals as long as the departures from the assumptions are not too great. In this section we present some methods for determining whether the data indicate significant departures from the assumptions.

Because the assumptions all concern the random error component, ε, of the model, the first step is to estimate the random error. Since the actual random error associated with a particular value of y is the difference between the actual y value and its unknown

Figure 12.34

Actual Random Error ε and Regression Residual $\hat{\varepsilon}$

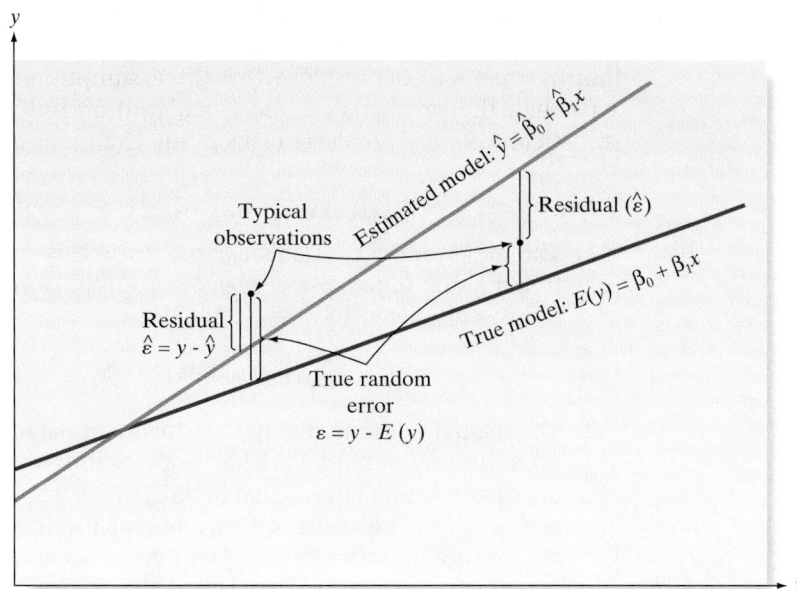

mean, we estimate the error by the difference between the actual y value and the *estimated* mean. This estimated error is called the *regression residual*, or simply the **residual**, and is denoted by $\hat{\varepsilon}$. The actual error ε and residual $\hat{\varepsilon}$ are shown in Figure 12.34.

> **DEFINITION 12.5**
>
> A **regression residual**, $\hat{\varepsilon}$, is defined as the difference between an observed y-value and its corresponding predicted value:
>
> $$\hat{\varepsilon} = (y - \hat{y}) = y - (\hat{\beta}_0 + \hat{\beta}_1 x_1 + \hat{\beta}_2 x_2 + \cdots + \hat{\beta}_k x_k)$$

Since the true mean of y (that is, the true regression model) is not known, the actual random error cannot be calculated. However, because the residual is based on the estimated mean (the least squares regression model), it can be calculated and used to estimate the random error and to check the regression assumptions. Such checks are generally referred to as **residual analyses**. Two useful properties of residuals are given in the box on p. 742.

Biography

FRANCIS J. ANSCOMBE (1918–2001)—Anscombe's Data

British citizen Frank Anscombe grew up in a small town near the English Channel. He attended Trinity College in Cambridge, England on a merit scholarship, graduating with first class honors in mathematics in 1939. He later earned his master's degree in 1943. During World War II, Anscombe worked for the British Ministry of Supply, developing a mathematical solution for aiming anti-aircraft rockets at German bombers and buzz bombs. Following the war, Anscombe worked at the Rothamsted Experimental Station, applying statistics to agriculture. There, he formed his appreciation for solving problems with social relevance. During his career as a professor of statistics, Anscombe served on the faculty of Cambridge, Princeton, and Yale Universities. He was a pioneer in the application of computers to statistical analysis and was one of the original developers of residual analysis in regression. Anscombe is famous for a paper he wrote in 1973 in which he showed that one regression model could be fit by four very different data sets ("Anscombe's data"). While Anscombe published 50 research articles on statistics, he also had serious interests in classical music, poetry, and art.

The following examples show how a graphical analysis of regression residuals can be used to verify the assumptions associated with the model and to support improvements to the model when the assumptions do not appear to be satisfied. Although the residuals can be calculated and plotted by hand, we rely on the statistical software for these tasks in the examples and exercises.

Properties of Regression Residuals

1. The mean of the residuals is equal to 0. This property follows from the fact that the sum of the differences between the observed y values and their least squares predicted $\hat{y}$ values is equal to 0.

$$\sum(\text{Residuals}) = \sum(y - \hat{y}) = 0$$

2. The standard deviation of the residuals is equal to the standard deviation of the fitted regression model, s. This property follows from the fact that the sum of the squared residuals is equal to SSE, which when divided by the error degrees of freedom is equal to the variance of the fitted regression model, s^2. The square root of the variance is both the standard deviation of the residuals and the standard deviation of the regression model.

$$\sum(\text{Residuals})^2 = \sum(y - \hat{y})^2 = \text{SSE}$$

$$s = \sqrt{\frac{\sum(\text{Residuals})^2}{n - (k + 1)}} = \sqrt{\frac{\text{SSE}}{n - (k + 1)}} = \sqrt{\text{MSE}}$$

First, we demonstrate how a residual plot can detect a model in which the hypothesized relationship between $E(y)$ and an independent variable x is misspecified. The assumption of mean error of 0 is violated in these types of models.*

EXAMPLE 12.14   ANALYZING RESIDUALS

Problem   Refer to the problem of modeling the relationship between home size (x) and electrical usage (y) in Example 12.7 (p. 698). The data for $n = 10$ homes are repeated in Table 12.8. MINITAB printouts for a straight-line model and a quadratic model fitted to the data are shown in Figures 12.35a and 12.35b, respectively. The residuals from these models are highlighted in the printouts. The residuals are then plotted

⊚ **ELECTRIC**

TABLE 12.8   Home Size–Electrical Usage Data

Size of Home x (sq. ft)	Monthly Usage y (kilowatt-hours)
1,290	1,182
1,350	1,172
1,470	1,264
1,600	1,493
1,710	1,571
1,840	1,711
1,980	1,804
2,230	1,840
2,400	1,956
2,930	1,954

*For a misspecified model, the hypothesized mean of y, denoted by $E_h(y)$, will not equal the true mean of y, $E(y)$. Since $y = E_h(y) + \varepsilon$, then $\varepsilon = y - E_h(y)$ and $E(\varepsilon) = E\{y - E_h(y)\} = E(y) - E_h(y) \neq 0$.

Figure 12.35a

MINITAB Printout for
Straight-Line Model of
Electrical Usage

```
The regression equation is
USAGE = 579 + 0.540 SIZE


Predictor        Coef   SE Coef      T       P
Constant        578.9     167.0   3.47   0.008
SIZE          0.54030   0.08593   6.29   0.000


S = 133.438    R-Sq = 83.2%    R-Sq(adj) = 81.1%


Analysis of Variance

Source          DF       SS      MS       F       P
Regression       1   703957  703957   39.54   0.000
Residual Error   8   142445   17806
Total            9   846402


Obs   SIZE   USAGE     Fit   SE Fit   Residual   St Resid
 1    1290  1182.0  1275.9    66.0      -93.9      -0.81
 2    1350  1172.0  1308.3    62.1     -136.3      -1.15
 3    1470  1264.0  1373.2    55.0     -109.2      -0.90
 4    1600  1493.0  1443.4    48.6       49.6       0.40
 5    1710  1571.0  1502.8    44.7       68.2       0.54
 6    1840  1711.0  1573.1    42.3      137.9       1.09
 7    1980  1804.0  1648.7    43.1      155.3       1.23
 8    2230  1840.0  1783.8    51.8       56.2       0.46
 9    2400  1956.0  1875.7    61.5       80.3       0.68
10    2930  1954.0  2162.0    99.6     -208.0      -2.34R

R denotes an observation with a large standardized residual.
```

on the vertical axis against the variable x, size of home, on the horizontal axis in Figures 12.36a and 12.36b, respectively.

a. Verify that each residual is equal to the difference between the observed y-value and the estimated mean value, $\hat{y}$.

b. Analyze the residual plots.

Solution **a.** Consider the first usage value in the data set, $y = 1{,}182$. For the straight-line model the predicted value for this observation is $\hat{y} = 1{,}275.92$ (highlighted on Figure 12.35a). Consequently, the residual is

$$\hat{\varepsilon} = (y - \hat{y}) = 1{,}182 - 1{,}275.9 = -93.9$$

Similarly, the residual for the first y value using the quadratic model (Figure 12.35b) is

$$\hat{\varepsilon} = 1{,}182 - 1{,}129.6 = 52.4$$

Both residuals agree (after rounding) with the first values given in the column labeled "**Residual**" in Figures 12.35a and 12.35b, respectively. Although the residuals both correspond to the same observed y value, 1,182, they differ because the predicted mean value changes depending on whether the straight-line model or quadratic model is used. Similar calculations produce the remaining residuals.

b. The MINITAB plot of the residuals for the straight-line model (Figure 12.36a) reveals a nonrandom pattern. The residuals exhibit a curved shape, with the residuals for the small values of x below the horizontal 0 (mean of the residuals) line, the residuals corresponding to the middle values of x above the 0 line, and

Figure 12.35b
MINITAB Printout
for Quadratic Model
of Electrical Usage

```
The regression equation is
USAGE = - 1216 + 2.40 SIZE - 0.000450 SIZESQ


Predictor           Coef      SE Coef        T       P
Constant         -1216.1        242.8    -5.01   0.002
SIZE              2.3989       0.2458     9.76   0.000
SIZESQ       -0.00045004   0.00005908    -7.62   0.000


S = 46.8013    R-Sq = 98.2%    R-Sq(adj) = 97.7%


Analysis of Variance

Source          DF        SS        MS        F       P
Regression       2    831070    415535   189.71   0.000
Residual Error   7     15333      2190
Total            9    846402


Obs   SIZE   USAGE      Fit   SE Fit   Residual   St Resid
  1   1290  1182.0   1129.6     30.1       52.4       1.46
  2   1350  1172.0   1202.2     25.9      -30.2      -0.77
  3   1470  1264.0   1337.8     19.8      -73.8      -1.74
  4   1600  1493.0   1470.0     17.4       23.0       0.53
  5   1710  1571.0   1570.1     18.0        0.9       0.02
  6   1840  1711.0   1674.2     19.9       36.8       0.87
  7   1980  1804.0   1769.4     21.9       34.6       0.84
  8   2230  1840.0   1895.5     23.3      -55.5      -1.37
  9   2400  1956.0   1949.1     23.6        6.9       0.17
 10   2930  1954.0   1949.2     44.7        4.8       0.35 X

X denotes an observation whose X value gives it large influence.
```

Figure 12.36a
MINITAB Residual Plot
for Straight-Line Model
of Electrical Usage

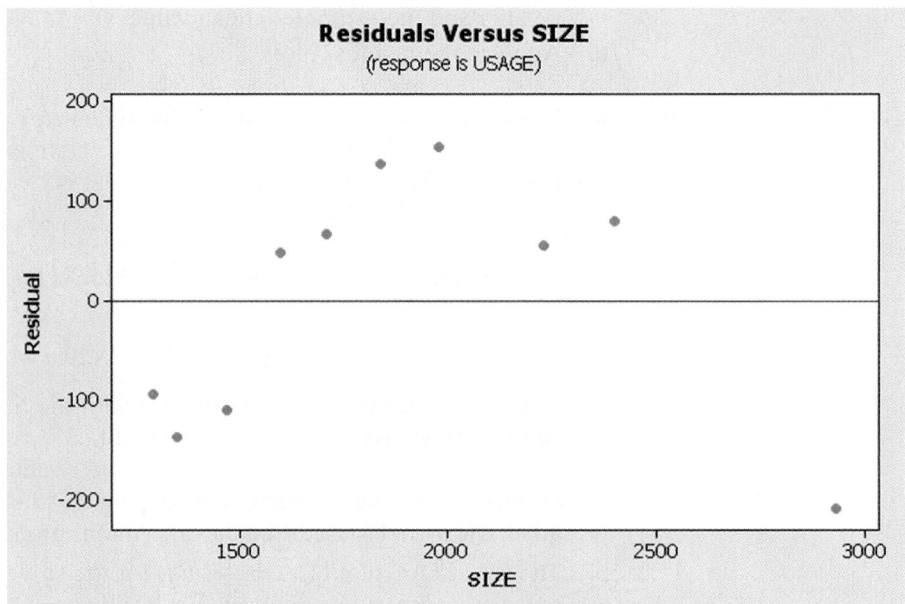

Figure 12.36b
MINITAB Residual Plot
for Quadratic Model of
Electrical Usage

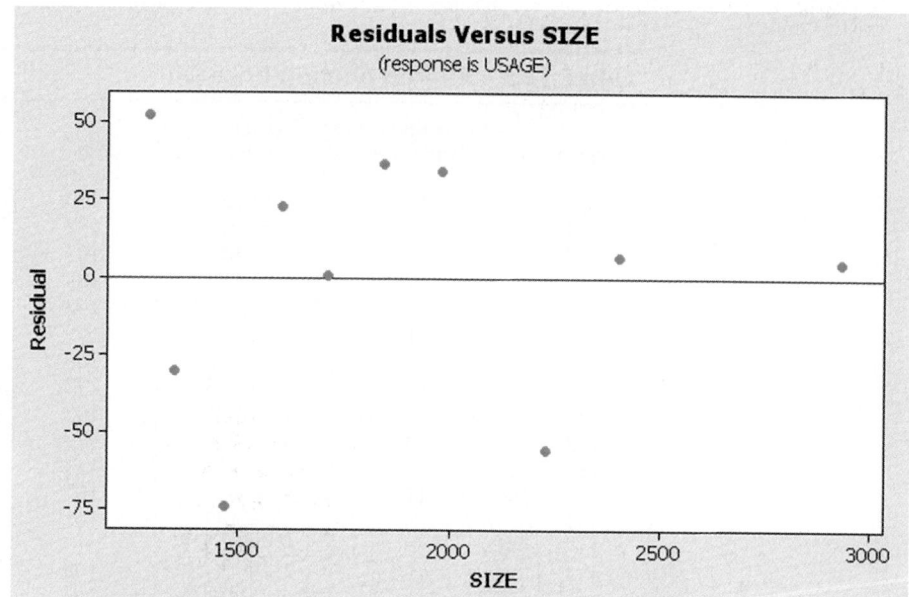

the residual for the largest value of x again below the 0 line. The indication is
that the mean value of the random error, *within* each of these ranges of x (small,
medium, large) may not be equal to 0. Such a pattern usually indicates that
curvature needs to be added to the model.

When the second-order term is added to the model, the nonrandom
pattern disappears. In Figure 12.36b, the residuals appear to be randomly distributed around the 0 line, as expected. Also, the variation of the residuals
around the 0 line appears to be much smaller for the quadratic model. The
implication is that the quadratic model provides a considerably better model
for predicting electrical usage.

Look Back The residual analysis verifies our conclusions from Example 12.7, where
we found the quadratic term, $\beta_2 x^2$, to be statistically significant.

■ ■ ■

Residual analyses are also useful for detecting one or more observations that
deviate significantly from the regression model. We expect approximately 95% of the
residuals to fall within 2 standard deviations of the 0 line, and all or almost all of them
to lie within 3 standard deviations of their mean of 0. Residuals that are extremely far
from the 0 line, and disconnected from the bulk of the other residuals, are called
outliers, and should receive special attention from the regression analyst.

DEFINITION 12.6

A residual that is larger than $3s$ (in absolute value) is considered to be an **outlier**.

EXAMPLE 12.15   IDENTIFYING OUTLIERS

Problem   Refer to Example 12.6 (p. 691) in which we modeled the auction price y of a grandfather
clock as a function of age x_1 and number of bidders x_2. The data for this example are
repeated in Table 12.9, with one important difference: The auction price of the clock at
the top of the second column has been changed from \$2,131 to \$1,131 (highlighted in
Table 12.9). The interaction model

$$E(y) = \beta_0 + \beta_1 x_1 + \beta_2 x_2 + \beta_3 x_1 x_2$$

⊙ **GFCLOCKALT**

TABLE 12.9 Altered Auction Price Data

Age x_1	Number of Bidders x_2	Auction Price y	Age x_1	Number of Bidders x_2	Auction Price y
127	13	$1,235	170	14	$1,131
115	12	1,080	182	8	1,550
127	7	845	162	11	1,884
150	9	1,522	184	10	2,041
156	6	1,047	143	6	845
182	11	1,979	159	9	1,483
156	12	1,822	108	14	1,055
132	10	1,253	175	8	1,545
137	9	1,297	108	6	729
113	9	946	179	9	1,792
137	15	1,713	111	15	1,175
117	11	1,024	187	8	1,593
137	8	1,147	111	7	785
153	6	1,092	115	7	744
117	13	1,152	194	5	1,356
126	10	1,336	168	7	1,262

is again fit to these (modified) data, with the MINITAB printout shown in Figure 12.37. The residuals are shown highlighted in the printout and then plotted against the number of bidders, x_2, in Figure 12.38. Analyze the residual plot.

Solution The residual plot dramatically reveals the one altered measurement. Note that one of the two residuals at $x_2 = 14$ bidders falls more than 3 standard deviations below 0. Note that no other residual falls more than 2 standard deviations from 0.

What do we do with outliers once we identify them? First, we try to determine the cause. Were the data entered into the computer incorrectly? Was the observation recorded incorrectly when the data were collected? If so, we correct the observation and rerun the analysis. Another possibility is that the observation is not representative of the conditions we are trying to model. For example, in this case the low price may be attributable to extreme damage to the clock, or to a clock of inferior quality compared to the others. In these cases we probably would exclude the observation from the analysis. In many cases you may not be able to determine the cause of the outlier. Even so, you may want to rerun the regression analysis excluding the outlier in order to assess the effect of that observation on the results of the analysis.

Figure 12.39 shows the printout when the outlier observation is excluded from the grandfather clock analysis, and Figure 12.40 shows the new plot of the residuals against the number of bidders. Now only one of the residuals lies beyond 2 standard deviations from 0, and none of them lies beyond 3 standard deviations. Also, the model statistics indicate a much better model without the outlier. Most notably, the standard deviation (s) has decreased from 200.6 to 85.83, indicating a model that will provide more precise estimates and predictions (narrower confidence and prediction intervals) for clocks that are similar to those in the reduced sample.

Look Back Remember that if the outlier is removed from the analysis when in fact it belongs to the same population as the rest of the sample, the resulting model may provide misleading estimates and predictions.

Now Work	*Exercise 12.119*

■ ■ ■

Figure 12.37

MINITAB Regression
Printout for Altered
Grandfather Clock Data

```
The regression equation is
PRICE = - 513 + 8.17 AGE + 19.9 NUMBIDS + 0.320 AGE_BIDS


Predictor      Coef   SE Coef       T      P
Constant     -512.8     665.9   -0.77  0.448
AGE           8.165     4.585    1.78  0.086
NUMBIDS       19.89     67.44    0.29  0.770
AGE_BIDS     0.3196    0.4790    0.67  0.510


S = 200.598   R-Sq = 72.9%   R-Sq(adj) = 70.0%


Analysis of Variance

Source           DF        SS        MS      F      P
Regression        3   3033587   1011196  25.13  0.000
Residual Error   28   1126703     40239
Total            31   4160290


Obs   AGE    PRICE      Fit   SE Fit   Residual   St Resid
  1   127   1235.0   1310.4     59.3      -75.4      -0.39
  2   115   1080.0   1105.9     62.1      -25.9      -0.14
  3   127    845.0    947.5     61.1     -102.5      -0.54
  4   150   1522.0   1322.5     37.1      199.5       1.01
  5   156   1047.0   1179.5     60.3     -132.5      -0.69
  6   182   1979.0   1831.9     82.9      147.1       0.81
  7   156   1822.0   1598.0     61.9      224.0       1.17
  8   132   1253.0   1185.8     39.7       67.2       0.34
  9   137   1297.0   1178.9     39.0      118.1       0.60
 10   113    946.0    913.9     58.6       32.1       0.17
 11   137   1713.0   1561.0     78.4      152.0       0.82
 12   117   1024.0   1072.6     53.1      -48.6      -0.25
 13   137   1147.0   1115.2     44.3       31.8       0.16
 14   153   1092.0   1149.2     59.0      -57.2      -0.30
 15   117   1152.0   1187.2     69.7      -35.2      -0.19
 16   126   1336.0   1117.6     43.4      218.4       1.12
 17   170   1131.0   1914.4    116.7     -783.4      -4.80R
 18   182   1550.0   1597.7     62.8      -47.7      -0.25
 19   162   1884.0   1598.3     57.0      285.7       1.49
 20   184   2041.0   1776.6     70.7      264.4       1.41
 21   143    845.0   1048.4     58.9     -203.4      -1.06
 22   159   1483.0   1421.8     40.6       61.2       0.31
 23   108   1055.0   1130.7     97.9      -75.7      -0.43
 24   175   1545.0   1522.7     55.4       22.3       0.12
 25   108    729.0    695.5     99.6       33.5       0.19
 26   179   1792.0   1642.7     57.6      149.3       0.78
 27   111   1175.0   1224.0    107.2      -49.0      -0.29
 28   187   1593.0   1651.3     68.6      -58.3      -0.31
 29   111    785.0    781.1     80.9        3.9       0.02
 30   115    744.0    822.7     75.5      -78.7      -0.42
 31   194   1356.0   1480.7    133.6     -124.7      -0.83 X
 32   168   1262.0   1374.0     57.7     -112.0      -0.58

R denotes an observation with a large standardized residual.
X denotes an observation whose X value gives it large influence.
```

Figure 12.38
MINITAB Residual Plot
for Altered Grandfather
Clock Data

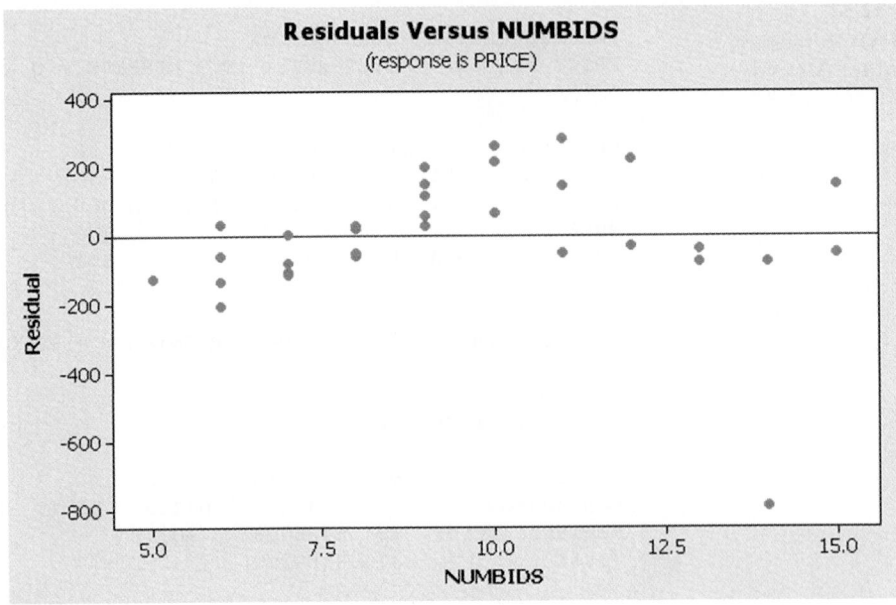

Figure 12.39
MINITAB Regression
Printout when Outlier
Is Deleted

```
The regression equation is
PRICE = 474 - 0.46 AGE - 114 NUMBIDS + 1.48 AGE_BIDS


Predictor      Coef   SE Coef      T       P
Constant      474.0     298.2    1.59   0.124
AGE          -0.465     2.107   -0.22   0.827
NUMBIDS      -114.12    31.23   -3.65   0.001
AGE_BIDS      1.4781   0.2295    6.44   0.000


S = 85.8286    R-Sq = 95.2%    R-Sq(adj) = 94.7%


Analysis of Variance

Source         DF        SS        MS       F       P
Regression      3   3933417   1311139  177.99   0.000
Residual Error 27    198897      7367
Total          30   4132314
```

Outlier analysis is another example of testing the assumption that the expected (mean) value of the random error ε is 0, since this assumption is in doubt for the error terms corresponding to the outliers. The next example in this section checks the assumption of the normality of the random error component.

EXAMPLE 12.16   CHECKING FOR NORMAL ERRORS

Problem   Refer to Example 12.15. Analyze the distribution of the residuals in the grandfather clock example, both before and after the outlier residual is removed. Determine whether the assumption of a normally distributed error term is reasonable.

Solution   A histogram and normal probability plot for the two sets of residuals are constructed using MINITAB and are shown in Figures 12.41 and 12.42. Note that the outlier appears to skew the histogram in Figure 12.41, whereas the histogram in Figure 12.42 appears to

Figure 12.40
MINITAB Residual Plot
when Outlier Is Deleted

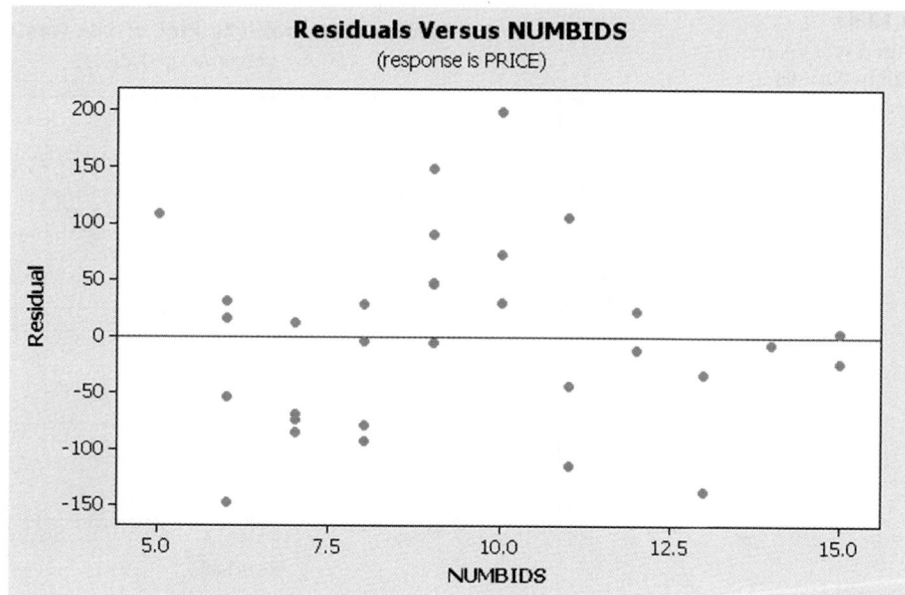

Figure 12.41
MINITAB Graphs of
Regression Residuals for
Grandfather Clock Model
(Outlier Included)

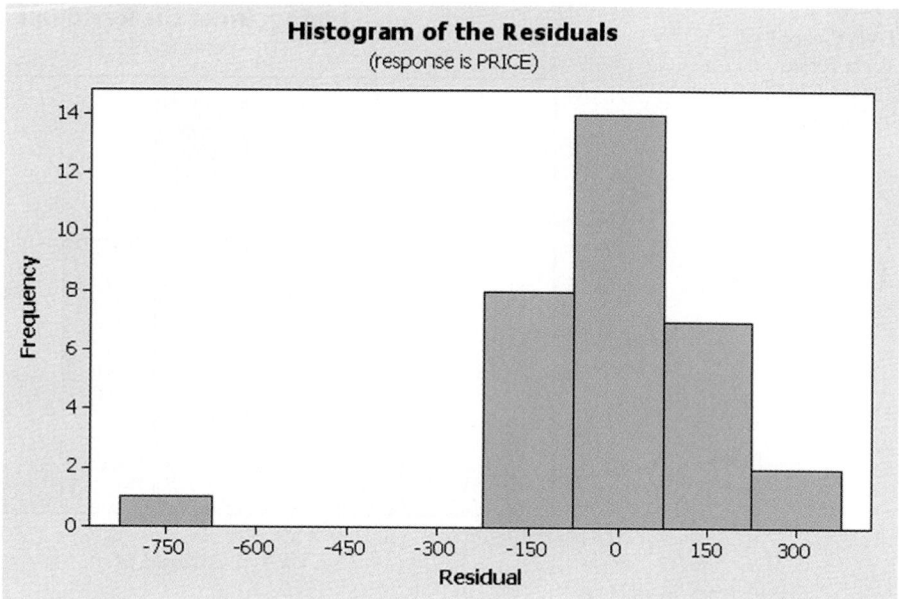

be more mound shaped. Similarly, the pattern of the residuals in the normal probability plot in Figure 12.42 (outlier deleted) is more nearly a straight line than the pattern in Figure 12.41 (outlier included). Although the graphs do not provide formal statistical tests of normality, they do provide a descriptive display. In this example the normality assumption appears to be more plausible after the outlier is removed. Consult the references for methods to conduct statistical tests of normality using the residuals.

Now Work   *Exercise 12.116d*

━━━━━━━━━━   ■ ■ ■   ━━━━━━━━━━

Of the four assumptions in Section 12.1, the assumption that the random error is normally distributed is the least restrictive when we apply regression analysis in practice. That is, moderate departures from a normal distribution have very little

Figure 12.41
Continued
(Outlier Included)

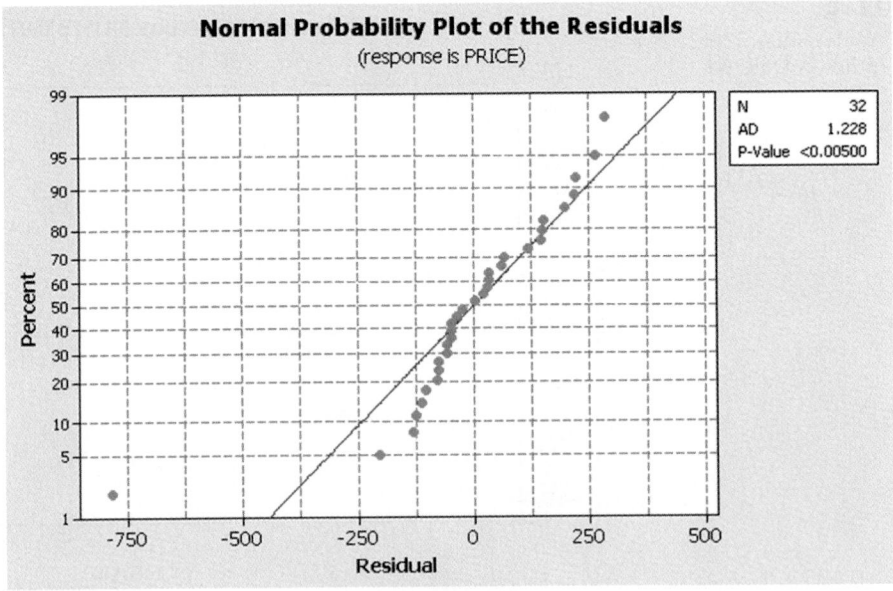

Figure 12.42
MINITAB Graphs of
Regression Residuals for
Grandfather Clock Model
(Outlier Deleted)

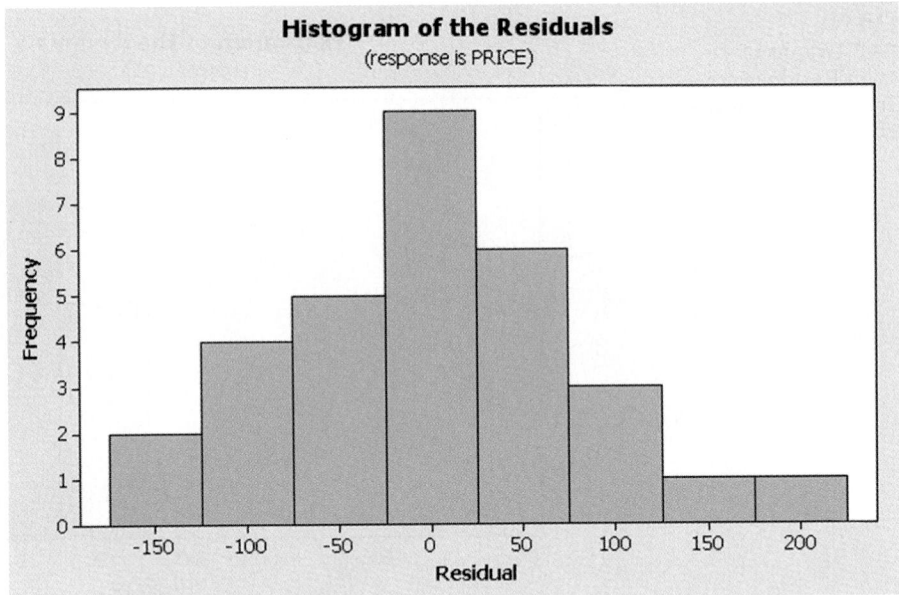

effect on the validity of the statistical tests, confidence intervals, and prediction intervals presented in this chapter. In this case, we say that regression analysis is **robust**
with respect to nonnormal errors. However, great departures from normality cast
doubt on any inferences derived from the regression analysis.

Residual plots can also be used to detect violations of the assumption of constant
error variance. For example, a plot of the residuals versus the predicted value $\hat{y}$ may
display one of the patterns shown in Figure 12.43. In these figures the range in values of
the residuals increases (or decreases) as $\hat{y}$ increases thus indicating that the variance of
the random error, ε, becomes larger or smaller as the estimate of $E(y)$ increases in
value. Since $E(y)$ depends on the x-values in the model, this implies that the variance of
ε is not constant for all settings of the x's.

In the final example of this section, we demonstrate how to use this plot to
detect a nonconstant variance and suggest a useful remedy.

Figure 12.42

Continued
(Outlier Deleted)

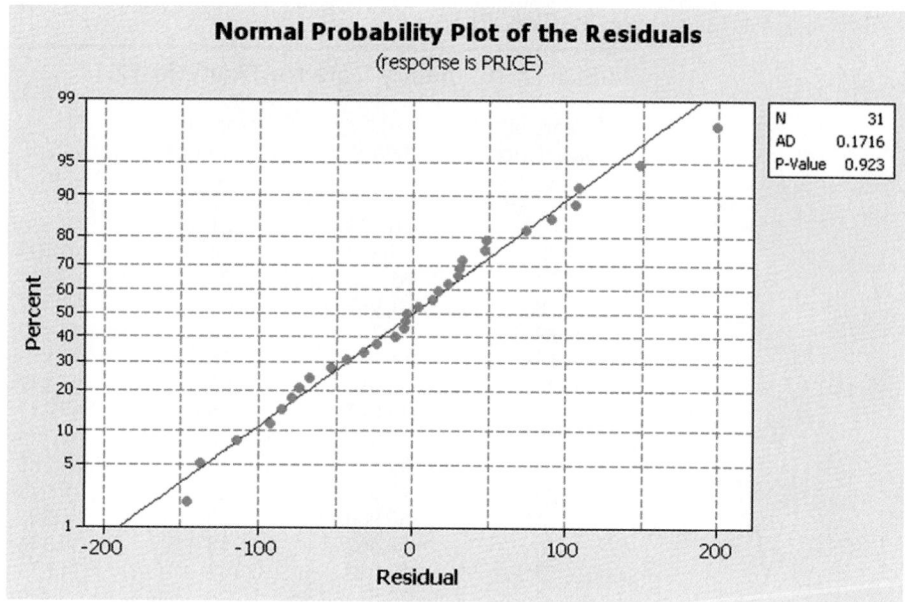

Figure 12.43

Residual plots showing
changes in the variance of ε

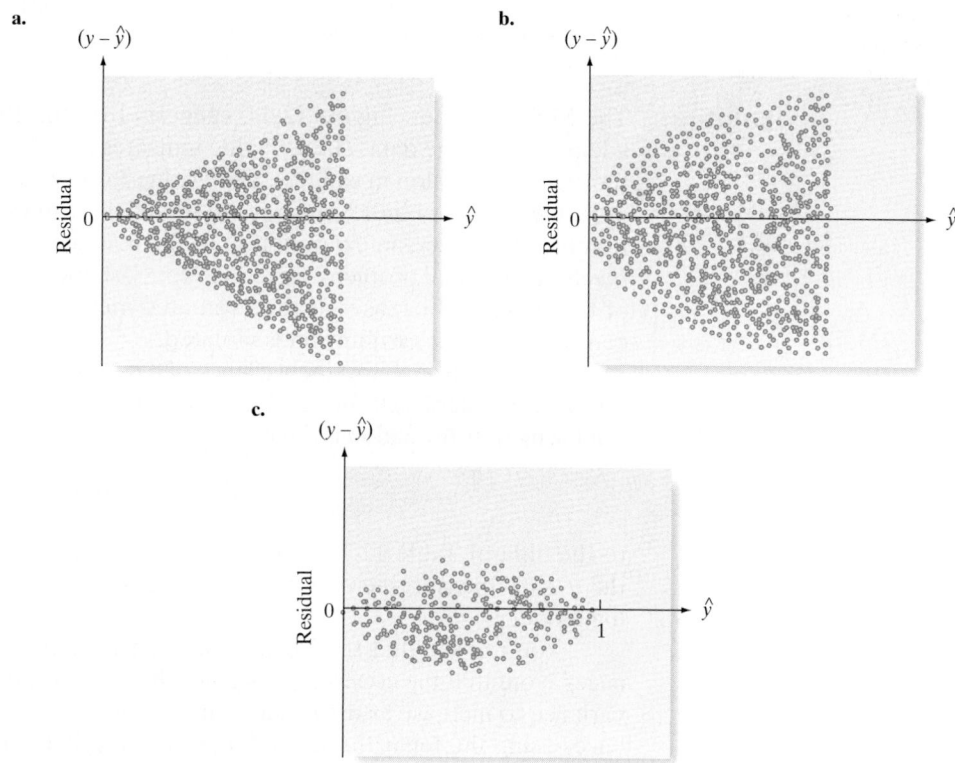

EXAMPLE 12.17   CHECKING FOR CONSTANT ERROR VARIANCE

Problem   The data in Table 12.10 are the salaries, y, and years of experience, x, for a sample of 50 social workers. The first-order model $E(y) = \beta_0 + \beta_1 x$ was fit to the data using SPSS. The printout is shown in Figure 12.44, followed by a plot of the residuals versus $\hat{y}$ in Figure 12.45. Interpret the results. Make model modifications, if necessary.

⊘ **SOCWORK**

TABLE 12.10   Salary Data for Example 12.17

Years of Experience x	Salary y	Years of Experience x	Salary y	Years of Experience x	Salary y
7s	$26,075	21	$43,628	28	$99,139
28	79,370	4	16,105	23	52,624
23	65,726	24	65,644	17	50,594
18	41,983	20	63,022	25	53,272
19	62,308	20	47,780	26	65,343
15	41,154	15	38,853	19	46,216
24	53,610	25	66,537	16	54,288
13	33,697	25	67,447	3	20,844
2	22,444	28	64,785	12	32,586
8	32,562	26	61,581	23	71,235
20	43,076	27	70,678	20	36,530
21	56,000	20	51,301	19	52,745
18	58,667	18	39,346	27	67,282
7	22,210	1	24,833	25	80,931
2	20,521	26	65,929	12	32,303
18	49,727	20	41,721	11	38,371
11	33,233	26	82,641		

Solution   The SPSS printout, Figure 12.44, suggests that the first-order model provides an adequate fit to the data. The R^2 value indicates that the model explains about 79% of the sample variation in salaries. The t value for testing β_1, 13.31, is highly significant (p-value ≈ 0) and indicates that the model contributes information for the prediction of y. However, an examination of the residuals plotted against $\hat{y}$ (Figure 12.45) reveals a potential problem. Note the "cone" shape of the residual variability; the size of the residuals increases as the estimated mean salary increases, implying that the constant variance assumption is violated.

One way to stabilize the variance of ε is to refit the model using a transformation on the dependent variable y. With economic data (e.g., salaries) a useful **variance-stabilizing transformation** is the natural logarithm of y, denoted $\ln(y)$.* We fit the model

$$\ln(y) = \beta_0 + \beta_1 x + \varepsilon$$

to the data of Table 12.10. Figure 12.46 shows the regression analysis printout for the $n = 50$ measurements, while Figure 12.47 shows a plot of the residuals from the log model.

You can see that the logarithmic transformation has stabilized the error variances. Note that the cone shape is gone; there is no apparent tendency of the residual variance to increase as mean salary increases. We therefore are confident that inferences using the logarithmic model are more reliable than those using the untransformed model.

Now Work   *Exercise 12.116b*

■ ■ ■

---

*Other variance-stabilizing transformations that are used successfully in practice are $\sqrt{y}$ and $\sin^{-1}\sqrt{y}$. Consult the references for more details on these transformations.

Figure 12.44

SPSS Regression Printout for First-Order Model of Salary

Model Summary

Model	R	R Square	Adjusted R Square	Std. Error of the Estimate
1	.887[a]	.787	.782	8642.441

a. Predictors: (Constant), EXP

ANOVA[b]

Model		Sum of Squares	df	Mean Square	F	Sig.
1	Regression	1.3E+10	1	1.324E+10	177.257	.000[a]
	Residual	3.6E+09	48	74691793.28		
	Total	1.7E+10	49			

a. Predictors: (Constant), EXP

b. Dependent Variable: SALARY

Coefficients[a]

Model		Unstandardized Coefficients		Standardized Coefficients	t	Sig.
		B	Std. Error	Beta		
1	(Constant)	11368.72	3160.317		3.597	.001
	EXP	2141.381	160.839	.887	13.314	.000

a. Dependent Variable: SALARY

Figure 12.45

SPSS Residual Plot for First-Order Model of Salary

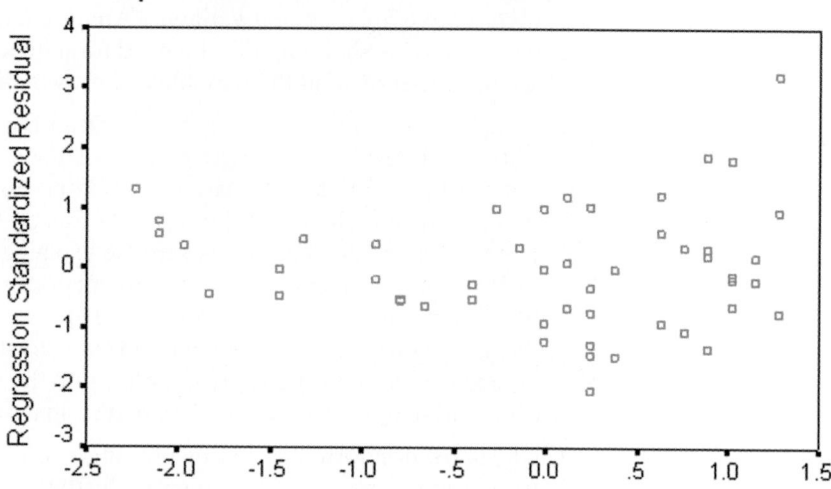

Scatterplot

Dependent Variable: SALARY

Figure 12.46

SPSS Regression Printout for Model of Log Salary

Model Summary

Model	R	R Square	Adjusted R Square	Std. Error of the Estimate
1	.929[a]	.864	.861	.1541127

a. Predictors: (Constant), EXP

ANOVA[b]

Model		Sum of Squares	df	Mean Square	F	Sig.
1	Regression	7.212	1	7.212	303.660	.000[a]
	Residual	1.140	48	.024		
	Total	8.352	49			

a. Predictors: (Constant), EXP

b. Dependent Variable: LNSALARY

Coefficients[a]

Model		Unstandardized Coefficients		Standardized Coefficients	t	Sig.
		B	Std. Error	Beta		
1	(Constant)	9.841	.056		174.631	.000
	EXP	.050	.003	.929	17.426	.000

a. Dependent Variable: LNSALARY

Residual analysis is a useful tool for the regression analyst, not only to check the assumptions, but also to provide information about how the model can be improved. A summary of the residual analysis presented in this section to check the assumption that the random error ε is normally distributed with mean 0 and constant variance is presented in the next box.

Steps in a Residual Analysis

1. Check for a misspecified model by plotting the residuals against each of the quantitative independent variables. Analyze each plot, looking for a curvilinear trend. This shape signals the need for a quadratic term in the model. Try a second-order term in the variable against which the residuals are plotted.

2. Examine the residual plots for outliers. Draw lines on the residual plots at 2-and 3-standard-deviation distances below and above the 0 line. Examine residuals outside the 3-standard-deviation lines as potential outliers, and check to see that approximately 5% of the residuals exceed the 2-standard-deviation lines. Determine whether each outlier can be explained as an error in data collection or transcription, or corresponds to a member of a population different from that of the remainder of the sample, or simply represents an unusual observation. If the observation is determined to be an error, fix it or remove it. Even if you can't determine the cause, you may want to rerun the regression analysis without the observation to determine its effect on the analysis.

3. Check for nonnormal errors by plotting a frequency distribution of the residuals, using a stem-and-leaf display, histogram, or normal probability plot. Check to see if obvious departures from normality exist. Extreme skewness of the frequency distribution may be due to outliers or could indicate the need

for a transformation of the dependent variable. (Normalizing transformations are beyond the scope of this book, but you can find it in the references.)

4. Check for unequal error variances by plotting the residuals against the predicted values, $\hat{y}$. If you detect a cone-shaped pattern or some other pattern that indicates that the variance of ε is not constant, refit the model using an appropriate variance-stabilizing transformation on y, such as $\ln(y)$. (Consult the references for other useful variance-stabilizing transformations.)

Figure 12.47
SPSS Residual Plot
for Model of Log Salary

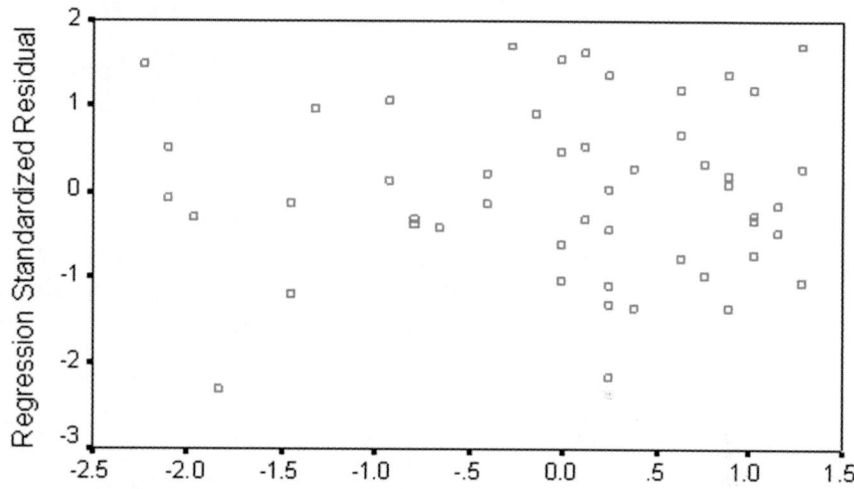

Scatterplot

Dependent Variable: LNSALARY

Regression Standardized Predicted Value

Residuals

Using the TI-83 Graphing Calculator

Plotting Residuals

When you compute a regression equation on the TI-83, the residuals are automatically computed and saved to a list called **RESID**. **RESID** can be found under the **LIST menu** (**2nd STAT**).

To make a scatterplot of the residuals,

Step 1   *Enter the data*
Press **STAT** and select **1:Edit**
Note: If the list already contains data, clear the old data. Use the up arrow to highlight '**L1**' or '**L2**'.
Press **CLEAR ENTER**.
Use the arrow and **ENTER** keys to enter the data set into **L1** and **L2**.

Step 2   *Compute the regression equation*
Press **STAT** and highlight **CALC**
Press **4** for **LinReg(ax + b)**
Press **ENTER**

Step 3 *Set up the data plot*

Press **2nd Y =** for **STATPLOT**

Press **1** for **Plot1**

Set the cursor so that **ON** is flashing.

For **Type**, use the arrow and Enter keys to highlight and select the scatterplot (first icon in the first row).

For **Xlist**, choose the column containing the *x*-data.

For **Ylist**, choose the column containing the residuals.

To enter the **RESID** as your **Ylist**:

    1. Use the arrow keys to move the cursor after Ylist:

    2. Press 2nd STAT for LIST

    3. Use the down arrow to highlight the listname RESID and press ENTER

Step 4 *View the scatterplot of the residuals*

Press **ZOOM 9** for **ZoomStat**

Example The figures below show a table of data entered on the TI-83 and the scatterplot of the residuals obtained using the steps given above.

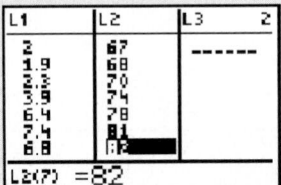
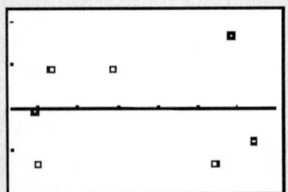

Statistics in Action Revisited

A Residual Analysis for the Condominium Sales Price Model

In the previous Statistics in Action Revisited section (p. 730), we found a second-order model to be both a statistically and practically useful model for predicting the auction price (*y*) of a condominium unit. Before actually using the model in practice, we need to examine the residuals to be sure that the standard regression assumptions are reasonably satisfied.

Figures SIA12.6 and SIA12.7 are MINITAB graphs of the residuals from the model. Except for a few outliers, the histogram shown in Figure SIA12.6 appears to be approximately normally distributed; consequently the assumption of normal errors is reasonably satisfied. The scatterplot of the residuals against $\hat{y}$ shown in Figure SIA 12.7 reveals no distinct pattern; thus, the assumption of a constant error variance is reasonably satisfied.

To find the outliers, we used MINITAB to compute the residuals of the model. A list of the unusual residual values is shown in the MINITAB printout, Figure SIA12.8. Four condo units (#111, #184, #185, and #188, highlighted on the printout) have residual values that are greater than 3 in absolute value. (Research reveals that all of these units are located on the fifth floor.) If we delete these observations from the data set and refit the model, we obtain the results shown in the MINITAB printout, Figure SIA12.9. You can see that the adjusted-R^2 value is now .793, an increase of about 9% from the previous value. Also, the standard deviation of the model is now $s = 15.1$, a reduction of about four hundred dollars from the previous value. Obviously, removing the outliers from the analysis yields a "better" prediction equation. If we can justify removing these four outliers (except for #111, they were all undersold units), then we can use the model shown in Figure SIA12.9 to obtain more accurate predictions of future condo sale prices.

Figure SIA12.6
MINITAB Histogram
of Residuals from Second-
Order Model of Condo
Sales Price

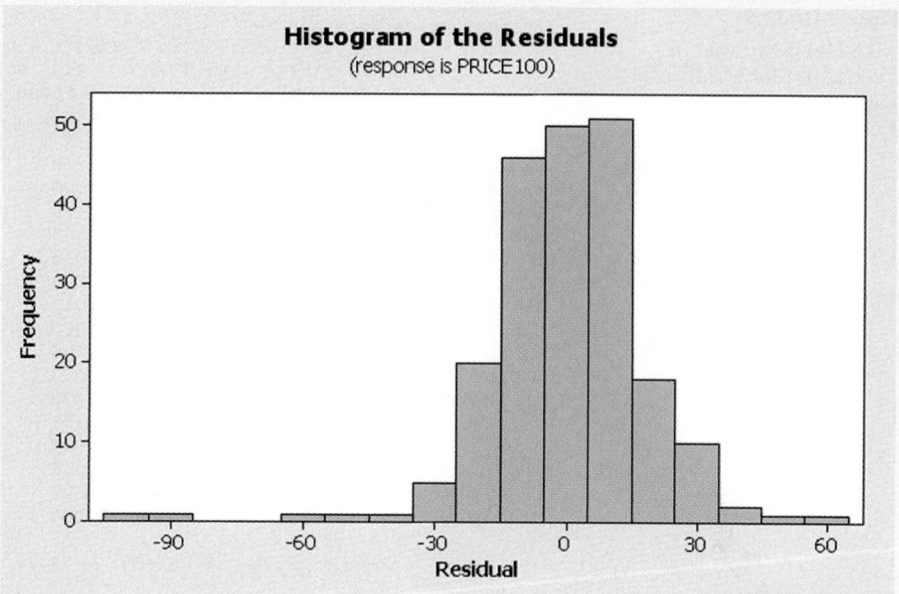

Figure SIA12.7
MINITAB Plot of Residuals
versus Predicted Values from
Second-Order Model of
Condo Sales Price

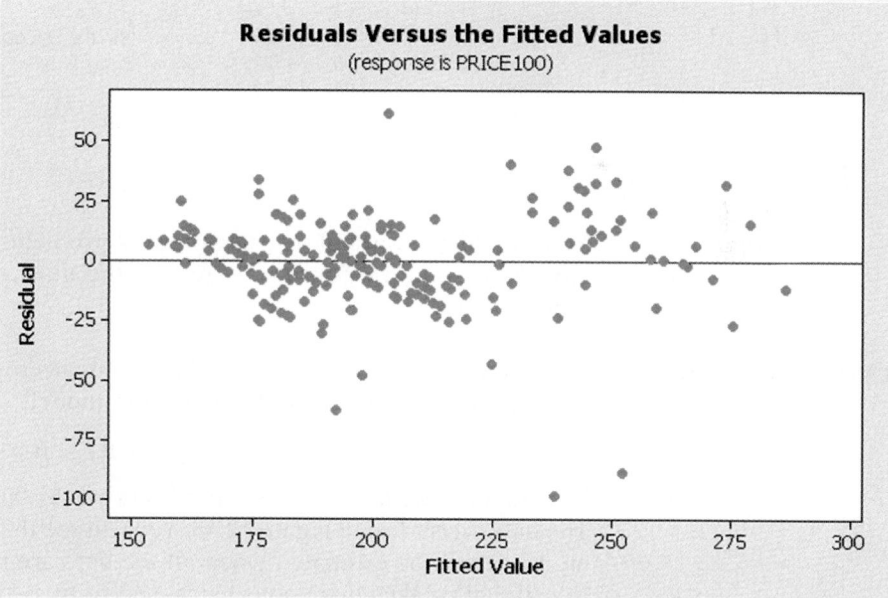

Figure SIA12.8
MINITAB List of Outliers
for the Second-Order Model
of Condo Sales Price

```
Unusual Observations

Obs   FLOOR   PRICE100      Fit   SE Fit   Residual   St Resid
 11    8.00    279.00    240.71    7.62      38.29       2.18R
 26    3.00    269.00    228.48    4.75      40.52       2.19R
 37    6.00    294.00    246.22    5.34      47.78       2.60R
111    5.00    265.00    202.98    3.69      62.02       3.30R
183    3.00    150.00    197.54    4.36     -47.54      -2.55R
184    5.00    140.00    238.04    6.02     -98.04      -5.40R
185    5.00    130.00    192.19    4.37     -62.19      -3.34R
186    6.00    182.00    224.70    5.17     -42.70      -2.32R
188    5.00    164.00    252.16    4.66     -88.16      -4.75R

R denotes an observation with a large standardized residual.
```

Figure SIA12.9

MINITAB Printout of Second-Order Model of Condo Sales Price with Outliers Deleted

```
The regression equation is
PRICE100 = 185 - 2.59 FLOOR - 1.98 DISTANCE - 0.387 FLR_DST + 0.912 FLOORSQ
           + 0.264 DISTSQ + 83.9 VIEW - 4.01 VU_FLR - 4.61 VU_DIST
           + 0.154 VU_FLR_DST - 0.426 VU_FLRSQ + 0.304 VU_DISTSQ - 13.1 END
           + 11.5 FURNISH - 31.8 AUCTION

Predictor       Coef   SE Coef       T      P
Constant      184.93     10.39   17.81  0.000
FLOOR         -2.591     3.570   -0.73  0.469
DISTANCE      -1.977     1.923   -1.03  0.305
FLR_DST      -0.3871    0.1593   -2.43  0.016
FLOORSQ       0.9120    0.3384    2.70  0.008
DISTSQ        0.2644    0.1115    2.37  0.019
VIEW           83.88     14.03    5.98  0.000
VU_FLR        -4.007     4.705   -0.85  0.396
VU_DIST       -4.608     2.529   -1.82  0.070
VU_FLR_DST    0.1535    0.2166    0.71  0.479
VU_FLRSQ     -0.4261    0.4624   -0.92  0.358
VU_DISTSQ     0.3041    0.1465    2.08  0.039
END          -13.142     6.244   -2.10  0.037
FURNISH       11.549     2.405    4.80  0.000
AUCTION      -31.816     2.954  -10.77  0.000

S = 15.0925   R-Sq = 80.7%   R-Sq(adj) = 79.3%

Analysis of Variance

Source           DF      SS     MS      F      P
Regression       14  181271  12948  56.84  0.000
Residual Error  190   43279    228
Total           204  224550
```

12.12 Some Pitfalls: Estimability, Multicollinearity, and Extrapolation

Problem 1   You should be aware of several potential problems when constructing a prediction model for some response y. A few of the most important are discussed in this final section.

Parameter Estimability   Suppose you want to fit a model relating annual crop yield y to the total expenditure for fertilizer, x. We propose the first-order model

$$E(y) = \beta_0 + \beta_1 x$$

Now suppose we have three years of data and $1,000 is spent on fertilizer each year. The data are shown in Figure 12.48. You can see the problem: The parameters of the model cannot be estimated when all the data are concentrated at a single x value. Recall that it takes two points (x-values) to fit a straight line. Thus, the parameters are not estimable when only one x is observed.

A similar problem would occur if we attempted to fit the quadratic model

$$E(y) = \beta_0 + \beta_1 x + \beta_2 x^2$$

Figure 12.48

Yield and Fertilizer Expenditure Data: Three Years

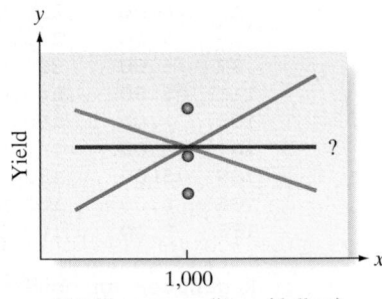

1,000

Fertilizer expenditure (dollars)

Figure 12.49

Only Two x Values
Observed: Quadratic Model
Is Not Estimable

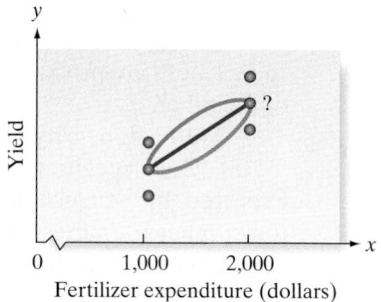

to a set of data for which only one or two different x-values were observed (see Figure 12.49). At least three different x-values must be observed before a quadratic model can be fit to a set of data (that is, before all three parameters are estimable).

In general, the number of levels of observed x-values must be one more than the order of the polynomial in x that you want to fit.

For controlled experiments, the researcher can select experimental designs that will permit estimation of the model parameters. Even when the values of the independent variables cannot be controlled by the researcher, the independent variables are almost always observed at a sufficient number of levels to permit estimation of the model parameters. When the statistical software you use suddenly refuses to fit a model, however, the problem is probably inestimable parameters.

Problem 2
Multicollinearity

Often, two or more of the independent variables used in a regression model contribute redundant information. That is, the independent variables are correlated with each other. For example, suppose we want to construct a model to predict the gas mileage rating of a truck as a function of its load, x_1 (in tons), and the horsepower, x_2 (in foot-pounds per second), of its engine. We would expect heavy loads to require greater horsepower and to result in lower mileage ratings. Thus, although both x_1 and x_2 contribute information for the prediction of mileage rating, y, some of the information is overlapping because x_1 and x_2 are correlated.

When the independent variables are correlated, we say *multicollinearity* exists. In practice, it is not uncommon to observe correlations among the independent variables. However, a few problems arise when serious multicollinearity is present in the regression variables.

> **DEFINITION 12.7**
>
> **Multicollinearity** exists when two or more of the independent variables used in regression are correlated.

First, high correlations among the independent variables increase the likelihood of rounding errors in the calculations of the β estimates, standard errors, and so forth. Second, and more important, the regression results may be confusing and misleading. Consider the model for gasoline mileage rating (y) of a truck,

$$E(y) = \beta_0 + \beta_1 x_1 + \beta_2 x_2$$

where x_1 = load and x_2 = horsepower. Fitting the model to a sample data set, we might find that the t-tests for testing β_1 and β_2 are both nonsignificant at the $\alpha = .05$ level, while the F-test for H_0: $\beta_1 = \beta_2 = 0$ is highly significant ($p = .001$). The tests may seem to be contradictory, but really they are not. The t-tests indicate that the contribution of one variable, say x_1 = load, is not significant after the effect of x_2 = horsepower has been accounted for (because x_2 is also in the model). The significant F-test, on the other hand, tells us that at least one of the two variables is making a contribution to the prediction of y (i.e., either β_1 or β_2 or both differ from 0). In fact, both are probably contributing, but the contribution of one overlaps with that of the other.

Multicollinearity can also have an effect on the signs of the parameter estimates. More specifically, a value of $\hat{\beta}_i$ may have the opposite sign from what is expected. In the truck gasoline mileage example, we expect heavy loads to result in lower mileage ratings and we expect higher horsepowers to result in lower mileage ratings; consequently, we expect the signs of both the parameter estimates to be negative. Yet, we may actually see a positive value of $\hat{\beta}_1$ and be tempted to claim that heavy loads result in *higher* mileage ratings. This is the danger of interpreting a β coefficient when the independent variables are correlated. Because the variables contribute redundant information, the effect of $x_1 =$ load on $y =$ mileage rating is measured only partially by β_1.

How can you avoid the problems of multicollinearity in regression analysis? One way is to conduct a designed experiment (Chapter 10) so that the levels of the x variables are uncorrelated. Unfortunately, time and cost constraints may prevent you from collecting data in this manner. Consequently, most data are collected observationally. Since observational data frequently consist of correlated independent variables, you will need to recognize when multicollinearity is present and, if necessary, make modifications in the regression analysis.

Several methods are available for detecting multicollinearity in regression. A simple technique is to calculate the coefficient of correlation, r, between each pair of independent variables in the model and use the procedure outlined in Section 11.6 to test for significantly correlated variables. If one or more of the r values is statistically different from 0, the variables in question are correlated and a multicollinearity problem may exist.* Other indications of the presence of multicollinearity include those mentioned above—namely, nonsignificant t-tests for the individual parameter estimates when the F-test for overall model adequacy is significant, and estimates with opposite signs from what is expected.†

Detecting Multicollinearity in the Regression Model

1. Significant correlations between pairs of independent variables

2. Nonsignificant t-tests for all (or nearly all) of the individual β parameters when the F-test for overall model adequacy is significant

3. Signs opposite from what is expected in the estimated β parameters

---

EXAMPLE 12.18

DETECTING MULTICOLLINEARITY

Problem   The Federal Trade Commission (FTC) annually ranks varieties of domestic cigarettes according to their tar, nicotine, and carbon monoxide contents. The U.S. surgeon general considers each of these three substances hazardous to a smoker's health. Past studies have shown that increases in the tar and nicotine contents of a cigarette are accompanied by an increase in the carbon monoxide emitted from the cigarette smoke. Table 12.11 presents data on tar, nicotine, and carbon monoxide contents (in milligrams) and weight (in grams) for a sample of 25 (filter) brands

---

*Remember that r measures only the pairwise correlation between x values. Three variables, x_1, x_2, and x_3, may be highly correlated as a group but may not exhibit large pairwise correlations. Thus, multicollinearity may be present even when all pairwise correlations are not significantly different from 0.
†More formal methods for detecting multicollinearity, such as variance-inflation factors (VIFs), are available. Independent variables with a VIF of 10 or above are usually considered to be highly correlated with one or more of the other independent variables in the model. Calculation of VIFs are beyond the scope of this introductory text. Consult the chapter references for a discussion of VIFs and other formal methods of detecting multicollinearity.

⊙ FTC

TABLE 12.11   FTC Cigarette Data for Example 12.18

Tar (x_1)	Nicotine (x_2)	Weight (x_3)	Carbon Monoxide (y)
14.1	.86	.9853	13.6
16.0	1.06	1.0938	16.6
29.8	2.03	1.1650	23.5
8.0	.67	.9280	10.2
4.1	.40	.9462	5.4
15.0	1.04	.8885	15.0
8.8	.76	1.0267	9.0
12.4	.95	.9225	12.3
16.6	1.12	.9372	16.3
14.9	1.02	.8858	15.4
13.7	1.01	.9643	13.0
15.1	.90	.9316	14.4
7.8	.57	.9705	10.0
11.4	.78	1.1240	10.2
9.0	.74	.8517	9.5
1.0	.13	.7851	1.5
17.0	1.26	.9186	18.5
12.8	1.08	1.0395	12.6
15.8	.96	.9573	17.5
4.5	.42	.9106	4.9
14.5	1.01	1.0070	15.9
7.3	.61	.9806	8.5
8.6	.69	.9693	10.6
15.2	1.02	.9496	13.9
12.0	.82	1.1184	14.9

Source: Federal Trade Commission

tested in a recent year. Suppose we want to model carbon monoxide content, y, as a function of tar content, x_1, nicotine content, x_2, and weight, x_3, using the model

$$E(y) = \beta_0 + \beta_1 x_1 + \beta_2 x_2 + \beta_3 x_3$$

The model is fit to the 25 data points in Table 12.11, and a portion of the SAS printout is shown in Figure 12.50. Examine the printout. Do you detect any signs of multicollinearity?

Solution   First, note that the F-test for overall model utility is highly significant. The test statistic ($F = 78.98$) and observed significance level (p-value $< .0001$) are highlighted on the SAS printout, Figure 12.50. Therefore, we can conclude at, say, $\alpha = .01$, that at least one of the parameters, β_1, β_2, or β_3, in the model is nonzero. The t-tests for two of three individual β's, however, are nonsignificant. (The p-values for these tests are highlighted on the printout.) Unless tar (x_1) is the only one of the three variables useful for predicting carbon monoxide content, these results are the first indication of a potential multicollinearity problem.

 The negative values for $\hat{\beta}_2$ and $\hat{\beta}_3$ (highlighted on the printout) are a second clue to the presence of multicollinearity. From past studies, the FTC expects carbon monoxide content (y) to increase when either nicotine content (x_2) or weight (x_3) increases—that is, the FTC expects *positive* relationships between y and x_2 and between y and x_3, not negative ones.

 All signs indicate that a serious multicollinearity problem exists.*

---

*Note also that the variance-inflation factors (VIFs) for both tar and nicotine, given on the STATISTIX printout, Figure 12.50, exceed 10.

Figure 12.50

SAS Printout for Model of
CO Content, Example 12.18

```
                             Dependent Variable: CO

                      Number of Observations Read        25
                      Number of Observations Used        25

                             Analysis of Variance

                                   Sum of        Mean
          Source            DF     Squares       Square     F Value    Pr > F

          Model              3    495.25781    165.08594      78.98    <.0001
          Error             21     43.89259      2.09012
          Corrected Total   24    539.15040


                  Root MSE            1.44573    R-Square    0.9186
                  Dependent Mean     12.52800    Adj R-Sq    0.9070
                  Coeff Var          11.53996


                             Parameter Estimates

                         Parameter     Standard                          Variance
          Variable    DF   Estimate      Error     t Value   Pr > |t|    Inflation

          Intercept    1    3.20219      3.46175      0.93     0.3655           0
          TAR          1    0.96257      0.24224      3.97     0.0007    21.63071
          NICOTINE     1   -2.63166      3.90056     -0.67     0.5072    21.89992
          WEIGHT       1   -0.13048      3.88534     -0.03     0.9735     1.33386


                    Pearson Correlation Coefficients, N = 25
                        Prob > |r| under H0: Rho=0

                               TAR        NICOTINE      WEIGHT

          TAR              1.00000        0.97661      0.49077
                                          <.0001       0.0127

          NICOTINE         0.97661        1.00000      0.50018
                           <.0001                      0.0109

          WEIGHT           0.49077        0.50018      1.00000
                           0.0127         0.0109
```

Look Back To confirm our suspicions, we had SAS produce the coefficient of correlation, r, for each of the three pairs of independent variables in the model. The resulting output is shown (highlighted) at the bottom of Figure 12.50. You can see that tar (x_1) and nicotine (x_2) are highly correlated $(r = .9766)$ while weight (x_3) is moderately correlated with the other two x's $(r \approx .5)$. All three correlations have p-values less than .05; consequently, all three are significantly different from 0 at $\alpha = .05$.

Now Work   *Exercise 12.118*

▬▬▬▬▬▬   ■ ■ ■   ▬▬▬▬▬▬

Once you have detected that multicollinearity exists, there are several alternative measures available for solving the problem. The appropriate measure to take depends on the severity of the multicollinearity and the ultimate goal of the regression analysis. Some researchers, when confronted with highly correlated independent variables, choose to include only one of the correlated variables in the final model. If you are interested only in using the model for estimation and prediction (step 6), you may decide not to drop any of the independent variables from the model. In the presence of multicollinearity, we have seen that it is dangerous to interpret the individual β parameters. However, confidence intervals for $E(y)$ and prediction intervals for y generally remain unaffected *as long as the values of the x's used to predict y follow the same pattern of multicollinearity exhibited in the sample data.* That is, you must take strict care to ensure that the values of the x-variables fall within the range of the sample data.

Problem 3
Prediction Outside the
Experimental Region

Many research economists have developed highly technical models to relate the state of the economy to various economic indices and other independent variables. Many of these models are multiple regression models, where, for example, the dependent variable y might be next year's growth in GDP and the independent

> **Solutions to Some Problems Created by Multicollinearity in Regression***
>
> 1. Drop one or more of the correlated independent variables from the model. One way to decide which variables to keep in the model is to employ step-wise regression (Section 12.10).
>
> 2. If you decide to keep all the independent variables in the model:
>   (a) Avoid making inferences about the individual β parameters based on the t-tests.
>   (b) Restrict inferences about $E(y)$ and future y values to values of the x's that fall within the range of the sample data.

variables might include this year's rate of inflation, this year's Consumer Price Index (CPI), etc. In other words, the model might be constructed to predict next year's economy using this year's knowledge.

Unfortunately, models such as these were almost all unsuccessful in predicting the recession in the early 1970s and the late 1990s. What went wrong? One of the problems was that many of the regression models were used to **extrapolate** (i.e., predict y for values of the independent variables that were outside the region in which the model was developed). For example, the inflation rate in the late 1960s, when many of the models were developed, ranged from 6% to 8%. When the double-digit inflation of the early 1970s became a reality, some researchers attempted to use the same models to predict future growth in GDP. As you can see in Figure 12.51, the model may be very accurate for predicting y when x is in the range of experimentation, but the use of the model outside that range is a danger-ous practice.

Problem 4
Correlated Errors

Another problem associated with using a regression model to predict a variable y based on independent variables $x_1, x_2, \ldots, x_k$ arises from the fact that the data are frequently *time series*. That is, the values of both the dependent and independent variables are observed sequentially over a period of time. The observations tend to be correlated over time, which in turn often causes the prediction errors of the regression model to be correlated. Thus, the assumption of independent errors is violated, and the model tests and prediction intervals are no longer valid. One solu-tion to this problem is to construct a **time series model**; the interested reader should consult the references for details of time series analysis.

Figure 12.51

Using a Regression Model Outside the Experimental Region

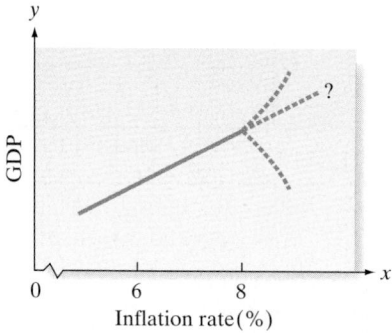

*Several other solutions are available. For example, in the case where higher-order regression models are fit, the analyst may want to code the independent variables so that higher-order terms (e.g., x^2) for a particular x-variable are not highly correlated with x. One transformation that works is $z = (x - \bar{x})/s$. Other, more sophisticated procedures for addressing multicollinearity (such as *ridge regression*) are beyond the scope of this text. Consult the references at the end of this chapter.

Exercises 12.108–12.123

Understanding the Principles

12.108 Define a regression residual.

12.109 Give two properties of the regression residuals from a model.

12.110 Define an outlier.

12.111 Define multicollinearity in regression.

12.112 Give three indicators of a multicollinearity problem.

12.113 Define extrapolation.

12.114 *True or False*. Regression models fit to time series data typically result in uncorrelated errors.

Learning the Mechanics

12.115 Consider fitting the multiple regression model

$$E(y) = \beta_0 + \beta_1 x_1 + \beta_2 x_2 + \beta_3 x_3 + \beta_4 x_4 + \beta_5 x_5$$

A matrix of correlations for all pairs of independent variables is given in the next column. Do you detect a multicollinearity problem? Explain.

	x_1	x_2	x_3	x_4	x_5
x_1	—	.17	.02	−.23	.19
x_2		—	.45	.93	.02
x_3			—	.22	−.01
x_4				—	.86
x_5					—

12.116 Identify the problem(s) in each of the residual plots shown below.

Applying the Concepts—Basic

12.117 Global warming and foreign investments. Refer to the *Journal of World-Systems Research* (Summer 2003) study of the link between foreign investments and carbon dioxide emissions, Exercise 12.16 (p. 679). Recall that the researchers modeled the level (y) of CO_2 emissions in 1996 as a function of seven independent variables for $n = 66$ developing countries. A matrix

Residual plots for Exercise 12.116

a.

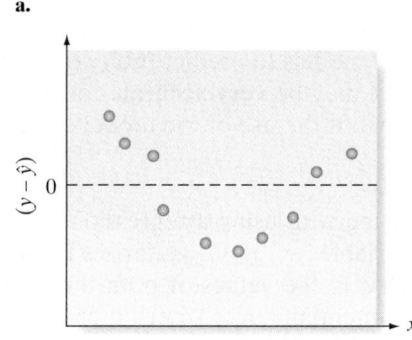

b.

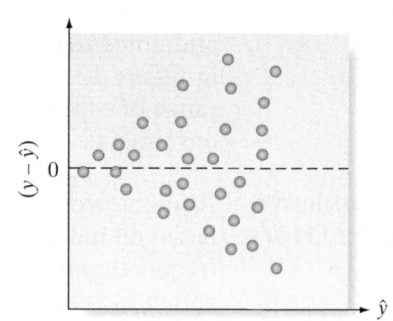

c.

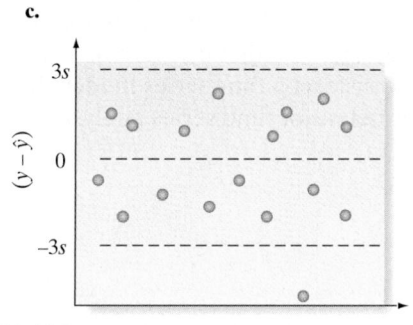

d.

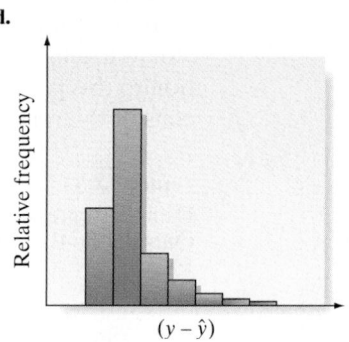

Correlations for Exercise 12.117

Independent Variable	x_2	x_3	x_4	x_5	x_6	$x_7 = \ln$ (level of CO_2 emissions in 1980)
$x_1 = \ln$ (foreign investments in 1980)	.13	.57	.30	−.38	.14	−.14
$x_2 =$ gross domestic investment in 1980)		.49	.36	−.47	−.14	.25
$x_3 =$ trade exports in 1980			.43	−.47	−.06	−.07
$x_4 = \ln$ (GNP in 1980)				−.84	−.53	.42
$x_5 =$ agricultural production in 1980					.45	−.50
$x_6 = 1$ if African country, 0 if not						−.47

Source: Grimes, P., & Kentor, J. "Exporting the greenhouse: Foreign capital penetration and CO_2 emissions 1980–1996," *Journal of World-Systems Research*, Vol. IX, No. 2, Summer 2003 (Appendix B).

given the correlation (r) for each pair of independent variables is shown at the bottom of p. 764. Identify the independent variables that are highly correlated. What problems may result from including these highly correlated variables in the regression model?

12.118 Factors identifying urban counties. Refer to the
NW *Professional Geographer* (Feb. 2000) study of urban and rural counties in the western United States, Exercise 12.18 (p. 680). Recall that six independent variables—total county population (x_1), population density (x_2), population concentration (x_3), population growth (x_4), proportion of county land in farms (x_5), and 5-year change in agricultural land base (x_6)—were used to model the urban/rural rating (y) of a county. Prior to running the multiple regression analysis, the researchers were concerned about possible multicollinearity in the data. Below is a correlation matrix, that is, a table of correlations between all pairs of the independent variables.

Independent Variable		x_1	x_2	x_3	x_4	x_5
x_1	Total population					
x_2	Population density	.20				
x_3	Population concentration	.45	.43			
x_4	Population growth	−.05	−.14	−.01		
x_5	Farm land	−.16	−.15	−.07	−.20	
x_6	Agricultural change	−.12	−.12	−.22	−.06	−.06

Source: Berry, K. A., et al. "Interpreting what is rural and urban for western U.S. counties." *Professional Geographer*, Vol. 52, No. 1, Feb. 2000 (Table 2).

a. Based on the correlation matrix, is there any evidence of extreme multicollinearity?

b. Refer to the multiple regression results in the table given in Exercise 12.18 (p. 000). Based on the reported tests, is there any evidence of extreme multicollinearity?

12.119 Passive exposure to smoke. Passive exposure to
NW environmental tobacco smoke has been associated with growth suppression and an increased frequency of respiratory tract infections in normal children. Is this association more pronounced in children with cystic fibrosis? To answer this question, 43 children (18 girls and 25 boys) attending a 2-week summer camp for cystic fibrosis patients were studied. (*The New England Journal of Medicine*, Sept. 20, 1990.) Researchers investigated the correlation between a child's weight percentile (y) and the number of cigarettes smoked per day in the child's home (x). The table at the bottom of the page lists the data for the 25 boys.

a. Using simple linear regression, the researchers predicted the weight percentile for the last observation ($x = 44$ cigarettes) to be $\hat{y} = 29.63$. Given that the standard deviation of the model is $s = 24.68$, is this observation an outlier? Explain.

b. Fit the model to the data and obtain the regression residuals. Show that the sum of the residuals is 0.

c. Refer to part **b.** Identify any other outliers in the data.

Applying the Concepts—Intermediate

12.120 Physical characteristics of boys. A physiologist wishes to investigate the relationship between the physical characteristics of preadolescent boys and their maximum oxygen uptake (measured in milliliters of oxygen per kilogram of body weight). The data shown on p. 766 were collected on a random sample of ten preadolescent boys.

a. Fit the regression model

$$y = \beta_0 + \beta_1 x_1 + \beta_2 x_2 + \beta_3 x_3 + \beta_4 x_4 + \varepsilon$$

to the data and give the least squares prediction equation.

b. It seems reasonable to assume that the greater a child's weight, the greater should be the maximum oxygen uptake. Is β_3, the estimated coefficient of weight, x_3, positive as expected? Give an explanation for this result.

⊙ CFSMOKE

Weight Percentile y	No. of Cigarettes Smoked per Day x	Weight Percentile y	No. of Cigarettes Smoked per Day x
6	0	43	0
6	15	49	0
2	40	50	0
8	23	49	22
11	20	46	30
17	7	54	0
24	3	58	0
25	0	62	0
17	25	66	0
25	20	66	23
25	15	83	0
31	23	87	44
35	10		

Source: Rubin, B. K. "Exposure of children with cystic fibrosis to environmental tobacco smoke." *The New England Journal of Medicine*, Sept. 20, 1990. Vol. 323, No. 12, p. 85 (data extracted from Figure 3).

BOYS10

Maximum Oxygen Uptake y	Age x_1 (years)	Height x_2 (centimeters)	Weight x_3 (kilograms)	Chest Depth x_4 (centimeters)
1.54	8.4	132.0	29.1	14.4
1.74	8.7	135.5	29.7	14.5
1.32	8.9	127.7	28.4	14.0
1.50	9.9	131.1	28.8	14.2
1.46	9.0	130.0	25.9	13.6
1.35	7.7	127.6	27.6	13.9
1.53	7.3	129.9	29.0	14.0
1.71	9.9	138.1	33.6	14.6
1.27	9.3	126.6	27.7	13.9
1.50	8.1	131.8	30.8	14.5

c. It would seem that the chest depth of a child should be positively correlated to lung volume and hence to maximum oxygen uptake. Is $\hat{\beta}_4$ significantly different from 0, as expected? If not, explain why.

d. Calculate the correlation coefficients between all pairs of the independent variables x_1–x_4. Do these correlations provide an explanation for the confusing signs and small t values associated with the estimated regression coefficients of the model?

DDT

12.121 Contamination of fish in the Tennessee River. Refer to the U.S. Army Corps of Engineers data on fish contaminated from the toxic discharges of a chemical plant located on the banks of the Tennessee River in Alabama. In Exercise 12.20 (p. 680) you fit the first-order model, $E(y) = \beta_0 + \beta_1 x_1 + \beta_2 x_2 + \beta_3 x_3$, where y = DDT level in captured fish, x_1 = miles captured upstream, x_2 = fish length, and x_3 = fish weight. Conduct a complete residual analysis for the model. Do you recommend any model modifications be made? Explain.

12.122 Analysis of Tokyo urban air. Chemical engineers at Tokyo Metropolitan University analyzed urban air specimens for the presence of low-molecular-weight dicarboxylic acid. (*Environmental Science & Engineering*, Oct. 1993.) The dicarboxylic acid (as a percentage of total carbon) and oxidant concentrations for 19 air specimens collected from urban Tokyo are listed in the next table. Consider the straight-line model relating dicarboxylic acid percentage (y) to oxidant concentration (x). Conduct a complete residual analysis.

12.123 Socialization of graduate students. *Teaching Sociology* (July 1995) developed a model for the professional socialization of graduate students working toward a Ph.D. in sociology. One of the dependent variables modeled was professional confidence, y, measured on

URBANAIR

Dicarboxylic Acid (%)	Oxidant (ppm)
.85	78
1.45	80
1.80	74
1.80	78
1.60	60
1.20	62
1.30	57
.20	49
.22	34
.40	36
.50	32
.38	28
.30	25
.70	45
.80	40
.90	45
1.22	41
1.00	34
1.00	25

Source: Kawamura, K., and Ikushima, K. "Seasonal changes in the distribution of dicarboxylic acids in the urban atmosphere." *Environmental Science & Technology*, Vol. 27. No. 10, Oct. 1993, p. 2232 (data extracted from Figure 4).

a 5-point scale. The model included over 20 independent variables and was fit to data collected for a sample of 309 sociology graduate students. One concern is whether multicollinearity exists in the data. A matrix of Pearson product moment correlations for ten of the independent variables is shown on p. 767. [*Note:* Each entry in the table is the correlation coefficient r between the variable in the corresponding row and corresponding column.]

a. Examine the correlation matrix and find the independent variables that are moderately or highly correlated.

b. What modeling problems may occur if the variables, part **a**, are left in the model? Explain.

Matrix of correlations for Exercise 12.123

Independent Variable	(1)	(2)	(3)	(4)	(5)	(6)	(7)	(8)	(9)	(10)
(1) Father's occupation	1.000	.363	.099	−.110	−.047	−.053	−.111	.178	.078	.049
(2) Mother's education	.363	1.000	.228	−.139	−.216	.084	−.118	.192	.125	.068
(3) Race	.099	.228	1.000	.036	−.515	.014	−.120	.112	.117	.337
(4) Sex	−.110	−.139	.036	1.000	.165	−.256	.173	−.106	−.117	.073
(5) Foreign status	−.047	−.216	−.515	.165	1.000	−.041	.159	−.130	−.165	−.171
(6) Undergraduate GPA	−.053	.084	.014	−.256	−.041	1.000	.032	.028	−.034	.092
(7) Year GRE taken	−.111	−.118	−.120	.173	.159	.032	1.000	−.086	−.602	.016
(8) Verbal GRE score	.178	.192	.112	−.106	−.130	.028	−.086	1.000	.132	.087
(9) Years in graduate program	.078	.125	.117	−.117	−.165	−.034	−.602	.132	1.000	−.071
(10) First-year graduate GPA	.049	.068	.337	.073	−.171	.092	.016	.087	−.071	1.000

Source: Keith, B., and Moore, H. A. "Training sociologists: An assessment of professional socialization and the emergence of career aspirations." *Teaching Sociology*, Vol. 23, No. 3, July 1995, p. 205 (Table 1).

Quick Review

Key Terms

[Note: Starred (*) items are from the optional sections in this chapter.]

Adjusted multiple coefficient of determination   674
Base level   710
Categorical variable   709
Coded variable   663
*Complete (nested) model   725
Complete second-order model   703
Correlated errors   763
Dummy variables   709
Extrapolation   763
First-order model   664
Global *F*-test   675
Higher-order term   663
Indicator variable   709

Interaction   691
Least squares prediction equation   665
Level of a variable   709
*Main effect terms   717
Mean square for error   667
Model building   664
Multicollinearity   759
Multiple coefficient of determination   674
Multiple regression model   662
*Nested model   725
*Nested model *F*-test   725
Outlier   745
Paraboloid   703
Parameter estimability   758

*Parsimonious model   730
Quadratic model   698
Quadratic term   698
Qualitative variable   709
*Reduced (nested) model   725
Residual   741
Residual analysis   741
*Response surface   728
Robust method   750
Saddle-shaped surface   703
Second-order model   698
Second-order term   698
*Stepwise regression   735
Time series model   763
Variance-stabilizing transformation   751

Key Formulas

$E(y) = \beta_0 + \beta_1 x_1 + \beta_2 x_2$ First-order model with two quantitative independent variables   664

$s^2 = \text{MSE} = \dfrac{\text{SSE}}{n - (k + 1)}$ Estimator of σ^2 for a model with k independent variables   667

$t = \dfrac{\hat{\beta}_i}{s_{\hat{\beta}_i}}$ Test statistic for testing H_0: $\beta_i = 0$ 670

$\hat{\beta}_i \pm (t_{\alpha/2}) s_{\hat{\beta}_i}$, where t depends on $n - (k + 1)$ df $100(1 - \alpha)\%$ confidence interval for $\beta_i = 0$ 670

$R^2 = \dfrac{\text{SS}_{yy} - \text{SSE}}{\text{SS}_{yy}}$ Multiple coefficient of determination   674

$R_a^2 = 1 - \left[\dfrac{(n - 1)}{n - (k + 1)}\right](1 - R^2)$ Adjusted multiple coefficient of determination   674

$F = \dfrac{\text{MS(Model)}}{\text{MSE}} = \dfrac{R^2/k}{(1 - R^2)/[n - (k + 1)]}$ Test statistic for testing H_0: $\beta_i = \beta_2 = \cdots = \beta_k = 0$ 675

$E(y) = \beta_0 + \beta_1 x_1 + \beta_2 x_2 + \beta_3 x_1 x_2$ Interaction model with two quantitative independent variables   691

$y - \hat{y}$ Regression residual   741

$E(y) = \beta_0 + \beta_1 x + \beta_2 x^2$ Quadratic model   698

$E(y) = \beta_0 + \beta_1 x_1 + \beta_2 x_2 + \beta_3 x_1 x_2 + \beta_4 x_1^2 + \beta_5 x_2^2$ Complete second-order model with two quantitative independent variables   703

$E(y) = \beta_0 + \beta_1 x_1 + \beta_2 x_2 + \cdots + \beta_{k-1} x_{k-1},$

where $x_i = \begin{cases} 1 & \text{if level } i+1 \\ 0 & \text{if not} \end{cases}$ Model with one qualitative variable at k levels   710

$F = \dfrac{(\text{SSE}_R - \text{SSE}_C)/\text{number of } \beta\text{'s tested}}{\text{MSE}_C}$ *Test statistic for comparing reduced and complete models   726

Language Lab

Symbol	Pronunciation	Description
x_1^2	x-1 squared	Quadratic term that allows for curvature in the relationship between y and x
$x_1 x_2$	x-1 x-2	Interaction term
MSE	M-S-E	Mean square for error (estimates σ^2)
β_i	beta-i	Coefficient of x_i in the model
$\hat{\beta}_i$	beta-i-hat	Least squares estimate of β_i
$s_{\hat{\beta}_i}$	s of beta-i-hat	Estimated standard error of $\hat{\beta}_i$
R^2	R-squared	Multiple coefficient of determination
R_a^2	R-squared adjusted	Adjusted multiple coefficient of determination
F		Test statistic for testing global usefulness of model
$\hat{\varepsilon}$	epsilon-hat	Estimated random error, or residual
SSE_R		Sum of squared errors for reduced model
SSE_C		Sum of squared errors for complete model
MSE_C		Mean square error for complete model
$\ln(y)$	Natural log of y	Natural logarithm of dependent variable

---

Chapter Summary Notes

- **Steps in multiple regression:** (1) Hypothesize the deterministic form of the model, (2) use the method of least squares to estimate the unknown β's, (3) assumptions on the random error (ε), (4) check the assumptions and make model modifications, (5) statistically evaluate the adequacy of the model, (6) if deemed useful, use the model for estimation and prediction.

- **Four assumptions for ε:** (1) mean of ε is 0, (2) variance of ε is constant for all x-values, (3) distribution of ε is normal, (4) values of ε are independent.

- **First-order model in k quantitative x's:** $E(y) = \beta_0 + \beta_1 x_1 + \beta_2 x_2 + \beta_3 x_3 + \cdots + \beta_k x_k$, where each β_i represents the change in y for every 1-unit increase in x_i, holding the other x's fixed.

- **Adjusted coefficient of determination (R_{adj}^2)** cannot be "forced" to 1 by adding independent variables to the model.

- **Recommendation for checking statistical utility of the model:** (1) Conduct the **global F-test**, (2) if test is significant, conduct t-tests on the "most important" β's only, (3) interpret the value of $2s$, (4) interpret the value of R_{adj}^2.

- **Interaction model in 2 quantitative x's:** $E(y) = \beta_0 + \beta_1 x_1 + \beta_2 x_2 + \beta_3 x_1 x_2$, where $(\beta_1 + \beta_3 x_2)$ represents the change in y for every 1-unit increase in x_1 for fixed x_2, and $(\beta_2 + \beta_3 x_1)$ represents the change in y for every 1-unit increase in x_2 for fixed x_1.

- Once interaction is tested and deemed important, avoid conducting t-tests on the first-order terms in the model.

- **Quadratic model in 1 quantitative x:** $E(y) = \beta_0 + \beta_1 x + \beta_2 x^2$, where $\beta_2 > 0$ implies *upward* curvature and $\beta_2 < 0$ implies *downward* curvature.

- Once curvature is tested and deemed important, avoid conducting a t-test on the first-order term in the model.

- **Complete second-order model in 2 quantitative x's:** $E(y) = \beta_0 + \beta_1 x_1 + \beta_2 x_2 + \beta_3 x_1 x_2 + \beta_4 (x_1)^2 + \beta_5 (x_2)^2$.

- **Dummy variable model for 1 qualitative x at three levels (A, B, C):** $E(y) = \beta_0 + \beta_1 x_1 + \beta_2 x_2$, where $x_1 = \{1$ if A, 0 if not$\}$ and $x_2 = \{1$ if B, 0 if not$\}$. Then $\beta_0 = \mu_C, \beta_1 = \mu_A - \mu_C$ and $\beta_2 = \mu_B - \mu_C$, where $\mu_i = E(y)$ at level i.
- **Complete second-order model in 1 quantitative x and 1 qualitative x at two levels (A, B):** $E(y) = \beta_0 + \beta_1 x_1 + \beta_2 (x_1)^2 + \beta_3 x_2 + \beta_4 x_1 x_2 + \beta_5 (x_1)^2 x_2$ where $x_2 = \{1$ if A, 0 if B$\}$.
- Two models are **nested** if one model (called the **complete** model) contains all the terms of another model (called the **reduced** model) plus at least additional term.
- A **parsimonious model** is a model with a small number of β parameters.
- Problems with using **stepwise regression** as the "final" model for $E(y)$: (1) Extremely large number of t-tests inflate the probabilities of Type I and Type II errors, (2) no higher-order terms (interactions and squared terms) included in the stepwise final model.

- Properties of **regression residuals**: (1) sum of residuals = 0, (2) sum of squared residuals = SSE.
- To detect a **misspecified model**: Plot residuals against each quantitative x in the model—look for trends (e.g., curvilinear trend).
- To identify **outliers**: Find residuals that are greater than $3s$ in absolute value.
- To detect **non-normal errors**: Graph residuals in a histogram, stem-and-leaf plot, or normal probability plot—look for strong departures from normality.
- To detect a **nonconstant error variance**: Plot residuals against y—look for patterns (e.g., cone-shaped pattern).
- **Multicollinearity** occurs when two or more of the x's in the model are correlated.
- Indicators of multicollinearity: (1) highly correlated x's, (2) significant global F-test, but all t-tests nonsignificant, (3) signs on the β estimates opposite from expected.
- **Extrapolation** occurs when you predict y for values of the x's that are outside of the range of the sample data.

Supplementary Exercises 12.124–12.154

Understanding the Principles

12.124 Why is model building the key to the success or failure of a regression analysis?

12.125 Write a model relating $E(y)$ to one qualitative independent variable that is at four levels. Define all the terms in your model.

12.126 Explain why stepwise regression is used. What is its value in the model-building process?

12.127 It is desired to relate $E(y)$ to a quantitative variable x_1 and a qualitative variable at three levels.
   a. Write a first-order model.
   b. Write a model that will graph as three different second-order curves—one for each level of the qualitative variable.

12.128 a. Write a first-order model relating $E(y)$ to two quantitative independent variables, x_1 and x_2.
   b. Write a complete second-order model.

Learning the Mechanics

12.129 Suppose you fit the model

$$y = \beta_0 + \beta_1 x_1 + \beta_2 x_1^2 + \beta_3 x_2 + \beta_4 x_1 x_2 + \varepsilon$$

to $n = 25$ data points and find that

$\hat{\beta}_0 = 1.26 \quad \hat{\beta}_1 = -2.43 \quad \hat{\beta}_2 = .05 \quad \hat{\beta}_3 = .62 \quad \hat{\beta}_4 = 1.81$
$\qquad s_{\hat{\beta}_1} = 1.21 \quad s_{\hat{\beta}_2} = .16 \quad s_{\hat{\beta}_3} = .26 \quad s_{\hat{\beta}_4} = 1.49$
SSE = .41 $\quad R^2 = .83$

   a. Is there sufficient evidence to conclude that at least one of the parameters β_1, β_2, β_3, or β_4 is nonzero? Test using $\alpha = .05$.
   b. Test $H_0: \beta_1 = 0$ against $H_a: \beta_1 < 0$. Use $\alpha = .05$.

   c. Test $H_0: \beta_2 = 0$ against $H_a: \beta_2 > 0$. Use $\alpha = .05$.
   d. Test $H_0: \beta_3 = 0$ against $H_a: \beta_3 \neq 0$. Use $\alpha = .05$.

12.130 Suppose you used MINITAB to fit the model

$$y = \beta_0 + \beta_1 x_1 + \beta_2 x_2 + \varepsilon$$

to $n = 15$ data points and you obtained the printout shown at the top of p. 770.
   a. What is the least squares prediction equation?
   b. Find R^2 and interpret its value.
   c. Is there sufficient evidence to indicate that the model is useful for predicting y? Conduct an F-test using $\alpha = .05$.
   d. Test the null hypothesis $H_0: \beta_1 = 0$ against the alternative hypothesis $H_a: \beta_1 \neq 0$. Test using $\alpha = .05$. Draw the appropriate conclusions.
   e. Find the standard deviation of the regression model and interpret it.

12.131 Suppose you have developed a regression model to explain the relationship between y and x_1, x_2, and x_3. The ranges of the variables you observed were as follows: $10 \leq y \leq 100, 5 \leq x_1 \leq 55, .5 \leq x_2 \leq 1$, and $1,000 \leq x_3 \leq 2,000$. Will the error of prediction be smaller when you use the least squares equation to predict y when $x_1 = 30$, $x_2 = .6$, and $x_3 = 1,300$, or when $x_1 = 60$, $x_2 = .4$, and $x_3 = 900$? Why?

12.132 The first-order model $E(y) = \beta_0 + \beta_1 x_1$ was fit to $n = 19$ data points. A residual plot for the model is shown on p. 770. Is the need for a quadratic term in the model evident from the residual plot? Explain.

12.133 To model the relationship between y, a dependent variable, and x, an independent variable, a researcher has taken one measurement on y at each of three different x

MINITAB Output for
Exercise 12.130

```
The regression equation is
Y = 90.1 - 1.84 X1 + .285 X2


Predictor    Coef   SE Coef      T      P
Constant    90.10     23.10   3.90  0.002
X1          -1.836    0.367  -5.01  0.001
X2           0.285    0.231   1.24  0.465


S = 10.68       R-Sq = 91.6%    R-Sq(adj) = 90.2%

Analysis of Variance

Source          DF     SS     MS      F      P
Regression       2  14801   7400  64.91  0.001
Residual Error  12   1364    114
Total           14  16165
```

Residual Plot for Exercise 12.132

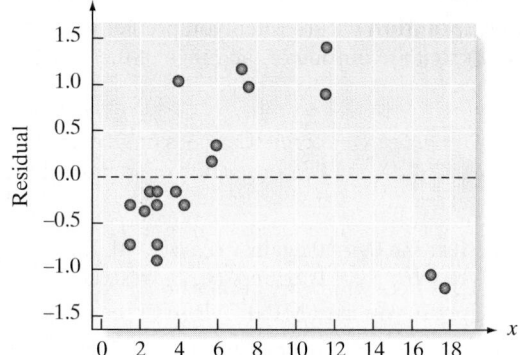

values. Drawing on his mathematical expertise, the researcher realizes that he can fit the second-order model

$$E(y) = \beta_0 + \beta_1 x + \beta_2 x^2$$

and it will pass exactly through all three points, yielding SSE = 0. The researcher, delighted with the "excellent" fit of the model, eagerly sets out to use it to make inferences. What problems will he encounter in attempting to make inferences?

12.134 Suppose you fit the regression model

$$E(y) = \beta_0 + \beta_1 x_1 + \beta_2 x_2 + \beta_3 x_2^2 + \beta_4 x_1 x_2 + \beta_5 x_1 x_2^2$$

to $n = 35$ data points and wish to test the null hypothesis $H_0: \beta_4 = \beta_5 = 0$.

a. State the alternative hypothesis.
b. Explain in detail how to compute the F statistic needed to test the null hypothesis.
c. What are the numerator and denominator degrees of freedom associated with the F statistic in part **b**?
d. Give the rejection region for the test if $\alpha = .05$.

Applying the Concepts—Basic

12.135 Acceptance of disability study. A disabled person's acceptance of a disability is critical to the rehabilitation process. The Journal of Rehabilitation (Sept. 1989) published a study that investigated the relationship between assertive behavior level and acceptance of disability in 160 disabled adults. The dependent variable, assertive-

ness (y), was measured using the Adult Self Expression Scale (ASES). Scores on the ASES range from 0 (no assertiveness) to 192 (extreme assertiveness). The model analyzed was $E(y) = \beta_0 + \beta_1 x_1 + \beta_2 x_2 + \beta_3 x_3$, where

$$x_1 = \text{Acceptance of disability (AD) score}$$
$$x_2 = \text{Age (years)}$$
$$x_3 = \text{Length of disability (years)}$$

The regression results are shown in the table.

Independent Variable	t	Two-Tailed p-Value
AD score (x_1)	5.96	.0001
Age (x_2)	0.01	.9620
Length (x_3)	1.91	.0576

a. Is there sufficient evidence to indicate that AD score is positively linearly related to assertiveness level, once age and length of disability are accounted for? Test using $\alpha = .05$.
b. Test the hypothesis $H_0: \beta_2 = 0$ against $H_a: \beta_2 \neq 0$. Use $\alpha = .05$. Give the conclusion in the words of the problem.
c. Test the hypothesis $H_0: \beta_3 = 0$ against $H_a: \beta_3 > 0$. Use $\alpha = .05$. Give the conclusion in the words of the problem.

12.136 ETS study of doctoral students. An Educational Testing Service (ETS) research scientist used multiple regression analysis to model y, the final grade point average (GPA) of business and management doctoral students. (*Journal of Educational Statistics*, Spring 1993.) A list of the potential independent variables measured for each doctoral student in the study is given below.

1. Quantitative Graduate Management Aptitude Test (GMAT) score
2. Verbal GMAT score
3. Undergraduate GPA
4. First-year graduate GPA
5. Student cohort (i.e., year in which student entered doctoral program: 1988, 1990, or 1992)

a. Identify the variables as quantitative or qualitative.

b. For each quantitative variable, give your opinion on whether the variable is positively or negatively related to final GPA.

c. For each of the qualitative variables, set up the appropriate dummy variable.

d. Write a first-order, main-effects model relating final GPA, y, to the five independent variables.

e. Interpret the β's in the model, part **d**.

f. Write a first-order model for final GPA, y, that allows for a different slope for each student cohort.

g. For each quantitative independent variable in the model, part **f**, give the slope of the line (in terms of the β's) for the 1988 cohort.

12.137 Distress in EMS workers. The *Journal of Consulting and Clinical Psychology* (June 1995) reported on a study of emergency service (EMS) rescue workers who responded to the I-880 freeway collapse during the 1989 San Francisco earthquake. The goal of the study was to identify the predictors of symptomatic distress in the EMS workers. One of the distress variables studied was the Global Symptom Index (GSI). Several models for GSI, y, were considered based on the following independent variables:

x_1 = Critical Incident Exposure scale (CIE)

x_2 = Hogan Personality Inventory-Adjustment scale (HPI-A)

x_3 = Years of experience (EXP)

x_4 = Locus of Control scale (LOC)

x_5 = Social Support scale (SS)

x_6 = Dissociative Experiences scale (DES)

x_7 = Peritraumatic Dissociation Experiences Questionnaire, self-report (PDEQ-SR)

a. Write a first-order model for $E(y)$ as a function of the first five independent variables, x_1–x_5.

b. The model of part **a**, fitted to data collected for $n = 147$ EMS workers, yielded the following results: $R^2 = .469$, $F = 34.47$, p-value $<.001$. Interpret these results.

c. Write a first-order model for $E(y)$ as a function of all seven independent variables, x_1–x_7.

d. The model, part **c**, yielded $R^2 = .603$. Interpret this result.

e. The t-tests for testing the DES and PDEQ-SR variables both yielded a p-value of .001. Interpret these results.

12.138 Proposition 48 study. *Sociology of Sport Journal* (Spring 1992) published research on Proposition 48, the NCAA regulation that requires a minimum of 700 on the Scholastic Assessment Test (SAT) for eligibility for college athletics. Critics have claimed that the SAT is biased against black student athletes and is not a valid predictor of academic success. To investigate the validity of the SAT as a predictor of academic success, high school GPA and study time were also considered as potential predictors of college GPA. The following variables are defined:

y = College GPA (the best measure of academic success)

x_1 = High school GPA

x_2 = Hours of studying in a season

x_3 = SAT score

The study conducted separate regression analyses for black and white student athletes. Some of the results are shown in the table.

	Parameter Estimate	Standard Error
Blacks		
Intercept	2.245	.09
High School GPA	−.040	.01
Study Hours	.14	.10
SAT Score	.001	.0025
Whites		
Intercept	1.970	.16
High School GPA	−.089	.01
Study Hours	.20	.23
SAT Score	.001	.0002

a. Write the model relating the mean value of the college GPA as a linear function of the three identified explanatory variables. Explain the meaning of each of the β parameters in your model.

b. Note that the estimate of the β parameter for SAT score for both models is .001. Interpret this estimate.

c. The reported p-value for testing $H_0: \beta_1 = 0$ is .734 for the black athletes' model and .000 for the white athletes' model. Interpret these values.

12.139 Algae growth study. The *Journal of Applied Psychology* (Dec. 1994) published research on the seasonal growth activity of algae in an indoor environment. The daily growth rate y was regressed against temperature x using the quadratic model $E(y) = \beta_0 + \beta_1 x + \beta_2 x^2$. A particular algal strain was grown in a glass dish indoors at temperatures ranging from $10°$ to $32°$ Celsius. The data for $n = 33$ such experiments were used to fit the quadratic model, with the following results:

$$\hat{y} = -2.51 + .55x - .01x^2, R^2 = .67$$

a. Sketch the least squares prediction equation. Interpret the graph.

b. Interpret the value of R^2.

c. Is the model useful for predicting algae growth rate, y? Test using $\alpha = .05$.

12.140 Emphysema and lung damage. A medical researcher wants to model the extent of lung damage in emphysema patients as a function of two variables: the number of years the patient has smoked (x_1) and the sex of the patient (x_2). Fifty emphysema patients are used in the study, and the response y for each patient is a subjective score ranging from 0 to 50. (High scores indicate extensive lung damage.)

a. Identify the independent variables as qualitative or quantitative.

b. Write the main effects (first-order) model for predicting the mean lung damage score, $E(y)$, from the two variables identified in part **a**.

c. Using a female patient as the base level, the researcher obtained the prediction equation

$$\hat{y} = -.85 + .65x_1 + .09x_2$$

What is the estimated difference between the mean lung damage scores for a male and female who have both smoked for 20 years?

d. Find the estimated mean lung damage score for a female patient who has smoked for 15 years.

e. It would not be surprising if the functional relation between lung damage and years of smoking were second order. Further, it is conceivable that years of smoking might affect men differently than women. Write a model that allows for these conditions.

12.141 Fructose diet experiment. The amount of fructose in an athlete's bloodstream is critical to performance. A researcher ran an experiment to determine whether diet had any effect on the level of fructose in the blood after 10 minutes of running on a treadmill. Twenty-one subjects were selected, and each subject's fructose level was measured after 10 minutes on a treadmill. Each subject was randomly assigned to one of three diets. One diet was high in protein, one high in carbohydrates, and one high in fruits. After one month on the diet, each subject was again asked to run on the treadmill for 10 minutes, and the fructose level in the bloodstream was measured. The dependent variable was the difference in the level of fructose in the blood between the second and the first runs on the treadmill.

a. Identify the independent variables in the experiment.

b. Write an appropriate regression model relating mean difference in fructose in the blood, $E(y)$, to the independent variables. Identify and code all dummy variables.

12.142 Crime in the UK. The *Journal of Quantitative Criminology* (Vol. 8, 1992) published a paper on the determinants of area property crime levels in the United Kingdom. Several multiple regression models for property crime prevalence, y, measured as the percentage of residents in a geographical area who were victims of at least one property crime, were examined. The results for one of the models, based on a sample of $n = 313$ responses collected for the British Crime Survey, are shown in the table below. [*Note:* All variables except Density are expressed as a percentage of the base area.]

a. Test the hypothesis that the density (x_1) of a region is positively linearly related to crime prevalence (y), holding the other independent variables constant.

b. Do you advise conducting t-tests on each of the 18 independent variables in the model to determine which variables are important predictors of crime prevalence? Explain.

c. The model yielded $R^2 = .411$. Use this information to conduct a test of the global utility of the model. Use $\alpha = .05$.

12.143 Gender and salary study. Much research—and much litigation—has been conducted on the disparity between the salary levels of men and women. Research reported in *Work and Occupations* (Nov. 1992) analyzes the salaries for a sample of 191 Illinois managers using a regression analysis with the following independent variables:

$$x_1 = \text{Gender of manager} = \begin{cases} 1 & \text{if male} \\ 0 & \text{if not} \end{cases}$$

Results for Exercise 12.142

Variable	$\hat{\beta}$	t	p-value
x_1 = Density (population per hectare)	.331	3.88	$p < .01$
x_2 = Unemployed male population	−.121	−1.17	$p > .10$
x_3 = Professional population	−.187	−1.90	$.01 < p < .10$
x_4 = Population aged less than 5	−.151	−1.51	$p > .10$
x_5 = Population aged between 5 and 15	.353	3.42	$p < .01$
x_6 = Female population	.095	1.31	$p > .10$
x_7 = 10-year change in population	.130	1.40	$p > .10$
x_8 = Minority population	−.122	−1.51	$p > .10$
x_9 = Young adult population	.163	5.62	$p < .01$
x_{10} = 1 if North region, 0 if not	.369	1.72	$.01 < p < .10$
x_{11} = 1 if Yorkshire region, 0 if not	−.210	−1.39	$p > .10$
x_{12} = 1 if East Midlands region, 0 if not	−.192	−0.78	$p > .10$
x_{13} = 1 if East Anglia region, 0 if not	−.548	−2.22	$.01 < p < .10$
x_{14} = 1 if South East region, 0 if not	.152	1.37	$p > .10$
x_{15} = 1 if South West region, 0 if not	−.151	−0.88	$p > .10$
x_{16} = 1 if West Midlands region, 0 if not	−.308	−1.93	$.01 < p < .10$
x_{17} = 1 if North West region, 0 if not	.311	2.13	$.01 < p < .10$
x_{18} = 1 if Wales region, 0 if not	−.019	−0.08	$p > .10$

Source: Osborn, D. R., Tickett, Al., and Elder, R. "Area characteristics and regional variates as determinants of area property crime." *Journal of Quantitative Criminology*, Vol. 8, No. 3, 1992, Plenum Publishing Corp.

$$x_2 = \text{Race of manager} = \begin{cases} 1 & \text{if white} \\ 0 & \text{if not} \end{cases}$$

$x_3 = $ Education level (in years)

$x_4 = $ Tenure with firm (in years)

$x_5 = $ Number of hours worked per week

The regression results are shown in the table as they were reported in the article.

Variable	$\hat{\beta}$	p-Value
x_1	12.774	<.05
x_2	.713	>.10
x_3	1.519	<.05
x_4	.320	<.05
x_5	.205	<.05
Constant	15.491	—
	$R^2 = .240$ $n = 191$	

a. Write the hypothesized model that was used, and interpret each of the β parameters in the model.

b. Write the least squares equation that estimates the model in part **a**, and interpret each of the β estimates.

c. Interpret the value of R^2. Test to determine whether the model is useful for predicting annual salary. Test using $\alpha = .05$.

d. Test to determine whether the gender variable indicates that male managers are paid more than female managers, even after adjusting for and holding constant the other four factors in the model. Test using $\alpha = .05$. [*Note:* The p-values given in the table are two-tailed.]

e. Why would one want to adjust for these other factors before conducting a test for salary discrimination?

Applying the Concepts—Intermediate

12.144 Carp excretion experiment. *Fisheries Science* (Feb. 1995) reported on a study of the variables that affect endogenous nitrogen excretion (ENE) in carp raised in Japan. Carp were divided into groups of 2 to 15 fish each according to body weight and each group placed in a separate tank. The carp were then fed a protein-free diet three times daily for a period of 20 days. One day after terminating the feeding experiment, the amount of ENE in each tank was measured. The next table gives the mean body weight (in grams) and ENE amount (in milligrams per 100 grams of body weight per day) for each carp group.

a. Plot the data in a scattergram. Do you detect a pattern?

b. Fit the quadratic model $E(y) = \beta_0 + \beta_1 x + \beta_2 x^2$ to the data. Use the resulting printout to test $H_0: \beta_2 = 0$ against $H_a: \beta_2 \neq 0$ using $\alpha = .10$. Give the conclusion in the words of the problem.

CARP

Tank	Body Weight x	ENE y
1	11.7	15.3
2	25.3	9.3
3	90.2	6.5
4	213.0	6.0
5	10.2	15.7
6	17.6	10.0
7	32.6	8.6
8	81.3	6.4
9	141.5	5.6
10	285.7	6.0

Source: Watanabe, T., and Ohta, M. "Endogenous nitrogen excretion and non-fecal energy losses in carp and rainbow trout." *Fisheries Science*, Vol. 61, No. 1, Feb. 1995, p. 56 (Table 5).

12.145 Depression among perfectionists. Perfectionists are persons who set themselves standards and goals that cannot be reasonably met or accomplished. One theory suggests that those individuals who are depressed have a tendency toward perfectionism. To study this phenomenon, 76 members of an introductory psychology class completed four questionnaires: (1) the ASO scale, designed to measure self-acceptance, (2) the Burns scale, designed to measure perfectionism, (3) the Zung scale, designed to measure depression, and (4) the Rotter scale, designed to measure perceptions between actions and reinforcement. (*The Journal of Adlerian Theory, Research, and Practice*, Mar. 1986.)

a. Write a first-order model relating depression (Zung scale) to self-acceptance (ASO scale), perfectionism (Burns scale), and reinforcement (Rotter scale).

b. The model, part **a**, was fitted to the $n = 76$ points and resulted in a coefficient of determination of $R^2 = .70$. Interpret this value.

c. Is there sufficient evidence to indicate that the model is useful for predicting depression (Zung scale) score? Test using $\alpha = .05$.

d. A t-test for the perfectionism (Burns scale) variable resulted in a (two-tailed) p-value of .87. Interpret this value.

12.146 Prototyping new software. To meet the increasing demand for new software products, many systems development experts have adopted a prototyping methodology. The effects of prototyping on the system development life cycle (SDLC) was investigated in the *Journal of Computer Information Systems* (Spring 1993). A survey of 500 randomly selected corporate level MIS managers was conducted. Three potential independent variables were: (1) *importance* of prototyping to each phase of the SDLC; (2) degree of *support* prototyping provides for the SDLC; and (3) degree to which prototyping *replaces* each phase of the SDLC. The next table gives the pairwise correlations of the three variables in the survey data for one particular phase of the SDLC. Use this information to assess the degree of multicollinearity in the survey data. Would you recommend using all three independent variables in a regression analysis? Explain.

Variable Pairs	Correlation Coefficient, r
Importance-Replace	.2682
Importance-Support	.6991
Replace-Support	−.0531

Source: Hardgrave, B. C., Doke, E. R., and Swanson, N. E. "Prototyping effects of the system development life cycle: An empirical study." *Journal of Computer Information Systems,* Vol. 33, No. 3, Spring 1993, p. 16 (Table 1).

12.147 Rating funny cartoons. Newspaper cartoons, although designed to be funny, often invoke hostility, pain, and/or aggression in readers, especially when those cartoons depict violence. A study was undertaken to determine how violence in cartoons is related to aggression or pain. (*Motivation and Emotion,* Vol. 10, 1986.) A group of volunteers (psychology students) rated each of 32 violent newspaper cartoons (16 "Herman" and 16 "Far Side" cartoons) on three dimensions:

y = Funniness (0 = not funny, ..., 9 = very funny)
x_1 = Pain (0 = none, ..., 9 = a very great deal)
x_2 = Aggression/hostility (0 = none, ..., 9 = a very great deal)

The ratings of the students on each dimension were averaged and the resulting $n = 32$ observations were subjected to a multiple regression analysis. Based on the underlying theory (called the inverted-U theory) that the funniness of a joke will increase at low levels of aggression or pain, level off, and then decrease at high levels of aggression or pain, the following quadratic models were proposed:

Model 1:
$E(y) = \beta_0 + \beta_1 x_1 + \beta_2 x_1^2, R^2 = .099, F = 1.60$
Model 2:
$E(y) = \beta_0 + \beta_1 x_2 + \beta_2 x_2^2, R^2 = .100, F = 1.61$

a. According to the theory, what is the expected sign of β_2 in either model?
b. Is there sufficient evidence to indicate that the quadratic model relating pain to funniness rating is useful? Test at $\alpha = .05$.
c. Is there sufficient evidence to indicate that the quadratic model relating aggression/hostility to funniness rating is useful? Test at $\alpha = .05$.

12.148 Machine downtime study. An operations manager is interested in modeling $E(y)$, the expected length of time per month (in hours) that a machine will be shut down for repairs, as a function of the type of machine (001 or 002) and the age of the machine (in years). The manager has proposed the following model:

$$E(y) = \beta_0 + \beta_1 x_1 + \beta_2 x_1^2 + \beta_3 x_2$$

where

x_1 = Age of machine
x_2 = 1 if machine type 001, 0 if machine type 002

a. Use the data obtained on $n = 20$ machine breakdowns, shown here, to estimate the parameters of this model.

SHUTDOWN

Downtime (hours per month)	Machine Age x_1 (years)	Machine Type	x_2
10	1.0	001	1
20	2.0	001	1
30	2.7	001	1
40	4.1	001	1
9	1.2	001	1
25	2.5	001	1
19	1.9	001	1
41	5.0	001	1
22	2.1	001	1
12	1.1	001	1
10	2.0	002	0
20	4.0	002	0
30	5.0	002	0
44	8.0	002	0
9	2.4	002	0
25	5.1	002	0
20	3.5	002	0
42	7.0	002	0
20	4.0	002	0
13	2.1	002	0

b. Do these data provide sufficient evidence to conclude that the second-order term (x_1^2) in the model proposed by the operations manager is necessary? Test using $\alpha = .05$.
c. Test the null hypothesis that $\beta_1 = \beta_2 = 0$ using $\alpha = .10$. Interpret the results of the test in the context of the problem.

12.149 Soil loss during rainfall. Phosphorus used in soil fertilizers can contaminate freshwater sources during rainfall runoff. Consequently, it is important for water quality engineers to estimate the amount of dissolved phosphorus in the water. *Geoderma* (June 1995) presented an investigation of the relationship between soil loss and percentage of dissolved phosphorus in water samples collected at 20 fertilized watersheds in Oklahoma. The data are given in the table on p. 775.

a. Plot the data in a scattergram. Do you detect a linear or curvilinear trend?
b. Fit the quadratic model $E(y) = \beta_0 + \beta_1 x + \beta_2 x^2$ to the data.
c. Conduct a test to determine if a curvilinear relationship exists between dissolved phosphorus percentage (y) and soil loss (x). Test using $\alpha = .05$.

PHOSPHOR

Watershed	Soil Loss x (kilometers per half-acre)	Dissolved Phosphorus Percentage y
1	18	42.3
2	17	50.2
3	35	52.7
4	16	77.1
5	14	36.8
6	54	17.5
7	153	66.4
8	81	67.5
9	183	28.9
10	284	15.1
11	767	20.1
12	148	38.3
13	649	5.6
14	479	8.6
15	1,371	5.5
16	9,150	4.6
17	15,022	2.2
18	69	77.9
19	4,392	7.8
20	312	42.9

Source: Sharpley, A. N., Robinson, J. S., and Smith, S. J. "Bioavailable phosphorus dynamics in agricultural soils and effects on water quality." *Geoderma*, Vol. 67, No. 1–2, June 1995, p. 11 (Table 4).

12.150 High school academic achievement. The *American Sociological Review* (Summer 1993) published a study of the variables that affect academic achievement in high school. Two variables hypothesized to have an impact on achievement were track (college program of courses or not) and sector (Catholic school or public school). One of the models considered in the study was

$$E(y) = \beta_0 + \beta_1 x_1 + \beta_2 x_2 + \beta_3 x_1 x_2$$

where

y = Verbal achievement score
x_1 = 1 if college track, 0 if not
x_2 = 1 if Catholic school, 0 if not

a. How would you determine whether the impact of track (x_1) on verbal achievement score (y) depends on sector (x_2)? Set up the null and alternative hypotheses for the test.

b. In terms of the β's, give the mean verbal achievement score, $E(y)$, for public school students not on a college program.

c. Repeat part **b** for public school students on a college program.

d. Subtract the results, parts **b** and **c**, to obtain the difference between the means of college-track and regular-track students in public schools.

e. In terms of the β's, give the mean verbal achievement score, $E(y)$, for Catholic school students not on a college program.

f. Repeat part **e** for Catholic school students on a college program.

g. Subtract the results, parts **e** and **f**, to obtain the difference between the means of college-track and regular-track students in Catholic schools.

12.151 Sale prices of apartments. A Minneapolis, Minnesota, real estate appraiser used regression analysis to explore the relationship between the sale prices of apartment buildings sold in Minneapolis and various characteristics of the properties. Twenty-five apartment buildings were randomly sampled from all apartment buildings that were sold during a recent year. The table on p. 776 lists the data collected by the appraiser. *Note*: Physical condition of each apartment building is coded E (excellent), G (good), F (fair).

a. Write a model that describes the relationship between sale price and number of apartment units as three parallel lines, one for each level of physical condition. Be sure to specify the dummy variable coding scheme you use.

b. Plot y against x_1 (number of apartment units) for all buildings in excellent condition. On the same graph, plot y against x_1 for all buildings in good condition. Do this again for all buildings in fair condition. Does it appear that the model you specified in part **a** is appropriate? Explain.

c. Fit the model from part **a** to the data. Report the least squares prediction equation for each of the three building condition levels.

d. Plot the three prediction equations of part **c** on a scattergram of the data.

e. Do the data provide sufficient evidence to conclude that the relationship between sale price and number of units differs depending on the physical condition of the apartments? Test using $\alpha = .05$.

f. Check the data set for multicollinearity. How does this impact your choice of independent variables to use in a model for sale price?

g. Consider the first-order model $E(y) = \beta_0 + \beta_1 x_1 + \cdots + \beta_5 x_5$. Conduct a complete residual analysis for the model to check the assumptions on ε.

Applying the Concepts—Advanced

12.152 Abundance of bird species. Multiple regression analysis was used to model the abundance y of an individual bird species in transects in the United Kingdom. (*Journal of Applied Ecology*, Vol. 32, 1995.) Three of the independent variables used in the model, all field boundary attributes, are

(1) Transect location (small pasture field, small arable field, or large arable field)

(2) Land use (pasture or arable) adjacent to the transect

(3) Total number of trees in transect

a. Identify each of the independent variables as quantitative or qualitative variables.

b. Write a first-order model for $E(y)$ as a function of total number of trees.

MNSALES

Code No.	Sale Price y ($)	No. of Apartments x_1	Age of Structure x_2 (years)	Lot Size x_3 (sq. ft.)	No. of On-Site Parking Spaces x_4	Gross Building Area x_5 (sq. ft.)	Condition of Apartment Building
0229	90,300	4	82	4,635	0	4,266	F
0094	384,000	20	13	17,798	0	14,391	G
0043	157,500	5	66	5,913	0	6,615	G
0079	676,200	26	64	7,750	6	34,144	E
0134	165,000	5	55	5,150	0	6,120	G
0179	300,000	10	65	12,506	0	14,552	G
0087	108,750	4	82	7,160	0	3,040	G
0120	276,538	11	23	5,120	0	7,881	G
0246	420,000	20	18	11,745	20	12,600	G
0025	950,000	62	71	21,000	3	39,448	G
0015	560,000	26	74	11,221	0	30,000	G
0131	268,000	13	56	7,818	13	8,088	F
0172	290,000	9	76	4,900	0	11,315	E
0095	173,200	6	21	5,424	6	4,461	G
0121	323,650	11	24	11,834	8	9,000	G
0077	162,500	5	19	5,246	5	3,828	G
0060	353,500	20	62	11,223	2	13,680	F
0174	134,400	4	70	5,834	0	4,680	E
0084	187,000	8	19	9,075	0	7,392	G
0031	155,700	4	57	5,280	0	6,030	E
0019	93,600	4	82	6,864	0	3,840	F
0074	110,000	4	50	4,510	0	3,092	G
0057	573,200	14	10	11,192	0	23,704	E
0104	79,300	4	82	7,425	0	3,876	F
0024	272,000	5	82	7,500	0	9,542	E

Source: Robinson Appraisal Co., Inc., Mankato, Minnesota.

c. Add main effect terms for transect location to the model, part **b**. Graph the hypothesized relationships of the new model.

d. Add main effect terms for land use to the model, part **c**. In terms of the β's of the new model, what is the slope of the relationship between $E(y)$ and number of trees for any combination of transect location and land use?

e. Add terms for interaction between transect location and land use to the model, part **d**. Do these interaction terms affect the slope of the relationship between $E(y)$ and number of trees? Explain.

f. Add terms for interaction between number of trees and all coded dummy variables to the model, part **e**. In terms of the β's of the new model, give the slope of the relationship between $E(y)$ and number of trees for each combination of transect location and land use.

Critical Thinking Challenges

12.153 FLAG study of bid collusion. Road construction contracts in the state of Florida are awarded on the basis of competitive, sealed bids; the contractor who submits the lowest bid price wins the contract. During the 1980s, the Office of the Florida Attorney General (FLAG) suspected numerous contractors of practicing bid collusion (i.e., setting the winning bid price above the fair, or competitive, price in order to increase profit margin). By comparing the bid prices (and other important bid variables) of the fixed (or rigged) contracts to the competitively bid contracts, FLAG was able to establish invaluable benchmarks for detecting future bid-rigging. FLAG collected data for 279 road construction contracts. For each contract, the following variables were measured. The data are saved in the **FLAG** file.

FLAG

1. Price of contract ($) bid by lowest bidder
2. Department of Transportation (DOT) engineer's estimate of fair contract price ($)
3. Ratio of low (winning) bid price to DOT engineer's estimate of fair price
4. Status of contract (1 if fixed, 0 if competitive)
5. District (1, 2, 3, 4, or 5) in which construction project is located
6. Number of bidders on contract
7. Estimated number of days to complete work
8. Length of road project (miles)
9. Percentage of costs allocated to liquid asphalt
10. Percentage of costs allocated to base material
11. Percentage of costs allocated to excavation
12. Percentage of costs allocated to mobilization
13. Percentage of costs allocated to structures
14. Percentage of costs allocated to traffic control
15. Subcontractor utilization (1 if yes, 0 if no)

Use the methodology of this chapter to build a model for low-bid contract price (y). Comment on how bid status impacts the price.

12.154 In Exercise 5.135 (p. 285), we introduced *The Bell Curve* (Free Press, 1994) by Richard Herrnstein and Charles Murray (H&M), a controversial book about race, genes, IQ, and economic mobility. The book heavily employs statistics and statistical methodology in an attempt to support the authors' positions on the relationships among these variables and their social consequences. The main theme of *The Bell Curve* can be summarized as follows:

1. Measured intelligence (IQ) is largely genetically inherited.
2. IQ is correlated positively with a variety of socioeconomic status success measures, such as prestigious job, high annual income, and high educational attainment.
3. From 1 and 2, it follows that socioeconomic successes are largely genetically caused and therefore resistant to educational and environmental interventions (such as affirmative action).

The statistical methodology (regression) employed by the authors and the inferences derived from the statistics were critiqued in *Chance* (Summer 1995) and *The Journal of the American Statistical Association* (Dec. 1995). The following are just a few of the problems with H&M's use of regression that are identified:

Problem 1   H&M consistently use a trio of independent variables—IQ, socioeconomic status, and age—in a series of first-order models designed to predict dependent social outcome variables such as income and unemployment. (Only on a single occasion are interaction terms incorporated.) Consider, for example, the model

$$E(y) = \beta_0 + \beta_1 x_1 + \beta_2 x_2 + \beta_3 x_3$$

where y = income, x_1 = IQ, x_2 = socioeconomic status, and x_3 = age. H&M utilize *t*-tests on the individual β parameters to assess the importance of the independent variables. As with most of the models considered in *The Bell Curve*, the estimate of β_1 in the income model is positive and statistically significant at $\alpha = .05$, and the associated *t* value is larger (in absolute value) than the *t*-values associated with the other independent variables. Consequently, *H&M claim that IQ is a better predictor of income than the other two independent variables*. No attempt was made to determine whether the model was properly speci-

fied or whether the model provides an adequate fit to the data.

Problem 2   In an appendix, the authors describe multiple regression as a "mathematical procedure that yields coefficients for each of [the independent variables], indicating how much of a change in [the dependent variable] can be anticipated for a given change in any particular [independent] variable, with all the others held constant." Armed with this information and the fact that the estimate of β_1 in the model above is positive, *H&M infer that a high IQ necessarily implies (or causes) a high income, and a low IQ inevitably leads to a low income.* (Cause-and-effect inferences like this are made repeatedly throughout the book.)

Problem 3   The title of the book refers to the normal distribution and its well-known "bell-shaped" curve. There is a misconception among the general public that scores on intelligence tests (IQ) are normally distributed. In fact, most IQ scores have distributions that are decidedly skewed. Traditionally, psychologists and psychometricians have transformed these scores so that the resulting numbers have a precise normal distribution. H&M make a special point to do this. Consequently, *the measure of IQ used in all the regression models is normalized (i.e., transformed so that the resulting distribution is normal), despite the fact that regression methodology does not require predictor (independent) variables to be normally distributed.*

Problem 4   A variable that is not used as a predictor of social outcome in any of the models in *The Bell Curve* is level of education. H&M purposely omit education from the models, arguing that IQ causes education, not the other way around. Other researchers who have examined H&M's data report that *when education is included as an independent variable in the model, the effect of IQ on the dependent variable (say, income) is diminished.*

a. Comment on each of the problems identified. Why do each of these problems cast a shadow on the inferences made by the authors?

b. Using the variables specified in the model above, describe how you would conduct the multiple regression analysis. (Propose a more complex model and describe the appropriate model tests, including a residual analysis.)

Student Projects

Note: The use of statistical software is required for this project.

This is a continuation of the Student Projects section in Chapter 11, in which you selected three independent variables as predictors of a dependent variable of your choice and obtained at least 10 data values. Now fit the multiple regression model using an available software package:

$$y = \beta_0 + \beta_1 x_1 + \beta_2 x_2 + \beta_3 x_3 + \varepsilon$$

where

y = Dependent variable you chose

x_1 = First independent variable you chose

x_2 = Second independent variable you chose

x_3 = Third independent variable you chose

a. Compare the coefficients $\hat{\beta}_1$, $\hat{\beta}_2$, and $\hat{\beta}_3$ to their corresponding slope coefficients in the Chapter 11 Student Projects, where you fit three separate straight-line models. How do you account for the differences?

b. Calculate the coefficient of determination R^2, and conduct the F-test of the null hypothesis H_0: $\beta_1 = \beta_2 = \beta_3 = 0$. What is your conclusion?

c. Check the data for multicollinearity. If multicollinearity exists, how should you proceed?

d. Now increase your list of three variables to include approximately 10 that you think would be useful in predicting the dependent variable. With the aid of statistical software, employ a stepwise regression program to choose the important variables among those you have listed. To test your intuition, list the variables in the order you think they will be selected before you conduct the analysis. How does your list compare with the stepwise regression results?

e. After the group of 10 variables has been narrowed to a smaller group of variables by the stepwise analysis, try to improve the model by including interactions and quadratic terms. Be sure to consider the meaning of each interaction or quadratic term before adding it to the model—a quick sketch can be very helpful. See if you can systematically construct a useful model for prediction. If you have a large data set, you might want to hold out the last observations to test the predictive ability of your model after it is constructed. (As noted in Section 12.10, using the same data to construct *and* to evaluate predictive ability can lead to invalid statistical tests and a false sense of security.)

REFERENCES

Barnett, V., and Lewis, T. *Outliers in Statistical Data*. New York: Wiley, 1978.

Belsley, D. A., Kuh, E., and Welsch, R. E. *Regression Diagnostics: Identifying Influential Data and Sources of Collinearity*. New York: Wiley, 1980.

Chatterjee, S., and Price, B. *Regression Analysis by Example*, 2nd ed. New York: Wiley, 1991.

Draper, N., and Smith, H. *Applied Regression Analysis*, 2nd ed. New York: Wiley, 1981.

Graybill, F. *Theory and Application of the Linear Model*. North Scituate, Mass.: Duxbury, 1976.

Kelting, H. "Investigation of Condominium sale prices in three market scenarios: Utility of stepwise, interactive, multiple regression analysis and implications for design and appraisal methodology." Unpublished paper, University of Florida, Gainesville, 1979.

Mendenhall, W. *Introduction to Linear Models and the Design and Analysis of Experiments*. Belmont, Calif.: Wadsworth, 1968.

Mendenhall, W., and Sincich, T. *A Second Course in Statistics: Regression Analysis*, 6th ed. Upper Saddle River, N.J.: Prentice Hall, 2003.

Mosteller, F., and Tukey, J. W. *Data Analysis and Regression: A Second Course in Statistics*. Reading, Mass.: Addison-Wesley, 1977.

Neter, J., Kutner, M., Nachtsheim, C., and Wasserman, W. *Applied Linear Statistical Models*, 4th ed. Homewood, Ill.: Richard Irwin, 1996.

Rousseeuw, P. J., and Leroy, A. M. *Robust Regression and Outlier Detection*. New York: Wiley, 1987.

Weisberg, S. *Applied Linear Regression*, 2nd ed. New York: Wiley, 1985.

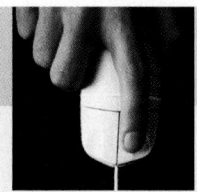

Using Technology

Multiple Regression Using MINITAB

To conduct a multiple regression analysis, first access the MINITAB worksheet file that contains the dependent and independent variables. Next, click on the "Stat" button on the MINITAB menu bar, then click on "Regression" and "Regression" again, as shown in Figure 12.M.1.

The resulting dialog box appears as shown in Figure 12.M.2. Specify the dependent variable in the "Response" box and the independent variables in the "Predictors" box.

[*Note:* If your model includes interaction and/or squared terms, you must create and add these higher-order variables to the MINITAB worksheet *prior to* running a regression analysis. You can do this by clicking the "Calc" button on the MINITAB main menu and selecting the "Calculator" option.]

Figure 12.M.1
MINITAB Menu Options for Regression

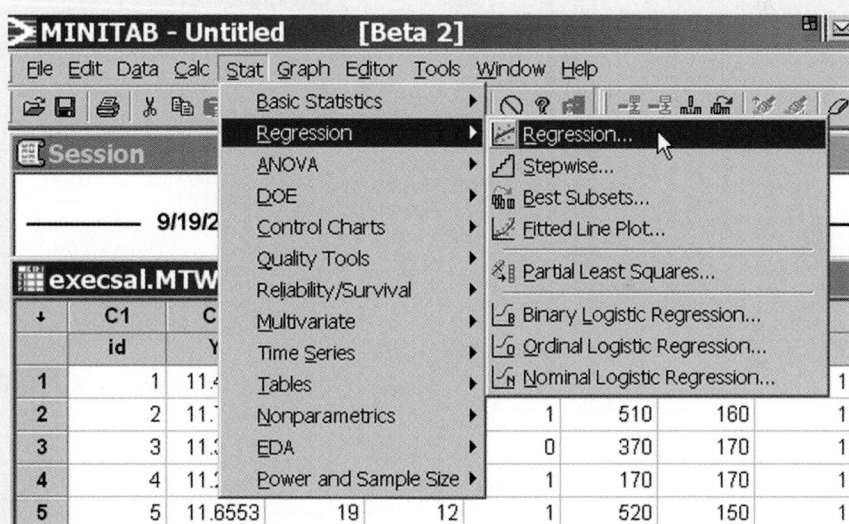

To produce prediction intervals for y and confidence intervals for $E(y)$, click the "Options" button and select the appropriate menu items in the resulting menu list. Residual plots are obtained by clicking the "Graphs" button and making the appropriate selections on the resulting menu. To return to the main Regression dialog box from any of these optional screens, click "OK. When you have made all your selections, click "OK" on the main Regression dialog box to produce the MINITAB multiple regression printout.

To conduct a stepwise regression analysis, click on the "Stat" button on the main menu bar, then click on "Regression", and click on "Stepwise" (see Figure 12.M.1). The resulting dialog box appears in Figure 12.M.3. Specify the dependent variable in the "Response" box and the independent variables in the stepwise model in the "Predictors" box. As an option, you can select the value of α to use in the analysis by clicking on the "Methods" button and specifying the value. (The default is $\alpha = .15$.) Click "OK" to view the stepwise regression results.

Figure 12.M.2
MINITAB Regression
Dialog Box

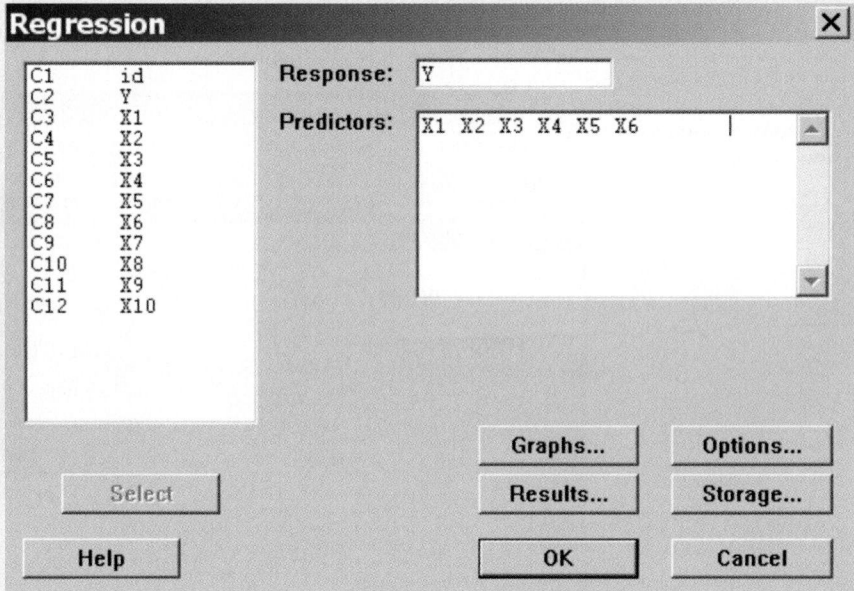

Figure 12.M.3
MINITAB Stepwise
Regression Dialog Box

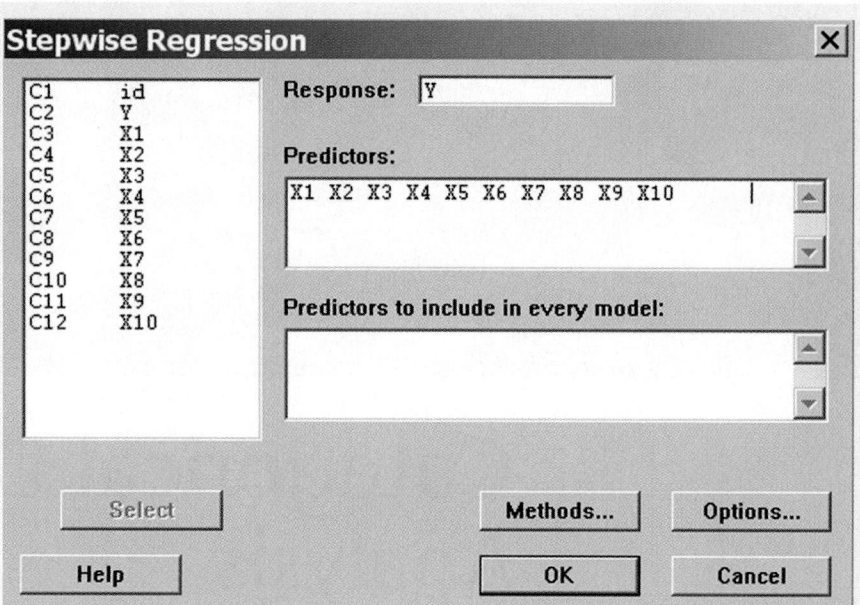

13

Categorical Data Analysis

Contents

13.1 Categorical Data and the Multinomial Experiment
13.2 Testing Categorical Probabilities: One-Way Table
13.3 Testing Categorical Probabilities: Two-Way (Contingency) Table
13.4 A Word of Caution about Chi-Square Tests

Statistics in Action

A Study of Coupon Users—Mail versus the Internet

Using Technology

Chi-Square Analyses Using MINITAB

Where We've Been

- Presented methods for making inferences about the population proportion associated with a two-level qualitative variable (i.e., a binomial variable)
- Presented methods for making inferences about the difference between two binomial proportions

Where We're Going

- Discuss qualitative (i.e., categorical) data with more than two outcomes.
- Present a *chi-square* hypothesis test for comparing the category proportions associated with a single qualitative variable—called a *one-way analysis*.
- Present a chi-square hypothesis test for relating two qualitative variables—called a *two-way analysis*.

Statistics in *ACTION*

A Study of Coupon Users—Mail versus the Internet

The service encounter is the critical interaction between a customer and the firm. In this encounter, the firm attempts to sell its services, to reinforce its offerings, and to satisfy the customer. A hot topic in marketing research is the exploration of a technology-based self-service (TBSS) encounter, where various technologies allow the customer to perform all or part of the service encounter. Examples of TBSS systems include ATMs, automated hotel checkout, banking by phone, self-scanning at retail stores, and transactions via the Internet such as Federal Express's package tracking and Charles Schwab's online brokerage services.

Marketing professor Dan Ladik (University of Suffolk) investigated a customer's motivation to use a TBSS and compared two customer segments—one that does not require any electronic technology use for service delivery and another that relies on the Internet for service delivery. The self-service delivery system studied was one that distributes discount coupons via both the mail and the Internet. Ladik investigated whether there were differences in customer characteristics and customer satisfaction between the mail (nontechnology) coupon users and the Internet (TBSS) coupon users.

The data for the study were obtained from a national services firm that specializes in discount coupons. (For reasons of confidentiality, the firm is not named.) Both customers who use the nontechnology mail delivery method and customers who use the firm's Internet Web site to access the coupons were sampled. For this *Statistics in Action* problem, we will focus on a subset of the full data set—a sample of 440 coupon users. Using a questionnaire, several qualitative variables were measured for each user. These are listed in Table SIA13.1. The data are saved in the **COUPONS** file.

In an attempt to answer the researcher's questions, we apply the statistical methodology presented in this chapter to this data set in two Statistics in Action Revisited examples.

Statistics in Action Revisited

- Testing Category Proportions for Customer Type in the Coupon Study (p. 790)
- Testing Whether the Coupon Customer Characteristics Are Related to User Type (p. 803)

◉ COUPONS

TABLE SIA13.1   Qualitative Variables Measured in the Coupon Study

Variable Name	Levels (Possible Values)
Coupon User Type	Mail, Internet, or Both
Gender	Male or Female
Education	High School, Vo-Tech/College, 4-year College Degree, or Graduate School
Work Status	Full Time, Part Time, Not Working, Retired
Coupon Satisfaction	Satisfied, Unsatisfied, Indifferent

13.1 Categorical Data and the Multinomial Experiment

Recall from Section 1.4 (p. 12) that observations on a qualitative variable can only be categorized. For example, consider the highest level of education attained by a professional hockey player. Level of education is a qualitative variable with several categories, including some high school, high school diploma, some college, college undergraduate degree, and graduate degree. If we were to record education level for all professional hockey players, the result of the categorization would be a count of the numbers of players falling in the respective categories.

When the qualitative variable of interest results in one of two responses (e.g., yes or no, success or failure, favor or do not favor), the data—called *counts*—can be analyzed using the binomial probability distribution discussed in Section 4.4. However, qualitative variables, such as level of education, that allow for more than two categories for a response are much more common, and these must be analyzed using a different method.

Qualitative data with more than two levels often result from a **multinomial experiment**. The characteristics for a multinomial experiment with k outcomes are described in the box. You can see that the binomial experiment of Chapter 4 is a multinomial experiment with $k = 2$.

> ### Properties of the Multinomial Experiment
>
> **1.** The experiment consists of n identical trials.
>
> **2.** There are k possible outcomes to each trial. These outcomes are sometimes called **classes**, **categories**, or **cells**.
>
> **3.** The probabilities of the k outcomes, denoted by $p_1, p_2, \ldots, p_k$, remain the same from trial to trial, where $p_1 + p_2 + \cdots + p_k = 1$.
>
> **4.** The trials are independent.
>
> **5.** The random variables of interest are the **cell counts**, $n_1, n_2, \ldots, n_k$, of the number of observations that fall in each of the k categories.

EXAMPLE 13.1   IDENTIFYING A MULTINOMIAL EXPERIMENT

Problem   Consider the problem of determining the highest level of education attained by each of a sample of $n = 40$ National Hockey League (NHL) players. Suppose we categorize level of education into one of five categories—some high school, high school diploma, some college, college undergraduate degree, and graduate degree—and count the number of the 40 players that fall into each category. Is this a multinomial experiment to a reasonable degree of approximation?

Solution   Checking the five properties of a multinomial experiment shown in the box, we have the following:

1. The experiment consists of $n = 40$ identical trials, where each trial is to determine the education level of an NHL player.

2. There are $k = 5$ possible outcomes to each trial corresponding to the five education-level responses.

3. The probabilities of the $k = 5$ outcomes, p_1, p_2, p_3, p_4, and p_5, remain the same from trial to trial (to a reasonable degree of approximation), where p_i represents the true probability that an NHL player attains level-of-education category i.

4. The trials are independent; that is, the education level attained by one NHL player does not affect the level attained by any other player.

5. We are interested in the count of the number of hockey players who fall into each of the five education-level categories. These five cell counts are denoted n_1, n_2, n_3, n_4, and n_5.

Thus, the properties of a multinomial experiment are satisfied.

■ ■ ■

In this chapter, we are concerned with the analysis of categorical data—specifically, the data that represent the counts for each category of a multinomial experiment. In Section 13.2, we learn how to make inferences about category probabilities for data classified according to a single qualitative (or categorical) variable. Then, in Section 13.3, we consider inferences about categorical probabilities for data classified according to two qualitative variables. The statistic used for these inferences is one that possesses, approximately, the familiar chi-square distribution.

13.2 Testing Categorical Probabilities: One-Way Table

In this section, we consider a multinomial experiment with k outcomes that corresponds to the categories of a *single* qualitative variable. The results of such an experiment are summarized in a **one-way table**. The term *one-way* is used because only one variable is classified. Typically, we want to make inferences about the true percentages that occur in the k categories based on the sample information in the one-way table.

To illustrate, suppose three political candidates are running for the same elective position. Prior to the election, we conduct a survey to determine the voting preferences of a random sample of 150 eligible voters. The qualitative variable of interest is *preferred candidate*, which has three possible outcomes: candidate 1, candidate 2, and candidate 3. Suppose the number of voters preferring each candidate is tabulated and the resulting count data appear as in Table 13.1.

TABLE 13.1   Results of Voter-Preference Survey

Candidate		
1	**2**	**3**
61	53	36

Note that our voter-preference survey satisfies the properties of a multinomial experiment for the qualitative variable, preferred candidate. The experiment consists of randomly sampling $n = 150$ voters from a large population of voters containing an unknown proportion p_1 who favor candidate 1, a proportion p_2 who favor candidate 2, and a proportion p_3 who favor candidate 3. Each voter sampled represents a single trial that can result in one of three outcomes: The voter will favor candidate 1, 2, or 3 with probabilities p_1, p_2, and p_3, respectively. (Assume that all voters will have a preference.) The voting preference of any single voter in the sample does not affect the preference of another; consequently, the trials are independent. And, finally, you can see that the recorded data are the numbers of voters in each of the three voter-preference categories. Thus, the voter-preference survey satisfies the five properties of a multinomial experiment.

In the voter-preference survey, and in most practical applications of the multinomial experiment, the k outcome probabilities $p_1, p_2, \ldots, p_k$ are unknown and we want to use the survey data to make inferences about their values. The unknown probabilities in the voter-preference survey are

$$p_1 = \text{Proportion of all voters who favor candidate 1}$$
$$p_2 = \text{Proportion of all voters who favor candidate 2}$$
$$p_3 = \text{Proportion of all voters who favor candidate 3}$$

To decide whether the voters have a preference for any of the candidates, we will want to test the null hypothesis that the candidates are equally preferred (that is, $p_1 = p_2 = p_3 = \frac{1}{3}$) against the alternative hypothesis that one candidate is preferred (that is, at least one of the probabilities p_1, p_2, and p_3 exceeds $\frac{1}{3}$). Thus, we want to test

H_0: $p_1 = p_2 = p_3 = \frac{1}{3}$ (no preference)

H_a: At least one of the proportions exceeds $\frac{1}{3}$ (a preference exists)

If the null hypothesis is true and $p_1 = p_2 = p_3 = \frac{1}{3}$, the expected value (mean value) of the number of voters who prefer candidate 1 is given by

$$E_1 = np_1 = (n)\frac{1}{3} = (150)\frac{1}{3} = 50$$

Similarly, $E_2 = E_3 = 50$ if the null hypothesis is true and no preference exists.

The following test statistic—the **chi-square test**—measures the degree of disagreement between the data and the null hypothesis:

$$\chi^2 = \frac{[n_1 - E_1]^2}{E_1} + \frac{[n_2 - E_2]^2}{E_2} + \frac{[n_3 - E_3]^2}{E_3}$$
$$= \frac{(n_1 - 50)^2}{50} + \frac{(n_2 - 50)^2}{50} + \frac{(n_3 - 50)^2}{50}$$

Note that the farther the observed numbers n_1, n_2, and n_3 are from their expected value (50), the larger χ^2 will become. That is, large values of χ^2 imply that the null hypothesis is false.

Biography

KARL PEARSON (1895–1980)—The Father of Statistics

While attending college, London-born Karl Pearson exhibited a wide range of interests, including mathematics, physics, religion, history, socialism, and Darwinism. After earning a law degree at Cambridge University and a Ph.D. in political science at the University of Heidelberg (Germany), Pearson became a professor of applied mathematics at University College in London. His 1892 book, *The Grammer of Science*, illustrated his conviction that statistical data analysis lies at the foundation of all knowledge; consequently, many consider Pearson to be the "father of statistics." A few of Pearson's many contributions to the field include introducing the term *standard deviation* and its associated symbol (σ); developing the distribution of the correlation coefficient; cofounding and editing the prestigious statistics journal *Biometrika*; and (what many consider his greatest achievement) creating the first chi-square "goodness-of-fit" test. Pearson inspired his students (including his son, Egon, and William Gossett) with his wonderful lectures and enthusiasm for statistics.

We have to know the distribution of χ^2 in repeated sampling before we can decide whether the data indicate that a preference exists. When H_0 is true, χ^2 can be shown to have (approximately) the familiar chi-square distribution of Sections 8.7 and 11.4. For this one-way classification, the χ^2 distribution has $(k - 1)$ degrees of freedom.* The rejection region for the voter-preference survey for $\alpha = .05$ and $k - 1 = 3 - 1 = 2$ df is

Rejection region: $\chi^2 > \chi^2_{.05}$

This value of $\chi^2_{.05}$ (found in Table VII) is 5.99147. (See Figure 13.1.) The computed value of the test statistic is

$$\chi^2 = \frac{(n_1 - 50)^2}{50} + \frac{(n_2 - 50)^2}{50} + \frac{(n_3 - 50)^2}{50}$$
$$= \frac{(61 - 50)^2}{50} + \frac{(53 - 50)^2}{50} + \frac{(36 - 50)^2}{50} = 6.52$$

*The derivation of the degrees of freedom for χ^2 involves the number of linear restrictions imposed on the count data. In the present case, the only constraint is that $\sum n_i = n$, where n (the sample size) is fixed in advance. Therefore, df $= k - 1$. For other cases, we will give the degrees of freedom for each usage of x^2 and refer the interested reader to the references for more detail.

Figure 13.1

Rejection Region for
Voter-Preference Survey

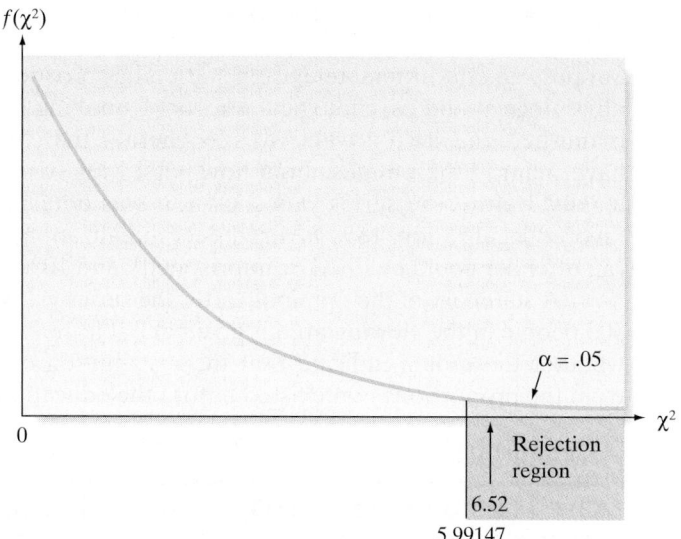

Since the computed $\chi^2 = 6.52$ exceeds the critical value of 5.99147, we conclude at the $\alpha = .05$ level of significance that there does exist a voter preference for one or more of the candidates.

Now that we have evidence to indicate that the proportions p_1, p_2, and p_3 are unequal, we can make inferences concerning their individual values using the methods of Section 7.3. [*Note:* We cannot use the methods of Section 9.3 to compare two proportions because the cell counts are dependent random variables.] The general form for a test of a hypothesis concerning multinomial probabilities is shown in the next box.

A Test of a Hypothesis about Multinomial Probabilities: One-Way Table

H_0: $p_1 = p_{1,0}$, $p_2 = p_{2,0}$, ..., $p_k = p_{k,0}$

where $p_{1,0}, p_{2,0}, \ldots, p_{k,0}$ represent the hypothesized values of the multinomial probabilities

H_a: At least one of the multinomial probabilities does not equal its hypothesized value

Test statistic: $\chi^2 = \sum \dfrac{[n_i - E_i]^2}{E_i}$

where $E_i = np_{i,0}$ is the **expected cell count**, that is, the expected number of outcomes of type i assuming that H_0 is true. The total sample size is n.

Rejection region: $\chi^2 > \chi_\alpha^2$, where χ_α^2 has $(k - 1)$ df

---

Conditions Required for a Valid χ^2 Test: One-Way Table

1. A multinomial experiment has been conducted. This is generally satisfied by taking a random sample from the population of interest.

2. The sample size n will be large enough so that for every cell, the expected cell count $E(n_i)$ will be equal to 5 or more.*

---

*The assumption that all expected cell counts are at least 5 is necessary in order to ensure that the χ^2 approximation is appropriate. Exact methods for conducting the test of a hypothesis exist and may be used for small expected cell counts, but these methods are beyond the scope of this text.

EXAMPLE 13.2   CONDUCTING A ONE-WAY χ^2 TEST

Problem   Suppose an educational television station has broadcast a series of programs on the physiological and psychological effects of smoking marijuana. Now that the series is finished, the station wants to see whether the citizens within the viewing area have changed their minds about how possession of marijuana should be considered legally. Before the series was shown, it was determined that 7% of the citizens favored legalization, 18% favored decriminalization, 65% favored the existing law (an offender could be fined or imprisoned), and 10% had no opinion.

A summary of the opinions (after the series was shown) of a random sample of 500 people in the viewing area is given in Table 13.2. Test at the $\alpha = .01$ level to see whether these data indicate that the distribution of opinions differs significantly from the proportions that existed before the educational series was aired.

⊙ **MARIJUANA**

TABLE 13.2   Distribution of Opinions about Marijuana Possession

Legalization	Decriminalization	Existing Laws	No Opinion
39	99	336	26

Solution   Define the proportions after the airing to be

p_1 = Proportion of citizens favoring legalization
p_2 = Proportion of citizens favoring decriminalization
p_3 = Proportion of citizens favoring existing laws
p_4 = Proportion of citizens with no opinion

Then the null hypothesis representing no change in the distribution of percentages is

H_0: $p_1 = .07$, $p_2 = .18$, $p_3 = .65$, $p_4 = .10$

and the alternative is

H_a: At least one of the proportions differs from its null hypothesized value

Test statistic: $\chi^2 = \sum \dfrac{[n_i - E_i]^2}{E_i}$

where

$$E_1 = np_{1,0} = 500(.07) = 35$$
$$E_2 = np_{2,0} = 500(.18) = 90$$
$$E_3 = np_{3,0} = 500(.65) = 325$$
$$E_4 = np_{4,0} = 500(.10) = 50$$

Since all these values are larger than 5, the χ^2 approximation is appropriate. Also, if the citizens in the sample were randomly selected, the properties of the multinomial probability distribution are satisfied.

Rejection region: For $\alpha = .01$ and df = $k - 1 = 3$, reject H_0 if $\chi^2 > \chi^2_{.01}$, where (from Table VII in Appendix A) $\chi^2_{.01} = 11.3449$.

We now calculate the test statistic:

$$\chi^2 = \frac{(39 - 35)^2}{35} + \frac{(99 - 90)^2}{90} + \frac{(336 - 325)^2}{325} + \frac{(26 - 50)^2}{50} = 13.249$$

Since this value exceeds the table value of χ^2 (11.3449), the data provide sufficient evidence ($\alpha = .01$) that the opinions on legalization of marijuana have changed since the series was aired.

Figure 13.2
SPSS Analysis of Data
in Table 13.2

OPINION

	Observed N	Expected N	Residual
LEGAL	39	35.0	4.0
DECRIM	99	90.0	9.0
EXISTLAW	336	325.0	11.0
NONE	26	50.0	-24.0
Total	500		

Test Statistics

	OPINION
Chi-Square[a]	13.249
df	3
Asymp. Sig.	.004

a. 0 cells (.0%) have expected frequencies less than
5. The minimum expected cell frequency is 35.0.

The χ^2-test can also be conducted using an available statistical software package. Figure 13.2 is an SPSS printout of the analysis of the data in Table 13.2. The test statistic and p-value of the test are highlighted on the printout. Since $\alpha = .01$ exceeds $p = .004$, there is sufficient evidence to reject H_0.

Look Back If the conclusion for the χ^2-test is "fail to reject H_0," then there is insufficient evidence to conclude that the distribution of opinions differs from the proportions stated in H_0. Be careful not to "accept H_0" and conclude that $p_1 = .07$, $p_2 = .18$, $p_3 = .65$, and $p_4 = .10$. The probability (β) of a Type II error is unknown.

| Now Work | *Exercise 13.9* |

■ ■ ■

If we focus on one particular outcome of a multinomial experiment, we can use the methods developed in Section 7.4 for a binomial proportion to establish a confidence interval for any one of the multinomial probabilities.* For example, if we want a 95% confidence interval for the proportion of citizens in the viewing area who have no opinion about the issue, we calculate

$$\hat{p}_4 \pm 1.96\sigma_{\hat{p}_4}$$

where

$$\hat{p}_4 = \frac{n_4}{n} = \frac{26}{500} = .052 \quad \text{and} \quad \sigma_{\hat{p}_4} \approx \sqrt{\frac{\hat{p}_4(1 - \hat{p}_4)}{n}}$$

Thus, we get

$$.052 \pm 1.96\sqrt{\frac{(.052)(.948)}{500}} = .052 \pm .019$$

or (.033, .071). Thus, we estimate that between 3.3% and 7.1% of the citizens now have no opinion on the issue of marijuana legalization. The series of programs may have helped citizens who formerly had no opinion on the issue to form an opinion, since it appears that the proportion of "no opinions" is now less than 10%.

*Note that focusing on one outcome has the effect of lumping the other $(k - 1)$ outcomes into a single group. Thus, we obtain, in effect, two outcomes—or a binomial experiment.

Statistics in Action Revisited

Testing Category Proportions for Customer Type in the Coupon Study

In the research on a technology-based self-service (TBSS) encounter (p. 783), 440 users of a firm's discount coupons were sampled and given a questionnaire to complete. One of the variables of interest to the researcher was type of coupon user. Recall (from Table SIA13.1) that the customers received the coupons in one of three ways: only through the mail (nontechnology user), only via the Internet (TBSS user), and via both the mail and Internet. What are the proportions of mail-only, Internet-only, and both users, and are these proportions statistically different?

To answer this question, we used SPSS to analyze the type of user variable in the **COUPONS** file. Figure SIA13.1 shows summary statistics and a graph to describe the three categories. From the summary table at the top of the printout, you can see that 262 (or 59.5%) of the customers are mail-only coupon users, 43 (or 9.8%) are Internet-only users, and the remainder (30.7%) use both mail and the

Internet. These sample percentages are illustrated in the bar graph at the bottom of Figure SIA13.1. In this sample of customers, the majority (almost 60%) obtain their coupons strictly through the mail.

Is this sufficient evidence to indicate that the true proportions in the population of customers are different? Letting p_1, p_2, and p_3 represent the true proportions for the mail-only, Internet-only, and both categories, respectively, we tested H_0: $p_1 = p_2 = p_3 = 1/3$ using SPSS. The printout is displayed in Figure SIA13.2. The cell frequencies and expected numbers are shown in the top table of the figure, while the chi-square test statistic (164.895) and p-value (.000) are shown in the bottom table. At any reasonably selected α-level (say, $\alpha = .01$), the small p-value indicates that there is sufficient evidence to reject the null hypothesis and conclude that the true proportions associated with the three user type categories are indeed statistically different.

Figure SIA13.1

SPSS Descriptive Statistics and Graph for Type of Coupon User

USER

		Frequency	Percent	Valid Percent	Cumulative Percent
Valid	Mail	262	59.5	59.5	59.5
	Net	43	9.8	9.8	69.3
	Both	135	30.7	30.7	100.0
	Total	440	100.0	100.0	

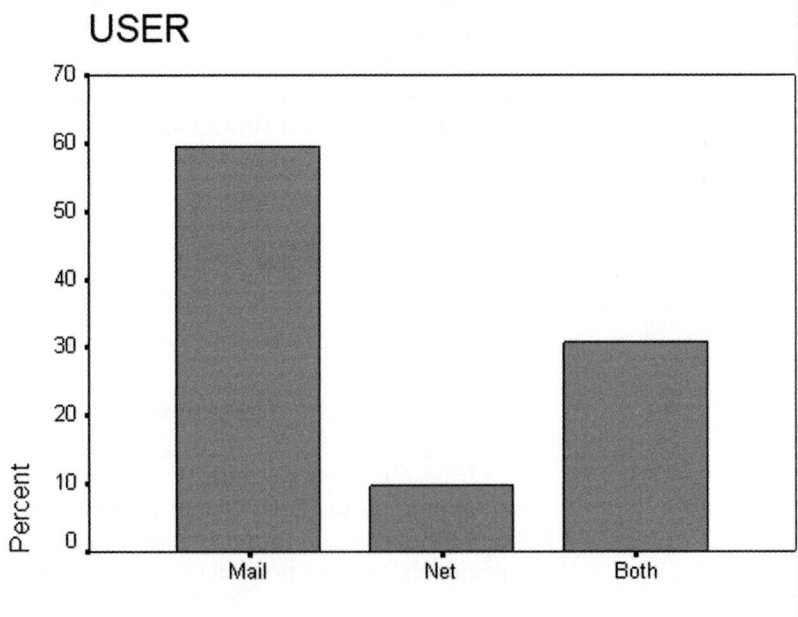

Figure SIA13.2
SPSS Chi-Square Test
User Type Categories

Chi-Square Test

Frequencies

USER

	Observed N	Expected N	Residual
Mail	262	146.7	115.3
Net	43	146.7	-103.7
Both	135	146.7	-11.7
Total	440		

Test Statistics

	USER
Chi-Square[a]	164.895
df	2
Asymp. Sig.	.000

a. 0 cells (.0%) have expected frequencies less than
5. The minimum expected cell frequency is 146.7.

Exercises 13.1–13.17

Understanding the Principles

13.1 What are the characteristics of a multinomial experiment? Compare the characteristics to those of a binomial experiment.

13.2 What conditions must n satisfy to make the χ^2-test for a one way table valid?

Learning the Mechanics

13.3 Use Table VII of Appendix A to find each of the following χ^2 values:
a. $\chi^2_{.05}$ for df = 10
b. $\chi^2_{.990}$ for df = 50
c. $\chi^2_{.10}$ for df = 16
d. $\chi^2_{.005}$ for df = 50

13.4 Use Table VII of Appendix A to find the following probabilities:
a. $P(\chi^2 \leq 1.063623)$ for df = 4
b. $P(\chi^2 > 30.5779)$ for df = 15
c. $P(\chi^2 \geq 82.3581)$ for df = 100
d. $P(\chi^2 < 18.4926)$ for df = 30

13.5 Find the rejection region for a one-dimensional χ^2-test of a null hypothesis concerning $p_1, p_2, \ldots, p_k$ if
a. $k = 3; \alpha = .05$
b. $k = 5; \alpha = .10$
c. $k = 4; \alpha = .01$

13.6 A multinomial experiment with $k = 3$ cells and $n = 320$ produced the data shown in the following table. Do these data provide sufficient evidence to contradict the null hypothesis that $p_1 = .25$, $p_2 = .25$, and $p_3 = .50$? Test using $\alpha = .05$.

	Cell		
	1	2	3
n_i	78	60	182

13.7 A multinomial experiment with $k = 4$ cells and $n = 205$ produced the data shown in the table.

	Cell			
	1	2	3	4
n_i	43	56	59	47

a. Do these data provide sufficient evidence to conclude that the multinomial probabilities differ? Test using $\alpha = .05$.
b. What are the Type I and Type II errors associated with the test of part **a**?
c. Construct a 95% confidence interval for the multinomial probability associated with cell 3.

13.8 A multinomial experiment with $k = 4$ cells and $n = 400$ produced the data shown in the following table. Do these data provide sufficient evidence to contradict the null hypothesis that $p_1 = .2$, $p_2 = .4$, $p_3 = .1$, and $p_4 = .3$? Test using $\alpha = .05$.

	Cell			
	1	2	3	4
n_i	70	196	46	88

Applying the Concepts—Basic

13.9 Jaw dysfunction study. A report on dental patients with temporomandibular (jaw) joint dysfunction (TMD) was published in *General Dentistry* (Jan/Feb. 2004). A random sample of 60 patients was selected for an experimental treatment of TMD. Prior to treatment, the patients filled out a survey on two nonfunctional jaw habits—bruxism (teeth grinding) and teeth clenching—that have been linked to TMD. Of the 60 patients, 3 admitted to bruxism, 11 admitted to teeth clenching, 30 admitted to both habits, and 16 claimed they had neither habit.

  a. Describe the qualitative variable of interest in the study. Give the levels (categories) associated with the variable.

  b. Construct a one-way table for the sample data.

  c. Give the null and alternative hypotheses for testing whether the percentages associated with the admitted habits are the same.

  d. Calculate the expected numbers for each cell of the one-way table.

  e. Calculate the appropriate test statistic.

  f. Give the rejection region for the test at $\alpha = .05$

  g. Give the appropriate conclusion in the words of the problem.

  h. Find and interpret a 95% confidence interval for the true proportion of dental patients who admit to both habits.

13.10 Location of major sports venues. There has been a recent trend for professional sports franchises in Major League Baseball (MLB), the National Football League (NFL), the National Basketball Association (NBA), and the National Hockey League (NHL) to build new stadiums and ballparks in urban, downtown venues. An article in *Professional Geographer* (Feb. 2000) investigated whether there has been a significant suburban-to-urban shift in the location of major sport facilities. In 1985, 40% of all major sport facilities were located downtown, 30% in central city, and 30% in suburban areas. In contrast, of the 113 major sports franchises that existed in 1997, 58 were built downtown, 26 in central city, and 29 in a suburban area.

  a. Describe the qualitative variable of interest in the study. Give the levels (categories) associated with the variable.

  b. Give the null hypothesis for a test to determine whether the proportions of major sports facilities in downtown, central city, and suburban areas in 1997 are the same as in 1985.

  c. If the null hypothesis, part **b**, is true, how many of the 113 sports facilities in 1997 would you expect to be located in downtown, central city, and suburban areas, respectively?

  d. Find the value of the chi-square statistic for testing the null hypothesis, part **b**.

  e. Find the (approximate) p-value of the test and give the appropriate conclusion in the words of the problem. Assume $\alpha = .05$.

13.11 Excavating ancient pottery. Refer to the *Chance* (Fall 2000) study of ancient Greek pottery, Exercise 2.12 (p. 38). Recall that 837 pottery pieces were uncovered at the excavation site. The table describing the types of pottery found is reproduced here.

Pot Category	Number Found
Burnished	133
Monochrome	460
Painted	183
Other	61
Total	837

Source: Berg, I., and Bliedon, S. "The pots of Phyiakopi: Applying statistical techniques to archaeology." *Chance,* Vol. 13, No. 4, Fall 2000.

  a. Describe the qualitative variable of interest in the study. Give the levels (categories) associated with the variable.

  b. Assume the four types of pottery occur with equal probability at the excavation site. What are the values of p_1, p_2, p_3, and p_4, the probabilities associated with the four pottery types?

  c. Give the null and alternative hypotheses for testing whether one type of pottery is more likely to occur at the site than any of the other types.

  d. Find the test statistic for testing the hypotheses, part **c**.

  e. Find and interpret the p-value of the test. State the conclusion in the words of the problem if you use $\alpha = .10$.

13.12 "Made in the USA" survey. Refer to the *Journal of Global Business* (Spring 2002) study of what "Made in the USA" on product labels means to the typical consumer, Exercise 2.13 (p. 38). Recall that 106 shoppers participated in the survey. Their responses, given as a percentage of U.S. labor and materials in four categories, are summarized in the table on p. 793. Suppose a consumer advocate group claims that half all consumers believe that "Made in the USA" means "100%" of labor and materials are produced in the U.S., one-fourth believe that "75 to 99%" are produced in the U.S. one-fifth believe that "50 to 74%" are produced in the U.S. and five percent believe that "less than 50%" are produced in the U.S.

Response to "Made in the USA"	Number of shoppers
100%	64
75 to 99%	20
50 to 74%	18
Less than 50%	4

Source: " 'Made in the USA': Consumer perceptions, deception and policy alternatives," *Journal of Global Business,* Vol. 13, No. 24, Spring 2002 (Table 3).

a. Describe the qualitative variable of interest in the study. Give the levels (categories) associated with the variable.

b. What are the values of p_1, p_2, p_3, and p_4, the probabilities associated with the four response categories hypothesized by the consumer advocate group?

c. Give the null and alternative hypotheses for testing the consumer advocate group's claim.

d. Compute the test statistic for testing the hypotheses, part **c**.

e. Find the rejection region of the test at $\alpha = .10$.

f. State the conclusion in the words of the problem.

g. Find and interpret a 90% confidence interval for the true proportion of consumers who believe "Made in the USA" means "100%" of labor and materials are produced in the U.S.

Applying the Concepts—Intermediate

13.13 Dosing errors at hospitals. Each year, approximately 1.3 million people in the United States suffer adverse drug effects (ADEs), that is, unintended injuries caused by prescribed medication. A study in the *Journal of the American Medical Association* (July 5, 1995) identified the cause of 247 ADEs that occurred at two Boston hospitals. The researchers found that dosing errors (that is, wrong dosage prescribed and/or dispensed) were the most common. The table summarizes the proximate cause of 95 ADEs that resulted from a dosing error. Conduct a test (at $\alpha = .10$) to determine whether the true percentages of ADEs in the five "cause" categories are different.

Wrong Dosage Cause	Number of ADEs
(1) Lack of knowledge of drug	29
(2) Rule violation	17
(3) Faulty dose checking	13
(4) Slips	9
(5) Other	27

13.14 Using instructional technology. In education, the term *instructional technology* refers to products such as computers, spreadsheets, CD-ROMs, videos, and presentation software. How frequently do professors use instructional technology in the classroom? To answer this question, researchers at Western Michigan University surveyed 306 of their fellow faculty. (*Educational Technology*, Mar.–Apr. 1995.) Responses to the frequency-of-technology use in teaching were recorded as "weekly to every class," "once a semester to monthly," or "never." The faculty responses (number in each response category) for the three technologies are summarized in the table.

Technology	Weekly	Once a Semester/ Monthly	Never
Computer spreadsheets	58	67	181
Word processing	168	61	77
Statistical software	37	82	187

a. Determine whether the percentages in the three frequency-of-use response categories differ for computer spreadsheets. Use $\alpha = .01$.

b. Repeat part **a** for word processing.

c. Repeat part **a** for statistical software.

d. Construct a 99% confidence interval for the true percentage of faculty who never use computer spreadsheets in the classroom. Interpret the interval.

13.15 Multiple sclerosis drug. Interferons are proteins produced naturally by the human body that help fight infections and regulate the immune system. A drug developed from interferons, called Avonex, is now available for treating patients with multiple sclerosis (MS). In a clinical study, 85 MS patients received weekly injections of Avonex over a 2-year period. The number of exacerbations (i.e., flare-ups of symptoms) was recorded for each patient and is summarized in the next table. For MS patients who take a placebo (no drug) over a similar 2-year period, it is known from previous studies that 26% will experience no exacerbations, 30% one exacerbation, 11% two exacerbations, 14% three exacerbations, and 19% four or more exacerbations.

Number of Exacerbations	Number of Patients
0	32
1	26
2	15
3	6
4 or more	6

Source: Biogen, Inc., 1997.

a. Conduct a test to determine whether the exacerbation distribution of MS patients who take Avonex differs from the percentages reported for placebo patients. Test using $\alpha = .05$.

b. Find a 95% confidence interval for the true percentage of Avonex MS patients who are exacerbation-free during a 2-year period.

c. Refer to part **b**. Is there evidence that Avonex patients are more likely to have no exacerbations than placebo patients? Explain.

13.16 Butterfly hotspots. *Nature* (Sept. 1993) reported on a study of animal and plant species "hotspots" in Great Britain. A hotspot is defined as a 10-km^2 area that is species-rich, that is, is heavily populated by the species of interest. Analogously, a coldspot is a 10-km^2 area that is species-poor. The table gives the number of butterfly hotspots and the number of butterfly coldspots in a sample of 2,588 10-km^2 areas. In theory, 5% of the areas should be butterfly hotspots and 5% should be butterfly coldspots, while the remaining areas (90%) are neutral. Test the theory using $\alpha = .01$.

Butterfly hotspots	123
Butterfly coldspots	147
Neutral areas	2,318
Total	**2,588**

Source: Prendergast, J. R., et al. "Rare species, the coincidence of diversity hotspots and conservation strategies." *Nature,* Vol. 365, No. 6444, Sept. 23, 1993, p. 335 (Table 1).

Applying the Concepts—Advanced

13.17 Analysis of a Scrabble game. In the board game Scrabble™, a player initially draws a "hand" of seven tiles at random from 100 tiles. Each tile has a letter of the alphabet and the player attempts to form a word

from the letters in his/her hand. In *Chance* (Winter 2002), scientist C. J. Robinove investigated whether a handheld electronic version of the game, called ScrabbleExpress™, produces too few vowels in the 7-letter draws. For each of the 26 letters (and "blank" for any letter), the accompanying table gives the true relative frequency of the letter in the board game as well as the frequency of occurrence of the letter in a sample of 700 tiles (i.e., 100 "hands") randomly drawn using the electronic game.

a. Do the data support the scientist's contention that ScrabbleExpress™ "presents the player with unfair word selection opportunities" that are not the same as the Scrabble™ board game? Test using $\alpha = .05$.

b. Estimate the true proportion of letters drawn in the electronic game that are vowels using a 95% confidence interval. Compare the results to the true relative frequency of a vowel in the board game.

SCRABBLE

Letter	Relative Frequency in Board Game	Frequency in Electronic Game	Letter	Relative Frequency in Board Game	Frequency in Electronic Game
A	.09	39	O	.08	20
B	.02	18	P	.02	27
C	.02	30	Q	.01	13
D	.04	30	R	.06	27
E	.12	31	S	.04	29
F	.02	21	T	.06	27
G	.03	35	U	.04	21
H	.02	21	V	.02	33
I	.09	25	W	.02	29
J	.01	17	X	.01	15
K	.01	27	Y	.02	32
L	.04	18	Z	.01	14
M	.02	31	# (blank)	.02	34
N	.06	36	Total		700

Source: Robinove, C. J. "Letter-frequency Bias in an Electronic Scrabble Game," *Chance,* Vol. 15, No. 1, Winter 2002, p. 31 (Table 3).

13.3 Testing Categorical Probabilities: Two-Way (Contingency) Table

In Section 13.1, we introduced the multinomial probability distribution and considered data classified according to a single criterion. We now consider multinomial experiments in which the data are classified according to two criteria, that is, *classification with respect to two qualitative factors.*

For example, consider a study in the *Journal of Marketing* (Fall 1992) on the impact of using celebrities in television advertisements. The researchers investigated the relationship between gender of a viewer and the viewer's brand awareness. Three hundred TV viewers were asked to identify products advertised by male celebrity spokespersons. The data are summarized in the **two-way table** shown in Table 13.3. This table is called a **contingency table**; it presents multinomial count data classified on two scales, or **dimensions, of classification**—namely, gender of viewer and brand awareness.

TABLE 13.3   Contingency Table for Marketing Example

		Gender Male	Female	Totals
Brand Awareness	**Could Identify Product**	95	41	136
	Could Not Identify Product	55	109	164
	Totals	150	150	300

TABLE 13.4a   Observed Counts for Contingency Table 13.3

		Gender Male	Female	Totals
Brand Awareness	**Could Identify Product**	n_{11}	n_{12}	r_1
	Could Not Identify Product	n_{21}	n_{22}	r_2
	Totals	c_1	c_2	n

TABLE 13.4b   Probabilities for Contingency Table 13.3

		Gender Male	Female	Totals
Brand Awareness	**Could Identify Product**	p_{11}	p_{12}	p_{r1}
	Could Not Identify Product	p_{21}	p_{22}	p_{r2}
	Totals	p_{c1}	p_{c2}	1

The symbols representing the cell counts for the multinomial experiment in Table 13.3 are shown in Table 13.4a; and the corresponding cell, row, and column probabilities are shown in Table 13.4b. Thus, n_{11} represents the number of viewers who are male and could identify the brand and p_{11} represents the corresponding cell probability. Note the symbols for the row and column totals and also the symbols for the probability totals. The latter are called **marginal probabilities** for each row and column. The marginal probability p_{r1} is the probability that a TV viewer identifies the product; the marginal probability p_{c1} is the probability that the TV viewer is male. Thus,

$$p_{r1} = p_{11} + p_{12} \text{ and } p_{c1} = p_{11} + p_{21}$$

Thus, we can see that this really is a multinomial experiment with a total of 300 trials, $(2)(2) = 4$ cells or possible outcomes, and probabilities for each cell as shown in Table 13.4b. If the 300 TV viewers are randomly chosen, the trials are considered independent and the probabilities are viewed as remaining constant from trial to trial.

Suppose we want to know whether the two classifications, gender and brand awareness, are dependent. That is, if we know the gender of the TV viewer, does that information give us a clue about the viewer's brand awareness? In a probabilistic sense we know (Chapter 3) that independence of events A and B implies $P(AB) = P(A)P(B)$. Similarly, in the contingency table analysis, if the **two classifications are independent**, the probability that an item is classified in any particular cell of the table is the product of the corresponding marginal probabilities. Thus, under the hypothesis of independence, in Table 13.4b, we must have

$$p_{11} = p_{r1}p_{c1} \quad p_{12} = p_{r1}p_{c2}$$
$$p_{21} = p_{r2}p_{c1} \quad p_{22} = p_{r2}p_{c2}$$

To test the hypothesis of independence, we use the same reasoning employed in the one-dimensional tests of Section 13.2. First, we calculate the *expected*, or *mean, count in each cell* assuming that the null hypothesis of independence is true. We do this by noting that the expected count in a cell of the table is just the total number of multinomial trials, n, times the cell probability. Recall that n_{ij} represents the **observed count** in the cell located in the ith row and jth column. Then the expected cell count for the upper-left-hand cell (first row, first column) is

$$E_{11} = np_{11}$$

or, when the null hypothesis (the classifications are independent) is true,

$$E_{11} = np_{r1}p_{c1}$$

Since these true probabilities are not known, we estimate p_{r1} and p_{c1} by the same proportions $\hat{p}_{r1} = r_1/n$ and $\hat{p}_{c1} = c_1/n$. Thus, the estimate of the expected value $E(n_{11})$ is

$$\hat{E}_{11} = n\left(\frac{r_1}{n}\right)\left(\frac{c_1}{n}\right) = \frac{r_1 c_1}{n}$$

Similarly, for each i, j,

$$\hat{E}_{ij} = \frac{(\text{Row total})(\text{Column total})}{\text{Total sample size}}$$

Thus,

$$\hat{E}_{12} = \frac{r_1 c_2}{n}$$

$$\hat{E}_{21} = \frac{r_2 c_1}{n}$$

$$\hat{E}_{22} = \frac{r_2 c_2}{n}$$

Finding Expected Cell Counts for a Two-Way Contingency Table

The estimate of the expected number of observations falling into the cell in row i and column j is given by

$$\hat{E}_{ij} = \frac{R_i C_j}{n}$$

where R_i = total for row i, C_j = total for column j, and n = sample size.

Using the data in Table 13.3, we find

$$\hat{E}_{11} = \frac{r_1 c_1}{n} = \frac{(136)(150)}{300} = 68$$

$$\hat{E}_{12} = \frac{r_1 c_2}{n} = \frac{(136)(150)}{300} = 68$$

$$\hat{E}_{21} = \frac{r_2 c_1}{n} = \frac{(164)(150)}{300} = 82$$

$$\hat{E}_{22} = \frac{r_2 c_2}{n} = \frac{(164)(150)}{300} = 82$$

The observed data and the estimated expected values (in parentheses) are shown in Table 13.5.

TABLE 13.5   Observed and Estimated Expected (in Parentheses) Counts

		Gender		
		Male	Female	Totals
Brand Awareness	**Could Identify Product**	95 (68)	41 (68)	136
	Could Not Identify Product	55 (82)	109 (82)	164
	Totals	150	150	300

We now use the χ^2 statistic to compare the observed and expected (estimated) counts in each cell of the contingency table:

$$\chi^2 = \frac{[n_{11} - \hat{E}_{11}]^2}{\hat{E}_{11}} + \frac{[n_{12} - \hat{E}_{12}]^2}{\hat{E}_{12}} + \frac{[n_{21} - \hat{E}_{21}]^2}{\hat{E}_{21}} + \frac{[n_{22} - \hat{E}_{22}]^2}{\hat{E}_{22}}$$

$$= \sum \frac{[n_{ij} - \hat{E}_{ij}]^2}{\hat{E}_{ij}}$$

Note: The use of $\sum$ in the context of a contingency table analysis refers to a sum over all cells in the table.

Substituting the data of Table 13.5 into this expression, we get

$$\chi^2 = \frac{(95 - 68)^2}{68} + \frac{(41 - 68)^2}{68} + \frac{(55 - 82)^2}{82} + \frac{(109 - 82)^2}{82} = 39.22$$

Large values of χ^2 imply that the observed counts do not closely agree and hence that the hypothesis of independence is false. To determine how large χ^2 must be before it is too large to be attributed to chance, we make use of the fact that the sampling distribution of χ^2 is approximately a χ^2 probability distribution when the classifications are independent.

When testing the null hypothesis of independence in a two-way contingency table, the appropriate degrees of freedom will be $(r - 1)(c - 1)$, where r is the number of rows and c is the number of columns in the table.

For the brand awareness example, the degrees of freedom for χ^2 is $(r - 1)(c - 1) = (2 - 1)(2 - 1) = 1$. Then, for $\alpha = .05$, we reject the hypothesis of independence when

$$\chi^2 > \chi^2_{.05} = 3.84146$$

Since the computed $\chi^2 = 39.22$ exceeds the value 3.84146, we conclude that viewer gender and brand awareness are dependent events.

The pattern of **dependence** can be seen more clearly by expressing the data as percentages. We first select one of the two classifications to be used as the base variable. In the above example, suppose we select gender of the TV viewer as the classificatory variable to be the base. Next, we represent the responses for each level of the second categorical variable (brand awareness in our example) as a percentage of the subtotal for the base variable. For example, from Table 13.5 we convert the response for males who identify the brand (95) to a percentage of the total number of male viewers (150). That is,

$$\left(^{95}/_{150}\right)100\% = 63.3\%$$

The conversions of all Table 13.5 entries are similarly computed, and the values are shown in Table 13.6. The value shown at the right of each row is the row's total

TABLE 13.6 Percentage of TV Viewers Who Identify Brand, by Gender

		Gender		
		Male	Female	Totals
Brand Awareness	**Could Identify Product**	63.3	27.3	45.3
	Could Not Identify Product	36.7	72.7	54.7
	Totals	100	100	100

expressed as a percentage of the total number of responses in the entire table. Thus, the percentage of TV viewers who identify the product is: $\left(\frac{136}{300}\right)100\% = 45.3\%$ (rounded to the nearest percent).

If the gender and brand awareness variables are independent, then the percentages in the cells of the table are expected to be approximately equal to the corresponding row percentages. Thus, we would expect the percentage who identify the brand for each gender to be approximately 45% if the two variables are independent. The extent to which each gender's percentage departs from this value determines the dependence of the two classifications, with greater variability of the row percentages meaning a greater degree of dependence. A plot of the percentages helps summarize the observed pattern. In the SPSS bar graph in Figure 13.3, we show the gender of the viewer (the base variable) on the horizontal axis, and the percentage of TV viewers who identify the brand on the vertical axis. The "expected" percentage under the assumption of independence is shown as a dotted horizontal line.

Figure 13.3 clearly indicates the reason that the test resulted in the conclusion that the two classifications in the contingency table are dependent. The percentage of male TV viewers who identify the brand promoted by a male celebrity is more than twice as high as the percentage of female TV viewers who identify the brand. Statistical measures of the degree of dependence and procedures for making comparisons of pairs of levels for classifications are available. They are beyond the scope of this text,

Figure 13.3
SPSS Bar Graph Showing
Percentage of Viewers Who
Identified the TV Product

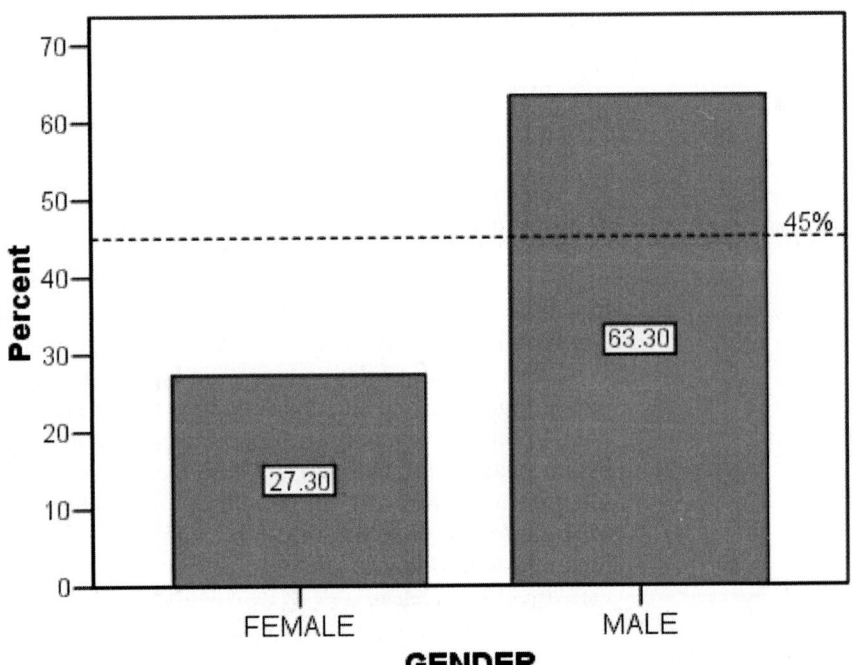

TABLE 13.7   General $r \times c$ Contingency Table

		Column				
		1	2	. . .	c	Row Totals
Row	**1**	n_{11}	n_{12}	. . .	n_{1c}	r_1
	2	n_{21}	n_{22}	. . .	n_{2c}	r_2
	⋮	⋮	⋮		⋮	⋮
	r	n_{r1}	n_{r2}	. . .	n_{rc}	r_r
Column Totals		c_1	c_2	. . .	c_c	n

but can be found in the references. We will, however, utilize descriptive summaries such as Figure 13.3 to examine the degree of dependence exhibited by the sample data.

The general form of a two-way contingency table containing r rows and c columns (called an $r \times c$ contingency table) is shown in Table 13.7. Note that the observed count in the *(ij)* cell is denoted by n_{ij}, the ith row total is r_i, the jth column total is c_j, and the total sample size is n. Using this notation, we give the general form of the contingency table test for independent classifications in the next box.

General Form of a Two-way (Contingency) Table Analysis: A Test for Independence

H_0: The two classifications are independent

H_a: The two classifications are dependent

Test statistic: $\chi^2 = \sum \dfrac{[n_{ij} - \hat{E}_{ij}]^2}{\hat{E}_{ij}}$

where $\hat{E}_{ij} = \dfrac{R_i C_j}{n}$

Rejection region: $\chi^2 > \chi^2_\alpha$, where χ^2_α has $(r - 1)(c - 1)$ df

---

Conditions Required for a Valid χ^2 Test: Contingency Tables

1. The n observed counts are a random sample from the population of interest. We may then consider this to be a multinomial experiment with $r \times c$ possible outcomes.

2. The sample size, n, will be large enough so that, for every cell, the expected count, $E(n_{ij})$, will be equal to 5 or more.

EXAMPLE 13.3

CONDUCTING A TWO-WAY ANALYSIS

Problem   A social scientist wants to determine whether the marital status (divorced or not divorced) of U.S. men is independent of their religious affiliation (or lack thereof). A sample of 500 U.S. men is surveyed and the results are tabulated as shown in Table 13.8.

  a. Test to see whether there is sufficient evidence to indicate that the marital status of men who have been or are currently married is dependent on religious affiliation. Test at $\alpha = .01$.

  b. Graph the data and describe the patterns revealed. Is the result of the test supported by the graph?

Solution   **a.** The first step is to calculate estimated expected cell frequencies under the assumption that the classifications are independent. Rather than compute

TABLE 13.8   Survey Results (Observed Counts), Example 13.2

| | | \multicolumn{5}{c}{Religious Affiliation} | |
		A	B	C	D	None	Totals
Marital Status	**Divorced**	39	19	12	28	18	116
	Never divorced	172	61	44	70	37	384
	Totals	211	80	56	98	55	500

these values by hand, we resort to a computer. The SAS printout of the analysis of Table 13.8 is displayed in Figure 13.4. Each cell in Figure 13.4 contains the observed (top) and expected (bottom) frequency in that cell. Note that $\hat{E}_{11}$, the estimated expected count for the Divorced, A cell, is 48.952. Similarly, the estimated expected count for the Divorced, B cell, is $\hat{E}_{12} = 18.56$. Since all the estimated expected cell frequencies are greater than 5, the χ^2 approximation for the test statistic is appropriate. Assuming the men chosen were randomly selected from all married or previously married American men, the characteristics of the multinomial probability distribution are satisfied.

The null and alternative hypotheses we want to test are

H_0: The marital status of U.S. men and their religious affiliation are independent

H_a: The marital status of U.S. men and their religious affiliation are dependent

The test statistic, $\chi^2 = 7.135$, is highlighted at the bottom of the printout, as is the observed significance level (p-value) of the test. Since $\alpha = .01$ is less than $p = .129$, we fail to reject H_0; that is, we cannot conclude that the marital status of U.S. men depends on their religious affiliation. (Note that we could not reject H_0 even with $\alpha = .10$.)

b. The marital status frequencies are expressed as percentages of the number of men in each religious affiliation category and highlighted in the SAS printout,

Figure 13.4

SAS Contingency Table
Printout for Example 13.3

```
                          The FREQ Procedure

                     Table of MARITAL by RELIGION

    MARITAL      RELIGION

    Frequency|
    Expected |A      |B      |C      |D      |NONE   | Total

    DIVORCED |    39 |    19 |    12 |    28 |    18 |   116
             |48.952 | 18.56 |12.992 |22.736 | 12.76 |

    NEVER    |   172 |    61 |    44 |    70 |    37 |   384
             |162.05 | 61.44 |43.008 |75.264 | 42.24 |

    Total        211      80      56      98      55     500


            Statistics for Table of MARITAL by RELIGION

    Statistic                      DF        Value        Prob

    Chi-Square                      4        7.1355      0.1289
    Likelihood Ratio Chi-Square     4        6.9854      0.1367
    Mantel-Haenszel Chi-Square      1        6.4943      0.0108
    Phi Coefficient                          0.1195
    Contingency Coefficient                  0.1186
    Cramer's V                               0.1195


                    Fisher's Exact Test

         Table Probability (P)       6.936E-06
         Pr <= P                       0.1251

                 Sample Size = 500
```

Figure 13.5

SAS Side-by-Side Bar Graphs Showing Percentage of Divorced Males by Religion

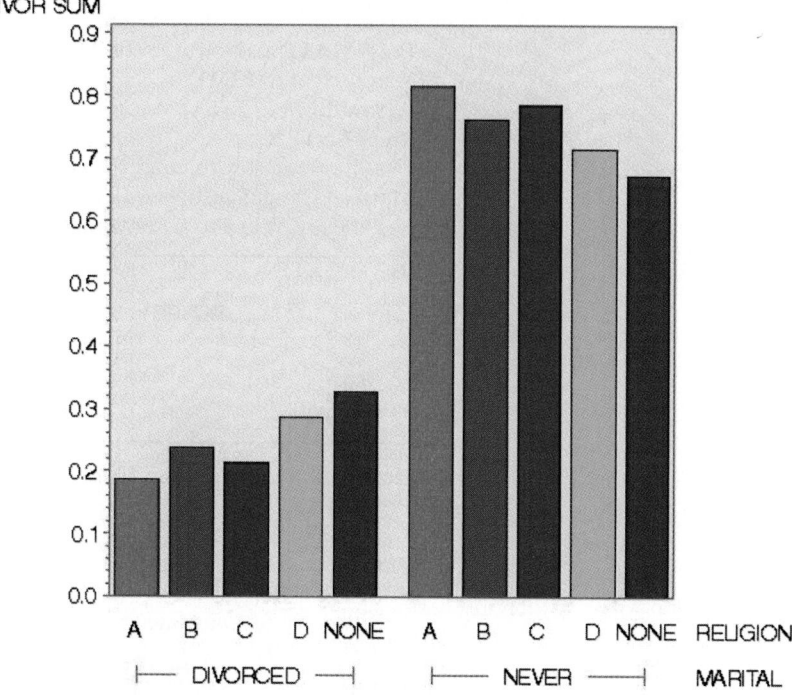

Figure 13.4. The expected percentage of divorced men under the assumption of independence is $(^{116}/_{500})100\% = 23\%$. A SAS graph of the percentages is shown in Figure 13.5. Note that the percentages of divorced men (see the bars in the "DIVORCED" block of the SAS graph) deviate only slightly from that expected under the assumption of independence, supporting the result of the test in part a. That is, neither the descriptive bar graph nor the statistical test provides evidence that the male divorce rate depends on (varies with) religious affiliation.

Now Work *Exercise 13.26*

■ ■ ■

Contingency Tables

Using the TI-83 Graphing Calculator

Finding *p*-values for Contingency Tables

Step 1 *Access the Matrix Menu to enter the observed values*
Press **MATRX** (Note: On the TI-83 Plus, press **2nd x⁻¹** for **MATRX**)
Arrow right to **EDIT**
Press **ENTER**
Use the arrow key to enter the row and column dimensions of your observed Matrix
Use the arrow key to enter your observed values into Matrix [A]

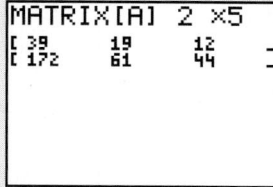

Step 2 *Access the Matrix Menu to enter the expected values*
Press **MATRX** (Note: On the TI-83 Plus, press **2nd x⁻¹** for **MATRX**)
Arrow right to **EDIT**
Arrow down to **2:[B]**
Press **ENTER**
Use the arrow key to enter the row and column dimensions of your expected
Matrix. (The dimensions will be the same as in Matrix A)
Use the arrow key to enter your expected values into Matrix [B]

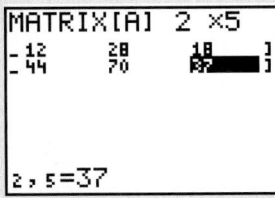

Step 3 *Access the Statistical Tests Menu and perform the Chi-square Test.*
Press **STAT**
Arrow right to **TESTS**
Arrow down to **C: χ^2 Test.**
Press **ENTER**
Arrow down to **Calculate**
Press **ENTER**

Step 4 *Reject H_0 if the p-value $< \alpha$*

Example Our Observed Matrix is $[A] = \begin{bmatrix} 39 & 19 & 12 & 28 & 18 \\ 172 & 61 & 44 & 70 & 37 \end{bmatrix}$

Our Expected Matrix is

$$[B] = \begin{bmatrix} 48.952 & 18.56 & 12.992 & 22.736 & 12.76 \\ 162.05 & 61.44 & 43.008 & 75.264 & 42.24 \end{bmatrix}$$

Use $\alpha = .05$ to test the following hypotheses:

H_0: The Matrix entries represent Independent events.
H_A: The Matrix entries represent events that **are not** independent.

The screens for this example are shown below.

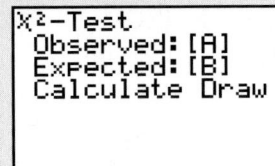

```
X²-Test
  X²=7.135464193
  P=.128900399
  df=4
```

As you can see from the screen the p-value is 0.1289. Since the p-value is **greater** than $\alpha = .05$, we **do not** reject H_0.

Statistics in Action Revisited

Testing Whether Coupon Customer Characteristics Are Related to User Type

In his study of a firm's discount coupon users (p. 783), marketing professor Dan Ladik wanted to know whether there are differences in customer characteristics (i.e., gender, education, work status, and satisfaction) among the three types of coupon users: mail-only (nontechnology) users, Internet-only (TBSS) users, and users of both mail and Internet coupons. One approach to analyzing these data is to determine whether each of the four customer characteristic variables are related to coupon user type. Since all the variables measured on the sample of 440 coupon users are qualitative, a contingency table analysis is appropriate.

Figures SIA 13.3–SIA13.6 show the SPSS contingency table analyses relating each customer characteristic to coupon user type. The p-values for the chi-square tests for the variables gender, education, satisfaction, and work status are .033, .361, .000, and .069, respectively. If we conduct each test at $\alpha = .01$ (we purposely choose a small α to minimize the chance of making a Type I error), the only significant result is for customer satisfaction. That is, the data provide evidence to indicate that the level of customer satisfaction depends on type of coupon user; however, there is not sufficient evidence to say that any of the other customer characteristics (gender, education, or work status) are related to type of coupon user.

The column percentages highlighted in the contingency table of Figure SIA13.5 reveal the differences in satisfaction levels of the three user types. The percentages of satisfied customers for mail-only and Internet-only coupon users are 65.6% and 69.8%, respectively. However, for users of both mail and the Internet, 91.1% are satisfied. This information was used by the coupon firm to develop a marketing strategy aimed at mail-only and Internet-only coupon users.

Figure SIA13.3
SPSS Contingency Table
Analysis—Gender versus
User Type

GENDER * USER Crosstabulation

			USER			
			Mail	Net	Both	Total
GENDER	Male	Count	84	7	31	122
		Expected Count	72.6	11.9	37.4	122.0
	Female	Count	178	36	104	318
		Expected Count	189.4	31.1	97.6	318.0
Total		Count	262	43	135	440
		Expected Count	262.0	43.0	135.0	440.0

Chi-Square Tests

	Value	df	Asymp. Sig. (2-sided)
Pearson Chi-Square	6.797[a]	2	.033
Likelihood Ratio	7.105	2	.029
Linear-by-Linear Association	4.371	1	.037
N of Valid Cases	440		

a. 0 cells (.0%) have expected count less than 5. The minimum expected count is 11.92.

Figure SIA13.4
SPSS Contingency Table Analysis—Education versus User Type

EDUC * USER Crosstabulation

			USER			
			Mail	Net	Both	Total
EDUC	HS	Count	34	7	19	60
		Expected Count	35.7	5.9	18.4	60.0
	VT/COLL	Count	96	20	62	178
		Expected Count	106.0	17.4	54.6	178.0
	COLL4	Count	85	9	38	132
		Expected Count	78.6	12.9	40.5	132.0
	GRAD	Count	47	7	16	70
		Expected Count	41.7	6.8	21.5	70.0
Total		Count	262	43	135	440
		Expected Count	262.0	43.0	135.0	440.0

Chi-Square Tests

	Value	df	Asymp. Sig. (2-sided)
Pearson Chi-Square	6.587[a]	6	.361
Likelihood Ratio	6.786	6	.341
Linear-by-Linear Association	3.546	1	.060
N of Valid Cases	440		

a. 0 cells (.0%) have expected count less than 5. The minimum expected count is 5.86.

Figure SIA13.5
SPSS Contingency Table Analysis—Satisfaction versus User Type

COUPSAT * USER Crosstabulation

			USER			
			Mail	Net	Both	Total
COUPSAT	Satisfied	Count	172	30	123	325
		Expected Count	193.5	31.8	99.7	325.0
		% within USER	65.6%	69.8%	91.1%	73.9%
	Indiff	Count	62	9	9	80
		Expected Count	47.6	7.8	24.5	80.0
		% within USER	23.7%	20.9%	6.7%	18.2%
	Unsatis	Count	28	4	3	35
		Expected Count	20.8	3.4	10.7	35.0
		% within USER	10.7%	9.3%	2.2%	8.0%
Total		Count	262	43	135	440
		Expected Count	262.0	43.0	135.0	440.0
		% within USER	100.0%	100.0%	100.0%	100.0%

Chi-Square Tests

	Value	df	Asymp. Sig. (2-sided)
Pearson Chi-Square	30.418[a]	4	.000
Likelihood Ratio	34.934	4	.000
Linear-by-Linear Association	25.717	1	.000
N of Valid Cases	440		

a. 1 cells (11.1%) have expected count less than 5. The minimum expected count is 3.42.

Figure SIA13.6
SPSS Contingency Table
Analysis—Work Status
versus User Type

WORK * USER Crosstabulation

			USER Mail	USER Net	USER Both	Total
WORK	FULL	Count	148	29	90	267
		Expected Count	159.0	26.1	81.9	267.0
	PART	Count	31	8	13	52
		Expected Count	31.0	5.1	16.0	52.0
	NONE	Count	31	3	17	51
		Expected Count	30.4	5.0	15.6	51.0
	RETIRED	Count	52	3	15	70
		Expected Count	41.7	6.8	21.5	70.0
Total		Count	262	43	135	440
		Expected Count	262.0	43.0	135.0	440.0

Chi-Square Tests

	Value	df	Asymp. Sig. (2-sided)
Pearson Chi-Square	11.687[a]	6	.069
Likelihood Ratio	12.208	6	.057
Linear-by-Linear Association	5.619	1	.018
N of Valid Cases	440		

a. 1 cells (8.3%) have expected count less than 5. The minimum expected count is 4.98.

Exercises 13.18–13.37

Understanding the Principles

13.18 What is a two-way (contingency) table?

13.19 *True or False.* One goal of a contingency table analysis is to determine if the two classifications are dependent.

13.20 What conditions are required for a valid chi-square test of data from a contingency table?

Learning the Mechanics

13.21 Find the rejection region for a test of independence of two classifications where the contingency table contains r rows and c columns and
 a. $r = 5$, $c = 5$, $\alpha = .05$
 b. $r = 3$, $c = 6$, $\alpha = .10$
 c. $r = 2$, $c = 3$, $\alpha = .01$

13.22 Consider the accompanying 2×3 (i.e., $r = 2$ and $c = 3$) contingency table.

		Column 1	Column 2	Column 3
Row	1	9	34	53
	2	16	30	25

 a. Specify the null and alternative hypotheses that should be used in testing the independence of the row and column classifications.
 b. Specify the test statistic and the rejection region that should be used in conducting the hypothesis test of part **a**. Use $\alpha = .01$.

 c. Assuming the row classification and the column classification are independent, find estimates for the expected cell counts.
 d. Conduct the hypothesis test of part **a**. Interpret your result.

13.23 Refer to Exercise 13.22.
 a. Convert the frequency responses to percentages by calculating the percentage of each column total falling in each row. Also convert the row totals to percentages of the total number of responses. Display the percentages in a table.
 b. Create a bar graph with row 1 percentage on the vertical axis and column number on the horizontal axis. Show the row 1 total percentage as a horizontal line on the graph.
 c. What pattern do you expect to see if the rows and columns are independent? Does the plot support the result of the test of independence in Exercise 13.22?

13.24 Test the null hypothesis of independence of the two classifications, A and B, of the 3×3 contingency table shown here. Test using $\alpha = .05$.

		B B_1	B B_2	B B_3
A	A_1	40	72	42
	A_2	63	53	70
	A_3	31	38	30

13.25 Refer to Exercise 13.24. Convert the responses to percentages by calculating the percentage of each B class total falling into each A classification. Also, calculate the percentage of the total number of responses that constitute each of the A classification totals.

a. Create a bar graph with row A_1 percentage on the vertical axis and B classification on the horizontal axis. Does the graph support the result of the test of hypothesis in Exercise 13.24? Explain.

b. Repeat part **a** for the row A_2 percentages.

c. Repeat part **a** for the row A_3 percentages.

Applying the Concepts—Basic

13.26 JAMA study of heart patients. The *Journal of the American Medical Association* (Apr. 18, 2001) published the results of a study of alcohol consumption in patients suffering from acute myocardial infarction (AMI). The patients were classified according to average number of alcoholic drinks per week and whether or not they had congestive heart failure. A summary of the results for 1,913 AMI patients is shown in the table.

AMI

		Alcohol Consumption		
		Abstainers	Less than 7 drinks/week	7 or more drinks/week
Congestive	**Yes**	146	106	29
Heart Failure	**No**	750	590	292
	Totals	896	696	321

Source: Mukamal, K. J., et al. "Prior alcohol consumption and mortality following acute myocardial infarction." *Journal of the American Medical Association,* Vol. 285, No. 15, Apr. 18, 2001 (Table 1).

a. Find the sample proportion of abstainers with congestive heart failure.

b. Find the sample proportion of moderate drinkers (patients who have less than 7 drinks per week) with congestive heart failure.

c. Find the sample proportion of heavy drinkers (patients who have 7 or more drinks per week) with congestive heart failure.

d. Compare the sample proportions, parts **a–c**. Does it appear that the proportion of AMI patients with congestive heart failure depends on alcohol consumption?

e. Give the null hypothesis for testing whether the proportion of AMI patients with congestive heart failure depends on alcohol consumption.

f. Use the MINITAB printout at the bottom of the page to conduct the test, part **e**. Test at $\alpha = .05$.

13.27 Late-emerging reading disabilities. Studies of children with reading disabilities typically focus on "early-emerging" difficulties identified prior to the 4th grade. In the *Journal of Educational Psychology* (June 2003), psychologists at Haskins Laboratories recently studied children with "late-emerging" reading difficulties (i.e., children who appeared to undergo a 4th-grade "slump" in reading achievement). A sample of 161 children was selected from 4th and 5th graders at elementary schools in Philadelphia. In addition to grade, the researchers determined whether each child had a previously undetected reading disability. A total of 66 children were diagnosed with a reading disability. Of these, 32 were 4th graders and 34 were 5th graders. Similarly, of the 95 children with normal reading achievement, 55 were 4th graders and 40 were 5th graders.

a. Identify the two qualitative variables (and corresponding levels) measured in the study.

b. From the information provided, form a contingency table.

c. Assuming the two variables are independent, calculate the expected cell counts.

d. Find the test statistic for determining whether the proportions of 4th and 5th graders with

```
Rows: FAILURE    Columns: ALCOHOL

           7ORMORE   ABSTAIN   LESS7     All

NO             292       750     590     1632
             273.8     764.4   593.8   1632.0

YES             29       146     106      281
              47.2     131.6   102.2    281.0

All            321       896     696     1913
             321.0     896.0   696.0   1913.0

Cell Contents:        Count
                      Expected count


Pearson Chi-Square = 10.197, DF = 2, P-Value = 0.006
Likelihood Ratio Chi-Square = 11.240, DF = 2, P-Value = 0.004
```

reading disabilities differs from the proportions of 4th and 5th graders with normal reading skills.

e. Find the rejection region for the test if $\alpha = .10$.

f. Is there a link between reading disability and grade level? Give the appropriate conclusion of the test.

13.28 Seatbelt use study. The *American Journal of Public Health* (July 1995) reported on a population-based study of trauma in Hispanic children. One of the objectives of the study was to compare the use of protective devices in motor vehicles used to transport Hispanic and non-Hispanic white children. On the basis of data collected from the San Diego County Regionalized Trauma System, 792 children treated for injuries sustained in vehicular accidents were classified according to ethnic status (Hispanic or non-Hispanic white) and seatbelt usage (worn or not worn) during the accident. The data are summarized in the table.

TRAUMA

	Hispanic	Non-Hispanic White	Totals
Seatbelts worn	31	148	179
Seatbelts not worn	283	330	613
Totals	314	478	792

Source: Matteneci, R.M., et al. "Trauma among Hispanic children: A population-based study in a regionalized system of trauma care." *American Journal of Public Health*, Vol. 85, No. 7, July 1995, p. 1007 (Table 2).

a. Calculate the sample proportion of injured Hispanic children who were not wearing seatbelts during the accident.

b. Calculate the sample proportion of injured non-Hispanic white children who were not wearing seatbelts during the accident.

c. Compare the two sample proportions, parts **a** and **b**. Do you think the true population proportions differ?

d. Conduct a test to determine whether seatbelt usage in motor vehicle accidents depends on ethnic status in the San Diego County Regionalized Trauma System. Use $\alpha = .01$.

e. Construct a 99% confidence interval for the difference between the proportions, parts **a** and **b**. Interpret the interval.

13.29 Masculinity and crime. Refer to the *Journal of Sociology* (July 2003) study on the link between the level of masculinity and criminal behavior in men, Exercise 9.25 (p. 452). The researcher identified events that a sample of newly incarcerated men were involved in and classified each event as "violent" (use of a weapon, throwing of objects, punching, choking, or kicking) or "avoided-violent" (pushing, shoving, grabbing, or threats of violence that did not escalate into a violent event). Each man (and corresponding event) was also classified as "high risk masculinity" (scored high on the Masculinity-Femininity Scale test and low on the Traditional Outlets of Masculinity Scale test) or "low risk masculinity." The data for 1,507 events are summarized in the table.

HRM

	Violent Events	Avoided-Violent Events	Totals
High Risk Masculinity	236	143	379
Low Risk Masculinity	801	327	1,128
Totals	1,037	470	1,507

Source: Krienert, J. L. "Masculinity and crime: A quantitative exploration of Messerschmidt's hypothesis," *Journal of Sociology*, Vol. 7, No. 2, July 2003 (Table 4).

a. Identify the two categorical variables measured (and their levels) in the study.

b. Identify the experimental units.

c. If the type of event (violent or avoided-violent) is independent of high/low risk masculinity, how many of the 1,507 events would you expect to be violent and involve a high-risk-masculine man?

d. Repeat part **c** for the other combinations of event type and high/low risk masculinity.

e. Calculate the χ^2 statistic for testing whether event type depends on high/low risk masculinity.

f. Give the appropriate conclusion of the test, part **e**, using $\alpha = .05$.

Applying the Concepts—Intermediate

13.30 Creating menus to influence others. Refer to the *Journal of Consumer Research* (Mar. 2003) study on influencing the choices of others by offering undesirable alternatives, Exercise 8.85 (p. 403). In another experiment conducted by the researcher, 96 subjects were asked to imagine that they had just moved to an apartment with two others and that they were shopping for a new appliance (e.g., television, microwave oven). Each subject was asked to create a menu of three brand choices for their roommates; then subjects were randomly assigned (in equal numbers) to one of three different "goal" conditions—(1) create the menu in order to influence roommates to buy a preselected brand, (2) create the menu in order to influence roommates to buy a brand of your choice, and (3) create the menu with no intent to influence roommates. The researcher theorized that the menus created to influence others will likely include undesirable alternative brands. Consequently, the number of menus in each goal condition that was consistent with the theory was determined. The data are summarized in the table on p. 808. Analyze the data for the purpose of determining whether the proportion of subjects who select menus consistent with the theory depends on goal condition. Use $\alpha = .01$.

MENU3

Goal Condition	Number consistent with theory	Number not consistent with theory	Totals
Influence/preselected brand	15	17	32
Influence/own brand	14	18	32
No influence	3	29	32

Source: Hamilton, R. W. "Why do people suggest what they do not want? Using context effects to influence others' choices," *Journal of Consumer Research,* Vol. 29, March 2003 (Table 2).

13.31 Politics and religion. University of Maryland professor Ted R. Gurr examined the political strategies used by ethnic groups worldwide in their fight for minority rights. (*Political Science & Politics,* June 2000.) Each in a sample of 275 ethnic groups was classified according to world region and highest level of political action reported. The data are summarized in the contingency table at the bottom of the page. Conduct a test at $\alpha = .10$ to determine whether political strategy of ethnic groups depends on world region. Support your answer with a graph.

13.32 Trapping grain moths. In an experiment described in the *Journal of Agricultural, Biological, and Environmental Statistics* (Dec. 2000), bins of corn were stocked with various parasites (e.g., grain moths) in late winter. In early summer (June), three bowl-shaped traps were placed on the grain surface in order to capture the moths. All three traps were baited with a sex pheromone lure; however, one trap used an unmarked sticky adhesive, one was marked with a fluorescent red powder, and one was marked with a fluorescent blue powder. The traps were set on a Wednesday and the catch collected the following Thursday and Friday. The table shows the number of moths captured in each trap on each day. Conduct a test (at $\alpha = .10$) to determine if the percentages of moths caught by the three traps depends on day of the week.

IRAQWAR

13.33 Iraq War survey. The Pew Internet & American Life Project commissioned Princeton Survey Research Associates to develop and carry out a survey of what Americans think about the recent War in Iraq. Some of the results of the March 2003 survey of over 1,400

MOTHTRAP

	Adhesive—No Mark	Red Mark	Blue Mark
Thursday	136	41	17
Friday	101	50	18

Source: Wileyto, E. P. et al. "Self-marking recapture models for estimating closed insect populations," *Journal of Agricultural, Biological, and Environmental Statistics,* Vol. 5, No. 4, December 2000 (Table 5A).

American adults are saved in the **IRAQWAR** file. Responses to the following questions were recorded:

1. Do you support or oppose the Iraq War? (1 = Support, 2 = Oppose)
2. Do you ever go online to access the Internet or World Wide Web? (1 = Yes, 2 = No)
3. Do you consider yourself a Republican, Democrat, or Independent? (1 = Rep., 2 = Dem., 3 = Ind.)
4. Have you or anyone in your household served in the U.S. military? (1 = Yes, I have; 2 = Yes, other; 3 = Yes, both; 4 = No)
5. In general, would you describe your political views as very conservative, conservative, moderate, liberal, or very liberal? (1 = Very conservative, 2 = Conservative, 3 = Moderate, 4 = Liberal, 5 = Very liberal)
6. What is your race? (1 = White, 2 = African-American, 3 = Asian, 4 = Mixed, 5 = Native-American, 6 = Other)
7. What is your income range? (1 = <10K, 2 = 10–20K, 3 = 20–30K, 4 = 30–40K, 5 = 40–50K, 6 = 50–75K, 7 = 75–100K, 8 = >100K)
8. Do you live in a suburban, rural, or urban community? (1 = urban, 2 = suburban, 3 = rural)

Conduct a series of contingency table analyses to determine if support for the Iraq War depends on one or more of the other categorical variables measured in the March 2003 survey.

13.34 Battle simulation trials. In order to evaluate their situational awareness, fighter aircraft pilots participate in battle simulations. At a random point in the trial, the simulator is frozen and data on situation awareness are immediately collected. The simulation is then continued until, ultimately, performance (e.g., number of kills) is measured. A study reported in *Human Factors* (Mar. 1995) investigated whether temporarily stopping the simulation results in any change in pilot performance. Trials were designed so

ETHNIC

		Political Strategy		
		No Political Action	Mobilization, Mass Action	Terrorism, Rebellion, Civil War
World Region	**Latin American**	24	31	7
	Post-Communist	32	23	4
	South, Southeast, East Asia	11	22	26
	Africa/Middle East	39	36	20

Source: Gurr, T. R. "Nonviolence in ethnopolitics: Strategies for the attainment of group rights and autonomy." *Political Science & Politics,* Vol. 33, No. 2, June 2000 (Table 1).

that some simulations were stopped to collect situation awareness data while others were not stopped. Each trial was then classified according to the number of kills made by the pilot. The data for 180 trials are summarized in the contingency table below. Conduct a contingency table analysis and fully interpret the results.

SIMKILLS

| | \multicolumn{5}{c}{Number of Kills} | |
	0	1	2	3	4	Totals
Stops	32	33	19	5	2	91
No Stops	24	36	18	8	3	89
Totals	56	69	37	13	5	180

13.35 Susceptibility to hypnosis. A standardized procedure for determining a person's susceptibility to hypnosis is the Stanford Hypnotic Susceptibility Scale, Form C (SHSS:C). Recently, a new method called the Computer-Assisted Hypnosis Scale (CAHS), which uses a computer as a facilitator of hypnosis, has been developed. Each scale classifies a person's hypnotic susceptibility as low, medium, high, and very high. Researchers at the University of Tennessee compared the two scales by administering both tests to each of 130 undergraduate volunteers. (*Psychological Assessment*, Mar. 1995.) The hypnotic classifications are summarized in the next table. A contingency table analysis will be performed to determine if CAHS level and SHSS level are independent.

a. Check to see if the assumption of expected cell counts of 5 or more is satisfied. Should you proceed with the analysis? Explain.

b. One way to satisfy the assumption, part **a**, is to combine the data for two or more categories in the contingency table (e.g., high and very high). Form a new contingency table by combining the data for the high and very high categories in both the rows and columns.

c. Calculate the expected cell counts in the new contingency table, part **c**. Is the assumption now satisfied?

d. Perform the chi-square test on the new contingency table. Use $\alpha = .05$. Interpret the results.

13.36 Gangs and homemade weapons. The National Gang Crime Research Center (NGCRC) has developed a 6-level gang classification system for both adults and juveniles. The six categories are shown in the table below. The classification system was developed as a potential predictor of a gang member's propensity for violence when in prison, jail, or a correctional facility. To test this theory, the NGCRC collected data on approximately 10,000 confined offenders and assigned each a score using the gang classification system. (*Journal of Gang Research*, Winter 1997.) One of several other variables measured by the NGCRC was whether or not the offender has ever carried a homemade weapon (e.g., knife) while in custody. The data on gang score and homemade weapon are summarized in the table at the bottom of the page. Conduct a test to determine if carrying a homemade weapon in custody depends on gang classification score. (Use $\alpha = .01$.) Support your conclusion with a graph.

HYPNOSIS

| | | \multicolumn{5}{c}{CAHS Level} |
		Low	Medium	High	Very High	Totals
SHSS: C Level	**Low**	32	14	2	0	48
	Medium	11	14	6	0	31
	High	6	14	19	3	42
	Very High	0	2	4	3	9
	Totals	49	44	31	6	130

Source: Grant, C. D., and Nash, M. R. "The Computer-Assisted Hypnosis Scale: Standardization and norming of a computer-administered measure of hypnotic ability." *Psychological Assessment,* Vol. 7, No. 1, Mar. 1995, p. 53 (Table 4).

GANGS

| | \multicolumn{2}{c}{Homemade Weapon Carried} |
Gang Classification Score	Yes	No
0 (Never joined a gang, no close friends in a gang)	255	2,551
1 (Never joined a gang, 1–4 close friends in a gang)	110	560
2 (Never joined a gang, 5 or more friends in a gang)	151	636
3 (Inactive gang member)	271	959
4 (Active gang member, no position of rank)	175	513
5 (Active gang member, holds position of rank)	476	831

Source: Knox, G. W., et al. "A gang classification system for corrections." *Journal of Gang Research,* Vol. 4, No. 2, Winter 1997, p. 54 (Table 4).

Applying the Concepts—Advanced

13.37 Efficacy of an HIV vaccine. New, effective AIDS vaccines are now being developed using the process of "sieving," that is, sifting out infections with some strains of HIV. Harvard School of Public Health statistician Peter Gilbert demonstrated how to test the efficacy of an HIV vaccine in *Chance* (Fall 2000). As an example, Gilbert reported the results of VaxGen's preliminary HIV vaccine trial using the 2×2 table on p. 000. The vaccine was designed to eliminate a particular strain of the virus, called the "MN strain." The trial consisted of 7 AIDS patients vaccinated with the new drug and 31 AIDS patients who were treated with a placebo (no vaccination). The table below shows the number of patients who tested positive and negative for the MN strain in the trial follow-up period.

a. Conduct a test to determine whether the vaccine is effective in treating the MN strain of HIV. Use $\alpha = .05$.

b. Are the assumptions for the test, part **a**, satisfied? What are the consequences if the assumptions are violated?

c. In the case of a 2×2 contingency table, R. A. Fisher (1935) developed a procedure for computing the exact *p*-value for the test (called *Fishers exact test*). The method utilizes the hypergeometric probability distribution of Chapter 4 (p. 225). Consider the hypergeometric probability

$$\frac{\binom{7}{2}\binom{31}{22}}{\binom{38}{24}}$$

This represents the probability that 2 out of 7 vaccinated AIDS patients test positive and 22 out of 31 unvaccinated patients test positive, that is, the probability of the table result given that the null hypothesis of independence is true. Compute this probability (called the *probability of the contingency table*).

⊙ **HIVVAC1**

		MN Strain		
		Positive	Negative	Totals
Patient Group	**Unvaccinated**	22	9	31
	Vaccinated	2	5	7
	Totals	24	14	38

Source: Gilbert, P. "Developing an AIDS vaccine by sieving." *Chance,* Vol. 13, No. 4, Fall 2000.

```
                    The FREQ Procedure

               Table of GROUP by MNSTRAIN

          GROUP      MNSTRAIN

          Frequency|
          Expected |NEG      |POS      |   Total

          UN       |      9  |     22  |     31
                   | 11.421  | 19.579  |

          V        |      5  |      2  |      7
                   | 2.5789  | 4.4211  |

          Total          14        24        38


        Statistics for Table of GROUP by MNSTRAIN

    Statistic                      DF      Value      Prob

    Chi-Square                      1     4.4112     0.0357
    Likelihood Ratio Chi-Square     1     4.2893     0.0384
    Continuity Adj. Chi-Square      1     2.7773     0.0956
    Mantel-Haenszel Chi-Square      1     4.2952     0.0382
    Phi Coefficient                       -0.3407
    Contingency Coefficient                0.3225
    Cramer's V                            -0.3407

    WARNING: 50% of the cells have expected counts less
             than 5. Chi-Square may not be a valid test.


                  Fisher's Exact Test

        Cell (1,1) Frequency (F)            9
        Left-sided Pr <= F            0.0498
        Right-sided Pr >= F           0.9940

        Table Probability (P)        0.0438
        Two-sided Pr <= P            0.0772

            Sample Size = 38
```

d. Refer to part **c**. Two contingency tables (with the same marginal totals as the original table) that are more contradictory to the null hypothesis of independence than the observed table are shown at right. First, explain why these tables provide more evidence to reject H_0 than the original table. Then, compute the probability of each table using the hypergeometric formula.

e. The p-value of Fisher's exact test is the probability of observing a result at least as contradictory to the null hypothesis as the observed contingency table, given the same marginal totals. Sum the probabilities of parts **c** and **d** to obtain the p-value of Fisher's exact test. (To verify your calculations, check the p-value at the bottom of the SAS printout, p. 810, labeled **Left-sided Pr < = F.**) Interpret this value in the context of the vaccine trial.

HIVVAC2

		MN Strain		
		Positive	Negative	Totals
Patient Group	**Unvaccinated**	23	8	31
	Vaccinated	1	6	7
	Totals	24	14	38

HIVVAC3

		MN Strain		
		Positive	Negative	Totals
Patient Group	**Unvaccinated**	24	7	31
	Vaccinated	0	7	7
	Totals	24	14	38

13.4 A Word of Caution About Chi-Square Tests

Because the χ^2 statistic for testing hypotheses about multinomial probabilities is one of the most widely applied statistical tools, it is also one of the most abused statistical procedures. Consequently, the user should always be certain that the experiment satisfies the assumptions given with each procedure. Furthermore, the user should be certain that the sample is drawn from the correct population—that is, from the population about which the inference is to be made.

The use of the χ^2 probability distribution as an approximation to the sampling distribution for χ^2 should be avoided when the expected counts are very small. The approximation can become very poor when these expected counts are small, and thus the true α level may be quite different from the tabled value. As a rule of thumb, an expected cell count of at least 5 means that the χ^2 probability distribution can be used to determine an approximate critical value.

If the χ^2 value does not exceed the established critical value of χ^2, *do not accept the hypothesis of independence.* You would be risking a Type II error (accepting H_0 if it is false), and the probability β of committing such an error is unknown. The usual alternative hypothesis is that the classifications are dependent. Because the number of ways in which two classifications can be dependent is virtually infinite, it is difficult to calculate one or even several values of β to represent such a broad alternative hypothesis. Therefore, we avoid concluding that two classifications are independent, even when χ^2 is small.

Finally, if a contingency table χ^2 value does exceed the critical value, we must be careful to avoid inferring that a *causal* relationship exists between the classifications. Our alternative hypothesis states that the two classifications are statistically dependent—and a statistical dependence does not imply causality. Therefore, *the existence of a causal relationship cannot be established by a contingency table analysis.*

Quick Review

Key Terms

Key Formulas

$$\chi^2 = \sum \frac{[n_i - E_i]^2}{E_i} \qquad \text{One-way table} \quad 787$$

where n_i = count for cell i

$E_i = np_{i,0}$

$p_{i,0}$ = hypothesized value of p_i in H_0

$$\chi^2 = \sum \frac{[n_{ij} - \hat{E}_{ij}]^2}{\hat{E}_{ij}} \qquad \text{Two-way table} \quad 799$$

where n_{ij} = count for cell in row i, column j

$\hat{E}_{ij} = R_i C_j / n$ 796

R_i = total for row i

C_j = total for column j

n = total sample size

Language Lab

Symbol	Pronunciation	Description
$p_{i,0}$	p-i-zero	Value of multinomial probability p_i hypothesized in H_0
χ^2	Chi-square	Test statistic used in analysis of count data
n_i	n-i	Number of observed outcomes in cell i of one-way table
E_i	e-i	Expected number of outcomes in cell i of one-way table when H_0 is true
p_{ij}	p-i-j	Probability of an outcome in row i and column j of a two-way contingency table
n_{ij}	n-i-j	Number of observed outcomes in row i and column j of a two-way contingency table
$\hat{E}_{ij}$	Estimated e-i-j	Estimated expected number of outcomes in row i and column j of a two-way contingency table
r_i	r-i	Total number of outcomes in row i of a contingency table
c_j	c-j	Total number of outcomes in column j of a contingency table

Chapter Summary Notes

- **Multinomial data** are qualitative data that fall into more than two *categories, classes*, or *cells*.
- Properties of a **multinomial experiment**: (1) n identical trials, (2) k possible outcomes to each trial, (3) probabilities of the k outcomes remain the same from trial to trial, (4) trials are independent, (5) the variables of interest are the *cell counts*.
- A **one-way table** is a summary table for a single qualitative variable.
- A **two-way table**, or **contingency table**, is a summary table for two qualitative variables.

- The **chi-square** (χ^2) **statistic** is used to test probabilities associated with one-way and two-way tables.
- Conditions required for a valid χ^2**-test:** (1) multinomial experiment, (2) sample size n is large — satisfied when expected cell counts are all greater than or equal to 5.
- A significant χ^2 test for a *two-way table* implies that the **two qualitative variables are dependent**.
- Chi-square tests for independence *cannot be used to infer that a causal relationship* exists between the two qualitative variables.

Supplementary Exercises 13.38–13.59

Understanding the Principles

13.38 What is the difference between a one-way chi-square analysis and a two-way chi-square analysis?

13.39 *True or False.* Rejecting the null hypothesis in a chi-square test for independence implies that a causal relationship exists between the two categorical variables.

Learning the Mechanics

13.40 A random sample of 250 observations was classified according to the row and column categories shown in the table.

		Column		
		1	2	3
	1	20	20	10
Row	2	10	20	70
	3	20	50	30

  a. Do the data provide sufficient evidence to conclude that the rows and columns are dependent? Test using $\alpha = .05$.
  b. Would the analysis change if the row totals were fixed before the data were collected?
  c. Do the assumptions required for the analysis to be valid differ according to whether the row (or column) totals are fixed? Explain.
  d. Convert the table entries to percentages by using each column total as a base and calculating each row response as a percentage of the corresponding column total. In addition, calculate the row totals and convert them to percentages of all 250 observations.
  e. Create a bar graph with row 1 percentages on the vertical axis and the column number on the horizontal axis. Draw a horizontal line corresponding to the row 1 total percentage. Does the graph support the result of the test conducted in part **a**?
  f. Repeat part **e** for the row 2 percentages.
  g. Repeat part **e** for the row 3 percentages.

13.41 A random sample of 150 observations was classified into the categories shown in the table.

			Category		
	1	2	3	4	5
n_i	28	35	33	25	29

  a. Do the data provide sufficient evidence that the categories are not equally likely? Use $\alpha = .10$.
  b. Form a 90% confidence interval for p_2, the probability that an observation will fall in category 2.

Applying the Concepts—Basic

13.42 Scanning Internet messages. *Inc. Technology* (Mar. 18, 1997) reported the results of an Equifax/Harris Consumer Privacy Survey in which 328 Internet users indicated their level of agreement with the following statement: "The government needs to be able to scan Internet messages and user communications to prevent fraud and other crimes." The number of users in each response category is summarized below.

Agree Strongly	Agree Somewhat	Disagree Somewhat	Disagree Strongly
59	108	82	79

  a. Specify the null and alternative hypotheses you would use to determine if the opinions of Internet users are evenly divided among the four categories.
  b. Conduct the test of part **a** using $\alpha = .05$.
  c. In the context of this exercise, what is a Type I error? A Type II error?
  d. What assumptions must hold in order to ensure the validity of the test you conducted in part **b**?

13.43 Risk factor for lumbar disease. One of the most common musculoskeletal disorders is lumbar disk disease (LDD). Medical researchers reported finding a common genetic risk factor for LDD (*Journal of the American Medical Association*, Apr. 11, 2001). The study included 171 Finnish patients diagnosed with LDD (the patient group) and 321 Finnish individuals without LDD (the control group). Of the 171 LDD patients, 21 were discovered to have the genetic trait. Of the 321 people in the control group, 15 had the genetic trait.
  a. Consider the two categorical variables, group and presence/absence of genetic trait. Form a 2×2 contingency table for these variables.
  b. Conduct a test to determine whether the genetic trait occurs at a higher rate in LDD patients than in the controls without LDD. Use $\alpha = .01$.
  c. Construct a bar graph that will visually support your conclusion in part **b**.

13.44 Performance of solder joint inspectors. Westinghouse Electric Company has experimented with different means of evaluating the performance of solder joint inspectors. One approach involves comparing an individual inspector's classifications with those of the group of experts that comprise Westinghouse's Work Standards Committee. In one experiment, 153 solder connections were evaluated by the committee and 111 were classified as acceptable. An inspector evaluated the same 153 connections and classified 124 as acceptable. Of the items rejected by the inspector, the committee agreed with 19.

a. Construct a contingency table that summarizes the classifications of the committee and the inspector.

b. Based on a visual examination of the table you constructed in part **a**, does it appear that there is a relationship between the inspector's classifications and the committee's? Explain. (A bar graph of the percentage rejected by committee and inspector will aid your examination.)

c. Conduct a chi-square test of independence for these data. Use $\alpha = .05$. Carefully interpret the results of your test in the context of the problem.

13.45 High-fiber food and cancer. According to research reported in the *Journal of the National Cancer Institute* (Apr. 1991), eating foods high in fiber may help protect against breast cancer. The researchers randomly divided 120 laboratory rats into four groups of 30 each. All rats were injected with a drug that causes breast cancer, then each rat was fed a diet of fat and fiber for 15 weeks. However, the levels of fat and fiber varied from group to group. At the end of the feeding period, the number of rats with cancer tumors was determined for each group. The data are summarized in the table at the bottom of the page.

a. Does the sampling appear to satisfy the assumptions for a multinomial experiment (see Section 13.1)? Explain.

b. Calculate the expected cell counts for the contingency table.

c. Calculate the χ^2 statistic.

d. Is there evidence to indicate that diet and presence/absence of cancer are independent? Test using $\alpha = 05$.

e. Compare the percentage of rats on a high-fat/no-fiber diet with cancer to the percentage of rats on a high-fat/fiber diet with cancer using a 95% confidence interval. Interpret the result.

13.46 Travel habits of retirees. A study in the *Annals of Tourism Research* (Vol. 19, 1992) investigates the relationship of retirement status (pre- and postretirement) to various items related to the travel industry. One part of the study investigated the differences in the length of stay of a trip for pre- and postretirees. A sample of 703 travelers were asked how long they stayed on a typical trip. The results are shown in the table, top right. Use the information in the table to determine whether the retirement status of a traveler and the duration of a typical trip are dependent. Test using $\alpha = .05$.

⊚ TRAVEL

Number of Nights	Preretirement	Postretirement
4–7	247	172
8–13	82	67
14–21	35	52
22 or more	16	32
Total	**380**	**323**

13.47 Hearing impairment study. The *Journal of Intellectual Disability Research* (Feb. 1995) published a longitudinal study of hearing impairment in a group of elderly patients with intellectual disability. The hearing function of each patient was screened each year over a 10-year period. At the study's conclusion, the hearing loss of each patient was categorized as severe, moderate, mild, or none. The classifications of the 28 surviving patients are summarized in the table below.

⊚ HEARIMP

Hearing Loss	Number of Patients
None	7
Mild	7
Moderate	9
Severe	5
Total	**28**

a. Conduct a test to determine whether the true proportions of intellectually disabled elderly patients in each of the hearing-loss categories differ. Use $\alpha = .05$.

b. Use a 90% confidence interval to estimate the proportion of disabled elderly patients with severe hearing loss.

13.48 Genetic crossing of snapdragons. According to one geneticist's theory, a crossing of red and white snapdragons should produce offspring that are 25% red, 50% pink, and 25% white. An experiment conducted to test the theory produced 30 red, 78 pink, and 36 white offspring in 144 crossings. Do the data provide sufficient evidence to contradict the geneticist's theory?

Applying the Concepts—Intermediate

13.49 Thumbs up or thumbs down. For over 20 years, movie critics Gene Siskel (formerly of the *Chicago Tribune*)

⊚ TUMORS

		Diet				
		High-Fat/No-Fiber	High-Fat/Fiber	Low-Fat/No-Fiber	Low-Fat/Fiber	Totals
Cancer Tumors	Yes	27	20	19	14	80
	No	3	10	11	16	40
	Totals	30	30	30	30	120

Source: Tampa Tribune, Apr. 3, 1991.

and Roger Ebert (*Chicago Sun-Times*) rated the latest film releases on national television's *Sneak Previews*. Since Siskel's death in 1999, Ebert has teamed with Richard Roeper (*New York Times Syndicate*) to rate movies on the syndicated TV-show *Ebert & Roeper At the Movies*. Ebert's and Roeper's ratings ("thumbs up" or "thumbs down") for 218 films released in 2003 are saved in the **THUMBSUP** file. Data for the first and last five movies in the file are shown in the table. Conduct a test to determine whether the movie reviews of the two critics are independent. Use $\alpha = .05$.

◉ THUMBSUP (First and last 5 movies listed)

Movie	Ebert Rating	Roeper Rating
Terminator 3	DOWN	UP
Finding Nemo	UP	UP
2 Fast 2 Furious	UP	DOWN
Hulk	UP	UP
Matrix Reloaded	UP	UP
.	.	.
.	.	.
Daredevil	UP	UP
Just Married	DOWN	DOWN
Sonny	DOWN	DOWN
Narc	UP	DOWN
City of God	UP	UP

Source: www.movies.com.

13.50 **Pig farm study.** An article in *Sociological Methods & Research* (May 2001) analyzed the data presented in the table. A sample of 262 Kansas pig farmers were classified according to their education level (college or not) and size of their pig farm (number of pigs). Conduct a test to determine whether a pig farmer's education level has an impact on the size of the pig farm. Use $\alpha = .05$ and support your answer with a graph.

◉ PIGFARM

		Education Level		
		No College	College	Totals
	<1,000 pigs	42	53	95
	1,000–2,000 pigs	27	42	69
Farm Size	2,001–5,000 pigs	22	20	42
	>5,000 pigs	27	29	56
	Totals	118	144	262

Source: Agresti, A., and Liu, I. "Strategies for modeling a categorical variable allowing multiple category choices." *Sociological Methods & Research,* Vol. 29, No. 4, May 2001 (Table I).

13.51 **Chess opening strategies.** In chess, the first few moves often play a determining role in the final outcome. Five different opening strategies are highly favored by chess experts. To determine whether one or more of these strategies is most preferred by grand masters in international competition, a random sample of 100 grand masters is taken, and each is asked which of the strategies he or she would prefer to employ. A summary of their responses is shown below. Do these data present sufficient evidence to indicate a preference for one or more of the strategies? Use $\alpha = .05$.

◉ CHESS

Strategy	A	B	C	D	E
Frequency	17	27	22	15	19

13.52 **Peanut butter market shares.** Marketing researchers often use the purchase of a sample of households, called a *scanner panel,* to understand the purchasing patterns and preferences of consumers. When shopping, these households present a magnetic identification card that permits their purchase data to be identified and aggregated. Researchers recently studied the extent to which panel households' purchase behavior is representative of the population of households shopping at the same stores. (*Journal of Marketing Research,* Nov. 1996.) The table below reports the peanut butter purchase data collected by A. C. Nielson Company for the panel of 2,500 households in Sioux Falls, South Dakota, over a 102-week period. The market share percentages in the right-hand column are derived from all peanut butter purchases at the same 15 stores at which the panel shopped during the same 102-week period.

a. Do the data provide sufficient evidence to conclude that the purchases of the household panel are representative of the population of households? Test using $\alpha = .05$.

b. What assumptions must hold to ensure the validity of the testing procedure you used in part **a**?

c. Find the approximate *p*-value for the test of part **a** and interpret it in the context of the problem.

◉ SCANNER

Brand	Size	Number of Purchases by Household Panel	Market Shares
Jif	18 oz.	3,165	20.10%
Jif	28	1,892	10.10
Jif	40	726	5.42
Peter Pan	10	4,079	16.01
Skippy	18	6,206	28.65
Skippy	28	1,627	12.38
Skippy	40	1,420	7.32
Total		19,115	

Source: Gupta, S., Chintagunta, P., Kaul, A., and Wittink, D. "Do household scanner data provide representative inferences from brand choices: A comparison with store data." *Journal of Marketing Research,* Vol. 33, Nov. 1996, p. 393 (Table 6).

13.53 **Birds feeding on gypsy moths.** A field study was conducted to identify the natural predators of the gypsy moth. (*Environmental Entomology,* June 1995.) For one part of the study, 24 black-capped chickadees (common wintering birds) were captured in mist nets and individually caged. Each bird was offered a mass of gypsy moth eggs attached to a piece of bark. Half the birds were offered no other food (no choice) and half were offered a variety of other naturally occurring foods such as spruce and pine seeds (choice). The numbers of birds that did and did not feed on the gypsy moth egg mass are given in the table. Analyze the data in the table to determine if a relationship exists between food choice and whether or not the chickadees fed on gypsy moth eggs. Use $\alpha = .10$.

MOTH

	Fed on Egg Mass	
	Yes	No
Choice of foods	2	10
No choice	8	4

13.54 Social play of children. The *American Journal on Mental Retardation* (Jan. 1992) published a study of the social interactions of two groups of children. Independent random samples of 15 children with and 15 children without developmental delays (i.e., mild mental retardation) were the subjects of the experiment. After observing the children during "freeplay," the number of children who exhibited disruptive behavior (e.g., ignoring or rejecting other children, taking toys from another child) was recorded for each group. The data are summarized in the two-way table below. Analyze the data and interpret the results.

13.55 Lifestyles of husbands and wives. The *Journal of Adlerian Theory, Research and Practice* (Mar. 1986) investigated the role lifestyles play in forming a marriage. Adlerian psychologists group people into one of the following five lifestyle types:
1. *Pleasing priority:* a strong desire to make others happy and win their approval
2. *Achieving priority:* industrial, responsible, orderly, with an active approach to life (particularly in the workplace)
3. *Outdoing priority:* a strong desire to be on top, in a position of superiority over others; likes to be considered the best at whatever he/she does
4. *Suppressing (control) priority:* maintains control over his/her emotions; discloses very little of himself/herself
5. *Avoiding priority:* is easily hurt; avoids any potentially painful (physical or emotional) situation.
   Data on both wife's and husband's lifestyles were obtained for 202 married couples living in family housing at the University of Georgia. The results are summarized in the table at the bottom of the page.
a. The key question of the study is: "Do the personality profiles contribute to marriage pairings?" In other words, is there a relationship between the personality priorities of the husband and wife pairs? Analyze the data for the researchers.
b. If no relationship between the personality priorities of the husband and wife pairs is found, part **a**, then analyze the data for husbands and wives separately. Are there differences in the proportions of husbands who fall in the five personality priorities? Are there differences in the proportions of wives who fall in the five personality profiles?

DISRUPT

	Disruptive Behavior	Nondisruptive Behavior	Totals
With Developmental Delays	12	3	15
Without Developmental Delays	5	10	15
Total	17	13	30

Source: Kopp, C. B., Baker, B., and Brown, K. W. "Social skills and their correlates: Preschoolers with developmental delays." *American Journal on Mental Retardation,* Vol. 96, No. 4, Jan. 1992.

LIFESTYLE

		Wife's Lifestyle					
		Pleasing	Outdoing	Avoiding	Control	Achieving	Totals
Husband's Lifestyle	**Pleasing**	9	6	6	3	9	33
	Outdoing	7	11	12	11	5	46
	Avoiding	8	8	6	11	5	38
	Control	8	10	7	15	11	51
	Achieving	4	6	7	10	7	34
	Totals	36	41	38	50	37	202

Source: Evans, T. D., and Bozarth, J. "Pairing of personality profiles in marriage." *Journal of Adlerian Theory, Research and Practice,* Vol. 42, No. 1, March 1986, pp. 59–64.

Interval	$x < -2$	$-2 \leq x < -1$	$-1 \leq x < 0$	$0 \leq x < 1$	$1 \leq x < 2$	$x \geq 2$
Frequency	7	20	61	77	26	9

13.56 Orientation clue experiment. *Human Factors* (Dec. 1988) published a study of color brightness as a body orientation clue. Ninety college students, reclining on their backs in the dark, were disoriented when positioned on a rotating platform under a slowly rotating disk that blocked their field of vision. The subjects were asked to say "Stop" when they felt as if they were right-side up. The position of the brightness pattern on the disk in relation to each student's body orientation was then recorded. Subjects selected only three disk brightness patterns as subjective vertical clues: (1) brighter side up, (2) darker side up, and (3) brighter and darker sides aligned on either side of the subjects' heads. The frequency counts for the experiment are given in the accompanying table. Conduct a test to compare the proportions of subjects that fall in the three disk-orientation categories. Assume you want to determine whether the three proportions differ. Use $\alpha = .05$.

◎ BODYCLUE

Disk Orientation		
Brighter Side Up	Darker Side Up	Bright and Dark Sides Aligned
58	15	17

Applying the Concepts—Advanced

13.57 Goodness of fit test. A statistical analysis is to be done on a set of data consisting of 1,000 monthly salaries. The analysis requires the assumption that the sample was drawn from a normal distribution. A preliminary test, called the χ^2 *goodness-of-fit test*, can be used to help determine whether it is reasonable to assume that the sample is from a normal distribution. Suppose the mean and standard deviation of the 1,000 salaries are hypothesized to be $1,200 and $200, respectively. Using the standard normal table, we can approximate the probability of a salary being in the intervals listed in the next table. The third column represents the expected number of the 1,000 salaries to be found in each interval if the sample was drawn from a normal distribution with $\mu = \$1,200$ and $\sigma = \$200$. Suppose the last column contains the actual observed frequencies in the sample. Large differences between the observed and expected frequencies cast doubt on the normality assumption.

Interval	Probability	Expected Frequency	Observed Frequency
Less than $800	.023	23	26
$800 < $1,000	.136	136	146
$1,000 < $1,200	.341	341	361
$1,200 < $1,400	.341	341	311
$1,400 < $1,600	.136	136	143
$1,600 or above	.023	23	13

a. Compute the χ^2 statistic based on the observed and expected frequencies.
b. Find the tabulated χ^2 value when $\alpha = .05$ and there are 5 degrees of freedom. (There are $k - 1 = 5$ df associated with this χ^2 statistic.)
c. Based on the χ^2 statistic and the tabulated χ^2 value, is there evidence that the salary distribution is nonnormal?
d. Find the approximate observed significance level for the test in part c.

13.58 Testing normality. Suppose a random variable is hypothesized to be normally distributed with mean 0 and standard deviation 1. A random sample of 200 observations on the variable yields frequencies in the intervals listed in the table at the top of the page. Do the data provide sufficient evidence to contradict the hypothesis that x is normally distributed with $\mu = 0$ and $\sigma = 1$? Use the technique developed in Exercise 13.57.

Critical Thinking Challenge

13.59 A "rigged" election? *Chance* (Spring 2004) presented data from a recent election held to determine the Board of Directors of a local community. There were 27 candidates for the Board, and each of 5,553 voters was allowed to choose 6 candidates. The claim was that "a fixed vote with fixed percentages [was] assigned to each and every candidate making it impossible to participate in an honest election." Votes were tallied in six time periods: after 600 total votes were in, after 1200, after 2,444, after 3,444, after 4,444, and, finally, after 5,553 votes. The data for three of the candidates (Smith, Coppin, and Montes) are shown in the accompanying table. A residential organization believes that "there was nothing random about the count and tallies each time period and specific unnatural or rigged percentages were being assigned to each and every candidate." Give your opinion. Is the probability of a candidate receiving votes independent of the time period? And, if so, does this imply a rigged election?

◎ RIGVOTE

Time Period	1	2	3	4	5	6
Votes for Smith	208	208	451	392	351	410
Votes for Coppin	55	51	109	98	88	104
Votes for Montes	133	117	255	211	186	227
Total Votes	600	600	1,244	1,000	1,000	1,109

Source: Gelman, A. "55,000 residents desperately need your help!", *Chance*, Vol. 17, No. 2, Spring 2004 (Figures 1 and 5).

Student Projects

Many researchers rely on surveys to estimate the proportions of experimental units in populations that possess certain specified characteristics. A political scientist may want to estimate the proportion of an electorate in favor of a certain legislative bill. A social scientist may be interested in the proportions of people in a geographical region who fall in certain socioeconomic classifications. A psychologist might want to compare the proportions of patients who have different psychological disorders.

Choose a specific topic, similar to those described above, that interests you. Clearly define the population of interest, identify data categories of specific interest, and identify the proportions associated with them. Now *guesstimate* the proportions of the population that you think fall in each of the categories. For example, you might guess that all the proportions are equal, or that the first proportion is twice as large as the second but equal to the third, etc.

You are now ready to collect the data by obtaining a random sample from your population of interest. Select a sample size so that all expected cell counts are at least 5 (preferably larger), then collect the data.

Use the count data you have obtained to test the null hypothesis that the true proportions in the population are equal to your presampling guesstimates. Would failure to reject this null hypothesis imply that your guesstimates are correct?

REFERENCES

Agresti, A. *Categorical Data Analysis*. New York: Wiley, 1990.

Cochran, W. G. "The χ^2 test of goodness of fit." *Annals of Mathematical Statistics*, 1952, 23.

Conover, W. J. *Practical Nonparametric Statistics*, 2nd ed. New York: Wiley, 1980.

Fisher, R. A. "The logic of inductive inference (with discussion)." *Journal of the Royal Statistical Society*, Vol. 98, 1935, pp. 39–82.

Hollander, M., and Wolfe, D. A. *Nonparametric Statistical Methods*. New York: Wiley, 1973.

Savage, I. R. "Bibliography of nonparametric statistics and related topics." *Journal of the American Statistical Association*, 1953, 48.

Using Technology

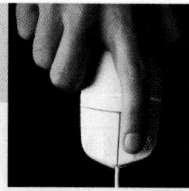

Chi-Square Analyses Using MINITAB

MINITAB can conduct chi-square tests for two-way (contingency) tables but cannot currently produce a chi-square test for a one-way table.

To conduct a chi-square test for a two-way table, first access the MINITAB worksheet file that contains the sample data. The data file should contain two qualitative variables, with category values for each of the n observations in the data set. Alternative, the worksheet can contain the cell counts for each of the categories of the two qualitative variables. Next, click on the "Stat" button on the MINITAB menu bar, then click on "Tables" and "Cross Tabulation and Chi-Square", as shown in Figure 13.M.1.

The resulting dialog box appears as shown in Figure 13.M.2. Specify one qualitative variable in the "For rows" box and the other qualitative variable in the "For columns" box. [*Note:* If your worksheet contains cell counts for the categories, enter the variable with the cell counts in the "Frequencies are in" box.] Next, select the summary statistics (e.g., counts, percentages) you want to display in the contingency table. Then click the "Chi-square" button. The resulting dialog box is shown in Figure 13.M.3. Select "Chi-Square analysis" and "Expected cell counts" and click "OK". When you return to the "Cross Tabulation" menu screen, click "OK" to generate the MINITAB printout.

Figure 13.M.1
MINITAB Menu Options
for Two-Way Chi-Square
Analysis

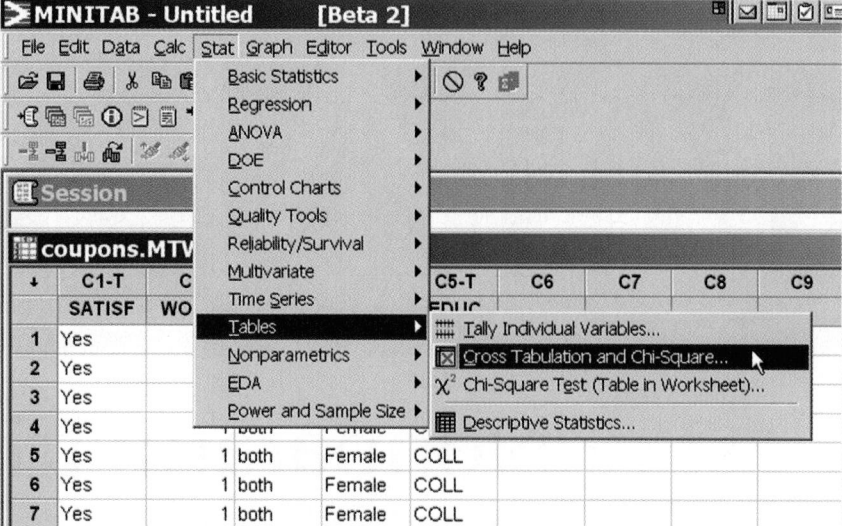

Figure 13.M.2
MINITAB Cross Tabulation
Dialog Box

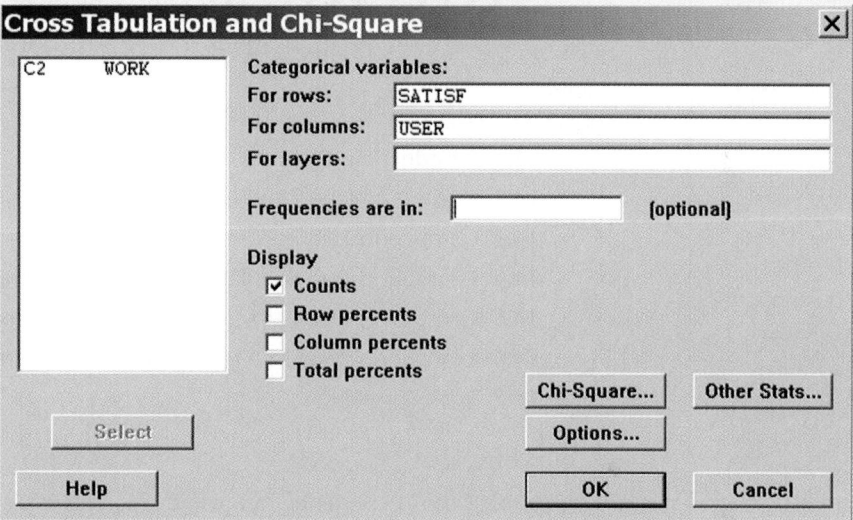

Figure 13.M.3
MINITAB Chi-Square
Dialog Box

[*Note:* If your MINITAB worksheet contains only the cell counts for the contingency table in columns, click the "Chi-Square Test (Table in Worksheet)" menu option (see Figure 13.M.1) and specify the columns in the "Columns containing the table" box. Click "OK" to produce the MINITAB printout.]

14

Nonparametric Statistics

Contents

Statistics in Action

Deadly Exposure: Agent Orange and Vietnam Vets

Using Technology

Nonparametric Analyses Using MINITAB

Where We've Been

- Presented methods for making inferences about means (Chapters 7–10) and for making inferences about the correlation between two quantitative variables (Chapter 11)

- These methods required that the data be normally distributed or that the sampling distributions of the relevant statistics be normally distributed.

Where We're Going

- Develop the need for inferential techniques that require fewer, or less stringent, assumptions than the methods of Chapters 7–10 and 11

- Introduce *nonparametric* tests that are based on ranks (i.e., on an ordering of the sample measurements according to their relative magnitudes)

Statistics in *ACTION*

Deadly Exposure: Agent Orange and Vietnam Vets

Agent Orange was the code name for a herbicide developed for the U.S. Armed Forces, primarily for use in tropical climates. The purpose of the product was to deny an enemy cover and concealment in dense terrain by defoliating trees and shrubbery where the enemy could hide. (The code name comes from the orange band that was used to mark the drums the herbicide was stored in.) Agent Orange was tested in Southeast Asia in the early 1960s and brought into ever widening use during the height of the Vietnam War (1967–1968); its use was diminished and eventually discontinued in 1971.

Agent Orange was a 50-50 mix of two chemicals, known conventionally as 2,4,D and 2,4,5,T. The combined product was mixed with kerosene or diesel fuel and dispersed by aircraft, vehicle, and hand spraying. As an unwanted by-product of the chemical manufacturing process, Agent Orange was found to be extremely contaminated with TCDD, or dioxin. In laboratory tests on animals, TCDD has caused a wide variety of diseases (including cancer), many of them fatal.

During the Vietnam War, an estimated 19 million gallons of Agent Orange was used to destroy the dense plant and tree cover of the Asian jungle. As a result of this exposure, many Vietnam veterans have dangerously high levels of TCDD in their blood and adipose (fatty) tissue. A study published in *Chemosphere* (Vol. 20, 1990) reported on the TCDD levels of 20 Massachusetts Vietnam vets who were possibly exposed to Agent Orange. The TCDD amounts (measured in parts per trillion) in both plasma and fat tissue of the 20 vets are listed in Table SIA14.1. These data are saved in the **TCDD** file.

What do the data tell us about the levels of TCDD in fat and plasma of Vietnam veterans? Is there a relationship between the fat and plasma TCDD levels? In the Statistics in Action Revisited sections listed below, we apply the nonparametric tests of this chapter to answer these questions.

TCDD

TABLE SIA14.1 TCDD Measurements for 20 Vietnam Vets

Vet	Fat	Plasma
1	4.9	2.5
2	6.9	3.5
3	10.0	6.8
4	4.4	4.7
5	4.6	4.6
6	1.1	1.8
7	2.3	2.5
8	5.9	3.1
9	7.0	3.1
10	5.5	3.0
11	7.0	6.9
12	1.4	1.6
13	11.0	20.0
14	2.5	4.1
15	4.4	2.1
16	4.2	1.8
17	41.0	36.0
18	2.9	3.3
19	7.7	7.2
20	2.5	2.0

Source: Schecter, A., et al. "Partitioning of 2,3,7,8-chlorinated dibenzo-*p*-dioxins and dibenzofurans between adipose tissue and plasma lipid of 20 Massachusetts Vietnam veterans." *Chemosphere,* Vol. 20, Nos. 7–9, 1990, pp. 954–955 (Tables I and II).

Statistics in Action Revisited

- Testing the Median TCDD Level of Vietnam Vets (p. 827)
- Comparing the TCDD Levels in Fat and Plasma of Vietnam Vets (p. 843)
- Testing whether the TCDD Levels in Fat and Plasma of Vietnam Vets Are Correlated (p. 865)

14.1 Introduction: Distribution-Free Tests

The confidence interval and testing procedures developed in Chapters 7–10 all involve making inferences about population parameters. Consequently, they are often referred to as **parametric statistical tests**. Many of these parametric methods (e.g., the small sample *t*-test of Chapter 8 or the ANOVA *F*-test of Chapter 10) rely on the assumption that the data are sampled from a normally distributed population. When the data are normal, these tests are *most powerful*. That is, the use of these parametric tests maximizes power—the probability of the researcher correctly rejecting the null hypothesis.

Consider a population of data that is decidedly nonnormal. For example, the distribution might be very flat, peaked, or strongly skewed to the right or left (see Figure 14.1). Applying the small sample *t*-test to such a data set may result in serious consequences. Since the normality assumption is clearly violated, the results of the *t*-test are unreliable: (1) the probability of a Type I error (i.e., rejecting H_0 when it is true) may be larger than the value of α selected; and (2) the power of the test, $1 - \beta$, is not maximized.

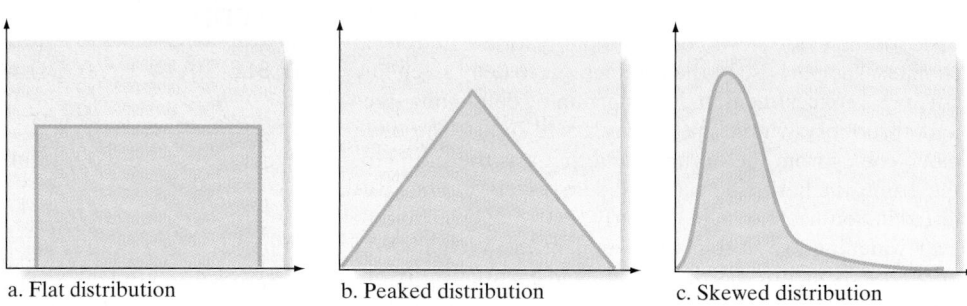

a. Flat distribution   b. Peaked distribution   c. Skewed distribution

Figure 14.1

Some nonnormal distributions for which the *t*-statistic is invalid

A host of *nonparametric* techniques are available for analyzing data that do not follow a normal distribution. Nonparametric tests do not depend on the distribution of the sampled population; thus, they are called *distribution-free tests*. Also, nonparametric methods focus on the location of the probability distribution of the population, rather than on specific parameters of the population, such as the mean (hence, the name "nonparametrics").

DEFINITION 14.1

Distribution-free tests are statistical tests that do not rely on any underlying assumptions about the probability distribution of the sampled population.

DEFINITION 14.2

The branch of inferential statistics devoted to distribution-free tests is called **nonparametrics**.

Nonparametric tests are also appropriate when the data are nonnumerical in nature but can be ranked.* For example, when taste-testing foods or in other types of consumer product evaluations, we can say we like product A better than product B, and B better than C, but we cannot obtain exact quantitative values for the respective measurements. Nonparametric tests based on the ranks of measurements are called *rank tests*.

DEFINITION 14.3

Nonparametric statistics (or tests) based on the ranks of measurements are called **rank statistics** (or **rank tests**).

---

*Qualitative data that can be ranked in order of magnitude are called *ordinal* data.

In this chapter, we present several useful nonparametric methods. Keep in mind that these nonparametric tests are more powerful than their corresponding parametric counterparts in those situations where either the data are nonnormal or the data are ranked.

In Section 14.2, we develop a test to make inferences about the central tendency of a single population. In Sections 14.3 and 14.5, we present rank statistics for comparing two or more probability distributions using independent samples. In Sections 14.4 and 14.6, the matched-pairs and randomized block designs are used to make nonparametric comparisons of populations. Finally, in Section 14.7, we present a nonparametric measure of correlation between two variables.

14.2 Single Population Inferences

In Chapter 8 we utilized the z- and t-statistics for testing hypotheses about a population mean. The z-statistic is appropriate for large random samples selected from "general" populations—that is, with few limitations on the probability distribution of the underlying population. The t-statistic was developed for small-sample tests in which the sample is selected at random from a *normal* distribution. The question is: How can we conduct a test of hypothesis when we have a small sample from a *nonnormal* distribution?

The **sign test** is a relatively simple nonparametric procedure for testing hypotheses about the central tendency of a nonnormal probability distribution. Note that we used the phrase *central tendency* rather than *population mean*. This is because the sign test, like many nonparametric procedures, provides inferences about the population *median* rather than the population mean μ. Denoting the population median by the Greek letter η, we know (Chapter 2) that η is the 50th percentile of the distribution (Figure 14.2) and as such is less affected by the skewness of the distribution and the presence of outliers (extreme observations). Since the nonparametric test must be suitable for all distributions, not just the normal, it is reasonable for nonparametric tests to focus on the more robust (less sensitive to extreme values) measure of central tendency, the median.

Figure 14.2

Location of the Population Median, η

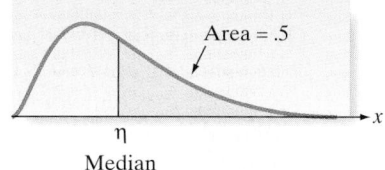

Area = .5

η
Median

For example, increasing numbers of both private and public agencies are requiring their employees to submit to tests for substance abuse. One laboratory that conducts such testing has developed a system with a normalized measurement scale, in which values less than 1.00 indicate "normal" ranges and values equal to or greater than 1.00 are indicative of potential substance abuse. The lab reports a normal result as long as the median level for an individual is less than 1.00. Eight independent measurements of each individual's sample are made. One individual's results are shown in Table 14.1.

⊚ SUBABUSE

TABLE 14.1 Substance Abuse Test Results

.78	.51	3.79	.23	.77	.98	.96	.89

If the objective is to determine whether the *population* median (that is, the true median level if an infinitely large number of measurements were made on the same individual sample) is less than 1.00, we establish that as our alternative hypothesis and test

$$H_0: \eta = 1.00$$
$$H_a: \eta < 1.00$$

The one-tailed sign test is conducted by counting the number of sample measurements that "favor" the alternative hypothesis—in this case, the number that are less than 1.00. If the null hypothesis is true, we expect approximately half of the measurements to fall on each side of the hypothesized median and if the alternative is true, we expect significantly more than half to favor the alternative—that is, to be less than 1.00. Thus,

Test statistic: S = Number of measurements less than 1.00, the null hypothesized median

If we wish to conduct the test at the $\alpha = .05$ level of significance, the rejection region can be expressed in terms of the observed significance level, or *p*-value, of the test:

Rejection region: p-value $\leq .05$

In this example, $S = 7$ of the 8 measurements are less than 1.00. To determine the observed significance level associated with this outcome, we note that the number of measurements less than 1.00 is a binomial random variable (check the binomial characteristics presented in Chapter 4), and *if H_0 is true*, the binomial probability *p* that a measurement lies below (or above) the median 1.00 is equal to .5 (Figure 14.2). What is the probability that a result is *as contrary to or more contrary to H_0* than the one observed? That is, what is the probability that 7 *or more* of 8 binomial measurements will result in Success (be less than 1.00) if the probability of Success is .5? Binomial Table II in Appendix A (using $n = 8$ and $p = .5$) indicates that

$$P(x \geq 7) = 1 - P(x \leq 6) = 1 - .965 = .035$$

Thus, the probability that at least 7 of 8 measurements would be less than 1.00 *if the true median were 1.00* is only .035. The *p*-value of the test is therefore .035.

This *p*-value can also be obtained by using a statistical software package. The MINITAB printout of the analysis is shown in Figure 14.3, with the *p*-value highlighted on the printout. Since $p = .035$ is less than $\alpha = .05$, we conclude that this sample provides sufficient evidence to reject the null hypothesis. The implication of this rejection is that the laboratory can conclude at the $\alpha = .05$ level of significance that the true median level for the tested individual is less than 1.00. However, we note that one of the measurements greatly exceeds the others, with a value of 3.79, and deserves special attention. Note that this large measurement is an outlier that would make the use of a *t*-test and its concomitant assumption of normality dubious.

Figure 14.3
MINITAB Printout
of Sign Test

Sign Test for Median: READING

Sign test of median =   1.000 versus <   1.000

	N	Below	Equal	Above	P	Median
READING	8	7	0	1	0.0352	0.8350

The only assumption necessary to ensure the validity of the sign test is that the probability distribution of measurements is continuous.

The use of the sign test for testing hypotheses about population medians is summarized in the next box.

Sign Test for a Population Median η

One-Tailed Test

H_0: $\eta = \eta_0$

H_a: $\eta > \eta_0$ [or H_a: $\eta < \eta_0$]

Test statistic:

S = Number of sample measurements greater than η_0 [or S = number of measurements less than η_0]

Two-Tailed Test

H_0: $\eta = \eta_0$

H_a: $\eta \neq \eta_0$

Test statistic:

S = Larger of S_1 and S_2, where S_1 is the number of measurements less than η_0 and S_2 is the number of measurements greater than η_0

[*Note:* Eliminate observations from the analysis that are exactly equal to the hypothesized median, η_0.]

Observed significance level:

p-value = $P(x \geq S)$

Observed significance level:

p-value = $2P(x \geq S)$

where x has a binomial distribution with parameters n and $p = .5$. (Use Table II, Appendix A.)

Rejection region: Reject H_0 if p-value $\leq \alpha$

Conditions Required for a Valid Application of the Sign Test

The sample is selected randomly from a continuous probability distribution. [*Note:* No assumptions need to be made about the shape of the probability distribution.]

Recall that the normal probability distribution provides a good approximation for the binomial distribution when the sample size is large. For tests about the median of a distribution, the null hypothesis implies that $p = .5$, and the normal distribution provides a good approximation if $n \geq 10$. (Samples with $n \geq 10$ satisfy the condition that $np \pm 2\sqrt{npq}$ is contained in the interval 0 to n.) Thus, we can use the standard normal z-distribution to conduct the sign test for large samples. The large-sample sign test is summarized in the next box.

EXAMPLE 14.1

APPLICATION OF THE SIGN TEST

Problem A manufacturer of compact disk (CD) players has established that the median time to failure for its players is 5,250 hours of utilization. A sample of 20 CDs from a competitor is obtained, and they are continuously tested until each fails. The 20 failure times range from 5 hours (a "defective" player) to 6,575 hours, and 14 of the 20 exceed 5,250 hours. Is there evidence that the median failure time of the competitor differs from 5,250 hours? Use $\alpha = .10$.

Solution The null and alternative hypotheses of interest are

$$H_0: \eta = 5{,}250 \text{ hours}$$
$$H_a: \eta \neq 5{,}250 \text{ hours}$$

Large-Sample Sign Test for a Population Median η

One-Tailed Test	**Two-Tailed Test**
$H_0: \eta = \eta_0$	$H_0: \eta = \eta_0$
$H_a: \eta > \eta_0$ [or $H_a: \eta < \eta_0$]	$H_a: \eta \neq \eta_0$

$$\text{Test statistic:} \quad z = \frac{(S - .5) - .5n}{.5\sqrt{n}}$$

[*Note:* S is calculated as shown in the previous box. We subtract .5 from S as the "correction for continuity." The null hypothesized mean value is $np = .5n$, and the standard deviation is

$$\sqrt{npq} = \sqrt{n(.5)(.5)} = .5\sqrt{n}$$

See Chapter 5 for details on the normal approximation to the binomial distribution.]

Rejection region: $z > z_\alpha$ Rejection region: $z > z_{\alpha/2}$

where tabulated z values can be found in Table IV, Appendix A.

Since $n \geq 10$, we use the standard normal z-statistic:

$$\text{Test statistic:} \quad z = \frac{(S - .5) - .5n}{.5\sqrt{n}}$$

where S is the maximum of S_1, the number of measurements greater than 5,250, and S_2, the number of measurements less than 5,250.

$$\text{Rejection region:} \quad z > 1.645, \text{ where } z_{\alpha/2} = z_{.05} = 1.645$$

Assumptions: The distribution of the failure times is continuous (time is a continuous variable), but nothing is assumed about the shape of its probability distribution.

Since the number of measurements exceeding 5,250 is $S_2 = 14$, then the number of measurements less than 5,250 is $S_1 = 6$. Consequently, $S = 14$, the greater of S_1 and S_2. The calculated z statistic is therefore

$$z = \frac{(S - .5) - .5n}{.5\sqrt{n}} = \frac{13.5 - 10}{.5\sqrt{20}} = \frac{3.5}{2.236} = 1.565$$

The value of z is not in the rejection region, so we cannot reject the null hypothesis at the $\alpha = .10$ level of significance.

Look Back The CD manufacturer should not conclude, on the basis of this sample, that its competitor's CDs have a median failure time that differs from 5,250 hours. The manufacturer will not "accept H_0" however, since the probability of a Type II error is unknown.

> **Now Work** *Exercise 14.5*

      ■ ■ ■

The one-sample nonparametric sign test for a median provides an alternative to the *t*-test for small samples from nonnormal distributions. However, if the distribution is approximately normal, the *t*-test provides a more powerful test about the central tendency of the distribution.

Statistics in Action Revisited

Testing the Median TCDD Level of Vietnam Vets

Recall that during the Vietnam War, U.S. soldiers were exposed to the herbicide, Agent Orange (p. 821). As a result of this exposure, many Vietnam veterans have dangerously high levels of the dioxin TCDD in their plasma and fat tissue. Some medical researchers consider a TCDD level of 3 parts per trillion (ppt) to be dangerously high. Do the data in Table SIA14.1 provide evidence to indicate that the median level of TCDD in both plasma and fat tissue of Vietnam vets exceeds 3 ppt? To answer this question, we applied the sign test to the data saved in the **TCDD** file. The MINITAB printout is shown in Figure SIA14.1.

We want to test $H_0: \eta = 3$ versus $H_a: \eta > 3$ for both variables, TCDD in fat and TCDD in plasma. According to the printout, 14 of the 20 Vietnam vets had

TCDD levels in fat above 3 ppt, and 12 of the 20 Vietnam vets had TCDD levels in plasma above 3 ppt. Consequently, the two test statistic values are $S = 14$ and $S = 12$, respectively. The one tailed p-values for the tests (highlighted on the printout) are .0577 and .1796, respectively.

Both tests are nonsignificant at $\alpha = .05$; at $\alpha = .10$, the test for TCDD in fat is significant. Therefore, the data provide sufficient evidence (at $\alpha = .10$) to say that the median TCDD in fat exceeds 3 ppt; however, the evidence is not as strong for TCDD in plasma.

Why apply the nonparametric sign test to the data rather than the more familiar Student's t-test? The MINITAB histograms in Figure SIA14.2 illustrate the problem. Clearly, the sample TCDD levels are not normally distributed. Consequently, the assumption required for the t-test to yield valid inferences is violated.

Figure SIA14.1
MINITAB Sign Tests for TCDD Data

Sign Test for Median: FAT, PLASMA

```
Sign test of median =   3.000 versus > 3.000

              N   Below   Equal   Above        P   Median
FAT          20       6       0      14   0.0577    4.750
PLASMA       20       7       1      12   0.1796    3.200
```

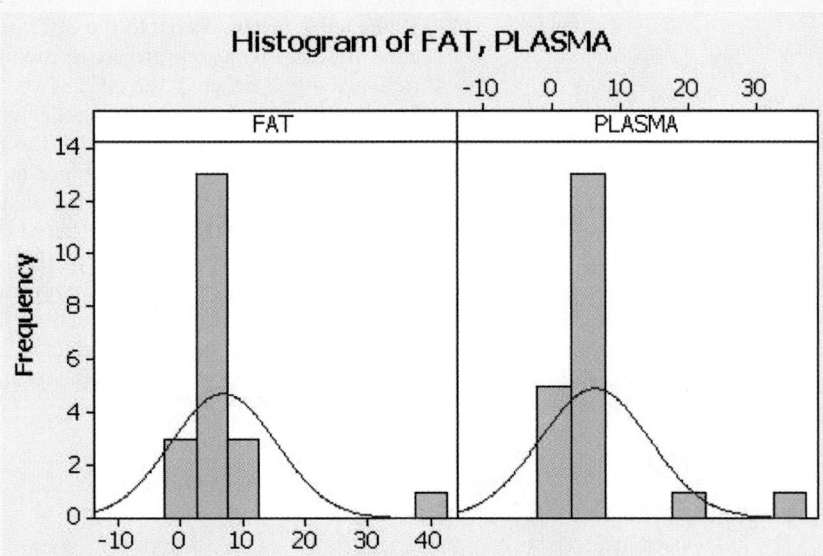

Figure SIA14.2
MINITAB Histograms for TCDD Data

Exercises 14.1–14.13

Understanding the Principles

14.1 Under what circumstances is the sign test preferred to the t-test for making inferences about the central tendency of a population?

14.2 What is the probability that a randomly selected observation exceeds the
  a. Mean of a normal distribution?
  b. Median of a normal distribution?
  c. Mean of a nonnormal distribution?
  d. Median of a nonnormal distribution?

Learning the Mechanics

14.3 Use Table II of Appendix A to calculate the following binomial probabilities:
  a. $P(x \geq 6)$ when $n = 7$ and $p = .5$
  b. $P(x \geq 5)$ when $n = 9$ and $p = .5$
  c. $P(x \geq 8)$ when $n = 8$ and $p = .5$
  d. $P(x \geq 10)$ when $n = 15$ and $p = .5$. Also use the normal approximation to calculate this probability, then compare the approximation with the exact value.
  e. $P(x \geq 15)$ when $n = 25$ and $p = .5$. Also use the normal approximation to calculate this probability, then compare the approximation with the exact value.

14.4 Consider the following sample of 10 measurements:

🌀 **LM14_4**

| 8.4 | 16.9 | 15.8 | 12.5 | 10.3 | 4.9 | 12.9 | 9.8 | 23.7 | 7.3 |

Use these data to conduct each of the following sign tests using the binomial tables (Table II, Appendix A) and $\alpha = .05$.
  a. $H_0: \eta = 9$ versus $H_a: \eta > 9$
  b. $H_0: \eta = 9$ versus $H_a: \eta \neq 9$
  c. $H_0: \eta = 20$ versus $H_a: \eta < 20$
  d. $H_0: \eta = 20$ versus $H_a: \eta \neq 20$
  e. Repeat each of the preceding tests using the normal approximation to the binomial probabilities. Compare the results.
  f. What assumptions are necessary to ensure the validity of each of the preceding tests?

14.5 Suppose you wish to conduct a test of the research hypothesis that the median of a population is greater than **NW** 80. You randomly sample 25 measurements from the population and determine that 16 of them exceed 80. Set up and conduct the appropriate test of hypothesis at the .10 level of significance. Be sure to specify all necessary assumptions.

Applying the Concepts—Basic

14.6 Caffeine in Starbucks coffee. Scientists at the University of Florida College of Medicine investigated the level of caffeine in 16-ounce cups of Starbucks coffee. (*Journal of Analytical Toxicology*, Oct. 2003.) In one phase of the experiment, cups of Starbucks Breakfast Blend (a mix of Latin American coffees) were purchased on six consecutive days from a single specialty coffee shop. The amount of caffeine in each of the six cups (measured in milligrams) is provided in the table.

🌀 **STARBUCKS**

| 564 | 498 | 259 | 303 | 300 | 307 |

  a. Suppose the scientists are interested in determining whether the median amount of caffeine in Breakfast Blend coffee exceeds 300 milligrams. Set up the null and alternative hypotheses of interest.
  b. How many of the cups in the sample have a caffeine content that exceeds 300 milligrams?
  c. Assuming $p = .5$, use the binomial table in Appendix A to find the probability that at least 4 of the 6 cups have caffeine amounts that exceed 300 milligrams.
  d. Based on the probability, part **c**, what do you conclude about H_0 and H_a? (Use $\alpha = .05$.)

🌀 **LICHEN**

14.7 Radioactive lichen. Refer to the 2003 Lichen Radionuclide Baseline Research project to monitor the level of radioactivity in lichen, Exercise 2.34 (p. 51). Recall that University of Alaska researchers collected 9 lichen specimens and measured the amount of the radioactive element cesium-137 (in microcuries per milliliter) in each specimen. (The natural logarithms of the data values, saved in the **LICHEN** file, are listed in the table on p. 829.) In Exercise 8.65 (p. 396), you used the t-statistic to test whether the mean cesium amount in lichen differs from $\mu = .003$ microcuries per milliliter. Use the MINITAB printout below to conduct an alternative nonparametric test at $\alpha = .10$. Does the result agree with the t-test in Exercise 8.65?

Sign Test for Median: CESIUM

```
Sign test of median = 0.00300 versus not = 0.00300

          N  Below  Equal  Above     P   Median
CESIUM    9     1      0      8  0.0391  0.00783
```

⊚ LICHEN

Location			
Bethel	−5.50	−5.00	
Eagle Summit	−4.15	−4.85	
Moose Pass	−6.05		
Turnagain Pass	−5.00		
Wickersham Dome	−4.10	−4.50	−4.60

Source: Lichen Radionuclide Baseline Research project, 2003.

14.8 Quality of white shrimp. In *The American Statistician* (May 2001), the nonparametric sign test was used to analyze data on the quality of white shrimp. One measure of shrimp quality is cohesiveness. Since freshly caught shrimp are usually stored on ice, there is concern that cohesiveness will deteriorate after storage. For a sample of 20 newly caught white shrimp, cohesiveness was measured both before storage and after storage on ice for two weeks. The difference in the cohesiveness measurements (before minus after) was obtained for each shrimp. If storage has no effect on cohesiveness, the population median of the differences will be 0. If cohesiveness deteriorates after storage, the population median of the differences will be positive.
   a. Set up the null and alternative hypotheses to test whether cohesiveness will deteriorate after storage.
   b. In the sample of 20 shrimp, there were 13 positive differences. Use this value to find the *p*-value of the test.
   c. Make the appropriate conclusion (in the words of the problem) if $\alpha = .05$.

14.9 Ammonia in car exhaust. Refer to the *Environmental Science & Technology* (Sept. 1, 2000) study of ammonia levels near the exit ramp of a San Francisco highway tunnel, Exercise 2.59 (p. 64). The daily ammonia concentrations (parts per million) on eight randomly selected days during afternoon drive-time are reproduced in the table. Suppose you want to determine if the median daily ammonia concentration for all afternoon drive-time days exceeds 1.5 ppm.

⊚ AMMONIA

1.53	1.50	1.37	1.51	1.55	1.42	1.41	1.48

   a. Set up the null and alternative hypotheses for the test.
   b. Find the value of the test statistic.
   c. Find the *p*-value of the test.
   d. Give the appropriate conclusion (in the words of the problem) if $\alpha = .05$.

Applying the Concepts—Intermediate

14.10 Hemotology tests on workers. The next table lists the lymphocyte count results from hematology tests administered to a sample of 50 West Indian or African workers. Test (at $\alpha = .05$) the hypothesis that the median lymphocyte count of all West Indian/African workers exceeds 20.

⊚ LYMPHO

14	28	11
15	17	25
19	14	30
23	8	32
17	25	17
20	37	22
21	20	20
16	15	20
27	9	20
34	16	26
26	18	40
28	17	22
24	23	61
26	43	12
23	17	20
9	23	35
18	31	

Source: Royston, J. P. "Some techniques for assessing multivariate normality based on the Shapiro-Wilk W." *Applied Statistics,* Vol. 32, No. 2, pp. 121–133.

14.11 Biting rates of flies. The biting rate of a particular species of fly was investigated in a study reported in the *Journal of the American Mosquito Control Association* (Mar. 1995). Biting rate was defined as the number of flies biting a volunteer during 15 minutes of exposure. This species of fly is known to have a median biting rate of 5 bites per 15 minutes on Stanbury Island, Utah. However, it is theorized that the median biting rate is higher in bright, sunny weather. To test this theory, 122 volunteers were exposed to the flies during a sunny day on Stanbury Island. Of these volunteers, 95 experienced biting rates greater than 5.
   a. Set up the null and alternative hypotheses for the test.
   b. Calculate the approximate *p*-value of the test. [*Hint:* Use the normal approximation for a binomial probability.]
   c. Make the appropriate conclusion at $\alpha = .01$.

14.12 Study of guppy migration. In a study of the excessive transitory migration of guppy populations, 40 adult female guppies were placed in the left compartment of an experimental aquarium tank which was divided in half by a glass plate. After the plate was removed, the numbers of fish passing through the slit from the left compartment to the right one, and vice versa, were monitored every minute for 30 minutes. (*Zoological Science,* Vol. 6, 1989.) If an equilibrium is reached, the researchers would expect the median number of fish remaining in the left compartment to be 20. The data for the 30 observations (that is, numbers of fish in the left compartment at the end of each one-minute interval) are shown in the table on p. 830. Use the large-sample sign test to determine whether the median is less than 20. Test using $\alpha = .05$.

14.13 Lunch at McDonald's. According to the National Restaurant Association, hamburgers are the number-one selling fast-food item in the United States. An economist studying the fast-food buying habits of

⊙ GUPPY

16	11	12	15	14	16	18	15	13	15
14	14	16	13	17	17	14	22	18	19
17	17	20	23	18	19	21	17	21	17

Source: Terami, H., and Watanabe, M. "Excessive transitory migration of guppy populations. III. Analysis of perception of swimming space and a mirror effect." *Zoological Science*, Vol. 6, 1989, p. 977 (Figure 2).

Americans paid graduate students to stand outside two suburban McDonald's restaurants near Boston and ask departing customers whether they spent more than $2.25 on hamburger products for their lunch. Twenty answered yes; 50 said no; and 10 refused to answer the question. (*Newark Star-Ledger*, Mar. 17, 1997.)

a. Is there sufficient evidence to conclude that the median amount spent for hamburgers at lunch at McDonald's is less than $2.25?

b. Does your conclusion apply to all Americans who eat lunch at McDonald's? Justify your answer.

c. What assumptions must hold to ensure the validity of your test in part **a**?

14.3 Comparing Two Populations: Independent Samples

Suppose two independent random samples are to be used to compare two populations and the *t*-test of Chapter 9 is inappropriate for making the comparison. We may be unwilling to make assumptions about the form of the underlying population probability distributions or we may be unable to obtain exact values of the sample measurements. If the data can be ranked in order of magnitude for either of these situations, the **Wilcoxon rank sum test** (developed by Frank Wilcoxon) can be used to test the hypothesis that the probability distributions associated with the two populations are equivalent.

Biography

FRANK WILCOXON (1892–1965)—Wilcoxon Rank Tests

Frank Wilcoxon was born in Ireland, where his wealthy American parents were vacationing. He grew up in the family home in Catskill, New York, then spent time working as an oil worker and tree surgeon in the back country of West Virginia. At age 25, Wilcoxon's parents sent him to Pennsylvania Military College, but he dropped out due to the death of his twin sister. Later, Wilcoxon earned degrees in chemistry from Rutgers (master's) and Cornell University (Ph.D.). After receiving his doctorate, Wilcoxon began work as a chemical researcher at the Boyce Thompson Institute for Plant Research. There, he began studying R. A. Fisher's (p. 516) newly issued *Statistical Methods for Research Workers*. In a now famous 1945 paper, Wilcoxon presented the idea of replacing the actual sample data in Fisher's tests by their ranks and called the tests the rank sum test and signed-rank test. These tests proved to be inspirational to the further development of nonparametrics. After retiring from industry, Wilcoxon accepted a Distinguished Lectureship position at newly created Department of Statistics at Florida State University.

Suppose, for example, an experimental psychologist wants to compare reaction times for adult males under the influence of drug A to those under the influence of drug B. Experience has shown that the populations of reaction time measurements often possess probability distributions that are skewed to the right, as shown in Figure 14.4. Consequently, a *t*-test should not be used to compare the mean reaction times for the two drugs because the normality assumption that is required for the *t*-test may not be valid.

Suppose the psychologist randomly assigns seven subjects to each of two groups, one group to receive drug A and the other to receive drug B. The reaction time for each subject is measured at the completion of the experiment. These data (with the exception of the measurement for one subject in group A who was eliminated from the experiment for personal reasons) are shown in Table 14.2.

Figure 14.4

Typical Probability Distribution of Reaction Times

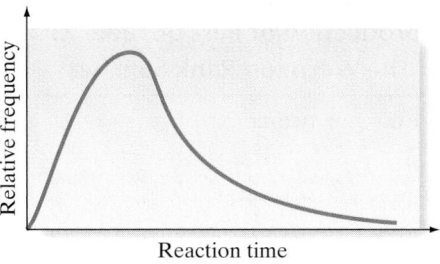

DRUGS

TABLE 14.2 Reaction Times of Subjects Under the Influence of Drug A or B

Drug A		Drug B	
Reaction Time (seconds)	Rank	Reaction Time (seconds)	Rank
1.96	4	2.11	6
2.24	7	2.43	9
1.71	2	2.07	5
2.41	8	2.71	11
1.62	1	2.50	10
1.93	3	2.84	12
		2.88	13

The population of reaction times for either of the drugs, say drug A, is that which could conceptually be obtained by giving drug A to all adult males. To compare the probability distributions for populations A and B, *we first rank the sample observations as though they were all drawn from the same population.* That is, we pool the measurements from both samples and then rank all the measurements from the smallest (a rank of 1) to the largest (a rank of 13). The results of this ranking process are also shown in Table 14.2.

If the two populations were identical, we would expect the ranks to be *randomly mixed* between the two samples. If, on the other hand, one population tends to have longer reaction times than the other, we would expect the larger ranks to be mostly in one sample and the smaller ranks mostly in the other. Thus, the test statistic for the Wilcoxon test is based on the totals of the ranks for each of the two samples—that is, on the **rank sums**. When the sample sizes are equal, for example, the greater the difference in the rank sums, the greater will be the weight of evidence to indicate a difference between the probability distributions of the populations. In the reaction times example, we denote the rank sum for drug A by T_1 and that for drug B by T_2. Then

$$T_1 = 4 + 7 + 2 + 8 + 1 + 3 = 25$$
$$T_2 = 6 + 9 + 5 + 11 + 10 + 12 + 13 = 66$$

The sum of T_1 and T_2 will always equal $n(n + 1)/2$, where $n = n_1 + n_2$. So, for this example, $n_1 = 6, n_2 = 7$, and

$$T_1 + T_2 = \frac{13(13 + 1)}{2} = 91$$

Since $T_1 + T_2$ is fixed, a small value for T_1 implies a large value for T_2 (and vice versa) and a large difference between T_1 and T_2. Therefore, the smaller the value of one of the rank sums, the greater the evidence to indicate that the samples were selected from different populations.

The test statistic for this test is the rank sum for the smaller sample; or, in the case where $n_1 = n_2$, either rank sum can be used. Values that locate the rejection region for this rank sum are given in Table XII of Appendix A. A partial reproduction of this

TABLE 14.3   Reproduction of Part of Table XII in Appendix A: Critical Values for the Wilcoxon Rank Sum Test

$\alpha = .025$ one-tailed; $\alpha = .05$ two-tailed

n_2 \ n_1	3		4		5		6		7		8		9		10	
	T_L	T_U	T_L	T_U	T_L	T_U	T_L	T_U	T_L	T_U	T_L	T_U	T_L	T_U	T_L	T_U
3	5	16	6	18	6	21	7	23	7	26	8	28	8	31	9	33
4	6	18	11	25	12	28	12	32	13	35	14	38	15	41	16	44
5	6	21	12	28	18	37	19	41	20	45	21	49	22	53	24	56
6	7	23	12	32	19	41	26	52	28	56	29	61	31	65	32	70
7	7	26	13	35	20	45	28	56	37	68	39	73	41	78	43	83
8	8	28	14	38	21	49	29	61	39	73	49	87	51	93	54	98
9	8	31	15	41	22	53	31	65	41	78	51	93	63	108	66	114
10	9	33	16	44	24	56	32	70	43	83	54	98	66	114	79	131

table is shown in Table 14.3. The columns of the table represent n_1, the first sample size, and the rows represent n_2, the second sample size. *The T_L and T_U entries in the table are the boundaries of the lower and upper regions, respectively, for the rank sum associated with the sample that has fewer measurements.* If the sample sizes n_1 and n_2 are the same, either rank sum may be used as the test statistic. To illustrate, suppose $n_1 = 8$ and $n_2 = 10$. For a two-tailed test with $\alpha = .05$, we consult part a of the table and find that the null hypothesis will be rejected if the rank sum of sample 1 (the sample with fewer measurements), T, is less than or equal to $T_L = 54$ *or* greater than or equal to $T_U = 98$. The Wilcoxon rank sum test is summarized in the next boxes.

Wilcoxon Rank Sum Test: Independent Samples*

Let D_1 and D_2 represent the probability distributions for populations 1 and 2, respectively.

One-Tailed Test

H_0: D_1 and D_2 are identical

H_a: D_1 is shifted to the right of D_2
[or H_a: D_1 is shifted to the left of D_2]

Test statistic:

T_1, if $n_1 < n_2$; T_2, if $n_2 < n_1$
(Either rank sum can be used if $n_1 = n_2$.)

Rejection region:

T_1: $T_1 \geq T_U$ [or $T_1 \leq T_L$]
T_2: $T_2 \leq T_L$ [or $T_2 \geq T_U$]

Two-Tailed Test

H_0: D_1 and D_2 are identical

H_a: D_1 is shifted either to the left or to the right of D_2

Test statistic:

T_1, if $n_1 < n_2$; T_2, if $n_2 < n_1$
(Either rank sum can be used if $n_1 = n_2$.) We will denote this rank sum as T.

Rejection region:

$T \leq T_L$ or $T \geq T_U$

where T_L and T_U are obtained from Table XII of Appendix A.

Ties: Assign tied measurements the average of the ranks they would receive if they were unequal but occurred in successive order. For example, if the third-ranked and fourth-ranked measurements are tied, assign each a rank of $(3 + 4)/2 = 3.5$.

---

*Another statistic used for comparing two populations based on independent random samples is the *Mann-Whitney U-statistic.* The U-statistic is a simple function of the rank sums. It can be shown that the Wilcoxon rank sum test and the Mann-Whitney U-test are equivalent.

> **Conditions Required for a Valid Rank Sum Test:**
> 1. The two samples are random and independent.
> 2. The two probability distributions from which the samples are drawn are continuous.

Note that the assumptions necessary for the validity of the Wilcoxon rank sum test do not specify the shape or type of probability distribution. However, the distributions are assumed to be continuous so that the probability of tied measurements is 0 (see Chapter 5), and each measurement can be assigned a unique rank. In practice, however, rounding of continuous measurements will sometimes produce ties. As long as the number of ties is small relative to the sample sizes, the Wilcoxon test procedure will still have approximate significance level α. The test is not recommended to compare discrete distributions for which many ties are expected.

EXAMPLE 14.2   APPLYING THE RANK SUM TEST

Problem Do the data given in Table 14.2 provide sufficient evidence to indicate a shift in the probability distributions for drugs A and B, that is, that the probability distribution corresponding to drug A lies either to the right or left of the probability distribution corresponding to drug B? Test at the .05 level of significance.

Solution

H_0: The two populations of reaction times corresponding to drug A and drug B have the same probability distribution.

H_a: The probability distribution for drug A is shifted to the right or left of the probability distribution corresponding to drug B.*

Test statistic: Since drug A has fewer subjects than drug B, the test statistic is T_1, the rank sum of drug A's reaction times.

Rejection region: Since the test is two-sided, we consult part a of Table XII for the rejection region corresponding to $\alpha = .05$. We will reject H_0 for $T_1 \leq T_L$ or $T_1 \geq T_U$. Thus, we will reject H_0 if $T_1 \leq 28$ or $T_1 \geq 56$.

Since T_1, the rank sum of drug A's reaction times in Table 14.2, is 25, it is in the rejection region (see Figure 14.5).† Therefore, there is sufficient evidence to reject H_0. This same conclusion can be reached using a statistical software package. The SAS printout of the analysis is shown in Figure 14.6. Both the test statistic ($T_1 = 25$) and one-tailed p-value ($p = .007$) are highlighted on the printout. The one-tailed p-value is less than $\alpha = .05$, leading us to reject H_0.

Look Back Our conclusion is that the probability distributions for drugs A and B are not identical. In fact, it appears that drug B tends to be associated with reaction times that are larger than those associated with drug A (because T_1 falls in the lower tail of the rejection region).

---

*The alternative hypotheses in this chapter will be stated in terms of a difference in the *location* of the distributions. However, since the shapes of the distributions may also differ under H_a, some of the figures (e.g., Figure 14.5) depicting the alternative hypothesis will show probability distributions with different shapes.

†Figure 14.5 depicts only one side of the two-sided alternative hypothesis. The other would show the distribution for drug A shifted to the right of the distribution for drug B.

Figure 14.5

Alternative Hypothesis
and Rejection Region for
Example 14.2.

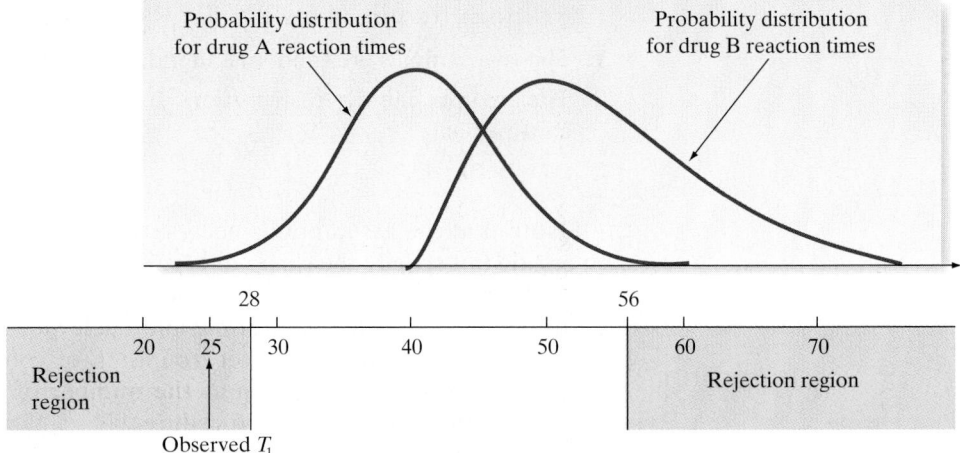

Figure 14.6

SAS Printout for
Example 14.2

The NPAR1WAY Procedure

Wilcoxon Scores (Rank Sums) for Variable REACTIME
Classified by Variable DRUG

DRUG	N	Sum of Scores	Expected Under H0	Std Dev Under H0	Mean Score
A	6	25.0	42.0	7.0	4.166667
B	7	66.0	49.0	7.0	9.428571

Wilcoxon Two-Sample Test

Statistic (S)	25.0000

Normal Approximation
Z	-2.4286
One-Sided Pr < Z	0.0076
Two-Sided Pr > |Z|	0.0152

t Approximation
| One-Sided Pr < Z | 0.0159 |
| Two-Sided Pr > |Z| | 0.0318 |

Exact Test
| One-Sided Pr <= S | 0.0070 |
| Two-Sided Pr >= |S - Mean| | 0.0140 |

Kruskal-Wallis Test

Chi-Square	5.8980
DF	1
Pr > Chi-Square	0.0152

Now Work *Exercise 14.18*

■ ■ ■

Table XII in Appendix A gives values of T_L and T_U for values of n_1 and n_2 less than or equal to 10. When both sample sizes, n_1 and n_2, are 10 or larger, the sampling distribution of T_1 can be approximated by a normal distribution with mean and variance

$$E(T_1) = \frac{n_1(n_1 + n_2 + 1)}{2} \quad \text{and} \quad \sigma_{T_1}^2 = \frac{n_1 n_2(n_1 + n_2 + 1)}{12}$$

Therefore, for $n_1 \geq 10$ and $n_2 \geq 10$ we can conduct the Wilcoxon rank sum test using the familiar z-test of Chapters 9 and 10. The test is summarized in the next box.

The Wilcoxon Rank Sum Test for Large Samples
($n_1 \geq 10$ and $n_2 \geq 10$)

Let D_1 and D_2 represent the probability distributions for populations 1 and 2, respectively.

One-Tailed Test

H_0: D_1 and D_2 are identical
H_a: D_1 is shifted to the right of D_2
   (or H_a: D_1 is shifted to the left of D_2)

Two-Tailed Test

H_0: D_1 and D_2 are identical
H_a: D_1 is shifted to the right or to the left of D_2

$$\text{Test statistic:}\quad z = \frac{T_1 - \dfrac{n_1(n_1 + n_2 + 1)}{2}}{\sqrt{\dfrac{n_1 n_2(n_1 + n_2 + 1)}{12}}}$$

Rejection region:

$z > z_\alpha$ (or $z < -z_\alpha$)

Rejection region:

$|z| > z_{\alpha/2}$

Exercises 14.14–14.29

Understanding the Principles

14.14 What is a rank sum?

14.15 *True or False.* If the rank sum for sample 1 is much larger than the rank sum for sample 2 when $n_1 = n_2$, then the distribution of population 1 is likely to be shifted to the right of the distribution of population 2.

14.16 What conditions are required for a valid application of the Wilcoxon rank sum test?

Learning the Mechanics

14.17 Specify the test statistic and the rejection region for the Wilcoxon rank sum test for independent samples in each of the following situations:
  a. H_0: Two probability distributions, 1 and 2, are identical
   H_a: Probability distribution for population 1 is shifted to the right or left of the probability distribution for population 2
   $n_1 = 7, n_2 = 8, \alpha = .10$
  b. H_0: Two probability distributions, 1 and 2, are identical
   H_a: Probability distribution for population 1 is shifted to the right of the probability distribution for population 2
   $n_1 = 6, n_2 = 6, \alpha = .05$
  c. H_0: Two probability distributions, 1 and 2, are identical H_a: Probability distribution for population 1 is shifted to the left of the probability distribution for population 2
   $n_1 = 7, n_2 = 10, \alpha = .025$
  d. H_0: Two probability distributions, 1 and 2, are identical H_a: Probability distribution for population 1 is shifted to the right or left of the probability distribution for population 2
   $n_1 = 20, n_2 = 20, \alpha = .05$

14.18 Suppose you want to compare two treatments, A and
NW B. In particular, you wish to determine whether the distribution for population B is shifted to the right of the distribution for population A. You plan to use the Wilcoxon rank sum test.
  a. Specify the null and alternative hypotheses you would test.
  b. Suppose you obtained the following independent random samples of observations on experimental units subjected to the two treatments:

🔹 **LM14_18**

| Sample A | 37, | 40, | 33, | 29, | 42, | 33, | 35, | 28, | 34 |
| Sample B | 65, | 35, | 47, | 52 | | | | | |

Conduct a test of the hypotheses described in part **a**. Test using $\alpha = .05$.

14.19 Suppose you wish to compare two treatments, A and B, based on independent random samples of 15 observations selected from each of the two populations. If $T_1 = 173$, do the data provide sufficient evidence to indicate that distribution A is shifted to the left of distribution B? Test using $\alpha = .05$.

14.20 Random samples of sizes $n_1 = 16$ and $n_2 = 12$ were drawn from populations 1 and 2, respectively. The measurements obtained are listed in the table.

🔹 **LM14_20**

Sample 1				Sample 2		
9.0	15.6	25.6	31.1	10.1	11.1	13.5
21.1	26.9	24.6	20.0	12.0	18.2	10.3
24.8	16.5	26.0	25.1	9.2	7.0	14.2
17.2	30.1	18.7	26.1	15.8	13.6	13.2

a. Conduct a hypothesis test to determine whether the probability distribution for population 2 is shifted to the left of the probability distribution for population 1. Use $\alpha = .05$.

b. What is the approximate p-value of the test of part **a**?

14.21 Independent random samples are selected from two populations. The data are shown in the table.

⊛ **LM14_21**

Sample 1		Sample 2		
15	16	5	9	5
10	13	12	8	10
12	8	9	4	

a. Use the Wilcoxon rank sum test to determine whether the data provide sufficient evidence to indicate a shift in the locations of the probability distributions of the sampled populations. Test using $\alpha = .05$.

b. Do the data provide sufficient evidence to indicate that the probability distribution for population 1 is shifted to the right of the probability distribution for population 2? Use the Wilcoxon rank sum test with $\alpha = .05$.

Applying the Concepts—Basic

14.22 Bursting strength of bottles. Polyethylene terephthalate (PET) bottles are used for carbonated beverages. A critical property of PET bottles is their bursting strength (i.e., the pressure at which bottles filled with water burst when pressurized). In the *Journal of Data Science* (May 2003), researchers measured the bursting strength of PET bottles made from two different designs—an old design and a new design. The data (pounds per square inch) for 10 bottles of each design are shown in the table. Suppose you want to compare the distributions of bursting strengths for the two designs.

⊛ **PET**

Old Design	210	212	211	211	190	213	212	211	164	209
New Design	216	217	162	137	219	216	179	153	152	217

a. Rank all 20 observed pressures from smallest to largest, and assign ranks from 1 to 20.

b. Sum the ranks of the observations from the old design.

c. Sum the ranks of the observations from the new design.

d. Compute the Wilcoxon rank sum statistic

e. Carry out a nonparametric test (at $\alpha = .05$) to compare the distribution of bursting strengths for the two designs.

14.23 Research on eating disorders. The "fear of negative evaluation" (FNE) scores for 11 female students known to suffer from the eating disorder bulimia and 14 female students with normal eating habits, first presented in Exercise 2.38 (p. 53), are reproduced at the bottom of the page. (Recall that the higher the score, the greater the fear of negative evaluation.) Suppose you want to determine if the distribution of the FNE scores for bulimic female students is shifted above the corresponding distribution for female students with normal eating habits.

a. Specify H_0 and H_a for the test.

b. Rank all 25 FNE scores in the data set from smallest to largest.

c. Sum the ranks of the 11 FNE scores for bulimic students.

d. Sum the ranks of the 14 FNE scores for students with normal eating habits.

e. Give the rejection region for a nonparametric test of the data if $\alpha = .10$.

f. Conduct the test and give the conclusion in the words of the problem.

14.24 Reading Japanese books. Refer to the *Reading in a Foreign Language* (Apr. 2004) experiment to improve the Japanese reading comprehension levels of University of Hawaii students, Exercise 9.13 (p. 449). Recall that 14 students participated in a ten-week extensive reading program in a second semester Japanese course. The number of books read by each student and the student's course grade are repeated in the table. Consider a comparison of the distribution of number of books read by students who earn an "A" grade and those who earn a "B" or "C" grade.

⊛ **JAPANESE**

Number of Books	Course Grade	Number of Books	Course Grade
53	A	30	A
42	A	28	B
40	A	24	A
40	B	22	C
39	A	21	B
34	A	20	B
34	A	16	B

Source: Hitosugi, C. I., and Day, R. R. "Extensive reading in Japanese," *Reading in a Foreign Language,* Vol. 16, No. 1, Apr. 2004 (Table 4).

a. Rank all 14 observations from smallest to largest, and assign ranks from 1 to 14.

⊛ **BULIMIA**

Bulimic students	21	13	10	20	25	19	16	21	24	13	14			
Normal students	13	6	16	13	8	19	23	18	11	19	7	10	15	20

Source: Randles, R. H. "On neutral responses (zeros) in the sign test and ties in the Wilcoxon-Mann-Whitney test." *The American Statistician*, Vol. 55, No. 2, May 2001 (Figure 3).

b. Sum the ranks of the observations for students with an "A" grade.

c. Sum the ranks of the observations for students with either a "B" or "C" grade.

d. Compute the Wilcoxon rank sum statistic.

e. Carry out a nonparametric test (at $\alpha = .10$) to compare the distribution of the number of books read by the two populations of students.

14.25 Visual acuity of children. In a comparison of visual acuity of deaf and hearing children, eye movement rates are taken on 10 deaf and 10 hearing children. The data are shown in the table. A clinical psychologist believes that deaf children have greater visual acuity than hearing children. (The larger a child's eye movement rate, the more visual acuity the child possesses.)

EYEMOVE

Deaf Children		Hearing Children	
2.75	1.95	1.15	1.23
3.14	2.17	1.65	2.03
3.23	2.45	1.43	1.64
2.30	1.83	1.83	1.96
2.64	2.23	1.75	1.37

a. Use the Wilcoxon rank sum procedure to test the psychologist's claim at $\alpha = .05$.

b. Conduct the test by using the large-sample approximation for the Wilcoxon rank sum test. Compare the results with those found in part **a**.

Applying the Concepts—Intermediate

14.26 Rain in Colorado. The data in the table, extracted from *Technometrics* (Feb. 1986), represent daily accumulated stream flow and precipitation (in inches) for two U.S. Geological Survey stations in Colorado. Conduct a test to determine whether the distributions of daily accumulated stream flow and precipitation for the two stations differ in location. Use $\alpha = .10$. Why is a nonparametric test appropriate for these data?

COLORAIN

Station 1			Station 2		
127.96	108.91	100.85	114.79	85.54	280.55
210.07	178.21	85.89	109.11	117.64	145.11
203.24	285.37		330.33	302.74	95.36

Source: Gastwirth, J. L., and Mahmoud, H. "An efficient robust nonparametric test for scale change for data from a gamma distribution." *Technometrics,* Vol. 28, No. 1, Feb. 1986, p. 83 (Table 2).

Applying the Concepts—Intermediate

14.27 Refer to the *Chance* (Fall 2002) study of a patent infringement case brought against Intel Corp., Exercise 9.20 (p. 450). Recall that the case rested on whether a patent witness's signature was written on top of key text in a patent notebook or under the key text. Using an X-ray beam, zinc measurements were taken at several spots on the notebook page. The zinc measurements for

three notebook locations—on a text line, on a witness line, and on the intersection of the witness and text line—are reproduced in the table.

PATENT

Text line:	.335	.374	.440			
Witness line:	.210	.262	.188	.329	.439	.397
Intersection:	.393	.353	.285	.295	.319	

a. Why might the Student's t-procedure you applied in Exercise 9.20 be inappropriate for analyzing this data?

b. Use a nonparametric test (at $\alpha = .05$) to compare the distribution of zinc measurements for the text line with the distribution for the intersection.

c. Use a nonparametric test (at $\alpha = .05$) to compare the distribution of zinc measurements for the witness line with the distribution for the intersection.

d. From the results, parts **b** and **c**, what can you infer about the mean zinc measurements at the three notebook locations?

14.28 Brood-parasitic birds. The term *brood-parasitic intruder* is used to describe birds that search for and lay eggs in the nests built by another bird species. For example, the brown-headed cowbird is known to be a brood parasite of the smaller-sized willow flycatcher. Ornithologists theorize that those flycatchers that recognize but do not vocally react to cowbird calls are more apt to defend their nests and less likely to be found and parasitized. In a study published in *The Condor* (May 1995), each of 13 active flycatcher nests was categorized as parasitized (if at least one cowbird egg was present) or nonparasitized. Cowbird songs were taped and played back while the flycatcher pairs were sitting in the nest prior to incubation. The vocalization rate (number of calls per minute) of each flycatcher pair was recorded. The data for the two groups of flycatchers are given in the table. Do the data suggest (at $\alpha = .05$) that the vocalization rates of parasitized flycatchers are higher than those of nonparasitized flycatchers?

COWBIRD

Parasitized	Not Parasitized
2.00	1.00
1.25	1.00
8.50	0
1.10	3.25
1.25	1.00
3.75	.25
5.50	

Source: Uyehara, J. C., and Narins, P. M. "Nest defense by Willow Flycatchers to brood-parasitic intruders." *The Condor,* Vol. 97, No. 2, May 1995, p. 364 (Figure 1).

14.29 Family involvement in homework. Refer to the impact study of the interactive Teachers Involve Parents in

Schoolwork (TIPS) program, Exercise 9.21 (p. 451). Recall that 128 middle school students were assigned to complete TIPS homework assignments, while 98 students were assigned traditional, non-interactive, homework assignments (ATIPS). At the end of the study, all students reported on the level of family involvement in their homework on a 5-point scale (0 = Never, 1 = Rarely, 2 = Sometimes, 3 = Frequently, 4 = Always). The data for the science, math, and language arts homework are saved in the **HWSTUDY** file. (The first five and last five observations in the data set are reproduced in the table.)

a. Why might a nonparametric test be the most appropriate test to apply in order to compare the levels of family involvement in homework assignments of TIPS and ATIPS students?

b. Conduct a nonparametric analysis to compare the science homework assignments of TIPS and ATIPS students. Use $\alpha = .05$.

c. Repeat part **a** for mathematics homework assignments.

d. Repeat part **a** for language arts homework assignments.

HWSTUDY

Homework Condition	Science	Math	Language
ATIPS	1	0	0
ATIPS	0	1	1
ATIPS	0	1	0
ATIPS	1	2	0
ATIPS	1	1	2
⋮	⋮	⋮	⋮
TIPS	2	3	2
TIPS	1	4	2
TIPS	2	4	2
TIPS	4	0	3
TIPS	2	0	1

Source: Van Voorhis, F. L. "Teachers' use of interactive homework and its effects on family involvement and science achievement of middle grade students." Paper presented at the annual meeting of the American Educational Research Association, Seattle, April 2001.

14.4 Comparing Two Populations: Paired Difference Experiment

Nonparametric techniques may also be employed to compare two probability distributions when a paired difference design is used. For example, consumer preferences for two competing products are often compared by having each of a sample of consumers rate both products. Thus, the ratings have been paired on each consumer. Here is an example of this type of experiment.

For some paper products, softness is an important consideration in determining consumer acceptance. One method of determining softness is to have judges give a sample of the products a softness rating. Suppose each of 10 judges is given a sample of two products that a company wants to compare. Each judge rates the softness of each product on a scale from 1 to 10, with higher ratings implying a softer product. The results of the experiment are shown in Table 14.4.

Since this is a paired difference experiment, we analyze the differences between the measurements (see Section 9.2). However, a nonparametric approach developed

SOFTPAPER

TABLE 14.4   Softness Ratings of Paper

Judge	Product A	Product B	Difference (A − B)	Absolute Value of Difference	Rank of Absolute Value
1	6	4	2	2	5
2	8	5	3	3	7.5
3	4	5	−1	1	2
4	9	8	1	1	2
5	4	1	3	3	7.5
6	7	9	−2	2	5
7	6	2	4	4	9
8	5	3	2	2	5
9	6	7	−1	1	2
10	8	2	6	6	10

$T_+ = \text{Sum of positive ranks} = 46$
$T_- = \text{Sum of negative ranks} = 9$

by Wilcoxon requires that we calculate the ranks of the absolute values of the differences between the measurements, that is, the ranks of the differences after removing any minus signs. *Note that tied absolute differences are assigned the average of the ranks they would receive if they were unequal but successive measurements.* After the absolute differences are ranked, the sum of the ranks of the positive differences of the original measurements, T_+, and the sum of the ranks of the negative differences of the original measurements, T_-, are computed.

We are now prepared to test the nonparametric hypotheses:

H_0: The probability distributions of the ratings for products A and B are identical.

H_a: The probability distributions of the ratings differ (in location) for the two products. (Note that this is a two-sided alternative and that it implies a two-tailed test.)

Test statistic: T = Smaller of the positive and negative rank sums T_+ and T_-

The smaller the value of T, the greater the evidence to indicate that the two probability distributions differ in location. The rejection region for T can be determined by consulting Table XIII in Appendix A (part of the table is shown in Table 14.5). This table gives a value T_0 for both one-tailed and two-tailed tests for each value of n, the number of matched pairs. For a two-tailed test with $\alpha = .05$, we will reject H_0 if $T \leq T_0$. You can see in Table 14.5 that the value of T_0 that locates the boundary of the rejection region for the judges' ratings for $\alpha = .05$ and $n = 10$ pairs of observations is 8. Thus, the rejection region for the test (see Figure 14.7) is

$$\text{Rejection region:} \quad T \leq 8 \quad \text{for } \alpha = .05$$

Since the smaller rank sum for the paper data, $T_- = 9$, does not fall within the rejection region, the experiment has not provided sufficient evidence to indicate

TABLE 14.5 **Reproduction of Part of Table XIII of Appendix A: Critical Values for the Wilcoxon Paired Difference Signed Rank Test**

One-Tailed	Two-Tailed	$n = 5$	$n = 6$	$n = 7$	$n = 8$	$n = 9$	$n = 10$
$\alpha = .05$	$\alpha = .10$	1	2	4	6	8	11
$\alpha = .025$	$\alpha = .05$		1	2	4	6	8
$\alpha = .01$	$\alpha = .02$			0	2	3	5
$\alpha = .005$	$\alpha = .01$				0	2	3
		$n = 11$	$n = 12$	$n = 13$	$n = 14$	$n = 15$	$n = 16$
$\alpha = .05$	$\alpha = .10$	14	17	21	26	30	36
$\alpha = .025$	$\alpha = .05$	11	14	17	21	25	30
$\alpha = .01$	$\alpha = .02$	7	10	13	16	20	24
$\alpha = .005$	$\alpha = .01$	5	7	10	13	16	19
		$n = 17$	$n = 18$	$n = 19$	$n = 20$	$n = 21$	$n = 22$
$\alpha = .05$	$\alpha = .10$	41	47	54	60	68	75
$\alpha = .025$	$\alpha = .05$	35	40	46	52	59	66
$\alpha = .01$	$\alpha = .02$	28	33	38	43	49	56
$\alpha = .005$	$\alpha = .01$	23	28	32	37	43	49
		$n = 23$	$n = 24$	$n = 25$	$n = 26$	$n = 27$	$n = 28$
$\alpha = .05$	$\alpha = .10$	83	92	101	110	120	130
$\alpha = .025$	$\alpha = .05$	73	81	90	98	107	117
$\alpha = .01$	$\alpha = .02$	62	69	77	85	93	102
$\alpha = .005$	$\alpha = .01$	55	61	68	76	84	92

Figure 14.7

Rejection Region for Paired
Difference Experiment

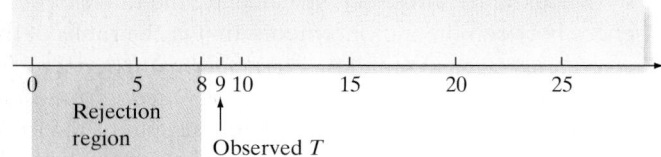

that the two paper products differ with respect to their softness ratings at the $\alpha = .05$ level.

Note that if a significance level of $\alpha = .10$ had been used, the rejection region would have been $T \leq 11$ and we would have rejected H_0. In other words, the samples do provide evidence that the probability distributions of the softness ratings differ at the $\alpha = .10$ significance level.

The **Wilcoxon signed rank test** is summarized in the following box. Note that the difference measurements are assumed to have a continuous probability distribution so that the absolute differences will have unique ranks. Although tied (absolute) differences can be assigned ranks by averaging, the number of ties should be small relative to the number of observations to ensure the validity of the test.

Wilcoxon Signed Rank Test for a Paired Difference Experiment

Let D_1 and D_2 represent the probability distributions for populations 1 and 2, respectively.

One-Tailed Test

H_0: D_1 and D_2 are identical

H_a: D_1 is shifted to the right of D_2 [or
  H_a: D_1 is shifted to the left of D_2]

Two-Tailed Test

H_0: D_1 and D_2 are identical

H_a: D_1 is shifted either to the left
  or to the right of D_2

Calculate the difference within each of the n matched pairs of observations. Then rank the absolute value of the n differences from the smallest (rank 1) to the highest (rank n) and calculate the rank sum T_- of the negative differences and the rank sum T_+ of the positive differences. [*Note:* Differences equal to 0 are eliminated, and the number n of differences is reduced accordingly.]

Test statistic:

T_-, the rank sum of the negative
differences [or T_+, the rank sum
of the positive differences]

Test statistic:

T, the smaller of T_+ or T_-

Rejection region:

$T_- \leq T_0$ [or $T_+ \leq T_0$]

Rejection region:

$T \leq T_0$

where T_0 is given in Table XIII in Appendix A.

Ties: Assign tied absolute differences the average of the ranks they would receive if they were unequal but occurred in successive order. For example, if the third-ranked and fourth-ranked differences are tied, assign both a rank of $(3 + 4)/2 = 3.5$.

Conditions Required for a Valid Signed Rank Test:

1. The sample of differences is randomly selected from the population of differences.

2. The probability distribution from which the sample of paired differences is drawn is continuous.

EXAMPLE 14.3

APPLYING THE SIGNED RANK TEST

Problem Suppose the police commissioner in a small community must chose between two plans for patrolling the town's streets. Plan A, the less expensive plan, uses voluntary citizen groups to patrol certain high-risk neighborhoods. In contrast, plan B would utilize police patrols. As an aid in reaching a decision, both plans are examined by 10 trained criminologists, each of whom is asked to rate the plans on a scale from 1 to 10. (High ratings imply a more effective crime prevention plan.) The city will adopt plan B (and hire extra police) only if the data provide sufficient evidence that criminologists tend to rate plan B more effective than plan A. The results of the survey are shown in Table 14.6. Do the data provide evidence at the $\alpha = .05$ level that the distribution of ratings for plan B lies above that for plan A?

⊙ **CRIMEPLAN**

TABLE 14.6 Effectiveness Ratings by 10 Qualified Crime Prevention Experts

| Crime Prevention Expert | Plan | | Difference | Rank of Absolute Difference |
	A	B	(A − B)	
1	7	9	−2	4.5
2	4	5	−1	2
3	8	8	0	(Eliminated)
4	9	8	1	2
5	3	6	−3	6
6	6	10	−4	7.5
7	8	9	−1	2
8	10	8	2	4.5
9	9	4	5	9
10	5	9	−4	7.5
			Positive rank sum = $T_+ = 15.5$	

Solution The null and alternative hypotheses are

H_0: The two probability distributions of effectiveness ratings are identical

H_a: The effectiveness ratings of the more expensive plan (B) tend to exceed those of plan A

Observe that the alternative hypothesis is one-sided (that is, we only wish to detect a shift in the distribution of the B ratings to the right of the distribution of A ratings) and therefore it implies a one-tailed test of the null hypothesis (see Figure 14.8). If the alternative hypothesis is true, the B ratings will tend to be larger than the paired A ratings, more negative differences in pairs will occur, T_- will be large, and T_+ will be small. Because Table XIII is constructed to give lower-tail values of T_0, we will use T_+ as the test statistic and reject H_0 for $T_+ \leq T_0$.

Figure 14.8

The Alternative Hypothesis for Example 14.3

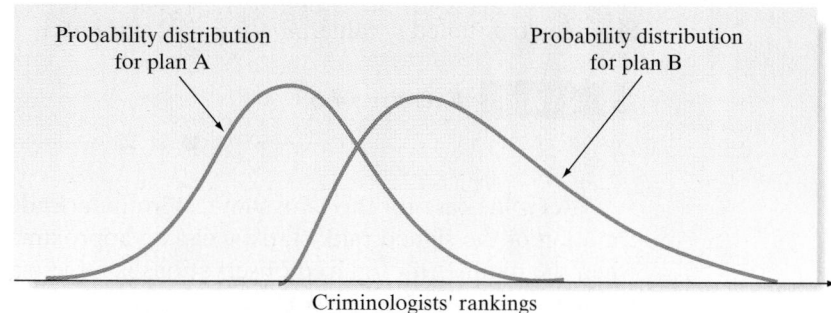

Probability distribution for plan A

Probability distribution for plan B

Criminologists' rankings

Figure 14.9

SPSS Printout for
Example 14.3

Wilcoxon Signed Ranks Test

Ranks

		N	Mean Rank	Sum of Ranks
A - B	Negative Ranks	6[a]	4.92	29.50
	Positive Ranks	3[b]	5.17	15.50
	Ties	1[c]		
	Total	10		

a. A < B

b. A > B

c. A = B

Test Statistics[b]

	A - B
Z	-.834[a]
Asymp. Sig. (2-tailed)	.404

a. Based on positive ranks.

b. Wilcoxon Signed Ranks Test

The differences in ratings for the pairs $(A - B)$ are shown in Table 14.6. Note that one of the differences equals 0. Consequently, we eliminate this pair from the ranking and reduce the number of pairs to $n = 9$. Looking in Table XIII, we have $T_0 = 8$ for a one-tailed test with $\alpha = .05$ and $n = 9$. Therefore, the test statistic and rejection region for the test are

$$\textit{Test statistic:} \quad T_+, \text{ the positive rank sum}$$

$$\textit{Rejection region:} \quad T_+ \leq 8$$

Summing the ranks of the positive differences from Table 14.6, we find $T_+ = 15.5$. Since this value exceeds the critical value, $T_0 = 8$, we conclude that this sample provides insufficient evidence at the $\alpha = .05$ level to support the alternative hypothesis. The commissioner *cannot* conclude that the plan utilizing police patrols tends to be rated higher than the plan using citizen volunteers. That is, on the basis of this study, extra police will not be hired.

An SPSS printout of the analysis, shown in Figure 14.9, confirms this conclusion. Both the test statistic and two-tailed *p*-value are highlighted on the printout. Since the one-tailed *p*-value, $.404/2 = .202$, exceeds $\alpha = .05$, we fail to reject H_0.

Now Work *Exercise 14.33*

■ ■ ■

As is the case for the rank sum test for independent samples, the sampling distribution of the signed rank statistic can be approximated by a normal distribution when the number *n* of paired observations is large (say $n \geq 25$). The large-sample *z*-test is summarized in the next box.

Wilcoxon Signed Ranks Test for Large Samples ($n \geq 25$)

Let D_1 and D_2 represent the probability distributions for populations 1 and 2, respectively.

One-Tailed Test

H_0: D_1 and D_2 are identical

H_a: D_1 is shifted to the right of D_2
   [or H_a: D_1 is shifted to the left of D_2]

Two-Tailed Test

H_0: D_1 and D_2 are identical

H_a: D_1 is shifted either to the
   left or to the right of D_2

$$\text{Test statistic:} \quad z = \frac{T_+ - [n(n + 1)/4]}{\sqrt{[n(n + 1)(2n + 1)]/24}}$$

Rejection region:

$z > z_\alpha$ [or $z < -z_\alpha$]

Rejection region:

$|z| > z_{\alpha/2}$

Assumptions: The sample size n is greater than or equal to 25. Differences equal to 0 are eliminated and the number n of differences is reduced accordingly. Tied absolute differences receive ranks equal to the average of the ranks they would have received had they not been tied.

Statistics in Action Revisited

Comparing the TCDD Levels in Fat and Plasma of Vietnam Vets

Medical researchers used the sample data in Table SIA14.1 to compare the TCDD levels in fat tissue and plasma for Vietnam veterans. Specifically, they wanted to determine if the distribution of TCDD levels in fat is shifted above or below the distribution of TCDD levels in plasma. To answer this question, first note that the data are collected as matched pairs—for each Vietnam vet in the sample the researchers recorded both the TCDD level in fat and in plasma. Therefore, the correct test to apply is the Wilcoxon signed rank test. The MINITAB printout for this analysis is shown in Figure SIA14.3.

Both the test statistic and p-value are highlighted on the printout. Since p-value = .073, at $\alpha = .05$ there is insufficient evidence to conclude that the distribution of TCDD levels in fat differs from the distribution of TCDD levels in plasma.

[*Note:* A histogram of the differences between the TCDD levels in fat and plasma, shown in Figure SIA14.4, illustrates that the sample differences are unlikely to have come from a normal population.]

Figure SIA14.3

MINITAB Signed Rank Test for TCDD Data

Wilcoxon Signed Rank Test: DIFF

```
Test of median = 0.000000 versus median not = 0.000000

                   N
                 for    Wilcoxon           Estimated
           N    Test   Statistic       P     Median
DIFF      20     19       140.0   0.073      1.175
```

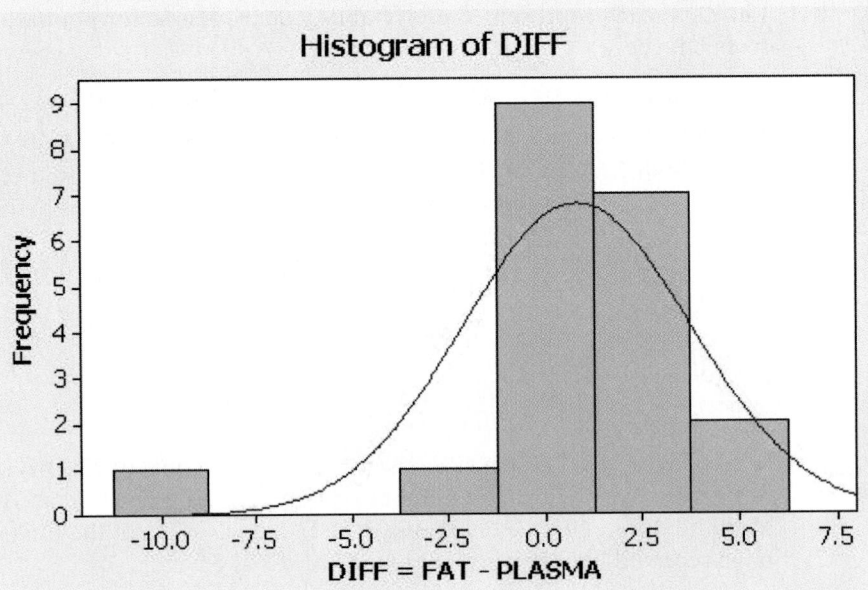

Figure SIA14.4
MINITAB Histogram for Differences in TCDD Levels

Exercises 14.30–14.43

Understanding the Principles

14.30 Explain the difference between the one- and two-tailed versions of the Wilcoxon signed rank test for the paired difference experiment.

14.31 In order to conduct the Wilcoxon signed rank test, why do we need to assume the probability distribution of differences is continuous?

Learning the Mechanics

14.32 Specify the test statistic and the rejection region for the Wilcoxon signed rank test for the paired difference design in each of the following situations:

a. H_0: Two probability distributions, A and B, are identical
H_a: Probability distribution for population A is shifted to the right or left of the probability distribution for population B
$n = 20, \alpha = .10$

b. H_0: Two probability distributions, A and B, are identical
H_a: Probability distribution for population A is shifted to the right of the probability distribution for population B
$n = 39, \alpha = .05$

c. H_0: Two probability distributions, A and B, are identical

H_a: Probability distribution for population A is shifted to the left of the probability distribution for population B
$n = 7, \alpha = .005$

14.33 Suppose you want to test a hypothesis that two treatments, A and B, are equivalent against the alternative hypothesis that the responses for A tend to be larger than those for B. You plan to use a paired difference experiment and to analyze the resulting data using the Wilcoxon signed rank test.

a. Specify the null and alternative hypotheses you would test.

b. Suppose the paired difference experiment yielded the data in the table. Conduct the test of part **a**. Test using $\alpha = .025$.

LM14_33

Pair	A	B	Pair	A	B
1	54	45	6	77	75
2	60	45	7	74	63
3	98	87	8	29	30
4	43	31	9	63	59
5	82	71	10	80	82

14.34 Suppose you wish to test a hypothesis that two treatments, A and B, are equivalent against the alternative that the responses for A tend to be larger than those for B.

a. If the number of pairs equals 25, give the rejection region for the large-sample Wilcoxon signed rank test for $\alpha = .05$.

b. Suppose that $T_+ = 273$. State your test conclusions.

c. Find the p-value for the test and interpret it.

14.35 A paired difference experiment with $n = 30$ pairs yielded $T_+ = 354$.

a. Specify the null and alternative hypotheses that should be used in conducting a hypothesis test to determine whether the probability distribution for population A is located to the right of that for population B.

b. Conduct the test of part **a** using $\alpha = .05$.

c. What is the approximate p-value of the test of part **b**?

d. What assumptions are necessary to ensure the validity of the test you performed in part **b**?

14.36 A random sample of nine pairs of measurements is shown in the table.

⊙ **LM14_36**

Pair	Sample Data from Population 1	Sample Data from Population 2
1	8	7
2	10	1
3	6	4
4	10	10
5	7	4
6	8	3
7	4	6
8	9	2
9	8	4

a. Use the Wilcoxon signed rank test to determine whether the data provide sufficient evidence to indicate that the probability distribution for population 1 is shifted to the right of the probability distribution for population 2. Test using $\alpha = .05$.

b. Use the Wilcoxon signed rank test to determine whether the data provide sufficient evidence to indicate that the probability distribution for population 1 is shifted either to the right or to the left of the probability distribution for population 2. Test using $\alpha = .05$.

Applying the Concepts—Basic

14.37 Students' attitudes towards parents. Refer to the *Journal of Genetic Psychology* (March 1998) study of attitudes of male college students toward their parents, Exercise 9.42 (p. 466). Recall that each of 13 students was asked to complete the following statement: My relationship with my father (mother) can best be described as (1) Awful, (2) Poor, (3) Average, (4) Good, or (5) Great. The study data are reproduced in the next table. The researchers want to compare male students' attitudes toward their fathers with their attitudes toward their mothers.

a. Why is a nonparametric test applicable for analyzing this data set?

b. Specify H_0 and H_a for the test.

c. Compute and rank the differences between the father and mother attitudes.

d. Sum the ranks of the positive and negative differences.

e. Find the rejection region for the test at $\alpha = .05$.

f. Give the conclusion of the test in the context of the problem. Compare your answer to the test results of Exercise 9.42b.

⊙ **FMATTITUDES**

Student	Attitude toward Father	Attitude toward Mother
1	2	3
2	5	5
3	4	3
4	4	5
5	3	4
6	5	4
7	4	5
8	2	4
9	4	5
10	5	4
11	4	5
12	5	4
13	3	3

Source: Adapted from Vitulli, W. F., and Richardson, D. K. "College student's attitudes toward relationships with parents: A five-year comparative analysis." *Journal of Genetic Psychology*, Vol. 159, No. 1, (March 1998), pp. 45–52.

⊙ **CRASH**

14.38 NHTSA new car crash tests. Refer to the National Highway Traffic Safety Administration (NHTSA) crash test data for new cars saved in the **CRASH** file. In Exercise 9.40 (p. 465) you compared the chest injury ratings of drivers and front-seat passengers using the Student's t-procedure for matched pairs. Suppose you want to make the comparison for only those cars that have a driver's star rating of 5 stars (the highest rating). The data for these 18 cars are listed in the table. Now consider analyzing these data using the Wilcoxon signed ranks test.

a. State the null and alternative hypothesis.

b. Use a statistical software package to find the signed ranks test statistic.

c. Give the rejection region for the test using $\alpha = .01$.

⊙ **CRASH5**

	Chest Injury Rating			Chest Injury Rating	
Car	Driver	Passenger	Car	Driver	Passenger
1	42	35	10	36	37
2	42	35	11	36	37
3	34	45	12	43	58
4	34	45	13	40	42
5	45	45	14	43	58
6	40	42	15	37	41
7	42	46	16	37	41
8	43	58	17	44	57
9	45	43	18	42	42

d. State the conclusion in practical terms. Report the *p*-value of the test.

14.39 Aggressiveness of twins. Twelve sets of identical twins are given psychological tests to determine whether the firstborn of the twins tends to be more aggressive than the secondborn. The test scores are shown in the table, where the higher score indicates greater aggressiveness.

a. For each twin set, compute the difference between the aggressiveness scores.

b. Rank the absolute values of the differences.

c. Use the ranks to find the signed rank statistic.

d. Do the data provide sufficient evidence (at $\alpha = .05$) to indicate that the firstborn of a pair of twins is more aggressive than the other?

AGGTWINS

Set	Firstborn	Secondborn
1	86	88
2	71	77
3	77	76
4	68	64
5	91	96
6	72	72
7	77	65
8	91	90
9	70	65
10	71	80
11	88	81
12	87	72

Applying the Concepts—Intermediate

14.40 Thematic atlas topics. The regional atlas is an important educational resource that is updated on a periodic basis. One of the most critical aspects of a new atlas design is its thematic content. In a survey of atlas

ATLAS

Theme	Rankings	
	High School Teachers	Geography Alumni
Tourism	10	2
Physical	2	1
Transportation	7	3
People	1	6
History	2	5
Climate	6	4
Forestry	5	8
Agriculture	7	10
Fishing	9	7
Energy	2	8
Mining	10	11
Manufacturing	12	12

Source: Keller, C. P., et al. "Planning the next generation of regional atlases: Input from educators." *Journal of Geography,* Vol. 94, No. 3, May/June 1995, p. 413 (Table 1).

users (*Journal of Geography*, May/June 1995), a large sample of high school teachers in British Columbia ranked 12 thematic atlas topics for usefulness. The consensus rankings of the teachers (based on the percentage of teachers who responded they "would definitely use" the topic) are given in the accompanying table. These teacher rankings were compared to the rankings a group of university geography alumni made three years earlier. Compare the distributions of theme rankings for the two groups with an appropriate nonparametric test. Use $\alpha = .05$. Interpret the results practically.

14.41 Treatment for tendon pain. Refer to the *British Journal of Sports Medicine* (Feb. 1, 2004) study of Chronic Achilles tendon pain, Exercise 10.62 (p. 557). Recall that each in a sample of 25 patients with chronic Achilles tendinosis was treated with heavy load eccentric calf muscle training. Tendon thickness (in millimeters) was measured both before and following treatment of each patient. These experimental data are reproduced in the table. Use a nonparametric test to determine if the treatment for tendonitis tends to reduce the thickness of tendons? Test using $\alpha = .10$.

TENDON

Patient	Before Thickness (millimeters)	After Thickness (millimeters)
1	11.0	11.5
2	4.0	6.4
3	6.3	6.1
4	12.0	10.0
5	18.2	14.7
6	9.2	7.3
7	7.5	6.1
8	7.1	6.4
9	7.2	5.7
10	6.7	6.5
11	14.2	13.2
12	7.3	7.5
13	9.7	7.4
14	9.5	7.2
15	5.6	6.3
16	8.7	6.0
17	6.7	7.3
18	10.2	7.0
19	6.6	5.3
20	11.2	9.0
21	8.6	6.6
22	6.1	6.3
23	10.3	7.2
24	7.0	7.2
25	12.0	8.0

Source: Ohberg, L. et al. "Eccentric training in patients with chronic Achilles tendinosis: normalized tendon structure and decreased thickness at follow up," *British Journal of Sports Medicine,* Vol. 38, No. 1, Feb. 1, 2004 (Table 2).

14.42 Neurological impairment of POWs. Eleven prisoners of war during the war in Croatia were evaluated for neurological impairment after their release from a Serbian detention camp. (*Collegium Antropologicum*, June 1997.) All 11 released POWs received blows to the head and neck and/or loss of consciousness during imprisonment. Neurological impairment was assessed by measuring the amplitude of the visual evoked potential (VEP) in both eyes at two points in time: 157 days and 379 days after release from prison. (The higher the VEP value, the greater the neurological impairment.) The data for the 11 POWs are shown in the table. Determine whether the VEP measurements of POWs 379 days after their release tend to be greater than the VEP measurements of POWs 157 days after their release. Test using $\alpha = .05$.

⚙ **POWVEP**

POW	157 Days After Release	379 Days After Release
1	2.46	3.73
2	4.11	5.46
3	3.93	7.04
4	4.51	4.73
5	4.96	4.71
6	4.42	6.19
7	1.02	1.42
8	4.30	8.70
9	7.56	7.37
10	7.07	8.46
11	8.00	7.16

Source: Vrca, A., et al. "The use of visual evoked potentials to follow-up prisoners of war after release from detention camps." *Collegium Antropologicum*, Vol. 21, No. 1, June 1997, p. 232. (Data simulated from information provided in Table 3.)

Applying the Concepts—Advanced

14.43 Bowlers' hot hand. Is the probability of a bowler rolling a strike higher after he has thrown four consecutive strikes? An investigation into the phenomenon of a "hot hand" in bowling was published in *The American Statistician* (Feb. 2004). Frame-by-frame results were collected on 43 professional bowlers from the 2002–2003 Professional Bowlers Association (PBA) season. For each bowler, the researchers calculated the proportion of strikes rolled after bowling four consecutive strikes and the strike proportion after bowling four consecutive nonstrikes. The data for four of the 43 bowlers, saved in the **HOTBOWLER** file, are shown below.

⚙ **HOTBOWLER**

Bowler	Proportion of Strikes	
	After 4 Strikes	After 4 Nonstrikes
Paul Fleming	.683	.432
Bryon Smith	.684	.400
Mike DeVaney	.632	.421
Dave D'Entremont	.610	.529

Source: Dorsey-Palmateer, R., & Smith, G. "Bowlers' Hot Hands," *The American Statistician*, Vol. 58, No. 1, Feb. 2004 (Table 3).

a. Do the data for the sample of four bowlers provide support for the "hot hand" theory in bowling? Explain.

b. When the data for all 43 bowlers is used, the *p*-value for the hypothesis test is approximately 0. Interpret this result.

14.5 Comparing Three or More Populations: Completely Randomized Design

In Chapter 10 we used an analysis of variance and the *F*-test to compare the means of *k* populations (treatments) based on random sampling from populations that were normally distributed with a common variance σ^2. We now present a nonparametric technique—the **Kruskal-Wallis *H*-test**—for comparing the populations that requires no assumptions concerning the population probability distributions.

Suppose a health administrator wants to compare the unoccupied bed space for three hospitals located in the same city. She randomly selects 10 different days from the records of each hospital and lists the number of unoccupied beds for each day (see Table 14.7). Because the number of unoccupied beds per day may occasionally be quite large, it is conceivable that the population distributions of data may be skewed to the right and that this type of data may not satisfy the assumptions necessary for a parametric comparison of the population means. We therefore use a nonparametric analysis and base our comparison on the rank sums for the three sets of sample data. Just as with two independent samples (Section 14.2), the ranks are computed for each observation according to the relative magnitude of the measurements *when the data for all the samples are combined* (see Table 14.7). Ties are treated as they were for the Wilcoxon rank sum and signed rank tests by assigning the average value of the ranks to each of the tied observations.

⊙ **HOSPBEDS**

TABLE 14.7 Number of Available Beds

Hospital 1		Hospital 2		Hospital 3	
Beds	Rank	Beds	Rank	Beds	Rank
6	5	34	25	13	9.5
38	27	28	19	35	26
3	2	42	30	19	15
17	13	13	9.5	4	3
11	8	40	29	29	20
30	21	31	22	0	1
15	11	9	7	7	6
16	12	32	23	33	24
25	17	39	28	18	14
5	4	27	18	24	16
$R_1 = 120$		$R_2 = 210.5$		$R_3 = 134.5$	

We test

H_0: The probability distributions of the number of unoccupied beds are the same for all three hospitals

H_a: At least two of the three hospitals have probability distributions of number of unoccupied beds that differ in location

If we denote the three sample rank sums by R_1, R_2 and R_3, the test statistic is given by

$$H = \frac{12}{n(n + 1)} \sum \frac{R_j^2}{n_j} - 3(n + 1)$$

where n_j is the number of measurements in the jth sample and n is the total sample size $(n = n_1 + n_2 + \cdots + n_k)$. For the data in Table 14.7, we have $n_1 = n_2 = n_3 = 10$ and $n = 30$. The rank sums are $R_1 = 120$, $R_2 = 210.5$, and $R_3 = 134.5$. Thus,

$$H = \frac{12}{30(31)} \left[\frac{(120)^2}{10} + \frac{(210.5)^2}{10} + \frac{(134.5)^2}{10} \right] - 3(31)$$
$$= 99.097 - 93 = 6.097$$

The H-statistic measures the extent to which the k samples differ with respect to their relative ranks. This is more easily seen by writing H in an alternative but equivalent form:

$$H = \frac{12}{n(n + 1)} \sum n_j (\overline{R}_j - \overline{R})^2$$

where $\overline{R}_j$ is the mean rank corresponding to sample j and $\overline{R}$ is the mean of all the ranks [that is, $\overline{R} = \frac{1}{2}(n + 1)$]. Thus, the H-statistic is 0 if all samples have the same mean rank and becomes increasingly large as the distance between the sample mean ranks grows.

If the null hypothesis is true, the distribution of H in repeated sampling is approximately a χ^2 (chi-square) distribution. This approximation for the sampling distribution of H is adequate as long as one of the k sample sizes exceeds 5. (See the references for more detail.) The degrees of freedom corresponding to the approximate sampling distribution of H will always be $(k - 1)$—one less than the number of probability distributions being compared. Because large values of H support the

Figure 14.10

Rejection Region for the Comparison of Three Probability Distributions

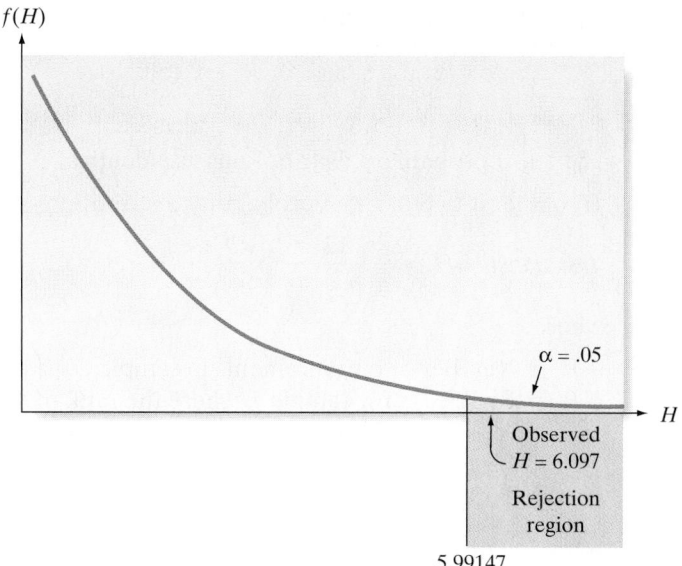

alternative hypothesis that the populations have different probability distributions, the rejection region for the test is located in the upper tail of the χ^2 distribution.

For the data of Table 14.7, the approximate distribution of the test statistic H is χ^2 with $(k - 1) = 2$ df. To determine how large H must be before we will reject the null hypothesis, we consult Table VII in Appendix A. For $\alpha = .05$ and df $= 2$, $\chi^2_{.05} = 5.99147$. Therefore, we can reject the null hypothesis that the three probability distributions are the same if

$$H > 5.99147$$

The rejection region is pictured in Figure 14.10. Since the calculated $H = 6.097$ exceeds the critical value of 5.99147, we conclude that at least one of the three hospitals tends to have a larger number of unoccupied beds than the others.

Now Work *Exercise 14.46*

The same conclusion can be reached from a computer printout of the analysis. The average ranks test statistic, and p-value of the nonparametric test are highlighted on the MINITAB printout in Figure 14.11. Since $\alpha = .05$ exceeds p-value $= .047$, there is sufficient evidence to reject H_0.

The Kruskal-Wallis H-test for comparing more than two probability distributions is summarized in the next box. Note that we can use the Wilcoxon rank sum test of Section 14.2 to compare a pair of populations (selected a priori) if the

Figure 14.11

MINITAB Kruskal-Wallis Test Comparing Three Hospitals

Kruskal-Wallis Test: BEDS versus HOSPITAL

```
Kruskal-Wallis Test on BEDS

HOSPITAL   N   Median   Ave Rank      Z
1         10    15.50      12.0   -1.54
2         10    31.50      21.1    2.44
3         10    18.50      13.5   -0.90
Overall   30               15.5

H = 6.10   DF = 2   P = 0.047
H = 6.10   DF = 2   P = 0.047   (adjusted for ties)
```

Kruskal-Wallis H-test supports the alternative hypothesis that at least two of the probability distributions differ.*

Kruskal-Wallis H-Test for Comparing k Probability Distributions

H_0: The p probability distributions are identical

H_a: At least two of the k probability distributions differ in location

Test statistic: $H = \dfrac{12}{n(n+1)} \sum \dfrac{R_j^2}{n_j} - 3(n+1)$

where

n_j = Number of measurements in sample j
R_j = Rank sum for sample j, where the rank of each measurement is computed according to its relative magnitude in the totality of data for the p samples
n = Total sample size = $n_1 + n_2 + \cdots + n_k$

Rejection region: $H > \chi_\alpha^2$ with $(k-1)$ degrees of freedom.

Ties: Assign tied measurements the average of the ranks they would receive if they were unequal but occurred in successive order. For example, if the third-ranked and fourth ranked measurements are tied, assign both a rank of $(3+4)/2 = 3.5$. The number of ties should be small relative to the total number of observations.

Conditions Required for the Valid Application of the Kruskal-Wallis *Test*

1. The k samples are random and independent.

2. There are 5 or more measurements in each sample.

3. The k probability distributions from which the samples are drawn are continuous.

Exercises 14.44–14.54

Understanding the Principles

14.44 Which of the following would lead you to conclude that the treatments in a balanced completely randomized design have distributions that differ in location?
a. Rank sums for all treatments are about equal.
b. Rank sum for one treatment is much larger than the rank sum for all other treatments.

14.45 Under what circumstances does the χ^2 distribution provide an appropriate characterization of the sampling distribution of the Kruskal-Wallis H statistic?

Learning the Mechanics

14.46 Data were collected from three populations, A, B, and
NW C, using a completely randomized design. The following describes the sample data:

$$n_A = n_B = n_C = 15$$
$$R_A = 235 \quad R_B = 439 \quad R_C = 361$$

a. Specify the null and alternative hypotheses that should be used in conducting a test of hypothesis to determine whether the probability distributions of populations A, B, and C differ in location.
b. Conduct the test of part **a**. Use $\alpha = .05$.
c. What is the approximate p-value of the test of part **b**?
d. Calculate the mean rank for each sample and compute H according to the formula that utilizes these means. Verify that this formula yields the same value of H that you obtained in part **b**.

14.47 Suppose you want to use the Kruskal-Wallis H-test to compare the probability distributions of three populations. The following are independent random samples selected from the three populations:

I:	34, 56, 65, 59, 82, 70, 45
II:	24, 18, 27, 41, 34, 42, 33
III:	72, 101, 91, 76, 80, 75

---

*A method similar to the multiple comparisons procedure of Chapter 10 can be used to rank the treatment medians. This nonparametric multiple comparisons of medians will control the experimentwise error rate selected by the analyst. Consult the references [Daniel (1990) and Dunn (1964)] for details.

a. What experimental design was used?
b. Specify the null and alternative hypotheses you would test.
c. Specify the rejection region you would use for your hypothesis test at $\alpha = .01$.
d. Conduct the test at $\alpha = .01$.

Applying the Concepts—Basic

14.48 Office rental growth rates. Real estate market cycles are commonly divided into four phases that are based on the rate of change of the demand and supply of properties: Phase I, Recovery; Phase II, Expansion; Phase III, Hypersupply; and Phase IV, Recession. Glenn Mueller of Johns Hopkins University studied the office market cycles of U.S. real estate markets. (*Journal of Real Estate Research* (Jul/Aug 1999.) For each of the four market cycles, office rental growth rates (i.e., growth rates for asking rents) were measured for a sample of six different real estate markets. These data (in percentages) are presented in the table. A MINITAB printout of a Kruskal-Wallis analysis of the data is also provided at the bottom of the page.

MKTCYCLE

Phase I	Phase II	Phase III	Phase IV
2.7	10.5	6.1	−1.0
−1.0	11.5	1.2	6.2
1.1	9.4	11.4	−10.8
3.4	12.2	4.4	2.0
4.2	8.6	6.2	−1.1
3.5	10.9	7.6	−2.3

Source: Adapted from Mueller, G. R. "Real estate rental growth rates at different points in the physical market cycle." *Journal of Real Estate Research*, Vol. 18, No. 1 (Jul/Aug 1999), pp. 131–150.

a. Specify the null hypothesis for the Kruskal-Wallis test.
b. Find the mean rank for each market cycle phase on the printout. Multiply this value by the sample size, 6, to obtain the rank sum for each phase. Verify these rank sums by ranking the data.
c. Find the Kruskal-Wallis test statistic on the printout. Substitute the rank sums from part **a** into the appropriate formula to verify this value.
d. Give the rejection region for the test at $\alpha = .05$.
e. Is there sufficient evidence to conclude that the distributions of office rental growth rates differ among the four market cycle phases?

f. Locate the *p*-value of the test on the printout and interpret this result.
g. What are the advantages and disadvantages of applying the Kruskal-Wallis *H*-test rather than the ANOVA *F*-test of Chapter 10?

14.49 Effect of scopolamine on memory. Refer to the *Behavioral Neuroscience* (Feb. 2004) study of the drug scopolamine's effects on memory for word-pair associates, Exercise 10.33 (p. 536). Recall that a completely randomized design with three groups was used— Group 1 subjects were injected with scopolamine, group 2 subjects were injected with a placebo, and group 3 subjects were not given any drug. The response variable was number of word pairs recalled. The data for all 28 subjects are reproduced below.

SCOPOLAMINE

Group 1 (Scopolamine):	5 8 8 6 6 6 6 8 6 4 5 6
Group 2 (Placebo):	8 10 12 10 9 7 9 10
Group 3 (No drug):	8 9 11 12 11 10 12 12

a. Rank the data for all 28 observations from smallest to largest.
b. Sum the ranks of the observations from group 1.
c. Sum the ranks of the observations from group 2.
d. Sum the ranks of the observations from group 3.
e. Use the rank sums, parts **b–d**, to compute the Kruskal-Wallis *H*-statistic.
f. Carry out the Kruskal-Wallis nonparametric test (at $\alpha = .05$) to compare the distributions of number of word pairs recalled for the three groups.
g. Recall from Exercise 10.33 that the researchers theorized that group 1 subjects will tend to recall the fewest number of words. Use the Wilcoxon rank sum test to compare the word recall distributions of group 1 and group 2. (Use $\alpha = .05$.)

14.50 Biting rates of flies. Refer to the *Journal of the American Mosquito Control Association* study of biting flies, Exercise 14.11 (p. 829). The effect of wind speeds in kilometers per hour (kph) on the biting rate of the fly on Stanbury Island, Utah, was investigated by exposing samples of volunteers to one of six wind speed conditions. The distributions of the biting rates for the six wind speeds were compared using the Kruskal-Wallis test. The rank sums of the biting rates for the six conditions are shown in the table on p. 852.

```
Kruskal-Wallis Test on GRWRATE

PHASE     N    Median   Ave Rank      Z
1         6    3.050        8.8    -1.50
2         6   10.700       20.8     3.33
3         6    6.150       14.8     0.90
4         6   -1.050        5.7    -2.73
Overall  24                12.5

H = 16.23   DF = 3   P = 0.001
H = 16.25   DF = 3   P = 0.001   (adjusted for ties)
```

Wind Speed (kph)	Number of Volunteers (n_j)	Rank Sum of Biting Rates (R_j)
<1	11	1,804
1–2.9	49	6,398
3–4.9	62	7,328
5–6.9	39	4,075
7–8.9	35	2,660
9–20	21	1,388
Totals	**217**	**23,653**

Source: Strickman, D., et al. "Meteorological effects on the biting activity of *Leptoconops americanus* (Diptera: Ceratopogonidae)." *Journal of the American Mosquito Control Association*, Vol. II, No. 1, Mar. 1995, p. 17 (Table 1).

    a. The researchers reported the test statistic as $H = 35.2$. Verify this value.

    b. Find the rejection region for the test using $\alpha = .01$.

    c. Make the proper conclusions.

    d. The researchers reported that the *p*-value of the test is less than .01. Does this value support your inference in part **c**? Explain.

Applying the Concepts—Intermediate

14.51 Social reinforcement of exercise. Two University of Georgia researchers studied the effect of social reinforcement on exercise duration in adolescents with moderate mental retardation. (*Clinical Kinesiology*, Spring 1995.) Eleven adolescents with IQs ranging from 32 to 61 were divided into two groups. All participated in a 6-week exercise program. Group A (4 subjects) received verbal and social reinforcement during the program, while Group B (7 subjects) received verbal and social reinforcement and kept a self-record of their individual performances. The researchers theorized that Group B subjects would exercise for longer periods than Group A. Upon completion of the exercise program, all 11 subjects participated in a run/walk "race" in which the goal was to complete as many laps as possible during a 15-minute period. The number of laps completed (to the nearest quarter lap) was used as a measure of exercise duration.

    a. Specify the null and alternative hypotheses for a nonparametric analysis of the data.

    b. The Kruskal-Wallis H-test was applied to the data. The researchers reported the test statistic as $H = 5.1429$ and the observed significance level of the test as *p*-value $= .0233$. Interpret these results.

    c. Are the assumptions for the test, part **b**, satisfied? If not, propose an alternative nonparametric method for the analysis.

14.52 Identifying mental disturbances. An experiment was conducted to determine whether a test designed to identify a certain form of mental disturbance could be easily interpreted by a person with little psychological training. Thirty judges were selected to review the results of 100 tests, half of which were given to disturbed patients and half to nondisturbed people. Of the 30 judges chosen, 10 were staff members of a mental hospital, 10 were

trainees at the hospital, and 10 were undergraduate psychology majors. The results in the table give the number of the 100 tests correctly classified by each judge. Do the data provide sufficient evidence (at $\alpha = .05$) of a difference in the probability distributions of the number of correct identifications among the three types of judges?

MENTALID

Staff		Trainees		Undergraduates	
78	76	80	69	65	74
79	86	75	81	70	80
85	88	72	76	74	73
93	84	68	72	78	75
90	81	75	76	68	73

14.53 The "name game." Refer to the *Journal of Experimental Psychology-Applied* (June 2000) study of different methods of learning names, Exercise 10.34 (p. 536). Recall that three groups of students used a different method to learn the names of the other students in their group. Group 1 used the "simple name game," Group 2 used the "elaborate name game," and Group 3 used "pairwise introductions." The accompanying table lists the percentage of names recalled (after one year) for each student respondent.

NAMEGAME

Simple Name Game:

24	43	38	65	35	15	44	44	18	27	0	38	50	31
7	46	33	31	0	29	0	0	52	0	29	42	39	26
51	0	42	20	37	51	0	30	43	30	99	39	35	19
24	34	3	60	0	29	40	40						

Elaborate Name Game:

39	71	9	86	26	45	0	38	5	53	29	0	62	0
1	35	10	6	33	48	9	26	83	33	12	5	0	0
25	36	39	1	37	2	13	26	7	35	3	8	55	50

Pairwise Intro:

5	21	22	3	32	29	32	0	4	41	0	27	5	9
66	54	1	15	0	26	1	30	2	13	0	2	17	14
5	29	0	45	35	7	11	4	9	23	4	0	8	2
18	0	5	21	14									

Source: Morris, P. E., and Fritz, C. O. "The name game: Using retrieval practice to improve the learning of names." *Journal of Experimental Psychology-Applied*, Vol. 6, No. 2, June 2000 (data simulated from Figure 1).

    a. Consider an analysis of variance *F*-test to determine whether the mean percentages of names recalled differ for the three name retrieval methods. Demonstrate that the ANOVA assumptions are likely to be violated.

    b. Use an alternative nonparametric test to compare the distributions of the percentages of names recalled for the three name retrieval methods. Use $\alpha = .05$.

14.54 Forums for tax litigation. In disagreements between the Internal Revenue Service (IRS) and the taxpayer that end up in litigation, taxpayers are permitted by law to choose the court forum. Three trial courts are available: (1) U.S. Tax Court, (2) Federal District Court, and (3) U.S. Claims Court. Each court possesses different requirements and restrictions that make the choice an important one for the taxpayer. A study of taxpayers' choice of forum in litigating tax issues was published in the *Journal of Applied Business Research* (Fall 1996). In a random sample of 161 litigated tax disputes, the researchers measured the taxpayer's choice of forum (Tax, District, or Claims Court) and tax deficiency (i.e., the disputed amount, in dollars). One of the objectives of the study was to determine those factors taxpayers consider important in their choice of forum. If tax deficiency (called DEF by the researchers) is an important factor, then the mean DEF values for the three tax courts should be significantly different.

a. The researchers applied a nonparametric test rather than a parametric test to compare the DEF distributions of the three tax litigation forums. Give a plausible reason for their choice.

b. What nonparametric test is appropriate for this analysis? Explain.

c. The table below summarizes the data analyzed by the researchers. Use the information in the table to compute the appropriate test statistic.

d. The observed significance level of the test was reported as *p*-value = .0037. Fully interpret this result.

Court Selected by Taxpayer	Sample Size	Sample Mean DEF	Rank Sum of DEF Values
Tax	67	$ 80,357	5,335
District	57	74,213	3,937
Claims	37	184,648	3,769

Source: Billings, B. A., Green, B. P., and Volz, W. H. "Selection of forum for litigated tax issues." *Journal of Applied Business Research*, Vol. 12, No. 4, Fall, 1996, p. 38 (Table 2).

14.6 Comparing Three or More Populations: Randomized Block Design

In Section 10.4 we employed an analysis of variance to compare p population (treatment) means when the data were collected using a randomized block design. The *Friedman F_r-test* provides another method for testing to detect a shift in location of a set of p populations.* Like other nonparametric tests, it requires no assumptions concerning the nature of the populations other than the capacity of individual observations to be ranked.

In Section 14.2, we gave an example in which a completely randomized design was used to compare the reaction times of subjects under the influence of one of two drugs. When the effect of a drug is short-lived (there is no carryover effect) and when the drug effect varies greatly from person to person, it may be useful to employ a *randomized block design*. Using the subjects as blocks, we would hope to eliminate the variability among subjects and thereby increase the amount of information in the experiment. Suppose that three drugs, A, B, and C, are to be compared using a randomized block design. Each of the three drugs is administered to the *same subject* with suitable time lags between the three doses. The order in which the drugs are administered is randomly determined for each subject. Thus, one drug would be administered to a subject and its reaction time would be noted; then after a sufficient length of time, the second drug administered; etc.

Suppose six subjects are chosen and that the reaction times for each drug are as shown in Table 14.8. To compare the three drugs, we rank the observations within each subject (block) and then compute the rank sums for each of the drugs (treatments). Tied observations within blocks are handled in the usual manner by assigning the average value of the ranks to each of the tied observations.

*The Friedman F_r-test was developed by the Nobel prize–winning economist Milton Friedman.

◉ **REACTION2**

TABLE 14.8　Reaction Time for Three Drugs

Subject	Drug A	Rank	Drug B	Rank	Drug C	Rank
1	1.21	1	1.48	2	1.56	3
2	1.63	1	1.85	2	2.01	3
3	1.42	1	2.06	3	1.70	2
4	2.43	2	1.98	1	2.64	3
5	1.16	1	1.27	2	1.48	3
6	1.94	1	2.44	2	2.81	3
		$R_1 = 7$		$R_2 = 12$		$R_3 = 17$

The null and alternative hypotheses are

H_0: The populations of reaction times are identically distributed for all three drugs

H_a: At least two of the drugs have probability distributions of reaction times that differ in location

The **Friedman F_r-statistic,** which is based on the rank sums for each treatment, is

$$F_r = \frac{12}{bk(k+1)} \sum R_j^2 - 3b(k+1)$$

where b is the number of blocks, k is the number of treatments, and R_j is the jth rank sum. For the data in Table 14.8,

$$F_r = \frac{12}{(6)(3)(4)}[(7)^2 + (12)^2 + (17)^2] - 3(6)(4) = 80.33 - 72 = 8.33$$

The Friedman F_r-statistic measures the extent to which the p samples differ with respect to their relative ranks within the blocks. This is more easily seen by writing F_r in an alternative, but equivalent, form:

$$F_r = \frac{12}{bk(k+1)} \sum b(\overline{R}_j - \overline{R})^2$$

where $\overline{R}_j$ is the mean rank corresponding to treatment j and $\overline{R}$ is the mean of all the ranks (i.e., $\overline{R} = \frac{1}{2}(k+1)$]. Thus, the F_r-statistic is 0 if all treatments have the same mean rank and becomes increasingly large as the distance between the sample mean ranks grows.

As for the Kruskal-Wallis H statistic, the Friedman F_r-statistic has approximately a χ^2 sampling distribution with $(k-1)$ degrees of freedom. Empirical results show the approximation to be adequate if either b (the number of blocks) or k (the number of treatments) exceeds 5. The Friedman F_r-test for a randomized block design is summarized in the next box.

> **Friedman F_r-*Test* for a Randomized Block Design**
>
> H_0: The probability distributions for the p treatments are identical
>
> H_a: At least two of the probability distributions differ in location
>
> *Test statistic:*　$F_r = \dfrac{12}{bk(k+1)} \sum R_j^2 - 3b(k+1)$
>
> where
>
> b = Number of blocks
> k = Number of treatments

> R_j = Rank sum of the jth treatment, where the rank
> of each measurement is computed relative to its position
> *within its own block*
>
> *Rejection region:* $F_r > \chi_\alpha^2$ with $(k - 1)$ degrees of freedom
>
> *Ties:* Assign tied measurements within a block the average of the ranks they
> would receive if they were unequal but occurred in successive order. For example,
> if the third-ranked and fourth-ranked measurements are tied, assign each a rank
> of $(3 + 4)/2 = 3.5$. The number of ties should be small relative to the total
> number of observations.

Conditions Required for a Valid Friedman F_r-Test

1. The treatments are randomly assigned to experimental units within the blocks.

2. The measurements can be ranked within blocks.

3. The p probability distributions from which the samples within each block
   are drawn are continuous.

For the drug example, we will use $\alpha = .05$ to form the rejection region:

$$F_r > \chi_{.05}^2 = 5.99147 \quad \text{(see Figure 14.12)}$$

where $\chi_{.05}^2$ is based on $(k - 1) = 2$ degrees of freedom. Consequently, because the
observed value, $F_r = 8.33$, exceeds 5.99147, we conclude that at least two of the
three drugs have probability distributions of reaction times that differ in location.

An SPSS printout of the nonparametric analysis, shown in Figure 14.13, confirms
our inference. Both the test statistic and p-value are highlighted on the printout. Since
p-value $= .0155$ is less than our selected $\alpha = .05$, there is evidence to reject H_0.

> **Now Work**   *Exercise 14.58*

Clearly, the assumptions for this test—that the measurements are ranked within
blocks and that the number of blocks (subjects) is greater than 5—are satisfied.
However, we must be sure that the treatments are randomly assigned to blocks. For the
procedure to be valid, we assume that the three drugs are administered in a random
order to each subject. If this were not true, the difference in the reaction times for the
three drugs might be due to the order in which the drugs are given.

Figure 14.12

Rejection Region for
Reaction Time Example

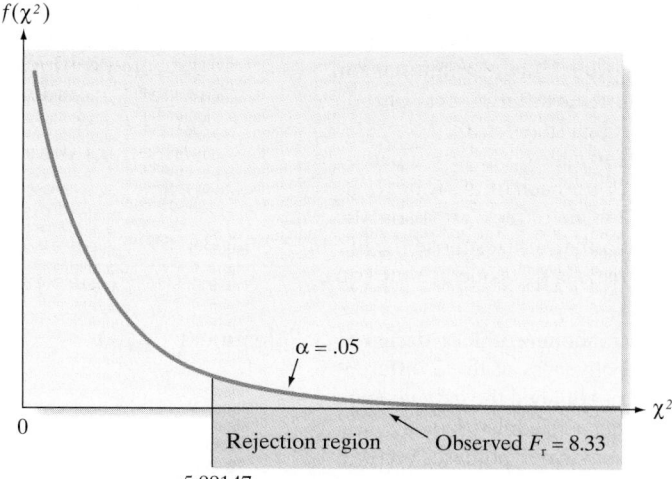

Figure 14.13
SPSS Friedman Test Printout

Friedman Test

Ranks

	Mean Rank
A	1.17
B	2.00
C	2.83

Test Statistics[a]

N	6
Chi-Square	8.333
df	2
Asymp. Sig.	.016

a. Friedman Test

Exercises 14.55–14.67

Understanding the Principles

14.55 Which of the following correctly describes how to rank the data in a randomized block design?
 a. For each treatment, rank the data across the blocks from smallest to largest.
 b. For each block, rank the data across the treatments from smallest to largest.

14.56 What conditions are required for a valid application of the Friedman F_r-test?

Learning the Mechanics

14.57 Data were collected using a randomized block design with four treatments (A, B, C, and D) and $b = 6$. The following rank sums were obtained:

$$R_A = 11 \quad R_B = 21 \quad R_C = 21 \quad R_D = 7$$

 a. How many blocks were used in the experimental design?
 b. Specify the null and alternative hypotheses that should be used in conducting a hypothesis test to determine whether the probability distributions for at least two of the treatments differ in location.
 c. Conduct the test of part **b**. Use $\alpha = .10$.
 d. What is the approximate p-value of the test of part **c**?
 e. Calculate the mean rank for each of the four treatments and compute the value of the F_r-test statistic according to the formula that utilizes those means. Verify that the test statistic is the same as that you obtained in part **c**.

14.58 Suppose you have used a randomized block design to
 NW help you compare the effectiveness of three different treatments, A, B, and C. You obtained the data given in the table and plan to conduct a Friedman F_r-test.
 a. Specify the null and alternative hypotheses you will test.

LM14_58

	Treatment		
Block	A	B	C
1	9	11	18
2	13	13	13
3	11	12	12
4	10	15	16
5	9	8	10
6	14	12	16
7	10	12	15

 b. Specify the rejection region for the test. Use $\alpha = .10$.
 c. Conduct the test and interpret the results.

14.59 An experiment was conducted using a randomized block design with four treatments and six blocks. The ranks of the measurements within each block are shown in the table. Use the Friedman F_r-test for a randomized block design to determine whether the data provide sufficient evidence to indicate that at least two of the treatment probability distributions differ in location. Test using $\alpha = .05$.

LM14_59

	Block					
Treatment	1	2	3	4	5	6
1	3	3	2	3	2	3
2	1	1	1	2	1	1
3	4	4	3	4	4	4
4	2	2	4	1	3	2

Applying the Concepts—Basic

14.60 Conditions impeding farm production. A review of farmer involvement in agricultural research was presented in the *Journal of Agricultural, Biological, and Environmental Statistics* (Mar. 2001). In one study, each of six farmers ranked the level of farm production constraint imposed by five conditions: drought, pest damage, weed interference, farming costs, and labor shortage. The rankings, ranging from 1 (least severe) to 5 (most severe), and rank sums for the five conditions are listed in the table below.

a. Use the rank sums shown in the table to compute the Friedman F_r-statistic.

b. At $\alpha = .05$, find the rejection region for a test to compare the farmer opinion distributions for the five conditions.

c. Make the proper conclusion, in the words of the problem.

14.61 Thematic atlas topics. Refer to the *Journal of Geography*'s published rankings of regional atlas theme topics, Exercise 14.40 (p. 846). In addition to high school teachers and university geography alumni, university geography students and representatives of the general public also ranked the 12 thematic topics. The rankings of all four groups are shown in the table on p. 858. A MINITAB analysis is provided (bottom of the page) to compare the atlas theme ranking distributions of the four groups.

a. Locate the rank sums on the printout.

b. Use the rank sums to find the Friedman F_r statistic.

c. Locate the test statistic and associated *p*-value on the printout.

d. Conduct the test and give the conclusion in the words of the problem.

14.62 Rotary oil rigs. Refer to Exercise 10.61 (p. 556) and the *World Oil* (Jan. 2002) study of rotary oil rigs. Three months were randomly selected and the number of oil rigs running in each of three states—California, Utah, and Alaska—was recorded. The data for the randomized block design are reproduced below. Consider a nonparametric test to compare the distributions of rotary oil rigs running in the three states.

OILRIGS

Month/Year	California	Utah	Alaska
Nov. 2000	27	17	11
Oct. 2001	34	20	14
Nov. 2001	36	15	14

a. State the null and alternative hypothesis for the test.

b. Rank the data within each month, then sum the ranks for each state.

c. Use the rank sums from part **b** to find the value of the test statistic.

FARM6

				Condition		
		Drought	Pest Damage	Weed Interference	Farming Costs	Labor Shortage
	1	5	4	3	2	1
	2	5	3	4	1	2
	3	3	5	4	2	1
Farmer	4	5	4	1	2	3
	5	4	5	3	2	1
	6	5	4	3	2	1
	Rank sum	27	25	18	11	9

Source: Riley, J., and Fielding, W. J. "An illustrated review of some farmer participatory research techniques." *Journal of Agricultural, Biological, and Environmental Statistics*, Vol. 6, No. 1, Mar. 2001 (Table 1).

Friedman Test: RANK versus GROUP blocked by THEME

```
S = 0.93   DF = 3   P = 0.819
S = 1.08   DF = 3   P = 0.782 (adjusted for ties)

                        Sum
                         of
GROUP   N   Est Median  Ranks
1       12     5.1250   27.0
2       12     6.1250   32.5
3       12     5.6250   29.0
4       12     6.1250   31.5

Grand median = 5.7500
```

ⓐ **ATLAS2**

	Rankings			
Theme	High School Teachers	Geography Alumni	Geography Students	General Public
Tourism	10	2	5	1
Physical	2	1	1	5
Transportation	7	3	7	2
People	1	6	2	3
History	2	5	9	4
Climate	6	4	4	8
Forestry	5	8	2	7
Agriculture	7	10	6	9
Fishing	9	7	10	6
Energy	2	8	7	10
Mining	10	11	11	11
Manufacturing	12	12	12	12

Source: Keller, C. P., et al. "Planning the next generation of regional atlases: Input from educators." *Journal of Geography,* Vol. 94, No. 3, May/June 1995, p. 413 (Table 1).

    d. Give the rejection region of the test at $\alpha = .05$.

    e. State the conclusion in the words of the problem. Does the conclusion agree with that for the ANOVA *F*-test conducted in Exercise 10.61?

14.63 Impact study of distractions while driving. The consequences of performing verbal and spatial-imagery tasks on visual search while driving were studied and the results published in the *Journal of Experimental Psychology: Applied* (Mar. 2000). Twelve drivers were recruited to drive on a highway in Madrid, Spain. During the drive, each subject was asked to perform three different tasks—a verbal task (repeating words that begin with a certain letter), a spatial-imagery task (imagining letters rotated a certain way), and no mental task. Since each driver performed all three tasks, the design is a randomized block with 12 blocks (drivers) and 3 treatments (tasks). Using a computerized head-free eye tracking system, the researchers kept track of the eye fixations of each driver on three different objects—the interior mirror, the off-side mirror, and the speedometer—and determined the proportion of eye fixations on the object. The researchers used the Friedman nonparametric test to compare the distributions of the eye fixation proportions for the three tasks.

    a. Using $\alpha = .01$, find the rejection region for the Friedman test.

    b. For the response variable, proportion of eye fixations on the interior mirror, the researchers determined the Friedman test statistic to be $\chi^2 = 19.16$. Give the appropriate conclusion.

    c. For the response variable, proportion of eye fixations on the off-side mirror, the researchers determined the Friedman test statistic to be $\chi^2 = 7.80$. Give the appropriate conclusion.

    d. For the response variable, proportion of eye fixations on the speedometer, the researchers determined the Friedman test statistic to be $\chi^2 = 20.67$. Give the appropriate conclusion.

Applying the Concepts—Intermediate

14.64 Effect of massage on boxers. Refer to the *British Journal of Sports Medicine* (Apr. 2000) experiment to investigate the effect of massage on boxing performance, Exercise 10.66 (p. 559). Recall that the punching power of each of eight amateur boxers was measured (in Newtons) after each of four rounds: (M1), round 1 following a pre-bout sports massage; (R1), round 1 following a pre-bout period of rest; (M5), round 5 following a sports massage between rounds; and (R5), round 5 following a period of rest between rounds. The data are reproduced in the table. Use the appropriate nonparametric test to compare the punching power means of the four interventions. Compare the results to those of Exercise 10.66.

ⓐ **BOXING**

		Intervention			
		M1	R1	M5	R5
	1	1243	1244	1291	1262
	2	1147	1053	1169	1177
	3	1247	1375	1309	1321
Boxer	4	1274	1235	1290	1285
	5	1177	1139	1233	1238
	6	1336	1313	1366	1362
	7	1238	1279	1275	1261
	8	1261	1152	1289	1266

Source: Hemmings, B., Smith, M., Graydon, J., and Dyson, R. "Effects of massage on physiological restoration, perceived recovery, and repeated sports performance." *British Journal of Sports Medicine,* Vol. 34, No. 2, Apr. 2000 (adapted from Table 3).

14.65 Plants and stress reduction. Refer to the Kansas State study designed to investigate the effects of plants on human stress levels, Exercise 10.67 (p. 559). The data on p. 859 are given as finger temperatures for each of ten students in a dimly lit room under three experimental conditions: presence of a live plant, presence of a plant photo, and absence of a plant (either live or photo). Analyze the data using a nonparametric procedure. Do the students' finger temperatures depend on the experimental condition?

PLANTS

Student	Live Plant	Plant Photo	No Plant (control)
1	91.4	93.5	96.6
2	94.9	96.6	90.5
3	97.0	95.8	95.4
4	93.7	96.2	96.7
5	96.0	96.6	93.5
6	96.7	95.5	94.8
7	95.2	94.6	95.7
8	96.0	97.2	96.2
9	95.6	94.8	96.0
10	95.6	92.6	96.6

Source: Elizabeth Schreiber, Department of Statistics, Kansas State University, Manhattan, Kansas

14.66 Absentee rates at a jeans plant. Refer to Exercise 10.68 (p. 559) and the *New Technology, Work, and Employment* (July 2001) study of daily worker absentee rates at a jeans plant. Nine weeks were randomly selected and the absentee rate (percentage of workers absent) determined for each day (Monday through Friday) of the work week. The data are reproduced in the table. Conduct a nonparametric analysis of the data to compare the distributions of absentee rates for the five days of the work week.

JEANS

Week	Monday	Tuesday	Wednesday	Thursday	Friday
1	5.3	0.6	1.9	1.3	1.6
2	12.9	9.4	2.6	0.4	0.5
3	0.8	0.8	5.7	0.4	1.4
4	2.6	0.0	4.5	10.2	4.5
5	23.5	9.6	11.3	13.6	14.1
6	9.1	4.5	7.5	2.1	9.3
7	11.1	4.2	4.1	4.2	4.1
8	9.5	7.1	4.5	9.1	12.9
9	4.8	5.2	10.0	6.9	9.0

Source: Boggis, J. J. "The eradication of leisure," *New Technology, Work, and Employment*, Volume 16, Number 2, July 2001 (Table 3).

14.67 Mosquito insecticide study. Refer to the *Journal of the American Mosquito Control Association* (Mar. 1995) study of the effectiveness of five different types of insecticides in controlling a species of Caribbean mosquito, Exercise 10.110 (p. 584). The resistance ratios (i.e., the dosage of insecticide required to kill 50% of the larvae divided by the known dosage for a susceptible mosquito strain) of the insecticides at each of seven Caribbean locations are reproduced in the table at the bottom of the page. Compare the resistance ratio distributions of the five insecticides using a nonparametric procedure. Are any of the insecticides more effective than any of the others?

MOSQUITO

		Insecticide				
		Temephos	Malathion	Fenitrothion	Fenthion	Chlorpyrifos
	Anguilla	4.6	1.2	1.5	1.8	1.5
	Antigua	9.2	2.9	2.0	7.0	2.0
	Dominica	7.8	1.4	2.4	4.2	4.1
Location	**Guyana**	1.7	1.9	2.2	1.5	1.8
	Jamaica	3.4	3.7	2.0	1.5	7.1
	St. Lucia	6.7	2.7	2.7	4.8	8.7
	Suriname	1.4	1.9	2.0	2.1	1.7

Source: Rawlins, S. C., and Oh Hing Wan, J. "Resistance in some Caribbean populations of *Aedes aegypti* to several insecticides." *Journal of the American Mosquito Control Association*, Vol. 11, No. 1, Mar. 1995 (Table 1).

14.7 Rank Correlation

Suppose 10 new paintings are shown to two art critics and each critic ranks the paintings from 1 (best) to 10 (worst). We want to determine whether the critics' ranks are related. Does a correspondence exist between their ratings? If a painting is ranked high by critic 1, is it likely to be ranked high by critic 2? Or do high rankings by one critic correspond to low rankings by the other? That is, are the rankings of the critics *correlated*?

If the rankings are as shown in the "Perfect Agreement" columns of Table 14.9, we immediately notice that the critics agree on the rank of every painting. High ranks correspond to high ranks and low ranks to low ranks. This is an example of *perfect positive correlation* between the ranks. In contrast, if the rankings appear as shown in the "Perfect Disagreement" columns of Table 14.9, high ranks for one critic correspond to low ranks for the other. This is an example of *perfect negative correlation*.

TABLE 14.9 Rankings of 10 Paintings by Two Critics

Painting	Perfect Agreement		Perfect Disagreement	
	Critic 1	Critic 2	Critic 1	Critic 2
1	4	4	9	2
2	1	1	3	8
3	7	7	5	6
4	5	5	1	10
5	2	2	2	9
6	6	6	10	1
7	8	8	6	5
8	3	3	4	7
9	10	10	8	3
10	9	9	7	4

TABLE 14.10 Rankings of Paintings: Less Than Perfect Agreement

Painting	Critic		Difference Between Rank 1 and Rank 2	
	1	2	d	d^2
1	4	5	-1	1
2	1	2	-1	1
3	9	10	-1	1
4	5	6	-1	1
5	2	1	1	1
6	10	9	1	1
7	7	7	0	0
8	3	3	0	0
9	6	4	2	4
10	8	8	0	0
				$\Sigma d^2 = 10$

In practice, you will rarely see perfect positive or negative correlation between the ranks. In fact, it is quite possible for the critics' ranks to appear as shown in Table 14.10. You will note that these rankings indicate some agreement between the critics, but not perfect agreement, thus indicating a need for a measure of rank correlation.

Spearman's rank correlation coefficient, r_s, provides a measure of correlation between ranks. The formula for this measure of correlation is given in the next box. We also give a formula that is identical to r_s when there are no ties in rankings; this provides a good approximation to r_s when the number of ties is small relative to the number of pairs.

Note that if the ranks for the two critics are identical, as in the second and third columns of Table 14.9, the differences between the ranks, d, will all be 0. Thus,

$$r_s = 1 - \frac{6 \sum d^2}{n(n^2 - 1)} = 1 - \frac{6(0)}{10(99)} = 1$$

That is, *perfect positive correlation* between the pairs of ranks is characterized by a Spearman correlation coefficient of $r_s = 1$. When the ranks indicate perfect disagreement, as in the fourth and fifth columns of Table 14.9, $\Sigma d_i^2 = 330$ and

$$r_s = 1 - \frac{6(330)}{10(99)} = -1.$$

Thus, *perfect negative correlation* is indicated by $r_s = -1$.

Biography

CHARLES E. SPEARMAN (1863–1945)—Spearman's Correlation

London-born Charles Spearman was educated at Leamington College before joining the British Army. After 20 years as a highly decorated officer, Spearman retired from the army and moved to Germany to begin his study of experimental psychology at the University of Liepzig. At the age of 41 he earned his Ph.D. and ultimately become one of the most influential figures the field of psychology. Spearman was the originator of the classical theory of mental tests and developed the "two-factor" theory of intelligence. These theories were used to develop and support the "Plus-Elevens" tests in England—exams administered to British 11-year-olds that predict whether they should attend a university or technical school. Spearman was greatly influenced by the works of Francis Galton (p. 593); consequently, he developed a strong statistical background. While conducting his research on intelligence, he proposed the rank order correlation coefficient—now called "Spearman's correlation coefficient." During his career, Spearman spent time at various universities, including University College (London), Colombia University, Catholic University, and the University of Cairo (Egypt).

Spearman's Rank Correlation Coefficient

$$r_s = \frac{SS_{uv}}{\sqrt{SS_{uu}SS_{vv}}}$$

where

$$SS_{uv} = \sum (u_i - \bar{u})(v_i - \bar{v}) = \sum u_i v_i - \frac{\left(\sum u_i\right)\left(\sum v_i\right)}{n}$$

$$SS_{uu} = \sum (u_i - \bar{u})^2 = \sum u_i^2 - \frac{\left(\sum u_i\right)^2}{n}$$

$$SS_{vv} = \sum (v_i - \bar{v})^2 = \sum v_i^2 - \frac{\left(\sum v_i\right)^2}{n}$$

u_i = Rank of the ith observation in sample 1
v_i = Rank of the ith observation in sample 2
n = Number of pairs of observations (number of observations in each sample)

Shortcut Formula for r_s*

$$r_s = 1 - \frac{6\sum d_i^2}{n(n^2 - 1)}$$

where

$d_i = u_i - v_i$ (difference in the ranks of the ith observations and 2)

For the data of Table 14.10,

$$r_s = 1 - \frac{6\sum d^2}{n(n^2 - 1)} = 1 - \frac{6(10)}{10(99)} = 1 - \frac{6}{99} = .94$$

*The shortcut formula is not exact when there are tied measurements, but it is a good approximation when the total number of ties is not large relative to n.

The fact that r_s is *close* to 1 indicates that the critics tend to agree, but the agreement is not perfect.

The value of r_s always falls between −1 and +1, with +1 indicating perfect positive correlation and −1 indicating perfect negative correlation. The closer r_s falls to +1 or −1, the greater the correlation between the ranks. Conversely, the nearer r_s is to 0, the less the correlation.

Note that the concept of correlation implies that two responses are obtained for each experimental unit. In the art critics example, each painting received two ranks (one for each critic) and the objective of the study was to determine the degree of positive correlation between the two rankings. Rank correlation methods can be used to measure the correlation between any pair of variables. If two variables are measured on each of n experimental units, we rank the measurements associated with each variable separately. Ties receive the average of the ranks of the tied observations. Then we calculate the value of r_s for the two rankings. This value measures the rank correlation between the two variables. We illustrate the procedure in Example 14.4.

EXAMPLE 14.4   APPLYING SPEARMAN'S RANK CORRELATION TEST

Problem   A study is conducted to investigate the relationship between cigarette smoking during pregnancy and the weights of newborn infants. A sample of 15 women smokers kept accurate records of the number of cigarettes smoked during their pregnancies, and the weights of their children were recorded at birth. The data are given in Table 14.11.

a. Calculate and interpret Spearman's rank correlation coefficient for the data.

b. Use a nonparametric test to determine whether level of cigarette smoking and weights of newborns are negatively correlated for all smoking mothers. Use $\alpha = .05$.

Solution   **a.** We first rank the number of cigarettes smoked per day, assigning a 1 to the smallest number (12) and a 15 to the largest (46). Note that the two ties receive the averages of their respective ranks. Similarly, we assign ranks to the 15 babies' weights. Since the number of ties is relatively small, we will use the

⊙ **NEWBORN**

TABLE 14.11   Data and Calculations for Example 14.4

Woman	Cigarettes per Day	Rank	Baby's Weight (pounds)	Rank	d	d^2
1	12	1	7.7	5	−4	16
2	15	2	8.1	9	−7	49
3	35	13	6.9	4	9	81
4	21	7	8.2	10	−3	9
5	20	5.5	8.6	13.5	−8	64
6	17	3	8.3	11.5	−8.5	72.25
7	19	4	9.4	15	−11	121
8	46	15	7.8	6	9	81
9	20	5.5	8.3	11.5	−6	36
10	25	8.5	5.2	1	7.5	56.25
11	39	14	6.4	3	11	121
12	25	8.5	7.9	7	1.5	2.25
13	30	12	8.0	8	4	16
14	27	10	6.1	2	8	64
15	29	11	8.6	13.5	−2.5	6.25
					Total =	795

shortcut formula to calculate r_s. The differences d between the ranks of the babies' weights and the ranks of the number of cigarettes smoked per day are shown in Table 14.11. The squares of the differences, d^2, are also given. Thus,

$$r_s = 1 - \frac{6 \sum d_i^2}{n(n^2 - 1)} = 1 - \frac{6(795)}{15(15^2 - 1)} = 1 - 1.42 = -.42$$

The value of r_s can also be obtained using a computer. A SAS printout of the analysis is shown in Figure 14.14. The value of r_s, highlighted on the printout, agrees (except for rounding) with our hand-calculated value of $-.42$.

Figure 14.14

SAS Spearman Correlation Printout for Example 14.4

```
              The CORR Procedure

   2  Variables:    CIGARETTES WEIGHT


   Spearman Correlation Coefficients, N = 15
          Prob > |r| under H0: Rho=0

                    CIGARETTES        WEIGHT

     CIGARETTES      1.00000        -0.42473
                                     0.1145

     WEIGHT         -0.42473         1.00000
                     0.1145
```

This negative correlation coefficient indicates that in this sample an increase in the number of cigarettes smoked per day is *associated with* (but is not necessarily the *cause of*) a decrease in the weight of the newborn infant.

b. If we define ρ as the **population rank correlation coefficient** [i.e., the rank correlation coefficient that could be calculated from all (x, y) values in the population], this question can be answered by conducting the test

H_0: $\rho = 0$ (no population correlation between ranks)

H_a: $\rho < 0$ (negative population correlation between ranks)

Test statistic: r_s (the *sample* Spearman rank correlation coefficient)

To determine a rejection region, we consult Table XIV in Appendix A, which is partially reproduced in Table 14.12. Note that the left-hand column gives values of n, the number of pairs of observations. The entries in the table are values for an upper-tail rejection region, since only positive values are given. Thus, for $n = 15$ and $\alpha = .05$, the value .441 is the boundary of the upper-tailed rejection region, so that $P(r_s > .441) = .05$ if H_0: $\rho = 0$ is true. Similarly, for negative values of r_s, we have $P(r_s < -.441) = .05$ if $\rho = 0$. That is, we expect to see $r_s < -.441$ only 5% of the time if there is really no relationship between the ranks of the variables.

The lower-tailed rejection region is therefore

Rejection region $(\alpha = .05)$: $r_s < -.441$

Since the calculated $r_s = -.42$ is not less than $-.441$, we cannot reject H_0 at the $\alpha = .05$ level of significance. That is, this sample of 15 smoking mothers provides insufficient evidence to conclude that a negative correlation exists between number of cigarettes smoked and the weight of newborns for the populations of measurements corresponding to all smoking mothers. This does not, of course, mean that no relationship exists. A study using a larger sample of smokers and taking other factors into account (father's weight, sex of newborn child, etc.) would be more likely to reveal whether smoking and the weight of a newborn child are related.

TABLE 14.12 Reproduction of Part of Table XIV in Appendix A: Critical Values of Spearman's Rank Correlation Coefficient

n	$\alpha = .05$	$\alpha = .025$	$\alpha = .01$	$\alpha = .005$
5	.900	—	—	—
6	.829	.886	.943	—
7	.714	.786	.893	—
8	.643	.738	.833	.881
9	.600	.683	.783	.833
10	.564	.648	.745	.794
11	.523	.623	.736	.818
12	.497	.591	.703	.780
13	.475	.566	.673	.745
14	.457	.545	.646	.716
15	.441	.525	.623	.689
16	.425	.507	.601	.666
17	.412	.490	.582	.645
18	.399	.476	.564	.625
19	.388	.462	.549	.608
20	.377	.450	.534	.591

Look Back The two-tailed p-value of the test (.1145) is highlighted on the SAS printout, Figure 14.14. Since the lower-tailed p-value, $.1145/2 = .05725$, exceeds $\alpha = .05$, our conclusion is the same: Do not reject H_0.

Now Work *Exercise 14.73*

■ ■ ■

A summary of Spearman's nonparametric test for correlation is given in the next box.

Spearman's Nonparametric Test for Rank Correlation

One-Tailed Test	**Two-Tailed Test**
H_0: $\rho = 0$	H_0: $\rho = 0$
H_a: $\rho > 0$ [or H_a: $\rho < 0$]	H_a: $\rho \neq 0$

Test statistic: r_s the sample rank correlation (see the formulas for calculating r_s)

Rejection region: $r_s > r_{s,\alpha}$	*Rejection region:* $|r_s| > r_{s,\alpha/2}$
[or $r_s < -r_{s,\alpha}$ when H_a: $\rho_s < 0$]	
where $r_{s,\alpha}$ is the value from Table XIV corresponding to the upper-tail area α and n pairs of observations	where $r_{s,\alpha/2}$ is the value from Table XIV corresponding to the upper-tail area $\alpha/2$ and n pairs of observations

Ties: Assign tied measurements the average of the ranks they would receive if they were unequal but occurred in successive order. For example, if the third-ranked and fourth-ranked measurements are tied, assign each a rank of $(3 + 4)/2 = 3.5$. The number of ties should be small relative to the total number of observations.

Conditions Required for a Valid Spearman's Test:

1. The sample of experimental units on which the two variables are measured is randomly selected.

2. The probability distributions of the two variables are continuous.

Statistics in Action Revisited

Testing whether the TCDD Levels in Fat and Plasma of Vietnam Vets Are Correlated

The medical researchers who examined the Vietnam War veterans exposed to Agent Orange also wanted an estimate of the correlation between the TCDD level in fat tissue and the TCDD level in plasma. Since the two variables were not normally distributed, they employed Spearman's rank correlation method. The SPSS printout for this analysis is shown in Figure SIA14.5.

The value of r_s (highlighted on the printout) is .774; thus, there appears to be a fairly strong positive correlation between fat and plasma TCDD levels in the sample of Vietnam vets. The one-tailed p-value of the test (also highlighted) is .000, indicating that there is sufficient evidence (at $\alpha = .01$) of a positive association between the two TCDD measures.

Figure SIA14.5

SPSS Spearman Rank Correlation Test for TCDD Data

Correlations

			FAT	PLASMA
Spearman's rho	FAT	Correlation Coefficient	1.000	.774**
		Sig. (1-tailed)	.	.000
		N	20	20
	PLASMA	Correlation Coefficient	.774**	1.000
		Sig. (1-tailed)	.000	.
		N	20	20

**. Correlation is significant at the 0.01 level (1-tailed).

Exercises 14.68–14.82

Understanding the Principles

14.68 What is the value of r_s when there is perfect negative rank correlation between two variables? Perfect positive rank correlation?

14.69 What conditions are required for a valid Spearman's test?

Learning the Mechanics

14.70 Use Table XIV of Appendix A to find each of the following probabilities:
  a. $P(r_s > .508)$ when $n = 22$
  b. $P(r_s > .448)$ when $n = 28$
  c. $P(r_s \leq .648)$ when $n = 10$
  d. $P(r_s < -.738$ or $r_s > .738)$ when $n = 8$

14.71 Specify the rejection region for Spearman's nonparametric test for rank correlation in each of the following situations:
  a. H_0: $\rho = 0$, H_a: $\rho \neq 0$, $n = 10$, $\alpha = .05$
  b. H_0: $\rho = 0$, H_a: $\rho > 0$, $n = 20$, $\alpha = .025$
  c. H_0: $\rho = 0$, H_a: $\rho < 0$, $n = 30$, $\alpha = .01$

14.72 Compute Spearman's rank correlation coefficient for each of the following pairs of sample observations:

a.
x	33	61	20	19	40
y	26	36	65	25	35

b.
x	89	102	120	137	41
y	81	94	75	52	136

c.
x	2	15	4	10
y	11	2	15	21

d.
x	5	20	15	10	3
y	80	83	91	82	87

14.73 The following sample data were collected on variables x and y:
NW

🔵 **LM14_73**

x	0	3	0	−4	3	0	4
y	0	2	2	0	3	1	2

  a. Specify the null and alternative hypotheses that should be used in conducting a hypothesis test to determine whether the variables x and y are correlated.
  b. Conduct the test of part **a** using $\alpha = .05$.
  c. What is the approximate p-value of the test of part **b**?
  d. What assumptions are necessary to ensure the validity of the test of part **b**?

Applying the Concepts—Basic

14.74 Mongolian desert ants. Refer to the *Journal of Biogeography* (Dec. 2003) study of ants in Mongolia, Exercise 11.22 (p. 608). Data on annual rainfall, maximum daily temperature and number of ant species recorded at each of 11 study sites are reproduced in the table.

   a. Consider the data for the 5 sites in the Dry Steppe region only. Rank the 5 annual rainfall amounts. Then rank the 5 maximum daily temperature values.

   b. Use the ranks, part **a**, to find and interpret the rank correlation between annual rainfall (y) and maximum daily temperature (x).

   c. Repeat parts **a** and **b** for the 6 sites in the Gobi Desert region.

   d. Now consider the rank correlation between the number of ant species (y) and annual rainfall (x). Using all the data, compute and interpret Spearman's rank correlation statistic.

GOBIANTS

Site	Region	Annual Rainfall (mm)	Max. Daily Temp. (°C)	Number of Ant Species
1	Dry Steppe	196	5.7	3
2	Dry Steppe	196	5.7	3
3	Dry Steppe	179	7.0	52
4	Dry Steppe	197	8.0	7
5	Dry Steppe	149	8.5	5
6	Gobi Desert	112	10.7	49
7	Gobi Desert	125	11.4	5
8	Gobi Desert	99	10.9	4
9	Gobi Desert	125	11.4	4
10	Gobi Desert	84	11.4	5
11	Gobi Desert	115	11.4	4

Source: Pfeiffer, M., et al. "Community organization and species richness of ants in Mongolia along an ecological gradient from steppe to Gobi desert," *Journal of Biogeography,* Vol. 30, No. 12, Dec. 2003 (Tables 1 and 2).

14.75 Effect of massage on boxers. Refer to the *British Journal of Sports Medicine* (Apr. 2000) study of the effect of massaging boxers between rounds, Exercise 11.56 (p. 623). Two variables measured on the boxers were blood lactate level (y) and the boxer's perceived recovery (x). The data for 16 five-round boxing performances are reproduced in the next table.

   a. Rank the values of the 16 blood lactate levels.

   b. Rank the values of the 16 perceived recovery values.

   c. Use the ranks, parts **a** and **b**, to compute Spearman's rank correlation coefficient. Give a practical interpretation of the result.

   d. Find the rejection region for a test to determine whether y and x are rank correlated. Use $\alpha = .10$.

   e. What is the conclusion of the test, part **d**? State your answer in the words of the problem.

BOXING2

Blood Lactate Level	Perceived Recovery
3.8	7
4.2	7
4.8	11
4.1	12
5.0	12
5.3	12
4.2	13
2.4	17
3.7	17
5.3	17
5.8	18
6.0	18
5.9	21
6.3	21
5.5	20
6.5	24

Source: Hemmings, B., Smith, M., Graydon, J., and Dyson, R. "Effects of massage on physiological restoration, perceived recovery, and repeated sports performance." *British Journal of Sports Medicine,* Vol. 34, No. 2, Apr. 2000 (data adapted from Figure 3).

14.76 Assessing dementia. The Milan Overall Dementia Assessment (MODA) is a neuropsychologically oriented test that provides a measure of an individual's cognitive deterioration. MODA scores range from 0 to 100, with higher scores indicating a greater degree of deterioration. A team of psychologists and neurologists administered the MODA to a sample of 30 patients with Alzheimer's disease. (*Neuropsychologia,* June 1995.) In addition, all patients were given a "face matching" test in which they were requested to match photographs of unknown faces. Scores on this test were recorded as the number of matching errors, with a maximum score of 27.

   a. The researchers reported Spearman's rank correlation between the MODA and face-matching scores as $r_s = .48$. Interpret this result.

   b. Test the hypothesis of a positive correlation between MODA score and face-matching score in Alzheimer's patients. Use $\alpha = .05$.

14.77 Exercise in MS patients. The metabolic and cardiopulmonary responses during maximal effort exercise in persons with multiple sclerosis (MS) was studied. (*Clinical Kinesiology,* Spring 1995.) The following variables were measured for each of 10 MS patients:

   1. Expanded Disability Status Scale (EDSS)—Ratings range from 1 to 4.5

   2. Peak oxygen uptake (liters per minute) during an arm cranking exercise (ARM)

   3. Peak oxygen uptake (liters per minute) during a leg cycling exercise (LEG)

   4. Peak oxygen uptake (liters per minute) during a combined leg cycling and arm cranking exercise (LEG/ARM)

Spearman's rank correlation was calculated for each pair of variables. The next table gives the correlation r_s between the variables in the corresponding row and column.

	EDSS	ARM	LEG/ARM	LEG
EDSS	1.000	−.439	−.655	−.512
ARM		1.000	.634	.722
LEG/ARM			1.000	.890
LEG				1.000

Source: Ponichtera-Mulcare, J. A., et al. "Maximal aerobic exercise of individuals with multiple sclerosis using three modes of energy." *Clinical Kinesiology,* Vol. 49, No. 1, Spring 1995, p. 10 (Table 2).

a. Explain why the table shows rank correlations of 1.000 in the diagonal.

b. Practically interpret each of the other rank correlations in the table.

c. Is there sufficient evidence (at $\alpha = .01$) to conclude that EDSS and peak oxygen uptake are negatively correlated? Perform the test for each of the three exercises.

Applying the Concepts—Intermediate

14.78 The "name game" Refer to the *Journal of Experimental Psychology—Applied* (June 2000) study in which the "name game" was used to help groups of students learn the names of other students in the group, Exercise 11.28 (p. 611). Recall that one goal of the study was to investigate the relationship between proportion of names recalled (y) by a student and position (order) of the student during the game (x). The data for 144 students in the first eight positions are saved in the **NAMEGAME2** file. (The first five and last five observations in the data set are listed here.)

a. To properly apply the parametric test for correlation based on the Pearson coefficient of correlation, r (Section 11.6), both the x and y variables must be normally distributed. Demonstrate that this assumption is violated for this data. What are the consequences of this violation?

b. Find Spearman's rank correlation coefficient on the SAS printout provided below and interpret its value.

c. Find the observed significance level for testing for zero rank correlation on the SAS printout and interpret its value.

d. At $\alpha = .05$, is there sufficient evidence of rank correlation between proportion of names recalled (y) by a student and position (order) of the student during the game (x)?

NAMEGAME2

Position	Recall
2	0.04
2	0.37
2	1.00
2	0.99
2	0.79
⋮	⋮
9	0.72
9	0.88
9	0.46
9	0.54
9	0.99

Source: Morris, P. E., and Fritz, C. O. "The name game: Using retrieval practice to improve the learning of names." *Journal of Experimental Psychology-Applied,* Vol. 6, No. 2, June 2000 (data simulated from Figure 2).

FCATs

14.79 FCAT scores and poverty. Refer to the *Journal of Educational and Behavioral Statistics* (Spring 2004) analysis of the link between Florida Comprehensive Assessment Test (FCAT) scores and sociodemographic factors, Exercise 11.24 (p. 608). Data on average math and reading FCAT scores of third graders, as well as the percentage of students below the poverty level, for a sample of 22 Florida elementary schools are saved in the **FCAT** file.

a. Compute and interpret Spearman's rank correlation between FCAT math score (y) and percentage (x) of students below the poverty level.

b. Compute and interpret Spearman's rank correlation between FCAT reading score (y) and percentage (x) of students below the poverty level.

c. Determine whether the value of r_s in part **a** would lead you to conclude that FCAT math score and percent below poverty level are negatively rank correlated in the population of all Florida elementary schools. Use $\alpha = .01$ to make your decision.

d. Determine whether the value of r_s in part **b** would lead you to conclude that FCAT reading score and percent below poverty level are negatively rank correlated in the population of all Florida elementary schools. Use $\alpha = .01$ to make your decision.

```
                    The CORR Procedure

              2   Variables:    POSITION RECALL


            Spearman Correlation Coefficients, N = 144
                   Prob > |r| under H0: Rho=0

                          POSITION         RECALL

            POSITION       1.00000         0.20652
                                           0.0130

            RECALL         0.20652         1.00000
                           0.0130
```

14.80 Company reputations. *The Wall Street Journal* (Feb. 7, 2001) reported on a Harris Interactive, Inc. survey of consumers to rate the reputations of America's most visible companies. The 1999 and 2000 ranks of 15 randomly selected companies are listed in the accompanying table. Conduct a test to determine if the 1999 and 2000 reputation ranks are positively correlated. Use $\alpha = .05$.

COREP

Company	1999 Rank	2000 Rank
Johnson & Johnson	1	1
Anheuser-Busch	14	6
Disney	10	8
Microsoft	15	9
FedEx	21	13
Wal-Mart	6	14
Coca-Cola	2	16
McDonalds	24	24
Yahoo!	19	27
Sears	29	35
America Online	26	39
Kmart	37	41
Toyota	28	19
Home Depot	8	4
IBM	17	7

Source: Harris Interactive, the Reputation Institute.

14.81 Study of child bipolar disorders. Psychiatric researchers at the University of Pittsburgh Medical Center have developed a new test for measuring manic symptoms in pediatric bipolar patients. (*Journal of Child and Adolescent Psychopharmacology*, Dec., 2003.) The new test is called the Kiddie Schedule for Affective Disorders and Schizophrenia Mania Rating Scale (KSADS-MRS). The new test was compared to the standard test, the Clinical Global Impressions-Bipolar Scale (CGI-BP). Both tests were administered to a sample of 19 pediatric patients before and after the patients were treated for manic symptoms. The changes in the test scores are recorded in the table below.

a. The researchers used Spearman's statistic to measure the correlation between the changes in the two test scores. Compute the value of r_s.

b. Is there sufficient evidence (at $\alpha = .05$) of positive rank correlation between the two test score changes in the population of all pediatric patients with manic symptoms?

14.82 Pain empathy and brain activity. Refer to the *Science* (Feb. 20, 2004) study on the relationship between brain activity and pain-related empathy in persons who watch others in pain, Exercise 11.58 (p. 623). Recall that 16 female partners watched while painful stimulation was applied to the finger of their respective male partners. The two variables of interest were y = female's pain-related brain activity (measured on a scale ranging from -2 to 2) and x = female's score on the Empathic Concern Scale (0 to 25 points). The data are reproduced in the accompanying table. Use Spearman's rank correlation test to answer the research question: "Do people scoring higher in empathy show higher pain-related brain activity?"

BRAINPAIN

Couple	Brain Activity (y)	Empathic Concern (x)
1	.05	12
2	−.03	13
3	.12	14
4	.20	16
5	.35	16
6	0	17
7	.26	17
8	.50	18
9	.20	18
10	.21	18
11	.45	19
12	.30	20
13	.20	21
14	.22	22
15	.76	23
16	.35	24

Source: Singer, T. et al. "Empathy for pain involves the affective but not sensory components of pain," *Science,* Vol. 303, Feb. 20, 2004. (Adapted from Figure 4.)

MANIA

Patient	Change in KSADS-MRS (%)	Improvement in CGI-BP	Patient	Change in KSADS-MRS (%)	Improvement in CGI-BP
1	80	6	10	−25	2
2	65	5	11	−35	2
3	20	4	12	−65	2
4	−15	4	13	−65	2
5	−50	4	14	−70	2
6	20	3	15	−80	2
7	−30	3	16	−90	2
8	−70	3	17	−95	2
9	−10	2	18	−90	1

Source: Axelson, D. et al. "A preliminary study of the Kiddie Schedule for Affective Disorders and Schizophrenia for School-Age Children Mania Rating Scale for children and adolescents," *Journal of Child and Adolescent Psychopharmacology,* Vol. 13, No. 4, Dec. 2003 (adapted from Figure 2).

Quick Review

Key Terms

Key Formulas

Test	Test Statistic	Large Sample Approximation
Sign	S = number of sample measurements greater than (or less than) hypothesized mean, η_0	$z = \dfrac{(S - .5) - .5n}{.5\sqrt{n}}$ 826
Wilcoxon rank sum	T_1 = rank sum of sample 1 or T_2 = rank sum of sample 2	$z = \dfrac{T_1 - \dfrac{n_1(n_1 + n_2 + 1)}{2}}{\sqrt{\dfrac{n_1 n_2 (n_1 + n_2 + 1)}{12}}}$ 835
Wilcoxon signed ranks	T_- = negative rank sum or T_+ = positive rank sum	$z = \dfrac{T_+ - \dfrac{n(n + 1)}{4}}{\sqrt{\dfrac{n(n + 1)(2n + 1)}{24}}}$ 843
Kuskal-Wallis	$H = \dfrac{12}{n(n + 1)}\sum\dfrac{R_j^2}{n_j} - 3(n + 1)$ 850	
Friedman	$F_r = \dfrac{12}{bk(k + 1)}\sum R_j^2 - 3b(k + 1)$ 854	
Spearman rank correlation (shortcut formula)	$r_s = 1 - \dfrac{6\sum d_i^2}{n(n^2 - 1)}$ 861	

where d_i = difference in ranks of ith observations for samples 1 and 2

Language Lab

Symbol	Description
η (eta)	Population median
S	Test statistic for sign test (see Key Formulas)
T_1	Sum of ranks of observations in sample 1
T_2	Sum of ranks of observations in sample 2
T_L	Critical lower Wilcoxon rank sum value
T_U	Critical upper Wilcoxon rank sum value
T_+	Sum of ranks of positive differences of paired observations
T_-	Sum of ranks of negative differences of paired observations
T_0	Critical value of Wilcoxon signed rank test
R_j	Rank sum of observations in sample j

H	Test statistic for Kruskal-Wallis test (see Key Formulas)
F_r	Test statistic for Friedman test (see Key Formulas)
r_s	Spearman's rank correlation coefficient (see Key Formulas)
ρ (rho)	Population correlation coefficient

Chapter Summary Notes

- **Distribution-free tests** — do not rely on assumptions about the probability distribution of the sampled population
- **Nonparametrics** — distribution-free tests that are based on **rank statistics**
- *One-sample* nonparametric test for the population median — **sign test**

- Nonparametric test for *matched pairs* — **Wilcoxon rank test**
- Nonparametric test for a *completely randomized design* — **Kruskal-Wallis test**
- Nonparametric test for a *randomized block design* — **Friedman test**
- Nonparametric test for *rank correlation* — **Spearman's test**

Supplementary Exercises 14.83–14.108

Understanding the Principles

14.83 When is it appropriate to use the *t*- and *F*-tests of Chapters 9 and 10 for comparing two or more population means?

14.84 How does a nonparametric test differ from the parametric tests of Chapters 8–10?

14.85 For each of the following, give the appropriate nonparametric test to apply
  a. Comparing two populations with independent samples
  b. Making an inference about a population median
  c. Comparing three or more populations with independent samples
  d. Making an inference about rank correlation
  e. Comparing two populations with matched pairs
  f. Comparing three or more populations with a block design

Learning the Mechanics

14.86 The data for three independent random samples are shown in the table. It is known that the sampled populations are not normally distributed. Use an appropriate test to determine whether the data provide sufficient evidence to indicate that at least two of the populations differ in location. Test using $\alpha = .05$.

☺ LM14_86

Sample 1		Sample 2		Sample 3	
18	15	12	34	87	50
32	63	33	18	53	64
43		10		65	77

14.87 A random sample of nine pairs of observations are recorded on two variables, *x* and *y*. The data are shown in the table.

☺ LM14_87

Pair	*x*	*y*	Pair	*x*	*y*
1	19	12	6	29	10
2	27	19	7	16	16
3	15	7	8	22	10
4	35	25	9	16	18
5	13	11			

  a. Do the data provide sufficient evidence to indicate that ρ, the rank correlation between *x* and *y*, differs from 0? Test using $\alpha = .05$.
  b. Do the data provide sufficient evidence to indicate that the probability distribution for *x* is shifted to the right of that for *y*? Test using $\alpha = .05$.

14.88 Two independent random samples produced the measurements listed in the table. Do the data provide sufficient evidence to conclude that there is a difference between the locations of the probability distributions for the sampled populations? Test using $\alpha = .05$.

☺ LM14_88

Sample 1		Sample 2	
1.2	1.0	1.5	1.9
1.9	1.8	1.3	2.7
.7	1.1	2.9	3.5
2.5			

14.89 An experiment was conducted using a randomized block design with five treatments and four blocks. The data are shown in the table. Do the data provide sufficient evidence to conclude that at least two of the treatment probability distributions differ in location? Test using $\alpha = .05$.

⊙ **LM14_89**

Treatment	Block			
	1	2	3	4
1	75	77	70	80
2	65	69	63	69
3	74	78	69	80
4	80	80	75	86
5	69	72	63	77

Applying the Concepts—Basic

14.90 Cancer remission time. The American Cancer Society funds medical research focused on finding treatments for various forms of cancer. One new treatment is being tested to determine whether the average remission time for a particularly aggressive form of cancer is extended by the treatment. Assume that current treatments have been shown to provide a median remission time of 4.5 years. Seven patients with this cancer are given the new treatment, and their remission times (in years) are as follows:

⊙ **REMISSION**

5.3	7.3	3.6	5.2	6.1	4.8	8.4

Use the sign test to determine if the median remission time is increased by the new treatment. Test at $\alpha = .05$.

14.91 Ordering exam questions. An educational psychologist claims that the order in which test questions are asked affects a student's ability to answer correctly. To investigate this assertion, a professor randomly divides a class of 13 students into two groups—7 in one group and 6 in the other. The professor prepares one set of test questions but arranges the questions in two different orders. On test A the questions are arranged in order of increasing difficulty (that is, from easiest to most difficult), while on test B the order is reversed. One group of students is given test A, the other test B, and the test score is recorded for each student. The results are shown in the table.

⊙ **TESTORDER**

Test A	90	71	83	82	75	91	65
Test B	66	78	50	68	80	60	

Use the Wilcoxon rank sum procedure to test for a difference (a shift in location) in the probability distributions of student scores on the two tests. Test using $\alpha = .05$.

14.92 River flood crests. Suppose we want to compare the mean flood crests on the Suwannee and Santa Fe rivers for four flood level years in Florida. The data (in feet), matched by year at six locations on the rivers, are shown in the table at the bottom of the page. Do the data indicate that the flood crests on the rivers were higher or lower in any one of the flood level years than any other? Test using the Friedman F_r-test (at $\alpha = .05$)

14.93 Methods of teaching reading. Sixth graders in an elementary school are taught reading by three different methods. Students were randomly assigned to three classes. One class used programmed instruction, a second used standard memorization techniques, and the third used an open classroom approach. The increases in reading levels attained by five students randomly selected from each of the three classes are shown in the table.

⊙ **READ3**

Programmed	Standard	Open
.9	1.0	1.7
1.5	.8	.5
.7	.9	1.6
1.1	1.2	1.4
.5	1.4	1.0

a. Rank the 15 data points in the table.
b. Obtain the rank sums for the three methods.
c. Use the rank sums to find the Kruskal-Wallis test statistic.
d. Do the data provide sufficient evidence to indicate that the probability distributions of increases in reading level differ for at least two of the methods? Use $\alpha = .05$.

14.94 Video teleconferencing study A study published in the *Journal of Business Communications* (Fall 1985) found that video teleconferencing may be a more effective method of dealing with complex group problem-solving tasks than face-to-face meetings. Ten groups of four people each were randomly assigned both to a specific communication setting (face-to-face or video

⊙ **FLOOD**

Station	Flood Stage	Flood Year			
		1	2	3	4
Suwannee River					
White Springs	77.0	79.9	74.4	88.5	85.2
Ellaville	54.0	59.8	52.8	65.0	68.1
Branford	29.0	29.3	29.2	35.6	38.9
Wilcox	14.0	11.6	13.0	18.6	22.3
Santa Fe River					
Three Rivers	19.0	24.1	24.3	27.2	30.6
U.S. 129 Bridge	21.0	23.8	23.8	30.7	34.2

teleconferencing) and to one of two specific complex problems. Upon completion of the problem-solving task, the same groups were placed in the alternative communication setting and asked to complete the second problem-solving task. The percentage of each problem task correctly completed was recorded for each group, with the results given in the accompanying table.

FACEVT

Group	Face-to-Face	Video Teleconferencing
1	65%	75%
2	82	80
3	54	60
4	69	65
5	40	55
6	85	90
7	98	98
8	35	40
9	85	89
10	70	80

The wilcoxon signed ranks test for matched pairs will be applied to the data.
- **a.** Specify the null and alternative hypotheses that should be used in determining whether the data provide sufficient evidence to conclude that the problem-solving performance of video teleconferencing groups is superior to that of groups that interact face-to-face.
- **b.** Compute the signed rank test statistic.
- **c.** Give the appropriate conclusion at $\alpha = .05$.

14.95 Feeding habits of fish Refer to the *Brain and Behavior Evolution* (Apr. 2000) study of the feeding behavior of blackbream fish, Exercises 2.142 (p. 99) and 11.105 (p. 99). Recall that the zoologists recorded the number of aggressive strikes of two blackbream fish feeding at the bottom of an aquarium in the 10-minute period following the addition of food. The table listing the weekly number of strikes and age of the fish (in days) is reproduced here.

BLACKBREAM

Week	Number of Strikes	Age of Fish (days)
1	85	120
2	63	136
3	34	150
4	39	155
5	58	162
6	35	169
7	57	178
8	12	184
9	15	190

Source: Shand, J., et al. "Variability in the location of the retinal ganglion cell area centralis is correlated with ontogenetic changes in feeding behavior in the Blackbream, Acanthopagrus 'butcher'." *Brain and Behavior,* Vol. 55, No. 4, Apr. 2000 (Figure H).

- **a.** Find Spearman's correlation coefficient relating number of strikes (y) to age of fish (x).

- **b.** Conduct a nonparametric test to determine whether number of strikes (y) and age (x) are negatively correlated. Test using $\alpha = .01$.

Applying the Concepts—Intermediate

14.96 Comparison of disposable razors. A major razor blade manufacturer advertises that its twin-blade disposable razor will "get you more shaves" than any single-blade disposable razor on the market. A rival blade company that has been very successful in selling single-blade razors wishes to test this claim. Marketing managers in this company randomly sampled eight single-blade shavers and eight twin-blade shavers, and counted the number of shaves that each got before a change of blades was indicated. The results are shown in the table.

RAZORS

Twin Blades		Single Blade	
8	15	10	13
17	10	6	14
9	6	3	5
11	12	7	7

- **a.** Do the data support the twin-blade manufacturer's claim? Use $\alpha = .05$.
- **b.** Do you think this experiment was designed in the best possible way? If not, what design might have been better?
- **c.** What assumptions are necessary for the test performed in part **a** to be valid? Do the assumptions seem reasonable for this application?

14.97 Spending on IS technology. University of Queensland researchers sampled private sector and public sector organizations in Australia to study the planning undertaken by their information systems departments. (*Management Science,* July 1996.) As part of the process they asked each sample organization how much it had spent on information systems and technology in the previous fiscal year as a percentage of the organization's total revenues. The results are reported in the table.

INFOSYS

Private Sector	Public Sector
2.58%	5.40%
5.05	2.55
.05	9.00
2.10	10.55
4.30	1.02
2.25	5.11
2.50	24.42
1.94	1.67
2.33	3.33

Adapted from Hann, J., and Weber, R. "Information systems planning: A model and empirical tests." *Management Science,* Vol. 42, No. 2 (July, 1996), pp. 1043–1064.

a. Do the two sampled populations have identical probability distributions or is the distribution for public sector organizations in Australia located to the right of Australia's private sector firms? Test using $\alpha = .05$.

b. Is the p-value for the test less than or greater than .05? Justify your answer.

c. What assumption must be met to ensure the validity of the test you conducted in part **a**?

14.98 Math literacy test. Many states now require that mathematics teachers pass a basic math literacy test in order to qualify to teach math in the state. An important part of the process is the development of a fair test. Suppose that one state board of education has developed a test and that a sample of 20 math teachers are asked to take it. The test will be considered too hard if at least 50% of *all* math teachers in the state score less than 60 (on a scale of 0 to 100).

a. Set up the appropriate null and alternative hypotheses to test whether the test is too hard.

b. Establish the appropriate test statistic and rejection region for the test using $\alpha = .05$.

c. What assumptions are necessary to ensure the validity of the test?

d. Suppose that 14 of the 20 sampled teachers score less than 60. What is the appropriate conclusion?

e. Calculate the observed significance level of the test. Interpret it.

f. Considering your answers to parts **d** and **e**, what would your recommendation be concerning the adoption of the test statewide?

14.99 Preventing metal corrosion. Corrosion of different metals is a problem in many mechanical devices. Three sealers used to help retard the corrosion of metals were tested to see whether there were any differences among them. Samples of 10 different metal compositions were treated with each of the three sealers, and the amount of corrosion was measured after exposure to the same environmental conditions for 1 month. The data are given in the next table. Is there any evidence of a difference in the probability distributions of the amounts of corrosion among the three types of sealer? Use $\alpha = .05$.

CORRODE

Metal	Sealer 1	Sealer 2	Sealer 3
1	4.6	4.2	4.9
2	7.2	6.4	7.0
3	3.4	3.5	3.4
4	6.2	5.3	5.9
5	8.4	6.8	7.8
6	5.6	4.8	5.7
7	3.7	3.7	4.1
8	6.1	6.2	6.4
9	4.9	4.1	4.2
10	5.2	5.0	5.1

14.100 Treating hypoglycemia. Hypoglycemia is a condition in which blood sugar is below normal limits. To compare the effectiveness of two compounds, X and Y, for treating hypoglycemia, each compound is applied to half the diaphragms of each of seven white mice. Blood glucose uptake in milligrams per gram of tissue is measured for each half, producing the results listed in the table below. Do the data provide sufficient evidence to indicate that one of the compounds tends to produce higher blood sugar uptake readings than the other? Test using $\alpha = .10$.

HYPOGLY

Mouse	X	Y	Mouse	X	Y
1	4.7	5.1	5	7.0	6.1
2	3.3	4.6	6	4.7	4.1
3	8.5	8.7	7	5.2	5.1
4	3.9	3.6			

14.101 Word association study. Three lists of words, representing three levels of abstractness, are randomly assigned to 21 experimental subjects so that seven subjects receive each list. The subjects are asked to respond to each word on their list with as many associated words as possible within a given period of time. A subject's score is the total number of word associates, summing over all words in the list. Scores for each list are given in the accompanying table. Do the data provide sufficient evidence to indicate a difference (shift in location) between at least two of the probability distributions of the numbers of word associates that subjects can name for the three lists? Use $\alpha = .05$.

WORDLIST

List 1	List 2	List 3
48	41	18
43	36	42
39	29	28
57	40	38
21	35	15
47	45	33
58	32	31

14.102 Eye pupil site and deception. An experiment was designed to study whether eye pupil size is related to a person's attempt at deception. Eight students were asked to respond verbally to a series of questions. Before the questioning began, the pupil size of each student was noted and the students were instructed to answer some of the questions dishonestly. (The number of questions answered dishonestly was left to individual choice.) During questioning, the percentage increase in pupil size was recorded. Each student was then given a deception score based on the proportion of questions answered dishonestly.

(High scores indicate a large number of deceptive responses.) The results are given in the next table. Can you conclude that the percentage increase in eye pupil size is positively correlated with deception score? Use $\alpha = .05$.

⊚ DECEPEYE

Student	Deception Score	Percentage Increase in Pupil Size
1	87	10
2	63	6
3	95	11
4	50	7
5	43	0
6	89	15
7	33	4
8	55	5

14.103 Monitoring highway speeders. A state highway patrol was interested in knowing whether frequent patrolling of highways substantially reduces the number of speeders. Two similar interstate highways were selected for the study—one heavily patrolled and the other only occasionally patrolled. After 1 month, random samples of 100 cars were chosen on each highway, and the number of cars exceeding the speed limit was recorded. This process was repeated on 5 randomly selected days. The data are shown in the table.

⊚ HWPATROL

Day	Highway 1 Heavily Patrolled	Highway 2 Occasionally Patrolled
1	35	60
2	40	36
3	25	48
4	38	54
5	47	63

a. Do the data provide evidence to indicate that the heavily patrolled highway tends to have fewer speeders per 100 cars than the occasionally patrolled highway? Test using $\alpha = .05$.

b. Use the paired t-test with $\alpha = .05$ to compare the population mean number of speeders per 100 cars for the two highways. What assumptions are necessary for this procedure to be valid?

14.104 Fluoride in drinking water. Many water treatment facilities supplement the natural fluoride concentration with hydrofluosilicic acid in order to reach a target concentration of fluoride in drinking water. Certain levels are thought to enhance dental health, but very high concentrations can be dangerous. Suppose that one such treatment plant targets .75 milligrams per liter (mg/L) for their water. The plant tests 25 samples each day to determine whether the median level differs from the target.

a. Set up the null and alternative hypotheses.

b. Set up the test statistic and rejection region using $\alpha = .10$.

c. Explain the implication of a Type I error in the context of this application. A Type II error.

d. Suppose that one day's samples result in 18 values that exceed .75 mg/L. Conduct the test and state the appropriate conclusion in the context of this application.

e. When it was suggested to the plant's supervisor that a t-test should be used to conduct the daily test, she replied that the probability distribution of the fluoride concentrations was "heavily skewed to the right." Show graphically what she meant by this, and explain why this is a reason to prefer the sign test to the t-test.

14.105 Defect rates over time. A manufacturer wants to determine whether the number of defectives produced by its employees tends to increase as the day progresses. Unknown to the employees, a complete inspection is made of every item that was produced on one day, and the hourly fraction defective is recorded. The resulting data are given in the table. Do they provide evidence that the fraction defective increases as the day progresses? Test at the $\alpha = .05$ level.

⊚ DEFECTS

Hour	Fraction Defective
1	.02
2	.05
3	.03
4	.08
5	.06
6	.09
7	.11
8	.10

14.106 Ranking wines. Two expert wine tasters were asked to rank six brands of wine. Their rankings are shown below.

⊚ WINETASTE

Brand	Expert 1	Expert 2
A	6	5
B	5	6
C	1	2
D	3	1
E	2	4
F	4	3

Do the data indicate a positive correlation in the rankings of the two experts? Test Using $\alpha = .10$.

14.107 Controlling for corn mold. A serious drought-related problem for farmers is the spread of aflatoxin, a highly toxic substance generated by mold, which contaminates field corn. Three sprays, A, B, and C, have been developed to control aflatoxin in field corn. To determine whether differences exist among the sprays, 10 ears of corn are randomly chosen from a contaminated corn field and each is divided into three pieces of equal size. The sprays are then randomly assigned to the pieces for each ear of corn, thus setting up a randomized block design. The table gives the amount (in parts per billion) of aflatoxin present in the corn samples after spraying. Determine whether there is evidence that the distributions of the levels of aflatoxin in corn differ for at least two of the three sprays. Test at $\alpha = .05$.

AFLATOX

Ear	Spray		
	A	B	C
1	21	23	15
2	29	30	21
3	16	19	18
4	20	19	18
5	13	10	14
6	5	12	6
7	18	18	12
8	26	32	21
9	17	20	9
10	4	10	2

Critical Thinking Challenges

14.108 Self-managed work teams and family life. Refer to the *Quality Management Journal* (Summer 1995) study of self-managed work teams (SMWTs), Exercise 9.135 (p. 506). Recall that the researchers investigated the connection between SMWT work characteristics and workers' perceptions of positive spillover into family life (one group of workers reported positive spillover of work skills to family life while another group did not report positive work spillover). The data collected on 114 AT&T employees, saved in the **SPILLOVER** file, are described below. In Exercise 9.135, you compared the two groups of workers on each characteristic using the parametric methods of Chapter 9. Reanalyze the data using nonparametrics. Are the job-related characteristics most highly associated with positive work spillover the same as those identified in Exercise 9.135? Comment on the validity of the parametric and nonparametric results.

SPILLOVER
Variables Measured in the SMWT Survey

Characteristic	Variable
Information Flow	Use of creative ideas (7-point scale)
Information Flow	Utilization of information (7-point scale)
Decision Making	Participation in decision regarding personnel matters (7-point scale)
Job	Good use of skills (7-point scale)
Job	Task identity (7-point scale)
Demographic	Age (years)
Demographic	Education (years)
Demographic	Gender (male or female)

Student Projects

In Chapters 10 and 14 we discussed two methods of analyzing a randomized block design. When the populations have normal probability distributions and their variances are equal, we can employ the analysis of variance described in Chapter 10. Otherwise, we can use the Friedman F_r-test.

In the Student Projects section of Chapter 10, we asked you to conduct a randomized block design to compare supermarket prices, and to use an analysis of variance to interpret the data. Now use the Friedman F_r-test to compare the supermarket prices.

How do the results of the two analyses compare? Explain the similarity (or lack of similarity) between the two results.

REFERENCES

Agresti, A., and Agresti, B. F. *Statistical Methods for the Social Sciences*, 2nd ed. San Francisco: Dellen, 1986.

Conover, W. J. *Practical Nonparametric Statistics*, 2nd ed. New York: Wiley, 1980.

Daniel, W. W. *Applied Nonparametric Statistics*, 2nd ed. Boston: PWS-Kent, 1990.

Dunn, O. J. "Multiple comparisons using rank sums." *Technometrics*, Vol. 6, 1964.

Friedman, M. "The use of ranks to avoid the assumption of normality implicit in the analysis of variance." *Journal of the American Statistical Association*, Vol. 32, 1937.

Gibbons, J. D. *Nonparametric Statistical Inference*, 2nd ed. New York: McGraw-Hill, 1985.

Hollander, M., and Wolfe, D. A. *Nonparametric Statistical Methods*. New York: Wiley, 1973.

Kruskal, W. H., and Wallis, W. A. "Use of ranks in one-criterion variance analysis." *Journal of the American Statistical Association*, Vol. 47, 1952.

Lehmann, E. L. *Nonparametrics: Statistical Methods Based on Ranks*. San Francisco: Holden-Day, 1975.

Marascuilo, L. A., and McSweeney, M. *Nonparametric and Distribution-Free Methods for the Social Sciences*. Monterey, Ca.: Brooks/Cole, 1977.

Wilcoxon, F., and Wilcox, R. A. "Some rapid approximate statistical procedures." The American Cyanamid Co., 1964.

Using Technology

Nonparametric Tests Using MINITAB

Sign Test

The MINITAB worksheet file with the sample data should contain a single quantitative variable. Click on the "Stat" button on the MINITAB menu bar, then click on "Nonparametrics" and "1-Sample Sign", as shown in Figure 14.M.1.

The resulting dialog box appears as shown in Figure 14.M.2. Enter the quantitative variable to be analyzed in the "Variables" box. Select the "Test median" option and specify the hypothesized value of the median and the form of the

Figure 14.M.1
MINITAB Nonparametric Menu Options

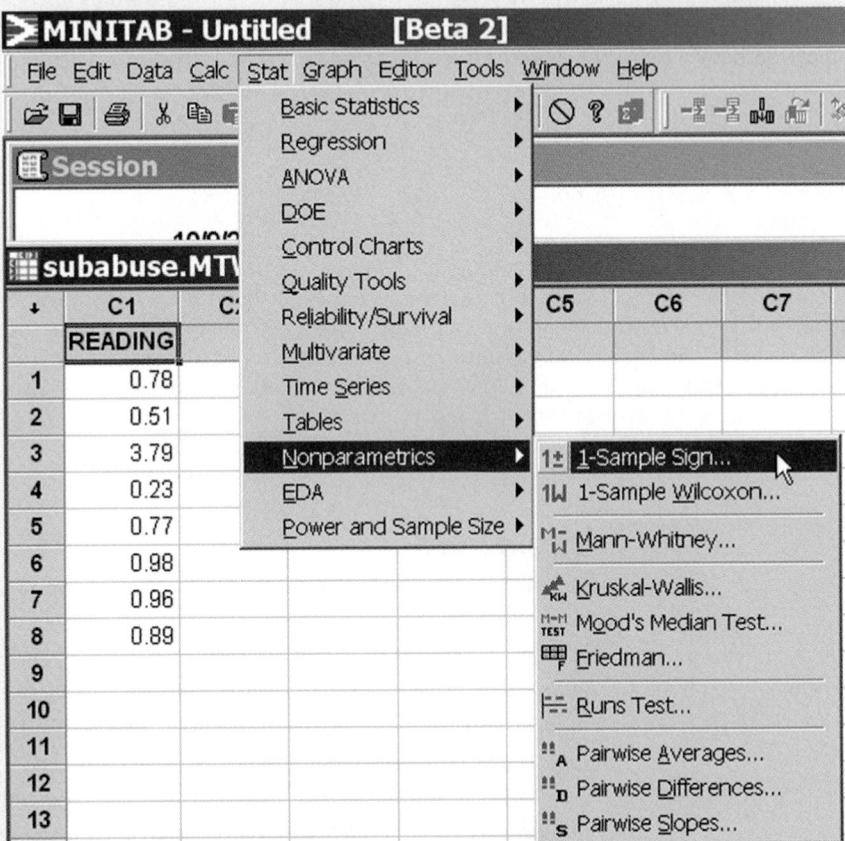

Figure 14.M.2
MINITAB 1-Sample Sign
Dialog Box

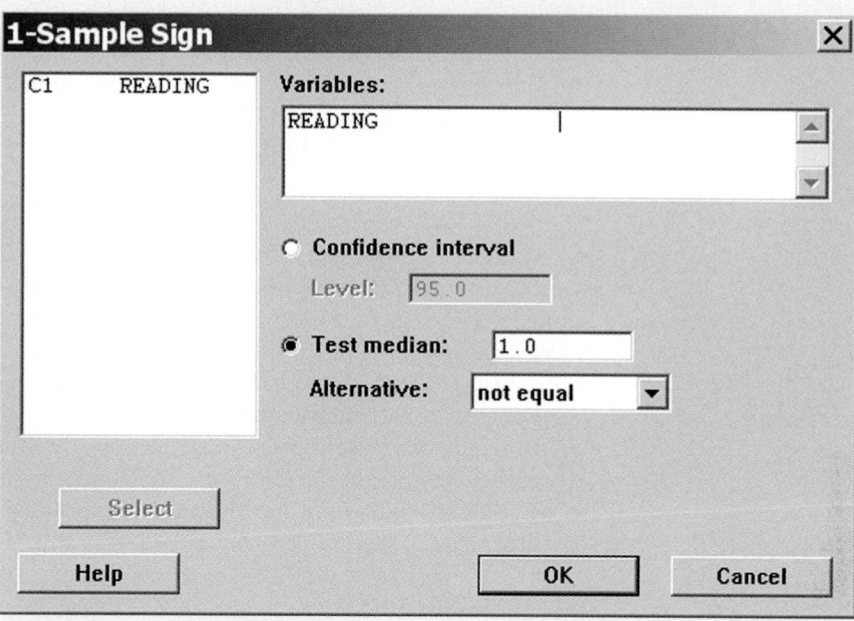

alternative hypothesis ("not equal", "less than", or "greater than"). Click "OK" to generate the MINITAB printout.

Rank Sum Test

The MINITAB worksheet file with the sample data should contain two quantitative variables, one for each of the two samples being compared. Click on the "Stat" button on the MINITAB menu bar, then click on "Nonparametrics" and "Mann-Whitney" (see Figure 14.M.1). The resulting dialog box appears as shown in Figure 14.M.3.

Figure 14.M.3
MINITAB Mann-Whitney
Dialog Box

Specify the variable for the first sample in the "First Sample" box and the variable for the second sample in the "Second Sample" box. Specify the form of the alternative hypothesis ("not equal", "less than", or "greater than"), then click "OK" to generate the MINITAB printout.

Signed-Rank Test

The MINITAB worksheet file with the matched-pairs data should contain two quantitative variables, one for each of the two groups being compared. You will need to compute the difference between these two variables and save it in a column on the worksheet. (Use the "Calc" button on the MINITAB menu bar.) Next, click on the "Stat" button on the MINITAB menu bar, then click on "Nonparametrics" and "1-Sample Wilcoxon" (see Figure 14.M.1). The resulting dialog box appears as shown in Figure 14.M.4.

Enter the variable representing the paired differences in the "Variables" box. Select the "Test median" option and specify the hypothesized value of the median as "0". Select the form of the alternative hypothesis ("not equal", "less than", or "greater than"), then click "OK" to generate the MINITAB printout.

Kruskal-Wallis Test

The MINITAB worksheet file that contains the completely randomized design data should contain one quantitative variable (the response, or dependent, variable) and one factor variable with at least two levels. Click on the "Stat" button on the MINITAB menu bar, then click on "Nonparametrics" and "Kruskal-Wallis" (see Figure 14.M.1). The resulting dialog box appears as shown in Figure 14.M.5. Specify the response variable in the "Response" box and the factor variable in the "Factor" box. Click "OK" to generate the MINITAB printout.

Figure 14.M.4
MINITAB 1-Sample
Wilcoxon Dialog Box

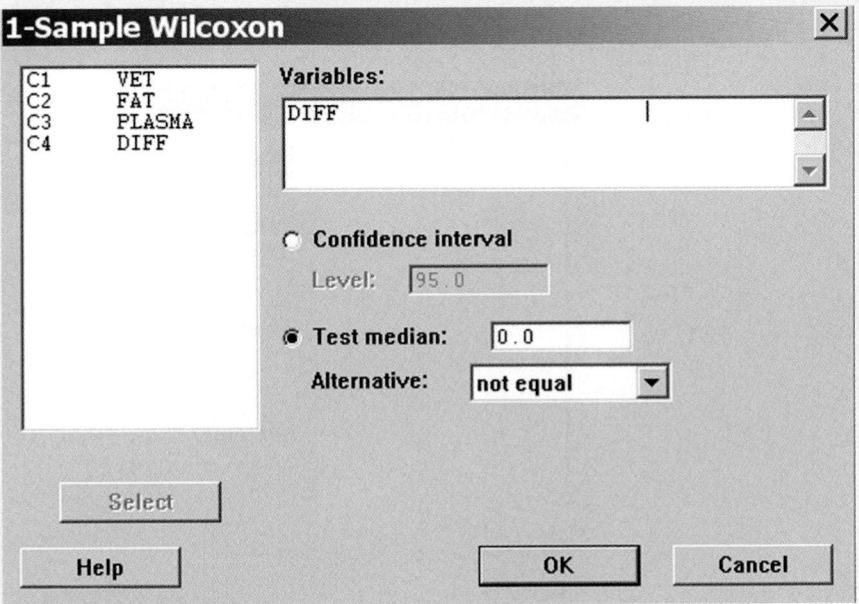

Figure 14.M.5
MINITAB Kruskal-Wallis
Dialog Box

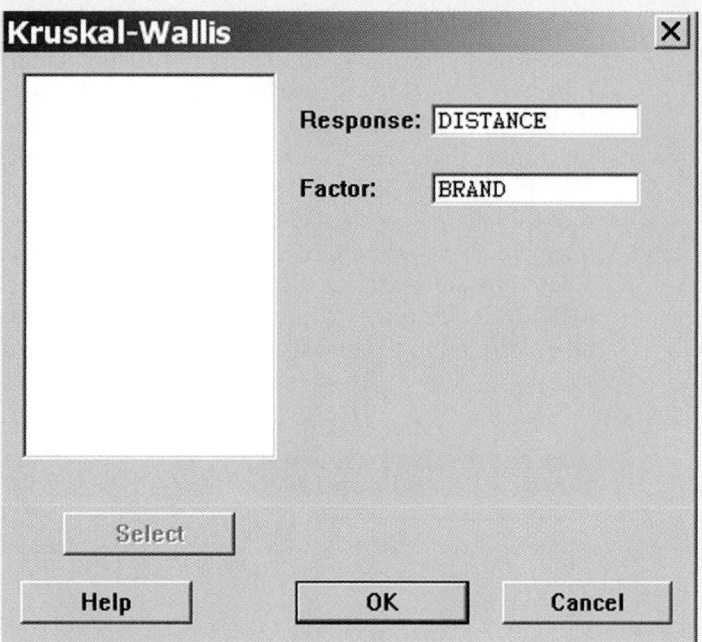

Friedman Test

The MINITAB spreadsheet file that contains the randomized block design data should contain one quantitative variable (the response, or dependent, variable) and one factor variable and one blocking variable. Click on the "Stat" button on the MINITAB menu bar, then click on "Nonparametrics" and "Friedman" (see Figure 14.M.1). The resulting dialog box appears as shown in Figure 14.M.6. Specify the response, treatment, and blocking variables in the appropriate boxes, then click "OK" to generate the MINITAB printout.

Figure 14.M.6
MINITAB Friedman
Dialog Box

Rank Correlation

To obtain Spearman's rank correlation coefficient in MINITAB, you must first rank the values of the two quantitative variables of interest. Click the "Calc" button on the MINITAB menu bar, and create two additional columns, one for the ranks of the *x*-variable and one for the ranks of the *y*-variable. (Use the "Rank" function on the MINITAB calculator.) Next, click on the "Stat" button on the main menu bar, then click on "Basic Statistics" and "Correlation". The resulting dialog box appears in Figure 14.M.7. Enter the ranked variables in the "Variables" box and unselect the "Display p-values" option. Click "OK" to obtain the MINITAB printout. (You will need to look up the critical value of Spearman's rank correlation to conduct the test.)

Figure 14.M.7
MINITAB Correlation
Dialog Box

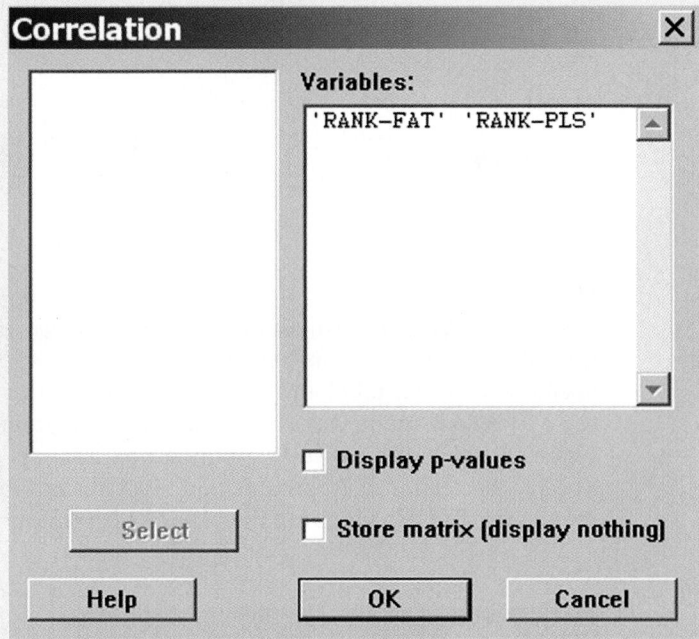

Appendix A
Tables

TABLE I Random Numbers

Row \ Column	1	2	3	4	5	6	7	8	9	10	11	12	13	14
1	10480	15011	01536	02011	81647	91646	69179	14194	62590	36207	20969	99570	91291	90700
2	22368	46573	25595	85393	30995	89198	27982	53402	93965	34095	52666	19174	39615	99505
3	24130	48360	22527	97265	76393	64809	15179	24830	49340	32081	30680	19655	63348	58629
4	42167	93093	06243	61680	07856	16376	39440	53537	71341	57004	00849	74917	97758	16379
5	37570	39975	81837	16656	06121	91782	60468	81305	49684	60672	14110	06927	01263	54613
6	77921	06907	11008	42751	27756	53498	18602	70659	90655	15053	21916	81825	44394	42880
7	99562	72905	56420	69994	98872	31016	71194	18738	44013	48840	63213	21069	10634	12952
8	96301	91977	05463	07972	18876	20922	94595	56869	69014	60045	18425	84903	42508	32307
9	89579	14342	63661	10281	17453	18103	57740	84378	25331	12566	58678	44947	05585	56941
10	85475	36857	53342	53988	53060	59533	38867	62300	08158	17983	16439	11458	18593	64952
11	28918	69578	88231	33276	70997	79936	56865	05859	90106	31595	01547	85590	91610	78188
12	63553	40961	48235	03427	49626	69445	18663	72695	52180	20847	12234	90511	33703	90322
13	09429	93969	52636	92737	88974	33488	36320	17617	30015	08272	84115	27156	30613	74952
14	10365	61129	87529	85689	48237	52267	67689	93394	01511	26358	85104	20285	29975	89868
15	07119	97336	71048	08178	77233	13916	47564	81056	97735	85977	29372	74461	28551	90707
16	51085	12765	51821	51259	77452	16308	60756	92144	49442	53900	70960	63990	75601	40719
17	02368	21382	52404	60268	89368	19885	55322	44819	01188	65255	64835	44919	05944	55157
18	01011	54092	33362	94904	31273	04146	18594	29852	71585	85030	51132	01915	92747	64951
19	52162	53916	46369	58586	23216	14513	83149	98736	23495	64350	94738	17752	35156	35749
20	07056	97628	33787	09998	42698	06691	76988	13602	51851	46104	88916	19509	25625	58104
21	48663	91245	85828	14346	09172	30168	90229	04734	59193	22178	30421	61666	99904	32812
22	54164	58492	22421	74103	47070	25306	76468	26384	58151	06646	21524	15227	96909	44592
23	32639	32363	05597	24200	13363	38005	94342	28728	35806	06912	17012	64161	18296	22851
24	29334	27001	87637	87308	58731	00256	45834	15398	46557	41135	10367	07684	36188	18510
25	02488	33062	28834	07351	19731	92420	60952	61280	50001	67658	32586	86679	50720	94953
26	81525	72295	04839	96423	24878	82651	66566	14778	76797	14780	13300	87074	79666	95725
27	29676	20591	68086	26432	46901	20849	89768	81536	86645	12659	92259	57102	80428	25280
28	00742	57392	39064	66432	84673	40027	32832	61362	98947	96067	64760	64584	96096	98253
29	05366	04213	25669	26422	44407	44048	37937	63904	45766	66134	75470	66520	34693	90449
30	91921	26418	64117	94305	26766	25940	39972	22209	71500	64568	91402	42416	07844	69618
31	00582	04711	87917	77341	42206	35126	74087	99547	81817	42607	43808	76655	62028	76630
32	00725	69884	62797	56170	86324	88072	76222	36086	84637	93161	76038	65855	77919	88006
33	69011	65795	95876	55293	18988	27354	26575	08625	40801	59920	29841	80150	12777	48501
34	25976	57948	29888	88604	67917	48708	18912	82271	65424	69774	33611	54262	85963	03547
35	09763	83473	73577	12908	30883	18317	28290	35797	05998	41688	34952	37888	38917	88050

(continued)

TABLE I Continued

Row	1	2	3	4	5	6	7	8	9	10	11	12	13	14
36	91576	42595	27958	30134	04024	86385	29880	99730	55536	84855	29080	09250	79656	73211
37	17955	56349	90999	49127	20044	59931	06115	20542	18059	02008	73708	83517	36103	42791
38	46503	18584	18845	49618	02304	51038	20655	58727	28168	15475	56942	53389	20562	87338
39	92157	89634	94824	78171	84610	82834	09922	25417	44137	48413	25555	21246	35509	20468
40	14577	62765	35605	81263	39667	47358	56873	56307	61607	49518	89656	20103	77490	18062
41	98427	07523	33362	64270	01638	92477	66969	98420	04880	45585	46565	04102	46880	45709
42	34914	63976	88720	82765	34476	17032	87589	40836	32427	70002	70663	88863	77775	69348
43	70060	28277	39475	46473	23219	53416	94970	25832	69975	94884	19661	72828	00102	66794
44	53976	54914	06990	67245	68350	82948	11398	42878	80287	88267	47363	46634	06541	97809
45	76072	29515	40980	07391	58745	25774	22987	80059	39911	96189	41151	14222	60697	59583
46	90725	52210	83974	29992	65831	38857	50490	83765	55657	14361	31720	57375	56228	41546
47	64364	67412	33339	31926	14883	24413	59744	92351	97473	89286	35931	04110	23726	51900
48	08962	00358	31662	25388	61642	34072	81249	35648	56891	69352	48373	45578	78547	81788
49	95012	68379	93526	70765	10592	04542	76463	54328	02349	17247	28865	14777	62730	92277
50	15664	10493	20492	38391	91132	21999	59516	81652	27195	48223	46751	22923	32261	85653
51	16408	81899	04153	53381	79401	21438	83035	92350	36693	31238	59649	91754	72772	02338
52	18629	81953	05520	91962	04739	13092	97662	24822	94730	06496	35090	04822	86774	98289
53	73115	35101	47498	87637	99016	71060	88824	71013	18735	20286	23153	72924	35165	43040
54	57491	16703	23167	49323	45021	33132	12544	41035	80780	45393	44812	12512	98931	91202
55	30405	83946	23792	14422	15059	45799	22716	19792	09983	74353	68668	30429	70735	25499
56	16631	35006	85900	98275	32388	52390	16815	69290	82732	38480	73817	32523	41961	44437
57	96773	20206	42559	78985	05300	22164	24369	54224	35083	19687	11052	91491	60383	19746
58	38935	64202	14349	82674	66523	44133	00697	35552	35970	19124	63318	29686	03387	59846
59	31624	76384	17403	53363	44167	64486	64758	75366	76554	31601	12614	33072	60332	92325
60	78919	19474	23632	27889	47914	02584	37680	20801	72152	39339	34806	08930	85001	87820
61	03931	33309	57047	74211	63445	17361	62825	39908	05607	91284	68833	25570	38818	46920
62	74426	33278	43972	10110	89917	15665	52872	73823	73144	88662	88970	74492	51805	99378
63	09066	00903	20795	95452	92648	45454	09552	88815	16553	51125	79375	97596	16296	66092
64	42238	12426	87025	14267	20979	04508	64535	31355	86064	29472	47689	05974	52468	16834
65	16153	08002	26504	41744	81959	65642	74240	56302	00033	67107	77510	70625	28725	34191
66	21457	40742	29820	96783	29400	21840	15035	34537	33310	06116	95240	15957	16572	06004
67	21581	57802	02050	89728	17937	37621	47075	42080	97403	48626	68995	43805	33386	21597
68	55612	78095	83197	33732	05810	24813	86902	60397	16489	03264	88525	42786	05269	92532
69	44657	66999	99324	51281	84463	60563	79312	93454	68876	25471	93911	25650	12682	73572
70	91340	84979	46949	81973	37949	61023	43997	15263	80644	43942	89203	71795	99533	50501

(continued)

TABLE I Continued

Row \ Column	1	2	3	4	5	6	7	8	9	10	11	12	13	14
71	91227	21199	31935	27022	84067	05462	35216	14486	29891	68607	41867	14951	91696	85065
72	50001	38140	66321	19924	72163	09538	12151	06878	91903	18749	34405	56087	82790	70925
73	65390	05224	72958	28609	81406	39147	25549	48542	42627	45233	57202	94617	23772	07896
74	27504	96131	83944	41575	10573	08619	64482	73923	36152	05184	94142	25299	84387	34925
75	37169	94851	39117	89632	00959	16487	65536	49071	39782	17095	02330	74301	00275	48280
76	11508	70225	51111	38351	19444	66499	71945	05422	13442	78675	84081	66938	93654	59894
77	37449	30362	06694	54690	04052	53115	62757	95348	78662	11163	81651	50245	34971	52924
78	46515	70331	85922	38329	57015	15765	97161	17869	45349	61796	66345	81073	49106	79860
79	30986	81223	42416	58353	21532	30502	32305	86482	05174	07901	54339	58861	74818	46942
80	63798	64995	46583	09785	44160	78128	83991	42865	92520	83531	80377	35909	81250	54238
81	82486	84846	99254	67632	43218	50076	21361	64816	51202	88124	41870	52689	51275	83556
82	21885	32906	92431	09060	64297	51674	64126	62570	26123	05155	59194	52799	28225	85762
83	60336	98782	07408	53458	13564	59089	26445	29789	85205	41001	12535	12133	14645	23541
84	43937	46891	24010	25560	86355	33941	25786	54990	71899	15475	95434	98227	21824	19585
85	97656	63175	89303	16275	07100	92063	21942	18611	47348	20203	18534	03862	78095	50136
86	03299	01221	05418	38982	55758	92237	26759	86367	21216	98442	08303	56613	91511	75928
87	79626	06486	03574	17668	07785	76020	79924	25651	83325	88428	85076	72811	22717	50585
88	85636	68335	47539	03129	65651	11977	02510	26113	99447	68645	34327	15152	55230	93448
89	18039	14367	64337	06177	12143	46609	32989	74014	64708	00533	35398	58408	13261	47908
90	08362	15656	60627	36478	65648	16764	53412	09013	07832	41574	17639	82163	60859	75567
91	79556	29068	04142	16268	15387	12856	66227	38358	22478	73373	88732	09443	82558	05250
92	92608	82674	27072	32534	17075	27698	98204	63863	11951	34648	88022	56148	34925	57031
93	23982	25835	40055	67006	12293	02753	14827	23235	35071	99704	37543	11601	35503	85171
94	09915	96306	05908	97901	28395	14186	00821	80703	70426	75647	76310	88717	37890	40129
95	59037	33300	26695	62247	69927	76123	50842	43834	86654	70959	79725	93872	28117	19233
96	42488	78077	69882	61657	34136	79180	97526	43092	04098	73571	80799	76536	71255	64239
97	46764	86273	63003	93017	31204	36692	40202	35275	57306	55543	53203	18098	47625	88684
98	03237	45430	55417	63282	90816	17349	88298	90183	36600	78406	06216	95787	42579	90730
99	86591	81482	52667	61582	14972	90053	89534	76036	49199	43716	97548	04379	46370	28672
100	38534	01715	94964	87288	65680	43772	39560	12918	86537	62738	19636	51132	25739	56947

Source: Abridged from W. H. Beyer (ed.). *CRC Standard Mathematical Tables*, 24th edition. (Cleveland: The Chemical Rubber Company), 1976. Reproduced by permission of the publisher.

TABLE II Binomial Probabilities

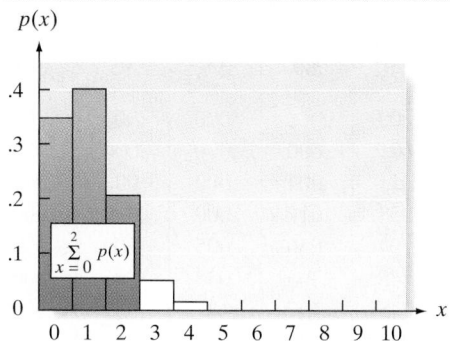

Tabulated values are $\sum_{x=0}^{k} p(x)$. *(Computations are rounded at the third decimal place.)*

a. $n = 5$

k \ p	.01	.05	.10	.20	.30	.40	.50	.60	.70	.80	.90	.95	.99
0	.951	.774	.590	.328	.168	.078	.031	.010	.002	.000	.000	.000	.000
1	.999	.977	.919	.737	.528	.337	.188	.087	.031	.007	.000	.000	.000
2	1.000	.999	.991	.942	.837	.683	.500	.317	.163	.058	.009	.001	.000
3	1.000	1.000	1.000	.993	.969	.913	.812	.663	.472	.263	.081	.023	.001
4	1.000	1.000	1.000	1.000	.998	.990	.969	.922	.832	.672	.410	.226	.049

b. $n = 6$

k \ p	.01	.05	.10	.20	.30	.40	.50	.60	.70	.80	.90	.95	.99
0	.941	.735	.531	.262	.118	.047	.016	.004	.001	.000	.000	.000	.000
1	.999	.967	.886	.655	.420	.233	.109	.041	.011	.002	.000	.000	.000
2	1.000	.998	.984	.901	.744	.544	.344	.179	.070	.017	.001	.000	.000
3	1.000	1.000	.999	.983	.930	.821	.656	.456	.256	.099	.016	.002	.000
4	1.000	1.000	1.000	.998	.989	.959	.891	.767	.580	.345	.114	.033	.001
5	1.000	1.000	1.000	1.000	.999	.996	.984	.953	.882	.738	.469	.265	.059

c. $n = 7$

k \ p	.01	.05	.10	.20	.30	.40	.50	.60	.70	.80	.90	.95	.99
0	.932	.698	.478	.210	.082	.028	.008	.002	.000	.000	.000	.000	.000
1	.998	.956	.850	.577	.329	.159	.063	.019	.004	.000	.000	.000	.000
2	1.000	.996	.974	.852	.647	.420	.227	.096	.029	.005	.000	.000	.000
3	1.000	1.000	.997	.967	.874	.710	.500	.290	.126	.033	.003	.000	.000
4	1.000	1.000	1.000	.995	.971	.904	.773	.580	.353	.148	.026	.004	.000
5	1.000	1.000	1.000	1.000	.996	.981	.937	.841	.671	.423	.150	.044	.002
6	1.000	1.000	1.000	1.000	1.000	.998	.992	.972	.918	.790	.522	.302	.068

(continued)

TABLE II Continued

d. n = 8

k \ p	.01	.05	.10	.20	.30	.40	.50	.60	.70	.80	.90	.95	.99
0	.923	.663	.430	.168	.058	.017	.004	.001	.000	.000	.000	.000	.000
1	.997	.943	.813	.503	.255	.106	.035	.009	.001	.000	.000	.000	.000
2	1.000	.994	.962	.797	.552	.315	.145	.050	.011	.001	.000	.000	.000
3	1.000	1.000	.995	.944	.806	.594	.363	.174	.058	.010	.000	.000	.000
4	1.000	1.000	1.000	.990	.942	.826	.637	.406	.194	.056	.005	.000	.000
5	1.000	1.000	1.000	.999	.989	.950	.855	.685	.448	.203	.038	.006	.000
6	1.000	1.000	1.000	1.000	.999	.991	.965	.894	.745	.497	.187	.057	.003
7	1.000	1.000	1.000	1.000	1.000	.999	.996	.983	.942	.832	.570	.337	.077

e. n = 9

k \ p	.01	.05	.10	.20	.30	.40	.50	.60	.70	.80	.90	.95	.99
0	.914	.630	.387	.134	.040	.010	.002	.000	.000	.000	.000	.000	.000
1	.997	.929	.775	.436	.196	.071	.020	.004	.000	.000	.000	.000	.000
2	1.000	.992	.947	.738	.463	.232	.090	.025	.004	.000	.000	.000	.000
3	1.000	.999	.992	.914	.730	.483	.254	.099	.025	.003	.000	.000	.000
4	1.000	1.000	.999	.980	.901	.733	.500	.267	.099	.020	.001	.000	.000
5	1.000	1.000	1.000	.997	.975	.901	.746	.517	.270	.086	.008	.001	.000
6	1.000	1.000	1.000	1.000	.996	.975	.910	.768	.537	.262	.053	.008	.000
7	1.000	1.000	1.000	1.000	1.000	.996	.980	.929	.804	.564	.225	.071	.003
8	1.000	1.000	1.000	1.000	1.000	1.000	.998	.990	.960	.866	.613	.370	.086

f. n = 10

k \ p	.01	.05	.10	.20	.30	.40	.50	.60	.70	.80	.90	.95	.99
0	.904	.599	.349	.107	.028	.006	.001	.000	.000	.000	.000	.000	.000
1	.996	.914	.736	.736	.149	.046	.011	.002	.000	.000	.000	.000	.000
2	1.000	.988	.930	.678	.383	.167	.055	.012	.002	.000	.000	.000	.000
3	1.000	.999	.987	.879	.650	.382	.172	.055	.011	.001	.000	.000	.000
4	1.000	1.000	.998	.967	.850	.633	.377	.166	.047	.006	.000	.000	.000
5	1.000	1.000	1.000	.999	.953	.834	.623	.367	.150	.033	.002	.000	.000
6	1.000	1.000	1.000	.999	.989	.945	.828	.618	.350	.121	.013	.001	.000
7	1.000	1.000	1.000	1.000	.998	.988	.945	.833	.617	.322	.070	.012	.000
8	1.000	1.000	1.000	1.000	1.000	.998	.989	.954	.851	.624	.264	.086	.004
9	1.000	1.000	1.000	1.000	1.000	1.000	.999	.994	.972	.893	.651	.401	.096

(continued)

TABLE II Continued

g. n = 15

k \ p	.01	.05	.10	.20	.30	.40	.50	.60	.70	.80	.90	.95	.99
0	.860	.463	.206	.035	.005	.000	.000	.000	.000	.000	.000	.000	.000
1	.990	.829	.549	.167	.035	.005	.000	.000	.000	.000	.000	.000	.000
2	1.000	.964	.816	.398	.127	.027	.004	.000	.000	.000	.000	.000	.000
3	1.000	.995	.944	.648	.297	.091	.018	.002	.000	.000	.000	.000	.000
4	1.000	.999	.987	.838	.515	.217	.059	.009	.001	.000	.000	.000	.000
5	1.000	1.000	.998	.939	.722	.403	.151	.034	.004	.000	.000	.000	.000
6	1.000	1.000	1.000	.982	.869	.610	.304	.095	.015	.001	.000	.000	.000
7	1.000	1.000	1.000	.996	.950	.787	.500	.213	.050	.004	.000	.000	.000
8	1.000	1.000	1.000	.999	.985	.905	.696	.390	.131	.018	.000	.000	.000
9	1.000	1.000	1.000	1.000	.996	.966	.849	.597	.278	.061	.002	.000	.000
10	1.000	1.000	1.000	1.000	.999	.991	.941	.783	.485	.164	.013	.001	.000
11	1.000	1.000	1.000	1.000	1.000	.998	.982	.909	.703	.352	.056	.005	.000
12	1.000	1.000	1.000	1.000	1.000	1.000	.996	.973	.873	.602	.184	.036	.000
13	1.000	1.000	1.000	1.000	1.000	1.000	1.000	.995	.965	.833	.451	.171	.010
14	1.000	1.000	1.000	1.000	1.000	1.000	1.000	1.000	.995	.965	.794	.537	.140

h. n = 20

k \ p	.01	.05	.10	.20	.30	.40	.50	.60	.70	.80	.90	.95	.99
0	.818	.358	.122	.012	.001	.000	.000	.000	.000	.000	.000	.000	.000
1	.983	.736	.392	.069	.008	.001	.000	.000	.000	.000	.000	.000	.000
2	.999	.925	.677	.206	.035	.004	.000	.000	.000	.000	.000	.000	.000
3	1.000	.984	.867	.411	.107	.016	.001	.000	.000	.000	.000	.000	.000
4	1.000	.997	.957	.630	.238	.051	.006	.000	.000	.000	.000	.000	.000
5	1.000	1.000	.989	.804	.416	.126	.021	.002	.000	.000	.000	.000	.000
6	1.000	1.000	.998	.913	.608	.250	.058	.006	.000	.000	.000	.000	.000
7	1.000	1.000	1.000	.968	.772	.416	.132	.021	.001	.000	.000	.000	.000
8	1.000	1.000	1.000	.990	.887	.596	.252	.057	.005	.000	.000	.000	.000
9	1.000	1.000	1.000	.997	.952	.755	.412	.128	.017	.001	.000	.000	.000
10	1.000	1.000	1.000	.999	.983	.872	.588	.245	.048	.003	.000	.000	.000
11	1.000	1.000	1.000	1.000	.995	.943	.748	.404	.113	.010	.000	.000	.000
12	1.000	1.000	1.000	1.000	.999	.979	.868	.584	.228	.032	.000	.000	.000
13	1.000	1.000	1.000	1.000	1.000	.994	.942	.750	.392	.087	.002	.000	.000
14	1.000	1.000	1.000	1.000	1.000	.998	.979	.874	.584	.196	.011	.000	.000
15	1.000	1.000	1.000	1.000	1.000	1.000	.994	.949	.762	.370	.043	.003	.000
16	1.000	1.000	1.000	1.000	1.000	1.000	.999	.984	.893	.589	.133	.016	.000
17	1.000	1.000	1.000	1.000	1.000	1.000	1.000	.996	.965	.794	.323	.075	.001
18	1.000	1.000	1.000	1.000	1.000	1.000	1.000	.999	.992	.931	.608	.264	.017
19	1.000	1.000	1.000	1.000	1.000	1.000	1.000	1.000	.999	.988	.878	.642	.182

(continued)

TABLE II Continued

i. $n = 25$

k \ p	.01	.05	.10	.20	.30	.40	.50	.60	.70	.80	.90	.95	.99
0	.778	.277	.072	.004	.000	.000	.000	.000	.000	.000	.000	.000	.000
1	.974	.642	.271	.027	.002	.000	.000	.000	.000	.000	.000	.000	.000
2	.998	.873	.537	.098	.009	.000	.000	.000	.000	.000	.000	.000	.000
3	1.000	.966	.764	.234	.033	.002	.000	.000	.000	.000	.000	.000	.000
4	1.000	.993	.902	.421	.090	.009	.000	.000	.000	.000	.000	.000	.000
5	1.000	.999	.967	.617	.193	.029	.002	.000	.000	.000	.000	.000	.000
6	1.000	1.000	.991	.780	.341	.074	.007	.000	.000	.000	.000	.000	.000
7	1.000	1.000	.998	.891	.512	.154	.022	.001	.000	.000	.000	.000	.000
8	1.000	1.000	1.000	.953	.677	.274	.054	.004	.000	.000	.000	.000	.000
9	1.000	1.000	1.000	.983	.811	.425	.115	.013	.000	.000	.000	.000	.000
10	1.000	1.000	1.000	.994	.902	.586	.212	.034	.002	.000	.000	.000	.000
11	1.000	1.000	1.000	.998	.956	.732	.345	.078	.006	.000	.000	.000	.000
12	1.000	1.000	1.000	1.000	.983	.846	.500	.154	.017	.000	.000	.000	.000
13	1.000	1.000	1.000	1.000	.994	.922	.655	.268	.044	.002	.000	.000	.000
14	1.000	1.000	1.000	1.000	.998	.966	.788	.414	.098	.006	.000	.000	.000
15	1.000	1.000	1.000	1.000	1.000	.987	.885	.575	.189	.017	.000	.000	.000
16	1.000	1.000	1.000	1.000	1.000	.996	.946	.726	.323	.047	.000	.000	.000
17	1.000	1.000	1.000	1.000	1.000	.999	.978	.846	.488	.109	.002	.000	.000
18	1.000	1.000	1.000	1.000	1.000	1.000	.993	.926	.659	.220	.009	.000	.000
19	1.000	1.000	1.000	1.000	1.000	1.000	.998	.971	.807	.383	.033	.001	.000
20	1.000	1.000	1.000	1.000	1.000	1.000	1.000	.991	.910	.579	.098	.007	.000
21	1.000	1.000	1.000	1.000	1.000	1.000	1.000	.998	.967	.766	.236	.034	.000
22	1.000	1.000	1.000	1.000	1.000	1.000	1.000	1.000	.991	.902	.463	.127	.002
23	1.000	1.000	1.000	1.000	1.000	1.000	1.000	1.000	.998	.973	.729	.358	.026
24	1.000	1.000	1.000	1.000	1.000	1.000	1.000	1.000	1.000	.996	.928	.723	.222

TABLE III   Poisson Probabilities

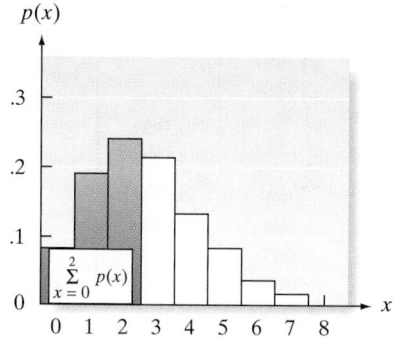

Tabulated values are $\sum_{x=0}^{k} p(x)$. *(Computations are rounded at the third decimal place.)*

λ \ k	0	1	2	3	4	5	6	7	8	9
.02	.980	1.000								
.04	.961	.999	1.000							
.06	.942	.998	1.000							
.08	.923	.997	1.000							
.10	.905	.995	1.000							
.15	.861	.990	.999	1.000						
.20	.819	.982	.999	1.000						
.25	.779	.974	.998	1.000						
.30	.741	.963	.996	1.000						
.35	.705	.951	.994	1.000						
.40	.670	.938	.992	.999	1.000					
.45	.638	.925	.989	.999	1.000					
.50	.607	.910	.986	.998	1.000					
.55	.577	.894	.982	.998	1.000					
.60	.549	.878	.977	.997	1.000					
.65	.522	.861	.972	.996	.999	1.000				
.70	.497	.844	.966	.994	.999	1.000				
.75	.472	.827	.959	.993	.999	1.000				
.80	.449	.809	.953	.991	.999	1.000				
.85	.427	.791	.945	.989	.998	1.000				
.90	.407	.772	.937	.987	.998	1.000				
.95	.387	.754	.929	.981	.997	1.000				
1.00	.368	.736	.920	.981	.996	.999	1.000			
1.1	.333	.699	.900	.974	.995	.999	1.000			
1.2	.301	.663	.879	.966	.992	.998	1.000			
1.3	.273	.627	.857	.957	.989	.998	1.000			
1.4	.247	.592	.833	.946	.986	.997	.999	1.000		
1.5	.223	.558	.809	.934	.981	.996	.999	1.000		

(continued)

TABLE III   Continued

λ \ k	0	1	2	3	4	5	6	7	8	9
1.6	.202	.525	.783	.921	.976	.994	.999	1.000		
1.7	.183	.493	.757	.907	.970	.992	.998	1.000		
1.8	.165	.463	.731	.891	.964	.990	.997	.999	1.000	
1.9	.150	.434	.704	.875	.956	.987	.997	.999	1.000	
2.0	.135	.406	.677	.857	.947	.983	.995	.999	1.000	
2.2	.111	.355	.623	.819	.928	.975	.993	.998	1.000	
2.4	.091	.308	.570	.779	.904	.964	.988	.997	.999	1.000
2.6	.074	.267	.518	.736	.877	.951	.983	.995	.999	1.000
2.8	.061	.231	.469	.692	.848	.935	.976	.992	.998	.999
3.0	.050	.199	.423	.647	.815	.916	.966	.988	.996	.999
3.2	.041	.171	.380	.603	.781	.895	.955	.983	.994	.998
3.4	.033	.147	.340	.558	.744	.871	.942	.977	.992	.997
3.6	.027	.126	.303	.515	.706	.844	.927	.969	.988	.996
3.8	.022	.107	.269	.473	.668	.816	.909	.960	.984	.994
4.0	.018	.092	.238	.433	.629	.785	.889	.949	.979	.992
4.2	.015	.078	.210	.395	.590	.753	.867	.936	.972	.989
4.4	.012	.066	.185	.359	.551	.720	.844	.921	.964	.985
4.6	.010	.056	.163	.326	.513	.686	.818	.905	.955	.980
4.8	.008	.048	.143	.294	.476	.651	.791	.887	.944	.975
5.0	.007	.040	.125	.265	.440	.616	.762	.867	.932	.968
5.2	.006	.034	.109	.238	.406	.581	.732	.845	.918	.960
5.4	.005	.029	.095	.213	.373	.546	.702	.822	.903	.951
5.6	.004	.024	.082	.191	.342	.512	.670	.797	.886	.941
5.8	.003	.021	.072	.170	.313	.478	.638	.771	.867	.929
6.0	.002	.017	.062	.151	.285	.446	.606	.744	.847	.916

λ	10	11	12	13	14	15	16
2.8	1.000						
3.0	1.000						
3.2	1.000						
3.4	.999	1.000					
3.6	.999	1.000					
3.8	.998	.999	1.000				
4.0	.997	.999	1.000				
4.2	.996	.999	1.000				
4.4	.994	.998	.999	1.000			
4.6	.992	.997	.999	1.000			
4.8	.990	.996	.999	1.000			
5.0	.986	.995	.998	.999	1.000		
5.2	.982	.993	.997	.999	1.000		
5.4	.977	.990	.996	.999	1.000		
5.6	.972	.988	.995	.998	.999	1.000	
5.8	.965	.984	.993	.997	.999	1.000	
6.0	.957	.980	.991	.996	.999	.999	1.000

(continued)

TABLE III Continued

λ \ k	0	1	2	3	4	5	6	7	8	9
6.2	.002	.015	.054	.134	.259	.414	.574	.716	.826	.902
6.4	.002	.012	.046	.119	.235	.384	.542	.687	.803	.886
6.6	.001	.010	.040	.105	.213	.355	.511	.658	.780	.869
6.8	.001	.009	.034	.093	.192	.327	.480	.628	.755	.850
7.0	.001	.007	.030	.082	.173	.301	.450	.599	.729	.830
7.2	.001	.006	.025	.072	.156	.276	.420	.569	.703	.810
7.4	.001	.005	.022	.063	.140	.253	.392	.539	.676	.788
7.6	.001	.004	.019	.055	.125	.231	.365	.510	.648	.765
7.8	.000	.004	.016	.048	.112	.210	.338	.481	.620	.741
8.0	.000	.003	.014	.042	.100	.191	.313	.453	.593	.717
8.5	.000	.002	.009	.030	.074	.150	.256	.386	.523	.653
9.0	.000	.001	.006	.021	.055	.116	.207	.324	.456	.587
9.5	.000	.001	.004	.015	.040	.089	.165	.269	.392	.522
10.0	.000	.000	.003	.010	.029	.067	.130	.220	.333	.458

	10	11	12	13	14	15	16	17	18	19
6.2	.949	.975	.989	.995	.998	.999	1.000			
6.4	.939	.969	.986	.994	.997	.999	1.000			
6.6	.927	.963	.982	.992	.997	.999	.999	1.000		
6.8	.915	.955	.978	.990	.996	.998	.999	1.000		
7.0	.901	.947	.973	.987	.994	.998	.999	1.000		
7.2	.887	.937	.967	.984	.993	.997	.999	.999	1.000	
7.4	.871	.926	.961	.980	.991	.996	.998	.999	1.000	
7.6	.854	.915	.954	.976	.989	.995	.998	.999	1.000	
7.8	.835	.902	.945	.971	.986	.993	.997	.999	1.000	
8.0	.816	.888	.936	.966	.983	.992	.996	.998	.999	1.000
8.5	.763	.849	.909	.949	.973	.986	.993	.997	.999	.999
9.0	.706	.803	.876	.926	.959	.978	.989	.995	.998	.999
9.5	.645	.752	.836	.898	.940	.967	.982	.991	.996	.998
10.0	.583	.697	.792	.864	.917	.951	.973	.986	.993	.997

	20	21	22
8.5	1.000		
9.0	1.000		
9.5	.999	1.000	
10.0	.998	.999	1.000

(continued)

TABLE III Continued

λ \ k	0	1	2	3	4	5	6	7	8	9
10.5	.000	.000	.002	.007	.021	.050	.102	.179	.279	.397
11.0	.000	.000	.001	.005	.015	.038	.079	.143	.232	.341
11.5	.000	.000	.001	.003	.011	.028	.060	.114	.191	.289
12.0	.000	.000	.001	.002	.008	.020	.046	.090	.155	.242
12.5	.000	.000	.000	.002	.005	.015	.035	.070	.125	.201
13.0	.000	.000	.000	.001	.004	.011	.026	.054	.100	.166
13.5	.000	.000	.000	.001	.003	.008	.019	.041	.079	.135
14.0	.000	.000	.000	.000	.002	.006	.014	.032	.062	.109
14.5	.000	.000	.000	.000	.001	.004	.010	.024	.048	.088
15.0	.000	.000	.000	.000	.001	.003	.008	.018	.037	.070

	10	11	12	13	14	15	16	17	18	19
10.5	.521	.639	.742	.825	.888	.932	.960	.978	.988	.994
11.0	.460	.579	.689	.781	.854	.907	.944	.968	.982	.991
11.5	.402	.520	.633	.733	.815	.878	.924	.954	.974	.986
12.0	.347	.462	.576	.682	.772	.844	.899	.937	.963	.979
12.5	.297	.406	.519	.628	.725	.806	.869	.916	.948	.969
13.0	.252	.353	.463	.573	.675	.764	.835	.890	.930	.957
13.5	.211	.304	.409	.518	.623	.718	.798	.861	.908	.942
14.0	.176	.260	.358	.464	.570	.669	.756	.827	.883	.923
14.5	.145	.220	.311	.413	.518	.619	.711	.790	.853	.901
15.0	.118	.185	.268	.363	.466	.568	.664	.749	.819	.875

	20	21	22	23	24	25	26	27	28	29
10.5	.997	.999	.999	1.000						
11.0	.995	.998	.999	1.000						
11.5	.992	.996	.998	.999	1.000					
12.0	.988	.994	.987	.999	.999	1.000				
12.5	.983	.991	.995	.998	.999	.999	1.000			
13.0	.975	.986	.992	.996	.998	.999	1.000			
13.5	.965	.980	.989	.994	.997	.998	.999	1.000		
14.0	.952	.971	.983	.991	.995	.997	.999	.999	1.000	
14.5	.936	.960	.976	.986	.992	.996	.998	.999	.999	1.000
15.0	.917	.947	.967	.981	.989	.994	.997	.998	.999	1.000

(continued)

TABLE III Continued

λ \ k	0	1	2	3	4	5	6	7	8	9
16	.000	.001	.004	.010	.022	.043	.077	.127	.193	.275
17	.000	.001	.002	.005	.013	.026	.049	.085	.135	.201
18	.000	.000	.001	.003	.007	.015	.030	.055	.092	.143
19	.000	.000	.001	.002	.004	.009	.018	.035	.061	.098
20	.000	.000	.000	.001	.002	.005	.011	.021	.039	.066
21	.000	.000	.000	.000	.001	.003	.006	.013	.025	.043
22	.000	.000	.000	.000	.001	.002	.004	.008	.015	.028
23	.000	.000	.000	.000	.000	.001	.002	.004	.009	.017
24	.000	.000	.000	.000	.000	.000	.001	.003	.005	.011
25	.000	.000	.000	.000	.000	.000	.001	.001	.003	.006

	14	15	16	17	18	19	20	21	22	23
16	.368	.467	.566	.659	.742	.812	.868	.911	.942	.963
17	.281	.371	.468	.564	.655	.736	.805	.861	.905	.937
18	.208	.287	.375	.469	.562	.651	.731	.799	.855	.899
19	.150	.215	.292	.378	.469	.561	.647	.725	.793	.849
20	.105	.157	.221	.297	.381	.470	.559	.644	.721	.787
21	.072	.111	.163	.227	.302	.384	.471	.558	.640	.716
22	.048	.077	.117	.169	.232	.306	.387	.472	.556	.637
23	.031	.052	.082	.123	.175	.238	.310	.389	.472	.555
24	.020	.034	.056	.087	.128	.180	.243	.314	.392	.473
25	.012	.022	.038	.060	.092	.134	.185	.247	.318	.394

	24	25	26	27	28	29	30	31	32	33
16	.978	.987	.993	.996	.998	.999	.999	1.000		
17	.959	.975	.985	.991	.995	.997	.999	.999	1.000	
18	.932	.955	.972	.983	.990	.994	.997	.998	.999	1.000
19	.893	.927	.951	.969	.980	.988	.993	.996	.998	.999
20	.843	.888	.922	.948	.966	.978	.987	.992	.995	.997
21	.782	.838	.883	.917	.944	.963	.976	.985	.991	.994
22	.712	.777	.832	.877	.913	.940	.959	.973	.983	.989
23	.635	.708	.772	.827	.873	.908	.936	.956	.971	.981
24	.554	.632	.704	.768	.823	.868	.904	.932	.953	.969
25	.473	.553	.629	.700	.763	.818	.863	.900	.929	.950

	34	35	36	37	38	39	40	41	42	43
19	.999	1.000								
20	.999	.999	1.000							
21	.997	.998	.999	.999	1.000					
22	.994	.996	.998	.999	.999	1.000				
23	.988	.993	.996	.997	.999	.999	1.000			
24	.979	.987	.992	.995	.997	.998	.999	.999	1.000	
25	.966	.978	.985	.991	.991	.997	.998	.999	.999	1.000

TABLE IV   Normal Curve Areas

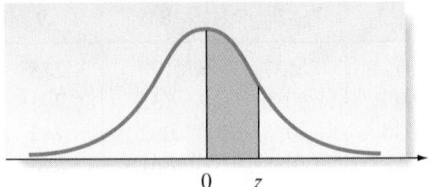

z	.00	.01	.02	.03	.04	.05	.06	.07	.08	.09
.0	.0000	.0040	.0080	.0120	.0160	.0199	.0239	.0279	.0319	.0359
.1	.0398	.0438	.0478	.0517	.0557	.0596	.0636	.0675	.0714	.0753
.2	.0793	.0832	.0871	.0910	.0948	.0987	.1026	.1064	.1103	.1141
.3	.1179	.1217	.1255	.1293	.1331	.1368	.1406	.1443	.1480	.1517
.4	.1554	.1591	.1628	.1664	.1700	.1736	.1772	.1808	.1844	.1879
.5	.1915	.1950	.1985	.2019	.2054	.2088	.2123	.2157	.2190	.2224
.6	.2257	.2291	.2324	.2357	.2389	.2422	.2454	.2486	.2517	.2549
.7	.2580	.2611	.2642	.2673	.2704	.2734	.2764	.2794	.2823	.2852
.8	.2881	.2910	.2939	.2967	.2995	.3023	.3051	.3078	.3106	.3133
.9	.3159	.3186	.3212	.3238	.3264	.3289	.3315	.3340	.3365	.3389
1.0	.3413	.3438	.3461	.3485	.3508	.3531	.3554	.3577	.3599	.3621
1.1	.3643	.3665	.3686	.3708	.3729	.3749	.3770	.3790	.3810	.3830
1.2	.3849	.3869	.3888	.3907	.3925	.3944	.3962	.3980	.3997	.4015
1.3	.4032	.4049	.4066	.4082	.4099	.4115	.4131	.4147	.4162	.4177
1.4	.4192	.4207	.4222	.4236	.4251	.4265	.4279	.4292	.4306	.4319
1.5	.4332	.4345	.4357	.4370	.4382	.4394	.4406	.4418	.4429	.4441
1.6	.4452	.4463	.4474	.4484	.4495	.4505	.4515	.4525	.4535	.4545
1.7	.4554	.4564	.4573	.4582	.4591	.4599	.4608	.4616	.4625	.4633
1.8	.4641	.4649	.4656	.4664	.4671	.4678	.4686	.4693	.4699	.4706
1.9	.4713	.4719	.4726	.4732	.4738	.4744	.4750	.4756	.4761	.4767
2.0	.4772	.4778	.4783	.4788	.4793	.4798	.4803	.4808	.4812	.4817
2.1	.4821	.4826	.4830	.4834	.4838	.4842	.4846	.4850	.4854	.4857
2.2	.4861	.4864	.4868	.4871	.4875	.4878	.4881	.4884	.4887	.4890
2.3	.4893	.4896	.4898	.4901	.4904	.4906	.4909	.4911	.4913	.4916
2.4	.4918	.4920	.4922	.4925	.4927	.4929	.4931	.4932	.4934	.4936
2.5	.4938	.4940	.4941	.4943	.4945	.4946	.4948	.4949	.4951	.4952
2.6	.4953	.4955	.4956	.4957	.4959	.4960	.4961	.4962	.4963	.4964
2.7	.4965	.4966	.4967	.4968	.4969	.4970	.4971	.4972	.4973	.4974
2.8	.4974	.4975	.4976	.4977	.4977	.4978	.4979	.4979	.4980	.4981
2.9	.4981	.4982	.4982	.4983	.4984	.4984	.4985	.4985	.4986	.4986
3.0	.4987	.4987	.4987	.4988	.4988	.4989	.4989	.4989	.4990	.4990

Source: Abridged from Table I of A. Hald, *Statistical Tables and Formulas* (New York: Wiley), 1952. Reproduced by permission of A. Hald.

TABLE V   Exponentials

λ	$e^{-\lambda}$	λ	$e^{-\lambda}$	λ	$e^{-\lambda}$	λ	$e^{-\lambda}$	λ	$e^{-\lambda}$
.00	1.000000	2.05	.128735	4.05	.017422	6.05	.002358	8.05	.000319
.05	.951229	2.10	.122456	4.10	.016573	6.10	.002243	8.10	.000304
.10	.904837	2.15	.116484	4.15	.015764	6.15	.002133	8.15	.000289
.15	.860708	2.20	.110803	4.20	.014996	6.20	.002029	8.20	.000275
.20	.818731	2.25	.105399	4.25	.014264	6.25	.001930	8.25	.000261
.25	.778801	2.30	.100259	4.30	.013569	6.30	.001836	8.30	.000249
.30	.740818	2.35	.095369	4.35	.012907	6.35	.001747	8.35	.000236
.35	.704688	2.40	.090718	4.40	.012277	6.40	.001661	8.40	.000225
.40	.670320	2.45	.086294	4.45	.011679	6.45	.001581	8.45	.000214
.45	.637628	2.50	.082085	4.50	.011109	6.50	.001503	8.50	.000204
.50	.606531	2.55	.078082	4.55	.010567	6.55	.001430	8.55	.000194
.55	.576950	2.60	.074274	4.60	.010052	6.60	.001360	8.60	.000184
.60	.548812	2.65	.070651	4.65	.009562	6.65	.001294	8.65	.000175
.65	.522046	2.70	.067206	4.70	.009095	6.70	.001231	8.70	.000167
.70	.496585	2.75	.063928	4.75	.008652	6.75	.001171	8.75	.000158
.75	.472367	2.80	.060810	4.80	.008230	6.80	.001114	8.80	.000151
.80	.449329	2.85	.057844	4.85	.007828	6.85	.001059	8.85	.000143
.85	.427415	2.90	.055023	4.90	.007447	6.90	.001008	8.90	.000136
.90	.406570	2.95	.052340	4.95	.007083	6.95	.000959	8.95	.000130
.95	.386741	3.00	.049787	5.00	.006738	7.00	.000912	9.00	.000123
1.00	.367879	3.05	.047359	5.05	.006409	7.05	.000867	9.05	.000117
1.05	.349938	3.10	.045049	5.10	.006097	7.10	.000825	9.10	.000112
1.10	.332871	3.15	.042852	5.15	.005799	7.15	.000785	9.15	.000106
1.15	.316637	3.20	.040762	5.20	.005517	7.20	.000747	9.20	.000101
1.20	.301194	3.25	.038774	5.25	.005248	7.25	.000710	9.25	.000096
1.25	.286505	3.30	.036883	5.30	.004992	7.30	.000676	9.30	.000091
1.30	.272532	3.35	.035084	5.35	.004748	7.35	.000643	9.35	.000087
1.35	.259240	3.40	.033373	5.40	.004517	7.40	.000611	9.40	.000083
1.40	.246597	3.45	.031746	5.45	.004296	7.45	.000581	9.45	.000079
1.45	.234570	3.50	.030197	5.50	.004087	7.50	.000553	9.50	.000075
1.50	.223130	3.55	.028725	5.55	.003887	7.55	.000526	9.55	.000071
1.55	.212248	3.60	.027324	5.60	.003698	7.60	.000501	9.60	.000068
1.60	.201897	3.65	.025991	5.65	.003518	7.65	.000476	9.65	.000064
1.65	.192050	3.70	.024724	5.70	.003346	7.70	.000453	9.70	.000061
1.70	.182684	3.75	.023518	5.75	.003183	7.75	.000431	9.75	.000058
1.75	.173774	3.80	.022371	5.80	.003028	7.80	.000410	9.80	.000056
1.80	.165299	3.85	.021280	5.85	.002880	7.85	.000390	9.85	.000053
1.85	.157237	3.90	.020242	5.90	.002739	7.90	.000371	9.90	.000050
1.90	.149569	3.95	.019255	5.95	.002606	7.95	.000353	9.95	.000048
1.95	.142274	4.00	.018316	6.00	.002479	8.00	.000336	10.00	.000045
2.00	.135335								

TABLE VI   Critical Values of t

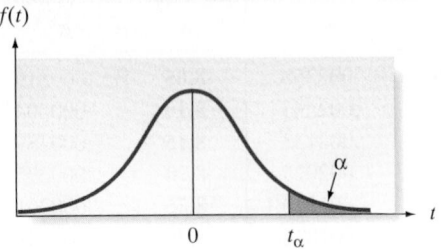

Degrees of Freedom	$t_{.100}$	$t_{.050}$	$t_{.025}$	$t_{.010}$	$t_{.005}$	$t_{.001}$	$t_{.0005}$
1	3.078	6.314	12.706	31.821	63.657	318.31	636.62
2	1.886	2.920	4.303	6.965	9.925	22.326	31.598
3	1.638	2.353	3.182	4.541	5.841	10.213	12.924
4	1.533	2.132	2.776	3.747	4.604	7.173	8.610
5	1.476	2.015	2.571	3.365	4.032	5.893	6.869
6	1.440	1.943	2.447	3.143	3.707	5.208	5.959
7	1.415	1.895	2.365	2.998	3.499	4.785	5.408
8	1.397	1.860	2.306	2.896	3.355	4.501	5.041
9	1.383	1.833	2.262	2.821	3.250	4.297	4.781
10	1.372	1.812	2.228	2.764	3.169	4.144	4.587
11	1.363	1.796	2.201	2.718	3.106	4.025	4.437
12	1.356	1.782	2.179	2.681	3.055	3.930	4.318
13	1.350	1.771	2.160	2.650	3.012	3.852	4.221
14	1.345	1.761	2.145	2.624	2.977	3.787	4.140
15	1.341	1.753	2.131	2.602	2.947	3.733	4.073
16	1.337	1.746	2.120	2.583	2.921	3.686	4.015
17	1.333	1.740	2.110	2.567	2.898	3.646	3.965
18	1.330	1.734	2.101	2.552	2.878	3.610	3.922
19	1.328	1.729	2.093	2.539	2.861	3.579	3.883
20	1.325	1.725	2.086	2.528	2.845	3.552	3.850
21	1.323	1.721	2.080	2.518	2.831	3.527	3.819
22	1.321	1.717	2.074	2.508	2.819	3.505	3.792
23	1.319	1.714	2.069	2.500	2.807	3.485	3.767
24	1.318	1.711	2.064	2.492	2.797	3.467	3.745
25	1.316	1.708	2.060	2.485	2.787	3.450	3.725
26	1.315	1.706	2.056	2.479	2.779	3.435	3.707
27	1.314	1.703	2.052	2.473	2.771	3.421	3.690
28	1.313	1.701	2.048	2.467	2.763	3.408	3.674
29	1.311	1.699	2.045	2.462	2.756	3.396	3.659
30	1.310	1.697	2.042	2.457	2.750	3.385	3.646
40	1.303	1.684	2.021	2.423	2.704	3.307	3.551
60	1.296	1.671	2.000	2.390	2.660	3.232	3.460
120	1.289	1.658	1.980	2.358	2.617	3.160	3.373
∞	1.282	1.645	1.960	2.326	2.576	3.090	3.291

Source: This table is reproduced with the kind permission of the Trustees of Biometrika from E. S. Pearson and H. O. Hartley (eds.), *The Biometrika Tables for Statisticians,* Vol. 1, 3d ed., Biometrika, 1966.

TABLE VII   Critical Values of χ^2

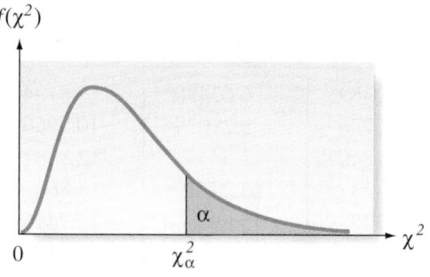

Degrees of Freedom	$\chi^2_{.995}$	$\chi^2_{.990}$	$\chi^2_{.975}$	$\chi^2_{.950}$	$\chi^2_{.900}$
1	.0000393	.0001571	.0009821	.0039321	.0157908
2	.0100251	.0201007	.0506356	.102587	.210720
3	.0717212	.114832	.215795	.351846	.584375
4	.206990	.297110	.484419	.710721	1.063623
5	.411740	.554300	.831211	1.145476	1.61031
6	.675727	.872085	1.237347	1.63539	2.20413
7	.989265	1.239043	1.68987	2.16735	2.83311
8	1.344419	1.646482	2.17973	2.73264	3.48954
9	1.734926	2.087912	2.70039	3.32511	4.16816
10	2.15585	2.55821	3.24697	3.94030	4.86518
11	2.60321	3.05347	3.81575	4.57481	5.57779
12	3.07382	3.57056	4.40379	5.22603	6.30380
13	3.56503	4.10691	5.00874	5.89186	7.04150
14	4.07468	4.66043	5.62872	6.57063	7.78953
15	4.60094	5.22935	6.26214	7.26094	8.54675
16	5.14224	5.81221	6.90766	7.96164	9.31223
17	5.69724	6.40776	7.56418	8.67176	10.0852
18	6.26481	7.01491	8.23075	9.39046	10.8649
19	6.84398	7.63273	8.90655	10.1170	11.6509
20	7.43386	8.26040	9.59083	10.8508	12.4426
21	8.03366	8.89720	10.28293	11.5913	13.2396
22	8.64272	9.54249	10.9823	12.3380	14.0415
23	9.26042	10.19567	11.6885	13.0905	14.8479
24	9.88623	10.8564	12.4011	13.8484	15.6587
25	10.5197	11.5240	13.1197	14.6114	16.4734
26	11.1603	12.1981	13.8439	15.3791	17.2919
27	11.8076	12.8786	14.5733	16.1513	18.1138
28	12.4613	13.5648	15.3079	16.9279	18.9392
29	13.1211	14.2565	16.0471	17.7083	19.7677
30	13.7867	14.9535	16.7908	18.4926	20.5992
40	20.7065	22.1643	24.4331	26.5093	29.0505
50	27.9907	29.7067	32.3574	34.7642	37.6886
60	35.5346	37.4848	40.4817	43.1879	46.4589
70	43.2752	45.4418	48.7576	51.7393	55.3290
80	51.1720	53.5400	57.1532	60.3915	64.2778
90	59.1963	61.7541	65.6466	69.1260	73.2912
100	67.3276	70.0648	74.2219	77.9295	82.3581

(continued)

TABLE VII Continued

Degrees of Freedom	$\chi^2_{.100}$	$\chi^2_{.050}$	$\chi^2_{.025}$	$\chi^2_{.010}$	$\chi^2_{.005}$
1	2.70554	3.84146	5.02389	6.63490	7.87944
2	4.60517	5.99147	7.37776	9.21034	10.5966
3	6.25139	7.81473	9.34840	11.3449	12.8381
4	7.77944	9.48773	11.1433	13.2767	14.8602
5	9.23635	11.0705	12.8325	15.0863	16.7496
6	10.6446	12.5916	14.4494	16.8119	18.5476
7	12.0170	14.0671	16.0128	18.4753	20.2777
8	13.3616	15.5073	17.5346	20.0902	21.9550
9	14.6837	16.9190	19.0228	21.6660	23.5893
10	15.9871	18.3070	20.4831	23.2093	25.1882
11	17.2750	19.6751	21.9200	24.7250	26.7569
12	18.5494	21.0261	23.3367	26.2170	28.2995
13	19.8119	22.3621	24.7356	27.6883	29.8194
14	21.0642	23.6848	26.1190	29.1413	31.3193
15	22.3072	24.9958	27.4884	30.5779	32.8013
16	23.5418	26.2962	28.8454	31.9999	34.2672
17	24.7690	27.5871	30.1910	33.4087	35.7185
18	25.9894	28.8693	31.5264	34.8053	37.1564
19	27.2036	30.1435	32.8523	36.1908	38.5822
20	28.4120	31.4104	34.1696	37.5662	39.9968
21	29.6151	32.6705	35.4789	38.9321	41.4010
22	30.8133	33.9244	36.7807	40.2894	42.7956
23	32.0069	35.1725	38.0757	41.6384	44.1813
24	33.1963	36.4151	39.3641	42.9798	45.5585
25	34.3816	37.6525	40.6465	44.3141	46.9278
26	35.5631	38.8852	41.9232	45.6417	48.2899
27	36.7412	40.1133	43.1944	46.9630	49.6449
28	37.9159	41.3372	44.4607	48.2782	50.9933
29	39.0875	42.5569	45.7222	49.5879	52.3356
30	40.2560	43.7729	46.9792	50.8922	53.6720
40	51.8050	55.7585	59.3417	63.6907	66.7659
50	63.1671	67.5048	71.4202	76.1539	79.4900
60	74.3970	79.0819	83.2976	88.3794	91.9517
70	85.5271	90.5312	95.0231	100.425	104.215
80	96.5782	101.879	106.629	112.329	116.321
90	107.565	113.145	118.136	124.116	128.299
100	118.498	124.342	129.561	135.807	140.169

TABLE VIII   Percentage Points of the *F*-distribution, $\alpha = .10$

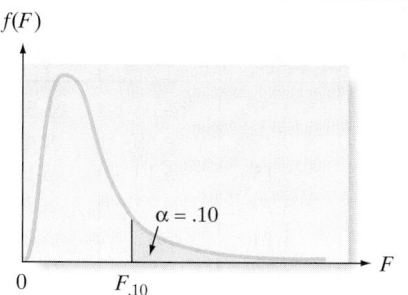

		v_1	NUMERATOR DEGREES OF FREEDOM							
	v_2	1	2	3	4	5	6	7	8	9
	1	39.86	49.50	53.59	55.83	57.24	58.20	58.91	59.44	59.86
	2	8.53	9.00	9.16	9.24	9.29	9.33	9.35	9.37	9.38
	3	5.54	5.46	5.39	5.34	5.31	5.28	5.27	5.25	5.24
	4	4.54	4.32	4.19	4.11	4.05	4.01	3.98	3.95	3.94
	5	4.06	3.78	3.62	3.52	3.45	3.40	3.37	3.34	3.32
	6	3.78	3.46	3.29	3.18	3.11	3.05	3.01	2.98	2.96
	7	3.59	3.26	3.07	2.96	2.88	2.83	2.78	2.75	2.72
	8	3.46	3.11	2.92	2.81	2.73	2.67	2.62	2.59	2.56
	9	3.36	3.01	2.81	2.69	2.61	2.55	2.51	2.47	2.44
	10	3.29	2.92	2.73	2.61	2.52	2.46	2.41	2.38	2.35
	11	3.23	2.86	2.66	2.54	2.45	2.39	2.34	2.30	2.27
	12	3.18	2.81	2.61	2.48	2.39	2.33	2.28	2.24	2.21
	13	3.14	2.76	2.56	2.43	2.35	2.28	2.23	2.20	2.16
	14	3.10	2.73	2.52	2.39	2.31	2.24	2.19	2.15	2.12
	15	3.07	2.70	2.49	2.36	2.27	2.21	2.16	2.12	2.09
	16	3.05	2.67	2.46	2.33	2.24	2.18	2.13	2.09	2.06
	17	3.03	2.64	2.44	2.31	2.22	2.15	2.10	2.06	2.03
	18	3.01	2.62	2.42	2.29	2.20	2.13	2.08	2.04	2.00
	19	2.99	2.61	2.40	2.27	2.18	2.11	2.06	2.02	1.98
	20	2.97	2.59	2.38	2.25	2.16	2.09	2.04	2.00	1.96
	21	2.96	2.57	2.36	2.23	2.14	2.08	2.02	1.98	1.95
	22	2.95	2.56	2.35	2.22	2.13	2.06	2.01	1.97	1.93
	23	2.94	2.55	2.34	2.21	2.11	2.05	1.99	1.95	1.92
	24	2.93	2.54	2.33	2.19	2.10	2.04	1.98	1.94	1.91
	25	2.92	2.53	2.32	2.18	2.09	2.02	1.97	1.93	1.89
	26	2.91	2.52	2.31	2.17	2.08	2.01	1.96	1.92	1.88
	27	2.90	2.51	2.30	2.17	2.07	2.00	1.95	1.91	1.87
	28	2.89	2.50	2.29	2.16	2.06	2.00	1.94	1.90	1.87
	29	2.89	2.50	2.28	2.15	2.06	1.99	1.93	1.89	1.86
	30	2.88	2.49	2.28	2.14	2.05	1.98	1.93	1.88	1.85
	40	2.84	2.44	2.23	2.09	2.00	1.93	1.87	1.83	1.79
	60	2.79	2.39	2.18	2.04	1.95	1.87	1.82	1.77	1.74
	120	2.75	2.35	2.13	1.99	1.90	1.82	1.77	1.72	1.68
	∞	2.71	2.30	2.08	1.94	1.85	1.77	1.72	1.67	1.63

DENOMINATOR DEGREES OF FREEDOM

Source: From M. Merrington and C. M. Thompson, "Tables of Percentage Points of the Inverted Beta (*F*)-Distribution," *Biometrika,* 1943, 33, 73–88.

(continued)

TABLE VIII   Continued

ν_1 / ν_2	10	12	15	20	24	30	40	60	120	∞
	NUMERATOR DEGREES OF FREEDOM									
1	60.19	60.71	61.22	61.74	62.00	62.26	62.53	62.79	63.06	63.33
2	9.39	9.41	9.42	9.44	9.45	9.46	9.47	9.47	9.48	9.49
3	5.23	5.22	5.20	5.18	5.18	5.17	5.16	5.15	5.14	5.13
4	3.92	3.90	3.87	3.84	3.83	3.82	3.80	3.79	3.78	3.76
5	3.30	3.27	3.24	3.21	3.19	3.17	3.16	3.14	3.12	3.10
6	2.94	2.90	2.87	2.84	2.82	2.80	2.78	2.76	2.74	2.72
7	2.70	2.67	2.63	2.59	2.58	2.56	2.54	2.51	2.49	2.47
8	2.54	2.50	2.46	2.42	2.40	2.38	2.36	2.34	2.32	2.29
9	2.42	2.38	2.34	2.30	2.28	2.25	2.23	2.21	2.18	2.16
10	2.32	2.28	2.24	2.20	2.18	2.16	2.13	2.11	2.08	2.06
11	2.25	2.21	2.17	2.12	2.10	2.08	2.05	2.03	2.00	1.97
12	2.19	2.15	2.10	2.06	2.04	2.01	1.99	1.96	1.93	1.90
13	2.14	2.10	2.05	2.01	1.98	1.96	1.93	1.90	1.88	1.85
14	2.10	2.05	2.01	1.96	1.94	1.91	1.89	1.86	1.83	1.80
15	2.06	2.02	1.97	1.92	1.90	1.87	1.85	1.82	1.79	1.76
16	2.03	1.99	1.94	1.89	1.87	1.84	1.81	1.78	1.75	1.72
17	2.00	1.96	1.91	1.86	1.84	1.81	1.78	1.75	1.72	1.69
18	1.98	1.93	1.89	1.84	1.81	1.78	1.75	1.72	1.69	1.66
19	1.96	1.91	1.86	1.81	1.79	1.76	1.73	1.70	1.67	1.63
20	1.94	1.89	1.84	1.79	1.77	1.74	1.71	1.68	1.64	1.61
21	1.92	1.87	1.83	1.78	1.75	1.72	1.69	1.66	1.62	1.59
22	1.90	1.86	1.81	1.76	1.73	1.70	1.67	1.64	1.60	1.57
23	1.89	1.84	1.80	1.74	1.72	1.69	1.66	1.62	1.59	1.55
24	1.88	1.83	1.78	1.73	1.70	1.67	1.64	1.61	1.57	1.53
25	1.87	1.82	1.77	1.72	1.69	1.66	1.63	1.59	1.56	1.52
26	1.86	1.81	1.76	1.71	1.68	1.65	1.61	1.58	1.54	1.50
27	1.85	1.80	1.75	1.70	1.67	1.64	1.60	1.57	1.53	1.49
28	1.84	1.79	1.74	1.69	1.66	1.63	1.59	1.56	1.52	1.48
29	1.83	1.78	1.73	1.68	1.65	1.62	1.58	1.55	1.51	1.47
30	1.82	1.77	1.72	1.67	1.64	1.61	1.57	1.54	1.50	1.46
40	1.76	1.71	1.66	1.61	1.57	1.54	1.51	1.47	1.42	1.38
60	1.71	1.66	1.60	1.54	1.51	1.48	1.44	1.40	1.35	1.29
120	1.65	1.60	1.55	1.48	1.45	1.41	1.37	1.32	1.26	1.19
∞	1.60	1.55	1.49	1.42	1.38	1.34	1.30	1.24	1.17	1.00

DENOMINATOR DEGREES OF FREEDOM

TABLE IX   Percentage Points of the F-distribution, $\alpha = .05$

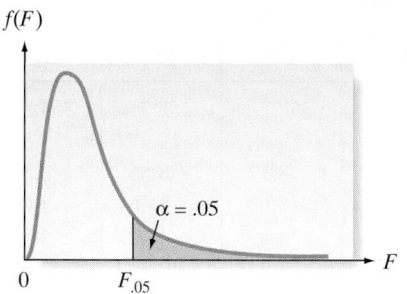

ν_2 \ ν_1	NUMERATOR DEGREES OF FREEDOM								
	1	**2**	**3**	**4**	**5**	**6**	**7**	**8**	**9**
1	161.4	199.5	215.7	224.6	230.2	234.0	236.8	238.9	240.5
2	18.51	19.00	19.16	19.25	19.30	19.33	19.35	19.37	19.38
3	10.13	9.55	9.28	9.12	9.01	8.94	8.89	8.85	8.81
4	7.71	6.94	6.59	6.39	6.26	6.16	6.09	6.04	6.00
5	6.61	5.79	5.41	5.19	5.05	4.95	4.88	4.82	4.77
6	5.99	5.14	4.76	4.53	4.39	4.28	4.21	4.15	4.10
7	5.59	4.74	4.35	4.12	3.97	3.87	3.79	3.73	3.68
8	5.32	4.46	4.07	3.84	3.69	3.58	3.50	3.44	3.39
9	5.12	4.26	3.86	3.63	3.48	3.37	3.29	3.23	3.18
10	4.96	4.10	3.71	3.48	3.33	3.22	3.14	3.07	3.02
11	4.84	3.98	3.59	3.36	3.20	3.09	3.01	2.95	2.90
12	4.75	3.89	3.49	3.26	3.11	3.00	2.91	2.85	2.80
13	4.67	3.81	3.41	3.18	3.03	2.92	2.83	2.77	2.71
14	4.60	3.74	3.34	3.11	2.96	2.85	2.76	2.70	2.65
15	4.54	3.68	3.29	3.06	2.90	2.79	2.71	2.64	2.59
16	4.49	3.63	3.24	3.01	2.85	2.74	2.66	2.59	2.54
17	4.45	3.59	3.20	2.96	2.81	2.70	2.61	2.55	2.49
18	4.41	3.55	3.16	2.93	2.77	2.66	2.58	2.51	2.46
19	4.38	3.52	3.13	2.90	2.74	2.63	2.54	2.48	2.42
20	4.35	3.49	3.10	2.87	2.71	2.60	2.51	2.45	2.39
21	4.32	3.47	3.07	2.84	2.68	2.57	2.49	2.42	2.37
22	4.30	3.44	3.05	2.82	2.66	2.55	2.46	2.40	2.34
23	4.28	3.42	3.03	2.80	2.64	2.53	2.44	2.37	2.32
24	4.26	3.40	3.01	2.78	2.62	2.51	2.42	2.36	2.30
25	4.24	3.39	2.99	2.76	2.60	2.49	2.40	2.34	2.28
26	4.23	3.37	2.98	2.74	2.59	2.47	2.39	2.32	2.77
27	4.21	3.35	2.96	2.73	2.57	2.46	2.37	2.31	2.25
28	4.20	3.34	2.95	2.71	2.56	2.45	2.36	2.29	2.24
29	4.18	3.33	2.93	2.70	2.55	2.43	2.35	2.28	2.22
30	4.17	3.32	2.92	2.69	2.53	2.42	2.33	2.27	2.21
40	4.08	3.23	2.84	2.61	2.45	2.34	2.25	2.18	2.12
60	4.00	3.15	2.76	2.53	2.37	2.25	2.17	2.10	2.04
120	3.92	3.07	2.68	2.45	2.29	2.17	2.09	2.02	1.96
∞	3.84	3.00	2.60	2.37	2.21	2.10	2.01	1.94	1.88

(Left vertical label: DENOMINATOR DEGREES OF FREEDOM)

Source: From M. Merrington and C. M. Thompson, "Tables of Percentage Points of the Inverted Beta (*F*)-Distribution," *Biometrika,* 1943, 33, 73–88.

(continued)

TABLE IX   Continued

ν_2 \ ν_1	NUMERATOR DEGREES OF FREEDOM									
	10	12	15	20	24	30	40	60	120	∞
1	241.9	243.9	245.9	248.0	249.1	250.1	251.1	252.2	253.3	254.3
2	19.40	19.41	19.43	19.45	19.45	19.46	19.47	19.48	19.49	19.50
3	8.79	8.74	8.70	8.66	8.64	8.62	8.59	8.57	8.55	8.53
4	5.96	5.91	5.86	5.80	5.77	5.75	5.72	5.69	5.66	5.63
5	4.74	4.68	4.62	4.56	4.53	4.50	4.46	4.43	4.40	4.36
6	4.06	4.00	3.94	3.87	3.84	3.81	3.77	3.74	3.70	3.67
7	3.64	3.57	3.51	3.44	3.41	3.38	3.34	3.30	3.27	3.23
8	3.35	3.28	3.22	3.15	3.12	3.08	3.04	3.01	2.97	2.93
9	3.14	3.07	3.01	2.94	2.90	2.86	2.83	2.79	2.75	2.71
10	2.98	2.91	2.85	2.77	2.74	2.70	2.66	2.62	2.58	2.54
11	2.85	2.79	2.72	2.65	2.61	2.57	2.53	2.49	2.45	2.40
12	2.75	2.69	2.62	2.54	2.51	2.47	2.43	2.38	2.34	2.30
13	2.67	2.60	2.53	2.46	2.42	2.38	2.34	2.30	2.25	2.21
14	2.60	2.53	2.46	2.39	2.35	2.31	2.27	2.22	2.18	2.13
15	2.54	2.48	2.40	2.33	2.29	2.25	2.20	2.16	2.11	2.07
16	2.49	2.42	2.35	2.28	2.24	2.19	2.15	2.11	2.06	2.01
17	2.45	2.38	2.31	2.23	2.19	2.15	2.10	2.06	2.01	1.96
18	2.41	2.34	2.27	2.19	2.15	2.11	2.06	2.02	1.97	1.92
19	2.38	2.31	2.23	2.16	2.11	2.07	2.03	1.98	1.93	1.88
20	2.35	2.28	2.20	2.12	2.08	2.04	1.99	1.95	1.90	1.84
21	2.32	2.25	2.18	2.10	2.05	2.01	1.96	1.92	1.87	1.81
22	2.30	2.23	2.15	2.07	2.03	1.98	1.94	1.89	1.84	1.78
23	2.27	2.20	2.13	2.05	2.01	1.96	1.91	1.86	1.81	1.76
24	2.25	2.18	2.11	2.03	1.98	1.94	1.89	1.84	1.79	1.73
25	2.24	2.16	2.09	2.01	1.96	1.92	1.87	1.82	1.77	1.71
26	2.22	2.15	2.07	1.99	1.95	1.90	1.85	1.80	1.75	1.69
27	2.20	2.13	2.06	1.97	1.93	1.88	1.84	1.79	1.73	1.67
28	2.19	2.12	2.04	1.96	1.91	1.87	1.82	1.77	1.71	1.65
29	2.18	2.10	2.03	1.94	1.90	1.85	1.81	1.75	1.70	1.64
30	2.16	2.09	2.01	1.93	1.89	1.84	1.79	1.74	1.68	1.62
40	2.08	2.00	1.92	1.84	1.79	1.74	1.69	1.64	1.58	1.51
60	1.99	1.92	1.84	1.75	1.70	1.65	1.59	1.53	1.47	1.39
120	1.91	1.83	1.75	1.66	1.61	1.55	1.50	1.43	1.35	1.25
∞	1.83	1.75	1.67	1.57	1.52	1.46	1.39	1.32	1.22	1.00

DENOMINATOR DEGREES OF FREEDOM

TABLE X   Percentage Points of the *F*-distribution, $\alpha = .025$

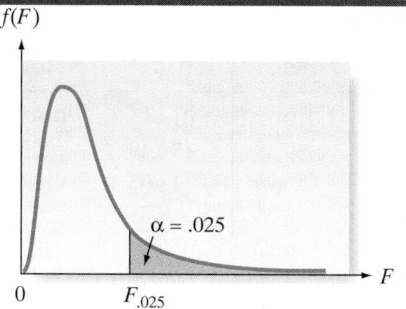

$f(F)$

$\alpha = .025$

$F_{.025}$

0

F

v_2 \ v_1	NUMERATOR DEGREES OF FREEDOM								
	1	**2**	**3**	**4**	**5**	**6**	**7**	**8**	**9**
1	647.8	799.5	864.2	899.6	921.8	937.1	948.2	956.7	963.3
2	38.51	39.00	39.17	39.25	39.30	39.33	39.36	39.37	39.39
3	17.44	16.04	15.44	15.10	14.88	14.73	14.62	14.54	14.47
4	12.22	10.65	9.98	9.60	9.36	9.20	9.07	8.98	8.90
5	10.01	8.43	7.76	7.39	7.15	6.98	6.85	6.76	6.68
6	8.81	7.26	6.60	6.23	5.99	5.82	5.70	5.60	5.52
7	8.07	6.54	5.89	5.52	5.29	5.12	4.99	4.90	4.82
8	7.57	6.06	5.42	5.05	4.82	4.65	4.53	4.43	4.36
9	7.21	5.71	5.08	4.72	4.48	4.32	4.20	4.10	4.03
10	6.94	5.46	4.83	4.47	4.24	4.07	3.95	3.85	3.78
11	6.72	5.26	4.63	4.28	4.04	3.88	3.76	3.66	3.59
12	6.55	5.10	4.47	4.12	3.89	3.73	3.61	3.51	3.44
13	6.41	4.97	4.35	4.00	3.77	3.60	3.48	3.39	3.31
14	6.30	4.86	4.24	3.89	3.66	3.50	3.38	3.29	3.21
15	6.20	4.77	4.15	3.80	3.58	3.41	3.29	3.20	3.12
16	6.12	4.69	4.08	3.73	3.50	3.34	3.22	3.12	3.05
17	6.04	4.62	4.01	3.66	3.44	3.28	3.16	3.06	2.98
18	5.98	4.56	3.95	3.61	3.38	3.22	3.10	3.01	2.93
19	5.92	4.51	3.90	3.56	3.33	3.17	3.05	2.96	2.88
20	5.87	4.46	3.86	3.51	3.29	3.13	3.01	2.91	2.84
21	5.83	4.42	3.82	3.48	3.25	3.09	2.97	2.87	2.80
22	5.79	4.38	3.78	3.44	3.22	3.05	2.93	2.84	2.76
23	5.75	4.35	3.75	3.41	3.18	3.02	2.90	2.81	2.73
24	5.72	4.32	3.72	3.38	3.15	2.99	2.87	2.78	2.70
25	5.69	4.29	3.69	3.35	3.13	2.97	2.85	2.75	2.68
26	5.66	4.27	3.67	3.33	3.10	2.94	2.82	2.73	2.65
27	5.63	4.24	3.65	3.31	3.08	2.92	2.80	2.71	2.63
28	5.61	4.22	3.63	3.29	3.06	2.90	2.78	2.69	2.61
29	5.59	4.20	3.61	3.27	3.04	2.88	2.76	2.67	2.59
30	5.57	4.18	3.59	3.25	3.03	2.87	2.75	2.65	2.57
40	5.42	4.05	3.46	3.13	2.90	2.74	2.62	2.53	2.45
60	5.29	3.93	3.34	3.01	2.79	2.63	2.51	2.41	2.33
120	5.15	3.80	3.23	2.89	2.67	2.52	2.39	2.30	2.22
∞	5.02	3.69	3.12	2.79	2.57	2.41	2.29	2.19	2.11

DENOMINATOR DEGREES OF FREEDOM

Source: From M. Merrington and C. M. Thompson, "Tables of Percentage Points of the Inverted Beta (*F*)-Distribution," *Biometrika*, 1943, 33, 73–88.

(continued)

TABLE X   Continued

v_2 \ v_1	NUMERATOR DEGREES OF FREEDOM									
	10	**12**	**15**	**20**	**24**	**30**	**40**	**60**	**120**	**∞**
1	968.6	976.7	984.9	993.1	997.2	1,001	1,006	1,010	1,014	1,018
2	39.40	39.41	39.43	39.45	39.46	39.46	39.47	39.48	39.49	39.50
3	14.42	14.34	14.25	14.17	14.12	14.08	14.04	13.99	13.95	13.90
4	8.84	8.75	8.66	8.56	8.51	8.46	8.41	8.36	8.31	8.26
5	6.62	6.52	6.43	6.33	6.28	6.23	6.18	6.12	6.07	6.02
6	5.46	5.37	5.27	5.17	5.12	5.07	5.01	4.96	4.90	4.85
7	4.76	4.67	4.57	4.47	4.42	4.36	4.31	4.25	4.20	4.14
8	4.30	4.20	4.10	4.00	3.95	3.89	3.84	3.78	3.73	3.67
9	3.96	3.87	3.77	3.67	3.61	3.56	3.51	3.45	3.39	3.33
10	3.72	3.62	3.52	3.42	3.37	3.31	3.26	3.20	3.14	3.08
11	3.53	3.43	3.33	3.23	3.17	3.12	3.06	3.00	2.94	2.88
12	3.37	3.28	3.18	3.07	3.02	2.96	2.91	2.85	2.79	2.72
13	3.25	3.15	3.05	2.95	2.89	2.84	2.78	2.72	2.66	2.60
14	3.15	3.05	2.95	2.84	2.79	2.73	2.67	2.61	2.55	2.49
15	3.06	2.96	2.86	2.76	2.70	2.64	2.59	2.52	2.46	2.40
16	2.99	2.89	2.79	2.68	2.63	2.57	2.51	2.45	2.38	2.32
17	2.92	2.82	2.72	2.62	2.56	2.50	2.44	2.38	2.32	2.25
18	2.87	2.77	2.67	2.56	2.50	2.44	2.38	2.32	2.26	2.19
19	2.82	2.72	2.62	2.51	2.45	2.39	2.33	2.27	2.20	2.13
20	2.77	2.68	2.57	2.46	2.41	2.35	2.29	2.22	2.16	2.09
21	2.73	2.64	2.53	2.42	2.37	2.31	2.25	2.18	2.11	2.04
22	2.70	2.60	2.50	2.39	2.33	2.27	2.21	2.14	2.08	2.00
23	2.67	2.57	2.47	2.36	2.30	2.24	2.18	2.11	2.04	1.97
24	2.64	2.54	2.44	2.33	2.27	2.21	2.15	2.08	2.01	1.94
25	2.61	2.51	2.41	2.30	2.24	2.18	2.12	2.05	1.98	1.91
26	2.59	2.49	2.39	2.28	2.22	2.16	2.09	2.03	1.95	1.88
27	2.57	2.47	2.36	2.25	2.19	2.13	2.07	2.00	1.93	1.85
28	2.55	2.45	2.34	2.23	2.17	2.11	2.05	1.98	1.91	1.83
29	2.53	2.43	2.32	2.21	2.15	2.09	2.03	1.96	1.89	1.81
30	2.51	2.41	2.31	2.20	2.14	2.07	2.01	1.94	1.87	1.79
40	2.39	2.29	2.18	2.07	2.01	1.94	1.88	1.80	1.72	1.64
60	2.27	2.17	2.06	1.94	1.88	1.82	1.74	1.67	1.58	1.48
120	2.16	2.05	1.94	1.82	1.76	1.69	1.61	1.53	1.43	1.31
∞	2.05	1.94	1.83	1.71	1.64	1.57	1.48	1.39	1.27	1.00

DENOMINATOR DEGREES OF FREEDOM

TABLE XI Percentage Points of the *F*-distribution, $\alpha = .01$

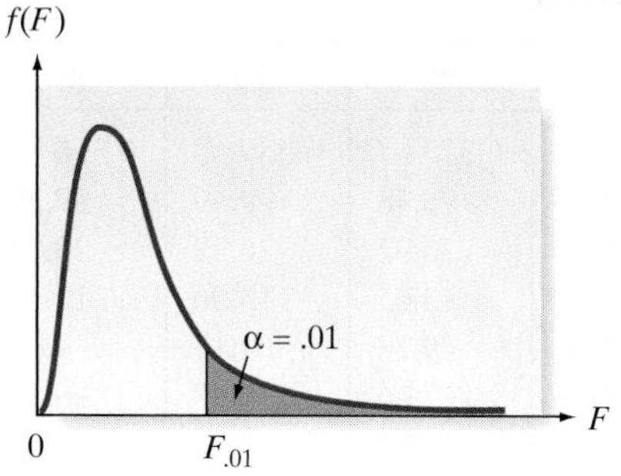

ν_2 \ ν_1	NUMERATOR DEGREES OF FREEDOM								
DENOMINATOR DEGREES OF FREEDOM	**1**	**2**	**3**	**4**	**5**	**6**	**7**	**8**	**9**
1	4,052	4,999.5	5,403	5,625	5,764	5,859	5,928	5,982	6,022
2	98.50	99.00	99.17	99.25	99.30	99.33	99.36	99.37	99.39
3	34.12	30.82	29.46	28.71	28.24	27.91	27.67	27.49	27.35
4	21.20	18.00	16.69	15.98	15.52	15.21	14.98	14.80	14.66
5	16.26	13.27	12.06	11.39	10.97	10.67	10.46	10.29	10.16
6	13.75	10.92	9.78	9.15	8.75	8.47	8.26	8.10	7.98
7	12.25	9.55	8.45	7.85	7.46	7.19	6.99	6.84	6.72
8	11.26	8.65	7.59	7.01	6.63	6.37	6.18	6.03	5.91
9	10.56	8.02	6.99	6.42	6.06	5.80	5.61	5.47	5.35
10	10.04	7.56	6.55	5.99	5.64	5.39	5.20	5.06	4.94
11	9.65	7.21	6.22	5.67	5.32	5.07	4.89	4.74	4.63
12	9.33	6.93	5.95	5.41	5.06	4.82	4.64	4.50	4.39
13	9.07	6.70	5.74	5.21	4.86	4.62	4.44	4.30	4.19
14	8.86	6.51	5.56	5.04	4.69	4.46	4.28	4.14	4.03
15	8.68	6.36	5.42	4.89	4.56	4.32	4.14	4.00	3.89
16	8.53	6.23	5.29	4.77	4.44	4.20	4.03	3.89	3.78
17	8.40	6.11	5.18	4.67	4.34	4.10	3.93	3.79	3.68
18	8.29	6.01	5.09	4.58	4.25	4.01	3.84	3.71	3.60
19	8.18	5.93	5.01	4.50	4.17	3.94	3.77	3.63	3.52
20	8.10	5.85	4.94	4.43	4.10	3.87	3.70	3.56	3.46
21	8.02	5.78	4.87	4.37	4.04	3.81	3.64	3.51	3.40
22	7.95	5.72	4.82	4.31	3.99	3.76	3.59	3.45	3.35
23	7.88	5.66	4.76	4.26	3.94	3.71	3.54	3.41	3.30
24	7.82	5.61	4.72	4.22	3.90	3.67	3.50	3.36	3.26
25	7.77	5.57	4.68	4.18	3.85	3.63	3.46	3.32	3.22
26	7.72	5.53	4.64	4.14	3.82	3.59	3.42	3.29	3.18
27	7.68	5.49	4.60	4.11	3.78	3.56	3.39	3.26	3.15
28	7.64	5.45	4.57	4.07	3.75	3.53	3.36	3.23	3.12
29	7.60	5.42	4.54	4.04	3.73	3.50	3.33	3.20	3.09
30	7.56	5.39	4.51	4.02	3.70	3.47	3.30	3.17	3.07
40	7.31	5.18	4.31	3.83	3.51	3.29	3.12	2.99	2.89
60	7.08	4.98	4.13	3.65	3.34	3.12	2.95	2.82	2.72
120	6.85	4.79	3.95	3.48	3.17	2.96	2.79	2.66	2.56
∞	6.63	4.61	3.78	3.32	3.02	2.80	2.64	2.51	2.41

Source: From M. Merrington and C. M. Thompson, "Tables of Percentage Points of the Inverted Beta (*F*)-Distribution," *Biometrika,* 1943, 33, 73–88.

(continued)

TABLE XI   Continued

v_2 \ v_1	NUMERATOR DEGREES OF FREEDOM									
	10	12	15	20	24	30	40	60	120	∞
1	6,056	6,106	6,157	6,209	6,235	6,261	6,287	6,313	6,339	6,366
2	99.40	99.42	99.43	99.45	99.46	99.47	99.47	99.48	99.49	99.50
3	27.23	27.05	26.87	26.69	26.60	26.50	26.41	26.32	26.22	26.13
4	14.55	14.37	14.20	14.02	13.93	13.84	13.75	13.65	13.56	13.46
5	10.05	9.89	9.72	9.55	9.47	9.38	9.29	9.20	9.11	9.02
6	7.87	7.72	7.56	7.40	7.31	7.23	7.14	7.06	6.97	6.88
7	6.62	6.47	6.31	6.16	6.07	5.99	5.91	5.82	5.74	5.65
8	5.81	5.67	5.52	5.36	5.28	5.20	5.12	5.03	4.95	4.86
9	5.26	5.11	4.96	4.81	4.73	4.65	4.57	4.48	4.40	4.31
10	4.85	4.71	4.56	4.41	4.33	4.25	4.17	4.08	4.00	3.91
11	4.54	4.40	4.25	4.10	4.02	3.94	3.86	3.78	3.69	3.60
12	4.30	4.16	4.01	3.86	3.78	3.70	3.62	3.54	3.45	3.36
13	4.10	3.96	3.82	3.66	3.59	3.51	3.43	3.34	3.25	3.17
14	3.94	3.80	3.66	3.51	3.43	3.35	3.27	3.18	3.09	3.00
15	3.80	3.67	3.52	3.37	3.29	3.21	3.13	3.05	2.96	2.87
16	3.69	3.55	3.41	3.26	3.18	3.10	3.02	2.93	2.84	2.75
17	3.59	3.46	3.31	3.16	3.08	3.00	2.92	2.83	2.75	2.65
18	3.51	3.37	3.23	3.08	3.00	2.92	2.84	2.75	2.66	2.57
19	3.43	3.30	3.15	3.00	2.92	2.84	2.76	2.67	2.58	2.49
20	3.37	3.23	3.09	2.94	2.86	2.78	2.69	2.61	2.52	2.42
21	3.31	3.17	3.03	2.88	2.80	2.72	2.64	2.55	2.46	2.36
22	3.26	3.12	2.98	2.83	2.75	2.67	2.58	2.50	2.40	2.31
23	3.21	3.07	2.93	2.78	2.70	2.62	2.54	2.45	2.35	2.26
24	3.17	3.03	2.89	2.74	2.66	2.58	2.49	2.40	2.31	2.21
25	3.13	2.99	2.85	2.70	2.62	2.54	2.45	2.36	2.27	2.17
26	3.09	2.96	2.81	2.66	2.58	2.50	2.42	2.33	2.23	2.13
27	3.06	2.93	2.78	2.63	2.55	2.47	2.38	2.29	2.20	2.10
28	3.03	2.90	2.75	2.60	2.52	2.44	2.35	2.26	2.17	2.06
29	3.00	2.87	2.73	2.57	2.49	2.41	2.33	2.23	2.14	2.03
30	2.98	2.84	2.70	2.55	2.47	2.39	2.30	2.21	2.11	2.01
40	2.80	2.66	2.52	2.37	2.29	2.20	2.11	2.02	1.92	1.80
60	2.63	2.50	2.35	2.20	2.12	2.03	1.94	1.84	1.73	1.60
120	2.47	2.34	2.19	2.03	1.95	1.86	1.76	1.66	1.53	1.38
∞	2.32	2.18	2.04	1.88	1.79	1.70	1.59	1.47	1.32	1.00

DENOMINATOR DEGREES OF FREEDOM

TABLE XII   Critical Values of T_L and T_U for the Wilcoxon Rank Sum Test: Independent Samples

Test statistic is the rank sum associated with the smaller sample (if equal sample sizes, either rank sum can be used).

a. $\alpha = .025$ one-tailed; $\alpha = .05$ two-tailed

n_2 \ n_1	3		4		5		6		7		8		9		10	
	T_L	T_U	T_L	T_U	T_L	T_U	T_L	T_U	T_L	T_U	T_L	T_U	T_L	T_U	T_L	T_U
3	5	16	6	18	6	21	7	23	7	26	8	28	8	31	9	33
4	6	18	11	25	12	28	12	32	13	35	14	38	15	41	16	44
5	6	21	12	28	18	37	19	41	20	45	21	49	22	53	24	56
6	7	23	12	32	19	41	26	52	28	56	29	61	31	65	32	70
7	7	26	13	35	20	45	28	56	37	68	39	73	41	78	43	83
8	8	28	14	38	21	49	29	61	39	73	49	87	51	93	54	98
9	8	31	15	41	22	53	31	65	41	78	51	93	63	108	66	114
10	9	33	16	44	24	56	32	70	43	83	54	98	66	114	79	131

b. $\alpha = .05$ one-tailed; $\alpha = .10$ two-tailed

n_2 \ n_1	3		4		5		6		7		8		9		10	
	T_L	T_U	T_L	T_U	T_L	T_U	T_L	T_U	T_L	T_U	T_L	T_U	T_L	T_U	T_L	T_U
3	6	15	7	17	7	20	8	22	9	24	9	27	10	29	11	31
4	7	17	12	24	13	27	14	30	15	33	16	36	17	39	18	42
5	7	20	13	27	19	36	20	40	22	43	24	46	25	50	26	54
6	8	22	14	30	20	40	28	50	30	54	32	58	33	63	35	67
7	9	24	15	33	22	43	30	54	39	66	41	71	43	76	46	80
8	9	27	16	36	24	46	32	58	41	71	52	84	54	90	57	95
9	10	29	17	39	25	50	33	63	43	76	54	90	66	105	69	111
10	11	31	18	42	26	54	35	67	46	80	57	95	69	111	83	127

Source: From F. Wilcoxon and R. A. Wilcox, "Some Rapid Approximate Statistical Procedures," 1964, 20–23.

TABLE XIII Critical Values of T_0 in the Wilcoxon Paired Difference Signed Rank Test

One-Tailed	Two-Tailed	$n = 5$	$n = 6$	$n = 7$	$n = 8$	$n = 9$	$n = 10$
$\alpha = .05$	$\alpha = .10$	1	2	4	6	8	11
$\alpha = .025$	$\alpha = .05$		1	2	4	6	8
$\alpha = .01$	$\alpha = .02$			0	2	3	5
$\alpha = .005$	$\alpha = .01$				0	2	3
		$n = 11$	$n = 12$	$n = 13$	$n = 14$	$n = 15$	$n = 16$
$\alpha = .05$	$\alpha = .10$	14	17	21	26	30	36
$\alpha = .025$	$\alpha = .05$	11	14	17	21	25	30
$\alpha = .01$	$\alpha = .02$	7	10	13	16	20	24
$\alpha = .005$	$\alpha = .01$	5	7	10	13	16	19
		$n = 17$	$n = 18$	$n = 19$	$n = 20$	$n = 21$	$n = 22$
$\alpha = .05$	$\alpha = .10$	41	47	54	60	68	75
$\alpha = .025$	$\alpha = .05$	35	40	46	52	59	66
$\alpha = .01$	$\alpha = .02$	28	33	38	43	49	56
$\alpha = .005$	$\alpha = .01$	23	28	32	37	43	49
		$n = 23$	$n = 24$	$n = 25$	$n = 26$	$n = 27$	$n = 28$
$\alpha = .05$	$\alpha = .10$	83	92	101	110	120	130
$\alpha = .025$	$\alpha = .05$	73	81	90	98	107	117
$\alpha = .01$	$\alpha = .02$	62	69	77	85	93	102
$\alpha = .005$	$\alpha = .01$	55	61	68	76	84	92
		$n = 29$	$n = 30$	$n = 31$	$n = 32$	$n = 33$	$n = 34$
$\alpha = .05$	$\alpha = .10$	141	152	163	175	188	201
$\alpha = .025$	$\alpha = .05$	127	137	148	159	171	183
$\alpha = .01$	$\alpha = .02$	111	120	130	141	151	162
$\alpha = .005$	$\alpha = .01$	100	109	118	128	138	149
		$n = 35$	$n = 36$	$n = 37$	$n = 38$	$n = 39$	
$\alpha = .05$	$\alpha = .10$	214	228	242	256	271	
$\alpha = .025$	$\alpha = .05$	195	208	222	235	250	
$\alpha = .01$	$\alpha = .02$	174	186	198	211	224	
$\alpha = .005$	$\alpha = .01$	160	171	183	195	208	
		$n = 40$	$n = 41$	$n = 42$	$n = 43$	$n = 44$	$n = 45$
$\alpha = .05$	$\alpha = .10$	287	303	319	336	353	371
$\alpha = .025$	$\alpha = .05$	264	279	295	311	327	344
$\alpha = .01$	$\alpha = .02$	238	252	267	281	297	313
$\alpha = .005$	$\alpha = .01$	221	234	248	262	277	292
		$n = 46$	$n = 47$	$n = 48$	$n = 49$	$n = 50$	
$\alpha = .05$	$\alpha = .10$	389	408	427	446	466	
$\alpha = .025$	$\alpha = .05$	361	379	397	415	434	
$\alpha = .01$	$\alpha = .02$	329	345	362	380	398	
$\alpha = .005$	$\alpha = .01$	307	323	339	356	373	

Source: From F. Wilcoxon and R. A. Wilcox, "Some Rapid Approximate Statistical Procedures," 1964, p. 28.

TABLE XIV   Critical Values of Spearman's Rank Correlation Coefficient

The α values correspond to a one-tailed test of $H_0: \rho = 0$. The value should be doubled for two-tailed tests.

n	$\alpha = .05$	$\alpha = .025$	$\alpha = .01$	$\alpha = .005$	n	$\alpha = .05$	$\alpha = .025$	$\alpha = .01$	$\alpha = .005$
5	.900	—	—	—	18	.399	.476	.564	.625
6	.829	.886	.943	—	19	.388	.462	.549	.608
7	.714	.786	.893	—	20	.377	.450	.534	.591
8	.643	.738	.833	.881	21	.368	.438	.521	.576
9	.600	.683	.783	.833	22	.359	.428	.508	.562
10	.564	.648	.745	.794	23	.351	.418	.496	.549
11	.523	.623	.736	.818	24	.343	.409	.485	.537
12	.497	.591	.703	.780	25	.336	.400	.475	.526
13	.475	.566	.673	.745	26	.329	.392	.465	.515
14	.457	.545	.646	.716	27	.323	.385	.456	.505
15	.441	.525	.623	.689	28	.317	.377	.448	.496
16	.425	.507	.601	.666	29	.311	.370	.440	.487
17	.412	.490	.582	.645	30	.305	.364	.432	.478

Source: From E. G. Olds, "Distribution of Sums of Squares of Rank Differences for Small Samples," *Annals of Mathematical Statistics,* 1938, 9.

TABLE XV   Critical Values of the Studentized Range, $\alpha = .05$

k ν	2	3	4	5	6	7	8	9	10	11
1	17.97	26.98	32.82	37.08	40.41	43.12	45.40	47.36	49.07	50.59
2	6.08	8.33	9.80	10.88	11.74	12.44	13.03	13.54	13.99	14.39
3	4.50	5.91	6.82	7.50	8.04	8.48	8.85	9.18	9.46	9.72
4	3.93	5.04	5.76	6.29	6.71	7.05	7.35	7.60	7.83	8.03
5	3.64	4.60	5.22	5.67	6.03	6.33	6.58	6.80	6.99	7.17
6	3.46	4.34	4.90	5.30	5.63	5.90	6.12	6.32	6.49	6.65
7	3.34	4.16	4.68	5.06	5.36	5.61	5.82	6.00	6.16	6.30
8	3.26	4.04	4.53	4.89	5.17	5.40	5.60	5.77	5.92	6.05
9	3.20	3.95	4.41	4.76	5.02	5.24	5.43	5.59	5.74	5.87
10	3.15	3.88	4.33	4.65	4.91	5.12	5.30	5.46	5.60	5.72
11	3.11	3.82	4.26	4.57	4.82	5.03	5.20	5.35	5.49	5.61
12	3.08	3.77	4.20	4.51	4.75	4.95	5.12	5.27	5.39	5.51
13	3.06	3.73	4.15	4.45	4.69	4.88	5.05	5.19	5.32	5.43
14	3.03	3.70	4.11	4.41	4.64	4.83	4.99	5.13	5.25	5.36
15	3.01	3.67	4.08	4.37	4.60	4.78	4.94	5.08	5.20	5.31
16	3.00	3.65	4.05	4.33	4.56	4.74	4.90	5.03	5.15	5.26
17	2.98	3.63	4.02	4.30	4.52	4.70	4.86	4.99	5.11	5.21
18	2.97	3.61	4.00	4.28	4.49	4.67	4.82	4.96	5.07	5.17
19	2.96	3.59	3.98	4.25	4.47	4.65	4.79	4.92	5.04	5.14
20	2.95	3.58	3.96	4.23	4.45	4.62	4.77	4.90	5.01	5.11
24	2.92	3.53	3.90	4.17	4.37	4.54	4.68	4.81	4.92	5.01
30	2.89	3.49	3.85	4.10	4.30	4.46	4.60	4.72	4.82	4.92
40	2.86	3.44	3.79	4.04	4.23	4.39	4.52	4.63	4.73	4.82
60	2.83	3.40	3.74	3.98	4.16	4.31	4.44	4.55	4.65	4.73
120	2.80	3.36	3.68	3.92	4.10	4.24	4.36	4.47	4.56	4.64
∞	2.77	3.31	3.63	3.86	4.03	4.17	4.29	4.39	4.47	4.55

(continued)

TABLE XV Continued

k ν	12	13	14	15	16	17	18	19	20
1	51.96	53.20	54.33	55.36	56.32	57.22	58.04	58.83	59.56
2	14.75	15.08	15.38	15.65	15.91	16.14	16.37	16.57	16.77
3	9.95	10.15	10.35	10.52	10.69	10.84	10.98	11.11	11.24
4	8.21	8.37	8.52	8.66	8.79	8.91	9.03	9.13	9.23
5	7.32	7.47	7.60	7.72	7.83	7.93	8.03	8.12	8.21
6	6.79	6.92	7.03	7.14	7.24	7.34	7.43	7.51	7.59
7	6.43	6.55	6.66	6.76	6.85	6.94	7.02	7.10	7.17
8	6.18	6.29	6.39	6.48	6.57	6.65	6.73	6.80	6.87
9	5.98	6.09	6.19	6.28	6.36	6.44	6.51	6.58	6.64
10	5.83	5.93	6.03	6.11	6.19	6.27	6.34	6.40	6.47
11	5.71	5.81	5.90	5.98	6.06	6.13	6.20	6.27	6.33
12	5.61	5.71	5.80	5.88	5.95	6.02	6.09	6.15	6.21
13	5.53	5.63	5.71	5.79	5.86	5.93	5.99	6.05	6.11
14	5.46	5.55	5.64	5.71	5.79	5.85	5.91	5.97	6.03
15	5.40	5.49	5.57	5.65	5.72	5.78	5.85	5.90	5.96
16	5.35	5.44	5.52	5.59	5.66	5.73	5.79	5.84	5.90
17	5.31	5.39	5.47	5.54	5.61	5.67	5.73	5.79	5.84
18	5.27	5.35	5.43	5.50	5.57	5.63	5.69	5.74	5.79
19	5.23	5.31	5.39	5.46	5.53	5.59	5.65	5.70	5.75
20	5.20	5.28	5.36	5.43	5.49	5.55	5.61	5.66	5.71
24	5.10	5.18	5.25	5.32	5.38	5.44	5.49	5.55	5.59
30	5.00	5.08	5.15	5.21	5.27	5.33	5.38	5.43	5.47
40	4.90	4.98	5.04	5.11	5.16	5.22	5.27	5.31	5.36
60	4.81	4.88	4.94	5.00	5.06	5.11	5.15	5.20	5.24
120	4.71	4.78	4.84	4.90	4.95	5.00	5.04	5.09	5.13
∞	4.62	4.68	4.74	4.80	4.85	4.89	4.93	4.97	5.01

TABLE XVI   Critical Values of the Studentized Range, $\alpha = .01$

v \ k	2	3	4	5	6	7	8	9	10	11
1	90.03	135.0	164.3	185.6	202.2	215.8	227.2	237.0	245.6	253.2
2	14.04	19.02	22.29	24.72	26.63	28.20	29.53	30.68	31.69	32.59
3	8.26	10.62	12.17	13.33	14.24	15.00	15.64	16.20	16.69	17.13
4	6.51	8.12	9.17	9.96	10.58	11.10	11.55	11.93	12.27	12.57
5	5.70	6.98	7.80	8.42	8.91	9.32	9.67	9.97	10.24	10.48
6	5.24	6.33	7.03	7.56	7.97	8.32	8.61	8.87	9.10	9.30
7	4.95	5.92	6.54	7.01	7.37	7.68	7.94	8.17	8.37	8.55
8	4.75	5.64	6.20	6.62	6.96	7.24	7.47	7.68	7.86	8.03
9	4.60	5.43	5.96	6.35	6.66	6.91	7.13	7.33	7.49	7.65
10	4.48	5.27	5.77	6.14	6.43	6.67	6.87	7.05	7.21	7.36
11	4.39	5.15	5.62	5.97	6.25	6.48	6.67	6.84	6.99	7.13
12	4.32	5.05	5.50	5.84	6.10	6.32	6.51	6.67	6.81	6.94
13	4.26	4.96	5.40	5.73	5.98	6.19	6.37	6.53	6.67	6.79
14	4.21	4.89	5.32	5.63	5.88	6.08	6.26	6.41	6.54	6.66
15	4.17	4.84	5.25	5.56	5.80	5.99	6.16	6.31	6.44	6.55
16	4.13	4.79	5.19	5.49	5.72	5.92	6.08	6.22	6.35	6.46
17	4.10	4.74	5.14	5.43	5.66	5.85	6.01	6.15	6.27	6.38
18	4.07	4.70	5.09	5.38	5.60	5.79	5.94	6.08	6.20	6.31
19	4.05	4.67	5.05	5.33	5.55	5.73	5.89	6.02	6.14	6.25
20	4.02	4.64	5.02	5.29	5.51	5.69	5.84	5.97	6.09	6.19
24	3.96	4.55	4.91	5.17	5.37	5.54	5.69	5.81	5.92	6.02
30	3.89	4.45	4.80	5.05	5.24	5.40	5.54	5.65	5.76	5.85
40	3.82	4.37	4.70	4.93	5.11	5.26	5.39	5.50	5.60	5.69
60	3.76	4.28	4.59	4.82	4.99	5.13	5.25	5.36	5.45	5.53
120	3.70	4.20	4.50	4.71	4.87	5.01	5.12	5.21	5.30	5.37
∞	3.64	4.12	4.40	4.60	4.76	4.88	4.99	5.08	5.16	5.23

(continued)

TABLE XVI   Continued

ν \ k	12	13	14	15	16	17	18	19	20
1	260.0	266.2	271.8	277.0	281.8	286.3	290.0	294.3	298.0
2	33.40	34.13	34.81	35.43	36.00	36.53	37.03	37.50	37.95
3	17.53	17.89	18.22	18.52	18.81	19.07	19.32	19.55	19.77
4	12.84	13.09	13.32	13.53	13.73	13.91	14.08	14.24	14.40
5	10.70	10.89	11.08	11.24	11.40	11.55	11.68	11.81	11.93
6	9.48	9.65	9.81	9.95	10.08	10.21	10.32	10.43	10.54
7	8.71	8.86	9.00	9.12	9.24	9.35	9.46	9.55	9.65
8	8.18	8.31	8.44	8.55	8.66	8.76	8.85	8.94	9.03
9	7.78	7.91	8.03	8.13	8.23	8.33	8.41	8.49	8.57
10	7.49	7.60	7.71	7.81	7.91	7.99	8.08	8.15	8.23
11	7.25	7.36	7.46	7.56	7.65	7.73	7.81	7.88	7.95
12	7.06	7.17	7.26	7.36	7.44	7.52	7.59	7.66	7.73
13	6.90	7.01	7.10	7.19	7.27	7.35	7.42	7.48	7.55
14	6.77	6.87	6.96	7.05	7.13	7.20	7.27	7.33	7.39
15	6.66	6.76	6.84	6.93	7.00	7.07	7.14	7.20	7.26
16	6.56	6.66	6.74	6.82	6.90	6.97	7.03	7.09	7.15
17	6.48	6.57	6.66	6.73	6.81	6.87	6.94	7.00	7.05
18	6.41	6.50	6.58	6.65	6.72	6.79	6.85	6.91	6.97
19	6.34	6.43	6.51	6.58	6.65	6.72	6.78	6.84	6.89
20	6.28	6.37	6.45	6.52	6.59	6.65	6.71	6.77	6.82
24	6.11	6.19	6.26	6.33	6.39	6.45	6.51	6.56	6.61
30	5.93	6.01	6.08	6.14	6.20	6.26	6.31	6.36	6.41
40	5.76	5.83	5.90	5.96	6.02	6.07	6.12	6.16	6.21
60	5.60	5.67	5.73	5.78	5.84	5.89	5.93	5.97	6.01
120	5.44	5.50	5.56	5.61	5.66	5.71	5.75	5.79	5.83
∞	5.29	5.35	5.40	5.45	5.49	5.54	5.57	5.61	5.65

Source: Biometrika Tables for Statisticians, Vol. 1, 3rd ed., edited by E. S. Pearson and H. O. Hartley (Cambridge University Press, 1966). Reproduced by permission of Professor E. S. Pearson and the *Biometrika* Trustees.

Appendix B
Calculation Formulas for Analysis of Variance

B.1 FORMULAS FOR THE CALCULATIONS IN THE COMPLETELY RANDOMIZED DESIGN

$$CM = \text{Correction for mean}$$

$$= \frac{(\text{Total of all observations})^2}{\text{Total number of observations}} = \frac{\left(\sum y_i\right)^2}{n}$$

$$SS(\text{Total}) = \text{Total sum of squares}$$

$$= (\text{Sum of squares of all observations}) - CM = \sum y_i^2 - CM$$

$$SST = \text{Sum of squares for treatments}$$

$$= \left(\begin{array}{c}\text{Sum of squares of treatments totals with} \\ \text{each square divided by the number of} \\ \text{observations for that treatment}\end{array}\right) - CM$$

$$= \frac{T_1^2}{n_1} + \frac{T_2^2}{n_2} + \cdots + \frac{T_k^2}{n_k} - CM$$

$$SSE = \text{Sum of squares for error} = SS(\text{Total}) - SST$$

$$MST = \text{Mean square for treatments} = \frac{SST}{k-1}$$

$$MSE = \text{Mean square for error} = \frac{SSE}{n-k}$$

$$F = \text{Test statistic} = \frac{MST}{MSE}$$

where

$$n = \text{Total number of observations}$$

$$k = \text{Number of treatments}$$

$$T_i = \text{Total for treatment } i \ (i = 1, 2, \ldots, k)$$

B.2   FORMULAS FOR THE CALCULATIONS IN THE RANDOMIZED BLOCK DESIGN

$$CM = \text{Correction for mean}$$

$$= \frac{(\text{Total of all observations})^2}{\text{Total number of observations}} = \frac{\left(\sum y_i\right)^2}{n}$$

$$SS(\text{Total}) = \text{Total sum of squares}$$

$$= (\text{Sum of squares of all observations}) - CM = \sum y_i^2 - CM$$

$$SST = \text{Sum of squares for treatments}$$

$$= \left(\begin{array}{c}\text{Sum of squares of treatment totals with}\\ \text{each square divided by } b, \text{ the number of}\\ \text{observations for that treatment}\end{array}\right) - CM$$

$$= \frac{T_1^2}{b} + \frac{T_2^2}{b} + \cdots + \frac{T_k^2}{b} - CM$$

$$SST = \text{Sum of squares for blocks}$$

$$= \left(\begin{array}{c}\text{Sum of squares of block totals with}\\ \text{each square divided by } p, \text{ the number}\\ \text{of observations in that block}\end{array}\right) - CM$$

$$= \frac{B_1^2}{k} + \frac{B_2^2}{k} + \cdots + \frac{B_b^2}{k} - CM$$

$$SSE = \text{Sum of squares for error} = SS(\text{Total}) - SST - SSB$$

$$MST = \text{Mean square for treatments} = \frac{SST}{k-1}$$

$$MSB = \text{Mean square for blocks} = \frac{SSB}{b-1}$$

$$MSE = \text{Mean square for error} = \frac{SSE}{n-k-b+1}$$

$$F = \text{Test statistic} = \frac{MST}{MSE}$$

where

$$n = \text{Total number of observations}$$
$$b = \text{Number of blocks}$$
$$k = \text{Number of treatments}$$
$$T_i = \text{Total for treatment } i \ (i = 1, 2, \ldots, k)$$
$$B_i = \text{Total for block } i \ (i = 1, 2, \ldots, b)$$

B.3   FORMULAS FOR THE CALCULATIONS FOR A TWO-FACTOR FACTORIAL EXPERIMENT

$$CM = \text{Correction for mean}$$

$$= \frac{(\text{Total of all } n \text{ measurements})^2}{n} = \frac{\left(\sum_{i=1}^{n} y_i\right)^2}{n}$$

$$SS(\text{Total}) = \text{Total sum of squares}$$

$$= (\text{Sum of squares of all } n \text{ measurements}) - \text{CM} = \sum_{i=1}^{n} y_i^2 - \text{CM}$$

SS(A) = Sum of squares for main effects, factor A

$$= \left(\begin{array}{c} \text{Sum of squares of the totals } A_1, A_2, \ldots, A_a \\ \text{divided by the number of measurements} \\ \text{in a single total, namely } br \end{array} \right) - \text{CM}$$

$$= \frac{\sum_{i=1}^{a} A_i^2}{br} - \text{CM}$$

SS(B) = Sum of squares for main effects, factor B

$$= \left(\begin{array}{c} \text{Sum of squares of the totals } B_1, B_2, \ldots, B_b \\ \text{divided by the number of measurements} \\ \text{in a single total, namely } ar \end{array} \right) - \text{CM}$$

$$= \frac{\sum_{i=1}^{b} B_i^2}{ar} - \text{CM}$$

SS(AB) = Sum of squares for AB interaction

$$= \left(\begin{array}{c} \text{Sum of squares of the cell totals} \\ AB_{11}, AB_{12}, \ldots, AB_{ab} \text{ divided by} \\ \text{the number of measurements in} \\ \text{a single total, namely } r \end{array} \right) - \text{SS}(A) - \text{SS}(B) - \text{CM}$$

$$= \frac{\sum_{j=1}^{b} \sum_{i=1}^{a} AB_{ij}^2}{r} - \text{SS}(A) - \text{SS}(B) - \text{CM}$$

where

a = Number of levels of factor A

b = Number of levels of factor B

r = Number of replicates (observations per treatment)

A_i = Total for level i of factor A ($i = 1, 2, \ldots, a$)

B_i = Total for level i of factor B ($i = 1, 2, \ldots, b$)

AB_{ij} = Total for treatment (ij), i.e., for ith level of factor A and ith level of factor B

B.4   TUKEY'S MULTIPLE COMPARISONS PROCEDURE. (EQUAL SAMPLE SIZES)

Step 1 Select the desired experimentwise error rate, α.

Step 2 Calculate

$$\omega = q_\alpha(k, v) \frac{s}{\sqrt{n_t}}$$

where

k = Number of sample means (i.e., number of treatments)

$s = \sqrt{\text{MSE}}$

v = Number of degrees of freedom associated with MSE

n_t = Number of observations in each of the k samples (i.e., number of observations per treatment)

$q_\alpha(k, v)$ = Critical value of the Studentized range (Tables XV and XVI of Appendix A)

Step 3 Calculate and rank the k sample means.

Step 4 Place a bar over those pairs of treatment means that differ by less than ω. A pair of treatments not connected by an overbar (i.e., differing by more than ω) implies a difference in the corresponding population means.

Note: The confidence level associated with all inferences drawn from the analysis is $(1 - \alpha)$.

B.5 BONFERRONI MULTIPLE COMPARISONS PROCEDURE (PAIRWISE COMPARISONS)

Step 1 Calculate for each treatment pair (i, j)

$$B_{ij} = t_{\alpha/(2c)}s\sqrt{\frac{1}{n_i} + \frac{1}{n_j}}$$

where

k = Number of sample (treatment) means in the experiment

c = Number of pairwise comparisons

[*Note:* If all pairwise comparisons are to be made, then $c = k(k - 1)/2$]

$s = \sqrt{\text{MSE}}$

ν = Number of degrees of freedom associated with MSE

n_i = Number of observations in sample for treatment i

n_j = Number of observations in sample for treatment j

$t_{\alpha/(2c)}$ = Critical value of t distribution with ν df and tail area $\alpha/(2c)$ (Table VI in Appendix A)

Step 2 Rank the sample means and place a bar over any treatment pair (i, j) whose sample means differ by less than B_{ij}. Any pair of means not connected by an overbar implies a difference in the corresponding population means.

Note: The level of confidence associated with all inferences drawn from the analysis is at least $(1 - \alpha)$.

B.6 SCHEFFÉ'S MULTIPLE COMPARISONS PROCEDURE (PAIRWISE COMPARISONS)

Step 1 Calculate Scheffé's critical difference for each pair of treatments (i, j):

$$S_{ij} = \sqrt{(k - 1)(F_\alpha)(\text{MSE})\left(\frac{1}{n_i} + \frac{1}{n_j}\right)}$$

where

k = Number of sample (treatment) means

MSE = Mean squared error

n_i = Number of observations in sample for treatment i

n_j = Number of observations in sample for treatment j

F_α = Critical value of F distribution with $k - 1$ numerator df and ν denominator df (Tables XIII, IX, X, and XI of Appendix A)

ν = Number of degrees of freedom associated with MSE

Step 2 Rank the k sample means and place a bar over any treatment pair (i, j) that differs by less than S_{ij}. Any pair of sample means not connected by an overbar implies a difference in the corresponding population means.

Short Answers to Selected Odd-Numbered Exercises

Chapter 1

1.11 qualitative; qualitative **1.13 a.** quantitative **b.** qualitative **c.** qualitative **d.** quantitative **1.15 b.** qualitative **e.** survey data
1.17 a. CEOs of all U. S. companies **b.** (1) qualitative, (2) quantitative, (3) quantitative, (4) quantitative **c.** CEOs selected only from top-ranked companies **1.19 a.** qualitative; qualitative; qualitative; quantitative **b.** sample **1.21 a.** quantitative **b.** quantitative
c. qualitative **d.** quantitative **e.** qualitative **f.** quantitative **g.** qualitative **1.23 a.** designed experiment **b.** smokers
c. quantitative **d.** population: all smokers in the U.S.; sample: 50,000 smokers in trial **e.** the difference in mean age at which each of the scanning methods first detects a tumor. **1.25 a.** designed experiment **b.** amateur boxers **c.** heart rate (quantitative); blood lactate level (quantitative) **d.** no difference between the two groups of boxers **e.** no **1.27 a.** designed experiment **b.** inferential statistics
c. all possible burn patients **1.29 b.** number of headers per game and IQ **c.** both quantitative **1.31 b.** answer to question posed;
qualitative **d.** survey data

Chapter 2

2.5 a. $X - 8, Y - 9, Z - 3$ **b.** $X - .40, Y - .45, Z - .15$ **2.7 a.** $39/266 = .147$ **b.** level 2 $-.286$, level 3 $-.188$, level 4 $-.327$,
level 5 $-.041$, level 6 $-.011$ **e.** level 4 **2.9 a.** relative frequencies: Black $-.207$; White $-.635$; Sumatran $-.017$; Javan $-.003$;
Indian $-.137$ **c.** $.842; .157$ **2.11 a.** qualitative; infant, child, medical, infant & medical, child & medical, infant & child, infant &
child & medical **b.** infant $-.061$; child $-.565$; medical $-.276$; infant & medical $-.001$; child & medical $-.030$; infant & child $-.062$;
infant & child & medical $-.004$ **d.** $.312$ **2.13 a.** survey **b.** quantitative **c.** about 60% **2.15** 60% of sampled CEOs had advanced
degrees **2.19** Over half of the whistle types were "Type a." **2.25 a.** 23 **2.27** frequencies: 50, 75, 125, 100, 25, 50, 50, 25

2.31 a.

Stem	Leaf
1	6
2	0 1 2 4 8
3	0 4 4 9
4	0 0 2
5	3

b. A students tend to read the most books

2.35 a.

Stem	Leaf
1	0000000
2	0
3	00
4	000
5	
6	
7	
8	
9	0
10	0
11	0
12	00

b. Yes **2.37** most of the PMI's range from
3 to 7.5

2.39 a.

Stem	Leaf
0	1 1 2 3 4 5 5 9
1	1 2 3
2	0 0
3	9
4	6
5	6
6	1 5
7	
8	
9	
10	0

c. eclipses **2.41 a.** Response rate of the familiarity group is higher. **b.** Familiarity group has highest rate; control group has lowest rate
2.43 a. 23 **b.** 123 **c.** 18 **d.** 51 **e.** 529 **2.45 a.** 6 **b.** 50 **c.** 42.8 **2.47** mean, median, mode **2.49** sample size and variability of the data
2.51 a. mean $<$ median **b.** mean $>$ median **c.** mean $=$ median **2.53** mode $= 15$; mean $= 14.545$; median $= 15$ **2.55 a.** 8.5 **b.** 25
c. .78 **d.** 13.44 **2.57 a.** mean $= 31.6$; median $= 32$; mode $= 34$ and 40 **b.** little or no skewness **2.59 a.** 1.47 **b.** 1.49 **2.61 a.** mean $=$
-4.86; median $= -4.85$; mode $= -5.00$ **2.63 b.** probably none **2.65 a.** qualitative, qualitative, quantitative, quantitative **c.** mean $= 93.6$;
median $= 91.5$; mode $= 86$ **d.** mean $= 95.3$; median $= 92$; mode $= 92$ **e.** mean $= 101.9$; median $= 103$; mode $= 113$ **f.** three centers
g. YND: mean $= 46.88$, median $= 43.7$, mode $= 0$; SLI: mean $= 20.17$, median $= 32.5$, mode $= 0$; OND: mean $= 0$, median $= 0$, mode $= 0$
2.67 a. mean $= -.15$; median $= -.11$; no mode **c.** mean $= -.12$; median $= -.105$; no mode **2.69** largest value minus smallest value
2.73 more variable **2.75 a.** 5, 3.7, 1.92 **b.** 99, 1,949.25, 44.15 **c.** 98, 1,307.84, 36.16 **2.77** data set 1: 1, 1, 2, 2, 3, 3, 4, 4, 5, 5; data set 2: 1, 1, 1, 1, 1,

5, 5, 5, 5, 5 **2.79 a.** 3, 1.3, 1.1402 **b.** 3, 1.3, 1.1402 **c.** 3, 1.3, 1.1402 **2.81 a.** 29, 75.7, 8.70 **b.** 24, 72.7, 8.53 **c.** A students **2.83 a.** .18
b. .0041 **c.** .064 **d.** Morning **2.85 a.** 12, 9.37, 3.06 **b.** 8, 5.15, 2.27; data less variable **c.** 8, 5.06, 2.25; data less variable **2.87 a.** dollars;
quantitative **b.** at least 3/4; at least 8/9; nothing; nothing **2.89 a.** $\approx 68\%$ **b.** $\approx 95\%$ **c.** $\approx$ all **2.91** range/6 = 104.17, range/4 = 156.25; no
2.93 a. $\bar{x}$ = 94.42, s = 4.38 **b.** (90.04, 98.80); (85.66, 103.18); (81.28, 107.56) **c.** 68.4%, 96.6%, 97.1%; yes **2.95 a.** unknown **b.** $\approx 84\%$
2.97 a. (0, 300.5) **b.** (0, 546) **c.** handrubbing appears to be more effective **2.99 a.** at least 8/9 of the velocities will fall within $936 \pm 3(10)$
b. No **2.101 a.** 19 ± 195 **b.** 7 ± 147 **c.** SAT-Math **2.103 a.** 25%; 75% **b.** 50%; 50% **c.** 80%; 20% **d.** 16%; 84% **2.105** $\approx 95\%$
2.107 a. $z = 2$ **b.** $z = -3$ **c.** $z = -2$ **d.** $z = 1.67$ **2.109** 26.5th percentile **2.111 a.** $z = -4.66$ **b.** $z = -.32$ **2.113 a.** 23 **b.** $z = 3.28$
2.115 a. -0.2 **b.** -0.06 **c.** $z = -4.65$ **2.117 a.** $z = 2.0$: 3.7; $z = -1.0$: 2.2; $z = .5$: 2.95; $z = -2.5$: 1.45 **b.** 1.9 **c.** mound-shaped; symmetric **2.123 a.** no, $z = .73$ **b.** yes, $z = -3.27$ **c.** no, $z = 1.36$ **d.** yes, $z = 3.73$ **2.125 a.** 4 **b.** $Q_U \approx 6, Q_L \approx 3$ **c.** 3 **d.** skewed right
e. 50%; 75% **f.** 12, 13, and 16 **2.127 a.** -1.26 **b.** No **2.129 b.** variability for 2000 is slightly greater than 1990 **c.** No **2.131 b.** medians:
Familiar ≈ 45, Treatment ≈ 40, Control ≈ 30 **b.** c. IQR for Treatment appears to be largest **d.** yes **e.** no **2.133 a.** suspect outliers: 74, 79, 79,
81, and 81 **b.** levels with $|z| > 3$: 74, 79, 79, 81, and 81 **2.135** Group 1: M = 48, IQR = 12; group 2: M = 47.5, IQR = 12; group 3: M = 46,
IQR = 4 **2.139** slight positive linear trend **2.141** yes, positive linear trend **2.143 a.** nonlinear increasing **b.** nonlinear increasing
2.145 a. no **b.** yes, slight positive **c.** reliability is suspect (only 5 data points) **2.147 a.** negative trend **b.** positive trend **2.155 a.** $-1, 1, 2$
b. $-2, 2, 4$ **c.** 1, 3, 4 **d.** .1, .3, .4 **2.157 a.** 3.1234 **b.** 9.0233 **c.** 9.7857 **2.159 a.** $\bar{x} = 5.67, s^2 = 1.0667, s = 1.03$ **b.** $\bar{x} = -\$1.5, s^2 = 11.5$,
$s = \$3.39$ **c.** $\bar{x} = .4125\%, s^2 = .0883, s = .30\%$ **d.** 3; \$10; .7375% **2.161** yes, positive **2.163** yes **2.165** $z = -1.06$ **2.167 a.** current:
42.0%; retired: 58.0% **b.** Most (40 of 50) Beanie babies have values less than \$100. **c.** yes, positive **d.** unknown; at least 84%; at least 93.7%
e. 44%; 98%; 100% **f.** 90%; 98%; 98% **2.169 c.** 26.4 ± 21.2 **d.** 24.5 ± 22.4 **2.171 a.** "favorable/recommended"; .635 **b.** yes
2.173 a. 28.5% **b.** 71% **c.** female **2.175 a.** at most 25% **b.** $\approx 2.5\%$ **2.177 b.** customers 238, 268, 269, and 264 **c.** $z = 1.92, 2.06, 2.13,$
3.14 **2.179 a.** Seabirds—quantitative; length—quantitative; oil—qualitative **b.** transect **c.** oiled: 38%; unoiled: 62% **e.** distributions are
similar **f.** 3.27 ± 13.4 **g.** 3.495 ± 11.936 **h.** unoiled **2.181 b.** yes **c.** A1775A: $\bar{x}$ = 19,462.2, s = 532.29; A1775B: $\bar{x}$ = 22,838.5,
s = 560.98 **d.** cluster A1775A **2.183 a.** quantitative: days and year **b.** lesser crimes **c.** yes **d.** mean = 41.4 days; median = 15 days
e. no; $z = 2.38$ **f.** no **2.185 a.** median **b.** mean **2.187** survey results biased, not reliable

Chapter 3

3.9 a. .5 **b.** .3 **c.** .6 **3.11** $P(A) = .55; P(B) = .50; P(C) = .70$ **3.13 a.** 10 **b.** 20 **c.** 15,504 **3.15 a.** $(R_1R_2), (R_1R_3), (R_2R_3), (R_1B_1),$
$(R_1B_2), (R_2B_1), (R_2B_2), (R_3B_1), (R_3B_2), (B_1B_2)$ **b.** 1/10 for each sample point **c.** $P(A) = 1/10, P(B) = 6/10, P(C) = 3/10$ **3.17** .05
3.19 a. .01 **b.** yes **3.21 a.** .261 **b.** Trunk $-.85$, Leaves $-.10$, Branch $-.05$ **3.23 a.** a. 1, 2, 3, 4, 5, 6, 7, 8, 9 **b.** c. .147, .101, .104, .133, .097,
.157, .120, .083, .058 **d.** .248 **e.** .418 **3.25 a.** 6 **b.** .282, .0365, .339, .032, .008, .274 **c.** .686 **3.27 a.** 28 **b.** 1/28 **3.29** Probabilities are the
same **3.31 a.** 15 **b.** 20 **c.** 15 **d.** 6 **3.41 b.** $P(A) = 7/8, P(B) = 1/2, P(A \cup B) = 7/8, P(A^c) = 1/8, P(A \cap B) = 1/2$ **c.** 7/8 **d.** no
3.43 a. $^3/_4$ **b.** 13/20 **c.** 1 **d.** 2/5 **e.** $^1/_4$ **f.** 7/20 **g.** 1 **h.** $^1/_4$ **3.45 a.** .65 **b.** .72 **c.** .25 **d.** .08 **e.** .35 **f.** .72 **g.** .65 **g.** A and C, B and
C, C and D **3.47** A = {eighth-grader scores above 655 on mathematics assessment test}; $P(A^c) = .95$ **3.49 a.** {11, 13, 15, 17, 29, 31, 33, 35}
b. {2, 4, 6, 8, 10, 11, 13, 15, 17, 20, 22, 24, 26, 28, 29, 31, 33, 35, 1, 3, 5, 7, 9, 19, 21, 23, 25, 27} **c.** $P(A) = 9/19, P(B) = 9/19, P(A \cap B) =$
$4/19, P(A \cup B) = 14/19, P(C) = 9/19$ **d.** {11, 13, 15, 17} **e.** 14/19, no **f.** 2/19 **g.** {1, 2, 3, ..., 29, 31, 33, 35} **h.** 16/19 **3.51 a.** (PTW-R,
Jury), (PTW-R, Judge), (PTW-A/D, Jury), (PTW-A/D, Judge), (DTW-R, Jury), (DTW-R, Judge), (DTW-A/D, Jury), (DTW-A/D, Judge) **b.** .684
c. .124 **d.** no **e.** .316 **f.** .717 **g.** .091 **3.53 a.** .157 **b.** .029 **c.** .291 **3.55 a.** $^1/_2$ **b.** 5/6 **3.63 a.** .5 **b.** .25 **c.** no **3.65 a.** .08 **b.** .40
c. .52 **3.67 a.** $P(A) = .4; P(B) = .8; P(A \cap B) = .3$ **b.** $P(E_1|A) = .25, P(E_2|A) = .25, P(E_3|A) = .5$ **c.** .75 **3.69** no **3.71** 1/3, 0,
1/14, 1/7, 1 **3.73** .364 **3.75 a.** .1 **b.** .522 **c.** The person is guessing. **3.77 a.** .75 **b.** .5 **c.** .4 **3.79 a.** 38/132 = .2879 **b.** 29/123 = .236
3.81 a. $P(A|I) = .9, P(B|I) = .95, P(A|N) = .2, P(B|N) = .1$ **b.** .855 **c.** .02 **d.** .995 **3.83 a.** .333 **b.** .120
c. No **d.** probability of playing Center depends on race **3.85 b.** .147 **c.** .85 **d.** .65 **e.** yes **f.** African Americans are stopped for
speeding more often than expected **3.87 a.** .116 **b.** .728 **3.89 a.** .553 **b.** $P(W|CA) = 1; P(W|CB) = .873; P(W|CC) = .559;$
$P(W|BA) = .741; P(W|BB) = .547; P(W|BC) = .342; P(W|AA) = .536; P(W|AB) = .216; P(W|AC) = .077$ **c.** .856 **d.** .552
3.91 b. Worst $-.250$, 2nd worst $-.200$, 3rd worst $-.157$, 4th worst $-.120$, 5th worst $-.089$, 6th worst $-.064$, 7th worst $-.044$, 8th worst $-.029$,
9th worst $-.018$, 10th worst $-.011$, 11th worst $-.007$, 12th worst $-.006$, 13th worst $-.005$ **c.** 250/800 = .313 **d.** .297, .284, .274, .267, .262,
.257, .255, .253, .252, .252, .251 **e.** .288 **3.97 a.** 35,820,200 **b.** 1/35,820,200 **c.** highly unlikely **3.109 a.** 4 **b.** 8 **c.** 32 **d.** 2^n

3.111 a. 35 **b.** 15 **c.** 435 **d.** 45 **e.** $\binom{q}{r} = \dfrac{q!}{r!(q-r)!}$ **3.113 a.** 56 **b.** 1,680 **c.** 6,720 **3.115 a.** 24 **b.** 12 **3.117 a.** 30 **b.** 1,050

3.119 a. 6 **b.** 1/3 **3.121 a.** 3,003 **b.** .0186 **c.** .5734 **3.123 a.** 21/252 **b.** 21/252 **c.** 105/252 **3.125 a.** 2,598,960 **b.** .002 **c.** .00394
d. .0000154 **3.129 a.** .225 **b.** .125 **c.** .35 **d.** .643 **e.** .357 **3.131** .2 **3.133 a.** .5 **b.** .99 **c.** .847 **3.135** .966 **3.137 a.** #4 **b.** #4
or #6 **3.139 a.** $A \cup B$ **b.** B^c **c.** $A \cap B$ **d.** $A^c|B$ **3.141 a.** 0 **b.** no **3.143** .5 **3.145 c.** $P(A) = 1/4; P(B) = ^1/_2$ **e.** $P(A^c) = 3/4;$
$P(B^c) = 1/2; P(A \cap B) = 1/4; P(A \cup B) = 1/2; P(A|B) = 1/2; P(B|A) = 1$ **f.** no, no **3.147 a.** no **b.** .3, .1 **c.** .37 **3.149 a.** false
b. true **c.** true **d.** false **3.151** .545 **3.153 a.** .02 **b.** .08 **3.155 c.** .6637, .1815, .1263, .0214, .0071 **d.** .2029 **3.157 a.** Total population,
Agricultural change, Presence of industry, Growth, Population concentration **b.** .18, .05, .27, .05, .45 **c.** .63 **3.159 a.** .385 **b.** .378 **c.** .183
3.161 a. .64, .32, .04 **b.** .72, .22, .06 **c.** dependent **3.163 a.** .4096 **b.** .0016 **c.** .9984 **3.165 a.** .7127 **b.** .2873 **c.** .9639 **d.** .3078
e. .0361 **f.** at least 3 **3.167 a.** 38,798,760 **b.** .105 **c.** .237 **3.169 a.** 1 to 2 **b.** .5 **c.** .4 **3.171 a.** .25 **b.** .0156 **c.** .4219 **d.** .001,
.000000001, .9970 **3.173** .993 **3.175** 4.4739×10^{-28} **3.177** yes **3.179** Marilyn

Chapter 4

4.3 a. discrete **b.** continuous **c.** continuous **d.** discrete **e.** continuous **f.** continuous **4.5 a.** continuous **b.** discrete **c.** discrete
d. discrete **e.** discrete **f.** continuous **4.7** 0, 1, 2, 3, ...; discrete **4.9** table, graph, formula **4.11 a.** .25 **b.** .40 **c.** .75 **4.13 a.** .7 **b.** .3
c. 1 **d.** .2 **e.** .8 **4.15 a.** b. $P(0) = 1/8, P(1) = 3/8, P(2) = 3/8, P(3) = 1/8$ **d.** 1/2 **4.17 a.** 1 **b.** .24 **c.** .39 **4.19 a.** .9985, .0015, 0, 0
c. 1 **4.21** $P(-3) = .1, P(-1) = .105, P(5) = .795$ **4.23 b.** .85 **c.** .6, 0, 0, 0, .7 **4.25** 7/8 **4.29** $\approx .95$ **4.31 a.** 3.8 **b.** 10.56 **c.** 3.2496
e. no **f.** yes **4.33 a.** $\mu_x = 1, \mu_y = 1$ **b.** distribution of x **c.** $\mu_x = 1, \sigma = .6; \mu_y = 1, \sigma = .2$ **4.35 b.** 4.65 **4.37 a.** a_1: 2.4; a_2: 1.5;

$a_3 - a_5$: .90; a_6: 1.65   **b.** a_1: .86, (.68, 4.12);a_2: .67, (.16, 2.84); $a_3 - a_5$: .3, (.3, 1.5); a_6: .57, (.51, 2.79)   **4.39** $-\$0.263$ **4.41 a.** $p(x) = .05$ for all x-values   **b.** 52.5   **c.** $(-5.16, 110.16)$ **f.** 33.25, 38.36   **g.** .525   **i.** .20   **j.** .65   **4.43 a.** mean $= 2.9$; median $= 3$ **b.** 3, 4   **d.** 3, 3

4.45 $p(x) = \binom{7}{x}.2^x.8^{7-x}$ $(x = 0, 1, 2, \ldots, 7)$ **4.47 a.** 15   **b.** 10   **c.** 1   **d.** 1   **e.** 4   **4.49 a.** .4096   **b.** .3456   **c.** .027   **d.** .0081

e. .3456   **f.** .027   **4.51 a.** $\mu = 12.5, \sigma^2 = 6.25, \sigma = 2.5$ **b.** $\mu = 16, \sigma^2 = 12.8, \sigma = 3.578$ **c.** $\mu = 60, \sigma^2 = 24,\sigma = 4.899$ **d.** $\mu = 63,$ $\sigma^2 = 6.3, \sigma = 2.510$ **e.** $\mu = 48, \sigma^2 = 9.6, \sigma = 3.098$ **f.** $\mu = 40, \sigma^2 = 38.4, \sigma = 6.197$ **4.53 a.** 0   **b.** .998   **c.** .137   **4.55 a.** $p = .5$

b. $p < .5$ **c.** $p > .5$ **4.57 b.** .05   **c.** 10   **4.59 a.** .001   **b.** .322   **c.** .994   **4.61 a.** .9861   **b.** .000000384   **c.** 5   **4.63 a.** .7908   **b.** .0555

4.65 a. .1   **b.** .7   **c.** .4783   **d.** .0078   **e.** yes   **4.67 b.** $\mu = 2.4, \sigma = 1.47$ **c.** $p = .9, q = .1, n = 24, \mu = 21.6, \sigma = 1.47$ **4.71** 3

4.73 b. $\mu = 3, \sigma = 1.7321$ **4.75 a.** .934   **b.** .191   **c.** .125   **d.** .223   **e.** .777   **f.** .001   **4.77 b.** $\mu = 3, \sigma = 1.7321$ **c.** .966   **4.79 a.** .368

b. .264   **c.** .920   **4.81 a.** .0224   **b.** .0850   **c.** $\mu = 3.8, \sigma = 1.9494$ **4.83 a.** 2   **b.** no, $p = .003$ **4.85 a.** $p(2) = .039, p(6) = .160,$ $p(10) = .047$ **c.** $\mu = 6.2, \sigma = 2.490$ **d.** very unlikely   **4.87** yes, .96   **4.91 a.** .3   **b.** .119   **c.** .167   **d.** .167   **4.93 b.** $\mu = 4, \sigma = .853$ **d.** .939   **4.95 a.** hypergeometric   **b.** binomial   **4.97 a.** .383   **b.** .0002   **4.99** .2693   **4.101 a.** .0883   **b.** .1585   **4.103** $P(x = 1) = .25$

4.105 a. $\mu = 113.24, \sigma = 4.194$ **b.** $z = 6.38$ **4.107 a.** Poisson   **b.** binomial   **c.** binomial   **4.109 a.** .192   **b.** .228   **c.** .772   **d.** .987 **e.** .960   **f.** 14; 4.2; 2.05   **g.** .975   **4.111 a.** .243   **b.** .131   **c.** .36   **d.** .157   **e.** .128   **f.** .121   **4.113 a.** .180   **b.** .015   **c.** .076   **4.115 b.** 1/30 **c.** .1455   **d.** .1559   **4.117** binomial   **4.119 a.** yes   **b.** .051   **c.** .757   **d.** .192   **e.** 3.678   **4.121 a.** $\mu = 520, \sigma = 13.491$ **b.** no, $z = -8.895$

4.123 .642   **4.125 a.** .383   **b.** .983   **c.** 20   **4.127 a.** .006   **b.** insecticide is less effective than claimed.   **4.129 a.** A: 4.60; B: 3.70   **b.** A: $\$46,000$; B: $\$55,500$ **c.** A: 1.34, 1.16; B: 1.21, 1.10   **d.** A: .95; B: .95   **4.131** .0005   **4.133** NASA: .0004; Air Force: .5155; Air Force

Chapter 5

5.3 b. $\mu = 20, \sigma = 5.774$ **c.** $(8.452, 31.548)$ **5.5 b.** $\mu = 3, \sigma = .577$ **c.** .577   **d.** .61   **e.** .65   **f.** 0   **5.7 a.** 0   **b.** 1   **c.** 1   **5.9 a.** .1333, .5714   **b.** .2667, 0   **5.11** yes   **5.13 a.** continuous   **b.** $\mu = 7, \sigma = .2887$ **c.** .5   **d.** 0   **e.** .75   **f.** .0002   **5.15** .4444   **5.17** symmetric, bell-shaped curve   **5.19** normal with $\mu = 0$ and $\sigma = 1$ **5.21 a.** .4772   **b.** .3413   **c.** .4545   **d.** .2190   **5.23 a.** 0   **b.** .8413   **c.** .8413 **d.** .1587   **e.** .6826   **f.** .9544   **g.** .6934   **h.** .6378   **5.25 a.** 1.645   **b.** 1.96   **c.** -1.96 **d.** 1.28   **e.** 1.28   **5.27 a.** 0   **b.** 1   **c.** 2.5   **d.** -3 **e.** 5   **f.** 1.4   **5.29 a.** .0456, .0026   **b.** .6826, .9544   **c.** 325.2; 261.6   **5.31 a.** 19.76   **b.** 36.72   **c.** 48.64   **5.33** 182   **5.35 a.** .9406   **b.** .9406 **c.** .1140   **5.37 a.** .1558   **b.** .2586   **c.** .0062   **d.** .9525   **5.39 a.** .4107   **b.** .1508   **c.** .9066   **d.** .0162   **e.** 841.8   **5.41 a.** .0735   **b.** .3651 **c.** 7.29   **5.43** $d = 56.24$ **5.45 a.** .3745   **b.** .7553   **5.47 a.** $z_L = -.675, z_U = .675$ **b.** $-2.68, 2.68$ **c.** $-4.69, 4.69$ **d.** .0074, 0 **5.51 a.** .68   **b.** .95   **c.** .997   **5.53** plot c   **5.55 a.** not normal (skewed right)   **b.** $Q_L = 37.5, Q_U = 152.8, s = 95.8$ **c.** IQR/s $= 1.204$ **5.57 a.** 7   **b.** 6.12   **c.** IQR/s $= 1.14$ **5.59** approx. normal   **5.61** both distributions approx. normal   **5.63** IQR/s $= 1.3$, histogram approx. normal   **5.65** The lowest score of 0 is less than one standard deviation below the mean.   **5.69 a.** yes   **b.** $\mu = 10, \sigma^2 = 6$ **c.** .726 **d.** .7291   **5.71 a.** .1788   **b.** .5236   **c.** .6950   **5.73 a.** 40   **b.** 5.66   **c.** 1.86   **d.** .9686   **5.75 a.** 10   **b.** 3.082   **c.** .16   **d.** .4364   **5.77 a.** no **b.** yes   **c.** yes   **5.79** ≈ 0 **5.81 a.** 0   **b.** .4641   **c.** no   **5.83 a.** 300   **b.** 800   **c.** 1   **5.89 a.** .367879   **b.** .950213   **c.** .223130   **d.** .993262 **5.91** .950213   **5.93 a.** .449329   **b.** .864665   **5.95** .434598   **5.97 a.** .367879   **b.** .606531   **c.** .135335   **d.** .814046   **5.99 a.** 17   **b.** .5862 **5.103 a.** exponential   **b.** uniform   **c.** normal   **5.105 a.** .9821   **b.** .0179   **c.** .9505   **d.** .3243   **e.** .9107   **f.** .0764   **5.107 a.** .6915 **b.** .0228   **c.** .5328   **d.** .3085   **e.** 0   **f.** .9938   **5.109 a.** .3821   **b.** .5398   **c.** 0   **d.** .1395   **5.111** .8315   **5.113 a.** 0   **b.** 1   **5.115 a.** .5124 **b.** .2912   **c.** .0027   **5.117 a.** .667   **b.** .333   **c.** 82.5°F   **5.119** .3125   **5.121** .0154; very unlikely   **5.123 a.** .0764   **b.** (65.024, 88.576) **5.125** approx. normal   **5.127 a.** (i) .384, (ii) .49, (iii) .212, (iv) .84   **b.** (i) .3849, (ii) .4938, (iii) .2119, (iv) .8413   **c.** (i) .0029, (ii) .0038, (iii) .0001, (iv) .0013   **5.129** 52 minutes   **5.131** .2676; no   **5.133** 0; unlikely   **5.135 a.** .05, .20, .50, .20, .05   **b.** z-scores   **c.** identical

Chapter 6

6.3 c. 1/16   **6.5 c.** .05   **d.** no   **6.13** unbiased, minimum variance   **6.15 a.** 5   **b.** $E(\bar{x}) = 5$ **c.** $E(M) = 4.778$ **d.** $\bar{x}$ **6.19** **b.** 1.61   **c.** $E(s^2) = 1.61$ **e.** $E(s) = 1.004$ **6.21** mean and standard deviation of sampling distribution of $\bar{x}$ **6.23** smaller **6.27 a.** $\mu = 100, \sigma = 5$ **b.** $\mu = 100, \sigma = 2$ **c.** $\mu = 100, \sigma = 1$ **d.** $\mu = 100, \sigma = 1.414$ **e.** $\mu = 100, \sigma = .447$ **f.** $\mu = 100, \sigma = .316$ **6.29 a.** $\mu = 2.9, \sigma^2 = 3.29,\sigma = 1.814$ **c.** $\mu_{\bar{x}} = 2.9, \sigma_{\bar{x}} = 1.283$ **6.31 a.** $\mu_{\bar{x}} = 30, \sigma_{\bar{x}} = 1.6$ **b.** approx. normal   **c.** .8944   **d.** .0228 **e.** .1292   **f.** .9699   **6.35 a.** 79   **b.** 2.3   **c.** approx. normal   **d.** .43   **e.** .3336   **6.37** .0838   **6.39 a.** $\mu_{\bar{x}} = 5.1, \sigma_{\bar{x}} = .4981$ **b.** yes   **c.** .2119 **d.** .4071   **6.41 a.** .0013   **b.** program did decrease the mean number of sick days   **6.43** handrubbing: $P(\bar{x} < 30) = .2743$; handwashing: $P(\bar{x} < 30) = .0047$; sample used handrubbing   **6.45** false   **6.47** true   **6.49 b.** $E(A) = \alpha$ **c.** choose estimator with smallest variance **6.51 a.** .5   **b.** .0606   **c.** .0985   **d.** .8436   **6.57 a.** $\mu_{\bar{x}} = 89.34, \sigma_{\bar{x}} = 1.3083$ **c.** .8461   **e.** .0367   **6.59 a.** $\mu_{\bar{x}} = 1.3, \sigma_{\bar{x}} = .240$ **b.** yes **c.** .1056   **d.** .0062   **6.61** .9772   **6.63 a.** .008   **b.** n large   **6.65 a.** discrete   **b.** approximately normal with $\mu_{\bar{x}} = 1.5, \sigma_{\bar{x}} = .01764$ **c.** .0023 **d.** no   **6.67** .9332   **6.69 a.** .0031   **b.** more likely if $\mu = 156$; less likely if $\mu = 158$ **c.** less likely if $\sigma = 2$; more likely if $\sigma = 6$

Chapter 7

7.5 yes   **7.7 a.** 1.645   **b.** 2.58   **c.** 1.96   **d.** 1.28   **7.9 a.** $28 \pm .784$ **b.** $102 \pm .65$ **c.** $15 \pm .0588$ **d.** $4.05 \pm .163$ **e.** no **7.11 a.** 83.2 ± 1.25 **c.** 83.2 ± 1.65 **d.** increases   **e.** yes   **7.13 a.** 19.3   **b.** 19.3 ± 3.44 **d.** random sample, large n **7.15 a.** mean egg length for all New Zealand birds   **c.** $\bar{x} = 68.28, s = 55.65$ **d.** 68.28 ± 20.27 **7.17 a.** $1.13 \pm .672$ **b.** yes   **7.19 a.** 19 ± 7.826 **b.** 7 ± 5.90 **c.** SAT-Math   **7.21** $.422 \pm .067$ **7.23 a.** μ_Y: $4.17 \pm .095$; μ_{MA}: $4.04 \pm .057$; μ_O: $4.31 \pm .062$ **b.** more likely   **7.25** Central Limit Theorem no longer applies; σ unknown   **7.29 a.** 2.228   **b.** 2.567   **c.** -3.707 **d.** -1.771 **7.31 a.** 5 ± 1.876 **b.** 5 ± 2.394 **c.** 5 ± 3.754 **d.** (a) $5 \pm .780$, (b) $5 \pm .941$, (c) 5 ± 1.276; width decreases   **7.33 a.** .009   **b.** 2.306   **c.** $.009 \pm .0037$ **7.35 a.** 2.8856 ± 4.034 **b.** $.4083 \pm .2564$ **7.37 c.** $(1,633.05, 2,778.85)$ **e.** $(1,532.39, 2,658.81)$ **7.39 a.** 7.3 ± 1.41 **c.** PMI values normally distributed **7.41 a.** 37.3 ± 7.70 **c.** One hour before: $\mu > 25.5$ **7.45 a.** yes   **b.** $.64 \pm .067$ **7.47 a.** yes   **b.** no   **c.** yes   **d.** no   **7.49 a.** .03 **b.** $.03 \pm .015$ **7.51** $.219 \pm .024$ **7.53 a.** $.368 \pm .129$ **b.** $.605 \pm .130$ **7.55 a.** .636   **b.** $.636 \pm .167$ **7.57 a.** $.524 \pm .095$ **7.59** 95%

confident that proportion of health care workers with latex allergy who suspect he/she has allergy is between .327 and .541   **7.61** true   **7.63** 519   **7.65 a.** 482   **b.** 214   **7.67** 34   **7.69 a.** $n = 16$: $W = .98$; $n = 25$: $W = .784$; $n = 49$: $W = .56$; $n = 100$: $W = .392$; $n = 400$: $W = .196$ **7.71 a.** .90   **b.** .05   **c.** 259   **7.73** 97   **7.75** 14,735   **7.77** 129   **7.79** 271   **7.81 a.** μ **b.** μ **c.** p **d.** p **e.** p **7.83 a.** $t = 2.086$ **b.** $z = 1.96$ **c.** $z = 1.96$ **d.** $z = 1.96$ **e.** neither t nor z **7.85 a.** .57 ± .049   **b.** 2,358   **7.87** (1) p; (2) μ; (3) μ; (4) μ **7.89 b.** no assumptions needed   **c.** 95% confident claim is false   **7.91** 1.37 ± .76   **7.95 a.** .660 ± .029   **b.** .301 ± .035   **7.97 a.** .044 ± .162   **b.** no evidence to indicate that this species has tendency to inbreed   **7.99 a.** .81 ± .41   **b.** 140   **7.101 a.** .378 ± .156   **b.** .703 ± .147   **c.** 41.405 ± 21.493   **d.** .378 ± .156   **7.103 a.** .094   **b.** yes   **c.** .094 ± .037   **7.105 a.** 49.3 ± 8.60   **c.** Population is normal.   **d.** 60   **7.107 a.** 781   **7.109 a.** yes   **b.** missing measure of reliability   **c.** 95% CI for μ: .932 ± .037

Chapter 8

8.1 null hypothesis   **8.3** α **8.7** no   **8.9** H_0: $p = .75$, H_a: $p \neq .75$ **8.11 a.** H_0: $p = .45$ **b.** H_0: $\mu = 2.5$ **8.13 a.** H_0: $p = .75$, H_a: $p > .75$ **8.15 a.** H_0: $\mu = 15$, H_a: $\mu < 15$ **b.** conclude mean mercury level is less than 15 ppm when mean equals 15 ppm   **c.** conclude mean mercury level equals 15 ppm when mean is less than 15 ppm   **8.17 a.** H_0: No intrusion occurs   **b.** H_a: Intrusion occurs   **c.** $\alpha = .001$, $\beta = .5$ **8.21 g.** (a) .025, (b) .05, (c) $\approx$.005, (d) $\approx$.10, (e) .10, (f) $\approx$.01   **8.23 a.** $z = 1.67$, reject H_0 **b.** $z = 1.67$, fail to reject H_0 **8.25 a.** Type I: conclude mean response for all New York City public school children is not 3 when mean equals 3; Type II: conclude mean response for all New York City public school children equals 3 when mean is not equal to 3   **b.** $z = -85.52$, reject H_0 **c.** $z = -85.52$, reject H_0 **8.27** yes, $z = 12.36$ **8.29 a.** H_0: $\mu = 16$, H_a: $\mu < 16$ **b.** $z = -4.31$, reject H_0 **8.31 a.** $z = 4.03$, reject H_0 **8.33 a.** no   **b.** $z = .61$, do not reject H_0 **d.** no   **e.** $z = -.83$, do not reject H_0 **8.35 a.** $z = -3.72$, reject H_0 **8.37** small p-values   **8.39 a.** fail to reject H_0 **b.** reject H_0 **c.** reject H_0 **d.** fail to reject H_0 **e.** fail to reject H_0 **8.41** .0150   **8.43** p-value $= .9279$, fail to reject H_0 **8.45 a.** fail to reject H_0 **b.** fail to reject H_0 **c.** reject H_0 **d.** fail to reject H_0 **8.47 a.** .3446   **b.** do not reject H_0 **8.49** p-value $< .0001$; reject H_0 at $\alpha = .05$ **8.51 b.** reject H_0 at $\alpha = .05$ **c.** reject H_0 at $\alpha = .05$ **8.53** small n, normal population   **8.57 a.** $t < -2.160$ or $t > 2.160$ **b.** $t > 2.500$ **c.** $t > 1.397$ **d.** $t < -2.718$ **e.** $t < -1.729$ or $t > 1.729$ **f.** $t < -2.353$ **8.59 a.** $t = -2.064$; fail to reject H_0 **b.** $t = -2.064$; fail to reject H_0 **c.** (a) .05 $< p$-value $< .10$; (b) .10 $< p$-value $< .20$ **8.61 a.** H_0: $\mu = 5.70$, H_a: $\mu < 5.70$ **b.** $t < -2.821$ **c.** $t = -4.33$ **d.** reject H_0 **8.63** $t = -.43$; fail to reject H_0 **8.65** $t = 3.72$, p-value $= .006$, reject H_0 **8.67** $t = 2.97$, fail to reject H_0: $\mu = .100$ **8.69** yes, $t = 1.95$ **8.71** qualitative   **8.73 a.** yes   **b.** no   **c.** yes   **d.** no   **e.** no   **8.75 a.** $z = -2.00$ **c.** reject H_0 **d.** .0228   **8.77 a.** $z = 1.13$, fail to reject H_0 **b.** .1292   **8.79 a.** .043   **b.** H_0: $p = .07$, H_a: $p \neq .07$ **c.** $z = -1.35$ **d.** $|z| > 1.96$.   **e.** do not reject H_0 **f.** .177   **8.81** yes, $z = 1.74$ **8.83** yes, $z = -2.71$ **8.85** $z = 15.46$, reject H_0: $p = .167$ **8.87** no, $z = 1.49$ **8.89** $z = 33.47$, reject H_0: $p = .5$ **8.91** power $= 1 - \beta$ **8.93 b.** 1,032.9   **d.** .7422   **e.** .2578   **8.95 a.** approx. normal, $\mu_{\bar{x}} = 50$, $\sigma_{\bar{x}} = 2.5$ **b.** approx. normal, $\mu_{\bar{x}} = 45$, $\sigma_{\bar{x}} = 2.5$ **c.** .2358   **d.** .7642   **8.97 a.** approx. normal, $\mu_{\bar{x}} = 10$, $\sigma_{\bar{x}} = 0.1$ **b.** approx. normal, $\mu_{\bar{x}} = 9.9$, $\sigma_{\bar{x}} = 0.1$ **c.** .8300   **d.** .8300   **8.99** power increases   **8.101** .1075   **8.105** false   **8.107 a.** $\chi^2 < 6.26214$ or $\chi^2 > 27.4884$ **b.** $\chi^2 > 40.2894$ **c.** $\chi^2 > 21.0642$ **d.** $\chi^2 < 3.57056$ **e.** $\chi^2 < 1.63539$ or $\chi^2 > 12.5916$ **f.** $\chi^2 < 13.8484$ **8.109 a.** $\chi^2 = 182.16$, reject H_0 **8.111 a.** H_0: $\sigma^2 = 225$, H_a: $\sigma^2 > 225$ **b.** $\chi^2 = 187.896$ **c.** p-value $= .9938$, do not reject H_0 **8.113** $\chi^2 = 12.61$, fail to reject H_0 **8.115** $\chi^2 = 2.16$, fail to reject H_0 **8.117 a.** no, $\chi^2 = 2.09$ **b.** The population of CO_2 amounts are normally distributed.   **8.119** alternative   **8.121** H_0, H_a, α **8.123** null   **8.125 a.** $z = -1.78$, fail to reject H_0 **b.** $z = -1.78$, fail to reject H_0 **c.** .29 ± .063   **d.** .29 ± .083   **e.** 549   **8.127 a.** $\chi^2 = 63.48$, reject H_0 **b.** $\chi^2 = 63.48$, reject H_0 **8.129 a.** H_0: Drug is unsafe, H_a: Drug is safe   **b.** α **8.131 a.** $t = -3.46$, reject H_0 **b.** normal population   **8.133 a.** H_0: $p = .5$, H_a: $p \neq .5$ **b.** $z = 5.99$ **c.** yes   **d.** reject H_0 **8.135 a.** H_0: $\mu = 10$, H_a: $\mu > 10$; one-tailed   **b.** $z > 1.645$ **c.** $z = 1.29$, fail to reject H_0 **8.137 a.** $t = 1.44$, p-value $= .169$, fail to reject H_0 **8.139 a.** $z = .70$, fail to reject H_0 **b.** yes   **8.141 a.** no, $z = 1.41$ **b.** small α **8.143 a.** no   **b.** $\beta = .5910$, power $= .4090$ **c.** power increases   **8.145** $z = 21.66$, reject H_0 **8.147 b.** .37, .24   **8.149 a.** yes, $z = 4.55$ **b.** unknown; Type II error   **c.** .01; Type I error   **d.** .94   **e.** .352

Chapter 9

9.1 normally distributed with mean $\mu_1 - \mu_2$ and standard deviation $\sqrt{\dfrac{\sigma_1^2}{n_1} + \dfrac{\sigma_2^2}{n_2}}$ **9.3 a.** no   **b.** no   **c.** no   **d.** yes   **e.** no   **9.5** b   **9.7 a.** 14; .4   **b.** 10; .3   **c.** 4; .5   **d.** yes   **9.9 a.** .5989   **b.** $t = -2.39$, reject H_0 **c.** $-1.24 \pm .98$ **9.11 a.** fail to reject H_0 at $\alpha = .10$ **b.** p-value $= .0575$, reject H_0 at $\alpha = .10$ **9.13 a.** $\mu_1 - \mu_2$ **b.** 12.5 ± 10.2   **9.15 a.** $z = -2.64$, reject H_0 **b.** $z = .27$, fail to reject H_0 **9.17 a.** H_0: $\mu_1 - \mu_2 = 0$, H_a: $\mu_1 - \mu_2 \neq 0$ **b.** reject H_0 **c.** fail to reject H_0 **9.19 a.** $(-.60, 7.96)$ **b.** independent random samples from normal populations with equal variances   **9.21 a.** $t = -7.24$, p-value $= 0$; reject H_0 **b.** $t = -.50$, p-value $= .62$; fail to reject H_0 **c.** $t = -1.25$, p-value $= .21$; fail to reject H_0 **d.** independent random samples   **9.23** no, $t = .18$ **9.25 a.** $\mu_1 - \mu_2$ **b.** no   **c.** .2262   **d.** fail to reject H_0 **9.27** $t = -.46$, fail to reject H_0 **9.29 a.** no standard deviations reported   **b.** $s_1 = s_2 = 5$ **c.** $s_1 = s_2 = 6$ **9.31** before   **9.33 a.** $t > 1.833$ **b.** $t > 1.328$ **c.** $t > 2.776$ $t > 2.896$ **9.35 a.** $\bar{x}_D = 2$, $s^2_D = 2$ **b.** $\mu_D = \mu_1 - \mu_2$ **c.** 2 ± 1.484 **d.** $t = 3.46$, reject H_0 **9.37 a.** $z = 1.79$, fail to reject H_0 **b.** .0734   **9.39 a.** μ_D **b.** paired difference   **c.** H_0: $\mu_D = 0$, H_a: $\mu_D > 0$ **d.** $t = 2.19$ **e.** reject H_0 **9.41 a.** 749.5 ± 278.1   **9.43 a.** H_0: $\mu_D = 0$, H_a: $\mu_D \neq 0$ **b.** $z = 2.08$, p-value $= .0376$ **c.** reject H_0 **9.45 a.** b. 95% CI for μ_D: 1.95 ± 1.91; control group has larger mean   **9.47** $t = -2.306$, p-value $= .0163$, reject H_0 at $\alpha = .05$ **9.49 a.** both have binomial distributions   **b.** approx. normal with mean $p_1 - p_2$ and standard deviation $\sqrt{\dfrac{p_1 q_1}{n_1} + \dfrac{p_2 q_2}{n_2}}$ **9.53 a.** $z < -2.33$ **b.** $z < -1.96$ **c.** $z < -1.645$ **d.** $z < -1.28$ **9.55 a.** .07 ± .067   **b.** .06 ± .086   **c.** $-.15 \pm .131$ **9.57** $z = 1.16$, fail to reject H_0 **9.59 a.** .153   **b.** .215   **c.** $-.062 \pm .070$ **9.61 a.** .143   **b.** .049   **c.** $z = 2.27$, fail to reject   **d.** reject H_0 **9.63** $z = -1.201$, fail to reject H_0 **9.65** yes; $z = 2.13$, reject H_0 at $\alpha = .05$ **9.67** $z = 5.47$, reject H_0: $p_1 - p_2 = 0$ in favor of H_a: $p_1 - p_2 > 0$ **9.69 a.** at least 18   **b.** sample may not be representative of the population   **9.73** $n_1 = n_2 = 24$ **9.75 a.** $n_1 = n_2 = 29,954$ **b.** $n_1 = n_2 = 2,165$ **c.** $n_1 = n_2 = 1,113$ **9.77** $n_1 = n_2 = 49$ **9.79** $n_1 = n_2 = 130$ **9.81** $n_1 = n_2 = 1,729$ **9.83** $n_1 = n_2 = 136$ **9.87** false   **9.89 a.** .025   **b.** .90   **c.** .99   **d.** .05   **9.91 a.** $F > 2.36$ **b.** $F > 3.04$ **c.** $F > 3.84$ **d.** $F > 5.12$ **9.93 a.** $F = 3.43$, reject H_0 **b.** $F = 3.43$, reject H_0 **9.95 a.** H_0: $\sigma^2_M = \sigma^2_F$, H_a: $\sigma^2_M < \sigma^2_F$ **b.** $F = 1.06$ **c.** $F > 1.26$ **d.** p-value $> .10$ **e.** fail to reject H_0 **9.97** $F = 1.16$, fail to reject H_0 **9.99 a.** $F = 1.44$, fail to reject H_0 **b.** $F = 5.25$, fail to reject H_0 **c.** no differences   **9.101** $F = 1.045$, fail to reject H_0 **9.103 a.** $\mu_1 - \mu_2$ **b.** $\mu_1 - \mu_2$ **c.** $p_1 - p_2$ **d.** σ_1^2 / σ_2^2 **e.** $p_1 - p_2$ **9.105 a.** $t = .78$, fail to reject H_0 **b.** 2.5 ± 8.99   **c.** $n_1 = n_2 = 225$

9.107 a. $3.9 \pm .31$ **b.** $z = 20.60$, reject H_0 **c.** $n_1 = n_2 = 346$ **9.109 a.** $t = 5.73$, reject H_0 **b.** 3.8 ± 1.84 **9.111 a.** p-value $= .871$, fail to reject H_0: $\mu_{no} - \mu_{yes} = 0$ **9.113 a.** time needed   **b.** climbers   **c.** paired experiment   **9.115 a.** H_0: $\mu_D = 0$, H_a: $\mu_D > 0$,   **b.** paired difference test   **d.** leadership: fail to reject H_0 at $\alpha = .05$; popularity: reject H_0 at $\alpha = .05$; intellectual self-confidence: reject H_0 at $\alpha = .05$ **9.117** $z = 1.21$, fail to reject H_0 at $\alpha = .01$ **9.119 a.** H_0: $\mu_1 - \mu_2 = 0$, H_a: $\mu_1 - \mu_2 \neq 0$ **b.** hours of work: fail to reject H_0; free time: reject H_0 at $\alpha = .05$; breaks: reject H_0 at $\alpha = .01$ **9.121 a.** t test   **b.** yes   **c.** no   **d.** fail to reject H_0 **e.** reject H_0 at $\alpha = .05$ **9.123** $t = 1.77$, fail to reject H_0 at $\alpha = .01$ **9.125** $z = -2.55$, reject H_0 **9.127** $-.33 \pm .22$ **9.129 a.** yes, $F = 10$ **b.** both populations normal   **9.131** $z = -3.55$, reject H_0: $p_1 - p_2 = 0$ **9.133** yes, $t = -2.19$ **9.135** use of creative ideas ($z = 8.85$); good use of job skills ($z = 4.76$)

Chapter 10

10.1 A, B, C, D   **10.5 a.** observational   **b.** designed   **c.** designed   **d.** observational   **e.** observational   **f.** observational   **10.7 a.** age when cancer is first detected   **b.** smokers   **c.** type of screening method   **d.** CT scan, chest X-ray   **10.9 a.** cockatiel   **b.** yes   **c.** experimental group   **d.** 1, 2, 3   **e.** 3   **f.** total consumption   **10.11 a.** ethical behavior   **b.** job type (high tech, low tech); sales task (new, maintenance)   **c.** high/new, high/maintenance, low/new, low/maintenance   **d.** college students   **10.17 a.** 6.59   **b.** 16.69   **c.** 1.61   **d.** 3.87   **10.19** dot plot b   **10.21 a.** (a) MSE $= 2$; (b) MSE $= 14.4$ **b.** (a) $t = -6.12$; (b) $t = -2.28$ **c.** (a) $|t| > 2.228$; (b) $|t| > 2.228$ **d.** (a) reject H_0; (b) reject H_0 **e.** both populations normal with equal variances   **10.23 a.** $F = 1.5625$; do not reject H_0 **b.** $F = 6.25$; reject H_0 **c.** $F = 25$; reject H_0 **d.** increases **10.25 a.**

Source	df	SS	MS	F
Treatment	2	12.36	6.18	2.93
Error	9	18.98	2.11	
Total	11	31.34		

b. $F = 2.93$, fail to reject H_0 **10.27 a.** completely randomized   **b.** treatments: 3, 6, 9, 12 robots; dep. var.: energy expanded   **c.** H_0: $\mu_3 = \mu_6 = \mu_9 = \mu_{12}$, H_a: At least 2 μ's differ   **d.** reject H_0 **10.29 a.** H_0: $\mu_{young} = \mu_{middle} = \mu_{old}$ **b.** reject H_0 **c.** H_0: $\mu_{young} = \mu_{middle} = \mu_{old}$; fail to reject H_0 **10.31** yes, $F = 3.90$ **10.33 b.** treatments: scopolamine, glycopyrrolate, and no drug; response: number of pairs recalled. **10.35** $F = 3.96$; do not reject H_0 **10.39 a.** no significant difference   **b.** μ_2 **c.** μ_1 **10.41 a.** 3   **b.** 10   **c.** 6   **d.** 45   **10.43** $\mu_1 > \mu_2$, $\mu_1 > \mu_3$, $\mu_4 > \mu_2$, $\mu_4 > \mu_3$ **10.45 a.** 6   **b.** μ_{12} is the largest mean; μ_3, μ_6, and μ_9 are not significantly different   **10.47 a.** yes   **b.** yes   **c.** no   **d.** .05   **e.** girl means not significantly different   **10.49 a.** reject H_0 **b.** Control and Slide not significantly different   **10.51** yes   **10.53** $\mu_{Control} > \mu_{CBT\text{-}WLT}, \mu_{CBT\text{-}IPT} > \mu_{CBT\text{-}WLT}$ **10.57 a.** 3 blocks; 5 treatments   **b.** 15   **c.** H_0: $\mu_1 = \mu_2 = \mu_3 = \mu_4 = \mu_5$, H_a: At least 2 μ's differ   **d.** $F = 9.109$ **e.** $F > 7.01$ **f.** reject H_0 **10.59 a.**

Source	df	SS	MS	F
Treatment	2	12.032	6.016	50.96
Block	3	71.749	23.916	202.59
Error	6	.708	.118	
Total	11	84.489		

b. yes, p-value $= .000$ **c.** yes, p-value $= .000$ **d.** $\mu_C < \mu_A < \mu_B$ **10.61 b.** California, Utah, Alaska   **c.** Nov 2000, Oct 2001, Nov 2001   **d.** H_0: $\mu_{Cal} = \mu_{Utah} = \mu_{Alas}$ **e.** .002; reject H_0 **f.** $\mu_{Cal} > \mu_{Utah}, \mu_{Cal} > \mu_{Alas}$ **10.63 a.** randomized block design   **c.** reject H_0 at $\alpha > .009$ **d.** $\mu_{control} < \mu_{burning}$ **10.65 b.**

Source	df	SS	MS	F
Prompt	4	1185.00	296.25	39.87
Week	5	386.40	77.28	10.40
Error	20	148.60	7.43	
Total	29	1720.00		

c. yes, $F = 39.87$ **d.** $\mu_{Control} < (\mu_{Inf\text{-}Low}, \mu_{Inf\text{-}Hi}) < (\mu_{Freq\text{-}Low}, \mu_{Freq\text{-}Hi})$ **10.67** do not reject H_0, $F = 0.02$ **10.69 a.** yes   **b.** reject H_0 **10.73 a.** 2   **b.** no   **c.** yes; 3 and 5   **d.** 15   **e.** df(Error) $= 0$; replication **10.75 a.**

Source	df	SS	MS	F
A	2	.8	.4000	2.00
B	3	5.3	1.7667	8.83
AB	6	9.6	1.6000	8.00
Error	12	2.4	.2000	
Total	23	18.1		

b. $SSA, SSB, SSAB$; yes, $F = 7.14$ **c.** yes   **d.** effects of one factor on the dependent variable are not the same at different levels of the second factor   **e.** $F = 8.00$, reject H_0 **f.** no   **10.77 a.** $F(AB) = .75; F(A) = 3; F(B) = 1.5$ **b.** $F(AB) = 7.5; F(A) = 3; F(B) = 3$ **c.** $F(AB) = 3; F(A) = 12; F(B) = 3$ **d.** $F(AB) = 4.5; F(A) = 36; F(B) = 36$ **10.79 a.** 6×6 factorial   **b.** Coagulant (5, 10, 20, 50, 100, 200); pH (4, 5, 6, 7, 8, 9); 36 treatments   **10.81 a.** event (3 wash-ups), strata (coarse, medium, fine, hydroid)   **10.83 a.** 2×2 factorial; Color (blue, red), question (difficult, simple)   **b.** sufficient evidence of interaction at $\alpha = .05$ **10.85 b.** sufficient evidence of interaction at $\alpha = .01$ **c.** no   **10.87 a.**

Source	df	SS	MS	F	P
Diet(D)	1	0.0124	0.0124	0.22	.645
Size(S)	1	8.0679	8.0679	141.18	.000
D × S	1	0.0364	0.0364	0.64	0.432
Error	24	1.3715	0.0571		
Total	27	9.4883			

b. main effect Diet: fail to reject H_0; main effect Size: reject H_0; Diet-Size interaction: fail to reject H_0 **10.89 a.** 464.4, 116.0, 104.0, 98.0   **b.** 6,739.605   **c.** SS(Sex) = 114.005, SS(Weight) = 41.405, SS(Sex × Weight) = 18.605   **d.** 324.63, 60.09, 177.81, 227.43   **e.** 789.97   **f.** 963.983

g.

Source	df	SS	MS	F
Sex	1	114.005	114.005	4.04
Weight	1	41.405	41.405	1.47
SxW	1	18.605	18.605	0.66
Error	28	789.968	18.213	
Total	31	963.983		

h. $F = 2.06$; no difference among treatment means   **10.95 a.**

Source	df	SS	MS	F
Treatment	3	11.332	3.777	157.38
Block	4	10.688	2.672	111.33
Error	12	.288	.024	
Total	19	22.308		

b. yes, reject H_0 **c.** yes; 6   **d.** yes, reject H_0 **10.97 b.** yes   **c.** yes   **10.99** $F = 8.295$, reject H_0; $\mu_{OND} < (\mu_{SLI}, \mu_{YND})$ **10.101 a.** yes, $F = 39.1$ **b.** no   **c.** $(\mu_{NFP}, \mu_{FP}) < (\mu_{CSH}, \mu_{CSFB})$ **10.103 a.** $F = .74$, do not reject H_0 **b.** not appropriate   **10.105 a.**

Source	df	F	p-value
Time Period	3	11.25	.0001
Station	9		
Error	27		
Total	39		

b. reject H_0, $F = 11.25$ **c.** largest: either 7/9-7/23 or 7/24-8/8; smallest: 8/24-8/31   **10.107 a.** experimental units: zircaloy components; blocks: ingots; response: pressure required to separate the bonded components; factor: bonding agent or element; factor type: qualitative; treatments: nickel, iron, copper   **b.** designed

c.

Source	df	SS	MS	F
Agent	2	131.9	65.95	6.36
Ingot	6	268.3	44.71	4.31
Error	12	124.4	10.37	
Total	20			

d. yes, $F = 6.36$ **e.** μ_{iron} is largest mean   **10.109 b.** yes, $F = 1,336.85$ **c.** yes; reject H_0, $F = 258.07$ **d.** no   **e.** nine treatment means   **10.111 a.** df(Period) = 1, df(Gender) = 1, df(P × G) = 1, df(Error) = 120, df(Total) = 123   **10.113 a.** 2×2 factorial   **b.** factors: tent type and location; treatments: (treated, inside), (treated, outside), (untreated, inside), (untreated, outside)   **c.** number of mosquito bites   **d.** effect of tent type on mean number of bites depends on location

Chapter 11

11.7 $\beta_1 = 1/3$, $\beta_0 = 14/3$, $y = 14/3 + (1/3)x$ **11.11** difference between the observed and predicted   **11.13** true   **11.17 c.** $\hat{\beta}_1 = .918$, $\hat{\beta}_0 = .020$ **e.** -1 to 7 **11.19 a.** $y = \beta_0 + \beta_1 x + \varepsilon$ **b.** $\hat{y} = 2.522 + 7.261x$ **11.21 a.** positive   **11.23 a.** $y = \beta_0 + \beta_1 x + \varepsilon$; negative   **b.** yes   **c.** no   **11.25 a.** $y = \beta_0 + \beta_1 x + \varepsilon$ **b.** $\hat{y} = -.607 + 1.062x$ **c.** positive   **d.** $\hat{y} = -.148 + 1.022x$ **11.27 a.** yes; positive   **b.** $y = \beta_0 + \beta_1 x + \varepsilon$ **c.** $\hat{\beta}_0 = 20.1275$, $\hat{\beta}_1 = .62442$ **11.29** $\hat{y} = 5.35 + .53x$; reduced by .053 meter; model not useful   **11.33 a.** 57.5; 3.19444   **b.** 257.5; 6.7763   **c.** 9.288; 1.1610   **11.35** 11.14: SSE $= 1.22$, $s^2 = .244$, $s = .494$; 11.17: SSE $= 5.134$, $s^2 = 1.03$, $s = 1.01$ **11.37 a.** SSE $= 858.17$, $s^2 = 286.06$, $s = 16.91$ **11.39 a.** 5.36   **b.** 3.42   **c.** reading score   **11.41 a.** 36.13   **b.** 72.26   **11.43** 0   **11.45** divide the value in half   **11.47 a.** 95%: .31 $\pm$ 1.13; 90%: .31 $\pm$ .92   **b.** 95%: 64 $\pm$ 4.28; 90%: 64 $\pm$ 3.53   **c.** 95%: $-.84 \pm .67$; 90%: $-.84 \pm .55$ **11.49 b.** $\hat{y} = 2.554 + .246x$ **d.** $t = .627$ **e.** fail to reject H_0 **f.** .246 $\pm$ 1.81   **11.51 a.** $t = -6.41$, reject H_0 **b.** $-.305 \pm .135$ **11.53** $-.0023 \pm .0016$ **11.55 a.** $y = \beta_0 + \beta_1 x + \varepsilon$ **b.** $\hat{y} = -8.524 + 1.665x$ **d.** yes, $t = 7.25$ **e.** 1.67 $\pm$ .46   **11.57 a.** $t = 6.286$, reject H_0 **b.** .88 $\pm$ .236   **c.** no evidence that slope differs from 1   **11.59** yes, $t = 2.858$ **11.63** true   **11.65 a.** perfect positive linear   **b.** perfect negative linear   **c.** strong positive linear   **d.** weak positive linear   **e.** strong negative linear   **11.67** .877   **11.69 b.** .185   **11.71 b.** both positive   **c.** .706   **11.73 b.** reject H_0 **c.** .49   **11.75 b.** piano: $r^2 = .1998$; bench: $r^2 = .0032$; motorbike: $r^2 = .3832$, armchair: $r^2 = .0864$; teapot: $r^2 = .9006$ **c.** Reject H_0 for all objects except bench and armchair.   **11.77 a.** $y = \beta_0 + \beta_1 x + \varepsilon$ **b.** $\beta_1 > 0$ **c.** no, $t = -3.12$ **11.79** $r = .5702$, $r^2 = .3251$ **11.83** true   **11.85 a.** $\hat{y} = 1.375 + .87x$ **c.** 1.5   **d.** .1875   **e.** 3.56 $\pm$ .33   **f.** 4.875 $\pm$ 1.065   **11.87 c.** 4.644 $\pm$ 1.118   **d.** $-.414 \pm 1.717$ **11.89 a.** no   **b.** (275.86, 407.26)   **c.** (332.95, 350.17)   **11.91** (92.298, 125.104)   **11.93 a.** $y = \beta_0 + \beta_1 x + \varepsilon$ **b.** $\hat{y} = 155.912 - 1.086x$ **c.** no, $t = -2.05$ **d.** $(-21.56, 279.09)$ **11.95 a.** Brand A: 3.349 $\pm$ .587; Brand B: 4.464 $\pm$ .296   **b.** Brand A: 3.349 $\pm$ 2.224; Brand B: 4.464 $\pm$ 1.120   **c.** $-.65 \pm 3.606$ **11.97** $E(y) = \beta_0 + \beta_1 x$ **11.99** true   **11.101 b.** $\hat{y} = x$; $\hat{y} = 3$ **c.** $\hat{y} = x$ **d.** least squares line has the smallest SSE   **11.103 a.** positive   **b.** yes   **c.** $\hat{y} = -214 + 1{,}118.4x$ **f.** $t = 4.32$, p-value $= .001$, reject H_0 **g.** .609   **i.** 93.27   **j.** (67.38, 119.15)   **11.105 a.** $y = \beta_0 + \beta_1 x + \varepsilon$ **b.** $\hat{y} = 175.7033 - .8195x$ **e.** $t = -3.43$, reject H_0 **11.107 a.** $H_0: \rho = 0$, $H_a: \rho \neq 0$ **b.** do not reject H_0 at $\alpha = .05$ **11.109** $\hat{y} = -92.458 + 8.347x$; $t = 3.248$, reject H_0; $r = .42$ **11.111 a.** yes   **b.** $\hat{\beta}_0 = -3.05$, $\hat{\beta}_1 = .108$ **c.** $t = 4.00$, reject H_0 **d.** $r = .756$, $r^2 = .572$ **e.** 1.09   **11.113 a.** $\hat{\beta}_0 = -13.49$, $\hat{\beta}_1 = -.0528$ **b.** $-.0528 \pm .0178$; yes   **c.** $r^2 = .8538$ **d.** (.5987, 1.2653)   **11.115 b.** $\hat{y} = 78.516 - .2389x$ **d.** $t = -2.31$, do not reject H_0 **e.** (301, 8.5)   **f.** $\hat{y} = 139.759 - .4497x$; $t = -15.35$, reject H_0 **11.117 a.** $\hat{y} = 46.4x$ **b.** $\hat{y} = 478.44 + 45.15x$ **d.** no, $t = .91$ **11.119** $\hat{y} = 2.55 + 2.76x$

Chapter 12

12.1 a. $E(y) = \beta_0 + \beta_1 x_1 + \beta_2 x_2$ **b.** $E(y) = \beta_0 + \beta_1 x_1 + \beta_2 x_2 + \beta_3 x_3 + \beta_4 x_4$ **c.** $E(y) = \beta_0 + \beta_1 x_1 + \beta_2 x_2 + \beta_3 x_3 + \beta_4 x_4 + \beta_5 x_5$ **12.5** test the null hypothesis that all the beta parameters (except β_0) are equal to 0   **12.7 a.** $t = 1.45$, do not reject H_0 **b.** $t = 3.21$, reject H_0 **12.9** df $= n - (k + 1)$ **12.11 a.** yes   **b.** yes, $F = 55.2$ **12.13 a.** $\hat{y} = 3.70 + .34x_1 + .49x_2 + .72x_3 + 1.14x_4 + 1.51x_5 + .26x_6 - .14x_7 - .10x_8 - .10x_9$ **c.** $t = -1.00$, do not reject H_0 **d.** (1.412, 1.608)   **12.15 a.** $E(y) = \beta_0 + \beta_1 x_1 + \beta_2 x_2 + \beta_3 x_3$ **c.** reject H_0: $\beta_1 = \beta_2 = \beta_3 = 0$ **d.** reject H_0 **e.** fail to reject H_0 **f.** fail to reject H_0 **12.17 a.** $E(y) = \beta_0 + \beta_1 x_1 + \beta_2 x_2 + \beta_3 x_3 + \beta_4 x_4$ **b.** reject H_0 **c.** yes   **d.** $H_0: \beta_1 = \beta_2 = \beta_3 = \beta_4 = 0$ **e.** $F > 3.32$ **f.** Thrill: reject H_0; change from routine: fail to reject H_0; surprise: reject H_0 **12.19 a.** $E(y) = \beta_0 + \beta_1 x_1 + \beta_2 x_2$ **b.** $\hat{y} = -20.35 + 13.35x_1 + 243.71x_2$ **d.** no, $t = 1.74$ **e.** 243.71 $\pm$ 138.38   **12.21 a.** $\hat{y} = 12.1804 - .0265x_1 - .45783x_2$ **c.** $t = -.50$, do not reject H_0 **d.** $-.4578 \pm .3469$ **e.** $R^2 = .529$, $R^2_{\text{adj}} = .505$; R^2_{adj} **f.** yes, $F = 21.88$ **12.23 a.** $E(y) = \beta_0 + \beta_1 x_1 + \beta_2 x_2 + \beta_3 x_3 + \beta_4 x_4 + \beta_5 x_5 + \beta_6 x_6 + \beta_7 x_7$ **b.** $\hat{y} = .998 - .022x_1 + .156x_2 - .017x_3 - .0095x_4 + .421x_5 + .417x_6 - .155x_7$ **12.25 b.** $F = 5.11$, reject H_0 **12.27 a.** model 1: $t = 2.58$, reject H_0: $\beta_1 = 0$; model 2: $t = 3.32$, reject H_0: $\beta_1 = 0$, $t = 6.47$, reject H_0: $\beta_2 = 0$, $t = -4.77$, reject H_0: $\beta_3 = 0$, $t = 0.24$, do not reject H_0: $\beta_4 = 0$; model 3: $t = 3.21$, reject H_0: $\beta_1 = 0$, $t = 5.24$, reject H_0: $\beta_2 = 0$, $t = -4.00$, reject H_0: $\beta_3 = 0$, $t = 2.28$, reject H_0: $\beta_4 = 0$, $t = .014$, do not reject H_0: $\beta_5 = 0$ **c.** Model 2   **12.29** 95% CI for $E(y)$ **12.33** $(-183.76, 207.25)$ **12.35** $(-1.233, 1.038)$ **12.37 a.** $E(y) = \beta_0 + \beta_1 x_1 + \beta_2 x_2 + \beta_3 x_1 x_2$ **b.** $E(y) = \beta_0 + \beta_1 x_1 + \beta_2 x_2 + \beta_3 x_3 + \beta_4 x_1 x_2 + \beta_5 x_1 x_3 + \beta_6 x_2 x_3$ **12.39 c.** interaction is present   **12.41 a.** $\hat{y} = -2.550 + 3.815x_1 + 2.630x_2 - 1.285x_1 x_2$ **e.** $H_0: \beta_3 = 0$, $H_a: \beta_3 \neq 0$ **f.** p-value $= .000$, reject H_0 **12.43 b.** $H_0: \beta_3 = 0$, $H_a: \beta_3 \neq 0$ **c.** reject H_0 **12.45 a.** $E(y) = \beta_0 + \beta_1 x_1 + \beta_2 x_2 + \beta_3 x_3 + \beta_4 x_4 + \beta_5 x_5 + \beta_6 x_1 x_2$ **b.** $H_0: \beta_4 = 0$ **c.** reject H_0 **d.** yes   **12.47 a.** affect of client credibility on likelihood depends on the level of linguistic delivery style   **b.** $H_0: \beta_1 = \beta_2 = \beta_3 = 0$ **c.** $F = 55.35$, reject H_0 **d.** $H_0: \beta_3 = 0$ **e.** $t = 4.008$, reject H_0 **f.** .114   **g.** .978   **12.49 a.** $E(y) = \beta_0 + \beta_1 x + \beta_2 x^2$ **b.** $E(y) = \beta_0 + \beta_1 x_1 + \beta_2 x_2 + \beta_3 x_1 x_2 + \beta_4 (x_1)^2 + \beta_5 (x_2)^2$ **c.** $E(y) = \beta_0 + \beta_1 x_1 + \beta_2 x_2 + \beta_3 x_3 + \beta_4 x_1 x_2 + \beta_5 x_1 x_3 + \beta_6 x_2 x_3 + \beta_7 (x_1)^2 + \beta_8 (x_2)^2 + \beta_9 (x_3)^2$ **12.51 a.** $t = 3.133$, reject H_0 **b.** $t = 3.133$, $t > 1.717$, reject H_0 **12.53 b.** moves graph to right or left   **c.** controls whether graph opens up or down   **12.55 b.** 1st-order model, $\beta_1 > 0$; 1st-order model, $\beta_1 < 0$; 2nd-order model   **12.57 a.** $E(y) = \beta_0 + \beta_1 x_1 + \beta_2 x_2 + \beta_3 x_1 x_2 + \beta_4 (x_1)^2 + \beta_5 (x_2)^2$ **b.** $\beta_4 (x_1)^2$, $\beta_5 (x_2)^2$ **12.59 b.** $E(y) = \beta_0 + \beta_1 x + \beta_2 x^2$ **c.** yes, $F = 203.16$ **d.** yes, $t = -3.39$ **12.61 a.** $E(y) = \beta_0 + \beta_1 x + \beta_2 x^2$ **b.** positive   **c.** no; $E(y) = \beta_0 + \beta_1 x$ **12.65** $E(y) = \beta_0 + \beta_1 x_1 + \beta_2 x_2$; $x_1 = \{1$ if level 2, 0 if not$\}$; $x_2 = \{1$ if level 3, 0 if not$\}$ **12.67 a.** $\hat{y} = 80 + 16.8x_1 + 40.4x_2$ **b.** $H_0: \mu_1 = \mu_2 = \mu_3$ **c.** $F = 24.72$, reject H_0 **12.69 a.** $E(y) = \beta_0 + \beta_1 x_1 + \beta_2 x_2$; $x_1 = \{1$ if democratic, 0 otherwise$\}$; $x_2 = \{1$ if dictatorship, 0 otherwise$\}$ **b.** $\beta_0 = $ mean risk level for communist $= \mu_{\text{communist}}$; $\beta_1 = \mu_{\text{democrat}} - \mu_{\text{communist}}$; $\beta_2 = \mu_{\text{dictator}} - \mu_{\text{communist}}$ **12.71 a.** $E(y) = \beta_0 + \beta_1 x$, where $x = \{1$ if flightless, 0 otherwise$\}$ **b.** $E(y) = \beta_0 + \beta_1 x_1 + \beta_2 x_2 + \beta_3 x_3$, where $x_1 = \{1$ if vertebrates, 0 if not$\}$, $x_2 = \{1$ if vegetables, 0 if not$\}$, $x_3 = \{1$ if invertebrates, 0 if not$\}$ **c.** $E(y) = \beta_0 + \beta_1 x_1 + \beta_2 x_2 + \beta_3 x_3$, where $x_1 = \{1$ if cavity within ground, 0 if not$\}$, $x_2 = \{1$ if trees, 0 if not$\}$, $x_3 = \{1$ if cavity above ground, 0 if not$\}$ **d.** $\hat{y} = 641 + 30{,}647x$ **e.** $t = 5.75$, reject H_0 **f.** $\hat{y} = 903 + 2{,}997x_1 + 26{,}206x_2 - 660x_3$ **g.** $F = 8.43$, reject H_0 **h.** $\hat{y} = 73.732 - 9.132x_1 - 45.01x_2 - 39.51x_3$ **i.** $F = 8.07$, reject H_0 **12.73 a.** $E(y) = \beta_0 + \beta_1 x$, where $x = \{1$ if enriched pond, 0 if natural pond$\}$ **b.** $\beta_0 = $ mean larval density of natural pond $= \mu_{\text{natural}}$; $\beta_1 = \mu_{\text{enriched}} - \mu_{\text{natural}}$ **c.** $H_0: \beta_1 = 0$, $H_a: \beta_1 > 0$ **d.** reject H_0 **12.75 a.** $E(y) = \beta_0 + \beta_1 x_1 + \beta_2 x_2 + \beta_3 x_3 + \beta_4 x_4$, where x_1-x_4 are dummy variables for the five reproductive classes   **b.** $\hat{\beta}_0 = 38$, $\hat{\beta}_1 = -22$, $\hat{\beta}_2 = 51$, $\hat{\beta}_3 = 5$, $\hat{\beta}_4 = 20$ **c.** $H_0: \beta_1 = \beta_2 = \beta_3 = \beta_4 = 0$ **d.** do not reject H_0 **12.77**

a. $E(y) = \beta_0 + \beta_1 x_1$ **b.** $E(y) = \beta_0 + \beta_1 x_1 + \beta_2 x_2 + \beta_3 x_3$; $x_2 = \{1$ if level 2, 0 otherwise$\}$; $x_3 = \{1$ if level 3, 0 otherwise$\}$ **c.** $E(y) = \beta_0 + \beta_1 x_1 + \beta_2 x_2 + \beta_3 x_3 + \beta_4 x_1 x_2 + \beta_5 x_1 x_3$ **d.** $\beta_4 = \beta_5 = 0$ **e.** $\beta_2 = \beta_3 = \beta_4 = \beta_5 = 0$ **12.79** $E(y) = \beta_0 + \beta_1 x_1 + \beta_2 x + \beta_3 x_2 + \beta_4 x_3 + \beta_5 x_4$, where x_2-x_4 are dummy variables   **12.81 a.** $\hat{y} = 48.8 - 3.4x_1 + .07x_1^2$; $\hat{y} = 46.4 + .3x_1 + .05x_1^2$; $\hat{y} = 41.3 - .7x_1 + .03x_1^2$ **12.83 a.** $x_1 =$ weight, $x_2 = \{1$ if catfish, 0 if not$\}$, $x_3 = \{1$ if largemouth bass, 0 if not$\}$ **b.** $E(y) = \beta_0 + \beta_1 x_1 + \beta_2 x_2 + \beta_3 x_3$ **c.** $E(y) = \beta_0 + \beta_1 x_1 + \beta_2 x_2 + \beta_3 x_3 + \beta_4 x_1 x_2 + \beta_5 x_1 x_3$ **d.** $\hat{y} = 3.13 + .00371 x_1 + 26.51 x_2 - 4.09 x_3$ **f.** $\hat{y} = 3.5 + .00344 x_1 + 25.59 x_2 - 3.47 x_3 + .00082 x_1 x_2 - .00129 x_1 x_3$ **g.** .00426   **12.85 a.** $E(y) = \beta_0 + \beta_1 x_1 + \beta_2 x_2 + \beta_3 x_1 x_2$, where $x_2 = \{1$ if plant, 0 if duck chow$\}$ **b.** $\hat{y} = 8.14 - .016 x_1 - 10.4 x_2 + .095 x_1 x_2$ **c.** .079   **d.** $-.016$ **e.** $t = .67$, fail to reject H_0 **12.87 a.** $E(y) = \beta_0 + \beta_1 x_1 + \beta_2 x_2$ **b.** β_1 for both   **c.** $E(y) = \beta_0 + \beta_1 x_1 + \beta_2 x_2 + \beta_3 x_1 x_2$ **d.** β_1; $\beta_1 + \beta_3$ **12.89** (a and b); (a and d); (a and e); (b and c); (b and d); (b and e); (c and e); (d and e)   **12.91** model with a small number of β parameters   **12.93 a.** complete: $\hat{y} = 14.6 - .61 x_1 + .44 x_2 - .08 x_3 - .06 x_4$; reduced: $\hat{y} = 14.0 - .64 x_1 + .40 x_2$ **b.** $SSE_R = 160.44$, $SSE_C = 152.66$ **c.** 5; 3   **d.** H_0: $\beta_3 = \beta_4 = 0$ **e.** $F = .38$, do not reject H_0 **f.** p-value $> .10$ **12.95 a.** $E(y) = \beta_0 + \beta_1 x_1 + \beta_2 x_2 + \beta_3 x_3 + \beta_4 x_4 + \beta_5 x_5 + \beta_6 x_6 + \beta_7 x_7 + \beta_8 x_8 + \beta_9 x_9 + \beta_{10} x_{10} + \beta_{11} x_{11}$ **b.**
$E(y) = \beta_0 + \beta_1 x_1 + \beta_2 x_2 + \beta_3 x_3 + \beta_4 x_4 + \beta_5 x_5 + \beta_6 x_6 + \beta_7 x_7 + \beta_8 x_8 + \beta_9 x_9 + \beta_{10} x_{10} + \beta_{11} x_{11} + \beta_{12} x_1 x_9 + \beta_{13} x_1 x_{10} + \beta_{14} x_1 x_{11} + \beta_{15} x_2 x_9 + \beta_{16} x_2 x_{10} + \beta_{17} x_2 x_{11} + \beta_{18} x_3 x_9 + \beta_{19} x_3 x_{10} + \beta_{20} x_3 x_{11} + \beta_{21} x_4 x_9 + \beta_{22} x_4 x_{10} + \beta_{23} x_4 x_{11} + \beta_{24} x_5 x_9 + \beta_{25} x_5 x_{10} + \beta_{26} x_5 x_{11} + \beta_{27} x_6 x_9 + \beta_{28} x_6 x_{10} + \beta_{29} x_6 x_{11} + \beta_{30} x_7 x_9 + \beta_{31} x_7 x_{10} + \beta_{32} x_7 x_{11} + \beta_{33} x_8 x_9 + \beta_{34} x_8 x_{10} + \beta_{35} x_8 x_{11}$ **c.** H_0: $\beta_{12} = \beta_{13} = \cdots = \beta_{35} = 0$
12.97 b. $\hat{y} = .906 - .023 x_1 + .350 x_2 + .275 x_5 - .003 x_1 x_2 + .005 x_1 x_5$, $F = 5.505$, $R^2 = .6792$, $s = .505$ **d.** $F = .29$, do not reject H_0
12.99 a. $E(y) = \beta_0 + \beta_1 x_1 + \beta_2 x_2 + \beta_3 x_3 + \beta_4 x_4 + \beta_5 x_5 + \beta_6 x_6 + \beta_7 x_7 + \beta_8 x_8 + \beta_9 x_9 + \beta_{10} x_{10}$ **b.** H_0: $\beta_3 = \beta_4 = \cdots = \beta_{10} = 0$ **c.** at least one of the additional variables is important   **e.** (8.12, 19.88)   **f.** yes   **g.** $E(y) = \beta_0 + \beta_1 x_1 + \beta_2 x_2 + \beta_3 x_3 + \beta_4 x_4 + \beta_5 x_5 + \beta_6 x_6 + \beta_7 x_7 + \beta_8 x_8 + \beta_9 x_9 + \beta_{10} x_{10} + \beta_{11} x_1 x_2 + \beta_{12} x_2 x_3 + \beta_{13} x_2 x_4 + \beta_{14} x_2 x_5 + \beta_{15} x_2 x_6 + \beta_{16} x_2 x_7 + \beta_{17} x_2 x_8 + \beta_{18} x_2 x_9 + \beta_{19} x_2 x_{10}$ **h.** H_0: $\beta_{11} = \beta_{12} = \cdots = \beta_{19} = 0$; nested model F-test   **12.101 a.** H_0: $\beta_4 = \beta_5 = 0$ **b.** H_0: $\beta_3 = \beta_4 = \beta_5 = 0$ **12.103** large number of t-tests inflating the overall α; no higher-order or interaction terms   **12.105 a.** 6   **b.** 5   **c.** 4   **12.107 a.** 11   **b.** 10   **c.** 1   **d.** $E(y) = \beta_0 + \beta_1 x_{11} + \beta_2 x_4 + \beta_3 x_2 + \beta_4 x_7 + \beta_5 x_{10} + \beta_6 x_1 + \beta_7 x_9 + \beta_8 x_3$ **12.111** two or more independent variables are correlated   **12.113** predicting y for x's outside the range of the sample data   **12.115** yes; x_4 is highly correlated with both x_2 and x_5 **12.117** x_4 is highly correlated with x_5 **12.119 a.** yes, residual $= 57.37$ **c.** no other outliers   **12.121** outliers: #105 and #115; error variances unequal   **12.123 a.** race and foreign status, year GRE taken and years in graduate program   **12.125** $E(y) = \beta_0 + \beta_1 x_1 + \beta_2 x_2 + \beta_3 x_3$; $x_1 = \{1$ if level 2, 0 otherwise$\}$; $x_2 = \{1$ if level 3, 0 otherwise$\}$; $x_3 = \{1$ if level 4, 0 otherwise$\}$ **12.127 a.** $E(y) = \beta_0 + \beta_1 x_1 + \beta_2 x_2 + \beta_3 x_3$, where $x_1 =$ quantitative, $x_2 = \{1$ if level 2, 0 if not$\}$, $x_3 = \{1$ if level 3, 0 if not$\}$ **b.** $E(y) = \beta_0 + \beta_1 x_1 + \beta_2 x_1^2 + \beta_3 x_2 + \beta_4 x_3 + \beta_5 x_1 x_2 + \beta_6 x_1 x_3 + \beta_7 x_1^2 x_2 + \beta_8 x_1^2 x_3$ **12.129 a.** yes, $F = 24.21$ **b.** $t = -2.01$, reject H_0 **c.** $t = .31$, do not reject H_0 **d.** $t = 1.21$, do not reject H_0 **12.131** yes; $x_1 = 60$, $x_2 = .4$, $x_3 = 900$ are outside range of sample data   **12.133** df(Error) $= 0$ **12.135 a.** yes, $t = 5.96$, p-value $= .00005$ **b.** $t = 0.01$, p-value $= .9620$, do not reject H_0 **c.** $t = 1.91$, p-value $= .0288$, reject H_0 **12.137 a.** $E(y) = \beta_0 + \beta_1 x_1 + \beta_2 x_2 + \beta_3 x_3 + \beta_4 x_4 + \beta_5 x_5$ **b.** model is statistically useful   **c.** $E(y) = \beta_0 + \beta_1 x_1 + \beta_2 x_2 + \beta_3 x_3 + \beta_4 x_4 + \beta_5 x_5 + \beta_6 x_6 + \beta_7 x_7$ **12.139 c.** yes, $F = 30.45$ **12.141 a.** diet at three levels   **b.** $E(y) = \beta_0 + \beta_1 x_1 + \beta_2 x_2$, where $x_1 = \{1$ if high protein, 0 if not$\}$, $x_2 = \{1$ if high fruits, 0 if not$\}$ **12.143 a.** $E(y) = \beta_0 + \beta_1 x_1 + \beta_2 x_2 + \beta_3 x_3 + \beta_4 x_4 + \beta_5 x_5$ **b.** $\hat{y} = 15.491 + 12.774 x_1 + .713 x_2 + 1.519 x_3 + .32 x_4 + .205 x_5$ **c.** $R^2 = .240$, $F = 11.68$, reject H_0 **d.** p-value $= .025$, reject H_0 **12.145 a.** $E(y) = \beta_0 + \beta_1 x_1 + \beta_2 x_2 + \beta_3 x_3$ **c.** yes, $F = 56$ **d.** do not reject H_0 **12.147 a.** negative   **b.** no, $F = 1.60$ **c.** no, $F = 1.61$ **12.149 b.** $\hat{y} = 42.25 - .0114x + .000000608x^2$ **c.** no, $t = 1.66$ **12.151 a.** $E(y) = \beta_0 + \beta_1 x_1 + \beta_2 x_6 + \beta_3 x_7$, where $x_6 = \{1$ if good,0 if not$\}$, $x_7 = \{1$ if fair, 0 if not$\}$ **c.** excellent: $\hat{y} = 188,875 + 15,617 x_1$; good: $\hat{y} = 85,829 + 15,617 x_1$; fair: $\hat{y} = 36,388 + 15,617 x_1$ **e.** yes, $F = 8.43$ **f.** x_1, x_3 and x_5 are highly correlated   **g.** assumptions are satisfied   **12.153** model includes $x_1 =$ DOT estimate, $x_3 = \{1$ if fixed,0 if competitive$\}$, and $x_5 =$ estimated days to complete

Chapter 13
13.3 a. 18.3070   **b.** 29.7067   **c.** 23.5418   **d.** 79.4900   **13.5 a.** $\chi^2 > 5.99147$ **b.** $\chi^2 > 7.77944$ **c.** $\chi^2 > 11.3449$ **13.7 a.** no, $\chi^2 = 3.293$
13.9 a. jaw habits; bruxism, clenching, bruxism and clenching, neither   **c.** H_0: $p_1 = p_2 = p_3 = p_4 = .25$, H_a: At least one p_i differs from .25
d. 15   **e.** $\chi^2 = 25.73$ **f.** $\chi^2 > 7.81473$ **g.** reject H_0 **h.** (.37, .63)   **13.11 a.** Pottery type; burnished, monochrome, painted, other
b. $p_1 = p_2 = p_3 = p_4 = .25$ **c.** H_0: $p_1 = p_2 = p_3 = p_4 = .25$ **d.** $\chi^2 = 436.59$ **e.** p-value ≈ 0, reject H_0 **13.13** $\chi^2 = 16$, reject H_0
13.15 a. $\chi^2 = 17.16$, reject H_0 **b.** $.376 \pm .103$ **c.** yes   **13.17 a.** yes, $\chi^2 = 360.48$ **b.** (.165, .223)   **13.19** true   **13.21 a.** $\chi^2 > 26.2962$
b. $\chi^2 > 15.9871$ **c.** $\chi^2 > 9.21034$ **13.23 a.** Column 1: 36%, 64%; column 2: 53.1%, 46.9%; column 3: 67.9%, 32.1%; totals: 57.5%, 42.5%
13.25 a. B_1: 29.9%; B_2: 44.2%; B_3: 29.6%   **b.** B_1: 47.0%; B_2: 32.5%; B_3: 49.3%   **c.** B_1: 23.1%; B_2: 23.3%; B_3: 21.1%   **13.27 a.** grade (4th and 5th); reading ability (normal and disability)
b.

	Reading Disability	Normal	Totals
4th	32	55	87
5th	34	40	74
Totals	66	95	161

c. 35.66, 51.34, 30.34, 43.66   **d.** $\chi^2 = 1.38$ **e.** $\chi^2 > 2.70554$ **f.** no, do not reject H_0 **13.29 a.** masculinity risk (high and low); event (violent and avoided-violent)   **b.** 1,507 newly incarcerated men   **c.** 260.8   **d.** 118.2, 776.2, 351.8   **e.** $\chi^2 = 10.1$ **f.** reject H_0
13.31 $\chi^2 = 35.41$, reject H_0 **13.33** Internet: $\chi^2 = .512$, do not reject H_0; party: $\chi^2 = 164.76$, reject H_0; military: $\chi^2 = 8.3$, reject H_0; views: $\chi^2 = 174.39$, reject H_0; race: $\chi^2 = 69.18$, reject H_0; income: $\chi^2 = 16.39$, reject H_0; community: $\chi^2 = 17.31$, reject H_0 **13.35 a.** no

b.

SHSS:C	CAHS Low	CAHS Medium	CAHS High
Low	32	14	2
Medium	11	14	6
High	6	16	29

c. yes   **d.** $\chi^2 = 46.70$, reject H_0 **13.37 a.** $\chi^2 = 4.407$, reject H_0 **b.** no   **c.** .0438   **d.** .0057; .0003   **e.** p-value $= .0498$, reject H_0
13.39 false   **13.41 a.** no, $\chi^2 = 2.133$ **b.** $.233 \pm .057$
13.43 a.

Genetic Trait	LLD Yes	No	Total
Yes	21	15	36
No	150	306	456
Total	171	321	492

b. $\chi^2 = 9.52$, reject H_0 **13.45 a.** possibly not   **c.** $\chi^2 = 12.9$ **d.** yes, reject H_0 **e.** $.233 \pm .200$ **13.47 a.** $\chi^2 = 1.143$, do not reject H_0
b. $.179 \pm .119$ **13.49** $\chi^2 = 36.31$, reject H_0 **13.51** no, $\chi^2 = 4.4$ **13.53** $\chi^2 = 6.171$, reject H_0 **13.55 a.** no, $\chi^2 = 13.715$ **b.** wife: no,
$\chi^2 = 3.198$; husband: no, $\chi^2 = 6.069$ **13.57 a.** $\chi^2 = 9.647$ **b.** 11.0705   **c.** no   **d.** $.05 < p$-value $< .10$ **13.59** $\chi^2 = 2.28$; insufficient
evidence to reject the null hypothesis of independence

Chapter 14

14.1 population not normal   **14.3 a.** .063   **b.** .500   **c.** .004   **d.** .151; .1515   **e.** .212; .2119   **14.5** $S = 16$, p-value $= .115$, do not reject H_0
14.7 $S = 8$, p-value $= .0391$, reject H_0 **14.9 a.** $H_0: \eta = 1.5$, $H_a: \eta > 1.5$ **b.** 3   **c.** .855   **d.** do not reject H_0 **14.11 a.** $H_0: \eta = 5$ **b.** $z = 6.07$, p-value $= 0$ **c.** reject H_0 **14.13 a.** yes, $z = 2.12$ **b.** no   **14.15** true   **14.17 a.** $T_1 \leq 41, T_1 \geq 71$ **b.** $T_1 \geq 50$ **c.** $T_1 \leq 43$ **d.** $|z| > 1.96$ **14.19** yes, $z = -2.47$ **14.21 a.** $T_1 = 62.5$, reject H_0 **b.** yes, $T_1 = 62.5$ **14.23 a.** $H_0: D_{female} = D_{normal}$, $H_a: D_{female} > D_{normal}$ **c.** 174.5
d. 150.5   **e.** $z > 1.28$ **f.** $z = 1.72$, reject H_0 **14.25 a.** $T_1 = 150.5$, reject H_0 **b.** $z = 3.44$, reject H_0 **14.27 a.** zinc measurements non-
normal   **b.** $T_1 = 18$, do not reject H_0 **c.** $T_2 = 32$, do not reject H_0 **14.29 a.** data not normal   **b.** $z = -6.47$, reject H_0 **c.** $z = -.39$, do not
reject H_0 **d.** $z = -1.35$, do not reject H_0 **14.33 a.** $H_a: D_A > D_B$ **b.** $T_- = 3.5$, reject H_0 **14.35 a.** $H_a: D_A > D_B$ **b.** $z = 2.499$, reject H_0
c. .0062   **14.37 a.** data not normal   **b.** $H_0: D_{father} = D_{mother}$, $H_a: D_{father} \neq D_{mother}$ **c.** $T_+ = 22; T_- = 44$ **d.** $T_+ \leq 11$ **e.** do not reject H_0
14.39 c. $T_- = 24.5$ **d.** no   **14.41** yes, $T_- = 50.5$ **14.43 a.** no, $T_- = 0$ **b.** reject H_0; supports the "hot hand" theory   **14.45** samples size for
each distribution is more than 5   **14.47 a.** completely randomized design   **b.** H_0: Three probability distributions are identical   **c.** $H > 9.21034$
d. $H = 14.53$, reject H_0 **14.49 b.** 84   **c.** 145   **d.** 177   **e.** $H = 18.4$ **f.** $H = 13.66$ **g.** $z = -3.36$, reject H_0 **14.51 b.** reject H_0
14.53 a. normality assumption violated   **b.** $H = 13.66, H = 13.66$ **14.55** ranking b   **14.57 a.** 6   **b.** H_0: The probability distributions for
the four treatments are identical   **c.** $F_r = 15.2$, reject H_0 **d.** p-value $< .005$ **14.59** $F_r = 13$, reject H_0 **14.61** p-value $= .819$, do not
reject H_0 **14.63 a.** $F_r > 9.21034$ **b.** reject H_0 **c.** do not reject H_0 **d.** reject H_0 **14.65** no, $F_r = .20$ **14.67** no; $F_r = 2.09$, do not reject H_0
14.71 a. $|r_s| > .648$ **b.** $r_s > .450$ **c.** $r_s < -.432$ **14.73 b.** $r_s = .745$, do not reject H_0 **c.** $.05 < p$-value $< .10$ **14.75 c.** .713
d. $|r_s| > .425$ **e.** reject H_0 **14.77 e.** no, $r_s = -.439$ **14.79 a.** $-.877$ **b.** $-.907$ **c.** reject H_0 **d.** reject H_0 **14.81 a.** .714   **b.** reject H_0
14.85 a. rank sum test   **b.** sign test   **c.** Kruskal-Wallis test   **d.** Spearman's test   **e.** signed rank test   **f.** Friedman's test   **14.87 a.** no, $r_s = .40$
b. yes, $T_- = 1.5$ **14.89** yes, $F_r = 14.9$ **14.91** no, $T_2 = 29$ **14.93** no, $H = 1.565$ **14.95 a.** $-.733$ **b.** do not reject H_0 **14.97 a.** yes, $T_2 = 105$
b. less than   **14.99** yes, $F_r = 6.35$ **14.101** yes, $H = 7.154$ **14.103 a.** yes, $T_+ = 1$ **b.** $t = -2.96$, reject H_0 **14.105** yes, $r_s = .929$
14.107 $F_r = 7.85$, reject H_0

Subject Index

Photo Credits

Florence Nightingale, Corbis/Bettmann; H.G. Wells, Getty Images, Inc.-Hulton Archive Photos; Wilfredo Pareto, The Granger Collection; John Tukey, Alfred Eisenstaedt/Getty Images/Time-Life Pictures; Pafnuty L. Chebyshev, Sovfoto/Eastfoto; John Venn, Maull and Fox/Gonville and Caius College; Blaise Pascal, The Granger Collection; Thomas Bayes, Prof. Stephen M. Stigler; Simeon Denis Poisson, Photo Researchers, Inc.; Carl F. Gauss, The Granger Collection; Abraham de Moivre, Picture Desk, Inc./Kobal Collection; Pierre-Simon Marquis de LaPlace, Picture Desk, Inc./Kobal Collection; Jerzey Neyman, Paul Bishop/University of California-Berkeley; William Sealy Gosset, The Granger Collection; Egon Sharpe Pearson, Biometrika; Friedrich R. Helmert, Bibliothek des GeoForschungsZentrums GFZ; Bradley Efron, Stanford University; Sir Ronald A. Fisher, Library of Congress; Francis Galton, Corbis/Bettmann; Francis J. Anscombe, Phyllis R. Anscombe; Karl Pearson, University College, London; Frank Wilcoxon, Florida State University, Department of Statistics; Charles E. Spearman, University College, London Department of Psychology

Chapter Opening Photos: Chapter 1, Getty Images; Chapter 2, Left Lane Productions/Corbis/Bettmann; Chapter 3, Morton Beebe/Corbis/Bettmann; Chapter 4, Alan Schein Photography/Corbis/Bettmann; Chapter 5, DiMaggio/Kalish/Corbis/Bettmann; Chapter 6, Neil Rabinowitz/Corbis/Bettmann; Chapter 7, Bryan Pickering/Eye Ubiquitous/Corbis/Bettmann; Chapter 8, Alan Schein Photography/Corbis/Bettmann; Chapter 9, Duomo/Corbis/Bettmann; Chapter 10, Lester Lefkowitz/Getty Images, Inc.-Image Bank; Chapter 11, Corbis Digital Stock; Chapter 12, Freelance Consulting Services Pty. Ltd./Corbis/Bettmann; Chapter 13, Grafton Marshall Smith/Corbis/Bettmann; Chapter 14, Kelly/Mooney Photography/Corbis/Bettmann

McClave/Sincich
Statistics, 10th Edition, Student Data CD
0-13-149827-4
© 2006 Pearson Education, Inc.
Pearson Prentice Hall
Pearson Education, Inc.
Upper Saddle River, NJ 07458
Pearson Prentice Hall™ is a trademark of Pearson Education, Inc.